PATTON

PATTON
A GENIUS FOR WAR

Carlo D'Este

HarperPerennial
A Division of HarperCollinsPublishers

Copyright acknowledgments follow page 977.

A hardcover edition of this book was published in 1995 by HarperCollins Publishers.

HarperCollins books may be purchased for educational, business, or sales promotional use. For information please write: Special Markets Department, HarperCollins Publishers, Inc., 10 East 53rd Street, New York, NY 10022.

First HarperPerennial edition published 1996.

Designed by Alma Hochhauser Orenstein
Photo insert designed by Barbara DuPree Knowles
Maps by George Ward

The Library of Congress has catalogued the hardcover edition as follows:

D'Este, Carlo, 1936–
    Patton: a genius for war / Carlo D'Este.
      p.   cm.
    Includes bibliographical references and index.
    ISBN 0-06-016455-7
    1. Patton, George S. (George Smith), 1885–1945. 2. Generals—United States—Biography. 3. United States—Army—Biography. I. Title.
    E745.P3D46   1995
    355'.0092—dc20   95-38433
    [B]

ISBN 0-06-092762-3 (pbk.)

96 97 98 99 00 ❖/ RRD 10 9 8 7 6 5 4 3 2 1

For Shirley Ann, Elizabeth, Liane,
Christopher, and Danielle

And in loving memory of my parents

Eleanor D'Este (1897–1992)
Charles D'Este (1896–1958)

CONTENTS

Illustrations follow pages 370 and 658.

MAPS

NOTE TO THE READER

Patton's dyslexia generated a lifelong writing problem manifested by mis-spelled words and the frequent omission of punctuation and capitalization. Throughout *Patton* there are extracts from Patton's letters and from his observations about places he visited. In order to allow a fuller understanding of how this condition dominated his life, I have attempted faithfully to represent his writings, mistakes and all.

More often than not, in his early letters and school papers, Patton tended to omit periods at the ends of sentences. However, only for the sake of clarity has it occasionally been necessary to insert minor corrections in punctuation and capitalization and artificially to break off long disquisitions in order to create a new paragraph. With this exception, Patton's writings are cited as he wrote them.

PROLOGUE

Who Was George S. Patton?

Ask virtually any American born after World War II what immediately comes to mind when the name "Patton" is mentioned, and chances are they will conjure an image of a large, empty stage dominated by an enormous, oversize American flag. A tall, uniformed figure suddenly strides to its center, a large blue sash trimmed in yellow draped across his chest, an array of medals on his left breast pocket, two ivory-handled pistols strapped to his waist, and a highly polished helmet on his head on which are set the four silver stars of a full general of the U.S. Army. Standing ramrod straight, the general begins to address an unseen audience of soldiers in blunt, often colorful language. On what is clearly the eve of a battle, he explains what he expects of them and how they will survive if they follow his advice. He concludes with the admonition: "The object of war is not to die for your country. It is to make the other poor dumb bastard die for his."

As the scene fades, we begin to lose sight of the fact that the man who has just spoken is an actor named George C. Scott in his most famous role, which in 1970 earned him the Oscar for Best Actor for his portrayal of Gen. George S. Patton Jr. in the film *Patton*. We have come to think of him as Patton himself. As one writer has accurately observed, the film "turned Patton the legend finally into Patton the folk hero. In the shape of Scott, with his dark scowling face and rasping voice, Patton had now become the essence of America's World War II. Just like the cowboy hero of the Old West, he had stepped into American mythology . . . the symbol of an older, simplistic America, untouched by social change, political doubts, [and] the uncertainties of the seventies and eighties."[1]

Although the architects of this powerful film strove diligently to reveal Patton as he really was, there were the inevitable distortions. Nor was it possible fully to portray his complex character in a film devoted solely to his World War II exploits. Moreover, *Patton* was based on the bestselling memoir of another famous general, Omar N. Bradley, who served as the film's chief military adviser.[2] It was ironical that Bradley received a considerable sum of money, including a percentage of the gross receipts, for his professional consultation on a film about a comrade-in-arms he despised and never understood.[3]

What inevitably emerges in the film is the portrayal of a brash, swashbuckling, controversial warrior. Yet, as one critic noted, if the film glorified anyone, it was Omar Bradley, not Patton.[4] Thus, for nearly half of the fifty years since his death in 1945, the primary sources of our collective knowledge of Patton are, largely, a popular film and the opinions of a general who detested him but who owed him a giant debt for his support during the final months of World War II. Add to this the fact that the image the real Patton presented to the world was a many-layered facade, and there exists ample justification for the question, Who was George S. Patton?

Although our knowledge of him is incomplete and shrouded in myth, it is indisputable that, for a variety of reasons, Gen. George S. Patton Jr. has earned a place in the pantheon of authentic American heroes. Throughout our relatively brief history as a nation, Americans have not only come to admire (and sometimes even venerate) men and women who have attained national prominence, but we have developed our own special breed of hero, modeled on the warriors who founded and tamed this nation. The Vietnam War has spawned a present-day revulsion for war as an instrument of national policy. Nevertheless, most Americans remain captivated by wars and the men who fight them. Our warrior-heroes range across the spectrum of American history: George Washington, Ulysses S. Grant, Robert E. Lee, Theodore Roosevelt, and John J. Pershing are among those who come immediately to mind. To this distinguished list can be added the name of George S. Patton.

Yet how little we really know of this man. Was he the tough, humorless, bloodthirsty warrior depicted by George C. Scott, or was he a romantic who would have been far more at home in an earlier age? The life of Patton is not only that of a uniquely American warrior but, paradoxically, that of a soldier who was very much out of his element in the twentieth century.

Patton was an ancestor worshiper, as we shall see, whose veneration of his forefathers verged on obsession. He saw himself as the modern embodiment of his heroic Confederate antecedents, and because of the enormously successful facade he created, the tender, romantic side of Patton was virtually unknown in his lifetime outside his circle of friends and admirers. The real Patton was an emotional and often humble man who could weep one moment, and seconds later put on his public face and curse in the most scatological of terms.

Virtually unknown, too, was Patton's deeply religious nature. He prayed often and almost always in private. "On one occasion in Palermo, Sicily, feeling in dire need to re-establish his lines of communication with the Almighty, he went into the great Cathedral. There he knelt in prayer for a solid hour with hardly a motion of his body. George Patton was convinced that God was on his side," and that there was indeed a god of Battles who would protect him.[5] On another occasion his wife, Beatrice, found him kneeling in prayer before a polo match. "Afterward she asked what he'd been praying for. 'For help in the polo game,' he replied. 'Were you praying for a win?' she inquired. 'Hell no,' he said, 'I was praying to do my best.'"[6]

Patton's detractors, and there are many (among them historian Paul Fussell, who has characterized him as one of World War II's "masters of chickenshit" for his strict dress code in the Third Army), believe he was little more than a headline grabber, out to enhance his own reputation by expending the lives of his men in an obsessive quest for personal glory. Others simply loathed him for his harsh methods, his unbending personality, his arrogance, his profanity, and the sheer wrath of his notoriously volatile temper.

With one major exception near the end of World War II, this perception is part of the myth of Patton as a passionate believer in providence and a man whose ambition was fueled by the convictions that "It is my destiny to lead the biggest army ever assembled under one flag," and "God isn't going to let me be killed before I do."[7] The reality is that Patton accepted the inevitability of death in combat but strove mightily to save the lives of his men. While it is true that Patton loved war, it was only in the pragmatic sense that he considered conflict an inevitable part of man's nature. He detested the death and devastation it wrought. However, if there were wars to be fought, he believed they ought to be conducted by the best qualified men, such as himself.

What made Patton so remarkable was his willingness to take risks and to make crucial life-and-death decisions no one else would dare. For all his military accomplishments, George C. Scott was right when he asserted that what made Patton unique was his individualism, his understanding that "You live and you die alone—he knew it and he lived it. . . . But foremost about Patton, I believe this man was an individual in the deepest sense of the word."[8]

Patton was an authentic and flamboyant military genius whose entire life was spent in preparation for a fleeting opportunity to become one of the great captains of history. No soldier in the annals of the U.S. Army ever worked more diligently to prepare himself for high command than did Patton. However, it was not only his astonishing breadth of professional reading and writing that separated Patton from his peers, but that intangible, instinctive sense of what must be done in the heat and chaos of battle: in short, that special genius for war that has been granted to only a select few, such as Robert E. Lee and German Field Marshal Erwin Rommel. Who but Patton would have tramped the back roads of Normandy in 1913 with a Michelin map to study the terrain because he believed he would someday fight a major battle there?

Patton's great success on the battlefield did not come about by chance but rather from a lifetime of study and preparation. He was an authentic intellectual whose study of war, history, and the profession of arms was extraordinary. His memory was prodigious, as was his intellect. Patton not only believed in the Scriptures but could quote them at length. For hours on end, he could recite not only verses from the Bible, but from his great love, poetry. His favorites were Homer's *Iliad* and Kipling's verse. He read voraciously and not only learned from what he read but managed to remember virtually all of it. As a young child, his nephew recalls sitting engrossed while Patton recited from memory lines from such diverse sources as Shakespeare, the Bible, Macaulay, and Kipling's *Barrack-Room Ballads*.[9]

To his family he was an accomplished and imaginative storyteller, whose tales were usually based on ancient heroes, occasionally embellished by imaginary characters who performed impossible feats of derring-do. During the latter stages of World War II, with his aide acting as referee and researcher, Patton used his encyclopedic knowledge to best a noted Harvard professor on historical subjects, and the high prince of the Catholic Church, New York's Archbishop (later Cardinal) Francis Joseph Spellman, on the Bible.[10]

To his detriment, what little the public knew of Patton was only what he permitted them to know. Patton's reputation has been perpetually tarnished by the facade he himself created and the public effortlessly accepted: that he was a swashbuckling, brash, profane, impetuous soldier who wore two ivory-handled revolvers and loved war so much he was nicknamed "Old Blood and Guts"—the general who slapped two soldiers in Sicily in August 1943 and was almost sent home in disgrace, his destiny unfulfilled because of momentary, irrational acts of rage.

Unfortunately, because he blatantly perpetuated his own self-created image, the legacy of "Old Blood and Guts" has not only become commonplace but the accepted perception of Patton. In the end his self-invented personality nearly destroyed him and has severely distorted his place in history.

A great deal has been written about Patton's battles and campaigns and very little about his lifetime of preparation for what he passionately believed was his destiny to lead a great army into battle. This book attempts to cover both. George S. Patton had so many faces and created so many illusions about himself that a major role of the biographer is to disentangle, to separate, and, in the end, to impart, as Gerald Clarke has written, "that rarest of all human gifts: understanding."[11] Ultimately, the saga that was Patton's life is infinitely more fascinating than his own self-created myth. Whether one admires or detests Patton, his rise to high command and his World War II exploits form the theme of this book.

What helps to make the tapestry of Patton's life so rich is the extraordinary wealth of material he penned. From diaries, essays, notes, poetry, and lectures, to his hundreds of intimate and revealing letters to his beloved wife, Beatrice, nothing Patton wrote was *ever* thrown away. To the contrary, there is ample evidence that he sensed such material would become the raw material of a future biography. Whether one chooses to view this attitude as prescient or merely vain presumption, the vast collection provides a rich foundation for a biographer.

Although the family archive is the bedrock on which any biography of George S. Patton must be built, it is by no means the only source of knowledge about him. As a public figure he has inspired biographies (some of them little more than hagiography), articles, and anecdotes, the vast majority of which deal exclusively with his World War II exploits and thus encompass only the final four years of his life. Until the recent publication of an intimate and revealing memoir of the Patton family by his grandson,* very little has been written about George S. Patton's childhood and military career prior to 1939, all of which were vital ingredients in the shaping of the man and the general. What little has previously been written about his ancestry is likewise shrouded in myth and misinformation. Fortunately there exists a wealth of other important references in libraries, archives, and long-out-of-print secondary sources that both enrich our understanding and permit a full accounting of his extraordinary life.

Once asked why he made the film *Patton*, which required seventeen years of effort to accomplish, executive producer Frank McCarthy replied that it was "to study this unique man," not "to lionize him. Only to study him and to say: 'My God, what a fascinating character this was!'"[12]

In the more than two hundred years of U.S. history there has never been (and may never again be) another American quite like George Smith Patton. He was one of a kind. The year 1995 marks the fiftieth anniversary of Patton's death, thus making it an appropriate occasion for a contemporary reexamination of the life of perhaps the most famous and controversial American soldier of the twentieth century.

---

*Robert H. Patton, *The Pattons: A Personal History of an American Family* (New York, 1994).

PART I

An Ancestry of Heroes

(1750–1885)

Americans have an insatiable craving for heroes.
—NEW YORK TIMES BOOK REVIEW, MAY 14, 1989

CHAPTER 1

The Pattons of Virginia

We ne'er shall look upon his like again.

—TRIBUTE TO COL. GEORGE S. PATTON, VMI, CLASS OF 1852

The first Patton in America was an enormously successful merchant and trader named Robert, who emigrated from his native Scotland to Virginia around 1770. Although little is known of this great-great-grandfather of George S. Patton Jr., he is thought to have been a rebellious Scottish patriot who "fled Scotland apparently after opposing the Crown in the interminable conflict for Scottish independence"[1] and later emigrated to Fredericksburg, Virginia, around 1770, via Bermuda.

According to Patton family lore, he fought for Bonnie Prince Charlie, the Young Pretender, and his real name was not Patton. Robert Patton has been variously described as "a smallish man who was hot-tempered and something of a dandy" and "a mule-headed, fiery little man with a fondness for ruffled shirts."[2] He is also believed to have dropped hints from time to time that he was the son of a landed aristocrat, and that Patton was the name he had adopted and was known by in Virginia. Another story has it that before his arrival in Virginia Robert lived in Bermuda, where he got into serious trouble when he killed the governor with a pistol after the latter insulted him.[3] The only known painting of Robert, depicting a clear-eyed, well-dressed young man, gives no hint of his personality.

All this is myth. In fact a great deal is known of the first Patton. Probably born Robert Paton in Mauchline, Ayr, Scotland, on September 24, 1750, well after the Scottish revolution, he emigrated to Culpeper in 1769 or 1770 from Glasgow.[4] Apparently indentured for a period of (probably five) years to the great Scottish mercantile syndicate of William Cunninghame,[5] Robert was based for a time at the Cunninghame depots in Falmouth and Culpeper before moving permanently to Fredericksburg. Patton's steady move upward within the Cunninghame syndicate to positions of greater responsibility is well documented.[6] In 1773 he was placed in charge of the Cunninghame operation in Culpeper and appears to have been one of its rising stars.

Robert Patton prospered in Virginia as a businessman[7] and subsequently, in October 1792, married well by gaining the hand of Anne Gordon Mercer, whose late father was Brig. Gen. Hugh Mercer, also a Scottish patriot and a legendary Revolutionary War hero.[8] The wedding took place on October 16, 1792, and was duly reported in the next edition of the (Fredericksburg) *Virginia Herald*.[9]

That Robert Patton was well established as a merchant in Fredericksburg by 1774 is clear from the fact that the master of the sloop *Speedwell* assigned a debt of forty-two pounds (more than sixteen hundred dollars—a huge sum at the time) to be repaid to him for wages advanced between July 1774 and May 1776.[10] He is reputed to have "made a competent fortune in business."[11] As a merchant and trader, Patton dealt in highly sought-after goods of the time, advertising for sale in 1792 in the local paper, the *Virginia Herald,* shipments of coal, salt, queensware (a beige-colored earthenware, popular at the time, often made by Wedgwood), eight to ten thousand "good" bricks, claret and other wines from London, Antigua rum, Holland gin, molasses, coffee, cotton, pepper, and muscovado sugar.[12] Until 1805, when the partnership was dissolved by mutual consent and the business run solely by Patton, he was associated with another local merchant named Williamson.[13]

About 1800 Robert Patton used his wealth to build a stately mansion he named "White Plains," on five acres overlooking the Rappahannock River and the falls north of the town.[14] In 1802 Robert was elected a vestryman in St. George's Episcopal Church, but like many other citizens of Fredericksburg, he and his wife soon grew disenchanted with the church and turned to Presbyterianism. The Patton family name appears prominently in references to the organization in 1808 of Fredericksburg's First Presbyterian Church, which was erected on land donated by Robert's wife.[15]

There is no evidence to suggest that Robert Patton was anything more than a conservative, upstanding merchant and benefactor of Fredericksburg, where he apparently spent his entire adult life. There is no record of service in the Revolutionary War, and according to Robert H. Patton, he "declined to serve in the Continental Army."[16] He is known to have returned temporarily to Glasgow, via England, in the summer of 1777 for his employer.[17] One surviving description of him by a Fredericksburg native was that "Mr. Patton was one of the noblest, most upright, most generous men she had ever known," while another noted that "Mercer's daughter was as frail as her husband was majestic."[18]

Robert Patton died in Fredericksburg on November 3, 1828, at the age of seventy-eight, and a brief obituary, which appeared in the *Virginia Herald*, read: "On Monday morning last, ROBERT PATTON, Sen., Esq.—an old and worthy citizen, and for many years a highly respectable merchant of this town."[19] Robert's Fredericksburg epitaph seems to have been that "he was one of the many fine Scotch merchants who have by their splendid integrity and thrift added much lustre to the commercial, social and religious history of old Fredericksburg."[20]

The union of Robert Patton and Anne Mercer produced seven children.[21] Their third child, John Mercer Patton, a physician, lawyer, and politician, was born in 1797. Like his maternal grandfather, John Mercer Patton studied to become a physician and graduated from the University of Pennsylvania in 1818. However, he never practiced the profession that was apparently forced on him by his father. The two are said to have quarreled repeatedly; nevertheless, John Mercer Patton eventually returned to Fredericksburg to fulfill his ambition to obtain a law degree. Patton prospered as a lawyer and served as a Virginia congressman from 1829 to 1838, before settling in Richmond. He was elected to four consecutive terms on the Executive Council of Virginia, and when Gov. Thomas W. Gilmer resigned in 1841, Patton became acting governor of the Commonwealth for a period of thirteen days.

John Mercer Patton was an independent-minded Democrat who was never afraid to speak with honesty and candor. In 1832 a major controversy erupted over a bill in Congress, sponsored by President Andrew Jackson, to recharter the Bank of the United States, during which Patton publicly rebuked the governor of Virginia for attempting to intimidate him into changing his vote.[22]

However, John Mercer Patton's greatest achievement was his pioneering work in helping to revise the Virginia civil and criminal codes. The resulting *Code of Virginia* of 1849 remained in force for the next quarter century. To the end of his life, Patton spoke out against any interference in America's religious or civil affairs by another country.[23] Patton and his wife, Margaret French Williams, produced twelve children, including nine sons, seven of whom were to serve in Confederate gray during the Civil War.[24]

Four of John Mercer Patton's nine sons attended the Virginia Military Institute (VMI) in Lexington, including the first George Smith Patton, who was born in 1833 in Fredericksburg and matriculated at VMI at the age of sixteen. George Patton had been carefully groomed by his parents to qualify for admission to VMI, and he became one of a large class of twenty-four.[25] When he graduated in 1852, George Patton was second in his class and was rated first in tactics, French, mathematics, Latin, and geology and chemistry.[26]

After graduation Patton seemed destined for a career as a teacher, and in this he was assisted by the first superintendent of VMI, the legendary Col. Francis H. Smith.[27] Patton taught in Richmond for two years while also studying for the bar in his father's law office. In November 1855 he continued the Patton tradition of marrying well when he wed Susan Thornton Glassell. Their union produced four children, the eldest of whom was Patton's father, the second George Patton, born on September 30, 1856, and christened George William Patton.[28]

Susan Thornton Glassell was descended from a distinguished family that could trace itself to George Washington's great-grandfather, King Edward I of England, and France's King Philip III. "Even farther in the dim recesses of time were sixteen barons who signed the Magna Carta, all of whom the Pattons believed were their direct ancestors."[29]

The family settled in Charleston, Kanawha County, in what is now West Virginia, where Patton practiced law with considerable success. He soon became a well-liked citizen and acquired the nickname "Frenchy" because of his goatee. "He was arrogant, a smart dresser, and displayed classic chivalry toward ladies, making him a dashing, romantic figure."[30] A photograph of Patton taken around 1860 depicts a dashing young man of aristocratic good looks who conveyed the perfect image of a successful lawyer. He has been described as "graceful and elegant as a speaker [who] was the charm of the social circle, where his genial wit, sparkling humor, ready repartee, and ringing laugh made him ever welcome. He seemed never to forget what he had once learned and could at will produce the choicest sentiments of the poets for the young and gay."[31] Devoutly religious, Patton also encouraged his men to attend chapel and rarely missed an opportunity to pray on his knees before his God.

Despite his youth, Patton was a visionary who saw war clouds on the horizon and was determined to prepare for action. Soon after moving to Charleston, he organized and commanded a company of militia, which became the Kanawha Riflemen, to which he attracted young aristocrats of high standing in the community, like himself. "Its bright uniforms and sharp drill were well-known throughout the area. The Kanawha Riflemen were said to be the best-drilled company in the entire Confederate Army, the result of Patton's superb military training." Despite their ceremonial status as a unit that "could dance as well as, and maybe better, than they could fight," their commander was nevertheless well known as stern and authoritarian.[32]

John Brown's insurrection in 1859 was the spark that galvanized Patton to action, and the Kanawha Riflemen was one of the Virginia militia units that converged on Harper's Ferry in the aftermath. By April 1861 Patton's unit had become Company H of the 22d Virginia Regiment, and Patton himself had become an ardent advocate of secession.[33]

Anticipating war, Patton moved his family from Charleston to the ancestral home, Spring Farm, near Culpeper Court House, shortly before Virginia seceded. His six-year-old son remembered "the coach coming to the door and my indignation at the fact that my toy drum, of which I was very proud, being left on the mantelpiece in the nursery. I cried bitterly as I was put into the coach." At Culpeper the entire Patton clan had gathered, including a number of cousins.[34]

The Patton homestead became a beehive of activity as the family prepared for war en masse. While the women made ponchos and uniforms, the Patton men went about the grim business of preparing themselves for war. "My grandmother gave each of her sons a T.B. [thoroughbred]) horse for himself and a nigrow [sic] body servant, with a second less well bred horse for the nigrow."[35] The matriarch of the family and the widow of John Mercer Patton was Margaret French Williams Patton, a strong-willed, resolute Virginia woman. Many tales about her have been passed down through generations of Pattons, one of which is that when she learned of the wounding of the youngest of her eight surviving sons, "she cried for the first time, but added it was because she had no more boys to send to fight the Yankees." Margaret Patton never accepted the defeat of the Confederacy and Patton's father relates that,

After the war . . . [she] was riding back from church on horseback with a Confederate officer. As they rode along she asked him, "Colonel, did you say 'Amen' when the minister prayed for the 'President of the United States and all others in authority?'" When the colonel said that he had, Mrs. Patton struck him with her whip.[36]

With secession, all the Patton brothers went off to war, except the eldest brother, Robert, who lived with a bulldog in one of the back rooms of the Patton homestead, and was an alcoholic former naval officer. Rarely mentioned in the family history, Robert was found dead in a farmyard near Culpeper in 1876, apparently the victim of drink.[37] The next eldest, John Mercer, became a colonel in command of the 21st Virginia Infantry but served only until mid-August 1862, when complications from a stomach disorder forced his permanent return to civilian life. Isaac Williams Patton (1828–90), who had previously settled in New Orleans, returned there to command a regiment of Louisiana infantry and was captured at Vicksburg.

Waller Tazewell Patton was the sixth son of John Mercer Patton and in 1855 became the third Patton to graduate from VMI. After graduation Tazewell (or "Taz,"as he was known within the family) taught at VMI for two years before becoming a lawyer in Culpeper. Soon after settling in Culpeper he was chosen to command the Culpeper Minutemen, a militia company that had first been raised in 1776 by one of his ancestors. Two of his younger brothers, Hugh Mercer and James French Patton, enlisted as privates, while both were still in their teens. Both later became lieutenants and were wounded, one at Cold Harbor and the other at Bull Run.[38]

Tazewell was severely wounded at Second Bull Run in late August 1862. After a long recuperation, he returned to his regiment in the spring of 1863. Elected to command the 7th Virginia Infantry, in July 1863 he met his destiny at Gettysburg, in the debacle on the third day of the battle that has been immortalized as Pickett's Charge. It has been aptly described as "a magnificent mile-wide spectacle, a picture-book view of war that participants on both sides remembered with awe until their dying moment—which for many came within the next hour."[39]

Of the more than fourteen thousand men who began the attack, less than half would return to the safety of their own lines. Among the first to perish were the officers who led their men into the cauldron of fire. The men of Pickett's division suffered the worst losses, nearly two-thirds, including all three brigade commanders. Of the thirteen regimental commanders, every single one was either killed outright or wounded.

One of those commanders, lying mortally wounded near a stone wall that afternoon, was twenty-nine-year-old Col. Waller Tazewell Patton, whose 7th Virginia had advanced the farthest before it was repulsed. Terribly wounded in the mouth, he was eventually removed from the battlefield and taken to a nearby Union hospital in Gettysburg. He was treated with kindness by a nurse who ministered to him during the final days of his life. Before the battle he had been troubled by a premonition that he would die that day.[40]

The incident in which Tazewell was wounded was witnessed by an enemy artillery officer, Lt. Henry T. Lee, whose battery had been positioned just behind the stone wall. During the attack,

he saw the two officers jump on the wall holding hands and instantly fall. The act so impressed him that when the charge was repulsed he went to look for them. One, a boy of nineteen, was dead, the other had his jaw shattered and was dying from a ghastly wound. The wounded officer motioned to Lee for a pencil and paper and wrote as follows: "As we approached the wall my cousin and regimental adjutant, Captain (name forgotten) pressed to my side and said: 'Its our turn next, Tazewell.' We grasped hands and jumped on the wall. Send this to my mother so that she may know that her son has lived up to and died according to her ideals."[41]

Fortunately a close relative was present to offer consolation, and he noted that Tazewell's only method of communication was to write, painfully, on a slate board. Foremost in his mind were his God, his mother, and his country. Shortly before his death, in a poignant letter to his beloved mother, he reaffirmed his devotion to God and asked for her prayers. The young colonel ended by scribbling on his slate board: "Tell my mother that I am about to die in a foreign land; but I cherish the same intense affection for her as ever."[42] When Waller Tazewell Patton died, on July 23, 1863, he was the first—but not the last—member of his family to perish in the service of the Confederacy.

George Smith Patton fought his first battle in nearby western Virginia, at a place called Scary Creek, in July 1861. He narrowly escaped death when he was thrown from his horse by the impact of a spent minié ball containing an ounce of pointed lead one-half inch wide.[43] The bone in Patton's upper right arm was shattered, and he was taken prisoner when he could not be moved and his comrades reluctantly left him behind. At a Union hospital the doctors told him the arm required amputation, but Patton adamantly refused. He had somehow been permitted to retain his pistol and made it uncompromisingly clear that he would shoot anyone who attempted to try. The arm did not heal properly, and Patton never regained full use of it. His young son later remembered watching his father use a knitting needle to remove a piece of bone from the wound.[44] Patton was eventually paroled and permitted to return to his family. When he recovered sufficiently from his wound he returned to the 22d Virginia as its commander, with the rank of lieutenant colonel.[45]

Patton continued to gain experience under his former VMI professor, Stonewall Jackson, and once again barely escaped death in May 1862, during the battle of Giles Court House. According to his son:

Being struck in the belly with a minié ball he thought the wound was fatal and so dismissed the surgeon, telling him to spend his time on those he could save. Shortly after this General Wharton [Patton's division commander] rode by and, having heard of the wound, asked . . . [him] how he was. Col. Patton replied that the wound was fatal and that he was writing a letter to his wife but that he did not feel like a dead man. Gen. Wharton dismounted and asked if he could examine the wound. He stuck his unwashed finger in it and exclaimed, "What is this," as his finger hit something hard. He then fished around and pulled out a ten dollar gold piece. The bullet [had struck] this and driven it into his flesh, and glanced off.[46]

While he had escaped death, Patton nevertheless suffered from blood poisoning and returned to the family home, now in Richmond, to recuperate. Patton's regiment had been operating in western Virginia, and in 1863 he again moved his family, this time from Richmond to Lewisburg, a small town near White Sulphur Springs. Patton's regiment was in the thick of the fighting during the Battle of Droop Mountain, in November 1863, where the Confederates were defeated by the Union cavalry of Maj. Gen. William Averell, a onetime friend of Patton. His son vividly recollected the grim aftermath:

> I remember seeing them retreating through Lewisburg. . . . Father had sent an ambulance with a pair of mules to the house and told my mother to take it and follow the army. . . . Late in the night my father came by with the last of the rear guard and stopped to tell us goodbye and give my mother a letter for General Averill [sic] asking him to see that we were not bothered.[47]

This was not Patton's last encounter with Averell. One morning he was breakfasting at a house when an orderly suddenly yelled: "The Yankees are coming!" Patton and his staff barely had time to escape Averell's cavalry by jumping out the back window while the lady of the house rushed to hide his saber under a mattress.

Susan Thornton Patton helped to care for both Confederate and Union wounded who were brought to the hotel in White Sulphur Springs. Young George followed his mother around with a bucket and sponge, and recalled that the smell was so awful that she fainted and had to be carried from the room.[48]

The greatest triumph of the Patton family during the Civil War took place at the Battle of New Market. When a Union force threatened Staunton, Confederate units were hastily assembled at New Market under the command of Maj. Gen. John C. Breckinridge. Vastly outnumbered, the Confederates were so desperate for reinforcements that the 247 young men of the VMI Corps of Cadets (the eldest of whom was only seventeen) were rushed from nearby Lexington. Federal and Confederate forces met on the Valley Turnpike at New Market on May 15, 1864. Here the South won a famous victory that forever immortalized the VMI Corps of Cadets. Four Pattons and their kin participated in the Battle of New Market. During the battle George Patton's 22d Virginia came to the rescue of his close friend and first cousin, Col. George Hugh Smith, whose 62d Virginia was in dire straits after being trapped in a ravine and badly decimated by Union canister. As one historian notes, "In retrospect, it emerges as a Patton military picnic."[49]

New Market established once and for all George Patton's credentials as an outstanding and innovative commander. When, during the battle, Federal cavalry attempted to penetrate the battalion on his left flank, Patton quickly improvised a hasty defense that shattered the charge. His brigade commander was often absent, and when he was present proved an ineffectual leader. Patton filled the void so often that most considered him the real brigade commander.

By the spring of 1864 the Pattons had moved again, this time to John Mercer Patton Jr.'s house, the Meadows, in Albemarle County. Susan Thornton Patton received a letter from her husband that he had joined Lt. Gen. Jubal Early's army and would soon be on a train passing through on the railway line at the bottom of the garden. As his son recalls, "He got off and stayed with us for several hours . . . then the last train composed of flat cars loaded with artillery stopped for him. I remember seeing a soldier on a car give him a hand to get aboard and as the train moved out he was leaning against a gun and waved us goodbye. I never saw him again."[50]

In July 1864 Patton's 22d Virginia was one of the units leading Early's fifteen-thousand-man Confederate Army of the Valley, which had advanced to the outskirts of Washington, within five miles of the White House. Although hastily emplaced Union reinforcements prevented Early's raiders from capturing Washington, he had become a very dangerous thorn in the Union hide. The seriousness of the threat posed by this large Rebel army was not lost on Gen. Ulysses S. Grant, who sent Gen. Philip Sheridan's Army of the Shenandoah to deal with Early "to the death," and to plunder the Shenandoah Valley. The two sides met on September 19, 1864, in the Third Battle of Winchester.

Outnumbered by twelve thousand, Early's army could not withstand a whirlwind Union attack on the Confederate left flank. Patton was then in command of his own "Patton's Brigade," and, although he had survived three earlier wounds, this time his luck ran out. The circumstances surrounding his death remain vague. Jubal Early's postwar memoir says only that "Colonel G.W. Patton [sic], commanding a brigade, was mortally wounded, and fell into the hands of the enemy."[51] It is known that Patton's brigade was attempting to defend the left flank that was eventually crushed by Sheridan's cavalry, which captured two thousand Confederate soldiers, among them the mortally wounded George Smith Patton.

Patton was one of several senior Confederate officers killed at Winchester, and Early later referred to him as "a gallant and efficient officer" whose loss "was deeply felt."[52]

A VMI memorial tribute published some years later indicates that Patton lingered for several days after being wounded. The house to which he had been taken was that of his cousin, Mary Williams.

In this interval the hope of recovery was inspired and sustained by the opinion of his surgeon that his wound though serious, was not mortal. A part of the last day of his life he was alone in his chamber. Cheerful, even buoyant, no fears were felt that a few brief hours would close his earthly course. A later visit to his chamber disclosed a great change, and warned his friends that death had sealed him for his own. A few words, unintelligible to the kind ones who ministered to him, escaped his lips, and his voice was hushed forever.[53]

Colonel Patton's death, on September 25, 1864, was apparently from a combination of fever and gangrene. He was barely thirty-one years old.[54]

When she learned of Patton's death from a Union newspaper some four days later, his distraught widow traveled to Winchester but arrived too late to attend his burial service. In the 1870s, when George S. Patton II was a cadet at VMI, George and Waller Tazewell Patton were interred together in the same grave in Winchester. According to his son's recollection:

> He [George Patton] was dug up and taken to the railroad station. When the train arrived with the body of Tazewell the two coffins were placed on a gun limber . . . and covered with a Confederate flag. . . . Many old soldiers were at the station, including a band with their instruments all draped in black. No noise was made and in utter silence by moonlight the cortege moved to the cemetery, Papa in [his VMI cadet] uniform, walking behind the limber.[55]

Those who had come to honor the Patton brothers did so in Confederate uniform and at great risk of arrest and incarceration. They formed an honor guard around the grave site, and as the coffins were lowered into the twin graves, they struck one another, causing the corpse of Waller Tazewell to break free. Patton later recounted that his uncle looked little different in death than he had in life.

It was later said of the second Patton to die on the field of battle that,

> preferring the profession of law to any other business and the sanctities of the home and family to all other pleasures, he had nevertheless, peculiar aptitude for a soldier's duty and a soldier's life. He enforced discipline without exciting dislike, and commanded his men without diminishing self-respect. . . . Colonel Patton appreciated the soldiers of our army as volunteers fighting in a sacred cause, and commanded their admiration while he won their love . . . "we ne'er shall look upon his like again." "[56]

Few Virginia families could claim to have contributed more to their cause or shed more of their blood than the Pattons and the Mercers. All were men of honor and principle who did their duty as they saw it in defense of their way of life and for their beloved Virginia. The list of their accomplishments was as long as it was distinguished. In all, some sixteen members of the Patton family and their kin fought for the Confederacy, and three of them died in its service.

*   *   *

After the death of George Patton, his family suffered destitution from the effects of a war that had finally overwhelmed the Confederacy and devastated Virginia, whose economy was in ruins, its currency worthless and its people desperate for the bare necessities of existence. The Pattons spent the winter of 1864–65 at Goochland, near Richmond, like so many others, "in great want of food and clothing." The widow Patton's hardship included responsibility for her blind father, the care of her four young children, and soon after Lee's surrender, the additional burden of caring for her brother, the gallant Capt. William T. Glassell, a former Confederate naval officer who had arrived suffering from tuberculosis contracted during his imprisonment in a Union POW camp. If Uncle John Mercer Patton had not sent them a steer, young George Patton believed that they would have starved.[57] Small wonder that, as a Northerner observed not long after Lee surrendered at Appomattox, across the land there was

> "no sign of human industry, save here and there a sickly, half-cultivated corn field. The country for the most part consisted of fenceless fields abandoned to weeds, stump lots and undergrowth.". . . Some 20,000 to 30,000 Virginia soldiers were dead. Thousands of others hobbled along city streets and country roads with an arm or leg missing . . . two generations of Virginians were maimed beyond description. . . . The future held no promise.[58]

Shortly afterward young William Patton showed up, driving an old Confederate ambulance pulled by two horses. With his help the Pattons packed their belongings and moved to a colonial mansion near Orange, Virginia, that had once belonged to the brother of President Madison. Other members of the Patton family also moved in, including the family matriarch, Margaret French Williams Patton, Uncle Hugh Mercer Patton, and a brother, George Glassell. The family farmed a small patch of land in the nearby river bottom. The task of Colonel Patton's young son was to walk behind the plough, dropping corn seed into the furrows and covering it up with his bare feet.

Finally, in the autumn of 1866, a letter arrived from Susan Glassell Patton's elder brother Andrew, who had settled in Southern California before the Civil War. It contained six hundred dollars and a request that she bring her family west. Although it was a princely sum for the time, it was not enough for eight people. To raise the extra money required for their forthcoming journey, Susan Patton sold everything the family owned "except her husband's sword, saddle, gold watch and his Bible. Willie sold what he had, and old Mr. Glassell had already given his worldly goods to the Confederate cause. There was nothing left for them [in Virginia] in the ruins of their politics and their plantations—and their way of life."[59]

CHAPTER 2

Don Benito Wilson

Ma has determined to let me take my father's full name.
I only hope I may be worthy of it.

—GEORGE SMITH PATTON II

In November 1866 the impoverished family of Col. George S. Patton left its beloved Virginia for the long and difficult journey to California. The Pattons sailed on the SS *Arizona* to Panama and then traveled overland across the Isthmus to the Pacific coast, where another ship took them to San Francisco. At San Francisco the Pattons boarded yet another vessel for the final leg of their journey to Los Angeles.

The trip was an ordeal, and Susan Patton nearly died of a fever en route.[1] It was also marked by a confrontation between Mrs. Patton and several other Southern passengers and two former Union generals. The wife of one attempted to kiss ten-year-old George Patton, who disgustedly refused, saying "he would never kiss a Yankee."[2] The Pattons arrived in San Francisco on December 19, 1866, to a warm reception from the family of Uncle Isaac Williams. En route to Los Angeles there were such rough seas that their vessel was eleven hours overdue and feared lost.

At San Pedro the twenty-six-year-old widow was reunited with her brother Andrew, whom she had not seen in nearly fourteen years. They were made welcome in Glassell's large home in Los Angeles.[3] A lawyer with seven children, Glassell was considered well-to-do but had lost heavily in financial speculations during the war. Now, with so many extra mouths to feed, he was having trouble making ends meet.[4]

Although relieved to be freed of the oppressive burden of life in post-war Virginia, Susan Glassell Patton nevertheless found Southern California, with its bare hills, wide-open spaces, strange customs, and mostly Spanish and Indian population in stark contrast to the lush green hills and valleys of her native Virginia. "I can never feel so much at home anywhere as I did in old Virginia," she nostalgically wrote her sister in 1867. "The great need of the country is a proper society, and if a number of Virginians would come to give it tone it would be a most desirable location. . . . The State of the South is gloomy beyond degree. . . . May God soon make his face to shine on our beloved land."[5]

She also felt her family's presence to be an intolerable imposition on her brother, and before long was able to move to a small adobe house and support her family by opening her own private school for girls in a rented schoolroom nearby. Her first class consisted of only eleven students, who each paid $3.00 per month, but eventually Susan Patton gained a reputation as a superb, no-nonsense teacher who could handle even the most ill-behaved pupil.[6]

Nevertheless, her loneliness without her dashing husband was evident in those first years in California. In 1867, on the third anniversary of his death, she wrote to her sister:

> This is the saddest time of all the year. . . . I feel like a stricken, broken down woman when I remember the fell blow that came upon me just three years ago, blotting out the light of life for me, and sending me and mine forth in the world homeless wanderers. . . . May [God] strengthen me in fighting life's battles with an unfaltering heart.[7]

Despite Susan's growing school (it now had nearly fifty pupils), the Pattons continued to live on the thin edge of privation, and she begged her sister to persuade John Mercer Patton to sell her late husband's law books for whatever cash they would bring. However, it was not until March 1868, when she received five hundred dollars from the sale of the former Patton homestead in Kanawha, that the financial constraints were eased somewhat. However, Susan suffered from a throat infection that eventually forced her, in December 1868, to stop teaching for a time, and in the 1870s her health continued to decline.[8]

For young George Patton II, the move to California was a blessing. Among the Glassell clan he had newfound friends and the opportunity to be a young boy, get into mischief, play ball, and hunt. He was painfully thin from his ordeal in Virginia, and this may have contributed to his contacting typhoid, which ravaged him with fever for a month. However, in an environment in which food and milk were plentiful, the young man throve and soon recovered.[9] His mother was a strict disciplinarian, and as the eldest child, George was expected to help support their fatherless family, which he did by scrubbing and cleaning a public school on weekdays and a church on

Sunday. He later told his son that the experience left him with a deep aversion to poverty and a fixed purpose in life to better himself so that his wife and family would never experience the same hardships he had endured.[10] "He swore that if he ever became affluent, he would always keep a Virginia-baked ham on the family sideboard as a symbol of overcoming those bitter years. And he did."[11]

Nevertheless, he deeply missed Virginia and lamented not knowing if he would ever return. "I hope that some day it will please God to bring me back to Virginia." In 1868, with the encouragement and approval of his mother, George William Patton, changed his middle name from William to Smith, thus becoming the second George Smith Patton. To his cousin Mercer he wrote, "Ma has determined to let me take my father's full name. I only hope I may be worthy of it."[12]

In January 1869 Col. George S. Patton's friend and VMI classmate Col. George Hugh Smith, late of the CSA, settled in San Francisco after a postwar odyssey that had first taken him to Mexico. Smith soon relocated to Los Angeles, where he joined Andrew Glassell's law firm. The comforting presence of George Smith brought the first happiness Susan Patton had known since her husband's death.

George Hugh Smith was born in Philadelphia in 1834, the son of an Episcopalian minister. Although his family later settled in Virginia, Smith's place of birth always rankled him because, "My playmates took advantage of it to call me [a] 'Yankee.'"[13] A man of brilliant intellect, foremost in his life was his beloved Virginia and the Confederacy. Smith's warrior nature emerged whenever either was slighted. He matriculated at VMI in 1850 at age sixteen and graduated in 1853, sixth in a class of twenty-five. Like his Patton kin, Smith taught for four years, earned a law degree, and settled in western Virginia to practice law. However, Smith was far too adventuresome to settle down at such a young age and soon left for what is now Washington State. After only a few weeks practicing law, he seized the opportunity to participate in a government expedition to survey for the construction of a road from the Oregon Territory to Fort Benton on the Missouri River. When the news of John Brown's raid finally filtered to the West, Smith immediately returned east and on Virginia's secession was commissioned a first lieutenant in the Provisional Army of Virginia. His VMI training served him well, and Smith soon rose to the rank of captain and was elected to the command of a rifle company.

Like his first cousin George Patton, Smith was captured early in 1861 and was promptly paroled to return to his home. In the spring of 1862 he was freed from his pledge and ordered back to active duty. Smith's outstanding leadership qualities while serving with Stonewall Jackson in the

Shenandoah Valley campaign of 1862 soon propelled him to the rank of colonel and the command of two regiments, first the 62d Virginia and later the 25th Virginia. He was badly wounded twice and by war's end had earned a deserved reputation for bravery.[14]

Lee's surrender at Appomattox was a particularly bitter pill for Smith, and to the end of his life he refused to swear allegiance to the United States, as required of all former Confederate officers. Like other generations of Pattons, Smith experienced visions that were so clear that they could not be dismissed as mere fantasy. On one occasion Smith found himself in a large ballroom filled with officers attired in the gray uniforms of the Confederacy. Each came up to Smith and silently shook his hand. Suddenly Smith became aware that he knew each man, for all had served under his command and all had died in the Civil War.[15]

In 1866 Smith fled Virginia for Mexico, where he spent a year surveying and a second year attempting to grow cotton. The venture was a disaster, and in 1868 Smith emigrated again, this time to California. After a year in San Francisco, Smith was drawn to Los Angeles. His decision to settle there was no coincidence, for George Smith had been in love with Susan Glassell Patton for many years and had never married because of his love for her. Most nights Smith could be found courting the widow Patton in her parlor.

In 1870 his hopes were realized with their marriage. Smith adopted Susan Glassell Patton's children, who adored him, particularly young George Patton. A kind and gentle man, Smith was very protective of the family of his late first cousin and became the father George had never really known. In addition to raising Susan's four children, their union produced two children, Annie Ophelia, who was later to marry Hancock Banning, and Eltinge, who died of tetanus in childhood. To the Pattons Smith was a "'Knight in Shining Armor' . . . to all who ever knew him."[16]

The marriage lasted only thirteen years. In 1883 Susan Glassell Patton Smith died of the cancer that for some time had been consuming her body. His granddaughter writes that Smith brought up the Patton children as his own. "He was a noble and generous man, and he raised George Smith Patton II (and later his son, George Smith Patton, Jr.) on stories of the heroism of George Patton, the real father and grandfather whom they never knew."[17]

George Smith Patton II was an excellent student and a budding orator. At the grammar school of Dr. T. H. Rose he was third on the honor roll at age twelve and determined to do better. "I do not feel very proud of my standing . . . I hope next term to do better. You know Ma is one who never feels satisfied unless her children stand highest."[18]

After the Civil War the Commonwealth of Virginia reserved several appointments each year to the sons of VMI graduates killed in action. Determined to follow in his father's footsteps, George Patton left California about 1871 and returned to Virginia to the Meadows, the home of his uncle, John Mercer Patton. His tutoring paid off, and he was admitted to VMI in 1874. When he reported to the Institute tailor to have his uniform fitted, Patton was informed that his measurements were precisely the same as those of his father. Twenty-seven years later, in 1903, the third George S. Patton to attend VMI repeated the same scene with the same tailor.[19]

During his three years at VMI, Patton was an exceptional student. He became, in succession, third corporal, cadet sergeant major and in his final year in 1876–77, the commander of A Company, a position which left him the first-ranking cadet officer in his class. In a photograph of Patton in his cadet uniform, taken in 1876, he bears an uncanny resemblance to his father. Once, while he was out riding in his cadet uniform, a Confederate general stopped and asked him if he was not Col. George S. Patton.[20] Young Patton's striking good looks were a near-carbon copy of his esteemed father's. Years later one of his VMI classmates wrote that he was the "handsomest man that Gawd ever made."[21]

Patton was haunted by his poverty, and on one occasion, while escorting a young lady to an outdoor band concert, he gallantly spread his handkerchief on the ground for her to sit on. It was full of holes, and although nothing was said, the next day cadet Patton received a new handkerchief from the lady.[22]

Patton's greatest triumph at VMI came in 1876, when, as first captain, he led the corps of cadets in parade at the national centennial celebrations in Philadelphia, an event notable for the fact that it was the first time a military formation from a Southern state had ever been permitted to appear in a Northern city since the end of the Civil War.[23]

After graduation in 1877 Patton remained at VMI for a year as an assistant professor of French and tactics. Patton was a Virginian at heart, but he found himself drawn back to Los Angeles and his family. His mother was ill with cancer, and although he would have preferred to practice law in New York in order to be near a cousin, Maggie French, with whom he had fallen in love, Patton knew his duty and his future were in California. He returned to Los Angeles in 1878 to study law under the tutelage of his uncle and stepfather, and passed the California bar exam in 1880.

Los Angeles was a rapidly growing region, and it was there that Patton—one of the most eligible young bachelors in the city—laid the groundwork for his future in both the law and politics. There is evidence, however, that he would have preferred a more adventuresome life: He signed up for a commission in the mercenary army of the Egyptian pasha, commanded by a British officer named Hicks. When his mother's cancer worsened, Patton was forced to decline, which ultimately saved his life when most of Hicks's force was later killed.

Patton's considerable oratorical skills had first become evident at the age of thirteen and in the intervening years had been honed to the point at which the *Los Angeles Star* took note of his "great talent for oratory."[24] The explosion of commerce and trade on the new frontier of California brought with it the evils of monopoly, corruption, and power politics on a grand scale. The railroads had not only opened the West to the rest of the country but elevated the owners of the Southern Pacific Railroad to unprecedented heights of power as the greatest landowners of the West.

Into this minefield stepped George Smith Patton. In 1884 he brought his oratorical skills to the podium to campaign for Grover Cleveland, the first Democrat to win the presidency since secession. The experience brought Patton prominence in Los Angeles. The year 1884 was also memorable for the end of his bachelor days. On December 11 he married Miss Ruth Wilson, the daughter of Benjamin Wilson, one of southern California's founding fathers. The *Los Angeles Herald* proclaimed that the wedding "marks an era in the social history of our county."[25]

Patton's new father-in-law, Benjamin Davis Wilson, was a pioneer, beaver trapper and trader, grizzly bear hunter, Indian fighter, justice of the peace, farmer, rancher, politician, horticulturalist, vintner, real estate entrepreneur, and one of the great landholders in Southern California. Born in Nashville, Tennessee, in 1811, the son of a Tennessee pioneer who had fought in the Revolutionary War as a major, Wilson was orphaned at age eight, and by the time he was fifteen had become a fur trapper and trader at Yazoo City, Mississippi. He traded mainly with local Choctaw and Chickasaw Indians before bad health led him to seek his fortune in the West. In 1833 he joined the Rocky Mountain Fur Company in Missouri and crossed the plains to Santa Fe, in the New Mexico Territory. From the autumn of 1833 to the spring of 1835, Wilson participated in an expedition that trapped beaver in Apache country near the Gila River.

Wilson was much too independent to remain in the employ of others, and in the spring of 1835 he returned to Santa Fe to form his own fur-trapping enterprise. He spent much of his life in the company of Indians and got along especially well with the Apache and their chief, Juan Jose. However, intrigue by rival Mexican trappers led to the murder of Juan Jose and Apache retribution against the Americans. Wilson and two others were captured and were to be put to death when the new chief connived to let Wilson escape. Nearly naked and without food, Wilson managed to elude angry Apache warriors, who pursued him until he reached the sanctuary of Santa Fe, nearly one hundred miles away.[26]

In this lawless place there were seemingly endless problems with Mexican intriguers and Indian wars. After his wilderness ordeal, Wilson operated his own trading company, and in 1837, when Santa Fe was ravaged by riots and the governor and other Americans were butchered, he was again saved from death by an Indian chief. Wilson remained in Santa Fe as a successful trader until 1841, when a resurgence of hatred against gringos made it unwise to remain. In September 1841 Wilson helped to organize a wagon train of restless settlers like himself and headed for Southern California, thus becoming one of the first pioneers ever to have crossed the American continent.[27]

Wilson was one of the unique breed of hardy trappers and traders known as Mountain Men, who were among the first to break through the mountains and cross the desert—previously barriers to exploration of the Far West. California historian Robert Glass Cleland has observed that the Mountain Men were "the pioneers of all Far Western frontiersmen, the trail blazers for subsequent explorers, the pathfinders of the course of empire to the western sea."[28] They counted among their number Kit Carson, Jedediah Smith, and Benjamin D. Wilson.

Southern California was to have been merely a stopover en route to his intended destination of China, but after failing several times to locate a China-bound ship, Wilson elected to remain in Southern California.[29] In 1843, within two years of his arrival, he purchased for one thousand dollars the Rancho Jurupa, a three-thousand-acre tract of land on the site of the present-day city of Riverside, and the following year married Ramona Yorba, the winsome sixteen-year-old daughter of one of the great Mexican landowners. Don Bernardo Yorba owned the Rancho Santa Ana, most of which is now Orange County, and a number of other haciendas, one of which later became the town of Yorba Linda, the birthplace of a future president of the United States, Richard M. Nixon. Their marriage was cut tragically short after only five years when Ramona died in 1849. The only daughter of this marriage later married her father's secretary, J. De Barth Shorb. As a wedding present Wilson deeded to the newlyweds several thousand acres of land, part of which is now the city of San Marino.[30]

Wilson was a tenacious warrior who feared neither man nor beast. In 1844 he was seriously mauled while tracking a ferocious grizzly bear that had slain one of his cows. After recovering from his wounds, he resumed the hunt and cornered the bear, managing only to wound it. The grizzly was finally killed the following day, after the unrelenting Wilson once again barely escaped death during another savage encounter between himself and one of the most dangerous animals on earth.[31]

On another occasion he killed a notorious Indian bandit and murderer named Joaquín. During the fight Wilson was struck by a poisoned arrow and only survived because his Indian servant sucked the venom from his wound.[32] Once he headed an expedition to track down and punish two rene-gade mission Indians who had slaughtered some settlers. In his typically direct manner, Wilson captured and held the Indian chief hostage and in return for his release demanded the heads of the two renegades. True to his word, Wilson brought the bloody heads in wicker baskets as affirmation to the provincial governor that he had satisfactorily carried out his assign-ment.[33] In 1845 an expedition headed by Wilson discovered a large colony of bears in the San Bernadino Mountains, and he named the place Big Bear after he and his companions had lassoed and killed twenty-two grizzlies.[34]

A man of frightful temper who did not suffer fools at all, Wilson disci-plined himself in his later years never to carry a gun in case he might do something rash. To Don Benito a man's word was his bond. Once, when he went to collect a five-thousand-dollar loan from a member of the Spreckels family (later famed as a sugar dynasty in California), the man insisted on a receipt for the money. Wilson returned home to fetch his pistol and con-fronted the foolish young man, calmly inquiring if he had ever observed death at firsthand. Told no, Don Benito rejoined menacingly, "Well, just wait about one minute!" Fortunately this practical lesson in integrity pre-vailed over death when the debt was hastily repaid in cash—without need of a receipt.[35]

In 1853 Wilson married a second time—Alabama-born Margaret Short Hereford, a widow who had been his housekeeper before her husband's death. The marriage produced two daughters, Ruth and Annie. Ruth Here-ford Wilson was the mother of George S. Patton Jr. Annie, a lifelong spin-ster, became Patton's beloved aunt Nannie.

The pre-gold-rush 1840s was a period of immense turbulence in the Califor-nia Territory during the fight for statehood. Don Benito, as Wilson came to be known, found himself caught in the middle of the struggle between the American settlers, led by Commodore Robert P. Stockton and John C. Fré-mont, who were attempting to wrest California from the control of Mexico and the Californians. His status as one of the major landholders in Southern California inevitably embroiled Wilson in the Mexican War of 1846–48. Threatened with arrest if he failed to support the native Californians, Wilson nevertheless endorsed the rebel American cause, was given a temporary commission as a captain by Commodore Robert Stockton, and barely escaped a Mexican firing squad before being briefly jailed in 1846 after the Battle of Chino.[36] In the aftermath of the revolution, which led to statehood for California in 1850, it was said of Wilson's role that he had "aided, per-haps more than any other man in southern California, in restoring peace and good feeling between the Americans and natives."[37]

During the years after California acquired statehood, Wilson increased both his landholdings and his influence. In 1850 he was elected chief clerk

of Los Angeles County, and a year later the first elected mayor of the city of
Los Angeles. In 1852 he was appointed an Indian agent by President Millard
Fillmore, a position that included the duties now accomplished by a justice
of the peace. Beginning in 1855 Wilson served two terms in the new California Senate. During his later years, Wilson helped to found the city of
Pasadena, became a city councilman, and later gained fame as one of the
best-known horticulturists in California. His landholdings, totaling more
than fourteen thousand acres, were situated all over Los Angeles County,
including four thousand acres on what is now the campus of the University
of California at Los Angeles (UCLA), which became a sheep and cattle
ranch. But it was his farm, orchards, and vineyards, called Lake Vineyard, in
the southwestern corner of what is now Pasadena, the "City of Roses," that
became Wilson's pride and joy.[38]

Wilson had turned land once described as "where a respectable jackrabbit wouldn't be seen" into a veritable Garden of Eden. In 1874 the Wilson
estate at Lake Vineyard consisted of thirteen hundred acres containing
102,000 vines, 1,600 orange trees, 1,200 lemon trees, 250 lime trees, and
several hundred olive and walnut trees. The adjoining Shorb estate consisted
of five hundred acres and 129,000 vines and some 2,500 fruit trees. The
combined wine harvest for 1873 was 75,000 gallons, and 5,000 gallons of
brandy. In 1874 the orangeries produced nearly 600,000 oranges, and it was
estimated that nearly this number remained unharvested on the trees. The
lemon crop amounted to nearly 75,000.[39] By 1883 his San Gabriel Wine
Company was capable of turning out 1.5 million gallons and was the largest
in the world, eclipsing even the great wineries of France.[40]

Visitors to the Wilson homestead were treated like royalty:

B. D. Wilson's Lake Vineyard was, in the early days, a combination of
the Hacienda days in California and the traditions of the South with the
wide acres, the lavish hospitality and the devotion to the old standards. It
was the center of social life for Americans in the Los Angeles area, and
most prominent visitors to the area were entertained by the Wilsons.[41]

One of Wilson's business associates was another pioneer, Phineas Banning (1831–1885), who had emigrated to California from Wilmington,
Delaware, in 1850. Banning was a colorful character, a onetime state senator, a general in the California National Guard, and a great showman who
had developed the harbor of nearby Wilmington (which he named for his
hometown) and thus put Los Angeles on the map as a port of call for commerce and trade and an alternative to the great port of San Francisco. Dub-
bing himself "Admiral of the Port," Phineas Banning

would meet all the boats that arrived at Wilmington, and when there were
passengers of distinction, he would send a vaquero on a horse to notify

Don Benito Wilson of their arrival. A coach would be standing at the door of the Banning house in Wilmington, with a vaquero hanging onto the bridle of each blindfolded bronco in harness. When the passengers were in the coach, the blindfolds would be whipped off by the vaqueros, who would leap aside, and the horses would be off at a run across the unbroken country to Lake Vineyard, often with Phineas Banning as whip.[42]

Don Benito died in March 1878 and never lived to celebrate the birth of his daughter Ruth's first child, at his beloved Lake Vineyard. Perhaps it was prophetic that the child was born on the evening of November 11, 1885, a day that in the coming century would be remembered each year as a memorial to the First World War. The newborn child was a son, and in honor of his father and grandfather he was christened George Smith Patton Jr. His elated father instantly nicknamed him "the Boy."[43]

PART II

Childhood
(1885–1903)

He is a well bred and a well brought up fellow . . .
[who] has developed a great taste and aptitude for the
study of military history and sciences. . . . If blood
counts for anything, he certainly comes of fighting
stock.

—JUDGE HENRY T. LEE (FORMER UNION OFFICER)

CHAPTER 3

"The Boy"

Patton's Childhood in Los Angeles

I must be the happiest boy in the world.

—GEORGE SMITH PATTON JR.

The newborn Patton's first months of life were difficult, and his nurse, a devout Irish Catholic named Mary Scally, was so deeply concerned that the unhealthy child might even die of the croup that she secretly had him baptized to prevent his immortal soul from entering heaven unchristened.[1] Fortunately he recovered, and the threat soon passed. His younger sister, born in 1887 and christened Anne, was always known to family and friends as Nita. Her brother was affectionately dubbed Georgie, a name no one except his family ever dared call him as an adult.

George S. Patton's youth and that of his younger sister were memorably happy ones. He called his parents Mama and Papa, and his first remembrances were of his father and of horses:

I had a rubber doll called Billy and . . . I recall his riding up to the house and putting me on the saddle. Later at Lake Vineyard Papa . . . and I went up to the corral . . . to look at some Shetland ponies one of which I chose and called Peach Blossom. . . . I remember very vividly playing with Nita . . . and seeing Papa come up on a chestnut mare belonging to Aunt Nannie. I wanted to go with him but he told me to play. As he rode on up the canyon Mary Scally, our nurse, said "you ought to be proud to be the son of such a handsome western millionaire." When I asked her what a millionaire was she said a farmer.

Young Georgie's memories were largely of life at Lake Vineyard, even though the family had moved to Los Angeles in 1886 so that his father could better tend to his business affairs. After Don Benito's death his vast business empire was taken over by his son-in-law, De Barth Shorb. Unfortunately, under Shorb's misdirection the Wilson enterprises narrowly averted bankruptcy. Shortly after his marriage in 1884, Patton was elected district attorney of Los Angeles, but his periodic ill health forced his resignation after barely a year in office. Patton later served two more terms as district attorney, and in 1894 he ran unsuccessfully as the Democratic candidate from California's Sixth Congressional District.

In 1885 Mr. Patton gave up not only a promising legal career but his post as district attorney to run the affairs of the Wilson estate, which was failing badly due to a combination of Shorb's mismanagement and the deadly effects of disease and drought on the vineyards and fruit trees. De Barth Shorb had become a multimillionaire from his Wilson inheritance and enjoyed a lifestyle that would have been the envy of a potentate. But his talent for profligate spending seems far to have exceeded his business acumen, and it left his family destitute after his death. Patton's father took over the ailing business, and although he fought to sustain the Wilson fortune, which Shorb had foolishly mortgaged, by 1899 the battle was lost and the Farmers and Merchants Bank finally foreclosed on the mortgage of the San Gabriel Winery. Patton was retained as the manager until 1903, when the bank sold the property to Henry Huntington, who used the land to build what became his famous library and gallery. He and Huntington were friends, and Patton now managed the Huntington Land and Improvement Company.[2] Although the years at the winery were difficult ones for George S. Patton II, his son recalled them with fondness as a time when he and his sister would ride carefree on horseback across the vast estate, and other occasions when they would ride in their father's single-seat wagon, which was drawn by a fine horse.

Patton's reminiscences of his childhood were written when he was nearly forty-two, but the happy memories were undimmed by time.

> When I went to tell Papa and Mama goodnight I used to kiss Papa many times and Mama only once; this was childish and thoughtless. Papa used to tell me that he was worried to death trying to keep out of the poor house. I told him I was worried too and when he asked why I said it was from fear that he would sell Broken, a Standard Bred stallion he had.

Despite his ups and downs in both politics and business, Mr. Patton kept his vow to escape poverty, and during his lifetime accumulated sufficient wealth that his family would never want for anything.

The Pattons lived for the most part at Lake Vineyard, in the original

Spanish-style home built by Don Benito in the 1850s. Its walls were of adobe brick, and it had one of the first slate roofs in California. The front was dominated by a large porch and steep stairs leading to the driveway. Numerous skunks lived under the house and used a hole next to the water pipes in an inside bathroom so regularly that the Pattons conveniently provided holes in the bathroom door to give the kittens easy access to the rest of the house. On the first visit of Georgie's future wife, Beatrice, to the Pattons', she was cautioned not to pet any black-and-white kitties she might observe in the halls.[3]

In 1902 Patton wrote a school essay in which he described the splendor of the original house at Lake Vineyard and the two wings that were later added from wood cut from a forest that had once covered the hillside behind. In front was a wide green lawn, and to its sides were an apple orchard and a garden filled with roses and vegetables:

> Almost half the first floor of the main building is kitchen, and almost half the kitchen is fireplace. After looking at the fireplace you would not wonder that the forests of the country are being exhausted. The parlor, which is one of the wings, is large and now seldom used. The carpet is thick and brilliant in color, hiding a fine floor of oak. There are many huge pictures on the walls. . . . In the attic are canes, swords, trunks, saddles, guns, beds, chairs, clothes, papers, books and rats.[4]

If the house and its kitchen were exceptionally large, it was for good reason. At one time or another Lake Vineyard was home to a vast number of Pattons and their kin. In addition to the family of George S. Patton II, Lake Vineyard domiciled Don Benito's widow, Margaret, until her death in 1898; Patton's spinster aunt Nannie; Miss Susan Patton, his father's unmarried sister; Mary Scally, the Patton nurse and nanny *extraordinaire*; and, for a number of years, his father's widowed sister, Nellie Patton Brown, and her six children. They also had a great many guests, and whenever there was a new minister he would stay at Lake Vineyard until he got settled.

When Ruth Wilson married George Patton II, her sister, Annie, was devastated. She had fallen deeply in love with the dashing young Patton even before he had attended VMI, yet she had lost the love of her life to her sister. When "the Boy" was born in 1885, he became the focus of her life. As a child George was adored by his loving parents and thoroughly spoiled by his aunt Nannie (the name by which she was known in the family), with whom he established a lifelong bond that he seemed loath ever to admit publicly. Young George and Nita enjoyed a comfortable youth that seemed free of pressures. The picture that emerges is of a close-knit family, except for the bizarre presence of Nannie, who completely dominated the Patton household. Her nephew could do no wrong, and in her quiet but controlling way she forbade any sort of criticism of Georgie. Nita she all but ignored. Of the two sisters, Nannie was prettier and thought to be more intelligent, and she became a surrogate mother to her beloved Georgie and a surrogate wife, whose obsessive love and idealization of the man whose hand she had lost verged on the maniacal. The fact that her sister was the real wife and mother was of absolutely no consequence. Nannie tenaciously shared everything in their marriage except George Patton's bed. The day they were about to board a train for their honeymoon in New Orleans, Nannie arrived unexpectedly at the train station, baggage in hand, with every intention of accompanying the newlyweds. It is not recorded how the Pattons talked her out of becoming their chaperone, but she eventually left them to what would become one of the infrequent times when they were free of her presence. That night the diary Nannie had kept virtually her entire life was left blank and never resumed.[5]

As Ruth Ellen Totten writes:

[Georgie] was the be-all and end-all of her life, the child she never had. Aunt Nannie's story is a sad one . . . in the 1870s, both the Wilson girls . . . fell in love with . . . young George Patton . . . [who] chose Ruth Wilson. . . . To make it sadder, as a matter of course for those days, they all lived together in old Lake Vineyard. . . . When the Pattons finally built their own home in 1900, Aunt Nannie . . . moved right in with them. So, all her life she lived in the house with the only man she had ever loved, and she lived vicariously in his son Georgie. . . . I have often wondered how Georgie Patton grew up to be the man he was with two strong-minded women (his mother and his aunt) baby-sitting him until he married Ma. Fortunately, Ma, a member of a large family, loved them all and took their constant presence for granted.[6]

The strong-willed Nannie Wilson sorely tried the patience of Georgie's parents, not only by her refusal to permit him to be punished, but on the rare occasions when she did not get her way and perceived that her sister or brother-in-law was thwarting her resolve to pamper her beloved "son," by intimidation. Then Nannie would suddenly become "ill" or "faint," take to her bed, and demand a doctor. When Georgie turned sixteen she gave him a gold ring shaped like a coiled snake, with eyes made from ruby chips, and, according to his nephew, he wore the ring on the third finger of his left hand for the remainder of his life.[7]

Nannie's rule of the Patton household verged on the tyrannical, and although she gave the false impression of being fragile and shy, her wiles served her well. Whatever their private feelings about Nannie may have been, the Pattons tolerated her as one of their own, even though she was of sufficient financial means to afford her own full-time driver and carriage. It was she who made the real decisions regarding the raising of Georgie, and their effects were to prove profound and far reaching.[8]

Aunt Nannie stubbornly refused to travel alone. Consequently, whenever she left Southern California, her sister or another family member was obliged to accompany her. During the years Georgie was at West Point, Nannie spent considerable time in Highland Falls, the small village outside the gates, in order to be near him. Even Patton eventually wondered why it was necessary for his mother to undertake the long and arduous transcontinental train journey merely to escort Nannie to or from California.[9] In this, as in so many other situations, Nannie nearly always got her way.

Georgie's mischievous childhood pranks were rarely punished, merely smiled at indulgently and forgotten. However, on one occasion, when he practiced what must have been his first experiment in armored warfare, there was hell to pay from his mother. He and several cousins ran amok with a farm wagon they had converted into a make-believe armored vehicle, which Patton later said was like one employed by John the Blind, the king of Bohemia.* In any event, the wagon, with its youthful warriors ready to do battle from behind the security of the tops of old wine barrels, careened out of control down the hill behind the house, and wreaked a dreadful toll upon the "enemy"—which turned out to be the Pattons' flock of turkeys, many of which were killed or maimed in what was Patton's first recorded combat action.

Ruth Wilson Patton was not amused; this time Georgie had gone too far and would have to be punished—Aunt Nannie be damned. Papa waffled, saying boys will be boys, and Nannie predictably insisted that her beloved Georgie be spared the rod. Undeterred, Ruth Patton spanked her son, but before doing so she "summoned the doctor and turned down Nannie's bed in anticipation of the physical collapse her sister invariably suffered whenever Georgie was punished."[10] It was one of the few occasions when Ruth Patton got the better of her feisty sister in matters of Georgie's upbringing.

Both during his childhood and later, when he was an adult, mealtimes in the Patton household were not merely repasts but memorable events, particularly dinner. Georgie and Nita loathed mush, which was a breakfast staple, and years later Patton wrote: "I can hear [my father] say every morning:

---

*John, king of Bohemia (1296–1346), is regarded as one of history's archetypal heroic warrior-kings of chivalry. John's life was spent in a crusading, obsessive pursuit of glory on the field of battle. He fought many battles and, in a campaign in Lithuania in 1337, was blinded. Nevertheless, even blindness did not deter him from fighting (on the side of the French) during the Battle of Crécy in 1346. John died a heroic death suicidally charging British archers, who cut him down in a hail of arrows. It is fitting that Patton would pick for a model a warrior whose life and death personified the heroes of his youth.

'Georgie, eat your mush.' He used salt instead of sugar on his and I did the same which was probably the reason I did not like it."

As Patton later recalled:

We always had white wine for midday dinner on Sunday and Nita and I were always given a little. When I was between eight and ten, I was sitting in the office with Papa. As usual before dinner he poured himself a glass of whiskey from a cupboard where he kept liquor in decanters. He then poured me a drink and said, "Son, this is not locked and you can get a drink whenever you want one." I never took one without him and seldom then. I think he did this because the two Wilson boys who had been very strictly raised both became drunkards after they came of age. Papa's idea was to make me think drinking common place and so set less store by it. Also he used to impress upon me that an ambitious man could not afford to drink.

Both Nannie and Patton's father would probably fit the modern-day definition of an alcoholic. Papa's consumption was quiet excess; Nannie's usually resulted in the spilling of the embarrassing and often harsh words of a spinster embittered by the denial of her life's dream. However, with the exception of the 1920s and 1930s, when he suffered from severe midlife crisis and black depression, spawned by the belief that he would die with his destiny unfulfilled, and drank with almost suicidal excess, Patton's consumption of alcohol was generally judicious and heedful of Papa's advice.[11]

Like Patton's surrogate mother, Papa was a tolerant parent, who, perhaps remembering his own wretched childhood, was remarkably easygoing on both of his children. In return, they adored their father. Georgie remembered: "In 1892 Mama had to take Aunt Nannie east for an operation. While they were gone Papa read Nita and myself the *Iliad* and then the *Odyssey* aloud." The exposure to Homer "led young Patton to perceive human struggles against the implacable destiny imposed by the gods, men who worked out their fates in heroic or mean fashion and received their just and deserved rewards."[12]

Papa Patton's granddaughter wrote many years later that the man they loved as "Bamps" had longed for a military career after graduating from VMI, a place obsessed with chivalrous memories of the Civil War and

haunted by the dead graduates, cadets, professors, officers young and old. Bamps was a romantic, and remembered his own father who died of his wounds as a hero. Dead heroes are so much more memorable than living ones, and his stepfather, that Prince of men, Colonel George Hugh Smith . . . never ceased his tales of the promise and the prowess of the dead George Patton. But Bamps could not follow his star into the service. He had his family to help, and a sense of responsibility toward his mother and his younger brother and sisters, so he went back to California . . . [and] did the right thing as he saw the right. So, all his dreams of glory and the sword were handed on to Georgie, and he lived his dream life vicariously through the young man he either referred to as the boy or my hero son.

Aunt Nannie decided that young Georgie was "delicate," and she, too, began to read aloud to him from classics that included *Pilgrim's Progress*, Plutarch's *Lives*, *The March of Xenophon*, *Alexander the Great*, and "anything and everything about Napoleon." In fact it was Nannie who deeply influenced his early education, and in Georgie she had a willing participant who listened raptly and absorbed deeply.

"He had Siegfried and Beowulf for his heroes; and Robert E. Lee and Stonewall Jackson. The stories of the Civil War he heard right from the men who had fought it . . . Georgie lived and played in the company of heroes, dead and alive." But one text above all others towered above the rest in young Patton's education: the Bible. And it was Nannie who wielded this most classic text of all like a cudgel, hammering its themes into his head as he sat next to her for three or four hours, day in and day out. Patton's grandson would later observe that Nannie taught Georgie that the Bible was the most noble tale of man's survival in the face of the oppression of both gods and evil men, and that Jesus emerged from the New Testament as the quintessential example of human courage: "Nannie's religious reading made her nephew's head swirl with alluring myths and legends that coalesced like a planet from a gaseous cloud into a worldview all his own.[13]

Nannie was never certain if her efforts were having any effect upon her beloved nephew, and even came to the sad conclusion that he might actually be dim-witted.[14] Until he started school at the age of eleven, he was unable to write, and his constant twirling of his long golden ringlets of hair as he sat quietly next to her was unnerving. Aunt Nannie probably never knew just how well her Georgie had absorbed her teachings. As an adult, and throughout his army career, the Bible would become his most fundamental guide, and God a source of solace, as well as the basis of practically everything he did. The invocation of "God" dominated his speeches and, either directly or by implication, his writings. And in his many times of trouble he would turn to God, not as a religious fanatic would, but in the almost serene belief that a higher power would help him to endure. Church, prayer, God, the Bible: All became foundations on which his life was built. His ability to quote the Bible at length for almost any occasion was the fruit of Aunt Nannie's labors of love. As a child who got on his knees (another lifelong habit) to recite his nightly prayers for his mother, Patton thought two small portraits on the nearby wall were of Jesus and God. Only later did he learn that the two bearded men were Stonewall Jackson and the man who was revered as the "God" of the Confederacy, Robert E. Lee.[15] Patton's nephew would later write, "The Bible was his companion and the church his refuge."[16]

As for living heroes, Georgie came to know a frequent guest of the Pattons, the infamous Col. John Singleton Mosby, the Confederate guerrilla leader. A prewar lawyer who had learned the law in prison after being expelled from the University of Virginia and imprisoned for killing another student, Mosby had migrated to California to work for the Southern Pacific Railroad.[17] Mosby delighted the impressionable young Georgie Patton with tales of the Civil War, all of which the boy absorbed like a sponge.

Thus, by the time he entered his teens, Patton had not only learned firsthand of the heroics of the men of the Confederacy but had been indoctrinated in the classics: Shakespeare, Homer, Sir Walter Scott, and Kipling; books about heroes, kings, conquerors, villains, gods, explorers, and adventurers and, above all, to the great soldiers of history: Caesar, Belisarius, Scipio, Hannibal, Napoleon, Joan of Arc, and Joachim Murat. All engendered in Patton a sense that he was in this life a reincarnation of soldiers of the past, that he had served in bygone armies and fought in the famous battles of history.

Where Nannie left off, Papa would take over with vivid and lavish—and, in all probability, exaggerated—tales, in which exemplars of the Civil War like Stonewall Jackson and the Patton colonels of VMI, who died a warrior-hero's death, became symbols to cherish and to emulate. Hugh Mercer, the Revolutionary War icon, also received his due. And finally, there was a living exemplar, his beloved step-grandfather, Col. George Hugh Smith, whose quiet counsel and tales of the Civil War and its battles instilled a profound sense of destiny in the young man. Smith's presence during Patton's youth may well have been the most important influence on his decision to become a soldier and to serve the Patton name. As an adult, Patton would accumulate an immense library of books, and his professional reading became one of the cornerstones of his later success in the U.S. Army, and on foreign battlefields.

With Georgie cuddling most nights in the warm sanctuary of Papa's arms, soaking up the Patton family legends that were dispensed with almost evangelistic fervor, it is hardly surprising that by the age of perhaps seven he was hopelessly seduced into the conviction that his life and destiny lay in perpetuating the Patton family name and its even more valorous achievements. Old Virginia and the glory of its cause; great battles such as Bull Run and Gettysburg; the beauty and majesty of cavalrymen clothed in Confederate gray, who charged courageously into a hail of enemy fire, sabers flashing; the belief that dying for such a cause was honorable—all these images and more were indelibly carved into the psyche of the young man nightly in the living room of the Patton home. Patton was at too impressionable an age not to be deeply affected by the kaleidoscope that swirled through his mind's eye, of highly romanticized warriors of old, of Pattons in glorious battle, of Hugh Mercer heroically dying for a new country.

It was not until Saint-Mihiel, the Meuse-Argonne, and North Africa that Patton began to understand that real war is hell. But as a youth, how could could he have thought war and warriors anything but romantic when they were recounted before a roaring fire where there was no death or carnage, only dreams? Part and parcel of Patton's immersion in ancestor worship was the notion—which grew into an absolute article of faith—that a dishonorable death meant a life wasted. Death was something that had to be earned through fulfillment in life. Anything less was not to have lived at all. A dishonorable death was disgraceful. The brave and noble characters who peopled Patton's mind—both real and mythological—were the models on which he would seek to pattern his life.

It never occurred to him, either as a youth or as an adult, that the Pattons did not hold a monopoly on courage or chivalry. Throughout his life George Patton completely rejected his Wilson heritage, even though he not only more closely resembled the Wilson side of the family but, as his daughter has pointed out, owed his traits as a warrior as much (or more) to his grandfather Wilson as he did to his Patton forefathers.[18] Biographer Martin Blumenson has written:

> He denied his debt to Benjamin Davis Wilson . . . who seemed to have transmitted to him his physical hardihood, mental perseverance, personal charisma, and driving willpower. Patton never wished to hear of his resemblance to Wilson, for Wilson was a self-made man and quite unlike the Patton aristocrats.[19]

He seemed to go out of his way to heap scorn on anyone connected with the Wilsons. In his 1927 family memoir he idolized his Patton ancestors to the total exclusion of those on his mother's side. Wilson rated but a single line as "my mother's father." When he wrote that their valorous deeds were performed "by men of my blood and . . . [it is] they who inspired me . . . [and] it is my sincere desire that any of my blood who read these lines will be similarly inspired and ever be true to the heroic traditions of their race," he meant the Pattons, not the Wilsons. Don Benito's strength of character, honesty, and steely resolution were traits passed to the grandson who never acknowledged his debt.

And, as Robert Patton points out, young Georgie (indeed, the entire Patton clan) was so imbued with the deeds of his forefathers that to accept his Wilson heritage would have been synonymous with treason and a rejection of their "desperate faith in their former glories [which] fostered a sense of themselves as natural-born noblemen. The more uncertain their circumstances, the more the Pattons waved their tattered flag of precious bloodlines."[20] Aunt Nannie apparently made no effort to champion her own father, and the result was that young George S. Patton was blinded to any heritage but that of his Virginia forefathers and their now-idolized, almost mythological achievements in defense of their name and birthright. Papa was as relentlessly effective in fueling the flames of the Pattons as Aunt Nannie was with the Bible. Georgie seems to have absorbed every detail. Throughout the remainder of his life he would repeatedly demonstrate that the thousands of hours of learning and brainwashing had not been in vain. In the process he conveniently managed to overlook, and indeed even scorn, Don Benito's great wealth, which kept the Pattons solvent and ensured that he would never want for anything. The silver spoon of Wilson wealth and good living was something that blessed Patton his entire life.[21]

When Don Benito died, in March 1878, at the age of sixty-six, his cortege consisted of seventy-five carriages and the attendees were described by the *Los Angeles Star* as "the largest assemblage of people ever congregated on a like occasion in Southern California."[22]

Patton's near paranoia about the Wilsons stemmed in part from a streak of racial prejudice that can be traced to his childhood. He felt that Don Benito had somehow sullied the Patton name by marrying a Mexican and converting to Catholicism. The notion that Catholic-Mexican *Wilson* blood ran through his veins was not only appalling, it was unacceptable and hence disowned in toto. However, Patton's chief villain and the focus of his anti-Wilson antipathy was De Barth Shorb, who had squandered his wealth and left his family impoverished after his death. Patton would later scornfully write: "he was either a fool or a crook and in a little while more of his management Mama and Aunt Nannie would have been beggars." When his father took over management of the Wilson estate, Patton proudly quoted a friend of his father, who had proclaimed that "God sent that young man to save the widow and orphan."

It was predictable that young Patton's interests would be horses, guns, and swords—and soldiers. At an early age George and Nita would amuse each other by playing soldier. "Nita and I had two blue coats with brass buttons. . . . Nita used to say she was a major while I claimed to be a private, which I thought was superior. When Papa would drive away in the morning he would salute us and ask how the private and major were."

From time to time "Nita and I would wear each other's clothes to dinner. One night when I was wearing her clothes Papa began talking about Lee and I got all excited and when he told me that since I was dressed as a girl I should not get so bloodthirsty. I cried." Patton got the message: It was fun but it was not manly. Papa bought Georgie a .22-caliber rifle, and he soon pleased everyone with his marksmanship by knocking an orange off a fence. It was the start of a life-long love affair with guns.

> One Christmas I got a steam train and another time a stationary engine, both of which ran for me until I was old enough to do it myself. I also had a soldier suit with a black woolly hat and pompom . . . there was a sword with belt and a rifle with a bolt action . . . I used to . . . [carry] it and two empty .22 shells with which I religiously loaded it to shoot at lions and robbers. Sometimes he would take his father's sword and I my toy one; he would kneel down and we would fight. We used to do the same thing with some boxing gloves he gave me.

His grandson notes how this interplay between the tangible and the fanciful may have helped symbolically to form a link between Georgie and his Confederate grandfather, who was mortally wounded wielding his saber at Winchester and on whose saddle Patton learned to ride.[23]

Riding and swordsmanship thus became second nature to Patton at an early age. After several years of playing with make-believe swords, one of which bore the inscription "Lt. Gen. G. S. Patton," his father fashioned a sword and scabbard that his son proudly wore. Patton's first sword also brought him some minor grief. "A store in Los Angeles was having a sale of 1870 French sword bayonets and I asked for one. I remember lying on the grass when we got home admiring it. Later I attacked the cactus with it and got well stuck."

Later Patton would make his own swords, and "once while riding with Papa accoutured with my sword and mounted on Peach Blossom I decided to charge and the saddle turned [over]. . . . I had a bad fall but was not really hurt. The saddle I then rode was the McClellan saddle on which my grandfather had been killed. On the pommel was a stain which I thought was his blood." This was the first of many falls from a horse that Patton experienced, taking a fearsome toll on his body.

Horses and horsemanship became second nature to the young Patton. When he was about ten, after years of pretend riding on a saddle in the Patton household, his father gave him an English saddle and a double bridle, along with two horses of his own, Galahad and Marmion. "I always saddled my own horse and groomed it to some extent. . . . I had a dog named Polvo about this time and he slept by Marmion. I remember once going to the stable at night when I was supposed to be studying and lying by Polvo, looking at Marmion, and thinking that I must be the happiest boy in the world. I was probably right."

Mr. Patton never economized as far as his family was concerned. Whatever the time or cost, nothing was too good that Papa did not find some means of lavishing it on Georgie: "I think at this time we lived on about three hundred dollars a month but we never wanted [for] a thing we did not get." Mr. Patton introduced his son to hunting and fishing at an early age. He was presented with a sixteen-gauge hammer shotgun at the age of ten, and two years later his father had to borrow from the bank in order to purchase an expensive twelve-gauge Le Favre that his father told him would last him the rest of his life. In 1927 he would write: "It is as good as ever." Nevertheless, Mr. Patton would struggle throughout much of his life to maintain a grand lifestyle for his family. His poor health often left him unable to practice law, and despite their large landholdings, the Pattons were more frequently than not cash poor. From time to time Mr. Patton would sell small parcels of land to raise money, usually at bargain-basement prices.[24]

The Patton family owned a cottage on fashionable Catalina Island, a playground for the affluent of Southern California, and it was there that Georgie learned to hunt, sail, and swim. Patton later remembered with considerable warmth that "these various hunting and fishing trips which he accompanied me on were a great proof of his affection for me, as he hated both hunting and sea fishing, but he went even when I was a grown man."

Despite his success as a fisherman, Patton had a long-standing distaste for fish. His first catch was a catfish, which became his breakfast the following morning, and "even then I disliked fish." Off Catalina he once caught a forty-five-pound yellowtail that was nearly as tall as its adolescent captor. As the proud Patton posed for Aunt Nannie's camera, his father instructed him to move behind the fish so that it would seem even larger.

He became an avid hunter, and after one of his first hunting trips, in which everyone but him had killed a single goat, Patton boasted to an assembled crowd that he had killed five. Afterward his father gently reminded him: "Son, it would have been more like a sportsman not to have mentioned the extra goats."

The relationship between father and son was such that minor transgressions were willingly admitted and just as readily forgiven by his indulgent father. On one occasion when Papa left Catalina for the mainland, he made his son promise not to swim beyond the end of the nearby wharf. "Once I dived and went so deep that I came up outside the limit. I could not rest until he got back so I could confess to him. Of course, he was not displeased. He was very proud of my swimming." When Patton was quite small he observed some men attempting to uproot a dead orange tree that refused to budge, and suggested they employ a horse to pull the tackle. His proud father noted to the family at dinner: "If the boy had not been there we could not have moved the tree." The same day his father had caught him throwing potato bugs into a brush fire and told him not to be cruel. Patton never forgot the admonition.

*   *   *

His parents were not the only adults to spoil young Georgie. For his six-teenth birthday his wealthy cousin Hancock Banning provided a dazzling entertainment in Georgie's honor. Banning had a fireworks-laden ancient steamer towed into the channel between the mainland and Catalina Island, where it was set afire. The result was one of the most spectacular fireworks displays ever seen in Southern California.[25]

From the time of Robert Patton, who had served as a vestryman in St. George's Episcopal Church in Fredericksburg, the Pattons had been staunch Episcopalian churchmen.

> They went to church every Sunday; paid their tithe; and either built the first Protestant churches in that part of California—as Don Benito did—or were on the building committees. The fine men of the church were in and out of their houses along with the Confederate veterans and the retired Mexican bandits. One of their dearest friends was darling Bishop Johnson . . . the Bishop of California.

One night not long after they were married, Lt. George S. Patton's bride saw the bishop and her husband

> walking up and down the lawn in deep conversation. Bishop Johnson, who was five feet tall, had his arm draped around Georgie's waist, and Georgie had his arm draped across the Bishop's shoulders. They were so rapt in conversation that Ma sneaked up to listen, and she heard Bishop Johnson say; "Yes, indeed, Georgie, I quite agree with you. You would have made a wonderful Bishop if you had had the Call." One of Georgie's childhood dreads was that he would "get the Call" . . . every night when he was a little boy, saying his prayers on his knees, he would pray that Jesus would not call him, because he wanted to be a soldier.

Patton had no formal schooling outside the family home until he was eleven. Until then he was tutored at Lake Vineyard, both to inculcate the youngster with a classical education and to spare him the scorn the family believed would have been heaped on him by his classmates. The exact date has never been recorded, but at an early age his parents realized that their son had a learning disability that prevented him from reading. Although they were unable to give the problem a name, young George S. Patton unques-tionably suffered from dyslexia.

Dyslexia, a disorder that is currently believed to affect as many as forty mil-lion Americans and 20 percent of the world's population,[26] afflicted many prominent people, ranging from Leonardo da Vinci to Albert Einstein,

Thomas A. Edison, Woodrow Wilson, Nelson Rockefeller, actors Cher and Tom Cruise, and major-league baseball star Andy Van Slyke. First diagnosed in 1896 by two British physicians, dyslexia was virtually unknown in the United States until the 1920s, and although great strides have since been made in what has become an enormous and important field of study, dyslexia still remains a complex and frustrating problem for both its sufferers and those who treat it. As one noted clinical psychiatrist has written: "Dyslexia has remained a scientific enigma, defying most attempts at medical understanding, diagnosis, prediction, treatment and prevention."[27]

The usual definition of dyslexia as "a learning disorder characterized by reading, writing and spelling reversals" is highly misleading.[28] This description barely scratches the surface of a complex disorder that, in addition to creating difficulties with reading and writing, includes an inability to concentrate, sharp mood swings, hyperactivity, obsessiveness, impulsiveness, compulsiveness, and feelings of inferiority and stupidity.[29] A tendency to boast is also very common among dyslexics. Moreover, dyslexia often affects spelling, grammatical, and mathematical abilities. Like Patton, many dyslexics are eventually able to overcome the reading and writing aspects of the disorder and lead productive lives. What is often overlooked by those who perceive dyslexia as merely a reading problem is the lifelong traumatic emotional effect it has on its victims. Until recently those who studied dyslexia never seem to have grasped fully that its reading aspects were merely the tip of the iceberg. Renowned dyslexia expert Dr. Harold C. Levinson writes:

Most dyslexics feel dumb, despite being smart. . . . Most often a dyslexic's compulsion to succeed is motivated by an overwhelming desire to prove to himself and others that he is not really as stupid as he feels. Accordingly, the dyslexic disorder frequently serves as a potent stimulus to achieve, reflecting a desperate attempt to reverse the humiliating feelings of inferiority that are invariably present.

However, as Levinson notes:

Unfortunately, tangible success and peer recognition, even adulation, do not neutralize or eliminate a dyslexic's feeling dumb. All too often, accomplished, even famous, dyslexics merely feel that they have succeeded in fooling everyone around them, and that others are not truly aware of how inept they really are. They attribute their successes to chance, a lucky break, a fluke of nature.[30]

These words are as much a description of George S. Patton as of the average dyslexic. As will be seen as the story of his life unfolds, virtually every symptom of dyslexia described above applied to Patton. Throughout his life Patton would deprecatingly refer to himself as having been slow, lazy, and stupid as a student. During his plebe year at West Point he would write to his future wife, Beatrice Banning Ayer: "I am either very lazy or very stupid or both for it is beastly hard for me to learn and as a natural result I hate to study."[31]

Dyslexics experience a need to justify to themselves and those who have no grasp of the nature of their problem that they are as good or better than ordinary people. In many, it often becomes a near-obsessive driving force in which the dyslexic seems to be saying to him- or herself, but secretly hoping that others will notice: "I'm smart too! I'm just as good as you!" This feeling of inferiority, the need for the dyslexic to prove not only to himself but to others that he or she is a person of intelligence and ability is the key not only to understanding the source of Patton's drive to succeed, but of the authoritarian, macho, warrior personality he deliberately created for himself. As Patton grew to manhood it was the dyslexia that fueled the fires of the ancestor-hero worship lighted by Papa and Aunt Nannie. To prove himself worthy of his Patton heritage would not only drive Patton, it would obsess him so powerfully, so single-mindedly, so outlandishly, that few could comprehend how anyone's life could be dominated by such demons.

As an adult, Patton would lampoon his inability to spell, once advising his nephew: "Any idiot can spell a word the same way time after time. But it calls for imagination and is much more distinguished to be able to spell it several different ways as I do."[32]

Patton was eleven before he began learning to read and write. In September 1897, when he was nearly twelve, Patton's father "finally rebelled against the 'hands that rock the cradle' ruling the boy, and sent him to the local grammar school," the Classical School for Boys, located on South Euclid Avenue in nearby Pasadena. Nannie's usual neurotic ploy of feigning illness failed this time. His first day was a poignant one in Patton's life: "We drove up in the old surrey and . . . Papa turned to me and said very sadly: 'Son, henceforth our paths diverge forever.' I have never forgotten that but though we lived more and more apart our hearts and minds never separated."

The principal of the Classical School for Boys was Dr. Stephen Cutter Clark, a noted Latin scholar and historian who was assisted by his brother, Mr. G. M. Clark. At what was essentially a small high school, Patton joined twenty-five other children of Southern California gentry and spent the next six years undergoing his first formal schooling. A diligent student, Patton nevertheless faltered because of his dyslexia. Algebra, geometry, and arithmetic were among the subjects taught and virtually all proved a struggle. In 1902, for example, Patton's report cards reflect examination and recitation grades in the fifties and sixties and occasionally in the low seventies. His deportment was exemplary, as were his marks in ancient and modern history, which consistently were in the high nineties. Other subjects included French, English, Latin, German, geography, reading, spelling, drawing, and declamation.[33] As his parents had feared, his greatest problems stemmed from mistakes in reading, writing, punctuation, and spelling. Patton was occasionally taunted by his fellow students for his spelling at the blackboard and his glaring mistakes when called upon to read orally to the class. Where Patton excelled was in his amazing capacity to memorize and quote verbatim and at length from the Bible or any book he had been exposed to during those many years of home schooling. Although he was not born with it, Patton developed a photographic mind that compensated dramatically for his dyslexia.

Young Patton benefited from the curriculum offered by the Clark brothers and, although unable to overcome his dyslexia, he nevertheless persevered. By the time he entered VMI in 1903, he had acquired the rudiments of a first-class education. Hidden behind the negative image created by the dyslexia lay an incredibly vast storehouse of knowledge of biblical and military subjects. Later, as an adult, Patton would make light of his academic ordeal by joking, "I had trouble with my a's and b's—and what the hell is that other letter?"[34]

Not only were logic and patriotism essential, but history was a centerpiece that helped young Patton beyond measure. Many of his essays dealt with the ancient warriors about whom he had been tutored by his father and aunt at Lake Vineyard. In retrospect it seems equally clear that the decision to expose him to the Clarks was one of the foundations of his future success, for it was there that Patton was able to express his thoughts and ideas about the men whom he would spend his life studying and emulating.

It was one thing to listen to tales, but quite another to channel that interest into ideas and concepts. History, as taught by the Clarks, was in the classical tradition of good versus evil, of ancient men who fought for their civilizations, usually for noble purposes but occasionally for the same base reasons that human beings have waged war from the dawn of history. With their emphasis on patriotism and self-sacrifice, it was hardly surprising that—having been exposed to the classics and the Bible—Patton emerged from their tutelage fully indoctrinated in the fundamental belief that a person's character invariably determined whether his or her life would be a success or a failure.[35]

Little of Patton's early schoolwork seems to have survived, but there are numerous examples of his essays from the period. Among the places he wrote about was the island of Sicily, which he would one day come to know only too well. In a 1902 essay he excoriated one of the Athenian commanders of the Sicilian expedition as "unfitted" for his post.

Among the lessons learned at the Clarks' school was that the character of those he studied had a great deal to do with their achievements. He described Themistocles as "eggotistical and had a right to be," and, "Cleon un like Periclese, was a man of violent passions. A great baster [bastard]." Of the ancients, Patton took as his hero not Alexander the Great ("in a fit of drunkeness [he] took his own life and his empire fell to pieces"), Hannibal, Caesar, or Constantine, but Epaminondas, a notable fourth-century B.C. Theban general: "Epaminondas was with out a doubt the best and one of the greatest Greeks who ever lived, with out ambition, with great genius, great goodness, and great patriotism; he was for the age in which he lived almost a perfect man."

An extract from a December 1901 essay on siege warfare, depicted below exactly as written, illustrates the extent of Patton's battle with dyslexia:

The atack on the castle was begun by a heavy discharge of arrows: which kept the defendors under cuver, but from loop holes the Normans answercd the fire with their crosboes and killed many outlaws. Then the out laws lead by the black knight attacked and took Barbacon after a fierce fight in which the black knight and Fron de Berf met in a hand to hand combat; the Normon being mortaly wounded. . . . After a brief rest the atack was renued . . . and most of their number slain or taken prisoner, the Tempeler and a fiew of his men cut their way out, leaving the castle in the hands of the out-laws.[36]

Patton's powers of imagination and attention to detail are evident, although his written work would have earned failing marks from any but the most indulgent teachers. The Clarks seem to have gone out of their way not only to encourage the young man but to avoid the criticism and ridicule that other less tolerant or wise teachers might have expressed.

A brief letter Patton wrote to the *San Francisco Call* in 1902 illustrates how his dyslexia affected even the simplest task: "Not being at all satisfied with your paper, during the last year, and being very much discussed with many of the articles, I would be very much obriged if you would discontinue sending it to me. Very truly yours, George S. Patton, Jr."[37]

At one point the Pattons considered sending their son to a private boarding school, but decided against it, undoubtedly in the belief he was not yet ready to live away from them.[38] The Classical School for Boys was Patton's only formal preparation for the difficult college years at VMI and the United States Military Academy that lay ahead.

During the summer of 1902 Patton was a tall, gangling youth with a shock of corn-blond hair. His feet had grown so fast that earlier that year they had literally split open his shoes, "and they have given no signs of stopping." He had also revealed a sense of humor. To a distant relative named Katherine Ayer, he had written that, "All the horses, dogs, pigs, cats, chickens, children, and relatives are well and growing bigger every day. Write again soon, the dogs and children want to hear from you."[39]

When Patton was not quite seventeen years of age, the last thing on his mind was an attraction to members of the opposite sex. For weeks his parents had talked about little else than the forthcoming visit of their distant relatives from Boston, Massachusetts, the Ayers.* Georgie gave hardly a passing thought to their mention of the young woman who would be accompanying them.

As befitted the wealthy of that era, the Boston family traveled in the comfort of their own railroad cars. At the station to greet them were the Patton and Banning clans, including Georgie who was present because he was about the same age as the young woman and "everyone thought it would be nice for her to have a young man to introduce her to his friends (not that anyone with such a huge family really needed friends, of course)."

The visitors were greeted effusively by their hosts, with a great exchange of hugs and kisses. The young woman had auburn hair worn in such a long braid that it hung nearly to the hem of her skirt. Although only sixteen, she had already entertained three proposals of marriage. Cradled in her arm was her constant companion, a doll she had named Marguerite:

> The girl and the doll were dressed alike in smart suit dresses of crash linen, suitable to the California climate. But Georgie took one look at the much touted "Belle of Boston" and backed away in disgust. In California the "Young Ladies" of sixteen had their hair up and most certainly did *not* play with dolls. If anyone thought he was going to escort this little kid around, they were making a big mistake. His friends would laugh him off the face of the earth!

The young woman's name was Beatrice Banning Ayer, and she would forever alter the life of George S. Patton Jr.

---

*Frederick Ayer's wife, Ellen Banning Ayer, was a distant relative of the husband of Papa Patton's half-sister, Anne Ophelia Smith. (See RHP.)

CHAPTER 4

The Belle of Boston
Beatrice Banning Ayer

By the end of . . . the summer, Ma and Georgie were in
love for the rest of their lives.

—RUTH ELLEN PATTON TOTTEN

The Ayers were a wealthy Massachusetts family whose patriarch was
eighty-year-old Frederick Ayer, a multimillionaire businessman. Born in
1822, Frederick Ayer, like Benjamin Davis Wilson, had started out dirt poor
and had acquired a massive fortune, first from the sale of patent medicines
and later from banking, real estate, printing, and the textile industry, which
flourished in New England in the nineteenth century.[1] Ayer's first wife, Cornelia, died of cancer in 1878, and he remained a widower until 1883 when
he met a vivacious actress named Ellen Barrows Banning who earned her
living in St. Paul, Minnesota, giving readings of Shakespeare, mostly at teas
and ladies' socials. She had been invited to a dinner party to meet an older
man who was a widower and considered a great catch. Instead she chose to
attend a performance of *Hamlet* by Edwin Booth. As she left the theater,
"the handsomest man that I had ever seen walked up to me and said, 'Are
you Miss Ellen Banning?'" Complimenting her on her good sense for skipping a dinner party in favor of a performance by America's greatest living
actor, "I decided that [you] should have both treats, and I have come for you
in my carriage with a chaperone to take you to the party."

Thirty-year-old Ellen Barrows Banning was swept off her feet by the
gallant and handsome Ayer. As she later told her daughter, "Once I looked

into those piercing blue eyes, if he had said 'Ellen Banning, will you follow me to the world's end?' I would have gone right with him just as I was. . . . He was like a knight in armor."

Ellen Barrows Banning made Frederick Ayer forget all about his beloved first wife. Outgoing, dramatic, and bubbling with life, she was, as her granddaughter Ruth Ellen Totten writes, "the dessert course of his life and that he lived so long as he did was because she made him laugh so much." Thus, at the age of sixty-two Frederick Ayer took as his bride (who always addressed him as "Sir Frederick") a descendant of Danish stock whose ancestors had emigrated first to Holland and then to England before the first Banning came to America in the early 1700s. One of the Dutch Bannings appears in Rembrandt's *The Night Watch* (whose actual title is *The Militia Company of Captain Frans Banning Cocq*), and John, the son of the first Banning, settled in Delaware and earned his fortune. A staunch patriot, John Banning cast Delaware's vote for George Washington for president, and when veterans of the Revolutionary War were paid off in near-worthless scrip, Banning personally indemnified the Delaware veterans with hard currency from his personal fortune. A branch of the Banning family eventually ended up in California, in the person of Phineas Banning, who had the good sense to become a business partner of Benjamin Davis Wilson.

The Ayer family was of English descent and traced their lineage to Wiltshire, from which Ayer's grandfather, also named Frederick, emigrated first to Salisbury, and later to Haverhill, Massachusetts, in 1635. Eventually the Ayers settled in Connecticut. As Beatrice Ayer's daughter would later write, "Ma was very proud of the Ayers, and after listening to the interminable tales of the ancestor-worshipping Pattons, would strike forth with, 'But don't forget—every man in *my* family could write Esquire after his name!'"

Frederick Ayer's rise from rags to riches was a saga worthy of Horatio Alger and the Great American Dream. An orphan, at the age of eleven he slept under the counter of the store in which he was employed as a clerk. At thirty-five he owned the store. In 1855 he and his brother, James Cook Ayer, went into the patent medicine business, producing Ayer's Sarsaparilla, Ayer's Hair Vigor, Ayer's Vegetable Pills, and Ayer's Cherry Pectoral, which were marketed and became popular all over the United States. The Ayers were unique in that they published on each bottle its ingredients, which in the case of the Cherry Pectoral included 1/16 of a grain of heroin. To his death, Ayer proudly insisted he never made a dollar of which he was ashamed.

A man of innate common sense and compassion, Ayer was years ahead of his time. Told he was insane by his competitors, he was the first in New England to provide the women who worked for him one day off a month with pay. "He was told the same thing when he helped finance Alexander Graham Bell and the New York subway system."[2]

Ayer once proudly told his children of a visit to Washington, D.C., and his spur-of-the-moment decision to "drop in" on President Abraham Lincoln in the White House. (In the eras before the risk of assassination made American presidents virtually inaccessible, it was not unusual for strangers to appear at the White House and be received by the president.) Ayer found Lincoln "hard at work in his office, tieless, vest unbuttoned, and in his shirt sleeves, wet with perspiration. Mr. Lincoln received us standing and looking terribly tired, bored and lifeless until, after introducing myself and our friends, I said, 'Mr. President, I have called to pay my respects to our President, but none of us has a favor to ask, not even a country post office.' At this, he woke up, and rushed at me with both hands, took both of mine, and shaking them vigorously said, 'Gentlemen, I am glad to see you. You are the first men that I have seen since I have been here that didn't want something.' "

Ayer's second marriage, to Ellen Barrows Banning in 1884, produced three children, the eldest of whom was born on January 12, 1886, and christened Beatrice. Two years later her brother, Frederick junior, was born, followed in 1890 by a younger sister, Mary Katharine (called Kay by her friends and siblings). Despite a thirty-one-year age difference, "theirs was a true love story that lasted until the end of their lives."[3]

The Ayers had first visited California in 1892, and after ten years Frederick Ayer was eager for a return visit to, among other things, try his hand at tandem driving, a sport excelled in by Ellen's brother, William Banning. The Ayers were staunch Republicans and the Pattons even more ardent Democrats, which made for interesting dinner table conversation at Lake Vineyard. When the Bannings, Ayers, and Pattons got together, there would often be as many as thirty around the dinner table.

Rare was the evening when there was not a heated political debate. One occasion made a powerful impression on Beatrice. The usual protocol was for the ladies to withdraw after dinner so the men could continue their oratory over coffee, port, and cigars. This night the Pattons and Bannings couldn't wait for the usual feminine departure and, perhaps because an excess of Don Benito's finest port had been consumed, the debate got rather heated and tempers began to flare, threats were made, and rumblings of settling the matter at dawn with bare fists and seconds. There were lots of "By God, sirs," and the situation looked ominous to young Beatrice until, at the stroke of 9:00 P.M. from the Dick Whittington clock over Mr. Patton's fireplace, there was a sudden scramble to grab coats and "raised fists were lowered to arms that clasped shoulders; voices were suddenly calm; and arrangements were made for the same party . . . to meet at one of the homes the following Tuesday, where they would, as George Patton senior said, 'continue the damned argument to its natural conclusion, as gentlemen.'"[4]

The Pattons never knew what to make of Ellen Banning Ayer, who was always known to her family and friends as Ellie. The actress in her delighted in making a grand entrance after the other guests were seated. During one of the Ayer visits to Southern California, everyone was already at the dinner table waiting for Ellie. They were all talking politics—the Patton clan lived and breathed politics, as did so many of the families who had been uprooted by the Civil War. That they were raving, tearing Democrats and the Ayers rock-ribbed Republicans just made it more exciting.

Nita Patton, then a young girl, had never seen anything quite like Ellie before and couldn't take her eyes from her. She said that Ellie looked around the table in a rather calculating way, and when she had decided who was the furthest person seated from her, she rose most gracefully, and walked around the table to her intended victim, who happened to be Annie Banning. Ellie paused, and so did the conversation.

> She took Annie's chin in one be-ringed and dimpled hand, and turning up her face, said to her in tones of thrilling and low register, "Annie, dear, what do you think of life?" No one in the family ever got over this. All our lives, when things in the family come to an impasse, all anyone has to do is make the remark, "What do you think of life?"—and we all break down and return to normal . . . For all Ellie's drama, she had a genuinely loving heart, and it showed up at all times—sometimes to the intense embarrassment of the object.

> Ellie Ayer was plump, her hands covered with jewels, and on her left arm she wore a gold bangle for each year of her marriage. Eventually these reached thirty-five and turned her into a walking jewelry store. Her granddaughter recalls that whenever she moved, "these slipped up and down and tinkled. She wore a fresh rose in her hair at all times . . . smelled deliciously and was great on hugging and kissing."

> Her two granddaughters also remembered how in 1918 little Beatrice and Ruth Ellen Patton were given a small white kid goat by Ellie. When the goat grew up it got into the habit of butting people from behind. The children were afraid of the animal until one day when it butted Bee's tutor and knocked her down. "After that we admired his gall but always kept our faces toward him."

In 1896 Frederick Ayer was advised to retire by his doctors. Protesting that he was a mere seventy-four and in the prime of his life, Ayer nevertheless acceded, and at his wife's suggestion the family spent the next two years in Paris, where Beatrice learned classical and conversational French. Her doll Marguerite was a present for not having spoken a word of English for three months. One winter was spent in Egypt, on a houseboat on the Nile River that was pulled by mules and horses on a towpath on the embankment. The experience proved a wrenching eye-opener for young Beatrice. For a time, her half-brother, Dr. James Ayer, accompanied them. After he removed a palm thorn from the foot of an Egyptian laborer, word spread like wildfire that there was a physician in the area. Soon the houseboat was swarming with lepers, sick babies, gangrene cases, and wailing adults and children afflicted with every known disease and deformity. Overwhelmed and without the equipment or means to treat them, Dr. Ayer did what he could. Beatrice and the other Ayer children were sent belowdecks whenever the boat landed, but once there they fought for places where they could peek through the curtained windows.

During their Nile adventure, Beatrice became enamored of the tattoo on the back of their boatman and, with a ten-dollar birthday gift from her half-brother, Chilly (Charles Fanning Ayer), decided she too would visit the tattoo parlor and have a full-rigged ship tattooed across her chest. Her governess discovered the missing Beatrice in the nick of time.

Beatrice became a talented musician, and, according to her daughter, "her gifts in that direction were closer to genius than to just talent." In addition to playing the piano, mandolin, steel guitar, and musical saw, she also wrote music, and in her middle age published a number of songs in a small book which was given to members of her family. One of them was a moving tribute to her mother, Ellie. The Ayer children enthusiastically participated in theatricals "in a big way" and Beatrice often played piano in many "better-than-amateur" concerts. She was accorded all the trappings of the upper class, one of which was attendance at a fancy Boston dancing school.*

Beatrice and her siblings were capable of good-natured mischief. Some years earlier Frederick Ayer and his sister-in-law became embroiled in a family feud over money, and when the woman's daughter married a man named Pearson and settled in Boston, her son was instructed under no circumstances *ever* to speak to an Ayer, and if they spoke to him, he was to go straight home. The Ayer children delighted in baiting the young man at the dancing school and used to draw lots to determine which one would either speak to him or ask him for a dance. As soon as one did, the poor child would flee.

In 1899 the Ayers moved from Lowell to Boston, where Frederick Ayer built an enormous mansion on fashionable Commonwealth Avenue that con-

---

*Another attendee was a very unpopular young man named Ernest Simpson, whom all the girls assiduously avoided because his hands always sweated through his kid gloves. Later Simpson became the second husband of a Baltimore socialite by the name of Wallis Warfield, the future duchess of Windsor.

tained an elevator, a dumbwaiter, electric lights, a telephone, and considerable amounts of Tiffany glass, marble inlay, parquet flooring, and heavy velvet drapes. In the rear of the house was a mews where the stables were maintained in the capable hands of the Ayer coachman, Henny, who said when he was hired that he did not particularly care about having a regular day off as long as he had permission to attend all public hangings. It was common to see Ayer and his wife and girls riding along the center strip of Commonwealth Avenue. Ayer rode until he was well into his nineties and thought nothing of chopping wood for exercise. "He was insistent that his girls ride in divided skirts—very daring, indeed—as he felt that side-saddle riding was bad for their backs." As a descendant has written of the Ayer children, "They lacked nothing that money could buy, but they learned, by God, to do exactly as they were told."[5]

The Ayers also maintained a fine country home, called the Farm, containing extensive gardens and greenhouses, in nearby Newton. It had been an engagement present for Ellie, who was passionate about roses. However, after Beatrice caught malaria, Frederick Ayer sold it, and in 1905 built a new summer retreat called Avalon-by-the-Sea in the fashionable seaside town of Pride's Crossing, near Beverly, Massachusetts.

Although some of the "old money" Boston Brahmins turned up their patrician noses at the brash, *nouveau riche* Ayers, they could hardly ignore their great wealth. Pride's Crossing was an equally clannish New England town, populated by the descendants of whaling and seafaring families who were less than elated by the presence of the vulgar "new money" represented by Frederick Ayer and his ilk. When Ayer tore down the old home on his new land and replaced it with a large mansion in the Italian ducal style, dominated by a fireplace large enough to "roast an ox," the gentry of Pride's Crossing were appalled.[6]

Beatrice Ayer made her social debut in Boston at the age of eighteen. "She had a marvelous time. It made her feel very grownup to have her name on Ellie's calling cards." Thursday was "at home" day for the Ayers, when the upper crust of Boston came to call and to take afternoon tea in the salon. With her brothers and sisters away at school, Beatrice became the center of attention in the Ayer household. According to her daughter, Beatrice Ayer was "a real 'Pocket Venus.' She had very beautiful blue eyes and exquisitely marked brows, a softly rounded chin and long, rich dark auburn hair. All this was in addition to being talented, witty and, of course, a considerable heiress."

Beatrice also became a skilled racing sailor, and crewed in the summer in the Atlantic for her cousin, a noted skipper who owned two racing yawls. The Ayers also owned their own schooner, the *Tempest*, and were members of the two choicest yacht clubs in Marblehead. Autumn was reserved for fox hunting, polo, skeet shooting, tennis, horse shows, and hunt breakfasts at the nearby Myopia Hunt Club. In the winter there was skiing, tobogganing, and skating. Beatrice Ayer developed into a first-class rider and a sought-after guest at the house parties held in the opulent salons of the rich members of the most exclusive equestrian club in New England. Although badly near-sighted and unwilling to wear glasses to correct the problem, she neverthe-less rode to hounds and jumped fences and ditches throughout her life with reckless disregard for her safety.[7]

Beatrice and her sisters also performed volunteer work to assist immi-grant children in a Boston settlement house, and it was considered extremely daring by other Boston mothers that Ellen Banning Ayer would even consider letting her daughters work "where they might 'catch a dis-ease.'"

This combination of beauty, brains, and great wealth attracted a number of suitors well before Beatrice made her formal debut. One was a Russian count, and one day she teasingly observed that if one scratched a Russian, a Tartar would emerge. "Immediately, the count rolled up his sleeve and ran his long shiny fingernails down his arm until the blood flowed from the welts and cried, 'Scratch me, Meees Ayer, scratch me!'" However, for all her privileged upbringing, Beatrice Banning Ayer had matured into a confi-dent, independent, strong-willed woman, a fact that both young Georgie Pat-ton and her father would soon discern.

Beatrice was closest to her younger brother, Frederick Ayer Jr., whom she loved dearly and who would later become a lifelong friend of George S. Patton. The devotion of the entire Ayer family to one another was common knowledge in Boston, and when they got together they were "like birds on a telephone wire." The Ayer children kept in touch with one another through-out their lives. "Ma's brothers and sisters were truly her best friends, not just relatives. I know that to Ma they were always, next to Georgie, 'the closest kin there is.'"

Beatrice and Georgie Patton had nearly met in 1892 during the Ayers' first visit to California, which they spent with the Bannings in Wilmington. One day her parents left for a day trip to Los Angeles, but six-year-old Beatrice, who often became ill riding in a carriage, declined and remained behind to read a new book. Later the Ayers mentioned they had seen Colonel Smith and the Patton family, including

> the dearest little boy, whose name was "Georgie," and who was just a few months older [than Beatrice]. . . . He had big blue eyes and beautiful golden curls, and was such a *good* little boy that he would have never let his father and mother and brother and sister to go off alone just so he could read a book—he would have come with them. Ma decided right

there Georgie Patton must be a little prig and she told her parents she
hoped she would never meet him and if she did, she would not play with
him.

As was their custom in the summer, the Patton and Banning clans
moved from the mainland to Catalina Island. The sons of Phineas Banning
had continued the entrepreneurial genius of their father by purchasing
Catalina Island and establishing a ferry service to transport the upper crust
of Southern California who could afford summer homes there. By the time
the Ayers arrived in 1902 the trip from San Pedro to Catalina had been
reduced to only an hour and a half via steamboat. The three families had a
memorable summer of fun and frolic, culminating in early September in a
fantasy play called *Ondine*, staged by the Banning, Patton, and Ayer children. Beatrice was given the lead role of Ondine, and Georgie was one of
the water sprites. A photograph of the three families taken that summer is
dominated by the white-bearded figure of Frederick Ayer in the center. At
the right is a relaxed, boyish George Patton in a summer suit and bow tie,
while on the opposite side next to Nita Patton is a demure Beatrice Patton, a
half smile on her face.

For young Georgie Patton it was the most important summer of his
life. The two young people were drawn together for the first time and "by
the end of the rehearsals, the play and the summer, Ma and Georgie were in
love for the rest of their lives." Two more unlikely opposites could not
have been imagined: the wealthy, well-educated, New England Yankee
young woman and the rather unsophisticated, dyslexic, rough-hewn son of
Virginia-born lawyers and Confederate warriors, who had grown up in the
still untamed environment of frontier Southern California.

After the magical summer of 1902, the two went their separate ways,
Georgie back for what would be his final year at the Classical School for
Boys, and Beatrice to Boston, where she continued her studies and a whirlwind social life. The two began a sporadic long-distance correspondence,
often through one of the adults. A letter to "Aunt Ruth" in November 1902
noted, "I know that Georgie's birthday comes sometime this month, but not
the exact date. When it comes, please spank him seventeen times for me and
give him my very best birthday compliments." She also revealed she would
like nothing better than another trip to California in 1903.[8] For Christmas
Beatrice sent Georgie a tiepin that brought a thank-you note that this was
"the very thing I most wanted." Signing himself, "Your faithful friend,
Kuhlborn [the character he portrayed in the summer play] or George S. Patton."[9]

Until their deaths neither would ever again long be out of the thoughts
of the other.

PART III

The Making of an Officer
(1904–1909)

God willing . . . and given the chance I will carve my
name on some thing bigger than a section room bench.

—CADET GEORGE SMITH PATTON JR. (MARCH 1905)

CHAPTER 5

A Father's Influence

The name of your son is . . . upon my list of those who will be invited to compete for my recommendation.

—SEN. THOMAS R. BARD TO GEORGE S. PATTON II

By the end of the summer of 1902, George S. Patton had not only fallen in love but had decided that he would become an army officer. His decision hardly came as a surprise to his parents and was clearly prompted by his unwavering belief that it was his obligation as the heir to the Patton name to carry on the family tradition by becoming a great soldier. Given young George's many years of indoctrination in the Patton heritage, it would have been a shock to his father had his son chosen any other profession. Clearly, Papa was pleased by "the Boy's" decision and it was agreed that he would seek admission to West Point rather than VMI, where three gererations of Patton men had trained to become soldiers.

VMI remained an alternative, although graduation did not guarantee a commission in the Regular Army. Other options were the nearby University of Arizona, where Papa's cousin, Col. John Mercer Patton III, was a professor of modern languages. Given the Patton presence at VMI since its founding, admission was a virtual certainty; acceptance at West Point, however, could only occur by means of a presidential or Congressional appointment. His son's academic record was mediocre, and Mr. Patton realized it would be a long and difficult process to obtain an appointment for Georgie. Even in his final year at Dr. Clark's school, Patton continued to struggle, his unknown dyslexic condition an unending source of frustration. In January 1903 Patton had written in one of his earliest letters to Beatrice, "My hard study is making me fat and stupid so that I have come to the conclusion that the onl[y] way to pass an Ex.[amination] is to try not to. . . . Please excuse this long yet truthful excuse."[1]

Appointments to the U.S. Military Academy at the turn of the century were as difficult to obtain as they are today. Congressmen from each district and territory in the United States were entitled to appoint one cadet, while the two senators from each state had only two appointments between them. A third, "at large," category of appointment permitted the selection of thirty cadets from anywhere in the United States by the president. For each candidate nominated there could be two alternates. The difference between then and now is that appointments could only be made when a cadet previously appointed by the same individual either graduated or dropped out of West Point.

All candidates had to pass a strenuous mental and physical examination before a board of army officers, and the standards for admission required a candidate to be "well versed" in reading, writing, spelling, English grammar, English composition, arithmetic, algebra, plane geometry, geography, history, the principles of physiology, and hygiene.[2] Virtually all posed a serious problem for young Patton.

Papa wasted no time putting to practical use the savvy he had gained from years of experience in the rough-and-tumble of Southern California politics. Mr. Patton decided that the best source of an appointment was Sen. Thomas R. Bard, whose next vacancy would occur in June 1904. [3]

Bard was hardly the ideal choice for a number of reasons, not the least of which was the fact that he and Patton were of entirely different political persuasions. Even more ominous was the fact that during the Civil War, Bard had become "a good hater of Rebels."[4] However, with virtually no other recourse on the political front, Mr. Patton began an all-out effort to lobby Senator Bard on behalf of his son that was to consume him for the next eighteen months.

The first big political gun unlimbered by Mr. Patton was Judge Henry T. Lee, a former Union officer and a well-connected Los Angeles Republican who was asked to recommend Georgie, and who obliged by writing a number of letters of recommendation. Other prominent Californians who took up their pens on his behalf included the venerable Col. George Hugh Smith. A bank president, an associate justice of the California Supreme Court, the president of an oil company, several other judges, a number of well-known lawyers, the postmaster of Los Angeles, and the naval aide to the governor of California all sent glowing letters of recommendation, predicting a distinguished military career.[5]

The blitz of letters continued into March 1903 and nearly overwhelmed Bard's private secretary, who was obliged to reply to each writer. Nevertheless, Bard remained determinedly noncommittal to Patton's patrons, and would only concede that he would offer young Patton "the opportunity of competing with other applicants."[6]

Mr. Patton's problem was complicated by the fact that while his son would turn seventeen in November 1902 and thus be eligible for admission to West Point in the autumn of 1903, Bard's first available appointment would not be until 1904. No one believed that Patton could have passed a competitive examination in 1903. However, an additional year might just be sufficient time to prepare him for West Point.

In February 1903 Mr. Patton contacted Francis C. Woodman, the headmaster of the Morristown (New Jersey) School, which was noted for preparing students for entrance examinations. Woodman's reply was hardly encouraging ("your son's case is one of those from which we distinctly shrink"), but he did agree to enroll Patton as a special student to prepare him for West Point.[7] Although Mr. Patton hedged his bets by reserving a place for his son at the Morristown School in the autumn of 1903, he considered it a last resort and stepped up his campaign to convince Senator Bard to appoint Georgie to West Point. At Woodman's suggestion Dr. Clark had administered the Princeton University entrance examination to Georgie, and in June 1903 came some encouraging results. Although he had failed plane geometry, he had passed algebra and U.S. history and was granted admission to the Princeton class of 1907.[8]

Patton never entered Princeton. Instead in September 1903 he enrolled at the Virginia Military Institute, after Papa had concluded that VMI was the logical place to prepare him for West Point. If he failed to gain admission, a commission from VMI might yet lead to an appointment in the Regular Army.[9]

Unlike the carefree summer of 1902, when he had first met Beatrice Ayer, 1903 was a time of intense, last-minute study at Lake Vineyard and on Catalina Island. As the day approached when he would embark on the first leg of attaining his dream of becoming an officer in the U.S. Army, Patton began to question whether he was good enough to live up to the family name—doubts that would continue to plague him until the end of his life:

> Just before I went away to the VMI I was walking with Uncle Andrew Glassell and told him that I feared that I might be cowardly. He told me that no Patton could be a coward. He was a most recklessly brave man. I told this to Papa and he said that while ages of gentility might make a man of my breeding reluctant to engage in a fist fight, the same breeding made him perfectly willing to face death from weapons with a smile. I think that this is true.[10]

He also conveyed his doubts and fears to George Hugh Smith, who reassured him that he would be able to do his duty. Smith also told Patton that a great war would soon engulf the world and that he must prepare himself through diligent study to play an important role in it.[11]

In September 1903 George S. Patton left Lake Vineyard and traveled by train across the United States, via San Francisco and Salt Lake City, to Lexington, Virginia. Patton was accompanied not only by his parents and Nita, but also by the obsessively adoring Aunt Nannie, who remained nearby throughout most of the year. The ritual of Nannie following her "son" from place to place was to become one of the more bizarre aspects of Patton family life.[12]

The question of whether or not Georgie would gain admission to West Point was unresolved. Senator Bard continued to play a closed hand and had yet to reveal the names of those he would even consider to fill his 1904 vacancy. Mr. Patton could only renew his campaign to convince Bard that his son should receive the coveted appointment. But for young Patton it was his final chance to prepare himself to fulfill his ambition and to live up to the academic achievements of his family, all of whom had excelled at VMI. After Papa bade his son farewell, Patton would always remember, "I never felt lower in my life."[13]

In the years since the Civil War, the institute had been rebuilt after being nearly destroyed in a June 1864 punitive raid in retribution for New Market, a month earlier. The exploits of Stonewall Jackson and the heroics of the VMI Corps of Cadets at New Market had become the foundation on which the rich tradition of VMI had been built in the thirty-eight years since the Civil War. An amazing 92 percent of VMI alumni had fought in the Civil War, including eighteen for the Union. Nineteen became generals (one of them a Union general) and 261 of them died, including three generals. VMI cadets had not only fought at New Market but also "augmented the thin line of Confederates manning the trenches between Petersburg and Richmond during the terrible winter of 1864–65."[14]

The new cadet was delighted when the school tailor recognized him as a Patton and noted that his uniform measurements were identical to those of his father and grandfather. Despite his status as a lowly "rat," Patton felt comfortable at VMI—where he was among other sons of graduates and Southern gentlemen—and his grades immediately began to reflect considerable improvement over those of his final year at Dr. Clark's school. By February 1904, of the approximately ninety students in his VMI class, Patton stood sixth in drawing, ninth in mathematics, tenth in Latin, and twenty-eighth in both history and English. His deportment was perfect: no demerits, and a well-earned "Excellent" was handwritten on his report card.[15]

Patton arrived at VMI determined to make good. His father had prepared him extraordinarily well, and also provided simple but useful advice to be a good soldier, a good scholar, and on the nights before he was to march on guard duty, to clean and shine his gun and brass until they were spotless. If there was time left over he was to study. He heeded his father's words, and "the result was that I never walked but one tour of Quarters guard, on all other occasions getting [selected as] Orderly."[16]

Patton understood exactly what was expected of him, and thus escaped the pitfalls that befell most "rats." This included keeping his mouth shut and never—ever—talking back to upperclassmen or instructors. The spit-and-polish Patton was head and shoulders above his classmates. His carriage was ramrod straight and his uniform impeccable; he was "always a little nattier; he executed the drill movements with a bit more snap, and his equipment always seemed to have a higher polish than that of his fellows."[17] Now six feet tall and weighing approximately 150 pounds, Patton became an exemplary soldier at VMI and set a standard for himself (and others) that he was to carry the rest of his life. A VMI historian later remembered Patton as "tall, blond, a fine looking young man . . . well liked by his fellow cadets . . . Patton was a good soldier."[18]

Nevertheless his lowly status left him yearning for something better. Some years later, when Beatrice was visiting nearby Natural Bridge, he would write to her:

> It was when I was a rat and Col. Marr took Mama and me over there one Sunday. . . . There were some girls on the porch and when they saw a cadet in the carriage they became interested until I got out and with a look of disgust one of them said "Oh, its only a rat" and then I saw the first necessity of chevrons.[19]

The routine of life at VMI began early with reveille and breakfast. Classes were from 8:00 A.M. to 4:00 P.M., followed by an hour of close-order drill on the parade ground. Patton learned how to "brace" (a position of rigid attention, with the chin firmly tucked into the chest, the face turning red from the exertion), and to memorize the various items of useless information required of all "rats."

Patton's first term was abruptly cut short in late October when Lexington was swept by an epidemic of deadly typhoid fever. As a precaution the VMI authorities decided to furlough the entire corps until the threat passed, and for a month the institute remained eerily devoid of life. Patton returned to California and spent half his unexpected vacation on trains, which took six days each way.[20]

Patton played left tackle for the 1903–4 scrub football team. One of his teammates later described him as "a tall, thin, hot-tempered 'rat' from Los Angeles." In 1945 Patton would recall that he was "probably the world's worst football player, but I did begin to inherit there—or one might say 'inhale' the fighting spirit of that great institution."[21]

Patton managed to stay out of trouble even when some of his classmates did not. As the commandant wrote to his father in January 1904, some members of the fourth class "distinguished themselves the other night, but George had the good sense not to be in it, or the good luck not to be caught."[22]

Although Patton's academic marks were generally high, there were already indications of trouble ahead. He had slipped in Latin and worried about the erratic nature of his performance. "Last week my Latin average was a little over 60 while my other studies were between 92 and 95. At home I would think this pretty fair but here I am utterly heart-broken." Fearful that his dream could slip away if his grades faltered, Patton buckled down and studied harder than ever, particularly to bring up his marks in Latin. The effort paid off, as his grades steadily improved.

He remained single-minded about West Point: "I *must* get that appointment. . . . The only reason I am so ancious to get in [to West Point] next year is that the joys of cadet life are not so grate as to make me wish to spend six years in the enjoyment of them. Five years will be bad enough but six o lord."[23]

For the benefit of Georgie's morale, Mr. Patton continued to provide optimistic assurances regarding West Point, which misled his son to write in January 1904,

> I suppose my appointment is pretty sure and I am glad of it because I wont have to study Latin here next year. . . . Mama asked in her last letter if I would not like to go in the Army from here I would like to but since I am to be a military man it would be much better for me to go to W.P.[24]

In spite of all Mr. Patton's efforts, Bard repeatedly resisted taking the easy way out by appointing Georgie without more ado and thus saving himself considerable further pressure from the small army of influential men who were bombarding him with letters. Bard soon confirmed that he was appointing several referees to administer informal competitive examinations, and also telegraphed the VMI superintendent, Gen. Scott Shipp, to inquire if Patton would be released from VMI to take the West Point examination. Assured by Shipp that he would, Bard cabled Mr. Patton that his son must attend the forthcoming examination to be held in Los Angeles.

Papa wrote to his son that it was his decision whether to remain at VMI or stake his future on West Point, but before he could even mail it, a telegram arrived from Bard announcing that the dreaded examination would be held in mid-February in Los Angeles. Mr. Patton wrote at once that: "This settles the matter. . . . I do not think you should fear to subject yourself to the required test . . . In the meantime make no change in your present course of hard work, and do not allow yourself to be upset or disconcerted by this matter."[25] Privately Papa was uncertain if his son was "sufficiently set in his determination to go into the army as a permanent career," but nevertheless believed he must take his chances on West Point in lieu of the certainty of completing VMI, to avoid a lifetime of regret such as he himself had endured.[26]

Patton sent Bard's secretary the required letter containing his personal data, and a certificate from the VMI surgeon that he was "entirely sound physically and of excellent physical development," was six feet one inch in height, weighed 167 pounds (a gain of nearly seventeen pounds since his admission the previous autumn), and had a chest measurement of thirty-eight inches. He incorrectly listed his birthdate as November 11, 1886.[27] Among the required certificate and documents was one from Dr. Stephen Cutter Clark, advising that George S. Patton had passed examinations in an array of subjects and had "showed himself an earnest and conscientious student. He was always very gentlemanly in his behavior to all the teachers and to his fellow students; a thoroughly clean, pure, conscientious young man, deservedly a favorite with all."[28]

George Patton returned to California in early February 1904. He had studied diligently during the lengthy train journey and at Lake Vineyard. As the most important day of the young man's life approached, both father and son grew increasingly anxious. After all the effort, George S. Patton's future would be determined by how well he did during one fateful day. On February 15 Patton dutifully completed Senator Bard's competitive examination without incident. The next day he was on a train for the six-day journey back to VMI.

The uneasy wait ended on March 4, when Mr. Patton received a brief telegram from Bard (dated the previous day), announcing: I HAVE TODAY NOMINATED YOUR SON AS PRINCIPAL TO WEST POINT. The preparation had paid off: Patton had scored first in the competitive examination. In 1947 Bard's son wrote to Beatrice Ayer Patton that he had just come across his father's records of Patton's appointment, and although "I did not find the report of the committee, it is clear enough that they judged the future General to be definitely the best of the lot."[29]

Papa's letter of congratulations to his son hinted at the extent of the grueling two-year ordeal:

> It has been a long and tiresome quest, but in your success I am sure
> that you will be more than compensated. It is a serious step you have
> taken, thus fixing your future career for life and I am sure you have done
> so with a full appreciation of all that it means . . . that which a man
> desires most strongly to do in this world, if he has really given it careful
> consideration, is what he is generally most fitted to do.

After the enormity of Bard's announcement had fully sunk in, Mr. Patton sent a second, heartfelt letter to George on March 4: "You cannot know how proud we feel . . . you may look forward to an honorable career—as a soldier of your country." Prophetically he wrote that all signs pointed to a period of war in which he believed the United States would play a leading role. "You have in you good soldier blood—and the opportunity before you is one to inspire your darndest effort. Be honorable—brave—clean—and you will reap your merited reward. . . . A thousand blessings. . . ."[30]

Patton wrote to Bard expressing his "deep sense of gratitude for the honor you have done me. I believe that I realize the gravity of the position in which this appointment places me and I will try to do my duty both at West Point and afterwards in the army to the best of my ability."[31] Senator Bard died in 1915, his political career a mere footnote of history but for his momentous decision to appoint young George Smith Patton to West Point.

Patton permitted himself a rare moment of pleasure.

Dear Papa:

Well I guess I have got it. And I am beastly glad and am sure you are. As for Mr. Bard I rank him and the pope on an equal plane of hollyness. . . . At last after all these many years this thing is finally setteled. I have just at this moment received my certificate, and now it only remains for the government inspectors to examin this hundred and seventy pounds of meat (which forms the earthly cage of my imortal soul) and if they consider that I am sufficiently sound to be killed, I suppose that like the Christmas turkey I will be admitted to the mental fatning pen at the point . . . tonight is the first time that I . . . can at last stop worrying. . . . I am sorry that I have been such a nusance . . . but according to that paper I signed I guess you will be freed from care on my account for at least eight years. Please thank the California Club for the literary effort on their part in my behalf and tell them that when I become a dictator I will send them my picture and autograph to be hung up along with the moose head, fish and the bear. . . . And with the help of God and a vigorous use of your influence I have the appointment.

Your loving and greatful son,
George S. Patton, Jr.[32]

Patton also wrote to share his triumph with Beatrice Ayer, but his elation was short-lived during his final three months at VMI, when he was forcefully reminded by upperclassmen that for all his success, he was still a lowly "rat." He was hazed, at times unmercifully:

They made him memorize magazine and newspaper articles about West Point and recite them on call, tore his bed apart, and ran him ragged on countless fool's errands. He took it all in stride and worked harder than ever—so hard that Major Strother [the commandant] . . . made him take a twenty-day leave to rest. Patton's notion of a vacation was to spend three weeks tramping over the Civil War battlefields in the Shenandoah Valley, studying the terrain so that he could visualize more clearly every engagement in that phase of the war and see for himself where his grandfather had died.[33]

In May 1904 Patton passed the required physical examination with flying colors and three weeks later received a letter from the War Department announcing: "I have the honor to inform you that you have met the requirements for admission, and that you will be regularly admitted as a Cadet upon reporting in person to the Superintendent of the Academy on the 16th day of June, 1904."[34]

On June 1 Patton's resignation from VMI was accepted with reluctance by General Shipp, who would have been pleased to have him return. In fact, Patton learned that had he remained at VMI, he would have been appointed first corporal the following year, an honor accorded to the outstanding plebe.[35]

George S. Patton's dream of entering West Point was about to become a reality. Had he known what dyslexia was, he might have had even more cause for exhilaration at his appointment. In a scant nine months he had risen above a passable but hardly notable academic record at Dr. Clark's school to an impressive one at VMI, and a perfect record of deportment that included not a single demerit. It was the first real challenge of Patton's life, and he had met it successfully by a combination of hard work and sheer determination. However, the road to an army commission still had to pass through West Point, which would be a new and even more daunting challenge. Despite his youth, his sheltered and pampered childhood, and his fantasy of becoming a great and famous general, Patton understood full well that West Point would be "the Hell to come."[36]

CHAPTER 6

"The Military School of America"

West Point, 1904–1905

The Corps! The Corps! The Corps! . . .
The long gray line of us stretches
Thro' the years of a century told. . . .

—HERBERT S. SHIPMAN

West Point began its existence as a school for engineers, and its first graduates in 1802 consisted of exactly two officers, who were commissioned in the Corps of Engineers.[1] By 1818 West Point had graduated a total of 202 officers, but both its input and class size remained small and the attrition rate high.[2] In 1843 there were still a mere 38 graduates in the class that included Ulysses S. Grant. Sylvanus Thayer, the third superintendent (1817–33) is credited with establishing West Point as a first-class institution and for originating class rankings, daily classroom recitations, and deportment, all of which made up a cadet's final standings and ultimately determined in which branch of the army he was commissioned. The top graduates became engineer and artillery officers, while the middle and bottom ranks of each class found themselves commissioned in the infantry. Thus, when Robert E. Lee finished second in the class of 1829 he became an engineer, while Grant, who finished in the middle of his class, was relegated to the infantry.[3]

By the time of the Civil War, West Point had come of age. The armies of the Union and the Confederacy were dominated by its graduates, who, in addition to Grant and Lee, included George B. McClellan, Abner Doubleday, George Gordon Mcade, J. E. B. Stuart, Philip Sheridan, William Tecumseh Sherman, Thomas J. "Stonewall" Jackson, Joseph G. Johnston, Ambrose E. Burnside, Jefferson Davis, and the "goat" (lowest-ranking graduate) of the class of 1861, an intensely ambitious young officer named George Armstrong Custer. "Of the sixty major battles of the war, fifty-five of them were commanded *on both sides* by West Point graduates, and the other five battles had a West Point commander on at least one side."[4] Although the Civil War pitted West Point classmates against one another, it failed either to seriously disrupt its mission or to prevent those who fought for the Confederacy from eventually sending their sons to the academy. And while the postwar classes were dominated by Northerners, by the time of the Spanish-American War in 1898, several former West Point graduates who had served the Confederacy as generals had been recommissioned in the U.S. Army, thus finally signaling an end to the cleavage the war had wrought.[5]

After the Civil War, the practice of hazing became an integral part of the daily life of a plebe at West Point. Bracing, as well as petty harassment at meals, on the parade ground, in quarters, and in the quadrangle soon became a part of West Point tradition. The excesses included strenuous physical and often harmful exercise, liberal doses of Tabasco sauce in a plebe's food, and elaborate funeral ceremonies for dead rats. Hazing produced a code of silence on the part of the hapless plebes, and it became a matter of dishonor to expose the upperclassmen who perpetrated such mischief. Among those who was hazed unmercifully at the turn of the century was the valedictorian of the class of 1903, Douglas MacArthur, himself the son of a noted army officer.[6]

At the turn of the century the U.S. Military Academy was considerably smaller than today's imposing facility, but in June 1904, when George Smith Patton exchanged VMI gray for that of a West Point plebe, it was little changed from the remote outpost of the nineteenth century. The first thing he noticed about West Point was that the third classmen assigned to greet and train the newcomers included many whose shouting, abuse, and instructions differed only semantically from those who had tormented him at VMI for the past year. Newly liberated from their own year of hell as plebes, many were on hand to greet the incoming members of the class of 1908, for the commencement of what is still nicknamed "Beast Barracks" (also known as "Plebe Camp" in Patton's time).

After being measured for and receiving uniforms and the other impedimenta of a cadet, the plebes were assigned to companies A through F on the basis of their height. As one of the tallest, Patton ended up in A Company. Plebes were assigned to four-man squads, which were harassed at each and every formation, usually by two third classmen. For the first month, plebes were taught the rudiments of military drill and ceremonies in a nearby tent city adjacent to the Plain (the plateau area of West Point, overlooking the Hudson). Demerits were assigned for infractions of discipline and regulations and punishment tours walked with rifle and pack.

In mid-July the new cadets were integrated into the annual summer camp with upperclassmen, who eagerly awaited their arrival and warned of things to come with shouts of: "We have been waiting for you," and endless harassment in the form of chores for them, that included "folding bedding; cleaning spurs, sabers, guns, breastplates, and shoes; sweeping the company streets; and, on hop nights, arranging clothing, putting on clean collars and cuffs, and making out the upper classman's hop cards."[7]

In spite of the harassment, Patton relished the pomp and ceremonial aspects of military life and thought West Point was less oppressive than VMI. During the summer of 1904 two generals were buried at the academy, which spawned in him romantic visions of great warriors and death:

> It was very impressive and the muffeled drums were great. . . . I certainly think it is worth going in the army just to get a military funeral. I would like to get killed in a great victory and then have my body born between the ranks of my defeated enemy escorted by my own regiment and have my spirit come down and revil [revel] in hearing what people thought of me. But I am afraid that I have not got enough sence or persistance. . . .[8]

George S. Patton arrived at West Point exceedingly conscious of his social status, believing that most of his fellow cadets were not of the same social ilk as those at VMI. He never lost his aversion to those of alleged inferior social status, a snobbish trait his grandson ascribes to Patton's father, who "considered himself to be of better stock, therefore of better character than most other men."[9] Most were "nice fellows but very few indeed are born gentlemen . . . the only ones of that type are Southerners." Patton eventually elected to room with two former VMI men; "both are gentlemen."[10] In August he announced: "I believe that some of the upper class men have begun to respect me, if not to like me and I am glad to say that this feeling is only apparent among the gentlemen for the rest I dont care and they know it." At supper one night Patton was harassed by a yearling who began yelling at him:

> I payed no attention infact was impudent, when all at once a man named Harris from Texas who is second corp[oral] in "A" Co. spoke up and said.

"Henry havent you enough brains to see that you cant make a gentleman do any thing if you yell at him." He then proceeded to give Mr. Henry (who looks like a pes-ant) hell. It speaks pretty well for my choice of room mates that Harris has chosen us to be the three plebes at his table.[11]

In the summer of 1904 Patton confided to his father that "I belong to a different class a class perhaps almost extinct or one which may never have existed yet as far removed from these lazy, patriotic, or peace soldiers as heaven is from hell. I know that my ambition is selfish and cold yet it is not a selfishnes for instead of sparing me, it makes me exert my self. . . . Of course I may be a dreamer but I have a firm conviction I am not and in any case I will do my best to attain what I consider—wrongly perhaps—my destiny."[12]

His first days at West Point also precipitated what would become a lifetime penchant for saying the wrong thing in public. He wrote of "catching a good deal of *hell* lately because in an unguarded moment I said that we braced harder at VMI than here." Ever since, "All the corps have been trying to show me my error and they have succeeded."[13] Like all cadets, he complained endlessly of the grind of cadet life, of the lousy food (meat so tough "the more you chewed it the bigger it got"), too little sleep, and the quaint customs. "If General Sherman's definition of war be right west point *is* war."[14] Patton began marking off on a calendar the number of days left in his plebe year, writing hopefully to Beatrice: "I will only be a plebe for two hundred and nienty seven day's more."

From the beginning he was a loner with few friends and a great many detractors. Unable to hide his disdain, he was deemed arrogant and remote. Whether because of a vendetta or mere hazing, in mid-August Patton, on guard duty, was attacked by three cadets. When one lunged and attempted to seize his rifle, Patton threatened to bayonet the first person to attack him. Fortunately the catch on his bayonet slipped and retracted into its sheath, or he might have killed the cadet. The muzzle of his rifle knocked his attacker away, and thereafter Patton was left alone.[15] Patton was also homesick, and wistfully recalled that, "I am rather sorry in a way that I went to VMI. . . . I have gotten so I dont care whether I amount to any thing or not but I am trying to over-come this somnistic condition and work . . . two rat years in succession are very depressing."

Although military life at West Point was less demanding than he had anticipated, he found the frequent twenty-four-hour guard duty "very hard . . . though we are allowed to go to bed at night none of us can take off so much as a glove and of course it is not very much fun sleeping in full uniform. . . . I will be in confinement next week for not knowing an order on guard. . . . I hate to get reported especially as I knew the order but did not understand the O.D. [Officer of the Day] when he asked me."[16] On at least one occasion that summer his dyslexia left him unable to read an order posted on the bulletin board.

The only welcome diversion from the harassment and military routine were mandatory daily hour-long dancing classes, designed to enhance their education as future officers and gentlemen. Dancing instructors were brought to West Point to conduct classes, which usually featured roommates as partners, although from time to time some young ladies would also participate. As one cadet in the class of 1891 has written: "It was here and only here during the entire 'Plebe Camp' that we had any fun."[17] George S. Patton enjoyed dancing and in one of his first letters from West Point to Beatrice Banning Ayer wrote of being "simply perfect" in the role of the female partner.[18]

The summer of 1904 was a series of highs and lows. He experienced severe mood swings, oftentimes in the course of a single letter home. Even an innocuous present from Beatrice was enough to raise doubts in a persona so fragile that he could write: "Beatrice sent me a little silver soldier for my watch fob. I hope there was no hidden sarcasm in it and that she did not mean I was a tin soldier."[19]

Although his third class year was still nearly eleven months off, Patton was already worrying about promotion and his future standing in the Corps of Cadets. ". . . to get a high corp [promotion] here a man has to be in the first ten men or else be the sun of an officer and since I am afraid that I cant be the one and am not the other I am in rather a bad way. Still I hope that I get some office for I know with out a doubt that I will make a much better officer than any of the present third class do."

Patton was not particularly tolerant of the West Point system:

Our whole class will have more demerits than any preceding class for since the upper class-men are not allowed to speak to us or correct us, they naturally bone us [with demerits] and they are quite right. Indeed I think that the system which they have adopted here of absolute forbarance toward plebes, will ruin the academy in a very few years. I have not one fifth the respect for an upper class man here that I had at the institute, and with out respect it is impossible to have good discipline.[20]

His constant complaining notwithstanding, Patton soon realized that West Point was indeed special: "The absolute honor of this place is amazing yet so ever present that after a time it ceases to be noticible. There is nothing but truth here and even the worst of the rabble to whom the name 'plebean' is *most* fitly applied soon learn this and conform to it."[21]

There were early indications that Patton might be exceptional. During his first military exercise in the hills above West Point, the "enemy" attempted to infiltrate through a long skirmish line. Patton was one of the guards and instead of concealing himself at an obvious place, hid in foot-high grass and after patiently waiting in the torrid heat for several hours, was rewarded when he "killed" an infiltrator. "I was highly praised for my hiding capacity." Yet, by his own admission, he would often get into trouble others easily avoided. "I try not to get boned," he told his father, "but cant seem to manige it. I get skined for some foolish offense such as yawning in ranks . . . still I think that I am better off than the majority."[22]

Despite the severe restrictions, Patton and Beatrice began a courtship that was to encompass his years at the academy. In his first letter he wrote: "I am certainly glad to know that I am missed and that you would like to see me."[23] Although plebes had scant time for the luxury of daydreaming, Beatrice began to intrude more and more into his thoughts, even though it is doubtful that he was yet prepared to admit that she had become an increasingly important part of his life.

The intense training was physically exhausting and the hazing often infuriating, but what bothered Patton most during the summer of 1904 was his uncertain academic prospects. "We begin studying on the first of september. I shall be rather glad when we do this and I at last find out just what my chances of being able to stick are."[24]

Finally came the day when the training ended and the hazing abated. The final day of summer camp Patton proudly reported to his father that it was when they were marching back to their barracks that "I realised that all of those people were looking at me and that I was part of the corps I felt fine. The only draw back being that I was only part of the corps not the whole thing. Of a truth, I am too ambitious, too much of a dweller in the future."

The academic ordeal Patton had long dreaded was about to begin. "We drew our books to day," he wrote at the end of August, "and they dont look very hard but of course they are." As his first semester at West Point neared, he attempted to express to Papa the tempest that raged within, the loneliness, his compelling need to excel, and his remoteness from others. It was just over a year, he noted,

> since I started to learn the profession of killing my brothers. . . . And in this year of contact with the world my respect for man has dwindeled instead of increasing. For even among the best, and the best are, I take it, those who devote them selves to the service of Mars, there is not the self sacrificing love of fame or self denying selfishness which I feel and which I had expected in others but rather a languid lacitude, [a] car[e]less indifference or hazy uncertainty not becoming in my estimation [of] a soldier or a man. . . . And if my nature prove incapable of the task I have set for my self or if the opportunity never comes I can at

least die happy in my own vanity knowing that I stood alone and that alone I fell.[25]

George Patton began to discover just how difficult West Point would turn out to be in September 1904 when the Corps of Cadets settled into a routine that hardly ever varied:

Cadets marched to every event: classes, athletics, meals, chapel services, and parades. Plebes were even marched to bath formations in the bathhouse across the area from their barracks. . . . Life was one formation after another, from reveille at 5:30 a.m. to taps at 10:00 p.m.. Only on Sundays did the plebe have much if any free time.[26]

The cadet rooms were Spartan, devoid of any creature comforts, and equipped only with iron bedsteads, hair mattresses, a single blanket and pillow per cadet, a chair and table, individual metal washbasins, soap, towels, and a crude clothespress. With regularity Patton and his fellow cadets would complain of the numbing cold they were compelled to endure.[27]

Not only were the living conditions wretched but plebe year academics were especially difficult for Cadet George S. Patton, who lamented that English was pure memorization and: "pretty hard for me because it is simply grammar and I know nothing of it. . . . I don't believe that there is any possibility of my being found [flunked] at least this year for there are some absolute fools in the present third class who got through. You should see this place at night it is absolutely soundless yet there are five hundred men in it and every one of them studying like hell."[28]

Mathematics filled each weekday morning and included geometry, algebra, plane and spherical trigonometry, surveying, and analytical geometry. Afternoons were devoted to French or to ethics and history. Tactical instruction each afternoon consisted of artillery and infantry tactics, fencing, bayonet exercises, and military gymnastics.[29]

The strain on the cadets was sufficiently heavy that several members of Patton's plebe class snapped before the school year ended. One attempted to kill a first classman, prompting Patton to observe, "so you see we study pretty hard and this knowledge may enable you to excuse some of my letters 'the very stupid ones.'"[30]

Patton was torn between an ability to see future greatness for himself and his dyslexia, which served unceasingly to implant the notion that he was both ordinary and stupid. His first plebe year was an uneven struggle to overcome an affliction about which he had no conception. "I dont know whether you knew or not that I have always thought that I was a military genius or at least that I was or would be a great general," he wrote Papa soon after classes started. "Well . . . at present I see little in which to base such a belief. I am neither quicker nor brighter in any respect than other men nor do they look upon me as a leader as it is said Napolions class mates looked upon him. In fact the only difference between me and other people is that I have ideals with out strength of character enough to live up to them and they have not even got them."[31]

Another aspect of Patton's struggle centered on an illusion. He began to affect personality traits intended to deceive others into believing he was someone entirely different—in short, to reinvent himself in the guise of a rugged, macho male.[32] When Patton entered VMI in 1903 he had already begun to display unmistakable signs of a significant personality change. As a teenager he had perceived that a military leader must present an image of invincibility and toughness, traits then utterly alien to him. Determined to prepare himself for generalship, Patton began acting in a manner that bore scant resemblance to his true persona. As biographer Martin Blumenson accurately observes of Patton, he concocted his own personal perception of how a leader and a general ought to look and behave, and he spent the remainder of his life honing that image by becoming profane, ruthless, and aristocratic. His famous scowl became so successful a part of his persona it seemed as if he had been born with it permanently engraved on his face. What Patton never understood was that while he succeeded beyond measure, in so doing, "he killed much of his sensitivity and warmth and thereby turned a sweet-tempered and affectionate child into a seemingly hard-eyed and choleric adult."[33]

There is ample evidence of the evolution of young George Patton from the happy-go-lucky youth of Southern California into the VMI and West Point cadet possessed by a single-minded ambition to succeed—a transformation that his classmates perceived as naked ambition. There was nothing wrong with aspiring eventually to become the first general in his class, but it was tactless to let it become common knowledge in a boastful fashion. Patton also bragged that he would letter at West Point in football, a feat he was unable to accomplish. His belief that he was different from other cadets, that he possessed a unique sense of commitment they lacked, that he was special where they were simply ordinary, was bound to breed resentment—and it did. When the upper classmen learned he had been at VMI, the hazing intensified. Other military institutions were regarded as "tin schools," none more so than VMI. During the summer of 1904 Patton was frequently and often forcibly reminded that he was now at West Point. His classmates soon scornfully dubbed him with the nickname "Georgie."

Nevertheless, despite his boastful attempts to portray himself as a tough guy, the ultramacho image that Patton cultivated in later years was not fully present in the youthful West Point cadet. Inside he remained a tenderhearted young man, always anxious to please his father and requiring constant parental approval and encouragement for everything he did.[34]

Patton's letters to Beatrice and his parents in the autumn of 1904 focused almost exclusively on his academic difficulties. West Point required extensive memorization, which turned out to be the only means by which Patton could keep pace with the demands of his instructors. The technique of memorizing he had learned from Nannie and Papa at Lake Vineyard now benefited him.

Although Patton struggled with academic subjects, he had no such problems on the parade ground, where he was far more comfortable. "I have been perfect so far in drill regulations," he proudly informed Papa.[35]

The grind was interrupted in mid-September by a welcome visit from his mother, Aunt Nannie, and his sister, Nita. "They were all looking splendidly and Nita seems to be quite grown up. I don't believe I ever will be."[36] Little has been written about Patton's loving relationship with his sister, which was devoid of jealousy or envy. To the end of his life he was protective and gentlemanly, and once said of her that while some varnishes "can hide the flaws in base wood, it cannot improve that which is already perfect."[37] Nita Patton fully reciprocated her brother's love and admiration. She lived most of her life in the Patton ancestral home at Lake Vineyard, and later turned it into a shrine dedicated to her famous brother. Swords, pistols, rifles, and machine guns, and a large portrait dominated the main room of the house.[38]

Throughout most of Patton's years at VMI and West Point,

either his mother or his doting Aunt Nannie lived in nearby lodgings. They wanted to be near "the boy" in case he needed anything. There are some pathetic letters between Georgie's parents, written during that time, telling each other how they miss each other, and how someday, when the children are grown, they will be together, never to part. There are many references to "walking hand-in-hand into the sunset." But in the meantime, they encouraged each other to stay near "the boy" . . . and bear their mutual loneliness as best they could.[39]

There is, however, no evidence in any of Patton's voluminous correspondence that their presence either reassured or inspired him, and his innermost feelings continued to be revealed mainly in his intimate letters to his beloved Papa, who rarely left California.

Though Patton and his roommate, Henry Ayres, had both attended VMI, they rarely agreed on anything. Ayres had a penchant for finding trouble, as well as a propensity for settling problems with fisticuffs. Patton regarded anyone who tempted fate as stupid; in turn Ayres thought Georgie arrogant. Trouble erupted one cold night when the two fought over whether their window should remain open or shut. A brutal fight ensued that wrecked their room and earned Patton a swollen face that reduced his eyes to slits. Miraculously their brawl escaped the attention of the authorities, and within a week the two fought again, "with Ayres getting a little the best of it, though he . . . had to use a toothbrush handle to pry his lips from between his teeth, so badly was his mouth swollen."[40]

As he continued to struggle in the classroom, Patton spoke openly in his letters of the uncertainty of his surviving academically. "I have my old fault in studying I don't consentrate but daudle along . . . I am also getting too many demerits for foolish things."[41] One of the daily dilemmas faced by dyslexics is that others believe they are merely stupid. There are few torments worse than being publicly identified as "slow." The harder he tried the worse he seemed to do, and by early October 1904 Patton's slide in English had worsened:

> I am doing rottin . . . and unless I do much better will . . . not even stop at the bottom I got an instructor who in an evil moment found out my utter lack of knowledge about English Grammar so he has been questioning me on it with much regularity and I with equal exactness have flunked. . . . "I don't seem to give a dam" I only hope that I will shake it off as I did in camp but I am absolutely worthless I know that I should study and don't. I see my lack of preparation today but tomorrow will be in the same fix. . . . If I were only my self of a year ago I would ask nothing better for then I tried and took a vital interest but now, *o! hell.*[42]

One of his most anguished letters read: "I am a characterless, lazy, stupid, yet ambitious dreamer; who will degenerate into a third rate second lieutenant and never command anything more than a platoon."[43]

Patton tried out for the football team but was cut and played intramural football for his cadet company, vowing he would try harder than ever the following year to make the varsity team. Football also became an excuse to resume his courtship of Beatrice. Cadets were allotted tickets to Army home games, and he announced his delight if she would come to any or all of them.[44]

Patton dabbled in poetry throughout most of his adult life, and in one of the first efforts, in 1904, he describes the fall landscape: "The hills across the river are very pretty now with all the different colors of Autumn." But to an imaginative mind they might almost seem to exemplify "[e]arthly vanity which takes on like the trees its most gaudy clothing just before it is forever quenched by the chill winter of failure."[45]

As the time neared when his progress would be formally noted, he wrote: "The best thing—the only thing now left for me to do is to by doubly

hard work live down the effects of a poor start"[46] What he could not understand was that his problem lay not in his study habits but his dyslexia. His
academic report for October was a mixed blessing. Of 153 cadets in the
fourth class, he rated 55th in math, 14th in Drill Regulations, but a dismal
139th in English.[47] It seemed to bolster his confidence, however, that he was
passing English. "My old instructor was a scoundril and did not like me so
he consequently gave me low marks. You need not bother about my being
found in English."[48]

His enthusiasm was fleeting, and less than a week later he reverted to
the pessimistic Patton of old. He complained that he had wasted the first
nineteen years of his life:

> I amount to very little more than when I was a baby. . . . I am fare in
> every thing but good in nothing. It seems to me that for a person to
> amount to some thing they should be good at least in one thing. Other
> boys appear to make successes but though [I] want to I dont succeed. Per-
> haps it is just that I lack that small fraction of courage, will power, or
> what ever it is which makes them succeed. Or perhaps I dont fail any
> worse than any one else only my jealousy makes me think I do.
>
> Still when I look at even my class mates I don't fell [feel] that
> sense of superiority which seemes to me should be felt by a (not great)
> but by a successful man. I some times fear that I am one of these
> darned dreamers with a willing spirit but a weak flesh a man who is
> always going to succeed but who never does. Should I be such an one
> it would have been far more merciful had I died ten years ago for I at
> least can imagine no more infernal hell than to be forced to live—fail-
> ure. . . . I am not sure I will be a general. . . . Perhaps I show weakness
> to write this letter but for the past three weeks I have had such an over
> powering sense of my own worthlessness that I had to give expression
> to it.[49]

To Patton perception was everything and his low self-esteem left him convinced he had little to offer anyone. Other than Papa there was no one to tell
him he was dead wrong.

In November Beatrice wrote to announce that she would soon be making
her debut in Boston. Patton fumbled his reply:

> You cant imagine how funny it seems to me that you are "coming
> out." Don't get angry but . . . I don't hold you in half as much awe as I
> did . . . any of the other people whose comming outs I remember. Now
> this is sent because you havent got lots more sence than they had because
> you have . . . well you dont seem very old. . . . Now please dont be mad

with me for if you want me to tell you that I think you are very old, sedate, and all the rest of it I will.*

He was obliged to decline Ellen Banning Ayer's formal invitation to her daughter's debut, but wrote to Beatrice that he hoped "you will have the very best time in the world at your coming out. . . ."[50]

Patton's November report showed dramatic improvement in English to seventy-first, but he remained fearful that there was too little time left "to redeem my honor."[51] After his math grade plunged during a bad week, Patton predicted that it might be better if he were to fail his plebe year. By some tortured logic he thought it would enhance his chance for a promotion to cadet corporal, the highest rank attainable by a third classman. But:

> I actually think that if I don't get a corp I will die so from the present out look you had better bring a coffin east in the spring. It may be a method of developing my character to give me such hard luck . . . I fancy there is no one in my class who so hates to be last or who tries so hard to be first and who so utterly fails.[52]

In December 1904 Patton deliberately faked illness in order to postpone having to sit for a forthcoming written recitation, an act that in later years would have been a violation of the honor code—for which the penalty is dismissal.[53] The ploy backfired when the West Point surgeon refused to excuse him from the dreaded recitation and then denied his request to return to full duty for a week. Kept on a liquid diet, Patton returned to duty "in a condition little short of starvation."[54] He confided to Beatrice that he had been frightened. "I think that in this attempt I illustrated Scotts verse 'o what a tangled web we weave, When first we practice to deceive.' only mine was a perfect 'Gordion knot.' "[55]

The December exams were of crucial importance, and plebes were on academic probation until they passed them. "Those of the class who passed became cadets; those who failed did not receive their warrants."[56] Patton passed. Papa did his best to encourage his son by letter, but Patton's self-esteem needed such constant boosting that his encouragement lasted only until the next academic crisis began the cycle all over again. Papa tried hard, but his pampered son was impossible to please. Even a minor lull between letters brought immediate wails of protest to his long-suffering father. Pat-

---

*Patton saved virtually every scrap of paper he ever wrote, resulting in a massive collection of papers. However, after his death Beatrice Ayer Patton burned most of her letters to her husband, including those written during his West Point years, a fact confirmed during a 1991 interview with her daughter, Ruth Ellen Totten. Thus we can only glimpse Beatrice's inner feelings through the letters that survived the flames of her fireplace, from the observations of family and friends, and by inference from Patton's replies.

ton's lifeline was letters of encouragement from home containing news of life in Los Angeles and the Patton household. But if they did not arrive with regularity, he would become depressed and occasionally testy. "You had better wake up and write," he once chastised his father. "I haven't had a letter for over two weeks."[57] After passing his midyear exams in December 1904 he reacted churlishly to his father's praise by writing, "You said I did well my last report but I can hardly commend your judgement. . . . In fact in my own opinion I am pretty darned poor, not that I am in the least discouraged I am only ashamed; but sad to tell my shame only makes me cuss myself and does not make me work harder."[58]

He alternated between the humility of self-deprecation and the nagging of a spoiled son accustomed to immediate acquiescence to his every whim, whether it was for stamps, writing paper, clothing, money, his favorite candy, or to perform some social deed on his behalf. Although he barely tolerated having Nannie or Mama nearby, he unhesitatingly used them to his own ends. Robert points out that his grandfather wanted it both ways, and wonders why they tolerated his selfish behavior: how, for example, he would disdain Nannie's presence one moment as a damned nuisance, and the next take advantage of her obsessive devotion to wangle a favor.[59]

The answer is complex but is consistent with the fact that he was their very own creation: a youth raised in the long shadow of dead, heroic Pattons. Papa, skillfully assisted by Nannie, had deliberately cultivated in his son the hope that he might become the heroic figure he himself had not. Robert fittingly points out that there was a certain vulnerability in Patton that others who knew him well could clearly discern. In private he did not take himself seriously, displayed a chivalry that charmed, and a vulnerability often manifested by tears over the smallest thing. Moreover, despite his faults and his nagging manner, Patton genuinely loved his parents, and, although unable to comprehend his affliction, they did all in their power to protect him.[60]

Athletics became a valuable outlet from the ordeal of the classroom. During the Christmas holidays in 1904 he decided to make the best of winter and took up skating, but his early attempts brought about frequent contact between his posterior and the ice, leaving him assured that his future as an outdoor sportsman lay elsewhere. Acknowledging to Beatrice that he had, "just gotten to the stage where I look upon any one who can stand up on the nasty things, with feelings near to worship," Patton vowed to keep at it.[61]

Other than playacting at Lake Vineyard, Patton had never fenced before entering West Point. Although fencing was a formal part of the curriculum, he also began to practice with a broadsword in the anonymity of the gym, where he could give free expression to his perception that being a great swordsman was an essential trait of a great general. One day while practicing in the corner of the gym with another cadet in order to avoid attention, Alvin Barber, a first classman and West Point's crack fencer, asked Patton to spar with him. He performed so well that henceforth fencing became an essential part of his life. For the moment, however, Patton was content merely to learn the rudiments of this sport of gentlemen, and by the spring of 1905 was sufficiently confident of his ability to write, "If I never amount to anything else I can turn instructor with the broad sword, for I am the best of the best in the class. It is lots of fun and I practice it as much as possible."[62] Eventually, he began to refer to himself as "Master of the Sword," a prophecy that would eventually be fulfilled.

Patton also joined the track team and described himself as "turning into a gray hound" even though on one occasion, "I almost brought my fiery life to a sudden and tragic conclusion," after tripping over a hurdle at full speed, falling on his head, and badly skinning his knees.[63] A week later he severely strained a tendon in his left ankle. These were the first of what would be a lifetime of accidents that might have killed unluckier or less well-conditioned men. Not only was Patton accident prone, but his impatience with the everyday problems of life sometimes led to acts of folly, such as the time he was left "hurting like hell" after deliberately cutting his gum with a pocket knife in an attempt to "let the beast of a [wisdom] tooth through." [64]

In early March the Corps of Cadets entrained for Washington, D.C., where it proudly marched down Pennsylvania Avenue in a howling wind during Theodore Roosevelt's inauguration parade. The train ride to Washington had been a nightmare of delays, late meals, and an air of disorganization that generated "little electric currents of rage running [throughout the train] . . . men beat on the floor with gun butts and gave little short howls like mad dogs," before finally disembarking in foot-deep mud in the black of a cold, raw March night.[65]

Beatrice and her parents had journeyed to Washington for the great event, and at the president's inaugural ball the night of March 4, George and Beatrice danced together for the first time. Nineteen-year-old George S. Patton was clearly smitten, and "had the finest time in the world," he wrote to her afterward.[66] He confessed to Papa that he could have danced on a hot stove with the same eagerness he and Beatrice had danced on the cold stone floor of the ballroom. Obliged to meet a midnight cadet curfew, Patton felt like Cinderella and departed with great reluctance. "Comeing out certainly had a wonderfully good effect on Beatrice . . . she is the prettiest girl I ever saw. . . . I am now probably suffering from a bad attack of puppy love."[67]

Throughout his life Patton avidly employed any means at his disposal to help advance or influence his career. In the spring of 1905 the entire Patton family was making plans to come east to see Georgie, and remain in New York for the summer. Patton, however, viewed the visit as more of an opportunity to better his standing and urged his father not to "forget to cultivate the Tacks (U.S. Army officers assigned to West Point to teach military subjects and to administer discipline in the Corps of Cadets), the other officers don't count but if you can get on the good side of the tacks I might get a 'make' [promotion]."[68]

Bolstered by Beatrice's encouraging letters, and finally admitting to himself that he was in love, Patton found even the impending end-of-year examinations less threatening. However, his March report (103d in French) left him as discouraged as ever:

> At last I have found my true place . . . I am nearly hopeless. I don't know what is the matter for I certainly work . . . I hate to be so low ranking for I still think that I am smarter than most of the men who rank me . . . it is exasperating to see a lot of fools who don't care beat you out when you work hard . . . I cant think of any thing but my own worthlessness so will stop writing.

> Your goaty son,
> George S. Patton, Jr.[69]

Promising to "do my damdest," Patton approached the exams "confident of getting through" and "happy at the prospects of an end to study."[70] Dyslexia so often results in fleeting highs and prolonged lows that Patton's euphoria was destined not to last. A week later the pendulum swung when another accident brought his morale crashing (literally): He fell over the seventh hurdle during a track meet and finished last instead of second. It resulted in a tortured letter to his father:

Dear Papa:

> I seem to be destined to damnation. . . . For an hour I would have gladly died Infact I had hysteria in a mild form. . . . A[yres] thought I was crazy and patted me on the back and raised thunder over me. It was pretty hard for I hate to be beaten and try so hard and Fail . . . and unless I make a 2.7 out of three Monday will have to take the math. exam too.

> Pa I am stupid there is no use talking I am stupid. It is truly unfortunate that such earnestness and tenacity and so much ambition should have been put into a body incapable of doing any thing but wish. . . . Asside from my sprained ankle gotten this afternoon I am well and sad.

> With lots of love
> George S. Patton, Jr.[71]

For the first time in Patton's life his father did not respond with sooth-ing platitudes but instead bluntly but compassionately said that while ambi-tion and winning were admirable traits, "You must school yourself to meet defeat and failure without bitterness—and to take your comfort in having striven worthily and done your best. . . . If you do not get a 'Corp'—take it with a smile—and keep on trying—your reward will come." Moreover, wrote Papa, from that time on, "You have got to fight your battles alone—to meet victory or defeat as becomes a man and a gentleman. . . . I have no fears for you—I know you are doing your best—and that is all you can do. When you have done that—for me you have *won*.. . . God bless and keep you."[72]

Two days later Mr. Patton received a telegram that he had secretly been dreading.

THE WESTERN UNION TELEGRAPH COMPANY

June 12, 1905
Dated: West Point, N.Y.

To: George S. Patton
San Gabriel, Calif

DID NOT PASS MATH. TURNED BACK TO THE NEXT CLASS.
PROBABLY FURLOUGH THIS SUMMER. WILL WIRE DEFINITELY.

//S// G.S. PATTON[73]

Mr. Patton immediately cabled his distraught son: IT IS ALL RIGHT MY BOY AND ALL FOR THE BEST. GOD BLESS YOU. FATHER. He also cabled his wife, who had only just arrived at West Point: DON'T WORRY—ALL FOR BEST—WIRE IF NECESSARY I SHOULD COME—BUT I HOPE [FOR AN] IMMEDI-ATE FURLOUGH [FOR GEORGIE]—AND ALL HOME.

When Patton returned to West Point in the autumn of 1905 to begin his mili-tary career all over again, he brought with him a small black notebook in which he began to record his thoughts, goals, and the happenings in his life. Not surprisingly, one of the first notations was: "Do your damdest always." To his death, Patton never fully understood that during his first year at West Point he had indeed "done his damdest" and had fallen victim not to stupid-ity or laziness but to dyslexia. Words written in 1984 by Dr. Harold N. Levinson, about his aim in the diagnosis and treatment of dyslexia words, would certainly have applied to Cadet George S. Patton in 1905: "To let no current or future dyslexic continue to feel stupid, dumb, and ugly."[74]

CHAPTER 7

"If at First
You Don't Succeed . . ."

You must do your damdest and win. Remember that is
what you live for. Oh you must! You have got to do
some thing! Never stop until you have gained the top or
a grave.

—CADET GEORGE SMITH PATTON JR.

Patton spent most of the summer of 1905 in California at the family's summer home on Catalina Island, preparing for his second year at West Point, hunting wild goats, and fishing. He continued to be accident prone and in July took the first of what would be many hard falls from the back of a horse. He wrote to Beatrice that, "it took me an hour to get the cactus all out of the horse and a lot of it is still in me." He also killed a goat but likened it to the experience of hell, "when I go to join the spirits of the goats I have killed for a low ranking man at the point is called a goat also."[1]

For George S. Patton it was a bittersweet time. The low self-esteem of dyslexics is the bane of their existence, but in this crucial moment in their son's life, his parents never "showed by word or deed their disappointment at my failure."[2] Their intuition that his problems were outside his control may have done more to contribute to Patton's future success than any other single act, even though it was more likely attributable to their penchant for never discussing family embarrassments.

Free of recriminations from his parents, Patton looked forward to his second year at West Point with a semblance of hope that he might eventually graduate. "It is scarcely possible that I may ever again be so happy and so sad at the same time. . . ."[3] Despite his problems and the ache at being at the other end of the continent from Beatrice, Patton "had a peach of a time at home which . . . is better than I deserved."[4] Unable to fathom his dyslexia, Patton attributed his turnback to destiny.[5]

Beatrice wrote to sympathize, and although her words helped, he could not envision a successful future. "I am awfully glad you all understand how it was with me. . . . [but] Looking at it in cold blood I have pretty small chance of coming out [graduating]."[6] Patton's greatest fear was that Beatrice would give up on him and not wait for him to graduate before committing herself to another. "[I]t never occurred to Georgie that Ma was in love with him," wrote his daughter. "He was too humble at that point in his life—and his failure to pass had humbled him even more deeply. . . ."[7]

That summer Patton began filling his newly acquired black notebook. The first was a poem about love, war, and fair maidens:

> *Oh! here's to the snarl of the striving steel*
> *When eye met eye on the foughten fiel'*
> *And the life went out with the entering steel*
> *In the days when war was war.*
>
> *Oh here's to the men who fought and strove*
> *Who parried and hacked and thrust and clove*
> *Who fought for honor and fought for love*
> *In the days etc.*
>
> *Oh! here's to the maids for whom they fought*
> *For whom they strove of whom they thought*
> *The maids whos love they nobly sought*
> *In the days etc.*[8]

In the notebook Patton recorded a hodgepodge of thoughts, poetry, principles of war, diagrams, admonitions, and social notes that affirmed that his terrible first year at West Point had matured him. Patton inscribed five principles that would guide him:

Genius is an immense capacity for taking pains.
Always do more than is required of you. . .
What then of death? is not the taps of death but first call to the reveille
    of eternal life.
We live in deeds not years.
You can be what you *will* to be.[9]

*   *   *

Between notes on history and tactics were such gems as a quotation from Beatrice that Patton was, "one of the few people in the world who can be courteous without being idiotic." Patton's homilies were also about love, among them this couplet: "Here is to the man who kisses a girl and keeps the secret close / But damn the man who kisses a girl and then goes out to boast." Beatrice had also told him: "Don't argue with a man If you cant convince him lick him. If you cant lick him keep still."

Patton's courtship continued during the summer of 1905, and he became more ardent with each letter they exchanged. As he prepared to return to West Point he hoped that "I may see you just for a little while. You see I have to come east a little early so as to be sure of getting to the Point on time . . . would you mind writing . . . and telling me whether you will be at home and if I may come to see you."[10]

Patton spent several of the most exhilarating days of his life in Beverly, Massachusetts, that August. In his first letter from West Point to his father, he no longer bothered to conceal his feelings. "While in Beverly I had absolutely a perfect time the Ayers did every thing in the world that they could for me. . . . That Beat. is certainly the best thing in her line in the world and I swallowed her hook to the swivlc (as one says of a fish) I guess that I am a fool to have such a case at such an early age. . . ."[11]

Fourth classmen who are turned back at West Point must retake their fourth year academic courses but are exempt from the hazing and harassment of plebe life. In his second year at West Point Patton existed in a sort of limbo where he was neither plebe nor third classman. Of his first day back he recorded: "It seemed very funny the first night at parade not to have any one tell me to get my shoulders back you see it is the first one I ever went to where I was not a 'Plebe.' If I am not the meanest corp.[oral] in the world when I do become one it will be a wonder for I will then have been a private three years."[12]

He tried out again for the football team and although relegated to the third string as a left end, the experience helped his self-confidence. After one pileup during a scrimmage against the varsity team, Patton was the last to emerge from under twenty-one men, but threw the runner for a yard loss. Then he injured his right arm and was taken to the infirmary, his football career over for the year. Although his injury had "probably saved me from breaking my neck," he remained determined to one day earn a place on the varsity team and the coveted A letter, writing to Beatrice that "unless I am found or killed I shall make this team before I graduate."[13] After recovering he turned his attention to the broadsword and in the spring of 1906, he again ran the high hurdles on the track team.[14]

Unlike the hospitalization of the previous December during which Patton was miserable, this time he was in high spirits and to pass the time even composed a poem that he described as "a by product of my pen as I tried to write the siven ages of man." For a young man with dreams of glory, the poem is remarkable for its intuitive sense of the folly of war:

> *And now we sing not of the stage of life*
> *But of that stage of which there is no counterpart on earth*
> *The stages of the life of a cadet.*
> *First there's the boy*
> *Unapt by nature he for aught of hardship*
> *Yet his early mind perverted by untruthful literature*
> *He sees a picture of war glorified*
> *And longs to be a soldier.*
>
> *He dreams of blood, of glory and of strife*
> *And knows not blood is pain and glory but a bubble*
> *Which bursts when riper age has made his folly clear.*
> *But why alas does knowledge come too late*
> *That we who in our youth did know it not*
> *Have wrecked our lives by learning it too late.*
> . . . .[15]

After the turbulence of his plebe year, Patton's second year at West Point seemed like an oasis of calm. He studied hard and for the first time saw his efforts rewarded, noting that, if anything, his classes were too easy. Midway through his first semester Patton ranked fourteenth in math, thirty-seventh in English and first in drill regulations in a class of 152 cadets.[16]

His letters to Beatrice became more frequent and included a standing invitation to visit him any time she "happened" to be in New York. "To forestall the excuse that I did not invite you to any particular hop I here by ask you to every dance to be given at West Point from now until I graduate and ask only that you let me know three or four days in advance so that I may make out your dance card."[17] She came to West Point for Patton's twentieth birthday, an occasion marred only by Patton's lifelong tendency to overindulge his sweet tooth for candy and cake. Afterwards he wrote joyfully to his mother that, "I am twenty and still alive. I had a peach of a birthday . . . truely one is in a bad way to be 20 and as hard hit as I am."[18]

When he was discharged from the hospital Patton weighed 160 pounds "stripped." His confidence grew and his letters to his parents and Beatrice continued to reflect an upbeat mood so different from the traumatic letters of a few months earlier. In early December, President Theodore Roosevelt attended the Army-Navy game, played at Princeton. At half time Roosevelt was formally escorted by sixteen cadets from the navy to the army side of the field. Patton was one of the cadets chosen, "to my great surprise and greater joy. I was the only one of the bunch who was not an army man or in some other way had a pull. After all it pays to be military. . . . You should have seen me . . . I nearly burst. It is a pretty big thing to be an escort to a president. . . . I think that I must have looked quite impressive but [Beatrice] was not in the least awed." He signed the letter, "Your distinguished son, George S. Patton, Jr."[19]

A third straight Christmas away from home failed to dampen his spirits, even though he lamented at still being a cadet private, "for this Pleabism is hard on a patrician like me. . . ."[20] One of his roommates was "found" but Patton sailed through the semiannual examinations with flying colors.[21]

But after a stressful week in January 1906, during which he made poor marks in French and English, he reverted to calling himself stupid and lazy, and wrote such a depressing letter to Beatrice that he burned it.[22] His complaints about being "naturally stupid" were a reflection of a dyslexic's need constantly to prove himself worthy in the eyes of others:

> Darn it I am a goat and had just as well learn to be content with my lot. . . . I grow weary of the rear rank and . . . am not very popular not because there is any thing the matter except that I am "Too damed military." That is I am better than they are. Now no one is more unjust than he who feels himself an inferior but dam them let them keep on some day I will show and make them feel how infernally inferior they are.[23]

Although she could only visit West Point infrequently, Beatrice proved to be the perfect foil for the emotional downside of Patton's dyslexia. Her letters bolstered his morale and made him begin to feel good about himself. After his melancholy letter in early 1906, Beatrice wrote to praise him and the results were dramatic. "Thank you for what you wrote about me it was the finest thing I ever heard about my self."[24]

Toward the end of his second year Patton began to anticipate the delights of advancement to the third class and set his sights on promotion to corporal, not just any corporal, but first corporal, the most prestigious office in his class. "I think I shall die when I get it. . . ." He also had been giving a good deal of thought to soldiering and declared that, "given the chance I will carve my name on some thing biger than a section room bench." But first he had to pass his year-end examinations after faltering slightly in March.[25]

The year ended on a high note when Patton not only routinely passed the exams with grades in the top third of his class but was selected second corporal.[26] His promotion meant that he would be a cadreman (one of those in charge of hands-on supervision in the plebe summer camp). Unfortunately and perhaps predictably, Patton was overzealous and managed to irritate virtually everyone, from his classmates to the tactical officers and, of course, the poor plebes who ran afoul of him. He soon learned that harassing plebes did not "afford me [as] much amusement as I had hoped. At first I hated to get after them and felt like a bruit when ever I 'crawled' them but soon I began to feel angry when ever I saw a Plebe and have been mad for a bout three days and that is not a very pleasing condition of mind."[27]

He was excited by the opportunity to be in command but seemed to have no concept of when enough was enough. "I believe that I reported more men than any other officer of the Day this summer," he told Beatrice, who had admonished him not to become overzealous. In addition to a long-running squabble with several classmates, Patton's first taste of authority ended in shock and disappointment when he was demoted from second to sixth corporal in late August. "Why I don't know unless I was too d—— military. For I certainly am the only man who can march this class. It certainly is the biggest shock I have had for a long time for I have tried hard to be a good soldier. Please don't think me too worth less for I will be adjutant yet."[28]

Although angry and hurt, Patton displayed no inclination to change his basic precept of demanding very high standards of those in his charge. "It is true that they don't like me but when I get out in front of them the foolishness stops," he proclaimed. His grandson accurately notes that while Patton was certainly headstrong, the later perception of him as disobedient or insubordinate was simply untrue and part of the iconoclastic image he himself had perpetuated through a lifteime of practice. To the contrary, as his oldest friend in the U.S. Army, Gen. Dwight D. Eisenhower, said of him: "[Patton] is fundamentally so avid for recognition as a great military commander that he will ruthlessly suppress any habit of his that will tend to jeopardize it."[29]

Patton continued to fill his notebook with thoughts, from ideas and quotes from Clausewitz to notes about cavalry. Some reflected his isolation: "Let people talk and be damed. You do what leads to your ambition and when you get the power remember those who laughed"; and, "No sacrefice is too great if by it you can attain an end."[30] Not surprisingly, after his demotion he wrote: "*Never trust* a person who has or thinks he has a cause to dislike you. He will surely stick you in the back. . . . I think that there must be a Destiny. . . . Look well whether the game be worth the candle. . . . I hope and pray that *what ever* it cost I shall gane my desire."[31]

On his twenty-first birthday he delivered some stern reflections on his life. As usual the dyslexia enhanced his harsh judgments of himself: "Man lives by deeds not years. At least I hope so for if at 42 I have as little to cause me self respect as I now have at 21 I had better say with Hector 'gape earth and swallow me.'"[32]

*   *   *

Patton's third class year was relatively uneventful. For the third straight year he tried and failed to make the varsity football team, and was again relegated to the third string. Football in those days was a brutal game played with a bare minimum of protection and, as Patton dutifully reported to his father, "I have not been doing at all well . . . [yet] I have managed to get pretty well bruised up and have been so stiff that I could scarcely bend over enough to put on my shoes."[33] He seemed to have lost his confidence of the previous year but believed the better food the football team received at their training table was worth the pain. Putting it all in perspective, Patton observed that he hoped his sister Nita's coming-out party was nice and "she raizes hell in society a lot better than I do in athletics."[34]

In mid-March 1907 Patton regained his second corporal stripes and jovially wrote to his father, "I take the opportunity of telling you I am in matters vegetable, animal, and mineral, the very living model of a modern Second Corporal. I begin to think that when it comes to scientific boning of bootlick I am almost it. . . . Of course, I spiffed a lot and braced my self more than anyone in the class. . . . With any sort of luck now I should get either Sgt. Major or even 1st Sergeant in June, one of which puts me in line for 1st Captain; the other for Adjutant. Still, I must not count my chickens too soon."[35]

Academically, Patton remained a mediocre student and stood near the bottom of his class in French and in the middle in drawing and math. By January 1907 his class had been reduced to 114 cadets and although Patton remained a borderline student in both French and Spanish, he successfully passed his exams and, as he had predicted, was promoted to sergeant major, the highest cadet position in the second class.

He had not been home since the summer of 1905 and spent most of the summer of 1907 in Southern California, and ten days with Beatrice in Massachusetts prior to his second class year. Patton was a dutiful son who adored his parents and never consciously caused either of them embarrassment or distress. One of the lone exceptions occurred that summer when, "for the only time in my life so far as I know I hurt Papa's feelings," when he publicly insisted on wearing his father's silk opera hat with a tuxedo to a dance. "Papa caught up with us with my straw hat and said the silk one was ridiculous and that just the day before he had seen Mr. Huntington wearing a straw with a tuxedo. I said that Papa did nothing but copy Mr. H. which was not true and hurt his feelings. I wore the straw."[36]

As the West Point days passed with increasingly rapidity, Patton continued to dream of glory and triumph in his chosen profession. In November 1907, an entry in his notebook served as a vivid invocation of the fire to succeed that burned within:

George Patton you have seen what the enthusiasm of men can mean for things done. As God lives you must of your self merit and obtain such applause by your own efforts and remember that though at times of quiet this may not seem worth much yet at the last it is the only thing and to obtain it life and happiness are small sacrifices. You have done your damdest and failed now you must do your damdest and win. Remember that is what you live for. Oh you must! You have got to do some thing! Never stop until you have gained the top or a grave.[37]

Patton also reminded himself that, "By perserverence and study and eternal desire any man can be great." Later entries were equally passionate:

What ever may happen what ever may be the temptation to slump . . . remember that you are a soldier and ever seek command. . . . If you die not a soldier and having had a chance to be one I pray God to dam you George Patton. . . . Never Never Never stop being ambitious . . . you have but one life live it to the full glory and be willing to pay. . . . Nothing is too small to do to win. . . . If you infringe your honor you have sold your soul. . . . An imperious conscience is your greatest defense. . . . Daring is wisdom it is the highest part of war.

His words of advice to himself bore the stamp of maturity far beyond the years of one so young and inexperienced. They were later to become the essence of his military philosophy:

- There is but one time to do a thing that is the first.
- Do not console your self with the thought, "I can make a mess of this but next time I will do better," there is no next time . . . there is but one time to win a battle or a campaign. It must be won the first time. Dont . . . say oh well I will with draw and win next week such a course is ruin absolute.
- In making an attack make only one and carry it through to the last house holder. Make the men who have gained ground lay down and hold it. What folly to let them fall back to take part in a fresh assault. . . . Remember Frederick the Great [who said to his faltering troops] "Come on men do you want to live for ever?"

And this from Napoleon: "To command an army well a general must think of nothing else."[38] Even as Patton used his black notebook to devise the tenets by which he would later govern his military career, the dyslexia continued to ambush his self-esteem in what became a daily struggle for recognition. It continued to haunt even his dreams. Shortly before the February 1908 cadet promotion list was announced, Patton wrote to his father of a dream in which, "I was the adjutant and I was having a fine time then next night I dreamed I was found and I was having a hell of a time. Every body was pointing their fingers at me and calling me stupid. I was so scared that I woke up. . . . There is no use talking the only thing I am good at is military. I can't to save my life care about studies and if did care about them I have not got the head for them. I can not sit down and study because I like to as some of these fools do."[39]

He expounded that to become a great soldier entailed learning from history in order to be

> so thoroughly conversant with all sorts of military possibilities that when ever an occasion arises he has at hand with out effort on his part a parallel. To attain this end I think it is necessary for a man to begin to read military history in its earliest and hence crudest form and to follow it down in natural sequence permitting his mind to grow with his subject until he can grasp with out effort the most abstruce question of the science of war. . . .[40]

Patton never conceived of any other career for himself. The obsession with succeeding in emulating the deeds of his forefathers blinded him to thoughts of any other profession. When Beatrice asked him if he were "to take away heredity, and love of excitement and desire of reputation will I like the army life? . . . if you take away those three things what is left in life? If there is any thing to live for except those three things you have taken away as worthless all I have ever dreamed of. . . ."[41]

He hoped his final West Point standing would be high enough to earn him a place in the cavalry arm. He wrote passionately of his desire to become a successful soldier:

> I am fool enough to think that I am one of those who may teach the world its value. . . . Now that is a rash thing to say and if twenty years from now with no war and no promotion some one should say "I thought *you* were going to teach the world?" why it would hurt. But if there were no dreamers I honestly think there would be little advance and even dreams may, no must come true if a man gives his life for what he believes. Of course it is hard for any one particularly for me who have never done much to give reasons why he believes in my self but foolish as it seems I do believe in my self. I know that if there is war "which God grant" I will make a name or at worst an end . . . [perhaps] it is only the folly of a boy dreamer who has so long lived in a world of imaginary battles that they only seem real and every thing else unreal . . . is it not better for a person to stick to the profession he has always thought about than for him to do something for which he has no

particular desire or capacity and I certainly have none for any thing but the army. I have thought about it so long that all the other parts of ambition are dead.[42]

Patton's early writings substantiate that the West Point years were far more than an essential period of preparation for his army commission. When the time came, Patton put into practice the theories of how men should be led and battles fought.

Although Patton's early writings reflect brilliance, he had yet to demonstrate to his contemporaries that his fiery intensity was anything more than the ravings of a temperamental opportunist. To the end of his cadet days he remained a dogmatic and unpopular cadet, a young man on the make. Beatrice attended the West Point–Yale football game in the autumn of 1908, and pronounced it "great fun watching him prance up and down the field at inspection, chest bulging and chevrons shining, serenely unconscious of the two pairs of cousinly eyes anxiously fixed upon him. He seemed by far the most military person on the post that day; our only anxiety was that he might break in two at the waistline."[43]

Patton's belief in discipline was seen as excessive, but when he was promoted to regimental adjutant in February 1908 he exulted that he had at last begun to live up to the high standards he had set for himself. As cadet adjutant Patton was always center stage in front of the Corps of Cadets. Impeccably dressed and a master of military posture, Patton was now the focus of attention. The adjutant read the orders of the day each morning and led the corps wherever they marched. It was what he had long coveted, and he made the most of it. His new position also entitled him to move from the drab barracks into a tower room in the First Division.

Blumenson notes: "He was very busy that spring. His studies, his duties as adjutant, and his activities in track, polo, horsemanship, and the broad sword gave him little leisure time. What also consumed much of his time was his habit of changing his uniform, he told Beatrice, fifteen times a day in order to be clean and neat always."[44] He thanked Beatrice for her faith in him. "Do you remember long ago . . . I said I would like to be adjutant but feared I never would be and you said I would . . . you probably think me a fool for being so pleased with my self but realy I am not so teribly stuck up for when you come down to it I have only beaten about a hundred men and that is not so very much. I wish it were more lots more. . . ."

He also began to display the enigmatic traits that would later characterize the public perception of him as a general. One day on the range, as Patton toiled in the pits, raising, lowering, and marking targets during rifle practice, he inexplicably stood up during the firing. He faced the firing line, unflinching as bullets angrily splattered all around him, and later stated that he had done it to test his courage under fire.[45] More to the point was the miracle that he was not wounded or killed giving in to the urge to satisfy his curiosity and prove his courage.

On another occasion, in February 1908, Patton was attending a class in electricity when a cadet asked if the spark from an induction coil would kill him if it passed through his hand:

> The Prof invited him to try it and the man refused. I hardly liked to see the class so easily scared so after the lecture I went down and asked if I could try it for I realy was curious to see how it would feel. At first he did not want me to do it but at last he allowed me and it hardly hurt at all though my arm is still a little stiff. He did not like at all having his bluff called though.[46]

However, it was in quite another area of his life that Patton's courage was soon to be put to the test. It had been nearly six years since he had fallen in love with Beatrice Ayer. Although their love for one another remained unspoken, it was plain that neither could bear to be without the other. Their letters had grown in frequency and intensity, at the expense of letters to his parents, which decreased dramatically in number during his last two years at West Point. Blumenson notes that Beatrice caused Patton—for perhaps the first time in his life—to think seriously about himself and his ambitions. However, despite gentle admonitions, she failed to cure him of his habit of writing excessively vain, "I"-oriented letters: He remained hopelessly "self-centered and visualized the world as an extension of himself."[47]

As 1908 drew to a close, graduation was no longer a distant fantasy but a mere six months away. Patton finally came to the realization that it was not enough simply to be in love with Beatrice, and that his life would never be complete unless he were to wed this woman who so thoroughly dominated every facet of his life. To avoid losing her he must soon not only propose to Beatrice, but win her hand from her formidable father, Frederick Ayer. George Patton found the prospect more terrifying than anything he had yet done in his young life.

PART IV

Junior Cavalry Officer
(1909–1917)

You know, looking back on Patton, he has been a
general all his life.

—CAVALRY OFFICER (FORT RILEY, KANSAS, CA. 1914)

CHAPTER 8

Love and Marriage

I have loved Beatrice ever since the summer
in California.

—CADET GEORGE S. PATTON

For nearly six years Patton's courtship of Beatrice Banning Ayer was primarily by letters, which focused on his life at VMI or West Point, his academic difficulties, and matters of male vanity, such as his wish that she could have seen him performing on the parade ground or in athletics. An early manifestation of his lifelong fear of aging resulted in a near paranoia about his appearance, in particular, his rapidly thinning "beautiful golden curls [which] were disappearing and he worried a lot about losing his hair—which, of course, he eventually did, although he loyally used Ayer Hair Vigor for years and years."

Patton found the Ayer family

so awfully nice that it is positively oppressive. We ride, swim, sail, motor and see *Bee*. . . . I have not told her that she is the only girl I have ever loved but she is—though it would be fatal for me to mention that fact now or perhaps ever. She is very nice to me and I think she likes me for she has been wearing my favorite color dresses ever since I said I liked them. Gosh, those skirted bi-peds at Catalina, who pawn themselves off as girls, aren't in it with her shadow. But, O Lord what an ass I am!

Although she missed Georgie, her love for him did not prevent Beatrice Ayer from having the time of her life, exulting in the exuberance of youth in a whirlwind of "balls, parties, beaus, concerts, art exhibits, theatres—all the things young ladies did." Her love life also became a topic of great amusement for her siblings. When Beatrice required an emergency appendectomy, Ellen Banning Ayer took charge of her recuperation with her customary bustling efficiency. In the Ayer family there was no such thing as a minor illness or minor surgery. In honor of the occasion Freddie and Kay Ayer composed a ditty for Beatrice, and to the end of their lives, would hum it whenever she "acted up." It was sung to the tune of the then-popular "Reuben, Reuben."

> *Georgie Porgie, so they say*
> *Goes a-courting every day.*
>
> *Sword and pistol by his side*
> *Beatrice Ayer for his bride.*
>
> *Doctor, Doctor, can you tell*
> *What will make poor Beatrice well?*
>
> *She is sick and she might die*
> *That would make poor Georgie cry.*
>
> *Down in the valley where the green grass grows*
> *There sits Beatrice, sweet as a rose.*
>
> *And she sings and she sings and she sings all day*
> *And she sings for Georgie to pass that way.*

Patton spent his Christmas furlough of 1908 at the Ayers' Commonwealth Avenue mansion in Boston. In their six years of courtship, both had adroitly sidestepped the question of actually declaring their love for each other. As daughter Ruth Ellen observes: "How much more fun a courtship was in those days when 'a glance, a bird-like turn of the head, the pressure of a hand' was as much a thrill to the suffering lover as getting right into bed is nowadays. They could savor every moment. . . ." Finally, Patton's feelings overruled his inability to express to Beatrice that he was madly in love with her. Some idea of his anxiety can be inferred from the fact that he was sufficiently terrified that Beatrice might refuse his declaration to have composed a fake telegram, which he carried in his pocket, purporting to order him back to West Point immediately.

He needn't have worried, for Beatrice readily responded that she loved him. In one of his rare letters to his mother, Patton confessed that, "if I dont tell somebody I will bust . . . every time I saw B I wanted to kiss her and had a hell of a time not to every time I went driving or walking." Almost certain that Beatrice would spurn his declaration, Patton kept postponing the event until one afternoon when they found themselves alone in the family library:

> I am not a coward but this business of pointing a gun at yourself and pulling the trigger in order to prove it is not loaded is not particularly enjoyable. . . . Well I did it and it was a very empty gun. Thank God! Oh she was a dear . . . never have I spent such an afternoon . . . were I never to smile again I still had not lived for nothing. The strange part is I think she has known it for a long long time six years. She said I should have known it too what an ass I have been. . . .

Neither would commit to a formal engagement, and their future was left undecided, in part because Patton expected to be sent to the Philippines, a routine assignment for a newly commissioned second lieutenant in the U.S. Army of the day. Beatrice told her parents of her love for Georgie Patton, and they merely smiled and said that of course they already knew, and they liked him.

Nevertheless Beatrice's love for him "scares me to death," and he seemed incapable of actually proposing marriage, apparently satisfied merely to have announced what had so long been in his heart, and convinced that he was unworthy of her.[1] After reading one of Beatrice's love letters, Patton wrote: "I sat with out moving for an hour and then went out and ran around the hills like a loon."[2] To his mother he acknowledged, "Gosh I have the queerest feelings. I am actually afraid."

Despite his feelings for Beatrice, Patton occasionally dated other women, and in early 1909 he was briefly smitten with an attractive and sexually alluring Vassar coed named Kate, whose family was very wealthy. "Yet if you put [her] $40,000,000 against the B. I fear that I would take the B. ass that I am when with the money I could be a general in no time," apparently forgetting that Frederick Ayer was also a multimillionaire.[3] Even when it came to love, his powerful ambition always came first.

Patton wrote to Frederick Ayer to spell out his career intentions, but in reality to announce himself as Beatrice's suitor. In his reply the wily old man—who did not want a career soldier for a son-in-law—coyly suggested that Patton ought to seek civilian work after performing his military service. "Your plan of life," he wrote, "is alright if you can have a command for a year in God's country and not in the Philippines. Fighting malaria is not war." Ayer had stopped just short of issuing an outright refusal, only because Patton had yet to ask formally for Beatrice's hand in marriage.[4]

Patton again wrote to Ayer to explain that "I have loved Beatrice ever since the summer [of 1902] in California," and then respectfully declared that after careful self-examination he believed that "I am only capable of being a soldier."[5] It was an ever-so-polite-but-firm rebuttal, indicating that he had no intention of taking up some other profession. In no doubt that marriage between his daughter and young Patton was inevitable but accustomed to getting his way, Frederick Ayer nevertheless began a deliberate campaign to wean him from an army career of which "he had the typical New England view of the 'brutal and licentious mercenary' . . . the Yankee[s] always thought of the army as the refuge for thieves and murderers."

The dialogue between the benevolent tycoon and the young cadet resumed a few days later when Ayer wrote:

> A man in the army must . . . always feel unsettled—That his location and home life are subject to the dictation and possible freak of another whom he may despise or even hate. A man like you should be independent of such control—His own master—Free to act and develop in the open world. . . . Every independent man should choose his own course in life. . . . I believe that the qualities of a good soldier . . . will help a man win in whatever calling he may choose.[6]

Patton lost none of his admiration for Ayer, and for a time briefly considered resigning from the army "all for the love of you. . . . The 'all' I just used is in derision of the little I can do to be good enough for you."

In yet another attempt to explain himself to Mr. Ayer, Patton wrote that he had never found any logical reasons why he wanted to be an army officer: "I only feel it inside. It is as natural for me to be a soldier as it is to breathe. . . ."[7]

Patton was torn between his love for Beatrice, the magnetic lure of a military career, and the inevitability of a clash with Frederick Ayer. As was his custom, he vented his feelings to his parents in a letter that searingly reflects the conflict between his love for Beatrice Banning Ayer and his obsessive ambition to be a successful soldier:

January 17, 1909

Dear Mama and Papa:

> . . . All my life I have done every thing I could to be a soldier for I feel inside that it is my job and that war will come. . . . When however I proposed to Beatrice I did something from instinct and against reason. At least it seems unlogical because she does not like war . . . and because a soldier should not marry. This because money seems an excellent tool, not for my own use, but to buy success and if I were unmarried I could get more things by paying attention to daughters of prominent people if

necessary marrying one of them. Now these things are not nice but they are logical and I had carefully planned to climb the ladder and I had a pretty clear field. But when I see B. all logic goes to hell. . . .

Realy I have no strength of character. I know what is right but I think of Beat and straight way stop all sane mental manoevers and fall down and worship her and enjoy doing it better than any thing else in the world. Am I an ass or just human. Would an embrionic great man have acted as I did or do I show my self a *mess*. God knows I am worried to death. I have got to, do you understand got to be great it is no foolish child dream it is me as ever I will be. I am different from other men my age. All they want to do is live happily and die old. I would be willing to live in torture, die tomorrow if for one day I could be realy great . . .

There is inside me a burning something. . . . I wake up at night in a cold sweat imagining that I have lived and done nothing. Perhaps all people have it but I dont believe they do. Perhaps I am crazy. There is no use concealing things from you for you might help and ought to know. I want to be a dictator or a president. No small ambition for a goat yet why not some one has to be why not me At least I think a man can not do wrong to try. . . .

> Your devoted much perplexed ass of a son
> George Patton

At one point Patton had no sooner implied to Ayer that he would resign his commission after a year and seek a suitable civilian profession than he wrote to Beatrice that, "before everything else I am a soldier. . . . I am *not* a *patriot*. The only thing I care for are you and my self in that I may be worthy of you . . . I may loose ambition and become a cleark and sit by a fire but if with sanity comes contentment with the middle of life may I never be sane. I dont fear failure I only fear a slowing up of the engine which is pounding on the in side saying up—up—some one must be on top why not you . . . But my hopes and views are so insane that I dont think he understands them; no one does, not even you and Lord knows I bother you seven days in the week with them— poor B. . . . I begin to think I was a fool to tell you all I did at first in this letter but I would rather make you made [mad] now than dissapoint you later."[8]

Unable to resolve his anguished conflict between love and career, he wrote again ten days later: "I have tried hard for your sake as you dont like it to think of another profession but it is like pulling hair. I dont know or care about other things. . . . I dont want to make money and that is all success in business amounts to. . . . As far as will goes I could but it would be self murder . . . I dont want to boast but the chances seem to point to my being some thing in the army out of it nothing."[9]

The problem was that not only did Mr. Ayer oppose his daughter living under the grim conditions of army life but Beatrice was "in the hell of a fix," torn between her love for Patton and her perceived duty to her eighty-seven-year-old father. Patton complained that the Ayers "dont understand the army business at all. It is inconceavable to them that a man can have no desire to gain [a fortune] and can wish to kill a fellow being by any such coarse method as shooting."[10]

Beatrice's overwhelming fear was that Patton would defy her father and choose the army over her. Although she dreaded the prospect of becoming an army wife, and for a time attempted gently to persuade him that perhaps a civilian career was not so bad after all, what if he elected to put his career over marriage to her, as seemed likely? She decided *that* prospect was even worse and wrote: "You must decide alone and then I will go with you *any* where." Henceforth Patton would no longer apologize for a profession he was proud of, and he informed Mr. Ayer that he intended to remain in the army.[11]

By early 1909 Patton's letters to Beatrice had undergone an abrupt transition. For nearly six years they had always formally begun: "Dear Beatrice." Now that he had declared his love for her, they began, as they would for the rest of his life, with: "Darling Beatrice," "Darling Beat," or "Beatrice Darling." Previously content to relate the daily happenings in his life, Patton suddenly began to articulate his feelings. "If I were ordered to the north pole or Hell I would love you and be true to you just the same. . . . Dearest I am devoted to you I love and worship you and only you for ever and ever."[12]

Still, their future seemed irretrievably bogged down in Patton's uncertain prospects in the army and Beatrice's reluctance to leave her family. Patton remained torn between his army career and its inevitable hardships, which hardly befitted a young woman of Beatrice's social stature:

> Even if I did suggest to my self leaving the army for you I could not live in Boston or any other decent place I am not made that way I dont fit but get fits in cities if I live in them. . . . I fear I would not do much more than live like a squash or some such plant. Beat that sounds as though I thought more of my self than of you but I dont dearest can't you understand? please do . . . it still bothers me [that] it might be awful for you . . . living in deserts and swamps. . . . Hell d—— d—— d—— is what I feel like saying when I think very long over such things . . . Perhaps I ought not to write them to you but they do bother me a lot.[13]

Fortunately Beatrice never learned that Patton had also written to his father: "It seems ridiculous that I should have fallen in love with a girl so completely useless as a wife for an army officer and there is no use avoiding that fact she has not one redeeming feature for a wife aside from the fact that I am madly in love with her and she is a d—— sight worse in love with me."[14] But he did tell her:

> The path that I seem intended to follow is not one that any one else will enjoy particularly you who are not by nature intended for such a life being too grand and bright and well educated. A woman to like the army ought to be narrow minded not over bright and half educated and that is the truth how ever gloomy it may sound. If I ever get to what I want you will like it and perhaps forget the mud of the road. . . . Oh! Beaty I love you every second and think of you and long for and pray for you. I would rather have your love than the world and all. Darling dearest darlingest Beatrice.[15]

The question of marriage thus remained unresolved: Patton did not formally propose, and Beatrice was as yet unable to bring herself to leave the Ayer family nest.

Patton's fifth and final year at West Point was his most successful one. At the annual Field Day in June 1908 he established a new school record in the 220-yard hurdles, won the 120-yard hurdles, and rounded out the most triumphant day of his athletic career at West Point as the runner-up in the 220-yard dash.[16] His feat won him a place in the cadet yearbook, the *Howitzer,* with the fifteen other wearers of the coveted letter *A*.[17] Patton also shot "Expert" with the rifle and continued to excel in swordsmanship.

The text accompanying the photograph of those first classmen who had won their *A* included a notation that, "It is said that Georgie Patton has compiled for future generals, a rule for winning any battle under any combination of circumstances."[18]

Patton had indeed composed a list of the traits of a future general. Many years later, after his son donated part of his father's vast collection of books to the Friends of the West Point Library, the following notation was unearthed on the final page of a textbook called *Elements of Strategy*: "End of last lesson in Engineering. Last lesson as Cadet, Thank God." Inscribed in the back cover was:

QUALITIES OF A GREAT GENERAL

1. Tactically aggressive (loves a fight)
2. Strength of character
3. Steadiness of purpose
4. Acceptance of responsibility

5.   Energy
6.   Good health and strength
                    //signed// George Patton
                    Cadet
                    U.S.M.A.
                    April 29, 1909[19]

To the end of his West Point days he remained virtually friendless, and his reputation as a "quilloid" (a description coined by the cadets for one who puts others on report for any infraction of the rule book) endured.[20] In the summer of 1908 eight first classmen were expelled after being caught hazing. Although he saw no harm in what his classmates had done, Patton seemed disinclined to haze, preferring instead to enforce discipline. When the commandant, Lt. Col. Robert L. Howze, demoted most of the first class cadet officers during a shakeup in the summer of 1908, Patton was unaffected. After Howze sent for Patton and "said some very foolish yet very nice things about me. I went about inflated to the bursting point all day."[21]

Patton's stubborn independence was demonstrated one day at the noon meal when he led the Corps of Cadets into the mess hall and, as they awaited the "Take seats" command, an unpopular [army] officer entered. Cadet custom was to stand silently at attention until the officer got the message and left the room. Patton, however, believed that any officer, whatever his alleged misdeeds, was deserving of proper respect for his rank. On this occasion, when the corps began to give the officer the silent treatment, Patton became so disgusted that he marched them out of the mess hall.[22]

Martin Blumenson assesses Patton's often stormy relations with his classmates as made up of equal parts of affection, admiration, and exasperation at his obsessive quest for eventual greatness, which most thought better left unspoken instead of publicly articulated. Sadly, despite five years at West Point, by the time Patton graduated in 1909, he could claim none of them as close friends.[23]

The editors of the *Howitzer* lampooned Patton with this entry under his photograph:

Confusion reigned supreme. The barracks were being shaken by a violent earthquake, and men came tumbling out of their divisions in all stages of dishabille. Suddenly the Cadet Lieutenant and Adjutant appeared in the area, faultlessly attired, as usual. Walking with firm step across the area, he halted, executed a proper about face and the stentorian tones rang out, "Battalion Attention-n-n-n! Cadets will refrain from being unduly shaken up. There will be no yelling in the area. The earthquake will cease immediately. By order-r-r-r of Lieutenant Colonel Howze!"

... Two broken arms bear witness to his zeal, as well as his misfortune, on the football field—but misfortune could not run fast enough to overtake him on the track. We believe that George's heart, despite its armored exterior, has a big soft spot inside, and have heard that Cupid has penetrated with his dart where the explosive D might fail.[24]

After agonizing over what branch of the army to apply for, Patton decided on the cavalry. Concerned lest he make a choice that would not guarantee him the best opportunity to gain promotion and, eventually, fame, he asked one of the Regular Army officers in the Department of Tactics, Capt. Charles P. Summerall, what he should do. Summerall, who in later years would become not only a close friend but a trusted mentor, advised Patton to select the cavalry.[25] He daydreamed that he would be assigned to the 15th Cavalry, stationed at Fort Myer, Virginia, across the Potomac River from Washington, D.C., where, "if I have any capacity," he wrote his father, "I might . . . get a boot lick on people of note."

Patton presented Beatrice with a photograph of himself in the uniform of a first classman. She had it mounted in a small, oval antique silver frame and kept it for the rest of her life. On the back she inscribed: "An unwritten page. I wonder what will be written on it. June 1909."[26]

Reflecting on his five years at West Point, Patton recalled his first day as a plebe. "How scared I was that day how earnest in my desire to succeed. I failed a little and did not succeed in much . . . but I did my best as I found it at least I have tried to do that always." He also reaffirmed his love for Beatrice: "That night in the moon light at Avalon a boy loved a girl. And now a boy a little older loves a girl even more dear than then. How fortunate almost unthinkably fortunate I was to meet you. . . . To day when I wrote my name on the board in the section room for the last time it was your face that I saw and it came to me how when a plebe I had first written my name on a similar board accross the hall how it had been your face that had been there . . . your cadet is going you made him may the officer be worthy of you."[27]

After five grueling years the ordeal of West Point ended on June 11, 1909, when 103 first classmen marched proudly in honor of the commencement speaker, the recently appointed secretary of war, Jacob McGavock Dickinson, who spoke of both the illustrious history of West Point and how its graduates had distinguished themselves in earlier wars.

Patton's final class standing was forty-sixth.[28] If that seems average, it was not, for had it not been for his dyslexia he undoubtedly would have graduated near the top of his class. Aside from himself, three of Patton's classmates were destined to later wear the four stars of a full general: William Hood Simpson, Jacob Loucks Devers, and Robert Lawrence

Eichelberger. A fourth, John Clifford Hodges Lee, became a controversial three-star general, and a dropout from Patton's original class of 1908, Courtney Hicks Hodges, was also destined to command an army in Europe in 1944–45.[29]

The Patton family and Beatrice were in attendance at this seminal event in the life of newly commissioned 2d Lt. George Smith Patton, Cavalry, U.S. Army. In 1909 Patton's father had been appointed to the West Point Board of Visitors and, after the ceremony ended, Capt. Morton F. Smith, of the Department of Tactics, told him, "You need have no worry over the future of your boy. He always does more than is asked of him." As Patton would later write, "Papa was pleased and told me this. With some exceptions I have always lived up to the ideal."[30]

The Pattons left for New York City and the graduation dinner of the class of 1909 at the elegant Hotel Astor. The next day his parents took him to Tiffany's, where they purchased his graduation present, an expensive watch which Patton later carried with him during the Mexican expedition in 1916 and throughout World War I.[31]

With his commission came orders to report to his first duty station, Fort Sheridan, Illinois, in September 1909. As he left West Point for an uncertain future in the Lilliputian Regular Army of the time, Patton had good reason to be proud: He had not only overcome his dyslexia but had achieved the all-important first step on the long road toward fulfilling his destiny of becoming a celebrated general.

Patton spent most of the summer of 1909 in California, much of it fishing off Catalina Island. He wrote to Ellen Banning Ayer to ask if he would be welcome to visit at the end of August before reporting to Fort Sheridan. His letter was both solicitous (for he genuinely cherished Ellie Ayer) and unintentionally humorous, noting that, "All the family would send you their love were they awake yet perhaps it is as well they are asleep for it would be but useless to send their love when you already have it."[32]

Like his prospective in-laws, Patton could also be the master of the grand gesture. The entire Ayer family was gathered on the terrace at Avalon in their finest summer dress when 2d Lt. Patton suddenly appeared on horseback. But instead of dismounting in front of the house, he kept riding right up the stairs and onto the terrace, where he alighted in front of Beatrice and solemnly bowed at her feet.[33]

During their brief time together the lovers burned the candle at both ends in a spree of socializing and yachting. The visit was enlivened one morning by the appearance on the terrace of a brass band, hired for the occasion of Katharine Ayer's nineteenth birthday. It was typically Ayer: festive and splashy, and Patton relished every moment. Beatrice wrote to Aunt Nannie that Georgie "certainly did look handsome—you are right, 'beautiful' is the word—and we have had the happiest visit . . . how I appreciate your having spared him to us; we have been so happy. . . . I am using one of your hankies to sponge the tears off [this letter]. . . . "[34]

Aunt Nannie could not bear to be separated from her beloved nephew and had followed him to both VMI and West Point. Now it was Chicago, where she rented an apartment to remain near "the Boy" during the first six months of his assignment to Fort Sheridan. Patton never recorded his feelings about Nannie's bizarre behavior, but he did write and visit her occasionally. Her absence from California must have pleased Mr. Patton, who was temporarily freed of her oppressive presence at Lake Vineyard.

Located just outside Chicago, Fort Sheridan was a small cavalry post and the home of an element of the 15th Cavalry Regiment, which was also based at Fort Myer, Virginia, and Fort Leavenworth, Kansas. Army regulations decreed that the first official act of a newly commissioned officer was to inform the regimental adjutant where he could be contacted prior to reporting for duty. Patton was unaware that the regimental adjutant was based not at Fort Leavenworth but Fort Myer. His letter was intercepted by his new commanding officer, Capt. F. C. Marshall, and returned to him with a polite suggestion that he write again. Patton immediately penned a second letter to the adjutant, candidly acknowledging his mistake.

Capt. Francis Cutler Marshall, Patton's first commanding officer, was one of the few men whom it can be said that Patton revered. An 1890 West Point graduate, Marshall had served with the 8th Cavalry in the Sioux Indian War of 1890–91, participated in the China Relief Expedition in 1900–1901, and taught in the Department of Tactics during four of Patton's five years at West Point. A man of dignity, patience, and quiet but very effective leadership, Marshall was perhaps the ideal man to tutor Lieutenant Patton on his first duty assignment. An officer who led by example, Marshall genuinely liked the ambitious young Patton, who reciprocated by trying hard to model himself on this outstanding soldier. Within days of his arrival Patton would write glowingly: "I am certainly glad that I got into Capt. Marshall's troop as he teaches me things that the other two [newly assigned lieutenants] never hear about from their troop commanders."[35]

Patton thought of his new assignment as a fourth plebe year,[36] and in a sense he was quite right for, like any other inexperienced second lieutenant, he had a great deal to learn. To absorb and master the duties of a junior officer was one thing, but the most important and clearly the most difficult aspect of his new profession was to earn the respect of his men. This an officer must achieve on his own.

Although his quarters at West Point were crude, nothing could have prepared him for Fort Sheridan. The small Regular Army of 1910 was poorly paid, and its enlisted ranks were populated by men of scant education and

little ambition. Between 1895—when the figure was less than $52 million—
and 1916, the average military budget was barely $150 million, and funds to
improve the squalid living conditions in the army's remote outposts were
virtually nonexistent.[37] Patton's bachelor quarters were on the third floor
of what was little better than a slum. They were, he wrote Beatrice, "pretty
bad . . . empty and very dirty. Save for one mahoginy desk . . . and an iron
bed there is no furniture."[38]

The active army of 1909 totaled 84,971 (4,299 officers and 80,672
enlisted men), most of whom was dispersed in small military garrisons that,
like Fort Sheridan, were rarely of more than battalion size.[39] Referred to as
"hitching post" forts, these tiny outposts were a relic of eighteenth-century
frontier America that now failed to meet the vision of Theodore Roosevelt's
administration in Washington, which believed that the U.S. Army must be
modernized to fight any potential future war.[40] However, attempts at consoli-
dating permanent military forts by secretaries of war and chiefs of staff, dating
to 1880, had failed to overcome congressional pork-barreling in allocating
funds to their constituencies where they would earn votes for re-election.[41]

On his first day of duty in the U.S. Army Patton had yet to be assigned specific duties and merely observed the unfolding of the daily routine of life in
a peacetime cavalry troop:

> I have a horse who is not bad nor good and a saddle . . . also an orderly
> who calls me "the Lieutenant" [it was customary at that time to address
> lieutenants as "Mister"]. I have not yet got my feet [on the ground] and so
> am a bit upset in my mind I will be all right in a day or two.[42]

Fort Sheridan was better situated than most army installations, but even
at best, duty at such a place was dreary. Patton was detailed to stable duty at
4:30 P.M., and was thus rarely able to take advantage of the custom that permitted an officer to visit the city any time between noon and midnight. The
winters were frigid, and the heat and humidity in summer were oppressive.
Always prone to hay fever, he suffered "to beat hell."

Fort Sheridan also housed a military prison, and one of Patton's duties
whenever he served as officer of the guard was to ensure that no one
escaped. "When I went on guard I had to go into the cage and count
them. . . . I felt like a convict my self before I had finished . . . they were
awfully ignorant looking and lord how they 'stunk.'" Patton recorded that
he ought to wear a diving suit to ward off the smell. "In their sleep they look
like the people in Hell in a book of Dante I used to read. . . . Think of it a
hundred and twenty ruined souls it is to me a very sad sight."[43]

Captain Marshall greatly eased Patton's transition. He also congratulated him on his forthcoming marriage, which, of course, was news to Pat-

ton who had yet to propose formally to Beatrice. "It seems some paper had a full account of it," he wrote to her. "I suppose coming events cast their shadows before but what a long shadow. Some times it scares me."

Marshall believed in realistic training and, gradually, he entrusted Patton with more responsibility by assigning him a variety of duties. He learned lessons that would later be used to good advantage. Patton especially enjoyed the practice marches and bivouacs that were standard fare for cavalry units, and performed well when put in charge of the advance patrol of Troop K, which entailed setting up, securing, and manning campsites. It fed his dreams that "Some day I will have a big tent and a refrigerator and a stove and a trunk and a lot of men cussing me for having so much baggage."[44]

Patton was unimpressed with the quarters of some married officers, on whom he was obliged by army protocol to pay a social call. "I met some people who positively make me ill when I think of the effect they would have on—a girl I know." He observed that the Marshalls managed to afford three servants and a striker (enlisted orderly) and dressed and behaved better than others at Fort Sheridan. Patton was at this stage of his life perhaps unduly impressed by people's status and tended to make judgments on the basis of social standing. One night he was invited to dine with a very wealthy executive in a stately mansion in the affluent Chicago suburb of Lake Forest, and came away in awe of his wealth. That Patton may not have desired a career in business, where wealth was an objective, did not prevent him from feeling comfortable—and somewhat envious—amid such trappings.

Regardless of the monotony of duty at Fort Sheridan and the empty feeling of being without Beatrice, Patton led an active social life both on post and in Chicago. In his off-duty time he attended football games (once diagramming and sending plays back to the coach at West Point), dances, formal balls at which he proudly wore his new suit and silk hat, the theater, dining out and the opera ("I . . . really like it though I don't understand a hell of a lot about it."). He dated several women but apparently failed to develop any lasting attachments, although he wrote to his father that one of them, a certain Miss Bishop, appeared to be in love with him. "She is a nice girl and rather pretty and very useful. I have taken her to a lot of plays. More than Beat. would care to hear about." On Halloween, Captain Marshall provided him a costume, complete with mask, and to the noisy accompaniment of loudly clanging pots and pans, the Marshall entourage visited each of the quarters on post and "raised the duce generally."[45] As a bachelor, Patton was frequently invited to dinner at the homes of married junior officers. His active and satisfying life more than made up for the long, dreary hours of repetitive troop duties.

Although Patton had begun to make a name for himself at Fort Sheridan, there were many harsh lessons yet to be learned. The first dealt with leadership, and was an example of his occasional tendency to permit his emotions to overrule his good sense. One afternoon he found an untethered horse in the stables, cursed a nearby soldier, and ordered him to run, tie the horse, and then run back. "This makes the other men laugh at him and so is an excellent punishment. The man did not understand me or thought he would dead beat so he started to walk fast. I got mad and yelled 'Run dam you Run.'" Patton sensed that he had needlessly humiliated one of his soldiers, and after assembling those who witnessed the incident, he publicly apologized for his behavior. "It sounds easy to write about but was one of the hardest things I ever did I think but I am glad I did it now that it is done."[46]

It would not be the last time Patton would embarrass himself by an irrational act. The positive aspect was that he not only quickly understood that swearing at an enlisted man was a serious impropriety for an officer, but possessed the courage to humble himself by issuing an immediate public apology. It was the first of many such apologies that Patton would deliver during his lifetime, making the act—as his grandson notes—"an art form, though it would never get any easier."[47]

Patton was so dissatisfied by the idleness of garrison duty that he and a fellow officer undertook a course of military studies four afternoons a week. Whether or not he cared to admit it, Patton was rapidly learning what Mr. Ayer had pointed out: that a military career in the peace time army was a dead end. He had scarcely been at Fort Sheridan a month before complaining: "God but I wish there would be a war. Until there is I see no hope of my ever needing to buy any more furnature [for his bachelor quarters] for you cant fill an empty heart with chairs."[48]

There was no better example of the boredom of military life than the high desertion rate ("five thousand last year in the best paied army on earth"), and the high occupancy rate of the post stockade. Drunkenness remained a serious problem, and in many places, particularly the larger cities, enlisted men were regarded with utter contempt. Discipline was swift and harsh, the duty usually mind numbing, and, although American soldiers were better paid than those of other countries, the pay was nevertheless wretchedly low. Some soldiers enlisted without fully understanding what they were letting themselves in for; literacy was a major problem, and conditions, though improved from those of the post–Civil War period, were at best austere.

Lucian K. Truscott, who as one of the outstanding generals of World War II would one day be intimately associated with Patton, has left a superb account of life in the cavalry. He writes of an existence that began well before dawn and was governed by the call of the bugle:

Troop officers took turns standing Reveille with their troops and, after roll call, reported to the officer of the day. . . . The bugle blew on numerous other occasions during the day at prescribed intervals: Mess, Police, Sick, and Drill calls. No bells, PA systems, telephone calls, or radio messages. Just the bugle. And we followed its orders.[49]

In short, the U.S. Army 2d Lt. George S. Patton entered in 1909 was rooted in the cavalry, infantry, and Coast Artillery Corps, and woefully unprepared to enter the era of modern warfare ushered in by the new century. Although significant reforms were being undertaken in Washington, there was little evidence of it in the "hitching post" forts such as Sheridan.

Patton spent Christmas of 1909 on furlough with Beatrice and the Ayers. As they had the previous year, the two lovers discussed marriage, and once again the matter was left unresolved. Patton's version, as written to his father several weeks later was:

> B. for rasons if any best known to her self will not say definitely that she will marry me . . . but as she has no objection or appears to have none to my telling her family . . . She certainly is an awful ass but then I suppose it is hard to blame a person for clinging to the present happiness and being slightly scared at changing it . . . I think that the only thing that scares her is the thought of leaving her family not of leaving her wealth for she does not give a dam for that. Any way she has got to or never leave them for I would look like an ass hanging around much longer.[50]

Nevertheless it was another magical time for both. Patton compared Beatrice to a moon goddess coming down to

> dress the world in silver. . . . I can but think that you are she. . . . Beaty you were so perfectly lovly that it was sort of sacred. I think that that evening and new years were more perfect than hours on this earth are. I was so happy I feared to wake up and when I did it was even more wonderful being still true.[51]

Eventually the Ayers would have to approve if they were to marry. Patton wanted no part of facing them, and it was eventually decided that Beatrice would seek their permission. Mr. Ayer's obvious disapproval of Patton's chosen profession was evident, but as her daughter has written, Beatrice "was the most strong-minded of the Ayer children; she always got her way in the end." But not without a fight:

> Granfer Ayer was not going to allow his darling to be taken off to a God forsaken Army Post in the middle of nowhere. . . . If Georgie really cared for precious Beatrice, he would do the decent thing, resign his commission and take a job (one which was waiting for him) in or near Boston. . . . [He] had not had his own way for nearly 80 years without getting accustomed to the feeling. . . . Georgie retaliated with heated despair. . . . He said that if he could not marry Ma he would never marry anyone, but that the army was his chosen career and his profession and he was going to stick to it, and serve his country in the way he knew best.

A confrontation between the strong-willed Beatrice and her autocratic father was inevitable, with neither party apparently prepared to be outmaneuvered. Frederick Ayer liked George Patton, but he was adamant that his beloved daughter not be exposed to army life, and he refused to approve of the marriage. Beatrice retaliated by locking herself in her bedroom and staging a hunger strike, refusing all suggestions that she change her mind.

> Every night on the stroke of midnight, there would be a basket dangling down the stairwell on a string to be filled by [sister] Kay with the choicest pickings from the larder. . . . She seemed to grow paler and more fragile each day . . . [and] could be seen leaning sadly on her windowsill. After about a week of this, Granfer Ayer capitulated to his concerned women-folk, and gave permission for the wedding to proceed. . . .
>
> [Frederick Ayer] had the last word, of a sort. I have never seen the letter he wrote to Georgie, but it was quoted to me by Ma and by Nita Patton, who said that it literally changed Georgie's life. Granfer Ayer said that he was persuaded that Georgie's vocation lay in the military, and so they would henceforth each do the thing they did best; he would earn the money, and Georgie would earn the glory.

Frederick Ayer's change of heart may well have been in response to a letter Patton sent him in early 1910, detailing his financial worth, which seemed to impress the old man, no doubt because he believed his future son-in-law to be of dubious wealth. Ayer's letter was a complete concession and rather generous:

> We confide her to you with our love and fullest confidence. Her frequent return to us we shall look for with longing. Will you keep this ever

in mind? I know your accommodations are not what you would have them in private life, but think that Beatrice enjoys roughing it to some extent, as all good sailors and soldiers must; and you know she is a pretty good sailor.[52]

Uncertain even of obtaining quarters at Fort Sheridan, Patton wrote his mother, "Gosh I do hate to get the poor kid into a life so different from that to which she is accustomed but what can I do." However, Frederick Ayer was taking no chances and informed Patton:

It has been my custom when my children have married and left our home to give them a monthly income, and [I] shall do the same to Beatrice and the younger ones. This is without regard to their circumstances. . . . I admire your firmness of purpose in sticking to the army until more strongly tempted by another occupation, and with every good wish for your early and steady advancement, I am,

Sincerely, your friend,
F. Ayer.[33]

The Marshalls invited Beatrice to visit Fort Sheridan to see for herself what army life would be like. She was appalled, and "at some point during her visit Ma offered to break their engagement as she didn't think she could ever make a good army wife. Georgie kissed her out of that fancy." Mr. Ayer was right; Beatrice would ever be the good soldier, but she did not have to like it and, in fact, came to loathe the shabbiness of their quarters at Fort Sheridan and the other "hitching posts" where Patton was assigned. Both clans duly arrived for what was as much an inspection of the army as it was to view the future homestead of Lieutenant and Mrs. George S. Patton.

It turned into a family affair: The Ayers escorted Beatrice from Massachusetts, while Mama and Nita arrived from California. Patton's father had decided to throw his hat back into the political arena and was attending a political convention at which he hoped to be nominated by the Democrats for the U.S. Senate. Although he would fail in 1910, Patton did obtain the Democratic nomination in 1916.

Mama disapproved of the lodgings her son would eventually acquire, and the two quarreled, with Patton attempting to explain to his mother that money simply would not buy him rank or a decent set of quarters; in those days post quarters were assigned solely by rank. Moreover, there was also a policy called "ranking out," which meant that an officer with seniority could force a more junior officer to vacate his quarters. As one of the most junior officers at Fort Sheridan, Patton was at the bottom of the eligibility list and wrote his father: "I am absolutely at sea as to where in Hell I will live when I get back but the only thing to do is to trust to God and good luck."[54]

As the wedding day approached everyone began to show signs of strain. Beatrice cried a good deal, leading Patton to admonish, "[B]ut B. if you go on crying at all times you will hurt your looks so stop."[55] To help relieve the tension the members of the two families began writing letters to one another. Patton's genuine fondness for Ellie Ayer is evident in a charming letter he wrote to her shortly after the engagement became official.

> I will try to take the best care in the world of her and keep her happy and in that way I may perhaps be able to cheer you a little. . . . When I think of marrying Beatrice and of her leaving home for me it frightens and surprises me. Frightness, at the thought of the wonder of it and of the vastness of my responsibility. . . . Truly God is good to me. Yet I think that in all reverence you are equally so. . . .
>
> Devotidely yours,
> George S. Patton, Jr.[56]

Patton was granted five weeks' leave for the marriage and a honeymoon in England. As the wedding day approached, Mr. Patton wrote lovingly to his future daughter-in-law. Beginning, "My Dear Little Girl," he spoke of his loneliness at Lake Vineyard without his family and how his spirits were raised by a letter from Beatrice. He had sent her some orange blossoms from Lake Vineyard and, "In my memory I saw the little white headed kid who was 'Georgie' . . . and I thought Lake Vineyard must send its flowery greetings to the dear little girl . . . so you see my dear little girl—when you put those poor faded blossoms in the scent bag you put with them this joyful loving thought of this far off father—who surrenders without fear or misgiving—his only 'little boy' to your loving keeping 'for always.' "[57]

At the last minute, Patton wrote with evident relief that he had been assigned quarters and drew a map to show Beatrice where they would live. "You don't know how glad [I am] . . . Two weeks from now we will probably be in a machine somewhere between Prides [Crossing] and Boston . . . It will be so strange I hope you will not be too sad Beaty. I will try to comfort you and darling we love each other so much that we will be happy. God bless our house."[58]

In one of his final letters to Beatrice before their marriage, he still managed to reveal signs of the fire burning within that no amount of happiness was ever destined to quench. Barely ten days before the wedding he wrote, "Beaty we must ammount to some thing."[59] When the occasion demanded it, Patton could be eloquent, as he was in his last romantic letter before the wedding:

May 22, 1910

Darling Beatrice:

This is the last letter I shall write you as your lover only hereafter I shall still be your lover but also your husband. Darling since I wrote my first letter to you almost eight years ago I have grown older and wiser and have thus been enabled to better understand and more clearly see your infinite perfection. So that in a way I may be said to love you more now than then . . . for I have ever loved you to the fullest of my power. God grant that if I develop in no other way my capacity for loving you may increase for it is only by a divine love that I can express to you my gratitude for all you are, have been, and will be to me . . .

When I think of the excited happiness with which my "Rattish" and "Plebish" hands have opened your darling letters of the past, and the emotions of hope or fear and always of pride which those letters elicited . . . I can hardly comprehend that we are hence forth to be one . . .

I have prayed that you should love and marry me yet not at the expense of your happiness so now that my prayer is to be granted it seems certain that you will be happy. God grant it! May our love never be less than now And our ambition as fortunate and as great as our love. Amen.

George[60]

The wedding of George Smith Patton Jr. and Beatrice Banning Ayer took place amid the splendor of Avalon on May 26, 1910, and was one of the noteworthy social events of that year on the North Shore of Massachusetts. Beatrice's engagement ring, which she wore to the end of her life, was a gold miniature of Patton's West Point class ring, in the center of which was a topaz, his birthstone. Beatrice wore the same wedding dress that her mother had worn for her own wedding in 1884. It was trimmed with real orange blossoms from Lake Vineyard, brought on the train by the Pattons in a box of wet cotton. Patton and the five ushers wore their full dress blue uniforms.[61] Beatrice's bridesmaid was her sister, Katharine; Georgie's best man was her brother, Frederick Ayer, Jr.[62]

The only blemish on this otherwise festive occasion occurred the day before the wedding when Mrs. Patton fell ill with influenza and was obliged to remain in her Boston hotel room during the ceremony. Although a great disappointment for the Pattons, it left an enthralled Aunt Nannie in the limelight of what was perhaps the "shining hour" of her life. "She could for an hour live her dream as she stood in the receiving line, next to the only man she had ever loved, with her adored Georgie whom she always thought should have been her son, marrying a girl that she herself truly loved."

The ceremony took place in St. John's Episcopal Church in nearby Beverly Farms against a background of white and green spring blooms that decorated the chancel. It was the first military wedding on the North Shore since the Civil War and the guests were drawn from Boston aristocracy, and from places as far away as New York, Minnesota (Ellie's home state), Virginia, Washington, D.C., and California. The rector of Boston's Old South Church performed the service, after which the newlyweds passed beneath the traditional crossed swords of Patton's classmates and cousin. Mr. Ayer hired a special train to convey the invited guests to and from Boston and carriages ferried them from the station to Avalon which was awash with flowers. A full orchestra on the terrace provided the music, and after Beatrice Patton cut their wedding cake with George's sword, it solemnly played the "Star-Spangled Banner."[63]

After the ceremony, the newlyweds posed for a formal wedding photograph, George in full uniform, with his officer's cap in his right hand and Beatrice in her dazzling wedding gown looking every inch "a rarely lovely bride" as the local paper later described her.[64]

The entire wedding was a typical Ayer production: first-class and lavish, with loving attention paid to the smallest detail. Ellie Ayer dispensed with her multitude of bangles and on this occasion wore only a single strand of pearls and a pendant. As the guests continued to celebrate, Lieutenant Patton and his bride slipped away to Boston where they spent the night before departing the following morning for New York to occupy the bridal suite of the SS *Deutschland* which was to take them to England for their honeymoon.[65]

The following morning when the newlyweds rang for breakfast, instead of a waiter, into their suite marched Ellie Ayer, carrying a single white rose in a crystal vase, followed by several of Beatrice's brothers and sisters. All had risen early that morning for the train ride to Boston to "be there when 'the children awoke.' Ma thought it was terribly thoughtful, but it almost killed Georgie."[66] The public Patton may have been a firebrand but privately he was a shy man, and this Ayer display of affection was decidedly not his idea of togetherness.

By all accounts the honeymoon was delightful. Both were avid travelers and delighted in visiting England, which was George's first ever foreign trip. They landed in Plymouth on June 3 and spent several days exploring Cornwall. After a stop in London, where Patton bought the first of what would soon become a vast collection of military books, they returned home in late June to set about the business of moving to Illinois and establishing their first home.[67] And, as the newlywed George S. Patton would soon learn, his life would never be the same again.

CHAPTER 9

"... And Baby Makes Three"

A marriage without conflicts is almost as inconceivable as a nation without crises.

—ANDRÉ MAUROIS

He had never seen anything as awful and revolting to all his sensibilities as the birth of his first child, and he never got over it.

—RUTH ELLEN PATTON TOTTEN

Beatrice Patton arrived at Fort Sheridan apprehensive, her head filled with seemingly endless advice about how to deal with this "wild westerner" she had married. To many easterners anything west of New England was scarcely better than a foreign land. Chicago was merely a place where one changed trains en route to California. One woman warned her that she ought never to leave her quarters without wearing her hat, as the sight of her long auburn tresses "might rouse some Indian local to go on a scalping spree. . . . She was warned against drinking the water, and told to be sure to ask the butcher if the meat was fresh . . . [and] Ellie gave her some sound advice. . . . [N]ever get intimate with your next door neighbors; never borrow anything; never confide in anyone but your husband, your doctor, your pastor and, of course, your mother."[1]

Beatrice happily reported to Aunt Nannie that "during our short housekeeping experience together," her husband had "added to his other accomplishments that of champion furniture polisher, varnish-and-painter, cook, plumber, carpenter, gardener and heavy chaperone. He will be a piano tuner next."[2]

Their first home was a far cry from Commonwealth Avenue or Lake Vineyard. The rooms were so small that only four chairs could be used in the dining room, and the bedroom closet was so narrow that Patton's straw boater had to be stood on its edge. Shortly before they moved in, the interior of their quarters had been painted peacock blue, with paint the quartermaster had left over from the painting of the railway station. Kay Ayer accompanied them and helped Beatrice establish her first household. The Pattons had barely settled into Quarters 92A at Fort Sheridan when, in early August, Troop K left for an extended period of summer maneuvers in Wisconsin. Patton was placed in charge of transporting the mules and horses. Beatrice would have preferred to accompany her husband and take temporary accommodations in nearby Sparta, Wisconsin, but was dissuaded from doing so by Captain Marshall. Not only would Patton be too busy to see her for more than a few hours a week, but the other officers' wives were remaining at Fort Sheridan.[3] After a short time she and her sister left to spend the remainder of the summer at Pride's Crossing.

Although they lacked realism, mock battles taught Patton a good deal during his first cavalry maneuvers. Marshall placed him in charge of a patrol and assigned him other responsibilities that broadened his experience of life in the cavalry. Although Patton proclaimed it "thrilling" to cut down a fictitious guerrilla force with sabers, his wise old sergeant, who had been through three wars, thought make-believe war "beat them all."[4] It was at this time that Patton began a habit that was to endure throughout his life. He began to read military classics to extend his knowledge and challenge himself to excel, but often found them difficult. Clausewitz was "about as hard reading as any thing can well be and is as full of notes of equal abstruceness as a dog is of flees." Patton missed his bride but had plenty of time on his hands for reading and sleeping and wishing for an assignment elsewhere.

When they were reunited in the middle of September, an elated Beatrice announced that she was pregnant with their first child, which was expected the following March. Patton, however, displayed little enthusiasm:

> To Ma, this was the ultimate goal of a woman's life; to find love, to have it returned, and to be able to bear a child to sanctify that love. . . . Georgie, on the other hand, felt a shadow creeping in between himself and his golden girl. He wanted her to himself, and he was slightly resentful that they were to be joined so soon by another. Ma sensed this, of course.

While both families were equally thrilled with Beatrice's news, it was for different reasons.

[Beatrice] had been raised in a close-knit, loving family, and she always felt "the more, the merrier" . . . and the Pattons were always obsessed with the thought that "the Name" must be preserved. They never had and never did, recover from the physical and spiritual losses of the Civil War, when so many Pattons and their kin had died on the bloody fields of Gettysburg and Winchester . . . With all of this in mind, Ma said to Georgie; "Will you mind terribly if this one is a girl?" He gave her the perfect answer—because he loved Ma with all of his heart; "What do *you* think, Beaty? I married one, didn't I?"

Beatrice found the army an alien place and made few friends during her stay at Fort Sheridan. She had little in common with the other officers' wives and longed for the culture of Boston that even cosmopolitan Chicago could not replace. On the only occasion when she decided to "drop by" her next-door neighbor in the other half of their quarters, Beatrice was about to knock, when:

The officer who lived there came shooting out the door, half of his face still covered with shaving lather, his suspenders hanging about his knees, his razor gripped firmly in his right hand, and running as fast as he could around the corner of the building; and after him came his wife, in her dressing gown, with her nightcap still on and her rolling pin clutched firmly in her right hand. They were so intent on this silent pursuit that neither of them saw Ma.

It was the last time she ever "dropped by" unannounced.

Beatrice avoided the post doctor at Fort Sheridan, whom she found repulsive, partly for his beard but mainly for his habit of chewing tobacco. Unwilling to become an object of interest to soldiers using the post infirmary or to entrust herself and her unborn child to a physician in whom she had no confidence, Beatrice was referred to a doctor in Chicago by the Ayer family physician in Boston. Nor was Patton destined to keep Beatrice to himself during her pregnancy. One of the Patton or Ayer family women was always present to keep her company, and to ease the difficult transition from Boston society to the austerity of Quarters 92A.

The birth of his first child turned out to be a shattering experience for George Patton. As their daughter writes, Beatrice had not only longed for a child by her beloved Georgie, but she wanted its birth to be a special time of sharing and triumph. The tiny bedroom became crowded with Ruth Patton, a doctor, a nurse the Ayers brought from Boston and Patton himself when he could find space. "The Ayers were outside on the landing.

To everyone it was an occasion of supreme joy, except to Georgie. He had to stand there and see the beautiful girl he had been married to less than a year being torn to pieces (in his eyes), by a monstrous stranger that was not either pretty or appealing or very much wanted by him. He had seen plenty of kittens and puppies and calves and colts born on the ranch, but he had never seen anything as awful and revolting to all his sensibilities as the birth of his first child, and he never got over it. When they tried to put the baby in his arms, he rushed out of the room and ran downstairs, and was sick in the kitchen sink. The family, all crowding in to cheer, felt that he was showing very suitable and sensitive emotions. No one knew how he really felt.

Patton's child was a girl, born at sunset on the afternoon of March 11, 1911. She was christened Beatrice Smith Patton, which was later changed to Beatrice Ayer Patton. Throughout her life she was fondly called "Bee" by her family. When the Patton family arrived *en suite* a month later they hid their disappointment, professed that there remained ample time for a male Patton heir to arrive one fine day. Nevertheless they were heard to refer to baby Bee Patton as "Smithy" or "Smith." With a nurse in constant attendance and later with his parents, sister Nita, and aunt Nannie seemingly always present in their tiny quarters, Patton became despondent and lonely. Indeed, there were moments when Patton thought he would never again be alone with his bride, and he resented the intrusion of his beloved family, even while he relished the opportunity to have them present.[5]

Patton had been the center of his family's attention his entire life, and he was woefully unprepared abruptly to play second fiddle not only to an infant, but an infant *girl*. To make matters worse, Beatrice was obliged to pay more attention to the child than to her husband. Unable to cope with this shattering blow to his outsize ego, Patton grew sullen and suffered the first of what would be occasional bouts of depression. On March 22, for example, he wrote to Aunt Nannie that "the accursed infant has black hair is very ugly and is said by some dastardly people to slightly resemble me which it does not, since it is ugly."[6] For a time the nickname coined by the Pattons stuck. In a letter to Aunt Nannie written in July 1911, Beatrice referred to the baby as "Smithy," and it was not until her first birthday that she was referred to as "Little Bee."[7]

Patton's disinterest in and jealousy of his daughter lingered well into the summer of 1911. Mother and daughter had again returned to Pride's Crossing, and when Patton wrote to Aunt Nannie in July, he made only passing reference to the fact that the child had gained nearly a pound. "I suppose that is fine," he disclosed, with an obvious lack of enthusiasm.[8] Beatrice's reaction to her husband's apathy has never been recorded, but there is no reason to doubt that his behavior was a profound disappointment when the occasion ought to have been a joyous celebration of their marriage. Patton seemed incapable of identifying with essayist Lafcadio Hearn, who observed: "If you ever become a father, I think the strangest and strongest sensation of your life will be hearing for the first time the thin cry of your own child." To the contrary, he would more than likely have agreed with the humorist who wrote that babies are: "an alimentary canal with a loud voice at one end and no responsibility at the other."[9] Nevertheless, when it came to his family, there was a puritanical streak in Patton that would lead to the unleashing of his wrath for the slightest real or perceived disrespect to his daughters or his wife.

In 1918, when he was in France preparing for combat, Patton apologized for his behavior at Fort Sheridan. His loneliness led him to write of the time he had just returned from maneuvers "and you gave me hell (which I deserved) for some of the letters I had written you still I wish we were there again I would behave better though I was very jealous at the time You were not so mad as circumstances justified you in being."[10]

His doldrums may have been exacerbated by the news that the deteriorating relations between Mexico and the United States indicated the imminence of war. What upset Patton was the fact that there was only a remote possibility he would ever participate in such a war. "By time this reaches you I may be a heap of rotting carrion on the sand hills of Northern Mexico or I may not," he wrote to Aunt Nannie. "But we hope to be the next troops sent . . . [even though] there may be no war. God forbid such an eventuality however."[11] Clearly anxious for something more exciting than Fort Sheridan, Patton thirsted for any assignment that would enable him to test his courage in combat, and Mexico offered the only possible prospect.

Like any marriage, the Pattons' required a considerable period of adjustment. Not only had there been precious little time for George and Beatrice to accustom themselves to living together, but with a newborn infant, the strains on each were multiplied. One day not long after Little Bee's birth, the colonel's wife unexpectedly came to call on a day when Patton was teaching new recruits to shoot on the rifle range. After some polite conversation the woman finally got to the point of her visit. Had Bee's birth somehow made a difference in their marital relations? Were they happy? "Was all well 'in the bedroom?'" Embarrassed, and beginning to take offense at such very personal questions, Beatrice was startled when the colonel's wife said, "I know you must think me an interfering old woman, Mrs. Patton, but when Colonel Girard came home at noon for his lunch, he mentioned the fact that Mr. Patton had been standing on the rifle butts all morning, between the targets, and he wondered if some circumstance had occurred of such a nature that Mr. Patton was trying to take his own life . . . perhaps because of some misunderstanding at home."

When her husband arrived home that afternoon, he encountered an enraged Beatrice, with her bags packed, awaiting a taxi to take her, baby Bee, and a maid to Chicago to entrain for Pride's Crossing. It took every ounce of Patton's powers of persuasion to convince Beatrice that he had, as he had previously at West Point, merely been curious to learn what George Washington meant when he wrote of the merry sound of bullets whizzing past his ears. Patton was again testing his courage under fire, and, although Beatrice was eventually persuaded to unpack her bags, she was not amused by her husband's antics.

On another occasion, in France, Patton learned that the price of taking his wife for granted was costly. Whenever they moved—and they would do so often during their long married life—the responsibility for packing and attending to the details of the move devolved on Beatrice. The day before they were scheduled to depart, Patton came home to find Beatrice exhausted, and thoughtlessly said: "I hope you remembered to pack all those swords under the bed." When she found some thirty swords and scabbards, something snapped.

> The next thing she remembered was chasing Georgie through the rooms with a sword uplifted in her two hands, and Georgie running madly ahead of her, jumping over chairs and tables, and his hands clasped over his head to [ward] off her stroke, yelling; "Don't! Don't! Please Don't!" She almost caught him, bringing the sword down so hard that it struck into the edge of a table. Georgie helped her to pack them.

In other aspects of their married life, Patton would readily defer to Beatrice. For example, Beatrice was fluent in French, a subject he had struggled to master at West Point, and he willingly accepted her help and advice. After translating an article from a French military journal in the summer of 1911, Patton asked his wife to revise it, expecting that "I may find some startling changes in my ideas."[12]

Although his pay was paltry, with Beatrice's monthly stipend from her father, money was never a problem in the Patton household. Patton kept a meticulous record of his expenses during the first few months of their marriage but soon gave it up as a waste of time.[13] Patton purchased his first automobile and in letters to his father displayed a considerable knowledge that "marked the beginning of his considerable interest in and knowledge of motorcars and would lead him eventually to the tanks." Patton also purchased his first horse in 1910 and, after attending several horse shows in Chicago—believing it would enhance his career—decided that he would participate the following year. Writing to Aunt Nannie, he confided, "it is a fine advertisement for a man [in the Army]."[14]

Then, and in the years to come, Beatrice often returned to Pride's Crossing to visit her parents. Her mother would frequently write to remind her daughter how much her elderly father missed his "thinking flower." Ellie "constantly played on Ma's feelings about coming home for a visit. I don't think she did it to devil Georgie, I think she really felt what she wrote. Ma was always torn between her duty to Georgie and her duty to her parents."

Beneath his serious facade Patton possessed a bawdy sense of humor, as he was to demonstrate on many an occasion. When Patton was stationed in El Paso, Texas, shortly before World War I, Beatrice, Little Bee, and Ruth Ellen (born in 1915) returned to Pride's Crossing for one of their frequent visits. It was Ellie's custom to have the grandchildren recite for her and her friends in the great living room that overlooked the ocean. Shortly before they departed Texas, Patton took Little Bee and Ruth Ellen aside and inquired, "if Ellie still asked us if we had a new 'piece' to recite. The Ayers were great on recitation . . . so Georgie said he was going to teach us a piece that was a big secret, and we were to tell it to nobody until Ellie asked us to recite."

The great moment came to pass at teatime, in the presence of Ellie and some family and friends, including an elderly woman who used an ear trumpet. As Ruth Ellen writes:

We had been brought down to be admired. We were dressed alike in our best hand-made dresses with real lace insets, each of us with a huge hairbow. . . . Ellie saw our entrance and said, "Beatrice, dear, do the girls have a nice piece they could recite for our friends?" Ma said she was sure we did, so there was a hush while we said our new piece that Georgie had taught us—which was:

There was a goddam spider
Lived up a goddam spout
There came a helluva thunderstorm
And washed the bastard out
And when the sun came out again
And dried up all the rain
Damn, if the old son-of-a-bitch
Didn't climb up that spout again!

I was looking right at Ellie, and saw the bangle-covered arm holding the teapot suspend itself in mid-air. There was a great silence. My sister Bee thought an encore was indicated, and we started the piece again, but were gently removed.

*   *   *

The following morning Beatrice was summoned for coffee with her aged father, who attempted without success to suppress a smile while saying:

> Little daughter, I feel that I have had some sort of communication with George that he wants his little family around him, and that is quite understandable. You must remember that I have passed my three-score-years and ten, and that every day thereafter has been an added blessing with the love of your mother and our wonderful children. I cannot expect them to be all about me forever. Georgie needs you and his little daughters, and I have taken the liberty of purchasing these tickets for your return to Texas on the first of the week.

In the spring of 1911, Patton again sought reassignment to the Cavalry School at Fort Riley by writing directly to his regimental commander, who informed him that at least three years of commissioned service were required before he would become eligible. The Patton way was never to take no for an answer if some means could be found to influence others. It was blatant bootlicking, but Patton merely deemed it an indispensable means of going about the business of looking after himself and his career.

"I wish I saw the chance of war," he complained to his father, "I get horribly bored doing nothing at all."[15] He reminded Papa of his acquaintance with the adjutant general of the army, who might be persuaded to help. ". . . [H]e who knocks it shall be opened unto. The trouble is that we hate so to knock. Yet why?"[16] A posting as military attaché in South America, duty as a tactical officer at West Point or, ideally, an assignment in Washington were all desirable options. With the insurrection there now over, what Patton emphatically did *not* want was an assignment to the Philippines.

When Captain Marshall left Fort Sheridan in May 1911 for temporary duty at Fort Leavenworth, Patton was given temporary command of Troop K, at a time when very junior officers were rarely given such opportunities. It was a clear signal that Patton's success in carrying out his first assignment had made the desired impression on an officer whom he esteemed and emulated. Yet, as Martin Blumenson notes, the honor seems to have had a minimal impact on Patton, who would have otherwise been expected to trumpet his achievement to one and all.[17]

In reply to a letter from Patton, Captain Marshall wrote that it was War Department policy not to assign any officer to West Point who had not attended the French Cavalry School at Saumur; "I would advise you to work your rabbit's foot good and hard and get sent there. There's no use putting it off, either. If you can get your friends interested you will have no difficulty."[19] Patton immediately wrote to Beatrice:

Should you think favorably of it I would try for it . . . your father would work the Mass[achussets] people for me and papa would try to fix it in California and in Texas. . . . I was glad to get the letter . . . for it gave me a chance to inquire how to use influence . . . if nothing else comes of it we will at least know more on that important point than we do at present.[20]

However, before Saumur could become a reality it seemed essential for Patton to escape Fort Sheridan, and his immediate efforts were now focused on obtaining a transfer to Fort Myer. Once there he could bring his influence and that of his family to bear on a future assignment to Saumur. By September 1911 Patton was optimistic, in part because Maj. Willie Horton, a beau of his sister-in-law, Kay Ayer, was "doing all he can in Washington and he is quite influential."[21]

In October 1911, the Pattons made their first journey as a family to Southern California, where Papa had built an elegant new five-bedroom Lake Vineyard mansion to replace the outdated house of Benjamin Davis Wilson, constructed in the 1850s. Patton was greeted by the Patton-Banning clan as a conquering hero. Daughter Ruth Ellen writes:

Georgie was the fair-haired boy at Lake Vineyard. Nothing was too good for him or, later when they were married, for Ma. He was adored by all his cousins. He was the great adventurer who had cut loose and gone off into the wilds of the unknown east, where fame and glory awaited him— they all hoped and felt sure.[18]

Whether it was Horton's influence or that of a member of Patton's family that resulted in his reassignment remains unclear but, much to the delight of both George and Beatrice, in the autumn of 1911, Patton received orders to report to Fort Myer, Virginia, for duty with the 15th Cavalry.[22]

CHAPTER 10

"A Young Man on the Make"

Enthusiasm finds the opportunities, and energy makes
the most of them.

—ANONYMOUS

When Patton reported to Fort Myer, Virginia, he instantly found himself
propelled from the backwater of the army to its most prestigious post. Situated on Arlington Heights overlooking the Potomac River and Washington,
D.C., Fort Myer had originally been called Fort Whipple during the Civil
War. Once part of Robert E. Lee's vast Virginia estate, it was considered one
of the bulwarks of the defense of the Union's capital city. To avoid the
ignominy of losing Washington to the Confederate Army, President Lincoln
ordered Arlington Heights seized in July 1861, shortly before the Battle of
Bull Run. Only Fort Stevens to the north of Washington was ever besieged,
by Jubal Early's army in 1864 (ironically, a raid spearheaded by Patton's
grandfather). By the 1870s Fort Whipple had become a permanent garrison,
and in the early 1880s it was renamed Fort Myer in honor of Brig. Gen.
Albert J. Myer, the army's first chief signal officer. In the aftermath of the
Civil War two hundred acres, centered on Lee's stately Arlington House,
became Arlington National Cemetery, the best-known burial ground of the
nation's military dead.[1]

Although originally devoted to communications activities, Fort Myer
was designated a cavalry post in 1887, when General Philip Sheridan
became its senior officer, and

from then until 1942 some of the Army's most celebrated mounted regiments formed the garrison. Horsemanship was a central activity, especially in the period between the World Wars, when the Army had a leading role in Olympic equestrian activities. Fort Myer was also the site for the earliest developments in the field of Army aviation. The Wright brothers had contracted with the Signal Corps to build a biplane. . . . On 9 September 1908, Orville Wright made fifty-seven complete circles over the drill field.[2]

The gulf between the general staff in Washington, many of whom were in a position to cultivate promotion and favor, and the line officers who served in the far-flung outposts of the army, often without recognition or promotion, was epitomized by the difference between the elegance of Fort Myer and the primitive "hitching post" forts like Sheridan.

By 1912 Fort Myer was also the residence of the U.S. Army chief of staff and other senior officers stationed in Washington. As the headquarters of the 15th Cavalry, it was perhaps the ideal duty station for the ambitious young Patton and his wife. Not only was the rolling countryside south of Washington honeycombed with riding trails, there was foxhunting in Virginia and Maryland and prestigious race meets each spring. Of equal importance was the fact that those stationed at Fort Myer were automatically granted access to Washington's high society. The 15th Cavalry provided escorts for military funerals and visiting dignitaries at state ceremonies, and played what was widely regarded as some of best polo on the eastern seaboard.[3] Officers stationed at Fort Myer automatically gained admittance to participation in the elite and glittering social life of Washington, where they frequently hobnobbed with members of Congress, Executive Branch, and—most important—the War Department. For a young man on the make like George S. Patton, duty at Fort Myer was a heaven-sent opportunity to exercise his growing proficiency at self-promotion with those who could most help advance his career. Patton had learned early that no matter how effective and proficient an officer he became, the influence of guardian angels in powerful positions was vital if he was to attain his dream of generalship and high command.[4]

In addition to its ceremonial duties, the activities of the 15th Cavalry typically included endless rounds of equitation, mounted and dismounted drills, marksmanship, scouting, patrolling, and grooming and maintenance of their animals. Patton was assigned to Troop A and soon impressed his new commander, Capt. Julian R. Lindsey, as a studious, hardworking, competent officer. The first thing Patton noticed about Fort Myer was that "people work much harder here than at Sheridan . . . it is all together more military." Patton had begun to write professional military articles while at Fort Sheridan, and at Fort Myer he continued what would become a lifelong practice. In February 1912 he produced an imaginative and original monograph for use by his troop, which was written in the form of Napoleonic maxims (principles of warfare) and called *Principles of Scouting*.[5]

One of the many virtues of duty at Fort Myer was the quality and size of the government quarters, which, even for a junior officer, were an enormous improvement over the dreadful abode that passed for the first Patton homestead at Fort Sheridan. The Pattons employed a full-time maid named Hannah and a chauffeur to drive the family car for, as George wrote to his father, everyone else seemed to have one, and they "could not keep up" [their social standing] without following suit.[6] Beatrice had found Fort Sheridan intolerable, and with the move to Fort Myer she was overjoyed to be in the far more comfortable environment of Washington, with its glamorous social life and its important people: "Ma, a city girl, fitted right into the Fort Myer–Washington life; and Georgie was beginning to grow greater in his own esteem, and with Ma at his side, who knew all the mores of the so-called Sophisticated East, he became more self-confident. Ma was able to entertain and do the things she had been brought up to do with style and verve."[7]

Good fortune soon smiled on Patton. Secretary of War Henry L. Stimson was fond of a daily ride on his horse. One day, while out for an early morning jaunt, Stimson encountered Lieutenant Patton on one of Fort Myer's many equestrian trails, and both men lost no time cultivating a friendship that was to endure for the remainder of Patton's life. Stimson was impressed by the young officer's dash; thereafter they would often ride together, and occasionally the secretary would recruit Patton to serve as his aide at social events at Fort Myer.[8] In later years their friendship was to serve Patton exceedingly well and help to save his career at its nadir.

Patton's assignment to Troop A lasted a mere three months, and in March 1912 he was reassigned as the squadron quartermaster. This new arrangement probably enabled Patton to practice and play on the stellar Fort Myer polo team.[9] Although he had bought several horses while at Fort Sheridan, he now realized that they were no match for the well-bred horses owned by those stationed at Fort Myer. To compete successfully at polo and in the local steeplechase races, he visited the famed horse country of Virginia and Kentucky in search of new horseflesh. In Lexington, Kentucky, he bought a registered Thoroughbred to add to his stable, which soon numbered seven horses.[10]

Sixteen years earlier a young French educator and sportsman, Baron Pierre de Coubertin, had resurrected the ancient Olympic games, which had fallen into disrepute under the Romans in the fourth century. In modern form an effort to promote international goodwill by means of amateur competition on the athletic field, de Coubertin's vision of a modern Olympics became a reality in 1896, when he managed to induce nine nations to send one hundred of their sportsmen to Athens to compete in the first games with two hundred Greeks as a celebration of the highest in amateur athleticism.[11]

The Fifth Olympiad, to be held in 1912 in Stockholm, Sweden, would become the largest and most successful yet of the revived modern games. Twenty-eight nations sent a total of four thousand athletes, including the first women, who competed in swimming events. For the first time one of the main events of the games was the Modern Pentathlon, a new version of the original Greek competition in which soldier-athletes vied against one another in five events: swimming three hundred meters, pistol shooting on a twenty-five-meter range, running a four-thousand-meter (two-and-a-half-mile) course, fencing, and riding a five-thousand-meter steeplechase.[12] The competition was limited to military contestants and when U.S. Army officials began considering their choice of representative, Patton immediately became virtually the only candidate. At West Point he had been a runner, his riding and swordsmanship were well known, and since early childhood he had learned to shoot and to swim long distances in the waters off Catalina Island.

In that era there were no Olympic trials or biannual international competitions at which athletes could vie for a berth on their national team. Instead appropriate entrants were sought and invited to participate. Each athlete then devised and carried out his own training program to prepare for the games. Patton was the first U.S. Army officer to represent the United States in the Modern Pentathlon. However, he was not named to the U.S. Olympic Team until May 10, 1912, leaving precious little time for training for the games which were to commence in early July. As Patton later wrote of the experience, he "was in excellent physical condition but had not run for about two years nor done any fast swimming for three."[13]

His grueling training regimen began immediately and brought untold misery to both Patton and his family. As daughter Ruth Ellen relates, "It was hard on everyone. He went on a diet of raw steak and salad and was, according to Ma, unfit for human companionship. But he had to push himself as he had such a short time in which to get into shape."[14] When George Patton trained for an athletic event, he was, as nephew Fred Ayer later related, about as pleasant to be around as "a tiger from whose jaws a haunch of game has just been snatched."[15] He gave up alcohol and tobacco and punished himself brutally in both swimming and running, his two weakest events. Patton knew only one way to train and that was mercilessly and without regard for himself or his safety. It was all the more difficult because he was not a natural athlete, had shown himself to be accident prone, struggled to run well, and, in reality, loathed swimming, perpetually disdaining it as a sport.

Yet the fires of ambition burned as deeply within Patton as they had at

West Point, and the Olympics presented a splendid opportunity for him to show what he could do on a world stage. All else was secondary. There was no respite when the U.S. team and the Patton clan embarked for Antwerp, Belgium, aboard the steamship *Finland* on June 14. Accompanied by his wife, parents, and sister Nita, Patton continued his grueling training during the voyage.

He practiced swordsmanship and running with the rest of the team, a regimen that began at dawn and included runs of two miles around the decks of the *Finland,* and pistol practice at targets rigged off the fantail. To accommodate the swimming team a special twenty-foot-long canvas pool was installed on the deck. Patton swam in place with a rope tied around his waist that left raw chafe marks.[16]

The Pattons traveled from Belgium to Sweden, arriving on June 29. There they were well received and fell in love with Sweden and its hospitable people. Patton continued his rigorous training, but he and his family were immediately caught up in an exciting round of parties, some of which were attended by the royal family. It was at one of these events that Patton met the aide to King Gustavus Adolphus V, Colonel Björling. The two men became lifelong friends. (The last photograph of Patton taken before his death was with Björling in Sweden in 1945.)

Mr. Patton accompanied his son to each practice and was ever present to provide encouragement. He also quickly became a favorite of the Swedish officers, who adored him. One evening Mr. Patton unintentionally sat down at a table outside a hotel that was reserved for members of a select club. An angry member strode up and insultingly placed his cane on the table in front of the startled Patton. Instantly a Swedish officer who knew Mr. Patton leaped from a nearby chair, broke the cane in half, apologized, and brought him to his own table.[17]

Besides himself there were forty-two other contestants in the Pentathlon, eight of whom were Swedish officers. Patton was the only American entered who actually participated.[18] The day before the games opened Patton had ill-advisedly rested instead of engaging in some form of workout to keep himself loose, and had only practiced pistol shooting, firing a near-perfect score (197 of 200), which boded well for the first day's competition. However, he was restless and apprehensive and barely slept that night, in part due to the long summer days, which brought scarcely an hour of darkness.

As a result Patton did poorly. He had shot well enough, and might have stood high in the pistol competition, but the oft-related tale is that one of his bullet holes could not be located in the target. It was thought that perhaps it had passed through the same hole from a previous round, and his generous Swedish competitors insisted this obviously must have been the case, but the missing bullet could not be located. Consequently the judges were obliged to penalize Patton ten points, and he wound up a dismal twenty-first. Beatrice blamed herself for his poor showing, later telling her daughter that, "her Georgie could not have failed on his own," and if she had taken him back to their hotel earlier he would have won, a chivalrous gesture but hardly a probable reason.[19] The tale of the so-called "lost round" is apocryphal. In reality not one but *two* of Patton's rounds missed the target, thus making it impossible for him to have won or even placed high in the pistol competition.[20]

On the second day the number of competitors had dropped to thirty-seven. Patton did well in the three-hundred-meter swimming event, finishing a very respectable sixth, although he was so exhausted at the finish that he had to be helped from the pool with a boathook. The third and fourth days were devoted to fencing, on the courts of the Royal Swedish Tennis Club, in which Patton finished third of the remaining twenty-nine. The event was perhaps the most demanding of the five Pentathlon events, and required each man to fence with a dueling rapier that weighed nearly one and one-half pounds, for three touches against every other competitor. Patton had good reason to be pleased with himself. "I was fortunate enough to give the French victor, Lieutenant Mas de la Tree, the only defeat he had."[21] Patton's pugnacious, slashing, give-no-quarter attacking style easily made him a crowd favorite but tactically often left him vulnerable to the finesse of his competitors, most of whom were far more experienced. Remarkably, of the twenty-nine opponents he met, Patton defeated twenty.[22] It was a noteworthy achievement, especially for an American of limited experience who had never been tutored by a world-class teacher.

Patton's offensive-mindedness with the sword was a harbinger of his future generalship on the battlefield. Throughout his career disdain for defense was a Patton trademark. To attack was to succeed, to defend was to invite defeat. In 1912, barely three years out of West Point, George S. Patton attacked the Pentathlon as later he would the German Army in World War II.

Now, only two events remained, and, except for his disastrous placement in the pistol shoot, Patton would have been in bona fide contention for a medal. His chances were greatly enhanced by another third-place finish in his best event, the five-thousand-meter steeplechase, where, on a borrowed Swedish cavalry horse, Patton and two Swedes registered a perfect score over the formidable course. The winners were then determined by the best time, and Patton's third place finish could only have been accomplished by a rider of championship caliber.[23]

The final day of the Modern Pentathlon was the dreaded four-thousand-meter cross-country run. By this time a mere fifteen competitors were left to line up in the Olympic stadium before the royal box to begin the race over a treacherous course that (under Olympic rules of the time) none of them had been permitted to view beforehand. The course wound through a twisting, hilly, forested path replete with mud-filled swamps. Even worse, it was an exceptionally hot day, with high humidity that soon drained the energy of the runners. As was his custom, Patton simply ran as fast as he could for as long as he could, without regard for pacing himself. It was yet another example of his refusal to settle for anything less than an all-out performance. Pacing oneself was for others; he would give it his best shot and damn the consequences.

Before the race his trainer had given him a shot of opium to provide additional stamina to help him through the event. Another name for opium is "hop," and its use was still legal in 1912. Although performance-enhancing drugs have been outlawed in present-day Olympic competition, the "hop" given Patton probably did him little good—other than inducing a feeling of well-being and spurring Patton to "run like hell" for the finish line. Dressed in a white shirt and knickers, Patton was the first to reenter the stadium for the final dash. Close behind him was a Swede, Gösta Asbrink, whom Patton later described as "a very hard and energetic sportsman." As he neared the finish line Patton began to stagger, and when he could no longer make his legs run, he began walking the final fifty meters. Asbrink passed him to win the race, as did another Swede who finished second. Patton somehow managed to cross the finish line in third place before he collapsed in a dead faint. Two of the other fifteen runners had also fainted, and one died in the torrid heat and humidity.[24]

Patton might well have died. How long he remained unconscious is not known, but he later wrote that it was several hours. "Once I came to I could not move or open my eyes and felt them give me a shot of more hop. I feared that it would be an overdose and kill me. Then I heard Papa say in a calm voice, 'Will the boy live?' And Murphy [the trainer] reply, 'I think he will but cant tell.' "[25] Patton was severely dehydrated from his ordeal but the effects of opium under such conditions undoubtedly worsened the situation and might easily have killed a less-well-conditioned athlete. Fortunately, he was young and at the peak of physical fitness, and thus survived the first of many close brushes with death.

Patton ultimately finished fifth in the final Pentathlon standings and, had he done better in his best event, the pistol, might have qualified for an Olympic medal, all of which were won by Swedes who had trained hard for eight months. Even so, his performance was superb; he defeated twenty of the twenty-nine fencers; and in swimming seventeen of the twenty-three competitors—this by a man whose only practice came in a canvas tank aboard ship. Small wonder the Swedish newspapers said of him, "His energy is incredible. In the distance running races, he returned to the stadium completely exhausted but did not falter. . . . In the fencing, his calm was unusual and calculated. He was skillful in exploiting his opponent's every weakness."[26]

When he left Sweden his accomplishments were little known outside Los Angeles. The limelight in Stockholm belonged to another American, Jim Thorpe, the man whom the king called "the most wonderful athlete in the world."[27] Thorpe—a Native American—was the first man to win the Modern Decathlon and the track-and-field version of the Pentathlon, only to be stripped later of his gold medals by the International Olympic Committee, for having once played semipro baseball.[28]

Whether on the playing field, the drill field, or the battlefield, Patton viewed as an adversary anyone who posed a threat to his aspirations, be he another athlete or a fellow army officer. The 1912 Olympics were a conspicuous exception, in which he generously praised his competitors, noting that the Modern Pentathlon was in reality

> an officers' competition, and certainly the high spirit of sportsmanship speaks volumes for the character of the officers of the present day. There was not a single . . . protest or any unsportsmanlike quibbling or fighting for points. . . . Each man did his best and took what fortune sent like a true soldier, and at the end we all felt more like good friends and comrades than rivals in a severe competition, yet this spirit of friendship in no manner detracted from the zeal with which all strove for success.[29]

The senior officer representing the U.S. Army in Stockholm was Lt. Col. Frederick S. Foltz, and in his report to the War Department he praised Patton's excellent showing and observed that, "he deserves great credit for the enthusiastic and exhaustive way in which he prepared himself for this very difficult, all around competition."[30]

For Patton, who had been raised from childhood on the purity and ethics of the ancient warriors, the 1912 Olympic Games were perhaps the closest approximation in his lifetime to that heroic ideal that he had fantasized about. Like himself, men who lived up to that model of perfection were to be admired, indeed even venerated. Throughout his life Patton disapproved of most of his contemporaries because he believed they lacked the essential qualities of a warrior, and he rarely had anything laudable to say about them. In Stockholm, Patton had only praise for his rivals, even though their goal was to win and thus to deny him the glory he sought. Each had earned the ultimate tribute that he could pay them. At the height of his fame during World War II, only an enemy officer, Field Marshal Erwin Rommel, would command such esteem.

Patton's relationship with the Swedish Pentathlon team would endure, and after World War II ended in 1945 the entire eight-man pistol team invited him to Stockholm, where, amid cigars, fine wine, and nostalgic reminiscences, George S. Patton once again fired his pistol in a symbolic reenactment of the Olympic competition and bettered his score of 1912, finishing second.[31]

News traveled slowly in 1912, and it was not until early August that the front page of *Los Angeles Times* proclaimed: "Young Patton has carried off honors in fencing, shooting, riding and swimming and according to the reports that are coming from Stockholm his athletic versatility is attracting considerable attention. . . ."[32]

Patton had arrived in Stockholm determined to improve his swordsmanship, and during the Olympics he inquired of his newfound friends the name of the finest teacher in Europe. Without exception he was told that the greatest swordsman was the "*beau sabreur*" of the French Army and the master of arms and instructor of fencing at the Cavalry School at Saumur, Adjutant M. Cléry.[33] From Stockholm the entire Patton family had embarked on a whirlwind European tour and had already visited Berlin, Dresden, and Nuremberg, where Beatrice and his father sampled donkey-meat sandwiches and drank German beer in small restaurants. While the remainder of the family continued their tour, Patton and Beatrice journeyed to Saumur, where he began a crash course in swordsmanship from the great master. For nearly two weeks Patton was tutored in the fine art of the sword. His brief stint merely whetted his appetite for further instruction from Cléry, with whom he had begun to form a close bond. He left to return to the United States on August 10, determined to use his influence to get himself detailed to Saumur at the first opportunity.

Although he lacked the notoriety the age of mass communications might have given him, his overseas adventure had been the first step toward fulfilling his destiny of living up to the Patton name by becoming a great soldier. The glow of his triumph in Stockholm and the exhilaration of Saumur would endure to the end of his life.

While Beatrice was on an extended visit with her parents at Pride's Crossing, Patton returned to his army duties at Fort Myer, from which he had departed three months earlier. The difference was that now he had been noticed. He was summoned to dinner with the army chief of staff, Maj. Gen. Leonard Wood, and his new friend, Henry Stimson, at which he was able to recount both his Olympic experience and his training under Cléry to sympathetic and important ears. His morning rides now occasionally included the company of General Wood. In the future Patton would take advantage of his acquaintance with Wood to write him letters suggesting various improvements in cavalry drill or procedure.

During his temporary bachelorhood, Patton threw himself into an endless round of polo and equestrian events. His duties with Troop A included teaching his men to shoot. He groused to Beatrice that, "the more I see of target practice the less I think of it. Our great trouble is that men do not do what they are told. They think too much! this talk about the independence of the American soldier will cost a lot of lives, if we would teach them to obay we would do much better than teaching them to shoot."[34] He also continued cultivating his contacts among the Washington elite. He finished third in a race at the Maryland State Fair at Timonium, returning a few days later to place first and win one hundred dollars and a silver plate.

Patton relied on Beatrice to pay the household bills and keep his life orderly. During her absence his frustration manifested itself in frequent complaints about unpaid bills and the manner in which she was handling the family finances. In early September he wrote a shamefully patronizing letter. "Your finances are perfectly ridiculous," he griped:

> To put all your money at practically no interest and so tied up that it takes an act of god to get it is pretty foolish. . . . I don't see how you could have let so many bills run from April or May the way you did. I think you had better let me run your bills here after or you will go to prison. . . . Inspite of your lack of brains I love you more all the time and miss you even here where I am very busy. I love you.[35]

Yet if more than a few days passed without a letter from Beatrice, his attempts at humor sometimes disguised a biting sarcasm. In 1915 when she was in California recovering from the birth of their second child, Patton churlishly complained, "I have had no news for over a week and while I don't care much for you still I like to hear at times."[36]

Although loath to admit it, Patton relied on Beatrice to maintain their affluent lifestyle, which included a growing stable of fine horses, tailored uniforms, dogs, and an automobile. However, in an age when men ruled the family roost, Patton never viewed his occasional wretched behavior toward his wife as male chauvinism. Although he was demanding, single-minded, and highly opinionated, as their marriage deepened over time, Beatrice somehow managed to assert her independence while at the same time instilling in her husband the illusion that he was the absolute master of the Patton household. Whenever she was absent in Massachusetts, he missed her deeply and sometimes his resentment boiled over in the form of hurtful letters. The longer she remained away the less tidy his life. Even the family dog, a bloodhound named Flip, caused Patton grief. One day he ran away and was later found in Washington's red-light district by a soldier.[37]

When Frederick Ayer wrote to question his workaholic lifestyle and mildly criticize his continued military career, Patton turned the tables by observing that there was a perfectly good reason for both:

> If you had not done more work than other people when you were my age you would not be now what you are. . . . I quite understand that what

I am doing looks like play to you but in my business it is in the best sort of advertising. It makes people talk and that is a sign they are noticing. And you know that the notice of others has been the start of many successful men."[38]

In October Beatrice and Little Bee returned to Fort Myer. The child was growing rapidly and to the delight of the Ayer clan had begun to talk. Not surprisingly her first word was "Dada," and whenever they passed the train station the child would loudly call "Dada" and once cried when her father did not come. "Georgie is perfectly crazy over her and she over him. Dear old things!" Beatrice proudly wrote to Aunt Nannie.[39]

In December 1912 Patton's budding friendship with General Wood paid off when he was detailed to the chief of staff's office as a staff officer and occasional aide to Wood and Stimson. In modern jargon Patton was an action officer but although the duties varied, usually involving the preparation of letters for Wood or staff studies on a variety of subjects, it put him at the right hand of the seat of power in the army. Neither his very junior rank nor his duties entitled him to the coveted title of general staff officer but this did not stop Patton, who wrote several important papers that were undoubtedly read by the chief of staff. No second lieutenant could have wished for more. His study of military history was beginning to pay off in high ratings and the notice of those who counted.

None counted more than Leonard Wood, who was something of a maverick. A doctor and a graduate of Harvard Medical School rather than West Point, Wood had taken over as chief of staff in 1910 determined to reform the creaking machinery of the army, which was mired more in tradition than in the realities of the twentieth century. Historian Russell Weigley has called Wood "a military evangelist," who, in concert with Stimson, fought for universal military training and a host of other reforms designed to increase the preparedness of the army for war.[40]

What got Patton the most recognition was a portion of his report on the Olympics that dealt with his fencing lessons with Cléry. His experience at Saumur not only brought him to the attention of the army's most senior officers but also gained him some valuable publicity, when a revamped version of the report appeared in the *Army and Navy Journal*.[41] Life was good, and Patton was making the most of his opportunities. Beatrice wrote to Aunt Nannie that her husband "has said so much about his charmed life that I half believe in it sometimes."

When Woodrow Wilson was inaugurated in February 1913 Patton was one of Wood's aides and rode in the great parade down Pennsylvania Avenue. By the time his tour in Washington and Fort Myer ended, the name "Patton" had been heard and read about with increasing frequency. Socially Beatrice used her talents to good effect when she and George hosted a dinner for Papa Patton, as well as other affairs all designed to bring Lieutenant Patton's name to the forefront and keep it there. In the March 12 issue of the *Army and Navy Journal* there were five separate articles featuring the Patton family.[42] And in the same month the first of many articles by Patton appeared in the prestigious *Cavalry Journal*. For some months he had been advising the Ordnance Department on the design of a new sword for the cavalry arm that was manufactured in 1913 to his precise specifications at the Springfield Armory. Technically it was called the U.S. Saber, M-1913; in reality it came to be known as the "Patton sword."[43]

In March 1913, too, Patton's enormously successful detail in the War Department ended with a warm letter from Wood in which he noted his "appreciation of the satisfactory manner in which you have discharged the duties assigned to you." Second lieutenants do not normally work for the army's most senior officer and rarely receive such letters of praise. For Second Lieutenant Patton it was a feather in his cap and a considerable boost to his self-esteem. His good work with the saber also enabled him to apply for a return to Saumur in the summer of 1913.

Patton participated in a flurry of horse shows, three-day events, flat races, steeplechases, and polo, leading to his experience in April of another of the many accidents connected with horses that he would suffer during his lifetime. This time he fell from his mount and slashed his scalp open in two places. Although he was only laid up for two days, Beatrice worried about her husband's impetuous riding style, and wrote with resignation to Aunt Nannie: "There isn't a bit of use in worrying about him. . . . I sort of hate to have him race, but the best way to keep him to his senses seems to be not to oppose him in any way."[44]

Patton was soon back in the saddle, competing furiously at some of the elite eastern courses, and playing polo. He found polo enthralling, writing to Beatrice that it was "wonderful," and that he "had never dreamed that either man or ponies could be so fine."[45] Later, he would declare that polo was like "a good war" and essential to the training of a commander for combat.[46] His nephew rated him as a good but certainly not great polo player, noting that he made up for his lack of skill by the ferocious manner in which he played, replete with profanities that were highly unsuitable for tender ears.[47]

Training his horses was another important aspect of Patton's equestrian life, and he once spent two days in a stable at Pimlico, Maryland, to learn how the experts did it. Fred Ayer remembers that Patton "did not tolerate disobedience from horses any more than he did from subordinates," citing an incident in which Patton decided that Hilda Ayer's [Fred's mother] horse, a stubborn gray named Gun Metal, required curing of his habit of rearing on its hind legs: "Goddammit, Hilda, I can stop the bastard from rearing." His prescription was to suggest breaking water balloons against the sides of Gun

Metal's head whenever he reared, making the animal believe it was blood rather than water running down his head and flanks. Hilda Ayer declined, saying, "Maybe so, but you're not doing it to my horse."

Not to be dissuaded, Patton replied, "All right, dammit, there's another way." Vaulting into the saddle, he yanked on Gun Metal's reins until he reared up. He then deliberately leaned backward, and the horse lost its balance and fell to the ground. Only an accomplished horseman could have landed out of harm's way, as Patton did. Before the startled horse could recover, he sat on its head to keep it from scrambling to its feet. "I guess that'll teach the dirty son-of-a-bitch a lesson." As brutal as they may have appeared, both of Patton's prescriptions for Gun Metal were accepted remedies for rearing. The latter, however, is exceptionally dangerous and potentially lethal to both horse and rider. It is practiced today only by expert riders in extreme cases. This time Patton failed. Gun Metal was perilously spooked for some time, and Hilda Ayer was left outraged.[48]

In the cavalry no officer could rightly call himself either an officer or a gentleman without being an expert horseman. On a more practical note, however, Patton also "saw in the horse a creature which he must dominate, by brute strength if need be."[49] Small wonder then, that Patton's equine relationships were as stormy as his human ones.

In June 1913 Patton received orders not only reassigning him to the Mounted Service School at Fort Riley, Kansas, on October 1, 1913, but authorizing his return to Saumur that summer, at his own expense, "for the purpose of perfecting yourself in swordsmanship." Both assignments were the payoff from his intense lobbying campaign to obtain a posting to Fort Riley. Nevertheless, Patton had no intention of being a mere student at the Mounted Service School. For months he had promoted the idea that a master of the sword be assigned there to teach swordsmanship. Not only did he ingeniously prepare the way by securing unofficial approval of his idea, but he even managed to persuade the commandant of the Mounted Service School to recommend it to the War Department, which then approved it. Naturally the ideal candidate to teach this new course was Lt. G. S. Patton, and his attendance at Saumur would earn him the right to the newly created title of master of the sword. His mastery of swordsmanship under Cléry would qualify Patton to become the army's first-ever master of the sword.[50]

For the second summer in a row the Pattons sailed for Europe, and once again the great Ayer wealth permitted them the luxury of shipping the family auto to France. (Patton's monthly salary as a second lieutenant was a lowly $157.50.) The $300.00 it cost to ship their car to France for a mere six weeks was the equivalent of $4,125.00 in 1991 dollars.[51]

They arrived at Saumur in late July after a leisurely drive from Cherbourg that took them through the hedgerow (*bocage*) country of Normandy.

It was countryside with which Patton would become intimately reacquainted in the summer of 1944.[52]

With baby Bee safe at Pride's Crossing, their sojourn at Saumur was perhaps the finest time of their marriage.

> They were alone at last, with no family whatsoever. Ma spoke perfect French . . . loved the French people and understood them. . . . Georgie was associating with some of the greatest warriors of the 19th and 20th centuries—the heroes of his youth in the flesh, the "beaux gallants.", . . The flower of the French Army, so soon to be mowed down by the trampling Bosch hordes. These were men of legend. There will never be their like again. They reminded Georgie of the descriptions of the Southern beaux and braves who had fought in the Civil War, and had been immortalized for him by his step-grandfather, Colonel George Hugh Smith.[53]

Patton toiled studiously under Cléry's tutelage and not only became an expert swordsman but also learned how to be a teacher. In Stockholm the previous year, his defense had been "the despair of his teachers, for the aggressive Patton was interested only in offense. His method of parrying was to counterattack."[54] The only word in Patton's vocabulary of swordsmanship was "thrust." He found the word "parry" as repugnant as he did "defense." In addition to intense fencing sessions and discussions with Cléry, Patton also attended classes at the Cavalry School. He learned to speak French, although he had to be assisted by Beatrice, who attended classes with him and took notes that she translated at night.

Another expert swordsman from a nearby military school, who came to Saumur for several days to assist Cléry, later wrote him a warm letter of friendship and praise, noting that Patton had the glorious task of teaching to others "the beauty and love of arms. I know it will be easy for you, for even in the short time I knew you I felt that you were a master."[55] He was Lt. Jean Houdemon, later to become a general and a French national hero in the two world wars. In 1947 Houdemon recalled the brief encounter: "I still see his proud and elegant figure, accompanied by Mrs. Patton, brightening the threshold of my house—a tall, thin cavalry man with a keen eye and a firm hand either to wield a sword or guide a horse . . ."[56] Houdemon found Patton a keen student of war, and for the brief period they were together they were inseparable, their riding, shooting, and fencing augmented by a study of battles and wars, particularly those of Napoleon. Houdemon would later write that Patton's historic campaigns of the Third Army in 1944–45 reminded him of Napoleon at his best, "from [the] overall strategy right to the psychological approach of the commander himself."[57]

When George was not fencing or studying, he and Beatrice toured the historic château region of the Loire Valley on foot, horseback, and automobile, using the efficient maps of Michelin. Caesar's legions and Attila the Hun's ravaging hordes had all been there before him, and Patton was convinced that he too had fought there in a previous life, and one day was destined to fight again in a modern war. "His intention was to study it 'for the next time around,'" said Ruth Ellen. "He said that the battles had been lost and won in these fields and hills through knowledge of the country; that history had already picked the battlefields, and that history was the greatest teacher. One of his mottoes was 'there are no practice games in life.'"[58]

It was as they prepared to return to Fort Myer that there occurred the incident when his angry wife chased him with one of his own swords, leaving Patton thankful Beatrice had not learned swordsmanship under Cléry. Among their many purchases while in France was a fine leather saddle, a gift for Leonard Wood. The unassuming Cléry had presented Patton with a photo of himself inscribed, "to my best pupil."[59] When the Pattons departed Saumur at the end of August they left behind many friends, and a full vessel of goodwill.

They sailed from Cherbourg in early September and returned to Fort Myer to prepare for George's immediate transfer to Fort Riley. The culmination of his triumphant tour at Fort Myer was a high rating on his final efficiency report from his regimental commander. For both George and Beatrice their brief tenure in the heartland of the army had been an unforgettable time. More important, it was a milestone for the career of George S. Patton. In less than two years he had represented the army and his country in the 1912 Olympics; gained the attention and respect of the army's top officials and the secretary of war; had been acknowledged as the army's foremost expert on swordsmanship; and, as a result of his five weeks at Saumur, the prized title of master of the sword in the U.S. Army was bestowed on him. It was quite an achievement for a mere second lieutenant of cavalry.

CHAPTER 11

"... A Home Where the Buffalo Roam"

... A very zealous and ambitious young officer.
—FROM 1915 EFFICIENCY REPORT

Fort Riley in 1913 was a throwback to the frontier army of the American West. Located in the desolate, rolling prairie country of central Kansas, adjacent to Junction City, an aptly named railroad interchange where two tributaries merge to form the Kansas River, Fort Riley was the home of the Mounted Service School (in 1920 it would be redesignated the Cavalry School).

First established in 1852 to protect westward-migrating settlers from the many large Indian tribes that inhabited the region, Fort Riley was situated at the crossroads of the Oregon Trail and a second westward trail to New Mexico, Arizona, and California. Fort Riley's twenty thousand acres are primarily rich alluvial plain but are bordered by white limestone escarpments nearly two hundred feet high, known to natives of the area as "The Rimrock."[1] For much of the year the grass is burned to a straw color, but in the spring the prairie turns such a luxuriant shade of green that it reminded Patton of France.

Freezing cold in winter and unbearably hot and humid in the summer, Fort Riley was, like Fort Sheridan before, the very antithesis of urbane Fort Myer. For all its lack of sophistication, the Fort Riley garrison nevertheless epitomized the Old Army of 1913, which was no more popular with the American public then than it had been a century earlier, when the army had become a permanent institution. If the Old Army was the western frontier, then Fort Riley was its heart and soul. There was not much interest in soldiers and soldiering, but the cavalry had long represented the most dynamic image the army possessed. To this day there remains an indefinable mystique about men who risked life and limb to ride hell bent for leather in uniforms of blue and gold across the great open spaces of the American West, as had Custer's 7th Cavalry. Since 1892 the Mounted Service School had been instructing aspiring young cavalry officers in the fundamentals of their profession, and, as Lucian Truscott has written, "The reservation at Fort Riley was a horseman's paradise . . . a detail to the Cavalry School was in every sense 'leading the life of Riley'!" It was also "a wonderful training ground for fighting men and for fighting combat leaders."[2]

Traces of the old frontier army were still in evidence at Fort Riley, including some grizzled cavalry veterans of the Indian wars of the nineteenth century. When Patton arrived in late September 1913, there was still a sign on the parade ground reading:

> Officers will not shoot buffalo on the parade ground from the
> windows of their quarters.

BY ORDER OF THE COMMANDING OFFICER[3]

The highlight of an average day at Fort Riley was not a parade, a horse show, or a foxhunt but rather the daily ceremony late each afternoon when every activity ceased for a few brief moments as the bugler blew retreat and the American flag was lowered from its place atop the post flagpole and reverently folded away until reveille the next morning. After Beatrice arrived in October 1913, she would often place Little Bee in her carriage and push her toward the flagpole, where both would observe the end of another day while waiting for Georgie to make his way up from the stables to join them.

Although Fort Riley was hardly glamorous, an assignment there was prized by cavalry officers. The Mounted Service School was highly regarded, and attendance was a prerequisite to the successful career of any cavalry officer. In 1913 the school was devoted "almost entirely to equitation, horsemanship, and the various arts and crafts associated with animal management."[4] Classes were held Monday through Friday, and there were off-duty studies required of each student. "This is the most strictly army place I have ever been in and also the most strictly business," Patton wrote to his father. "We start at eight o'clock and get through at three thirty which is more work than I have ever done in the army."[5]

As usual when the Pattons moved, Beatrice had returned to her family while her husband attended to the business of establishing the family homestead and commencing his new duties. His first letter to her was somewhat ominous: "I have not done much to day except get a house," he wrote. "It is fully as ugly as the one at Sheridan but a little larger. . . . The rooms are finished in yellow pine like sheridan [sic] and it is not allowed to paint it. . . . [There] is only one bath tub and that very small." There was no room for a live-in servant. "I hired a colored woman to come and clean it. It is the dirtiest thing you ever saw though clean dirt. . . ."[6] He got to work varnishing the floors and doors, turned the back room into a storage area, and built shelves in the kitchen. In subsequent letters there was more unfavorable news. "I think if you are to survive this place at all you will have to ride horse back as there is not another thing to do. There is not even a place to go in an automobile. . . . You have no idea how deserted this post is. It is out in the plains all by itself with nothing near it." For once he began to demonstrate thoughtfulness. "I love you and miss you," he wrote in early October, "but don't want you to come until every thing is all right here. You certainly have to give up a lot on my account."[7]

At the Mounted Service School Patton was both pupil and teacher. As the newly anointed master of the sword he taught three separate classes in swordsmanship to both his fellow classmates and members of the faculty, virtually all of whom outranked him. Some resented the notion of being taught by such a junior officer, and Patton was aware that he must do something to gain their respect and undivided attention. His remedy was typically Pattonesque. One day he arrived with a package that he unwrapped but did not open. Then he began his lecture:

> Now, gentlemen, I know many of you outrank me, some of you by many grades, and I realize how hard it must be to take instruction from a man you must regard as still a little damp behind the ears. But gentlemen, I am about to demonstrate to you that I have been an expert with the sword, if in nothing else, for at least fifteen years, and in that respect I am your senior.

Patton then withdrew from the package (which had been sent to him by his mother) the two wooden swords he and Nita had used as children in Lake Vineyard, and waved them in the air. There was a brief moment of silence, and then the class broke up in gales of laughter. Although Lieutenant Patton's problems were now mostly a thing of the past, he found that keeping them interested "is the hardest job I ever tried and I certainly am tired at night."

His efforts attracted the attention of the school commandant, who not only wrote "excellent" on his first efficiency report but praised Patton for "his great zeal and proficiency in his work." An officer who knew him at Fort Riley said of him during World War II: "You know, looking back on Patton, he has been a general all his life."[8]

In addition to his other school activities, Patton also wrote the drill regulations for the new cavalry sword, as well as the introduction to a pamphlet titled *Army Racing and Records for 1913,* which the War Department hoped would serve to enhance the image of the cavalry by publicizing and encouraging army officers to participate in racing and polo. Patton's introductory essay emphasized—indeed it was more of an exhortation than an introduction—the value of competition and knowledge of horsemanship.[9]

The Pattons visited Lake Vineyard during the 1913 Christmas break, and on their return the reality of life at Fort Riley nearly overwhelmed Beatrice. As Georgie had warned her, there was very little to occupy the interest of a Boston aristocrat in the middle of Kansas:

> Ma didn't speak the same language that was spoken by the other Army wives. Her interests had always been music, the theatre, racing boats, her family, the life of a cultivated Eastern heiress. She had never had to worry about money, or "making do" and she didn't understand the gossip or wistful references to "olden days." In those days the "old army" was a club, with an inner circle of people who were the sons and grandsons and daughters and granddaughters of Army officers and who knew each other from the cradle to the grave. . . . Having a great deal of her mother Ellie's acting ability, she could put on a good show, but her heart really wasn't in it, and she was lonely—and even a little bit bored.

Nita Patton and Kay Ayer were frequent visitors. At an isolated post like Fort Riley attractive young women were prized, and during these times the Patton household was a busy and exciting place.

With so little to stimulate her active imagination and equally little of common interest to share with her husband, Beatrice began again to question seriously if she was cut out to be an army wife. She had virtually nothing in common with the other wives, a great many of whom were southern, which Beatrice associated, however snobbish it may have seemed, with the servant class. A hired couple came in to take care of the household duties. "She was beginning to feel she was a terrible failure as an Army wife and mother. . . . It all seemed very wild and crude and savage."

Nevertheless Beatrice always referred to her experience at Fort Riley as her "waking up" period, a time when she reasserted her identity and accepted the reality that she was married to a career officer, and thus there would be many more Fort Rileys in their immediate future. Porch sitting was a way of life at Fort Riley and on the afternoon of her "awakening," the wife of one of Georgie's classmates who lived down the street was over-heard to reply in a loud voice to a series of "What says?" from her deaf mother-in-law: "I said that with that pretty little Mrs. Patton gone so much to see her folks, Mrs. Merchant seems to be getting her hooks into young Mr. Patton."

A distressed Beatrice grabbed Little Bee and ventured to the Rimrock, which was about as alien as a place could be from New England, and while there experienced what can only be termed a unique encounter with her inner self. She began to examine the path surrounding the escarpment and suddenly:

> She was not standing on the lone prairie in the Middle West . . . [but] on the shores of a dead sea, millions of years gone dry. She ran as fast as she could push the carriage, handed the baby to Hannah, and sat right down and wrote to Lauriats book store in Boston . . . to send her immediately all the books they had on American marine fossils.

In them she learned that Kansas had been the bottom of the Permian Sea for 350 million years during the Paleozoic Era.

> No longer were the plains of Kansas bleak to her. They were a treasure house to be explained, explored and exploited. That one minute she told us, changed her whole life. Her inner eye had been opened. She never again in her life had a dull moment, or a single regret for the fun and games of her childhood. She had discovered the whole world.

Not only did this unique experience change Beatrice's life forever, but she suddenly began to view her neighbors in a kindly spirit. The wife of Captain Eli DuBose Hoyle (another future general), who was something of a legend in her time and a charter member of the Old Army's inner circle, became one of Beatrice's closest friends. A woman of Amazonian propor-tions who was "as wide as she was high" and played the piano "like an angel," Mrs. Hoyle had beautiful eyes, and her charm and hospitality made Beatrice feel completely at ease for the first time. With Mrs. Hoyle to guide and befriend her, Beatrice began to feel very much at home in the company of the Old Army wives. It was much like family life in the Ayer household. To her intense joy, many years later Beatrice Patton's only son would marry the Hoyles' great-granddaughter.[10]

In May 1914 Patton graduated from the Mounted Service School and was rated "proficient" in each of the various subjects taught. He was also com-mended as the first master of the sword and rated suitable for duty as an instructor in swordsmanship.[11]

Even though Patton was far removed from Washington, he managed to find means of ensuring he would not be forgotten by those whose influence counted. When Leonard Wood was replaced as chief of staff by an officer whom Patton did not know, he unhesitatingly wrote to the deputy chief of staff, Maj. Gen. Hugh L. Scott, to deplore a scheme to dispose of a section of the Fort Riley reservation: "In writing this letter I fully realize that I run a grave risk of overstepping the bounds of military decorum. I trust however, that you will excuse any presumption on my part on the grounds that I am personally, perfectly disinterested in the matter." Although writing directly to an officer who would soon become the new army chief of staff was about as far out of bounds as a second lieutenant could get, Patton knew full well that if he had written through military channels, the letter would never have arrived in Washington with his name associated with a recommendation to retain the land. As it would on numerous other occasions, Patton's audacity paid off when he received a favorable reply from Scott.[12] During World War II Patton would employ the same method to outfox his superiors. Until someone forced him to stop, he would do things his unorthodox way. *"L'audace, l'audace, toujours l'audace!"* became his motto, dating back to his earliest days in the army. What he had learned at Fort Myer was continuously refined to ensure that no one ever forgot who he was.

Patton's prestige was further enhanced in June 1914, when he received a letter from the American Olympic Committee announcing that he had been unanimously elected a member of the U.S. team for the Sixth Olympiad, to be held in Berlin. His selection had come about on the recommendation of the president of the committee, Col. Robert M. Thompson, whom the Pattons had met in December 1912 at a hunt breakfast in Washington, D.C.[13] His self-promotion had once again paid off handsomely.

He also thoroughly enjoyed the distinction of being the only master of the sword and wanted to continue teaching for another year. Moreover, he had no intention of leaving Fort Riley until he completed a second-year course of instruction for promising company-grade officers. He was selected as one of only ten students and spent the summer break of 1914 in his usual flurry of activity, which included a leave with the Ayers in Massachusetts and a trip to Columbia, Missouri, in search of additional horses for his growing stable.[14]

His visit to Pride's Crossing coincided with one by his father, and the two men traveled together back to Fort Riley, where father and son enjoyed an extended reunion. The two rode the open plains, and for the first time Mr. Patton was able to view his son at work and at play, unimpeded by a family event such as Little Bee's birth. His memoir of his father recalled with obvious fondness these all-too-rare occasions. Yet Patton also used them to enhance his image, noting that whenever his father visited him, "it did me a lot of good as his intelligence, character and learning impressed favorably the better class of officers whom he met."[15] Beatrice wrote to her father-in-law: "You don't know what a comfort it is to me to have you with Georgie! He loves you so much that you can keep him in order better than anyone else in the world."[16]

Yet, despite his unique accomplishments as an Olympic athlete and master of the sword, Patton was often restless and dissatisfied, naively criticizing himself to his father for failing to live up to his stated goal of becoming a brigadier general by the time he was twenty-seven: "Now I am twenty-nine and not [even] a first Lieutenant."

On June 28, 1914, the assassination in Sarajevo of the Austrian Archduke Franz Ferdinand soon inflamed Europe and led to the outbreak of World War I. George Patton followed the momentous events in Europe with more than passing interest. Here—at last—was what seemed to offer the first opportunity since his commissioning to experience war firsthand, but dismay followed when Woodrow Wilson declared the United States neutral a day after Germany had launched its offensive in the West. He wrote at once to Leonard Wood for advice and assistance, and to request a year's leave of absence "on some pretext" that would enable him to join the war in France. Lest Wood mistake his intentions, Patton pointedly noted that his objective was to participate in combat, not observe, for "it is only by doing things others have not done that one can advance."[17]

It was an audacious gambit—and one that provides a clear example of Patton's burning passion to experience war. As he wrote Wood:

> If I can get the leave I can manage the rest. Of course, with the understanding that I will never apply to the United States for help if I get in trouble or captured. I can turn over the Swordsmanship at Riley to an officer who did well under me last year and will continue my method. As my family does not rely on me for support I would only be risking my self. Please do not think this a spontaneous folly. I have contemplated it for years. I would not bother you except that I am encouraged by the interest you have already taken in me and because I value your opinion above that of any one . . . "[18]

Wood's reply dashed Patton's hopes: "Don't think of attempting anything of the kind, at present. If you can get a leave, all right; but go to look on. We don't want to waste youngsters of your sort in the service of foreign nations unless they need you more than appears to be the case now. Stick to the present job. . . . I know how you feel, but there is nothing to be done."[19] It was no coincidence that his next efficiency rating noted Patton as "a very zealous and ambitious young officer."

It was in early 1915 that Maj. Charles D. Rhodes, a veteran of the Sioux War of 1890–91, and Leonard Wood's chief aide during Patton's brief tenure in the War Department in 1912, rated Patton "excellent" all around. However, Rhodes was also the first officer to put on record what many regarded as his overzealousness. Although Patton lacked experience with troops, Rhodes found the lieutenant a "most promising young officer of high ideals, devotion to duty, and marked industry. He is somewhat impulsive and intolerant of the opinions of others, and needs a period of severe duty with troops to counter-balance his protracted duty away from troops and to round out his efficiency as an all around officer."[20]

In June 1914 Beatrice had become pregnant for the second time, but this event appeared to have no bearing on Patton's consuming desire to help the French fight World War I. Since his experience with the birth of Little Bee, Patton had made clear his reluctance to have a second child. But whenever he inquired what she would like for an anniversary, Christmas, or birthday present, Beatrice would always reply: "I want a baby." When she announced her pregnancy there was an unshakable conviction, at least in the minds of the Patton family, that this time around Beatrice would certainly bear for them a second "Boy." The child was due in February 1915, and "he" would be born at Lake Vineyard, in the same "extraordinarily ugly bed, [that had been] bought in 1856 for the second Mrs. Wilson, when Don Benito was furnishing the new ranch house he had built for her"—the same bed in which Ruth Patton and her son had been born.

For weeks beforehand the Pattons dashed about in a frenzy of activity that even included planting gardens that the child could see from "his" newly decorated bedroom with its pink-and-white wallpaper. Aunt Nannie was summarily exiled, and her bedroom was turned into a sort of family command post, a "production room for the first grandson, to be the proud bearer of the name George Smith Patton, IV." Shortly before her due date, Ruth and Nita Patton escorted Beatrice from Fort Riley to Lake Vineyard, where she was made to sit in leisure on the veranda, with little more to do than smell the pungent aroma of orange blossoms and drink cow's milk from the Lake Vineyard herd.

Although Patton was obliged to remain at Fort Riley, he too became fully involved in the drama, sending a warning to his father to "tell the doctor if there is any question between her life and the life of the child, the child must go. This is probably an unnecessary caution but I insist on it. If he will not subscribe to it get another doctor who will. . . . She is a brave woman and I hope she will have no trouble."[21] Alas, when the momentous event took place on February 28 after another long and difficult labor, Beatrice had produced a second daughter, who remained nameless for a time.

So certain had everyone been that the child would be a boy that little thought had been given to a name for a girl.

For the second time an atmosphere of gloom pervaded the Patton household. Painfully aware that she had "failed" again, the plucky Beatrice informed her mother-in-law: "Well, Aunt Ruth, better luck next time!" With a shocked expression, Ruth Patton exclaimed, "Beatrice dear, please don't mention 'next time' to your Uncle George. He has had a *very* hard day!" Patton was awakened in the middle of the night and handed a telegram by his frightened housekeeper, who was certain Beatrice had died. In his first letter after the big event, Patton wrote "D-E-L-I-G-H-T-E-D!!" He also admitted:

> I am very glad from a selfish point of view that I was not there. . . . Though had it been possible I would have been for I think it might have comforted you a little. . . . You had better have it named out there where you can get more advice. All I know is that I don't like the sound of either Ruth or Ellen. You might call it Beatrice Second like a race horse. I certainly like the sound of that name the best of any. I love you with all my heart and hope you have not suffered or are not suffering more than necessary.[22]

Quite naturally such tongue-in-cheek advice on naming the baby was ignored. Instead, to honor her mother and mother-in-law, Beatrice had the baby christened Ruth Ellen.

The attending physician had turned out to be inept, not only leaving Beatrice in extreme discomfort but decreeing that the infant could nurse at each breast for no more than ten minutes. By the time mother and new daughter returned to Fort Riley, Ruth Ellen had become sickly, cried incessantly, and daily became thinner. "I was not a bundle of joy," she related many years later. "I had everyone mad at everyone else because I never stopped my crying. Finally in desperation, Ma took me to a local civilian doctor in Junction City, a great-hearted man of genius, Dr. Fred O'Donnell. He took one look at me and said, cheerfully, 'Well, Mrs. Georgie, I think we can save this little lady if we start right now. She's starving to death.'" Put on a formula of cream laced with a small dab of brandy, Ruth Ellen soon became a model infant, and everyone began sleeping at night. Once again George Patton swallowed his disappointment in a busy schedule of soldiering.

Shortly before the birth of his second daughter, Patton's beloved step-grandfather, George Hugh Smith, died at the age of eighty-one. In the final years of his life, Smith had risen to prominence as a California state senator and a commissioner of the state supreme court and in particular for his writings about jurisprudence, which brought him recognition not only in the United

States but also in Britain and in Europe. "He was a great mind wasted and I am sorry," Patton wrote to his father. "I was very fond of him and wish I could have seen more of him. He did not have the military mind in its highest development because he was swayed by ideas of right and wrong rather than those of policy. Still he was probably more noble for his fault."[23] George Hugh Smith had epitomized the ideal of the citizen-soldier who fought for Virginia and its way of life, which makes Patton's rather scornful reaction to his death even more incomprehensible.

On May 7, 1915, a German submarine torpedoed and sank the Cunard liner *Lusitania* off the coast of Ireland, with a loss of 1,198 lives, 128 of them American. Patton was incensed, blaming what he regarded as American cowardice on President Wilson and his policy of neutrality. "Anyone but a woman can see that the loss of life is a question of indifference to Germany," he complained to his father. "Who dares . . . to say that one can be 'Too proud to fight.' In any other country or age that pride has always been called by another name. . . . I think that we ought to declare war if Germany failes as she should to pay heed to our foolish talk [Wilson's diplomatic protests]. If Wilson had as much blood in him as the liver of a louse is commonly thought to contain he would do this." There was more than a kernel of truth in Patton's anti-Wilson diatribe, for in less than two years the United States would declare war on Germany. As Patton saw it, "By the time Germany beats the allies we will have time if we start now to get an army."[24]

Patton graduated from the second-year course at the Mounted Service School in June 1915. It had been a busy year, apportioned between being a student, teaching, and writing about swordsmanship. Whenever Beatrice was absent he began to increase his reading of military history. It was not uncommon for Patton to immerse himself in books for twelve hours at a time, and the mental uplift he derived encouraged him to continue to study even more. "My mind is less like a potato than has been the case with it since we left Sheridan . . . the more I read the more I see the necessity for reading. War now will not be gained by a highly educated 'bottom' but by a well developed 'top,'" of which he fully intended to be a part.[25]

By 1915 Patton had established himself as the army's leading expert on the sword, had developed the criteria for a new cavalry saber badge, and had written an illustrated pamphlet, really an instructor's manual, titled *The Diary of the Instructor in Swordsmanship*. The pamphlet also contained some of Patton's usual strong opinions, among them that anyone who refused to admit to a saber "touch" ought to be tried by court-martial. Naturally any notion of using the saber defensively was scorned.[26]

In the spring of 1915 Patton was again injured in a fall, this time when his horse stepped in a hole while he was riding on the prairie. "She rolled on me," he wrote Beatrice, who was still in California, "and in getting up she kicked me in the head with her hind foot and cut quite a hole in which I had five stitches taken. . . . When I get less hair than I now have I will look just like a German duelist."[27] Yet the cut was nothing to jest about and was in fact considerably more serious than the one he had sustained at Fort Myer in 1913, keeping Patton on sick call for nine straight days.[28]

After graduation he was granted two and one-half months leave and the Pattons returned to Massachusetts for a leisurely summer of relaxation and fun, a vacation that was terminated one day when his car overturned as he was returning from a polo match. Another driver discovered him, managed to lift the car, and removed an unconscious stranger, dressed in polo clothes, whose head had been battered by gravel and was covered in crankcase oil. Patton was taken to the nearby home of Mrs. Charles G. Rice (the mother of Hilda Ayer, née Rice, who was married to Beatrice's brother, Frederick), from which his savior had departed only moments earlier. He was placed on a bed more dead than alive. When Mrs. Rice opened Patton's mouth, she found him strangling from the oily gravel and sludge he had swallowed. The stranger's clothes contained the labels of her son-in-law, Frederick Ayer Jr., who at that moment was on his honeymoon. It was only after Mrs. Rice cleaned him up and summoned a doctor that she discovered the man wearing her son-in-law's polo clothes was George S. Patton, whose life she had just saved. Needless, to say Patton was eternally grateful, and when Mrs. Rice died in 1933 after a fall from a horse, he came from Fort Myer to attend her funeral, saying that he would have come no matter if he had been halfway around the world.[29]

During the early part of 1915 a rumor, which soon proved true, began sweeping the army that in October the 15th Cavalry would be reassigned to constabulary duty in the Philippines to replace the 8th Cavalry. Once his schooling ended Patton, was due to return to the regiment.[30] Although God could do little to avert this perceived disaster to his career, George Patton could and would help himself.

After chaperoning Beatrice and the children to Pride's Crossing, he immediately journeyed to Washington. Patton never recorded just how he pulled strings, but once again he managed to persuade someone of influence in the War Department, possibly his West Point mentor, Charles P. Summerall. He was pleased when the black doorman at a theater recognized him. "No one else did. Some day I will make them all know me."[31]

Patton's new posting was to be on the Mexican border with the 8th Cavalry, which was then in the process of relocating to Fort Bliss, Texas, from the Philippines.[32]

Before George S. Patton left Fort Riley he presented the Mounted Service School with a cup to be awarded annually to the winner of a mounted saber competition for the Troop Officer's Class. Known as the Patton Cup, his legacy soon became a highly sought-after prize in which each contestant had one and one-half minutes to complete a series of jumps and use his saber to penetrate twenty dummies placed along a difficult course.[33]

In mid-September Patton reported to Fort Bliss, which was located outside El Paso. After a summer holiday in Pride's Crossing, Beatrice had accompanied him back to Fort Riley to help pack and ship what by now had become a very large household. Without asking his father he consigned everything— horses, three dogs, and his household goods—to Lake Vineyard, collect. Patton's stable of horses had grown to eleven, and shipping this menagerie was no small chore. Some years later he recalled with evident glee having sent his father a telegram that "he was to pay the freight. He thought this a great joke and used to tell about it at the California Club."[34] He and Beatrice then drove from Fort Riley to Southern California. The trip was grueling and averaged barely twenty miles per hour on the mostly unpaved, dusty highways. It was a feat of endurance, and George and Beatrice arrived weary and covered in dust and road grime. When they left Lake Vineyard several weeks later by train, Beatrice accompanied her husband as far as Fort Bliss, where the two parted company for the latest of their many painful separations. It would be some months before the family was reunited.

When Patton reported for duty he learned that most of the 8th Cavalry had not yet arrived from the Philippines, only a small advance party being present. Shocked when informed he was to be examined for promotion the very day of his arrival, Patton explained his plight to the regimental executive officer who advised him to wire the War Department, which immediately authorized a five-week postponement. Patton had a brief reunion with Major Marshall, who was visiting Fort Bliss en route to the Philippines and was delighted at the accomplishments of his young protégé. Patton, in turn, went all-out to impress Marshall's host, another future general, who coincidentally happened to be a member of his forthcoming promotion board, Capt. Howard R. Hickock. Patton also took pains to cultivate the president of the board, his new squadron commander, Maj. George T. Langhorne.[35]

With only a few duties to distract him during the first weeks, he studied hard and, although he had worried about passing the examination, which covered tactics, cavalry drill, and field service regulations, apparently did so with ease. The resulting certification from the board meant that Patton's name was placed on the War Department promotion list.[36] Although he was no longer assigned to the 15th Cavalry, Patton learned he was still subject to a transfer to his former unit whenever he was promoted and a first lieutenant's slot opened up in the Philippines.[37]

*   *   *

El Paso, originally founded in 1682 as a Spanish mission, was the oldest settlement in Texas. By 1915 it had become a small, Wild West border town on the Rio Grande. A "refuge for Mexican jetsam, and hideout for western gunmen, the town relied heavily for existence on the railroad and on the army garrison at Fort Bliss," a treeless, arid military reservation on the northern outskirts of El Paso, overlooked by the bare and forbidding Franklin Mountains.[38] The recently built fort contained stables that were firetraps and tiny, poorly constructed quarters, news that was hardly music to Beatrice's ears.

Patton was lonely and frequently wrote how much he missed Beatrice. He ended one letter by declaring, "can't send any kisses as we had onions for dinner." Unassigned for several weeks, he assisted Major Langhorne in forming a polo team. To his utter disgust, Patton fell off his borrowed cow pony during his first match. For once the only injury he suffered was to his pride.

For a short time he was acting commander of Troop D and wrote Beatrice that it seemed strange to command a cavalry troop once again. Throughout his life Patton never failed to be moved by military reviews, and when his troop was on guard detail he attended a regimental parade as a spectator. "It was a fine sight all with sabers drawn and all my [Patton] sabers [being used]. It gives you a thrill and my eyes filled with tears . . . it is the call of ones ancestors and the glory of combat. It seems to me that at the head of a regiment of cavalry any thing would be possible."[39]

Patton's assignment to border duty in 1915 coincided with a period of increasingly turbulent relations between the United States and Mexico. Many times during his military career Patton found himself in the right place at the right time. The southwestern United States in late 1915 was such a place. Although most of the western United States had been settled and tamed by the turn of the century, the Wild West—where colorful, independent, violent men still lived by the law of the gun—would continue to survive for another twenty years in such locales as southwest Texas.

Into this still-untamed land came George Smith Patton. It was to prove a match made in heaven. He had always fancied himself a warrior-hero in the mold of King Arthur, standing alone in majestic opposition to an evil force. On the Mexican border during the next year he would again be provided with a unique opportunity to become that hero, and in the process to solidify his claim as a "comer" in the U.S. Army officer corps.

CHAPTER 12

Pershing and the Punitive Expedition

I want to go more than anyone else.

—LT. GEORGE S. PATTON

Since 1910 Mexico had been wracked by instability, and by 1914 there was increasing hostility in Mexico toward Woodrow Wilson's interference in Mexican affairs and inept handling of U.S.-Mexican relations. When the thirty-year dictatorship of Porfirio Díaz ended in 1910 with his overthrow, there followed a period of revolution in which the fledgling democratic government of President Francisco Madero was overthrown in 1913 by a military coup that installed General Victoriano Huerta as president. Huerta was believed to have murdered Madero and the vice president. Wilson refused to recognize the Huerta regime, and an incident in Tampico and the occupation of Vera Cruz in 1914 by an American military force was an attempt by the president to persuade the Mexican people to replace yet another in a long line of despots. Wilson's well-intentioned crusade to remove Huerta backfired and instead exacerbated the growing anti-Americanism in Mexico. The United States withheld recognition of the Huerta regime, and sent military supplies to his opponent, Venustiano Carranza, the governor of Coahuila, a former senator and wealthy landowner, who called for the Mexican people to rise up and topple the new dictator. Unfortunately Carranza was a ruthless revolutionary who was hardly better than the man he was striving to replace.

Among Carranza's early supporters was a charismatic renegade from Durango named Francisco Villa, who was better known to his legion of followers as "Pancho." A notorious bandit leader, Villa epitomized the Mexicans' love of *macho* and had become the Latino version of Robin Hood, looting the rich, rustling their cattle, and giving to the poor. Villa may have been a folk hero but he was also a cold-blooded killer who "could shoot down a man point-blank, showing no more emotion than if he were stepping on a bug."[1] As a military commander Villa was daring and, after learning the hard way the folly of the cavalry charge, possessed of a tactical boldness that Patton himself might have admired.[2]

The bloody Vera Cruz incident brought about the resignation of Huerta and his replacement as president by Carranza. Breaking with Carranza, the ambitious Villa formed his own rival Conventionalist party and began opposing his former cohort. The Mexican economy was in disarray and instability grew as intrigue and lawlessness swept the nation. Pancho Villa had counted on American support to obtain the presidency. Instead, when Wilson recognized the new Carranza government in October 1915, an irate Villa swore revenge against the United States.[3]

By the end of 1915 not only had Villa and his *pistoleros* launched a series of raids along the U.S.-Mexican border that frightened Americans living in Texas, New Mexico, and Arizona border towns, but Carranza's forces were also engaged in similar burning and looting. Mexico was swept by violence as Villistas, Carranzistas, and several other factions turned the political climate into a state of virtual anarchy. Wilson, who had once supported Pancho Villa, now regarded him as little more than a bandit who threatened the security of the southwestern United States. Fighting appeared imminent, and the War Department began deploying troops to Texas and New Mexico to meet this threat.

One of those alerted was Brig. Gen. John J. Pershing, then commanding the 8th [Infantry] Brigade at the Presidio of San Francisco. Anticipating further trouble from marauding Villistas and Carranzistas and the need for a possible punitive expedition against Mexico, the War Department hastily sent Pershing and his troops to Fort Bliss in April 1914. Pershing was placed in command of some five thousand troops guarding the U.S.-Mexican border from Arizona to a bleak outpost in the Sierra Blanca mountains ninety miles southeast of El Paso.[4]

Pershing's career had advanced dramatically since he had been tagged with the derisive nickname "Black Jack" in 1897 at West Point, where he was one of the most unpopular tactical officers ever to serve at the military academy. (Pershing had served for two years in Montana with the all-black 10th Cavalry Regiment. At West Point he was sneeringly referred to behind his back as "Nigger Jack," later modified to "Black Jack.")[5] Now one of the most highly respected officers in the army, Pershing quickly earned the confidence and respect of the inhabitants of El Paso, who viewed him as their protector from Villa. A tough, experienced veteran of the Indian wars and

the Moro uprising in the Philippines, Pershing was a no-nonsense disciplinarian whose glacial stare when he was angry could instantly instill fear
into even the most veteran trooper. In his new assignment Pershing was constantly in motion, "sometimes by car, more often mounted, he trekked his
domain with regularity—a general [who] never lost control of his farthest
flock."[6] He trained his men hard, sent them on maneuvers, and confidently
prepared to accede at once to an order to do battle with the Mexicans.

On August 27, 1915, while he was in Texas, a terrible misfortune struck
when Pershing's wife, Frankie, and three of their daughters died from suffo-
cation when a fire ravaged their quarters at the Presidio of San Francisco
shortly before they were to depart for Fort Bliss. When informed of the
tragedy, an anguished Pershing cried, "My God! My God! Can it be true?"
Among the outpouring of messages of condolence was one signed "Francisco Villa."[7]

In mid-October 1915 Patton had been at Fort Bliss barely a month when
Troops A and D were ordered to the Sierra Blanca mountains, where the
army had established a chain of outposts to guard the border sector southeast of El Paso. The trip took four days by horseback and wagon train under
a blazing hot sun. The tiny town of Sierra Blanca was a whistle stop on the
railway line from El Paso and a violent holdover from the Old West, populated by cowboys and gunfighters of fearsome reputation. Situated at an elevation of 4,500 feet in the rocky Sierra Blancas, the entire town consisted of
approximately fifty people, twenty houses, a saloon, and a hotel. Troops A
and D were based at Sierra Blanca, with one manning the outposts and the
other in reserve to protect the railroad and respond to any trouble along the
border. Once a month they rotated with each other.[8]

Patton soon became acquainted with the locals, one of whom was the
elderly Sierra Blanca town marshal, Dave Allison, whose white hair and
cherubic face were the facade of a renowned gunman who had once slain a
notorious bandit named Orasco and his gang. Impressed to be in the company of an illustrious gunfighter, Patton wrote to Beatrice. "He kills several
Mexicans each month. He shot Orasco and his four men each in the head at
sixty yards. He seemed much taken with me."[9] The whiskey flowed copi-
ously in the saloon, and on any given Saturday night the local *pistoleros* and
cowhands entertained themselves by firing their weapons at targets both real
and imagined. Patton's profanity, his ability to shoot a pistol, and his free-
spending ways at the saloon, where he delighted in buying beer for them,
quickly made him a popular figure with the rough-hewn Texans, who sensed
a kindred spirit.

The first task of Patton and his superior, 1st Lt. Daniel D. Tompkins, the
troop executive officer, was to inspect the two outposts situated at either end
of the Sierra Blanca mountain chain, a 100-mile, three-day trip by horseback

over some of the most rugged terrain in the United States. There were few roads but "miles and miles of loose stone. . . . It is the most desolate country you ever saw. Rocks and these thorny bushes."[10] The area was virtually devoid of any human habitation. Patton slept rough on the ground and availed himself of the opportunity to shoot game. "Americans can't live there or if they do they don't live long." He boasted to Beatrice that, "I made the darndest shot with a pistol you ever saw I hit a jack rabbit running at about fifteen yards while riding at a trot. My reputation as a gun man is made."[11] His elation at becoming a crack shot would soon suffer a major reversal.

Patton ingenuously reported to Beatrice his pleasure at meeting a local cowgirl, "which shows that I am a social success though from the talk here she is easy of conquest. Very easy!"[12] One night, at the farthest cavalry outpost, called Love's Ranch, he escaped potential injury that left him "very scared" when a "crazy drunk" trooper pointed a pistol at him before being tackled from behind and disarmed by one of the NCOs.

His adventures included guarding more than thirty miles of railroad track. Although the duty was hard and involved no fighting, "I am hopelessly dirty [but] I feel contented. . . ."[13] He passed his thirtieth birthday in the saddle, inspecting his outposts.

In Hot Springs he bested a man with shoulder-length hair in a rifle match but lost in a pistol competition. He also met a remarkable panther hunter who regaled Patton with tales of his adventures, "which others said were true. He was very dark [skinned] and commented on it. Saying 'Dam it a fellow took me for a Mex and I had to shoot him three times before he believed I was white.' This impressed me very much and I assured him that he was the whitest man I had ever seen."[14]

On another occasion Patton was the senior officer in Sierra Blanca, when he received an urgent telegram from Fort Bliss that a Mexican bandit named Chico Chano was en route, with a force of 200 men, to raid Sierra Blanca. Somewhat skeptical, Patton did nothing other than order his men to sleep with their weapons handy. At 11:00 P.M. three more telegrams arrived. One was signed "John J. Pershing," and it shook Patton's complacency and made him pay serious heed to the threat. Over the next several days there was evidence of both Villistas and Carranzistas on both sides of the border, but to his disappointment, there was no clash between the Mexicans and the U.S. cavalry.[15]

A few days later he received another urgent telegram with an unconfirmed intelligence report that a large Mexican force was on the loose near Fort Quitman, a border outpost. Patton was ordered to "act with vigor," which he interpreted as an order to "attack first and ask questions next. So I decided that if possible I would make a saber charge. . . . I thought I had a medal of honor sewed up and laid awake planning my report until one a.m." He and his men were spoiling for a fight, but after a thiry-two-mile trek across the wasteland in bright moonlight they arrived at Fort Quitman and found that if there had been a force of Carranzistas, it had long since left. In all, a disappointed Patton was in the saddle for eleven hours and covered between sixty and seventy miles. "The last 17 miles was awful. The dust was so thick that you could not see the fence at the side of the road."[16]

With only two hours' sleep in the previous two days, Patton felt fine but tired. He stolidly accepted that his first encounter with an enemy would have to await another day: "I had great hopes of seeing how my sabers would work but better luck next time."[17] Nevertheless, the incident provided small but unmistakable clues to Patton's future aggressive behavior on the battlefield.

Soon after this Beatrice decided to visit her husband. She left the children in the care of her parents and journeyed to El Paso, where Patton met her and they drove in the family auto to Sierra Blanca. For a brief time Major Langhorne shared their house. Patton admired Langhorne and his splendid automobile. Like the Pattons, Langhorne came from wealth and had brought to Sierra Blanca an elegant eight-cylinder Cadillac that was kept in an adobe garage behind the house.[18]

Beatrice quickly became as well liked as her husband by the townsfolk, who appreciated having in their midst an elegant lady who knew all about horses but was not pretentious. One of Beatrice's chief admirers was the sheriff (the town had both a sheriff and a marshal), who encouraged Patton to quit the army and become his partner in running a spa.

Patton's first attempt at emulating a Wild Westerner was catastrophic. One evening the Pattons were having dinner with some local businessmen at the town's only hotel when suddenly a gun went off, the lights went out, and a pair of strong hands grabbed Beatrice and unceremoniously dragged her under the table. Unable to determine the source of the gunfire, the group then nervously dispersed. The moonlight was exceptionally bright, but that did not prevent Patton from slamming the family auto head-on into a cattle gate that was plainly visible. Beginning to sob, with tears running down his cheeks, he said to Beatrice accusingly: "God dammit, you don't give a damn about me! That was my pistol that went off; I might have been killed, and you didn't even say anything or ask me if I was alright!" When he finally calmed down, Patton explained he had been emulating local custom on dress-up occasions by wearing his pistol in his trouser fly, and that somehow it had fired and shot a hole through his trouser leg into the floor.[19] Patton's future attempts to imitate Wyatt Earp would be with his pistol holstered in plain sight on his waist.

Beatrice was the glue that kept their long marriage together. Military wives who survived the frequent moves to distant posts, the low pay, the often-squalid quarters, and being both mother and father to their children during prolonged separations from their husbands (who were in the field training or off fighting a war in some godforsaken place) were indeed a special breed of women. It must have been especially difficult for Beatrice to make the transition from socialite to the wife of a lowly second lieutenant. And while their unlimited income from Mr. Ayer enabled them to enjoy many of the trappings of wealth, places like Fort Sheridan and Sierra Blanca were unavoidable for an ambitious young cavalry officer. Although Beatrice rarely verbalized her private feelings about her husband's chosen career, on at least this one occasion her frustration erupted. It was a measure of the character of this remarkable woman that she willingly endured years of such hardships out of love and respect for her husband's passion to be a career soldier. Nevertheless, after one of the frequent, brutal windstorms that plagued Sierra Blanca, Beatrice cried and told her husband she wished he would resign from the army.

Despite their brief sojourn there, the Pattons were so popular in Sierra Blanca that when George was ordered back to Fort Bliss, most of the town turned out for a farewell gala evening of barbecue, music, and square dancing. Guns and liquor were surrendered at the door and babies were wrapped in blankets and stacked like cordwood on a large brass bed in a bedroom. The Pattons danced with everyone, and the regret at their departure was mutual.[20] When they returned to El Paso, Beatrice found that she liked Fort Bliss and decided it was time for the family to be reunited. Patton applied for post quarters, and Beatrice returned to Massachusetts to bring the family west.

For a few all-too-brief months Beatrice's and George's lives were unexceptional, and it seems to have made for a welcome respite in their otherwise hectic existence. Beatrice became a cavalryman's wife again, and George played polo, hunted, and defended the saber to the Cavalry Equipment Board, now headed by his former superior officer, Maj. Charles D. Rhodes.

One of the usual number of visitors to the Patton household at Fort Bliss was his sister, Nita. Twenty-nine, very attractive, and as yet unmarried, Nita was "a tall blonde Amazon with enormous capabilities of love and loyalty and great good sense. In every way she was, like her only brother, slightly larger than life-size." At one of Fort Bliss's frequent social events, Nita was introduced to Black Jack Pershing. Although the death of his wife and daughters was still a raw wound, Pershing was immediately attracted to Nita Patton and she to him. Pershing's biographer writes that Nita possessed the same fine facial features as his late wife, and, sensitive to his hurt and sadness, she was instantly captivated by him. "She encouraged the general and they grew closer than friends."[21] Pershing was a man who needed women in his life, particularly for their attention and flattery. Nita and Black Jack Pershing were strongly attracted to each other. Although their relationship would eventually develop into a full-fledged love affair, its early progress was abruptly cut short in early March by events outside their control.

In early January 1916, Villa had begun exacting bloody revenge against the United States because of its support of the Carranza regime. At Santa Ysabel sixteen American mining engineers were kidnapped from a train and summarily executed. Even though the United States was on a virtual war footing, and despite considerable outrage in Washington over this atrocity, there was as yet no order for intervention. Two months later, however, Villa and his band of between four and five hundred men were on the move north from their bastion in the state of Sonora toward the U.S.-Mexican border. In their wake they left a trail of pillage, kidnapping, and murder. The Villistas kidnapped and held an American woman for nine days after slaughtering her husband; others were raped and strangled.[22] As Villa's force moved toward the United States, word of its advance spread north of the border, and any lingering doubts about their intentions were now removed.

Villa's target was the small border town of Columbus, New Mexico, and in the early morning hours of March 9 he and his raiders struck Columbus and began indiscriminate burning, looting, and killing. The raid left eighteen Americans dead. Although Mexican losses were very high, Villa had achieved his aim of arousing the United States.[23]

News of the raid did not reach Fort Bliss until March 10, and before long a distraught Patton learned of the rumor that the 8th Cavalry was not included in Pershing's plans for a retaliatory raid against Villa. The Columbus raid occurred just as Patton had been preparing to travel at his own expense to Rock Island, Illinois, to defend his saber to the "damned fools" on the Cavalry Board, who were considering reverting back to a curved saber. In 1916 Patton's vision of the army of the future did not extend beyond the cavalry and his passionate belief in the importance of the cavalry sword. He had again written to Leonard Wood, this time to oppose any change in the cavalry saber. Wood again counseled patience: "You are quite right. . . . It will be a long step backward if we revert to the old drill regulations. However, there is no use in being discouraged. The instruction you gave at Riley is very valuable and will eventually count."[24] His fiery defense of the saber was published in the *Cavalry Journal*.

Pancho Villa's raid immediately altered Patton's plans. A punitive expedition was clearly imminent, and he had no intention of missing his first big opportunity to see action. But how to gain admission? On March 12 Patton learned that the 8th Cavalry would not participate in the punitive expedition, and silently cursed the rotundity of his commanding officer, which he held responsible for the exclusion. Pershing insisted on a high state of physical fitness, and the 8th Cavalry commander was the very antithesis. "There should be a law killing fat colonels on sight," he complained.[25] From that day Patton would distrust any fat officer.

Patton concluded that being detailed to Pershing's staff was the best means of participating in the punitive expedition. The fact that there were no vacancies did not dissuade him from pleading with his regimental adjutant for a recommendation. He also appealed to both Maj. John L. Hines, the expedition's adjutant general, and to one of Pershing's aides, Lt. Martin C. Shallenberger. When Pershing learned of Patton's inquiries, he telephoned to ask if it was true that he wanted to accompany the expedition. To Patton's excited yes, Pershing replied that he would see what he could do. Sensing that this was merely lip service and not a genuine commitment, Patton decided that he must personally persuade Pershing to include him.

That evening he arrived unannounced at Pershing's quarters and told the general that he would perform any job, no matter how menial, and that he was good at handling newspaper correspondents. Pershing retorted: "Every one wants to go. Why should I favor you?"

"Because I want to go more than anyone else," replied Patton. As Pershing's biographer records, there was "a cold look from steely eyes, no flicker of a smile, no thawing of official pose, just a short last sentence: 'That will do.'" Pershing telephoned him the following morning and said, "Lieutenant Patton, how long will it take you to get ready?" Patton had already packed his gear and informed the startled Pershing that right away would be fine: "I'll be God Damned. You are appointed Aide."[26]

Thus began the most important and rewarding professional relationship of Patton's life. In 1924 he composed a small memoir of Pershing in which he noted:

> It was three years before I learned from him why he took me. It seems that in '98 Lieut. Pershing was an instructor at West Point. The policy was that no instructors should go to the [Spanish American] war. Lieut. Pershing used every normal means to secure an exception and finally went A.W.O.L. to Washington where, by a line of talk similar to the one I employed on him in 1916, he secured the detail to Cuba.[27]

Undoubtedly Patton had made a favorable impression, but it would be misleading not to consider that Pershing's growing attraction to his sister, Nita, may also have contributed to his decision to employ an extra (unauthorized) aide-de-camp. For the time being Patton replaced Lt. James L. Collins, Pershing's other aide, who was absent but due to rejoin the headquarters shortly.

Despite the euphoria that at last something was being done about Villa, Patton correctly perceived that chasing the Mexican would prove far more difficult than some thought:

I think that we will have much more of a party than many think as Vil-
las men at Columbus fought well and . . . [Mexico is] very bad [terrain]
for regular troops. There are no roads and no maps and no water for the
first 100 miles. If we can induce him to fight it will be all right but if he
breaks up [his force] it will be bad, especially if we have Carranza on
our rear. They can't beat us but they will kill a lot of us. Not me
though.[28]

Under Pershing's whip hand things happened in a hurry. Patton was
responsible for making the logistic arrangements to move Pershing and his
staff to Columbus, where thousands of troops were converging to form the
Punitive Expedition. His multitude of duties included organizing the daily
business of the headquarters, arranging the general's visit to units and
accompanying him to take notes, bearing messages, drafting letters and
messages dictated to him, establishing a censorship program for the expedition, developing supply estimates, and generally making himself such an
indispensable asset to Pershing that he unhesitatingly retained him after
Lieutenant Collins returned to duty, even though there was no official billet
for a third aide.[29]

Occasionally, during rare interludes when Pershing found time for
horseback rides into the wilderness, Patton would accompany him. The
feisty general saw a great deal of himself in the eager young lieutenant:

Jack enjoyed his enthusiasm, his quiet adoration, his almost comical tries
at emulation. A burning professionalism touched a kindred current in Pershing; he read Patton's frequent papers on tactics with interest and criticized them carefully. Patton's effervescent nature brightened headquarters
considerably, and his eagerness lightened the work of inspections. Inspections were the key to soldiering, and Jack taught Patton the virtues of
close troop knowledge by example.[30]

Patton thought Pershing "likes me almost too much for I have volunteered
to take several messages which he has refused to let me do for fear of my
getting hurt."[31]

Pershing's influence on young Patton cannot be overemphasized. He
was the very model of a military commander, whose ideas of duty and discipline meshed perfectly with Patton's own conception. Pershing would not
brook disorder or sloppiness of mind or person or billet, and he was a
superb organizer of troops. He even possessed the same short-fused temper
as Patton. In his memoir of Pershing, Patton praised his professionalism. In
Pershing he had at last found the perfect example of a senior commander,
whom he would later successfully emulate, refining to his own lofty standards what he had learned:

Under the personal supervision of the General every unit . . . every horse and every man was fit; weaklings had gone; baggage was still at the minimum, and discipline was perfect. When I speak of supervision I do not mean that nebulous staff control so frequently connected with the work. . . . General Pershing knew to the minutest detail each of the subjects in which he demanded practice and by his physical presence and personal example and explanation insured himself that they were correctly carried out.[32]

The expedition headquarters ran like a well-oiled machine, and to the impressionable Patton, who now saw his own ideas confirmed by the personal example of an officer he came to idolize, the kind of leadership Pershing brought to the command of the Punitive Expedition and later to the AEF in France embodied many of his own strongest ideals of generalship. It was Pershing's "personal care which gets the results and only this *personal* care will," he inscribed in a small notebook.[33] Here was clear affirmation of the merits of strong leadership. The fact that Pershing was an autocratic leader only served to reinforce the graphic lessons Patton was gratefully absorbing. Years later, at the commencement of World War II, Pershing would proudly remind Patton: "I can always pick a fighting man and God knows there are few of them."[34]

One of the best examples of Pershing's leadership occurred when he arrived to find Columbus in a state of utter chaos. Trains, supplies, troops, and trucks were pouring into the tiny town, and there was no one in charge. There was neither chief quartermaster nor ordnance officer, and medical supplies, tents, trucks, automotive parts, tires, radios, and other matériel were piling up and creating a major logistical nightmare. By the sheer weight of his personality, Pershing quickly asserted his authority at Columbus by assigning duties and bringing order out of what had been one hell of a mess when he arrived. As his biographer writes, "General Pershing's glacial presence brought some order from confusion," and he took steps to ensure that "the expedition would not be stalled by disorder at the start."[35]

Pershing's arrival at Columbus came not a moment too soon. An order from Washington ordered the Punitive Expedition to cross the border no later than March 15, barely forty-eight hours later. The president's orders to Pershing were typical of the guileless Wilsonian method: Pursue and punish Villa, but do not upset the Carranza government by firing on any of his troops (as if Villistas and Carranzistas would be wearing signs identifying themselves). The futility of Wilson's edict was plain even before the expedition commenced, when the local Carranzista commander at nearby Palomas threatened to attack the Americans and only Pershing's insightful decision to hire the man as a guide averted a major incident.[36]

It would soon get worse. Not only would Pershing pursue Villa, but

Carranza and his men would take advantage of Wilson's terms to make life miserable for the Punitive Expedition. Within a matter of days the list of Carranzista outrages grew, ranging from harassment to firefights with American troops. Attempts to set up negotiations with the Carranza government were rebuffed with contempt, and the two countries edged closer than ever to war.[37]

Pershing's mandate to catch Villa was doomed to failure from its inception. Northern Mexico is a vast wasteland of desert with few towns, and is dominated by the barren and rugged Sierra Madres, whose peaks average ten to twelve thousand feet and are honeycombed with deep canyons that offered Villa and his men excellent hiding places. The few roads were little better than dirt trails, which threw up huge clouds of dust in dry weather and turned into quagmires in the rain. Villa's men had the advantage of being on familiar ground, leaving Pershing with the problem not only of entrapping the bandit and his followers but of coping with the harsh desert, where food and water were at a premium. In military terms Pershing was confronted with the enormous, almost insurmountable, logistical problem of resupplying a large force that each day would advance farther from its source of supply in the United States. Moreover, since the Punitive Expedition employed a number of independent forces that eventually operated on both sides of the Sierra Madres, the resupply problem became even more severe.

Added to Pershing's woes were the poor communications between himself and his forces, with ineffective radios that constantly seemed to break down at crucial moments. Since the first flights at Fort Myer in 1908, the untested Signal Corps Aviation Service had possessed only a few crude aircraft. Six were sent to Pershing and proved invaluable as the only means of linking his widely scattered force. Although aviation was becoming popular across the country and the army had identified it in 1913 as "a vital necessity," parsimonious congressional appropriations had left the United States a dismal fourteenth among the nations possessing aviation capability.[38]

The inexperienced 1st Aero Squadron was equipped with the dangerously unstable Curtiss JN-2 "Jennies." The gallant men who flew these deathtraps were all pioneer aviators. Three of them—Carl Spaatz, Millard Harmon, and Ralph Royce—would become prominent commanders in the Army Air Corps during World War II. A number of other young lieutenants and captains assigned to the Punitive Expedition were also destined for high rank: Courtney H. Hodges, William H. Simpson (Patton's fun-loving West Point classmate), Kenyon A. Joyce (Patton's future superior officer and mentor), Lesley J. McNair, and Brehon B. Somervell.[39]

The six Jennies lasted barely a month before all had crashed. Two were lost within the first week of the expedition. Despite the planes' brief partici-

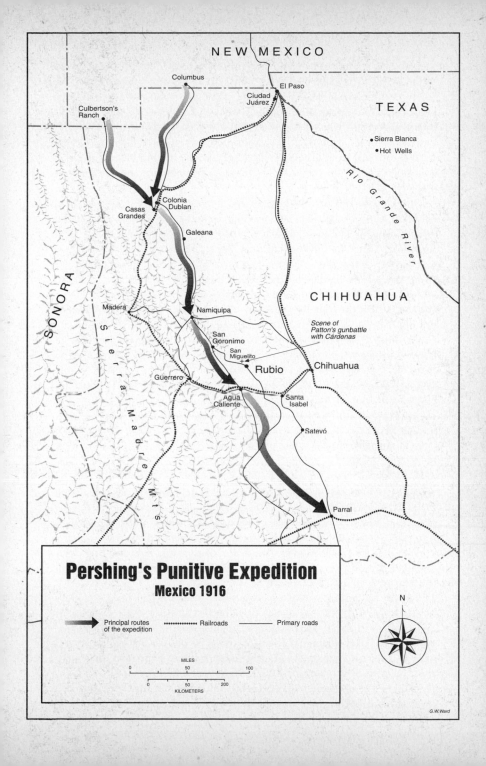

NEW MEXICO

Columbus

El Paso

Ciudad
Juárez

TEXAS

Culbertson's
Ranch

• Sierra Blanca
• Hot Wells

Colonia
Dublan

Casas
Grandes

Galeana

Rio Grande River

CHIHUAHUA

Madera

Namiquipa

San
Goronimo

*Scene of
Patton's gunbattle
with Cárdenas*

SONORA

San
Miguelito

Rubio

Chihuahua

Guerrero

Agua
Caliente

Santa
Isabel

• Satevó

S
i
e
r
r
a
M
a
d
r
e
M
t
s
.

Parral

Pershing's Punitive Expedition
Mexico 1916

➤ Principal routes
of the expedition

•••••• Railroads

—— Primary roads

N

MILES
0   50   100

0   50   200
KILOMETERS

G.W.Ward

pation, the need for aviation in a modern war was affirmed. Although used in the Punitive Expedition to carry mail and dispatches, their potential for aerial reconnaissance and intelligence was not lost on the army leadership or Congress, which soon raised the appropriation from three to eight hundred thousand dollars.[40] Among the interested observers of this first ill-fated but gallant attempt to use aircraft in support of a ground force was George Patton, who would himself employ modern versions of small reconnaissance aircraft during the great drive by his Third Army in France in the summer of 1944.[41]

In the new twentieth century of mechanization and weapons of mass destruction, the days of the U.S. Cavalry were, by 1916, numbered. In the great war raging in Europe, the cavalry played only a very small role, but it was ideally suited for the Punitive Expedition. In fact, the cavalry was better able to operate in the desolate mountains of Mexico, and horses turned out to be far more reliable than the trucks used to resupply the expedition, which frequently broke down; all of which contributed to the inescapable conclusion that the United States was woefully unprepared to fight a modern war.

In April Patton finally persuaded Pershing to use him as a courier to deliver an urgent message to the 11th Cavalry, whose exact whereabouts to the south were only vaguely known. "Almost a needle in a haystack," wrote Patton. "As I started the General shook me warmly by the hand saying 'Be careful, there are lots of Villiastas.' Then still holding my hand he said, 'But remember, Patton, if you don't deliver that message don't come back.' It was delivered."[42]

On another occasion, after a motorcycle courier had turned back after being fired on, Patton volunteered to carry an order to Maj. Frank Tompkins, the commander of a provisional squadron of the 13th Cavalry. Taking advantage of an opportunity to be where there might be action, Patton remained temporarily. When it became evident that Tompkins seemed to be misinterpreting Pershing's orders, Patton audaciously contradicted him. "I told him I would take the responsibility for moving in the way which I thought the order intended we should go." Patton was right and Tompkins complied. It was an example of Patton's willingness to take risks that might result in great harm to his career if he were wrong.

At the commencement of the expedition, Patton began keeping a diary that provides an important record of what he did as well as his thoughts about military matters, and what he was absorbing by Pershing's example. Pershing established his command post with Col. George A. Dodd's 2d Cavalry Brigade, based at Culbertson's Ranch, one hundred miles west of El Paso, near the New Mexico–Arizona–Mexico border.[43] Patton accompanied Pershing when Colonel Dodd's cavalry crossed the border and advanced fifty miles into Mexico. Pershing was noted for eschewing the trappings of rank, and the cold, rain, sleet, and wind of the Sierra Madres did not prevent him from sleeping on the ground from March to May without a tent, doubling up with one of his aides for the additional warmth secured by two blankets: "No frost or snow prevented his daily shave."[44]

At least once Pershing went without sleep for two days. During his daily motor trips in an open-topped Dodge touring car, which often took him thirty to fifty miles ahead, Pershing never carried anything more than a single blanket and a toilet kit. His automobile became his command post, "Here he wrote his dispatches—this was G.H.Q."[45] The tiny staff, the crude rations, and the hardship voluntarily endured were unprecedented for a modern general.[46] Euphoric at being with Pershing, Patton so relished the prospect of distinguishing himself in the pursuit of Villa that he did not even notice the hardships. His only disappointment was that the cavalry had not been permitted to bring their (Patton) sabers on the expedition.

The first night in Mexico, Patton's saddle blanket was stolen while he was eating dinner. Pershing lent him one of his, and Patton's diary records: "I stole another one for him."[47] The following day the expedition moved another fifty-eight miles south to Colonia Dublán, where Pershing established his permanent command post. For once, the cavalry was the best means of chasing Villa, and at Dublán Pershing began to plot the means by which he intended to catch the Mexican.

CHAPTER 13

The Bandit Killer

GEORGE S. PATTON SHOOTS VILLISTA CAPTAIN
—*PASADENA NEWS* (MAY 25, 1916)

Pershing's penchant for motoring around Mexico with minimal security and even less concern for his own safety led to more than one hair-raising close call in the spring of 1916. His entourage consisted of three open automobiles manned by fifteen men armed only with nine rifles. Patton was in the lead car, which was traversing unmapped and "semi-hostile mountain and desert," when an armed Mexican suddenly appeared in the headlights. Nearby, what seemed to Patton "a veritable army seemed to lurk." With "halting Spanish and beating heart," Patton rushed forward, unsure whether the Mexicans were friend or foe, and "prejudiced my hope of eternal salvation by a valuable description of ourselves as the advance guard of an automobile regiment." Suddenly Pershing appeared, identified himself, and demanded to know why in hell he was being stopped. Patton had visions of a massacre, "but the commanding presence of the General and his utter disregard of danger over-awed the Mexicans and we went on, though personally it was more than a mile before I ceased feeling bullets entering my back." Two hours later a convoy of three trucks carrying airplane spare parts and gas was attacked by the same Mexicans, leaving Patton to muse over the maxim attributed to Caesar that "Fortune favors the bold."[1]

Rumors of Villa's whereabouts abounded, and though scattered elements of his band were found and engaged, there was very little action, and Villa himself remained at large. To paraphrase a term used in the Vietnam era, the Punitive Expedition was mostly "search" and very little "destroy." Patton soon became bored with the inactivity. Other than a violent firefight at Guerrero on March 30, when some thirty Villistas were killed by the 7th Cavalry, led by the redoubtable Colonel Dodd, search operations continued in the area around Rubio, where the people were unfriendly and unwilling to betray Villa. Unknown to the Americans, two days earlier, during a battle with Carranza forces, Villa's leg had been shattered by a bullet from a Remington .44-caliber rifle, and the bandit leader's primary concern was now to survive a ghastly wound that would have killed a lesser man.

The routine and the boredom were alleviated only by occasional duck hunting with pistols, and Patton's habit of taking target practice at rabbits or the green glass insulators used on telephone poles.[2] He expressed sympathy for the plight of the Mexican peasants and viewed the intervention as a futile gesture, but "if we leave ruin total and complete will follow. These people pay 50% Taxes to the state and the other 50% to the ranch owner. . . . If we take the country we could settle it and these people would be happier and better off."[3] Several weeks earlier he had concluded, "I realy think that Villa bad as he is, and he is unspeakable, was the man for us to have backed; he was the French Revolution gone wrong . . . but old Villa is Damed hard to find."

Despite these views, Patton also gave vent to what would become an almost violent streak of disdain and prejudice against what he clearly regarded as inferior peoples. Today it would be a blatantly racist attitude, but in Patton's time it was considered neither unusual nor unacceptable. In the case of the Mexican peasants, Patton was put off by their living conditions and the behavior that poverty traditionally spawns. "Now there is nothing left but for us to take the country and exterminate the present inhabitants they are so far behind that they will never catch up they are much lower than the Indians. They have absolutely no morals."[4] Throughout his life Patton's response to the sort of poverty he saw in Mexico, North Africa, and Sicily was defensive and antagonistic, as if he could not bear to witness squalor and destitution firsthand.

The officer corps of Patton's era was virtually all white, strongly upper middle class, overwhelmingly Anglo-Saxon Protestant, conservative in its political views, and tainted by an institutional anti-Semitism and racial bias. Officers did not necessarily dislike Jews or blacks, they were simply *different* and therefore suspect. Like Patton, many grew up barely knowing a black man or a Jew—and those who did had formed opinions that hardened as they became adults. The word "nigger" was an integral part of Patton's vocabulary, and even though such terms are unacceptable in the late twentieth century, in 1916—and even much later—the letters and diaries of many officers who later attained high command in World War II contain frequent such references to blacks and Jews. Their authors considered them to be routine vernacular, and would have been surprised had they been informed that they were racist epithets. The armed forces remained segregated until after World War II, when President Harry S. Truman ordered integration. His biographer, David McCullough, writes that even Truman, whose civil rights record was admirable, privately "could still speak of 'niggers,' as if that were the way one naturally referred to blacks."[5]

That Patton expressed anti-Semitic and antiblack values is beyond question; that he was a racist is less certain. Like Truman, he was a product of his times and clearly distrusted both blacks and Jews; the former were simply considered inferior, whereas Jews were distrusted and often despised for their success. President Theodore Roosevelt once invited Booker T. Washington to the White House, but privately viewed blacks "as a race" as "altogether inferior to the whites." Among the failures of the liberal, reformminded Woodrow Wilson was the reinstitution of segregation in the federal government—which had been integrated for nearly fifty years.[6]

The military profession was socially isolated from the outside world, encrusted in tradition, and extremely slow to adapt to or accept change. Many career officers were the second or third generation to attend West Point, to which appointments came from white Anglo-Saxon Protestant congressmen and senators. It was not until after World War II that the social base of the Regular Army broadened to include a larger percentage of officers with lower-middle-class and working-class backgrounds. The outlook of the 1980s and 1990s, which has produced a black chairman of the Joint Chiefs of Staff and many high-ranking black officers, simply did not exist in the early part of this century. Jews and blacks were viewed as outsiders, even those few who managed to gain entrance to West Point. Moreover, the early years of the twentieth century were a particularly virulent period of worldwide anti-Semitism, fueled and exemplified by the Dreyfus affair in France. In short, despite its isolation from the civilian populace, the officer corps represented the views of the society from which its members came, and that society was largely segregationist and anti-Semitic. After fifty years the wounds of the great Civil War were only beginning to heal, and George S. Patton, whose only experience of blacks was limited to those he had seen performing menial labor, tended to view them with dispassionate curiosity. On the one hand he could and did admire the toughness and courage of the men of the segregated 10th Cavalry, while on the other disdaining them and their officers because they were not part of his social order.

One of Pancho Villa's most trusted subordinates was Gen. Julio Cárdenas. The commander of Villa's personal bodyguard, the Dorados, Cárdenas was believed to be in hiding in the vicinity of Rubio. If Pershing could not run Villa to earth, then at least a big fish like Cárdenas, who was believed to have participated in the raid on Columbus, might be snared. Search operations west of Rubio were intensified. Naturally Patton craved a role and pestered Pershing for an opportunity to participate in the manhunt for Cárdenas. Pershing finally relented, doubtless aware that there would be no peace and quiet in his headquarters unless he acquiesced to his aide's incessant exhortations. Patton was temporarily attached to 1st Lt. Innis Palmer Swift's Troop C, 13th Cavalry.[7]

The search inevitably led them to San Miguelito Ranch, where Cárdenas's family was thought to be residing. The rancho had just been searched to no avail by the 16th Infantry, but the commander reported that several armed Mexicans had been seen departing in great haste for the sanctuary of the nearby mountains. A subsequent search by Troop C of both the rancho and the surrounding area found no trace of Cárdenas, but his wife and baby were discovered, and his uncle was located at a nearby rancho. According to Patton: "The uncle was a very brave man and nearly died before he would tell me anything." Although there is no way to determine what his role might have been, the wording of his letter suggests that the uncle was unsuccessfully tortured for information about his nephew.[8] For some days afterward Patton remained privately suspicious that Cárdenas was still in the area and convinced that the presence of his family would eventually draw him back to San Miguelito.[9]

On May 14 Pershing placed Patton in command of a foraging expedition to obtain a fresh supply of maize. The party, consisting of Patton, ten soldiers from the 6th Infantry Regiment, and two civilian guides, was dispersed in three Dodge touring automobiles.[10] In Rubio he spotted a suspiciously large but unarmed group of some sixty Mexicans. One of the guides, an ex-Villista named E. L. Holmdahl, knew several of the men, whom Patton later described as "a bad lot." After purchasing a large supply of grain for the horses, Patton decided to test his premise about Cárdenas by launching a raid on San Miguelito before there was an opportunity for anyone in Rubio to warn the bandit leader.

Expecting trouble, the entourage approached the uncle's rancho at Saltillo but found no trace of Cárdenas.[11] Before approaching the Cárdenas hacienda at San Miguelito, six miles farther north, Patton stopped and outlined his plan, based on a surprise attack, to surround and flush out the bandit leader, trapping anyone inside.

Patton later wrote several accounts of what transpired at San Miguelito, one to Beatrice that same day, another for Pershing, and in 1928 yet another. They vary not only in the level of detail but also in the sequence of events. They are here combined to recount, as closely as possible in Patton's own words, what occurred (appropriately enough) at high noon on May 14, 1916, in what became a U.S.-Mexican version of the infamous shootout at the OK Corral in Tombstone, Arizona. Patton wrote:

About a mile and a half south of the house the ground is lower than the house. And one cannot be seen until topping this rise. As soon as I came

over this, I made my car go at full speed and went on past the house . . .
four men were seen skinning a cow in the front. One of these men ran to
the house and at once returned and went on with his work. I stopped my
car northwest of the house and the other two southwest of it. I jumped out
carrying my rifle in my left hand [and] hurried around to the big arched
door leading into the patio. . . . I rounded the corner and walked about
half way to the gate. When I was fifteen yards from the gate three armed
men dashed out on horseback, and started around the southeast corner.

So schooled was I not to shoot, that I merely drew my pistol and
waited to see what would happen. . . . When they got to the corner they
saw my men coming that way and turned back and all three shot at me.
One bullet threw gravel on me. I fired back with my new pistol five
times. Then my men came around the corner and started to shoot. I did
not know who was in the house. There were a lot of windows only a few
feet from our right side. Just as I got around the corner three bullets hit
about seven feet from the ground and put adobe [chips] all over me.

Patton had approached the hacienda at a run, bobbing and weaving in
case someone inside fired on him, and when the three riders dashed out the
gate he had hollered "Halt!" to no avail.[12] As Patton, Holmdahl, and two soldiers were reloading behind the safety of the north wall of the hacienda, the
three Mexicans, one of whom was wounded, were trapped inside the courtyard, cut off by the presence of the troopers Patton had ordered to cover the
only other two avenues of escape to the southeast and southwest: "I
reloaded my pistol and started back when I saw a man on a horse come right
in front of me. I started to shoot at him but remembered that Dave Allison
had always said to shoot at the horse of an escaping man and I did so, and
broke the horse's hip. He fell on his rider and as it was only about ten yards,
we all hit him. He crumpled up."

In his 1928 version, Patton noted that, "impelled by misplaced notions
of chivalry," he "did not fire on the Mexican who was down until he had
disentangled himself and rose to fire." During the confusion a second Mexican had somehow eluded the Americans. By the time he was detected, the
Villista was nearly a hundred yards east of the hacienda and about to make
good his escape when, in a hail of rifle fire, he pitched forward dead in the
sand near a stone wall. Patton had shot at him three times with his rifle, as
had four or five of the soldiers. Two of the three Mexicans were now dead.

Still uncertain how many Villistas were present, Patton, "thought there
were some men in the patio and as the flat roof had a parapet I was afraid
they would climb up there and shoot us. I hated to climb up but hated worse
not to, so took two men and told two others to watch the roof." Two soldiers
propped a dead tree against the wall while Patton climbed onto the dirt roof
of the hacienda. Suddenly it gave way under his feet, plunging him through

up to his armpits. He might have been cut in half if there had been anyone inside the house with a saber, and with considerable urgency he quickly managed to pull himself back atop the roof.

Meanwhile, the ex-Villista Holmdahl, who had been covering the front door, spotted a man running from a gate in the southwest corner toward the nearby fields. "He was dropped at about two hundred yards and held up his hand in a token of surrender but as Holmdahl approached him he drew his pistol and fired at Holmdahl who then killed him."

According to Patton's account: "All this time there had been four men out in front skinning a cow. They never looked at us at all," as if such events were perfectly commonplace. In reality they hoped by their passive reaction to avert being killed by what were clearly some very sinister American *hombres* who shot first and asked questions afterward. Patton was uncertain if further danger lurked inside the hacienda, and the four skinners were rounded up as "we each got behind a Mex and went in" to conduct a room-by-room search.

In one room they encountered Cárdenas's mother and wife, who was rocking her infant daughter in her arms. With hatred in their eyes, they stood unmoving and in stony silence. The search of the hacienda yielded no further Villistas, but Patton did discover one immovable heavy wooden door, whose lock he promptly shot off. Inside they found several elderly women cowering in the corner, fully expecting to be killed by these fearsome intruders. One of them finally began to intone a prayer to God the Father and the saints above to save their souls and to bring his wrath down upon these evil Americanos.[13]

During the gun battle, it had been impossible to determine if Cárdenas was one of the three killed.[14] In the aftermath, however, Patton was able to determine that one of the horses was that of Julio Cárdenas, complete with a silver saddle and a saber. His corpse was identified by the skinners.[15] Wounded by Patton as the three riders were driven back into the courtyard, he was the third man killed several minutes later by Holmdahl as he attempted to flee the hacienda. An examination of his cartridge belts revealed that the Villista leader had been wounded four times before Holmdahl's *coup de grâce* and had fired thirty-five rounds before he died. The other two dead Mexicans were an unnamed Villista captain and a private.

After nearly doing himself significant anatomical damage with a "hair trigger" Colt .45 automatic pistol at Sierra Blanca the previous year, Patton had exchanged it for an ivory-handled Colt 1873 single-action .45-caliber revolver.[16] To ensure that there would be no embarrassing repetition, Patton kept only five shells in the gun and left the chamber opposite the hammer blank.

As they were preparing to depart, Patton spotted some fifty Villistas

heading toward them on horseback at full speed, no doubt attempting to rescue Cárdenas. Some shots were exchanged before the Americans beat a
hasty retreat toward Rubio, the three bloody corpses strapped across the
blistering-hot hoods of the automobiles like trophies of a hunt.

No one at Pershing's headquarters had any idea where Patton and his
men were, and had a bullet in the gas tank put his autos out of action, the
result might well have been extremely unpleasant. Thus, as Patton put it,
"we withdrew gracefully" from San Miguelito ranch. Patton ordered the
telephone wires cut to prevent an ambush ahead. The convoy created considerable excitement but met with no adversity as it passed through Rubio
with its grisly trophies. It was nearly 4:00 P.M. when the convoy rolled into
Pershing's field headquarters. It was a bizarre scene. Never before had Pershing been presented with the corpses of his dead enemies. (For that matter,
the experience proved to be unique in his career.) Nevertheless he was
pleased that at least someone had enlivened the hunt for Villa and taken out
a key member of his band.

Pershing permitted Patton to keep the saddle and saber. But something
had to be done about burying the three bandits who were beginning to
decompose disagreeably in the steamy late afternoon heat. It was decided to
hold a quick, impromptu funeral. Against the backdrop of a blood-red sunset, graves were hastily dug, but no one seemed to know the words to the
burial service. Finally a veteran sergeant spoke up and said he knew what to
do. Raising one hand, he intoned: "Ashes to ashes and dust to dust, / If Villa
won't bury you, Uncle Sam must."[17]

Patton's feat created an instant sensation in the press across the United
States. A Boston newspaper featured a photograph of Patton taken outside his
tent, his pipe in his mouth, under the headline MEXICAN BANDIT-KILLER
WELL KNOWN IN BOSTON. The accompanying article proclaimed that "Patton
and his men left the camp in their autos and fought the bandits from their
autos, that is to say, they sprang directly from their cars into the fight, putting
the encounter in a class by itself." The report was actually based on a graphic
account in the *New York Times* by correspondent Frank Elser, whom Patton
had befriended. Pershing's biographer has written: "Newsmen caught some
yearning touch of glory in the thin, reedy-voiced lieutenant who strove to
look like John J. Pershing."[18] Overnight Patton became nationally acclaimed
as the "Bandit Killer." Perhaps more important was that, unwittingly, "Patton
had initiated motorized warfare in the U.S. Army."[19]

Whether or not Patton actually killed anyone has never been clearly
established. What seems indisputable is that he was responsible for wounding
Cárdenas before he was killed by Holmdahl. Patton's rifle shots *may* have
killed a second bandit but were part of a fusillade fired by himself and four or
five others at virtually the same instant. And, finally, Patton and several of his
soldiers shot the first Mexican, whose horse he had killed in the initial

encounter. Again, his shots *may* have been the ones to have killed the man. What is certain is that Patton's quick thinking and sound plan of attack prevented all three Villistas from escaping into the sanctuary of the nearby hills.

His first letter to Beatrice after the Rubio affair was surprisingly subdued. "As you have probably seen by the papers, I have at last succeeded in getting into a fight. . . . I have always expected to be scared but was not nor was I excited. I was afraid they would get away. I never heard a bullet but some say that you do not at such close range. I wondered a little at first that I was not hit, they were so close."[20] Patton reveled in his newfound notoriety. After returning to what he termed "the windiest place in the world" where he killed two snakes outside his tent, Patton was teased

> because I used a pistol instead of a saber the other day, but it simply goes to show that an officer should be able to use all arms, for being on foot I could not have used a saber. The Gen. has been very complimentary telling some officers that I did more in half a day than the 13 Cav. did in a week. He calls me the "Bandit," there is another bandit near here that I wanted to take a try at, but he would not let me. It is just as well as my luck might change. . . . You are probably wondering if my conscience hurts me for killing a man it does not. I feel about it just as I did when I got my sword fish, surprised at my luck.[21]

The incident at San Miguelito was one of the few highlights of the Punitive Expedition. For the remainder of 1916 the hunt for Villa began to wane and was replaced by the tedious routine of life in a temporary bivouac. Boredom spawned drunken shoot-outs between troops or with local Mexicans. Pershing initiated a tough new training regimen that included cavalry maneuvers to keep his men occupied and sharp, but Pancho Villa had gone to ground, and between the Carranza regime's growing intransigence and Wilson's restrictive guidance, nothing could change the fact that the Punitive Expedition was doomed to failure.

On May 23, 1916, Patton was at last promoted to first lieutenant, and after nearly seven years was no longer at the bottom of the army hierarchy. And while he daydreamed of going off to war, Beatrice was relieved that with his mandatory foreign service now being fulfilled in Mexico, Patton would avoid duty in the Philippines, and the family would remain together in El Paso, perhaps indefinitely.

But no end to the expedition was in sight when, in June, Dublán was transformed into an enormous military encampment complete with a railhead that disgorged tons of supplies. A thousand civilian workers built large ordnance and aircraft repair shops by day and brawled by night in the saloons and whorehouses of what had once been a sleepy Mexican town.[22]

Although Pershing kept up a perpetually torrid pace, it was garrison work and lacked excitement. Despite constant training and inspections, in which Patton participated, the monotony continued. "We are all rapidly going crazy from lack of occupation and there is no help in sight," Patton wrote Papa in July. The Cavalry Equipment Board came to Mexico, and Patton spent some of his idle time preparing to testify. He spent many hours in animated conversation with Pershing, discussing the saber, but the general showed little interest in it as a weapon. Patton also continued the habit begun at West Point of writing poetry on a variety of subjects to pass the time. "I inclose to [two] disgusting poems I have composed," he wrote to Beatrice in July; "Tear them up."[23] One, a naughty and amusing attempt at humor titled "The Turds of the Scouts," was intended as tribute to the hardy Texans attached to the Punitive Expedition.[24]

Life became so routine that even Pershing craved diversion, and he would occasionally accompany Patton and other members of his staff to hunt in the nearby hills. For the unsuspecting the general's presence often turned a leisurely hunt into a test of will and perseverance. Once, when their car became mired in mud, Pershing dismounted and strode off in the direction of Dublán, while behind him Patton and a guide struggled to keep up with the older man. "We did four miles in 50 minutes," Patton later complained in his diary. It was the "hardest walking [I] ever did. . . . [I was] stiff for several days."[25]

To remain with Pershing (who was not authorized to have a third aide until he was promoted to major general later in 1916), Patton was attached "on paper" to the 10th Cavalry, one of the oldest and proudest cavalry regiments in the U.S. Army.[26] One of several black regiments authorized by Congress in 1866, and segregated until 1948, the 10th Cavalry was one of the army's most decorated units. Its men had been given the nickname of "Buffalo Soldiers" by the Cheyenne Indians, who admired and respected their courage on the battlefield.

Most of the officers were white, and although he never formally served with the 10th Cavalry, Patton was invited to attend a celebratory dinner marking its fifty-eighth anniversary. He was impressed with an elaborate ceremony re-creating their battles and noted in his diary that they were staged by Maj. Charles Young, the third black officer to graduate from West Point, "who refused to sit down at the table on the pretext he was not feeling well." Young's sudden "illness" was a deliberate ploy to avoid offending his white counterparts in a segregated army.[27] Patton believed Young's act to be chivalrous.

As tensions continued to rise between Mexico and the United States, Patton expected war and wrote Mr. Ayer that Beatrice ought to return to their care. "If I am wounded she could get to the border before I could and if I am killed—which I shant be—she would be better at home."[28] Instead of war, there was more of the same: dust, rain, wind, mud, flies, rats, tarantulas, centipedes, rattlesnakes, and frequent intestinal problems from bad food and water. He toyed briefly with the idea of resigning from the army to accept a commission at the higher rank of major in a Southern California volunteer unit being raised to fight in Mexico, but for a number of reasons decided against it. Among them was that his request would have to merit Pershing's approval, and Patton was loath to bite the hand that "was good enough to bring me down here." Although the idea of becoming a major had appeal, common sense won out, and after some hesitation the offer was rejected.[29]

Mr. Patton had decided to make a run for the U.S. Senate and won the Democratic nomination. George repeatedly encouraged his father and offered the sort of fatherly advice he himself had received for so many years. "I am glad you decided to run. . . . Don't go at it in any half way but whoop it up and tell them all sorts of lies. Especially how much Wilson has helped the army—which he has not." Patton despised Wilson and observed: "I would like to go to hell so that I might be able to shovel a few extra coals on that unspeakable ass Wilson how you can support him is beyond me."[30]

Patton was not only frustrated with the president's clumsy handling of American relations with Mexico, and his failure to declare war in the summer of 1916, he was also irritated by Wilson's claims that the United States was militarily prepared. "He has . . . not the backbone of a jelly fish. This alledged preparadness is a lie," he wrote Papa. "We have no army and will never have until we have universal service not [just universal] training. . . . Wilson has preserved peace. Peace of the jackall feasting on what the lions have killed. Peace with the American name a by word and a hissing. Oh! he is a great representative of a fine people."[31]

Although the Punitive Expedition spurred passage of the National Defense Act of 1916, which, among other provisions, gave the president authority to federalize the National Guard, Patton was right: The United States remained woefully unprepared for either economic or military mobilization in the now likely event that it would be drawn into the war raging in Europe. However, the act succeeded in crippling the War Department General Staff, which by the spring of 1917 comprised a mere nineteen officers.[32]

At the end of August Pershing decided to take a brief leave in Columbus and, perhaps to ensure that Nita would come, invited Patton to accompany him. For a week they forgot the hardships of Mexico in a flurry of riding, dining, dancing, and conversation during which the knowledgeable Nita offered practical advice to her beau about the problems he faced in Mexico. It was ironic that Pershing and Nita found themselves chaperoned by Patton and his wife. On the return trip to Dublán, Patton noticed that his commanding general was suddenly much more lively and peppering him with none-too-subtle questions about his sister. It was now abundantly clear that Black Jack Pershing had begun to fall hard for Nita Patton. "A surprised staff found him witty, humorous, laughing, a warm companion. What caused the change? Patton knew but kept quiet. Daily he heard Jack's recitation of Nita's fascinations and reckoned friendship close to love. 'He's all the time talking about Miss Anne,' George wrote Beatrice, and added with some envy, 'Nita may rank us yet.'"[33]

With little to do in El Paso without her husband, Beatrice left for Lake Vineyard in September to assist her father-in-law's campaign. Addressing him as "Dear Senator," Patton's letters aggressively exhorted his father to hold nothing back in the campaign against Hiram Johnson, the Republican governor of California. "Don't hesitate at rough stuff . . . He will probably sling mud. If he does you sling rocks . . . Go after his private life. That will get the Sufferage vote. *Remember* this is no practice game but the whole show the *finals* . . . You must win. All your life has been a preparation for this so you *must* land it." Patton saw his father's destiny wrapped up in this political battle, and he continued to press him as he later would his troops.[34]

In early October disaster again struck the accident-prone Patton. The gasoline lamp in his tent caught fire, severely burning his face and hair. "I ran out side and put my self out." By the time Patton reached a nearby field hospital his face was "hurting like hell." He remained in great pain for some days and could only subsist on a liquid diet ingested through a tube. Miraculously his eyes were undamaged, and eventually the doctor predicted that his face would heal without permanent scarring. When Patton was finally fit enough to pen a letter to Beatrice, he described himself as looking "like an old after-birth of a Mexican cow. . . . [I] would have hated worst to have been blinded because I could not have seen you."[35]

Patton's doctors advised sick leave, and he was granted two weeks. His face and hands were swathed in bandages when Beatrice met him in Columbus. Together they traveled by train to Lake Vineyard. She was obliged to re-dress his mangled face several times. "This made her sick to her stomach, which embarrassed her terribly. She felt it was wrong to have such a reaction—not a bit like Florence Nightingale . . . poor Georgie had to tell her what to do while she seemed to be 'pulling yards of skin off his face.' "[36]

In Los Angeles he was treated by an uncle by marriage, Dr. Billy Wills. Papa was in the final stages of his campaign for the Senate, and after being granted a two-week extension of his leave by Pershing, Patton accompanied him on a trip to California's Imperial Valley.[37] Even though his head was bandaged like a mummy's, Papa took his "hero son" to his clubs and favorite haunts, extolling him as the "bandit killer."[38]

Patton was at his father's side on election night when he was soundly defeated by Governor Johnson, who ran a strongly antibusiness campaign, in the era of Upton Sinclair and his powerful muckraking novels. "He never flinched and took it with a smile. Papa's efforts carried California for Wilson [sic] and secured the president's reelection. On the strength of this I tried to get Papa to push himself for secretary of war, but he was too high souled to be a good advocate for him self and lost out."[39]

George S. Patton II was a kind, decent, and honorable man, but as a campaigner he was too much the gentleman, lacking the passion and glib tongue of a successful politician. For the remainder of his life his son would never forgive Woodrow Wilson for failing to pay what Patton believed was a debt owed his father. "This was a great calamity as he would have made a magnificent secretary [of war]."[40] The selfish side of Patton bemoaned what Papa might have done for his career had Wilson appointed him.

Although Pershing lauded him for his role in the Cárdenas shoot-out, and had lost none of his high regard for Patton, he was disturbed by his young protégé's single-mindedness and his intolerant views of others. During his convalescence Patton received a thoughtful letter from Pershing that, in addition to wishing him a speedy recovery, contained some forthright advice:

> Do not be too insistent upon your own personal views. You must remember that when we enter the army we do so with the full knowledge that our first duty is toward our government, entirely regardless of our own views under any given circumstances. We are at liberty to express our personal views only when called upon to do so or else confidentially to our friends, but always confidentially and with the complete understanding that they are in no sense to govern our actions.[41]

Patton's future actions would confirm his failure to heed Pershing's advice.

He began composing an article commissioned by Pershing about cavalry training. Their discussions resumed, and one night Pershing said, "Why of course you are one of the broadest and best cavalrymen I know." A delighted Patton returned the favor by writing to Beatrice, "the more I see of the man the better my opinion of his brains becomes."[42] However, they quarreled at least once over the content of Patton's article.[43] The bone of contention was Pershing's disdain for the saber and Patton's contention that it had a place in mounted action, along with the pistol. Both men stubbornly refused to budge until Patton became offended, picked up his papers saying, "Very good, sir," and began to leave the room—but not too quickly, in case Pershing changed his mind. "Just as I left he called out to put in the saber. . . . To day he said, 'well you got your way didn't you.' I said why 'no sir not at all' and we both laughed."[44] Pershing rarely gave in to anyone, and regardless of its seemingly trivial nature, it was an enormous moral victory for the master of the sword on a point of honor.

Despite the intellectual stimulation he received from Pershing, the long, dreary days in Mexico left Patton questioning his circumstances and as uncertain as ever whether he would ever realize his dreams. He became very lonely and insecure, and unburdened himself to Beatrice.

> If I could only be sure of the future I would get out. That is if I was sure that I would never be above the average army officer I would for I don't like the dirt and all except as a means to fame. If I knew that I would never be famous I would settle down and raise horses and have a good time. It is a great gamble to spoil your and my own happiness for the hope of greatness. I wish I was less ambitious, then too some times I think that I am not ambitious at all only a dreamer. That I don't realy do my damdest even when I think I do. This job I now have is not good as I have not enough to do and get lazy. Well this is not much of a Christmas letter. I hope you have a nice time and stay young. I wish I had married you when you were eighteen except that baby B. would now be fifteen which would be a bore. I love you with all my heart and wish that I were with you this and all Christmasses. I love you. I love you. I love you. George.[45]

Patton had ample time to take stock of himself and his military career and concluded that his father-in-law's prediction had come true. He was in a dead-end career. He was now thirty-two years old and required reading glasses, which he was to use for the remainder of his life. It had been a year since Beatrice had joined him at Sierra Blanca and in a fit of despair told him she wanted him to resign. As he did during times of stress, Patton turned to poetry in an attempt to atone for what Beatrice was obliged to endure in his quest to fulfill his destiny:

To Beatrice

O! Loveliest of women
What e'er I gain or do
Is naught if in achieving
I bring not joy to you.

I know I often grieve you—
All earthly folk are frail!
But if this grief I knowing wraught
My life's desire would fail!

The mandates of stern duty
Oft take us far apart
But space is impotent to check
The heart which calls to heart.

Perhaps by future hidden
Some greatness waits in store
If so, the hopes your praise to gain
Shall make my efforts more.

For victory, apart from you,
Would be an empty gain
A laurel crown you could not share
Would be reward in vain.

You are my inspiration
Light of my brain and soul
Your guiding light by night and day
Will keep my valor whole.[46]

Patton's poetry was the source of some embarrassment within the family. Beatrice had showed his 1916 poems to Mr. Patton, who wrote back, "As for the 'Poems' the only possible excuse seems to be that they were written in the hospital—I was afraid to show them to the family."[47]

Patton both feared and detested the prospect of aging. He interpreted his need for glasses as merely one of the signs. "I hate to get old and also for you to get old," he wrote Beatrice in January 1917. "It is true you look just as young as you did when I went to West Point, but I hate to have us out of the twenties. Since we have lost a year of each other, it almost seems we should not age." Ruth Ellen noted later:

Getting older was always a worry to Georgie. He worried about losing his hair. He worried about losing his figure, and used to try on his cadet uniform to see if he could still get into it. On his 50th birthday, he refused to get out of bed, as he said his chances of being a hero were over; Caesar had conquered Gaul when he was in his 30's; Alexander had conquered the known world in his 30's; Napoleon was finished at 50; the only possible one he had left was the Duke of Wellington who was about 53 when he won at Waterloo. Ma finally got him out of bed by persuading him that he had been 50 for a whole year without knowing it.

Beatrice was far too busy to worry much about aging. "I only remember once when she was in her late 40's," writes Ruth Ellen, "I found her weeping bitterly in front of her dressing table mirror. When I asked her what was wrong, she wailed, 'Oh, no one that ever sees me now would ever know what a pretty girl I was.' She didn't wear makeup, and she refused to tint her hair because she had a great, if concealed, contempt for women whom she referred to as 'mutton dressed as lamb.'"[48]

*   *   *

In January 1917 the ill-fated attempt to capture Pancho Villa ended with the recall from Mexico of the Punitive Expedition. The relief felt by everyone was palpable. "We are all dead tired of it," Patton wrote Beatrice. "There is much excitement here as to our withdrawal."[49]

The only other optimistic news was that he was ordered to take the examination for promotion to captain in March. Pershing asked Patton to pass the word that Nita should be told of the probable dates of their return to Columbus. On January 27 the first of 10,690 men and 9,307 horses embarked for Columbus. It took a week to assemble the full expeditionary force, which proudly rode across the border with Black Jack Pershing at its head. Not far behind him was 1st Lt. George S. Patton, Jr.[50]

The budding romance between his sister and his commanding general was a troubling dilemma for Patton:

> Georgie loved his sister dearly, and General Pershing was his idol, but the way things were building up between them disturbed him very much. He had visions of what would be said about himself, if the romance blossomed; that Georgie Patton had climbed to rank and influence on the coattails of his Commanding Officer. He wanted Nita to be happy, but, oh, he did want to be his own man!

In the aftermath of the Punitive Expedition, Pershing's courtship of Nita Patton continued with newfound ardor. When Maj. Gen. Frederick Funston died suddenly in February 1917, Pershing was chosen to succeed him as commanding general of the Southern Department, based at Fort Sam Houston, in San Antonio. In March, Pershing journeyed to Los Angeles for a joyful week with the Patton clan at Lake Vineyard, and to ask for Nita's hand in marriage. Patton had been reassigned to the 7th Cavalry at Fort Bliss, and when Pershing stopped for a visit en route to San Antonio, he remained ambivalent about whether or not he wanted his idol to become his brother-in-law. Equally plagued by serious doubts was Mr. Patton, who kept his misgivings largely to himself. His was the natural protective distrust of a suitor who was twenty-seven years older than his beloved daughter. Yet Nita was a mature twenty-nine, very much in love, and not to be dissuaded in her determination to wed John J. Pershing. Beatrice, ever the romantic, was in favor of love conquering all, and by the time Pershing departed for Texas there was a tacit understanding that they would one day wed. However, their dreams of a life together were soon to be dashed in the quagmire of a world war.

PART V

World War I
(1917–1918)

The most colossal, murderous, mismanaged butchery
that has ever taken place on earth.

— ERNEST HEMINGWAY

Flanders was such a carnage and waste: a million and
a half young men died, and they have statues in Whitehall . . . to the fuckwits who engineered it. Haig and all
those old bastards up on their horses.

—LIAM NEESON

CHAPTER 14

"Over There: The Yanks Are Coming"

The United States Enters World War I

I am a sort of "Pooh-Bah" and do everything no one else does.

—PATTON

In the spring of 1917 the United States was drawn into what was destined to become the bloodiest conflict in the history of mankind. Both Wilson and his secretary of state, William Jennings Bryan, had subscribed to the naive belief that "a nation could remain aloof from war by refusing to prepare for it."[1] Although officially neutral, the United States had for some time been supporting the Allied powers—Britain, France, and Russia—with military hardware and loans that totaled nearly a billion dollars. America's so-called neutrality was a charade that fooled no one, least of all the Germans.

What had begun with a minor event in June 1914—the assassination in Sarajevo of Archduke Franz Ferdinand of Austria—had, by virtue of the complicated linkage of treaties and alliances, escalated into a war of such dimensions that it engulfed not only Europe but the Middle East. By waiting until 1917 the United States avoided bloodbaths such as Gallipoli, Ypres,

Arras, Passchendaele, and the Somme—horrific testimony to the carnage that characterized World War I. When one side or the other did launch an offensive, the results had become predictable. On a single day—July 1, 1916—the British Army sustained fifty-seven thousand casualties, including more than nineteen thousand dead, in the Battle of the Somme, while at Gallipoli British and French forces fought the Turks in an eight-month killing frenzy.[2] As appalling as these battles were, they were dwarfed by the 1916 siege of Verdun, in which the French and German armies racked up nearly one million casualties during the bloodiest battle in the history of warfare. By 1917 the Western Front had become a grotesque scar that ran from the North Sea across northern France to the Swiss border, and the war had evolved into a stalemate—a war of attrition in which trench warfare had become the norm.

At the outset of the war there had been genuine attempts by the belligerents to win a military victory on the battlefield, but by the end of 1915 it had become evident that the tactical use of their armies could not be achieved. On the Western Front there emerged new tactics, in which victory was sought by attrition, by wearing out the other side's troops and its military assets faster than one's own were being consumed.[3] The result was that the infantry literally became cannon fodder for their generals.[4]

The Central Powers (Germany, Austria, and Turkey) were wary of the United States but did not refrain from the provocation of unlimited submarine warfare, which by early 1917 culminated in the indiscriminate sinking of Allied and neutral ships wherever they were found. In the end it was this policy more than anything else that goaded Woodrow Wilson into seeking a declaration of war from Congress against the Central Powers, and Congress duly complied.[5]

Having entered the war, the United States faced the daunting prospect of mobilization on a massive scale. In December 1916 the strength of the army was a paltry 108,399 officers and men, whose fundamental weaknesses had been exposed during the Punitive Expedition. Not only was America a slumbering giant, but neither its military establishment, its people, nor its civilian industry was prepared for a major war to be fought more than three thousand miles from its shores. Thus, as historian Russell F. Weigley notes, "the help that America could offer in 1917 was mostly a promise."[6]

A token force of four infantry regiments was hastily formed into the 1st Division, which was to earn immortality in this and subsequent wars as perhaps the best known of all U.S. Army formations, the Big Red One. A token show of American commitment occurred in Paris on July 4, when a single battalion of the 16th Infantry Regiment, fresh from Mexico and still in their campaign hats, marched proudly through the streets of Paris to the cheers of *"les hommes au chapeau de cow-boy."*[7]

Among the many varied and complex problems to be resolved were the production and supply of weapons, uniforms, and vehicles of all sorts; the purchase of more than three hundred thousand horses and mules; the creation of training facilities; and the development and manufacture of modern new weapons that would permit the U.S. Army to compete with the German Army.[8]

The decision to institute a Selective Service System to draft men aged eighteen to thirty-five (later raised to forty-five) had by the end of the war in 1918 registered 24 million American men and caused the drafting of 2.75 million.[9] War fever swiftly gripped the United States. Men rushed to volunteer in droves and posters appeared, the best known of which depicted "Uncle Sam" pointing and proclaiming: I WANT YOU FOR U.S. ARMY. Patriotism in the form of slogans and songs swept the nation. In addition to forming a large military force, millions of Americans would be required to man the factories needed to produce war matériel.

The decision to send an American Expeditionary Force to France required the ablest commander-in-chief who could be found, an officer of proven courage and resolution to carry out the exceptionally difficult task ahead. The selection fell to the new secretary of war, Newton D. Baker, and from the outset there was only one bona fide candidate for the appointment: Maj. Gen. John J. Pershing, the commanding general of the Southern Command, and the officer who had without public complaint faithfully implemented a policy of which he did not personally approve during the Punitive Expedition. Pershing was summoned to Washington and when he arrived, Secretary Baker was not disappointed. "At fifty-six, Pershing was an imposing figure, tanned, ramrod-straight, and meticulously groomed. This man had presence, and he was all soldier." [10]

Secretary Baker gave Pershing a signed order from Wilson and a virtual *carte blanche* to create an American Expeditionary Force. "I shall give you only two orders," said Baker. "One to go and one to return." Pershing was charged with one of the most challenging tasks ever given to an American military commander: to create and train a fighting force from a tiny peacetime army that was ill-equipped to fight any kind of war, much less one against the formidable German Army. In 1917 the United States Army possessed only 285,000 Springfield rifles, 544 three-inch field guns, and sufficient ammunition for a mere nine-hour bombardment. The shortage of basic weapons was so dire that the troops of the 89th Division were obliged to whittle mock weapons from pieces of wood. Of the fifty-five planes in the fledgling Flying Service, 93 percent were obsolete, and the remaining four were obsolescent.[11] *Punch,* the renowned humor magazine, would shortly provide a British version of the arrival of the AEF. In a cartoon meant to refer to the slurs cast upon the British Expeditionary Force in 1914, the crown prince was shown beseeching Kaiser Wilhelm: "For Gott's sake, Father, be careful this time—don't call the American army 'contemptible!'" [12]

                    *   *   *

Thanks to Pershing, George S. Patton remained at Fort Bliss when his former unit departed for duty in New Mexico. He was given command of
Troop A, 7th Cavalry, an excellent assignment for a junior first lieutenant
filling a captain's slot. It was a hectic time for Patton, with not only his new
duties but studying for and passing, in March, the competitive examination
for promotion to captain. He also found time to write about cavalry matters,
including his opinion on a proposed new cavalry saddle requested by Pershing. In April he was again featured in the *Cavalry Journal* in an article
disputing those who wished to introduce a curved saber. The argument was
finally settled when the commandant of the Mounted Service School recommended to the War Department the retention of the Patton saber.[13] Clearly it
was Patton's passionate defense of his saber that saved the day before the
Cavalry Board, which had shown every intention of abolishing it. To culmi-
nate a brief but satisfying interval, Patton's rating officer cited him as "progressive, active, zealous, and an excellent troop commander."[14]

    Once again he had anticipated events, and several weeks before war
was formally declared on Germany, Patton had solicited three letters of recommendation for a commission in what was certain to be a large volunteer
army raised for service in France. Again, there was the prospect of attaining
the rank of major or even lieutenant colonel in such a force for an officer of
his reputation and experience. One of the letters was from Pershing.[15] However, the passage of the new Selective Service Act ruled out the raising of
volunteer forces; thus, if Patton intended to participate in the war he would
have to remain in the army.

    Then came the news that Mr. Ayer was ill with pneumonia and an
urgent summons for Beatrice to return to Pride's Crossing. Patton was
granted a thirty-day leave and wrote Pershing that he was sorry to give up
his new command, "as I was doing well with it and had the horses in fine
shape . . . [however] as the Seventh is now full of captains I doubt if I get
my promotion in it and if I do not I shall try to get some regiment nearer
Boston so Mrs. Patton can be near her parents. . . . I have just had a letter
from Nita. She talks in a most warlike way and speaks of fighting in Flanders as if it were a thing assured."[16] At Pershing's request Patton enclosed a
copy of his diary of the Punitive Expedition.

    When the Pattons arrived at Pride's Crossing, they not only found
ninety-year-old Frederick Ayer gravely ill and unlikely to recover fully, but
Ellie also in bed with two nurses in attendance. Then Beatrice, too, fell ill,
and with Kay Ayer about to be married, Patton found the chaos almost more
than he could bear. Ostensibly the Ayer family troubles soon brought Patton
to Washington in search of a new assignment. "Beatrice will have to be
here," he wrote Pershing, "as her parents are too sick to be left alone. . . . All
the people here are war mad and every one I know is either becoming a

reserve officer or explaining why he cant." However, it was the prospect of obtaining a war assignment for himself that really lured Patton to Washington. "Of course, if we go to France it will be all right as in that case she cant be with me any way and could stay here as well as any where else."[17]

Patton learned unofficially from a friend (and a future chief of staff of the U.S. Army), Capt. Malin Craig, that Pershing had directed that the matter of any assignment for him be placed in temporary abeyance. He should not have been surprised, for Pershing had hinted in a recent letter that something was brewing. Initially Pershing had only been directed to form and command a division—the AEF appointment would come a few days later—and Patton had indeed been placed on a select list of officers he had chosen to form his headquarters.

The declaration of war profoundly depressed Mr. Patton, who was unwell and still anguished over his stinging loss to Hiram Johnson the previous November. He missed Georgie dreadfully, and on April 30, on his return from Washington, where it had been made clear that he would not be offered an important presidential appointment in the second Wilson administration, he wrote: "I do not feel at all desirous to take some second class job." He understood that his son wanted to be a part of the forthcoming AEF. "I hate to think of your going—but I would hate you to be passed over if you want to go.[18]

On May 18 a telegram arrived from the adjutant general ordering Patton to report immediately to Pershing in Washington.[19] Patton telephoned Papa, who was in Washington, and was told to have Beatrice—now well enough to travel—accompany him. While changing trains in New York the following morning, Patton and Beatrice learned in a newspaper of Pershing's appointment to command the AEF. Mr. Patton met them at Washington's Union Station and informed his son that Pershing would soon be embarking for France. Just how soon, Patton learned from Pershing's new aide, Capt. Nelson Margetts; they would be leaving the following Wednesday, a mere four days hence. Beatrice telegraphed Fort Bliss for her husband's striker to hasten to Washington with his uniforms. In the interim Patton ordered a new tailor-made uniform and wore a borrowed one belonging to a West Point classmate who would become one of his closest friends, Everett S. Hughes.[20]

Patton was also delighted to learn that he had been promoted to captain in mid-May. However, as he soon discovered, his new captain's bars had, at least for the time being, earned him the pedestrian task of supervising sixty-five enlisted orderlies, chauffeurs, engineers, medical corpsmen, and signal personnel of Pershing's advance headquarters, who were to accompany him to France. It was his job to see that these men were properly clothed and smartly turned out each day and carried out their duties efficiently. It was detail work and if it was somewhat unglamorous, Patton was not bothered. He was content merely to be a playing a part in this great adventure.[21] He had long ago ascertained that the first step on the path to success was to get one's foot in the door. There would be ample time to arrange for an assignment that would actually get him into combat. Officially Patton would for some time be carried on the rolls of the AEF Headquarters as "attached"; unofficially, he was the commander of the headquarters troop.

Beatrice and his parents remained in Washington, and Patton saw them only infrequently. Papa introduced him to his friend, Secretary of the Interior Franklin K. Lane, who observed: "That boy of yours is all wool and a yard wide or I am no judge," which greatly pleased the elder Patton.[22] When Pershing delayed his departure by five days, both men used the precious extra time to visit their loved ones. Mr. and Mrs. Patton hosted a large dinner in honor of Pershing, and the caption under a photograph that later appeared in a Washington newspaper publicized the prevailing rumor that Black Jack and Nita would one day marry.[23] Some weeks earlier news of the engagement had been leaked to the Los Angeles papers. Although their engagement was now public knowledge, the wedding was put on hold until after the war. Daughter Ruth Ellen never learned the identity of the culprit but believes it was not a member of the immediate family, "as there were still grave doubts in several minds."[24]

Then, all too quickly, came the moment of bittersweet parting. Patton and some of his men journeyed to New York by train, accompanied by his family, where they joined Aunt Nannie. He had started a new diary that he would keep for the remainder of the war, and in it he recorded that the mood of his family was one of near despair at the forthcoming departure of Pershing's entourage on the steamer HMS *Baltic* for Liverpool.

Patton bade farewell to Beatrice on the morning of May 28, 1917. Married for seven years, they had already been separated for more than two of them. Neither left a record of their parting, but it cannot have been anything less than very painful. Both knew and understood the reality of what lay ahead, and neither was under any illusion that American intervention would somehow magically bring about an end to a bitter war that had been raging since 1914.

Mr. Patton accompanied his son to Lower Manhattan, where "he told me good by with a smile," as he boarded a ferry for Governor's Island.[25] Mr. and Mrs. Patton returned to California; Nita joined the Red Cross, and Beatrice returned to Avalon, where she would remain for the duration of the war. She would not see her husband again for nearly two years.

In a dense fog the *Baltic* slipped out of New York Harbor and into the Atlantic for the ten-day trip to Liverpool. Boat drills, French lessons, inoculations for the many diseases that plagued soldiers in France, lectures on the threat of venereal disease, and grim jokes about the very real U-boat threat occupied Patton's time during the voyage.

Pershing and his party of nearly two hundred men were given a gala dockside reception in Liverpool by the Lord Mayor, the admiral of the port, the local army commander and a company of Royal Welch Fusiliers, complete with regimental mascot, "a formidable-looking goat." All stood stiffly at attention and saluted as the regimental band played both "The Star-Spangled Banner" and "God Save the King." When the Americans arrived by train at London's Euston Station, they were again formally greeted by the Lord Mayor of London and the commander of the British Home Forces, Field Marshal Viscount John French. Patton was taken to the Tower of London for a grand reception by the members of the 1st Battalion, the Honourable Artillery Company (called a "company," it was in fact a regiment), who lined the walks of the historic courtyard and cheered.

Patton's first days in England were filled with excitement, as he met a host of dignitaries, signed the king's guest book at Buckingham Palace, was wined and dined at the exclusive White's Club—the world's oldest—and taken to the theater, after which, "I was stopped 20 times by women of the street." In the Honourable Artillery Company mess in the Tower, Patton dined with and was toasted by the regimental commander, who noted that the regiment "would take pride in our success and look upon us as adopted children. . . . There was much cheering."[26]

During their five days in England, Pershing met everyone from King George V to Prime Minister David Lloyd George; South Africa's Gen. Jan Christian Smuts; the first sea lord, Admiral John Jellicoe; and even the munitions minister, a man named Winston S. Churchill. One of Pershing's young staff officers wrote: "The two groups of men who would have to work together in war looked each other over and tried to determine who knew what, who could be trusted and how far one could extend that trust."[27]

At Paris's Gare du Nord, a group of dignitaries headed by the barrel-chested commander in chief and hero of the Marne, Marshal Joseph Joffre, and the minister of war, Paul Painlevé, repeated the by-now standard warm reception given Pershing. He established a small headquarters on the rue Constantine, near Les Invalides, and quickly made himself popular with the French by visiting Napoleon's tomb and kissing his sword. The French saw the AEF as the saviors of France and, as its head, Pershing was awarded the Grand Cross of the Legion of Honor in the solemn atmosphere of Les Invalides. Americans were suddenly enormously popular in France, with everyone from delegates, who stood and cheered Patton in the Chamber of Deputies, to the average Parisian, who showered the men of the AEF with flowers and enthusiastically waved small American flags. For Patton the journey through the streets from the Gare du Nord to the Hotel Continental where he was billeted, "was the most inspiring drive I ever took."[28]

Patton's diary and his frequent letters to Beatrice recorded the gala events of life in one of the most exciting cities of the world. He was immediately caught up in the social life of the French capital, but complained that he had too little to do. He accompanied Pershing during a visit to an aerodrome where four hundred planes were assembled, and was awed by the sight. "At last I have been up in an aeroplane," he wrote excitedly to Beatrice. "I had always expected to be scared but it is quite different from what you would suppose . . . though I was going about a hundred miles an hour I hardly seemed to be moving at all. The country looks beautiful just like a huge map and one can distinguish things much better than I had thought possible."[29] Several days later Patton again flew, this time with his friend the flamboyant Col. Billy Mitchell, whom he had met in Mexico the previous year.

When George and Beatrice parted in New York, their plan had been for Patton to get settled in France and, when the time was right, for Beatrice to take up residence in Paris, where he would join her whenever he could get leave or a pass from the front. "We are just as safe here as you are at Avalon," he assured her. "It is hard to think that war is going on so near. . . ." Perhaps she might come in August and bring daughter Bea, but he thought Paris too difficult a place for little Ruth Ellen. It meant that they would have to run the dangerous gauntlet of the Atlantic. Patton naively suggested purchasing "some sort of rubber garment which is a life pre-server" because of the U-boat threat. The very real danger to his family does not seem to have entered into Patton's thinking. It should have, for his own diary on June 24 records that Pershing told him that for that week alone, German U-boats had sunk more than four hundred thousand tons of Allied shipping and that unless a way was found to stop such losses, "we would never get over 500,000 men to France."[30]

Despite his whirlwind schedule, Pershing occasionally found time for Patton and inevitably their talk turned to their loved ones. "It is certainly the most intense case I have ever seen," Patton dutifully reported to Beatrice. Black Jack deeply missed Nita and was desperate for her company but accepted that his duties made it impossible for her to join him in Paris, which would have made a precedent-setting bad impression.[31] Patton's plans to reunite with Beatrice in Paris were soon dashed when Pershing persuaded the War Department to issue an order banning wives from France. Pershing had done so because of the chaos it would create and the morale factor for the majority, who were unable to afford it. "You see the British had to send back 60,000 women who came over," George informed Beatrice. It was time to pull strings:

Now the only thing to do is to put pressure to work on the secretary of state so you can come. Not as a nurse but straight out. I know it can be done. . . . I am sure that with [brother-in-law] Freddie [Ayer] . . . Pa and every one else the thing can be done. The sooner you do it the better as the restrictions will probably get more and more severe. I disapprove your coming as a red cross nurse for . . . we would be so far apart we could never meet. . . . There is not the *least* doubt that with proper influence you can get pas[s]ports. Use the influence.[32]

Patton used his own contacts in France to help bring this about. He spoke with Maj. Robert Bacon, a former secretary of state (Bacon served briefly in 1909, under Theodore Roosevelt, succeeding Elihu Root), who assured him that, "If Pa gets a diplomatic job you could come either as his secretary or his interpreter and he could leave you when he went home . . . Or you could come as Paris correspondent for the [Boston] Transcript."[33]

Patton considered trying to get Beatrice to England but gave up the idea as impractical. She attempted to gain approval from the War Department and was confident enough of success to have booked passage on a steamer, only to be informed that a passport would not be issued. The rejection left her in tears of frustration. "This disappointment was so unexpected & cruel," she wrote Aunt Nannie. "And poor Georgie—with his little apartment for us two in Paris. I wouldn't care so much if I didn't know how he is feeling. And I tried so hard to go. Well—I will try to behave and fill my place here as well as I can, so help me God. This is a hard time on all of us."[34]

Daughter Ruth Ellen equates Patton's motivation to that of the early–Civil War mentality that infected the North. "People still remembered how ladies and gentlemen from Washington, D.C. had driven out in their carriages to watch the first battle of Bull Run. General Pershing, aware that this was no party war, but the war to change the course of world history, ordered that no women were to come to Europe."[35] Patton could only respond helplessly that, "Poor Beat! I feel as badly for your sorrow as for my own dissapointment. . . . I am so sorry."[36]

Typically neither gave up hope and continued to seek some loophole that would empower Beatrice to join her husband, including using his father's influence to gain the ear of Wilson's confidant, Col. Edward M. House, and ultimately even Wilson himself. Patton admitted that Pershing would be "awfully angry" when he learned of his duplicity, but "It will wear off in time. I have as you know a rather unfair advantage over J. which I have never used but I could in that particular case."[37]

In the end their attempts to pull strings came to naught. "Ma was finally persuaded that her arriving in Europe would look like favoritism and influence, and all sorts of innuendoes would be started that might hurt Georgie's career, so, with an aching heart, she abandoned her plan."

Even with his imperfect command of French, Patton found himself very much at home in Paris. Although he was quite comfortable in the Hotel Continental, where he had cultivated the headwaiter, Patton soon arranged to share a spacious three-bedroom apartment off the Champs-Elysées with a French interpreter and an American ambulance driver. The idea of living where only French was spoken appealed to him. His letters were a steady stream of reassurances that he was in no danger. "One thing is sure with my present job I am bound to live long even if I am not happy. Don't picture me going to a drunkards grave for I have drunk very little and never buy over half a bottle [of wine]." Nevertheless, the good life of Paris soon bored Patton, who chafed for something more exciting. "I am not having very much to do infact if a sergeant could not do all I have to do I would bust him but I am trying to do the best I can. . . . Perhaps some time I shall get a real job . . . Still if I were not here I would be a raving lunatic by now so should be patient."[38] To rate the use of an automobile became a badge of prestige in the AEF headquarters, and there were never enough to go around. Occasionally Patton became the butt of anger, when he could not produce one for this or that major or colonel. "The work is very confining. I am a sort of 'Pooh-Bah' and do everything no one else does," he lamented.[39]

When Beatrice worried that he might be killed, Patton wrote to reassure her he was "a lot safer here than I usually am at home because I don't play polo or race or jump or do any other interesting thing."[40]

Life in Paris was deceptively unreal, when less than sixty miles away men were living and dying in the trenches of the Western Front. For Patton it was a time to enjoy the social life of a gentleman to which he was accustomed, even as he thirsted to find a suitable posting that would test him as a soldier. His desire for a challenging assignment was moderated only by the knowledge that it would be many months before the AEF was committed to battle. One of his many duties was to investigate how a military police system might be instituted in France to help control a massive number of American doughboys when the AEF expanded to what would eventually be a strength of some 2 million.[41]

Patton may not have cared for his job, but most of the time it kept him busy. In late August he wrote of "catching hell" attempting to arrange transport for VIPs and staff officers on short notice. "If it does not make me quite crazy it will perhaps develop in me a placid disposition. At least it is getting me out of the habit which you so justly object to of my saying things can't be done. For here there is no such word. The day before yesterday I disrupted three governments and was damned by all three."[42]

Although he continued to describe his job as "a rotter" and complain that, "I only don't resign it because I think I will get a better one some day," Patton took advantage of his wife's great wealth to somehow purchase a twelve-cylinder, five-passenger Packard automobile. He paid the astounding sum of $4,386, the 1992 equivalent of more than $50,000. "I hope you approve," he told Beatrice when he wrote asking her to deposit the funds for his use in his Los Angeles bank. Those in the AEF headquarters less monetarily endowed undoubtedly resented a very junior captain parading about in such an extravagant luxury merely to commute the few blocks between his apartment and his office.[43]

Mail was censored, but this too was one of Patton's many mundane jobs, and he acted as his own censor. However, "I have been leaving the kissing marks off for fear that they might be interpreted as cipher but you know I mean them anyway. . . . Don't think that I am having a roaring time and not thinking of you for it is not so. Paris is a stupid place with out [you] just as heaven would be under the same conditions."[44]

Patton's love of pomp and ceremony was rewarded on Bastille Day when thirty thousand French *poilus* representing 260 regiments paraded down the Champs-Elysées. From fine seats in the stands, "it was a very impressive sight for one knew he was looking at men who had been through the test. I did not see ten officers with out a wound chevron and many of them were beardless boys."[45]

From time to time, Patton would fill in as Pershing's aide and on July 20 was fortunate enough to be in this capacity when Black Jack visited the British commander in chief, Field Marshal Sir Douglas Haig. Patton and a bright young military attaché, George Quekemeyer, were ordered to plan the route Pershing would take from Paris to Haig's headquarters at Montreuil. Late on the afternoon of July 7, Patton and Quekemeyer set out on a journey that took them through Picardy and Flanders and bleak scenes of what war had done to western France. With their headlights reduced for security purposes to tiny slits that emitted scant light, they barely crawled along a nightmarish landscape of muddy tracks alive with marching troops, vehicles, and animals. When they arrived at Haig's command post nearly six hours later, they were met by his chief of intelligence, Gen. John Charteris, who was impressed by his first contact with American officers. When the day came for Pershing to visit Haig, Patton had installed a plate on the front of his staff car that read "U.S. No 1."[46]

At Montreuil and later at his spacious villa in Saint-Omer, Patton was introduced not only to Haig but to a number of other prominent officers, including Gen. Hugh Trenchard, a pioneer of British aviation. He was thrilled to have dined in such august company and discussed the virtues of the saber with Haig, fishing with Maj. Sir Philip Sassoon, and tactics with Gen. Sir Hubert Gough, the British Fifth Army commander. As he proudly reported to Beatrice, "Gough walked around with his hand on my shoulder for some time. . . . I went to five lunches and four dinners and never sat next to less than a Brigadier General, usually a major General."[47] For the first time Patton was exposed to the details of the enormity of waging war on a massive scale. And he clearly made an impression on Sir Douglas Haig, who wrote in his diary of Patton that "The A.D.C. is a fire-eater! [who] longs for the fray."[48]

To an Ayer family friend Patton reported that he had just returned "from a most interesting trip where I saw the working of over a million men from the inside. It is stupendous and fine. The more one sees of war the better it is. Of course there are a few deaths but all of us must 'pay the piper' sooner or later and the party is worth the price of admission."[49] Their inspection of British facilities and units included observation balloons. "George Patton happily tagging along . . . gawked at the balloons awhile, then marveled at the 'parishoots' used by the intrepid men who swung in those stationary gondolas that so attracted German fighters."[50]

In July the U.S. 1st Division began arriving in France. Several members of the division staff stopped off in Paris en route to their new training area at Gondrecourt, in Lorraine, and were met at the Gare d'Orsay by Patton. One of them was the division operations officer (G-3), Capt. George C. Marshall, and though there was nothing special about their first meeting, many years later Marshall's would be the hand that helped guide Patton to high command and the fulfillment of his lifelong dream.[51]

As the summer of 1917 wore on, Patton continued to chafe at being a glorified errand boy for the headquarters staff, many of whom were reserve officers and thus to be scorned as unworthy of their uniforms. "They ought never make them higher than lieutenants the majors are insufferable." He began to fence every morning. "My fencing is putting me in fine shape. . . . I enjoy it hugely . . . and it is about the only thing I like doing much. . . . If I go broke, or rather if you do, I can always teach fencing."[52]

"To look at Paris one would never know that it was at war," he wrote to Mr. Ayer. "All the people seem gay and spend a large part of their time being run over by our automobiles or else running over us." His job was "stinking" but "at least it is devoid of monotony . . . and so far I have made a go of it."[53] Patton worked long hours at unfulfilling work that left him feeling "like a rat chewing an oak tree" but in a letter to Papa, Pershing assured him that, "George is eager to get to the front when the time comes, and I shall of course give him his chance."[54] Patton and two others jokingly formed a club; "to be a member you must be a cavalry man who never rides and who never goes with in fourteen miles of the trenches." Finally, there came hints that he would soon be moving with Pershing to a new, unspecified location. Even this prospect did little to cheer him. "I can't see for the life of me where I am going to do much in this war personally but my luck will hold I suppose. . . ."[55]

Pershing realized he could no longer command the AEF from Paris when most of his troops were training in eastern France, and in early September he relocated his headquarters from Paris, with its distractions and intrigue, to Chaumont, the capital of the Haute-Marne department, a small, pleasant city of sixteen thousand in the mountainous, forested southeastern corner of Lorraine.[56] Patton accompanied him, and in the months that followed he would suddenly find himself catapulted from obscure taxi overseer and de facto headquarters commandant into the limelight of command.

CHAPTER 15

Tank Officer

I will have to grow and grow a lot. But I *will*. Here is my chance.

—PATTON

From the time of his arrival in France, Pershing's most pressing problem was the incessant attempts by both the British and the French to gain operational control of the AEF. Wilson had previously informed his new allies that the AEF would fight only as an independent army under American command. However, Pershing soon found himself pressured from all sides to assign American troops to British and French units as replacements. The French claimed the situation was so desperate that at least a token replacement force was required. Pershing not only adamantly refused such entreaties but reiterated that the AEF would be an independent force whose size must be a minimum of one million men by May 1918, and that it would simply not fight until it was properly equipped and trained. To the end of the war, Pershing remained unbending, although at times the intrigue to sway him reached the level of opéra bouffe.[1]

The troops of the AEF arriving in France were raw and untrained, and Pershing's first priority was to prepare them for the ordeal ahead. He had already observed firsthand the appalling conditions at the front, which soon resulted in outright mutiny in the dispirited French Army. Assisted by the French, Pershing instituted a tough training program by establishing a number of schools to train American officers and men in virtually every aspect of warfare, and to redress what a private in the 9th Division called their woeful ignorance of the basic principles of the soldier.[2] A General Staff College was established at Langres, and scattered throughout eastern France were other schools that taught everyone from chaplains to artillerymen, quartermasters to cooks.[3]

After touring the French and British battlefields and consulting with the other Allied commanders, Pershing came to the conclusion that for an American army to succeed, intensive training would be required, and he decreed: "The standards for the American Army will be those of West Point."[4] In addition to the large number of specialty schools, there were schools at army, corps, and division level, all designed to weld the AEF into a cohesive fighting force that would not be committed to battle until he deemed the entire army ready to fight. Although Pershing's training methods have been criticized as too elaborate, the decision to turn much of Lorraine into a vast staging and training area was soon to benefit Patton.[5]

As he had during the Punitive Expedition, Pershing relentlessly inspected every corner of his growing domain. He would turn up unexpectedly anywhere American troops were based, and nothing seemed to escape his personal attention. An artillery commander of the 42d Rainbow Division was descending from a hayloft when he encountered the commander in chief, who demanded to know what he had done to prepare himself to be a colonel commanding a regiment of artillery.[6] No one was immune from Pershing's sharp eye and scathing condemnation if he was found wanting. The complacent or neglectful were soon dismissed or court-martialed.

While Pershing drove himself at a pace that would kill an ordinary man, Patton remained in charge of the headquarters enlisted men, and was also the post adjutant and, temporarily, the provost marshal, "and any other little thing that people think about giving me. . . . I am nothing but a hired flunkey. I shall be glad to get back to the line again and will try to do so in the spring." The early days of the AEF at Chaumont were chaotic, "hundreds of clearks rushing about and officers shouting for a place to stay. we could do nothing so let them yell. . . . I can think of about one million things I would rather be. . . . I am also in charge of Passes and anti air craft defenses. . . . You know how I hate to telephone. Well I live at the end of one now."[7]

Claiming he was, "too much of a savage for city life," Patton found that he actually preferred Chaumont to Paris, yet the place "bores me to death." His billet and mess were at Pershing's large nearby house. "My room is small but cheerful over looking the garden and park," and the countryside offered numerous interesting places where he could ride his horses.[8] Social life in Chaumont consisted of a weekly movie and an occasional dance at the nearby American hospital. Patton attended one and wrote of the nurses: "I have never seen such a lot of horrors in my life . . . and they dance like tons of brick. I don't think I shall go again it is too much work dancing with people out of ones own class who are not dressed up."[9]

His frustration mounted when officers less qualified than himself began receiving promotions. Patton was number 113 on the list of captains of cavalry eligible for promotion, and he was outraged that unqualified senior officers were being promoted after three months' training, while men with nine years' experience, like himself, were still captains.[10] Nevertheless he admitted that he should be thankful for his good fortune and despite missing Beatrice dreadfully, "if I were there [there] would probably be no living with me. Every one says I have the best position of any young Cavalry man. . . ."[11] He wrote proudly to Mr. Ayer (who had recovered from his bout of pneumonia but was increasingly frail):

> We run railroads and build them. Build docks, charter ships, houses, hotels, factories. We buy coal, wood, movable houses. Horses, Automobiles. Aeroplanes. Clothes. Move troops. Make telegraph lines and almost every other sort of human occupation. We have a hell of a time finding office space for all our clearks and officers. Each day we think that the town will not hold another man then ten more come and we tuck them in it is very exciting.[12]

Although Patton performed his duties with great efficiency, it was simply that, for a soldier who craved action, they were of the wrong sort. "This war would be a lot more interesting if we could have some fighting but . . . I fear we will be at it a long time yet before we do any killing."[13]

Patton's separation from Beatrice was painful and seemed to remind him of his own mortality and his dread of growing old. "B you don't know how much I love you . . . try not to worry too much and get grey hair. I don't like them. I put tonic on my head every day and take exercises so as to keep my youthful appearance."[14] He hoped to write each day and usually managed to average one every second day. The loneliness and his dyslexia still produced miserable days. "I would give a lot to have you consol me and tell me that I amounted to a lot even when I know I don't," he wrote on October 8.

Toward the end of September, Patton's interest was aroused by talk around the headquarters of the creation of a new branch of the army:

> There is a lot of talk about "Tanks" here now and I am interested as I can see no future in my present job. The casualties in the Tanks is high that is lots of them get smashed but the people in them are pretty safe as safe as we can be in this war. It will be a long long time yet before we have any so don't get worried. We will see each other and talk it over before I will even have a chance to apply. I love you too much to try to get killed but also too much to be willing to sit on my tail and do nothing.[15]

*   *   *

Early in the war the stalemate in France brought about attempts by the British to create an armored vehicle that could be used to crush the deadly barbed wire, climb up and over trenches, and advance across no-man's land and into the German trench works, thereby creating a breach that could be exploited by the infantry. If such a vehicle could be developed, it just might turn the tide in favor of the Allies. The idea for an armored vehicle was first brought to the attention of the War Office in October 1914 by a war correspondent with the BEF, Lt. Col. Ernest D. Swinton, who had learned of an American Holt caterpillar tractor "which could climb like the devil."

Among the traditionalists in the British military there was scant enthusiasm for any form of original thinking, particularly for a mechanical monstrosity that then did not even exist. One exception, however, was the independent-minded Winston Churchill. In 1915 Churchill had been the first lord of the Admiralty, and the idea held spontaneous appeal to his resourceful imagination, particularly if the development of an armored vehicle might enable the navy to get in on the action in France by piloting armored "landcruisers." Churchill ordered his staff to get busy.[16]

The War Office was only spurred to action in 1915, when it learned that the Admiralty was independently engaged in developing a land cruiser. A joint army-navy committee was formed, and when Swinton returned to duty in the War Office, he emerged from an interview with the secretary, calling the effort, "the most extraordinary thing I have ever seen. The Director of Naval Construction appears to be making land battle-ships for the Army, who have never asked for them and are doing nothing to help. You have nothing but naval ratings doing all your work. What are you? Are you a mechanic or a chauffeur?"[17]

By 1916, a design breakthrough occurred, and a tank called "Mother" (later "Big Willie") became the first operational armored vehicle produced by the British.[18] The machine was equipped with a six-pounder gun inside a small turret on each side, and four machine guns.[19]

Officially designated the Mark I, on September 15, 1916, forty-nine "Big Willies" participated in the Battle of the Somme, with mixed results. There was as yet no doctrine for their employment, and instead of massing, they were employed individually across the front. Moving forward in the black of night, seventeen broke down or became hopelessly stuck in the mud before arriving at the start line for the attack, which commenced at dawn. Although mud and mechanical breakdown claimed most of the "Big Willies," nevertheless, several of these ungainly iron monsters were spectacularly successful and terrorized the Germans. One led the assault on the town of Flers, south of Arras, which was taken without a single loss, to the cheers of the New Zealand infantry who followed it. Another straddled a trench work and captured three hundred German prisoners, while still another attacked a German artillery battery before being knocked out.

According to the German press, the panic created in the front lines by these first tanks on the battlefield was reflected in cries of: "The devil is coming!"

The British painted their tanks in a variety of rainbow colors, and although these incredible machines may have looked comical, they created very real alarm in the German High Command. "Secret and urgent orders were issued to the German troops to hold their ground at all costs and fight to the last man against these new and monstrous engines of war, which they complained were both cruel and effective."[20] Flawed though it was, the use of the first armored vehicles on a battlefield forever changed the course of modern warfare.

During this time the French had also been developing their own version of the tank, which was designed as an infantry personnel carrier. But after observing the success of the British, they redesigned their machine to become an artillery weapon to support the advancing infantry.[21] The French Renaults were lightly armored and far more mobile than the heavier British "Big Willies" and were employed for the first time in the spring of 1917.

U.S. Army officers from the American Military Mission had been interested spectators to these events, and although their reports to the War Department were disparaging of the future of the tank as a weapon of war, they generated considerable interest in Washington. It was inevitable that one of these reports would eventually land on the desk of John J. Pershing. A joint Anglo-French tank board had been created but, not surprisingly, its members failed to agree on either tactics or equipment. The British continued to propound the doctrine of using their heavy tanks independently, while the French were in favor of close cooperation between the light Renault tanks and the infantry.[22]

Despite reports of design deficiencies and rampant mechanical breakdowns, Pershing was quick to grasp that here was a new weapon of considerable potential that was worth another look. He established an AEF tank board to investigate further. The board's report favored American use of tanks, which were "destined to become an important element in this war," and recommended the immediate creation of a Tank Department to implement a plan to manufacture and field a force of two hundred heavy and two thousand light tanks, modeled on the British Mark VI and the Renault.[23] The officer initially appointed by Pershing was Patton's mentor, Lt. Col. LeRoy Eltinge.

In mid-October Patton contracted jaundice after what he later termed "an attack of excessive fish-eating" and was hospitalized with "a fine cavalry yellow all over."[24] His stomach was pumped daily and he was placed on an all-milk diet. To his daughter Bee he wrote, "I am like you now I drink milk and as I don't like it I take a long time over it the same as you do but I hope now that you eat better."[25] Patton remained a sickly shade of yellow for several weeks and was kept in the hospital bemoaning his condition as "the most stupid disease imaginable. . . ." He was cheered by the expectation of being promoted to major in the near future. "I am counting my chickens before they are hatched but I feel sure of being a major very soon."[26]

During his hospitalization Patton began to realize that he could not continue indefinitely using his influence with Pershing. "I am going to get away from him and stand on my own feet then if I can't make a go of it it is my fault."[27] Patton shared a hospital room with Col. Fox Conner and although the two officers discussed tanks, Patton had about decided to seek a new assignment as an Infantry major when Colonel Eltinge[28] visited to say that a tank school was soon to open in the nearby town of Langres and "would I take it. Inspite of my resolution to the contrary I said yes. But I kept discussing it pro and con with . . . Connor and again decided on Inft."[29] Another version of this event, written in 1928, was that Eltinge had put it to him: "Patton, we want to start a Tank School, to get anything out of tanks one must be reckless and take risks, I think you are the sort of darned fool who will do it."[30]

In early October Patton entered in his diary: "Some time about the end of September Col. Eltinge asked me if I wanted to be a Tank officer. I said yes and also talked the matter over with Col. McCoy who advised me to write a letter asking that in the event of Tanks being organized that my name be considered I did so."[31] Patton had briefly toyed with the idea of asking for an assignment to a new bayonet school, which "is exactly like what I did at Riley" and when a British colonel came to lecture on the bayonet, Patton wrote: "His chief point was physocoligy of war. His words were almost verbatim what I used to say at Riley and for saying which people thought me brutal."[32]

The idea of tanks proved more appealing, and he wasted no time writing directly to Pershing that he wished to be considered for command in the new tank force. Referring to the duties of tanks as analogous to those performed by cavalry in war, for which he had considerable experience, he wrote:

> I have run Gas Engines . . . and have used and repaired Gas Automobiles since 1905. . . . I speak and read French better than 95% of American officers so could get information from the French Direct. I have also been to school in France and have always gotten on well with frenchmen. I believe I have quick judgement and that I am willing to take chances. Also I have always believed in getting close to the enemy and have taught this for two years at the Mounted Service School where I had success in arousing the aggressive spirit in the students.

> He also reminded Pershing that he was "the only American who has ever made an attack in a motor vehicle."[33]

The more he thought about it, the more Patton came to the realization that Colonel Eltinge had offered him a unique opportunity. "I believe that with my usual luck I have again fallen on my feet," he wrote to Papa in early November.

> It is so apparently a thing of destiny that . . . my name is in and [I] will start in before I am thirty two. Here is the sporting side of it. There will be hundred[s] of Majors of Infantry but only one of Light T[anks]. The T. are only used in attacks so all the rest of the time you are comfortable. Of course there is about a fifty percent chance that they wont work at all but if they do they will work like hell. Here is the golden dream."

As Patton began to envision it: "1st. I will run the school. 2 then they will organize a battalion and I will command it 3. Then if I make good and the T. do and the war lasts I will get the first regiment. 4. With the same "IF" as before they will make a brigade and I will get the star."[34] Patton's prophecy was, with one minor exception, amazingly accurate.

A decade later Patton admitted his enthusiasm for tanks had not existed in Paris the previous summer when he was introduced

> to a French Officer . . . [who] was a Tank enthusiast who regaled me for several hours with lurid tales of the value of his pet hobby as a certain means of winning the war. In the report I submitted . . . I said, couching my remarks in the euphemistic jargon appropriate to official correspondence, that the Frenchman was crazy and the Tank not worth a damn. On the 17th of November following I was detailed as the first officer in the Tank Corps.[35]

Patton passed muster with Pershing and on November 10, 1917, became the first U.S. Army soldier to be assigned to the new Tank Corps. Patton's orders directed him "to proceed to Langres, France, and report to the Commandant of the Army Schools for the purpose of establishing the First Army Tank School."[36] He was provided with one assistant, a young artillery officer, 1st Lt. Elgin Braine, previously assigned to the 1st Division. On November 18 the entire U.S. Army Tank Corps consisted of Patton and Braine.

The news cannot have reassured Beatrice, whose recent letters were filled with sadness and apprehension that perhaps George was at the front and had not informed her. "I hate to have you feel sad all the time for it makes you worry and worrying makes you feel old, etc. Hence don't worry."[37] Shortly before his thirty-second birthday, on November 11, he wrote, "It is sixteen years ago next July since I decided to marry you and we have been together about five."[38] Nor were her spirits raised by a letter a

month later in which Patton wrote that he had just had "my usual yearly
accident," returning to Paris from a visit to the British front near Amiens,
where he had met Colonel J. F. C. Fuller, the brilliant and eccentric military
thinker and second in command of the British tank force. Patton had viewed
"the flash of guns and the trench rockets going up." The automobile in
which he was riding rammed a closed railroad gate, and the impact thrust
Patton's head through the windshield, cutting open an artery in his temple
and gouging a hole in the point of his jaw an inch long and an inch deep,
which required five stitches at the American hospital at Neuilly, "missed the
carroted [carotid] artery and jugular and facial nerve about an 1/8 of an inch
if it had gotten them I would probably have cashed in," he wrote his
shocked wife.[39]

In early November Patton told her of his transfer to tanks, reiterating
that it was a gamble he had been mulling over for a month. "The light tank
is a new invention and may not work at all. If it does not I can still go to an
infantry Battalion and would have lost only my time . . . [but] if it works I
will have pulled one of the biggest coups of my life so far."[40]

To the end of his assignment as the AEF headquarters troop commander, Patton remained a stickler for discipline. Although he was to hand over
his command the following day, he had observed a lapse of saluting and
sloppy dress by officers assigned to Chaumont, and unhesitatingly wrote
directly to the commander in chief to point out what he considered was a
failure of good order and discipline within Pershing's own headquarters.[41]
Patton ruthlessly enforced uniform regulations and there is little doubt that
Pershing's influence had hardened his resolve not to tolerate sloppiness,
even if meant correcting his superiors. There were many sins an officer
could commit, but from the time he was a cadet at West Point handing out
demerits for such infractions, very high on his list was anyone who failed to
display the highest standards of military dress and demeanor.

In his diary he noted that he had been obliged to "cuss out a lot of captains and majors for not saluting." He invited Pershing to review his troop
on that final morning and two hours before the general's appearance Patton
inspected his men. Six were found to have dirty uniforms and were summarily banished to KP and fatigue duty.[42] Two hours later Patton was en route to
Paris and a future that would take him to the heights of his profession.

Before the newly designated chief could organize the AEF Light Tank
School he first had to learn something about tanks. By his own admission,
Patton's knowledge prior to November 1917 was limited to his dismissive
and cursory investigation the previous July. He had already visited Langres,
where the new school would be, but could not begin his own training and
familiarization until permission was obtained from the French for an orien-
tation visit to their light tank training center at Chamlieu, in the forest of

Compiègne. There, for two weeks, he familiarized himself with the French light tank. Patton drove a Renault, fired its gun, and pronounced himself thrilled by the experience. "It is easy to do after an auto and quite comfortable though you can see nothing at all. . . . It is funny to hit small trees and see them go down. They are noisy . . . [and] rear up like a horse or stand on their head with perfect immunity. . . . The thing will do the damdest things imaginable."[43] Patton discussed tactics in the mess with French cavalry, infantry, and artillery officers ("they fight all the time as to whether or not they should use trenches."), inspected the maintenance facility, began translating lesson plans and met with the head of the French tank force. In the evening Patton crawled into a Renault to learn how it worked and inundated his hosts with so many questions that they were obliged to bring in an authority to help them satisfy his thirst for information.[44] During the second week he was joined by Lieutenant Braine, and the two officers toured the Renault tank factory to learn how the tank was designed and made.[45]

In Braine, Patton had been assigned an officer of outstanding mechanical ability and innate common sense whose contributions as one of the architects of the U.S. tank force have never received their full due. In 1918, while Patton spearheaded the establishment and training of the first tank brigade in France, Braine acted as his liaison officer in the United States, and helped to guide the production of the American light tank through a bureaucratic minefield.

While Patton was undergoing his crash course in tanks, the British launched the largest tank attack yet mounted, at Cambrai, on November 20, 1917. Five infantry divisions spearheaded by most of the 324 tanks of the British Tank Corps attacked the Hindenburg Line (the principal front line, so named for Field Marshal Paul von Hindenburg, chief of the German General Staff) en masse across a seven-mile-wide front and caught the Germans flatfooted. In six hours they penetrated ten thousand yards to the fourth level of the German defenses and overran two divisions, capturing eight thousand prisoners. In a mere six hours British tanks had taken more ground than had the one hundred thousand troops of Gen. Sir Hubert Gough's Fifth Army during the massive and costly Third Battle of Ypres in July 1917.

The problem with Cambrai was that there was no plan for exploiting a breakthrough of such proportions. Nearly 180 tanks broke down during the attack, most to mechanical failure, and without them the infantry failed to seize the initiative, thus enabling the Germans to rush reinforcements to plug the gaping hole in their line. Cambrai begged the question of who was more astounded—the British, who never envisioned or prepared for a decisive penetration, or the Germans, who were stunned by the British offensive.[46] Tactics for tank use had yet to be devised, and at the time only a few men grasped the significance of Cambrai, which had conclusively proven that tanks were a deadly weapon that could play a vital role on the modern battlefield and were fully capable of penetrating and crushing an entrenched enemy position. Among these men were J. F. C. Fuller and, even though he was not present, George S. Patton, who informed Beatrice that his timing in joining the Tank Corps could not have been better. "Lots of people have suddenly discovered that in the tanks they have always had faith and now express a desire to accept the command of them but fortunately I beat them to it by about four days . . . the job . . . [ahead] is huge . . . Some times I wonder if I can do all there is to do but I suppose I can. I always have so far."[47]

After Chamlieu, Patton visited Colonel Fuller to learn firsthand about the Battle of Cambrai.[48] For a solid month Patton had immersed himself in tanks, and the extent of what he learned was promptly put to the test. After rendering a verbal report of his experience to Colonel Eltinge, Patton's first real challenge was to consolidate his newfound knowledge into specific recommendations for the new American tank force. With Braine's assistance, Patton began to distill his thoughts into what became a fifty-eight-page memo to the Chief of the Tank Service, titled *Light Tanks*. The paper outlined in clear language his proposed organization and rationale for what Patton himself many years later wrote on his copy, "was and is the Basis of the U.S. Tank Corps. I think it is the best Technical Paper I ever wrote." Patton's self-appraisal was unerring—the paper outlined in clear language his proposed organization of and rationale for what became the U.S. Tank Corps. Patton's report was a masterpiece of originality and clear thinking that spelled out his recommendations for the organization, tactics, equipment and training of the new American tank force. It presented both a historical and technical perspective of the Renault tank, and efficiently analyzed the successes and failures of both French and British tanks and tactics to the present moment.

Tanks failed, he noted, whenever they got ahead of the infantry, thus losing the benefit of their natural and mutual support. "The proper conception of the light tank is as a heavily armored infantry soldier, with equal activity and greater destructive and resistant powers." Patton also provided a forecast of the future role of armor and of the philosophy by which he was to become famous when he added, "If resistance is broken and the line pierced the tank must and will assume the role of pursuit cavalry and 'ride the enemy to death.' "[49]

Coming from someone whose knowledge of tanks was little more than a month old, his memo was an astonishing reflection of the visionary aspect of Patton's mind. How difficult was the effort? Very. For every single detail

had to be worked out to fulfill all conceivable conditions. Even a list of tools and spare parts down to and including extra wire and string . . . and nothing to base it on but a general knowledge of soldiering. Honestly I think not many men could have combined the exact mechanical knowledge with the general Tactical and organizational knowledge to do it. But I think I did a good job. . . . I hope I can make a success of this business but starting with nothing is hard. Now I feel helpless and almost beaten but I will make a go of it or bust.

To Aunt Nannie he wrote that he had "worked up a pretty good report on tanks. It was [as] interesting as it was original. . . . The proof is in the eating and we are getting ready for dinner."[50]

During the final days before taking up his new post he explained that he felt, "sort of like a Rat without a tail just now running tanks when there are none. I don't know where to go or what to do yet feel that I should be doing something fast and furious."[51] Patton's anxiety was partly the natural apprehension of one about to sail into uncharted waters. But there was also the ever-present dyslexia to complicate what was already a formidable challenge. To his beloved confidante, Beatrice, he confided his fears and his determination to succeed:

> Tomorrow I start on my way again. All alone to go to a new place and organize the Light Tank Service. I feel unusually small in self esteem. I have been so long a small but important cog in a machine . . . it is hard to go off and be the last word all by myself. . . . Actually I am in quite a "Funk" for there is nothing but me to do it all. Starting the Fencing School was a similar experience but vastly smaller and then too I had a model to copy. Here it is all original and all to be conceived and accomplished. The most cheering thing is that Gen. Harbord, Col. Eltinge and Col. Malone all seem confident I can do it. I wish I were as sanguine. I am sure I will do it but just at this moment I don't see how. I will have to grow and grow a lot. But I will. Here is my chance. If I fail it will be only my fault. I won't even have you to pick on.[52]

As 1917 drew to a close, Patton could look back on a year that had taken him from the hot, dusty plains of Mexico to the freezing cold of France, where he was destined to play an important role in the Great War. In a blizzard he drove from his new quarters in Langres to Chaumont for Christmas dinner with Pershing and several of his closest staff officers. Pershing presented Patton with a cigarette holder. "We all drank to being Together Christmas 1918."[53]

CHAPTER 16

"Great Oaks from Little Acorns Grow"

I am getting a hell of a reputation.

—PATTON

In December 1917 Lt. Col. LeRoy Eltinge was replaced as the temporary head of the AEF Tank Corps by Col. Samuel D. Rockenbach, a cavalry officer and 1889 VMI graduate who had been given the thankless task of overseeing the creation of the Tank Corps. He later recalled reporting to Eltinge, who opened his desk and presented him with a bundle of papers, saying: "Here's all we know about Tanks, go after them."

For a time the entire Tank Corps consisted of Patton, Braine, and Rockenbach. Patton and Rockenbach were complete opposites. From the time of their first meeting in December 1917, Patton was uneasy around Rockenbach and conveyed to Beatrice that "I guess he does not care a whole lot for me but my theory that if you do your best no one can hurt you will be put to the proof."[1]

Whereas Patton was aggressive and restless, Rockenbach "brought a wealth of experience and maturity to his post as commander of the tank force and principal tank adviser to General Pershing. Hardly an original thinker, even-tempered, lacking a sense of humor, and displaying a tendency to fixed, narrow opinions, Rockenbach was able to balance Patton's headstrong enthusiasm and channel his creativity."[2] Patton's daughter, who knew Rockenbach after the war at Camp Meade, Maryland, has described him as "famous for his razor-edged tongue and his martinet-ship. His wife, Emma, was a breezy aristocrat, and a notable horsewoman. When she was asked why on earth she had married 'Rocky,' she replied, 'I married him for his conformation, of course. Did you ever see a finer piece of man-flesh?"[3]

Rockenbach and Patton managed to coexist remarkably well because of their mutual cavalry connection and VMI roots. Eventually Rockenbach even came to admire Patton, but the young captain still found it difficult to deal with his superior and never fully trusted him.

> Col. R. is the most contrary old cuss I ever worked with as soon as you suggest any thing he opposes [it] but after about an hours argument comes round and proposes the same thing him self. So in the long run I get my way but at a great waste of breath. It is good discipline however for me for I have to keep my temper. At the end of each argument I feel completely done up. I guess he does too. Still he was trying to have me made a Lieut Colonel so I ought not be too hard on him.[4]

Rockenbach was not only responsible for everything to do with tanks and the Tank Corps at AEF headquarters, but also acted as Patton's commanding officer. In the still ambiguous and primitive state of the tank in France, Patton soon found that virtually every aspect of his new assignment came under Rockenbach's purview. Thus, even though they were uncomfortable bedfellows, both officers were exemplars of their profession who put the task at hand ahead of their personal feelings.[5]

Patton's reaction to Rockenbach was mercurial. At times he would praise him, but when frustrated would roundly criticize Rockenbach in his letters to Beatrice. When, by the summer of 1918, Patton had yet to receive his first American tank, he wrote in annoyance, "The more I see of Gen. R. the less I think of him he is nothing but a good hearted wind bag. I truly believe others would have pushed this show along much better."[6]

What Patton failed to appreciate was how well Rockenbach managed to isolate him from the red tape and the endless wrangling with the British and French, whose cooperation and assistance were essential. Rockenbach ensured that Patton was left free to concentrate on organizing the new tank center and deliberately kept him outside the policy battles that raged in the corridors of the AEF and in the Supreme War Council, where Rockenbach spoke for Pershing on tank matters.[7] For some time British and French cooperation on tank matters did not extend to furnishing tanks to the new American force. If the AEF wanted tanks, the United States would have to design and manufacture its own. The task was monumental and in hindsight seems to have been all but impossible in the short time. With many fingers in the pie in both France and the United States, for the army to have agreed on a design, geared up the necessary manufacturing capability, and produced and shipped to France the hundreds of tanks the AEF required was fantasy.

   As was his longtime custom, Patton used his frequent letters to his wife and, occasionally, his father as a means of expressing his ideas, fears, jealousies, frustrations, and impatience. His criticism of others—be they contemporaries or superior officers who were either too slow or timid or, in his opinion, had achieved an undeserved promotion or coveted assignment—was usually scathing. Sometimes in a fit of pique he would condemn another, only to retract or modify his opinion in a subsequent letter. Mostly he was driven by frustration. In 1918 the most frequent object of his ire was Rockenbach.

   In mid-December Patton arrived in the ancient walled Roman city of Langres, high in the Haute Marne, and rented the first floor of an abandoned château-hotel, which he and Braine temporarily called home. The owner, the Countess d'Aulan, preferred living in Paris and the place had not seen use as a hotel for nearly three hundred years.[8] It had no gas or electricity, and the only light came from oil lamps. When Patton asked the rental agent about the owner, he was told the count was a hero who had died in the war. Assuming it to have been in World War I, Patton asked what battle. Oh no, replied the agent, the count had died in the twelfth century during the Second Crusade![9]

   An old woman and her two daughters were hired as cook and maids. After Patton visited another officer's quarters and commented on how chilly the place was, one of the maids offered to sleep with him to keep him warm. Patton demurred, saying he was too old and would be quite happy to have the fire stoked up instead.[10]

   Langres was a dark and cold place in December, and there was no suitable location in the town for the tank school. With Braine in Paris working on tank procurement, the dismal winter days and the daunting prospect of ever getting the school successfully launched left him moody and uncertain. He had no staff, no site, no equipment, and none of the other myriad components required to start such an enterprise from scratch. "Sometimes I feel depressed over ever getting a school started. If I ever get the staff I know I can deliver the goods but . . . it is a hell of a job. Fortunately I have no students so far."[11]

   He wrote notes about the school and several "bum" poems, which he sent to Beatrice.

   Another day is done and yet no tank school is going. I am getting nearvous very much so feeling that I am lazy and worthless and yet not seeing how to procede. I hate the feeling. Perhaps if I had more brains I would see something to do but the roads are so slipperly that one can not even explor the d—— country. . . . Having been on a staff where some one else does your thinking is quite plesant but bad for one's originality . . . may be I shall get some back bone. It is one of the few times

I ever felt I lacked it but I seem to or maybe it is just the situation. But it is most unpleasant.[12]

Patton continued both to reassure Beatrice and justify to her his reasons for leaving Pershing's staff. To continue there:

> I would have been simply an office boy. . . . I have always talked blood and murder and am looked on as an advocate of close up fighting. I could never look my self in the face if I was a staff officer and comparatively safe. . . . The Tanks were I truly believe a great opportunity for me. I ought to be one of the high ranking men one of the two or three at the top. I am fitted for it as I have imagination and daring and exceptional mechanical knowledge. . . . Tanks will be much more important than aviation and the man on the ground floor will reap the benefit. It would not have been right either to J. [Pershing] or myself to have hung on any longer[. B]esides I was loosing my independence of thought and a little more of it would have made a nothing of me.[13]

In the nearby village of Bourg, Patton and Braine found a site suitable for the tank school. Patton later described Bourg as "the dirtiest place in the world full of cows and chickens but it has a fine view being on the edge of a cliff over a long and beautiful valley."[14] Among its advantages were adequate terrain for training, two excellent access roads, and a railhead for his tanks. However, getting the French to provide the land proved difficult and required time-consuming negotiations. Patton thought the French were stalling, and when they refused outright a request for the site, he paid a visit to the French headquarters commandant, a colonel whom he politely called a fool. "He did not like it much." They were "d——fools," he complained to Beatrice. "You would think we were doing them a hell of a favor to fight for them. . . . I will get the ground or bust. . . ."[15] Eventually the site was acquired and became the official new home of the Army Tank School, Captain G. S. Patton, Jr., Director.

The days now became a blur of activity. When he wasn't developing course plans and requirements, Patton was on the move, one day escorting Rockenbach around the proposed site, another accompanying him to the Renault tank works or to Chamlieu for meetings with their French counterparts. Other journeys took him to the British Tank Corps at Bernecourt, where he held further discussions with Fuller and Gen. Hugh Elles, their celebrated commander, who had personally led the Cambrai attack. Organizing the school consumed hours of his time. He and Braine spent two days drawing up a list of spare parts for the tank. "It takes an awful lot of stuff to kill a German," he mused.[16]

The temporary return of Braine, himself constantly on the move as a sort of roving ambassador of the Tank Corps, brought badly needed administrative help, but for the most part Patton was on his own. He spent a great deal of time in Chaumont, reviewing with Rockenbach plans for the school and the organization of the Tank Corps. With the first American tank barely on the drawing board, many months away from construction and delivery, the primary problem was the acquisition of tanks with which to train Patton's students.

As word of the new tank force spread throughout the AEF, men had begun volunteering. The first to arrive in early January were ten Coast Artillery lieutenants, and by midmonth Patton had eighteen officers assigned to the school, who were immediately sent to other AEF schools for training in other military skills. At once Patton began weeding out those who were unsuitable. When two of the new lieutenants (both former NCOs) did not perform well and thus had to be reduced to their original ranks, "They broke down and cried like babies. To see old strong men cry is not pleasant but there was nothing to do. War is not run on sentiment."

From the first day Patton became notorious as an unbending disciplinarian who set high standards and would brook no nonsense. Woe to the officer or NCO who was found deficient in dress or military courtesy: "I am getting a hell of a reputation for a skunk when officers don't salute me I stop them and make them do it. I also reported a reserve lieutenant to day for profanity." Noting that King Louis XI of France ate only eggs to avoid being poisoned, Patton wrote with some pride: "I expect some of them would like to poison me."[17] His first memorandum dealt with soldierly appearance and deportment. Shoes and brass would be highly polished and hair cut short, "so that they look like soldiers and not like poets. . . .* There is a wide spread and regrettable habit in our service of ducking the head to meet the hand in rendering a salute. This will not be tolerated." Saluting *would* be carried out with precision and smartness.[18]

Patton himself spared no expense to ensure that *his* dress exceeded that of his charges. "I have just gotten a new pair of very nice boots at a very high cost [in Paris] but one must look well in order to hold peoples attention."[19] On another occasion, Patton, in a harbinger of an incident that nearly destroyed his career a quarter of a century later, "cussed a reserve officer for saluting me with his hands in his pockets . . . he said that he demanded to be treated like an officer. I almost hit him but compromised by taking him to the General who cussed him good. Some of these officers are the end of the limit. I bet the Tank Corps will have discipline if nothing else . . . We have a long way to go to make an army."[20] The vengeful lieutenant was the billeting

---

*For one who wrote poetry himself most of his life, Patton's choice of simile in this instance was decidedly poor.

officer in Langres, and he subsequently hounded Patton by twice ordering him out of quarters he had just moved into, until the commanding general intervened.[21]

During a visit to the 82d Infantry Division, Patton had noted with great interest their shoulder patches and he enjoined his own officers to create a new patch for the tanks: "I want you officers to devote one evening to something constructive. I want a shoulder insignia. We claim to have the firepower of artillery, the mobility of cavalry, and the ability to hold ground of the infantry, so whatever you come up with it must have red, yellow, and blue in it."[22] So it was that 1st Lt. Will G. Robinson and another officer spent that night with paper and a set of crayons, attempting to comply with Patton's wishes. After many hours of trial and error they finally settled on a red-yellow-and-blue triangle that to this day remains the shape of the insignia of the armored divisions of the U.S. Army. Their "pyramid of power" design delighted Patton, who approved it on the spot. Pulling a one-hundred-dollar bill from his pocket, he handed it to Lieutenant Robinson with orders to find someone in Langres to make the patches that day. The enterprising Robinson soon returned with three hundred patches. Even at this early stage, Patton was attempting to provide the Tank Corps with an identity. As Robinson later wrote: "If there was anything he wanted, it was to make the Tank Corps tougher than the Marines and more spectacular than the Matterhorn. That triangle was the first step."[23]

No detail was too small to escape Patton's attention. He drew up designs for a Tank Corps brassard (a distinctive band worn around the upper sleeve of the uniform) and an ornament to be worn on the uniform collar.[24] During the summer of 1918, when he delivered a series of lectures to officers of the AEF General Staff School, Patton offered reassurance that the mission of the new Tank Corps was to assist, not to supplant, the infantry. "Tanks in common with all other auxiliary arms are but a means of aiding infantry, on whom the fate of battle ever rests, to drive their bayonets into the bellies of the enemy. Hence tanks when operating in numbers must always conform to the needs and frailties of the infantry," he declared.[25]

No one escaped his critical eye. All officers were ordered to synchronize their watches with those of the adjutant at each morning's mess. Even the medical officer endured the wrath of Patton's temper for not keeping his heels together while at attention during an inspection. "The poor Dr. is so scared of me that he giggles when I speak to him and is almost incoherent it is quite amusing If I ever get to be an 'old general' I fear that B[ea] Jr & R[uth] E[llen] will have fiew suitors."[26]

Patton's insistence on discipline and attention to detail extended to his family, and even from a distance of 3,500 miles he could not resist giving Beatrice advice that often sounded like a command. After receiving some Christmas photographs, he complained that baby Ruth Ellen's hair "looks awful. Why don't you cut it." He also informed Beatrice that he expected her to be "the same age when I get back as when I left," and instructed her to "die your hair for I don't like gray hair at all."[27] (When an enlisted man mistook him for a second lieutenant, Patton was secretly delighted because junior lieutenants were not "old looking.") On another occasion he noted: "The letters from the children are very cute but I fear that [little] B is a little given to posing and I don't think much of it. Of course she loves you every one does but her style of expression seems a little forced. Don't get mad at me for saying it for I suppose I am wrong."[28]

Without the necessary site and tanks to train with, it was hard and frustrating work, but "by degrees I grow." Yet, even when the head of the AEF schools complimented him for "taking hold better than any man he had," Patton thought that "the rest must be pretty poor for I certainly have not done much to be proud of so far. I don't feel that I am doing my best. . . . I never do. If I ever do feel that I am earning my pay I must realy begin to get some where."

He traveled to Chaumont, where he delivered a superb lecture on tanks to a group of AEF colonels and generals, still without benefit of an actual tank. But Patton seemed incapable of giving himself credit or accepting that there were circumstances over which he had no control, such as the absent tanks. "I am feeling very low," he wrote only hours after his lecture. "I am disgusted with the whole business."[29]

Despite a nonstop schedule of appointments, visits, and the writing of plans and lectures, Patton's desire for action and his high level of frustration all too often left him angry and disillusioned. "Unless I get some Tanks soon I will go crazy for I have done nothing since november and it is getting on my nerves."[30] Patton's mood alternated between acute doldrums and near euphoria. A few days after complaining of going "crazy" he proudly informed Beatrice that, "I think I am more or less of a mechanical genius for I simply know by looking at an engine all about it."[31]

When Patton learned that he had been promoted to major on January 26, 1918, he decided to pin on the insignia of his new rank, even though he had yet to be formally notified of his promotion. "I feel sort of like a thief but that does not bother me at all as there are so many militia majors around here that one must have leaves to keep ones self respect." Never bothered by such protocol as official orders, Patton did the same thing in 1943.[32]

In early March, Pershing invited Patton to accompany him on an inspection trip to the 1st Division in a sector of the front line that had seen little recent action. The division commander was his former West Point tactical officer, Charles P. Summerall, and, as usual, Patton was anxious to get as close to the front at possible to observe an artillery barrage scheduled for shortly after midnight. Although Summerall had given his blessing for Patton to visit an observation post close to the front, Pershing noticed him strapping on his helmet and growled: "Where in hell are you going?" Patton had to tell him, and Pershing "forbade it."[33] Although normally not so protective of his officers, in Patton's case Pershing seemed in no hurry to expose his protégé to unnecessary danger.

Patton's limited social life was never dull, however. His dinner guests ranged from the local *sous-préfet,* "a funny old chap, who regaled us till half past eleven with the most shocking stories about ladies [which] no American of his age and rank would dare to talk about as he did," to visiting senior officers. Unable to spell correctly the name of a French baroness who invited him to tea in her nearby château, he wrote that it was "Madame La Baronne Pig and Sheep," who was, he reassured Beatrice, "quite nice and safely old."[34]

Rockenbach's obvious choice to command the first two tank battalions to be organized was Patton, who was recommended for promotion to lieutenant colonel: "I will have about 1400 men under me by May if every thing works out well." For a time former secretary of war Henry L. Stimson, now a lieutenant colonel, joined his mess, and the two renewed the friendship begun at Fort Myer. Their reunion, while insignificant at the time, would prove to be of untold benefit to Patton in World War II.[35]

While awaiting his first shipment of training tanks from the French, Patton put his new officers to work in a variety of tasks ranging from laying water lines to converting farm stables and lofts almost overnight into the living accommodations and shops that would make up the new school. When his first two companies of troops arrived on February 17, they found a hot meal awaiting them and "a nice latrine all dug. I think the villagers thought we were digging for gold as they came out and watched the operation." Patton expressed pleasure with his "good and efficient" officers and praised the quality of his new troops, all of whom were draftees. "They are really a very fine bunch of men much above the ordinary."[36] For the first time in his military career, "I am the absolute boss and it seems strange at first not to have to ask any one any thing at all."[37]

Drills and exercises commenced only after the men removed the heaps of manure that littered the new training area. Foot drills simulated what they would later do aboard their tanks, and various forms of command by arm signal, touch, or sound were learned and practiced. Patton taught them how tanks would be employed on the battlefield and personally met the high standards he demanded. They learned how to troubleshoot gasoline engines. They were lectured on a variety of military subjects, including camouflage, reconnaissance, map reading, and reading aerial photographs. There were short courses at other AEF schools on machine guns and other weapons, to ensure "that the use of these weapons would be so instilled in [their] minds . . . that under all conditions their use would be automatic and deadly."[38] There were also unpleasant but mandatory gas drills, necessitated by the German use of deadly chlorine and mustard gas, starting in 1915 at Ypres. Patton was unconcerned, believing gas would prove mainly a nuisance for his tankers.

Although he was generally pleased with the progress of both his officers and men, Patton returned tired and angry from a trip to England with Rockenbach to inspect a new British tank, complaining that things seemed to go well only when he was present. "We have a stupendous job and little time and none of my officers are worth a damn. I have to instruct all of them in every thing under heaven . . . maping, Visual training, Aiming, gas Engines, signaling, reconnaissance, Intelligence, and some other things I cant recall."[39]

He still found time to miss Beatrice. "When this war is over I am going to insist on using a single bed for *both* of us at the same time. There is perhaps more than one reason for this, but the only one which the censor and modisty will allow me to mention is that I am tired of being cold and especially of getting into a large and empty bed full of cold sheets. Hence you will have to go to bed first."[40]

Patton was aware that his letters were sometimes hurtful and that he was not the same man without her calming influence and common sense, which usually managed to keep his volatile personality in check. He confessed:

> Without your advice I am apt to make mistakes of judgment. All of which shows that I did a good thing by marrying you even if I don't treat you with 'respect' as you say or rather infer I don't in one of your letters. . . . Well I am sorry I have treated you that way for you are one of the few people man or woman for whom I have any. . . . When I realize that I am with out question a very superior soldier and yet realize as I do my many shortcomings I can but feel sorry for Gen P. served by men worse than I.[41]

In March 1918 Frederick Ayer died after a long illness, thus ending a flamboyant life filled with both material and spiritual success. Notified by a telegram from Ellie Ayer that "our great Commander has gone," Patton at once penned one of the most compassionate and loving letters he ever wrote his wife. Mr. Ayer was "the most perfect mortal I know. . . . Beatrice Jr and Ruth Ellen should be wonderful children with such a grandfather. It is futile to attempt to comfort you. Words, especially written words, are totally inadequate to consol for such a loss. . . . I know Darling that you are suffering all that the human soul can suffer. . . . Beaty my whole heart is at your feet. . . . My poor Darling Please take comfort if you can. May God help and strengthen you."

Beatrice Patton had been suffering for some time from an unspecified ailment, and, as it turned out, her ordeal was only beginning. Two weeks later Ellen Banning Ayer died and yet another sad telegram arrived. (It is not mere conjecture to suggest that Ellen's death, despite her ill health, was due to a broken heart.)

The Ayers' death was the final straw to destroy the Pattons' efforts to reunite in France or England. Patton wrote that he had always hoped to be present to help Beatrice when her parents died, but "it could not be. What has happened has quite reconciled me to your not having come to France. You would have felt so terribly to have been away. It seems a heartless thing to say but I think that Ellie is happier than she would have been to have continued on with out your father. They were as nearly one as is possible to be—as nearly as one as we are. I do not think I would care much about keeping on if you were gone. Because if you were not around to admire what I did what the rest thought would make little difference."[42]

Patton would later write in a brief history of the tank school that his objective had been thorough training "in the highest ideals of discipline, neatness, devotion to duty and esprit de corps. These results were produced by vigorous attention to close order drill, by the enforcement of great personal neatness on the part of the officers and men and by lectures pointing out to the men the necessity for the ends sought."[43] The key was discipline. Patton called it "instant, cheerful, unhesitating obedience. . . . Discipline is not a foolish thing, it is not a demeaning thing, it is a vital thing." Like a coach exhorting a football team, he cautioned his officers:

> Lack of discipline at play means the loss of a few yards. Lack of discipline in war means death or defeat, which is worse than death. The prize for a game is nothing. The prize for this war is the greatest of all prizes— freedom. . . . The reason the Bosch has survived so long against a world in arms is because he is disciplined. Since 1805 he had bred this quality as we breed speed in horses; but he is neither the inventor nor the patentee . . . We cannot wait until A.D. 2018 to breed discipline as they have done. But we are as intelligent as football players, far more intelligent than the Greeks or the Romans or the Persians or the Gauls of two thousand years ago . . . when we the quarterbacks give the signal of life or death in the near day of battle, you will not think and then act, but will act and if you will, think later—after the war. It is by discipline alone that all your efforts, all your patriotism, shall not have been in vain. Without it heroism is futile. You will die for nothing. With DISCIPLINE you are IRRESISTIBLE.[44]

For Patton a major factor in attaining discipline was his near-obsessive insistence on spit and polish. It was the *sine qua non* of his long-held conviction that to attain success on the battlefield a soldier had to be focused, and only by establishing and enforcing the very highest standards of discipline could he be taught to react efficiently and instinctively in the midst of chaos.

What had begun without a single tank and with only himself and Braine would, in little more than a year, evolve by March 1919 into to an enormous training center of four hundred tanks and five thousand officers and men. The comment that "great oaks from little acorns grow" also applied to Patton's philosophy of training. Men's lives often depended on split-second reactions. Only a trained and well-disciplined soldier was capable of this. Patton's "field of dreams" was built on the bedrock of hard, thorough training and discipline that began the first day and never let up. Throughout his long military career Patton never once lowered his standards or his microscopic attention to the smallest detail. Many did not like it, and more epithets were directed at Patton than perhaps any soldier in modern military history, but most of the men in his charge flourished, and it was the secret to his success as a commander and trainer of troops.

Patton's tankers soon began to evince a sense of pride in their new-found, razor-sharp military bearing. Saluting became so smartly executed that the byword became, "Give 'em a George Patton."[45] He himself was somewhat in awe of what he was accomplishing. "I don't see why they like me as I curse them freely on all occasions." Someone wrote a song about Patton "which is most complimentary and says that 'We will follow the Colonel through hell and out the other side.'"[46] One corporal recalled how Patton had instructed the students, "Why you God damned sons of bitches, do you think the Marines are tough? Well you just wait until I get through with you. Being tough will save lives." Another enlisted man noted that in the Tank Corps "we had officers to be proud of."[47]

Patton's drills for both officers and enlisted men included saluting, close-order drill, and athletics, games, and calisthenics from which no officer under the age of thirty-five was exempt. The area was kept scrupulously policed. Bourg may have been manure-ridden but under his aegis the cleanliness of the town itself became part of the "police" details carried out by his troops. The mud notwithstanding, "I make the chauffeurs wash their machines after the last trip each night no matter what the hour."[48]

Through two world wars, Patton was never seduced by the lure of weapons of war as cure-alls that would defeat a resilient enemy such as the German army. The fighting man was the key to winning, and fifteen years later he wrote:

When Samson took the fresh jawbone of an ass and slew a thousand men therewith he probably started such a vogue for the weapon . . . that for years no prudent donkey dared to bray. . . . History is replete with count-less other instances of military implements each in its day heralded as the last word—the key to victory—yet each in its turn subsiding to its useful but inconspicuous niche. Today [in 1933] machines hold the place for-merly occupied by the jawbone, the elephant, armor, the long bow, gun powder, and latterly, the submarine. They too shall pass.[49]

The secret he emphasized was not new weaponry, no matter how useful "be they the tank or the tomahawk . . . wars may be fought with weapons, but they are won by men. It is the spirit of the men who follow and the man who leads that gains the victory."[50] Bourg was the first place in which Patton was able to articulate his philosophy to a large and receptive audience that became ingrained with his philosophy of war.

Since his commissioning, Patton's problems had been not philosophical but physical, in the form of his unfortunate tendency to attempt new and innovative methods of accidentally killing himself. In February 1918, for example, he was in an accident when his automobile skidded on a rain-slicked highway as he was driving to a nearby railway station. This time only the vehicle sustained damage, but Patton had to walk three miles in the rain and missed his train.[51] In March 1918 he had yet another fender-bender, this time with his own car, which had to be sent to Paris for repairs. His horse Sylvia Green was hit by a truck, but neither horse nor rider sustained injury.[52] "We are now wearing a gold V on our left sleeves to show that we have been in the zone of the armies for six months. . . . If we get wounded we wear the same on the right arm for each wound. I would like to be hit in some nice fat part so I could get one."[53] When an acquaintance received a flesh wound in his leg, Patton wrote to Aunt Nannie how he envied any man permitted to wear a wound chevron and "pose as a hero." Beatrice undoubt-edly shuddered at the prospect.

The days were long and strenuous. Because of the limited equipment on which to train, the students worked in relays from six A.M. to six P.M., Monday through Saturday. Sundays were reserved for inspections.[54] To harden the troops each company ran double-time in formation each morn-ing for one kilometer. A newly opened YMCA canteen was about the only amusement available. "The only advantage is that they are working so hard that they go right to bed. I have no [venereal] 'disease' here now and hope to keep it out . . . it is easy if the men will only take the available precau-tions." In response to what he knew would have been Beatrice's disap-proval of anything but abstinence, Patton noted, "Of course I know what your remidy would be but I don't approve of that as men who are apt to be killed are entitled to what pleasures they can get."[55]

One of the reasons Patton was respected by virtually all of his subordi-nates at Bourg was that he taught by example and was never afraid to get his own hands dirty. One morning he was inspecting the underside of a tank when "about a pint of oil got in my face." He shrugged it off. During a class in map reading when some of his students could not visualize a contour line on a map, Patton made a miniature hill out of a potato and cut slices out of it to make his point. It was something Captain Marshall had taught him at Fort Sheridan and, "then as now it worked." His secret was preparation and a superb memory. "Some times it seems to me that all I have ever done has been in preparation for my present job." Still, he thought that "genius as Napolion put it is simply a memory of detail. I have a hell of a memory for poetry and war."[56]

At Bourg there was no shortage of mud, of which "there is certainly a magnificent supply." Patton was issued a truck, motorcycle and an automobile. Although he would now have to "hoof" it less, he worried that walking might "unduly develop my legs and hence [I will] not look so well in boots." He looked forward to the longer days, noting that "I have always thought and still more strongly think that our chief fault as an army has been that of taking things too easily when the days get longer I can work them more." He managed to find time to ride one of his horses. "Riding ten miles on Miss Green daily has improved her disposition and my health vastly."[57]

The school was running smoothly, which was not lost on Rockenbach, who was duly impressed during a visit to Bourg. The average commander, on learning of the imminent arrival of a superior officer to inspect, is likely to overreact with a flurry of activity designed to present a near-perfect impression. Not Patton. When Rockenbach made an unannounced visit in mid-March and declared himself well pleased, Patton, in an expression of supreme confidence, wrote that he was "glad as I had not known he was coming so things were just as they usually are."[58] Secretary of War Newton Baker also visited in mid-March, and Patton observed: "He is a little rat but very smart."[59]

Applications were pouring in from quality volunteers throughout the AEF. Still, "a Tank Corps without tanks is quite as exciting as a dance without girls."[60] He knew they were en route but did not know when they would actually arrive and champed at the bit in anticipation, grumbling that he felt "like an expectant mother waiting the advent of a child in the shape of a train load of tanks but thus far there have been no premonitory symptoms."[61]

When the first ten of twenty-five Renault light tanks finally arrived by train on March 23, Patton was awakened by his orderly, Pfc. Joseph Angelo, who excitedly informed him that he was needed at the railroad siding, where he soon drove each tank off the flatcars. He was the only one who knew how to drive a tank.[62] This handful of tanks were the only training vehicles used by the American tank force until shortly before their first combat six months later.

The Renault was a crude machine with a two-man crew, the driver and the tank commander, who also doubled as the gunner for the 37-mm cannon. It was capable of speeds of only four to five miles per hour. Nevertheless their arrival thoroughly excited Patton, who now had something better than a beat-up old truck to simulate a tank.[63]

The tanks of 1918 contained none of today's hi-tech instruments for communication between tanks or between members of the crew. Since it had no lights inside, the crews became proficient by learning to operate the tank and its guns blindfolded. The commander could only signal instructions to his driver by means of a series of kicks. As one of the original tankers (and Patton's lifelong friend), Harry Semmes, relates: "A kick in the back told him to start forward; a kick on the right or left shoulder told the driver to turn; a kick on the head was the signal to stop; repeated kicks on the head was the signal to back."[64] Thus, striking an enlisted man, normally a court-martial offense in the U.S. Army, became an accepted means of operating a tank.

Not until mid-September were there sufficient tanks for both Patton's tank battalions to train in simultaneously. He arranged with the French for his crews to practice driving at their trench-mortar school. The firing range was littered with shell craters and his crews practiced driving across this terrain to simulate what they would soon encounter on a real battlefield. The training was made more difficult by what seemed to be incessant rains and snow that lasted until well into spring. Rarely was there a day of sunshine.

Before long the school had outgrown the facility at Bourg and Patton was obliged to seek an overflow facility at nearby Brennes. And still new students continued to arrive in droves. "I am having a hard time putting them away but so far have managed to do it but it is like a sardine factory. Still I got a compliment out of it for Col R. told his adjutant that he could send them to me as I had never kicked yet. It is the old thing of the willing horse being ridden to death."[65] To his intense relief, Patton began to acquire assistance from men of engineering and mechanical experience, such as Capt. G. D. Sturdavent, who was the president of the Grant Six Automobile Company. Although not permanently assigned to Bourg, Sturdavent was of considerable help to Patton during his brief visit to the school in April. After several attempts Patton also succeeding in arranging for the transfer to the Tank Corps of Capt. Serano Brett, a Regular Infantry officer, whose specialty was the machine gun. Brett would soon become one of Patton's key subordinates and the commander of one of his tank companies, and later of a battalion.

Rare is the military unit that can claim it never bartered with another organization or facility and never made a "midnight requisition" for needed supplies or equipment that were unavailable through the supply system. The rate at which the tank center was burgeoning made bartering and "midnight requisitions" a virtual necessity, and, if he did not actively encourage initiative of this sort, the pragmatic Patton turned a blind eye to such activities by his staff. In mid-June he noted that he was "in a little trouble my self over some Pipe I 'stole' for the Center here. The Engineers are very mad at me and the Inspector General is coming down to investigate the affair I will probably get repramanded for cutting red tape but it ought not to hurt me as I am only guilty of too much initiative. Which is a quality often missing over here. Don't worry about me. The inspector is Olmstead who lived next to us at [Fort] Myer."[66]

Patton perceived that if the new Tank Corps was to succeed on the battlefield, new tactics would have to be developed, observing:

The British tactics of Cambrai were no longer applicable . . . we rightly contended that since Tanks were a supporting arm they should conform to the normal formations of Infantry and not demand that the Infantry should conform to theirs. . . . An investigation of French tanks also failed to appeal. The method they advocated placed the tanks behind the infantry reserve battalions in which position they followed placidly until the necessity for their intervention arose. In theory this was all right but in practice it demonstrated that a period of from one to two hours often elapsed between the need for the employment of the tanks and the time of their arrival. . . . Under the circumstances it seemed best to devise a system of our own. This we did and while it was far from perfect it worked.[67]

What Patton did was to create and teach a new doctrine whereby two of the three companies of tanks in the new light tank battalion were placed in the assault echelon and the remaining company in reserve. During the spring and summer of 1918 he taught these new tactics until each tanker thoroughly understood what the tanks were to do and how they were to function with the infantry.[68]

By April, Patton felt confident enough to risk a series of field exercises and demonstrations to test both the training of his officers and men and the operation of tanks in direct support of the infantry. On April 16, to test his theories, he held the first of a series of maneuvers, each of which he personally choreographed from directives he himself had written. Five days later ten Renault tanks participated in a simulated battle in support of an infantry battalion on a mock battlefield. He invited the commandant of the Army General Staff School at nearby Langres to send his students to observe the exercise. To ensure they understood what was occurring, he assembled a package of maps, diagrams, and a fact sheet about the tank for each of nearly two hundred officers. Patton checked himself out of the hospital in order to attend his own exercise. (He had developed a skin rash from wearing a gold chain sent to him by his mother, and had infected the rash with blisters by overmedicating it with iodine. Later he thought that the rash might have been due to nervousness.[69]

Despite a driving rain the exercise was a stunning success. Although one tank fell into a shell hole, Patton had kept one in reserve, which was immediately thrust into the attack and "everything went on fine." He had passed his first major hurdle with flying colors, and the only casualties were some of the visiting officers, who were thrown from their horses, which reacted with terror to the noise of artillery fire and simulated grenades. "They certainly are rotten riders," Patton chortled. He later inscribed a note on a memorandum: "I ran this show. It was the first Tank Maneuver ever held in US Army." It is difficult to argue with his conclusion that the demonstration had convinced a great many AEF officers of the importance and future value of the light tank on what would soon be an American battlefield. To Beatrice he wrote with justifiable pride: "I was realy more than pleased."[70]

He had been promoted to lieutenant colonel at the beginning of April, and five days after the successful exercise he was appointed to command the first U.S. light tank battalion, appropriately designated the 1st Light Tank Battalion. The advance from junior first lieutenant the previous May to lieutenant colonel was a remarkable one. "I feel more or less of a fool being a colonel. . . . How do you feel being a Mrs. Colonel," he asked Beatrice.[71] Behind the promotion was the hand of Pershing. "General P. had a hell of a time getting me promoted as they said I was too young but he finally put it over." Even before his promotion came through, Patton's dyslexia-driven need to put himself down clashed with what he knew to be his virtues. "If I am a lt Col. I have surely gone some and feel like an imposter though dangerous modesty is not one of my many faults."[72]

But becoming a lieutenant colonel was by no means the end of Patton's dramatic rise. By the time of his first battle some months later, Patton would be commanding more than a mere battalion in his date with destiny in the Meuse-Argonne. At long last the Patton name was to be put to the ultimate test.

CHAPTER 17

Baptism of Fire
The Saint-Mihiel Offensive

*Do ever in all things our damdest
And never oh never retreat.*

—PATTON[1]

The great thing about war is that it shows up character.
—JOHN MORTIMER[2]

By the summer of 1918 Patton despaired of ever receiving enough tanks in time to fight with the AEF when it was at last committed to battle. "One year ago to day," he wrote on June 13,

> we reached Paris full of desire to kill Germans. We are still full of desire but . . . some times I deeply regret that I did not take the infantry last November instead of the tanks. The regiment I had the chance to join has been at it now for five months. Of course I have done a lot but I keep dreading lest the war should finish before I can realy do any fighting. That would destroy my military career or at least give it a great set back . . . the unknown is always full of terrors and I wake up at night in a sweat fearing that the d—— show is over. . . . I trust that it is doing my character a lot of good for I keep at it inspite of constant difficulties and discouragements. But unless I get into a fight or two it is all wasted effort.[3]

On May 19 he was given permission by Rockenbach to visit the French front, and although there was no prospect of imminent combat, Patton had been contemplating his own mortality. In the event anything did happen to him, Patton penned a letter Beatrice was to open only in the event of his death. He entrusted it to Capt. Joseph W. Viner, one of his first (and finest) tank commanders.[4] It read:

May 20, 1918

Darling Beat:

I am leaving this letter with Capt Viner who will send it to you if he feels well assured I have been killed if I am not you will never see it. Of course if I am reported killed I may still have been Captured so don't be too worried.

I have not the least premonition that I am going to be hurt and feel foolish writing you this letter but perhaps if the thing happened you would like it. . . . Beatrice there is no advice I can give you and nothing that I could suggest that you would not know better than I. Few men can be so fortunate as to have such a wife.

All my property is yours though it is not much. My sword is yours also my pistol the silver one   I will give [the horses] Sylvia to Gen. Pershing and Simalarity to Viner.

I think that if you should fall in love you should marry again I would approve. . . . The only regret I have in our marriage is that it was not sooner and that I was mean to you at first. . . . It is futile to tell you how much I love you. Words are as inadequate as is love for a person like you.

If I go I trust that it will be in a manner such as to be worthy of you and of my ideals.

Kiss Beatrice Jr and Ruth Ellen for me and tell them that I love them very much and that I know they will be good.

Beat I love you infinitely.
George[5]

For nearly two weeks Patton sampled life in a French tank unit near the front lines, learning how tanks were employed. "I wrote it all down."[6] Even six miles away "you can hear the guns all the time but it is simply a con-stant roar as there are so many guns you can't distinguish any seperate explosion. . . . I hope to get up closer for a day or two and can then tell you how it feels when it is hitting nearer."[7] His French had improved dramatically, and he voiced his admiration for Frenchmen as soldiers. "Personally I like them much better than the British possibly because they do not drink Tea," which Patton described as "a most hellish and wasteful practice."[8]

With increasing frequency his letters articulated the void he felt had been created by his separation from Beatrice. "When we are together again I am going to drink a gallon of black coffee so I will stay awake and can make love to you. Oh! Beaty I miss you terribly and feel it more when I am idle and have more time to think. . . . I love you with all my heart."[9] In another he admitted that "all I try to do is for the effect it will have on you. In fact my attitude towards you is more that of a lover uncertain of his chances than of a husband."[10]

After several days leave in Paris in early June he wrote of his disgust at being so safe and sound: "When I see all the officers hanging round Paris doing little, I hate my self that I am not in the inft. for now we would be in action and at least doing something to stop the 'Bosch.'" He recovered the fragment from a German 150-mm shell that had fallen nearby and "had a piece of it cut out by a soldier in a Tank repair unit. He put your name on one side and the date May 30, 18 on the back. To day I took it to 'Cartier' and am having a gold chain put on it for a bracelet. It will be realy unique and is my own idea. I hope you will like it for it might have made you a widow."[11]

Patton concluded his letter with a hilarious account of a British officer and his wife in the room next to his in the Hotel Meurice.

> I know I should not have listened but I felt lonesome so I did. It was most amusing. They talked in a most impersonal way and every few minutes would stop and kiss. At last curiosity got the better of me and I looked through the key hole. They were both very properly clad in gray dressing gowns . . . and were sitting quite a ways apart eating crackers. . . . They had been separated only five months but she said it was years. I could not help thinking of us.
>
> Well they went to bed she in a nitie as thick as sail cloth he in canton flannel and in twin beds. At this point the thought occured to me, 'No wonder the Bosch beat the British if they are that cold blooded.' Soon however he asked her if she was tired. Foolish question! she said no but that he must be and that he should go to sleep at once.
>
> I feared for a moment that the cracker dirt had clogged his soul but was reasured on hearing him say that he was not at all tired at the same time there was much squeaking of beds and I felt assured he had gotten into hers. . . . Then all grew quiet again and I feared that they had gone to sleep. But soon my ears told me that I was in error. . . . Here I went to bed. It is sort of funny but verges on the tragic. . . . I hope you are not too shocked at my conduct. . . . I love you with all my heart and will never use twin beds or canton flannel pajamas when I see you.[12]

By August 1918 the staff gods in Washington had twice changed the designation of the school. Patton now commanded the 1st Brigade, Tank Corps, consisting of the 344th and 345th Tank Battalions.[13]

Patton's twenty-five training Renaults were mechanical nightmares, and his maintenance shop worked twenty-four-hour days in an often vain attempt to keep them running. The staff could count their blessings if there were as many as ten tanks in service, enough to train two platoons at one time.[14] Patton began emphasizing night training, which proved highly successful. He had taught his men well, and they responded by conducting night marches exactly as he had laid them out in his directives. But, as historian Dale Wilson points out in his landmark study of the Tank Corps, training had reached the point where it had become so repetitive that the men were in danger of losing the edge Patton had so carefully been honing. The training had hardened his 950 troops (50 officers and 900 enlisted men), who were now spoiling to put it to the test of battle.[15] In a sense Patton was the victim of his own success, for his men were "all dressed up with no place to go" and no tanks to take them there. When four of his officers were arrested for drinking in public with women, Patton grumbled: "We are getting full of virtue here. I don't think much of it. The French do as they please so why not we."[16]

Patton submitted to Rockenbach a paper he thought so "revolutionary" that it would modify U.S. tank tactics, but although it *was* innovative it was proffered at precisely the wrong moment. This was not the time to introduce another new set of tactics that would require more training, but rather a time to try out what had already been well learned. Whether or not it was intended as such, Rockenbach's reply—which Patton had the good sense to accept—was a politely worded reproof to leave well enough alone.[17]

Somehow Patton and three of his senior officers found time to attend the twelve-week course at the General Staff School in nearby Langres. The course was particularly demanding on Patton, who was obliged to return at night to Bourg to supervise his own school. Although he did not have the slightest interest in becoming a staff officer, he correctly perceived the General Staff School as a training ground that might enrich his experience. Among those attending were his West Point classmate Maj. William H. Simpson; Maj. John Shirley Wood, a 1912 graduate who was destined to become one of Patton's closest friends; and Capt. Joseph Stilwell, who had graduated from the Military Academy the year Patton entered and would later gain fame in Burma as "Vinegar Joe" Stilwell.[18] The school staff and the visiting speakers were all men who would later become famous for their achievements; among them were: Maj. Adna R. Chaffee, Gen. Hugh Trenchard, Major Alexander M. "Sandy" Patch, and Lt. Col. George Catlett Marshall.[19]

Despite the rigors of burning the candle at both ends, Patton miraculously found time to continue pouring out notes, directives, and lectures. One delivered to the AEF Line School was so well received that one officer informed him that "it was the best lecture he had ever heard by soldier or civilian. . . . I feel quite elated."[20]

Without warning the great wait suddenly came to an end. On the morning of August 20 Patton was still at the General Staff School when he was handed a tersely worded note reading: "You will report at once to the Chief of the Tank Corps accompanied by your Reconnaissance officer and equipped for field service."[21]

For more than a year Pershing had been forging the AEF into a fighting force that by the end of August 1918 exceeded 1,300,000 men. Troops were pouring into France at the rate of 10,000 per day. During the summer of 1918, Pershing had been injecting American units into the front lines to gain combat experience. The great German offensive in late May was thwarted barely fifty miles from Paris, at Château-Thierry on the Marne, by the U.S. 2d and 3d Divisions. At Cantigny the 1st Division won a small but important victory.

The Marine Brigade fought magnificently at Belleau Wood, where the French were beginning to fall back under intense German pressure. It was there that several immortal phrases became part of military lore. When a French officer ordered a marine officer to withdraw his unit he was bluntly informed: "Retreat, hell, we just got here." Instead the marines attacked several days later on June 6, 1918. One platoon was led by two-time Medal of Honor winner Gunnery Sgt. Dan Daly, who exhorted his men with words that would be repeated twenty-six years later at another place in France, called Omaha Beach: "Come on, you sons of bitches. Do you want to live forever?"[22] Belleau Wood and Château-Thierry became landmarks in the history of the U.S. Army and the U.S. Marine Corps.

The First Battle of the Marne in the summer of 1914 had left a large bulge in the form of a triangle some twenty-four miles wide and fourteen miles deep in the Allied front line between the Meuse and Moselle Rivers. It was centered on Saint-Mihiel, a town situated on a bend in the Meuse, about twenty-five miles southeast of Verdun. The bulge commenced eight miles east of Verdun, ran south to Saint-Mihiel, and then northeast to a point on the Moselle near Pont-á-Mousson. It was known as the Saint-Mihiel salient. It was the ideal place to baptize by fire an untested American army and in the process eliminate a long-standing threat to Verdun and a nagging and dangerous thorn in the Allied side.

On August 10 Pershing assumed command of the U.S. First Army, with its headquarters at Neufchâteau. Pershing and the French chief of staff, Marshal Ferdinand Foch, agreed that the proper place to employ First Army was in the Meuse-Argonne sector of western Lorraine, located to the west of

Verdun. They also concurred that the Saint-Mihiel salient must first be eliminated. Where they violently disagreed was over Foch's insistence that the
First Army be fragmented and much of it placed under French command.
Once again Pershing vehemently declined to break up an American force to
support a French attack in the Aisne sector, west of Verdun, but did accept
Foch's taunt that the time had come for the Americans to make good on
their promises. In early September he would prove it by launching an offensive to crush the Saint-Mihiel salient.

At the end of August the Saint-Mihiel sector was to become Pershing's
responsibility. By World War I standards it had been relatively quiet since
April 1915, when the French had suffered nearly sixty thousand casualties
in a failed attempt to recapture the salient. The front lines were about a half
mile apart, and although there was little infantry action, there were frequent
gas attacks and artillery bombardments. American troops had replaced the
French and early on learned the perils of the deadly poison gas. The salient
itself consisted of forested heights along the eastern banks of the Meuse that
descended into the Woëvre Plain, which contained a number of small lakes,
swamps, and woods. Through the salient flowed only one river of any consequence, the Rupt de Mad, a tributary of the Moselle.[23]

Among the units scheduled to participate in the offensive was George S.
Patton's newly formed tank brigade. The trouble was, the American tank
program was so bogged down in a maze of bureaucracy that no tanks had
yet been manufactured to fill the brigade's needs. Instead, in June the
French had promised but had yet to deliver 144 Renaults, enough to fill Patton's requirements. In addition, three American heavy tank battalions (using
150 British Mark Vs), then training in England, were being relocated to
France to add punch to the offensive, along with a French light tank regiment. In all three Allied tank brigades would be participating in the battle:
one heavy U.S. brigade, one French light brigade, and Patton's 1st Tank
Brigade. If all went according to plan, there would be more than 555 Allied
tanks in support of the Saint-Mihiel offensive.[24]

Patton left the tank center in the capable hands of newly promoted
Major Viner, while he and his reconnaissance officer, 1st Lt. Maurice H.
Knowles, traveled with Rockenbach and his chief of staff, Lt. Col. Daniel D.
Pullen, to Neufchâteau. Pullen would command the 3d Tank Brigade, a
mixed Anglo-French force, while Patton's 1st Tank Brigade was to consist
of the 344th and 345th Tank Battalions, plus twenty-four Schneider medium
tanks of the French IV Groupement [regiment]. At the last minute the
British backed out, leaving an enormous void in Rockenbach's plans.[25]

Because the Saint-Mihiel salient consisted mainly of a lowland plain of
clay soil that, after prolonged rains, turned into a virtual swamp that would
seriously inhibit the movement of tanks, Patton went forward to learn for
himself the nature of the terrain. During the night of August 21–22 he

accompanied a French patrol into no-man's land, describing the experience as "most interesting" but "not at all exciting." After crawling for well over a mile they came to the German outer wires where "the Bosch whistled at us and we whistled back and having seen what we wanted went home." For Beatrice, Patton "picked some dasies for you in the bosch wire and will send them when I get back. . . . I rather hoped we would have a patrol encounter but nothing happened."[26]

His firsthand report contradicted Rockenbach, who had reconnoitered the area in early 1918 and concluded it was suitable only for very limited tank use. Patton disagreed, arguing that unless there were heavy rains, he was certain that his tanks could get the job done. Patton's findings on his trip to the front evolved into a cardinal principle he was to employ for the rest of his military career: the "absolute necessity for a tank officer to personally see the ground" on which he was to fight.[27] He again made forays beyond the front line to survey the German front on which he would operate, and returned satisfied that the ground would support his tanks—*if* the rains held off.

The original plan was for Patton's brigade to support the V Corps attack from the northern corner of the salient. However, after days of detailed planning, he was suddenly summoned to the headquarters of the IV Corps and informed that his brigade would now be in support of the 1st Division (the Big Red One) and the 42d (Rainbow) Division.

Time was desperately short, and Patton had to scrap his plans and not only develop a new tactical battle plan but devise new logistical arrangements to get his tanks to railheads near the IV Corps front and establish fuel dumps near the front line. Although Patton managed to establish a ten-thousand-gallon fuel dump, he was denied vital oil and lubricants. Unfamiliarity with the Tank Corps resulted in extensive problems with staff officers in higher headquarters who had no concept of the unusual needs of the fledgling armored force. "One fatuous staff officer said that the French mud would lubricate the tanks' tracks."[28] The 42d Division had never trained with tanks; its commanders, while receptive to the idea of tank support, had received only the barest of briefings. When Patton requested that smoke be laid on as part of the preparatory artillery barrages to protect his tanks from German antitank guns, the 42d Division G-3 refused because it would have meant amending and reissuing the division fire plan. A furious Patton wrote in his diary: "The biggest fool remark I ever heard showing just what an S.O.B. [he is]." It took Rockenbach's personal intervention to compel the 42d Division to honor Patton's request.[29]

Patton's Renaults did not begin arriving at Bourg until August 24, and had to be serviced and prepared for the forthcoming offensive. With communications between tanks so tenuous, Patton designed an ingenious system of identifying each tank with markings in the form of playing-card suits:

spades, hearts, diamonds, and clubs. There were four platoons in each tank company, and each platoon was identified by one of the four suits. Each of the five tanks within a platoon was given both a suit and a number. Thus, the fifth tank in the "hearts" platoon of C Company, 344th Tank Battalion, became known as the Five of Hearts.[30]

To prepare, move, and join men and tanks at the right time and place is exceptionally difficult under the best of conditions. It had never been done before, and Patton's problems were increased by the necessity of moving his tanks the relatively short distance from Bourg to the front via the French railways. Although the offensive was postponed until September 12, the last tank did not detrain until two *hours* before H hour. With barely ten minutes to spare they moved into position, where they were greeted by Patton. Even though most had been without sleep for two nights, the tankers' concern was not sleep but whether or not they would miss out on the coming battle. Some cried out, "Oh, Colonel, can't you make them wait for us?" And, "Will we be in time?"[31]

Patton's custom of writing letters to raise the morale of his troops began at Saint-Mihiel, where he issued his final order, in which he exhorted his tankers to champion the future of the Tank Corps by accomplishing their mission on the battlefield. To those familiar with Patton's later career, his words bore a close resemblance to his speeches to his troops during World War II. "No tank is to be surrendered or abandoned to the enemy," he reminded his men:

> If you are left alone in the midst of the enemy keep shooting. If your gun is disabled use your pistols and squash the enemy with your tracks . . . remember that you are the first American tanks. You must establish the fact that AMERICAN TANKS DO NOT SURRENDER. . . . As long as one tank is able to move it must go forward. Its presence will save the lives of hundreds of infantry and kill many Germans. . . . This is our BIG CHANCE; WHAT WE HAVE WORKED FOR. . . . MAKE IT WORTH-WHILE.[32]

The dreaded rainy season had commenced on September 8, and by D day it had been raining incessantly for nearly four days. Beginning at 1:00 A.M., four hours before H hour, 2,800 Allied guns began hurling shells into the German lines and lighting up the skies—just as the rains turned into a torrent. When the ground attack commenced at 5:00 A.M., Patton was sitting on a hillside overlooking the front line. "When the shelling first started I had some doubts about the advisability of sticking my head over the parapet, but it is just like taking a cold bath, once you get in it it is all right. And I soon got out and sat on the parapet."[33] Ahead of him lay the pandemonium of the battlefield.

Although Patton had planned down to the smallest detail what each of his tanks would do on September 12, Murphy's law—the first rule of warfare—quickly came into play. At Saint-Mihiel the lack of training and cooperation between tanks and infantry led to untold problems. Less than thirty minutes after H hour, Major Brett's tanks, whose mission was to lead the 1st Division infantry, had advanced so fast that the infantry was nowhere to be seen. On the right flank, Capt. Ranulf Compton's 345th Tank Battalion was to follow closely behind the infantry of the 42d Division and then lead the attack on the villages of Essey and Pannes. Instead Compton's tanks encountered great difficulty breaching the main German trench line. The trench works were fourteen feet wide in places and the rain-soaked muddy banks surrounding them were insurmountable barriers for the majority of the battalion's tanks. Two tanks were knocked out by direct hits from the heavy German shell fire that pounded the 345th Tank Battalion and the lead brigade of the Rainbow Division. By approximately 9:30 A.M., when Patton arrived, only five had managed to reach the outskirts of Essey.[34]

Patton was intensely disappointed when the German infantry failed to put up much of a fight, but the battlefield nevertheless remained a very dangerous place from ground and artillery fire. In anticipation of an Allied offensive, the Germans had decided to withdraw to the area behind the Woëvre Plain and permit the Allies to advance into the plain before counterattacking. However, three days before the offensive the Germans learned that the salient was to be attacked from two flanks, and they decided to begin a phased withdrawal. By H hour two divisions were already withdrawing.[35]

Despite the rain and foggy conditions that pervaded the battlefield the morning of September 12, what Patton could see from the relative safety of his hillside perch was that most of his tanks seemed to be stalling in the trenches that crisscrossed the German front lines. "It was a most irritating sight." Patton's telephone wire had run out, and at this point he faced a dilemma.[36] Should he violate Rockenbach's explicit orders to remain in radio contact with his headquarters or move ahead into the unknown reaches of the battlefield? At 7:00 A.M. Patton simply could not resist the lure of the battlefield any longer. He left his adjutant in charge of the telephone and, calmly smoking his pipe, entered the battlefield on foot with Lieutenant Knowles and four runners. He passed some dead and wounded, but when he saw one American sitting in a shell hole holding his rifle, Patton thought the soldier was malingering and went to "cuss him out," only to discover he was dead from a bullet in the head.[37]

In his letters and after-action reports Patton described in detail how

the whole country was alive with [tanks] crawling over trench[es] and into the woods. It was fine but I could not see my right battalion so went to look for it, in doing so we passed through several town[s] under shell fire but none did more than throw dust on us. I admit that I wanted to duck and probably did at first but soon saw the futility of dodging fate, besides I was the only officer around who had left on his shoulder straps and I had to live up to them. It was much easier than you would think and the feeling, foolish probably, of being admired by the men lying down is a great stimulus."[38]

Near the village of Maiserais, Patton encountered his French battalion commander, Maj. C. M. M. Chanoine, who was directing repair work on one of his tanks. As Patton began walking away, a German 150-mm shell struck the tank, killing or wounding the entire crew. Major Chanoine was knocked unconscious but miraculously recovered and was able to carry on.

Patton's journey next took him to the outskirts of the village of Essey, where the 84th Brigade of the 42d (Rainbow) Division seemed to be stalled. The brigade commander was Brig. Gen. Douglas MacArthur, who was personally leading his infantrymen. "I walked right along the firing line . . . they were all in shell holes except the general . . . who was standing on a little hill. I joined him and the creeping barrage came along toward us, but it was very thin and not dangerous."[39] Neither officer had much to say to the other. Patton recorded: "Each one [of us] wanted to leave but each hated to say so, so we let it come over us." Such is the nonsense written about Patton that one of MacArthur's biographers describes their meeting as an inevitable "macho duel," during which "Patton flinched at one point and then looked annoyed with himself, whereupon the brigadier said dryly, 'Don't worry, major [sic]; you never hear the one that gets you.'"[40] Their meeting that morning was wholly by chance, and the only thing on Patton's mind was to uphold the honor of the Tank Corps in the same manner in which he had ordered his officers to behave. Upstaging MacArthur was not on his agenda. However, as one writer has observed: "One properly placed German shell at this moment in World War I would have eliminated two major, inspiring and controversial figures of World War II."[41]

It was now nearly 10:00 A.M., and on the road on the northern edge of the village of Essey there came the first real test of Patton's courage since the shoot-out in Mexico more than two years earlier. The bridge spanning the Rupt de Mad was reported to have been mined by the Germans, even though American engineers were unable to detect any evidence of telltale wires. In his official report, Patton described his brief discussion with MacArthur, in which he had asked "if we could move the tanks forward across the bridge at Essey which contrary to expectations was found intact. He gave his consent in [that] the bridge was not mined. We walked over the bridge in a most cat-like manner expecting to be blown to heaven any moment but to our great relief found that the bridge had not been tampered with."[42]

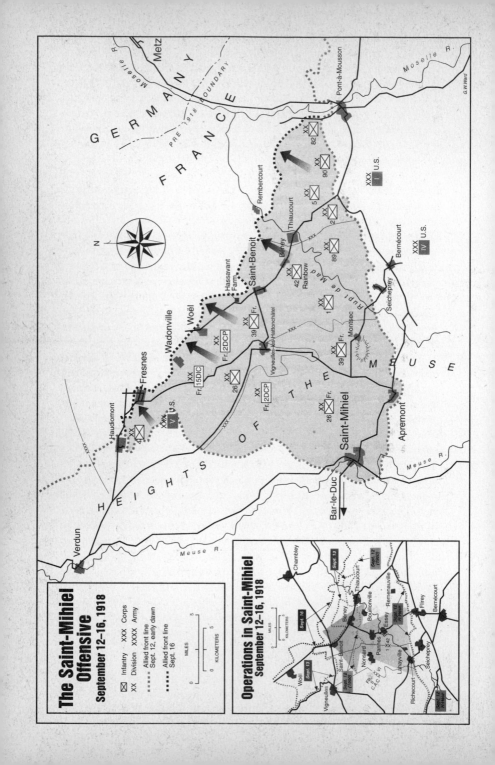

The Saint-Mihiel Offensive
September 12–16, 1918

⊠ Infantry	XXX Corps
XX Division	XXXX Army

•••••• Allied front line Sept. 12, early dawn

- - - - Allied front line Sept. 16

MILES
0   5

KILOMETERS
0   5

Operations in Saint-Mihiel
September 12–16, 1918

MILES
0   5

KILOMETERS
0   5

It is another measure of the Patton myth that an accomplished historian and biographer, William Manchester, would write of how deeply the scenes at Essey affected MacArthur, saying that "it was that vein of compassion which set him apart from the Pattons of the army. He could be ostentatious and ruthless, and . . . a killer. Yet his attitudes toward war would always be highly ambivalent, exulting in triumph while pitying the victims of battle."[43]

Lack of compassion was not part of Patton's personality, and he was as deeply affected by the horror of war as MacArthur, or any other battlefield commander. Some time prior to Saint-Mihiel, Patton confided to Beatrice his first impressions of war. Daughter Ruth Ellen writes:

> The war was all around him when he wrote Ma a letter, which shows a side of him that she always saw, but that few others, outside his immediate family, ever knew existed. He wrote to her that he had been inspecting a battlefield at night, and that the dead soldiers, as yet unclaimed by the burial teams, were lying there in the moonlight. He said it was hard to tell the Americans and British from the Germans, and they all looked alike—very young and very dead—and he began to think how often their mothers had changed their diapers and wiped their noses, and suddenly the whole concept seemed unbearable, and he decided that the only way to survive under such a stress was to try to think of soldiers as numbers, not as individuals, and that the sooner the allies won, the sooner the slaughter of the innocents would cease. However, no matter what he said, he could never quite do that. To him his men were individuals, people and responsibilities, always.[44]

For Patton, poetry was both a means of cheering himself up and of inspiring himself. "He was always worried—as he had been that long-ago day at Fort Sheridan when he had stood on the target butts—that he would not be able to face the song of the bullet that had his name on it. He told us . . . that you never heard the bullet that killed you as the missile travels faster than the sound."[45] At such times, it was not uncommon for Patton to express himself through poetry, as he did in a poem called "The Moon and the Dead."

The roar of the battle languished
The hate from the guns grew still,
While the moon rose up from a smoke cloud
And looked at the dead on the hill.

Pale was her face with anguish
Wet were her eyes with tears,
As she gazed at the twisted corpses
Cut off in their earliest years.

Some were bit by the bullet,
Some were kissed by the steel,
Some were crushed by the cannon
But all were still, how still!

The gas wreaths hung in the hollows
The blood stink rose in the air,
And the moon looked down in pity
At the poor dead lying there.

Light of their childhood's wonder,
Moon of their puppy love,
Goal of their first ambition
She watched them from above.

Yet not with regret she mourned them
Fair slane [slain] on the field of strife,
Fools only lament the hero
Who gives for faith his life.

She sied [sighed] for the lives extinguished
She wept for the loves that grieve,
But she glowed with pride at seeing
That manhood still doth live.

For though the moon is winsome
In wisdome she is old
Nor grieves she for the fallen
Nor grudges she the bold.

Her years are for the hero
Her hate is for the cur
Her utter loathing for the hound
Who shrinkes from righteous war.

The moon sailed on contented
Above the heapes of slane,
For she saw that honor liveth
And manhood breathes again.[46]

In Essey, MacArthur saw "a sight I shall never quite forget." The rapid
American advance forced the Germans to evacuate the town in such a hurry

that they had abandoned a battery of guns, the complete instrumentation and music of a band, and an officer's horse, which was found saddled in a barn. A number of French civilians, mostly elderly men, also emerged from cellars. They had been there during the four years of German occupation and had no idea that the United States had even entered the war. MacArthur had "to explain to them that we were Americans."[47]

Although several Germans surrendered to both Patton and MacArthur, Patton himself did not remain in Essey but after receiving MacArthur's permission, moved his tanks forward toward the next village, called Pannes. After more than three hours on the move, Patton, Lieutenant Knowles, and one of his runners, a sergeant named Graham, were too exhausted to walk any farther and hitched a ride on one of the three remaining tanks. The road between Essey and Pannes was littered with dead Germans and horses, the remains of what had once been a German artillery battery. Two of the tanks soon ran out of gas, and as Patton's tank reached the outskirts of Pannes, the infantry, which had been following them, apparently frightened by the horrific sights of dead men and animals, refused to enter the town.[48]

To reassure the nervous infantrymen Patton explained that his tank would lead the way into Pannes. The sergeant commanding the tank was equally apprehensive, and Patton told him he would accompany the tank. He perched himself atop the Renault, while Lieutenant Knowles and Sergeant Graham sat on the tail of the tank. "I watched one side of the street and they the other."

Patton's official after-action account describes what happened:

> We continually expected to be shot off our precarious perch. At the north end of the town we saw one German and Lieut. Knowles and Sgt. Graham, the runner, got off the tank to affect the capture. To their great surprise they found thirty instead of one but using their pistols they captured the entire crew. The tank continued out the northern end of the town . . . Colonel Patton, who was still sitting on the top of the tank, here had the most horrible experience; he could hear machine gun fire but could not locate them until glancing down the left side of the tank about six inches below his hand he saw the paint flying from the side of the tank as the result of numerous machine gun bullets striking against the tank. Owing to his heroic desire to make the tank a less enticing target he leaped from the tank and landed in a shell hole a great distance away. This shell hole however was exceedingly small and the Germans took an unpleasant delight in shooting at its upper rim.[49]

Not only had German gunners zeroed in on him, but the tank on which he had been riding continued on, unaware that Patton was no longer a pas-

senger. The nearest infantry were about two hundred meters away, at the
edge of the village. As Patton wrote:

> He was in a great state of perplexity . . . if he moved backwards and con-
> ducted a strategical withdrawal the Infantry would think a tank officer
> was running away; should he move forward he would become a distinct
> target of the four machine guns which he was now able to see about 500
> meters to his front. He finally solved the problem by moving sideways
> until he regained the Infantry. During the course of this movement he was
> repeatedly forced to seek shelter in small shell holes.[50]

The infantry commander adamantly refused Patton's request to move
his troops forward. Patton then asked if he would send a runner to the tank
which was cruising about in the open, some 500 yards to their front. "To this
request the heroic Infantry made this reply, 'Hell no, it aint my tank.' " Patton's only choice was to once again expose himself.

To his father he wrote:

> I drew a long breath and went after the tank on foot as I could not let
> it be going against the whole town alone. It is strange but quite true that at
> this time I was not the least scared, as I had the idea of getting the tank
> fixed in my head. I did not even fear the bullets though I could see the
> guns spitting at me. I did however, run like H———. On reaching the tank
> about four hundred yards out in the field I tapped on the back door with
> my stick, and thank God it was a long one. The sergeant looked out and
> saluted and said "what do you want now Colonel." I told him to turn and
> come back   he was much depressed. I walked just ahead of him on the
> return trip and was quite safe.

The Germans could be seen in the distance retreating to the north,
apparently scared off by the presence in Pannes of Patton's five tanks. The
tanks again moved forward, this time accompanied by the infantry, until one
of the tanks began mistakenly firing on an American machine-gun emplace-
ment. When Patton asked the machine-gun officer to apprise the tank of its
mistake, the officer told him off. Again Patton had to act as a runner, but
"this time however he did so with less speed since there was very little fire.
This third advance of the tanks finally proved successful."[51]

Patton had barely slept in four days, and after losing his rations was so hungry that he ate some crackers taken from a dead German. "They were very
good but I would have given a lot for a drink of the brandy I had had in my
sack." (His rations had been stolen by some German POWs. When Knowles
and Graham captured the Germans in Pannes, Patton was giving instructions

for the next attack, on the town of Bény. He handed his haversack to Sgt. Graham, who was guarding the POWs. When Graham left for a few moments to capture another nearby German, the Germans emptied Patton's haversack and filled it with stones. Patton only discovered the loss some hours later.)[52]

With things well in hand at Pannes, Patton began walking toward Brett's battalion on the left flank, where he found the tanks of the 344th Battalion out of gas and Brett in tears from exhaustion and bleeding from a slight bullet wound in his nose. "I comforted him and started home alone to get some gas. It was most interesting over the battle field like the books but much less dramatic. The dead were about mostly hit in the head. There were a lot of our men stripping off buttons and other things but they always covered the faces of the dead in a nice way." As for fear, "Vanity is stronger than fear . . . [in] war as now waged there is little of the element of fear," he informed Papa.[53]

A wet, muddy, and exhausted Patton encountered an officer in a staff car, apparently rubbernecking. He gave Patton a lift, but the road was filled with infantrymen trudging forward, and their automobile was soon stuck behind them. German airplanes had been active in the Saint-Mihiel sector all day, and when one spotted the congestion it dropped a bomb that killed two soldiers walking directly behind the car in which Patton and the officer were riding. His luck continued to hold, and he had proved "to my own satisfaction that I have nerve. I was the only man on the front line except gen mcArthur who never ducked a shell. I wanted to but it is foolish as it does no good. If they are going to hit you they will."[54]

The results of Patton's training methods were immediately evident in the teamwork and heroics of many members of his brigade, among them Capt. Harry Semmes. As his tank was crossing the Rupt de Mad, it suddenly sank, becoming completely submerged. Semmes managed to escape through the turret and was swimming ashore when he remembered that his driver was trapped inside the tank. Although under fire from German infantry in a nearby trench, Semmes swam back to the tank, dived into it, and dragged the driver out by his ears. The two men then swam ashore and killed one of the Germans who had been shooting at them.

First Lt. Julian K. Morrison, a platoon leader in Company A, came under fire from a German machine-gun nest situated in a nearby woods. Unable either to attack it with his tanks or reach it with his guns, Morrison dismounted from his tank and attacked the German gunners on foot with only his pistol. Although twice wounded, he persisted and captured the guns. Morrison later observed that it was made clear during his training that a Tank Corps officer was expected to die, if necessary, to accomplish the mission. During his lectures Patton had instructed his officers that they must "go forward, go forward. If your tank breaks down go forward with the Infantry. There will be no excuse for your failure in this, and if I find any tank officer behind the front line of infantry I will—." Given Patton's penchant for colorful language, no one who heard him was in any doubt as to his meaning.

Early on many of Major Brett's tanks had broken down or stalled in the rubble and barbed wire of the German trench works. However, the deadliest enemy of the tank was the boglike ground, which one of Patton's repair and salvage officers described as, "sticky, soggy, awful mud in which the tanks wallowed belly deep."[55] After three tanks had broken down, including the one he was commanding, an annoyed and frustrated Brett got out of his tank and, on foot, led those that remained for several kilometers, all the time under machine-gun and rifle fire. Brett had, noted Patton, set "a fine example of coolness and courage to all his command."[56] Later Brett shot a German machine-gun crew from the steeple of a church in the village of Nonsard. His three company commanders had all distinguished themselves by directing their tanks' advance, even though constantly exposed to enemy fire. Patton recommended both Brett and Semmes for the Distinguished Service Cross for their bravery on September 12.

As for Patton, his exploits on September 12 went unrecognized. Douglas MacArthur won his fifth Silver Star for gallantry at Essey; George S. Patton was severely admonished by his commanding officer when an irate Rockenbach learned where he had been that day. Rockenbach told Patton the night before the offensive: "There is no question of personal courage in this war; it is a business proposition where every man must be in his place and performing his part. Keep control of your reserve and supply, you have no business in a Tank and I give you the order not to go into this fight in a tank." As Rockenbach told a postwar audience: "Patton obeyed his order, but saw his duty to go in the fight on *top* of a tank."[57]

Although Patton could claim to have carried out his order, Rockenbach was understandably furious that he had obviously violated its intent. A brigade commander's place, he scathingly informed Patton, was not in the front lines but at his command post, where he could direct the battle and, not incidentally, be reached by his higher headquarters. For Patton to have been wandering around on the battlefield was so grossly irresponsible that he was seriously considering relieving him of command for insubordinate behavior. Rockenbach reinforced his displeasure with a letter spelling out a number of points about what tank officers were expected to do in combat. For all his aggressiveness and bluster, on those occasions when he was called on the carpet and knew he was in severe trouble for some serious infraction, Patton inevitably became calm and apologetic and usually managed to defuse the situation with eloquent *mea culpa*s and promises to behave himself in the future. He was diplomatic enough to sense when contrition, not belligerence, would soothe troubled waters. It usually sufficed. In this instance it did not hurt when, several days later, Pershing wrote to commend Rockenbach for the performance of his tanks at Saint-Mihiel. Rockenbach had little choice but to endorse the letter to Patton and his men with similar congratulations on their good work.[58]

It should be noted that during World War I, command and control were more theoretical than practical. Radio communications between commanders and units were nonexistent, and the use of pigeons to communicate was hardly a prescription for any commander to keep a successful grip on a battle. It was problematical how much control Patton could have exercised from a command post in the rear. Thus, in spite of of the worst tonguelashing he had ever received, Patton was unmoved and utterly unrepentant, even though Rockenbach was correct in pointing out that by abandoning his post in the rear he had effectively cut himself out of the chain of command.[59] "General R. gave me hell for going up but it had to be done. At least I will not sit in a dug out and have my men out fighting." In future operations Rockenbach ordered Patton to remain at his brigade command post. Although Patton took steps to ensure that there were better communications with Rockenbach during the next offensive, he still had no intention whatsoever of remaining in the rear while his brigade was in action.

Patton had thoroughly indoctrinated the Tank Corps in his brand of aggressive leadership. To a man his officers were resolved never to fall behind their infantry, no matter what. In his account, Lieutenant Morrison later explained:

> In the Saint-Mihiel drive Tankers could be seen any where from one to seven kilometers in front of the infantry. Everyone fought—cooks, company clerks, mess sergeants, runners and mechanics. So closely was the order that the Tank Corps nearly starved for two or three days afterward. Needless to say before the next fight orders came out to the effect that anyone leaving the post assigned to him would be dealt with by Court Martial ... [however] the courage of the Tank Corps ... [was] sufficiently proved.

Although Patton was momentarily in Rockenbach's doghouse, he had succeeded beyond expectation in inspiring the officers and men of the Tank Corps to the very highest standards of behavior on the battlefield. He taught them; they listened and achieved. Rockenbach's point notwithstanding, inasmuch as this was the first-ever day of combat for the untested U.S. Army Tank Corps, the example Patton set by being seen on the battlefield was of far greater importance than anything he could have done had he remained behind the lines. In terms of inspiration alone the value of his presence was incalculable, and sent the message that Patton practiced what he preached.

At Cambrai the British commander, Gen. Hugh Elles, believed strongly that it was important for him to set the example in their first tank battle, which he did by riding in the lead tank with his battle flag displayed from the turret. Colonel Fuller, his chief of staff, thought it a bad idea but soon changed his mind. "Tankers belong in tanks," he said.[60]

Although the offensive did not officially end until September 14, the Saint-Mihiel salient had been eliminated by dawn on September 13. The Germans had withdrawn to the Hindenburg Line, and although the First Army had to repulse several counterattacks during the night of September 13–14, for all practical purposes the Saint-Mihiel offensive had succeeded on its first day. The 1st Tank Brigade played only a minor role in the offensive after September 12. Most of Brett's tanks were still out of gas, and the only resupply was from sleds towed by some of the tanks. Gas trucks bearing the precious fuel were mired in traffic jams on roads leading to the front. A few tanks were refueled by draining the tanks of others for whatever gas could be obtained. By midday both battalions were refueled and ready to roll. Compton's battalion advanced to St. Benoit, while fifty of Brett's tanks advanced toward the village of Vigneulles, four miles to the west. For the second day in a row, Patton did a good deal of walking around the Saint-Mihiel battlefield.

At 5:00 A.M. on September 14, Brett personally reconnoitered the area around Wöel on a captured German motorcycle in hopes of locating elements of the missing 1st Division. He returned empty-handed and with the knowledge that his battalion was in no-man's land, where there were no troops to be seen from either side. Patton arrived at about 6:00 A.M. and with no infantry to support, decided to use Brett's tanks to make a reconnaissance into no-man's land. On the road to Wöel, Patton encountered Pershing's intelligence officer, Brig. Gen. Dennis Nolan, and informed him that Brett's battalion was "looking for a fight and the First Division." Nolan, who had just come from Wöel, told him that the town had been newly evacuated by a German battalion.

Four men on captured horses were sent out in different directions in an attempt to locate friendly infantry, to no avail. At noon Patton ordered Captain Semmes to send a reconnaissance patrol toward Wöel to again search for any sign of American troops. Near Jonville, the patrol, consisting of three tanks and five infantrymen, under the command of Lt. Ted McClure, ran into a retreating German infantry unit and a 77-mm artillery battery and became embroiled in a fierce firefight. When the German artillery began to turn their guns into direct-fire weapons in an attempt to destroy the tanks, McClure led a cavalrylike charge and succeeded in capturing not only the German 77s but a quantity of machine guns. The infantry were routed.

The tanks soon came under German artillery fire, and McClure and two of his men were wounded by shrapnel. When two of his tanks broke down, the enterprising young lieutenant refused to abandon them. Instead he coupled all three together, and the lone serviceable tank began towing the other two toward friendly lines. He sent one of the wounded men to the rear to request assistance, and five tanks were dispatched to reinforce what today would have been called Task Force McClure. All returned safely.[61]

McClure's courageous performance that day helped establish the credibility of the Tank Corps and a tradition of excellence that has characterized the performance of army tankers in the years following its re-creation as the Armored Force in 1941. It was the valor of men like McClure, Brett, and Semmes that more than justified their new nickname, the "Treat 'Em Rough" boys.[62]

This final tank action of the Saint-Mihiel offensive earned high praise from Patton for being the leader of "the only [known] successful operation of tanks absolutely unaided by other troops in attacking and routing an enemy."[63] More important, it sowed the seeds of what was to become Patton's trademark employment of armor in World War II—the deep penetration. Lieutenant McClure's exploits so impressed Patton that he later told Semmes that he believed his tanks could have slashed into the German rear if only McClure's force had been larger. He vowed that the next time such an opportunity presented itself he would employ a larger number of tanks and aim for a genuine breakthrough.[64] What Patton had begun to envision went far beyond the officially prescribed mission of the Tank Corps as a support arm of the infantry. He had grasped the enormous potential of the tank as a potentially decisive factor on the modern battlefield. It was the concept of mobile warfare, and it had begun with what became barely a footnote in the history of the battle of Saint-Mihiel.[65]

During the interwar years Patton would refine his ideas and by 1928 would state that with the advent of the tank and the airplane there was now a solution to the problem of delivering the coup de grace to an enemy force. "Such a [tank and aerial] force could be used in a manner analagous to that employed by Napoleon with his heavy cavalry. The tanks and attack planes or a large proportion of them should be held as a reserve to be used after a general battle had developed the enemies plans and sucked in his reserves. Then at the predetermined time and place this force should be launched ruthlessly and in mass."[66]

The battle of the Saint-Mihiel salient lasted barely thirty-six hours, but for the fledgling AEF, its first offensive was a stunning triumph that had in a single day erased what the Germans had held for four years, and in the process reaffirmed what had already been proven at Belleau Wood and Château-Thierry: that American troops were the equal of any who fought on the battlefields of France. In his postwar report Pershing would write that "the allies found that they had a formidable army to aid them, and the enemy learned finally that he had one to reckon with."[67]

MacArthur argued in vain to his superiors that the success in the Saint-Mihiel salient was a golden opportunity to have captured Metz and turned the German left flank, an argument later taken up by the British historian and military thinker Basil Liddell Hart. Others, like the I Corps commander, Maj. Gen. Hunter Liggett, thought that "the possibility to tak[e] Metz . . . existed only on the supposition that our Army was a well coordinated machine, which it was not, as yet!"[68] One of the German commanders defending the Saint-Mihiel salient said: "I have experienced a good many things in the five years of war and have not been poor in success but I must count the 12th of September among my few black days."[69] German casualties included fifteen thousand POWs and the loss of 257 guns; American losses numbered seven thousand.[70]

Patton was disappointed by the Battle of Saint-Mihiel. He felt that the minimal resistance of the Germans had not resulted in a true test of the ability of the tank as a fighting machine. Losses were minor, two tanks lost to direct hits by German fire, forty stalled in trenches and thirty out of gas. Five tankers were killed and nineteen wounded, four of them officers. "The great feat the tanks performed was getting through at all," he explained to Beatrice. "The conditions could not have been worse."[71]

Overall, "as a fight, this operation was not very decisive to the Tanks, but as an exploit in mechanics, driving and endurance it is unequalled. Tanks designed to cross six-foot trenches were made to cross trenches from ten to fourteen feet wide, and not only one but trench after trench."[72] Of the 174 tanks in action, 3 took direct hits and 43 others were out of action due to mechanical trouble or trapped in trench works. On September 16 Patton was able to report 131 tanks fit for action, a remarkable achievement, given the conditions encountered on the battlefield.[73]

Despite his own feelings and Rockenbach's annoyance with him, Patton had every reason to feel pleased with Saint-Mihiel. First, and most consequential, he had overcome the fear of combat that every soldier endures and had proved beyond any doubt his personal courage. And, while his tanks had not acquitted themselves particularly well in their first test of battle, his men had. Mechanical and tactical problems can be fixed; repairing leadership deficiencies is a lot harder. Patton had clearly imbued the officers and men of the Tank Corps with his brand of inspired leadership. Moreover, as Dale Wilson notes, until Saint-Mihiel: "Patton had adhered to the idea that tanks were strictly an infantry support asset." But the battle "provided him with a vision of what more mechanically advanced tanks might be able to accomplish on future battlefields operating as an independent combat arm. Patton also exhibited that rare ability to adjust quickly to a rapidly changing situation on the battlefield. This trait would later become a hallmark of his World War II operations."[74]

By the standards of World War I, the Saint-Mihiel offensive was both brief and relatively minor. Nevertheless sixteen American divisions and one French colonial division, consisting of more than 650,000 men, battled German Composite Army C and liberated more than two hundred square miles of the German-held salient.[75]

Now, in a mere ten days, the First Army was under orders from Marshal Foch to shift its operations from Saint-Mihiel to the Meuse-Argonne, where a fresh offensive was to be launched on September 26. Patton's 1st Tank Brigade was slated to play an important role in that offensive.

CHAPTER 18

Valor Before Dishonor

The Meuse-Argonne

His date with destiny, so long anticipated
and dreaded, came on September 26, 1918.

—RUTH ELLEN TOTTEN

The grand Allied design for the early autumn of 1918 called for a series of offensives across the entire front, to be kicked off on September 26 with the U.S. First Army and the French Fourth Army initiating attacks in the Meuse-Argonne. The First Army was to attack from the south across a twenty-four-mile front, from the Argonne Forest to the Meuse River, with the French launching their offensive to the west, on the American left flank. In subsequent days massive Allied attacks would commence, with British, Belgian, French, Australian, and American forces attacking to crush the Hindenburg Line in the west.

The problems facing Pershing between the end of the Saint-Mihiel battle and the commencement of the Meuse-Argonne offensive were staggering. First, 220,000 French troops had to be removed from the area before American troops could be positioned but, to deceive the Germans, only at the very last moment before the offensive began. The burden of planning fell on (now) Col. George C. Marshall, the First Army G-3, who was responsible for moving more than 500,000 American troops into position in the Argonne, along with more than 2,000 guns and 900,000 tons of ammunition and supplies, along only three roads.[1] To avoid detection, troop units had to march toward Bar-le-Duc, a town on the Marne some twenty-five miles southwest of Saint-Mihiel, before turning north on one of the few roads that led toward the sparsely populated forests of the Argonne.

Those units in the Saint-Mihiel sector, such as Patton's tank brigade, had to be moved sixty miles during the rainy season. Marshall's well-laid plans proved impossible to carry out: "They broke down almost from the beginning. Thousands of the 90,000 horses that were hauling supplies through the waterlogged country near the Meuse collapsed or died in their traces, causing monumental traffic jams. In the almost constant drizzle, engineers worked tirelessly with rocks and gravel, mud and logs to repair mired roads."[2] Night after night men sweated and labored in the mud and rain on roads scarcely fit for a horse and buggy.

A plan was concocted to deceive the Germans into believing that Pershing intended to continue the offensive in the former Saint-Mihiel salient to either capture Metz or drive eastward deep into Alsace. The deception worked well, leaving the Germans convinced that the movement of American units was to form a new army to be located between Saint-Mihiel and Verdun. Part of the plan included a demonstration near Pont-à-Mousson by fifteen tanks of Compton's 345th Tank Battalion. The obvious but false conclusion the Germans were to draw was that Patton's tanks were part of a large American force being assembled to launch an offensive in that sector. During the early evening hours of September 22, the tanks motored into no-man's land just long enough for the Germans to learn of their presence but not calculate their actual strength.[3]

The terrain over which the First Army would attack strongly favored the Germans, who had used it skillfully to establish three defensive belts in front of the main (fourth) Hindenburg Line, behind which was yet another line, the Freya Stellung. To successfully penetrate these defenses, an attacker would have to run a gauntlet of fire that was dominated in the center of the sector by the heights surrounding the town of Montfaucon. The Argonne Forest was a thousand feet above sea level and consisted of a series of deep ravines and bluffs that were heavily defended, while on the right flank the Meuse River provided an unfordable barrier. As the First Army chief of staff, Brig. Gen. Hugh A. Drum, would later state: "This was the most ideal defensive terrain I have ever seen or read about."[4] The only positive aspect was that the twenty-four-mile front was held by only five German divisions. Like Saint-Mihiel, the Meuse-Argonne sector had been largely inactive since the first year of the war and for some time had been considered a place where battle-weary troops from both sides were sent to rest and recuperate. Parties of Germans and French could frequently be seen squirrel hunting in no-man's land, where the corpses of soldiers killed in 1914 still lay where they had fallen, macabre skeletons covered by uniforms of blue or gray.[5]

To compensate for the miserable terrain and the German ability rapidly to reinforce the Argonne front with another fifteen divisions, Pershing and his staff drew up a bold plan that relied on surprise and a preponderance of forces. Numbering 250,000, the assault troops were to crush German resistance quickly in the first two defensive belts so that the First Army would advance the ten miles to the Hindenburg Line within twenty-four hours, thus preventing German reinforcements from being moved forward. Nine American divisions would simultaneously assault at H hour on September 26. Marshal Pétain, the French commander in chief, believed that if surprise and boldness did not pay off at once, the Americans faced the likely possibility of a winter stalemate before the second defensive line.[6] It was a risk that Pershing believed had to be taken, and, as historian John Toland has written, "probably only Pershing among Allied commanders would have dared to take the risks he faced," particularly since his only veteran troops were still disengaging from the Saint-Mihiel salient, and four of the nine assault divisions were untested in battle.[7] In short, the risks were enormous, but so too were the potential rewards.

The 1st Tank Brigade was spared the chaos of movement to the Argonne by road. Patton's tanks returned to the Saint-Mihiel railhead and entrained on flatcars for the journey to a railroad siding at Clermont-en-Argonne. They arrived in the dead of night and unloaded in the by-now-familiar drizzling rain. It was a difficult and hazardous undertaking, and Semmes credits Patton with making the impossible occur.[8]

The tanks were moved into a wooded area three kilometers north. With German planes active in the Meuse-Argonne sector during daylight, it had been impossible to remove the telltale tracks at night. Patton instinctively suspected danger and ordered his tanks moved again, this time obliterating any sign of their tracks by laying small branches over them. Not long after his tanks left the first bivouac area, German artillery began raining down on the site.

One of Patton's letters to Beatrice before the new offensive spoke in vague terms of going out "night owling" to accomplish "certain things not dangerous which must be done," a reference to conducting a personal reconnaissance of the German front from no-man's land. He described the countryside as "like a haunted forest . . . in the day it is nearly deserted but from dark to dawn it is alive with men and horses & guns you never dreamed of much less saw such numbers of guns. It is wonderful."[9] Up to the day before the offensive, the French manned the front lines to lull the Germans into believing that nothing would happen in this sector. American reconnaissance patrols, Patton's included, wore blue French uniforms and helmets to maintain the deception.

As it had been at Saint-Mihiel, the mission assigned Patton's tank brigade was to support two divisions of the I Corps. The 28th and 35th Divisions (both National Guard divisions, the 28th from Pennsylvania and the

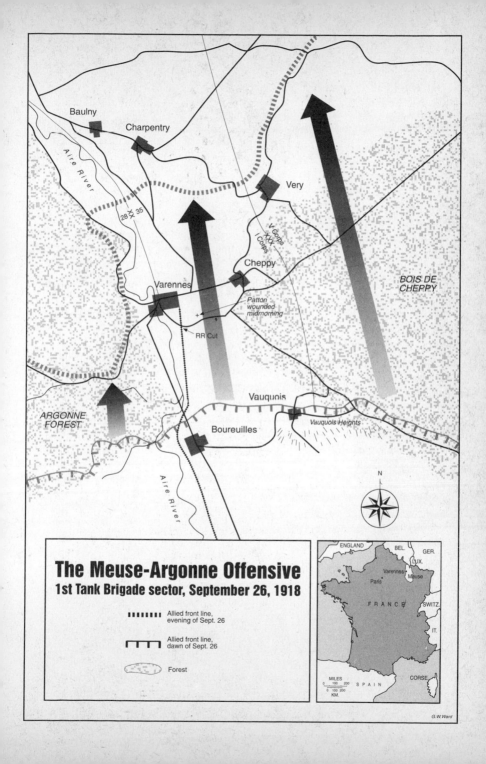

Baulny

Charpentry

Aire River

28 X 35

Very

V Corps
XXX
I Corps

Cheppy

Varennes

Patton
wounded
midmorning

RR Cut

BOIS DE
CHEPPY

ARGONNE
FOREST

Vauquois

Boureuilles

Vauquois Heights

Aire River

N

The Meuse-Argonne Offensive
1st Tank Brigade sector, September 26, 1918

▮▮▮▮▮▮▮ Allied front line,
         evening of Sept. 26

┌┬┬┐   Allied front line,
         dawn of Sept. 26

⬭   Forest

ENGLAND   BEL.
                    GER.
              LUX.
      Varennes   Meuse
  Paris
    F R A N C E   SWITZ.

                        IT.

MILES
0 100 200   CORSE
0 100 200   S P A I N
KM.

G. W. Ward

35th from Kansas-Missouri), were to attack side by side toward the villages of Varennes and Cheppy. The 28th Division would assault north along the eastern edge of the Argonne Forest, while the 35th Division attacked across the open terrain north of the Vauquois Heights.

Patton's reconnaissance made it obvious that his tanks could not operate in the Argonne Forest. In fact the entire sector was not well suited for the use of tanks, and Patton's report was couched in negative terms. There was only a narrow strip on the right flank of the 28th Division in which tanks could operate, and to further complicate matters, the Aire River (boundary line) flowed between the axis of advance of the two divisions, thus eliminating any possibility of mutual support.

Patton assigned the most difficult mission to the more experienced Brett and his 344th Battalion. As at Saint-Mihiel, the infantry would spearhead the assault, and in the 35th Division sector, be followed by two tank companies of the 344th. Although hardly ideal, it was nevertheless an imaginative solution to the difficulty posed by the inhospitable terrain. Even so Patton considered the Argonne better terrain for his tanks than the marshlands of Saint-Mihiel.

Accompanying his operations order was a memorandum that included a detailed examination of the terrain and how both the infantry and his tanks ought to be employed in the forthcoming offensive. Brig. Gen. Malin Craig, the I Corps chief of staff, was sufficiently impressed with what Patton had written that he ordered a copy furnished to each division commander.[10]

In the few hectic days prior to the offensive, Murphy's law inevitably applied. One hundred thousand gallons of gasoline arrived in tank cars without a single pump. "Now we can't get it out except by dippers!!!" stormed Patton. Of the many frustrations he experienced at Saint-Mihiel, undoubtedly the greatest were the numerous opportunities lost because there was no gasoline for his tanks. Resupply had been a time-consuming nightmare, and, having learned this lesson well, Patton approached the Argonne campaign determined to rectify the problem. "The Saint-Mihiel Offensive had taught the necessity for an immediate supply of extra gas, so each tank moving into action was required to carry two 20-liter cans tied to its tail." The two large fuel dumps Rockenbach established along Route Nationale 3, not far from the front, failed to satisfy Patton, who managed to deploy an additional twenty thousand gallons for the 344th in two forward gas dumps of his own, within a half mile of the front line.[11]

Patton became increasingly irritated when things went wrong. His brigade staff bore the brunt of his criticism and complaints that "I spoon fed these hounds so much that they are helpless and run to me every time they ought to go to the W.C. to see if it is all right." But Patton ultimately held himself responsible for, "reaping what I sewed. . . . Some times I think I am not such a great commander after all. Just a fighting animal. Still I will improve in time. At least if one learns by mistakes I ought to be wise. I have made all there are [to make]." As for his "rotten staff," he vowed he would "never do it again if I pull through this. But it is a big if. Hellish big."[12] Mostly his reaction was the result of fatigue, frazzled nerves, and his never-ending frustration, when, despite his powerful personality, his troops were unable to respond to the high (some would say, impossible) standards he set for them.

Somehow, before the offensive was launched, Patton managed to write several letters to Beatrice, but offered few clues as to where he was other than to note that: "I fancy our next show will be less easy than the first that is if the bosch fight and I think they will. . . . I will wire you after the next fight as I did this time."[13] He also reminded Beatrice how much he missed her, sometimes in ways that in more modern times might draw a return blast. "You are always young and fat, not too fat to me, in my thoughts of you. I wish I could squeeze you and pinch you and cuddle you and love you tonight instead of going out to work in the mud alone. I shall have forgotten how. Perhaps?"[14]

His final letter was dated September 25.

> Just a word to you before I leave to play a little part in what promises to be the biggest battle of the war or world so far. We kick off in the morning but this will not be mailed until after that. If the Bosch fights he will give us hell but I don't think he intends to fight very hard. . . . I will have two Battalions and a group of French tanks in the show in all about 140 Tanks. We go up a stinking river valley which will not be at all a comfortable place in a few hours. . . . I think that after this show we will have a rest I hope so for the men are tired and all the tanks need over hauling. . . . I have your picture with me so feel quite safe. I am always nearvous about this time just as at Polo or at Foot ball before the game starts but so far I have been all right after that I hope I keep on that way. . . . I love you you you always.[15]

In what was clearly an attempt to mollify Rockenbach, Patton included a notation in his written orders that his front-line command post would open at H + 1 hour at the site of the 35th Division field headquarters. The tip-off that Patton would not necessarily operate from there could be found in the reference that he would have from six to ten runners with him. In fact his entourage on the morning of September 26 consisted of himself, his recon-naissance officer, Captain Knowles; his signal officer, 1st Lt. Paul S. Edwards, twelve enlisted runners, a number of pigeons carried in baskets, field telephones, and a large quantity of telephone wire.[16] If Rockenbach intended to admonish him a second time, Patton's disobedience would not be for lack of preparation.

*   *   *

At 2:30 A.M. on the morning of September 26, some 2,800 guns began hammering the German front. From the fires of the smallest French 75-mm gun used by many of the American artillery batteries, to the mammoth fourteen-inch railroad guns situated far to the rear, the sight was, as air ace Capt. Eddie Rickenbacker observed from his Spad fighter, like "a giant switchboard which emanated thousands of electric flashes as invisible hands manipulated the plugs." An American corps commander recorded the event "as the sound of the collision of a million express trains."[17] During the three hours preceding H hour, the Allies expended more ammunition than both sides managed to fire throughout the four years of the Civil War. The cost was later calculated to have been $180 million, or $1 million per minute. Twenty-five miles away, the windows rattled in the house occupied by the German army commander, Gen. Max von Gallwitz.[18]

On Hill 290, half a mile west of Neuvilly and not far from Patton's crude dugout command post, Battery D, 2nd Battalion, 129th Field Artillery Regiment began firing one thousand rounds per hour from its four 75-mm guns in support of the 35th Division. The commanding officer was a thirty-four-year-old National Guardsman from Missouri, a recent graduate of the AEF Artillery School, Capt. Harry S. Truman. From Truman's vantage point the fireworks were as breathtaking as they were deadly. It appeared "as though every gun in France was turned loose. . . . My guns were so hot that they would boil [the] wet gunnysacks we put on them to keep them cool," Truman wrote to his wife, Bess, a month later.[19]

With H hour upon them, thousands of infantrymen awaited the signal to move forward into no-man's land. To a man they experienced the heart-pounding, suffocating, dry-mouthed, gut-wrenching feeling that inevitably precedes the first moments of combat. Most had never fought before. On September 26, and on subsequent days, Patton and his tankers would experience combat in the Argonne offensive that would make Saint-Mihiel seem like child's play.

During the march to the front lines the night of September 25–26, Brett's 344th Tank Battalion was preparing to cross a bridge over the Aire at Neuvilly when German 77-mm shellfire began raining down, killing two MPs guarding the bridge. Patton was present, and concern mounted as to how his tanks could be gotten safely across the bridge. Then Patton noticed there were predictable lulls between barrages, which he took advantage of to speed a tank company across before the shelling began again. In this way the tanks navigated this dangerous place without incident or loss of life.[20]

For much of the year the Argonne is shrouded in dense fog, and the morning of September 26, 1918, was no exception. At H hour, 5:30 A.M., American doughboys left the safety of their trenches and began advancing into the fog as rolling artillery barrages began chewing up the ground in front of them—and presumably any Germans unlucky enough to get in the way.

Initially all went well across the front, except the main attack centered on the Montfaucon heights, which proved every bit as difficult as predicted, despite an eight-to-one numerical superiority. Patton managed to resist the impulse to move forward and remained in the 1st Tank Brigade command post in a woods outside Neuvilly until 6:30 A.M., when he and his command group left and began following in the tracks of Major Brett's leading tank companies, astride Route Nationale 46. The tanks soon ran into a German minefield but, as Patton would later write, "thanks to the courtesy of the Germans in leaving up warning signs [ACHTUNG MINEN] the tanks avoided this danger."[21]

Patton's own description conveys a sense of the confusion that prevailed that morning. "It was terribly foggy and in addition they [American artillery] were shooting lots of smoke shells so we could not see ten feet. I started forward at 6:30 to see what was doing but could see little. Machine guns were going in every direction in front behind and on both sides. But no one could tell who they belonged to. I had six men—runners—with me and a compass," which became Patton's only means of navigation.[22]

By 10:00 A.M., Patton's party had advanced as far as a crossroads about 500 yards south of the village of Cheppy, where he sent his first and only pigeon message to inform Rockenbach of his location and what little he knew of the tactical situation. They sat down to await the arrival of Compton's reserve tanks.[23] What Patton did not know until the fog began lifting a few minutes later was that in the confusion and the fog he had (just as at Saint-Mihiel) actually advanced beyond his own tanks, many of which had become entangled in a trench works some 125 yards to his rear.

Trouble began as soon as the fog started to lift. A German communiqué characterized the fleet of tanks that suddenly appeared in their midst as "like the brothel of hell."[24] Nevertheless, what the German gunners could see, they could hit, and the Germans seemed to be everywhere. "The tanks, as well as the Infantry, were subjected to intense fire from the front, flanks and sometimes from the rear," Patton later wrote in his official report. When German artillery began ranging in on the crossroads and machine guns situated in and around Cheppy, along the road to Varennes, and in the nearby Bois de Cheppy, began raking the crossroads, Patton and his party hastily took refuge behind a nearby narrow-gauge railway cut. (It was later determined that there were at least twenty-five German machine-gun nests defending Cheppy.)

The capture of Cheppy was essential to the accomplishment of the 35th Division's mission on the first day of the offensive. However, it was at Cheppy that, instead of an orderly advance, bedlam reigned. As they had during the Saint-Mihiel offensive, Patton's tanks had advanced beyond most of the infantry they were supporting. Artillery fire from both sides and German machine-gun fire combined to create such chaos that many of the doughboys of the green 35th Division simply could not cope. Historian Dale Wilson observes: "The division's first-day mission would have given pause to veteran soldiers; for the National Guardsmen from Kansas and Missouri entering battle for the first time it proved to be too much. Enthusiasm can only carry a soldier so far."[25]

Some had become lost in the fog and were wandering around looking for their units; others had panicked and were fleeing toward friendly lines when they encountered Lt. Col. George S. Patton, who ordered them to remain with him. The lifting of the fog and the increasingly intense German machine-gun fire made remaining with Patton preferable to risking almost certain death by exposing themselves by dashing for the rear:[26] "So I collected all the soldiers I found who were lost and brought them along. At times I had several hundred."[27]

In the next several hours the legend of George S. Patton the warrior was born. Here is his account, written two days later to Beatrice: "All at once we were shot at to beat hell with shells and machine guns. Twice the inft started to run but we hollored at them and called them all sorts of names so they staied. But they were scared and some acted badly, some put on gas masks, some covered their face[s] with their hands but none did a damed thing to kill Bosch. There were no officers there but me."[28]

German fire raining down around the railway cut became so intense that Patton and his ad hoc infantry force were forced to seek sanctuary on the reverse slope of a small rise about a hundred yards to the south. A nearby German trench works had become a troublesome bottleneck. The leading tanks of Brett's Company C were entangled in a very wide and deep trench and the remainder of Capt. Math L. English's Renaults were halted and unable to advance. The situation quickly worsened when Captain Compton's reserve companies and Major Chanoine's two battalions of heavy Schneider tanks began arriving, creating a massive traffic jam. When the fog began lifting, the Germans sent spotter aircraft aloft, enabling them to direct increasingly heavy artillery fire on this lucrative target.

A disaster was in the making unless the trench works was breached—which from Patton's vantage point appeared increasingly unlikely by the minute. Patton wrote that their presence presented "a dangerously large target to the enemy. However, before they registered on the spot, the tanks were scattered behind various cover. Two French Schneider tanks persisted in pushing forward, and were stalled in the only [other] crossing over the trench system."[29]

When Patton noticed that the two Schneiders and English's Renaults remained stuck, he sent Lieutenant Knowles and later, his batman, Pfc.

Joseph T. Angelo, with orders to get them freed and moving forward at once to attack and eliminate the German machine guns. When nothing happened he sent Lieutenant Edwards to tell Captain English, "to have five tanks come up over the hill and attack the machine guns in front." Still nothing happened. Finally, exasperated and enraged, Patton went himself. "I went to the bottom of the slope and found that the tanks were being held up on account of the two trenches. A group of French tank men were sitting in the trench with shovels."

According to Edwards, in the face of very heavy machine-gun fire and increasingly effective artillery fire, Patton immediately

> went over to the tanks which were being splattered with machine gun fire and removed the shovels and picks and put the men to work. . . . In spite of the repeated requests that he step down in the trench from his exposed position the Colonel steadfastly refused to do so saying "To Hell, with them—they can't hit me." There were a number of casualties among those who were tearing down the sides of the trenches for the passage of the tanks but the Colonel refused to budge.[30]

Patton's own version was that when there was no sign of progress in freeing the tanks:

> I decided to do business. . . . So I went back and made some Americans hiding in the trenches dig a passage. I think I killed one man here he would not work so I hit him over the head with a shovel. It was exciting for they shot at us all the time but I got mad and walked on the parapet. . . . At last we got five tanks accross and I started them forward and yelled and cussed and waved my [walking] stick and said come on. about 150 doughboys started but when we got to the crest of the hill the fire got fierce right along the ground. We all lay down.[31]*

As later described by his daughter, the moment of his date with destiny, "so long anticipated and dreaded," now occurred. As he crouched at the foot of the low hill and began sending hand signals to his tanks, the Germans had gotten the range and their fire intensified, reminding Patton of a lawn mower cutting the grass at Lake Vineyard.

He was afraid. His hands were sweating and his mouth was dry. There was a low bank of clouds behind the rising ground, and he looked up and saw, among the clouds, his ancestors. The ones he had seen in pictures

---

*Patton began using a walking stick as a result of his experience at Saint-Mihiel, mainly to enable him to tap it on the side of a tank to gain the attention of its crew.

looked like the daguerreotypes and their paintings; there was General Hugh Mercer, mortally wounded at the Battle of Princeton; there was his grandfather, Colonel George Patton . . . there was his great-uncle, Colonel Waller Tazewell Patton . . . there were other faces, different uniforms, dimmer in the distance, but all with a family look. They were all looking at him, impersonally, but as if they were waiting for him. He knew what he had to do, and continued the tank action.[32]*

Patton's batman, Pfc. Angelo, records that, "In a few minutes the tanks began to move forward over the hill where small groups of Inft. had taken cover in shell craters. The Col. asked an Inft. Sgt. if there were any of his Officers present, to which question he answered no. The Sgt. then asked Col. Patton what they should do, [and] he replied 'follow me' to which they consented and followed." Patton had arisen, waved his walking stick over his head, and shouted, "Let's go get them, who's with me?" and begun to move toward the top of the hill, with about one hundred infantrymen in his wake.[33] However, as soon as they reached the exposed ground at the crest of the hill, their presence attracted heavy machine-gun fire, and all sought sanctuary on the reverse slope, as men on both sides of them were brutally cut down.

Like it or not, Patton was suddenly the infantry commander he had occasionally thought about becoming. He and he alone had held the frightened infantrymen together in the railway cut and later, behind the hill. In fact, one of the reasons why he had sent three runners before going himself to organize the tanks was his instinctive understanding that if he left, the infantry would panic and bolt. The motto of the infantry is "Follow Me," and in its finest tradition Patton led the way, after first ridding himself of what he perceived was momentary cowardice. "I saw that we must go forward or back and I could not go back so I yelled "who comes with me."[34] There was considerable yelling, "but only six of us started. My striker, me and 4 doughs. I hoped the rest would follow but they would not and soon there were only three [of us] but we could see the machine guns right ahead so we yelled to keep up our courage and went on. Then the third man went

---

*Nearly thirty-three years later Patton's own son would have the same experience in the Korean War. Captain George S. Patton IV and a South Korean colonel were on foot on a road under artillery bombardment, unable to reach his jeep and its radio, to warn his unit of the situation. The colonel advised young Patton that it was too dangerous to cross the road. "I looked up at the sky and there he was. He said, 'Get your ass across the road.' That was the message I got. I took a deep breath and took off and as I crossed the road the shelling stopped—or relaxed. I think my sense of duty and obligation to my men would have made me cross that road anyway. But his appearance in the clouds helped to spur me on." (Maj. Gen. George S. Patton, USA [Ret.], quoted in Jeffrey St. John, "Reflections on a Fighting Father," *The New American*, Dec. 16, 1985.)

down."[35] Patton and Angelo were now alone and exposed to every German gunner who chose to shoot at them. Patton was armed with a pistol in a holster strapped to his waist, but even unsheathed it was as useless against machine guns as his walking stick. Blumenson likens them to Don Quixote and Sancho Panza wandering alone in the wilderness.[36]

By his own admission, Patton was terrified and, as he later wrote briefly of the event in 1927:

> Just before I was wounded I felt a great desire to run, I was trembling with fear when suddenly I thought of my progenitors and seemed to see them in a cloud over the German lines looking at me. I became calm at once and saying out loud "It is time for another Patton to die" called for volunteers and went forward to what I honestly believed to be certain death. Six men went with me; five were killed and I was wounded so I was not much in error.[37]

A machine-gun bullet struck Patton with terrific force, and he toppled to the ground, blood seeping from a serious wound. The lone American who was neither wounded nor killed was Pfc. Angelo.

However one opts to interpret Patton's brief experience, the fact remains that it profoundly influenced him to risk almost certain death. In retrospect, he would certainly have agreed with Shakespeare's observation that "Our doubts are traitors,/And make us lose the good we oft might win/By fearing to attempt."[38]

Although both men had some protection from a tank they had been walking next to, it was not enough to protect them from the deadly German machine guns. The bullet that found Patton struck him in the left upper thigh, "and came out just at the crack of my bottom about two inches to the left of my rectum. It was fired at about 50 m [meters] so made a hole about the size of a dollar where it came out."[39] When Angelo asked if he had been hit, Patton whispered, yes. He immediately lapsed into shock but managed to remain conscious throughout his ordeal. Angelo managed to get Patton into "a small shell hole," where he then sliced open his trousers and applied a bandage to stem the bleeding. However, nothing could be done about evacuating Patton to the safety of an aid station in the rear, where the wound could be properly treated. German infantry had moved into the railroad cut about forty yards away, previously vacated by Patton and his band of infantry. Any attempt to move from the tenuous sanctuary of the shell hole was now impossible.

The miracle was that both men were not killed outright. Patton has described how "I felt a blow in the leg but at first I could walk so I went about 40 feet when my leg gave way My striker, the only man left yelled 'oh god the colonels hit and there aint no one left.' He helped me to a small shell hole and we lay down and the Bosch shot over the top as fast as he could and he was very close."[40] Patton thought he was wounded about 11:15 A.M., while others placed the time at between 10:30 and 11:00.[41] Patton himself thought he was there about an hour; however, in his condition, time is frequently distorted. The time he lay in the shell hole has not been accurately established, but it was probably closer to two hours.

Before Patton could be moved the German machine guns had to be silenced, and that took quite some time. Throughout his ordeal, "one of my tanks [sat] guarding me like a watch dog."[42] On Patton's order, a runner was sent to find and inform Major Brett that he was now in command of the brigade, but in the chaotic conditions he was not located until midafternoon. Patton also insisted that no attempt be made to rescue him until the situation was stabilized.[43]

Following Patton's wounding, Captain Compton finally managed to maneuver two platoons of Company B around the west side of the troublesome hill and all of Company C around the eastern side. German infantry in nearby trenches to the west were erased at about the time approximately one hundred troops of the 35th Division's 138th Regiment arrived on the scene, under the command of an infantry major. Together Compton and the major launched a joint attack and while one tank-infantry force outflanked Cheppy, the other managed to enter and secure the village. The time was approximately 1:30 P.M.[44]

This joint action by Patton's tankers and the 138th Regiment may well have been the first-ever example of tank-infantry cooperation in an offensive situation. Nothing recorded about the earlier Saint-Mihiel campaign even approached the capture of Cheppy by tanks and infantry working together as a team. If there was to be evidence that the tank had a future, it was this small but important tank-infantry action. Unfortunately, whatever useful lessons might have been learned were to be lost in the aftermath of the war.

While Patton awaited evacuation, a 35th Division medic happened by the shell hole and changed his bandage. "Patton thanked him courteously."[45] Even though he claimed to have experienced no pain from his wound, Patton reluctantly began to accept the fact that, at least for the moment, his war was over. Patton's thoughts and state of mind were later recorded by his daughter. Although numb with shock, he recalled feeling "a great calmness of mind and of spirit," and "kept thinking that he was nearly thirty-three years old, and that his grandfather Patton had been thirty-three years old when he had taken the shell fragment . . . at Cedar Creek; and how young he had been; and what a waste it all was."

Patton knew that he was alive but that "part of him had died; he was a little bit in both worlds. In his own words: 'I was overwhelmed by a deep feeling of warmth and peace and comfort, and of love. I knew profoundly death was related to life; how unimportant the change-over was; how ever-lasting the soul—and the love was all around me, like a subdued light.' "[46]

It was not merely his ancestors who provided Patton with inspiration. Many nights his beloved father would suddenly appear, sit down, and talk with him and reassure him that he would act honorably and bravely in battle. To Patton, this apparition, "was just as real as [being] in his study at home at 'Lake Vineyard.'"[47]

With the deadly German machine guns silenced at last, Patton was finally rescued about 1:30 that afternoon. Four of his enlisted men removed their commander from the shell hole to sanctuary behind the nearby hill, where he had lain that morning. Major Brett still had not yet been located, and Patton's final order before being evacuated was to send Lieutenant Edwards to search for him. The two-mile journey on a litter carried by five men was "not at all plesant." Accompanied by the faithful Angelo, who refused to leave his side, Patton ordered the ambulance driver to detour via the command post of the 35th Division, where he personally rendered a report of the situation at the front to a division staff officer. His duty done, Patton finally permitted himself to be delivered to a nearby evacuation hos-pital, where he slipped into the comfort of anesthesia as a surgeon named Elliot operated on him.[48]

When Capt. Harry Truman's battery reached the crossroads near Cheppy on the morning of September 27 they discovered savage testimony to the fury of the battle that had raged there the day before. "Heaped in a pile, were seventeen American dead, infantrymen, while down the road a dozen more were lying 'head to heel,' all shot in the back after they'd gone by. . . ."[49] It is not difficult to imagine this being Patton's fate had not the god of battles (in whom he devoutly believed) decided to spare his life.

That Patton believed there was a higher power protecting him is evident in a letter he wrote nearly a month later to his wife, stating that he would not have had the courage to expose himself had he not thought of her and his ancestors. "I felt that I could not be false to my 'cast' and your opinion." Nor did he actually believe *he* would be hit. "One has a sort of involuntary fear of the bullets but not a concrete fear of being hit." Patton recalled "some story by Kipling where the officers smoked to reassure the men. So I smoked like a factory. We were then being shelled heavily from in front and were under rifle fire from both flanks and in front. But I kept saying to myself I am not to be hit I know it so I felt better but it was quite bad men were falling or rather being blown to bits all around."[50] The experience taught Patton that even he was mortal.

Patton awoke the morning of September 27 to find two of his company commanders in the beds next to his—Harry Semmes and Dean Gilfillan.

Semmes had been wounded several times, including by a sniper's bullet in his head, as he attempted to find a safe route for his tanks through a bog near the Vauquois heights; Gilfillan's heroism included taking two direct hits on his tank, and being wounded by machine guns and shrapnel, none of which deterred him from knocking out two machine-gun nests and mowing down a coterie of fleeing German infantry. Captain Gilfillan was later awarded the Distinguished Service Cross (DSC) for valor, and Semmes earned two DSCs for his bravery at Saint-Mihiel and Vauquois ridge.[51]

Patton managed to write his account of September 26, but, weak and exhausted, he mostly "slept a lot." Although the bullet had not struck a vital organ, his luck was little short of miraculous. "The Dr. says that he can't see how the bullet went where it did with out crippeling me for life. He says he could not have run a probe without getting either the hip joint, sciatic nerve or the big artery yet none of these were touched. 'Fate' again. I have never had any pain and can walk perfectly."[52] A French colonel who came to call thought differently, and Patton reported that he said: "I am so glad you were only wounded you are one of those gallant men who always get killed But you will get it yet."[53]

Before being evacuated on September 29, he cabled Beatrice: SLIGHTLY WOUNDED NO DANGER. In a pouring rain he and Semmes were placed on stretchers aboard "a cattle train" and taken on a very uncomfortable twenty-hour journey to a base hospital near Dijon. They traveled in boxcars fitted with racks on which the wounded were triple-stacked. Patton was in a top rack, where "the iron bars of the stretcher hurt my back and I could not move," and was fed only once, a meal of coffee and bread spread with molasses. Three days later he was able to write to Beatrice that he was "missing half my bottom but other wise all right. . . . The hole in my hip is about as big as a tea cup and they have to leave it open" to be drained. "I suffer none at all except when they dress the wounds. I look as if I had just had a baby or was unwell. Still we broke the Prussian guard with the tanks so it is all fine. This is a stupid letter but it is hard to write."[54]

Patton was immediately placed in quarantine for a week in case he was carrying the highly contagious meningitis virus. He was the senior of fifty officers in a large ward serviced by only two nurses. When the officer in the next bed died from a broken back, it served as a visible reminder that the war was never very far away.

Patton had Semmes for companionship, and they spent hour upon hour discussing tactics, the performance of their tanks at Saint-Mihiel and in the Argonne, and how they might employ them in the future. When Patton was not talking he continued sleeping a great deal. It would take many more days of "'cultures' of my bottom to see if there are any bugs" before Patton's wound was eventually sewed up. He assured Beatrice that his scar would not show "unless the styles change. I surely am a lucky fellow." He hoped that they would have enough money after the war for a second honeymoon. But even if they didn't, "Lets do it any how. We will have to get used to each other" all over again.[55]

The boredom and inactivity left him restless at being confined in "a rotten place with a cemetary just out side where they bury people all day long."[56] Patton began to feel better after being moved to a new ward. When the weather permitted he was allowed to sit outside in a wheelchair and smoke his pipe and read. It left him time to reflect on the battle that had nearly cost him his life.

Outwardly Patton seemed undaunted by his close brush with death:

> I feel terribly to have missed all the fighting it seems too bad but I had to go in when I did or the whole line might have broken. Perhaps I was mistaken but any way I believe I have been sited for decoration either the Medal of Honor or the military cross. I hope I get one of them. . . . Peace looks possible but I rather hope not for I would like to have a few more fights. They are awfully thrilling like steeple chasing only more so.[57]

Even the pleasant news that Rockenbach had recommended his promotion to full colonel left Patton with mixed feelings. "I would like it in a way but the more rank one gets the harder it is to get into a fight and fights certainly are fun. This is not a pose either it is actually so and one of the few things I could enjoy as you know I like most things solely for the results."[58]

Patton's wound proved to be slow healing, and he reported with disgust that, "my d-wound is still full of bugs so they can't sew me up. It is most Provoking. I have just been in to cuss out the surgeon but it does no good . . . it is impossible to give special attention to anyone here. I am feeling fine and want to get out."[59]

The following day he proudly wrote: "What do you think of me. I just got my colonelcy over the wire and am not yet 33. That is not so bad is it. Of course I have class mates in the engineers who are colonels but none others. So I feel quite elated though as a matter of fact I don't believe I deserve it very much. . . . I do hope I get the decoration I would prefer it to the promotion." He addressed the envelope to "Mrs. Colonel G.S. Patton, Jr."[60] Several days later he wrote to Aunt Nannie that his promotion "is not bad for 32. Though I had always intended to be a general at 26."[61]

Patton may have been only days shy of his thirty-third birthday, but his appearance belied his age. Gaunt from the loss of nearly thirty pounds and often unshaven, he was not the immaculately dressed and groomed soldier his men knew so well. His doctors thought him about forty-five years old: "I am a lot older in some things, he admitted." His earlier soldiering in the United States and Mexico seemed like an eternity ago. Patton had once wondered if he could even successfully command a battalion of militia. "I wondered at the time if I could have done it. Now I know that I could command a division. Things are realy much easier than they appear."[62]

On October 19 Braine arrived with a fistful of letters from Beatrice, whom he had visited before returning to France. One of Patton's first questions was how his wife looked, "if you had any gray hair. . . . I always think of you as Undine so I don't want you to look 33, even if I do." Toward the end of October he wrangled a transfer to a hospital at Langres, which for Patton was like coming home.[63] Within days he became an outpatient and was able to return to his quarters at Bourg, where he immediately resumed command of the tank center. Major Brett remained the commander of the 1st Tank Brigade until the war ended.

His first day back Patton issued a sharply worded decree on all aspects of personal deportment and discipline. Anyone who might foolishly have fantasized that their colonel had mellowed while in the hospital was soon set straight. He may have had "a whole bath towel stuffed in my bottom and [been] bleeding like a stuck pig," but the fire-breathing George S. Patton was back.

Pershing's tactics of massing overwhelming numbers of troops in the Meuse-Argonne had backfired when the First Army was unable to prevent a massive German reinforcement followed by strong counterattacks. The work of the men of the graves registration service was never ending. Next to Patton, Major Brett was the most aggressive tank commander in the AEF, but even he could do little in the face of such resistance. By the end of September 26, forty-three of the original one-hundred forty tanks had either been knocked out by the Germans or had failed mechanically. Two days later there were only fifty-three tanks still in action.[64]

The World War I battles fought in the Argonne Forest bore an uncanny resemblance to the Battle of the Hürtgen Forest in World War II*; both had few roads, were honeycombed with gullies and ravines, and were heavily forested, which restricted vision to the range of a hand grenade and favored the defender, who covered every avenue of approach with machine-gun fire. It was the worst of all possible places to fight and a bloody killing ground nonpareil. It was in the Argonne where a hillbilly rifleman from Tennessee

---

*By November 1944, Allied forces were stalled along the German border. In mid-November Lieut. Gen. Omar Bradley launched a major offensive, with the U.S. First, Third, and Ninth Armies aimed at driving to the Rhine and encircling the Ruhr. The spearhead of the First Army attack was VII Corps, whose mission was to attemot to duplicate the Saint-Lô breakout through the Hürtgen Forest toward Cologne. Despite massive aerial bombardment, what became known as the Battle of the Hürtgen Forest was a bitterly fought colossal failure, with very high American casualties in what Bradley himself termed "sheer butchery on all sides."

named Alvin C. York won the Medal of Honor and immortality as one of the greatest marksmen in the history of the U.S. Army.

The Argonne Forest was not secured until nearly mid-October, by which time the First Army was more than 1 million men strong. The campaign ended with the Armistice. American casualties exceeded 122,000 men (26,277 killed and 95,786 wounded), while the Germans lost 100,000. Losses in the Tank Corps accurately reflected the high casualty rate in the AEF. By late October the 1st Tank Brigade had only 80 of 834 men fit for duty.[65]

Patton's tankers acquitted themselves with distinction in the Argonne. On September 26, outside Varennes, the heroic rescue of a wounded officer trapped in his tank resulted in the first award of the Medal of Honor to a member of the Tank Corps, and Joe Angelo earned a Distinguished Service Cross for saving Patton's life.[66] "The tank corps established its reputation for not giving ground. They only went forward. And they are the only troops in the attack of whom that can be said."[67]

The army hierarchy may not have fully appreciated the newfound importance of the tank, but the men of the 35th Division certainly did. A tank lieutenant wrote of being greeted as a savior. "Thank God," said one captain, whose infantry had been pinned down by machine-gun fire. Even the commanding general and his staff expressed their appreciation whenever elements of Patton's brigade passed by.[68] Patton has related how the only way one German machine-gun could be silenced was by literally running over it with a tank; "but even in death they were holding to their gun. My men buried them and put up a cross 'To two brave men though S.O.B.s.' "

At the eleventh hour of the eleventh day of November 1918, the guns on the Western Front fell silent, and the most terrible war in the history of mankind had ended. Coincidentally, it was George S. Patton's thirty-third birthday, and for the Armistice to occur on this day was a good omen indeed.[69] His diary recorded: "Peace was signed and Langres was very exited [excited]. Many flags. Got rid of my bandage. Wrote a poem on peace."[70]

Four days earlier, at Avalon, Beatrice Patton was awash in memories of her beloved parents as she sorted and packed their belongings. Suddenly church bells began ringing in unison. Beatrice burst into tears, and Ruth Ellen remembered her mother exclaiming: "The war is over! The war is over! Your father will be coming home!" Like that of many Americans, Beatrice's celebration was premature. The United Press Paris correspondent had evaded the censors with a dispatch that erroneously reported that an armistice had been signed on November 7. Immediately reprinted in American newspapers, it became known as the "false" armistice. [71]

Some 350 miles east of Bourg, in the deep snows of southeastern Bavaria, an eccentric German corporal named Adolf Hitler spent the night of Novem-

ber 11, 1918, on sentry duty in a wooden watchtower guarding Russian prisoners of war.[72] The war may have been over for Germany, but within months
of the disastrous Treaty of Versailles Adolf Hitler founded an organization
that twenty years later would avenge Versailles and produce the war that
would fulfill the destiny of George S. Patton, Jr.

CHAPTER 19

Bitter Aftermath

I saw battle-corpses, myriads of them,
. . . the slain soldiers of the war,
They themselves were fully at rest, they suffer'd not,
The living remained and suffer'd, . . .
And the armies that remained suffered.

—WALT WHITMAN, *WHEN LILACS LAST IN THE DOORYARD BLOOM'D*

Well this is a hellish stupid world now and life has lost
its zest.

—PATTON

From the dawn of time men have been driven by an unfathomable need to prove their courage and masculinity in some bizarre rite of passage that defines their lives. Conversely, the notion of laying down one's life in a foreign land is as frightening as it is repellent. The great British poet of World War I, Siegfried Sassoon, was right when he wrote: "Soldiers are dreamers; / when the guns begin / They think of firelit homes, clean beds, and wives."[1]

Thus it is one of the great paradoxes of war that while most soldiers have little affinity for what they are obliged to do on the battlefield, they suffer withdrawal symptoms when it ends. Whether it was a caveman discovering the power of dominance with a simple club or a contemporary Scot displaying his physical prowess in the grueling Highland games, man's compulsion to prove his masculinity is as fundamental as life itself.

Yet such is the enigma of war that it still manages to induce a sense of euphoria in defying death or maiming injury, and walking away intact. For the soldier who is ruthlessly thrust into months or even years of living on the edge, the abrupt cessation of war more often than not brings with it an inchoate but acute sense of letdown. The human mind, accustomed as it is to a certain discipline and order, is incapable of processing sudden change without trauma. The end of war rips the fabric of that special bond men in combat share with one another—a bond so compelling that neither subsequent family happiness nor material success can ever replace it.[2]

The end of war is, in short, a sort of massive hangover, a culture shock that often manifests itself in antisocial behavior, alcoholism, and severe depression. The narcotic-like effect of being in combat cannot suddenly be replaced by a return to a normal lifestyle. Men honed to a fine edge to kill other men in a variety of ruthless ways can hardly be expected to become instant pillars of their community.

Moreover, there is an exceptionally fine line between bravery and cowardice. The subject of why men risk their lives in the face of almost certain death has long been a puzzling one. Why, for example, did Patton finally overcome his admitted fear of death the morning of September 26 at Cheppy? He admitted to an almost overwhelming terror that, in the end, was overcome only by the even greater fear that his men would think him cowardly for not risking his life. If his troops viewed him badly, that would last only until his death; but if he failed to act, his dishonor to his family heritage would last for eternity. If it can be said that Patton was driven to an act of extreme bravery, then the underlying reason is at least partially delineated by Patton's friend and mentor, Gen. J. G. Harbord:

Every soldier took into action a confused panorama of hastily prepared ammunition dumps, gun positions, tractors, telephone wires, artillery pulled by emaciated and exhausted horses, tractors and trucks, maps enough to carpet the battlefield—all the thousand products of a mighty industrial age which we had attempted to adapt to military uses. As a background for this panorama he remembered faint legends of old frontier days, traditions of other Americans in other wars, his particular conception of patriotism, the farewells of his mother and sister, and perhaps another, his local pride in his own neighborhood and its interest to him. All these things added to *nervous force, energy without limit, confidence, youth and optimism—we were substituting for experience.* To doubt audibly was to be a traitor.[3]

Above all it was the peer pressure of not being seen to be a coward that drove men like Patton to take such appalling risks. If, as the great Euripides wrote, around 412 B.C., "a coward turns away, but a brave man's choice is danger," then Patton's case is that of a reluctant and frightened hero, a warrior so anxious to prove himself that he risked death to avoid in death the wrath of his kinsmen who died before him on other battlefields.[4]

Patton's poem "Fear" provides some additional clues to what drove him:

I am that dreadful, blighting thing,
Like rat-holes to the flood,
Like rust that gnaws the faultless blade,
Like microbes to the blood.

I know no mercy and no truth,
The young I blight the old I slay.
Regret stalks darkly in my wake
And Ignominy dogs my way.

Sometimes in virtuous garb I rove,
With facile talk of easier way,
Seducing, where I dare not rape
Young manhood from its honour's sway.

Again in awesome guise I rush
Stupendous, through the ranks of war,
Turning to water with my gaze
Hearts that before no foe could awe.

The maiden who has strayed from right,
To me must pay the mead of shame,
The patriot who betrayed his trust,
To me must own his tarnished name.

I spare no class, or cult, or creed,
My course is endless through the year.
I bow all heads, and break all hearts,
All owe me homage—I am FEAR![5]

Unfortunately, part of the Patton myth has grown from his actions on the first day of the Meuse-Argonne campaign. One biographer had Patton "uttering war cries and waving his saber . . . looking vaguely like Ben Hur in a chariot. . . . Everything about the man now resulted in two reactions: War cries and Rallying. . . . So he war cried and rallied, reorganizing enough of the demoralized infantry to cover the advance of his tank, leading them forward until most of them were killed by relentless hammering of machine guns and a neat semicircle of shrapnel around his midriff cut short his path to glory." [6]

On November 11, 1918, men who moments before had been killing one another now laid down their arms and for the first time left the battlefield without fear of having to return to fight another nameless battle. Lawrence Stallings, a World War I marine, wrote, "Where veterans in the Meuse line were halted in their tracks, forbidden to fraternize with the enemy, mainly they stretched on the ground, thanked God they were still alive, and built their first campfires—feeling ill at ease because no guns were firing anywhere. . . . There was no stillness on the Yank firing lines, as at Appomattox. It was a matter of noisy laughter, of men too weary to shed tears."[7]

Capt. Harry S. Truman was both relieved and elated that he had survived, writing that, no matter how horrific, it had still been "the most terrific experience of my life."[8] Truman was proud that his leadership had helped his troops to survive; for Patton it was more personal: He had been robbed of any further chances to grasp the brass ring. Men like Truman were driven by honor and the will to survive; Patton was driven by a personal compulsion to earn acceptance from the ghosts of his warrior ancestors through feats of bravery. At times the emotional baggage of the need to overcome fear was almost unbearable. More often than not at such times, Patton turned to poetry as an outlet for his emotions.

His poem, "Peace—November 11, 1918" is a paean to the soldier, not of joy but of sadness, confusion, and anger. Above all else, it released his pent-up emotions and the feeling, so common among his fellow soldiers, that there was no more war to fight—that for the first time peace had replaced the daily sight of death. In short, the "high" of war was over and only the hangover remained.

I stood in the flag-decked cheering crowd
Where all but I were gay,
And gazing on their extecy,
My heart shrank in dismay.

For theires was the joy of the "little folk"
The cruel glee of the weak,
Who, banded together, have slain the strong
Which none alone dared seak.

The Bosch we know was a hideous beast
Beyond our era's ban,
But soldiers still must honor the Hun
As a mighty fighting man.

The vice he had was strong and real
Of virtue he had none,

Yet he fought the world remorselessly
And very nearly won. . . .

And looking forward I could see
Like a festering sewer,
Full of the fecal Pacafists
Which peace makes us endure. . . .

None of the bold and blatant sin
The disregard of pain,
The glorious deeds of sacrefice
which follow in wars train.

Instead of these the little lives
Will blossom as before,
Pale bloom of creatures all too weak
To bear the light of war.

While we whose spirits wider range
Can grasp the joys of strife,
Will moulder in the virtuous vice
Of futile peaceful life.

We can but hope that e're we drown
'Neath treacle floods of grace,
The tuneless horns of mighty Mars
Once more shall rouse the Race

When such times come, Oh! God of War
Grant that we pass midst strife,
Knowing once more the whitehot joy
Of taking human life.

Then pass in peace, blood-glutted Bosch
And when we too shall fall,
We'll clasp in yours our gory hands
In High Valhallas' Hall.[9]

This poem was the first of many that flowed like a dirge from Patton's pen in the months during and after the war. Those written before the Armistice tended to be more upbeat, often self-assertive reminders not to capitulate to the god of fear in the battles to come. Shortly after returning to the United States in 1919, Patton attended a play in Washington about the war and pensively wrote to Pershing: "The noise of the shells and the machine guns made me feel very homesick. War is the only place where a man realy lives."[10] Although he would have denied it, the effects of the November 1918 armistice were to be visible in Patton's behavior during the twenty-one-year hiatus between the two world wars.

The end of the war also brought about the undoing of Nita Patton's engagement to Black Jack Pershing. She had spent the war in Washington but in the late spring of 1919 was in London with her sister-in-law, Kay Merrill, when a letter arrived from Pershing to the effect that "the feeling" was gone, and that they ought not to proceed with announcing their engagement or of marrying until "the feeling" returned. To make his point, Pershing had not bothered to send her a ticket to the great victory gala held in Paris the night of July 3, 1919, at which he was the guest of honor.[11] A phone call to AEF headquarters by Patton's brother-in-law, Keith Merrill, brought the response that of course the Merrills were expected to escort Nita to the ball.[12] Humiliated, Nita broke her engagement to Pershing by returning the diamond ring he had presented her during their whirlwind courtship in 1917.

Nita Patton was a woman of great pride who rebuffed numerous entreaties to patch things up with Pershing.[13] An obvious reason for the breakup of their romance was that Pershing had found companionship and perhaps "the feeling" with another woman, named Elizabeth Hoyt.[14] Beatrice Patton offered another explanation:

> General Pershing had been under a terrible strain for the war years and had done a fantastic job. As the war drew to its successful close he was wined and dined and flattered and praised by the great and the near great, and some of the most beautiful women in Europe who were not above falling at his feet to gain something for their heart's interests. He had a Caesar's triumph. Nita with her blond Viking good looks and carriage and her predominantly good sense, was just there and could more or less be propped in a corner until he had time to regroup and reconsider. Only, Nita removed herself with all flags flying.[15]

Despite the shattered romance, Pershing was an occasional guest in the Patton household during the postwar years, when George was again stationed at Fort Myer. His two daughters were now old enough to know about the romance between the still "arrestingly handsome" Pershing and their aunt Tinta (the Patton children's nickname for Nita). Once, when Bee Patton asked her mother how Aunt Nita could ever have been in love with "that silly old man," Beatrice replied: "The John J. Pershing you children know is not the Black Jack Pershing that Tinta fell in love with. Lots of men die in wars, but some of them who have very strong bodies go on living long after the person inside of them, the real them, is dead. They are dead because they used themselves all up in the war. That's one of the most terrible things about war."[16]

When Pershing visited, he and Patton would spend hours reminiscing about their experiences, aided by liberal doses of alcohol. One night Pershing began to cry and said, "Georgie, Georgie, if I hadn't been such a damned egotistic fool, my children would have been just a little younger than your children with the same beautiful blond hair, and the same true blue eyes." On several occasions before returning from Europe, Patton had written his sister to suggest a "dignified reunion" with Black Jack in Europe. Well before the final snub in July 1919, Nita had known that their relationship was at an end. She was too proud to revive it and rejected her brother's attempts to arrange a reconciliation. The only one truly pleased by their breakup was Papa Patton, who "did not consider General Pershing good enough for Nita, snobbishly observing that Pershing's father had been a brakeman on the railroad. Georgie was relieved too at having the taint of favoritism or nepotism removed from whatever his future might hold."[17] Their breakup haunted Pershing the rest of his life. Nita Patton never married, and Black Jack remained a widower.[18]

During Patton's lengthy recuperation he and Beatrice wrote to one another virtually every day. Beatrice still badly wanted another child, and in one of her letters mentioned to George that she was an excellent mother and loved being one. Denied any further role in the war, and in the doldrums of recovery, he was not amused and fired off one of the most hurtful letters he ever sent to Beatrice: "Your childish proclivities, of which you boast, do not interest me at all. I love you too much and am jealous, or something of the children. Your only chance to have another child is accident or Immaculate Conception. You ought to be complimented. But being pig-headed, I suppose you are not. I love you too much."[19]

With the urgency of war removed, Patton found it more difficult to face each day. "I fear that laziness which has ever pursued me is closing in on me at last. It will be funny to command 74 men in a troop of cavalry after having commanded a thousand and more in battle and to be through by noon each day." To avoid what he termed "the devil of idleness," Patton decided to write a book he planned to title "War as She Is," for "in prose it is the pen which makes the sword great in peace. So if I write a good book I might get to be a general before the next war. If I start the next war as a Brig. Gen. and hit the same pace I gained in this I will make three grades or end up as a full general."[20] It turned out to be a remarkably accurate prediction. The book, however, never materialized beyond the twenty-six pages Patton wrote in late 1918 and early 1919.

To avert even the specter of idleness, he wrote a series of lectures on tank tactics to the General Staff College, the outline of thoughts that he planned to use to form his book, and an after-action report on the exploits of the 1st (renamed the 304th) Tank Brigade, which he and Major Brett (who would himself be a colonel in the next war) wrote for Rockenbach. After citing the problems encountered in the two campaigns, the report concluded that "the value of tanks as attacking units and as a fighting arm had been demonstrated."[21] After one lecture to a group of generals Patton groused that it was "a rotten affair as they all went to sleep." He thought them all second-rate.[22]

Despite his discontent, Patton used the time to absorb valuable lessons that he would later put to excellent use in World War II. "I wish I had known as much when I was fighting as I do now but there was no one who knew and we had to learn by experience. I have been reading some German documents about tanks and they furnish the greatest compliments we could have received. They under estimated the tank and it cost them the war perhaps at least it hurried the end."[23]

Letters praising the Tank Corps arrived, and Patton was quick to note that "lots of them don't belong to me by right as I was out oft but I trained them and taught them all they know."[24] He had left his brigade in Major Brett's highly capable hands. During his recuperation, Brett had written to congratulate Patton on his promotion, to note that he was "damned sorry to see you leave the 1st Brigade," and to reassure him. "Don't worry about the old Brigade. I fought them until we had no personnel left and then we organized the remnants into a Provisional Company and gave them another whirl for their money. Just now the company is laying back at Exermont waiting to tear into them again. . . . Its a miracle you weren't killed."[25]

Patton was never known for his adulation of brother officers, who might be construed as threats to his professional advancement. But he was effusive in his praise of Brett, writing to him in late November 1918 that he wanted to put in writing "what I have long felt in my heart." [The Tank Corps'] enviable record

> both in peace and war, has been due more to your earnest and constant efforts in training and valorous conduct in battle than to that of any other man or officer. Not only did you work here when we had nothing, not even hope, without a murmur, but, in battle you fought the Brigade until there was nothing left and even after that you fought on. As far as I know no officer of the A.E.F. has given more faithful, loyal, and gallant service.[26]

Patton was confident that he would soon receive the coveted Distinguished Service Cross, but on November 17:

The most terrible thing happened to me. I heard . . . I will not get the DSC. Why I don't know as one is not even supposed to know that one has been recomended. I think that R. was in too big a hurry and put in without sufficient data. Or else some one got me from behind. The worst part of it is that once rejected you cannot again be recomended. I woke up all last night feeling that I was dying and then it would occur to me what had happened. I cannot realize it yet. It was the whole war to me. All I can ever get out of two years away from you.

But I will be G.D. [goddamned] if I am beat yet. I don't know what I will do but I will do something. If not I will resign and join the French army as a Captain or something. Gen. R. thinks my colonelcy is a compensation but it is nothing. I would rather be a second Lt. with the D.S.C. than a general with out it. It means more than an "A" and it would be of vast value in [the] future.[27]

For anyone else a promotion to full colonel might well have been more than adequate compensation, but not for Patton, who was devastated by the news and began lobbying to correct what he believed was a major miscarriage of justice before it was too late. He was somewhat mollified the following day to learn that Brig. Gen. Harry A. Smith, the commandant of the AEF schools, was recommending him for the Distinguished Service Medal. Patton had gone to Smith for advice and learned (even though it was against regulations for Smith to tell him) that " 'I have this day recomended you for the D.S.M. for having had the finest spirit and discipline in your command that I have ever seen'. . . . Any how I feel less alone in the world than I did. I just said my prayers for them both. I have a crude religion. But an everlasting love for you."[28]

At one point he learned that a colonel who knew Beatrice and Nita was the president of the AEF awards board and known to be "very fair." When the board considered upgrading DSC recommendations for six of his officers and men to the Medal of Honor, Patton thought it possible that he too had a chance of receiving the highest decoration America could bestow on its military heroes. "All I want is fairness not partiality. I would surely like to have the blue ribbon with the white stars." However, of his tank men, only Cpl. Donald M. Call and another corporal eventually received the Medal of Honor.[29]

As for his DSC, Patton learned that Rockenbach's recommendation had been favorably endorsed by the First Army chief of staff, his old friend Hugh Drum. However, before Pershing ever saw it, the AEF adjutant general disapproved it, noting that no further action was to be taken in the matter.[30] Although Rockenbach's efficiency rating noted that Patton had been "recommended for the Distinguished Service Cross for conspicuous courage, coolness, energy and intelligence in handling troops in battle," it was of no particular help in influencing the decision to grant or not to grant him the DSC.[31] A subsequent report rendered in December was slightly less effusive, noting that Patton was "very efficient, but youthful. He will, I believe, sober into one of highest value."[32]

At Chaumont, Patton called on LeRoy Eltinge, now a general, who sent him to Rockenbach. Once again Patton's glib tongue succeeded in impressing a superior to bend to his wishes. With the aid of a letter drafted by Patton, Rockenbach agreed to reopen the matter, and within days a fresh recommendation that included eleven firsthand accounts of his valor was wending its way through the military bureaucracy.

He informed Beatrice of the disappointing news that it seemed unlikely that he would be coming home in the near future. There was equal uncertainty over the future of the Tank Corps. "I don't know whether there will be a regular tank corps or not and if there is I don't know if I will stay in it. I would rather be a capt of cav[alry] than a major of Tanks but a Lt. Col of tanks might be different. Although tanks in peace time would be very much like coast artillery with a lot of machinery which never works."[33]

Too late for the war, a number of American-built Renault tanks had finally arrived in France. They were faster and better built than the French version and Patton lamented that with the ground now hard from the cold weather, "if only we could have had a few hundred of them during the war it would have been something . . . if there was only a war on. . . ." Instead writing, lectures, dances, and intramural football occupied his time. During his first football game Patton kept intact his record for new and innovative ways of injuring himself when his foot was punctured by a nail, "but not far. As it was in my left foot I still limp on the same side." Even Sundays were no longer a day of duty, a feeling Patton found "quite strange" after so many months of nonstop drilling. At a dinner dance a woman told Patton that his men revered him for his feat of bravery in crossing the bridge at Essey. "I was pretty sure it was not mined. If it had been it would not have hurt me at all as there would have been nothing left to hurt. It is funny that this small thing should stick in the minds of the soldiers."[34]

When she received his letter about the Essey bridge, Beatrice immediately wrote back, "as I read . . . about the 'little affair of the Essey bridge' which you say you 'forgot to mention;' I am dumb—If only you were near enough for me to whisper it to you. . . . Georgie, you are the fulfillment of all the ideals of manliness and high courage & bravery I have always held for you, ever since I have known you. And I have expected more of you than any one else in the world ever has or will."[35]

After clearing out Avalon, Beatrice and the children moved to Washington, D.C. It was a demonstration of her independence that she did so without bothering to ask her husband's advice or consent. However, when Beatrice wrote suggesting that she might do volunteer work, Patton admonished her, "I wish you would get over this fool idea of war work. And attend only to your self. Your hair, your chin and your tummie. I have done plenty of war work for the family."[36]

In early December word filtered through Patton's old-boy network that he would receive the DSC. He was overjoyed. When it was published later that month, the official War Department General Order for the award read "for extraordinary heroism" on September 26, 1918.[37] He wrote to thank Rockenbach. "I shall always prize it more than any thing I could have gotten in the war . . . with out your earnest effort I should not have gotten it. Thank you again."[38]

Patton was granted a seven-day leave and left for Paris, which had once again become the City of Light and was crowded with soldiers and filled with gaiety. A week earlier Truman had been granted a similar leave and found Paris "as wild as any place I saw." Despite the revelry and upbeat mood that pervaded the city, Patton was lonely without Beatrice and wrote her that "the most melancholy thing I have ever tried is amusing my self alone. I doubt if I stay the entire seven days . . ."[39] He left Paris in the possession of a police dog he named Char, who "has a long pedigree. Since marrying you I have never been satisfied with any thing but the best in dogs."[40]

When Woodrow Wilson arrived in Paris to crown his crusade for peace at the forthcoming peace conference, Patton was in the crowd that greeted him. He was more impressed with the fierce smell of bad tobacco than he was with the president, who soon signed his name to the most ruinous peace treaty in modern history.

On December 17 Patton was one of twenty-four officers and enlisted men who were awarded DSCs at a review ceremony at Bourg in the presence of the entire complement of the 1st Tank Brigade and the tank center. Patton sent the DSC ribbon and the citation to his father as his Christmas present for 1918. Patton again spent Christmas with Pershing, who presented him with a scarf. "I was the only D.S.C. there so was well pleased with my self." He cut a dashing figure with three service chevrons and a wound stripe on his sleeve and the DSC over his breast pocket. "There are all sorts of rumors about our going home soon I hope they are correct as I would like to rest up a while before the next war when ever it may be. I hope I do as well next time as I did this [one] . . . I surely am some soldier if I say so my self. . . ."[41]

The year 1918 ended as quietly as it had begun. Patton received two more efficiency ratings from Rockenbach and Gen. Harry Smith, who assessed him as "one of our very best." For Patton it was the "end of a fine year full of interest. I hope it will be the only one in which I am away from B."[42]

The advent of 1919 brought with it the twin problems of boredom and low morale for his men, whom Patton kept busy with training and housekeeping work. Still there were limits to what even he could accomplish, particularly with the men of the AEF returning home in huge numbers. During the first two months after the Armistice more than eight hundred thousand troops were discharged, but the preponderance of the AEF was not slated to return home until the spring of 1919. Many of them lived in crude encampments where mud was king and apathy its queen, and as the time passed, their morale plummeted to new lows. The mighty AEF war machine had taken a year to create and it could not be dismantled overnight. Moreover, more than two hundred thousand of the newly created U.S. Third Army were on occupation duty in and around Coblenz, and along the Rhine River. They would remain until the following summer, when Germany signed the Treaty of Versailles and it was finally determined that they would not be required to fight the German army again.[43]

Pershing's order to conduct prescribed drills was a total failure, and eventually sports and a wide range of educational programs replaced gun drills and road marches. The morale of the Regular Army officer corps suffered right along with conscripts awaiting discharge. Patton was fortunate to have become a colonel before November 11, when all promotions were frozen by the War Department. It was clear that a return to a small peacetime army would bring with it demotions from wartime ranks to permanent regular grades.[44] In Patton's case that meant captain. Officers who were wearing stars or eagles one day suddenly appeared in captain's bars or the gold oak leaves of a major on their uniforms.

On January 3 Patton was ordered to be prepared to close down his operations and move his tankers to the United States on short notice. Meanwhile Patton traveled in France and Luxembourg to give lectures and demonstrations on tanks. Not only did the experience keep him busy, but most of it was in the province of Lorraine, over which he was destined to fight a series of bitter battles a quarter of a century later. In Luxembourg he was billeted with the mother-in-law of the army commander, a major whose army was discharged after it went on strike. Patton was appalled. "This is the first country in the world to have no army. It is a horrible example of what not to do."[45]

Toward the end of January, Rockenbach learned that Patton was slated for occupation duty in the Rhineland. The growing shortage of tank officers led him to request Patton's retention for at least an additional month, by which time it was expected that his brigade would depart for the United States. The approval of Rockenbach's request was merely a deferral of the duty. Although Patton could remain at Bourg, when his troops departed he would still be transferred to occupation duty. As gung-ho as Patton was, this news arrived like a bombshell. Even he had wearied of the seemingly endless military duties that had little to do with advancing his career. Still recovering from his wound, the days on the road visiting one unit after another, and nights spent in cold, crude billets where drunkenness and wenching were commonplace, had become wearisome. He wanted only to be with his wife and family.

In addition to losing both her parents in 1918, Beatrice had not been in good health for some months. "God-Dam," he penned in his diary in a fit of rage and dismay. Enough was enough. It was time to go home.

During his final weeks in France Patton was not always pleasant to be around. "I have seen JJP [Pershing] make people cry but to day is the first time I ever did it. I surely gave one of my captains hell and he howled but it did him no good. It is a great accomplishment and I set out to do it."[46]

When he was not traveling, Patton was composing the brief treatise that was all that was ever written of his proposed book. One day he went to Chaumont and found all the AEF Medal of Honor recipients at a luncheon in their honor, hosted by Pershing and a number of high ranking generals. The generals, assisted by Patton, took the occasion to serve them. "All of them were young except one captain and one corporal. The rest were just boys but all had fine clear eyes. It struck me as a splendid contrast the Brains of the army and the brawn." One of them was former corporal Donald Call, now a lieutenant, whom Patton introduced to General Summerall. "I wish I had gotten an M.H. . . . I will get an M.H. in the next war. I hope."[47]

His time was enlivened when his presence at Chaumont coincided with a visit by the Prince of Wales (the future duke of Windsor), who was visiting Pershing and several American units. After inspecting the VIII Corps, Patton wrote Nita:

> On the way back I rode with the Prince and he told me a lot of stories supposed to be bad   some were. He said "Bein a dashed prince rather cramps ones style What?" There was a reception and later a dinner. After the dinner the prince and several of us danced to a phonograph and then he wanted to play poker but none of us knew how so we shot craps sitting on the floor. The H.R.H. got a hundred and fifty of my francs and then went to bed. He did not have much money and had to borrow to start the game . . . the next day we left for Commercy to inspect the 35th Division. There were twenty thousand men in ranks and we walked about seven miles to inspect every man   every one with a wound stripe was talked to by the Gen. and the Prince . . . On the way back I rode . . . with J. and we

talked for about three hours he told me all sorts of secret history. . . .
When I left after dinner the Prince said 'I should like awfully to nock
about with you in america on the border.' He possibly says that to every
one.[48]

The departure date of Patton's brigade was now fixed for March 1, but
until the morning of February 24 Patton remained uncertain of his status.
When his request to remain with his troops was disapproved, he used his
influence to lobby key members of the AEF staff. "My orders to accompany
the brigade home came this morning so at last I am going," he wrote to
Rockenbach. To Pershing he sent a heartfelt letter of thanks. "I have
attempted in a small way to model my self on you and what ever success I
have had has been due to you as an inspiration."[49]

Not long before he left France, Patton's father admonished him for
some of his poetry and his tendency to let his "gift of gab" get him into dangerous trouble. "You are now 34—and a Col and the dignity going with
your rank invests what you say with more importance so I hope in your
speeches you will be very careful & self restrained—for your own good &
for your future—Another gift you have developed I really regret—and that
is the ability to write verse upon vulgar & smutty subjects," which he found
both undignified and potentially hurtful. "You may some day want to enter
public life. . . . All the really big men I have known—abstained from repeating vulgar stories. . . . I don't want to preach and will say no more but I am
sure your own judgment—upon reflection will agree with mine."[50]

In March Patton and his men arrived in Marseilles after a lengthy train journey. There they boarded the SS *Patria* for the voyage to New York via
Gibraltar. Patton was pleased when the port commander informed him that
of the units that had come through the port of Marseilles, his brigade was
the best disciplined.[51] As the senior officer, he was troop commander of the
2,103 officers and men aboard the ship. Rough seas and worm-infested meat
made the ten-day trip from Gibraltar to New York miserable.

Several days before landfall, it was learned that publisher William Randolph Hearst was to be among the members of the official welcoming committee when the ship docked in Brooklyn. While the *Patria* was still in New
York Harbor, it was met by a police patrol boat and some members of
Mayor John F. Hylan's welcoming committee. A soldier yelled from the
deck, "Is William Randolph Hearst on board?" Another trooper threw a
packet tied with string into the boat. It contained a resolution condemning
Hearst for his pro-German sentiments, signed by some fifty officers and 450
men.[52] When the ship docked several officers aboard told reporters where to
obtain copies of the protest letter.[53]

Although Patton and his tankers received considerable newspaper coverage, both in New York and throughout the country, their homecoming was the subject of controversy when the *New York Herald Tribune* reported the affair as headline news. In a front-page article under a provocative headline, the *Herald* declared that Patton's men believed Hearst was "un-American, pro-German and inhumanitarian . . . [and that] Colonel George Smith Patton, jr., an officer of the regular army, whose home is in San Gabriel, California, did not sign the document. It was said that as an army officer in command of troops he did not feel that it was proper for him to align himself with any factional protest."[54]

Patton was thoroughly dismayed and momentarily convinced that the protest would seriously mar what otherwise turned out to be a triumphant return, perhaps even resulting in a court-martial. His role in the Hearst affair remains one of denial that he knew anything about it before the unwelcome newspaper coverage. In letters to Pershing and his father, Patton disclaimed any advance knowledge of this action, claiming that even if they disagreed with Hearst, they ought to have done nothing. In a fit of unjustified paranoia, he told Papa: "Some fools must have . . . done it to get me in trouble . . . I hope I don't get hit for something of which I was perfectly inoscent."[55]

He need not have worried, for once the *Patria* docked, the protest was soon lost in the euphoria of homecoming. Patton became the darling of the press, whose reporters quoted him and headlined him in articles chronicling the exploits of the tank corps. In the *New York Herald* his photograph appeared under a caption reading: TANK FIGHTERS OF NEW YORK AMONG 2,110 BACK HOME, COLONEL PATTON TELLS HOW BIG MACHINES BY HUNDREDS ATTACKED GERMANS.[56]

And in newspapers from the *Washington Post* to the *Richmond Times-Despatch,* Patton was the man of the hour. His hometown *Los Angeles Herald* reported, TANK VICTORY OF YANKS IS DESCRIBED BY PATTON. [57]

For a man who two years earlier had been apprehensive over his future, Patton had come a long way. After leaving New York for France in May 1917 as a junior captain, he now returned as the leading battlefield commander and expert of the Tank Corps. He wore the eagles of a full colonel and on his breast was the DSC, four battle stars and the French croix de guerre. Soon he would add the Distinguished Service Medal (DSM) to his ribbons. As the voyage ended, Brett presented Patton with a letter signed by himself and forty-nine of the sixty-five tank officers aboard the *Patria* that read: "A testimonial of personal affection for Colonel Patton for his energy, his leadership, his courage, his constant attention to the welfare of his officers and men, his understanding and foresight, his sterling sense of justice and fairplay."[58] It was a touching and unexpected tribute from his loyal officers.

There were festive reunions of hugs, kisses, and tears, and euphoria swept the Brooklyn docks as the returning soldiers were reunited with their loved ones. Among those present to greet her man was Beatrice Patton. The only record of their reunion is daughter Ruth Ellen's recollection that, "I have been told that he was walking with a cane because of his wounded leg, but that when he saw Ma standing on the dock, he laid down his cane and walked down the gang-plank unaided."[59]

Whatever the exact circumstances of his return, the war was finally over for Patton. His moment in the spotlight was to prove fleeting. Ahead lay the inevitable reductions in rank and the barren interwar years, during which Patton would need every ounce of perseverance to survive in an army that was to prove itself as woeful as its pre-1917 predecessor.

PART VI

The Interwar Years
(1919–1939)

Where the hell am I?

—PATTON

CHAPTER 20

Eisenhower, Patton, and the Demise of the Tank Corps

(1919–1920)

The "war to end all wars" had been fought; the swords had been beaten into ploughshares; the military appropriations had been cut—all was to be well.

—RUTH ELLEN PATTON TOTTEN

Patton's brigade was assigned to the new permanent home of the Tank Corps at Camp Meade, Maryland. In 1919 Meade's primary function was to help demobilize the massive military force created to fight the war. Patton had barely arrived when, in mid-April, he was ordered to temporary duty in Washington, where he and several other veteran tank officers constituted a tank board to examine and recommend the basic doctrine of the peacetime Tank Corps, and how it ought to be organized, trained, and employed in future wars.[1] For several months Patton and his team pored over records and reports and traveled extensively to visit the Springfield (Massachusetts) Armory, the Rock Island (Illinois) Arsenal, the Holt Caterpillar Tractor factory in Illinois, and a tank works in Davenport, Iowa, where his face became swollen. He was diagnosed with chicken pox and laid up for several days until he was able "to get looking decent again." His ever-present fear of growing old led him to observe: "I was delighted to find that I was still young enough to have it."[2]

Patton's final efficiency ratings from France arrived later in 1919, including one from Pershing. All were laudable, especially one by Brig. Gen. Harry Smith, who rated him "one of the most active and forceful officers in the service . . . [and] one of the strongest officers in the Army."[3] The AEF had also approved Patton's Distinguished Service Medal, and he proudly informed his father that he thought only he and MacArthur held both the DSC and DSM. Later, however, he put himself down by writing that it was common knowledge the only reason he won the medal was because of the splendid manner in which the Tank Corps carried out its mission: "They won the medal and fortune pinned it on him."[4]

Despite the long separation from his wife and family, the task of re-establishing his brigade at Camp Meade meant little time for a leisurely reunion. It was midsummer of 1919 before he was granted an extended leave to visit his family at Lake Vineyard. The only record of that visit is Patton's brief remark: "Papa and I had long talks about the war. . . . He was so proud of me that he embarrassed me when he presented me to his friends. He always said: 'Mr. so and so you remember my son Colonel Patton.' Neither he or mama ever made an actual fuss. The nearest I can recall is when one day Mama called me 'Her hero son.'"[5]

Beatrice bluntly informed her husband that she refused to remain apart from him while he was stationed at Camp Meade, even though it was within commuting distance of their leased home in Washington, D.C. Although there were few officer's quarters at Camp Meade, Beatrice insisted that they find a means to create their first home together since Fort Bliss. Ruth Ellen remembers that her father managed to requisition an entire disused wooden barracks, covered by tarpaper, in the middle of a sandlot, "which Ma turned into an unforgettable home. The only paint available at the Quartermaster stores was blue and yellow. So, the whole part of the barracks we lived in was painted blue, yellow, blue and yellow, yellow and blue. . . . The latrine presented a problem in decor, but she solved this by planting trailing ivy in the urinals."

Their barracks home was such a dangerous fire hazard that cooking inside was banned, and the Pattons had to eat in the mess hall, where most meals came out of a can. Beatrice soon issued a fresh—and nonnegotiable—ultimatum: "She *would* live at Camp Meade and she *would* have a kitchen. . . . A day or so after her 'pronounciamento,' a much harassed Georgie appeared in a tank, hauling a timber sled on which was a small signal house he had found abandoned on the range. 'Here's your goddam kitchen,'" he said. With the assistance of military prisoner labor, Patton managed to erect a foundation for the new family kitchen and connected it to their barracks with a covered boardwalk. Beatrice was thrilled and thought it reminiscent of Mount Vernon, where the food was also cooked in a kitchen located outside the main house. It too received copious applications of the ubiquitous blue and yellow paint. For once Beatrice had a large complement of hired help, which included a housekeeper, a governess, an English cook, and six Mexican servants. The Patton stable was soon bursting with a dozen horses. George had also purchased a Pierce-Arrow, noting that, "I can afford it and believe in enjoying my self between wars."[6] The Pattons were now a two-automobile family.

Since 1918 the Tank Corps in the United States had been headed by Col. Ira C. Welborn, an infantryman and Medal of Honor winner during the Spanish-American War. Ever loyal to Rockenbach and suspicious of Welborn, whom he did not know, Patton wrote to his former commander that "Col. W. is dead from the neck up" and urged Rockenbach to hurry home to claim the job rightfully for himself, which he did in the summer of 1919.[7] When the Rockenbachs came to dinner one evening, Beatrice had neglected to remove two signs she had placed in the dining room for the benefit of her two daughters. One behind Bee said, SWALLOW, and another behind Ruth Ellen implored, CHEW. Mrs. Rockenbach smiled and said, "We have the same trouble at our house too, Mrs. Patton. The General never stops chewing and I never stop talking. So you see, we can still learn from the young."[8]

Although they barely knew him, Patton's two young daughters were thrilled actually to have him around, but it did not last. "Bee and I had been led to expect so much—a knight in shining armor, a playmate, a fearless killer of the dreaded Hun, a teller of tales. What we got was far below our expectations." Patton found as much difficulty adjusting to his own family as he did to the peacetime army. Ruth Ellen later wrote:

> I realize now that he was in considerable pain at the time; worried about his future in the tank corps of his creation; and having a hangover from the war, which is a very real thing. A man goes from the command of thousands of men where his judgement means victory or defeat, life or death, to the shrinking command of a handful of men, and the narrowing horizons of peacetime duty with not enough money and not enough troops, and the tender trap of home and family—and, it is a let-down. I guess things didn't come up to Georgie's expectations either.

A stern and uncompromising father, Patton found it difficult to relate to his two daughters, both of whom he perpetually resented for not having been born male. As the youngest, Ruth Ellen had a particularly tempestuous relationship with her father. Now four years old and, unlike her older sister, too young to have had any recollection of George Patton until he stormed back into their lives in the spring of 1919, she thought he was an ogre. Everything I did was wrong. I will never forget the first time he ever spoke directly to me. I had rushed into the house not knowing he was there, and he was sitting on the living-room floor among a lot of dismembered guns which he was cleaning and re-assembling. He looked up at me as I hovered in the doorway, and with his really charming smile said; "Hello, little girl." I was so overwhelmed with the attention that I burst into tears and howls, and he began to yell at Ma to "come and take the baby away," that she was "making a goddam awful noise," and that he hadn't touched her.

Shortly after Beatrice admonished him to try harder, Patton bought his daughters a white bull terrier puppy named Tank, the first of several bull terriers, the most famous of which was his World War II companion, Willie. Tank turned out to be stone deaf, and the Patton children communicated with Tank by banging on the floor. Despite his handicap, the dog somehow managed always to be at the door to greet his masters when they arrived home. The major difference between Tank and Willie was that even deaf, Tank was a fighter who never gave ground to invading dogs who frequently attacked him but almost always lost. Much to Patton's dismay, Willie turned out to be a coward.

Eventually Patton relaxed in the presence of his family and "began to tell us the wonderful stories that Ma had promised us. He was a great raconteur." The children's favorite story was a lengthy fable he made up about Reginald (Shark) Fulke, the Black Count of Anjou. Beatrice would chide him for the "raw history" he imparted, but Patton would respond, "It's history, Bee, history! You dug all this out of the archives at Saumur yourself! Do you want the girls to grow up ignorant?" The Patton girls loved it. "As the stuff of history is often cruel and bawdy and bloody and unfair, that's the way Georgie told it. Bee and I were the age where we particularly enjoyed the Rabelaisian bits." Using his great knowledge of history, Patton created an epic tale of medieval life and times with no detail left to the imagination. Later Patton's half-aunt Ophelia Smith hired a genealogist, who traced the family line to an illegitimate son of Edward III. "Georgie was simply charmed to think that somewhere in the misty tangles of his background there was a genuine no-good, hell-blasted, fascinating Fulke."

During her childhood the strong-willed Ruth Ellen would struggle with her father, who approached fatherhood much as he did disciplining a military unit. In 1927 Beatrice and Bee were in the United States, and twelve-year-old Ruth Ellen and three-year-old George (Beatrice had finally had her much-desired third child in 1924) were left in the care of their father in Hawaii. As a pampered child, Patton had never been forced to eat anything he disliked, and, on the rare occasions he was pressed, Aunt Nannie would always manage to get him off the hook with her patented hysterics. When they were first married, Beatrice was appalled at his wasteful practice of eating only what he pleased. To appease his wife, Patton supervised his children's eating habits with an iron hand and began to revere a clean plate, which he knew she rated second only to godliness. One day the family servant served breadfruit for lunch, and, never having heard of this Hawaiian delicacy, Ruth Ellen refused to consume it. Her equally stubborn father insisted she would eat it or "take a licking." Ruth Ellen chose the licking.

Patton ordered her to pick out her own whip. "Scared to death, and mad as hell, I picked out what I thought was the worst whip on the rack. It was a hunting whip with a long thong with a red cracker on the end of it. . . . I took the hunting whip back to the dining room, and nonplussed, he cracked it and then laid it along my . . . legs four times." Patton's young son, George, sat at a nearby table goggle-eyed at what had just happened to his sister, who, "swollen with rage and self-pity," defiantly proclaimed: "THAT didn't hurt." Some years later Beatrice admitted to Ruth Ellen that her father had written to confess the incident and offered her a divorce for having mistreated her child.

While Patton was stationed at Camp Meade an incident took place that well illustrates Beatrice Patton's devotion to her husband. The Pattons were attending an elegant white-tie dinner in Washington one evening. Patton left Beatrice at the door while he parked their car. As Beatrice waited in the foyer, an older, overweight (obviously deskbound) officer observed Patton enter, resplendent in his dress uniform and medals and muttered, "Just look at the little boys they are promoting to Colonel these days; look at that young chicken still wet behind the ears, wearing a Colonel's eagle." The next thing Beatrice remembered was sitting astride the officer's shoulders, banging his head on the black-and-white-marble floor tiles "while Georgie and several others were trying to pull her off."[9]

While Patton was on detached service in Washington, he was temporarily replaced as commanding officer of the 304th by a lieutenant colonel named Dwight David Eisenhower, who was better known to his friends as Ike. However, Eisenhower was soon posted to a detail that would take him across the United States and neither officer returned to Camp Meade until the autumn of 1919, when they met for the first time.[10] Patton commanded the light tanks of the 304th Brigade, while Eisenhower commanded a battalion of newly manufactured Mark VIII Liberty tanks that had come off the assembly line too late to be used in France.

Dwight Eisenhower was an infantry officer who had been detailed to the Tank Corps the previous year. To his intense disappointment he was sent to Camp Colt, in Gettysburg, instead of France, in March 1918. Near the site of Pickett's Charge, where Waller Tazewell Patton had died, Eisenhower established and ran the largest tank training center in the United States so impressively that he was promoted from captain to lieutenant colonel in seven months.[11]

The two men were vastly different. Patton was a combat veteran and five years older than Eisenhower, who graduated from West Point in 1915, and would become the crown jewel of "the class the stars fell on."[12] Patton seemed monumentally egotistical next to the more self-effacing Eisenhower who was particularly struck by the incongruity of this impatient, self-confident, elegant soldier with his high, squeaky voice. Patton loved horses, riding, and polo; Eisenhower suffered from a serious knee injury sustained playing football at West Point and had no interest in polo. Both men were avid riders, hunters, and pistol shooters. One was wealthy and could afford an extravagant lifestyle; the other was dirt poor and had difficulty making ends meet. For a time the two played poker together twice a week. One could afford to lose; the other managed to supplement his meager pay by a shrewdness at the poker table that would later serve him well. One of Eisenhower's biographers has noted another significant difference: "Everybody thought from the beginning of Georgie's career that there were no limits to the heights he might achieve. For most of his life very few people thought that Eisenhower would achieve anything much."[13]

On balance they seemed a genuinely mismatched pair: the brash, outspoken cavalryman-tanker and the fun-loving midwesterner whose roots were in the infantry. Yet the two soon forged an enduring friendship that lasted until shortly before Patton's death. As Eisenhower would later write: "From the beginning he and I got along famously. Both of us were students of current military doctrine. Part of our passion was our belief in tanks—a belief derided at the time by others."[14] It was also a relationship that "was to delight and dismay [Eisenhower] for the rest of his life."[15] The only other thing they had in common was that each was losing his hair at an alarming rate, a "competition" that Eisenhower would win hands down.

They shared a knowledge of tanks and a passionate belief that the tank had a greater role in the army of the future than as a mere subordinate arm of the infantry. Both thirsted for something more exciting than the bland peacetime existence of Camp Meade, where Eisenhower spent his leisure time gardening and perusing a Burpee seed catalog for entertainment. The highway to the main encampment had been plagued by a number of holdups, and their boredom led Eisenhower and Patton to undertake their own version of "cops and robbers." Armed with half a dozen pistols, they drove very slowly down the road in the hope they would be attacked. According to Ike's son, John, "Dad always said, 'We wanted to see what a fellow's face looked like when he's looking into the other end of a gun.' They were both going to pull guns in different directions and fix this guy. They thought they were going to be a two-man posse on the blacktop road there, but nobody stopped them. They were both disappointed," especially

Patton, who would have relished a repeat of the Rubio shoot-out.[16]

In 1919 the temperance movement achieved passage of the Eighteenth Amendment, ushering in Prohibition. Both men avidly joined in the new all-American pastime of distilling their own bootleg alcohol. Eisenhower concocted his gin in an unused bathtub, while Patton brewed beer in a woodshed, storing the bottles in the covered walkway between his quarters and the kitchen shack. One evening there was a sudden noise that sounded like a machine gun, followed by a series of soft booms. Patton instantly dived to the floor, and the cook began screaming in the kitchen. The beer had exploded. "Georgie got up, rather shamefacedly, and explained that it had sounded so much like hostile fire that he instinctively had taken cover." Beatrice "laughed and laughed and called him 'her hero,' and he got very red."[17]

Many were misled by Eisenhower's easygoing manner and charming smile, a disarming facade that masked a brilliant, ambitious officer who thirsted as deeply as Patton did to advance his chosen career by answering the call of war. He was also a strict disciplinarian with a terrible temper, which, when triggered, often caused its recipients to quake in their boots. In his own less obvious manner, the fires within Dwight Eisenhower burned as passionately as those of George S. Patton. When World War 1 ended just as he was about to deploy his unit to France, a frustrated Eisenhower lamented: "I suppose we'll spend the rest of our lives explaining why we didn't get into this war. By God, from now on I am cutting myself a swath and will make up for this."[18]

Eisenhower and his wife, Mamie, were neighbors of the Pattons and like them, expended considerable money and effort in remodeling their barracks into a suitable home (including a lawn and a white picket fence), the only difference being that Dwight Eisenhower could barely afford the expense. Patton's daughters became enchanted by Mamie Eisenhower, whom they found "the most glamorous creature that ever appeared in our lives. She insisted, very daringly we thought, that we call her Mamie and not Mrs. Eisenhower, as she said she wasn't that old yet." Beatrice Patton disapproved of her children calling any adult by their first name but could do little when Mamie insisted. She "wore filmy negligees most of the day, and drank a lot of iced tea, which she stirred by swirling it around in her glass. We thought this the ultimate in sophistication, and tried to do the same with our milk, with bad results."

Despite the budding friendship between their husbands, Beatrice Patton and Mamie Eisenhower were never close and seemed to share only the common trait of frailty, although Beatrice's occasional bouts of ill health never deterred her from leading a vigorous, athletic life. Whereas the urbane Beatrice was an avid participant in life and a major influence in promoting her husband's career, Mamie was mostly an unsophisticated and untraveled spectator. Although a woman of wit and charm, she could barely cook, sew, or make a bed, and

> worried about her health all the time . . . suffered from claustrophobia but could not walk any distance in the open . . . hated the seaside, disliked going abroad, was afraid of flying, was prostrated by heat and cold, was terrified of insects, and became victim of a host of complaints, physical and psychological . . . Nevertheless, Mamie was tougher than she seemed and she outlived Ike by nearly a decade.[19]

Although Woodrow Wilson's personal utopian vision of peace was a fantasy, it was his specter of a postwar world in which peace reigned supreme that gripped the United States in the aftermath of the Armistice. Wilson had taken Europe by storm as the living embodiment of a golden age in which war would be relegated to the history books. The creation of the League of Nations was to be Wilson's eternal contribution to the world. However noble its intentions and potential role as the world's peacekeeper, the League instead became a toothless tiger, relegated by the concept of persuasion without force to the status of an inconsequential body of diplomats and politicians. As historian Robert Leckie observes: "Not even Woodrow Wilson could have envisioned . . . [that] the death of the dynasties and the end of empires was to be so quickly succeeded by the day of the dictator."[20] The surviving veterans of the German Army were bitter at what they believed was the perfidy of the Jews and the politicians; many, including Corporal Hitler, vowed one day to avenge their betrayal at Versailles, and from within.

In the United States, the aversion to all things military became virtually a national mania. Militarism in any form was replaced by a militant pacifism that took seriously the notion that America had indeed fought "the war to end all wars." Veterans who months earlier had been wined and dined as national heroes suddenly found themselves shunned. The nation had lost interest in war and in the symbol of war, the military establishment. "Wrenched by fears of radicalism, economic depression, and renewed entanglement in Europe's distresses, the American people yearned for conditions of tranquility, innocence, and isolation."[21]

Demobilization occured on such a massive scale that it managed to overshadow the purge of the Union Army after the Civil War. By the end of June 1919 the army had discharged more than 2.6 million enlisted men and 128,436 officers, and by January 1920 the Regular Army had been reduced to only 130,000, including the troops still on occupation duty in Germany.[22] After the terrible bloodshed in Europe, demobilization bred disinterest in future war or in such mundane subjects as the role of the tank. Interest in developing tactics and learning the possible future uses of the tank had largely evaporated. To all but a handful of officers, the army seemed to care only about reverting to its prewar size. The last thing anyone wanted to hear was talk of another war from professional soldiers. Men like Patton and Eisenhower thus became anachronisms, even within their own army.

Rockenbach displayed little inclination to buck the tide, which he clearly saw was running against him. When Brig. Gen. Fox Conner took command of an infantry brigade in Panama, he asked for Eisenhower as his executive officer. Even though Ike was now an intimate part of his small circle of senior Tank Corps officers, Rockenbach refused to release him for another reason: Since 1919 he had successfully coached the Tank School football team. Retaining a top-notch football coach in the 1920s was considered as important as being successful in one's other duties.[23]

Isolationism, and an obsessive panic over the Russian revolution and the spread of communism dominated the Europe of the 1920s. While Europe went its own disastrous way, there was no enthusiasm for either a large American standing army or in spending badly needed appropriations to sustain a permanent military force. Thus, with barely a whimper, what was left of the Regular Army establishment settled into an interwar period of hibernation.

Although Patton would come full circle in his outlook and thinking during the 1920s and 1930s, his was (along with those of Bernard Law Montgomery, J. F. C. Fuller, and Basil Liddell Hart in Britain and Charles de Gaulle in France) one of the few military voices struggling to be heard in a wilderness of indifference. By contrast, Heinz Guderian and an emerging new breed of German officers were soon gaining attention for their innovative ideas.

When the National Defense Act became law in June 1920, the Regular Army was authorized 300,000 men, but only 200,000 were left from what had once been a gigantic force of 3.7 million. By January of the following year, Congress mandated a reduction in enlisted strength to 175,000, and in June 1921, a further reduction to 150,000 was ordered. The ax fell again the following year, when Congress decreed that the active army consist of no more than 12,000 commissioned officers and 125,000 enlisted men, a strength at which it remained until 1936.[24] The entire War Department appropriation was established at about $300 million annually, and with the navy considered the first line of defense, the army's share was woefully insufficient to meet its needs. In 1922 the United States ranked seventeenth among nations with standing armies, and in 1933, the chief of staff, Gen. Douglas MacArthur, noted that what few tanks the army possessed (except for a dozen or so experimental models) were "completely useless for employment against any modern unit on the battlefield."[25]

The roaring twenties in the United States have been described as a period of "unrestrained hedonism."[26] Woodrow Wilson, who had led the nation like the college professor he had once been, died, to be replaced in the White House by Warren G. Harding. Harding and his successor, Calvin Coolidge, were arguably the two most "hands-off" presidents ever to occupy the Oval Office. They gave America exactly what it wanted: do-nothing government. Harding's incompetence spawned high government corruption (the Teapot Dome scandal), and Coolidge's principal claim to fame was ensuring that he spent from two to four hours of his workday sleeping—and that the country knew it. Prohibition brought with it the speakeasy, racketeers, bootleggers, and a significant rise in crime and corruption. America seemed to live for the moment, with an unquenchable thirst for excitement, no matter how trivial or outrageous. Lurid headlines trumpeting murder, kidnapping, and sex scandals were the lifeblood of the tabloid newspapers.

The U.S. Army of the interwar years was all but forgotten and became so emasculated that at any given time only one-quarter of the Regular Army officers and one-half of the enlisted men were available to fill positions in fighting units, many of which existed simply on paper. (Of the nine authorized divisions only three actually existed.) The few Infantry units that did exist were scattered among as many as forty-five separate army posts. The sad truth was that the army had come full circle from its frontier days before World War I, and what had been one of the world's most modern armies was again a second-rate force kept afloat only by dedicated senior officers like Pershing, MacArthur, Summerall, Fox Conner, and junior officers such as Marshall, Eisenhower, and Patton, who stuck it out despite appalling living conditions, reductions in rank to prewar levels, miserable pay, and no incentives whatsoever to remain soldiers.

Patton was hardly alone in wondering why he did not resign his commission. Brilliant airmen like his friend Gen. Billy Mitchell were stymied in their belief in the future of air power. It was a measure of the times that Mitchell would be court-martialed in 1925 for insubordination for making public his quarrel with the War Department General Staff over his advocacy of air power and an independent air force. During these dismal years it was not uncommon in communities where military men were stationed to find signs ordering dogs and soldiers to keep off the grass.

Eisenhower has credited Patton with first introducing him to Fox Conner, Pershing's erstwhile chief of operations in the AEF and one of the most brilliant officers ever produced by the army. Over the years Conner became Eisenhower's teacher and a father figure whom he admired above all others: "In a lifetime of association with great and good men, he is the one . . . to whom I owe an incalculable debt." The two were introduced at a Sunday dinner at the Pattons', and Eisenhower would later note that perhaps the greatest reward of his friendship with George S. Patton was meeting Conner that fateful afternoon in the autumn of 1920.[27]

The dinner to which Conner and Eisenhower had been invited was typical of the social affairs hosted by George and Beatrice Patton. It was a custom they had learned well from their parents, of creating a setting of congeniality, in which good food and drink were the stimuli for meaningful conversation. In an era when radio was still the newest of inventions, such events were the cornerstones of the social life of families like the Pattons. After dinner on this particular day, Patton and Eisenhower spent the remainder of the afternoon touring Camp Meade, conversing with Conner about their ideas for the future of the tank, and attempting to answer his many questions. It was another indicator of the new era that Fox Conner was one of the few officers to even express an interest in what they were doing, and the only one ever to provide any encouragement.[28]

Their brief time at Camp Meade brought together two of the pivotal future leaders of World War II, just as the army was sliding to its nadir. Both might justifiably have walked away from any responsibility for the future of a spiritless army. Fortunately, however, its soul was held together by the dedication of these men and others like them who argued that the tank of the future ought to be more than merely an auxiliary arm of the infantry. As Eisenhower has explained: "George and I and a group of young officers thought this was wrong. Tanks could have a more valuable and more spectacular role. We believed . . . that they should attack by surprise and mass. . . . We wanted speed, reliability and firepower."[29]

By themselves they stripped a tank down to its last nut and bolt and managed to put it back together—and make it run. They tinkered with supporting weapons and endlessly debated tank employment and the tactics of surprise. In what both acknowledged was one of the most frightening moments of either man's life, one day they narrowly escaped death when a cable stretched taut between two tanks suddenly snapped and barely missed their heads as it cut a lethal path at lightning speed through the nearby brush and saplings. Patton likened it to his near-death wounding at Cheppy, while Eisenhower later wrote: "We were too startled at the moment to realize what had happened but then we looked at each other. I'm sure I was just as pale as George. That evening after dinner, he said, 'Ike, were you as scared as I was?' 'I was afraid to bring the subject up,' I said. We were certainly not more than five or six inches from sudden death."[30]

Another time, Patton was test-firing a .30-caliber machine gun while Eisenhower observed the trajectory of the bullets through field glasses. Without warning the weapon "cooked" and began spewing bullets everywhere. The two future generals raced off in panic but returned to disable the gun with sheepish expressions on their faces. After this second near-disaster both decided they had about exhausted their luck. Nevertheless, they kept at it within a small circle of converts, among them Serano Brett, learning lessons, analyzing tactics, and creating scenarios in which tanks were employed in a variety of situations and conditions. As Eisenhower later said of this period in their careers: "These were the beginnings of a comprehensive tank doctrine that in George Patton's case would make him a legend. Naturally, as enthusiasts, we tried to win converts. This wasn't easy but George and I had the enthusiasm of zealots."[31]

Patton penned a steady output of specialized reports and essays, many of which dealt with technical aspects of the tank—its design, maintenance, and operation. Always searching for new information and ideas, Patton even wrote to the Naval War College to inquire how the navy maneuvered and fought its warships because tank engagements would resemble somewhat "the nature of a sea fight."[32]

Patton, Brett, and several other officers traveled to Hoboken, New Jersey, to examine the Christie convertible, M1919, a revolutionary new tank designed by a former Ordnance Department technician, inventor, and race-car builder and driver named J. Walter Christie, whose Front Driver Motor Company had just designed and built the first tank to meet specific military requirements.[33] Despite some serious mechanical flaws, Patton and his team were enthused about what they perceived as its great virtues: speed (up to sixty miles per hour with tracks), and mobility (it could climb a two-and-a-half-foot wall and jump a seven-foot ditch). Their report led to a favorable endorsement of Christie's project by the Tank Corps and a contract from the Ordnance Department to design a new tank for the army. Historian Eric Larrabee notes that Christie was not only "the unsung hero of American military technology," but that Patton's vision enabled him to find in Christie a kindred soul who believed as he did in the necessity for mobility in armored vehicles of the future: "In the Christie we are buying a principle not a vehicle."[34] For the next decade Patton would continue to work with and support Christie's attempts to refine his tank. It has been suggested but never proved that Patton may have helped privately to finance Christie.[35]

Although the project would eventually flounder and be discontinued in 1924, Christie nevertheless remained an important link in the evolution of the modern tank and again displayed his genius in 1922, when he successfully demonstrated for Rockenbach his latest invention, an amphibious vehicle that swam across the Hudson River from Hoboken to New York.[36] This vehicle was a forerunner of the DD amphibious tank utilized in the invasion of Normandy on D Day, June 6, 1944.[37]

Although the U.S. Army never fully appreciated either Christie or his revolutionary tank, the Soviet Union saw its potential, and in 1928 two of his tanks were sold to the Russians, who eventually copied and refined what they called the "Christie-Russki" into what became perhaps the finest tank of World War II, the famous T-34 that broke the back of the great German panzer offensive in 1943 at Kursk, ensuring the eventual defeat of the Wehrmacht on the Eastern Front.[38]

In 1920 both Eisenhower and Patton published provocative articles in the *Infantry Journal*.[39] Eisenhower's stressed the need for a newly designed tank of the future, armed with two machine guns and a six-pounder main gun, able to cross nine-foot trenches and speed cross-country at twelve miles per hour. The clumsy, inefficient machines of Saint-Mihiel and the Meuse-Argonne were now relics of the past and "in their place we must picture this speedy, reliable and efficient engine of destruction. . . . There is no doubt . . . that in the future tanks will be called upon to use their ability of swift movement and great fire power . . . against the flanks of attacking forces."[40]

Patton's article contended: "The tank is new and for the fulfillment of its destiny, it must remain independent, not desiring or attempting to supplant infantry, cavalry, or artillery. . . . Like the air service, they are destined for a separate existence. The tank corps grafted on . . . [one of these] will be like a third leg to a duck—worthless for control, for combat impotent."[41] In his lectures, too, Patton continued to insist that the tank should not be incorporated into the infantry.

Patton was reassured that at least he was not alone when, in 1920, J. F. C. Fuller published an important book about tanks and their role in the Great War. Among Fuller's arguments was that if the British had been able to put two fresh brigades of medium tanks into the Battle of Cambrai, for less than one-fifth of a single day's cost of the war, "the greatest war in history might have closed on or near the field of Waterloo in a decisive victory ending in an unconditional surrender or an irretrievable rout."[42]

What Patton and Eisenhower had concluded was heresy to the leaders of the infantry, with its long tradition (and privilege) of being the decisive arm of war and the "Queen of Battle." What these two upstart tank officers were suggesting would alter the whole doctrine of land warfare. By combining maneuver with the traditional supporting artillery fire, they began to make the case that well-designed tanks could maneuver in mass and either outflank an enemy position or, as Fuller suggested, tear gaping holes in an enemy line and precipitate the collapse of an entire front. The possibilities of such tactics were nothing less than breathtaking. In future wars the infantry would close on the enemy forces and hold them in place, while tanks enveloped and either destroyed them or set the stage for the infantry to win a decisive victory at a fraction of the cost in blood of the linear battles of the past.

Some infantrymen naturally regarded Eisenhower's article as blasphemous, and in the autumn of 1920 he was summoned to Washington for an unpleasant audience with the chief of infantry and warned that his ideas "were not only wrong but dangerous and that henceforth I would keep them to myself. Particularly I was not to publish anything incompatible with solid infantry doctrine. If I did, I would be hauled before a court-martial. . . .

George, I think, was given the same message. This was a blow. One effect was to bring George and me even closer."[43] The two began spending even more time together, riding, studying, stalking bandits, and cementing a friendship that both prized. Eisenhower noted in his memoirs: "With George's temper and my own capacity for something more than mild irritation, there was surely more steam around the Officers Quarters than at the post laundry."[44]

Anticipating their eventual attendance at the prestigious School of the Line (as the future Command and General Staff School at Fort Leavenworth, Kansas, was then called), Patton obtained copies of previous tactical problems used there and said to Eisenhower, "Let's you and I solve these together. . . . I worked the problems with him . . . and got a lot of fun out of it. We'd go to his house or my house and the two of us would sit down, and while our wives talked for the evening, we would work the problems . . . and grade ourselves."[45] Patton thought Leavenworth's authors were overcautious and their solutions "much too timid. Time is waisted." The two officers also reworked each problem by adding tanks to the scenario and were pleased when the side with armor "won." However, it was not all work and no play. They made sure there was ample time set aside for their twice-weekly poker game.[46]

For all their compatibility, however, there was a fundamental disagreement between the two men, for, as Stephen Ambrose writes, "they began an argument that would last until Patton's death. Patton thought that the chief ingredient in modern war was inspired leadership on the battlefield. Eisenhower felt that leadership was just one factor. He believed that Patton was inclined to indulge his romantic nature, neglecting such matters as logistics, a proper world-wide strategy, and getting along with allies."[47] Although they would not serve together again until World War II, the passionate discussions of their Camp Meade days would continue during the interwar years by correspondence.[48]

However deep their commitment to the future of the tank, neither was actually willing to stick his neck out publicly in the manner of Billy Mitchell, whose court-martial for calling his superiors "incompetent, criminally negligent and almost treasonable" became a national *cause célèbre*. Patton may have been zealous in the practice of his profession, but, as one of his early biographers has observed, he simply could not envision a time when the nation "would open its purse to the tune of billions of dollars to equip not only our own armies but also those of our Allies with trucks and tanks and jeeps and mobile artillery," all of which were the essential ingredients of a mechanized army.[49]

Nor was he a revolutionary who was prepared to sacrifice his career for a cause that he deemed hopeless. Nevertheless, until his return to the cavalry, Patton used any excuse he could think of to induce continued interest in tanks by anyone in a position of influence. In 1920 he wrote in the *Cavalry Journal*: "The tank is a special, technical and vastly powerful weapon . . . give it half a chance, over suitable terrain and on proper missions and it will mean the difference between defeat and victory."[50]

In 1925 Eisenhower signaled that he had clearly understood the message from his chief. As a student at the Command and General Staff School, he authored a paper asserting that it was "apparent that tanks can never take over the mission of the infantry, no matter to what degree developed. Advancing infantry will continue to be the deciding factor."[51] Patton, too, retreated into the conventional ideas of the time and sought a different approach that would educate his fellow officers in other aspects of leadership and soldiering. In so doing he forfeited an opportunity to become the father of the American armored force of World War II, a role that was taken over by other visionary officers.

Nevertheless, both Patton and Eisenhower made significant contributions that became evident only during the Second World War. Ambrose credits both men:

> As World War II would prove, the two young officers had it exactly right, so right in fact that their conclusions seem commonplace today. But in 1920 they were two decades ahead of most military theorists, except for Fuller and Liddell Hart, whose works were just beginning to appear in any case. Eisenhower and Patton were true pioneers, original and creative in their thought. But the Army was not pleased. The Great War had been won by infantry, charging in mass. Future wars would be won the same way. The Army's postgraduate schools all taught that basic lesson, all scorned maneuvering and outflanking attempts and concentrated on the problem of how to get the infantry across no-man's land.[52]

What Patton and Eisenhower did together was what Patton would continue to do for nearly twenty years: to study and prepare for the day when his knowledge might actually be put to use. But with the lethargy of the interwar years, even Patton eventually concluded that the future of the tank service was indeed bleak, and without funds and commitment it had become a dead-end career.[53] Few of the old army cavalrymen had succumbed to the lure of the iron horse, and with its demise only a matter of time, these men remained adamant that one of the most elite arms of the U.S. Army not be permitted to die. Peacetime offered them a return to the nostalgic days of polo, mounted drill, and the routine of a simpler, romantic life that cavalrymen everywhere prized.

In 1919, during congressional hearings on a bill to reorganize the army, both Secretary of War Newton Baker and the chief of staff, Gen. Peyton C. March, argued for retaining the independence of the Tank Corps.[54] Patton's friend, Maj. Gen. Charles P. Summerall, the wartime commanding general of the 1st Division, spoke in favor of retaining an independent tank arm, but a week later whatever positive effect Summerall may have made was torpedoed by Pershing. "The Tank Corps," he told the committee, "should not be a large organization," and ought to be placed "under the Chief of Infantry as an adjunct of that arm."[55] Rockenbach's arguments for an independent corps were too conservative, and neither passionate nor persuasive enough at the most critical moment in the history of the tank service. In the end it hardly mattered. Pershing's towering reputation and his opposition sounded the death knell of the Tank Corps. A War Department mandate ensured that the corps would lose its identity and with it, its *raison d'être*. For Patton, Viner, Brett, Braine, Mitchell, Eltinge, Daniel D. Pullen, Ralph I. Sasse, Eisenhower, and the other intrepid pioneers of the Tank Corps who comprised the "Treat 'Em Rough" boys, it was a very bitter pill.

For many years branch chiefs in the War Department had administered the engineering, ordnance, medical, and quartermaster functions, but for the first time since the Civil War the National Defense Act reestablished these offices for the combat arms: the infantry, artillery, and cavalry. Each chief was a major general, and collectively the branch chiefs constituted what was known as the special staff of the War Department. Each of these powerful men spoke for his particular service and was a virtual law unto himself when it came to branch doctrine, research and development, and control of personnel, including assignments. The Tank Corps had no independence, few friends in court, and was now at the mercy of the chief of infantry, Maj. Gen. Charles Farnsworth, who had no fondness for either tanks or the officers who commanded them. Eisenhower's censure in 1920 for daring to advocate the study of tanks and the need to enlarge their role was a message that even the dimmest could appreciate. As late as 1938, when the war clouds of World War II were already visible, the chief of cavalry, Maj. Gen. John K. Herr, expressed sorrow that so many of his officers were transferring to mechanized units. He blamed the General Staff for deliberately attempting to destroy the cavalry, and refused to recognize that armor would ever replace the current missions performed by the cavalry. Herr, a gallant and highly regarded officer, was proof that during the interwar years outdated beliefs would die hard.[56]

The total authorization of the Tank Corps had been set in 1919 at 154 officers and 2,508 enlisted men, and in the autumn of that year the dreaded demotions finally caught up with Patton and Eisenhower. At the end of June 1920 Patton reverted to his permanent Regular Army grade of captain, but he was fortunate enough to be promoted to major the following day. Eisenhower was also demoted to captain a month later but he, too, was promoted to major several days later. Both officers were among only a handful to be so promoted. The army had indeed fallen on hard times.[57] Col. Henry E. Mitchell, who had commanded the other tank brigade in the AEF, overnight became Major Mitchell, and Rockenbach was demoted to colonel. Only Pershing, the army's highest-ranking officer, was untouched by the reductions.

Whatever his previous complaints about Rockenbach, Patton had every reason to be grateful to his superior, who kept him in command of the 304th Tank Brigade as a captain even though other former tank officers assigned to Camp Meade now outranked him. In what was an exceptionally generous gesture, Rockenbach assigned anyone superior in rank to Patton to other units or to his own headquarters.[58] By the time Patton's AEF veterans were discharged, the 304th Brigade had been reduced to a mere company. However, he was so successful in recruiting new officers and men to the unit that by mid-1920 the brigade was back to its authorized strength of thirteen companies.[59]

Less than a year after the war ended, Patton was nearly alone in his alarm over the nation's drift into indifference. The country and the army were "in a hell of a mess and there seems to be no end to it," he wrote Nita in October 1919:

> We are like the people in a boat floating down the beautiful river of fictitious prosperity and thinking that the moaning of the none too distant waterfall—which is going to engulf us—is but the song of the wind in the trees. We disregard the lessons of History . . . and we go on regardless of the VITAL necessity of trained patriotism—HIRING an army. Some day it too will strike and then the end. FINIS written in letters of Blood on the map of North America. Even the most enlightened of our politicians are blind and mad with self delusion. They believe what they wish may occur not what history teaches will happen . . . squeamishness is fatal to any race.[60]

But gradually the work of a peacetime army overtook even Patton, who settled into a routine and in his off-duty time began a steady round of competition in swordsmanship events, horse shows, steeplechasing, and polo. He hunted, socialized, and became the gentleman officer he had been in pre-war days. After months of reflection Patton finally came to the conclusion that he had no future in a corps that remained subservient to the infantry. Unlike France, where he routinely used thousands of gallons of fuel in his tanks, under the parsimonious peacetime appropriations his tanks at Camp Meade could barely be cranked up, much less employed for training. Instead they became useless relics that sat in his motor pool. The fate of the Tank Corps lay partly at his own feet. As one of the leading authorities on tanks he—and his insistence that armor be treated as part of the infantry—was responsible, at least in part, for the army's decision not to transfer the Tank Corps to the cavalry.[61]

Patton's demotion in the summer of 1920 was the turning point of his career, compelling him to decide where his future lay. He could stay in the infantry with the possibility of advancing to general under stifling conditions, or he could request a transfer back to the cavalry, where he was well known and respected. The choice seemed clear. Before he left France to return to the United States, Patton had been required to fill out a form listing his preferences for future duty. His list contained only two items: service in the Tank Corps and "fighting." A more realistic preference statement, completed in April 1919, requested either duty in England as the military attaché or admission to the School of the Line at Fort Leavenworth. By the summer of 1920 it was obvious that he would have neither, and he submitted his request for a transfer to the cavalry, which was approved in September.[62] He could not have been encouraged by the report of a War Department inspector general, who, in the spring of 1920, rated Patton "efficient" and noted his keen interest in tanks, but overall, he wrote, "I consider him average." It was hardly a ringing endorsement.[63]

The change would become official on October 1, 1920, the same month he was ordered to report for duty with the 3d Cavalry at Fort Myer. Leaving his beloved brigade proved very difficult. To the faithful Rockenbach he again offered appreciation for "your long suffering and great kindness to me during the past three years. While serving under you I have had the most vivid and interesting experiences of my life and shall always remember your considerate treatment of me and my various vagaries."[64] The same day, at a review in his honor, Patton addressed the seven hundred men of the 304th Brigade for the final time. He thanked them, even though they probably thought him "the meanest man in the world." But he never believed it necessary to

> apologize for my acts since what ever I have done has been the result of an honest effort to perform my duty as I saw it. When I have cussed out or corrected any of you, men or officers it has been because according to my lights you were wrong, but I have never remembered it against you. I have never asked any of you to Brace more, work more, fight more, than I have been willing to do my self; with the result that in keeping up with you in France I had to get shot.

As for his DSM, "It was the Brigade not I who won it. . . . God bless the 304th Brigade."[65]

The speech was followed by a ceremony in which the brigade sergeant major and two enlisted men advanced and presented Beatrice with a foot-high loving cup, with a light tank etched on one side and an inscription from the enlisted men to both Pattons on the other. Beatrice "made a little speech, & we all cried some . . . it was very touching." Then the brigade passed in review for Patton.

Never again would George S. Patton command on a scale where he could come to know the men individually and interact with them in a manner that is impossible with the command of large formations. Of course, not all of them appreciated their hard-boiled commander, but most were devoted to him, and twenty years later a great many would seek to serve again in some capacity under Patton's command. But, for the immediate future, it was back to the cavalry and the good times.

CHAPTER 21

"If You Want to Have a Good Time, Jine the Cavalry"

(1920–1922)

You must be: a horse master; a scholar; a high minded gentleman; a cold blooded hero; a hot blooded savage.

—PATTON[1]

While weighing his decision to leave the Tank Corps, Patton was reminded of a line from one of Gen. J. E. B. Stuart's favorite songs, "If you want to have a good time, jine the Cavalry."[2] His return to Fort Myer, Virginia, and the 3d Cavalry brought him back to his first love, where his proficiency in horsemanship and the sword could again be put to constructive use.

It seems doubtful that he ever saw the irony of a decision in which he had tried to have it both ways. In the early days of the Tank Corps in France he had little choice but to promote the tank as a weapon to support the infantry. Not only was too little then known or understood about tanks and their potential, but to have promoted independence in 1917 or 1918 would have jeopardized their acceptance by the same infantrymen they were sent into battle to support. By the time Patton and others began lobbying for an independent Tank Corps, it was too late and the die had been cast in favor of the infantry. It was a genuine catch-22, and Patton became one of its primary victims. Although he still kept ties to tank men and still sat on the tank board in the War Department, the former Tank Corps had lost its most eloquent spokesman.

Patton's return to the cavalry was a journey back in time to the glory days when America's heroes were cavalrymen who charged into battle with pennants flying and bugles blaring. With the advent of World War I, the cavalry had become an anachronism, whose "pony soldiers had little more to do than train and look pretty for ceremonial functions. With no Indians to fight, their officers kept fit by playing polo."[3] During the interwar years the only action seen by the cavalry was the disgraceful assault on veterans in Washington during the Bonus March in 1932, an event in which Patton was to play a major role. Unfortunately British general Edmund Allenby's stunning success with cavalry in the Middle East had helped to blind the old guard into the mistaken belief that horse cavalry still had a future.[4]

Nevertheless the closest the Patton family came to establishing roots during his thirty-six years in the nomadic military profession was at Fort Myer, which became the one place where they felt comfortable. Its peaceful setting well met their needs, from the social to the sporting. Moreover, Patton's return in the autumn of 1920 brought him command of the 3d Squadron, 3d Cavalry. Patton's rank and position entitled him to occupy Quarters No. 5 on Officers Row, a large Victorian brick home set amid neatly trimmed lawns and the leafy splendor of Fort Myer, with a superb view of the Potomac River and Washington. After living in so many second-rate quarters during the ten years of their married life, the Pattons (and their children) embraced Fort Myer.

Patton returned to cavalry duty as quietly as he had left it in 1917, and with barely a ripple he was soon (literally) back in the saddle. Although he would eventually become the best-known apostle of the tank and its employment on the battlefield, Patton's true love was horsemanship. Despite his unhappiness at leaving his beloved tank brigade, Fort Myer had its compensations.

Although his return to the 3d Cavalry coincided with the beginning of the roaring twenties, Patton remained the same tough taskmaster with his troops that he had always been. There was ample time set aside for fun and games, but serving under Patton's command remained a serious, challenging, no-nonsense business. He continued to impart the lessons he had learned so well at Bourg and refined at Camp Meade. As historian Roger Nye points out, what Patton did was to train "his officers to be trainers. He was very explicit about which paragraphs of Cavalry Drill Regulations he wanted them to teach to their troopers, and he later questioned the men to see if they had learned . . . [in] 'Patton's School for Professionals.'"[5]

The status quo was for others, and he constantly tested new innovations, techniques, drills, or field maneuvers. In one month alone Patton scheduled twenty-two lectures and delivered sixteen personally, using note cards on which he had carefully typed his teaching points.[6] By requiring them to practice as well as preach leadership and command, Patton was able to keep the mental level of his officers high. In one exercise he demonstrated how a general must exercise leadership in a desperate situation, noting that one should not always survive such a situation. In fact, "THERE IS NOTHING MORE PATHETIC AND FUTILE THAN A GENERAL WHO LIVES LONG [enough] TO EXPLAIN A DEFEAT."[7] He continued his practice of writing notes to himself, and in 1921 began keeping a field notebook, "prepared by my self." On the cover page Patton inscribed the words: "SUCCESS IN WAR DEPENDS UPON, THE GOLDEN RULE [OF] WAR. SPEED—SIMPLICITY—BOLDNESS." The notebook was a compendium of what he had learned about war in ten years in the army, some of which will be recognizable to those familiar with the film *Patton*:

> The "Fog of war" works both ways. The enemy is as much in the dark as you are. BE BOLD!!!!!

> When the enemy wavers throw caution to the winds. He may have a reserve that will stop your pursuit but it cannot restore the battle. A violent pursuit will finish the show. Caution leads to a new battle.

> Officers must not hesitate to lead. Before an attack is declared hopeless the senior officer must lead an attack in person.

> War means fighting. Fighting means killing, not digging trenches. Find the enemy, attack him, invade his land, raise hell while you are at it.

> Ride the enemy to death. Talk blood and death it is good for the men. Try to make fenatics of your men it is the only way to get great sacrefices.

> YOU ARE NOT BEATEN UNTIL YOU ADMIT IT. Hence DON'T.

> OFFICERS MUST BE MADE TO CARE for their men. That is the Sole Duty of All Officers.

> To move swiftley, strike vigorously and secure all the fruits of victory is the secret of successful war.[8]

The Pattons quickly settled into the pleasant routine of life in the army's most glamorous posting. Most days Beatrice took her daughters to nearby Arlington National Cemetery, where they walked and played amid the gravestones:

> It never seemed a sad place to us as Ma said that a cloud of glory hung over it. We had our special walk and our favorite places . . . There were

statues of mounted soldiers on horses. My sister Bee chose as her special hero the one-armed General Pat Kearney on his prancing horse . . . We were all used to the sound of the Dead March . . . booming out three or four times a day—and oftener during the flu season. . . . Bee and I attended all the funerals possible. Our quarters at that time were next to the hospital which was conveniently next to the cemetery, and across the street from the chapel. We would follow discreetly behind the cars and were once, according to Ma, found dancing solemnly behind the tomb-stones to the hymn,"Abide With Me," as played by the Army band. There was something basic and terrible and splendid about the flag-draped cof-fin on the caisson and, in the case of an officer's funeral, the curveting charger with the rider's empty boots turned backwards in the stirrups (to confuse the ghost, Georgie told us) led by a soldier, who, in the olden times, would have been the dead warrior's squire, leading his favorite horse to be sacrificed on the tomb so that he could join his dead master. The terrible throb of the music reached us as nothing else ever had—I know now that it was like the sound of irreconcilable grief.

Beatrice was an ardent lover of flowers, which her husband fondly pre-sented her on every occasion. Eventually her enterprising daughters con-cluded that a great many useful flowers were going to waste on the nearby graves of Arlington, and it was some time before she learned the source of the fresh flowers they began bringing home to her each day. Thinking they came from the post greenhouse (a fixture on army posts of that period), Beatrice believed that she had two very thoughtful children until one day when they dragged home in their little red wagon a large wreath (so big they had to remove it from the stand it had been mounted on) from the grave site of a recently buried Civil War hero. A horrified and pale Beatrice unsuccess-fully attempted to hide the evidence by stuffing it into a garbage can, and in desperation removed it to the cellar, where she hacked it to pieces and jammed it into their coal furnace.

When Patton was informed that evening, the shock was so great that he was obliged to leave the room, or so his daughters thought: "When he returned, he was wiping his eyes and his face was very red, so we thought he must have been crying." In fact he had left the room to keep from laughing:

Georgie got down on his knees and put his arms around us (by this time we were bawling like stuck pigs) and explained that flowers from funerals were "government property," a phrase we perfectly understood to mean "no-no," and that while it was a lovely gesture, he didn't think "they" would like it. We understood about "they" too; "they" said "Keep Off the Grass" and "No Parking" and "Keep Out," and no one in their right minds would cross "they."

The army of the interwar years was desperately poor, underpaid, and under-appreciated, and, worst of all, unfit to fight another war. The average enlisted man could not afford to be married and, unless—like Patton—they had an outside income, even senior noncoms and officers had considerable difficulty making ends meet. However, there were compensations to being a soldier, for as Patton's daughter writes:

> In those days nobody had any spending money, but they had everything that money could buy; polo, hunting on the border posts, racing, horse shows, skeet shooting, tennis, a hop at least once a month, prize fights for the enlisted men, football and baseball, even enlisted servants, called "orderlies" or "strikers" . . . and, of course, the post greenhouse. The poor old American taxpayer footed the bill for these goodies, but he certainly got an efficient army out if it! One that won him at least two wars.

The primary mission of the 3d Cavalry was to perform in ceremonies, funerals, parades, inaugurations, and special formalities for visiting heads of state. As such, the cavalrymen were highly trained and could perform a variety of trick rides. When Patton returned to Fort Myer, he began looking for some method of improving the lot of the enlisted men by earning additional income for the Soldier's Rest and Recreation Fund and other post charities. With the winter weather making outdoor drills impractical, most of the drilling took place in the large indoor post arena. Patton was the driving force in the inauguration of a series of indoor exhibition drills and riding events held at Fort Myer each Friday afternoon from January to early April.

Competition was exceptionally keen, and eventually each troop of the 3d Cavalry developed a series of spectacular and entertaining routines that they performed. One might perform acrobatics on horseback, while another developed a musical ride involving many riders with guidons (the identification flag of a military unit, carried on a long pole) executing various movements at different gaits or performing jumping exhibitions. The artillery used their gun carriages with teams of horses. The Fort Myer riding hall had two tiers of galleries that seated nearly eighteen hundred spectators. The exhibitions spawned what came to be known as the Society Circus, a series of exhibition rides and competitions held on Saturday and Sunday, an idea that immediately caught on and soon became one of the seasonal highlights of the Washington social scene. Congressmen and senators were given free tickets, and they often brought constituents and their families. Debutantes found Fort Myer, with its large numbers of eligible, dashing young bachelor officers, a marvelous place to snag a husband, and their families would buy boxes and have parties as the riding events took place. Beatrice held luncheons at Quarters No. 5 and helped organize teas at the officers' club. "It was terribly exciting," writes Ruth Ellen, "to watch Georgie 'doing his stuff.'" The girls always knew those whom their father wished to impress and

> whether he liked them or not—usually not—because he would smile what we privately called his "sissy baby smile" (sissy-baby was the worst word in sisterly language). It was a dead giveaway all his life. His smile would go so far around the corners of his face that his ears would lie flat, and he would crinkle his eyes and show his teeth, but his eyes inside were not smiling. It always amazed us that strangers did not recognize it because even the dogs did.

At Arlington, Beatrice taught Ruth Ellen to read by tracing out the letters on the gravestones. The religious experiences of the Pattons consisted of a series of experiments that his daughter notes were "rich and full." Ellie Ayer was a strict Protestant, but as a child Beatrice and her brother and sister had been permitted to attend the various churches in Boston, until they finally settled on the Congregational Church. Beatrice wanted her own children to have the same freedom of choice. "She had a true feeling for prayer and depended on it all her life. In her religion she was not a mystic, as was Georgie, [who was] a devout, tithing, church-going Episcopalian. Beatrice encouraged her daughters to try all the Sunday schools on post. Eventually the girls finally settled on the Baptist Sunday School; firstly because it served cocoa and cookies and, secondly, because the Baptist hymns were so great. We would come home singing them and Ma, who had never heard them before, was crazy about them and, in fact, took the 'Old Rugged Cross' as one of her lifelong favorites."

Beatrice had a fine ear for music and could easily pick up a tune and play it on the piano. By contrast, the Pattons were musically destitute.

> When Georgie was a cadet he had to learn the beat of the various bugle calls which ran the service life in the days before loudspeakers, so that he could tap them out with his fingers. It filled us with awe to learn that when he was a plebe at West Point he had stood up when the band played the "Dashing White Sergeant," and had sat down when it started the "Star Spangled Banner." We were as full of music as Ma was, although not so gifted and Georgie's lack was incredible to us.

Beatrice volunteered to teach Sunday School and to take on an unruly group of older boys who were the bane of the chaplain. Her classes were well attended week after week, and, unable to stand it any longer, the post chaplain finally asked the secret of her success. An embarrassed Beatrice admitted that she told her charges that if they behaved themselves, learned their lessons, and attended for four consecutive Sundays they would be permitted to come to the Patton home and put their finger in the trigger of Patton's revolver that had killed Julio Cárdenas in 1916.

Time and again Beatrice demonstrated what a terrific asset she was to her husband's career. For example, in the summer of 1920, he had arranged for a tank demonstration at Camp Meade and invited seven generals from the War Department. His tanks were sparkling and lined up for inspection in the motor pool. However, his distinguished visitors seemed singularly unimpressed and carried out their inspection in silence. Patton, who thought his charges little more than circus spectators viewing a line of elephants, mounted one of the tanks and invited someone to accompany him on a ride. Patton's eloquent briefing about its many virtues failed to induce anyone actually to drive the machine. All refused, and one general sarcastically remarked that tanks were known to blow up, topple over and even turn turtle on a steep hill.

> With a snort of disgust Patton turned, beckoned to his wife, and with a sweeping theatrical gesture invited her to enter a small whippet tank. Though dressed fit for a garden party, she disappeared nimbly into the turret. Patton let out a howl of triumph; howled to his visitors over the racket of thundering motors to wait—and watch—and was off. He handled the tank as he would a polo pony . . . spinning in lightning turns, jarring to abrupt stops, thudding down the field at break-neck speed. The onlookers saw a rare demonstration of the maneuverability of armored cavalry. His performance produced dry smiles. The top brass, having taken one look at bruised, begrimed and topsy-turvy Mrs. Patton, who had lost her hat, were still determined not to sample any of Patton's tankmanship.[9]

Patton then turned to the assemblage and observed, "You see, gentlemen, how easy it is to handle? Now, who would like to make a try?" No one did, and a silence ensued until one general said, "Thanks, Major Patton, we have seen enough."[10] Catty remarks about Beatrice's unladylike behavior began circulating around Camp Meade, though Mamie Eisenhower defended her, saying she had shown everyone her mettle but that generals were afraid of soiling their uniforms.[11] Even in a losing cause Beatrice had proved a willing and able participant in any venture that might enhance George's career.

Patton's commanding officer, with whom he had previously served at Fort Riley, was Col. W. C. Rivers who, along with his formidable wife, had a well-deserved reputation as a holy terror. Patton's children thought his name excruciatingly funny ("a joke of outstanding scatology as a toilet to us was a 'W.C.,' and having the added attraction of that last name was just too

much"). For daughter Bee, Arlington had not been the first flower "incident." While at Fort Riley, Beatrice had looked out her window one day and
seen little Bee picking the heads off all of Mrs. Rivers's tulips. "Ma went to
pieces and called Bee in, gave her a hearty scolding, and sat down to write a
note of apology to her neighbor with the terrible reputation." When Beatrice
went next door to deliver her note there were tears streaming down her face,
but instead of an angry woman, she found Mrs. Rivers standing on her front
porch in a fit of laughter.[12]

At Fort Myer, Colonel Rivers once demonstrated his sense of humor by
including the following item in the post daily bulletin: "Every dog on duty
at Fort Myer will be given an extra ration today because it was observed that
no dogs attended the funeral of Major General Smith today. By order of the
Commanding Officer, W. C. Rivers, Colonel, USA." Such was life in the
army of the interwar years.

One night a close friend of the Pattons, a colonel named Emile Cutréa,
dined chez Patton, accompanied by his wife and his mother-in-law. The
mother-in-law was a shrew who spent the evening complaining about how
her daughter was married to one of the barbarians that made up the officer
corps—the Pattons excepted, of course (they were impressive because they
"smelt good" and "spoke good English," and primarily because they were
known to be rich). Then the woman remarked that when her "poor little
girl" and her husband had been stationed at Camp Meade, "they had to
move into a set of quarters that was all painted blue and yellow inside!!!
Can you imagine what kind of people would do a thing like that!" Beatrice
tactfully changed the subject, while Patton began talking very fast to his
friend, who had turned an interesting shade of crimson.[13]

A neighbor of the Pattons was General Summerall, whose wife's very
elderly parents lived with them. Mrs. Summerall, an "army brat" from an
old military family, was known to spend a great deal of time taking care of
her parents. One day Beatrice offered to "sit" with them while Mrs. Summerall went out shopping. Before departing Mrs. Summerall said: "I forgot
to tell you, Mrs. Patton, that my parents are quite senile, and they love to set
little fires. You have to be careful." Beatrice found two tiny old people who
looked like "dried up crickets" in rocking chairs next to the fireplace. The
wife whispered to her husband, "Is SHE gone?" When he replied yes, they
both smiled at Beatrice and said, "Dear Mrs. Patton, do you have any
matches with you by chance?"

In March 1921 the 3d Cavalry participated in the inauguration day parade of
Warren G. Harding, during which Patton rode his horse beside the presidential limousine. Patton's first efficiency rating from Colonel Rivers noted him
"below average" in tact but "superior" in a category created especially for
him, called "mental energy."[14] When he was not on the drill field, Patton was

usually on the polo field, either at Fort Myer or as a member of the army team. When it did not perform well against other teams at the American Junior Polo Championship in October 1921, Patton wrote a letter offering suggestions that might as well have been rules for combat. While demeaning his own ability and experience, Patton proffered what he termed "self evident truths." He hoped to dispel "a wide spread falicy among us" that he and his fellow officers were rough, tough characters on the polo field. "Compared to good players we are as lambs to raging lions." Tactics were too meek and too much trust placed in God, who was "seldom on the job." Speed, discipline, and riding were all in dire need of practice by the best players the army could obtain.[15] What Patton did not—in fact, could not—say in his letter was that not winning was intolerable. However, according to Gen. Hamilton H. Howze, who entered the army in 1930, the problem was not so much the quality of the army's polo players but the fact that "our mounts were selected mostly from horses bought on the civilian market at an average price of just $156."[16] Patton's, of course, was one of the few exceptions.[17]

One of the army's finest polo players was a young officer whose later career encompassed both serving under Patton and eventually succeeding him as the Third Army commander. From the day he joined in 1917 as a young lieutenant, to his retirement in 1948 as a four-star general, Lucian K. Truscott was destined to join the ranks of the great military leaders of the U.S. Army of World War II. On the polo field Truscott was as successful as he was uncompromising. His philosophy of polo and war could just as well have been Patton's—and so could his profanity. "Listen, Son, goddamit," he once growled to his own son:

> Let me tell you something and don't ever forget it. You play games to *win*, not lose. And you fight wars to win! That's spelled W-I-N! And every good player in the game and every good commander in a war, and I mean really *good* player or *good* commander, every damn one of them has to have some sonofabitch in him. If he doesn't, he isn't a good player or commander. And he *will* never be a good commander. Polo games and wars aren't won by gentlemen. They're won by men who can be first-class sonsofbitches when they have to be. It's as simple as that. No son-ofabitch, no commander.[18]

On the "fields of friendly strife" of the interwar years there were no bigger sonsofbitches that Lucian K. Truscott and George S. Patton. As General Howze notes, polo provided invaluable lessons in inuring young officers to danger and taught them teamwork and to think and respond with split-second quickness. "Most of all, it was enormous fun."[19]

Polo is a physically demanding and often dangerous sport. Howze recalls two occasions on which a player was killed, and broken arms and legs were everyday occurrences. Patton played it with abandon and with his customary disregard for his body. Once while playing polo at the Myopia Hunt Club, his head was slashed open by a mallet. He stopped only long enough to jam an iodine-soaked cotton wad into the wound before resuming play. [20]

When he was not playing polo Patton was attending horse shows as far away as Syracuse, New York, and the ribbons and cups began to pile up. In November 1921 he injured his hip when he and his mount fell during a competition at Madison Square Garden. In January 1922 he became quite ill, narrowly averting cardiac arrest, from an allergic reaction to tainted shellfish that produced severe swelling and shock.

He also made a point of keeping in touch with Pershing and other senior officers he had come to know in France, such as Generals Smith, Harbord, and Summerall. In the summer of 1922 the army polo team trained on Long Island, where Patton met the likes of the Harrimans, the Belmonts, and others whose names graced the *New York Social Register.* "These are the very nicest rich people I have ever seen," he wrote to Beatrice.

In the spring of 1922 Beatrice's sister, Kay Merrill, nearly died of childbed fever in Madrid, where her husband was the American consul. Kay was removed to London on a stretcher and placed in the care of a noted British physician, who saved her life. Without consulting one another the Ayer kin began arriving in England to be near their sister. Beatrice was the first, and during her lengthy absence that year tending to her sister and niece and nephew, Mama and Papa Patton turned up in Washington to help supervise the Patton household, whose master was frequently traveling with the army polo team. Mr. Patton announced that he had always wanted to take his family to England for the summer, and off they went.

The George Patton of the interwar years displayed frequent traits of anger and outrageous behavior in ways that exasperated even his patient and adoring wife. With nothing meaningful to propel his career, it was almost as if he were openly rebelling against his fate by behaving outlandishly. There is no question that he was bored, and even less doubt that he returned from the war with his thirst for fulfilling his destiny barely quenched. Then, his demotion three grades from colonel to captain was a blow that left bitter frustration, which was not assuaged by his repromotion to major. Although polo, other sports, and participation in equestrian events had brought him to the peak of physical fitness—perhaps the best of his life—Patton never lost his obsessive fear of growing old and losing his physical and mental powers. While unafraid of dying, he greatly feared that he might not do so honorably. When a car salesman attempting to sell him a Cadillac noted that the doors opened and closed, "just like the gate of a vault in a graveyard," Patton stalked off. [21]

Those who suffered the most from Patton's frustrations and resentments were his family and friends, whom he insulted innovatively and without tact. Patton was briefly stationed in Boston in 1924–25, and at social events he occasionally delighted in shocking women he disdained by inviting them to view where he had been wounded in 1918, unbuckling his belt, and asking the victim to help him lower his pants. His worst behavior, however, was reserved for sister-in-law Kay Merrill and her husband. One such occasion was a formal dinner at the Merrill home in Washington, attended by a number of foreign dignitaries and their wives. Patton, attired in his dress blues, accepted a plate of steak and, after removing a bone, which he unobtrusively wrapped in a napkin, left the room. He returned momentarily, waving the bone and announcing in a loud voice heard by all that he had never been given enough to eat in the Merrill house, and anyone who cared to was welcome to join him later in gnawing one in front of the fireplace. Patton thought it merely a harmless prank that everyone ought to have found humorous but the deathly silence that followed left the Merrills hopping mad.

Patton reserved his most tactless and hurtful remarks for Keith Merrill, whom he regarded as little better than a coward for evading the war in the Foreign Service. For his repeatedly insulting remarks about her husband, Patton was eventually banned for a time from the Merrill home by his sister-in-law. Whenever he misbehaved in public, it was left to Beatrice to restore order from the wreckage her husband left in his wake. Publicly all was well, but her own private anger at his outrageous *enfant-terrible* antics can only be imagined.[22]

George Marshall would later say of Patton that "he would say outrageous things and then look at you to see how it registered; curse and then write a hymn."[23]

Much of Patton's anger in this, the darkest period of his life, was reflected in his poetry, which continued to flow from his pen. In "The War Horses" Patton lamented:

Where are they now, the horses who fought for us in France,
The horses of the limber and the gun?
They shared our toil and triumph and they shared our joy and pain,
But we left them, yes, we left them every one.

I can see them still, all shivering, with their rumps set to the rain,
On the muddy cheerless horse lines of the front.
They didn't make a murmur and they took the worst that came,
And when they died, they didn't even grunt.

During the period 1922–24 Patton wrote "The Soul in Battle":

In the valley of the slaughter where the winged Valkyrie dwell
And the souls of men go naked to their God

I have seen the curtain parted, I have glimpsed the flinty trail
The final road the spirits have all trod.

Yet in the awesome clearness of the future there made plane
The spirit loses something of its dread
And life with all its littleness is very very drab
While the living view the corpses as not dead.

The veneer of life is melted by the hot blast of the shell
And we behold our fellows very plain
Not as cur or fool or hero but like some poor flustered thing
A trembling beast reluctant to be slane.[24]

Although Patton wrote most of his war poems in 1917 and 1918, the blackest and most depressing verse began appearing in the early 1920s, reflecting his disillusionment at being on an endless merry-go-round. In "A Soldier's Burial," written in 1919, he lamented the dead but mused that perhaps they were just as well off. It was a far mellower version of a bitter poem, "The Moon and the Dead" (1918), in which Patton had written of twisted corpses: "Some were bit by the bullet, some were kissed by the steel, some were crushed by the cannon, but all were still how still!/The gas wreathes hung in the hollows, the blood stink rose in the air and the moon looked down in pity at the poor dead laying there."[25]

Perhaps his most melancholy poetry was composed in 1922 and included one called "Anti-Climax" in which he wrote:

'Tis hard to see the slobbering lips
And bleary lashless eye,
The firm set mouth and eagle gaze
They had in days gone by.

Have those scrawny necks like withered kelp
Ever boomed the deep hurrah?
Has the shufling tramp of those dragging feet
Ever filled a foe with awe?

'Twere better indeed for the sake of fame
To have died in the lust of youth,
Than have lived to wither and drool and blink
In ages pitiless truth.

Behold yon statue picturing
A hero slane in strife!

And then in pity gaze upon
That comrade still in life.
The one a glorious super-man,
Replete with life and fire,
The other a poor doddering wrathe,
A witch light in the mire.

And having looked who can but choose
To go while life is gay,
To go while glory guilds the road
And Victory crowns the day.
(May 30, 1922)

The same day Patton composed a second poem, "The Lament of the New Heroes":

Strange it is how all the glory of the splendor of our youth,
Is forgotten with the passing of the need.
While the toothless twisted creature of an era that is gone . . .
Yes the lank and scabby creatures who were drafted years ago,
Now are lauded and applauded by the throng.
Though the battles which they fought in—if they fought—
Which none may know, were as nothing to the fights
which we have won. . . .

Patton's behavior during the 1920s, his unpredictability and frequent anger, suggests a combination of his ever-present dyslexic mood swings and depression. Outwardly he was the same dynamic officer he had always been: demanding, emphatic, innovative, but inwardly his anger raged unabated. Patton's hyperactive, workaholic nature was evident in his unrestrained participation in riding events. And when he was not on duty, as if to prove that he was as good as ever, he was involved in playing polo, steeplechasing, skeet shooting, fox hunting, entering horse shows, and playing tennis, handball, or squash. Patton also began learning to fly, as well as adding hundreds of new books to his personal library. He read them all avidly, took his usual copious notes, and incorporated what he learned into his growing lexicon of knowledge. The Pattons remained socially active, and when he found the time to sleep is a complete mystery. The spiral of activity that dominated his life was hardly the stuff of a self-proclaimed man of leisure. His zeal and proficiency in horsemanship alone led to the accumulation of some four hundred ribbons and two hundred cups.[26]

Patton never seemed able to slow his frantic pace. It was as if he must drain the very last drop from the bottle of life while there was still time. It also served as a means of ensuring that he would not suffer the boredom of peacetime duty or have time to ponder his fate. If there was no war for him to fight in, he would substitute excellence in everything he did, accepting and mastering every challenge, daring to push himself (and his luck) to its outermost limits. Patton drove his car with the same recklessness he displayed in propelling one of his polo ponies toward the goal. He knew only two speeds: fast and faster.

Patton also continued to indulge his thirst for adventure. One night in 1922, as he was returning to his hotel from a nearby Long Island horse show, Patton spied three men and what he perceived was a damsel in obvious distress, who was being abducted into a truck parked at the curb. He slammed on his brakes, leaped from his Pierce-Arrow brandishing a revolver, and ordered the men to unhand the woman. Terrified at the sudden intrusion of an obvious madman waving a gun at them, they complied, but eventually managed to convince him that the lady was only being assisted into a truck by her fiancée. For a time the incident became one of Patton's favorite topics of conversation at the social events he attended on Long Island. When one woman asked why he carried a weapon to a horse show, Patton is said to have replied: "I always believe in being prepared, even when I'm dressed in white tie and tails."[27]

One of the most important functions performed by the 3d Cavalry was funeral escort detail at Arlington National Cemetery, which was rotated among the squadrons. Caskets of dead veterans would be met at Washington's Union Station, whence their flag-draped caissons would be escorted through the city and across the Potomac to a final resting place in Arlington. On May 23, 1922, Patton was routinely leading a detail from Union Station to Arlington, but on reaching the Virginia side of the Memorial Bridge, whether out of boredom or spring fever, Patton suddenly ordered the cortege to proceed at a gallop. In the ensuing chaos the horses pulling the caisson broke free, obliging Patton to chase after them "to recover both the corpse and the decorum."[28]

Patton's evenings were generally reserved for reading and study. He devoured what Fuller and others had to say about every aspect of war, from the theoretical to the practical. He studied wars of the past such as the Russo-Japanese War of 1905 and the little-known Franco-German War of 1866–67. It was his custom to scribble in the margins of his books and later type or inscribe the results—often on note cards—into a synopsis that became a permanent part of his personal papers. During the interwar years Patton consulted an eclectic list of the famous and the lesser known, ranging from Napoleon and Clausewitz to du Picq, Jomini, Cromwell, Xenophon, and Frederick the Great.

No writer or war was too obscure for Patton, who read, studied, and absorbed what he read into his own evolving concept of war and how to fight it. His marginal notes were sometimes divided into "how," "when," "F" (for future wars—probably a passage to reinforce his own ideas), and "Q" (worth quoting). In one instance he incorrectly suggested that he was the first in the United States to define discipline, which was actually first done in 1879 by a general addressing the Corps of Cadets at West Point.[29]

Patton paid particular attention to the writings of the German military thinkers and generals of World War I. He instinctively recognized that there was a great deal to be learned from his former enemy, and he read virtually everything that had been translated into English: Balck, von Seekt, Scharnhorst, von Falkenhayn, and others. To learn about future wars he took great interest in what Gen. Friedrich von Bernhardi had to say about *The War of the Future in the Light of the Lessons of the World War*.[30] He also studied such French leaders as Foch and Pétain. The men and the campaigns of the Civil War were examined as a labor of love, and over the years Patton became an expert on many of its battles. Everything that interested him was painstakingly typed on note cards and suitably annotated with additional ideas and comments. Virtually all of Bernhardi's book found its way into his notes.[31]

From the great and the unknowns he compiled quotations that interested or inspired him. A sampling:

> *Napoleon*: "Fortune is a woman who must be wooed while she is in the mood. . . . The greatest general is he who makes the fiewest mistakes— i.e. he who neither neglects an opportunity nor offers one. . . . The only right way of learning the science of war is to read and reread the campaigns of the great captains." [Patton added: "And think about what you read."]
>
> *General Hugh Elles* (at Cambrai): "Do your damdest."
>
> *Arab proverb*: "Victory is gained not by the number killed but by the number frightened."
>
> *Buddecke*: "No matter how high an officer goes as a teacher or administrator his true roll is a leader of troops. To be good at this he must study MILITARY HISTORY—Work Map PROBLEMS and Think War."
>
> *Frederick the Great*: "Ride the enemy to death. L'Audace—L'Audace—Tout jour L'Audace." [These words became the cornerstone of his future generalship.]

And finally, a quotation from Patton himself: "A leader is a man who can adapt principles to circumstances."[32]

Patton had no particular love for his former enemy, but his personal feelings did not prevent him from admiring the professionalism of their officer corps and the German General Staff, all of which reinforced his growing belief in the necessity for training troops to the highest possible level. He also detected that "between the Prussian concept of professionalism and the needs of the American democratic society" there existed "a linkage: conditioning young men for courage and sacrifice, drilling them in the knowledge of their calling, [and] motivating them to perform with dedication."[33]

Beatrice remained in England visiting her sister, Kay, until August 1922, while her husband worked tirelessly to ensure that there would be no such thing as an idle day in his life. By December 1922 Patton's brief tour of duty at Fort Myer ended. He had been accepted in the Advanced Officers' Course at the Cavalry School at Fort Riley, where—after spending Christmas in Los Angeles—he reported for duty in January 1923.

CHAPTER 22

Past and Future Warrior Reincarnate

Man's a damned liar. I know because I was there.
—MARGINAL NOTATION IN ONE OF PATTON'S HISTORY BOOKS

In 1943, before the Allied landings in Sicily, Gen. Sir Harold Alexander said to Patton, "You know, George, you would have made a great marshal for Napoleon if you had lived in the 19th Century." Patton immediately countered, "But I did."[1] His response was not bravado, for Patton genuinely believed that he had lived before and would live again. Nor was Alexander's observation unique. One of Bee Patton's suitors thought that Patton "should have been a Virginia dragoon, mounted on a big black horse with a saber in his hand," and his World War II aide, Charles Codman, proclaimed that he was "not a contemporary figure," but possessed of the "virtues of the classic warrior."[2]

From childhood, Patton believed that he had lived in earlier times as, among others, a Viking warrior and a Roman legionnaire. Except to a few close friends and family intimates, he never publicly articulated his belief in reincarnation, but the evidence that he believed in a prelife and an afterlife is overwhelming, as were his varying levels of what he believed to be memory of these experiences.[3]

According to his daughter, privately Patton

believed implicitly in reincarnation, and brought all of us up to believe in it. He said it was the only acceptable explanation for injustice and inequality. He believed that we were put on earth to better our souls and bring them closer to the perfection of God, and he quoted Revelations Chapter III, verse 12, "Him that overcometh shall I make a pillar in the temple of my God and he shall go no more forth." . . . He had read the Bhagavad-Gita—the Bible of the Brahmin sect in India—many times. He used to recite it to us, especially, "As a man, casting off worn out bodies, entereth into others that are new. For sure is the death of him that is born, and sure the birth of him that is dead, therefore, over the inevitable thou shalt not grieve."[4]

Patton proclaimed himself unafraid of death, but as Ruth Ellen notes:

He said he was always afraid in battle and anyone who said he was not was either a liar or a moron. My mother asked him why, if he wasn't afraid of death, he was afraid in battle and he told her he was afraid of being maimed and was ashamed of such a fear, and he also said he was a little bit afraid of dying because he knew, by experience, it could be long, drawn out and painful. . . . "Believe me, God is not a hypocrite. He didn't make a great world like this one, and go to make one on the other side that isn't just as great."[5]

History is replete with vivid and inexplicable examples of men and women whose accounts of former lives cannot be dismissed as the mere ravings of a lunatic fringe. How, then, to explain the genius of Patton, his attraction to history, and his extraordinary insight into both the past and those who lived and recorded it? Was his destiny guided by the hand of the past, making his karma inevitable? Were his frequent brushes with death merely the marks of a brave and occasionally foolish man? Was he being protected—as he believed—by the unseen hand of his ancestors? Were his insightful and often inexplicable observations the accurate recollections of one who had lived multiple lives?

The answers are, of course, hypothetical and circumstantial. Perhaps his knowledge and sense of former lives was mere coincidence, but what is irrefutable is that for both George and Beatrice Patton it was as natural to believe in an afterlife as it was to conceive of fulfilling one's destiny on earth. In short, George Patton had lived many lives, always in the mantle of a warrior, and would do so again after the end of this present life. What others have described as the experience of having existed before, of being cognizant of things they simply could not have known in their present life, he called déjà-vu.

His nephew Fred Ayer once asked Patton if he believed in reincarnation and was told, "I don't know about other people; but for myself there has never been any question. I just don't think it, I damn well know there are places I've been before, and not in this life."[6] Patton spoke eloquently of his experience in France in 1918 on meeting a young French liaison officer who offered to show him around Langres. Patton declined, saying it was not necessary as he knew the place quite well. He then proceeded instinctively to guide his driver, as if directed by an unseen spirit that whispered in his ear. Unerringly Patton went to the site of a Roman amphitheater and two temples dedicated to Mars and Apollo, pointed to the spot where Caesar had pitched his tent, and noted that he knew it to be the right place because he had been there before.[7]

Once when his family was heard discussing the death of an obscure relative, Patton interjected: "What is so strange or unusual about dying? You just walk back out the same door you came in. When you came in, you had no complaints about where you'd come from."[8] As he lay near death on the battlefield near Cheppy on September 26, 1918, he had understood then that when Saint Paul said (1 Cor. 15:26) "and the last enemy that shall be destroyed is death," he really meant "'the fear of death' because when you are so close to Thanatos, you cannot fear him; you know you have known him before."[9]

In Patton's case, belief in reincarnation offered a means of fulfilling his ultimate and eternal destiny of leading a great army in battle:

To him this was enough of a goal to be worth going through countless rebirths. He declared that he had once hunted for fresh mammoth, and then in other ages had died on the plains of Troy, battled in a phalanx against Cyrus the Persian, marched with Caesar's terrible Tenth Legion, fought with the Scottish Highlanders for the rights and hopes of the House of Stuart, [fallen] on Crecy's field in the Hundred Years' War, and [taken] part in all the great campaigns since then.[10]

He had also been a pirate and fought with both Alexander the Great and Napoleon as a cavalryman.[11] Nye notes: "He always suffered horrible deaths. God determined when he would return and fight again."[12] Curiously, although he believed the spirits of his ancestors were ever present to guide him as they had at Cheppy, Patton apparently did not believe that they themselves reincarnated, thus posing another inexplicable conundrum.[13]

Patton had occasional flashbacks to his former lives. He despised the cold, particularly after the accident in Mexico in 1916 when his fuel lamp exploded in his face, and he suffered extreme discomfort whenever he exposed his face to the sort of weather that existed one winter day in 1934. He and Beatrice were mounted on their horses awaiting the start of a fox-hunt, when Patton began to describe Napoleon's retreat from Moscow in the winter of 1812, how the dead were removed in carts in cold so severe that it turned blood brown on the frozen snows of Russia. Patton recalled trudging along in the column of retreating French soldiers, with one of his arms so numb it might have been lost in battle. Ahead rode Napoleon on his horse. Such events were no laughing matter to Patton, who believed that such horrific memories "scar your soul" and thus survive death and are reborn in a new life, to be relived in the mind and body of the current persona.[14]

Perhaps the most vivid of Patton's memories was that of his experiences as a Viking warrior. Daughter Ruth Ellen has recounted his descriptions of his Viking life, which are too convincing and graphic to be dismissed as the creations of an inventive mind. They include being carried from a battlefield on a great shield carried by four warriors who thought him dead, of riding in the great Viking ships, and of the nauseating, incredible stench of unwashed, hairy bodies, leather, fish, and oils.[15] As grandson Robert points out, "Georgie's descriptions of his past lives were always dreadful," and included a flashback to his life "in which, as a besieged Carthaginian dying of thirst in the second century B.C., he drank urine out of his helmet."[16]

Perplexed that Patton's great library of books contained nothing dealing with either reincarnation or Buddhism, daughter Ruth Ellen once asked him why. Patton replied that his oft-mentioned "I was there" answer had been "born into him, not induced by listening or reading."[17] Nevertheless, his interest in and study of other cultures is suggestive of his belief in an afterlife.

Patton studied neither reincarnation, science, nor the great religions of the world in which reincarnation is a basic doctrine, such as Buddhism. It was quite simply an article of faith that he had lived before as a warrior and would continue to do so in future lives. Undoubtedly Patton's strong sense of religion and his deep belief in God contributed to the certainty of his own eternal mortality. One of his favorite poems was Wordsworth's *Ode on Intimations of Immortality from Recollections of Early Childhood,* with its theme of the everlasting soul. Ruth Ellen relates that when their grandfather died, Patton made them memorize Sir Edwin Arnold's poem "After Death," in which, as historian Nye writes, "the everlasting soul speaks from afar that the temporary body has been abandoned, but the separation from those left behind will not be long."[18]

For Patton, death merely meant that his soul would once again change bodies—a philosophy with which the members of his family seem to have been thoroughly imbued. Ruth Ellen, for example, spoke of her belief that death is not to be feared but is merely one of the many lives we live through reincarnation.[19]

What concerned Patton was that he attain immortality by fulfilling his destiny of becoming a great military leader during *this* particular lifetime. Unlike those who worry about heaven and hell and good and evil, Patton never intellectualized or rationalized the issue. It was a straightforward proposition. He was presently living in the body of a man named Patton. In earlier lives he had always been a warrior and in future lives would continue as one. Nothing else mattered. He did not seek to gain wisdom or study the metaphysical reasons for the existence of an afterlife. Unlike composer Richard Wagner, and others, who included redemption as a fundamental part of reincarnation, Patton was not concerned with the ultimate end of the cycle of birth and death; he merely wanted to continue life as yet another warrior.

Religion played a large role in Patton's life. He not only believed in God and, in particular, a god of battles who served as his protector, but in a conception of religion in which God was personalized and not some obscure, invisible spirit to which one prayed in church. Nevertheless Patton found comfort and solace in churches. Former NATO commander Gen. John R. Galvin tells the tale of a German cleric who, as a young priest in 1945, came upon Patton in a medieval church in Bad Wimpfen am Berg, in Germany's Neckar Valley, which had survived destruction in the war. "Patton was doing a most unusual thing for a man of such reputation. The warrior, with notebook and pencil in hand, was calmly sketching the stained-glass windows."[20]

It was equally inevitable that signs of his beliefs would emerge from prayer and be reflected both in the poetry he wrote and the poets he read and admired. They included "After Death" in which Edwin Arnold had written, "He made life—and He takes it—but instead Gives more: praise the Restorer, Al-Mu'hid! . . . In light ye cannot see, Of unfulfilled felicity, And enlarging Paradise; Lives the life that never dies."[21] The best expression of his past lives appears in a lengthy poem written in 1922 during his bleak days at Fort Myer. Titled "Through a Glass Darkly," Patton demonstrates a powerful belief in God and alludes to earlier lives, the first of which may have been as a caveman. He even suggests that while Christ was on the cross

> *Perhaps I stabbed our Savior*
> *In His sacred helpless side.*
> *Yet I've called His name in blessing*
> *When in after times I died.*[22]

This most moving and complex of his poems concludes with the words:

> *So forever in the future*
> *Shall I battle as of yore,*
> *Dying to be born a fighter*
> *But to die again once more.*[23]

*   *   *

Occasionally in the interwar years Patton would speak of his former lives during the family's after-dinner conversation. His earliest remembrance of déjà-vu was as a youth at Lake Vineyard, where he and his cousins would climb into a wagon and, after covering themselves with shields made of barrelheads, send it careening down a hill. The young warriors were told to fire their arrows through the shields as it came abreast of the enemy. The "enemy" turned out to be turkeys that scattered in all directions, providing ample food for the Patton table in the following days. Asked where he got such a foolish idea, he replied, "in Bohemia centuries ago, John-the-Blind had done just that and won a great victory against the Turks. Where had he heard about it? 'Oh', he told them, 'I was there.'"[24]

In early 1918, one of Patton's newly assigned officers at the Tank School was Lt. Harry Semmes, who became a lifelong friend. Semmes was unique in that he was one of the few intimates ever permitted to learn of Patton's former life experiences, one of which Patton related to him in the autumn of 1918. According to Semmes, shortly after his arrival in Langres Patton was ordered to a secret destination in a part of France where he had never been. Although he had no idea where he was going, near the top of a hill Patton leaned forward and asked his driver "if the camp wasn't out of sight just over the hill and to the right. The driver replied, 'No sir, our camp where we are going is further ahead, but there is an old Roman camp over there to the right. I have seen it myself." As Patton was leaving the camp he asked an officer: "Your theater is over here straight ahead, isn't it?" The officer responded, "We have no theater here, but I do know that there is an old Roman theater only about three hundred yards away." Semmes notes: "This conversation took place well after dark and it was impossible for Patton to see the theater."[25]

During the interwar period Patton was able to indulge fully his interest in the Civil War and frequently took his family on outings to various battlefields, delivering detailed orations on the battle fought there. One of his friends of long-standing was the German military attaché, Gen. Friedrich von Boetticher, who was himself a well-versed student of the Civil War. Von Boetticher accompanied the Pattons on the day in the 1930s when they visited the sites of the Battle of the Wilderness. Ruth Ellen recounts that: "Georgie had us all lined up and assigned each of us a position, as was his custom." Beatrice was, as usual, the Union forces, "and young George and I were somebody's troops." General von Boetticher was designated time-keeper and critic:

> While we were lining up and receiving our orders, we noticed a group of
> tourists with a guide. Georgie was gesticulating and explaining the battle,
> and one old gentleman with "pork chop" whiskers was getting further and
> further from the guide, and closer and closer to our group, obviously

intrigued. Georgie finished his speech by saying that he was General Early and would be with his aides on a nearby rise of ground. General von Boetticher, who had the book they were using, pointed out that General Early had been in another part of the battlefield, and he and Georgie began to discuss it. The old gentleman could stand it no longer and broke in, saying, "The young gentleman is quite right; General Early was on that nearby rise. I was in that battle, [with a New York Regiment] and, gentlemen, we ran like hell!" Georgie looked at him calmly and said, "Of course, General Early was on that rise; I saw him there myself." General von Boetticher methodically corrected the text of the book with his pen.[26]

Fred Ayer vividly remembers accompanying the Patton family on a visit to the Gettysburg battlefield in 1932. Without notes or maps Patton led them unerringly to the place where his great-uncle had fallen. Although Patton was dressed in mufti and carrying a walking stick, it seemed to young Ayer that his uncle suddenly appeared in a gray uniform, with a sword in his hand, where he guided them over the most hallowed battlefield in America, describing and re-creating in such astonishing and accurate detail that to his family it was as if they had been transformed back to the hot summer of 1863 and were reliving the Battle of Gettysburg. The scene was so poignant that it left Ayer shivering. In his mind's eye it was not his uncle George before him but "a proud son of Virginia, defender of her honor. He *was* Waller Tazewell Patton, or George S. Patton of Winchester, and he was about to die for a glorious cause."[27]

It is not surprising that Patton was, according to Ruth Ellen,

one of the leading authorities on medieval armor in his day. He made my brother and all of his nephews suits of armor out of tin, aluminium, and for chain mail he used those awful pot scrubbers the army used to use, made of metal links. He was absolutely adamant that the armor be correct, but he deliberately mixed up the periods as he said no one went out and bought a brand new suit of armor, except a king or the like. They used what was hanging in the armory and made up bits and pieces to fill in.[28]

Patton also underwent what his daughter calls psychic experiences in which he encountered someone who had died. Twice his father suddenly appeared in a vision, once five days after his death, and again in the family cemetery in San Gabriel.[29] Patton was, of course, not alone in a belief in the ability to commune with the dead. Catherine Marshall has written of her late husband, Peter Marshall, who came to her and said she was not to think of him as dead. Sometimes these are in the form of psychic dreams, another phenomenon experienced by the Patton family.[30]

Patton's sister, Nita, had at least one such experience. She adored her mother and was very lonely after her death in 1928. "She awoke one night to the exquisite scent of Parma violets," the only scent Ruth Patton ever wore, "and felt someone rise from the edge of her bed and a hand pass across her cheek, and with that she felt a sense of profound happiness and peace that she never lost."[31]

Patton's many recollections of himself as a medieval warrior were perfectly in keeping with his vision of himself as a warrior-hero of historical fact and fiction. As Professor Carmine Prioli notes, from the time at Lake Vineyard that he pulled at an iron bolt in a wall and imagined himself as King Arthur pulling his sword Excalibur from the stone, to his visit to Tintagel castle in Cornwall in 1910 (during his honeymoon), where the legendary king was still thought to exist, Patton's visions of himself were those of a romantic, heroic warrior. That he and Beatrice appeared in costume as Arthur and Guinevere in 1935 at a Washington military pageant was no surprise.[32] Patton's love of history, his medievalism, and his poetry were all natural complements to his sense of déjà-vu.

Ruth Ellen believes that the family trait of remembrance was far more than an admirable ability. Waller Tazewell Patton was noted for his superb memory, and both of Patton's daughters could memorize and, years later, recite poetry with phenomenal accuracy. Several of the poems Patton required his daughters to memorize focused on death and rebirth.[33]

During Patton's first duty assignment in Hawaii in the 1920s, the Patton girls devoured continuing soap opera love stories in the *Honolulu Advertiser* despite warnings from Beatrice not to read such trash. Patton permitted them to continue the practice, provided they memorized and recited the first three paragraphs of each night's story. For years they had been made to memorize a poem each week—Patton called it a "brain exercise"—and this new effort proved relatively simple. Later he paid them ten cents a page to read items of his choosing, such as *Plutarch's Lives*. Ruth Ellen also memorized one of her father's poems about reincarnation, in which he recalled the sights, sounds, and smells of death during the Crusades:

MERCENARY'S SONG (A.D. 1600)
I am no callow Christian,
No pus-paunched prelate, I,
I hope not for salvation,
Nor fear the day I'll die

In wantonness of appetite,
In women, wine and war,
In fire and blood and rapine
In these my pleasures are.

I love the smell of horse dung,
The sight of corpse-strewn mud,
The sound of steel on armour
The feel of clotting blood.

The women I have ravished,
The infants I have slain,
The priests and nuns I've roasted,
They haunt me not again.

Priests talk of soul's salvation,
And shining lights afar,
But give me a harlot's laughter
And the battle flash of war.

Priests talk of soul's damnation
The white hot pits of hell;
I fear more wounds that fester
And gape and rot and smell

Then here's to blood and blasphemy!
And here's to whores and drink!
In life you know you're living
In death we only stink.

When Ruth Ellen recited the poem in school, her horrified teacher took her by the hand to the headmaster, who sent home a note. A distressed Beatrice explained to her husband that not all parents were so open with their children. Patton thought it hilarious.[34]

In North Africa in 1943, Patton spoke of dueling Rommel in the desert one on one, each in a tank by himself, winner take all. (The invitation to Rommel would have been engraved, and written in iambic pentameter.) Outsiders might have interpreted this as the fantasy of an aging general, but to those who understood him, like his young aide, Capt. Dick Jenson, Patton was indeed serious. In the film *Patton* Jenson observes, "Too bad jousting's gone out of style. It's like your poetry, General. It isn't part of the twentieth century."[35]

Whether our destiny is a form of immortality in another life or is merely "ashes to ashes and dust to dust," what can be said of Patton is that he believed passionately in both his own destiny and déjà-vu. Both were guiding tenets of his life and statements of personal philosophy. Given his extensive depression during the interwar years, his close brushes with death may well have sustained him until World War II, when he was at last presented with the opportunity to fulfill his life's ambition.

CHAPTER 23

Student Days, Boston Baked Beans, and Hawaiian Leis

(1923–1928)

As I approach 41 and there is no war I . . . fear that I
shall live to retire a useless soldier.

—PATTON

Patton spent five months at Fort Riley in 1923 attending the Advanced
Course at the Cavalry School, where he impressed his instructors with his
grasp of all facets of the military art. He finished high in his class and the
commandant, Brig. Gen. Malin Craig, noted on the combined rating of
twenty-eight instructors and four course directors that Patton did everything
"exceptionally well." Instruction included the machine gun, about which the
students were expected to learn the tactical employment, but not its mechanical aspects. One day Patton approached a young instructor, Lt. Paul Robinett, and said: "They say that the machine gun is the killingest weapon on
the battlefield. If that is so, I have got to know more about it. Will you give
me some personal instruction on Saturday afternoons?" Before he left Fort
Riley, Patton had mastered every detail of the weapon. In return he taught
fencing to Robinett and a group of bachelor lieutenants on Sunday afternoons in the attic of their quarters:

It was a gory business. With every thrust he would shout some such remark as: "I've knocked his eye out! See the blood spew! I've bored him through the heart! I've ripped his guts out!" One by one we stepped up to be dismembered. On and on Patton would go until almost exhausted. . . . Among ourselves we agreed that Major Patton carried a bucket of blood with him and liked to throw it around. We noted, however, that he was more circumspect among his elders, and indeed the assistant commandant . . . later agreed that he was surprised to learn this side of Patton, saying, "I had no idea he was like that."[1]

Patton's performance gained him a coveted selection to attend the General Service School at nearby Fort Leavenworth, Kansas.[2] When George reported to Leavenworth in September, it was without Beatrice, who was pregnant and remained in Massachusetts to await the birth of their third child.

Built astride the bluffs overlooking the Missouri River, on the border between Kansas and Missouri, Fort Leavenworth played an important role in the expansion of the United States. Well before the Midwest was settled, the river served as the principal highway for traders and explorers who sought the great untapped natural wealth of the young nation. Situated near the eastern end of the Oregon and Santa Fe Trails (and noted in the journal of Lewis and Clark in 1804), the fort was established as a permanent cantonment in 1827 by Col. Henry Leavenworth, commander of the 3d Infantry Regiment.[3]

Fort Leavenworth became a key outpost in keeping the peace, for the great exploration of the new frontiers of the United States, and for the three expeditions of the Mexican War of 1846.[4] During the 1850s the Kansas Territory saw bloody strife and lawlessness that by the outbreak of the Civil War left Leavenworth a Union oasis. In the 1870s Leavenworth was the site of a military prison, and years later became notorious as the site of a federal penitentiary built on the perimeter of the fort.

In 1881 Gen. William Tecumseh Sherman established Leavenworth as the site of what eventually became the most important military school for any professional army officer. In Patton's time it was called the Command & General Staff School, and later redesignated the U.S. Army Command & General Staff College (its present title).[5] Those who rise to the army's highest positions, including such celebrated officers as Douglas MacArthur and George C. Marshall, are all graduates of the school.[6] To become what Marshall referred to as "a Leavenworth" man was a prestigious mark of success without which no officer could aspire to future high command or compete for the few high positions in the peacetime army.[7]

The highly competitive, yearlong course at Leavenworth was taken very seriously by those fortunate enough to be selected.[8] Careers were made and broken, as were more than a few marriages. Nervous breakdowns were unexceptional, depression and insomnia were as common as colds, and there was even the occasional suicide. As Patton noted of his own experience, "I have been studying to beat hell." So intense was the competition that in one letter Patton unfeelingly noted, "So far I have done better than I have expected on almost all the problems though I doubt if I stand as high relatively as I did at Riley . . . [but] I think some of the others will crack—I hope so."[9]

The typical school day was long and highly structured and required considerable off-duty study. In addition to turning out qualified staff officers, the course at Leavenworth (the most challenging year of an officer's peacetime career) was designed to test the mettle of its students by placing them under great stress and obliging them to think and react even when exhausted, which was usually the case in war.[10] Using the case-study method, students were obliged to solve a variety of tactical problems for which there was generally an "approved" or "school solution."

After a day spent in conferences, lectures, and map and field exercises, students studied, often far into the night, which led to Patton's being summoned before an academic board for allegedly employing an unauthorized study aid, when a classmate "noticed [him] studying under a strange blue light and reported him. Upon questioning, an embarrassed Patton admitted he bought the light because it supposedly restored his hair."[11] To his continuing dismay, Patton had begun losing much of his hair to baldness, and the light was an example of "hope springs eternal" in his never-ending quest for something that would keep him forever young. "He stopped using the light—he went on getting bald."

Beatrice gave birth to their third and last child on Christmas Eve 1923 at Avalon, the Ayer family estate. On leave, Patton was present, as was the entire Patton clan, for the great day when Beatrice finally fulfilled her husband's passion for a son to carry on the Patton name. The child appropriately was christened George Smith Patton IV.[12] Beatrice journeyed to Leavenworth with her newborn son to be with George during the final months of the course. The Patton girls were left in the care of the Merrills and were, by Ruth Ellen's admission, "rather horrid."[13]

Unemployed, but armed with a letter of recommendation from his former commander, Joe Angelo wrote to congratulate Patton and to "hope [the baby] will some day be an officer like his father."[14]

The harsh demands of the school and the competition to become an honor graduate (one of the top twenty-five) remained intense right up to the end of the course. Patton's class graduated 248 officers. He stood twenty-fifth and was cited as an honor graduate, one of only two cavalry officers to attain this distinction.[15]

Social life was an important part of the Leavenworth experience, and prior to Beatrice's arrival, Patton's had been limited to riding and tennis. His wife's great wealth enabled them to entertain lavishly, on a scale inconceivable to the average officer, who was obliged to subsist on a miserly army income. An example of the Pattons' ability to live well appeared on the society page of the *Leavenworth Times* on June 1, 1924: "Major and Mrs. George S. Patton, of Fort Leavenworth entertained with one of the very handsome dinners of the season at the golf club last evening for one hundred and thirty guests. The decorations of the table were exquisite pink peonies."[16]

Patton kept a meticulous one-hundred-page notebook of the course and his part in it. He and Eisenhower had kept in touch, and when Ike attended the school in 1925–1926, Patton sent him the notebook.[17] It must have helped, for Eisenhower finished first in his class and earned a letter of praise from his friend, who wrote that he was pleased "to think my notes helped you" but that "I feel sure that you would have done as well without them."[18] Blumenson believes that the latter comment was merely politeness on the part of Patton, who was certain that his notebook was the *sole* reason Eisenhower had been so successful.[19]

Patton's letter to Eisenhower also provided clues to his future behavior. "We talk a hell of a lot about tactics and such and we never get to brass tacks. Namely what is it that makes the Poor S.O.B. who constitutes the casualtie lists fight and in what formation is he going to fight. The answer to the first is Leadership that to the second—I don't know." Patton did believe that one of the keys was superior artillery fire, without which "the solitary son of a bitch alone with God is going to skulk as he always has . . ." Patton also told his friend that, having now graduated from Leavenworth, he should "stop thinking about drafting orders and moving supplies and start thinking about some means of making the infantry move under fire." He prophesied that "victory in the next war will depend on EXECUTION not PLANS."[20]

Patton also commended the work of a French officer, Ardant du Picq, to Eisenhower. Du Picq's *Battle Studies*, a study of leadership and what makes men fight, had become a classic and was so highly regarded by Patton that he prepared 138 notes on soldiers and soldiering. Among them was a recommendation that skulkers—those who desert the front lines and their comrades without authority or who malinger to avoid combat—ought to be shot. "The execution of the skulker is necessary, not for his sin but for his betrayal of his comrades. Judas is execrated for the betrayal of One, should he who betrays hundreds escape?"[21]

Patton and Eisenhower emerged from Leavenworth with a discernible difference in philosophy, which would one day contribute to the rupture of their friendship. Although there was an obvious overlap, Eisenhower perhaps believed more firmly in the need for efficacious plans, while Patton placed more faith in their execution.

The commandant was his old friend, Gen. Harry Smith, who attempted to retain him as an instructor at Leavenworth, describing him as "a fine soldier . . . [and] one of our best students . . . [who] has demonstrated that he understands the theory, as well as the practice of war." Smith failed, however, and Patton was assigned to the First Corps Area Headquarters in Boston as the G-1 (personnel officer), where he reported for duty in July 1924.[22]

Patton's year at Fort Leavenworth was one of the better of the succession of unhappy years that comprised the two decades between the world wars. During a summer holiday before reporting for duty in nearby Boston, Beatrice wrote to Aunt Nannie that Georgie "is having a grand time here and is busy every minute. He seems like his old self again—he has been so changed since the war I feared it was permanent! But this summer he is just like a kid—every stern line has gone out of his face and . . . the kids . . . are having a grand time with him."

Beatrice was ecstatic over Patton's new assignment. The family rented a large homestead called Sunset Hill, near the beach in Beverly, complete with paddocks, stables, and a large contingent of English servants who had previously worked for the Merrills. Baby George's nanny was a Cockney and her husband was an ex-poacher who had been gassed at Ypres. "Ma was in her element. She had her family and old friends nearby—and Boston with the theatre and the symphony—all the little refinements she had missed. And she had, at last, her little son."

Ruth Ellen was passing through what she called the "mouse period" of her life, and had decided to raise thoroughbred mice, complete with a stud, females, and a studbook kept just as her father did with horses. All went well until the mice began straying from their appointed places in the barn and were soon everywhere, including the house. One morning as Major Patton pulled on his boots, one of Ruth Ellen's mice bit his toe. There were predictable bellows of outrage, and the crisis had barely passed when, as Patton went to put on his officer's cap, he let out a terrible howl and threw it to the floor, yet another mouse revealed scampering over his balding head. "After that, my mice lived in the stable."

Beatrice attempted to improve her horse jumping, and for her birthday Patton purchased a handsome new black mare named Dinah, informing the children they were to remain upstairs with their mother until he rang the doorbell. When it rang, they found Dinah standing in the middle of the dining room. With his flair for the dramatic, Patton had brought the horse in through the French windows. "Many things happen in life that get forgotten," wrote Ruth Ellen, "but a horse in the dining room is unforgettable!"

In September 1924 John J. Pershing retired, and an unsigned editorial in the *Boston Daily Transcript* was actually written by Patton, who told his mentor that he believed him to be "the Greatest American Soldier. . . . In my opinion the greatest honor a soldier can have is to have been prevaleged to serve under you as I was allowed to do. While you were on the active list I did not want to be thought guilty of bootlicking for my own benefit."[23]

After only eight months Patton received orders to Hawaii, ending the most idyllic period of their married life, and greatly disappointing Beatrice.[24] In one respect she knew that the new orders were a blessing. The children's tutor and piano teacher, a Miss Dennett, had fallen madly in love with Patton and, although Beatrice had no desire to fire a fine teacher, "Georgie was getting sticky about it . . . [and] finally gave out an ultimatum: either Miss Dennett went or he did."

During his brief stint in Boston, Patton wrote several important papers, one of which appeared in the *Cavalry Journal*. It was his vision of the future, in which a general observed the battlefield on a televisionlike monitor, filmed from a helicopter that captured images of tanks fighting each other and aircraft delivering tongues of flame that were precursors of napalm, the dreadful substance used in World War II flamethrowers and in aircraft ordnance. Patton was certain that many new mechanical devices would be invented in the not-too-distant future. His own recommendation as an interim measure was to create an armored car by surrounding a regular two-ton truck with armor plate, suitable for use with the horse cavalry.[25]

The Patton of old delivered a lecture to the 11th Cavalry in Boston in November 1924, titled "Cavalry Patrols," preaching the fire-and-brimstone tactics that became his trademark during World War II. During the interwar years Patton would state repeatedly that the next war would be fought where roads would be scarce. Six years later he would refine his comments by citing a remark once made by a Russian officer to Union general George McClellan, "In war all roads are bad." He would also quote Pershing that, "Future wars may begin in the air but they will end in the mud."[26]

Although Patton's remarks were meant to apply to the horse cavalry, they were also the forerunners of his own aggressive employment of tanks and infantry during World War II. "Get around if you can't get through," he exhorted. "IT IS THE DUTY OF CAVALRY AND SHOULD BE ITS PRIDE TO BE BOLD AND DASHING."[27]

In March 1925 Patton left for Hawaii without Beatrice, who had been hospitalized and remained at Avalon in ill health, with lingering complications from little George's difficult birth fifteen months earlier. He traveled by ship from New York to Panama, San Francisco, and back to Los Angeles for a brief reunion with his family, duplicating the journey undertaken after the

Civil War by his father and maternal grandmother. He even complained, as she had, of noisy children and squabbling women. During the trip from San Francisco to Honolulu, there was a fire in one of the holds containing Patton's household effects. Water from the fire hoses ruined Beatrice's piano, which was later found "floating happily" in eight feet of water. Many of his precious military books were lost. When their household effects were delivered to his new quarters, he sorrowfully reported that "practically everything is ruined." But, as he pointedly remarked at the end of one letter, "the bed is not *hurt at all*."[28]

Patton filled two positions at Schofield Barracks, acting G-1 and G-2 (intelligence officer) of the Hawaiian Division. He loathed the intense heat and humidity but eventually decided that "the beauty of the place grows on one." He was particularly impressed with Schofield Barracks, an old-fashioned military post with parade grounds, green lawns, neat barracks, eight bands, and plenty of the pomp he so dearly loved. As many as thirteen thousand men turned out smartly for division reviews.[29]

One of Patton's first efficiency reports suggested that he was "a man of energy and action . . . better qualified for active duty than the routine of office work."[30] In early 1926 Patton learned that the commandant of cadets at West Point was soon to be reassigned, and he cabled Generals Smith, Harbord, and Craig that he desired the appointment. Could they put in a good word for him? They tried, but the superintendent favored someone else, and there the matter died.[31]

In August 1926 Patton was decorated at a regimental ceremony with the civilian Treasury Life Saving Medal. Three years earlier he and Beatrice had been sailing in a twelve-foot catboat off Salem when a squall erupted. As they made for shore, they discovered a capsized boat. In a rough sea, the Pattons skillfully tacked their boat back and forth and managed to rescue three young boys from certain death. Beatrice later obtained statements from the grateful youths and sent them to the Treasury Department in the hope that her husband might receive a Life Saving Medal. The caption in a photograph in a Boston newspaper of Patton being decorated proclaimed: "Saved Three Boys from Drowning." Two medals ought to have been awarded, for the rescue could not have been carried out without Beatrice's skill and bravery, but only her husband received the coveted silver medal, engraved on the back with the words: "To George S. Patton, Jr., for bravely rescuing three boys from drowning, August 21, 1923." The award was a perfect example of how Beatrice helped to advance her husband's career.[32]

Patton spent Christmas of 1925 in California and not long after returning to the islands was joined by his family. They were greeted at dockside by a happy Patton, his arms filled with leis for each of them. The Pattons genuinely loved the Hawaiian Islands. The household soon grew to include a governess for baby George, two Japanese housekeepers, three grooms for the horses, and the usual large stable of polo ponies and horses. Beatrice had thoroughly immersed herself in Hawaiian history and culture before her arrival, and it was a case of love at first sight. "It was as if she had been waiting for Hawaii all her life." Later she made many Hawaiian friends and became an expert on the life and culture of the islands. Beatrice's study of Polynesian culture later resulted in the writing of a historical novel *The Blood of the Shark*, which was published in 1936. They found Honolulu society very much to their liking and were introduced to Walter F. Dillingham, one of the wealthiest and most influential men in the islands. "The Dillinghams took the Pattons to their collective bosoms and a lifelong friendship grew."[33]

Polo was mediocre when Patton arrived but improved dramatically when he organized an army team at Schofield Barracks. Before long the Pattons knew everyone in the islands who counted, and after polo matches frequently entertained the sailing and horse set. Beatrice thrived on music, singing, and dancing, while George preferred drinking with the men and swapping tales of polo, horses, hunting, and sailing. At first Patton was rather standoffish toward Beatrice's many new Hawaiian friends, but soon realized that their color and culture were not bars to friendship. "When Georgie became a friend," wrote Ruth Ellen, "he came with a whole heart and no reservations, and the Hawaiians who knew him all loved him and treated him with both laughter and respect."

One of Patton's additional duties was to purchase horses for the cavalry, and this entailed trips to the big island of Hawaii, where most were bought from the half-millon-acre Parker Ranch. The entire family would travel by overnight steamer and remain several days to explore some ancient ruins.

Beatrice continued to be his strongest asset. Her skills as a hostess, mother, and wife enhanced his standing in both the civilian and military communities. Somehow she managed to juggle her roles as provider-wife-hostess *extraordinaire*, always showing off her husband in the best possible light. The task was enhanced by her own personal interests in the Hawaiian people and their culture. During their first tour in Hawaii, Beatrice began writing a book of Hawaiian legends translated into French. (The book was later published in Paris, with superb illustrations by a Hawaiian artist.) An older Frenchwoman, who was giving Beatrice a refresher course in French, took a fancy to her husband. At this time the Pattons frequently spent their weekends in an exclusive Honolulu hotel. On one occasion Patton joined his social companions in the water in the wee hours of the morning after an evening of partying. He soon discovered the woman swimming after him, in the nude, bent on seduction on a nearby flat channel marker. A terrified Patton apparently broke records swimming away from this new species of barracuda. Out of breath, he staggered into their beach-front cottage and begged his wife to save him. Beatrice nearly choked with laughter.[34]

Beatrice learned the hula and taught it to Ruth Ellen, which made Patton jealous. Later, when the family visited Lake Vineyard, Beatrice put on a hula show that shocked Mother Patton. Beatrice played the guitar, her small son played the drum rattles, and Ruth Ellen did the hula.[35]

Their social life was complicated by Patton's superior, a stuffy colonel who scorned drinking and once ordered him to spy on a major suspected of having committed the unpardonable sin of serving bootleg liquor at his New Year's party. Patton adamantly refused and threatened to resign his commission before committing such a disgusting act. When the colonel was transferred to Boston, the snobbish Mrs. Colonel suggested to Beatrice that perhaps some letters of introduction to the local gentry might be in order. Beatrice replied that her family were all simple country folk who lived outside Boston. The woman demurred, as undoubtedly she and the good colonel would have little in common with such people.

Patton decided to get even and learned from Beatrice that the leaves of their beautiful oleander secreted a sap known to be highly irritating if it touched the skin. Wearing garden gloves, Patton crafted a lei of oleander blossoms, and on the day the colonel departed, went to his ship and draped the lei around his neck. The colonel burst into tears and said, "Oh, Georgie, I didn't know you cared!"

Patton's behavior still occasionally bordered on the bizarre. He often took Bee and Ruth Ellen to the cinema, but they missed the finales of a great many war films: Patton would angrily stalk out with his daughters if the uniforms or weapons were inaccurate. After Bee once misbehaved, Patton announced to his wife, "Beatrice Ayer Patton, how did a beautiful woman like yourself ever have two such ugly daughters?" Nor did Patton endear himself to the girls when Beatrice complained that they lacked "graces" by not knowing how to play tennis. He bought each of them rackets and balls, and during lunch hour they were required to practice in the broiling sun while he hovered nearby on horseback, yelling at them if they slacked off. After several weeks of this torture, the girls vowed never to play tennis again as long as they lived.

Patton's love-hate relationship with Ruth Ellen was never more in evidence than on the occasion when she was about twelve and being schooled to ride in a horse show. Virtually nothing she did was right, and Patton's criticism came in torrents. Finally he shouted in exasperation, "Get off that goddamn horse and let me show you how to do it." As Patton rode off to demonstrate jumping, Ruth Ellen was overheard to mutter: "Dear God, please let that son of a bitch break his neck."[36]

*   *   *

Patton had come to believe the United States would one day fight a war with Japan. His suspicions were aroused when one of his NCOs showed him that many Japanese migrant workers lived in small shacks near key inlets and bays. According to Ruth Ellen:

> When Japanese warships came into Honolulu to refuel or repair, there would always be parties of uniformed Japanese officers visiting these little farms. The word was that the farmers were relatives of the officers. Georgie said that as the caste system in Japan was still dominant, he wondered if there were not trained saboteurs, or signalmen, changing places with their so-called "cousins" which would have been easy. . . . All these things he reported to Military Intelligence from where it disappeared into the vast maw of Washington unnoted.

In the spring of 1926 Mr. and Mrs. Patton and Nita journeyed to Hawaii and spent several months at Schofield Barracks, doting on their new grandson who would now carry on "the Name." Convinced that she would catch leprosy from one of the natives, the neurotic Aunt Nannie remained in California. Although it was a happy time for the family, it was obvious that Papa was unwell, his health having been in serious decline for some years. It is believed that as a young man he contracted a case of latent tuberculosis from his uncle, William Glassell, the Confederate naval officer, who had acquired the disease in a Union POW camp during the Civil War.

Papa wrote that he hoped the family could all "come home" to California for "one more Christmas together." The Pattons spent Christmas 1926 at Lake Vineyard, and it was then Mr. Patton told his son he had TB in his left kidney but that the family was not to know, which of course they already did. His doctors had decided an operation was too life threatening.[37]

He told George that his career in the army was fate and that he was being prepared to fulfill some special role. "He used to assure me with the greatest confidence that I would yet be in the biggest war in history . . . and I believed him, particularly as I have always felt the same thing concerning my self." One evening Papa said, "Son you had more experiences and saw more of life in two years in France that I have in all my seventy years." Patton thought him not bitter but merely wistful. "He had a romantic and venturesome spirit . . . and a sense of duty to those he loved. I am sure he felt that death was the only adventure left for him and yearned for it with out fear and with great curiosity and anticipation."

The Pattons had barely returned to Hawaii when, in February 1927, Mr. Patton's condition worsened and a lifesaving operation became necessary. In great pain, he was wheeled into surgery; his son kissed him good-bye, not knowing if he would survive the ordeal. "The courage of that act would have won the Medal of Honor on a field of battle . . . he was very weak; yet going to what he thought was death he tried to cheer me with out thought of self." Patton remained for three weeks but finally could stay no longer. The night before he left for Hawaii, Patton "kissed him for the last time and said good by. He smiled and . . . the last words I ever heard him say were: 'Good by Old Man take care of yourself.'"[38]

In June 1927 a dreaded telegram announced Mr. Patton's death, apparently from tuberculosis, and cirrhosis of the liver from too many years of drinking more than was good for him. Patton was "absolutely undone" and "blamed himself for not having been at Bamps' bedside when he died; for not having been there when Bamps needed him; and for having spent so much of his life away from home; for all the little things he might have done for his father." Beatrice reminded him that he had been a good and dutiful son, "but he was still inconsolable."[39]

There was no ship to California for several days, and the family witnessed Patton's show of "almost unreasonable grief." He made his three children memorize Arnold's "After Death," of which Ruth Ellen notes, "This poem—this affirmation of faith and trust—has seen us through everything that has ever happened to us in the griefs of our later years." Although daughter Bee managed to arrive in time from Virginia, where she was attending an exclusive private school, Patton was too late for his father's funeral. The wake lasted three days, and Papa's corpse was never left alone for a single second. A steady wailing for the dead emanated from the back of the house where old Mary Scally and her Irish friends were joined by many Mexicans, whose families dated back to the time of Don Benito Wilson.

Upstairs Aunt Nannie had for days been screaming hysterically for her beloved George "to come and take her with him, and that she knew that it was she he had always loved and not Ruth, whom he had married." Her dreadful wailing pierced the ears of the mourners sitting in suffering silence, while hour upon hour could be heard the refrain that Georgie should have been *her* son. "Wait for me, George! Wait for me!" As they had for so long, Ruth and Nita simply ignored the crazed Nannie, and throughout the ordeal somehow managed to remain composed and retain the great dignity that was so characteristic of both women. For years Aunt Nannie had been insufferable, her moods exacerbated by alcohol, and the scenes that Patton fortuitously missed were frightful.[40]

The morning he arrived at Lake Vineyard in full uniform, Patton went directly to his father's flower-strewn grave in the family plot in nearby San Gabriel.[41] "Suddenly," he wrote a short time later, "I seemed to see him in the road wearing his checked overcoat and with his stick which he waved at me as he had been used to do when he was impatient and wanted to go somewhere." In the vision Mr. Patton was using his cane as a sword to slash off the heads of the flowers. "As he passed Georgie, he looked at him with his warm smile and a twinkle in his eye, and went on down the path." Patton "knelt and kissed the ground then put on my cap and saluted not Papa, but the last resting place of that beautiful body I had loved. His soul was with me and but for the density of my fleshly eyes I could have seen and talked with him."[42]

A few days later, while sitting at his father's desk examining the contents of his safe, which included the minié ball that killed great-uncle Waller Tazewell at Gettysburg and the shell fragment that had killed his grandfather at Winchester, Patton had a second vision. With tears streaming down his face Patton suddenly saw Papa standing in the doorway. He was frowning and seemed to be shaking his head. "Georgie felt a clear, distinct message that 'all was well; he was not to mourn.' He said that his mind seemed to answer, 'I understand. Alright, Poppa.' At that, his father gave him his radiant smile and turned and walked to the stairs and only when Georgie saw the trousered legs disappearing up to the door into the front hall did he remember that the Honorable George S. Patton had been dead and in his grave for five days."

The death of his father was the most traumatic event of Patton's life. He had lost the best friend he ever had, and the man who had sacrificed so much for a son he adored. For the next several days Patton wrote furiously to produce a touching memoir of his father, and of his own childhood:

> Oh! darling Papa I never called you that in life as both of us were too self contained but you were and are my darling. I have often thought that life for me was too easy but the loss of you has gone far to even my count with those whom before I have pitied. God grant that you see and appreciate my very piteous attempt to show here your lovely life. I never did much for you and you did all for me. Accept this as a slight offering of what I would have done.[43]

No son could have provided a more loving tribute to his father.

The last year of Patton's Hawaiian tour was troublesome. In November 1926 he had finally been appointed the permanent G-3 of the Hawaiian Division. After a brigade field exercise that autumn, Patton, who acted in the capacity of an official observer, severely criticized the unit's performance and presented the commander, a brigadier general, with a harshly worded written critique. However true his remarks, the paper was highly indiscreet and should never have been delivered without the concurrence of the commanding general.[44] It was an example of Patton's lapses of military protocol, combined with his impatience when he disagreed with others. Within the military hierarchy of a division staff, the G-3 is the first among equals. He is the division commander's principal tactical adviser, and is responsible for every aspect of training and operations. As he did with every position he ever occupied, Patton carried out his duties with a zeal that heightened his unpopularity. He issued a series of strict directives designed to smarten up military discipline. Their implicit message was that the brigade and regimental commanders were not doing their job. His acerbic imprint graced whatever displeased him, with little concern for the possible consequences. In the margin of a report by a board of officers examining defenses against low-flying aircraft, he wrote: "This is a poor paper . . ." The report's principal author was the future commander of the U.S. Army Air Corps in World War II, Maj. Henry H. "Hap" Arnold.[45]

When Patton recommended an exercise to demonstrate how to defend against an enemy air attack on infantry, Maj. Gen. William R. Smith, the division commander, agreed. The exercise went badly, and another blistering report emanated from Patton's pen, only this time it contained Smith's blessing. However, as Blumenson notes, everyone in the division knew that the author of the comments was not General Smith but George S. Patton. The old-boy network reacted with fury, and, in response to complaints from his senior officers, Smith relieved Patton, citing as the reason that he was "too positive in his thinking and too outspoken."[46]

Although Patton's new appointment as the G-2 was disguised as a routine reassignment (which was perfectly within the commander's prerogative), the relief was one of the bitterest moments of Patton's career. Unfortunately he had been thinking and acting more like the World War I colonel he had once been than a relatively junior major. Patton's yen for a war was inconsistent with the politics of peacetime soldiering, and his indiscretions and his inability to suffer fools had finally earned him one antagonist too many.[47]

Much of his outpouring of essays and lectures was the result of his frustration with peacetime duty. While in Hawaii he wrote about the war he knew and understood and discerned as few others did. Patton's social life had never been better, his writing was prolific, and eventually his bitterness at being relieved was partially mitigated by his final efficiency report. Maj. Gen. Fox Conner was the new commander, and he wrote, "I have known him for fifteen years, in both peace and war. I know of no one whom I would prefer as a subordinate officer."[48]

Shortly before they departed Hawaii in April 1928 for Washington, where Patton was assigned to the office of the Chief of Cavalry, George and Beatrice bought the only home either ever owned. After the death of his father, Lake Vineyard had lost its magical appeal. Patton favored Virginia, the land of his roots, but most of Beatrice's large family resided in Massachusetts, and they decided to purchase a home near Avalon. Still convinced that he would die in a foreign land, and that a war with Japan was imminent, he was content that Beatrice would be well taken care of in the bosom of her family.

Some months earlier, when they had first discussed the idea, Patton had written, partly tongue in cheek, that he had "always hoped I would secure supreme command and such fame that after the war I would be able to become President or dictator by ballot or by force. In that case we would have persuaded a grateful people to build us a marble Palace at the flag pole at Fort Myer. However as I approach 41 and there is no war I . . . fear that I shall live to retire a useless soldier in which case as we could still hunt it would be nice to have a place in the hunting country [of Massachusetts]."[49]

Set amid the beautiful Myopia Hunt country, the house, named Green Meadows, was a large colonial that had been bought by neighbors and Beatrice's brothers to prevent it from being subdivided after the owners had died. It was offered to the Pattons at cost, and they accepted by telegram from Hawaii. They spent the late spring of 1928 in Beverly, but Patton could not remain long and in the early summer reported for duty in Washington. Beatrice remained behind to take charge of restoring Green Meadows.[50]

CHAPTER 24

The Washington Years
(1928–1934)

I do . . . regard with horror a state of affairs which would
make our country both unready and unwilling to defend
its honor.

—PATTON

Patton rented a large house called Woodley, near the Washington Cathedral,
that had once been the summer White House of a president. Among his
pleasures was renewing his friendship with Henry L. Stimson during horseback rides through Rock Creek Park. Patton used these social occasions to
keep Stimson informed of what was transpiring in the War Department, and
to solicit his support.

In 1929 Stimson became secretary of state under new President Herbert
Hoover, and that same year Stimson bought the property Patton was renting,
and the family moved to a six-bedroom house in northwest Washington that
was thought to be haunted. The site of frequent parties, and often filled with
guests, the place took nine servants to run. Not long after moving in, Patton
made a striking impression on two very proper spinsters from Boston who
entered the house at the same moment he emerged stark naked from his
dressing room upstairs. Calmly the ladies walked outside, closing the front
door behind them. Patton fled in horror to the safety of his bedroom while
the spinsters rang the front bell and reentered as if nothing had occurred.

Dinner in the Patton household was a formal affair. Beatrice and the two
girls would appear dressed in long gowns, while Patton wore a dinner

jacket. Conversation was "light and elegant," as the family servants bustled about, serving a multicourse meal.

Patton's propensity for putting his foot in his mouth was demonstrated one evening when the phone rang and a gentleman with a cultured Harvard accent asked for Major Patton. Thinking it was an old family friend, he said: "Why Francis, you damned old nigger lover! What in hell are you doing in Washington, and who the hell let you out of jail?" The caller replied: "This is the secretary of state. I called to see if Major Patton would care to come over to Woodley for a game of squash rackets and a drink."

The Pattons quickly adapted to the Washington social scene. Their usual pursuits were foxhunting, steeplechasing, horse shows, and polo. Patton bought a boat that he kept berthed on the Potomac. Bee and Ruth Ellen wanted only to lie on deck in bathing suits and attract the attention of young males, but Patton was a stern "admiral" who permitted no lounging aboard his vessel and made them shine brass and perform other decidedly tedious shipboard duties.

Fall and winter were devoted to foxhunting in the lush Virginia country-side. Too vain to wear her glasses and consequently riding "blindly" behind her husband, hardly ever seeing the fox or what her horse jumped over, Beatrice soon gained a deserved reputation as a fearless rider. Beatrice also spent time attending to the establishment of Green Meadows as the family homestead. Patton was more interested in the outdoor aspects and, with fla-grant disregard for his own safety, built a dangerously high five-foot jump, when four-foot jumps were the norm. If he could not fight, then at least he could challenge himself by placing himself in mortal danger.[1]

Through the Stimsons, the Pattons socialized with and entertained everyone of note in Washington, but were disdained by the professional party givers, who resented their presence in what they regarded as their own private preserve. When it became known they were planning a large ball for Bee's debut, one of the biddies called to remind Beatrice that she had failed to clear the event with the clique and to tell her that her ball could *not* be held because it might establish a bad precedent. When Beatrice politely demurred, she was told that the clique would steal away all her guests by scheduling a number of grand balls for the same evening. They must have gagged when Pershing, the Stimsons, and everyone who was anyone in Washington society turned up at the Pattons' and had a grand time. The event even made the first page of the society section.

It was during his tour in Washington that Patton claimed to have made what his family referred to as his "great sacrifice." He reputedly turned down an assignment in London in the office of the military attaché, even though it would have meant a golden opportunity to play world-class polo, hunt with the bluebloods of England, and participate in internationally rec-ognized horse shows. He told a disappointed Beatrice:

We have two marriageable daughters who . . . will be rich someday. If we go to London it stands to reason that one or both of them will marry an Englishman. Englishmen, well-bred Englishmen, are the most attractive bastards in the world, and they always need all the money they can lay their hands on to keep up the castle, or the grouse moor, or the stud farm, or whatever it is they have inherited. I served with the British in the war, and I heard their talk. They are men's men, and they are totally inconsiderate of their wives and daughters; everything goes to their sons, nothing to the girls. I just can't see Little Bee, or Ruth Ellie in that role. Someday, just tell them what I did for them and maybe they won't think I'm such an old bastard after all.

Patton loved his daughters deeply and was extremely protective, particularly of his eldest, Bee, who had grown into a beautiful blond but very shy young woman. In 1929 Patton returned to West Point for his twentieth reunion, and, to ensure Bee had a good time, he contacted several old friends and asked that they select whomever they considered the "best cadet" to escort Bee to the hops and other social events. They chose John Knight Waters, a young third classman from Baltimore whom they considered the most outstanding cadet at the Academy, and from their first date in 1929 until they married five years later, the two were inseparable. Like his future father-in-law, Waters was destined to become a cavalry officer and later attain the exalted rank of four-star general.[2]

In October 1928 Ruth Wilson Patton died on a train in New Mexico while enroute to Washington with Nita. Ruth Patton's health had been declining for some years, and the death of her husband had left her with little incentive to continue living. Nita had always been her favorite child. Her Georgie had been given everything he ever wanted, but "Nita had had to take second best all her life . . . [and] had inherited her tremendous strength of character and her total integrity. . . . Georgie [later] turned over his entire share of the California estate to his sister so that she could live at Lake Vineyard in comfort for life."

Patton had been a major since 1920, and prospects for promotion in the peacetime army remained so bleak that even the death of a senior officer was welcomed. One night he came home from work and announced: "General so-and-so died today and every colonel in Washington is like a wolf bitch in heat, waiting to see who will step up into his slot."

His new duties left him joylessly advocating tanks without betraying the cavalry and seriously harming his career. In the ensuing years he would waver first one way, then another. Blumenson notes that Patton's problem was that subtlety and discretion were not his strong suit. He accepted that the cavalry of the future must be mechanized to replace the horse, but as a

representative in high standing of the chief of cavalry, he could hardly recommend the demise of his own service, and, in the end, supported the cavalry.[3] About as far as Patton was willing to venture was to agree that it was
desirable to attach armored cars to a cavalry division, "a vision not too far
removed from that of the mechanizationists."[4] His ambivalence left him at
odds with the pioneers of armor, headed by Adna Chaffee, who would eventually propel the army into the age of mechanization.

At conventions and other gatherings, ranging from marines to quartermasters to the Pennsylvania National Guard, he delivered a mixed message.
To the American Remount Association in the spring of 1930 Patton suggested that the horse and tank could coexist. "What is wanted is better types
of both run by men who know their powers and limitations." To the infantry
and his fellow cavalrymen he declared: "If the 14th Century Knight could
adapt himself to gunpowder, we should have no fear of oil, grease, and
motors."[5] In another speech he tartly warned those who thought the cavalry
was dead not to fight against it: "They will be corpses, not we." Such flip-
flopping appeased his superiors and enabled him to keep a foot in both
camps.

He studied the famous and the infamous and drew lessons from each.[6]
Patton thought Gen. George McClellan's downfall came about because he
had taken counsel of his fears, a conviction that had become a vital part of
his military principles.[7] He continued to learn from the Germans and
devoured the memoirs of Hans von Seekt (the World War I chief of staff of
the German Eleventh Army), writing "*Bull*" next to his comment that a
commander ought to delegate most routine duties to his chief of staff in
order to keep "his mind fresh and free for the great decisions." Patton disavowed this traditional concept, which he would modify in World War II to
fit his belief that a good chief of staff ought to play a more important role in
the day-to-day operations of an army.[8] In 1929 Patton vividly applied his
extraordinary knowledge of history when he served as an umpire during
cavalry maneuvers at Fort Bliss. "His final critique disconcerted his audience by equating the attacking Brown forces with the Etruscans and the
White forces with the defending Romans. He told of the Etruscans' crossing
the Apennines into the Po valley in the 6th century B.C."[9]

That year Dwight Eisenhower was assigned to the War Department and
renewed his friendship with George Patton. The Eisenhowers were frequent
guests of the Pattons. As their son, John, later remembered:

I always held the Pattons in considerable awe, because of their obvious
wealth and the fact that Patton was a lieutenant colonel while Dad was a
major. Silver horsemanship trophies covered a complete wall in the living
room of their quarters. Patton was a good-humored man who loved to
joke. His language was full of the purple expressions for which he later

became famous. I was astonished that he not only swore profusely around ladies but also encouraged all three of his children to do the same. When young George, a fine boy slightly younger than I, would come out with an appropriate piece of blasphemy, Patton would roar with pleasure.[10]

The only major difference in their relationship ten years after they first met at Camp Meade was that Eisenhower was no longer a part of the debate over mechanization.

During the late 1920s there was renewed interest in the creation of an armored force, and at the instigation of Secretary of War Dwight Davis, and Patton's friend General Charles P. Summerall (the army chief of staff from 1926 to 1930), an attempt was made in 1928 to create at Camp Meade a small experimental combined-arms force of tanks, artillery, cavalry, air corps, and engineers, but the project faltered for lack of funds.

When he arrived in France in 1917, Patton's romantic notion of combat had been dashed by the grotesque specter of trench warfare. As historian William Woolley notes, "Real war had meant to him movement and decisiveness. West Point and his experience in Mexico had reinforced that view, so that he saw the static and indecisive trench warfare that he found in France in 1917 as an unhealthy aberration. During the war and for the next dozen years afterward Patton mulled over the question of what had gone wrong, reaching the conclusion that the culprit was the mass army."[11]

Thus, despite his vivid imagination, vision, and pioneering accomplishments during World War I, Patton's ambivalence was a direct reflection not only of his own ideas but of the office in which he served from 1928 to 1932. As the chief of the Plans and Training Division he was squarely in the middle of what Woolley calls "the intellectual citadel of traditionalism within the cavalry"—those who published the cavalryman's bible, the *Cavalry Journal*, and lobbied for its retention. Although their mission was mandated by law to investigate and propose the weapons and doctrine of the future, these were the same officers who fought a rear-guard action to prevent the takeover of their service by mechanical machines.[12] A classic conflict of interest, for Patton it meant an unhappy obligation to serve two masters.

To retain his post, Patton distanced himself from those he termed "pure mechanizationists" and satisfied the traditionalists by frequently pointing out the limitations of armored vehicles. Philosophically Patton and the mechanizationists were as far apart as ever. "What the mechanizationists had proposed as a revolutionary means to overthrow a traditional system of warfare, Patton had finally come to accept as an evolutionary means to restore it. . . . Patton had joined the mechanizationists; yet he remained a traditionalist."[13]

However, even Chaffee, who toiled tirelessly to create and sell the notion of an armored force, never reckoned on the acrimonious opposition and deliberate obstructionism of the diehard cavalrymen during the decade of the 1930s to the idea of replacing their beloved horses with machines. Patton thus found himself at the center of a controversy he could not resolve. His head told him that the cavalry was a glorious anachronism and that the future lay in an armored force, but his heart led him to condemn what would ironically fulfill the destiny he had compromised his career to await. Had he joined his friend Adna Chaffee, Patton might have gained joint billing and a rightful place as the patriarch of the American armored force.[14] However, as late as 1932, while a student at the U.S. Army War College, he continued to support the notion of a future role for the horse cavalry. In short, although Patton eventually advocated an armored force, he arrived via the back door, and in so doing forfeited his place in history as one of its true pioneers. Only later in World War II would Patton take what Chaffee and the mechanizationists had created and turn it into an instrument that revolutionized modern war. *That* would be his enduring contribution.

One of Patton's pronouncements that he would later surely regret was a 1933 article in the *Cavalry Journal* in which he wrote that although in a future war armored vehicles and horses would operate together, his readers should not be swayed by history's many examples of new military implements that were heralded as the key to victory only to fail the test of battle. "When Samson took the fresh jawbone of an ass and slew a thousand men therewith he probably started such a vogue for the weapon, especially among the Philistines, that for years no prudent donkey dared to bray. History is replete with countless other instances. . . . Today machines hold the place formerly occupied by the jawbone. . . . They too shall pass."[15] Yet, while publicly condemning mechanization he was so attracted to the idea that he worked tirelessly to create an organization table for a mechanized cavalry regiment that might be employed with the traditional horse cavalry.[16] Beyond this he was not willing to venture.

It can be argued that the reasons for the continued obstinacy of the cavalrymen to remain at least one war behind reality were a last, desperate attempt to retain the traditions of the Old Army. "Mechanization's threat to horse cavalry involved more than military obsolescence; to let this change happen would destroy a world of social rituals based on the horse and the romance of the cavalry. As long as one could justify a military role for horse soldiers, the polo matches, the fox hunts, and the horse shows all fell into place as perfectly appropriate—good training—as if it would always be this way."[17]

To understand why Patton returned to the cavalry and dodged the controversy of mechanization by hedging his bets, one need look no further than the comfortable life afforded a cavalry officer, especially a *rich* cavalry officer of his social standing. The obsession to prove himself in a war continued to dominate his dreams, but until a war enabled him to fulfill his perceived destiny of becoming a famous warrior, the cavalry could at least satisfy his romantic nature and permit him to bide his time in comfort. Beatrice, however, was torn between a profound desire to have her mercurial husband happy and successful, which meant another war, further long separations (of which there had already been enough), and possibly even his death, or to have him safe but unhappy. The social and sporting life in Hawaii and Washington could hardly have been more agreeable, on the other hand.

In September 1931 Patton entered the U.S. Army War College. Until 1951, when it was moved to its present location at Carlisle Barracks, Pennsylvania, the War College was located in southeast Washington at the site of what is now Fort McNair. While Leavenworth had been the stepping-stone to higher staff duty, the elite War College existed to prepare its students for future high command and staff.[18]

As he had at Leavenworth, Patton worked doggedly to achieve recognition. He also again lobbied for assignment as the commandant of cadets to West Point after the course ended in June 1932. J. G. Harbord wrote a letter of recommendation to the commandant of the War College, Maj. Gen. William D. Connor, who was slated to become the new superintendent the following year. Another friend from the AEF, Hugh Drum, also put in a good word for Patton, but it all came to naught for the second time when the post remained unfilled until 1933 and then went to another.[19]

The most important part of the course was a research paper prepared by each student. Patton's outlined his vision of the next war to be fought by the United States.[20] He again returned to a familiar theme in which he argued the merits of both large conscripted citizen armies such as the AEF of World War I, and the much smaller professional armies of other nations, which he believed were the real key to winning rapidly and decisively. With his usual blend of historical example and attention to the most minute detail, Patton managed to turn his essay into a lesson in the history of warfare and of the warriors of many races and nations.

The paper was forwarded by the commandant to the War Department with praise that it was a "work of exceptional merit."[21] Patton also chaired a student committee that examined mechanization and reported its recommendations for consideration by the War Department General Staff. The committee's final report bore Patton's imprint and endorsed the notion that there was no place in the army for a major armored force operating independently, but that the infantry and cavalry ought to originate their own mechanized units to support each other in their particular role on the battlefield. In other words, there was no place in the army for the mechanizationists. The views of Patton's committee report coincided with those of the new chief of staff, Gen. Douglas MacArthur, who, in any event, had no funds to support the development of an independent armored force and decreed that each combat arm of the army should create and employ their own armored vehicles as they saw fit.[22]

In November 1931 Patton's studies were interrupted so that he could attend the funeral of Aunt Nannie, who had died at the age of seventy-three. "I never knew, until I saw her in the majesty of death, what a noble face she had," he told Ruth Ellen.[23]

Although he had positively worshiped his father, it was not until he returned to Lake Vineyard three years after she died that Patton finally acknowledged how deeply he missed his mother. He composed a touching letter that he left in her trinket box. Admitting that he had never showed her 'his love and admiration for a lifetime of courage', Patton wrote:

> Children are cruel things. Forgive me. I had always prayed to show my love by doing something famous for you, to justify what you called me when I got back from France, "My hero son." Perhaps I still may, but time grows short. I am 46. In a few moments we will bury the ashes of Aunt Nannie. All the three who I loved and who loved me so much are now gone. But you know that I still love you and in the presence of your soul I feel very new and very young and helpless even as I must have been 46 years ago. . . . I have no other memories of you but love and devotion. It is so sad that we must grow old and seperate. When we meet again I hope you will be lenient for my frailties. In most things I have been worthy. Perhaps this is foolish but I think you understand. I loved and love you very much. Your devoted son G.S. Patton, Jr.[24]

A 1932 charcoal drawing of Patton may well be the best depiction of him ever produced. Unlike most of Patton's later photos, especially those of the World War II period, in which he appears autocratic and severe, the artist captured the image of a thoughtful, somewhat pensive man in the uniform of a cavalry officer. There is about him an air of profound sadness.[25]

Patton graduated from the War College in June 1932 with a "superior" academic rating from the commandant and an outstanding evaluation that described him as "an aggressive and capable officer of strong convictions. An untiring student."[26]

He returned to Fort Myer in July 1932 as the executive officer of the 3d Cavalry. The family moved from a large rented house in Washington to one in Rosslyn, just outside the gates of Fort Myer, whose highlight was its all-black bathroom. The nation was in the depths of the Great Depression. World War I had been over for nearly fourteen years, yet there emerged in the pages of the *Cavalry Journal* an image that nothing had changed in the splendid, isolated sphere of the cavalry, whose leaders remained in total denial of the effects of the Great War, and of the obsolescence of the horse. "As the Chief of Cavalry begged Congress for more horsemen, he seemed plagued by a terrible amnesia that denied the machine gun, the gas barrage, and the totally obliterating power of modern artillery that had altered the geography of Belgium 20 years earlier."[27]

Although the army was as woefully behind the times, its men underpaid and unappreciated, it nevertheless was a sea of tranquillity in which life was structured and governed by rituals as old as the nation. Patton and other cavalrymen were living in what amounted to "a dream world accessible to very few. At Fort Myer, especially, a young cavalry officer close to the social whirl of Washington must have felt very secure, so close to the rich and famous and powerful that one might easily imagine he was one of them."[28]

The tranquillity of Patton's life in the cavalry was rudely interrupted in the summer of 1932 by the Bonus March on Washington, carried out by disgruntled and distraught veterans of World War I who believed their government had betrayed them. After the war, as a means of partially compensating soldiers for their war service, Congress had voted a bonus in the form of bonds due to mature in 1945. With the nation in the throes of the worst economic depression in its history, the 1928 Republican party slogans of "A Chicken in Every Pot" and "Two Cars in Every Garage" had turned to ashes. Many of the 3,500,000 veterans began demanding that in such hard times Congress ought to make good on their promise by authorizing payment now, when money was so desperately needed. During the winter of 1931–32 the agitation for payment was widely reported in the press, and various veterans' groups across the United States began supporting a march on Washington to lobby Congress.[29]

The previous year Patton's former batman, Joe Angelo, then an unemployed riveter about to lose his home because he could no longer pay his property taxes, had walked 160 miles to Washington from Camden, New Jersey, and had testified in full uniform complete with medals before a congressional committee on how he had won the DSC by saving Patton's life. "I could go right over here to this cavalry camp across the river and get all the money I want or need from Colonel Patton. But that ain't right. I don't want him to feel under any obligation to me for saving his life. He owes me nothing. I did my duty. All I ask is a chance to work or a chance to get my money on my certificate."[30]

Destitute veterans, many of them homeless, had already begun turning up at Fort Myer, where sympathetic soldiers paid small sums from their monthly wages to help support the hiring of some to perform menial kitchen and stable duties. By May 1932 what had begun as a trickle had reached as many as 25,000 men who converged on Washington. Although some brought their families, virtually all had no place to live and no money to afford food or accommodations. Many occupied condemned buildings, but the majority ended up creating an enormous shantytown in the mud flats of Anacostia, which most dubbed "the Flats"; others, "Hooverville." In the misguided belief that their protests would spark other groups to follow suit, the Hoover administration reacted by denying recognition and aid to the veterans.

Among the more than eleven thousand men who occupied the Anacostia Flats was Joe Angelo. The place had become a tourist attraction, and Angelo became one of its most colorful leaders. To demonstrate his disgust at their plight, he dug a hole in the ground and had himself covered up with dirt, using only a stovepipe to breathe. Calling himself a "living corpse," Angelo told everyone who paid ten cents admission to view his plight: "This is a swell country where you get to be buried alive if you want to eat." Vowing to remain until the bonus was paid, Angelo said: "If they don't vote the bonus they're going to have to erect a monument here where a wearer of the DSC starved to death."[31]

Each day veterans turned up on Capitol Hill and lobbied Congress, whose members, while sympathetic, failed to enact legislation that would have fed and sheltered the men and paid their bonus. Except for a few hundred reactionary veterans (led by two communists), the marchers had organized themselves into a paramilitary organization called the Bonus Expeditionary Force (BEF), which was led and policed by its own members, who were committed to a peaceful attainment of their goals.

As the weeks passed, tensions grew, Congress dithered, then adjourned in mid-July having authorized the veterans to borrow only against their certificates for transportation back to their homes. About ten thousand remained in Washington, vowing to "stay till 1945 to get the bonus," and chanting, "Bonus or a job." Tensions appeared to ease despite the distribution by the small number of hard-core communist veterans of handbills condemning the government and demanding action.

However, the bonus marchers failed to reckon on the paranoia of the Hoover administration, which completely overreacted. The afternoon of July 28 Secretary of War Patrick J. Hurley ordered mobilization to disperse the BEF. The 3d Cavalry was ordered into the city for riot duty.[32] The regiment quickly assembled and, with Patton at its head, galloped across the Memorial Bridge and onto the Ellipse, south of the White House, where it awaited the arrival of an infantry battalion being sent by steamship from nearby Fort Washington. Two hundred cavalry troopers on horseback, clad in steel helmets and with carbines, gas masks, and drawn sabers, were a sight never seen before in Washington.

As the executive officer, Patton did not exercise direct command of any element of the regiment (except, of course, in the absence of the commanding officer) and, although not required to participate, simply could not resist the lure of the event. While they awaited the infantry (who arrived later, with bayonet-tipped rifles and gas masks on), Patton decided to use the time to reconnoiter "by trotting stonily along Pennsylvania Avenue. Several thousand veterans along the route greeted him with mixed cheers and jeers."[33]

The affair might never have ended in tragedy had the authorities acted with the same compassion as District of Columbia chief of police, Pelham D. Glassford, a former army brigadier general, who had gone out of his way to assist rather than persecute the bonus marchers, who might well have departed of their own accord before long. Seizing on the alleged threat posed by the tiny communist element, MacArthur had for some days been exhorting Hurley to do something about the bonus marchers for whom he had little sympathy. In what has been condemned by historian Piers Brendon as "one of the most shameful episodes in American history," the bonus marchers were attacked by tanks, infantry, and Patton's cavalry in downtown Washington.

About four P.M., as the troops swept down Pennsylvania toward the Capitol, thousands of spectators witnessed the dispersal of the veterans, some of whom heaved rocks and bricks and were tear-gassed by the infantry. Some of the spectators were heard to curse the troops: "Shame! Shame!" and "You goddamned bums!"[34] Even though they did not resist, the veterans were tear-gassed and forcibly driven into Hooverville by about eight hundred federal troops. The cavalry advanced with sabers drawn and occasionally used the flats of their blades across the buttocks to disperse anyone who resisted. Patton used the flat of his saber at least once, and the veterans were soon driven across the Anacostia Bridge.

Officers stationed in Washington were required to wear civilian clothes. But on this fateful afternoon MacArthur appeared on horseback, in full dress uniform, to direct the operation personally, much to the disgust of Eisenhower, who was then one of the chief of staff's military assistants. Hoover had twice sent direct orders through Hurley to MacArthur that the military was *not* to pursue the BEF across the river to Hooverville. Using the flimsy pretense that he had not heard, and refusing to be bothered by "people coming down and pretending to bring orders," MacArthur deliberately ignored Hoover's instructions and, in so doing, disobeyed his first but certainly not his last president.[35]

Early that evening, as the troops began clearing "the Flats," someone set fire to one the shanties. As many of the veterans fled into the night toward Maryland, "Hooverville" was engulfed in flames and reduced to ashes. MacArthur fatuously defended his actions by claiming that the BEF was a communist-inspired plot to incite an uprising, and that the fire had been deliberately set by the veterans.[36] Although no record of the incident appears anywhere in Patton's personal papers, one account notes that he was hit by a brick that knocked him into the mud during the melee the night of July 28.[37]

The morning of July 29, the 3d Cavalry was still manning picket lines around Anacostia when an infantry sergeant approached a group of regimental officers and asked for Patton. In tow was Joe Angelo:

When Major Patton saw them, his face flushed with anger: "Sergeant, I do not know this man. Take him away, and under no circumstances permit him to return!" The sergeant led the downcast man away. Then Major Patton turned to the small group of officers and said: "That man was my orderly during the war. When I was wounded, he dragged me from a shell hole under fire. I got him a decoration for it. Since the war my mother and I have more than supported him. We have given him money. We have set him up in business several times. Can you imagine the headlines if the papers got wind of our meeting here this morning!" Then he added, "Of course, we'll take care of him anyway!"[38]

Although Patton described the Bonus March fiasco as "a most distasteful form of service," he nevertheless regarded the desperate men of the ragtag BEF as Bolsheviks who deserved their fate. Shortly after the incident Patton wrote a paper that was intended to explain what happened but instead managed to be generally unsympathetic. "In my opinion, the majority were poor, ignorant men, without hope and without really evil intent, but there were several thousand bad men [sic] among them and many weak sisters joined them." Although he argued that riot control duty was "onerous service," his description of the assault on the marchers smacked of self-satisfaction and included this observation: "Bricks flew, sabers rose and fell with a comforting smack, and the mob ran. We moved on after them, occasionally meeting serious resistance. . . . Two of us charged at a gallop, and had some nice work at close range with the occupants of the truck, most of whom could not sit down for some days."[39] With his penchant for detail, it may have merely been Patton's way of recounting what occured, nevertheless, despite his anguish at having to attack his own comrades, his unnecessary participation in the incident did him no credit. The ultimate humiliation was his embarrassing confrontation with Joe Angelo and the knowledge that the man who had saved his life would never forgive him.

At some point Angelo ended up in Walter Reed Army hospital, where the Pattons visited him. He is remembered as a sad little man whose eyes were disproportionately large. "We wondered how he could have dragged Georgie into the shell hole and saved him," recalls Ruth Ellen.

Several myths about Patton emerged from the Bonus March, the first of which was that he had sheared off the ear of a protester with his saber. In 1951 a congressman who apparently served under him during World War II falsely alleged that Patton had been responsible for planning the Bonus March attack for MacArthur.[40]

Two years later Patton was interviewed at Fort Myer about his role in the Bonus March. The interviewer found Patton a well-dressed, handsome, and athletic figure."What do *you* think of this New Deal and all those socialist schemes—and all the labor troubles? Where is it all going to lead?" he asked. Told that Roosevelt was merely attempting to save capitalism, Patton was incredulous. "Just two years ago I was busy protecting the conservative Republican President Hoover and now, think of it, here I am supposed to defend President Franklin D. Roosevelt with all the forces at my command! Well, anyway, we professional soldiers are supposed to be non-partisan. It is our duty to defend the party in power, whichever it is."[41]

On Memorial Day 1932 Patton delivered a passionate speech to the American Legion in Alexandria, Virginia, in which he warned that disarmament was actually a prescription for war, and if the United States did not remained prepared for war the deaths of those who fought in World War I might have been in vain. "Perpetual peace," he said, was "a futile dream" dominated by internationally minded pacifists (the "jellyfish of the world") who were "constantly working to change Armistice day into disarmament day."

"Do not misunderstand me for in spite of the fact that I am what is sometimes referred to as a brutal and licentious merciary I am not hunting trouble or advising others to do so. I do however regard with horror a state of affairs which would make our country both unready and unwilling to defend its honor." His summation of American unpreparedness still envisioned small professional armies fighting each other: "A permanent fire department is better than volunteer one," he proclaimed.[42]

After fourteen long years as a major, Patton was promoted to lieutenant colonel in March 1934. His duties at Fort Myer dealt with such mundane responsibilities as constructing squash courts, a skeet course, and riding herd on the regimental officers, all of whom he insisted learn to play polo. He also campaigned for every officer to attend the regular Saturday night hops. When attendance did not rise as he had hoped, those who did not respond found the regimental band playing in front of their quarters on a Saturday night. Attendance soon became near perfect.[43]

He built a twenty-foot cabin cruiser in his garage from a design in *Popular Mechanics,* but when it was completed the boat proved too big to get through the door. With the help of his ten-year-old son he eventually managed to get the boat to the Potomac, where it nearly overturned. "Patton learned a lot about boat (and garage) design from that venture."[44]

Much of the Pattons' time during the Fort Myer years was devoted to the Cobbler Hunt, of which he and Beatrice were the cosponsors for several years. The Pattons were extremely popular with the horse set, which included a number of prominent Washingtonians, among them congressmen and senior officers such as Generals Hugh Drum and Billy Mitchell. The Cobbler Hunt was considered so important that it was officially supported by Fort Myer. The 1934 season opened with 2,500 guests sampling a steer roasted on a huge grate as part of a gigantic feast. The *pièce de résistance* was a tournament in which the participants reenacted the trials of knights of old without the armor. "The Pattons really stage a show when they get under way," noted one of their admirers in the local paper. "To Colonel Patton was the responsibility of charging the knights about 30 in number. His words of advice resounded across the foothills of the Cobbler in no uncertain tone."[45] During the Cobbler Hunt Trials in April 1934, a judge's decision apparently went against the Pattons even though many thought they had won the event. It prompted a letter from a prominent member commending Patton and his family for graciously accepting the ruling. "After all the kicking and crabbing one generally sees and hears in events of this nature it *is* a pleasure and an *encouragement* to *occasionally see some real sportsmanship* show up."[46]

On June 27, 1934, Lt. John K. Waters and Bee Patton were married in the same church in Pride's Crossing as her parents twenty-four years earlier. It was only the second military wedding on the North Shore since the Civil War, the Pattons' in 1910 having been the first. Bee wore the same wedding dress her grandmother and mother had worn at their own weddings in 1885 and 1910. The entire affair emulated the 1910 wedding, down to the choir singing the wedding march from Wagner's *Lohengrin*. Patton was resplendent in his dress blues and medals. Publicly the photographs show only what appears to have been a proud father, with a semblance of a smile on his face, giving away his daughter to an outstanding young man who was everything one could ever wish in a son-in-law.

Behind the facade was a tormented human being. Ruth Ellen was the bridesmaid, and she would "never forget Georgie's face" as he led Bee down the aisle. "He looked just like a child who is having his favorite toy taken away. All his determination to remain forever young was being undermined by having a daughter getting married. He was forty-nine years old, and he still had not won a war or kept his part of the bargain with Grandfer Ayer about winning the glory. He looked stricken to the heart."

CHAPTER 25

War Clouds

(1935–1939)

Because of his innate dash and great physical courage and endurance he is a cavalry officer from whom extraordinary feats might be expected in war.

—BRIG. GEN. KENYON JOYCE

Beatrice was overjoyed when Patton received orders for a second tour in Hawaii in the spring of 1935. She had been writing a novel about Hawaii, and this tour would provide the impetus for her to complete and publish *Blood of the Shark*.[1] Patton had been powerfully influenced by British general Sir Edmund Allenby, whose exploits had made him one of his personal heroes. The two had met during World War I, and Allenby had told Patton that "for every Napoleon, Alexander, and Jesus Christ that made roles of [*sic*] history, there were several born. Only the lucky ones made it to the summit. He felt that in every age and time, men were born ready to serve their country and their god, but sometimes were not needed; you had to be at the right place at the right time—you had to be lucky."

With Allenby's admonition in mind, Patton was determined to bring some "action" back into what had been fourteen years in limbo amid mounting frustration with the peacetime army. He told Beatrice he preferred death to being a nobody. He would now test his mettle and his luck by soloing a schooner from California to Hawaii. The year before the Pattons had purchased a forty-foot schooner named *Arcturus*, and George began prepping for the journey by taking courses in celestial navigation several times a week.

Beatrice refused to leave her husband to his fate and informed him she was not about to let him "go and drown without her." Unable to handle the large, heavy sails, Beatrice signed on as the ship's cook, even though she had never in her life cooked so much as a meal. While her husband attempted to learn navigation, Beatrice undertook a crash course in cooking from her sister, Kay. Her brother, Fred, who was an expert sailor, provided a list of supplies. The *Arcturus* was shipped by rail from Washington to San Diego, where in May 1935 Patton embarked for Honolulu with an amateur crew of five and the experienced sailor who had come with the boat the previous year. Despite what he had managed to learn, the truth was that Patton knew almost nothing about navigation. As they left port, he turned to Beatrice and said: "We can learn, can't we?"[2]

The trip was a great adventure. Beatrice was constantly seasick and slept for most of the journey, and the cooking chores were handled by others. The Pattons kept a notebook they titled, "When and If We Ever Build a Boat." The 2,238-mile journey took fifteen days, and when they sighted Mauna Kea volcano there was a collective sigh of relief. They were greeted at the dock by friends and family, dancing girls in hula skirts, and a band.

Although outwardly another two years of sun and fun, Patton's second Hawaiian tour was miserable. He was fifty years old and wallowing in the depths of mental despair and midlife crisis. He could see that his career was at a dead end, and he was even more fearful that "Time's winged chariot [was] hurrying near." His behavior verged on the intolerable: he complained about Beatrice's gray hair, worried that his daughter Bee would make him a grandfather, and generally made himself "extremely disagreeable. Ma bought a book called, 'Change of Life In Men' which simply infuriated him . . . he burned it. . . . He started to seek the company of younger people than himself and Ma, and let himself be flattered by eternal harpies who are always standing in the wings of successful marriages, hoping the wife will falter and the man will be there to feast on."

Patton continued to play polo with a ferocity that was intended to compensate for his age, and he paid the price. During a scrimmage prior to the all-important Inter-Island Games he sustained a serious fall on his head when his pony threw him to the ground. Waving off assistance, Patton staggered to his feet and finished the scrimmage. That night he attended a cocktail party and spent the weekend on his boat. Although his doctor ruled it only a slight concussion, en route back to the yacht harbor Patton suddenly looked at Beatrice and inquired: "Where the hell am I? The last thing I remember is seeing the ground come up and hit my face!"

Patton had always been a modest social drinker who rarely ever showed the effects of alcohol consumption, but his latest fall changed him forever. As Ruth Ellen writes:

Whatever had happened to his head, it cut down on his capacity to carry alcohol and whereas he had never shown his liquor in his whole life, just a couple of drinks would make him quite tight. This upset Ma no end . . . and she told him so. Ma had never criticized Georgie in his life, and that upset him and he got very disagreeable. . . . Sometimes he would take an extra drink while she was looking at him; just to "show her."

Even worse, "He did not appear well when he got tight; he got tearful and sentimental; recited poetry out of place and context and picked on Ma. . . . It was horrible to see feet of clay where the winged heels once were."

Worse was to come when one of Ruth Ellen's closest friends arrived in Honolulu en route to the Far East. Jean Gordon was the twenty-one-year-old daughter of Louise Raynor Ayer, Beatrice's half-sister, and has been described as "a lovely young woman of great charm, intelligence, and sensitivity."[3] When Jean made a play for Patton, he was flattered that a beautiful young woman would find a balding, middle-aged man attractive, and seemed powerless to resist. He fell hard, made a damned fool of himself, and nearly wrecked his marriage. For some reason Beatrice seemed oblivious, even though, as her grandson would later write, Patton and Jean had scarcely troubled to conceal their attraction to each other. Shortly before the family was to have visited another of the Hawaiian Islands on a horse-buying expedition, Beatrice became ill. Beatrice somewhat naively permitted her husband and Jean to make the trip unchaperoned, in what later seemed to grandson Robert an ill-timed recipe for the heartache that ensued. When they returned neither Beatrice nor Ruth Ellen had the slightest doubt about what had occurred. "A powerful tension settled over the family" at the very moment when Beatrice's novel was being published, her achievement overshadowed; "Georgie's action had stolen her moment."[4]

When her ship sailed for Japan, Jean stood on deck, her neck laden with leis placed there by Patton, who waved eagerly to her from dockside. Beatrice was deeply wounded and tearfully said to Ruth Ellen: "It's lucky for us that I don't have a mother because, if I did, I'd pack up and go home to her now." But, she said:

Your father needs me. He doesn't know it right now, but he needs me. In fact, right now he needs me more than I need him. Perhaps there is a reason for all this. I want you to remember this; that even the best and truest of men can be be-dazzled and make fools of themselves. So, if your husband ever does this to you, you can remember that I didn't leave your father. I stuck with him because I am all that he really has, and I love him and he loves me.[5]

Beatrice retreated into her pursuit of music and the study of Hawaiian culture, and confined her anguish to a private thought book. Patton is thought to have presented her with some sort of reconciliation gift, but no one in the family ever knew for certain what passed between them. Her grandson writes: "Whatever peace offerings he brought her in apology, it was through Beatrice's will to forgive that the marriage endured."[6] While his behavior was inexcusable, this period of his life does reflect Patton's sorry mental state and just how low his self-esteem had fallen. For Beatrice it must have seemed an eternity since her husband had toasted all army wives at a 1924 banquet held at West Point: "May we live to make them happy, or, on the great day come, so die as to make them proud."[7]

In 1936 Patton learned that Simon Buckner was departing West Point and the post of commandant of cadets would once again become available. But for the fourth time Patton lost out, this time because he was serving "overseas" and thus could not be appointed.[8]

It was also during his second tour in Hawaii that Patton had his famous clash with his commanding general, Hugh Drum. During World War I Drum was a trusted member of Pershing's AEF staff and also his First Army chief of staff before being reduced to major during the postwar cutbacks. By 1935 Drum was a major general commanding the Hawaiian Department, and it was at his instigation that Patton was invited to become his G-2. Yet the ambitious Drum, whose background was humble, deeply resented Patton's wealth and lifestyle. Lieutenant Colonel Patton was a friend of the most important people in the islands and was invited to every social gathering of any consequence. By contrast, Drum, as the senior officer on the island of Oahu, was accorded only pro forma invitations, and was decidedly not a favorite of the socially important. That he came to resent George S. Patton was perhaps inevitable.

The Inter-Island Polo Championships, a weeklong series of matches, was one of the major social events of the year for the elite of Hawaii. In 1936 Patton's army team was playing the Oahu team, captained by his dear friend Walter Dillingham, in front of a large audience that included General Drum, who occupied a front-row seat in the flag-draped VIP box.

Patton's voice was high-pitched, and did not carry particularly well. He regarded it as a curse and compensated by a tendency to yell often. What he shouted during polo matches was a 1930s version of what is commonly called "trash talk" among modern-day athletes. On this afternoon horses and riders furiously banged into one another, and amid the banging of mallets and the pounding hoofbeats, many four-letter words could clearly be heard emanating from Patton as he attempted to ride off his friend Dillingham. After numerous expletives, Patton could be heard screaming: "Why you old son of a bitch, I'll ride you right down Front Street."

Without warning the humorless Drum ordered the match suspended and summoned Patton to his box. Patton rode up, dismounted, and in front of the large crowd was severely reprimanded by an irate Drum for using foul language and for conduct unbecoming an officer and a gentleman. "I'm relieving you of the captaincy of the Army team, for using offensive language in front of the ladies and insulting your competitors. You will leave the field at once."[9] Patton's daughter recalls: "There was a stunned silence. All you could hear were people and horses breathing." An equally shocked Patton drew himself stiffly to attention and, with a "Yes, sir," saluted Drum, and began to lead his mount from the field.

Walter Dillingham and Frank Baldwin, the captain of the Maui team, quickly conferred in the middle of the field. Dillingham then rode up to the general's box, followed by Baldwin, and inquired if it was true that Drum had just relieved Colonel Patton. When Drum replied that he had, Dillingham turned to Baldwin and announced in a loud voice heard by all that this meant the end of the matches for the year. If Patton was relieved his team would not retake the field, and the tournament would end right there. Baldwin informed Drum: "I have never heard Georgie Patton use foul language of any kind." There was another moment of silence as the two most powerful men on Oahu, Drum and Dillingham, stared each other down. Faced with the responsibility (and embarrassment) of ruining the matches and earning the permanent enmity of the islanders, Drum blinked. He summoned Patton back, and after warning him to watch his language, gathered his family and drove away.

Although Patton would later repay Drum in kind during the 1941 Louisiana maneuvers, there is no evidence to support the notion of a feud between the two officers. Certainly the polo incident severely strained their relationship, and neither ever forgot it, but to their credit, each behaved with professional decorum.[10]

Patton's focus as G-2 was the security of the Hawaiian Islands and their vulnerability to attack. He had closely followed the growing hostility and appetite for conquest of the militant Japanese leadership (which had invaded and occupied Manchuria in 1931 and invaded China in 1933), and believed war with Japan was likely. The brutal atrocities committed against the Chinese during the Sino-Japanese War led Patton to author a plan to intern Japanese living in the islands in the event of a surprise attack. In 1937 he wrote a paper entitled "Surprise," a chillingly accurate prediction of a Japanese attack on Hawaii. Although there had previously been concern about a possible Japanese invasion, naval planners had scoffed at the notion, arguing that an attacking enemy force would be vulnerable to the island-based aircraft and submarines, even with the main battle fleet as far away as the Atlantic.[11]

Patton apparently was not impressed by the navy's War Plan Orange (the strategy for fighting a war with Japan). Among his observations was that there might well be "the unheralded arrival during a period of profound peace of a Japanese expeditionary force within 200 miles of Oahu during darkness; this force to be preceded by submarines who will be in the immediate vicinity of Pearl Harbor." In addition to sabotage of key military installations and the assassination of commanding officers, he envisioned the possibility of "an air attack by [Japanese] navy fighters and carrier borne bombers on air stations and the submarine base using either gas or incendiary bombs."[12] It appears to have been the first-ever prophecy of the disaster later called Pearl Harbor, and was even more significant for having been made well before 1940, when Roosevelt ordered most of the Pacific Fleet to the Hawaiian Islands as a signal meant to deter Japanese aggression in Asia.[13]

Patton displayed no concern for the rights or treatment of the Japanese population of Hawaii, believing that most were undercover secret agents of the Japanese imperial army or navy sent there to carry out sabotage when so ordered. His suspicions had originated during his first tour in Hawaii and were intensified by Japan's aggression in the 1930s. Patton's plan for the mass arrest of suspects and their incarceration as virtual hostages was even more ruthless than the later internment of Japanese Americans in 1942.

Patton's virulent anti-Japanese sentiments were not only the result of his visceral distrust and dislike (like that of most Caucasians in Hawaii) of other races, but a reflection of an acute anxiety (bordering on terror) that his life and career were a dismal failure. His womanizing, his inability to handle alcohol, and his bad temper had all led to a growing alienation from his children, and from Beatrice, who loyally stuck by him when others would long since have decamped. Through Patton's pain and torment, Beatrice was convinced not only that he needed her strong presence, but that only she could save him from self-ruin.

Minus Ruth Ellen, who had a broken arm, the Pattons left Hawaii for San Diego in June 1937 aboard the *Arcturus*. The boat was caught in the same violent Pacific storm in which aviatrix Amelia Earhart disappeared. When the *Arcturus* eventually arrived, badly damaged, its crew reckoned themselves very fortunate to be alive.[14]

Patton was looking forward to a long leave before reporting to his new assignment at Fort Riley, but fell victim to another serious accident while riding with Beatrice and son-in-law, Johnnie Waters. On this day, "by his own admission," Patton was riding in what he called the "danger zone," in which the head of his horse was even with the stirrups of Beatrice's horse, Memorial, a position he had warned everyone was exceptionally dangerous, particularly during fly season. When the animal suddenly lashed out at a horsefly, its hoof broke Patton's leg with a sound "like a dry stick snap-ping."

He was rushed to nearby Beverly Hospital with an injury so serious that blood clots were released into his circulatory system, causing throm-bophlebitis. An embolism came within seconds of killing him and, accord-ing to his friend and physician, Dr. Peer P. Johnson, it was only Patton's excellent physical condition that saved his life. He later told Fred Ayer that he remembered passing out and then being aware of lying on a battlefield on a large Norse shield, when two Vikings cloaked in armor arrived and lifted it for the journey to Valhalla. Patton then recalled that one of them silently shook his head and he was lightly lowered to the ground. When he regained consciousness he was in a hospital bed. "I guess they're not ready to take me yet," he said. "I still have a job to do."[15]

In the 1930s the wonder drugs of today were unknown, and treatment generally consisted of immobilization. His doctors feared complications. For Patton this was a new low. "His frenetic tour under General Drum, his growing discouragement with lack of promotion, and his ever present worry that he would never have 'his war' piled up on him, and he was suicidally depressed," wrote Ruth Ellen. To make matters worse, the army sent a board of medical officers from Boston to determine if Patton had been drunk or on drugs at the time of the accident. Although he was cleared, he took it as a clear signal that the goddess of fortune was no longer smiling upon him.[16] Although Patton recovered normally, his injury was of sufficient seriousness that it might well have ended his career then and there, resulted in the ampu-tation of his leg, or even killed him. At least his life and his career had been saved.[17]

During his lengthy hospital stay Patton's friends decorated his room with streamers of colored toilet paper and someone placed a sign above his door reading HULA HULA NIGHT CLUB in large red letters. Liquor flowed freely, and his nurses fawned over him. When he got better and could roam the halls on crutches, Patton told obscene jokes and left laughter and chaos in his wake. To a woman friend who had recently given birth and remarked how nice it was to be able to lie on her stomach again, Patton replied in a loud voice for the benefit of other patients and hospital staff: "Why, my dear young lady, had you retained that position in the first place, you would probably not have been in your recent unenviable condition."[18]

Outwardly all was festive, but behind the facade Patton had fallen vic-tim to a deep depression, which the injury exacerbated. Only Beatrice was able to detect his growing despair. Ever since their first voyage to Hawaii in 1935, the Pattons had often spoken of designing and sailing their dream schooner and used to joke about "when and if" they ever had the opportu-nity, or as Patton put it, *When the next war is over, and if I live through it.* To prevent Patton from sliding any deeper into the well of depression, Beatrice invited noted boat designer John Alden to collaborate with her husband to design such a boat, using the notebook they had been cramming full of ideas for more than two years. Alden designed a schooner they christened the *When and If,* and when it was launched it became the family pride and joy. It was Patton's intention that one day he and Beatrice would sail the boat around the world, a dream that was never fulfilled.[19]

His recuperation at Green Meadows took another six months. Although the center of attention, Patton was determined to repay Memorial, and he hobbled on crutches to the stables and beat the horse with one of them, cursing and threatening to kill the animal. Beatrice was infuriated and told her husband he was childish to blame her horse for an accident that was clearly his own fault. Patton petulantly complained that Beatrice was denying him one of the few pleasures he had left, which was to kill her damned horse.

It was perhaps the low point of their marriage. Although she attempted to keep up her husband's flagging morale (which he called "a dirty French word"), she was exasperated and weary of his irrational behavior. To escape the confrontations she rode her horse daily and twice a week skippered her boat in races off Marblehead or Manchester, which only made Patton angrier because of what he insisted was her neglect of him. For her part Beatrice resented having to do everything: If marriage was a partnership, then the present state of affairs had degenerated to the point where she was doing all of the "getting along."

With prescribed exercise Patton rapidly improved, but his state of mind was blacker than ever. One day, on the edge of despair, Beatrice implored her daughters to go and cheer their father up. "One or both of you girls go down and talk to your father! He's lying brooding in a hammock about not knowing his children and the fact that no one loves him, and he's very depressed, and I can't do another thing with him!"

When faced with the unpleasant task of confronting their mercurial father, the Patton girls drew straws. Ruth Ellen lost and found herself tongue-tied in the presence of her glum and unresponsive father. In desperation she asked him to recite one of their favorite poems, a trick that had worked in the past. Patton responded and when he finished, arose and, squeezing Ruth Ellen affectionately, said: "My, I certainly enjoyed our talk. I don't see enough of you girls!"

Patton's bizarre behavior during the interwar years manifested itself in various ways. His sweet tooth led him to consume entire boxes of candy or a whole fruitcake, then have his stomach pumped at a hospital. He would eat the fillings from chocolates and then attempt to hide the evidence by carefully replacing the wrappings, and "he once caused an uproar by hollering at a military band during a solemn ceremony, 'Quit playing that goddamn music!' unaware that the song was the national anthem."[20]

At times it must have seemed to his family that he was hopelessly out of control. Grandson Robert writes of the times when his own father was exposed to Patton's erratic personality, "the tears at each passing birthday, the drinking, the time Georgie, meaning to honor a horse that had gone lame, decided to shoot the animal himself rather than allow some stranger— a veterinarian—to put it down with a needle." Patton botched the effort and only wounded the distraught animal. Son George witnessed his father frantically "'crying, weeping, and trying to hold the horse so he could shoot him again.' A groom ran out and finished the job. Georgie was devastated with guilt. He shut himself in his room and didn't emerge for a day."[21]

By February 1938 Patton had recovered sufficiently to report for duty at Fort Riley, where he became the executive officer of the 9th Cavalry and a member of the faculty of the Cavalry School.[22] A close friend and fellow instructor was Maj. Terry de la Mesa Allen, one of the army's most colorful, unconventional, and audacious officers, who had contrived to get himself sent to the AEF in 1917. Unable to find suitable employment in a front-line infantry unit, Allen quickly located a combat training center about to graduate a class, unobtrusively slipped into the line of officers receiving their diplomas, and was sent to command a battalion. He was severely wounded in the Meuse-Argonne.[23]

Allen and Patton developed a friendship based on their mutual cavalry backgrounds. Theirs was a robust but friendly rivalry, especially over a manual Allen was writing about cavalry tactics. Patton met his match in Allen, and the two argued furiously at all hours of the day and night, each convinced his ideas were correct. Five years later they would be reunited on the battlefields of North Africa.

Son-in-law Johnnie Waters had been stationed at Fort Riley during Patton's second tour in Hawaii. Beatrice Patton Waters soon learned firsthand of the frustration and resentment Patton's wealth generated in others less well off. She had become a superb rider, and her life in Kansas was centered on horses and horse shows. Despite flawless performances, however, Bee Waters never seemed to qualify for the ribbons. Nevertheless she declined to complain, having been indoctrinated by her parents with the principle of never protesting or questioning a judge's decision. "To be a good sport was as important as being a good Christian." At a cocktail party she learned the truth from one of her father's fair-weather friends, a drunken cavalry colonel, who angrily declared she had "damn well won" the final class: "I want to tell you here and now how we just love to give you the gate—you riding your goddam father's goddam thousand-dollar horses against these flea-bitten remounts. It just does us good to show up Georgie's goddam thoroughbreds with him off in Hawaii where he can't do a goddam thing about it!" Bee Waters did not bother telling the colonel that she had been riding a horse that cost only $170.[24]

Patton was hardly alone in his frustration with the current state of the U.S. Army. Senior officers who had spent their entire careers preparing for a war that never came were reduced by utter frustration to attempting to kill one another on the polo field or taking inordinate risks in the riding ring. "Scott, Chaffee, Wainwright, Herr, Henry, Patton, Richardson, Van Voorhis, Lear—all the officers who were within five years to become the builders and leaders of the greatest armored force the world has ever seen, were at Riley and Bliss and Leavenworth and Clark . . . ready . . . [but] with no place to go." Perhaps the only difference lay in the intensity of Patton's disquiet that he would never get a second chance at war.

Fort Riley turned out to be a tonic for Patton. He was back among the cavalrymen he was most comfortable with, and he began to relax and enjoy life again. The family again began doing things together, riding for recreation, participating in horse shows, and happily chasing jackrabbits across the prairie. Patton resumed writing military articles and was acting so much like his old self that—despite her delight—Beatrice was certain it was too good to be true and could not possibly last.

By the summer of 1938 Patton had fully regained his health, and after nearly twenty years finally regained his World War I rank, when he was promoted to full colonel on July 1. After the briefest assignment of his career he reported to Fort Clark, Texas, where he assumed command of the 5th Cavalry Regiment. An oasis on the site of an artesian well (from which flowed five million gallons of water a day) in the heart of the badlands of southwest Texas, near Eagle Pass and the Mexican border, Fort Clark was the army's ultimate backwater, a place where old cavalry soldiers were sent to finish out their careers in peace and quiet. Fort Clark had served an important purpose during the Civil War and as a cavalry outpost in the time of the Indian wars, but on the eve of World War II it was an anachronism whose only purpose seemed to be as a place to station an outdated regiment of cavalry.

Col. George S. Patton blew in like a Texas dust storm to disturb what had been years of tranquillity. He was reunited with Kenyon Joyce, who commanded the post and had been his superior at Fort Myer. Patton counted Joyce among his closest friends both socially and professionally. For his part Joyce saw in Patton an officer destined for greater things. In Patton's 1934 efficiency report he had written: "I believe this officer could be counted on for great feats of leadership in war." Although others gave Patton consistently high ratings in the 1930s, only Joyce went on record to predict what he might do in a war.[25]

Unused to commanding officers who actually took war seriously and demanded the ultimate in spit and polish, the officers and men of the 5th Cavalry had never seen the likes of George S. Patton. He conducted sand table exercises, made them *walk* through mock battles, instructing them that this was the way they would have to fight the next war, and preached that horses were obsolete.[26] At the annual Third Army war games he tested his ideas and tactics and drove his troops unrelentingly in the harsh terrain and heat so intense "it makes your blooming eyeballs crawl."[27] He delighted in outflanking the opposition and when an "enemy" artillery colonel refused to surrender to one of his captains, Patton rode up and "stuck my white pistol in his face."[28] In the 5th Cavalry the duty was strenuous and the socializing was not for the abstinent. Before Beatrice arrived Patton attended a regimental function in which he had his first experience since World War I "of being the old man at a party . . . along about midnight I found it expedient to go home."[29]

Fort Clark reminded Patton of his Southern California roots. One night the Pattons were dining in Piedras Negras, a Mexican border town opposite Eagle Pass, when a Yaqui Indian dressed in the fancy uniform of a Mexican general, with the cruelest, snakelike eyes Ruth Ellen had ever seen, began staring at Patton, who kept repeating that he knew the man from his past. The proprietor informed Patton that the bill had been taken care of by the general, who asked to be remembered to Teniente Patton, whom he had last met under very different circumstances. The penny finally dropped: "I *knew* I had seen that man before. He was one of Pancho Villa's officers! That goddam Yaqui shot at me and nearly winged me!"

The Pattons' otherwise serendipitous tour at Fort Clark was marred when a recruit accidentally poured an excess of chlorine into the post water supply the night before a regimental review. With the entire regiment lined up in full dress uniform on the parade ground, pandemonium suddenly erupted as everyone began breaking ranks and heading for the nearest bushes to vomit, "including the glorious Colonel of the Regiment, Saber George, himself!" Virtually all of Fort Clark had chlorine poisoning except a greatly amused Beatrice Patton, who had drunk only tea and was thus unaffected.

Patton's ebullient mood continued through the autumn, and Ruth Ellen remembered it as one of the best times of their lives. To her father's utter dismay, the chief of cavalry, Maj. Gen. John Herr, personally called Patton at home in December 1938 to inform him he was being reassigned to Fort Myer and the command of the 3d Cavalry, replacing Col. Jonathan Wainwright, who, unlike the Pattons, had no outside source of income and could no longer bear the heavy financial burden required of the officer commanding Fort Myer. Herr assured Patton that he had done a splendid job at Fort Clark but was badly needed at Fort Myer.[30]

As much as he loved Fort Myer, Patton was devastated and immediately blamed Beatrice, asserting that she had done a terrible thing by taking him away from real soldiering at Fort Clark for the task of being a glad-handing socialite in Washington. It would be the death knell of his career, he said, and it was all her fault because of *her* money. "A lot of very fancy language got thrown around as we packed, and Ma got her feelings hurt." The automobile trip to Washington was tense and quite unpleasant. When they arrived the chip on Patton's shoulder remained, and he now blamed the Wainwrights for his plight, complaining that they had "done it on purpose." When Beatrice challenged him as to what exactly they had done, Patton could not answer but argued that it had been to make him look bad. The tranquillity of Riley and Clark was indeed gone, and it was 1937 all over again. Tensions in the Patton family ran high. No one was happy, least of all Beatrice, who bore the brunt of her husband's increasingly fragile personality and his tendency to blame her for virtually everything that was wrong with his life. The miracle was that the marriage survived at all.

For the first time in their long marriage Beatrice fell apart from the tension and strain of constantly having to massage Georgie's massive ego, and she took to her bed with a form of kidney colic that left her utterly miserable. "When Ma got sick, Georgie fell to pieces" and was treated by the post physician, Dr. Albert Kenner, who was to become the Third Army surgeon several years hence. As usual, Patton was a terrible patient.

The winter was a nightmare for both, and to escape the tension, Beatrice created what she called her "sanity plan," which included a daily horseback ride. One of her companions was Eleanor Roosevelt, whose horse was stabled at Fort Myer. The two women became fast friends, and Mrs. Roosevelt was an occasional luncheon guest at the Pattons'. Beatrice found the president's wife shy and very lonely and thought she had the finest eyes she had ever seen in a woman.

Between the required, almost daily entertainment of Washington VIPs, sponsorship of the Society Circus (which Patton himself had originated many years earlier), and the frequent need to entertain friends and West Point classmates, this tour of duty at Fort Myer was no picnic.

Whenever Patton arrived at a new post, he lost no time asserting his influence. He summoned the chaplain and bluntly told him that his sermons were too long. "I don't yield to any man in my reverence to the Lord, but, Goddammit, no sermon needs to take longer than 10 minutes. I'm sure you can make your point in that time." The following Sunday morning Patton appeared in the front pew and ostentatiously took out his watch when the chaplain began his sermon. Exactly ten minutes later the chaplain concluded his sermon. The troops were delighted.[31]

Patton's troops had little money to spend after the middle of the month. Coincidentally, his dog would mysteriously disappear at the same time each month, and he would post a notice in the regimental stables offering a two-dollar reward for its return. A day or two later the dog would always turn up, and Patton would present the reward to a soldier.[32]

Patton's black moods continued, and he became so tyrannical that no one wanted to be around him. He lashed out at his loved ones and at the servants, constantly sending back food for not being cooked to his satisfaction or for being too hot or cold—and sometimes for no reason at all. Ruth Ellen was understandably disgusted by his behavior, but Beatrice would only reply that Oscar Wilde had been right when he wrote that "each man kills the thing he loves."[33] No one was spared his harsh tongue. Ruth Ellen believed that it was fate (and no doubt a just punishment) when Patton fractured his leg again when he fell into the cabin of the *When and If*.

Patton the taskmaster was very much in evidence. A young cavalry officer named James H. Polk, who had just been given command of his own troop, thought he and his unit had done quite well during a live ammunition attack problem and were about to be praised by Patton. "Some things you never learn," he wrote years later, as a retired four-star general. Patton had noticed a delay in opening fire and was told by Polk that it was against regulations to issue ammunition earlier. "I helped write those very same stupid regulations for damn fools like you," he roared. "Now, let's have a real exercise." In the realistic training that followed it was obvious to Polk that "our colonel was preparing us for the real thing."[34]

Yet there remained Patton the enigma. On the polo field he gave no quarter. Off it he lectured his officers how privileged they were to play a sport that was superb training for leaders. "If you are privileged to have a thing," he told them, "you must take care of it. Mrs. Patton, the children and I replaced the divots after the first game. Hereafter all players will go over the field and replace the divots after each game." He also directed that players report to the stables after each match to ensure that their ponies were looked after.[35]

One of Patton's officers was Paul Robinett, who had taught him the machine gun in 1923. Robinett recalls that one day after playing polo, Patton took his horse to the indoor riding hall to school the animal for a forthcoming exhibition ride. He set a three-foot six-inch post-and-rail jump in the center of the hall, "mounted his charger, adjusted the stirrup straps, galloped around the hall a few times and then down the center for the jump. The horse hesitated and started to refuse. He kicked the animal and hit it with his crop. The horse attempted to jump but struck the upper bar with his knee and fell." Patton landed on his head in the tanbark, saved from serious or fatal injury only by his polo helmet. Although badly shaken, Patton arose "swearing like a trooper, caught the horse, remounted, and took the jump a number of times in rapid succession. Then he came over to where I was standing . . . and queried, 'Do you know why I made my horse take those jumps?'" Robinett replied he supposed it was to discipline the horse. "Not at all," said Patton, "I did it just to prove to myself that I am not a coward!"[36]

That some of his contemporaries and classmates were now generals only deepened Patton's gloom. It was even more painful whenever he was obliged to entertain them. On such occasions he would trot out what Ruth Ellen called his "mirthless smile" and pretend to be happy for them, when instead, it ate at him like a cancer. Beatrice remained frustrated and seemingly powerless to rescue her husband once again from the depths of his despair. The tension was so palpable that Ruth Ellen was only happy when away from the Patton household, in the company of her current beau, a young artillery officer named James Totten.

Neither of her parents was pleased when Ruth Ellen announced she intended to marry Jim Totten. Patton realized that she was deadly serious when he discovered she had told him *before* informing Beatrice. With a look of incomprehension, he said: "Goddammit, you can't marry him! He's too short; he's a field artilleryman; and he's a Catholic!" But then, over a pitcher of martinis, he offered some pungent advice. Only Patton could have told his younger daughter that if she married a man of small stature (Totten was five feet five and a half inches and weighed about 135 lbs.) it would mean that each morning a small man left his home much like a rooster "leaving his own dunghill, crowing and flapping his wings and shouting to every rooster within hearing distance that he was ready for all comers; such a man must have a wife whose love and loyalty are unquestionable; he must be the head man in his own home." Small men, he said, were usually better than big men (their reactions were quicker because nerve impulses had less distance to travel!), and since such men were closer to the ground, they had less distance to fall. Above all "a small man had to be dead sure of himself, and when he was, he was usually meaner than hell, which made him a good Army officer." As bizarre as this advice sounded, Ruth Ellen later admitted that her father's counsel was invaluable and helped sustain her throughout her long and fruitful marriage.

Beatrice was displeased and even Nita had discouraged Ruth Ellen, who burst into tears and said, "Well, then, I guess I will never marry. I'll just stay home and take care of Ma." After a moment of silence, and with a look of horror on her face, Nita almost shouted, "What! And be like me? One sacrifice on the altar of family loyalty is enough. Go home and marry your young man, and God help us both. I'll come to your wedding."

The night Lieutenant Totten came to ask for her hand, Patton put on his coat, left the house, and went to the post theater, where he sat through the same film three times. As Ruth Ellen watched in horrified silence, her mother threw cold water on their marriage, arguing that he should at least be an Episcopalian. Although the two had known one another for two years, Beatrice did not believe it was long enough. She need not have worried. James Totten was a fine young officer and a dignified gentleman whose army lineage, like that of his prospective father-in-law, dated to the Revolutionary War, in which an ancestor served as George Washington's chief of engineers.[37]

Robert Patton (1750–1828).
(Photo courtesy of the Patton family)

A rare 1880 lithograph of Lake Vineyard.
(*From* History of Los Angeles County, California)

Margaret Hereford Wilson (1820–1898).
(*Photo courtesy of the Patton family*)

Benjamin Davis Wilson—
"Don Benito" (1811–1878).
(*Photo courtesy of the Patton family*)

Patton's grandfather,
Col. George S. Patton I
(1833–1864).

Susan Glassell
Thornton Patton
(1835–1883).

Mother,
Ruth Wilson
(1861–1928).

Anne Wilson
(Aunt Nannie)
(1858–1931).

Father, George S. Patton II
(1856–1927)

(Photos this page courtesy of the Patton family)

"The Boy"
George Smith Patton Jr.

Georgie and Aunt Nannie.

Georgie, age fifteen.

(Photos this page courtesy of the Patton family)

The Pattons, the Ayers, and assorted other kinfolk at the Patton estate on Catalina Island, summer 1902. George S. Patton is seated on the right, and his future wife, Beatrice Banning Ayer is on the left, seated next to Nita Patton. In the center are Beatrice's parents, Ellen and Frederick Ayer (white beard). Behind them is Nannie Wilson. On her right is Patton's mother, Ruth, and on her left is Patton's beloved step-grandfather, Col. George Hugh Smith. (*Photo courtesy of the General Phineas Banning Residence Museum*)

VMI cadet.
(*Photo courtesy of the Patton family*)

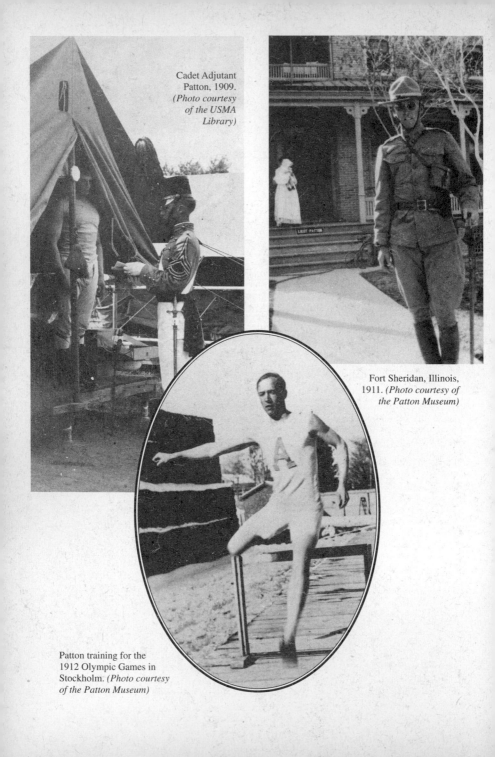

Cadet Adjutant Patton, 1909. *(Photo courtesy of the USMA Library)*

Fort Sheridan, Illinois, 1911. *(Photo courtesy of the Patton Museum)*

Patton training for the 1912 Olympic Games in Stockholm. *(Photo courtesy of the Patton Museum)*

At the Cavalry School, Fort Riley, Kansas, 1914–15, Patton was both a student and, as the army's first master of the sword, an instructor in fencing. *(Photo courtesy of the Patton Museum)*

Beatrice, "Little Bee," George, 1915. *(Photo courtesy of the Patton family)*

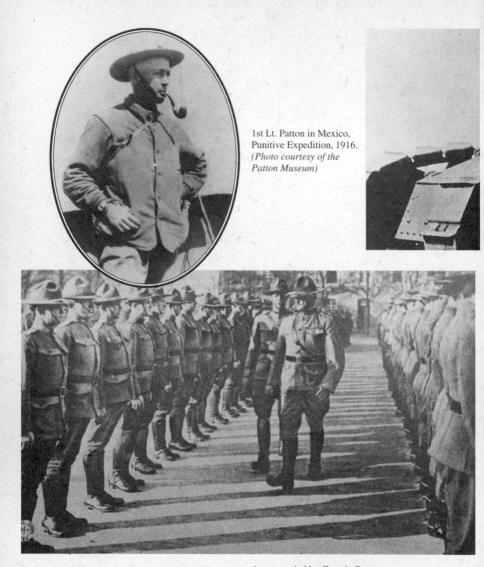

1st Lt. Patton in Mexico,
Punitive Expedition, 1916.
*(Photo courtesy of the
Patton Museum)*

Accompanied by Captain Patton,
Gen. Black Jack Pershing inspects
Patton's headquarters troop, AEF,
Chaumont, France, 1917.
(Official U.S. Army photograph)

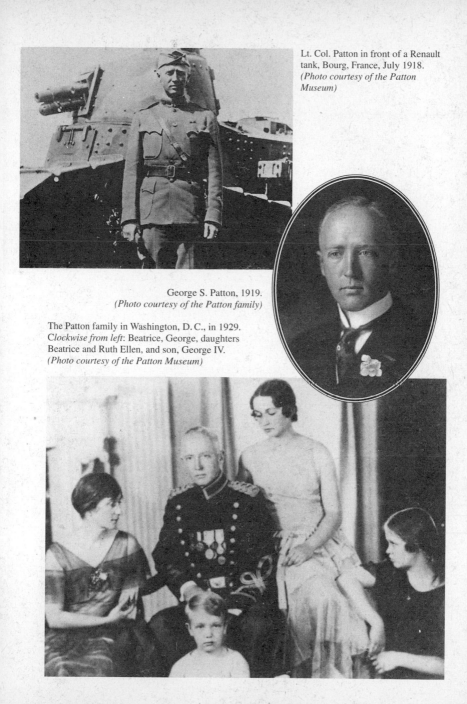

Lt. Col. Patton in front of a Renault
tank, Bourg, France, July 1918.
*(Photo courtesy of the Patton
Museum)*

George S. Patton, 1919.
(Photo courtesy of the Patton family)

The Patton family in Washington, D.C., in 1929.
C*lockwise from left*: Beatrice, George, daughters
Beatrice and Ruth Ellen, and son, George IV.
(Photo courtesy of the Patton Museum)

Commanding officer, 3d Squadron, 3d Cavalry, Fort Myer, Virginia, 1921.
(Photo courtesy of the Patton Museum)

Patton's polo team in Hawaii, 1927. *(Photo courtesy of the Patton family)*

Patton's career and family life during the interwar years revolved around horses and horsemanship. Son George, Beatrice, Ruth Ellen, Bee, Patton, 1931. *(Photo courtesy of the Patton Museum)*

Master of the Cobbler Hunt, Virginia, 1932. *(Photo courtesy of the USMA Library)*

One of several superb charcoal drawings of Patton by Donald Gordon Squier, 1932. *(Official U.S. Army photograph, National Archives)*

Despite the smiles of his family, Patton was a deeply unhappy man, as reflected in this scene at the family dinner table in 1939 that includes Nita Patton's two adopted sons, Peter and David. Ruth Ellen is on the left, son George is next to Beatrice. Daughter Bee is seated on the right, next to Nita. *(Photo courtesy of the Patton Museum)*

Patton, Lt. Jimmy Totten, and his bride, Ruth Ellen, 1940.
(Photo courtesy of the Patton Museum)

Commanding general, 2d Armored Division, at the Red River, Louisiana Maneuvers, 1941. *(Official U.S. Army photograph, National Archives)*

Patton the teacher: explaining the Louisiana Maneuvers to the 2d Armored Division. *(Photo courtesy of the Patton Museum)*

With his intimate friends Geoffrey Keyes and Col. Harry A. "Paddy" Flint *(right)*, Louisiana Maneuvers, 1941. *(Photo courtesy of the Patton Museum)*

Commanding general, I Armored Corps,
Desert Training Center, California, 1942.
(Official U.S. Army photograph, National Archives)

II Corps Commander,
Tunisia, 1943.
*(Photo courtesy of the
Patton Museum)*

Patton the diplomat:
with the sultan of
Morocco and his son.
*(Official U.S. Army
photograph, National
Archives)*

Air Vice Marshal Sir Arthur Coningham. *(Official U.S. Army photograph, National Archives)*

Maj. Gen. Ernest N. Harmon. *(Photo courtesy Norwich University)*

Maj. Gen. Terry de la Mesa Allen. *(Official U.S. Army photograph, Dwight D. Eisenhower Library*

Lt. Gen. Omar N. Bradley. *(Official U.S. Coast Guard photograph, National Archives)*

In one of their increasingly rare festive moments, Eisenhower officially promotes Patton to lieutenant general at the II Corps command post, March 1943. *(Official U.S. Army photograph, Dwight D. Eisenhower Library)*

Patton, Eisenhower, and Montgomery. *(Imperial War Museum)*

Patton had every reason to be proud of his children, each of whom married well. Although he never lived long enough to see his son marry, son George would wed Joanne Holbrook, the daughter of another distinguished family with a West Point lineage dating to 1812.[38]

Totten refused to compromise either his ethics or his religion. After listening politely to Beatrice for a half hour, he rose and asked what she thought of turncoats. Told she did not approve of them, Lieutenant Totten replied, "Neither do I. Goodnight." And he walked out. An hour later Patton returned to find the two women weeping, convinced that the romance was over. The formidable Pattons had met their match in Jimmy Totten. Although for several days Ruth Ellen believed that she would end up like Nita, Totten returned and bravely asked Patton for her hand in marriage. After fifteen minutes alone the deal was done, and they emerged with Patton's approval. Ruth Ellen and Jim Totten were officially engaged.[39]

The two men who became Patton's sons-in-law eventually wore six stars (John Waters later retired as a four-star and Jim Totten as a two-star general.) Although they never became close friends, George Patton and Jim Totten had a healthy respect for each other, and for Ruth Ellen Patton Totten that was good enough. They were married in July 1940 in St. John's Episcopal Church in Beverly Farms, and Ruth Ellen became the third Patton to wear Ellen Banning Ayer's 1884 wedding dress.

Patton was outfitted in his dress whites, replete with decorations. His daughter thought he looked "gorgeous." During the twenty-minute drive to the church, father and daughter found they had little to say to each other. At last Patton broke the silence and said: "I guess you know your mother and George and I will miss you." Ruth Ellen replied "yes" and fought back tears. "I hope we gave you a happy childhood," and the tears flowed. Her father's only advice was: "If you treat Jim the way your mother has treated me, you will be alright." Jimmy Totten was the light of Ruth Ellen's life and their marriage lasted twenty-seven years until his untimely death in 1967.

The ceremony was traditional, but the departure of the happy couple after their reception on the lawn of Great Meadows was as memorable as it was bizarre. Nearly two hundred people attended—family and friends—some of them on the eccentric side, including an elderly cousin named Annie Ruggles, who explained her preference for leather-covered upholstery in her automobile "because it's so much easier to clean when people throw up." After the traditional rice shower, their chauffeur-driven auto had begun to move away, when, from a nearby stone wall, Patton suddenly let out a piercing scream and leaped on to the roof, where he knelt and emptied two revolvers into the sky. The shaken chauffeur managed to stop the vehicle and Patton nonchalantly strolled away as if nothing at all had occured.[40]

*   *   *

During the 1930s Japanese and German aggression had gone unchecked, and by the summer of 1939 the world teetered on the brink of its second world war of the century. On September 1, 1939, the first shots of the greatest, most devastating war in history were fired on the border between Germany and Poland, and, although there was little immediate effect in the United States, this event was to change the lives of millions, including George S. Patton, who had waited more than twenty years for a second chance to achieve his destiny. This time he would not be denied.

PART VII

Prelude to War
(1939–1942)

We have tried since the brith of our nation to promote our love of peace by a display of weakness. This course has failed us utterly.

—GEN. GEORGE C. MARSHALL, 1945

CHAPTER 26

Division Commander

Hell on Wheels
—MOTTO OF THE 2D ARMORED DIVISION

The United States was unaffected by the first months of the Second World War. After Hitler's blitzkrieg trampled Poland in September 1939, Britain and France declared war on Germany, but America remained officially neutral. As Europe headed toward the most devastating conflagration in history, little had changed in Washington, and priorities were often pegged more to their social impact than their military necessity. Life at Fort Myer went on much as before. Although keenly aware of the unsettled state of the world, even Patton seemed to have been disarmed by more than two decades in the wilderness of the interwar years. In one instance he wrote the chief of cavalry in early September 1939 to recommend exclusion of an officer from appointment to the War College because "he is of more value to the Cavalry in his present position as riding companion to Mrs. Roosevelt."[1]

Patton soon became reacquainted with the army's new chief of staff, Gen. George C. Marshall, who was to have an inordinately large influence on his career. Although recognized during World War I as an outstanding young officer, Marshall too had labored in obscurity during the wilderness years. He was immensely influential during the late 1920s and early 1930s as the assistant commandant of the Infantry School at Fort Benning, Georgia, where more than two hundred future World War II generals passed through or were permanently assigned. Marshall began keeping a black notebook in which he recorded the names of officers he considered promising. Dwight Eisenhower, Omar Bradley, Walter Bedell Smith, and Joseph Stilwell were among those entered, with notations about their attributes.

Though Marshall barely knew him, he included the name of George S. Patton.

Marshall was one of Franklin Delano Roosevelt's most inspired appointments. A difficult and demanding officer to work for, Marshall was renowned for his bluntness and impatience. Those who served under him soon learned that they either produced or were cut from his team. Not even the president could intimidate Marshall, who never hesitated to disagree with Roosevelt when he thought his commander in chief was mistaken. It is said that only the president and crusty Joe Stilwell ever dared to call Marshall "George" to his face, and that Roosevelt only did so once. In the highly politicized atmosphere of Washington, the president valued Marshall's willingness to speak truthfully and candidly without fear of the consequences. It was the beginning of what historian Thomas Parrish has so aptly described as "the most triumphantly effective political-military team in American history—a team whose achievements rested on candor and hard-won mutual respect."[2]

Marshall has been called "the organizer of victory," and his official biographer has written that he was the "perfect blend of soldier and civilian."[3] This visionary leader guided the army through the immensely complex transition from a pathetically ill-equipped, inadequate peacetime force to the mighty war machine that eventually totaled nearly eight million men in the army and Army Air Forces. Air Corps Maj. Gen. Elwood Quesada later called Marshall "the aristocrat of our time . . . [and] the most selfless man that has reached public office in the last century in this country."[4]

When Marshall assumed office in September 1939, he also inherited Quarters One at Fort Myer, where the army chief of staff traditionally lives. Patton offered Marshall the hospitality of his quarters while Quarters One was being readied, and the new chief of staff responded that he would be happy to "batch" with him for a few days. Writing to Beatrice (who was at Green Meadows), Patton boasted that he had "just consumated a pretty snappy move. Gen. George C. Marshall is going to live at our house!!! He and I are batching it. I think that once I can get my natural charm working I wont need any letters from John J. P. or any one else."[5]

Patton did not need to earn Brownie points with Marshall. The chief of staff already understood the mercurial Patton and liked him both personally and professionally. Even when Patton got himself into trouble, Marshall never lost sight of his leadership qualities and his great value to the army. "Marshall thought him an extraordinary character. 'He would say outrageous things and then look at you to see how it registered.' He would 'curse and then write a hymn.'" Marshall's stepdaughter, Molly, liked Patton and thought him amusing, "like a little boy." She was Marshall's favorite riding companion during his daily early-morning outings, but Patton, who longed to accompany them, habitually arrived at the Fort Myer stables first. Marshall and Molly would arrive to find Patton mounted on his horse, clearly signaling his desire to accompany them. Behind Marshall's cold demeanor there lurked a sense of humor, and on such mornings the general delighted in frustrating Patton by ignoring him and riding off with Molly. To Patton's acute vexation, he was never permitted to accompany Marshall on his morning ride.[6]

Marshall's wife, Katherine, also had Patton's number, and one day at Fort Myer bluntly told him, "George, you musn't talk like that. You say these outrageous things and then you look at me to see if I'm going to smile. Now you could do that as a captain or a major, but you aspire to be a general, and a general cannot talk in any such wild way." Patton laughed.[7]

With the advent of war in Europe, Patton began to sense that perhaps he was not washed up after all. During the summer of 1939 he led part of the 3rd Cavalry in the annual First Army maneuvers with his usual tenacious flair and disregard for anything short of winning. He also began a letter-writing campaign to ensure that those in high places would not forget him. Already the U.S. Army was about to expand under the aegis of Gen. Lesley J. McNair, who would train the massive force that fought the Second World War.

One of Marshall's most important decisions was to streamline an army riddled with overage officers in key positions who had for years obstructed modernization. He began ruthlessly weeding out and retiring many of its aged officers. War was a young man's game, and few over the age of fifty-five survived or attained high troop command. To inject a youthful new image into an army so badly in need of overhaul, those over the age of fifty would not be tapped for high command positions. There were the inevitable exceptions, and one of them was Patton, who had made such a vivid impression on Marshall in France during World War I. Although others were promoted and reassigned to important positions, Patton continued to stagnate at Fort Myer, frustrated and convinced that Marshall considered him too old for high command, totally unaware that many years earlier he had been identified as one of the future leaders of the army.

Marshall sought the opinions of key senior officers, and in the summer of 1940 he wrote confidentially to solicit recommendations from the chief of cavalry of colonels who ought to be promoted to brigadier general. Patton headed a list of twenty cavalry lieutenant colonels and full colonels under fifty-five years of age.[8] In 1941 McNair rated Patton second of five armored division commanders, noting that he was "good" but "division [command] probably his ceiling." At the bottom of a list of eighteen potential division commanders appeared the name "Eisenhower."[9]

Despite McNair's lukewarm rating, Marshall had already made up his mind about Patton, having written years earlier in his black book: "George will take a unit through hell and high water." Below it he added: "But keep a tight rope round his neck." Also: "Give him an armored corps when one becomes available."[10]

Marshall apparently saw through Patton's pretense of supporting the horse cavalry advocates over the mechanizationists as a means of preserving his career, and told his executive officer, Lt. Col. Leonard T. Gerow (a future corps commander on June 6, 1944, and in the Normandy campaign): "Patton is by far the best tank man in the Army. I know this from the First World War. I watched him closely when he commanded the first tanks we ever had. I realize he is a difficult man but I know how to handle him."[11]

Intellectually and professionally Marshall and Patton shared a common bond that included learning the art of war from history. Patton's penchant for disparaging virtually every officer he ever served with never extended to Marshall, about whom there has never been recorded a single negative comment.[12] Marshall judged his officers solely on their performance, which was why Dwight Eisenhower and George S. Patton rated so high on his list of those to assume positions of great responsibility. Patton and Marshall had not served together during the entire interwar period, thus Marshall's judgment was based solely on his impressions and remembrance of Patton's performance of duty in France.[13]

Patton was certified eligible for promotion to brigadier general, but even with major maneuvers scheduled for 1940 at Fort Benning and in Louisiana, he nevertheless deemed his future prospects dim. That year he abruptly abstained from polo and hunting to avoid a crippling injury that might bar him from an active role—if anyone bothered to remember him, that is.[14] To help ensure that Marshall did, Patton had a jeweler in New York send him a set of eight silver stars, which he presented to the chief of staff in honor of his promotion to four-star general. Marshall replied with appreciation: "I will wear these stars with satisfaction and honor to the Army."[15]

In the spring of 1940 Patton participated as an umpire in the Third Army maneuvers in Louisiana, where Kenyon Joyce's (horse) cavalry division was thrashed by mechanized and tank forces combined into a provisional armored division. At the conclusion of the exercise Patton was one of a group of officers that included his friend and rival, Adna Chaffee, who met clandestinely in the basement of an Alexandria, Louisiana, high school to decide that the time had come to create an independent armored force, free from any control by the powerful cavalry and infantry chiefs. When the recommendations of the "basement conspirators" landed on Marshall's desk a few weeks later, he readily approved the creation of the first autonomous tank force and placed it in the capable hands of Chaffee, who set out to create an American panzer force.[16] There is no record of what Patton's contribution to that landmark meeting may have been.

Although he continued to command the 3d Cavalry, by the summer of 1940 it had become obvious even to Patton that the appointment of Chaffee to head the Armored Force signaled the death knell of the cavalry. Mechanization would be the future of Marshall's new U.S. Army. Yet Patton remained an outsider who seemed destined to play no part. Nevertheless, many diehard cavalrymen blamed their demise not on their own blindness to reality but rather on the spite of special interests and the jealousy of certain deskbound officers.[17]

Indeed, any lingering doubts about the necessity of developing a new armored force were removed once and for all by the sudden and devastating end of the so-called Phony War in Europe when, on May 10, 1940, German armored forces launched a blitzkrieg against France.[18] Within a matter of days the French army was defeated, Paris fell, and the French government sued for an armistice, signed on the site of the German surrender and humiliation in the Forêt de Compiègne in 1918. The British Expeditionary Force, likewise overwhelmed, was surrounded near the port city of Dunkirk. Most of the BEF escaped to England in a hastily organized operation by the Royal Navy, which the new prime minister, Winston Churchill, called "the deliverance of Dunkirk."

The blitzkrieg was largely the brainchild of Gen. Heinz Guderian, one of two impressive German military leaders whom Patton had studied closely in the late 1930s. The other was a previously obscure panzer general named Erwin Rommel. The writings of Guderian, Rommel, and a series of army G-2 translations of other German military works of the period left no doubt of the powerful promise of modern armor on the battlefields of Europe—a promise tragically fulfilled at the expense of the Polish cavalry in September 1939.[19]

Although still an outsider, Patton had seen enough during the 1940 maneuvers and in the newspapers reporting the tragic events in France to understand clearly where his future lay. Even so, he wrote yet another article for the *Cavalry Journal* in support of his beloved horse cavalry. Citing the lessons learned in the Third Army maneuvers, Patton stressed the positive capabilities of cavalry and how they ought to be employed in war. Among his pronouncements was one he had previously introduced: "GRAB THE ENEMY BY THE NOSE AND KICK HIM IN THE PANTS."[20] This principle would soon become George S. Patton's best-remembered byword. The only difference was that his cavalry would be mechanized.

In June 1940 Patton wrote to Chaffee and received a reply that he had been placed on Chaffee's preferred list of officers recommended to command the new armored brigades being formed in the two armored divisions (the 1st at Fort Knox and the 2d at Fort Benning) that initially comprised the Armored Force.[21] "It is a job which you could do to the queen's taste, and I need just such a man of your experience in command of an armored brigade. . . . We have an enormous job in front of us to get this thing organized, trained, and going in a minimum of time . . . I shall always be happy to know that you are around close in any capacity when there is fighting to be done."[22]

In July, Patton was attending Ruth Ellen's wedding when he learned from a Boston newspaper that he had been reassigned to Fort Benning and the 2d Armored Division, commanded by a friend, Brig. Gen. C. L. Scott, a bluff ex-cavalry officer, who had led the army polo team during the junior championships in 1922 and had managed the U.S. Olympic Equestrian Team in the early 1930s. Scott had immediately selected Patton from the list of potential brigade commanders sent him by Chaffee. Patton was astute enough to perceive that Marshall had also had something to do with his new assignment, and he wrote the chief of staff his thanks. Marshall wrote back that he thought "it would be just the sort of thing you would like most to do at the moment. Also, I felt that no one could do that particular job better."[23] Another positive signal that his luck had at long last changed came that same month, when Henry L. Stimson returned for a second stint as secretary of war.

Although Patton clearly deserved selection, it did him no harm that the army's "old boy network" had worked quickly to provide him a challenging new assignment just when his future looked bleak. As Roger Nye writes, although his long marriage to the cavalry had ended, "the heady ideas fostered by that union were not left behind. Patton retained the cavalry's notions of speed, maneuver, flank and rear attack, firepower, shock action, leadership, and the exacting care of men and equipment; now he would apply them to men in machines."[24]

On July 24, 1940, Patton formally relinquished command of the 3d Cavalry and, in a moving ceremony that left him in tears, he was escorted to the gates of Fort Myer by an honor guard.[25] The following month he assumed command of the 2d Armored Brigade, which consisted of 350 officers, 5,500 enlisted men, 383 tanks, and 202 armored cars. Although he relished returning to tanks, his pleasure was tempered by the absence of Beatrice during his first weeks at Fort Benning. His loneliness was visible. For the first time in their thirty-year marriage there seemed to be outright estrangement. The strains in their relationship during the interwar years had come to a head in the past several years, and his letters to Beatrice make it clear that their marriage was on the rocks. During previous separations Patton had compensated by working so hard he barely had time to be lonely. Now, however, even though he was soon to become a general, it was a hollow triumph without his beloved Beatrice at his side.

Patton had begun learning to fly and had another brush with death on the anniversary of his mother's birthday, which was also the first day of the Battle of Saint-Mihiel in 1918. "I was nearly killed that day but came nearer this afternoon when the motor on the airplane conked on the take off." Of his near death, he wrote, "fortunately or unfortunately as you happen to look at it we just got back to the field and no one was any the wiser except the pilot and me. But you almost got your wish that I die soon. . . . I have been expecting a letter all day from you about starting our lives anew. But it has not come so I suppose it wont. I dont blame you at all I simply am sorry for both of us."[26] A month earlier he had written painfully: "I hope some day you may forgive me but I will be damned if I see why you should. I love you anyhow."[27] In another letter he wrote with anguish that he had

> only my self to blame and you are the one I have hurt. I suppose the most charitable thing to this is that I was crazy. But I cant see how you or any other self respecting person could ever forgive me. . . . I love you and miss you terribly but can see no future I have hurt you too much. Perhaps when we are both dead you will forgive me. I hope it will be a comfort to us. You are the only person I have ever loved."[28]

Even in the face of circumstances that would have wrecked lesser marriages, Beatrice again stood by her husband, and later that year joined a desperately unhappy Georgie, who had missed her calming presence more deeply than he cared to admit. "For so long when I have done any thing worth while you have always been the gallery. It is hard to have no gallery any more and I feel quite sorry for my self but more sorry for you because I have shattered all your ideals. . . . I even hate to type that I love you because it sounds like such a lie. But I do." It was the most moving apology he would ever make.[29]

Beatrice's arrival seemed to calm the troubled waters of their marriage, and they again became the social couple of note, often appearing at costume balls dressed as King Arthur and Guinevere or Rhett Butler and Scarlett O'Hara.[30] Whatever turmoil existed in their private lives remained hidden from public view. Patton disdained the superb red-brick quarters of officer's row in favor of a house built by the post engineers in an unfashionable area of Fort Benning that was near his troops. Because Patton paid for the addition of a porch and closet with two hundred dollars of his own money, there were later unfounded charges and rumors circulated by his critics that there was something improper about what he had done.[31]

Patton's initial days in the "Hell on Wheels" division were anything but serene. Unlike France, where he had been the driving force of the tank training program, at Fort Benning Patton found himself playing a vastly different role. Perhaps it was because he had been away from tanks for twenty years, but those early days were filled with indecision and an inability to grasp his role. The 2d Armored was woefully ill equipped and understrength, its equipment both antiquated and often nonexistent. Even worse, morale in his brigade was at rock bottom. How was he to train these pampered young men? "Now we've got to make them attack and kill! God help the United States," he lamented with gloom.[32]

The men required motivation, and too many of the officers were not sufficiently fortified to provide the leadership required to organize and train an armored division, for which there were no precedents.[33] His would-be tankers possessed none of the motivation of the doughboys of World War I, and his tanks were obsolete pieces of scrap iron. *New York Times* correspondent Hanson W. Baldwin has described them as "a partly organized rabble of khaki-wearing civilians."[34]

Patton's inner turmoil was so great that despite wedding his future to tanks, he entertained one final temptation to return to his beloved horse cavalry when General Herr wrote to inquire if he desired to command Joyce's cavalry division the following spring. Without Beatrice's wise counsel he dithered. "Of course I would rather command it than any thing but on the other hand promotion will probably be better in the Tanks. I shall probably stay in the tanks. It makes little difference."[35]

Facing the greatest challenge of his career, Patton attempted to repeat the methods by which he had achieved such great success at Bourg, where his driving presence and personal intervention in every detail of tank operations and training had spelled the difference. He sped from place to place in a command car, an airplane or a light tank outfitted with a steamboat siren, which he boasted could be heard eight miles away on a clear day. Patton usually stood ramrod straight beside the driver, and operated the horn himself to clear the way and announce his imminent arrival. His towering appearance frightened most, and those who did not respond with alacrity were subjected to further piercing blasts that usually brought immediate compliance. Once during the Carolina maneuvers in 1941 two tank columns appeared at the same time at a crossroads. A dirty and weary captain was directing traffic and attempting to sort out the mess when Patton arrived in his scout car, horn blaring. The captain had his back turned and in response to Patton's obnoxious horn shouted: "Shut up there, you sonuvabitch." When he turned around and saw Patton towering over him, he fled in panic through some nearby poison sumac bushes. Patton was merely amused.[36]

Nevertheless, at Fort Benning in 1940, for the first time in his career, Patton's brand of personal leadership did not work well and, instead of producing results, fostered confusion and disorganization that were exacerbated by his constant intervention. Friends and associates observed the spectacle with trepidation. Where was the fire-breathing Patton of old? Was he finally washed up?[37]

Of greater concern at the moment was his anguish over the future state of his marriage; his personal troubles cannot have helped his state of mind. Nonetheless, as his frustrations multiplied, Patton finally began to perceive and correct his mistakes. Patton's longtime study of history began to pay off, and the first example he hammered into the officers and men of the 2d Armored was the wretched ordeal of the Poles the previous year. After barely a month in command, Patton addressed the officers of the entire division, and his theme was one he would observe for the remainder of his career. He began to expound his "ever victorious" philosophy, that the best defense was a damned good offense and that in the new armored division there existed the means to accomplish an American version of the German blitzkrieg.

Later calling an armored division "the most powerful organization ever devised by the mind of men," Patton cited Caesar, who, "in the winter time . . . so trained his legions in all that became soldiers and so habituated them to the proper performance of their duties, that when in the spring he committed them to battle against the Gauls it was not necessary to give them orders, for they knew what to do and how to do it." He concluded: "I know we shall attain it and when we do may God have mercy on our enemies; they will need it."[38]

Exactly one month later the ceremonial guns boomed a salute to the promotions of Scott to major general and Patton to brigadier general. Legend within the 2d Armored has it that Scott said to Patton: "Well, George, they just promoted the two most profane men in the army."[39] He wrote facetiously to the newly promoted Terry Allen, "The ARMY HAS CERTAINLY GONE TO HELL when both of us are made [promoted]. I guess we must be in for some serious fighting and we are the ones who can lead the way to hell," later writing, "all that is now needed is a nice juicy war."[40] To another friend, Robert L. Eichelberger, Patton wrote: "At last they have had the sense to promote the two best damn officers in the U.S. Army."[41] Patton's years in the wilderness were finally over. It had taken thirty-one years of service, but at last he was a general.

During the autumn of 1940, training intensified with exercises at all levels and reviews for visiting dignitaries, including a delighted Secretary of War Stimson. Chaffee's role was virtually identical to Rockenbach's in World War I and included not only command of the Armored Force but also of I Armored Corps, at Fort Knox, Kentucky. However, Chaffee was ill with cancer, and when Scott was selected to assume temporary command of I Corps in September, Patton became the acting division commander. In December 1940 both assignments became permanent. In the months following Patton's appointment he became a nationally known figure, and his mailbox began to fill with letters from his former tankers of the earlier war, and from others far and wide. After Pearl Harbor they intensified. Those who wrote ranged from Boston Brahmins (Henry Cabot Lodge Jr., a future United Nations ambassador, who served as a lieutenant colonel in the 2d Armored), to the barely literate. Supreme Court Justice Frank Murphy asked to be called to active duty in 1942 as a lieutenant colonel in the 2d Armored, only to be forbidden by Roosevelt. An ex-tank lieutenant wrote that his age was no barrier. "What can I do; where can I fit; How may I be of service to my country?" Another who pleaded for an assignment under Patton was Frank S. Leavitt, known professionally as the wrestler Man Mountain Dean.[42]

What the division lacked and Patton supplied was a will to fight. The many important lessons learned in thirty years of service became rules for the employment of his division: how reconnaissance units should deploy and advance, refueling techniques, road march discipline, the proper use and firing of weapons, tank gunnery, maintenance, camouflage, security, radio discipline, and the proper method of maneuvering tanks and armored cars in all types of situations. No detail was too small to escape Patton's inquisitive mind. This time, however, he had plenty of help, and he began delegating authority instead of attempting to attend to every detail himself.

Above all else he insisted on the highest standards of discipline and dress. When it was reported that 2d Armored troops had been seen in nearby Columbus, Georgia, in sloppy uniforms, often drunk and disorderly, an edict signed by Patton quickly appeared. Culprits paid a stiff price for ignoring his rules, and, as the historian of the 2d Armored has written, soldierly pride and discipline became "one of the division's hallmarks."[43]

As he began to stamp his personality and battle philosophy on the division, Patton was likely to turn up anywhere, at any time. There was no room for complacency. Patton gave true meaning to the division's motto. Traffic jams at crossroads enraged him, and when the division was spread over a large area during maneuvers, he frequently arrived by air. One day he came across just such a scene, and as his plane dived he was heard to scream: "Get those goddamn tanks off the roads and into the bushes." Compliance was instant.[44]

Patton's curt manner could convey a chilling message in few words, such as the day he sent for his G-3 and inspector general and, looking up briefly, said: "I want you two to know that I do not judge the efficiency of an officer by the calluses on his butt." Both readily grasped Patton's message and hastily deserted their offices to spend time in the field conducting staff visits.[45]

Although some were unaware of it at the time, those Patton identified as promising were often praised to other senior officers or recommended for important assignments. One was Maj. Isaac D. White, a graduate of Norwich University, who commanded the division reconnaissance battalion. Before Patton left the division for a new assignment, he summoned White to his office and invited him to join his staff. The feisty White wanted no part of a staff job and politely declined, fully expecting Patton's wrath to descend on him. Instead Patton said: "I don't blame you a damn bit," and later recommended him for promotion from major to brigadier general. During World War II White confirmed Patton's evaluation by advancing to the grade of major general and command of the 2d Armored.[46]

Before joining the division White encountered Patton at Fort Riley during the latter's brief visit. Patton had just learned of his promotion to brigadier general, and said, "Well, I'm going to get a lot more of these [stars]." Yet I. D. White was one of those who saw through the Patton facade. "It was showmanship and in many ways his image did not reflect his true character. People just don't believe you when you tell them that he was a very deeply religious man and that he was also a very emotional and a very thoughtful person in many ways." White credits Patton with thoroughly stamping his imprint on the division.

> General Patton was really the person who instilled the division with great pride in itself and developed a great esprit, as well as a great deal of the aggressiveness which characterized the division throughout its entire service. . . . I remember an incident . . . [when he visited] men on the rifle range. He asked a soldier what he was doing and the soldier said, "Just trying to hit the target." And the General said, "You are like hell! You are trying to kill some German son of a bitch before he can kill you."

White believed that what separated Patton from the rest was that

> he really inspired everybody with the idea that when you have gone just as far as you can go, you can still go a little bit further. He also I think instilled the division with the idea that no mission was too difficult to accomplish . . . You might not have loved him but you respected him and admired him and you wanted to put out for him. . . . Every unit in the division developed a very fierce and intense pride in its accomplishments.[47]

Colorful stories soon circulated about those who had run afoul of him. One morning his jeep drove past a sergeant whose back was turned and a lieutenant who was facing him but failed to notice or salute his commanding general. Some yards past, Patton suddenly shouted: "Back up this goddam car," the jeep screeching to a halt in front of the officer and the NCO. Excusing the sergeant, Patton demanded to know why the lieutenant had not saluted his car. "Never let that happen again," he ordered and drove off, leaving the two in a state of shock. To his friend Harry Semmes, who was riding in the jeep with him, Patton said: "Sounds cruel, doesn't it? But it saves lives—that'll spread." As the Patton legend similarly spread, many who served under him would later brag of the severity of their particular chewing out. To his officers he said that those who exchanged salutes did so as a mark of respect for one another as part of the brotherhood of the U.S. Army. His logic in being so strict was based on the simple premise: "If you can't get them to salute when they should salute, and wear the clothes you tell them to wear, how are you going to get them to die for their country?"[48] Some despised him and his methods, but most seemed to feel themselves part of a team whose commander knew what he was doing. It engendered a sense of confidence that must be experienced to be fully understood.

It took tremendous effort and skill to turn untrained men into a fighting division through endless drilling and field exercises and, in 1941, participation in several major maneuvers. The division improved but still suffered from such Pattonesque difficulties as "waffle ass"—a disease contracted by those who sat too much—and "student complex"—a malady brought about by too much schooling and insufficient field experience, which resulted in a tendency, found primarily in NCOs and junior officers, to wait for instructions rather than rely on initiative.[49] Under Patton initiative was a cardinal rule. Once, in 1944, while in command of the Third Army, Patton encountered the 90th Division commander, Maj. Gen. Raymond McLain, at a crossroads in Normandy. McLain informed him that a nearby town would be captured by 5:00 P.M. Patton replied, "Listen, Mac, we are going through to Germany, and we don't want to stop too long. I want that town by 3:00 o'clock." Planning all along to capture the town by 3:00, he told Patton the later time, knowing he would cut it in half.[50]

Patton's interest in all phases of soldiering was never more evident than in his attempt to design a suitable uniform for his tankers, a practice begun in France in 1918. In December 1940 a group of National Guard and Reserve generals arrived at Fort Benning to observe some exercises by the 2d Armored. One of these officers was McLain, and as he arrived he observed "a tall, stately officer leaning against a tree dressed in green padded jodhpurs, a pea green double-breasted jacket fitted at the waist with brass buttons running from the waist up over the right shoulder, with a pearl [sic] handled pistol on a peculiar sling under his left arm." His helmet, reputedly obtained from the Washington Redskins football team, had a gold raised band around it that looked very much like a halo. On his face was a look of amusement. "What do you think of my uniform," demanded Patton. "I don't quite understand it," replied a perplexed McLain.[51]

Patton explained that he was devising a uniform he hoped the army would adopt. To illustrate, he had McLain climb into a tank, whereupon the pistol strapped to his waist impeded his entry and exit. Noting that a split second might make the difference between life and death if the tank were hit and the crew forced to abandon it in haste, Patton said: "I am trying to find a way that will save [their] life." Although the army never adopted Patton's version, it was an example of his unfailing attention to detail in everything he ever did.[52]

Patton's sincere attempt to produce a new uniform for his armored troops was comical. When he posed in it for a group of photographers, he resembled a football player dressed as a bellboy. Behind his back Patton was often referred to as the "Green Hornet" (the title of a popular 1940s radio program), and although the uniform soon passed into history, the nickname stuck for a short time until his better-known sobriquet was coined.[53] Once, when some visiting senior officers came from Washington to observe division maneuvers, one of Patton's staff attempted to brief a general who waved him away. "No, don't bother about that. We just came down to have a look at Georgie's uniform."[54]

McLain later served under Patton in Sicily, and in 1944, when Patton was asked by Omar Bradley who ought to be selected from a long list of candidates to command the 90th Infantry Division, he replied: "McLain." As the discussion ended Bradley said: "George, you haven't commented on any of these officers. This Division is going to belong to you; you had better speak up." Patton replied, "Hell, damn it, I told you McLain and that is what I meant." McLain got the 90th Division.[55] He had earned Patton's trust, and that was good enough. McLain's performance later justified Patton's judgment.

Although its origins are still unclear, about this time Patton was christened with the nickname that was forever to identify him: "Old Blood and Guts." At the division officer school Patton frequently lectured his junior officers and one of his more colorful descriptive phrases was that in combat they must expect to be "up to their necks in blood and guts." In the bachelor officer quarters one night, a lieutenant is said to have looked at his watch and observed that "it was about time to go hear old Blood and Guts," a remark that was greeted with laughter.[56] Another version is that the nickname resulted from a speech given to the division in which Patton referred to the necessity for an armored division in combat to possess "blood and brains." When the widely discussed speech reached the ears of reporters, it was reported in the newspapers as "blood and guts."[57]

Whatever its origin, the nickname eventually stuck and although no one is known ever to have called him "Old Blood and Guts" to his face, it became a magnet for his critics, who cited it as an example of Patton's monumental ego and his wanton disregard for the lives of his troops. Within the 2d Armored, however, he was generally referred to by that most reverent of military titles, the "Old Man." Within the War Department, Patton was a subject of mirth and still referred to by many as the "Green Hornet."[58]

If Patton's uniform produced light relief, his training methods were anything but humorous. Over and over again he drilled his troops that they would form the spearhead of the coming war, the worthy modern-day successors to the original tank men of 1918.

In December 1940, 6,500 troops and nearly twelve hundred tanks and vehicles made a successful 270-mile road march from Fort Benning to Panama City, Florida, and back, testing their road discipline and ability to move. Patton's premise was that an armored division that lacked the ability to move speedily and efficiently was useless, and this had been their first real test.

The United States instituted the draft in 1940, and by early 1941 new trainees began arriving at Fort Benning to fill shortages of up to 50 percent that existed the previous summer. By April the 2d Armored paraded 14,000 men at a mounted review held in a driving rain on the day Patton was promoted to major general. He addressed his troops with the challenge that "armored and aerial warfare make higher demands on courage and discipline than have ever been experienced by the fighting men of our race." He reminded them that the greatness of and success of armies depended far more on the courage, discipline, and loyalty of the men composing them, than it did on the perfection of the machines which they operated. "Victory will be decided by those possessing the highest attributes of courage and loyalty." As for himself, "I can but promise that I shall never ask any man to undergo risks which I, myself, do not incur."[59] It the was same principle he would employ throughout the coming war, that well-led, courageous soldiers were the ultimate military machine.

Becoming a general failed to inhibit Patton's idiosyncrasies. In fact, as he became more and more visible to the public, his tendency toward impulsive and irrational acts led to precisely the trouble Katherine Marshall had predicted. Several of Patton's misdeeds during World War II led to the necessity for a public apology. The best known took place in 1943 in Sicily, but the first occurred at Fort Benning in 1941. The senior officer on post was Maj. General Lloyd R. Fredendall, a dour and imperious infantryman who also commanded the 4th Infantry Division. One morning a staff car passed a 4th Division motorcycle MP at a high rate of speed. The MP took off after the vehicle with sirens howling. Ignoring the siren, the staff car merely went faster. When the vehicle was finally pulled over in the 2d Armored area, the MP discovered Patton inside and said: "General, I am sorry but I have to arrest you for speeding in our division area. General Fredendall—."

At the mention of Fredendall's name Patton exclaimed in anatomical terms what the general could do, and ordered the MP to "get his ass out the Second Division area." There was no love lost between the 4th Division and the 2d Armored. Fredendall was fond of employing maintenance teams at surprise roadblocks on the roads of Fort Benning to inspect and cite vehicles found deficient.[60] There seemed to be particular relish when the vehicles belonged to Patton's command.

When the MP commander duly informed Fredendall that one of his MPs had just attempted to arrest Patton, he exclaimed with obvious relish:

"That son of a bitch. He's not going to get away with this."[61] Fredendall telephoned Patton and, after a tense conversation, ordered him to report to his
office to apologize to the MP. "And right NOW, George," he thundered. Patton complied and after a frigid exchange between the two generals, the MP
was summoned. Fredendall said, "Corporal, General Patton has something
to say to you." Patton said, "Soldier, I'd like to tell you that you did a good
job this morning in trying to apprehend somebody who was violating the
rules. And as for the remarks I made to you, I sincerely owe you my apologies and I hope that you will accept them."[62]

During his first maneuvers in 1941, Patton discovered a command car
parked in plain sight and ordered the driver located and arrested. His new
aide, 2d Lt. Richard N. Jenson, had the unenviable task of telling Patton that
the vehicle belonged to an umpire and was not subject to the usual rules of
concealment. When Patton encountered a captain, he informed him of what
he had done and where the driver could be found. The captain was to release
him from arrest and present the commanding general's apology for the mistake. It was typical of Patton that he would commit some indiscretion, usually the result of momentary rage, only to apologize for his behavior and
carry on, the matter all but forgotten. Unlike others, the mercurial Patton did
not carry grudges; he was a man of the moment and when that moment was
over, so too was the incident it triggered. Generals are not in the habit of
apologizing. Patton was an exception.[63]

Patton's driver was M/Sgt. John L. Mims, who had driven for his predecessor and had been retained. As Mims relates, Patton knew only one
speed—f-a-s-t—and it was not uncommon for him to race at speeds of up to
eighty-five miles per hour around Fort Benning. He delighted in defying the
MPs, whom he seemed to disdain, apparently after they had hassled him one
day as he was leaving the post in civilian clothes in his automobile.[64] Once,
as an MP vainly attempted to flag his staff car down as it sped through an
intersection, Mims began to comply, only to hear Patton shout that he
should "get the hell going."[65] The experience one morning soon after Patton's arrival was typical. Mims was driving at the posted speed limit of 25
mph when Patton said: "Sergeant, I am in a hurry, speed up." Mims duly
speeded up to 35 and had barely gone a hundred yards when Patton
screamed at him: "If you're scared to drive this God damn thing, get out and
walk and let me drive." Mims, who valued his plum duty assignment, hit the
accelerator and thereafter drove at unsafe speeds of 85–90 mph to avoid further outbursts from his volatile passenger. Patton once allowed Mims 120
minutes to drive the 119 miles from Fort Benning to the Atlanta airport.[66]
Patton's disregard for his personal safety had clearly not changed with his
advancement to general.

Patton's treatment of Mims was typical of his dealings with other soldiers. When angered by some transgression he would lash out furiously in

his colorful language, and moments later the matter would be forgotten, as if it had never occurred. "It would make you wonder how a man could raise so much hell and commotion and then be so calm."[67]

Yet behind the blasphemy there lay a moral man who could not abide lewd behavior. Patton's protégé, I. D. White, recounts an incident in November 1940 when

> There was what we used to call a "smoker" for the officers of the division. There was booze, music and dancing girls. As the evening wore on the girls got rid of most or all of their clothes, and put on some obscene dances. At this point General Patton got up and walked out. The next day he told us there would be no more such activities and gave us a lecture about obscenity, etc. A few hours later he was explaining a tactical maneuver in the most obscene language! How do you characterize such a person?[68]

By the summer of 1941 Patton's exploits at Fort Benning had captured the imagination of the press, and in July he appeared on the cover of *Life* magazine, posing sternly in his "Green Hornet" uniform in front of a tank. Inside, a lengthy article extolled the leadership of "Patton of the Armored Force" and the realistic training of the 2d Armored.[69]

In 1940 Dwight Eisenhower was a lieutenant colonel with an uncertain future that left him deeply concerned that he would never receive one of the important new assignments in the rapidly expanding army. He had continued serving MacArthur in the Philippines from 1936 to 1939, when he was posted to Fort Lewis, Washington, as the executive officer of an infantry regiment. He and Patton resumed their correspondence in the late summer of 1940, with Patton advising that he expected to command an armored division shortly. Would Eisenhower care to serve under him in some capacity? If so, "I shall ask for you either as Chief of Staff which I should prefer or as a regimental commander you can tell me which you want for no matter how we get together we will go PLACES. If you get a better offer in the meantime take it." When he concluded that he hoped "we are together in a long and BLOODY war," it was another of his typically outrageous statements that historians have cited to prove that he was a warmonger.[70] Although the United States would not formally enter the war for more than a year, Patton had long since concluded that American participation was inevitable. Such comments were typical of his propensity to say things for their shock value to old friends like Eisenhower.

Eisenhower was eager to serve under Patton and immediately replied: "I am flattered. . . . It would be great to be in the tanks once more, and even better to be associated with you again. I suppose it's too much to hope that I could have a regiment in your division, because I'm still almost three years away from my colonelcy, but I *think* I could do a damn good job of commanding a regiment."[71] Eisenhower's well-deserved reputation as an outstanding staff officer left him deeply apprehensive that he would miss an opportunity to command a tank regiment in combat under Patton.[72]

Patton urged him to use any pull he had to wrangle a transfer and assured him that seniority was not a factor, as "I dont bother with rank anyhow." Moreover, "there will be 10 new generals in this corps pretty damned soon."[73] Others desired Eisenhower's services as well, however, and, with Marshall's blessing, in June 1941, he was transferred to Fort Sam Houston, Texas, as the Third Army chief of staff. In August 1941, in Louisiana, there began the largest maneuvers ever held by the U.S. Army, during which both Eisenhower and Patton would solidly cement their reputations as potential senior leaders.

CHAPTER 27

The 1941 Tennessee, Louisiana, and Carolina Maneuvers

I want the mistakes [made] down in Louisiana, not over in Europe, and the only way to do this thing is to try it out, and if it doesn't work, find out what we need to make it work.

—GEN. GEORGE C. MARSHALL [1]

By the summer of 1941 Western Europe had become a Nazi-held dominion. In June Hitler unleashed Operation Barbarossa—the massive surprise attack against Russia, in which more than three million German ground troops supported by three thousand tanks and two thousand aircraft swarmed over a front extending from the Black Sea to the Arctic, catching the Red Army flat-footed. The British were under siege from a naval blockade, and although the Royal Air Force had staved off invasion by winning the Battle of Britain, the outlook remained grim, and was made more dismal still in the Far East, when Hong Kong fell in December. Despite the loss of the battleship *Bismarck* in May 1941 at sea, the Germans were winning the Battle of the Atlantic, and Adm. Karl Doenitz's German U-boats were exacting a mounting toll on convoys in the icy waters of the North Atlantic. Between September 1939 and the end of 1941, Allied shipping losses were an appalling 2,361 ships, totaling 8,545,606 tons. [2]

Although the United States remained officially neutral, Roosevelt responded to Churchill's pleas for help by means of Lend-Lease. In September 1940, the president arranged for delivery to the British of fifty destroyers in return for the granting of leases to the U.S. Navy in Bermuda, the Caribbean, and Newfoundland (popularly referred to as the destroyers-for-bases deal). In what came to be known as the "undeclared war" against German submarines, Roosevelt proclaimed that the U.S. Navy would protect American shipping over an area that covered half the Atlantic Ocean. As Germany and the United States edged closer to outright war, Roosevelt committed the nation to supplying a steady flow of war material vital to Britain's survival.

As American industry began gearing for war, George Marshall and Leslie McNair continued to mastermind the rebirth of the army, both in size and, perhaps more important, in its doctrines as a fighting force. Conflict continued to rage over the concept of the mobility of the lightly armed cavalry of the Old West and the newer belief in the use of massive force first employed by Ulysses S. Grant in the Civil War, and taken to new levels by Pershing during World War I. Conversely, Robert E. Lee had demonstrated that force combined with mobility could produce results. Marshall's dilemma was to enlarge and convert the army into a force that could successfully challenge the mobility and mechanization of the German army, whose success on the battlefields of Poland and France forewarned of the difficulties ahead.

When Germany invaded Poland, the United States was a third-rate power with an army that still ranked a miserable seventeenth in the world.[3] So sorry was the state of the U.S. Army in 1939 that—had Pancho Villa been alive to raid the southwestern United States—it would have been as ill prepared to repulse or punish him as it had been in 1916.

Marshall's mobilization efforts had brought about a more mobile, streamlined, triangular infantry division, the creation of the armored division, and the introduction of improved new weapons. By the end of 1940 the army's strength was 620,000 men, and on June 30, 1941, it stood at 1,460,998. Now the time had come to test the U.S. Army's preparedness to fight on a modern battlefield.[4] The year 1941 was monopolized by division-, corps-, and army-level war games. As a worried McNair noted, "We didn't know how soon war would come, but we knew it was coming . . . and we had to get together *something* of an Army pretty darn fast."[5] That assessment came to fruition in the late summer of 1941 during the Louisiana maneuvers, in what Marshall billed as "a combat college for troop leading" and as a field laboratory "for the new armored, anti-tank, and air forces that had come of age since 1918."[6]

Two months earlier the Second Army held its own maneuvers, which served as an important test for the 2d Armored Division, in mountainous, heavily forested terrain in Tennessee, ill-suited for mobile warfare. The opening phase of the exercise called for the 2d Armored, part of the Red force, to attack the Blue (enemy) VII Corps. As Patton attempted to move the 2d Armored undetected into the maneuver area, the spirit of the exercise was livened by a challenge laid down by the VII Corps commander, who offered a twenty-five-dollar reward for the capture of Patton. Not to be out-done, Patton reciprocated by doubling the prize for anyone who bagged his opponent (the VII Corps commander).

It was Patton's custom to brief his troops personally before an exercise or, later, a combat operation. The Tennessee maneuvers were no exception, and in May he articulated his essentials of discipline, which he insisted would one day save their lives and create a pattern of behavior that would permit them to react instinctively in the terrible heat and confusion of com-bat. Vintage Patton, it included the admonition that although it was only a war game, they must "think this is war . . . that is the only chance, men, that you are going to have to practice. The next time, maybe, there will be no umpires, and the bullets will be very real."

Patton also drilled home another of his principles of war:

> Remember, that one of the greatest qualities which we have, is the ability to produce in our enemy the fear of the unknown. Therefore, we must always keep on moving, do not sit down, do not say, "I have done enough," keep on, see what else you can do to raise the devil with the enemy. I remember once when I was trying to play football at West Point, I didn't do so well. The coach came up and said, "Mr. Patton, if you can't do anything else, throw a fit!" The same thing applies to armored forces, if you can't think of anything else to do, throw a fit, burn a town, do something.[7]

Such was the Promethean size of the new armored division that when the 2d Armored left Fort Benning for Tennessee, *each* of its two columns was sixty miles long.[8] While a friendly infantry division immobilized the enemy force, Patton's plan was to strike the flank of the Blue force with infantry and tanks. The scheme was the first instance of what became an American practice of using combined-arms task forces (infantry, armor, artillery, and combat engineer) in various combinations to suit the tactical situation. The Red force was awarded the victory, but of greater significance was the 2d Armored's feat of maneuvering into position in the dark, without lights and under radio silence. As the division historian notes: "Umpires with the tankers called the road march the most magnificent ever made by tanks."[9] Patton's tough training was beginning to show dramatic results.

In the remaining phases the 2d Armored played various roles. Patton spent most of his time prowling the front lines, alternately encouraging or cussing his officers and men. When they captured an "enemy" division commander and his command post, Patton gleefully paid each man the promised bounty. Another phase of the exercise expected to last an entire day ended after a mere three hours, when the 2d Armored's aggressive attacks overwhelmed their opponents.

Yet mistakes were made, leading Patton to remark in his critique that "if there were not mistakes, there would be no need for maneuvers."[10] After one of the 1941 maneuvers, Patton's critique included a plate on which reposed a wet noodle. Holding it up for all to see, he attempted to push it across the plate and, failing, said: "Gentlemen, you don't push the noodle, you pull it. In other words, you lead."[11] Patton's division emerged from the Tennessee maneuvers with an enviable reputation as an aggressive, exceptionally well-led division. Nevertheless his old friend Floyd Parks, who was then serving on McNair's staff, wrote to suggest that Patton ought to tone down his ostentatiousness, advice he studiously ignored. If he angered others by his methods, so be it. Patton that believed the results more than justified the means.[12]

Arguments still raged between the proponents of antitank weapons, (who predicted this weapon would herald the death knell of armor), and the tankers (who believed that the lethal new antitank weapons could be circumvented by developing new tactics of fire and maneuver). The Tennessee maneuvers were the first test of both, and in the wake of those who claimed victory for the antitank forces, Patton warned his men of the folly of attempting to employ what he called "the famous bull-fighting method of charging the enemy with the purpose of crushing him beneath our tracks." Although Patton's reputation as a successful commander was by now well established, it was also typical that he consistently gave credit for his achievements to his troops. After the Tennessee maneuvers he heaped praise upon the men of the 2d Armored.

The Tennessee maneuvers were corps-level exercises.[13] The autumn exercise in Louisiana and eastern Texas would be the first in which two armies battled each other. Under the overall control of McNair, the scenario pitted the 160,000 troops of Lt. Gen. Ben Lear's Second Army, whose mission was to invade Louisiana, against the 240,000 troops of the Third Army, commanded by Lt. Gen. Walter Krueger. Twenty-seven divisions participated, and the entire state of Louisiana became a gigantic maneuver area.

The Louisiana war games became one of the most watched and reported events of 1941. "War" between the Red and Blue forces kicked off at noon on September 15, with Lear's Red Army attacking across the Red River and driving toward Shreveport with two corps, one of which was the I Armored. Lear was a tough disciplinarian who was known for his abrasive manner. His plan was to use the mobility of his armored forces to outflank the Blue army, which was defending southern Louisiana. However, in Walter

Krueger and his brilliant chief of staff, Dwight Eisenhower, Lear was up against worthy, imaginative opponents with clear ideas of how to defeat the invader. Despite heavy rains, Krueger's airmen soon located the Red Force armor (including Patton's 2d Armored) crossing the Red River. The Blue forces were able to bottle up Lear's tanks, and within a short time the Red forces were in desperate trouble. Patton's tanks attempted to pry open the flank for a breakout, but Krueger and Eisenhower were too quick, and the 2d Armored suffered heavy losses in what soon became a rout. By the fourth day, Lear's force was nearly surrounded and facing annihilation when McNair mercifully ended the first phase.[14] Needless to say Patton was not pleased to be on the losing side.

When Phase II began a week later, the roles were reversed, with the 2d Armored part of Krueger's attacking Blue force, whose mission was to capture Shreveport and defeat Lear's Red force, defending the city. The original plan called for a Blue force frontal attack, but when Lear declined to do battle, Krueger elected to launch a bold flank attack with the I Armored Corps. It was spearheaded by the 2d Armored, which initiated a 350-mile end run to outflank the Red force. The journey took Patton back into Texas, where he was able to swing his tanks north and then east until he had gotten behind Lear's army and could attack Shreveport from the rear. When a key bridge collapsed due to flooding, the length of the march increased to nearly 400 miles. Although McNair later denied that Patton's audacity was the underlying reason, he nevertheless decided to end the final phase of the Louisiana maneuvers prematurely after barely five days. With Patton and the Blue force poised to deliver the *coup de grâce* to Lear, there seemed no reason to continue.[15]

Patton had driven his men with his usual combination of enthusiasm, whip-cracking, and personal example. When his tanks ran out of gas, the division paid in cash (possibly Patton's) to have them refueled at local filling stations.[16] Patton's journey actually took his division outside the designated maneuver area, resulting in howls of outrage that he was playing by his own rules. Patton merely grinned and retorted that he was unaware of the existence of any rules in war. Winning was all that mattered, and he did not give a damn what it took so long as the enemy was defeated. Whatever the technicalities of his violation of the rules for the maneuver, in the eyes of his superiors he had shown the sort of initiative and daring that had long been absent in the U.S. Army. Even if it smacked of grandstanding, George Patton had no intention of letting a little matter of fuel deter his advance or ruin a golden opportunity to rout his opponent. Three years later the same problem would confront Patton, but this time there would be no friendly filling stations with supplies of fuel he could tap.

Even though the 2d Armored was strung out over a wide area, Patton never hesitated in attacking Shreveport with only a fraction of his division.

He understood the risk but also realized that surprise was on his side. So great was the shock on the enemy side that he could afford to attack with a force of only regimental size and defeat a much larger defending force. It was a lesson he was to carry into the war and use well. The weather had been foul and had served as an excellent test of the ability of armor to move under even the poorest conditions.

The Louisiana maneuvers not only tested doctrine and tactics but also identified the army's most promising leaders—the men who would carry the burden of fighting during the coming war. Among the stars of the Louisiana maneuvers was Eisenhower, who won acclaim even as he grumbled that he would rather have been commanding a unit of Patton's tanks. Eisenhower, credited with devising the strategy by which Krueger's Third Army bested Lear's Second Army, caught the attention of the press, who extolled his virtues in print even as they misidentified him as "Lt. Col. D. D. Ersenbeing." Shortly after the Japanese attack on Pearl Harbor, Marshall ordered Eisenhower to Washington for duty with the War Department, the first step in his meteoric rise to supreme command.[17]

The Louisiana maneuvers also confirmed Marshall's belief that new, younger men were needed to command the army's combat units. Senior officers were ruthlessly purged: Thirty-one of the forty-two army, corps, and division commanders were either relieved or shunted aside to make way for a new generation of commanders, which included Omar N. Bradley, Terry Allen, Leonard T. Gerow, and Patton's classmate William H. Simpson. Twenty of twenty-seven division commanders were replaced in 1942. Only eleven (of forty-two) senior officers achieved higher command, among them Patton's onetime nemesis, Lloyd Fredendall.[18] Patton emerged as the star of the armored force. Some said his name alone was worth an armored division.

Despite the complaints of some congressmen that the army was wasting money conducting such exercises, the Louisiana maneuvers were, as Christopher Gabel concludes, "unprecedented in U.S. Army history and have never been duplicated in size or scope since." Under the tutelage of Marshall and McNair, "the Army was forced to make good two decades of virtual disarmament in two years' time." The result was that "the GHQ maneuvers of 1941 revealed both the penalties of military unpreparedness and the power of American resolve. The war that followed transformed both the Army and the nation forever."[19]

The Louisiana maneuvers had barely ended when the 2d Armored participated in its third major war game of 1941. Even as the lessons learned in Louisiana were being assessed to correct the many shortcomings in tactics, doctrine, logistics, and training, plans were being finalized for the final exercise of the year in the Carolinas, between Hugh Drum's First Army and

IV Corps, which included the I Armored Corps (the 1st and 2d Armored Divisions); in all, some three hundred thousand troops would participate.

Within the armored force there was a searching analysis over the employment of tanks and how the problems noted in Louisiana could be resolved short of the battlefield. Patton complained that the weapons of the division were being used piecemeal, thus forfeiting its great assemblage of firepower. "Each time we fight with only one weapon . . . we are not winning a battle, we are making fools of ourselves," he told his division.[20] The armored proponents also faced a growing sentiment within the army, led by McNair, for antitank weapons and doctrine whose premise was their superiority as tank killers. Devers, Scott, Patton, and other armor proponents had to demonstrate that the tank could survive on the battlefield by putting the lessons of Louisiana into practice during the Carolinas exercise.

Held in mid-November, the Carolina maneuvers became a test of the mobility of armor against the numerical superiority of conventional forces, augmented by heavy antitank defenses. The exercise also pitted Patton against Hugh Drum. If Patton was looking for a payback for his humiliation in 1936, it was not long in coming. Despite a huge numerical advantage, Drum was so intent on winning at all costs that he descended to committing flagrant violations of maneuver rules in order to gain an unfair advantage even before the exercise kicked off.[21]

In 1941 the army had yet to refine the means of supplying the enormous appetite of its tanks and vehicles. Tanker trucks would eventually be used to supply the fuel-guzzling Sherman tanks, but at the time of the Carolina maneuvers, Patton's most pressing problem was keeping his tanks gassed from hundreds of five-gallon jerry cans, a difficult and often backbreaking task. When he learned that the I Armored Corps was responsible for Patton's resupply, the corps G-4, a Regular Army colonel, hastily transferred, leaving an inexperienced young reserve lieutenant named Porter B. Williamson in charge.

Fuel was brought into a railroad siding in large tank cars, but Williamson had no way to transfer it to Patton and was unable to contact the 2d Armored. He was searching for the division headquarters when he encountered a Hell on Wheels tank submerged in a small stream, with what appeared to be a gray-haired sergeant with a high pitched, squeaky voice in command. When the "sergeant" put on his helmet, he saw that it had two silver stars on it. Williamson identified himself, and Patton barked: "Where the hell is my gas. I need gas!" and demanded to see the corps G-4. Informed that the colonel was en route to Fort Knox, Patton asked who the hell *was* the G-4? "I am," replied Williamson. Disgusted, Patton walked to his tank and returned with a pile of maps. "We need gas. Lots of gas. We've got a real problem here. Without gas then the war game is over. I cannot move ten miles without gas."[22]

Although Patton was notorious for taking liberties with the rules, in reality he attempted to play by them rather than with them. In his discussion with Lieutenant Williamson, he expressed concern that the use of the fuel in the railhead might violate the maneuver rules. Reassured it would not, Patton turned to the young officer and said: "You son-of-a-bitch! We can sure learn from each other! May I borrow your map?[23] Gen. Drum will not even have a map showing where I am going to attack! . . . He is an old friend, but I sure as hell would enjoy capturing him."[24]

One of the 1941 maneuvers produced another map story about Patton, in which he encountered a tank company commander and inquired if he knew his location. The captain replied that he did not, whereupon Patton exploded and asked if he had maps. "Yes, sir, but I haven't had time to check my position by the map because I have just finished capturing an infantry regiment." Patton reached over and asked the officer for his maps and ripped them to pieces, saying: "Captain, you don't need maps. Carry on," and sped off.[25]

To Hugh Drum's acute embarrassment, early on the first day of the final phase of the exercise, he was captured at a roadblock by I. D. White's 82d Reconnaissance Battalion. With great politeness a captain greeted Drum: "Good morning, general. Will you join me?" Drum was spared the humiliation of being taken to the rear (and no doubt an encounter with Patton) when McNair, the exercise director, ordered him released an hour later on the flimsy grounds that he could not have been spirited successfully through First Army lines.[26] Drum, who cried foul, complaining that Patton's action was a violation of the maneuver rules (obviously ignoring his own casual disregard of the same rules), was actually released so that the exercise could continue. A civilian telephone operator who spoke with him during his brief captivity observed: "My, but that man was sure having a hissy-fit!"[27] A great deal has been made of the incident as an example of the antagonism between two longtime enemies. While it is true that the maneuvers generated animosities and rivalries that played out during the course of the war, Patton himself never commented either publicly or privately on the matter, and the stories that circulated were more fantasy than *cause célèbre*. Nevertheless, although Drum was the senior officer, his embarrassment at Patton's hands during the Carolina maneuvers is thought to have crippled his chances for high command, relegating him to obscurity and retirement in 1943.[28]

Among the VIPs observing the final phase of the Carolina maneuvers was George C. Marshall, who noted Patton's impressive performance. Marshall was later criticized by a senator for leaving Washington at a critical moment to observe the Carolina maneuvers. "The Chief of Staff rested his defense on one episode of the trip. He recalled with relish the splendid performance of a certain George S. Patton."[29]

*   *   *

With his customary aggressiveness and flamboyance, Patton had settled once and for all that even if tactics and training required further refinement, good leadership could still win the day on the battlefield. As historian Gabel notes:

> A wide variety of operational styles emerged during the [three 1941] maneuvers, ranging from Patton's high-spirited armored raids to Drum's methodical general assaults. Patton's style seemed progressive at the time, but it actually owed more to J. E. B. Stuart than to Heinz Guderian. Drum's methods were enormously successful on maneuvers but held little appeal for Marshall and McNair, who wished to forge the Army into a rapier, not a battle-ax.[30]

Gabel's comparison of Patton to J. E. B. Stuart is accurate, for not only Patton's mastery of the art of surprise but his continual emphasis on the importance of reconnaissance were lessons he had learned from studying the Civil War. Time and again during 1940 and in the 1941 maneuvers, Patton stressed the necessity to secure information about the enemy and to ensure that it was properly disseminated.[31] To that end intelligence became the lifeblood of Patton's cut-and-slash tactics, and he relied on his reconnaissance forces, operating behind enemy lines, as his eyes and ears.

Patton never lost sight of the vital importance of his men's survival and his subsequent reputation as a bloodthirsty warrior ("our blood, his guts"), uncaring of his men and bent only on success, is unfounded. No commander ever devoted more time to training his troops to such a high standard that would *save*, not waste, their lives. He spent inordinate amounts of time preparing his lectures and then addressing his entire division personally in order to drive home his message. Again and again he warned of what they must do to survive, and how they must not needlessly waste their lives by some foolish act. While many division commanders commonly spent most of their time in their command post, Patton was hardly ever there. Instead, he continued the practice begun at Saint-Mihiel and refined during the 1941 maneuvers, of visiting his units in the front lines. He wrote to Floyd Parks of the importance of personal leadership: "We have many generals capable of keeping a G-2 [intelligence and] G-3 [operational] map, and very few capable of leading anything. . . . It is my convinced opinion that General Scott and I are the only general officers who participated in the [Tennessee] maneuvers who know by personal observation how our troops behaved."[32]

No one who experienced Patton's brand of leadership was in any doubt of his sincerity. Indeed, if there was a single, defining theme in his training methods, it was to repeat—again and again—how to stay alive and win on the battlefield. "Do as I have taught you," he would proclaim, "and you will stay alive." Patton understood, as few did, that tough, realistic training, combined with strong, effective leadership, spelled the difference between success and failure. Gen. Matthew B. Ridgway once said that it didn't hurt morale for the troops to see a dead general once in a while, and Patton would have agreed with him!

Patton's success in Louisiana and the Carolinas also seems to have spiked an attempt by the antitank cabal in Washington to sabotage the Armored Force by relegating it to an insignificant role. Among their complaints were comments that Patton's tactics were no way to fight a war.[33] McNair, who had kept his thumb firmly on Patton during the Louisiana and Carolina maneuvers and had exhibited evidence of a distinct bias against the Armored Force, continued to advocate a separate antitank arm that could mass its resources to defeat armor, a position that was never proved in the maneuvers.[34]

To the contrary, the 1941 maneuvers finally resolved the question of what form the new armored force would assume. When Adna Chaffee died of cancer in August 1941, his replacement was Patton's West Point classmate Jacob L. Devers, who carried on Chaffee's vision and argued for the formation of an all-armored corps that would employ enormous firepower to create breakthroughs on the battlefield—the same type of *corps de chasse* the British were then employing without success in North Africa. Eventually one armored division was integrated into the traditional corps of two or more infantry divisions, and instead of creating breakthroughs, the division would exploit whatever advantage the infantry could create.[35] It was hardly what the armor purists desired, but it was the best they could hope for and became the doctrine utilized throughout the war.

At the conclusion of the Carolina maneuvers in December 1941, the 2d Armored returned to Fort Benning after nearly six months of rigorous field training. Although as yet untested in combat, it was one of the best-trained divisions in the army. When Patton assembled his officers to critique the division's performance, he offered only one comment: "Keep off the roads," a reference to the vulnerability of vehicles to air attack and ambush. Of his own performance he was the first to note a need for improvement:

> Tactically I made mistakes, both in training and in operations, which I am now correcting through further education of myself and the officers of this division. Those of you who have been with us since the beginning know that at all times we have been preparing for war. Our first training goal as set by General Chaffee, was to be ready for war by September 30, 1940. Of course, we were not ready; we had neither the men nor the equipment, but we made a damned good try. I shall be delighted to lead you against any enemy, confident in the fact that your disciplined valor and high training will bring victory.

Patton's passionate speech was an attempt to motivate them into being warriors who must put their heart and soul into becoming expert killers. "The only good enemy is a dead enemy," he told them. "Misses do not kill, but a bullet in the heart or a bayonet in the guts do. Let every bullet find its billet—it is the body of your foes. . . . Battle is not a terrifying ordeal to be endured. It is a magnificent experience wherein all the elements that have made man superior to the beasts are present: courage, self-sacrifice, loyalty, help to others, devotion to duty."

Then the romantic warrior in Patton burst forth: "Remember that these enemies, whom we shall have the honor to destroy, are good soldiers and stark fighters. To beat such men, you must not despise their ability, but you must be confident in your own superiority. . . . Remember too that your God is with you." He concluded with this poem:

The God of our Fathers, known of old,
Lord of our farflung battle line.
Beneath who's awful hand we hold
Dominion over palm and pine.

The earth is full of anger,
The seas are dark with wrath;
The Nations in their harness
Go up against our path!
Ere yet we loose the legions—
Ere yet we draw the blade,
Jehova of the Thunders,
Lord God of Battles, aid!

E'en now their vanguard gathers,
E'en now we face the fray—
As Thou didst help our fathers,
Help Thou our host today!
Fulfilled of signs and wonders,
In life, in death made clear—
Jehova of the Thunders,
Lord God of Battles, hear![36]

It was another of his amazing performances, combining death, scripture, poetry, and the basic tenets of war. No other commander in American military history ever successfully invoked the fear of combat in words anyone could easily understand and, even more important, identify with. Of the terror of battle, Patton recalled the words of a French marshal whose visitor on the eve of a battle noted his knees trembling. The marshal, a veteran and hero of twenty years of war, replied: "My lord duke, if my knees but knew where I shall this day take them, they would tremble even more." The tremor they would experience was not fear but simply the excitement "every athlete feels just before the whistle blows . ... you will be inspired by magnificent hate."[37]

Therein lay the heart of the problem facing any military commander, whether Robert E. Lee in the Civil War, George S. Patton in World War II, or Norman Schwarzkopf in the Gulf War of 1991: how to motivate decent young men raised on the precepts of the Bible, the sanctity of human life, and the immorality of killing to become an efficient cog in a gigantic killing machine such as an armored division. While it was enough to make their mothers cringe, the only method whereby a Patton or a Schwarzkopf could succeed on the battlefield was to trespass on the inherent decency of Americans by training and motivating their men to survive by killing others whose task was to kill *them*. Patton did it as well or better than virtually anyone else.

The testing of the new army in the 1941 maneuvers had come none too soon. Shortly after the 2d Armored returned to Fort Benning came the shocking news flash and the black headlines revealing the surprise Japanese attack on the U.S. Pacific fleet at Pearl Harbor on December 7, 1941. The following day the United States and Britain declared war on Japan. Germany and Italy issued declarations of war on December 11, 1941. It was an indicator of U.S. unpreparedness that the Italian dictator, Benito Mussolini, would scorn the United States as a second-rate power whose industrial capacity was nothing more than a journalistic hoax, and whose military importance was so negligible that its involvement in the war would be of little consequence. From the insularity of isolationism the United States was now irrevocably thrust into a two-ocean, global war that would swiftly test the nation's military training on the battlefield.

In the wake of Chaffee's death Patton was promoted in January 1942 to command the I Armored Corps, and there came the sad day when he said farewell to his beloved 2d Armored Division. As he often did on such occasions, Patton was overcome by emotion and cut short his farewell speech. He would always consider the 2d Armored his, and even after his death its postwar commanders would keep in touch with Beatrice Patton to report news of Georgie's division.[38] Years later Gen. I. D. White would recall Marshall's words to Patton: "George, you didn't make the 2d Armored Division, the 2d Armored made you."[39]

Patton and his new boss, Jake Devers, were not only classmates but competitors. Patton continued his practice of writing informally and directly to whomever he pleased, and once bypassed Devers on some matter directly to Stimson. At a dinner party not long afterwards, Beatrice began to leave the room so the men could talk shop, but Devers asked her to remain. "We need a judge in here," he said:

George and I are going to settle some things right here. . . . Now, George, I have your recommendations here and we've given them all careful consideration. I don't give a damn who commands, but I'm the commanding officer right now, and I'm going to command, and I'm going to make the decisions. . . . You went this way and we want to go a little differently than you do. You're too much of a horse cavalryman. I'm a little more on fire power. Are you going to play ball?

A grinning Patton arose, saluted Devers, and said, "Yes, boss." And that was that. "He was a good soldier—always was," recalled Devers.[40]

Transferred from Fort Knox to Fort Benning shortly before Patton took command, the I Armored Corps was situated in a tent city in a remote corner of the post. The day of his arrival to assume command was as colorful as it was unforgettable. Patton sent written notice he would assume command at precisely 11:00 A.M. January 15, 1942. Some of the older officers who knew Patton by reputation scrambled to transfer out before he arrived. At 10:00 the corps staff heard the scream of sirens emanating from the main post. Some thought it a fire, while others gazed through binoculars to determine its location. Another announced it must be a grass fire.

At 10:30 A.M. sharp, a dozen motorcycle outriders, with sirens blaring, stormed into the corps cantonment. They wore gleaming helmets emblazoned with the emblem of the 2d Armored Division, and carried polished rifles. They surrounded the entire corps area, standing a silent, rigid vigil, speaking to no one. A short time later there again came the sound of sirens howling in the distance, the rumble of heavy vehicles, and huge clouds of dust. It was a second convoy consisting of two tanks and several trucks filled with soldiers and flags. The troops dismounted, and several placed two covered flags in stands they had brought with them. While five soldiers remained with the flags, the remainder formed a double line in front of the headquarters. As the corps staff wandered outside to observe these strange proceedings, it finally occurred to the chief of staff to suggest that perhaps they, too, ought to assemble. Dressed in a variety of uniforms and headgear, the staff sauntered into a ragged semblance of a military formation and began chatting among themselves, until intimidated into silence by the steely glares of Patton's honor guard.

As everyone stood at attention (or at least what passed for attention among the corps staff), an immaculately polished command car drove up and stopped in front of the headquarters. Standing erect in the back, his hand on the support bar, was Patton. For many this was their first glimpse of their new commander. Without so much as a glance in their direction, Patton dismounted and approached the 2d Armored honor guard and barked: "Sergeant! Post the colors!" He held a rigid hand salute as the color guard unfurled the American flag and a red general-officer flag with the two white stars of a major general.

A staff officer murmured, "Here comes a tear-jerking speech!" Instead, Patton ordered "Dismissed!" and the honor guard drove off, some of them in tears. Patton stood in silence for several minutes, occasionally looking at his watch. On the stroke of 11:00 , he announced: "I assume command of I Armored Corps! At ease!" Still at rigid attention himself, Patton addressed the assembled staff: "We are in for a long war against a tough enemy. We must train millions of men to be soldiers. We must make them tough in mind and body, and they must be trained to kill. As officers we must give leadership in becoming tough physically and mentally. Every man in this command must be able to run a mile in fifteen minutes with a full military pack." When an overweight senior officer guffawed, Patton angrily resumed: "I mean every man! Every officer and enlisted man, staff and command, every man will run a mile! We will start running from this point in exactly thirty minutes! I will lead!" Then, turning his head slowly so that he made eye contact with every officer, Patton said, "I have in my hand the orders for the transfer of every officer in I Armored Corps. Every order is signed and dated today. Every officer wanting a transfer or refusing to run a mile will leave this command before the sun goes down! Those officers wanting to remain in I Armored Corps be back at this point in thirty minutes. Dismissed!" One of those present was Lt. Porter Williamson, who recalled: "The silence was so total the sun seemed to stop in the sky. The two stray dogs, always running around our area, stopped and remained motionless as if at attention awaiting Gen. Patton's command."[41]

Several older officers elected to transfer, but the majority stuck it out. Thirty minutes later most assembled outside. When Patton appeared, his war face was gone, a smile in its place. "I do not see a full military pack among any of you," he said, "so we will not run today! But get in shape! We will be doing a mile run every day! Let's all go over to the Fort Benning Officers Club and have lunch." By that night no one in the I Armored Corps was in any doubt that they had a tough new commander. It was a scenario he was to repeat with the same chilling similarity little more than a year later on a battlefield in North Africa.

At his first staff meeting, Patton dropped a bombshell when he bluntly announced:

The war in Europe is over for us. England will probably fall this year. Our first chance to get at the enemy will be in North Africa. We cannot train troops to fight in the desert of North Africa by training in the swamps of Georgia. I sent a report to Washington requesting a desert training center in California. The California desert can kill quicker than the enemy. We will lose a lot of men from heat, but the training will save hundreds of lives when we get into combat. I want every officer and section to start planning on moving all our troops by rail to California.[42]

While Devers fought the bureaucracy in Washington to obtain the best available officers to staff ten new armored divisions, and his G-1 wore out his welcome in the officer assignment branch, Patton worked to distill the lessons learned during the eighteen months he had been back in tanks. To Marshall he sent heartfelt thanks for "the honor conferred upon me. . . . The best thanks I can offer you is to assure you that if I fail, it would not be for lack of trying." Marshall replied that Patton had made "a fine record," and I have been particularly glad that time and maneuvers permitted your thorough development as a division commander. You should now be able to make a great contribution to the armored force as [a] corps commander."[43]

In February 1942, when *Life* attempted to publish a second article on Patton, he wrote the editor to request that they desist, even though Devers had given his approval on the grounds that the publicity would enhance the prestige of the new Armored Force. Feeling that they had built him up to levels he could not possibly live up to, Patton thought its publication might ruin his career and was better suited to his eulogy should he be killed in battle.[44]

In March 1942 Patton was selected to command a new desert training center being formed in the arid wastelands of southeastern California. Patton had mixed emotions, for it meant saying good-bye to Fort Benning. But it was time to move on. Although the United States was at war and he was retaining command of the I Armored Corps, Patton nevertheless perceived his new assignment as a dead-end training job that was taking him too far from Washington, where he could use his influence for a better shot at a field command when the first combat units were deployed overseas. It apparently did not occur to him that his corps would be the obvious choice.

Patton's farewell party at Fort Benning was memorable. Although he had poor senses of taste and smell, he loved parties and fancied himself a connoisseur when it came to mixing drinks. His family had assembled at Fort Benning to give him a royal send-off, and even though the drinks had already been ordered from the officer's club, Patton canceled the order, proclaiming that *he* would mix the contents of the household liquor cabinet into one gigantic punch. Appalled, Beatrice begged him not to proceed, but Patton insisted that his concoction would be served so cold that no one would worry about the taste. Moreover, as a thrifty New Englander, surely she could recognize the value of using up expensive booze that would otherwise simply go to waste.

His concoction of Scotch, wine, and bourbon was poured into oilcans that had been converted into serving containers. The brew, which he dubbed "Armored Diesel," apparently had to be sampled to be believed. His officers all came and drank toasts to their commander. As a grim-faced Beatrice looked 'on, the party soon began grinding to a halt as one officer after another collapsed glassy-eyed into the arms of total intoxication from Patton's powerful concoction (it was said to be *worse* than real diesel). Knowing better than to drink this dreadful potion, his family consumed only beer and was spared the same fate. Patton was high on the adrenaline of playing the host and thought it a splendid evening. (Family lore has it that the recipe for "Armored Diesel" later mercifully vanished.)

The mood of the women present was hardly boosted when Patton raised his glass in a toast, and insensitively said, "Here's to the wives. My, what pretty widows you're going to make." Some of his 2d Armored officers presented him with a silver cigarette box engraved with their names and the words: "To a Gallant Soldier whose friendship we cherish. May you go on to further deeds of valor in your country's service." Patton was so moved that he turned his back on the gathering, the tears streaming down his face. The following morning, with more tears, and cheers ringing in their ears, the Pattons left for California.[45]

CHAPTER 28

Countdown to War

This is my last night in America. It may be for years and
it may be forever.

—PATTON

The War Department wisely concluded that it was essential to train American troops in desert warfare in the event U.S. forces were eventually committed to combat in North Africa. Patton was selected to create the first Desert Training Center, situated near the town of Indio, 200 miles east of Los Angeles, in 16,200 square miles of arid desert. During his brief tenure, which began in March 1942, more than sixty thousand troops trained at this vast center.[1]

Accompanied by some of the officers who would later become key players in the invasion of North Africa and in the Seventh and Third Armies, Patton selected the training site during a four-day reconnaissance of southeastern California and western Arizona. The party ranged over a huge desolate area without encountering a single other human. Patton used a Piper Cub aircraft to view his new domain from above, and later personally reconnoitered a large area on foot and on horseback. The 20,000-square-mile area of desert and mountains, rocks, crags, cactus, mesquite, and dry-salt-lake beds was declared "probably the largest and best training ground in the United States," and the new training center, dubbed "Little Libya," the world's largest, closely emulated the conditions that would later be encountered in North Africa. Daytime temperatures often climbed to 130 degrees in the shade. In winter they ranged from freezing at night to 100 degrees at midday.[2] However, the humidity of Louisiana, he noted, was far worse than the dry heat of the desert and the weather from October to the end of May "is what babies cry for and old, rich people pay large sums of money to obtain."[3]

Many of the first troops to arrive scorned it as "the place that God forgot."[4] With stunning rapidity tent cities began to sprout, and soldiers began learning the important art of improvisation. To simulate the harsh conditions of war there were few amenities. Until a permanent base camp was erected, few tents had wooden floors, and electricity, hot water, and clean sheets were not included in a stay at the "Hotel Patton."[5] With no time for bureaucratic studies or lengthy investigative reports, Patton drew on a variety of sources and candidly admitted: "I have a limited amount of knowledge about the desert, so do not hesitate to give me the most trivial details which, from your experience you might consider superfluous."[6] Patton gathered around him anyone who knew anything at all about the desert, including a noted explorer of the Gobi. "So-called experts—bearded and unbearded, natty and matted and tattered—attached themselves to General Patton and the base camp, all of them blessed with one attribute in common—a mouth that kept talking."[7]

Patton's training was primarily designed to toughen his troops physically and mentally to withstand the rigors of desert warfare for protracted periods, with only a minimum of food and water (only one canteen a day was permitted during maneuvers), but with no decrease in combat efficiency. When the Los Angeles water district suggested building water storage tanks, Patton demurred. "They have no time to do anything except learn to fight," he said. Most units had to walk or ride several miles to a quartermaster water point to take a cold shower. Training began at the lowest level and progressed to large combined-arms exercises, requiring movements of three hundred miles, involving the entire corps. Patton's message to his officers was straightforward: "If you can work successfully here, in this country, it will be no difficulty at all to kill the assorted sons of bitches you meet in any other country."[8]

Patton decreed that within a month after their arrival, everyone must be able to run a mile in ten minutes with a rifle and full field pack. He seemed to be everywhere, even off duty. Radio reception was so poor that Patton paid for out of his own pocket and installed his own equipment so the troops would have news and music. He also used it as a quick means of communicating with everyone, and had microphones installed both in his office and his tent so he could override the station anytime he chose, which he often did day and night. There would be a click, the music would stop and a gruff, high-pitched voice would announce: "This is General Patton!" He would remind them to stay alert for an exercise, be on their toes for visiting brass, or occasionally berate or relieve an officer, which usually drew howls of laughter from the enlisted men. Once he announced that colonel so-and-so was removed from command at once. "If you know what's good for you,

you will stay away from me for a week," he bawled. On other occasions he would use the radio to declare that he had found "a damn good soldier today," and then announce his name and his unit. Every company-size unit and higher was ordered to have a bugler, and each night some one hundred men in the various base camps would play taps at approximately the same moment, producing a mournful echo across miles of desert and off the nearby mountains. More important, it helped to create a sense of unity and high morale. For Patton it brought back a flood of warm memories of his service with Pershing, chasing Pancho Villa.[9]

Patton was uncompromising and spared no one, least of all himself, eschewing the usual luxuries found on a conventional military installation.[10] He was an uncompromising driver and disciplinarian, whether it was on the drill field or policing the bivouac area. "He stated that the greatest difficulty in the Army was the lack of initiative and sense of responsibility among the younger officers," and within two weeks of his arrival two officers found themselves under arrest for failing to carry out their duties properly. He was equally uncompromising with himself and lived in the base camp with his men. "He was with his men in whatever they did. He would be in the first of the tanks . . . from his cub plane or from his hill he would study his troops . . . he supervised them, he exhorted them, he taught. . . . His men and officers never knew what vehicle—jeep, Packard sedan, tank or half-track, cub plane or tractor—might suddenly erupt him." He informed Devers that "I propose to hold the housekeeping arrangements here to the minimum . . . to spend just as little time as possible on 'prettying up' and as much time as possible on tactical and technical instruction."[11]

Patton had his own personal hill, dubbed "The King's Throne," overlooking the training ground, from which he would closely scrutinize with binoculars everything that went on below him. When he observed anything requiring correction or admonishment, he would shout instructions into his radio.

Vehicles were not permitted on a paved highway during exercises, and one morning he spied a truck several miles away and gave chase. The driver was a sergeant who was asked if he couldn't take the desert. The sergeant replied he had no problem with the desert, but his transmission certainly did and his unit's trucks were constantly failing. To see for himself, Patton got down on his hands and knees to peer under the vehicle; then he ordered the sergeant to write him a report on the problem, which he promised to fix. Within earshot he noted that this was a damn good soldier who cared more for his truck than he did for his commanding officer's orders.[12]

Most corps commanders were little-known, remote figures, but what made Patton unique was that his troops knew him by constant daily personal contact. Patton would scold his troops and then say: "I'm doing this for your own good. If you'd done [in combat] what you did here, you'd be killed."[13]

But he never asked his men to do anything he would not do himself, and often would "bawl hell out of a man" for getting his tank stuck in the sand, and then proceed to pull it out himself with a tank or recovery vehicle. One day, when tanks were being unloaded from flat cars too slowly, Patton became impatient and personally demonstrated how and where to place support timbers. The tanks soon began rolling smoothly off the train.

Conditions at the desert training center were harsh, equipment failed routinely, and men fell from heat exhaustion and rattlesnakes. There were the inevitable training accidents and deaths, and when they occurred Patton was to be found in the hospital tent with the victim until he died.[14]

He readily confessed that he deliberately wore his "war face" because war was not a smiling business, and threatened to shoot anyone photographed smiling. Patton's profanity was equally calculated. "War is hell," and the language of war is never polite, he explained. To prepare young American men to kill, "swearing helps to get the point across."[15] Once he introduced a speaker as "the noblest work of God—a killer."[16] His frequent lectures were also often self-deprecating. "I do not know of a better way to die than to be facing the enemy. I pray that I will fall forward when I am shot. That way I can keep firing my pistols! I was shot in the behind in World War I! I do not want to be hit there again. I got a medal for charging at the enemy, but I have had to spend a lot of time explaining how I got shot in the behind! I want to fall forward!" He would conclude: "Every man is expendable—especially me."[17]

His exhortations were a series of "nevers": Never give up; never dig in; never defend, always attack; never worry about defeat, think of and plan only for victory; you win by never losing. He would caution that to win a battle a man had to make his mind run his body because the body will always give up from exhaustion. But when you are tired, the enemy is just as exhausted: "Never let the enemy rest." No detail was too small; one day he shouted that pillows were worthless baggage that ought to be discarded, and that by sleeping straight a soldier would get more oxygen to his brain and therefore fight better.[18]

Although it was common practice in war to cover up or subdue insignias of rank and organization to prevent the enemy from learning what unit they were fighting and what rank they held, Patton declared he would shoot the first person to do so. Officers must lead and let their rank be seen. "We want the enemy to *know* they are facing the toughest fighting men in the world!" He told his men to let their rank be seen, and, if they were captured, he advised them that their captors would treat Patton's men kindly. It was all part of his calculated efforts to prepare them for war. He insisted they should not fear being killed in combat, as the chances of death were greater driving an automobile or going to bed, where most people died. Again and again, he repeated: "Never take counsel of your fears!" Patton could not abide cowardice and preached: "A coward is always in hell because he will suffer a thousand deaths every day. A brave man will only die once."[19]

Another pet peeve was foxholes, which he called graves. Keep moving rapidly and there would be no need for them. Wars were won not by holding land, which he deemed worthless, but by killing the enemy. Man himself was the ultimate war machine.

He scorned clergymen as "pulpit killers" for preaching: "Thou shalt not kill," and argued that they knew less about the Bible and the way God worked than he did. Pointing to the story of David slaying Goliath, he insisted it was not a sin to kill if one served on the side of God. Kill or be killed, he exhorted.[20] Calling himself "a hell of a guy," he would remark: "I'm giving them hell one minute and crying over them the next."

While Patton's screaming presence could unnerve even the strongest individual, junior officers seemed to live in particular terror of incurring his wrath. The number "arrested" was legion. He wrote to retired Gen. Malin Craig that his main problem was young, inexperienced officers who lacked self-confidence. "I believe that the vigorous use of a polished toe against their hind ends may eventually induce them to do something besides sit on their asses!" While some inevitably grumbled over the harsh conditions and relentless training, most came to respect their tough commander and morale remained high. After narrowly missing some telephone poles near his headquarters while landing in his Piper Cub, Patton's signalmen, without orders (and to the dismay of the Southern California Telephone Co.) took them down and buried the wires underground. "The boys said they did not want their General killed."

If an officer fell asleep during one of his lectures, Patton would walk over, awaken him, and inquire how long he had been without sleep. If the reply was less than forty-eight hours, Patton would storm that a soldier must be capable of such sustained periods in combat. If the period was longer he would merely order the offender to bed.

To ensure that his troops remained alert, Patton frequently attempted to infiltrate his own heavily guarded base camp. One night he was prowling outside its perimeter with Lieutenant Williamson, whom he sent forward to see if the guard knew his instructions. When asked from what direction he expected trouble, the guard pointed directly at Patton's headquarters in the center of the camp, and replied *that* was where he expected trouble. Patton, who overheard the conversation, began laughing, and waved off Williamson. "That man understands his mission." The astonished guard exclaimed: "My God! You gotta expect General Patton from all directions." Patton, who believed in the maxim that good soldiers should dread their own commander more than the enemy, was pleased. "We are doing better," he observed. "We are [now] up to the level of the Roman legions."[21]

Patton did his best to arrange for entertainment and sports. Baseball teams sprouted, famed conductor Leopold Stokowski led one hundred musicians in a concert, and the versatile pianist Victor Borge also entertained the troops. The fifteen hundred inhabitants of Indio were nearly overwhelmed by the presence of the army. One night nine thousand troops stormed Indio's few restaurants and bars, and as many as seven hundred might be found on a single city block. The Indioans were friendly, but when Patton addressed the local women's club on Memorial Day, some complained that soldiers were whistling at them. Patton said his men suffered from many restrictions but if they wished, he would ban whistling. "But, if I were you women I wouldn't worry if they whistled at me. If I went into the street and they didn't whistle at me, then I'd worry." No more was heard on the subject.[22]

Patton installed Beatrice in a rented ranch house in Indio and visited on weekends, usually ordering Sergeant Mims to drive like hell. He had left his enlisted aide, M/Sgt. William Meeks, at Fort Benning, but in April he wrote his friend, James Ulio, the army adjutant general, to have Meeks reassigned to him because he was "the only man who can find anything."[23] Meeks soon rejoined Patton and remained with him until his death.

Worried he would be forgotten while he was in California, Patton deluged Devers and McNair with letters and memos citing his accomplishments, and wrote to Malin Craig that "I wish to God that we would start killing somebody somewhere soon, and trust that you will use your influence to see that I can take a hand in the killing. Just to keep my hand in for Marshal Rommel I have shot one or more jackrabbits every day that I have been here, with a pistol."[24]

With time precious, during the first twenty-three days of training Patton conducted thirteen major tactical exercises, not all of which were successful. Some VIPs were present to observe a mock attack, narrated personally by Patton. When there was no attack, a red-faced Patton turned to his visitors and exclaimed: "This is exactly how battles are lost in war. This maneuver is called off until tomorrow morning. I can assure everyone that this mix-up will not happen tomorrow." Patton assembled his commanders to determine what had gone wrong. Slapping his swagger stick against his thigh as he questioned the lieutenant in command of the tanks that had failed to attack, Patton learned that the radios had failed, and the officer never received the order to attack. Turning to the colonel in overall command, he said: "You should have dashed over there on your two damn feet. Radios will always be going dead! We must be ready." The colonel was invited to his office for a "chat," and to calm the trembling lieutenant, Patton put his arm around him and said reassuringly that he had acted in accordance with army regulations. However, he advised if the young officer wanted to be Napoleon he must think only of the mission. "Forget about Army Regulations. . . [which] are written by those who have never been in battle. . . . Our only mission in combat is to win. If we do not win, you can forget everything!"[25]

The distinctive mark of a soldier is his uniform and insignia, and, contrary to army policy, Patton lobbied hard for unit markings on all his vehicles and uniforms. Eventually he convinced his superiors that the benefits far outweighed the need for denying the enemy knowledge of what units he was opposing, correctly arguing that it was nonsense to believe that the Germans would not readily identify their opponents. To McNair he wrote passionately:

> There is a regrettable and widespread belief among civilians and in the Army that we will win this war through materiel. In my opinion we will only win this war through blood, sacrifice, and high courage. In order to get willing fighters we must develop the highest possible Esprit de Corps. Therefore, the removal of distinctive badges and insignia from the uniform is highly detrimental. To die willingly, as many of us must, we must have tremendous pride not only in our nation and in ourselves but in the unit in which we serve. . . . [It] is of vital moment to our ultimate victory.[26]

To the end of his life Patton argued against the stupidity of repressing the single most effective means of keeping high morale.

To remind McNair not to forget him, he wrote several weeks later to ask that "when serious fighting starts I be given a chance to prove in blood what I have learned in sweat," and was reassured when Devers informed him in mid-July that an armored corps was under serious consideration for deployment, and that Patton would command it.[27] "THANKS," he immediately wrote back. "I WONT LET YOU DOWN."[28]

In April, Marshall had sent Eisenhower to London as his special envoy to Churchill, and in June named him commander of all American troops in the newly designated European Theater of Operations (ETO). Now wearing the three stars of a lieutenant general, Eisenhower wrote to alert Patton that the time was rapidly approaching when his services would be urgently required.[29]

For the British the war had been a calamitous series of reversals ever since April 1940, including Norway, Dunkirk, Greece, Crete, Burma, Hong Kong, Dieppe, Tobruk, and most recently Singapore (the single worst defeat in British military history). In the Mediterranean, which they had heretofore regarded as their own private preserve, the British were reduced to the status of intruders. Less observable was the Battle of the Atlantic, where the Allies were losing the fight with German U-boats, which were sinking Allied shipping faster than it could be replaced.

In early 1941 Hitler sent an expeditionary force commanded by General

Erwin Rommel to North Africa to bolster the sagging fortunes of the Italians. Although Britain now had an ally, its army continued to fight a series of mostly losing battles across the far-flung Empire, as the United States began gearing to join the fray sometime in 1942. Until that day came, the war in the desert pitted the Eighth Army against Rommel's German-Italian Panzer Army. By July 1942 the Eighth Army was defending the final German obstacle to Egypt: Alam Halfa. Only with the arrival in August 1942 of Gen. Sir Harold R. L. G. Alexander as commander in chief, Middle East, and Lt. Gen. Bernard Montgomery as the new Eighth Army commander were the British able to begin reversing the humiliating defeats of 1941 and 1942.

British and American discussions concerning the opening of a second front somewhere in Europe or the Mediterranean (primarily to ease the pressure on the Red Army by drawing off portions of the Wehrmacht) had been ongoing since Pearl Harbor, resulting in agreement that the defeat of Germany would take priority over operations in the Pacific against Japan. The so-called soft underbelly strategy favored by Churchill, of attacking the Axis along the southern periphery of its empire, was adamantly opposed by Marshall, who argued that Allied military operations must result in an immediate and direct threat to Nazi Germany. This meant a cross-Channel invasion, not "hunt-and-peck" operations in the Mediterranean, which, as one British officer noted, the United States regarded "as a kind of dark hole, into which one entered at one's peril."[30]

The fall of Tobruk to Rommel in June 1942 was the heaviest blow struck against the British in the Middle East, and came during an official visit to Washington by Churchill. FDR offered whatever help the United States was capable of providing. The prime minister readily accepted and asked that a number of the new American Sherman tanks be sent at once to bolster the Eighth Army. Marshall countered with the offer of a fully equipped armored division, and immediately summoned Patton to Washington in June 1942 to prepare plans for its deployment.

With a stern admonition that he must make do with only a single division, the chief of staff sent Patton to the nearby War College to read the planning documents and be briefed. Patton, however, failed to heed his directive, and the following morning learned a harsh lesson in what it meant to cross Marshall, who was intensely surprised to receive a letter from Patton forcefully proposing that *two* divisions be deployed. "The decision [was] final. . . . I didn't want to hear from him on that," he recalled after the war. Disgusted that Patton had flouted his orders, Marshall said simply: "Order him back to Indio," and immediately sent "one of my secretaries of the General Staff" personally to put Patton on a plane to California "that morning, which they did."

From Indio a deeply shaken Patton repeatedly attempted to telephone

Marshall, who was never available to accept his call. In desperation Patton contacted the deputy chief of staff, Lt. Gen. Joseph T. McNarney, and announced that after "a lot of thinking," he had concluded "that maybe I could do the job after all with the forces your stupid staff is willing to give me." When informed of Patton's change of heart, Marshall smiled and said: "You see, McNarney, that's the way to handle Patton."[31]

When the 2d Armored could not be deployed in time, Churchill readily accepted Marshall's offer to send the British three hundred tanks and one hundred howitzers.[32] Although there was no further need for Patton's services in Washington, he spent several miserable weeks believing that he had ruined his chances for a combat command by angering Marshall. When he learned through the grapevine that his fears were misplaced, he wrote contritely to Floyd Parks that "I am willing to take anything to any place at any time regardless of consequences."

By the summer of 1942, the two Allies had yet to agree on a common strategy, and it had become evident that a cross-Channel invasion of France could not take place before mid-1943. The impasse ended only when Churchill succeeded in obtaining American agreement for an entirely new plan called Gymnast, a joint Anglo-American operation to seize French North Africa in November 1942. Despite Marshall's opposition, Gymnast (later renamed Torch by Churchill) fulfilled Roosevelt's insistence on the commitment of U.S. combat forces in 1942, rather than awaiting a joint invasion of Europe in 1943, which remained problematic at best. Although he would not have put it as crudely as Patton did to his troops, Roosevelt too was chafing for an enemy to kill.

After neutralizing the Vichy French regime in Morocco and Algeria, the Allies were to seize Tunisia, thereby trapping Rommel's German-Italian army between the Torch forces and Montgomery's Eighth Army, which was to break out of its defensive position near El Alamein and drive Rommel westward into Tunisia.

Three separate invasion forces from the United States and Britain were to rendezvous off North Africa, and land simultaneously to seize Casablanca, Oran, and Algiers. Torch was the first of many compromises between the Allies. In this instance Churchill agreed to a massive buildup of American forces in Britain for a cross-Channel operation in 1943, and conceded that an American officer should command Allied forces in the Mediterranean. That choice was Patton's friend and Marshall's protégé Dwight David Eisenhower, whose new command was designated Allied Force Headquarters (better known by its initials AFHQ). Not the least of the reasons that the British approved of Eisenhower was their mistaken belief that his inexperience would make him susceptible to the influence of Churchill and the British chiefs of staff. They expected to nominate a British deputy commander and were displeased when Eisenhower selected Mark Clark as his deputy. In British parlance the hard-nosed Clark was not a "gentleman," but it was his initiative and drive that brought together the Torch plan in a few short weeks.

The rise to high command and fame of the man already known simply as "Ike" to his friends was the most dramatic of any American officer of World War II. After the Louisiana maneuvers Marshall brought him to the War Department to head the War Plans Division. Patton wrote to suggest that after he finished developing war plans he ought to apply for the command of an armored division. However, it was too late for Eisenhower to serve under Patton. His star was already in the ascendant as a result of his brilliant performance in Louisiana, and glowing recommendations from friends and colleagues had caught Marshall's attention. Although he had recommended others for the job, Marshall sent Eisenhower and Mark Clark to London in April 1942 to coordinate U.S. planning with Churchill and the British chiefs of staff.[33] Although Eisenhower's concept of inter-Allied cooperation was the bedrock of his success as the Allied commander in chief in the Mediterranean, and later in northwest Europe, he was under no illusions as to the magnitude of his task. He later recalled: "The British and the Americans came together like a bulldog and a cat."[34]

Despite his gaffe with Marshall, by the summer of 1942 Patton's reputation had been greatly enhanced by his success at the Desert Training Center. He saw very little of his family and when his son, George, won an appointment to West Point as a member of the class of 1946, Patton sent him a blunt lecture on deportment that resembled the admonitions he gave his troops. "Do your damndest in an ostentatious manner all the time. Never make excuses. Never be late, always be well dressed. If some little fart hases you dont get mad do what he says and take it out on some one else next year . . . remember the Virginia adage 'that the best trainer is Old Doctor Work.'" Concluding, "We are realy proud of you for the first time in your life see to it that we stay that way," the letter bore scant resemblance to the tactful, caring advice his gentle father had given him in 1904.[35] Robert Patton describes Patton's relationship with his son as "amiable if distant," and if what was intended as a letter of congratulations seemed unduly harsh, it was nevertheless meant as praise.[36]

At the end of July Patton was again hastily summoned to Washington, where he learned that he was to command the Western Task Force of Torch: the invasion of Casablanca. He never returned to California and could only send a message to his troops, apologizing for failing to say a personal goodbye, and expressing his "unparalleled honor to have commanded such men." After Patton left, many of his troops who stayed behind expressed the wish that they could have accompanied him.

Patton was not Marshall's first choice. The chief of staff originally intended to appoint Maj. Gen. Robert L. Eichelberger, who instead was sent to the southwest Pacific, where he distinguished himself under MacArthur. Why Patton? As Marshall's official biographer notes:

> Perhaps it was a sign of some inner regret in Marshall that he would never have a chance to prove himself in conflict that he prized the eccentrics like Patton and Wingate who were difficult to live with but who exulted in the fray. Or it may have been the temper and the fury of his own nature, rigidly disciplined and long pent up, responding sympathetically to natures that were never curbed. Whatever his reason, he called Patton to fighting command and saw that he had his chance.[37]

Patton's headquarters was in the Munitions Building on Independence Avenue, which was the site of the War Department before the construction of the Pentagon. His first act was to summon from California several trusted members of his I Armored Corps staff. Patton's chief of staff was Col. Hobart R. "Hap" Gay, a former cavalry officer and confidant whose loyalty sometimes ran to blind faith in his chief.[38] Blumenson describes Gay as a plain man who was utterly devoted to Patton, "a splendid companion who liked to ride and hunt, a superb staff officer who ran the military details of the headquarters with exceptional efficiency." Gay acted effectively as Patton's foil under any and all conditions, and although his superiors would repeatedly urge Gay's replacement, Patton stubbornly refused. The problem was that Gay "lacked the breadth and depth of intellectual capacity. His prejudices and politics paralleled Patton's, and as a consequence reinforced instead of correcting them."

The two men first met in 1938, when, as a young cavalry officer, Lt. Hobart Gay was assigned to Patton's staff at Fort Myer. Patton's reputation throughout the cavalry as a tough taskmaster so thoroughly daunted Gay that he had attempted unsuccessfully to have himself assigned elsewhere. His first encounter with Patton merely reinforced that trepidation. Gay was a fine polo player, and he had no sooner arrived at Fort Myer than Patton stopped him as he was crossing the parade ground and said: "I bought this @#%%@$ horse and can't do anything with it. I want you to come ride it on a hunt." Gay blanched, said, "Yessir," and on the first jump the horse broke a leg and had to be destroyed. "I was scared to death." Ready to tell his wife not to bother unpacking their household goods, Gay was shocked when Patton put his arm around his shoulder and said, "Thank God, I will never have to ride that S.O.B. [again]."[39] From that time on Gay served with him, and a bond was forged between the two that lasted until Patton's death.

In Washington in 1942, Gay made his imprint on the Western Task Force staff during the planning for Torch. For example, whenever it became too noisy in the bullpen atmosphere of the Munitions Building, Patton would growl: "Tell 'em to stop that goddam racket. I can't hear myself think," and Gay would come out and announce: "The General wants QUIET." And that would be the end of the noise.[40]

Eisenhower was enthusiastic when he learned that Patton would be part of his new command and wrote to Marshall: "I am delighted you fixed upon him as your choice for leading the American venture."[41] Patton had barely unpacked his bags when he was ordered to England to assist Eisenhower in planning Torch. On August 5, accompanied by several other officers, including his friend Gen. Jimmy Doolittle, Patton boarded a four-engine PanAm Stratoliner for the long flight to England via Newfoundland and Ireland. In London he was billeted at the plush Claridge's Hotel, where he found the women in the dining room "hideous" and the wartime capital drab. He met with Brig. Gen. Lucian Truscott and Eisenhower about Torch. No one liked the plan. "We both feel the operation is bad and is mostly political. However, we are told to do it and intend to succeed or die in the attempt. If the worst we can see occurs, it is an impossible show, but, with a little luck, it can be done at a high price. . . . "[42]

He noted in his new diary that most American officers in London seemed too pro-British. "I am not, repeat not, Pro-British."[43] Despite a growing Anglophobia that eventually reached manic proportions, Patton rated Adm. Lord (Louis) Mountbatten and several of his senior staff officers "damn good fighting men."[44] He was not, however, impressed with the American and British staff officers planning Torch. At this early stage the operation was ambiguous and ill defined, and during two conferences chaired by Eisenhower, Patton thought the U.S. Navy planners were "certainly not on their toes," and exhibited the first signs of displeasure and distrust of his friend, Ike. He found Ike "not as rugged mentally as I thought; he vacillates and is not a realist." When Eisenhower asked what Patton thought of the Torch plan, he replied that it was much too complicated and needed serious revision. Like Patton, Eisenhower was committed to supporting Torch, even though he believed that the operation was "strategically unsound" because it impeded rather than enhanced the opening of a second front in Europe.[45]

Despite his private disdain for the British, Patton managed to impress an upper-class gathering at an intimate dinner party. Accompanied by Truscott, whom he promised a command if Eisenhower would release him, Patton was in rare conversational form and dazzled his audience with tales of Mexico, Pershing, and the great shoot-out at Rubio. "He told them of murderers he had known . . . of the rare California vintage he sampled and enjoyed only to discover, when the great cask was cleaned, the body of a drowned Mexican . . . his audience was duly appreciative, and registered the appropriate degrees of horror, astonishment, and doubt." After they departed, Patton charged Truscott with ensuring the hostess received flowers the following day.[46]

Patton also met with Ike's new deputy commander, Mark Clark, who had hitched his wagon to Eisenhower's star and himself enjoyed an equally meteoric rise, from lieutenant colonel to major general. Patton was wary of Clark, whom "I do not trust . . . yet." He thought that Clark seemed more preoccupied with enhancing his own future than with in winning the war. Patton's distrust of Clark undoubtedly stemmed from pique that an officer very junior to him was now also a two-star general. Before Eisenhower appointed him deputy commander in August 1942, with responsibility for planning Torch, Clark was slated to command the Center Task Force (the U.S. II Corps) in the invasion of Oran. His replacement was Patton's rival, the despised Lloyd Fredendall.

After a round-robin of meetings Patton judged himself "the only true gambler in the whole outfit," and when the plan began to jell in mid-August, Patton was permitted to return to Washington.[47] However, it was late September before he would admit that Torch seemed workable, although still risky. "I feel very calm and contented. It still can be a very desperate venture if the enemy does everything he should and we make a few mistakes. I have a sure feeling we will win."[48] Eisenhower duly reported to Marshall his satisfaction with Patton, and quoted both Clark and Patton as saying: "We unanimously want to assure you that regardless of academic calculations as to [its] eventual success, we have no other thought except that of carrying out this operation through to the utmost of our abilities."[49]

Another officer who reported for duty with Patton was Brig. Gen. Geoffrey Keyes, an ex-cavalryman and longtime friend of "firm character and level head," and one of the few men (besides Gay and Semmes) with whom Patton would share his innermost thoughts. In time other veteran officers were recruited and formed the nucleus of what later became the Seventh and Third Army staffs. He was authorized a second aide and sent for Alexander C. Stiller, a former Tank Corps sergeant and Arizona cowboy, who reported in his World War I helmet, uniform, and leggings. Lt. (later Maj.) Al Stiller was utterly devoted to Patton (perhaps to a fault) and served his master primarily as a bodyguard and general gofer. Capt. Richard Jenson, the son of one of Patton's Southern California childhood sweethearts, remained Patton's primary aide-de-camp. Patton never once said an unkind or harsh word about Jenson, an outstanding young officer whom he came to cherish as deeply as his own son.

After World War II, Omar Bradley and some of the original members of the II Corps staff perpetuated the myth that Patton was a mediocre commander, ill served by a poor staff that more often than not failed to do its job—a view, colored by deep prejudices against Patton, that was neither balanced nor accurate. Although Patton scorned details and left his staff to carry out the policies he announced, he was never dismissive of the staff process. To the contrary, he staffed his headquarters only with men he knew and trusted to do their jobs well. Anyone who failed was replaced. Eventually the key members of the staff consisted of his deputy, Geoffrey Keyes; his chief of staff, Gay; and Hugh Gaffey, who became the deputy when Keyes moved to a corps command after the Sicily campaign. Lt. Col. Paul D. Harkins (a four-star general in Vietnam in the early 1960s) was Gay's assistant chief of staff. Col. Oscar W. Koch, who had known Patton at Fort Riley and had been on his 2d Armored Division staff at Fort Benning, joined Patton after Torch as his extremely competent intelligence officer (G-2). Col. Kent C. Lambert was the operations officer (G-3), and the G-4 was Col. Walter J. Muller, who later earned distinction as perhaps the best staff logistician in the ETO. "Maud" Muller possessed such a keen knack for getting things done that Patton would deliberately refrain from inquiring about the means by which his G-4 did so.[50] As one of Patton's Third Army staff officers would later observe: "You either knew your job, or you didn't. If you didn't, or if you were the cause of any friction in the Headquarters, you were quickly and quietly gotten rid of . . . but if you knew your job you were allowed to perform it in your own way. . . . Results were the only thing that counted."[51]

When the Western Task Force began assembling and training on the East Coast, Patton's disenchantment with the navy and what he regarded as its undue pessimism and negative attitude during the preparations became a prime subject of Washington gossip and nearly led to his relief. It was his first brush with the senior service and he had yet to learn the hard way the perils of interservice rivalry with those schooled in the art of intrigue, and of criticizing sailors in the same dismissive manner as he was accustomed to treating his army contemporaries. The trouble began at one of the first meetings between the soft-spoken Torch naval commander, Adm. H. Kent Hewitt, and Patton. Irked by what he perceived as the navy's less than helpful attitude in planning the landing of his invasion force, Patton lost his temper and let fly with "a torrent of his most Rabelaisian abuse," causing those accompanying Hewitt to flee in "virtual panic, convinced they could never work with a general so crude and rude as Patton."[52]

Hewitt complained directly to Adm. Ernest J. King, the gruff chief of naval operations. Patton and the navy were sparring so frequently that King intervened with Marshall to seek Patton's relief. At Eisenhower's urging, Marshall refused and attempted to soothe the ruffled feathers of King and Hewitt by insisting that "Patton is indispensable to 'Torch,'" and that the very same qualities that made him so disagreeable to work with also made him successful in battle.[53] It was the first of many times when Eisenhower and Marshall would intervene to bail Patton out of trouble and save his career. Marshall later admonished: "Don't scare the Navy, they are plenty scared of you now, for they know you have sailed in more danger and can navigate better than any of them."[54]

The problem with Torch was inexperience and insufficient time to produce little better than haphazard plans and training, and it left Patton nervous and uncertain that even he would survive. Although a joint plan for amphibious operations had existed for some time, the army and navy were simply unused to working with each other, and the consequences ranged from farcical to desperate. Secret rehearsals were held along isolated places on Chesapeake Bay, and during one troops stormed the beaches only to be greeted by an ice-cream vendor anxious to sell his wares to a receptive clientele. In another, when only a single landing craft managed to find the correct beach, the new commander of the 2d Armored Division, Ernest N. Harmon, grumbled: "If they can't find an objective in peaceful Chesapeake Bay, with a lighthouse beacon for help, how are they going to find an objective on a foreign shore and under conditions of war?"[55]

Patton managed to lure Lucian Truscott from England to command a regimental combat team. Truscott had served first on Mountbatten's Combined Operations Staff, had (with William O. Darby) organized the U.S. Army Rangers, and had become one of the few American experts on amphibious warfare. When Truscott warned of the lack of night practice landings, Patton replied hotly: "Dammit, Lucian, I've already had enough trouble getting the Navy to agree to undertake this operation. All I want is to get them to sea and to take us to Africa. Don't you do a damn thing that will upset them in any way." Nevertheless, with the help of Admiral Hewitt, Truscott's regimental combat team was one of the few to practice night landings. "We could not afford to wait," wrote the official U.S. Navy historian. "Everyone was green at the game, and there was no more time for more training and rehearsals. If D-Day, 8 November, could not have been met, winter conditions on the beaches would have forced postponement until next year."[56]

Patton frequently visited Newport News, Virginia, to observe the amphibious training being conducted by the navy for the units of the Western Task Force, which were taught how to load ships, waterproof vehicles, and practice landings on beaches. Ernie Harmon always knew that Patton had visited when units reported that their officers had been ordered into arrest after incurring his wrath, more often than not without a clue as to what they had done wrong. Harmon would always release them from arrest, and Patton would usually awaken him in the middle of the night to inquire if "those bastards" had reported their arrest. When Harmon said yes, Patton's voice usually mellowed, and typically he would reply: "Well, Ernie, release them; they probably did as well as they knew, and we have to manage with them." Harmon thought Patton's quick reaction to his own unfairness "typical of a strange, brilliant, moody, sometimes ill-advised military leader."[57]

Shortly before their training ended, Patton returned to address the assembled fifteen thousand men of his beloved division. Harmon, who fully expected "a long Pattonesque harangue," described the speech:

The general, resplendent in riding breeches and shining boots, took over the microphone and after a few opening sentences was so overcome by emotion that he choked up and tears coursed down his face. The two of us then walked off the stage. The Division stood up spontaneously and cheered and cheered again. There was an electrical quality about him which, without even a formal speech, communicated itself to masses of men. I knew, without completely understanding why, that the officers and men of the Western Task Force would have confidence in their leaders during the coming invasion.[58]

After Patton's departure, Harmon told Oscar Koch, who soon became Patton's G-2, "The damned old fool. All the way over here he was telling me about how he was going to raise hell with the troops and so on, and instead of that he broke down and cried!"[59]

At Fort Bragg, North Carolina, Patton addressed the men of the 47th Regiment of the 9th Division. The commander later recalled:

The General's speech that sunny fall day was a little different from the movie version. For one thing it was bloodier. He suggested ways to make the enemy suffer. And he closed by quoting part of Kipling's 'Recessional.' At the end every man in the combat team cheered, a genuinely spontaneous cheer. And there were cries of, 'More! More!'. . . The general grinned and came down the steps, entered his car, and was gone. The troops were dismissed but the excitement lingered. They were talking and laughing and letting off steam. Never before had they heard a general talk like that. He had made a deep impression.[60]

Most of the sailors and naval officers with amphibious experience were already in the Pacific, and it took a major effort to learn the basics not only of amphibious warfare but of naval gunfire and minesweeping. As one officer commented sarcastically: "Having to use the incompetent to train the ignorant was not a method calculated to produce the best condition of readiness."[61] Nevertheless there was ample resentment of the army, "which did not seem to know what it was doing, [which] untidily cluttered up transport vessels during rehearsals . . . [and whose presence] came to be seen by navy personnel as a form of punishment."[62]

Interservice cooperation was not enhanced by the facts that Patton and his staff were in Washington, while Hewitt's was in Norfolk, and that there were fundamentally different approaches to the problem of landing the Western Task Force safely on the shores of Morocco. The army was understandably anxious to land its troops, equipment, and supplies quickly, while the navy wanted to limit its exposure to hostile gunfire by landing the invasion force and then returning to disembark everything else. Each thought the other ignorant, spoiled, and uncooperative.[63] Gradually, however, Hewitt and Patton came to appreciate each other, and with understanding came increased co-operation.

One of Patton's final acts before leaving for North Africa was a visit to his mentor, eighty-one-year-old Black Jack Pershing, who was in ill health in Walter Reed Army Hospital. "He did not recognize me until I spoke. Then his mind seemed quite clear. He looks very old. It is probably the last time I shall see him but he may outlive me." When Patton thanked him for giving him a chance in Mexico, the old general replied: "I can always pick a fighting man and God knows there are few of them. I am happy they are sending you to the front at once. I like Generals so bold that they are dangerous. I hope they give you a free hand . . . he said that at the start of the war he was hurt because no one consulted him, but was now resigned to [it]. . . . He almost cried. It is pathetic how little he knows of the war."[64]

In a final gesture of respect, Patton dropped to his knees to ask for Pershing's blessing, "which he gave me with great emotion. I kissed his hand; then I put on my cape and gave him the salute. Twenty years dropped from him."[65] Pershing "squeezed my hand and said, 'Goodbye, George, God bless and keep you and give you victory.'"[66]

Patton also attended to the traditional, important, last-minute business of soldiers going off to war, writing to his brother-in-law Frederick Ayer.

> By the time you get it you will have read in the papers where I am and if I still AM. In spite of my faults you have treated me as a real brother and I have felt that way towards you. I do appreciate all you have been and done to and for me. My admiration for you as a man is without limit. . . .
>
> The job I am going on is about as desperate a venture as has been undertaken by a force in the world's history. We will have to meet and defeat superior numbers on a coast where one can only land 60% of the time. So my proverbial luck will have to be working all out. However, I have a convinced belief I will succeed. If I don't I shall not survive . . . there is the off chance that political interests may help and we shall have, at least initially, a pushover. Personally I would rather have to fight—it would be good practice. However in any event we shall have to fight and fight hard and probably for years. . . . So far as B. and the children are concerned, I know that under your supervision they could not be better

off. I am enclosing a sealed letter to B. which you are only to give her when and if I am definitely reported dead. I expect you to keep it a long time. . . . This all sounds very gloomy, but is not really so bad. All my life I have wanted to lead a lot of men in a desperate battle; I am going to do it; and at fifty-six one can go with equanimity—there is not much one has not done. Thanks to you and B. I have had an exceptionally happy life. "Death is as light as a feather; reputation for valor is as heavy as a mountain."

Very affectionately,
/s/ G.S. Patton, Jr.[67]

To his beloved nanny, Mary Scally, Patton wrote that by the time she read his letter, "I will either be dead or not. If I am, please put on a good Irish wake; if I am not, get busy with the Pope."[68]

On October 21 Patton and Hewitt were summoned to the White House, where they were effusively greeted by Roosevelt, who said: "Come in Skipper and Old Cavalryman and give me the good news." The president later wrote: "I asked him whether he had his old Cavalry saddle to mount on the turret of a tank and if he went into action on the side with his saber drawn . . . Patton is a joy."[69]

In his diary Patton claimed credit for having instigated the meeting, "with the hope that he [FDR] would put some heat on Hewitt about the necessity of [the] landing. As nothing came of it, I said, 'The Admiral and I feel that we must get ashore regardless of cost, as the fate of the war hinges on our success." Roosevelt replied, "Certainly, you must," and wished them both godspeed.[70] According to another account of the meeting, a highly emotional Patton said: "Sir, all I want to tell you is this—I will leave the beaches either a conqueror or a corpse," a remark he also repeated during his final courtesy call on Marshall.[71]

Patton's departure from Washington left most of his civilian office staff in tears. And as Beatrice prepared to bid Georgie farewell at nearby Bolling Field, she was given a reprieve in the form of an invitation from an Air Corps general to accompany her husband to Norfolk, grateful that George did not remark on the redness of her eyes. "The waiting has been terrible," she later wrote. "He went to Mexico on one and a half hour's notice, and on five days' notice to France in 1917. But after all, eleven weeks is a short time to organize the greatest expeditionary force ever to sail at once from our U.S." At Norfolk Patton was so excited that he jumped out of the plane, only to be called back by Beatrice for their final good-bye. "I was glad, for the great strain was over for him at last. Proud to tell, for once I did not cry."[72]

Patton's final note to Beatrice before sailing must have been wrenching to read: "It will probably be some time before you get a letter from me but I will be thinking of you and loving you." For the third time in their lives war would force them into a painful period of separation and uncertainty. Whatever the troubles of their marriage, however, Beatrice's love for her husband and pride in his accomplishments were boundless: "What a man. He is very great."[73]

After observing the 3d Division embarkation, Patton reflected: "If everyone does his part these seemingly impossible tasks will get done. . . . All of us are relieved that the strain is over and the staff and all of the men I have seen are cheerful and full of confidence. . . . We have all done our full duty and will succeed." Of his own role in Torch, Patton wrote: "When I think of the greatness of my job . . . I am amazed, but on reflection, who is as good as I am? I know of no one."

On the eve of their sailing, Patton attended a conference of 150 army and navy officers, chaired by Admiral Hewitt, to review the invasion plan. For most this was the first time they were informed where they were bound. "At this shore conference, Admiral Hewitt gave a calm and reasoned statement of Operation 'Torch' and its purpose," and the transport commander noted that the navy's mission was "to serve the troops—to die for them if necessary." Patton, who remarked in his diary that his predecessor, a navy captain, had spoken "for three hours and said nothing," arose and "talked blood and guts for five minutes and got an ovation."[74] During his fiery, blunt speech Patton

> exhorted the Navy to remember Farragut, but predicted that all our elaborate landing plans would break down in the first five minutes, after which the Army would take over and win through. "Never in history," said he, "has the Navy landed an army at the planned time and place. If you land us anywhere within fifty miles of Fedhala and within one week of D-day, I'll go ahead and win. . . . We shall attack for sixty days, and then, if we have to, for sixty more. If we go forward with desperation, if we go forward with utmost speed and fight, these people cannot stand against us."[75]

Afterward Patton mingled with the navy officers and delivered this message: "Don't worry about it. We'll [make] do with what you can give us."[76]

Patton also had tough words for his invasion commanders, whom he summoned on October 14 for a final good-bye and pep talk. He told them, "If you don't succeed, I don't want to see you alive." After his death, Beatrice recalled, "These men were all friends of his and when he came home to lunch alone with me, he was more shaken than I could believe. He had a strong drink, a glass of beer and two cups of coffee, and then told me about it. He had asked General Marshall in to shake hands with them, but the General 'had to go to the Senate' and excused himself. I think G[eorgie] was very much hurt."[77]

Patton also sent each commander an identical letter noting that he was certain that their unit would cover itself in glory: "You must conquer! The success of the whole operation depends on you. . . . Do not attack without cause. If you must attack, spare no man's life. . . . I am looking to you and your officers . . . [for] fearless leadership." He expressed an intention to see each ashore, not to "usurp your deeds, but in order to share with you the dangers of our soldiers," and concluded: "I see no point in surviving defeat, and I am sure that if all of you enter into battle with equal resolution, we shall conquer, and live long, and gain more glory."[78]

Patton's reputation is built primarily on his success as a commander on the battlefield, but training troops was the foundation of those accomplishments. In World War I he had done it all, and as a division and later an armored corps commander, his legacy from Fort Benning and Indio was that of a superb organizer and trainer of troops. The Desert Training Center was Patton's final opportunity to refine tactics, maneuvers, and the employment of large formations, and he made the most of it, relentlessly emphasizing fundamentals and skills of survival: reconnaissance, tank, artillery, and small arms gunnery, radio and telephone communications, and the employment of air support and logistics.

By the time he went to war Patton had read and absorbed virtually every book in English of any significance about mobile warfare, including a 1941 War Department translation of an article in the *Frankfurter Zeitung*, which recounted that German tanks operated as much as one hundred kilometers ahead of the advancing armies. When Patton "went off to war he was America's most effective advocate of a daring armor doctrine."[79]

Now he would put those organizational and training skills to the ultimate test on the battlefields of North Africa. On October 24, 1942, Admiral Hewitt's flagship *Augusta* and the ships of the Western Task Force (Naval Task Force 34) slipped quietly from Norfolk into the open sea and formed into convoys before running the U-boat gauntlet across the South Atlantic. It was a solemn moment, as Americans left the shores of the United States to participate in their second world war of the twentieth century. Truscott recalls, "It was borne in upon me with an awesome finality that, for better or worse, the die was cast."[80] An equally somber Patton reflected in his diary: "This is my last night in America. It may be for years and it may be forever. God grant that I do my full duty to my men and myself."[81]

PART VIII

The War in
the Mediterranean

Casablanca to Messina
(1942–1943)

Hold 'em by the nose and kick 'em in the ass!

—PATTON

CHAPTER 29

The "Torch" Landings

The God of fair beginnings has prospered here my
hand.

—PATTON

Torch was a complex and difficult first test of the United States Army in
World War II, whose logistics aspects alone were a nightmare. With three
separate convoys of inadequately trained soldiers and sailors converging on
North Africa, there were endless opportunities for snafus. No one had even
the vaguest notion of the level of resistance of the Vichy French, whom no
one wanted to fight and risk turning into a German ally. The gambit the
Allies hoped to pull off was to invade North Africa with such swiftness that
the French would not be encouraged to offer resistance and instead would
quickly join their alliance against Germany. As Churchill had cabled Roosevelt: THE FIRST VICTORY WE HAVE TO WIN IS TO AVOID A BATTLE.[1]

There were to be three simultaneous landings on November 8, 1942:
task forces sailing from Great Britain were to land at Algiers (Eastern), and
Oran (Center), and Patton's (Western), in Morocco. Patton's objective was
to capture Casablanca, which was to become the principal railhead for the
resupply of the army in North Africa. In the main landing 19,000 troops of
the U.S. 3d Division were to assault the small fishing port of Fedala, about
fifteen miles north of Casablanca, and another 9,000 under Truscott north of
Rabat were to capture the important airfield at Port Lyautey. Finally, 6,500
troops under Harmon were to land at Safi, some one-hundred forty miles
south of Casablanca, to protect the task force against a counterattack from
Marrakech.[2]

Patton faced two major problems: the unpredictable weather in Novem-

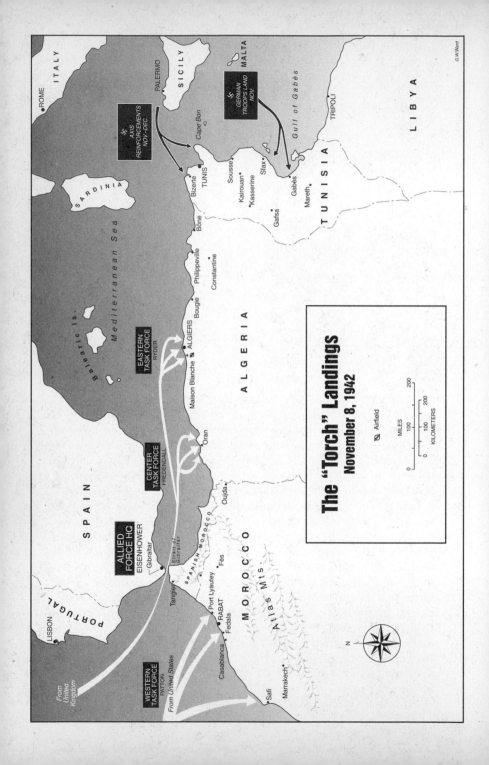

ber, which made landings from the Atlantic Ocean dangerous, and intelligence so ambiguous that he could not determine whether he would be greeted with open arms or by all-out resistance. Unaware that Mark Clark had brokered a last-minute but nevertheless shaky alliance with the French in Algiers, Patton was edgy over the prospect of fighting the French in Morocco with a decidedly inferior invasion force. He had written to Eisenhower that if his own G-2's intelligence estimate proved correct, "we will have quite a fight." Eisenhower himself only rated Torch's odds of success at a mere 50–50.[3]

During the voyage of Task Force 34 Patton prowled the *Augusta*, frustrated by his enforced inactivity. "I keep feeling that I should be doing something but there is nothing to do," he lamented in his new diary. A marine orderly followed him wherever he went, leaving Patton feeling "like a prisoner." However, the time was put to good use in cementing relations with Hewitt, of whom he now wrote, "I like him better all the time." To a friend he wrote: "I am delighted at the wholehearted spirit of cooperation they have evinced." Patton also noted that, "If this is my last letter—which I doubt—rest assured that for a would-be corpse, I feel very well."[4]

During the two-week voyage across the Atlantic the one-hundred-ship task force soon settled into a routine of frequent zigzag maneuvers, abandon-ship drills, signal drills, and practice alerts. Task Force 34 was protected from marauding U-boats by continuous air cover from escort carriers, some forty destroyers and cruisers, and three battleships, all of which deeply impressed Patton.[5] At one point the *Augusta* suddenly left the formation and made a complete circle of the convoy. On a nearby ship, Harmon joked: "There goes Georgie, getting in a little horse exercise."[6]

To fill the long hours and gain a perspective for his forthcoming role in Morocco, he read the Koran, although it "bored the hell out of him."[7] To kill time he and his staff shot carbines off the fantail. Patton also lectured endlessly. "I have been giving everyone a simplified directive of war. Use steamroller strategy. . . . Attack weakness. Hold them by the nose and kick them in the arse."[8] He sent a written exhortation to all officers of the Western Task Force that they were to "impress your men that fire wins battles . . . we are better in all respects than our enemies, but to win, the men must know this. It must be their absolute belief. We must have a superiority complex. . . . Always attack. Never surrender."[9]

Although he disdained any notion of fear, the days of idleness played on Patton's emotions. "Every once in a while the tremendous responsibility of this job lands on me like a ton of bricks, but mostly I am not in the least worried. I can't decide logically if I am a man of destiny or a lucky fool, but I think I am destined. . . . I feel my claim to greatness hangs on an ability to inspire. Perhaps when Napoleon said 'Je m'engage et puis je vois,' he was right. It is the only thing I can do in this case. . . . I have no personal fear of death or failure. This may sound like junk or prophecy, within a week."[10] Shortly before D day he ordered a proclamation read to his troops that will forever be identified with him:

> When the great day of battle comes, remember your training . . .
> you must succeed, for to retreat is as cowardly as it is fatal.
> Americans do not surrender. During the first days and nights
> ashore you must work unceasingly, regardless of sleep,
> regardless of food. A pint of sweat will save a gallon of blood.
> The eyes of the world are watching us. The heart of America
> beats for us. God is with us. On our victory depends the
> freedom or slavery of the human race. We shall surely win.[11]

When Patton was handed a propaganda leaflet in French, prepared by the Office of War Information, which was to have been air dropped on various Moroccan cities precisely at H hour, he quickly noted that accents were missing from several words. Exploding in anger at his hapless G-2, who had brought him the offending document for approval, Patton turned the air blue with expletives, refused to permit their distribution, and demanded each be corrected by hand. "Do you expect me to land on French soil introduced by such illiterate calling cards—Goddamnit?"[12] A hastily assembled team that included his personal secretary, Joseph D. Rosevich, was still correcting the leaflets at H-hour. "Oh, hell! It's just like the Old Man—a goddamn perfectionist!" he complained.[13]

For diversion he read a detective novel and still could not decide whether to worry or be confident. At 10:30 P.M. he went to bed fully dressed and slept until 2:00 A.M., when he went on deck to observe lights blazing ashore in Fedala and Casablanca. After a period of bad weather, the sea was dead calm on the eve of the invasion. "God is with us."

Several hours earlier, the Allies had broadcast a radio message to the French from Roosevelt, noting that they came as friends, but it had no effect at Casablanca when at 6:50 A.M. the invasion fleet began exchanging fire with French shore batteries and the uncompleted battleship *Jean Bart*, which lay in Casablanca harbor.[14] Shortly after 8:00 A.M. seven French destroyers steamed out of Casablanca harbor in an attempt to interfere with the Fedala landings by shelling the beaches and the helpless U.S. transports offshore. Around 10:00, just before Patton's Higgins landing craft was to have been lowered into the sea for his trip ashore, a French light cruiser and several destroyers launched the most serious attack yet.[15] The *Augusta*'s rear turret fired its three guns, and the blast blew the landing craft "to hell and we lost all our things except my pistols," temporarily delaying his journey ashore. The blast also blew out Patton's tactical radios and severed all communications with Eisenhower in Gibraltar. One of the shells fired by the French cruiser contained a deep yellow dye (to enable visual spotting by gunners). "I was on main deck just back of number two turret leaning on the rail when one hit so close that it splashed water all over me," liberally covering his leather jacket with yellow slime. Stiller rushed to clean Patton's jacket but was soundly rebuffed. As a naval officer who witnessed the incident later wrote, Patton said: "Leave it there. This will stay on the #$%&?# jacket as long as I am able to wear it."[16]

Unruffled, Patton stood on deck as an enemy warship again took aim at the *Augusta*. "When I was on the bridge later, one hit closer but I was too high to get wet." Patton remained aboard the *Augusta* for lunch ("naval war is nice and comfortable") as the shelling kept on. At noon a delayed message from Harmon reported the capture of Safi nearly seven hours earlier. At 1:20 P.M. a landing craft deposited Patton and his staff in the surf off Fedala, where everyone got soaked wading ashore. Initial attempts to persuade the French garrison to surrender had failed, but the 3d Division had secured the waterfront area, and that night Patton slept in a blacked-out hotel in Fedala, where he wrote in his diary that "God has been very good to me today."

At Safi, Harmon's force landed successfully with only minor difficulty, but at Mehdia and Fedala, Murphy's law applied. The Mehdia landings were bungled by the navy and were hours late and on the wrong beaches. A near disaster was averted only by Truscott's outstanding leadership and personal intervention, while at Fedala the beaches were chaotic and cluttered with landing craft, supplies, and mobs of curious Arabs whose presence only exacerbated the problems facing the invaders. In both places French naval gunfire and Vichy ground troops disrupted the landings. D day was perhaps best typified when the first American bazooka fired in combat hit a nearby tree instead of a French tank.[17]

The French resident general, August Noguès, was not inclined to stick his neck out by surrendering French Morocco to the Allies without orders from the Vichy leader, Marshal Pétain. The most sympathetic and anti-Vichy French officer was Gen. Emile Béthouart, who would have welcomed the Americans, but his preinvasion coup to take control of the French military failed, resulting in his arrest. Later it took Patton's personal intervention to save Béthouart from imprisonment and possible execution for treason.

The Allies had enlisted the aid of a senior French officer, Gen. Henri Giraud, who was proclaimed the French commander in chief, but few, including those in Casablanca, accepted his authority. Thus, while the drama played out between Béthouart and Noguès, the landings began and the French fought back furiously.

Torch was unlike any other battle Patton would ever fight. It was not a "hold them by the nose and kick 'em in the ass" type of battle, but rather an unconventional fight against "friends." The battle for Casablanca was really a series of skirmishes at three widely scattered locations, over which Patton had virtually no control or influence, except in a minor way at Fedala. To make matters worse, with his own radio link out, communications between the *Augusta* and AFHQ were so poor that it became necessary to employ one of Hewitt's speedy sea craft to act as a courier to Gibraltar. Patton was left to make decisions about the French largely on his own.

The seventy-four-hour battle for Casablanca was hardly Patton's finest hour. It had been decided not to invade Rabat in part because of faulty intelligence and fatuous political advice that it would be wise to avoid attacking the political and spiritual capital of Morocco. (Only after the fact did it become evident that the French would likely have surrendered at once had Rabat been invaded.) Instead, they elected to stand and fight for Casablanca, Fedala, and Mehdia.[18]

During the preliminary Torch planning, Patton's original directive was to invade Oran, but in early September it was changed to Casablanca, which left an impossibly short sixty-four days to train for the landings. The decision to surround Casablanca in the hope the French would immediately surrender left the Western Task Force to fight the most difficult of the Torch battles.[19]

Patton's political adviser was an ex-naval officer named Paul Frederick Gilbert, who foisted his personal views on Patton "in the guise of political advice." Both Gilbert and Patton's intelligence officer advised bypassing Rabat on religious grounds, a recommendation Patton accepted without question. Respecting the notion of avoiding confrontations with Morocco's Muslim population was a concept his historically trained mind could accept. Preinvasion intelligence was so dreadful that Patton's G-2 staff gleaned most of what they knew of Morocco from travel guides.[20]

Before dawn on November 9, Patton was prowling the beaches, which were still in such a state of utter chaos that landing craft were taking half an hour to unload and drift free of the sand. He jumped into the water and shouted at some nearby soldiers: "Come back here! Yes, I mean *you*! *All* of you! Drop that stuff and come back here. Faster than that, goddam it. On the double!" Waist deep in the water and with his shoulder against the bow of the landing craft, Patton showed them what he wanted done. He yelled for them to await the next wave: "Lift and push. Now! *Push*, goddammit, *push*! . . . Don't you realize that boat has other trips to make? How do you expect to fight a war without ammunition? Now go and take that stuff up to the dump. On the double."[21]

Those who had served under Patton were not surprised to see their commanding general in the surf; those who hadn't soon learned he was likely to turn up expectedly at any time. Cursing and prodding, Patton helped to restore a semblance of order when a boat capsized in the turbulent surf, drowning sixteen soldiers. Only three bodies ("a nasty blue color") were recovered, one of them by Patton himself. When a frightened soldier threw himself on the sand and began sobbing, Patton "kicked him in the fanny with all my might and he jumped right up and went to work. Some way to boost morale. As a whole the men were poor, the officers worse. I saw one Lieutenant let his men hesitate to jump into the water. I gave him hell. I hit another man who was too lazy to push a boat."[22] By late morning Patton was dehydrated and had been without food since the previous night, and on the afternoon of November 9 he returned to the *Augusta*, where he remained that night, mirthfully joking to his hosts "that the 'goddam Navy' had demolished the kitchen of the house he had selected for his headquarters."[23]

Overall Patton spent some eighteen hours on the beaches outside Casablanca, later reflecting: "People say that Army commanders should not indulge in such practices. My theory is that an Army commander does what is necessary to accomplish his mission and that nearly 80 per cent of his mission is to arouse morale in his men." He felt he had genuinely made a difference, even though the real hero on the Fedala beaches was the navy beachmaster, Cmdr. J. W. "Red" Jamison, whom Patton later praised for having "saved the whole goddam operation."[24]

By November 10 an increasingly restless Eisenhower cabled Patton, "The only tough nut left is in your hands. Crack it open quickly." To his chief of staff, Bedell Smith, he signaled, "If he [Patton] captures Casablanca by noon tomorrow, I will recommend both him and Fredendall for third stars."[25] Patton, however, had faithfully kept Eisenhower informed of his activities by means of lengthy letters. "I regret that you are mad with me over my failure to communicate; however, I cannot control interstellar space and our radio simply would not work. . . . [However] the only person that lost by it was myself, as by my failure to communicate with you, the press was probably unable to recount my heroic deeds." On a more serious note, he reminded his new boss, "I feel that in successfully accomplishing the job you handed me, this force achieved the impossible. . . . I am forced to believe that either my proverbial luck or more probably the direct intervention of the Lord was responsible."[26]

When Ernie Harmon reported to Patton in Casablanca, he was greeted gruffly: "Where in hell have you been all this time?" The tough, outspoken Harmon had learned to be wary of Patton's unpredictability, and later observed: "I thought we had done a remarkable job at Safi . . . perhaps the small boy still under my skin expected a slap on the shoulder, a gesture which meant 'well done.' Then I saw the slightly sardonic twinkle in his eye and the smile and realized once again that this was Patton's way."[27] His generals may not have received many laurels, but wherever Patton observed outstanding leadership or praiseworthy acts, he immediately cited them. After learning that an infantry platoon leader had held a lighthouse against an entire French battalion, Patton said: "Lieutenant, what is your rank?" When he replied: "Second Lieutenant, Sir," Patton responded: "You are a liar, Sir, you are now a First Lieutenant"—and promoted the officer on the spot.[28]

By the night of November 10 Noguès concluded that further resistance was pointless and he sent word to Patton that he was willing to negotiate a surrender. Patton ignored Noguès, arguing that it was a bad idea to change plans at the last minute, and instead ordered an unequivocal reply: cease hostilities at once or endure an all-out assault on Casablanca the following morning. He did, however, warn Hewitt to be ready to call off the naval guns, and decided to postpone the attack until after daylight.

The gambit worked when the French abruptly surrendered, and the next afternoon Gay stood on the steps of the Hotel Miramar in Fedala with an honor guard to greet the black limousine, flanked by motorcycle outriders, bearing Noguès and the naval commander, Adm. François Michelier. To avert further bloodshed, Patton had wisely ordered red-carpet treatment for the French in the belief that there was "no use kicking a man when he is down, particularly a potential ally."

Patton opened the meeting by congratulating the French for their gallantry, which seemed to partially mollify Michelier, who had lost most of his fleet. Then the stiff terms of surrender outlined in the draft document (one which Patton had been ordered to use if the French resisted) were read by Col. William H. Wilbur of Patton's staff, (who won a Medal of Honor at Casablanca for risking his life in an attempt to avoid bloodshed and arrange a cease-fire). The air suddenly turned frigid, and, proclaiming that Patton's terms would oblige the Americans to maintain order in Morocco without French assistance, Noguès clearly suggested that the United States would regret such an act. As the official historian writes: "No draft terms took adequate cognizance of the Allied dependence on continued French capacity to control the Sultan's native subjects."[29]

Patton quickly sensed the impossibility of attempting to govern seven million Arabs and Berbers while carrying out his military mission in French Morocco with only his tiny force. As an eyewitness, Patton's future aide Charles Codman, described it: "Communications with General Eisenhower in Gibraltar and with General Clark in Algiers are nil. General Patton is on his own. It did not take him long to decide. Rising to his full height, he picked up the familiar typescript of Treaty C and tore it into small strips." Addressing the French as officers and gentlemen with whom he had served honorably and studied under at Saumur, Patton noted his complete faith in their word as French officers. "'If each of you . . . gives me his word of honor that there will be no further firing on American troops and ships, you may retain your arms and carry on as before—but under my orders. . . . Agreed?' It was."[30] The war for Morocco was officially over.

Overall the benefits of trusting the French outweighed the risk, despite decidedly anti-Vichy American sentiment. Nevertheless, Patton's decision was inherently risky and would have backfired badly if Noguès elected to double-cross him: "French cooperation in protecting the hundreds of miles of railroad to Oran was essential. . . . Internal trouble in Morocco might have spelled defeat in Tunisia."[31]

Before the French delegation departed, Patton held up his hand and said, "Gentlemen, we have now settled everything, but there is one disagreeable formality which we should go through." Worried looks were quickly replaced by smiles as champagne bottles were opened and Patton offered a toast "to the happy termination of a fratricidal strife and to the resumption of the age-old friendship between France and America." (He subsequently wrote of this occasion: "They drank $40.00 worth of champagne, but it was worth it.")[32]

"Again the hand of God," wrote an elated Patton. "I said I would take Casa by D plus 3 and I did. A nice birthday present." He was fifty-seven years old and ended a memorable birthday with these words: "The God of fair beginnings has prospered here my hand. . . . To God be the praise."[33]

The battle for Casablanca had been won at a cost of 337 killed, 637 wounded, 122 missing, and 71 captured.[34] Near the scene of the fiercest battles, tears welled in Patton's eyes at a moving ceremony for the American dead of Torch. Patton also attended a ceremony for the French dead. "Now we were Allies united and fighting for the freedom of man," wrote Harry Semmes.[35]

After Stimson telephoned Beatrice on November 8 that the Torch landings had been successful, she wrote to him immediately:

> Darling Georgie: I realize that there are months and perhaps years of waiting and anxiety ahead of me, yet today all I can think of is your triumphs and the thought that rings through my mind like a peal of bells— that the first jump is taken. . . . God with us. . . . I wish I were with you, fighting in your army. I would be a better soldier than here at home; but there can be no separation between us except by amputation. . . . What hurts you hurts me too, and I share in your triumph, even thousands of miles away.[36]

Beatrice became a minor celebrity and was swamped with fan mail about her now famous husband, whose picture was splashed over the front pages of newspapers across America. Even son George began receiving a large volume of mail at West Point. When John Waters was decorated for bravery, Beatrice wrote: "What a family." For the remainder of the war, she worked tirelessly to support the war effort by speaking at war bond rallies,

doing radio interviews—anything that would enhance her husband's growing reputation. When she visited some wounded men from Patton's command at Walter Reed Army Hospital, one soldier said he had lain down after
being fired on and was hit. Beatrice unsympathetically remarked: "Well,
look what it did for you. Maybe if you'd stood up and kept going ahead, you
wouldn't have got hit."[37]

Beatrice had no interest in remaining at Green Meadows for the duration of the war and moved in with her sister, Kay, in Washington, where she
would be near those, like the Stimsons, who were at the center of the war
effort. Patton's rapid rise from obscure colonel to invasion commander
seemed to coincide with a rejuvenation of their relationship, and was clearly
reflected in her wartime letters. "Actually, I am so thrilled over your success
that I just don't worry or do anything but shine in your reflected light. I feel,
as you do, that all your life has pointed to this and that you still have many
big things ahead, and that God is with you and guiding your every
move. . . . I feel sure that you are marked by destiny and that I am willing to
wait on God for that." Their thoughts had become so closely harmonized
that one morning Beatrice awoke speaking Arabic, a language she knew a
smattering of but had not spoken since she was twelve years old.

The press had begun to use the term "Old Blood and Guts," moving
several of his World War I tankers to express their disgust in letters to the
editor. One wrote: "As one who . . . has seen the tears stream down his
cheeks when the boys died, [I] say that our boys in Africa are under the best
and most humane military leader I have ever known."[38]

To Beatrice's consternation, Patton would occasionally lapse into his
habit of reflecting upon his own mortality. When a friend was killed with a
shot through the heart, he mused: "I think a nice clean death like Doyle's
would be the easiest way out. Dont worry because if anything happens I will
be dead and you will have been notified long before you get this."[39] Old
friend Walter Dillingham wrote to recall that a general had once said: "You
were a difficult officer in time of peace, but a 'hell of a good soldier' in time
of war. If you are spared I am sure that the enemy will consider this an
understatement."[40]

Although there was considerable dissatisfaction with the navy's performance, Patton responded with gratitude that they had managed to get him
ashore and that their defects were based solely on lack of time to train. Patton sent Hewitt a dispatch that was transmitted to every ship in the fleet and
read: "It is my firm opinion that the great success attending the hazardous
operations carried out on sea and on land by the Western Task Force could
only have been possible through the intervention of Divine Providence manifested in many ways. Therefore, I shall be pleased if . . . our grateful thanks
be expressed today in appropriate religious services."[41] Some may have
interpreted Patton's message to mean that it was thanks to God rather than

the U.S. Navy that his force had gotten safely ashore. However, anyone who knew of Patton's deep reverence for a higher power would not have mistaken his meaning. He genuinely believed the success of Torch had been the will of God and during the battle for Casablanca conveyed his thanks virtually every day in his diary, and in letters and dispatches.

With the brief war in Morocco at an end, Patton found himself a rear-area soldier with no battles to fight, a point he made to everyone with whom he had contact. He sent Stimson a full and colorful report of the events, the country, and the people, and heaped lavish praise on the men of the Western Task Force and the navy, regretting only that "personally I did not have an opportunity of engaging in close combat. Had the fight lasted a few hours longer, I believe I would have achieved my ambition. . . . "[42]

One of Patton's most important functions during Torch was that of diplomat, a role most soldiers are not trained to undertake. Once military operations against the French ended, it was Patton's task to ensure that not only the French but the Moroccan leaders cooperated with the Allies. From invasion commander, Patton now became a proconsul. As the senior American in Morocco, he *was* the United States.

Patton excelled at diplomacy and was an excellent representative of the United States. He spoke French, knew the protocol expected of a diplomat and proconsul, and flawlessly carried out the niceties of polite exchange that he had learned to perfection from a lifetime of practice and study. The profane swashbuckler was replaced by the suave diplomat, a transition he accomplished with consummate ease. As biographer Blumenson notes, few comprehend Patton as a diplomat, but in Morocco he was in his element, with a political mission to "bring French Morocco under Allied control without disturbing the French control over a restless native population," which required "some rather nice manipulation of French and Moroccan sensibilities."[43]

Before Casablanca fell Patton wrote to the sultan of Morocco to express assurance that he and his men had not come as conquerors but as friends, at the behest of the president of the United States, who, "on his word of Honor" gave assurance that when their common Nazi enemy was vanquished, he and his troops would leave Morocco forever. In a respectful tone, backed up by a velvet-gloved fist, he implored the sultan to use his powers to help end "this fratricidal strife," noting that if the French continued to fight, "it is my military duty and purpose to attack by air, by sea, and by land, with the utmost violence known to modern war."[44]

Patton demonstrated his understanding of his mission during the brief but difficult negotiations with the French commanders on November 11. His independent decision to scrap the formal guidelines laid down by the State Department and act on his own had not only been accepted with alacrity by the French, but was never criticized within the Allied high command.

Patton's formal call on the sultan took place on November 16 and was to have included an escort to the palace by the 2d Armored Division. However, at the last minute he elected not to embarrass the French, whose escort troops looked paltry by comparison. "It would just rub it in on the French" and "would appear boastful on my part, so I dismissed them." Patton inspected a French honor guard, a battalion of Moroccan cavalry, resplendent in gleaming uniforms, and a bodyguard of the governor-general in white uniforms with red Moroccan leather trappings. As he passed through the ranks of the soldiers standing resplendent in his honor, Patton thought "how pathetic to think that one of the light tanks in the escort could easily have destroyed all of the splendid creatures standing at salute."[45]

He and Noguès then proceeded to the royal palace, where, inside the outer square of the palace grounds, some four hundred rifle-bearing black troops dressed in red coats and white gaiters were drawn up at attention. After dismounting, another ceremony took place in Patton's honor, as the music of drums, cymbals, and horns "played with great abandon." They passed through a second gate into a large courtyard "completely encircled by men dressed in white Biblical costumes." Inside the palace "we entered a long room . . . and on the right side were a large number of gold chairs . . . the floor was covered with the thickest and most beautiful rugs I have ever seen. At the end of the room, on a raised platform sat the Sultan, who is a very handsome young man, extremely fragile, with a highly sensitive face."

With his love of pomp and ceremony, Patton was treated to a unique event at which he was the centerpiece. During an elaborate exchange of courtesies in Arabic and French, Noguès rambled on for some ten minutes before it occurred to Patton "that the U.S. was getting scooped. So when Noguès sat down, I stepped into the middle of the floor [and] without asking anyone's permission began an extemporaneous speech," in which he presented the compliments of the United States, assuring the sultan that "with the help of God," they would "achieve certain victory against our common enemy, the Nazi." Evoking the name of "our great President, General Washington," Patton noted that "the friendship of America and France dates from the same period."[46]

It was an amazing performance that had a profound effect on both French and Arabs, who were immensely pleased by his forceful but diplomatic words. The sultan replied graciously that he was honored to accept Patton's invitation to visit his command. "He also said that my gracious presence would have a profound and salutary influence on the entire Moslem world, of which he was the spiritual head."[47]

In the coming months Patton would frequently entertain the sultan (whom he referred to privately as Sa Majesté) and escort him on inspection trips, along with other VIPs in the Moroccan hierarchy. He seemed genuinely relaxed and comfortable in the presence of royalty. "Perhaps I should have been a statesman," he wrote Beatrice several days later.

Hunting wild animals in the Atlas mountains, he passed through cedar forests of enormous trees twelve feet thick and one hundred or more feet high rarely seen by Westerners. During a wild pig hunt he impressed his hosts with his prowess by killing one running pig with a single shot through the brain at ninety yards and another in the heart at fifty yards, and a jackal with buckshot. Afterward he joined in a noisy feast of sheep, chicken, and stew. Aware of Arab tradition, Patton carefully observed the custom of eating with only three fingers of his right hand. He was regally received by the pasha of Marrakech, and immediately afterward wrote a detailed historical treatise on this little-known part of Africa, in which he described such subjects as the architecture, and the street wares, habits, and costumes of the people.

Few Europeans or Americans had ever seen such spectacles of grandeur, pageantry, pomp, and precision as the tributes accorded Patton. He was feted, saluted, cheered, chanted at, danced for, showered with gifts, and treated as royalty.[48] In Rabat, at a splendid, joint Franco-American parade one hundred thousand people cheered him and shouted "*Vive l'Amérique!!*" He was in his element; the child who had listened hour upon hour to tales of ancient warriors, of pomp and circumstance, and of strange exotic lands was now living the dream of a lifetime as a benign conqueror.

In mid-November Patton flew to Gibraltar and met with Eisenhower inside "the Rock," a place he found depressing, in no small part because of the disdainful attitude of many British staff officers on the AFHQ staff. He thought too many of Ike's words were British ("He speaks of lunch as 'Tiffin' and of gasoline as 'petrol.'") "I was disappointed in him. He talked of trivial things. We wasted a lot of time at lunch with the governor of the Rock, an old fart in shorts with skinny legs. I truly fear that London has conquered Abilene."[49] Eisenhower's reaction to such visits with his longtime friend were equally mixed: Although pleased to see him, and able to let his hair down about private matters he rarely spoke of with others, they nearly always meant a night of nonstop conversation.

Patton despised the inaction, and his diary began to fill with his foreboding at again being left behind as many of his troops were gradually being withdrawn from Morocco and sent to join the fray in Tunisia. Nevertheless he insisted on tough training and conducted maneuvers at least once a week, during which his troops went sleepless for a twenty-four-hour stretch. Still, Eisenhower had made no decision over the future of his force. For the time being his corps trained and organized itself for future combat. "I hate this organizing," he complained in his diary. "Keyes can do it better than I can. I am a fighter."[50]

A week later Patton flew in the nose of a B-25 bomber to Oran to visit Fredendall, whom he found very gloomy:

> He fears that he and I will hold the bag while our troops in small bunches are shipped to the British. I fear that the British have again pulled our leg. . . . This was *not* the initial plan. I seem to be the only one beating my wings against the age of inaction. The others simply say how much better off we are than the people at home. I don't want to be better off—I want to be Top Dog and only battle can give me that. Bea sent me a lot of stars but I fear that I shall have no occasion to put them on. . . . The waiting is hard. Perhaps I am being made perfect through suffering, for I do suffer when I cannot move. . . . I am pretty low today.[51]

His enemies dismissed Patton's complaints as the words of someone more interested in keeping American troops together than in fighting the Germans, yet his views were shared by most other senior American officers who believed it was a serious mistake to piecemeal American units. When word of Patton's concern reached Eisenhower's ears, however, "he dismissed it with a scoff."[52]

When Patton inspected two regiments of the 3d Division marching toward Rabat, he learned of two soldiers who had done a fine job during the landings, and ordered his car to return and offer them a ride. A drunken soldier sassed Patton, who pointed his whip at the man and told him to take his hand out of his pocket. The GI complained he had been hit and offered to fight his commanding general, until quieted by Stiller. Not long before such conduct would have earned a trip to the brig; now Patton merely shrugged and left others to deal with the matter.

Patton's driving ambition suffered a major setback when his rival, Mark Clark, was rewarded by promotion to lieutenant general after his secret mission by submarine to negotiate with the French prior to the Torch landings. Even more discouraging was the announcement that Clark was to command the newly forming Fifth U.S. Army, while, "I simply have a Corps. . . . I felt so awful that I could not sleep for a while, but I shall pass them yet."[53]

Inevitably Patton began to draw comparisons between Pershing, the supreme commander of the AEF, and Eisenhower, the supreme commander of Torch, and he found his friend sorely wanting. While he understood the need to coexist with the British, Patton thought Eisenhower had gone too far, a belief that Clark and Bradley eventually came to share. He found Algiers a snakepit of intrigue that was best avoided, and cited a remark by Eisenhower in February 1943: "George, you are my oldest friend, but if you or anyone else criticizes the British, by God I will reduce him to his permanent grade and send him home. The reason that I have not promoted you is that I want to promote three of you . . . [but Fredendall] is reported to have talked against the British. If he has, by God I'll bust him." A perplexed Patton later asked Mark Clark if he had been similarly accused. Clark replied that he had not, and "that Ike had talked to him the same way . . . thinks General Marshall told him to do it to all of us. 'Cromwell beware ambition, by it the angels fell.'"[54]

Patton's attitude toward Eisenhower varied with his mood of the moment. He once recorded how proud he was of Eisenhower during a conversation with a senior British general, Sir Bernard Paget. "Ike certainly makes a fine impression when he talks. . . . I think I could do better in the same job," but admitted that he seemed "to lack something which makes the politicians trust Ike."[55]

Several days after being dressed down, Patton received a note from Eisenhower cautioning him for sometimes making a bad impression by engaging his mouth before his brain, a habit he needed to curb. Eisenhower assured him of his friendship and expressed his "intense desire" to see Patton promoted as quickly as possible. Patton took the rebuke in stride, believing that his friend meant well by his criticism, but he was discouraged that "I certainly have thus far failed to sell myself in a big way to my seniors."[56] Patton's reply, which he never mailed, acknowledged that he did value Eisenhower's advice. "For years I have been accused of indulging in snap judgments. Honestly, this is not the case because, like yourself, I am a profound military student and the thoughts I express, perhaps too flippantly, are the result of years of thought and study. . . . It may be that I am not overawed in the presence of high personages and therefore speak too freely." He promised that "I shall never let you down." Patton never explained why the letter was not sent, and his reasons remain unknown.[57]

Although distraught over Eisenhower's apparent displeasure with him, Patton was even more disturbed by the growing belief that he was destined to miss the war in Tunisia altogether. The war was going so badly for the Allies there that Eisenhower summoned him to Algiers on December 9, but when Patton's aircraft attempted to land it received a rude welcome from nervous Allied antiaircraft gunners. As exploding shells burst all around, Patton felt "as if the whole ship was going to pieces." Eisenhower directed him to visit the front and report back why Allied tank losses were so high. He and Stiller drove 250 miles in the cold and rain in a topless scout car into Tunisia, where he arrived soaked and half frozen at the First Army headquarters of British Lt. Gen. Kenneth Anderson, the senior Allied commander in Tunisia until Gen. Sir Harold Alexander was appointed the commander in chief of Allied land forces in February 1943.

Torch was a period of great transition, during which the new Allies were sizing one another up and forming opinions and attitudes—many of them unfavorable—that would endure throughout the war. Patton's subsequent

Anglophobia clearly had its roots in this visit to the front, where he formed a negative opinion of British officers and their soldiers, just as a great many British officers, starting with Alexander, would soon form equally unflattering judgments of American fighting men and their commanders. Patton had held his British comrades-in-arms in high esteem during the First World War and, even though he had not been overly impressed with them during his brief sojourn in London the previous August, it appears to have been his observations in Tunisia that caused him to form an abiding negative impression. After a briefing at the British 5th Corps, Patton left convinced that a recent setback was the result of poor leadership, and at the 78th Division he was astonished and disgusted to arrive at 8:30 A.M., and find the commanding general just arising.[58]

His opinion of the Arabs was reflective of the attitude that they were an inferior race. The Arabs in Tunisia, he noted, "are lower—if this is possible—than the Arabs in Morocco. Most of their dwellings look like manure piles, and I believe they are." He observed many were cross-eyed and wondered why. At Béja he stopped to admire "one of the finest medieval castles I have ever seen," noting its defenses against ancient attackers. Everywhere he saw the evidence of successful Luftwaffe attacks on British and American vehicles, whose wrecks littered the road. On the main highway to the front, he observed that Arab herdsmen continually blocked their passage, the sheep making the roads slippery with large piles of manure. "If this road is to be used to its maximum, both these practices will have to be stopped."

He finally located his son-in-law, Lt. Col. John K. Waters, who commanded a tank battalion in the 1st Armored Division, and the two had a brief reunion and were photographed together. Earlier Waters had narrowly avoided being wounded when a bullet passed through his uniform without leaving a scratch. Patton was impressed, and the two discussed how Waters might extricate his unit from its precarious position if the Germans attacked. "I was very much pleased with his attitude and also the behavior of the men, who were very glad to see me."

Waters also explained how he had attempted to get a nearby Royal Artillery battery to fire on German guns shelling a nearby town, but had been refused and told he must obtain permission from the brigade artillery officer, which took some two and a half hours. Patton found the British brigade commander "trembling all over. He told me this was due to fatigue. From the smell of his breath I could see that it was due to something else."

The troops of the 1st Armored Division told Patton that he was the first general they had seen during their twenty-four days of combat, "a sad commentary on our idea of leadership," leaving him disgusted with both Eisenhower and Clark, neither of whom, he complained in his diary, had yet been to the front. "[They] have no knowledge of war. Too damned slick, especially Clark." American light tanks had already proved no match for the more heavily armored German panzers, with their high-velocity guns. (The siutation was soon to become even more dangerous with the introduction of the Tiger tank and its gun.) American tank losses were very high (Waters's unit alone had already lost thirty-nine, or nearly two-thirds of its tanks), and this deeply worried Patton.[59]

Patton also stopped at Combat Command B of the 1st Armored Division to visit his former cavalry instructor, Paul M. Robinett, now a brigadier general. Jumping from his command car, Patton "shouted in a high shrill voice, 'Where are the damned Germans, I want to get shot at!' He looked hard and fit and was full of enthusiasm and good cheer as he greeted everyone in the most friendly and comradely fashion. . . . Patton was really full of himself and regaled us with stories, anecdotes, and details of the Casablanca operation."[60] That night Patton dropped his showman's act and, extracting a flask from his pocket, said to Robinett: "Let's have a drink for old times' sake and for two old cavalrymen who have gone astray." Slapping Robinett on the back, Patton said: "I'm glad you are here. There is a lot to be done before we are through and I expect to be in it."[61]

His report to Eisenhower cited the inadequacy of the puny 37-mm Grant tank against the 75-mm guns of the German Mark IV panzer.[62] Although Patton's attitude toward Eisenhower was becoming increasingly critical, Eisenhower remained high on his friend, which was reflected in his assessment: "Among the American commanders, Patton I think comes close to meeting every requirement made on a commander."[63]

Patton's return journey from Algiers to Casablanca aboard a transport was a nightmare. Spurning the advice of the crew, he insisted that they undertake the journey, even though it meant landing in the dark. "I was nervous about being away so long and intended chancing it, although I knew . . . I might get shot down by my own anti-aircraft." When the plane's radio failed, Patton became genuinely alarmed and "very depressed." He was spared "friendly fire" when the plane entered a violent rainsquall and the pilot had barely located the field with a flare when the rains obliterated the runway. For nearly one and a half hours the pilot circled Casablanca and then, with gas running out, decided to chance a belly landing on any piece of open ground they could locate. "If we could not see the ground—which was most of the time—we would have [had] to jump." Finally the rain let up momentarily, and the pilot dived through a hole and made a perfect landing, thus successfully ending the latest in a long line of Patton's brushes with death. Grateful and relieved, he emerged from the plane and took the pilot and copilot to his quarters, where he "gave them dinner and kept them all night. I am also writing a letter of commendation."[64]

Air crashes were as much of a menace in North Africa as the was Luftwaffe, and the frequency of his close calls frightened Patton. Several weeks after the first incident he was returning to Morocco in poor visibility. "We nearly hit several mountains and I was scared till I thought of my destiny. That calmed me. I will not be killed in a crash."[65]

In Morocco his daily rounds included visiting the sick and wounded in hospitals. In one Patton encountered a young GI who had been twice rescued from the sea after both his landing craft were sunk, and said, "Now you can go back home justly feeling that you have done your bit and in time get well again." The soldier replied with a heartfelt plea: "I don't want to go back home; I came over here to see something but I haven't seen anything yet." On Patton's order, the soldier was taken on an aerial sightseeing trip of North Africa before being evacuated to the United States.[66]

With the neutralization of French North Africa, in mid-November 1942 the Allies opened the second phase of the Torch plan—a rapid thrust into Tunisia to capture the key ports of Bizerte and Tunis before Axis forces in North Africa could be augmented. Then, while Montgomery pursued Rommel into Tunisia, the Torch forces would drive east to Tripolitania and trap Panzer Armee Afrika. The plan and reality soon diverged.

After Alamein, Rommel had urged Hitler to remove German forces from North Africa while there was still time, arguing that prolonged operations there would serve no useful purpose, resulting only in their senseless annihilation. Hitler angrily spurned Rommel's advice. Fearing the loss of Tunisia's vital ports, which the Allies would certainly utilize to mount an invasion of southern Europe, Hitler ordered the immediate reinforcement of Tunisia by creating the Fifth Panzer Army under Gen. Hans-Jürgen von Arnim. More than 100,000 German and Italian reinforcements began pouring into Tunisia about the time the Allies were attempting to capture Tunis in early November. In contrast to the inexperienced Anglo-American Torch force, von Arnim's army consisted mostly of veteran troops, thus setting the stage for the first major confrontation of the war in the Mediterranean.

After the Battles of Alam Halfa and El Alamein, the Eighth Army was now extended across nearly a thousand miles of Libyan Desert. By mid-November, as the Torch forces were struggling to establish a fully operational front in western Tunisia, it reached Tobruk, the scene of its humiliating defeat six months earlier.

The five-hundred-mile supply line between Algeria and Tunisia, heavy winter rains, and uninspired Allied leadership combined to turn the overoptimistic timetable established for Torch into a shambles. Moreover, the Allies had not counted on Hitler's countermove and received two bloody noses attempting to capture Tunis in December 1942. The result was a winter stalemate. As Patton had seen for himself during his brief visit to the front, the Luftwaffe was active and effective for one of the few times during the war, and inexperienced Allied troops found themselves pitted against veteran German formations. Eisenhower confessed to a colleague in Washington that military operations in North Africa thus far had "violated every recognized principle of war, are in conflict with all operational and logistic methods laid down in text-books and will be condemned, in their entirety, by all Leavenworth and War College classes for the next twenty-five years."[67]

Churchill had correctly surmised that once the United States was fully committed in the Mediterranean, it would be a case of, "in for a penny; in for a pound." Torch was an excellent example of the opportunistic nature of Britain's Mediterranean policy, which was again demonstrated in January 1943, when the Allied leadership met at Casablanca to debate future strategy after the end of the North African campaign, and, as Churchill reminded Roosevelt, "because we have no [suitable] plan for 1943."

The British were committed to the removal from the war once and for all of Mussolini and the Italians, and Churchill insisted on playing out the Mediterranean option when it became evident that the campaign in Tunisia was badly behind schedule. Thus, while continuing to affirm the eventual necessity for a cross-Channel invasion of northwest France, Churchill sought to buy time for its planning and preparation by nibbling away at what he termed "the soft underbelly" of Germany in the Mediterranean, where the Allies had the Axis on the run and must not surrender the initiative.[68]

Stalin refused to attend Casablanca but nevertheless cast a long shadow over the deliberations of his two allies. Although they were soon to win the most important victory of the war at Stalingrad, the Russian dictator continued to badger both Churchill and Roosevelt during the second half of 1942 to open a second front without delay, and further relieve the pressure on the beleaguered Red Army by forcing Hitler to maintain formations defending western Europe.

An Anglo-American clash at Casablanca was inevitable when the Combined Chiefs of Staff[69] began the stormiest negotiations ever to occur between the two Allies. Sir Alan Brooke, the chief British strategist and military spokesman, came to Casablanca fully committed to an invasion of Sicily, following the Tunisian campaign, to reinforce Allied success in the Mediterranean and to gain a foothold in southern Europe.

The deadlock centered around the timing and form of the cross-Channel invasion,[70] even as Marshall continued to insist that "the Mediterranean was a blind alley to which American forces had only been committed because of the President's insistence that they should fight the Germans somewhere."[71] Deadlock ensued over Marshall's demand for the transfer of Allied forces in North Africa to the United Kingdom for the cross-Channel invasion, while the British chiefs of staff continued to insist upon Churchill's avowed goal of the "cleansing of North Africa to be followed by the capture of Sicily."[72] The compromise that Churchill loftily proclaimed was, "the most complete strategic plan for a world-wide war that had ever been conceived," deferred Operation Overlord, (the code name given the cross-Channel invasion), and instead authorized the invasion of Sicily in the summer of 1943.[73]

Patton's primary function at Casablanca was to provide security for the conference, and to act as a glorified majordomo and greeter to a glittering array of Allied leaders whose official hosts were Mark Clark and his Fifth Army. Patton was critical of Clark's arrangements and distressed that the French were excluded and that the British were "pulling hell out of our leg and no one knows it. I believe they want to discredit the French with the Arabs so that after this war they can 'acquire' French West and North Africa. The tragedy to me is that we will let them do it."[74]

Patton and Mark Clark failed to agree on much of anything. "His whole mind is on Clark . . . for one hour he spent his time cutting Ike's throat. And Ike, poor fool, sent him here. . . . It is most discouraging." When Clark insisted on replacing French troops guarding the Casablanca airfield with Americans (to ensure they did not fire on the conference site), Patton exploded, and the two quarreled. Patton's version is that he managed to dissuade Clark only by threatening to resign his command. "It would have been the crowning insult to the French and would have given the Nazis a wonderful propaganda weapon and roused the Arabs."[75]

Patton officially greeted Roosevelt at a ceremony to mark his arrival in Casablanca. At the airfield the president inspected an honor guard commanded by Lieutenant Colonel Semmes. Virtually the entire Western Task Force, some forty thousand spit-and-polished troops, lined the road for miles as Roosevelt reviewed them with obvious pleasure and considerable emotion. This grand scene was sullied by FDR's Secret Service bodyguards, who rode next to him with drawn pistols "to protect the Commander-in-Chief from being assassinated by his own loyal troops." Patton was not alone in his disgust at this gesture.[76]

Despite the heavy security, which included a battalion of infantry to guard the site of the conference at Anfa, a Casablanca suburb, Patton was visibly nervous about his responsibility for guarding the Allied leaders. He stormed at the president's personal physician: "I hope you'll hurry up and get the hell out of here. The Jerries occupied this place for two years and their bombers know how to hit it. They were around ten days ago and it's a cinch they'll be back."[77]

Patton had moved into a sumptuous villa and each evening lavishly entertained a group of VIPs. Among his many guests was a bevy of admirals, generals, and politicians, among them Giraud (the Allied-sponsored French commander in North Africa, of whom Patton said: "I fear he is too much a soldier to run his job as dictator—at least he wants to fight"); Marshall; Clark; Roosevelt's trusted adviser Harry Hopkins (whom Patton quite liked); and FDR himself, whom Patton thought "really appeared as a great statesman." Patton also entertained Churchill, noting only that "we got on well."[78] Churchill's personal assistant mentioned Patton in his postwar memoirs, complaining: "Not satisfied with the barbed wire, the Americans had brought in a number of guard dogs which prowled around biting every one impartially."[79]

When Churchill ran out of cigars and Marshall ordered that some be found at once, Marshall's staff was stymied, but Patton's efficient young aide, Capt. Dick Jenson, produced a box of White Owls sent by his wife. The five-cent price on the label was hurriedly obliterated and the cigars presented to Churchill. Later that night Marshall observed, "The strangest thing happened. The Prime Minister lit one of those cigars, took one puff and put it aside. I can't understand—he must be giving up smoking."[80]

At a small ceremony in his villa, Roosevelt presented Colonel Wilbur with the Medal of Honor for gallantry at Casablanca. Churchill's senior military assistant, General (later Lord) Ismay, stood next to Patton and later wrote: "There were tears in his eyes and a burning conviction in his voice, as he said: 'I'd love to get that medal posthumously.' Patton was emotional and impulsive . . . but I have not the slightest doubt that he meant what he said, and that his greatest ambition was to die on the field of battle."[81]

One of those unimpressed with Patton was Gen. Sir Alan Brooke, the dour chief of the Imperial General Staff, whose own diary noted: "A real fire-eater and definitely a character." Brooke found Patton unlike any British general he had met (including Montgomery), and after the war added:

> My meeting with Patton had been of great interest. I had already heard of him, but must confess that his swashbuckling personality exceeded my expectation. I did not form any high opinion of him, nor had I any reason to alter this view at any later date. A dashing, courageous, wild and unbalanced leader, good for operations requiring thrust and push but at a loss in any operation requiring skill and judgment.[82]

Another frequent guest was Brig. Gen. (later Maj. Gen.) Everett S. Hughes, a logistician, classmate, and longtime friend of both Patton and Eisenhower. In North Africa, Eisenhower wore several command hats, one of which was commander of all U.S. troops. Hughes was Eisenhower's deputy, and on the infrequent occasions he and Patton met, the two spent hours exchanging gossip and opinions, most of which Hughes recorded in an illegible diary.[83] Hughes, a harsh critic of those running the war, was one of the few with whom Patton could relax and speak in complete confidence. Unfortunately, Hughes's gossip usually sent Patton's frustration level to new heights. At Casablanca a major topic of conversation among the visiting

VIPs was the politics of dealing with the French, which to Patton's soldierly mind was secondary to the pursuit of the real war in Tunisia. Both men agreed that Torch was "still a General Staff war, not a soldier's war."

Anxious to impress Marshall, Patton saw little of him at Casablanca, and deplored the weather, which prevented the chief of staff from personally observing his highly disciplined troops and learning of his accomplishments in Morocco. Also present with Marshall was Patton's former boss, Jake Devers, who wrote to Beatrice that "George's troops were by far the best-disciplined and best-trained looking group of soldiers that I saw in all my travel." Although Devers thought Patton "seemed a little depressed" at not being able to fight Rommel, he was "demonstrating at all times that he is a real soldier's soldier."[84]

Patton and Eisenhower managed one night of lengthy conversation. Eisenhower said he was considering appointing him deputy commanding general to run the war in Tunisia, while he dealt with the politics and intrigue of AFHQ. The following morning Eisenhower pinned a second Distinguished Service Medal on Patton's uniform for Casablanca, and reiterated the notion of appointing him his deputy.[85]

In fact it was Marshall who, before leaving Casablanca, recommended that Eisenhower appoint a deputy commander and suggested Patton. Although he briefly considered moving Patton to Algiers to become his "Deputy Commander for Ground Forces," and utilizing "his great mental and physical energy in helping me through a critical period," the quid pro quo of Casablanca was American acceptance of a British suggestion (originated by Brooke and put to Roosevelt personally by Churchill) that Eisenhower ought to appoint Alexander his deputy.[86] The plan to utilize Patton thus died when Marshall cabled Eisenhower: ALEXANDER WILL BE YOUR MAN WHEN BRITISH EIGHTH ARMY JOINS YOU AFTER [CAPTURING] TRIPOLI.[87]

Eisenhower retained Patton in Morocco to plan the American component of the invasion of Sicily, and Marshall sent Maj. Gen. Omar N. Bradley to act as Eisenhower's "eyes and ears." Patton was relieved to be off the hook and confided in his diary: "I think I was fortunate in not being made Deputy Commander-in-Chief to Ike. I guess destiny is still on the job. God, I wish I could really command and lead as well as just fight."[88]

Partly to alleviate his boredom, in mid-February 1943 Patton accepted an invitation from the British hero of Alamein, Bernard Montgomery, to attend a study week to impart the lessons of the battlefield learned since 1939–40 to senior officers from England and the Middle East. Patton arrived in Tripoli in a B-17 bomber to find himself the only American officer present. Montgomery incorrectly interpreted this as a lack of professional interest on the part of the American leadership, when in fact the reason was the fighting in Tunisia and training for the forthcoming invasion of Sicily. Monty described Patton as "an old man of about 60," and the Eighth

Army commander's impression—however brief—of this "elderly" American general was hardly flattering. Patton himself candidly admitted that he was "certainly the oldest looking general here."[89]

With the exception of Monty, New Zealand's Gen. Bernard Freyberg (winner of a Victoria Cross in World War I), and one or two others, Patton thought the senior British officers he met unremarkable. Of Montgomery he wrote: "small, very alert, wonderfully conceited, and the best soldier—or so it seems—I have met in this war. My friend General Briggs says he is the best soldier and the most disagreeable man he knows." The rest "are the same non-committal clerical types as our generals." Unlike Eisenhower, who all but revered Alexander as the best British general he ever knew, Patton rated Alexander, soon to become his tactical boss on the battlefield, quiet and not very impressive looking. "The British say that Montgomery commands and Alexander supplies—this may be so."

Having been weaned under Pershing, whose greatest battle was to retain the AEF under American command and control, Patton was shocked when Fredendall's II Corps was placed under British command. "We have sold our birthright. . . . I am shocked and distressed. . . . One is inclined to think that fighting ability is at a discount."[90]

For the ordinary soldier, war is mostly "hurry up and wait," a continual boredom alleviated only by training, movement from place to place and, occasionally, battle. For George S. Patton it was no different. While Fredendall commanded the II Corps, nothing he did seemed to be leading to a confrontation with Rommel in Tunisia. His entire life had been in preparation to be a combat commander, but during World War I he had served under fire a mere five days. In Morocco there had been only four days of fighting, in which he had played only a minor role. And now it seemed that the damned war would once again go on without him. Driven by a lifetime of inner desire and an obsession to carve a meaningful military destiny for himself in the pages of history, it was a very frustrating time, made more so by the steady drain of Patton's units to Tunisia.[91]

The boredom was temporarily relieved in early February when Patton accompanied the pasha of Marrakech on a wild animal hunt in the Atlas Mountains. They arrived in a Rolls-Royce, awaited by a thousand beaters. The pasha rode a fine black mule draped with a great purple saddle, with Patton beside him on a gray stallion. Soon "the largest and blackest" wild boar Patton had ever seen broke cover, charged the party, and was shot by the Pasha, to no effect. Then, as the enraged beast headed straight for Patton, "with a nasty expression on his face, I put a solid slug in his right eye at about 3 yards and he fell dead so that the blood from his mouth wet my feet. It was quite exciting but I was not perturbed till afterwards. I never am at the time."[92] Word of Patton's exploit quickly spread through the Arab grapevine and was embellished to three boars shot dead, "Bang. Bang. Bang," literally at his feet. "He is formidable, your General Patton," the grand vizier noted to Major Codman.[93] In French Morocco at least, the legend of George S. Patton was secure.

He poured out his frustrations to Beatrice, who wrote back long gossipy letters designed to reassure him that he was "a better man in every way than you have ever been and are learning all the time the ways of the dove as well as those of the serpent and the fire dragon."[94] Patton was dismayed by the high-level backbiting "between soldiers who are primarily politicians. We have many commanders but no leaders. . . . Some times I wish I was retired but I guess I would not like that either. Probably I would only be content if I was god and probably some one ranks him."[95]

The inexperience of Eisenhower and Clark rankled, as did the fact that both men were his juniors in both time in service and experience. Yet most of what bothered him was simply frustration over his inaction. He had seen for himself how badly the fight was being lost in Tunisia, and it ate at him. "I cant sleep for thinking of what I know and can't say. . . . I feel just like a bird in a room who can see the out side through the window and beats himself to death trying to get out. Some day the glass will break or the window will open."[96] He was temporarily buoyed by the news that he was to command the American invasion element in Sicily: "It is an honor to be trusted with the American part of the plan. I feel I will win. I feel more and more that I have a mission; then I lose my confidence but get it back."[97]

Patton had gotten enormous publicity from the 1941 cover story in *Life* and now was on the verge of becoming a media darling for his colorful, unorthodox behavior. Robert St. John of NBC submitted a piece that cast Patton as "a combination of Buck Rogers, the Green Hornet, and the Man from Mars . . . the rootin', tootin', hip-shootin' commander of American Forces in Morocco . . . Major General George Patton, Junior . . . he has enough DASH and DYNAMITE to make a Hollywood adventure-hero look like a drugstore cowboy."[98] Such tales created an indelible image in the minds of the American public of a swashbuckling hero with whom they could identify. It did him no harm when he articulated his own vision of the romantic hero of old in which he and Rommel would duel one on one—to the death. "It would be like a combat between knights in the Old days. The two Armies could watch. I would shoot at him. He would shoot at me. If *I* killed him, I'd be the champ. America would win the war. If he killed me . . . well . . . but he WOULDN'T."[99]

Then, in early March, came the devastating news that his son-in-law was missing in action after the disastrous battle of Sidi-Bou-Zid. It would be several weeks before word was received that John Knight Waters was alive and a prisoner of war in Germany. The uncertainty of his fate was a severe blow to the entire Patton family. Patton wrote that if it had been his own son, George, "I could not feel worse. . . . I feel terribly sorry for [daughter] B[eatrice]."[100] When Beatrice telephoned her daughter with the unhappy news, "after a short silence, she replied, 'I am sure Johnnie's alive, and anyway, don't worry, Ma, the Regulars will show.'"[101] The campaign in Tunisia had become unraveled in disastrous fashion, and Patton was abruptly summoned to help clean up the mess.

CHAPTER 30

A Summons to Battle

When I was a little boy at home I used to wear a wooden sword and say to myself: "George S. Patton, Jr., Lieutenant General."

—PATTON

Hitler's decision to reinforce North Africa led to a winter stalemate in December 1942 and deferred the final showdown in Tunisia until after the heavy winter rains ended and full-scale offensive action could be resumed at the end of March 1943. The disarray in Tunisia was not confined to mud so sticky and lethal that it devoured vehicles, but extended to the new Anglo-American alliance and its new commander in chief, Dwight Eisenhower.

In North Africa the Anglo-American commonality of language could not obscure the fundamental differences between the traditions practiced in the British army for centuries, and the rawness of a peacetime, largely draftee American army suddenly thrust into a global war. Eisenhower had become mired in running AFHQ, fighting the war of the paper shufflers, and coping with the political fallout from Torch, instead of attending to his most important duty—that of running the battle in Tunisia—which could not be accomplished from Algiers, four hundred miles away. These were trying times for Eisenhower, and, as even Patton had learned since Torch, behind Eisenhower's infectious grin and sunny outer facade lay a fiery temper that occasionally erupted with awesome fury.[1]

The campaign in Tunisia required orchestration by a tactical commander—something Eisenhower had neither the time nor the experience to do. For all his intellectual brilliance, Eisenhower had never before commanded above battalion level. In Tunisia he left the battle in the hands of the senior

Allied commander, British Lt. Gen. Kenneth Anderson, an officer later shunted into obscurity for his alleged mishandling of the British First Army in Tunisia. A far better general than he has been portrayed, Anderson was in many ways the victim of Eisenhower's inexperience and incompetence as a battlefield commander. The result, as one biographer has charged, was that "from Algiers he was not able to exert a grip on the battle, but he also failed to forge an effective chain of command."[2]

Among the proliferating command problems in Tunisia was Lloyd Fredendall, the II Corps commander, whose anti-British, anti-French sentiments were further undermining an already messy campaign. It was Brooke's utter lack of confidence in Eisenhower as a commander that prompted him to choreograph the assignment of one of Britain's most admired battlefield generals, Harold Alexander, to take over the faltering ground campaign and the newly formed 18th Army Group.[3]

In the months since Alamein, Rommel's Afrika Korps had not only fought a skillful delaying action across North Africa but had demonstrated that it was still a deadly foe. Rommel and his superior, Field Marshal Albert Kesselring (the German commander in chief in the Mediterranean), saw an opportunity to strike a blow that could split the Allies in Tunisia in two and threaten the overextended lines of communication from Algeria into Tunisia. The German attack, delivered by the entire panzer strength of the Afrika Korps and von Arnim's Fifth Panzer Army, was aimed at the inexperienced, fragmented American forces guarding the southern flank of the Allied front: the U.S. II Corps.

Deployed across a large area, the corps was vulnerable to attack and defeat in detail by superior Axis forces. Contrary to Eisenhower's orders, which called for a large mobile reserve to be deployed behind a reconnaissance and screening force, "American infantry had been lumped on isolated djebels [hills] . . . and mobile reserves were scattered in bits and pieces along the line."[4]

At the front, American vulnerability was obvious to the lowest-ranking private soldiers, even if their senior commanders were too remote to grasp the situation. Soldiers possess a marvelous ability to reduce events to their simplest common denominator. And so it was in Tunisia, with an unnamed GI's pithy observation that, "Never were so few commanded by so many from so far away!"[5]

The penalty for these unsound dispositions was paid in February, when Axis units inflicted disastrous defeats at Sidi-Bou-Zid and Kasserine Pass. Two of von Arnim's veteran panzer divisions surprise-attacked with vastly superior firepower and quickly chewed up units of the 1st Armored Division at Sidi-Bou-Zid. American units were deployed in so-called penny-packet formations (independent, self-contained, self-supporting, brigade-size forces) that the British had used with disastrous consequences in 1941–42, before Montgomery took command of the Eighth Army.[6] With the Allies unable to contain the powerful panzer forces or to reinforce in time, Sidi-Bou-Zid quickly turned into a first-class military disaster as position after position was attacked and overrun. Farther south a German-Italian battle group of Rommel's Afrika Korps advanced with little opposition and attacked U.S. forces defending the Kasserine Pass, with equally grave consequences. There the American commander had not bothered to occupy the commanding terrain of the hillsides but instead had deployed his troops across the valley floor "as if to halt a herd of cattle."[7]

It was after Sidi-Bou-Zid that Lt. Col. John K. Waters was reported missing in action. For several agonizing weeks after the news, before word was received that he was alive and in a German POW camp, Patton was unsure if his son-in-law had survived. He was later reassured when a graves registration team failed to locate any sign of his corpse on the battlefield. He recovered an ammunition clip from a burned-out tank and sent it to his grandsons as a memento of their father in the event he had not survived.

Kasserine and Sidi-Bou-Zid were humiliating and disastrous defeats whose results reached far beyond those of a battlefield debacle. American armor and artillery were no match for superior German armaments. Not only were American tactics and dispositions unsound, but once attacked at Sidi-Bou-Zid some troops abandoned their positions and equipment and fled to the rear. It was so bad that Eisenhower's naval aide wrote in his diary: "proud and cocky Americans today stand humiliated by one of the greatest defeats in our history."[8]

In the United States the revelation of what occurred at Kasserine was public shock and disbelief that "shook the foundations of their faith, extinguished the glowing excitement that anticipated quick victory, and, worst of all, raised doubt that the righteous necessarily triumphed."[9] At the front the debacle was seen by many as far more than merely the failure of green troops against veteran soldiers. "To some of us, at least," wrote General Robinett, "a better explanation might have been green commanders and obsolete equipment."[10]

II Corps suffered approximately six thousand casualties at Kasserine and Sidi-Bou-Zid. Equally dismaying, the obvious problems of training and leadership made it evident that without quick and decisive action the possibility of strategic defeat loomed large. Of even greater harm was the impact of Kasserine on Anglo-American relations. The British viewed Kasserine as irrefutable evidence that American fighting ability was more talk than reality. Although Eisenhower attempted to put on a brave face in his communications to Marshall, the truth was that Kasserine and Sidi-Bou-Zid had left a bitter legacy. First impressions count, and it would take virtually the entire war to sway Alexander from his low opinion of American troops and their commanders.

*   *   *

The failures of American leadership began with Eisenhower himself, who continued to exhibit the uncertainties of inexperience of high command, manifested by a tendency to interfere on the battlefield and in his hesitation to address the growing problems within II Corps. "Eisenhower's willingness to interfere in the affairs of his subordinates," observed his biographer Stephen Ambrose, "ill became a man who often waxed eloquent on the subject of the sanctity of unity of command and the chain of command; his violation of his own principles was a reflection of his lack of confidence in Fredendall, whom he was either unwilling or afraid to remove."[11]

Having retained Fredendall after it was obvious that he was unfit to command II Corps, Eisenhower compounded his own problems. "And at the crucial moment, when Rommel was at his most vulnerable, he had failed to galvanize his commanders, which allowed Rommel to get away. Kasserine was Eisenhower's first real battle; taking it all in, his performance was miserable. Only American firepower, and German shortages, had saved him from a humiliating defeat."[12] The only faintly positive postscript to this fiasco came from Rommel, who was not deceived by the poor American performance and believed that once matured by combat, the U.S. Army would prove a formidable foe.[13]

What scant comfort could be taken from such a grim situation was found when Robinett's troops captured some Italians near Kasserine. The uniforms of Italian private soldiers were usually festooned with large and small aluminum stars. A GI presented Robinett with a helmetful, saying, "Sir! We have captured a whole flock of Italian brigadier generals!" An officer "picked out some very large stars and good-naturedly remarked: 'I suggest that we send these to General Patton. He would not approve of the Italians having larger stars than his.'"[14]

At the tactical level, American problems were compounded by Lloyd Fredendall, one of the most inept senior officers to hold a high command during World War II. He was completely out of touch with his command, had balked at cooperating with Anderson, and failed to make a positive impact on his troops or subordinate commanders, with whom he feuded constantly, particularly Orlando Ward of the 1st Armored, who despised him. At the time of Kasserine, the corps headquarters was located some sixty-five miles behind the front lines in an enormous underground bunker in a deep gorge. A battalion of combat engineers spent valuable weeks constructing this monstrosity when their skills were urgently needed elsewhere. Eisenhower, who timidly counseled Fredendall to get out of his command post more often, was among the visitors to emerge embarrassed and appalled from Fredendall's bunker, snidely nicknamed "Shangri-la, a million miles from nowhere," and "Lloyd's very last resort."[15] A British historian and former combat general has aptly described Fredendall as "a prime

specimen of the traditional over-ripe, over-bearing and explosive senior officer in whom the caricaturists have always delighted."[16] Small wonder, then,
that British opinion of American leadership and fighting ability was so low;
it could hardly have been otherwise.

Eisenhower urgently attempted to redress a situation spinning out of
control in Tunisia by summoning Ernie Harmon from Morocco, where he
was training the 2d Armored Division for Sicily. Having at last accepted that
Fredendall had lost control of II Corps, Eisenhower sent the tough, outspoken Harmon to the front as his personal representative to help reverse the
situation by whatever means necessary. While Fredendall sulked in his
bunker and took no active role, Harmon brought common sense and firm
leadership to the rapidly deteriorating front when it was most needed, and
when they were given a taste of determined leadership, American troops
responded by blunting the German offensive and convincing Rommel that it
was time to back off and regroup. When Patton learned that the 10th Panzer
Army had been stopped at Thala, he told Eisenhower: "Well, von Arnim
should have read about Lee's attack at Fort Stedman" (outside Petersburg,
Virginia, where stalwart Union reserves beat off a last-ditch Confederate
counterattack in 1865).[17]

In his after-action report to Eisenhower on Kasserine, Harmon condemned Fredendall as "a son-of-a-bitch" who was unfit for command, and
later told Patton that he was a moral and physical coward.[18] Alexander, too,
had visited Fredendall and, when asked his opinion by Eisenhower, replied:
"I'm sure you must have better men than that."[19] Stung by Alexander's and
Harmon's candid assessments, Eisenhower belatedly dismissed Fredendall
and first offered the command of II Corps to Harmon, who declined, saying
he could not accept the position after recommending the relief of its commander. The next obvious choice was Patton.[20]

On the afternoon of March 4, Patton had been out riding a horse lent him by
General Noguès when Eisenhower left an urgent message that he was to
depart the following morning for extended field service. The AFHQ chief of
staff, Maj. Gen. Walter Bedell Smith, would only reveal that Patton might
be relieving Fredendall. "Well, it is taking over rather a mess but I will
make a go of it. I think I will have more trouble with the British than with
the Boche. 'God favors the brave, victory is to the audacious.' "[21] Patton had
intended to slip quietly away the morning of March 5, but was given a royal
send-off by his staff, who lined up in the street outside his headquarters with
a band and a military police honor guard. Virtually the entire staff signed a
letter reading: "Good luck and God bless you. If this isn't a private fight we
would all like to go with you."[22]

During a brief stopover at the Maison Blanche airfield outside Algiers,
Patton conferred for a half hour on the tarmac with both Eisenhower and

Bedell Smith before flying on to Constantine to meet with Alexander. Patton was to replace Fredendall, on the flimsy grounds "that it was primarily a tank show and I know more about tanks. [Ike] stressed that criticism of the British must stop. I fear he has sold his soul to the devil on 'Cooperation,' which I think means we are pulling the chestnuts for our noble allies." The meeting was primarily a lecture by Eisenhower, who reversed the guidance he had given Fredendall by instructing Patton to control the battle from the corps command post. Cmdr. Harry C. Butcher witnessed the meeting and recorded that Patton, embittered over Johnnie Waters, "damned the Germans so violently and emotionally that tears came to his eyes three times. . . . Now he has the opportunity to chase Rommel to his heart's content." Eisenhower also lectured Patton about his reckless disregard for his own safety. "He doesn't need to prove his courage. General Ike wants him as a corps commander, not as a casualty."[23]

At Constantine, Alexander seemed genuinely glad to see him. "He was very friendly and complimentary in his remarks, stating he wanted the best Corps Commander he could get and had been informed that I was that man." That night he was thoroughly briefed by the American liaison officer to Alexander and came away firmly convinced that "that I too must 'cooperate' or get out."[24]

Patton arrived at II Corps at 10:00 A.M. on March 6 to find Fredendall still at breakfast. Although Patton's first impression of the corps staff was that their dress and discipline were poor, he was more generous toward Fredendall whom he thought had conducted himself "very well," given the circumstances. One of the first people Patton encountered at II Corps was Maj. Gen. Omar N. Bradley, newly arrived in North Africa and sent by Eisenhower as his "personal observer." When Bradley arrived in Tunisia in late February 1943 as an unassigned and very junior major general, untested in combat, he was quickly and rather rudely shocked when he arrived at II Corps about a week before Patton to be greeted by a hostile Fredendall, who banished him to a dingy, windowless, filthy hotel nearby.[25]

Like Harmon before him, Bradley was distressed by what he discovered as he toured the II Corps area of operations. The problems Bradley noted went far beyond II Corps: He observed a lack of urgency in Northwest Africa that went clear to Eisenhower who was, he thought, strangely unconcerned about the outcome of the campaign. "But no one seemed concerned about the ultimate way it would end, nor did they seem frightened that we might be unable to maintain the schedule for the Italian invasion." Unlike Fredendall, Patton would not countenance having "one of Ike's goddamn spies" in his command, and easily persuaded Bedell Smith to assign Bradley to the position of deputy corps commander.

Alexander and his chief of staff inspected II Corps three days after Patton had assumed command and seemed impressed by the red-carpet treatment accorded them. Patton was especially pleased with the 1st Division and the performance of his two longtime friends, Terry Allen and Brig. Gen. Theodore Roosevelt, the assistant division commander. Patton, in turn, was delighted that Alexander took the time to talk personally and at length to the officers of the II Corps staff. When the entourage visited a nearby phosphate mine, Patton thought the director looked liked a German and that an attempt might be made to kill Alexander, "so I walked in front of him."[26]

Although Alexander found Patton a great improvement over Fredendall, he remained skeptical about his American corps and how well it would fight. Even though II Corps would soon show marked improvement under Patton, the significant American recovery unfortunately made little impression on Alexander, whose first inspection of his new command left him dissatisfied with Anderson and pessimistic about American fighting ability. He wrote a confidential and decidedly gloomy assessment of the American soldier and his leaders to Brooke, observing that American troops were "soft, green and quite untrained. Is it surprising then that they lack the will to fight. . . . There is no doubt that they have little hatred for the Germans and Italians and show no eagerness to get in and kill them . . . unless we can do something about it, the American Army in the European theatre of operations will be quite useless and play no useful part whatsoever." What really concerned Alexander was Operation Husky, the forthcoming invasion of Sicily, and the need for a rapid turnabout in American performance.[27] Alexander reported his findings to Eisenhower in far more tactful terms, saying that although American troops were inexperienced in battle, they should soon be the equal of any fighting soldiers in the world.[28] The message, however politely couched, was that Alexander rated the American forces in Tunisia rank amateurs in the art of war.

Patton had only ten days to make his presence felt before the corps was to open a diversionary offensive ordered by Alexander to threaten Rommel's flank while Montgomery launched his long-awaited attack against the Mareth Line. Discipline in II Corps was nonexistent. "No salutes. Any sort of clothes and general hell."[29] He ruthlessly used shock tactics to cajole, bully, encourage, and excite his men into believing that they were capable of defeating their enemy. Patton seemed to be everywhere at once, using every leadership stratagem learned during his thirty-four-year military career. Uniform regulations were strictly enforced, and offenders punished with fines, many nabbed personally by Patton. Though seemingly trivial, even nonsensical, what Bradley called Patton's "spit and polish" reign had a badly needed positive effect.

> Each time a soldier knotted his necktie, threaded his leggings, and buckled on his heavy steel helmet, he was forcibly reminded that Patton had come to command the II Corps, that the pre-Kasserine days had ended, and that

a tough new era had begun . . . these Patton reforms promptly stamped his personality upon the corps. And while they did little to increase his popularity, they left no doubt in anyone's mind that Patton was to be the boss.[30]

On his first morning as II Corps commander, Patton ordered the mess to close at 7:30 A.M. Later, when the corps headquarters moved, the mess was closed at 6:00. The staff was accustomed to straggling in at all hours; henceforth they would either arrive on time or go hungry. One of Patton's first orders was that all his officers would wear their insignia of rank on their helmets, even though it would make them better targets. "That's part of your job of being an officer," he said. Patton also told his lieutenants and captains that they were expendable and must personally lead their men into battle and take the same risks as their lowest-ranking soldier.[31]

Although he had far less time to effect a change in the fighting spirit of II Corps, Patton's challenge was not unlike the one that faced Montgomery when he assumed command of the Eighth Army in August 1942. Paul Robinett describes Patton's impact on II Corps:

A rare sense of showmanship supported his qualities of leadership. . . . Patton raced through the countryside . . . he radiated action, glamor, determination, and hearty but reserved comradeship. He came with a Marsian speech and a song of hate; gross, vulgar, and profane, although touchingly beautiful and spiritual at times. . . . The old soldiers, who knew him as *Gorgeous Georgie* or *Flash Gordon*, rejoiced at his coming, even though they feared his rashness. They knew that he would demand much, but that there would be a pat on the back for every kick in the pants and that their interests would be his interests.[32]

Patton's inspections took the usual form of exhortation and speeches, sometimes to the troops, sometimes to their officers. Robinett recalls that Patton's speech to the officers of his regiment was

so fantastic that the best of us could not give a good report of just what was said. In any case, it would be unprintable. . . . Of all the senior commanders in World War II, General Patton understood best the teachings of one of the very greatest American soldiers, Gen. William T. Sherman: "No man can properly command an army from the rear, he must be at the front . . . at the very head of the army—[he] must be seen there, and the effect of his mind and personal energy must be felt by every officer and man present with it."

Robinett also thought that "Patton's undoubted success as a commander can, in large measure, be attributed to his unrelenting application of this principle.[33]

Patton's urgency was based as much on the fact that Fredendall had never issued any orders about proper dress and deportment as it was on impending combat operations. "It is absurd to believe that soldiers who cannot be made to wear the proper uniform can be induced to move forward in battle," he wrote in one of his first diary entries in Tunisia. Although he had clearly given Fredendall the benefit of the doubt, within a week Patton concluded that his predecessor had "just existed—he did not command, and with few exceptions, his staff was worthless due to youth and lack of leadership."[34]

The immediate crackdown on sloppy uniforms and discipline netted many offenders, most often for failing to wear a helmet. Harry Semmes remembers that among the front-line soldiers "there was a fable that Patton caught one of his staff officers on the latrine without his helmet and fined him $25." GI helmets were most often referred to as "pisspots," but in II Corps they also gained notoriety as "the $25 derby," the usual fine imposed by Patton.[35] (His fines were deliberately handled in a manner that would not reflect unfavorably on the official records of his troops.)

With no time to train II Corps, Patton believed that being seen by his troops and enforcing discipline were two things that would have an immediate impact: "Discipline consists in obeying orders. If men do not obey orders in small things, they are incapable of being [led] in battle. I *will* have discipline—to do otherwise is to commit murder. . . . I cannot see what Fredendall did to justify his existence."[36]

Patton's first impression of Tunisia was that it was "the coldest damn place I have ever seen and it has rained every day since I came . . . it is very hard on the men."[37] As he and Bradley began a whirlwind inspection of the corps, Patton found some units in which morale was high and discipline excellent, while others were obviously in need of quick improvement.

On the evening of March 12, Manton Eddy, the 9th Division commander, notified Patton that he had heard on the radio of his promotion to lieutenant general. Although the official orders would not reach him for another day or two, that night Patton slept proudly under a three-star flag that Dick Jenson had been carrying for a year for just this moment, along with appropriate sets of the three silver stars of a lieutenant general. That night, too, in his diary Patton recalled his childhood dream: "When I was a little boy at home I used to wear a wooden sword and say to myself 'George S. Patton, Jr., Lieutenant General.' At that time I did not know there were full generals. Now I want, and will get, four stars."[38] His pleasure was momentarily tempered, however, by his need to keep warm. "It is realy awful. Jimmy Doolittle is sending me a fleace lined coat. . . . Lots of love. George. *Lt*. Gen. (FIRST TIME)," he wrote to Beatrice.[39]

*   *   *

The pairing of Patton and Bradley set the stage for one of the most unusual relationships of military history: the mercurial Patton who might curse one moment and the next get down on his knees to ask for God's help; and the unflappable, businesslike, taciturn Bradley. If ever two military men were mismatched it was this duo. From the time that circumstance thrust them together in March 1943, it was evident that Patton and Bradley were *not* birds of a feather. They had never before served together, and their disparate personalities assured that there would be little beyond the formalities of a senior-subordinate relationship. The straitlaced Bradley never condoned Patton's use of profanity, although he undoubtedly understood Patton's motives. From the time of their earliest association in Tunisia, Bradley was privately critical of Patton. After the war he recorded his dismay at Patton's methods:

> Why does he use profanity? Certainly he thinks of himself as a destined war leader. Whenever he addressed men he lapsed into violent, obscene language. He always talked down to his troops. When . . . Patton talked to officers and men in the field, his language was studded with profanity and obscenity. I was shocked. He liked to be spectacular, he wanted men to talk about him and to think of him. "I'd rather be looked at than over-looked." Yet when Patton was hosting at the dinner table, his conversation was erudite and he was well-read, intellectual and cultured. Patton was two persons: a Jekyll and Hyde. He was living a role he had set for himself twenty or thirty years before. An amazing figure![40]

Bradley's abhorrence of Patton's profane manner colored his opinion forever. "At times I felt Patton, however successful he was as a corps commander, had not yet learned to command himself." While admitting that "those mannerisms achieved spectacular results . . . they were not calculated to win affection among his officers or men," a criticism in which Patton himself would readily have concurred, though many would just as likely have disputed it.[41]

An example of Patton's methods backfiring occurred in Morocco, as the 3d Division was training for the Sicily invasion. Patton drove up in his command car, stopped at the entrance to the division HQ company, and summoned the headquarters commandant, Maj. James K. Watts: "You, man, come over here." As Major Watts complied, Patton yelled, "You son-of-a-bitch! When I tell you to come, I want you to run. Where the— hell is your headquarters?" After informing Patton where he would find the commanding general, Major Watts boldly said: "Sir, I resent being called a son-of-a-bitch. I think you owe me an apology." Although Patton readily agreed and said he was sorry, it was an example of him at his absolute worst.[42]

An even more odious example occured when Patton shamelessly humiliated his friend Terry Allen during a visit to the 1st Division. When Patton assumed command of II Corps it was inevitable that these two charismatic and iron-willed men would clash, even though they were old friends. During the battle of El Guettar they squabbled endlessly over tactics. On this occasion, after being up all night, Allen and Roosevelt had just gone to bed when Patton arrived early one morning at the 1st Division command post. Roosevelt appeared in a bathrobe as Patton began emoting for Allen's benefit. When he discovered a series of slit trenches around the perimeter of the command post, Patton demanded: "What the hell are those for?" Terry Allen replied that they were for protection against air attack by the Luftwaffe. "Which one is yours?" When Allen pointed it out to him, Patton walked over, unzipped his fly, and urinated into it. "There," he said. "Now try to use it." Allen and Roosevelt both had burly bodyguards who carried stripped-down Thompson submachine guns. Bill Donley, Roosevelt's bodyguard, later told how all that could be heard was the loud click of safeties being deactivated in the deathly silence of the crisp morning air. Had either commander given the order, there is little doubt that they would gladly have shot Patton where he stood. Patton got the message and left as quickly as he had come.[43] It was an ugly incident, a humiliating experience for Terry Allen, and one that did no credit whatsoever to Patton, who always seemed to overreact when they came into contact.[44] Omar Bradley would later cite this as a justifiable example of why he despised Patton.

Bradley also criticized Patton's expression of frustration on learning that an ailing Rommel (who had returned to Germany after two grueling years in the desert) was no longer commanding Army Group Africa. "He was possessed of the idea that he, George Patton, was here to do battle with Rommel. 'Let me meet Rommel in a tank and I'll shoot it out with the son-of-a-bitch.' "[45] Although Patton would later criticize Bradley when the tables of command were turned in northwest Europe, he thought highly of him in North Africa, praising him both publicly and privately; "a swell fellow," he wrote Beatrice. After he left II Corps, Patton wrote to Bradley: "I want to repeat that I never enjoyed service with anyone as much as you and trust that someday we can complete our warlike operations."[46]

Bradley's dislike of Patton stemmed in no small part from his rigid upbringing, in which alcohol and profanity were forbidden. As his collaborator, Clay Blair, points out: "Bradley came from a strict, teetotaling Christian Church, small town, Midwest background. He did not even *taste* alcohol until he was thirty-three years old. [His wife] Mary Bradley did not drink at all . . . and *detested* anyone who was even a moderately heavy drinker."[47]

There was probably another reason for the disparity between Patton and Bradley. Patton and the armored division commanders of World War II were a special breed, quite different from their brother officers. Mostly ex-cavalrymen, they tended to be combative, high strung, flexible, and fiercely loyal to their troops. They led from the front, seized opportunities, created high morale in their units, argued against orders they thought wrong, and took responsibility for acting on their own. On the battlefield armored divisions habitually operated in uncertainty, often across vast areas, and always with the object of destroying the enemy.

By virtue of their mission and training, the infantry and their commanders functioned in narrower terms of terrain, position, and the seizing of specific objectives. There were other significant differences: Armored division commanders inspired, and, contrary to Bradley's characterization of Patton, infantrymen too often led by threats. Failure was punished by swift recrimination and, frequently, relief. Schooled in the rigid doctrines of Fort Benning's Infantry School, Bradley was uncomfortable with those who displayed independence and dash, preferring instead the company of more conservative infantrymen who thought in like terms. As for Patton, Bradley recognized his genius but thought him a dangerously loose cannon, too obsessed with the attainment of personal glory. In short, they were, as Blumenson has written, World War II's "odd couple," and "their problem of accommodating their distinctly different styles and outlooks was a contributory cause [to the faltering of the Allies]" during the latter stages of the Normandy campaign.[48]

Famed American war correspondent Ernie Pyle virtually canonized Bradley in his dispatches as the "G.I. General." Just as Patton's private persona bore no relation to his public image, so too did Bradley's, thanks to Pyle. The public was fed the image of a plain, soft-spoken general with whom the average civilian could readily identify. If truth is the first casualty of war, so was the pretense that Omar Bradley was a general of the masses, an image that Bradley himself gladly encouraged for the remainder of his life. The real Omar Bradley was somewhat narrow-minded and utterly intolerant of failure. As combat historian S. L. A. Marshall categorically asserts, Bradley "was played up by Ernie Pyle. . . . The GI's were not impressed with him. They scarcely knew him. He's not a flamboyant figure and he didn't get out much to troops. And the idea that he was idolized by the average soldier is just rot. He didn't make that much of an imprint . . . [and it] couldn't compare with Patton's on the minds of troops."[49]

Patton has earned an unjustified reputation for sacking commanders, but the truth is that Orlando Ward of the 1st Armored Division was the only senior combat commander he relieved during the *entire* war, and this only after repeated warnings and attempts to force an improvement in the division's performance.

By contrast, in northwest Europe, as the First Army, and later, 12th Army Group commander, Bradley was notoriously intolerant of failure, as the numerous division commanders he sacked could attest. Not surprisingly,

Bradley's favorites were fellow infantrymen in whose company he felt both comfort and kinship. In his war memoir and later autobiography, he praises career infantryman Lt. General Courtney Hodges, his stolid, dependable, and utterly uninspiring successor in command of the First Army, as a far better general than Patton. Noted historian Charles B. MacDonald, himself a decorated infantry company commander, is scathingly critical of both Bradley and Hodges for their handling of one of the bloodiest and most senseless battles fought by American troops in the war, the Hürtgen Forest in November 1944.[50]

One of the war's most decorated and respected heroes was paratroop commander James M. Gavin, who admired Bradley (for "his fine appreciation of the tools of his trade" and his "tactical hands-on knowledge"[51]), but deplored the American tendency to sack senior officers too quickly. The British tended to retain unsatisfactory commanders far too long before relieving them, while the exact opposite occurred in the U.S. Army, where senior commanders were frequently relieved for the slightest transgression, often without adequate time to grow into a job. This practice greatly disturbed Gavin, who has written: "I have a haunting memory that does not diminish with the passage of time of how unfairly and thoughtlessly we treated some of our senior officers," an implicit reference to Bradley, the most ruthless senior American commander in the European Theater of Operations.[52] If all this was well in the future, its roots lay in the untidy battles fought in Tunisia, which helped shape Bradley's approach to command.

Patton's presence in Tunisia was being so well publicized in the United States that he soon worried that he was receiving too much "cheap publicity" in the press at home, "and I fear I shall get a lot more. There are 49 correspondents and photographers here sniping at me."[53] In defense factories across the nation there began appearing large posters reading:

> Old Blood & Guts attacks Rommel
> "Go Forward"
> "Always Go Forward"
> "Go until the last shot is fired and the last drop of gasoline is
> gone and then go forward on foot!"
> —Lieut General George S. Patton, Jr.
> DARE WE WORKING HERE DO LESS?[54]

Patton was impressed with Alexander and his feats in two world wars: "What a man." Patton also thought him a snob "in the best sense of the word." When the two generals observed an Arab place a large vat of water on the shoulders of a tiny Arab woman, Patton said: "That is arab chivalry," and Alexander replied: "Let's stop and jolly well kick ass."[55] Alexander seemed equally impressed with Patton and the grand manner in which he was received with an honor guard during his first visit to II Corps, or so he thought. But as Patton would soon discover, Alexander's affinity for a kindred soul did not extend to entrusting a large American combat force with anything other than the most menial of missions. Nothing would change Alexander's deeply ingrained belief that American units did not fight well and therefore should not be entrusted with major military missions.

For all Patton's commitment to leadership by example, one of the first and most important lessons he learned in Tunisia was to let common sense overrule valor when it came to the Luftwaffe, which wreaked havoc by strafing Allied convoys and any vehicle caught in the open. "So far I have not been straffed from the air, but it is bound to come, as I have to drive miles every day. If you see the plane in time, you stop the car and run like hell for 50 yds off the road and lay down. It seems most undignified but all do it."[56]

He forthrightly admitted to Beatrice that combat frightened him just as it did everyone else. "I still get scared under fire. I guess I will never get used to it, but I still poke along. I dislike the straffing the most."[57] He constantly worried that fear might drive him to cowardice. "I do hope that I do my full duty and show the necessary guts," he wrote the day before his corps went on the offensive. "I rather dislike mines, and the whole damned country is full of them. We lose officers daily, mostly with legs blown off or broken. We have to have sand bags in the bottom of the cars. That helps some."[58]

On March 21, during the Battle of El Guettar, he barely escaped death when a hill at the front was hit with a salvo of 150-mm German shells, "almost on the spot [where he had been sitting]," moments after he had departed. He was also strafed twice during one of his daily excursions about the front. An Associated Press report noted that when attacked by three low-flying Messerschmitts, "the general did what any good soldier would do— dived to the ground. So did everybody else—damn fast. He wasn't the first or the last man to hit the deck."[59] Although Patton relished the aura of traveling about in an MP-escorted command car with sirens screaming, three-starred pennants and the II Corps flag flying, he secretly worried that they would attract the unwelcome attention of the Luftwaffe. "At first I was nervous for fear of air attack," he wrote on March 17, "but soon got used to it. Courage is largely habit and self-confidence. I thank God that He has again aided me."[60]

Patton could still be counted on to do the unexpected. Even his and Bradley's aides-de-camp were not immune from standing watch at night. As Bradley's aide, Chester Hansen, notes: "We were always glad to see those big generals come with aides because it simply meant another man and a shorter watch." To guard his headquarters, Patton had a number of half-tracks with .50-caliber machine guns, one of which was directly outside the room in which they slept. At the first sign of the Luftwaffe, they would open fire with a great chattering sound. Patton had instructed them to fire at German aircraft because he believed it improved the morale of his troops.[61]

In the weeks that followed, Patton's battles would be on both the battle-field and in the boardrooms of the Allied high command. His first task had been to rid II Corps of the legacy of Fredendall and somehow convince his troops that they could do far better. Now the results of his efforts would be put to the ultimate test. Before Patton could square accounts with Alexander and the British, II Corps would have to prove itself in the forthcoming bat-tle, to which history has given the name El Guettar.

CHAPTER 31

Allies

If we ever fight another war with allies, God forbid . . .
—FORMER AEF STAFF OFFICER

Despite the rotten weather, one of Alexander's first acts as the new Allied ground commander in chief was to launch an offensive designed to stabilize the situation and pave the way for the defeat of von Arnim's Axis army in Tunisia. While Montgomery's Eighth Army hammered away at the Mareth Line, Alexander launched the First Army in a successful offensive against von Arnim's panzer army in northern Tunisia. Patton's mission was both secondary and diversionary. II Corps, now four divisions strong, (1st, 9th and 34th Infantry and the 1st Armored), was to attack the Axis flank to draw off reserves that might otherwise interfere with Montgomery's offensive to drive along the coastal plain toward Tunis and crush the enemy between the British First and Eighth Armies. Alexander's directive that II Corps was not to advance beyond the Eastern Dorsal* was an unmistakable warning that Patton was not to interfere with Montgomery's offensive.

The apathy that existed under Fredendall was now a thing of the past, and the night before the offensive opened on March 16 Patton somberly informed his staff: "Gentlemen, tomorrow we attack. If we are not victorious, let no one come back alive." Bradley wrote: "With that, George excused himself and retired alone to his room to pray."[1] In his diary Patton wrote that it had been a "horrible" day,

*The Eastern and Western Dorsals are hill masses that form the chain of mountains that encompass most of Tunisia. Both were key terrain features during the campaign in Tunisia in 1942–43.

one of my pre-match [polo] days . . . the hardest thing a general has to do is wait for the battle to start after all the orders are given. . . . Everything there was time to do has been done. Not enough, but all there is time for. Now it is up to others and I have not too much confidence in any of them. Wish I were triplets and could personally command two divisions and the Corps. Bradley, Gaffey and Lambert are a great comfort. God help me and see to it that I do my duty, but I must have Your help. I am the best there is, but of myself I am not enough. "Give us the victory, Lord." Went to bed and slept well till 0600.[2]

Bradley found Patton's erratic, profane but reverent personality utterly bewildering. His "last appeal for victory even at the price of death was looked upon as a hammy gesture by his corps staff . . . [and] helped to make it more clearly apparent to them that to Patton war was a holy crusade."[3]

Patton conceived of the forthcoming offensive to capture the hill towns of Gafsa, Sened, and Maknassy in the Eastern Dorsal as a modern-day reenactment of the Second Battle of Manassas: Stonewall Jackson (II Corps) to make a flank attack so that Longstreet's corps (Montgomery's Eighth Army) could make a breakthrough. He also attempted to evade the restrictions imposed by Alexander. Patton was determined to begin the reversal of American fortunes by taking to the offensive. Although Alexander had ordered him to confine his actions to the mountains of the Eastern Dorsal, Patton had a far more ambitious scheme in mind and devised a two-phase plan. As the main elements of the 9th and 34th Divisions feinted toward Faid and Fondouk, the 1st Armored and an infantry regiment of the 9th Division were to recapture the area southeast of Kasserine, drive through the Maknassy Pass, and threaten to break out into the coastal plains on the other side. The 1st Division would then capture Gafsa and drive toward Gabès via El Guettar. If the operation was successful, the threat to the German-Italian army would be considerable and simply could not be ignored by Alexander.

Alexander's far more conservative prescription was that II Corps merely regain the ground lost the previous month, including airfields for the RAF, and Gafsa, which was to become a future logistical base for Eighth Army. By *threatening* to break out toward Gabès, Alexander's aim was to force von Arnim to commit his reserve at Gafsa, rather than Mareth, when Montgomery launched his breakthrough attempt.

Despite heavy rains the operation opened auspiciously the night of March 16–17 as the 1st Division seized Gafsa and moved reconnaissance elements toward El Guettar. Von Arnim took Alexander's bait and overreacted to the American threat, perceiving it to be part of a coordinated Allied plan to envelop his rear, trapping the Italian First Army and, if successful, the Afrika Korps as well. He committed the veteran 10th Panzer Division to

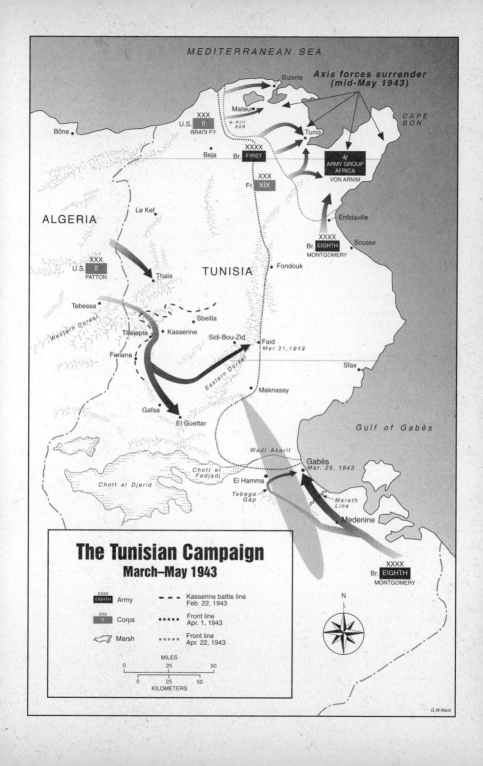

MEDITERRANEAN SEA

Axis forces surrender
(mid-May 1943)

Bizerte

Mateur

+ Hill
609

CAPE
BON

Bône

U.S. XXX II
BRADLEY

Tunis

Beja

Br. XXXX FIRST

ARMY GROUP
AFRICA

VON ARNIM

Fr. XXX XIX

ALGERIA

Le Kef

Enfidaville

XXXX
Br. EIGHTH
MONTGOMERY

Sousse

U.S. XXX II
PATTON

Thala

TUNISIA

Fondouk

Tebessa

Sbeitla

Western Dorsal

Thelepte

Kasserine

Sidi-Bou-Zid

Faid
Mar. 31, 1943

Feriana

Eastern Dorsal

Sfax

Gafsa

Maknassy

El Guettar

Gulf of Gabès

Wadi Akarit

Gabès
Mar. 29, 1943

Chott el
Fedjadj

El Hamma

Chott el Djerid

Tebega
Gap

Mareth
Line

Medenine

The Tunisian Campaign
March–May 1943

XXXX EIGHTH	Army	--- Kasserine battle line Feb. 22, 1943
XXX II	Corps	···· Front line Apr. 1, 1943
	Marsh	···· Front line Apr. 22, 1943

N

MILES
0   25   50

0   25   50
KILOMETERS

XXXX
Br. EIGHTH
MONTGOMERY

G.W. Ward

the II Corps sector to reinforce the Italian Centauro Division at El Guettar and the Maknassy Pass, which was the key to his defense of the Eastern Dorsal.

On March 22, 10th Panzer arrived at the very moment when the 1st Armored Division was attempting to breach Maknassy Pass, which had been virtually undefended twenty-four hours earlier. Aware of the limited American role dictated by Alexander, and perhaps overly concerned about the ability of the Luftwaffe to attack from nearby bases, Orlando Ward had elected to regroup and renew his attack the following day. This unfortunate decision resulted in another stinging setback for 1st Armored and infuriated Patton, who had counted on the success of this operation to carry out his expanded offensive. Patton's rage was not confined to Ward; the weather had turned foul, and everywhere it was "a sea of mud, just awful."

Patton's fortunes improved when an armored battle group of 10th Panzer, supported by infantry, assault guns, artillery, and Stuka dive-bombers, was encountered the morning of March 23 on the open plain along the Gabès-Gafsa highway east of El Guettar. The 1st Infantry Division, backed by massed artillery and tank destroyers, was disposed in the hills overlooking the highway. The battle raged throughout the day, as the ambushed 10th Panzer lost thirty-two tanks and suffered heavy infantry losses before withdrawing in disarray. Patton's elation was tempered by his disgust over the poor German tactics, which resulted in what he thought was the senseless slaughter of good infantry.

During the eight-day offensive Patton was constantly on the move to visit his front-line troops to motivate their commanders and influence the outcome in any way he could. En route to visit Ward on March 19, Patton was shot at by an Arab, the bullet coming close enough for him to hear its whine. The one day he was unable to leave his headquarters, he complained it was "a hell of a way to fight a war." His nightly diary entries contained pleas to the Almighty to stick with him. After he narrowly missed death on a hillside overlooking El Guettar, these seemed a reasonable request.

Unlike a modern-day corps commander, such as Lt. Gen. Frederick M. Franks, who commanded the U.S. VII Corps in the Gulf War in 1991, Patton's ability to influence the action on the battlefield was limited to the range of his radios and the distance he could cover in a day across a vast operational area. Franks stayed in touch with his commanders on the ground from a helicopter-borne command post equipped with laptop computers and sophisticated communications. Attack helicopters, missiles, the most modern and deadly tank in the world, and an array of lethal weapons completed the corps arsenal. In stark contrast, Patton's control over II Corps at El Guettar was so limited that he was obliged to sit virtually incommunicado on a hillside to observe the battle, during which he was nearly killed by German artillery.

By World War II standards, the Battle of El Guettar was a minor engagement, but for the U.S. Army it was a significant victory. Patton's corps had successfully carried out an important mission and had begun the process of redemption for Kasserine. As one participant would later observe: "Probably, the greatest training benefit of the Battle of El Guettar was learning that the opponent was not ten feet tall."[4]

After the battle Patton and Bradley examined the devastated German and Italian positions and came away impressed with the power of American artillery and appreciative of the hell their men had endured. "After seeing how strong the position was . . . I wonder that we ever drove them out," Patton wrote in his diary.[5] Although Rommel had already left Africa, Patton consoled himself with the knowledge that he had bested the Germans in battle, even though he would have preferred acting out his chivalrous dream of a one-on-one duel to the death.

The lesson for von Arnim and his commanders was equally clear: The U.S. Army was no longer to be taken lightly as an adversary. In less than three months American forces would demonstrate the accuracy of correspondent Drew Middleton's observation: "Armies never learn from other armies. They have to learn by themselves, and a lot of the tactics that we used were those the British had used disastrously two years earlier and discarded."[6]

In the aftermath of El Guettar, Eighth Army broke through the Mareth Line in early April, and Alexander offered Patton an opportunity to play a potentially spectacular role by thrusting his tanks to the coast at Gabès, thus severing the Afrika Korps from Fifth Panzer Army and trapping both.[7] Patton sent a 1st Armored task force, commanded by Col. C. C. Benson, one of his former AEF tankers, which encountered tough opposition and in three days made only scant progress, losing thirteen tanks. Patton's daily presence at the front made little impact, despite "a terrible temptation to interfere."

What no one except Beatrice knew was that his daily forays into the front lines that enabled him to see and be seen by his troops camouflaged an ever-present, gnawing fear, which he struggled constantly to overcome. He was shot at "quite a little. . . . My old fear of fear came up again today," he confessed to Beatrice. "There was a bad place on the road that they were shelling quite accurately. I began to find reasons why I need not go any further but of course I did and nothing happened."[8] During an air raid alert the previous night, Patton thought it was a ground attack and ran outside with his carbine in his hand, his "old lust of battle" as "hot as ever though I admit I hate shells and bombs."[9]

Several weeks later he again tested himself by deliberately walking through an uncleared minefield to restore what he perceived as flagging confidence in a 1st Armored unit. His obsession with forging ahead seemed secondary only to his insistence on defying death, as if to prove to himself over and over again that he did not lack courage—lending credence to Eisenhower's foresight in warning him not to act foolishly and become a casualty.

By early April the offensive that had begun so promisingly at El Guettar simply ran out of steam, as II Corps continued to struggle against the stiffened Axis defenses along the Eastern Dorsal. Patton's lone opportunity had come and gone, and with it his patience with Ward. He thought the 1st Armored commander lacked drive and had not done well. As Bradley later wrote:

> Patton pushed Ward constantly to break through at Maknassy and seemed unreasonable toward Ward for his failure to break through. This was climaxed when Patton called Ward one evening [March 24] and demanded, "Goddamn it! I want that hill in front of you. Get off your ass; get a pistol in your hand, and lead that attack yourself." Ward did lead a night attack at the head of his infantry . . . [and] was slightly wounded in the eye. He probably would have preferred to have been killed that night.[10]

Bradley's biased version discounts the situation Patton faced when he assumed command of II Corps. The 1st Armored Division had been von Arnim's prime victims at Sidi-Bou-Zid, and Rommel's at Kasserine Pass, and Patton rightly considered its leadership too timid and in need of a solid jolt to restore both morale and confidence. His decision to force Ward to lead the attack, while unprecedented and seemingly heartless, was less so in light of the circumstances. Nor was it made without considerable trepidation. In his diary Patton wrote: "Now my conscience hurts me for fear I have ordered him to his death, but I feel that it was my duty." Although the attack failed, Patton decorated Ward with a Silver Star. "I believe his action would have merited the DSC except for the fact that it was necessary for me to order him to do it."[11]

Bradley's opinion may have derived from the fact that Patton relieved Ward not long after this incident. However, Patton's loss of confidence in Ward was based on his belief that the highly popular artilleryman was simply not the right commander to lead the division. "I have little confidence in Ward or in the 1st Armored Division. Ward lacks force. The division has lost its nerve and is jumpy. I fear that our troops want to fight without getting killed," he wrote on March 28.[12] He needled Ward incessantly and on March 31 ordered him to attack even if he sustained losses of 25 percent. "I feel quite brutal in issuing orders to take such losses, especially when I personally am safe, but it must be done. Wars can only be won by killing, and the sooner we start the better."[13]

During the attack for the Maknassy heights, Keyes urged Patton to visit the 1st Armored to make clear his feelings about its importance, but he declined, believing a personal visit might "scare him [Ward] to death," and sent Gaffey in his stead, a decision he later regretted.[14] On April 4 Patton dispatched Bradley to inform Ward he was to be relieved. "While Alexander has written me a letter asking that I relieve him, I did not use this as a cloak for my act. I should have relieved him on the 22d or 23d [of March], but did not do so as I hate to change leaders in battle, but a new leader is better than a timid one."[15]

Ward's relief was greeted with extreme hostility by the men of the 1st Armored. Many never forgave Patton, particularly when his able replacement, Ernie Harmon, arrived and articulated the same message: that the division was in serious need of a good shaking-up and a stern new brand of leadership. In salty, Pattonesque language, Harmon informed his officers and NCOs that the casualties at Kasserine and Sidi-Bou-Zid had died unnecessarily. Some of those who did not, he remarked, were yellowbellies who were afraid to die. Harmon's aim, as Patton's had been in sacking Ward, was to revitalize a division that was feeling sorry for itself, and by the end of the Tunisian campaign Harmon had restored credibility to the 1st Armored, whose anger at him (and Patton) ebbed as they began to win battles.

However painful to both men, Patton's decision to replace the well-liked "Pinky" Ward with the bullheaded, fearless Ernie Harmon proved to be the right one.[16] Indeed, Patton's instincts rarely failed him and, as he once related to author John P. Marquand: "You've got to get the feel of troops to lead them. You've got to know how much is left in them and how much more they can take. I know just how much more by looking in their faces."[17] Patton had correctly sensed the need to shake up the 1st Armored and taken the unpleasant step of replacing a man he personally liked.

Despite commanding a front-line corps, there were too many moments of boredom and inaction to suit Patton's aggressive nature. As any veteran soldier will attest, combat is long periods of boredom, sandwiched between unmitigated terror, and it was no different for the commanding general of II Corps. "The hardest thing I have to do is to do nothing . . . the men on the ground have to do the fighting and by and large they are doing well." Some days he was obliged to remind himself that he was earning his pay.

Part of Patton's routine were the almost daily visits to the clearing stations and field hospitals, where "the men like to see me." Some he found "pretty gruesome," and he made a point of thanking God for helping him to keep his composure. At one hospital he noted: "It was strange how the men followed me with their eyes, fearing I would not speak to each one. I talked to all who were conscious. One little boy said, 'Are you General Patton?' I said, 'Yes,' and he said, 'Oh, God.' Another one said, 'You know *me*. You made a talk to my battalion at Casablanca.' I told him I remembered him well."[18] At another he saw one "poor devil" missing a leg. "I asked him how he felt and he said fine, since you came to see me. I suppose I do some good but it always makes me choke up. I have no personal feeling of responsibility for getting them hurt, as I took the same chances, but I hate to look at them."[19]

Omar Bradley and others have criticized Patton's style of command as too often demeaning and suggested that he was not well liked by his troops. If Patton's ruthlessness shocked some of the troops, the presence of a three-star general in their midst just as often had precisely the right effect. Typical was his unexpected appearance at the position of an engineer unit of the 9th Division. Pvt. Charles B. Hoffman observed the arrival of a jeep with a three-star flag. Its occupant dismounted and said his name was Patton, the new commander of II Corps. "I'll never forget his uniform, from his helmet to the riding boots he wore . . . and [the] binoculars hanging from his neck." As a lieutenant briefed him, Patton "kept looking at the hills ahead. I remember his last words were, 'We'll beat the sons of bitches,' and he left. I looked at Lt. Ware with amazement but we both smiled and sure felt good." After El Guettar the 9th Division was given a two-day rest period, during which Patton demonstrated his pleasure with their performance by arranging for everyone to have a steak dinner.[20]

Patton's detractors—and there were many—thought he was a royal pain in the ass, and were delighted that one of Bradley's first acts after he assumed command of II Corps was to revoke Patton's edict that everyone wear neckties, and permit shirtsleeves to be rolled up. In the minds of many, "it marked a basic difference between Patton and him and established Bradley forever as the 'soldier's soldier' who looked out for the little guy."[21]

Patton was often unfairly tough on the commanders. During the battle of El Guettar, Patton strode into the operations tent of the 47th Regiment where he encountered Maj. Gen. Manton S. Eddy, the division commander, conferring with the regimental commander, Col. Edwin H. Randle. Patton took Eddy aside and tongue-lashed him for failing to be forward with his troops. As Randle later recalled, Eddy said: "In all my career I've never been talked to as Patton talked to me this morning. I may be relieved of command." He was not, and in northwest Europe not only commanded a corps under Patton but also received a Distinguished Service Cross. Several days later, when Patton again charged into Randle's command post and ordered him to attack, Randle retorted, "Hell, General, I have been ordering my men forward against the German position and unless I can get some air power and heavy artillery support, I don't intend to order my men forward any more." Patton left without a further word. It was an example of his tendency to respect those with the courage to stand up to him, who were thereafter usually left alone.[22]

Randle also has more pleasant memories of Patton. At El Guettar, he appeared one morning and asked, "Have you got some place where I can see what's going on?" To get to a forward observation post the two officers were obliged to drive in plain sight of the enemy, which often fired its artillery at moving vehicles. Randle thought to himself, "You'll be in one big mess if you get the corps commander shot up." Arriving without incident, Patton looked out over the valley and the desolate ground as far as the eye could see and exclaimed: "Gee! I had no idea it looked like this up here!" Randle later wrote: "That exclamation, 'Gee!' doesn't sound like Patton language, but that is exactly what he said. Patton remained for an hour, and when they returned to his command post, Randle apologetically offered him a lunch of cold C rations. "He grinned and said, 'Sure!' and we sat on a bench each opening a can and eating with a spoon. I think he enjoyed that too. . . ."[23]

During the battle of El Guettar, Patton arrived at the 9th Division headquarters shortly after a Luftwaffe "bed-check Charlie" had harassed the division staff with his nightly bombing, and in a loud voice heard by all, said: "Manton, I want you to get these staff officers out of those holes and out here where they can be shot at." An overwrought staff colonel went berserk and had to be evacuated to a hospital for psychiatric examination. He never returned to his unit.[24]

On April 1, 1943, Patton's beloved aide, Capt. Dick Jenson, was killed when several five-hundred-pound bombs from Junker twin-engined bombers landed on Colonel Benson's forward command post. Jenson had been there at the behest of Patton, who wanted him to gain some experience and exposure to front-line conditions. Although there was enough advance warning for Bradley and the other officers to dive into nearby slit trenches, one bomb landed on the edge of Jenson's foxhole, and he was killed instantly by the concussion. Bradley, who was barely ten feet away from Jenson, was unscratched. Patton was utterly devastated; it was as if he had lost his own son. The man regarded as the most hardboiled American commander of the war got down on his knees and, tears streaming down his cheeks, wept unashamedly over Jenson's body. It was one of the saddest days of Patton's life.[25] In his diary, Patton recorded his sorrow:

I am terribly sorry as he was a fine boy, loyal, unselfish and efficient. As soon as he was brought in I went to the cemetery with Gaffey. He was on a stretcher rolled up in a shelter half. We uncovered his face and I got on my knees to say a prayer, and all the men did the same. . . . I kissed him on the brow and covered him up. At 1600 [hours—4 P.M.] Stiller, Sgt. Meeks, Sgt. Mims and I went to the cemetery. Dick . . . was wrapped in a white mattress cover. We had a squad and a trumpeter, but did not fire the volleys, as it would make people think an air raid was on. The Corps Chaplain read the Episcopal service and he was lowered in. There are no coffins here, as there is no wood . . . He was a fine man and officer. He had no vices. I can't see the reason that such fine young men get killed. I shall miss him a lot. C'est la guerre.[26]

Like nothing else he had ever experienced, Dick Jenson's death haunted Patton for the remainder of his life, and served as a constant reminder of the awesome responsibility of being a commanding general. "I am really more broken up over Dick than I can express," he wrote to Keyes. "I did not know how fond I was of him." He sent a moving letter to Jenson's widow and the day after his funeral returned to photograph the grave and place flowers on it. Correspondents calling on him in Gafsa after the funeral found him still in tears.

Patton's simmering rage at being ignored by Alexander boiled over in early April. Acting in no small part out of frustration at Jenson's death, Patton personally appended a note to the routine Situation Report (Sitrep) for April 1 that read: "Forward troops have been continuously bombed all morning [by the Luftwaffe]. Total lack of air cover for our units has allowed German air force to operate almost at will." To emphasize the point, Patton took the unusual step of sending the Sitrep over his personal signature.[27]

When it landed on the desk of the commander of Allied tactical air forces in North Africa, Air Vice Marshal Arthur "Mary" Coningham, the tough, outspoken airman exploded in anger. Although his American deputy, Brig. Gen. Lawrence S. Kuter, intended to urge Patton to retract his message, his hot-tempered boss, bent on avenging the slander of his air force, was in no mood to listen to reason. In retaliation Coningham deliberately sent a provocative signal to every commander in the Mediterranean, sarcastically suggesting that the Sitrep must be someone's idea of an April Fool's Day joke, and that the real problem was that II Corps was not battleworthy.[28]

Coningham's message nearly created an international incident. Eisenhower's air deputy, Air Chief Marshal Sir Arthur Tedder (who also commanded all Allied air forces in the Mediterranean), was handed a copy of Coningham's signal on April 1 when his aircraft landed at Constantine to refuel. Tedder thought it was "ill-judged" and later observed that "for some hours there was grave danger of very serious political and international repercussions . . . this was dynamite with a short, fast-burning fuse, and the situation could well have led to a major crisis in Anglo-American relations." Moving quickly to defuse the situation, Tedder telephoned Eisenhower that he had ordered Coningham to cancel the signal and personally apologize to Patton. One of Eisenhower's biographers has written that he "saw the whole thing as a personal failure. He was not doing his job. . . . "[29] Eisenhower nearly resigned, since it seemed clear to him that if he could not control his own subordinate commanders he had no business being the Allied commander in chief. Only timely intervention by the pragmatic Bedell Smith prevented Eisenhower from sending such a cable to Marshall.[30]

Tedder, Lt. Gen. Carl A. "Tooey" Spaatz (who commanded the Allied strategic bombers), and Kuter drove to Patton's headquarters in Gafsa on April 3 to engage in some hasty fence-mending. Kuter (whom Patton snidely referred to in his diary as "some boy wonder") later recalled how Patton greeted them "wearing his fiercest scowl," which never left his face. "We expressed regrets that Coningham's message had gone beyond the facts in the case, in[to] the area of acrimonious and controversial opinion. Patton maintained a belligerent posture which impressed me as the attitude of a small boy who knew he'd been bad but believed he would get away with it." The airmen left after assuring Patton that Coningham would be calling on him shortly.[31]

The Luftwaffe picked that precise moment to attack Patton's headquarters. Three or four Focke-Wulf 190 fighter-bombers made a strafing run straight down the street outside the room where the meeting was taking place, firing their machine guns and dropping a number of small bombs. The only casualty was a camel that dislocated its leg running away.[32] According to Bradley plaster flaked from the ceiling, and the door was wedged shut from the concussion of the German bombs. "Tedder packed his pipe, looked up mischievously from the table and smiled. Tooey looked out the window . . . turned to Patton and shook his head. 'Now how in hell did you ever manage to stage that?'" Tedder said: "I knew you were a good stage manager, but this takes the cake." With a huge grin on his face, Patton retorted: "I'll be damned if I know, but if I could find the sonsabitches who flew those planes I'd mail them each a medal."[33]

The next day there was an epic showdown between two of the war's foremost prima donnas. Coningham had telephoned to ask if he might call on Patton about noon, but the notion of dining with his adversary was so distasteful that Patton deliberately ate lunch early. Coningham arrived to find Patton "seated squarely behind the dead center of his flag-surrounded desk, wearing his polished helmet" and his ivory-handled revolvers. Patton deliberately snubbed Coningham by failing to rise or to shake his hand and began lecturing the airman, "extolling the 'unquestioned bravery of his 62,000 [sic]* men' and concluded with a vigorous fist on the desk."

Coningham replied: "I am so dreadfully sorry for that horrible signal. I want to apologize and do what I can to make amends. I would never have done it, but I had received a barrage of requests all day. I am proud of my air force and will not have them criticized." To emphasize the point, Coningham pounded his fist on Patton's desk. Patton snorted that if the occasion were to arise again he would prepare additional Sitreps exactly like the one

---

*Approximately 82,000 men were actually assigned to II Corps.

Coningham had challenged. However, Patton did concede that "personally
. . . I am willing to accept your apology and forgive you for indicating that I
was a fool, but I cannot accept an oral apology for your calling 60,000
American soldiers un-battleworthy and failing in their duty." Not to be out-
done, Coningham said that "he also enjoyed a good fight. . . ."

By now the two generals were practically shouting at each other. Patton
hotly replied that his men too had been under a barrage all day and it was
Coningham's fault. "Pardon my *also* shouting, but I too have pride and will
not stand for having Americans called cowards. I have asked for an official
investigation. If I had said half what you said I would be a Colonel and on
my way home." Finally, common sense (and perhaps the dreaded words
"official investigation") prevailed, and both men backed off. Coningham
then said: "I am awfully sorry. What can I do to make amends?"

Patton required a cable "specifically retracting your remarks about the
lack of battle-worthiness of our men," to be sent "to the same people to
whom you sent the first message," after which "I shall consider the incident
closed."

"Done," replied Coningham, and the two shook hands, whereupon Patton changed his mind and took the airman off to lunch "with much laughter
and good fellowship."[34] Coningham wrote his own version the next day,
noting that the tragedy of the incident was that "the message conveyed the
impression that I was suggesting that all of the II Corps were not battle-
worthy. This was not even thought about. . . . I thought of only very small
numbers."[35]

"I like him very much," wrote Coningham, "he is a gentleman and a
gallant warrior. But on the slightest provocation he breathes fire and battle,
and as I also like fighting I could not resist the challenge when he turned the
barrage on to me . . . we had a very friendly lunch and talked of work, and
we both now consider the matter closed."[36]

Despite their fiery confrontation, Patton, too, decided he liked his
adversary. When they parted Coningham said: "I can't thank you enough.
You have been very generous." Patton replied: "It is always easy to be generous to a gentleman who admits his mistakes." Privately Patton thought the
promised signal would never materialize because Eisenhower, in the interests of inter-Allied unity, would probably tell Coningham it was not necessary. Coningham, however, was true to his word and that night issued an
effusively sincere public apology that was distributed throughout Tunisia.[37]

The next morning Patton sent a letter that acknowledged "our most sincere appreciation of your more than generous signal. Personally, while I
regret the misunderstanding, for which I was partially responsible, I cannot
but take comfort and satisfaction from the fact that it gave me an opportunity of becoming better acquainted with you, because to me you exemplify
in their most perfect form all the characteristics of the fighting gentleman."[38]

In fact Patton was "mad and very disgusted" by Coningham's message and kept the issue alive by writing to Eisenhower that the apology was "altogether inadequate."

Eisenhower fired back a letter admonishing Patton for his "unwise distribution" of his original Sitrep, and for "demanding the last pound of flesh" from Coningham. Next time, said Eisenhower, Patton should notify him by "confidential report."[39] Patton received Eisenhower's letter with scarcely concealed anger, and after reading it said: "It is noteworthy that had I done what Conygham [sic] did, I would have been relieved. Ike told me later that he could not punish Conygham because he was a New Zealander and political reasons forbad. Unfortunately, I am neither a Democrat or a Republican—just a soldier."[40]

Publicly, however, the contretemps had been resolved, and for this both Coningham and Patton deserve credit for swallowing their oversize egos to settle what might easily have been the nastiest, most potentially damaging Anglo-American quarrel of the war. It could not have occurred at a worse moment. For all Eisenhower's talk of harmony and cooperation, the Anglo-American marriage was teetering on the brink of rebellion by the American ground commanders who, like Patton, believed that Eisenhower had sold them out to the British. The souring relations between the two parties threatened to poison the alliance at its most formative moment and claim Eisenhower as its best-known victim. In fact Patton's handling of the potentially explosive Coningham incident (ham-fisted as it was) nevertheless may have saved Eisenhower's job. At the very least Patton converted the strong-willed Coningham into an uneasy ally in the war against their real enemy, the Axis.

Alexander's continued distrust of the U.S. Army was again reflected in his plan for the decisive battle of the Tunisian campaign in which the British were to make the main effort, while II Corps was relegated to a minor role of protecting the British flank. The insignificance of the American role did not escape the attention of Marshall, who bluntly signaled his dissatisfaction to Eisenhower. What disturbed the chief of staff was the implication in the U.S. press that the 34th Division had spoiled Montgomery's chances of trapping the Afrika Korps, and that American troops were being given menial tasks on the battlefield.

Patton's highly charged emotions were hardly unique and certainly cannot be attributed to his thirst for personal recognition. What he penned in his diary on April 7 was a consensus felt by all American field commanders in Tunisia that

after having spent thousands of casualties making a breakthrough, we are not allowed to exploit it. The excuse is that we might interfere with the Eighth Army. . . . One can only conclude that when the Eighth Army is going well, we are to halt so as not to take any glory. It is an inspiring method of making war and shows rare qualities of leadership, and Ike falls for it. Oh! for a Pershing. . . . Sic Transit Gloria Mundi.[41]

The issue of the 34th became a *cause célèbre* in Tunisia and further deepened Anglo-American discord at the top. Although Bradley later took full credit in his memoirs for saving the 34th Division from Alexander's censure, it was actually Patton who responded when Alexander informed him on April 11 that one of his corps commanders, Lt. Gen. J. T. Crocker, had reported that the 34th Division was "no good," and that he planned to place II Corps under the control of Anderson's First British Army. Publicly Patton managed to control his outrage that the British had leaked their displeasure to the press and now planned to relegate II Corps to a demeaning role.[42]

Ever since Kasserine, American commanders had been seething with indignation over what they considered unfair British criticism, loose talk to war correspondents, and Eisenhower's order that there was to be no protesting of British leadership. Even the mild-mannered Bradley began to question the wisdom of Eisenhower's failure to restrain the British. When correspondents reported Crocker's public criticism of the 34th Division, Bradley was irate and later noted: "No words can describe Patton's rage and fury at Ike." In his diary Patton's prose was purple. "God damn all British and all so-called Americans who have their legs pulled by them. I will bet that Ike does nothing about it. I would rather be commanded by an Arab. I think less than nothing of Arabs." A day later Patton was still seething that "Ike is more British than the British and is putty in their hands. Oh, God, for John J. Pershing."[43]

Patton was invited to lunch with Alexander on April 11, where he encountered Coningham, who took him aside to say that he deeply appreciated "your most generous letter." Patton slyly noted in his diary that he lied when he replied that "it was from the heart," but thought his diplomatic gesture enhanced the likelihood of receiving the air support he required from Coningham. Less pleasing was his private conversation with Alexander concerning the 34th Division and II Corps.

That evening Patton penned a strongly worded letter to Alexander protesting the relegation of II Corps to the role of a bridesmaid. "Frankly, I am not happy about it, and I feel that I should be lacking in candor if I failed to again bring the matter to your attention," he wrote. It was a matter of national prestige that his corps should have an equal role with the British. A day later he sent a second letter, which Bradley hand-carried to Alexander, about retaining the 34th Division under II Corps control. "If we deny the 34th Division a chance, we will so besmirch its reputation as to render its future utilization of dubious value. . . . General Bradley . . . who will hand you this letter, can give you other cogent reasons."[44] Alexander agreed only reluctantly, and the incident deepened American skepticism of Alexander and the British.[45]

American commanders in Tunisia had become so paranoid over Eisenhower's edict about cooperation with the British that Patton thought his letter to Alexander might cost him his job. "I feel all the time there must be a showdown and that I may be one of its victims."[46] In the privacy of his diary he continued to rage, and might have been even more unforgiving had he learned that there was every likelihood Eisenhower would never have intervened had Marshall not all but directed him to ensure that U.S. forces were given a meaningful role in the final battles in Tunisia.

In mid-April Eisenhower felt that the situation in Tunisia had stabilized sufficiently for Patton to return to his permanent command in Morocco to complete planning for the invasion of Sicily. Even if the feeling was not reciprocated, during his brief command of II Corps, Patton had been impressed by Bradley. The principal combat element of the U.S. invasion force in the Sicily landings was to have been the VI Corps, commanded by Maj. Gen. Ernest J. Dawley. However, as Patton told Bradley: "I've worked with you and I've got confidence in you. On the other hand I don't know what in hell Dawley can do. If you've got no objection, I'm going to ask Ike to fix it up." At Patton's urging VI Corps was switched by Eisenhower to Mark Clark's Fifth Army, and II Corps slated for Sicily. Bradley was rewarded with his first corps command and, at the end of the campaign, a third star.[47]

Ironically, Patton eventually rated higher in Erwin Rommel's esteem than he did in Omar Bradley's. After Kasserine, Rommel had observed:

> In Tunisia the Americans had to pay a stiff price for their experience, but it brought rich dividends. Even at that time, the American generals showed themselves to be very advanced in the tactical handling of their forces, although we had to wait until the Patton Army in France to see the most astonishing achievements in mobile warfare. The Americans, it is fair to say, profited far more than the British from their experience in Africa, thus confirming the axiom that education is easier than re-education.[48]

Although Patton had warmly praised Bradley, the latter had virtually nothing positive to say about Patton in his two postwar memoirs. With only the faintest of praise, Bradley described Patton's command of II Corps as anything but his "finest hour." Although he credited Patton with restoring discipline and self-confidence, Bradley argued that II Corps had been so badly handcuffed by Alexander that "on the whole we had merely learned to walk."[49] In his war memoir, *Crusade in Europe*, Eisenhower was far more

generous to Patton.[50] Fifty years later an assessment of the Tunisian campaign would put his role in Tunisia into perspective: "Patton pushed his men
to fight and dress like the best soldiers in the world. Within days they knew
they were led by a commander who would not let them fail."[51]

Bradley later caustically asserted that his command of II Corps had
been far less ostentatious and that he was firm but more compassionate.
Bradley's first act was to ease or rescind several of what he considered to be
Patton's drastic regulations. "I coaxed rather than ordered. . . . II Corps was
soon working smoothly, as a good team should [wrongly implyng it had not
under Patton]."[52]

Bradley also scolded Patton for his criticism of Eisenhower, citing his
advice not to provoke Eisenhower further and deplored Patton's diary entry
of April 15 (excerpted in the press after an enlisted man pilfered and sold it
to a New York newspaper) in which he condemned Eisenhower as "an ass"
for his pro-British bias for telling Alexander that "he did not consider himself an American but as an ally."[53] Yet Bradley's own memoirs are scathingly
critical of Eisenhower, and make clear his disgust with his superior's apparent favoritism of the British. When Eisenhower wrote that as the new II
Corps commander he must be "tough" and produce results, Bradley thought
it so "patronizing" that he filed it without reply as "so much grist for the historians."[54] Bradley's attitude toward the British was formed in Tunisia, hardened into an implacable Anglophobia in Sicily, and flowered in full force in
Normandy in 1944. Despite the similarity of their views toward the British,
there is little evidence of warmth in what was primarily a professional relationship between the two West Point classmates. After the war they had little
to do with each other, and only after Eisenhower's death did Bradley reveal
that "I shared Patton's misgivings about Ike, though I was less harsh in my
private judgments and never criticized him before others. Ike was too weak,
much too prone to knuckle under to the British, often . . . at our expense."[55]

Eisenhower had become so concerned over the American role in Tunisia
that he met with Alexander, Patton, and Bradley on April 14 to clear the air.
Eisenhower's version is that the discussions were frank and centered on
expanding the Americans' role to permit them to fight as a national entity. If
he and Bradley were in full accord, as Eisenhower's diary suggests, it was
news to Patton, whose mood was at an all-time low in Tunisia. His diary
was a bitter diatribe of complaints that Eisenhower had never even mentioned his and Bradley's victory at El Guettar ("some leader"); he scorned
Ike's "we are all allies" pitch. In a stunning turnabout Eisenhower later criticized Patton for taking his orders about cooperation "so literally that they
had been meek in acquiescing without argument to [Alexander's] orders
from above." Having been sent two widely varying but well-meaning sets of
guidance by Eisenhower, Bradley and Patton both might have been forgiven
for wondering just what it was their boss expected of them.

During the final month of the Tunisian campaign II Corps proved equal to the challenge Marshall had insisted it be given. What Alexander had intended to be yet another minor diversionary action, at Bizerte, turned out to be decisive in sealing off the northern end of the Allied trap. The 34th Division was deliberately given an important and difficult task and performed brilliantly during the bloody battle for Hill 609. The Allied trap was sprung when the Eighth Army pushed the Axis army into the coastal plain around Tunis, where von Arnim surrendered in mid-May. About 100,000 of the approximately 250,000 Axis troops bagged near Tunis were German. Bradley sent Eisenhower a simple two-word message: "Mission accomplished." At about the same time Alexander was sending his now famous message to Churchill: "Sir, it is my duty to report that the Tunisian campaign is over. All enemy resistance has ceased. We are masters of the North African shores."[56]

With the Tunisian campaign at an end, the excellent performance by II Corps ought to have erased the British skepticism of American fighting ability. But Alexander's behavior during the forthcoming Sicily campaign was to suggest that the triumph at Bizerte had not altered his impression that American combat troops were still not up to the standard set by the Eighth Army. Despite first-class leadership on the part of Patton, Bradley, and Harmon, the deplorable example of Fredendall still lingered in the minds of the senior British commanders.[57]

For the Allies the Tunisian venture was a testing ground. As Bradley later noted: "In Africa we learned to crawl, to walk—then run."[58] The price of this experience for the fledgling U.S. Army was not cheap. American losses were 18,221, including 2,715 killed.[59]

With the Allies in complete control of North Africa, a turning point in the war had come. They, and not Hitler, would henceforth dictate the time and place of future engagements. That Tunisia was a battlefield laboratory was all too evident from the experience of Kasserine, which for all its bitter aspects, became a lesson that American commanders were doggedly determined not to repeat. It also marked the emergence of a new generation of American generals, who were to make their mark throughout the remainder of the war. The performance of these men was to provide powerful evidence that these lessons were never forgotten.

After forty days in command, and with decidedly mixed emotions, Patton left II Corps in the hands of Omar Bradley on April 15. He was pleased to be departing, satisfied that he had done a good job under trying conditions, but pessimistic over command arrangements that left the British in tactical control of his corps. Although Eisenhower had made it clear from the outset that he was merely a stand-in, he seemed uncertain of his status as late as April 13, two days prior to the change of command. "I would like to finish this fight," he wrote, "but shall not argue, as it seems to me that I am in the hands of fate, who is forging me for some future bigger role."

Although it is widely believed that Patton's growing criticism of Eisenhower was selfishly motivated, considerable evidence to the contrary suggests that the real reason was nationalism, and Patton and Bradley's insistence that the U.S Army be given a fair chance to prove itself:

> Bradley, Everett Hughes, General [Lowell W.] Rooks (the AFHQ deputy chief of staff), and I, and probably many more feel that America is being sold. I have been more than loyal to Ike and have talked to no one and have taken things from the British that I would never take from an American. If this trickery to America comes from above, it is utterly damnable. If it emanates from Ike, it is terrible. I seriously talked to Hughes of asking to be relieved as a protest. I feel like Judas. Hughes says that he and I and some others must stick it out to save the pieces. I am not sure, but I love fighting and if I asked to be relieved, I would not even be a good martyr.[60]

Patton implored the Almighty for continued divine guidance and remarked in his diary that he had learned a great deal in a short period of time. His conclusions about leadership and war, however seemingly self-important, were sound, and included the necessity for a single commander to run the war, not separate air, sea, and ground chiefs, an arrangement that lasted until the end of the war in 1945.

> The trouble is we lack leaders with sufficient strength of character. I could do it and possibly will. As I gain experience I do not think more of myself but less of others. Men, even so-called great men, are wonderfully weak and timid. They are too damned polite. War is very simple, direct, and ruthless. It takes a simple, direct, and ruthless man to wage war. Sometimes I wonder if I will have to laugh at myself for writing things like the above. But I think not. I have developed a lot and my never small self-confidence has vastly grown.[61]

En route to Constantine, Patton stopped to visit the ancient Roman city of Timgad, founded by the Emperor Trajan, where he employed his knowledge of ancient history to marvel at what men like Trajan had accomplished. "It was a wonderful sight." At Trajan's arch "I was tremendously impressed with this monument of a great and vanished race. Yet I have fought and won a bigger battle than Trajan ever heard of."[62] When Patton encountered Mark Clark in Constantine he wondered why he had bothered being so jealous. "I think I have passed him, and am amused at all the envy and hatred I wasted on him and many others. Looking back, men seem less vile."[63]

For all his savage criticism of Eisenhower, Patton still esteemed him as a friend, and found that they had a great deal in common, including the need for toughness "to build and run an effective army." During one of their marathon nights of conversation in Algiers, the two agreed that "utter ruthlessness, even to their best friends, was mandatory. Troops had a right to good leadership," with Patton concluding that what the army required most was "bravery plus brains." A fascinated Harry Butcher observed this exchange and concluded that "both gave every exterior indication of toughness but actually were chicken-hearted underneath."[64]

Patton was unable to curb his pen and lashed out at his friend for favoring the British. That "s.o.b." Crocker, he wrote, had "publicly called our troops cowards. Ike says that since they were serving in his Corps that was O.K. I told him that had I so spoken of the British under me, my head would have come off. He agreed, but does nothing to [Crocker]."[65] The morning he left for Casablanca Patton talked "very plainly to Ike . . . strange to say he took it—but he has a sophist argument, probably provided by the British for everything he has done. I told him he was the reverse of J. J. Pershing."[66]

Patton's Anglophobia was largely confined to senior British commanders. He admired and respected the British Tommy, and those who met him appear to have reciprocated his feelings. When he departed II Corps, one of the most complimentary letters Patton received was from the commander of a tank battalion of the British 6th Armored Division, whom Patton had decorated with a Silver Star.[67]

After the bruising intramural quarrels with Eisenhower and Alexander, Patton's mood brightened immeasurably when told of a message of praise from Marshall: "You have done a fine job and have justified our confidence in you." A tearful Patton said: "I owe this to you, Ike," to which Eisenhower curtly replied: "The hell you do."[68] Before returning to Casablanca on April 17, Patton gathered some nasturtiums and, with Hap Gay and Sergeant Meeks, visited the military cemetery at Gafsa to bid his final farewell to Dick Jenson. "There are more than 700 graves there now," he noted with sadness, on the same day that Jenson's widow wrote to thank Patton for "the beautiful tribute you paid Dick. You gave him the happiest years of his life. After his Father's death you took his place, and his admiration and affection for you was unbounded. The 'old man' could do no wrong."[69] Two nights later, still overcome with emotion, his voice quivering and the tears running down his cheeks, Patton exclaimed to Butcher: "I guess I really am a Goddam old fool."[70]

Patton returned to Morocco having learned the harsh lesson that in wars (particularly those conducted by coalitions), politics and command were inseparable yet uncomfortable bedfellows. For a time at least, the turmoil of the war between the generals in Tunisia could be forgotten. In Casablanca he was warmly greeted by his staff, and wrote in his diary his first night back: "I have been gone 43 days, fought several successful battles, commanded 95,800 men, lost about ten pounds, gained a third star and a hell of a lot of poise and confidence, and am otherwise the same."[71]

CHAPTER 32

"A Dog's Breakfast"

If I win I can't be stopped! If I lose I shall be dead.

—PATTON

In 1943 the Allies were obliged to fight one campaign while planning another—the invasion of the island of Sicily, a strategically important but remote outpost of Mussolini's modern-day Roman empire from which Axis air and naval forces seriously threatened to interdict Allied shipping in the Mediterranean.

By pursuing military action in Sicily the British hoped to knock Italy out of the war and, with Overlord still many months off, there was little Marshall could do to resist the next logical step, the invasion of Italy, where the Allies could establish air bases from which their air forces could aid the strategic bombing effort throughout the "soft underbelly" of Europe. Most important, by electing to carry the Mediterranean war into Italy, Churchill intended to compel Hitler to maintain large numbers of troops there and not in northwest France.

When Brooke orchestrated Alexander's appointment as the Allied land force commander in Tunisia, it meant that he would fill the same role in Husky. Allied naval and air forces were also commanded by British officers: Adm. Sir Andrew Browne Cunningham and Air Chief Marshal Sir Arthur Tedder. Brooke's brilliant coup meant that Eisenhower was scarcely more than a figurehead, with little control over the actions of his British subordinates. Not only did they all have to agree on the invasion plan, but if Eisenhower disagreed with their actions he either had to persuade them to change a decision or resign. He had no authority to replace them. Other than those filled by Eisenhower and Patton, every Allied senior command position in the Mediterranean was held by a British officer, thus stacking the deck in favor of the British, a fact which deeply rankled the American commanders, particularly Patton who for months had warned of the perils of giving the British a free hand.

From its inception as a strategic compromise at Casablanca, the planning of Husky was plagued by interminable problems of organization and command. The Combined Chiefs had decreed that Eisenhower create a separate headquarters to plan Husky, and by late January 1943 an inter-Allied planning group was formed in Algiers, which from its birth was ineffective because of Alexander's preoccupation with the faltering campaign in Tunisia.

The designated invasion commanders were Montgomery and Patton, whose I Armored Corps (now located in Rabat and designated the Western Task Force) was the American invasion element. Only Montgomery, whose Eighth Army would form the (British) Eastern Task Force, had any early involvement or active interest in the planning of Husky, and what he saw filled him with gloom.[1]

The initial plan devised by Alexander's staff called for a series of amphibious landings along the six-hundred-mile coast of Sicily by Allied task forces of varying sizes. When he first learned of the proposed plan, Montgomery was horrified by what in reality was a return to the "penny-packet" warfare that had been so disastrously employed under his predecessor, Gen. Sir Claude Auchinleck, and in Tunisia. Referring to the plan by another typically British expression, "a dog's breakfast," he denounced the entire Husky planning effort as "a hopeless mess." As time passed without resolution, Montgomery repeatedly warned his superiors in typically blunt language that unless a sensible plan was soon developed, the Allies were courting serious trouble.

The inter-Allied squabbling that had begun in Tunisia continued with a vengeance over Sicily, as Alexander's lack of grip on Husky exacerbated the already contentious interservice rivalries. Everyone had his own private ax to grind: Tedder, the air commander, insisted on the prompt acquisition of airfields in southeastern Sicily for his tactical aircraft, while Cunningham, the naval commander in chief, backed the multiple invasion plan, which provided maximum security and dispersion for the Allied fleet. Montgomery, acting in Alexander's stead, contended that the only acceptable plan that would guarantee the success of the invasion was to concentrate all the ground forces in the southeastern corner of Sicily and attack as a single unified force.

For three months the arguments and debates raged within the Allied high command, as a number of invasion plans were proposed, considered, and then scrapped because none could satisfy everyone. Finally a calculated act of insubordination by Montgomery brought about a compromise plan everyone could agree on. In late April 1943, frustrated by the inaction and deeply worried that the planning stalemate might never be broken in time for the invasion, Montgomery, at considerable risk to his reputation, deliberately precipitated a crisis by signaling Alexander that he intended to proceed with planning for the employment of the entire Eighth Army in southeastern Sicily.

Patton was largely a spectator on April 29, 1943, as the British wrangled among themselves during one of the most contentious Allied meetings of the war. Alexander failed miserably as a mediator, and after nearly three hours of furious argument the Allied leadership remained deadlocked. Patton deliberately stayed out of what was primarily an intra-British fight, and was surprised that no one was sacked.[2]

On May 2, Montgomery cornered Bedell Smith in the lavatory of AFHQ, where he outlined his proposed plan (writing on a steamed-up mirror with his finger) and insisted it be presented at once to Eisenhower. However ludicrous the setting, Montgomery's discussion with Bedell Smith actually broke the impasse. The Husky debates had been entirely a British internecine quarrel, the only major operation of the Second World War in which a crucially important decision was orchestrated in an Algerian privy.

Grounded by heavy rains in Mostaganem (165 miles west of Algiers), where he had recently moved his headquarters and "after the worst drive I have ever had," Patton reported to Eisenhower, saying: "I am sorry I am late for the meeting but I did the best I could." Eisenhower waved off the apology. "Oh, that's all right. I knew you would do what you were ordered without question and told them so. We had better get hold of Alexander and Hewitt and show you the new setup."[3]

Montgomery's plan was approved by Eisenhower and the Allied commanders in chief, and although it did not endear him to his associates, there was general relief all around. When he addressed his fellow commanders, Montgomery candidly admitted that he could be a "tiresome person," but few would quarrel with his logic: "I have seen so many mistakes made in this war, and so many disasters happen, that I am desperately anxious to try and see that we have no more."[4]

The newly approved invasion plan called for the Eighth Army to land four divisions and one independent infantry brigade along a fifty-mile front in southeast Sicily, from Syracuse to the Pachino Peninsula. At the same time, along the south coast Patton's forces would make their primary landings at Gela and Scoglitti with the 1st and 45th Divisions of Bradley's II Corps; while to the west, at Licata, newly promoted Maj. Gen. Lucian Truscott's 3d Division would land to protect the American left flank. The object of the assault landings was to seize an Allied bridgehead in southeastern Sicily and, in the process, capture the key ports of Syracuse and Licata and the airfields near Gela, from which Tedder's airmen would operate in support of the advancing ground forces. D day was designated as July 10, 1943.

Under Montgomery's plan, once the Gela airfields were captured, Patton's only mission was to protect the Eighth Army's left flank while the British made the main effort toward Messina. Although disappointed and disgusted at playing second fiddle to Montgomery and the British, Patton was uncharacteristically reticent. Privately he was furious at what he described as "war by committee," and thought that Husky was yet another betrayal of the U.S. Army by Eisenhower. When Alexander anxiously inquired if he was satisfied with the new plan, Patton tersely replied: "General, I don't plan—I only obey orders."[5] Before returning to Mostaganem, Patton met with Admiral Cunningham, who was still smarting over Monty's success and the changes made to Husky. The crusty admiral urged Patton to protest his limited role. Patton emphatically refused with the comment: "No, goddammit, I've been in this Army thirty years [sic] and when my superior gives me an order I say, 'Yes, Sir!' and then do my Goddamndest to carry it out."[6]

The senior British commanders all despised Montgomery, whom Tedder once snidely referred to as "a little fellow of average ability who has had such a build-up that he thinks of himself as Napoleon." Strangely, despite the disappointing change of mission and the domination of recent events by Montgomery, Patton's opinion remained favorable. "Monty is a forceful, selfish man," he remarked, but "I think he is a far better leader than Alexander. . . . " He wrote to Beatrice that "Monty and I had quite a conference and got on fine. . . . I should hate to be married to him in either meaning but as a partner we will be fine."[7]

The perception of Montgomery as an arrogant, insufferably pompous, glory-seeking egomaniac intent on capitalizing on his newfound fame as the vanquisher of Rommel and the vaunted Afrika Korps is untrue. A wealth of documentation effectively demolishes the allegation made by his enemies that his motive in seeking to alter the Husky plan and the role of the Eighth Army was self-advancement, at the expense of Patton and the U.S. Army. In fact the evidence suggests quite the opposite: that Montgomery acted courageously in a no-win situation, against powerful opposition that left him so far out on a limb that even his mentor, Alan Brooke, could not have saved him.

Although Patton publicly accepted his subordinate role with good grace, those closest to him knew that he was devastated. "The U.S. is getting gypped," he wrote in his diary.

Only an act of God or an accident can give us a run for our money. On a study of 'form,' especially in the higher command, we are licked. Churchill runs this war . . . the thing I must do is retain my SELF-CONFIDENCE. I have greater ability than these other people and it comes from, for lack of a better word, what we must call greatness of soul based on a belief—an unshakable belief—in my destiny. The U.S. must win—not as an ally, but as a conqueror. If I can find my duty I can do it. I must. This is one of the bad days.[8]

Back in Mostaganem Patton finally vented his anger, telling his staff: "This is what you get when your Commander-in-Chief ceases to be an American and becomes an Ally." Alexander later attempted to soften the blow by writing in his campaign despatch that Patton's attitude was "an impressive example of the spirit of complete loyalty and inter-Allied co-operation."[9]

As the date for the invasion grew near and there had yet to be any indication of what Alexander intended to do after the landings, Patton insisted that Alexander produce a signed document, "with the binding effects of a treaty," which would establish a clear-cut boundary between his forces and the Eighth Army, and provide for supplies and numerous other important details. Patton was backed by Bedell Smith, who invoked Field Marshal Lord Gort's remark about Montgomery: "In dealing with him one must remember that he is not quite a gentleman."[10]

In mid-May II Corps officially joined Force 343. Patton greeted Bradley at Mostaganem airfield with an honor guard and hosted a luncheon in his honor, at which he drank to the health of "the Conqueror of Bizerte." Patton's praise, "He grows on me as a very sound and extremely loyal soldier," seems not to have been reciprocated by any change of feeling in Bradley, whose dislike and distrust of his boss was to increase dramatically in the coming days.[11]

In May Patton suddenly found himself in trouble with Marshall for a foolish act by his G-3, Col. Kent Lambert, who broke censorship rules by sending home letters that detailed the Torch operations. The censors in Washington intercepted the letters, which were delivered to Marshall. The first had been sent back to Patton for action, and in the second Lambert had written, "my friend Patton said, 'Nuts, file it.' So I escaped." Not only did Lambert's actions cost him a promotion, but Marshall was furious that Patton had apparently turned a blind eye to the first letter. What angered the chief of staff was not so much the divulging of important information, but rather that Patton had willingly condoned a clear violation of War Department orders. Although the official correspondence made no mention of it, Eisenhower later related that Marshall "suggested that, as an example to others, I remove Patton from his command and send him home." Eisenhower asked Patton if he had made such a remark about Lambert. Patton replied "that 'nuts' is about the only expletive I do not use." Gay also reprimanded Lambert, and Patton "personally cussed him out."[12] Eisenhower took no further action, but the incident was the second occasion in six months that a senior officer had considered Patton's relief.[13]

Over the years Patton had grown so accustomed to playing the role he had created for himself more than forty years earlier that it never occurred to him that there were occasions when it could and did get him into trouble. The only two officers Patton had to keep happy were Eisenhower and Marshall. Yet time after time during World War II Patton jeopardized his standing with one or both. The *enfant terrible* within not only tended to confirm the public's perception of him, but more than once threatened to destroy his career. Thus it was that shortly before the invasion of Sicily, Patton's tendency to excess was extremely harmful when, in the presence of both Marshall and Eisenhower, he raged at a squad of 1st Division infantrymen who had just carried out a practice assault landing. In a paroxysm of anger he demanded: "And just where in hell are your goddamned bayonets?" As Bradley later wrote: "George blistered them with oaths.* Eisenhower and Marshall stood within earshot in embarrassed silence." A senior AFHQ officer whispered to Bradley: "Well, there goes Georgie's chance for a crack at higher command. That temper of his is going to finish him yet."[14] Despite his intemperate outburst and Bradley's pessimism, Patton soon learned that Marshall had been impressed. The chief of staff clearly thought Patton was worth the price tag he came with.[15]

Although this particular incident seemed to be one of Patton's acting performances (moments later he remarked to Bradley, "Chew them out and they'll remember it"), there were other occasions when his emotions and uncontrollable anger overtook common sense, with serious consequences. It was impossible for most people to distinguish between Patton's charades and his true emotions. Both Patton and Montgomery were consummate actors, and most of what they did in public was for the purpose of impressing their troops. Monty claimed with pride: "I rehearse it. I take immense trouble over what I say. I write it down before hand, I practice it in front of the mirror when I am shaving."[16] When stationed in Hawaii in 1927, Patton once sent his nephew a photograph of his "war face," which "I have been practicing before a mirror all my life. I'm going to use it again to scare hell out of the Germans."[17]

The result of Patton's impulsiveness was the nearly predictable tendency to self-destruct at the most inopportune moments. This particular incident only tended to confirm Bradley's increasingly negative opinion. Like

---

* Beatrice rarely swore, but one day, when Patton was complaining about some Irish soldiers in his command, she said they were only interested in the three F's. Ruth Ellen remembers her father asking "what the three F's were and without turning a hair, Ma replied, 'Fighting, funerals and fucking.' I have never seen anyone look as shocked as Georgie did!"

Monty's, Patton's lack of tact was the one character flaw that consistently landed him in hot water. Fortunately his most outrageous remarks were confined to his diary and letters to Beatrice. In June he was introduced to King George VI, whom he pronounced "just a grade above a moron. Poor little fellow." And, when the British Secretary of State for War, Sir James Grigg, told him that Alexander had confidently predicted that American troops would soon be the best in the world, Patton replied that they were *already* the best. In his diary he wrote of his impoliteness, "it is the only way to talk to an Englishman."[18]

On May 20 he attended an Allied victory parade in Tunis, an impressive and majestic affair that he hoped would be "only the first of many such triumphal processions that I shall participate in. Bradley and I also hope that in the next one, we will have a more conspicuous role."[19] The French, he thought, "have an innate capacity for ceremonial marching." The British also marched and performed well: "There was one Sergeant Major . . . who should be immortalized in a painting. He typifies all that is great in the British non-commissioned officer, and he certainly knew it. I have never seen a man strut more."[20]

A number of senior British officers were included in the official reviewing party with Eisenhower, but Bradley and Patton were relegated to a minor reviewing stand occupied by French civilians and inconsequential military officers. Patton cloaked his distress with anger, declaring it "a goddamned waste of time," and even Bradley was disgusted that it seemed "to give the British overwhelming credit for the victory in Tunisia. For Patton and me, the affair merely served to reinforce our belief that Ike was now so pro-British that he was blind to the slight he had paid to us and, by extension, the American troops who fought and died in Tunisia."[21]

As he had been for months, Patton remained deeply troubled over the effects that his criticism was having upon his twenty-four year friendship with Eisenhower.[22] Out of loyalty he wrote to Marshall "a letter which largely overstated his merits, but I felt that I owe him a lot and must stay in with him. I lied in a good cause. As a matter of fact I know of no one except myself who could do any better than Ike, and God knows I don't want his job." At a press conference the correspondents asked Eisenhower what had happened to Patton, who had officially disappeared. Eisenhower would only reply evasively: "I have had to pull General Patton out to plan a bigger operation, and at the moment he commands the I Armored Corps. Please lay off mentioning him."[23]

With sirens screaming and motorcycle escorts clearing the way, Patton sped from unit to unit and delivered a speech that was part pep talk and part lecture, in which he assured them that their fears were normal, and that as good soldiers they would overcome them as long as their officers provided top-notch leadership. "All men are afraid in battle," he noted. "The coward is the one who lets his fear overcome his sense of duty." He was keenly aware of the power of a commander to influence his troops. "Leadership is a funny thing. . . . I don't know how I do it," he confessed to Beatrice.[24] Repeatedly, he used his wealth of experience to deliver a simple message to his troops: "make the other dumb son-of-a-bitch die for *his* country!" At the headquarters of the Services of Supply a large crowd clapped and cheered him. Patton returned the favor when "I kissed my hand to them, which brought down the house. However, I am not sure they knew who I was— probably thought I was Ike."[25]

Inevitably, Patton's newfound fame had its price. Fan mail increased, but when a major story about him in *Time* left the unflattering impression that he was critical of his own troops, Patton raged at the injustice but accepted that there was little he could do about it. He wrote to remind Stimson that he had been misinterpreted:

> The nasty article . . . was most unfortunate in that it gave some of the soldiers an idea that I criticized them, whereas one of my many faults is that I am too pro-American and can see nothing wrong in anything that our soldiers do. . . . So long as you, General Marshall, and my family understand the situation, I have no complaint and look forward with considerable pleasure to the time when I will make all these all-wise analysts and newspaper correspondents eat a few words. I am mean enough to hope that this meal will give them an acute case of indigestion.[26]

It was the beginning of Patton's love-hate relationship with the press.

There was the usual outpouring of instructions and guidance, inspections, more brief pep talks to small groups, and constant corrections when mistakes in training or leadership were noted. "None of the lesser generals do it," he complained. Patton devoted a great deal of his time perfecting the small but important details of the forthcoming invasion by testing machine guns, rockets and antitank grenades against pillboxes and determining the best means of detonating Teller mines* with rifle fire, the optimum uses of smoke, and attempting to solve a myriad of other problems that would face the invaders. No detail was too small to escape his consideration and advice.

Patton also continued his long-standing habit of visiting hospitals. At one a 1st Division enlisted man said: "General, don't you remember me? I am in the 1st Division." Patton said he did, and the soldier escorted him through the hospital and introduced him to other II Corps soldiers. Another GI said that the other divisions were okay, but *his* 9th Division was the best

---

*A German land mine widely utilized during World War II, the Teller contained nineteen pounds of TNT. It was first used in the Western Desert in 1941 during the battle for Tobruk.

outfit. Patton shook hands with him for "standing up for his own," and left feeling confident: "We have now an army."[27]

Patton perpetually worried about what others thought of him, particularly Eisenhower and Mark Clark, whose relationship he viewed as "some sort of an unholy alliance. I should not worry as I seem to be doing nicely, but I do worry. I am a fool—those two cannot upset destiny. Besides I owe each of them a lot, but of course don't know to what extent they have undercut me? The next show must be a success or a funeral." During a dinner with Marshall at which Clark was present, Patton went out of his way to praise his rival, noting in his diary: "I am getting tactful as hell and in this case it is true. If you treat a skunk nicely he will not piss on you—as often."[28] Yet there was no reason for such paranoia. Marshall thought Patton ought to have equal billing with Monty, and in mid-May Eisenhower readily approved the elevation of I Armored Corps to army status. On D day it would officially become the Seventh United States Army, and Patton would be subordinate only to Alexander and Eisenhower.

When Patton, Keyes, and Brig. Gen. Maxwell D. Taylor of the 82d Airborne met to discuss the attack on May 5, developing the basic plan for the American invasion took about an hour. Patton reminisced about its simplicity: "Some day bemused students will try to see how we came to this decision and credit us with profound thought we never had. The thing as I see it is to get a definite, simple plan quickly, and win by execution and careful detailed study of the tactical operation of lesser units. Execution is the thing, that and leadership."[29]

Patton's hard work and contributions to Husky were never recognized by Bradley, whose relations with him were soon to be at their nadir. Later harshly criticizing him for allegedly failing to interest himself in the details of Husky, Bradley claimed Patton was all too willing to accept any plan so long as it contained a role in it for himself. "George is spectacular. Does not like drudgery. And planning is drudging work."[30] Bradley's denunciations notwithstanding, Patton's important contribution to the planning of the invasion was that he articulated a relatively uncomplicated, straightforward concept of the operation, and then left his staff to work out the details.[31]

On June 21 the Allied commanders met in Algiers to present their final plans to Eisenhower, and to one another. Patton recorded his impressions in his diary, beginning with Eisenhower, who he thought failed to assert his authority as the commander in chief during a ten-minute opening speech. A senior American officer called the entire afternoon "boring" except for Force 343's twenty-two-minute presentation, which bore Patton's imprint: "short, beautifully illustrated and carefully rehearsed." Patton himself spoke for six minutes and thought "we stole the show," noting that it was only thirty seconds longer than rehearsed. "Ike was pleased and, for a change, said so." After the briefing Patton shook the hand of and thanked each staff officer involved in the Force 343 presentation. His G-2 wrote: "If this were the kind of act which might be interpreted as showmanship, it certainly was not considered such by the recipients of their commander's regards."[32]

In the weeks prior to Husky, Patton spent considerable time with Alexander, and their conversations inevitably turned to the differences in their respective armies:

> Alex said that it was foolish to consider British and Americans as one people, as we are each foreigners to the other. I said that it was so and the sooner everyone recognized it the better. I told him that my boisterous method of command would not work with British no matter how successful with Americans, while his cold method would never work with Americans. He agreed. I found out that he has an exceptionally small head. That may explain things.[33]

Before the invasion Eisenhower sent Maj. Gen. John Porter Lucas to Patton as his "eyes and ears." Lucas was highly rated, had extensive experience in a variety of command and staff jobs, and had been sent by Marshall to Eisenhower as an officer of "military stature, prestige and experience." Had the new "spy" in his headquarters been anyone but Johnny Lucas, Patton would have complained bitterly to Eisenhower. However, the two were longtime friends, and Patton trusted him implicitly. Lucas was diplomatic, sensible, and persuasive, and Patton frequently sought his advice, even to the point of employing him as a conduit to Eisenhower.

In May 1943 Patton appointed Lt. Col. Charles R. Codman, a cultured Boston Brahmin, to replace Dick Jenson as his principal aide. A wine buyer for the elite firm of S. S. Pierce before the war, Charley Codman had distinguished himself in World War I as an aviator, and wore a chestful of medals that included the French croix de guerre with palm. He spoke fluent French and was sophisticated and wise in the ways of keeping a commander like Patton content. Patton, in turn, thought Codman was the epitome of a true gentleman, and relied heavily on him as much for his friendship as for his wise counsel and efficient management of his busy schedule.[34]

Shortly before the Husky landings Eisenhower prepared a confidential memo in which he analyzed the strengths and weaknesses of each senior member of his Allied team. He found Monty "a very able, dynamic type of army commander," who needed only a strong boss to keep him in line. He rated his longtime friend George S. Patton:

> A shrewd soldier who believes in showmanship to such an extent that he is almost flamboyant. He talks too much and too quickly and sometimes

creates a very bad impression. Moreover, I fear that he is not always a good example to subordinates, who may be guided by only his surface actions without understanding the deep sense of duty, courage, and service that make up his real personality. He has done well as a combat corps commander, and I expect him to do well in all future operations.[35]

Eisenhower's top rating was reserved for Omar Bradley, whom he called, "about the best rounded, well balanced senior officer that we have in the service. . . . I feel that there is no position in the army that he could not fill with success."[36]

Soldiers are especially preoccupied with their mortality before a battle. What most probably do not realize is that their generals are no different. Before the Husky landings Patton reflected on his own chances of survival, noting in his diary that "Bradley, Terry Allen and I were talking over Husky and guessing at the chances of at least one of us getting it. I doubt if it will be me."[37] Earlier he and Bradley had assessed their chances of survival as "not better than 50–50 but God or luck will tip the beam to us."[38]

He continued to write in both his diary and letters to Fred Ayer of his fears, both real and imagined:

> It is rather interesting how you get use[d] to death. I have had to go inspect the troops daily. In which case you run a very good chance . . . of being bombed and shot at from the air, and shelled or shot at from the ground. I had the same experience every day, which is for the first half-hour the palms of my hands sweat and I feel very depressed. Then, if one hits near you, it seems to break the spell and you don't notice them any more. Going back in the evenings over the same ground and at a time when the shelling and bombing is usually heavier, you become so used to it you never think about it . . . the same mental attitude I have in the New York Horseshow. The first time I ride over the jumps, I am scared. By the end of the week, it is simply routine.[39]

"It looks as if fate were fattening me up for something," he wrote on July 4. "I hope I perform when the time comes. Battles take years to get ready for, all one's life can be expressed in one little decision but that decision is the labor of uncounted years. It is not genius but memory— uncounted memory—and character, and Divine Wrath which does not hesitate nor count the cost." When his plane encountered rough weather after a visit to Oran, "I had to do a most distasteful thing. I asked Codman to tell the pilot that, if it got any worse, to turn back. This is . . . the only time I ever publicly showed timidity."[40]

Yet Patton could not seem to resist the lure of testing himself again and again. In May he wrote to Malin Craig of walking down a street where machine guns were firing only two feet away (presumably part of a training exercise). Calling it "good experience, even for me," Patton thought it would be helpful for young soldiers to see their commanding general under fire. Soon "I am going out in a tank and have them put high bursts just over me. I think I have something there, but I cannot explain it. I wish to surprise some of my friends."[41] Even the god of battles must have wondered what he had left to prove.

In Algiers on July 5 Patton paid his final respects to Eisenhower, who took the occasion to deliver a lecture on the poor discipline of Terry Allen's 1st Division. "I told him he was mistaken and that, anyhow, no one whips a dog just before putting him in a fight." Patton left disgusted with his friend:

> At no time did Ike wish us luck or say he was back of us—fool. . . . I told him that I was very appreciative of being selected. He said, 'You are a great leader but a poor planner.' I replied that, except for Torch, which I had planned and which was a huge success, I had never been given a chance to plan. He said that if 'Husky' turned into a slugging match he might recall me to get ready for the next operation, and let Bradley finish Husky. I protested that I would like to finish one show. I can't make out whether he thinks Bradley is a better close fighter than I am or whether he wants to keep in with General Marshall, who likes Bradley. I know that Bradley is completely loyal to me . . . I would not trade places with any-one I know. I am leading 90,000 men in a desperate attack and eventually it will be over 250,000. If I win I can't be stopped! If I lose I shall be dead."[42]

One of the army's most brilliant planners and strategists, Brig. Gen. Albert C. Wedemeyer, was temporarily assigned to Patton. Wedemeyer flat-tered him by suggesting that he was the one to run the army after the war. "I declined, being a liar," noted Patton, "but it is much too soon to make plans . . . one must end as a beloved, victorious leader and many men must die before I am that. To be that I must take chances above and beyond the call of duty. And luck may not hold!"[43] He thought often about the bullet meant for him. "I don't like the whine of bullets any more than I ever did but they attract me just the same." His remarks in July 1943 varied little from those made at West Point, when he had deliberately exposed himself to live fire on the rifle range. Only a select few knew of Patton's obsession with his own mortality, or that he was rarely free from an unsettling fear that he would fail and disgrace himself. His demons surfaced repeatedly in brood-ing letters sent to Fred Ayer and Beatrice.

Patton's obsession with fate (and death) was again evident when he wrote to Ayer: "It is not a goodbye, as the higher ranking I get, the less chance I have to do any real fighting. However, one can always take a long swim, and swimming in oily water, which is on fire is not healthy. . . . I have a great many lives and do not feel at all dead. . . . If we should not meet again until we get to the other side, I am assured on credible authority that the heavenly foxes are fast, the heavenly hounds keen, the fogbank fences high and soft, and the landings firm." He wrote to Beatrice that he did not expect to be killed, "but one can never tell. It is all a matter of destiny. . . . Well, when you get this you will either be a widow or a radio fan, I trust the latter. In any case I love you."[44]

CHAPTER 33

"Born at Sea, Baptized in Blood"
Seventh Army Commander

George, let me give you some advice. If you get an
order from Army Group that you don't like, why, ignore
it. That's what I do.

—MONTGOMERY

Accompanied by Lucas and Maj. Gen. Matthew B. Ridgway, the 82d Airborne Division commander, Patton boarded Admiral Hewitt's flagship,
Monrovia, on the morning of July 6. He marveled at the sight of the huge
fleet and ruminated: "I hope God and Navy do their stuff. . . . We will not be
stopped." In a brief but moving ceremony, Hewitt presented Patton with his
first Seventh Army flag. The motto of Seventh Army became: Born at sea,
Baptized in Blood.

An eyewitness, correspondent Hal Boyle, saw a "fire of pride" in Patton's eyes, writing later that "it was to him not a ship's deck he stood upon,
but a peak of glory."[1]

Patton's first message to his troops as an army commander read: "When
we land we will meet German and Italian soldiers whom it is our honor and
privilege to attack and destroy. . . . The glory of American arms, the honor of
our country, the future of the whole world rests in your individual hands. See
to it that you are worthy of this great trust. God is with us. We shall win."[2]

Patton shared a cabin with Lucas, and by the third day at sea noted that although "I have the usual shortness of breath I always have before a polo game, I would not change places with anyone I know right now." Only Beatrice knew of his uncertainties. "I doubt I will be killed or wounded, but one can never tell. It is all a question of destiny. . . . I have no premonitions and hope to live forever." On the eve of D day, as the fleet neared the coast of Sicily, Patton summoned the chaplain to his cabin after supper to join him in prayer. Then he slept briefly and restlessly, dreaming that a black kitten and then many cats were spitting at him. Forty minutes before H hour he appeared on deck to observe the landings, which took place throughout the early morning hours of July 10. He thought to himself that despite his own anxiety the Italians must be scared to death.[3]

The invasion of Sicily began disastrously when the British glider force that was to land near Syracuse ahead of the seaborne forces encountered danger-ously high winds, smoke from the island, and heavy flak from both enemy and friendly guns—which, contrary to orders, mistakenly fired on the aerial armada. Casualties were heavy. A paratroop regiment of the 82d Airborne fared little better when the winds and flak dissolved the neat aerial forma-tions into a confused jumble over a large area of southeastern Sicily. More than three thousand paratroopers who were to have been dropped northeast of Gela instead landed over a thousand-square-mile area.

Coordination and cooperation between the air, naval, and ground forces was extremely poor, and became one of the bitterest inter-Allied problems. The air plan was so vague that the naval and ground commanders did not have the foggiest notion *how* the airmen intended to support their particular operation. During May and June they raised these and other questions, but satisfactory answers were not forthcoming from the RAF and the U.S. Army Air Force chiefs, who had grown increasingly independent and reluctant to take orders from the army or navy. Patton's air counterpart, Col. Lawrence Hickey, was a firm advocate of close air-ground cooperation but paid dearly for his stance. When Hickey, whose boss was none other than "Mary" Con-ingham, elected to colocate with Patton's HQ, Coningham relieved him. This repudiation was not lost on Patton and his staff, who saw it as proof of the indifference of the airmen to their requirements. Coningham had a very long memory for those who slighted him, and it is not unreasonable to con-clude that this act signaled a payback to Patton for the humiliation of their earlier encounter.[4]

Patton became so concerned over the lack of air cooperation that he pleaded with Hewitt to circumvent the airmen in favor of naval carrier sup-port. "You can get your Navy planes to do anything you want; but we can't get the Air Force to do a goddam thing!" he declared.[5] But Hewitt had no carriers with which to oblige, and the invasion forces sailed from North

Africa without the slightest notion of what air support they would have on D day and during the campaign.

Unlike the ill-fated airborne and glider operations, the amphibious invasion of Sicily was a great success. Enemy resistance was mixed: In some places there was heavy fighting, while in others the Italian defenders surrendered with alacrity. The most important of the three American landings took place at Gela, where a Ranger force commanded by Lt. Col. William O. Darby created a powerful diversion by attacking the town while the main 1st Division landings occurred along the beaches of the Plain of Gela to the east. By midmorning, after a short but vicious battle, Darby's Rangers had cleared Gela, but only after fending off a counterattack by Italian tanks.[6]

His stormy friendship with Terry Allen in no way affected Patton's decision to assign Allen's 1st Division to carry out the Gela landings, which he correctly envisioned would be the most difficult of the three assault landings.[7] Omar Bradley, by contrast, had little use for Allen, whom he held responsible for the 1st Division's alleged notorious lack of discipline in North Africa. "He was possibly the most difficult commander I had to handle throughout the war," Bradley later wrote. The 36th (National Guard) Division was originally slated to assault Gela, but Patton pleaded with Eisenhower that "I want those [1st Division] sons of bitches. I won't go on without them!" Eisenhower acceded, and the 1st Division quickly demonstrated the astuteness of Patton's decision.

On the morning of July 11 German and Italian formations launched a counterattack to throw Patton's army back into the sea. Their primary target was the 1st Division, which came under heavy attack across its front. Battles raged all day, and the 1st Division was supported by elements of Col. James M. Gavin's paratroopers who made a heroic stand at Biazza Ridge, preventing the Hermann Göring Division from driving a deadly wedge between the 1st and 45th Divisions. By early afternoon German panzers had penetrated the center of the 1st Division's defenses and were loose on the Gela Plain, threatening even the division command post. It took a "Custer's last stand" to hold back the German panzers, which were destroyed or driven off by deadly naval gunfire.

Patton had remained aboard the *Monrovia* on D day feeling "like a cur, but I probably did better here." By the morning of July 11 he could no longer stand being cooped up and made his first appearance in Gela at precisely 9:30 A.M. As Signal Corps cameramen recorded the scene, Patton, with Hap Gay behind him, waded ashore, resplendent in an immaculate uniform complete with necktie neatly tucked into his pressed gabardine shirt, knee-length polished black leather boots, and his ever-present ivory-handled pistols strapped to his waist. Shells began falling into the water as close as thirty yards, but Patton ignored them, telling Gay: "It's all right,

Hap, the bastards can't hit us on account of the defilade afforded by the town."[8]

Patton arrived at Darby's command post in Gela just in time to witness a second counterattack by Italian tanks. From a rooftop he observed with fascination as the tanks advanced across the open plain north of Gela straight for the town—and him. He noticed a young naval officer standing in a nearby street with a radio, and shouted: "Hey, you with the radio!" Extending his arm in the direction of the Italians, he ordered: "If you can connect with your Goddamn Navy, tell them for God's sake to drop some shell fire on the road." Soon, the cruiser *Boise* obliged and began hammering the enemy tanks. Before departing the Ranger command post, Patton instructed a captain to "kill every one of the goddam bastards."[9]

During the afternoon Patton turned up at the 1st Division command post but found Allen away. When the dog-tired Allen returned, he found a euphoric Patton sitting with his feet propped up on a field desk, a big cigar in his mouth. How was the division doing? asked Patton. Allen replied that it was no tea party—that the 1st Division was doing okay but needed additional artillery support. With a wave of his hand Patton said: "I'm now an Army commander; take this up with Bradley."

Whether Patton failed to appreciate the dilemma of the 1st Division or was merely engaging in his long-standing habit of baiting his old friend, he took the occasion to express his displeasure that the 1st Division had failed to capture a key airfield in the hills north of Gela. Allen was one of the most attack-minded officers ever to command a U.S. division, but at that moment his one and only concern was to ensure that his battered division somehow survived the fury of the German counterattack.

July 11, 1943, marked the beginning of a chilling decline in Patton's relations with Omar Bradley, who accused him of meddling when he countermanded a tactical order Bradley had previously given Allen. An irate Bradley later confronted Patton, who apologized for interfering in what was clearly none of his business. Although obviously out of line, Patton was nevertheless nettled by his subordinate's protest, and when he conferred with Eisenhower the following day, he deliberately remarked that Bradley was "not aggressive enough." When he learned from Eisenhower what Patton had said, Bradley branded it an unforgivable slur.[10]

Aboard the *Monrovia* that night Hap Gay confided in his diary the opinion that Patton's personal presence at the front lines that day had much to do with the failure of the German counterattack. Patton himself thought it "the first day in this campaign that I think I earned my pay and actually I was not much scared. I am well satisfied with my conduct today. God certainly watched over me today."[11] In fact Patton's presence on July 11 had practically nothing to do with the successful American stand. The Rangers, infantry, paratroopers, artillerymen and tankers—virtually anyone with a weapon—contributed, along with the magnificent support of the navy, to withstanding the Axis counterattacks.

Patton's major achievement that dramatic day was not his presence on the battlefield but his initiative months earlier in persuading Eisenhower to switch the 1st Division. "In doing so," admitted Bradley, "[Patton] may have saved II Corps from a major disaster." Despite his personal antipathy to Allen, Bradley praised the 1st Division for its stand at Gela. "Only the perverse Big Red One with its no less perverse commander was both hard and experienced enough to take that assault in stride."[12]

The American public was given a wildly exaggerated perception of Patton in Sicily. The *New York Herald Tribune* trumpeted: PATTON LEAPED ASHORE TO HEAD TROOPS AT GELA . . . [TO] TAKE PERSONAL COMMAND of THE COUNTERATTACK ON GELA. The *Los Angeles Evening Herald-Express* proclaimed: REVEAL GEN. PATTON LED YANKS AGAINST NAZI TANKS IN SICILY. The account read in part: "Lieut. Gen. George S. Patton, commander of American fighting forces in Sicily, leaped into the surf from a landing boat and personally taking command, turned the tide in the fiercest fighting of the invasion at Gela."[13]

The triumphant American stand at Gela on July 11 was marred by tragedy that night, when the aerial convoy ferrying Col. Reuben H. Tucker's 504th Regimental Combat Team was shot to pieces by the so-called friendly fire of Allied naval forces. What ought to have been a milk run, in which the 2,300 paratroopers of Tucker's force were to be dropped as reinforcements into the Gela bridgehead, turned into a nightmare. The navy had been under aerial attack for most of the day, and less than an hour before the arrival of the 504th, Axis aircraft launched their largest strike yet. Only minutes after the attack had abated, the unsuspecting troop carrier planes appeared, and before anyone could stop it, indiscriminate fire began from virtually every antiaircraft gun ashore and afloat. "Friendly fire" killed sixty pilots and crewmen and eighty-one paratroopers.

Patton, Ridgway, and the naval commanders had all attempted to ensure that every unit was notified of the time and route of the airborne convoy, but some had clearly failed to get the word. Before the invasion Patton had directed all ground and naval commanders to be aware that there might be flights of friendly troops during any of the first six nights of Husky, and several hours earlier Patton, noticing that the antiaircraft gunners were jumpy, attempted unsuccessfully to postpone the drop. "Am terribly worried," he wrote in his diary. This tragic incident, in which a substantial element of the 504th was literally blown from the sky, not only cost precious lives but nearly became the death knell to the future employment of airborne forces. Eisenhower ordered a halt to further operations pending a formal inquiry. Air, ground, and naval commanders all blamed one another, but Patton bore the brunt of Eisenhower's outrage in a bitter and depressing end to the most crucial day of the Sicily campaign.

*   *   *

The next several days were to prove equally unhappy for Patton, when, in two separate incidents near Biscari, seventy-three Italian POWs were massacred by a captain and a sergeant of the 45th Division. Before D day Patton had personally addressed the entire division to present his traditional prebattle speech, which was witnessed by General Wedemeyer, who recorded:

> He admonished them to be very careful when the Germans or Italians raised their arms as if they wanted to surrender. He stated that sometimes the enemy would do this, throwing our men off guard. The enemy soldiers had on several occasions shot our unsuspecting men or had thrown grenades at them. Patton warned the members of the 45th Division to watch out for this treachery and to "kill the s.o.b.'s" unless they were certain of their real intention to surrender.[14]

Because of their inexperience Patton felt that they needed one of his most dynamic pep talks (of the sort seen in the film *Patton*), and on June 27, after intense preparation, he delivered the speech in two shifts to the men of the 45th Division.[15] Once ashore the inexperienced troops of the 45th Division discovered that there was considerable truth in Patton's admonition, which was later substantiated by Canadian reports of Germans faking surrender. Indeed, German atrocities in Sicily were so rampant during the first few days of the campaign that both sides soon abandoned taking prisoners. Truscott's 3d Division also had numerous unpleasant experiences, particularly with the Italians.[16]

When a horrified Bradley learned of the two incidents he promptly reported them to Patton, who, by his own diary admission, cavalierly dismissed the matter as "probably an exaggeration" and instructed him "to tell the officer responsible for the shootings to certify that the dead men were snipers or had attempted to escape or something, as it would make a stink in the press and also would make the civilians mad. Anyhow, they are dead so nothing can be done about it."[17] However, the notes of Chester Hansen reflect that when informed by Bradley of the incident, Patton replied: "Try the bastards."[18]

Whether on Patton's order or his own initiative, Bradley ordered the two men to face a general court-martial for the premeditated murder of the seventy-three POWs. Their defense attorneys argued what later came to be called "the Nuremberg defense"—namely, that they had only been following the orders given in Patton's June 27 speech to the 45th Division. Although this ploy was rejected, Patton did become the object of an official inspector general's investigation, and an investigating officer was sent from Washington to England in 1944 prior to the Normandy landings to look into the incidents. Patton was grilled by the inspector general, whose report concluded that there was no culpability on his part and that the defense lawyers had used "quite unethical methods" in attempting to blame Patton.

Nevertheless, an investigation of the Biscari incidents by an American historian later concluded that Patton's actions verged on a war crime, and that his speeches to the 45th Division, if not actually urging the shooting of prisoners, certainly implied that the fewer the better. However, what Patton said and what the officers of the 45th Division instructed their troops to do appear to have differed. One private testified: "We were told that General Patton said . . . 'Fuck them. No prisoners!' " An officer noted that Patton had said: "The more prisoners we took, the more we'd have to feed, and not to fool with prisoners."[19] His friend Everett Hughes wrote to the U.S. Army inspector general that he did not believe Patton had acted improperly:

> I am convinced that Patton is a fighter for he looks at war realistically and does what few men in our army have yet dared to do—talk openly about killing. George believes that the best way of shortening the war is to kill as many Germans as possible and as quickly as possible. I am so concerned over the success of this war and the need for fighting men and fighting leaders and I am so convinced that Patton is one of the men we need that I have written the above to you as an old friend and also the Inspector General to tell you what I must tell someone.

The investigation "went no further."[20]

When the incident arose again in 1944, Patton was irate that he had been accused of abetting the atrocities. "The fact that Johnnie Waters is a prisoner would, aside from anything else, have made me the last one to do a thing for which they could retaliate on him."[21] He wrote scornfully to Beatrice that "the friends of freedom" were attempting "to cook up another incident about some unnecessary killings—if killings in war are ever unnecessary."[22]

Whatever Patton's public stance, privately he had few if any qualms over the deaths of his enemy or, for that matter, anyone who interfered with his mission. In August 1943 he would write to Beatrice: "Some natives have the foolish trick of cutting our telephones. Many, very many, have sudden and fatal diseases incidental to lead poisoning." He also noted that although "these Boches and the Italians that are left are grand fighters, [they] have now pulled the white flag trick four times. We take few prisoners."[23] It is the nature and brutality of war that men in the heat of combat often become virtually immune to such acts. Once one side began to shoot POWs, the other inevitably retaliated. Sicily was in no way unique, nor was Patton's presence the difference. The following summer in Normandy, when the Canadian 3d Division and the elite 12th SS (*Hitler Jugend*) Panzer Division clashed near Caen, one side or the other began killing its prisoners, which brought instant retaliation.[24]

Sicily became a low point in Patton's long-standing friendship with Eisenhower, whose angry cable after the airborne tragedy demanded action against those responsible and implied that the fault lay within Patton's command. Eisenhower was still visibly irritated when he visited Patton aboard the *Monrovia* on July 12. According to Butcher: "Ike spoke vigorously to Patton about the inadequacy of his reports of progress reaching headquarters at Malta." Patton was also castigated for absenting himself ashore the previous day. Eisenhower contrasted the Seventh Army's reporting with that of the Eighth, which obviously satisfied the Allied commander. However, Lucas could not understand Eisenhower's displeasure, noting that he personally checked the reports and "they seemed to me to be as complete as they could well be under the circumstances."[25]

This unpleasant encounter illustrates the extent of the deterioration in relations between the two commanders. "Ike had stepped on him hard. There was an air of tenseness. I had a feeling that Ike was disappointed. He said previously that he would be happy if after about five days from D-Day, General Bradley were to take over because of his calm and matter-of-fact direction," noted Butcher.[26] Not once during his visit did Eisenhower ever compliment Patton on any aspect of Seventh Army operations. Moreover, his criticism of Patton's alleged reporting failures may have been misplaced, for the chain of command was through Alexander's headquarters, which had been redesignated 15th Army Group. What Alexander's headquarters did with Patton's reports was outside the control of the Seventh Army commander. As Lucas observed, "The C-in-C should have mentioned the fact that a most difficult military operation was being performed in a manner that reflected great credit on American arms."[27]

Butcher also wrote in his diary: "When we left General Patton I thought he was angry." Patton was indeed livid but resigned himself to the fact that the incident merely represented a further example of Eisenhower's pro-British bias. The following day Patton received a cable from Eisenhower "cussing me out" for the loss of the 504th. "As far as I can see, if anyone is blamable, it must be myself . . . perhaps Ike is looking for an excuse to relieve me. . . . If they want a goat, I am it."[28] Patton well understood the principle that in such instances a commander is always ultimately responsible. Still, he was deeply hurt by the cable, and although a formal investigation later concluded that no one was to blame, he remained frustrated that everyone's best efforts had failed to prevent a tragedy. He personally apologized to Ridgway, who, despite the formal report clearing everyone, still believed that "Ike, in his mind, continued to hold Patton responsible."[29]

Patton's confrontation with Eisenhower could not have come at a worse time, for it left him uncharacteristically reluctant to act at the very moment when he should have challenged the most misguided decision made by the Allies in Sicily.

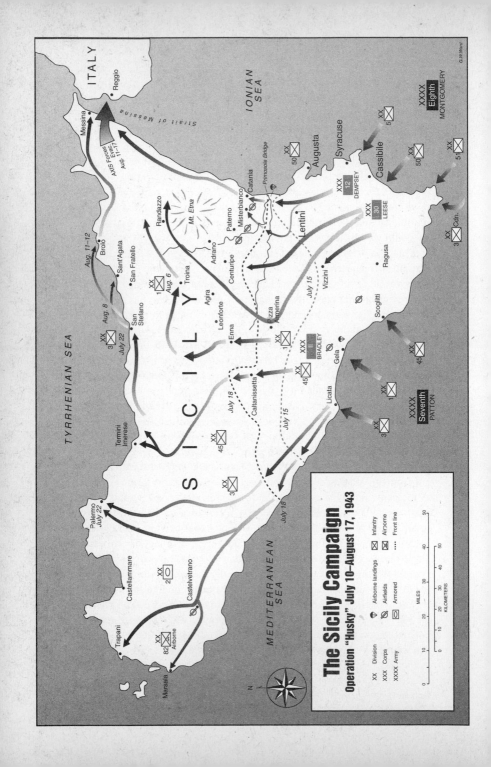

The Sicily Campaign
Operation "Husky" July 10–August 17, 1943

*   *   *

Within the first forty-eight hours the Allies were poised to exploit the absence of a coordinated Axis defense of the island. During the planning phase Patton and Montgomery never met to discuss strategy, nor had Alexander ever promulgated a master plan for the conquest of Sicily. The result was that among the three senior ground commanders there was no common agreement on campaign objectives. Alexander preferred to await actual developments before asserting himself, and after the war indicated that he never had any intention of doing so until the Seventh Army had seized the Gela airfields and the Eighth had control of the ports of Syracuse and Augusta and the Plain of Catania.

Alexander was powerfully influenced by the conviction that American fighting ability was still inferior to that of the more experienced and reliable British army. He refused to acknowledge that the U.S. Army now fighting in Sicily bore scant resemblance to the force that had been humiliated at Kasserine five months earlier.

What passed for strategy was Alexander's notion that Patton would act as the shield in his left hand while the Eighth Army served as the sword in his right. As one of Montgomery's senior staff officers has written: "The two armies were left largely to develop their operations in the manner which seemed most propitious in the prevailing circumstances." The crux of the matter was that Alexander was simply not prepared to entrust the Seventh Army with anything more than a secondary role in the campaign.

Bradley's later recollection of the strategy of the campaign was that the first objective was to push inland, take the high ground, protect the beaches, and get the airfields. Once securely established, the Allies were to cut Sicily in half, face east and roll up all Axis forces trapped between them and the great Mount Etna *massif* and the eastern coast. How this was to occur was left unanswered. Beyond this generalized purpose lay an expectation that the overall Allied objective was Messina. During the preinvasion planning AFHQ called Messina the key to the island, even though Alexander never saw fit prior to D day to provide even the vaguest guidance, nor did he anticipate the speed with which the two armies would secure such sizable bridgeheads.

In short, Alexander did nothing to prepare either himself or his two army commanders to fight the campaign in Sicily. The inevitable result was that his two strong-willed subordinates began to act independently of Alexander and each other. Thus, when Montgomery proposed on July 12 what Alexander had envisioned all along—that the Eighth Army make the main effort to cut Sicily in two—he found ready acceptance. The problem was that the proposed route of advance was across the interarmy boundary and directly across the route of advance of the U.S. 45th Division. Montgomery signaled Alexander that he wanted the troublesome boundary line near Vizzini moved farther to the west, despite the fact that the 45th Division was best positioned to carry out this task.

Alexander ordered Patton to hand over the Vizzini-Caltagirone highway (Highway 124) to Eighth Army, which in turn required Bradley to move the entire 45th Division back to the Gela beaches and then north to new positions on the left flank of the 1st Division. Although the order disgusted Patton, he complied immediately and without protest. Bradley was thunderstruck. "My God," he complained to Patton, "you can't let him do that," believing that he ought to have resisted Alexander's order.[30]

So ludicrous was this situation that once the orders were issued, transferring the disputed boundary and Highway 124 to Montgomery, the Canadians encountered stiff resistance while the 45th Division stood helplessly by, unable to come to their aid even though their artillery was within one mile of the highway. What the better-positioned American infantry, with the advantage of close artillery support, could have accomplished with relative ease became a costly ordeal for the Canadians. Whom to blame has become a matter of opinion. For example, a recent study of coalition war faults Bradley, who, "instead of negotiating the details of the hand-over with his British neighbor, Oliver Leese . . . cut off his nose to spite his face. . . . Thereafter, Bradley and Patton assumed what Montgomery always practiced, that under weak leadership senior commanders should interpret their orders to suit themselves. In North West Europe it led to tacit conspiracy to ignore Eisenhower."[31]

Leese, the 30th Corps commander, later conceded that it was "an unfortunate decision. We were still inclined to remember the slow American progress in the early stages in Tunisia, and I for one certainly did not realize the immense development in experience and technique which they had made in the last weeks of the North African campaign. I have a feeling now that if they [the 45th Division] could have driven straight up this road, we might have had a chance to end this frustrating campaign sooner."[32] Instead Alexander played directly into German hands by granting them precious time to impart their defensive genius and dictate the timetable for the campaign.

Memories of this frustrating incident haunted Bradley for the rest of his life. In his autobiography, *A General's Life*, he recalled often wondering "if I received such orders now, would I really obey them." Undoubtedly Bradley would have agreed with Montgomery's remark to Patton during the campaign. Referring to Alexander, whom he esteemed as a friend but disdained as a commander in chief, Montgomery said: "George, let me give you some advice. If you get an order from Army Group that you don't like, why, ignore it. That's what I do."[33]

This fateful decision was in reality proof that Alexander's original plan was to employ the Seventh Army in the role of a protective shield for the

Eighth. The U.S. Army historians would later write with a passion and bitterness unusual in an official history that Bradley "was ready and willing to take Highway 124 and Enna, thus encircling the German defenders facing Eighth Army.... For all practical purposes, Seventh Army could have stayed on the beaches; its brilliant assault achievements . . . completely nullified. . . . "[34]

The problem lay neither with Montgomery's impetuousness nor with Patton's apathy but with Alexander's unwillingness to take control of the ground campaign at its most critical moment. The decision shattered all pretense of cohesiveness and led to a situation in which the two army commanders virtually dictated conflicting and divisive courses of action for their respective armies and created an absurd and unnecessary personal rivalry.

Messina was the only real strategic objective in Sicily, and Bradley considered the decision to be the turning point of the campaign. For the remainder of his life Bradley never forgave Montgomery for relegating his corps to the same insignificant role of supporting the British it had had in Tunisia, even though the real culprit was Alexander. Patton merely perplexed him. Why, he wondered, had Patton agreed to such "an outrageous descision"? But "he would—and did. Like a lamb. No one has ever satisfactorily explained Patton's position on this decision. It was wholly out of character. By all rights, he could have been expected to roar like a lion. My guess is that at that point, Patton sincerely believed Ike was 'looking for an excuse' to relieve him."[35]

Patton was reeling from the airborne disaster and Eisenhower's stinging criticism, and Bradley's conclusion that he felt "certain that Patton believed that if he caused a ruckus on this day Ike would fire him," was undoubtedly correct.[36] However, what counted was what Patton himself *perceived* at that crucial moment. Apparently Patton had not grasped that, by tradition, the British considered orders from above subject to challenge and discussion. He had already made it clear before D day that, "when I get an order, I carry it out."

When Alexander visited the Seventh Army on July 13, he already had Montgomery's letter requesting the disputed highway in his pocket but failed to mention it to Patton, who was disgusted that Alexander "the Allied commander of a British and American Army, had no American with him. What fools we are."[37] Plainly Alexander had already made up his mind to relegate Seventh Army to a secondary role *before* visiting Patton and, even worse, declined the opportunity to inform Patton personally at a time when he might at least have smoothed what were certain to be ruffled American feathers. Alexander's visit left Patton seething with frustration and was a severe setback to mutual trust and goodwill within the Anglo-American leadership.

The boundary-line contretemps left Patton determined to alter the course of the campaign on his own. The only concession Patton had managed to wring from Alexander was permission to expand his operations to the west and to capture the ancient city of Agrigento, provided that on no account he become embroiled in a major engagement that might risk exposing the Eighth Army left flank. Having gotten his foot in the door, Patton now sought a means of capturing Agrigento without incurring Alexander's displeasure.

On July 14 Patton visited Lucian Truscott and told him that he needed to find some means of capturing the city of Porto Empedocle but could not do so without first capturing nearby Agrigento. Despite Alexander's admonition, Patton made it clear that he was "extremely anxious to have that port." Truscott immediately understood and replied that he not only sympathized but could undoubtedly help. Surely the 15th Army Group could have no objection to a reconnaissance in force undertaken on his own initiative? "General Patton, with something of the air of the cat that had swallowed the canary, agreed that he thought they would not."[38]

Still smarting over being treated as a second-class citizen, Patton decided that the time had come to take matters into his own hands. His real goal was to capture the great northern port city of Palermo, which, Truscott notes, "drew Patton like a lode star." But to accomplish this, he would have to outscheme both Monty and Alexander. On July 19 he wrote to Beatrice that "Monty is trying to steal the show and with the assistance of Devine Destiny (a snide reference to Eisenhower) may do so but to date we have captured three times as many men as our cousins. . . . If I succeed, Attila will have to take a back seat."[39] After conferring with Lucas and Wedemeyer, Patton decided that he must persuade Alexander personally to grant him more latitude in the employment of his army. "I am sure neither he nor any of his British staff has any conception of the power and mobility of the Seventh Army, nor are they aware of the political implications. . . . I shall explain to General Alexander on the basis that it would be inexpedient politically for the Seventh Army not to have equal glory in the final stage of the campaign."[40]

Patton and Wedemeyer arrived unexpectedly at Alexander's headquarters in North Africa on the afternoon of July 17 to make a personal plea for a greater American role. Alexander was clearly caught off guard, and their confrontation took place in a civilized but deadly serious manner. From the impressions of eyewitnesses, there seems to have been no doubt in Alexander's mind that Patton had arrived spoiling for a showdown.

Montgomery's separate offensives in eastern Sicily had been blunted by the tenacious Germans and were in serious trouble, and Patton knew it. In the rough, mountainous terrain of central Sicily, where the road net was poor and vehicular mobility severely limited, the Germans had given Leese's Anglo-Canadian corps the same bloody nose they had to Montgomery's other corps while defending the important Plain of Catania and virtually stalemating the British advance toward Messina. Moreover, Patton instinctively understood that if he could persuade Alexander to give him a green light, he could gain control of his own destiny. Perhaps sensing that he could no longer rein Patton in, Alexander quickly acceded to his plan to use II Corps for a drive to the north coast, near Termini, to cut Sicily in two, while the remainder of the Seventh Army was detached for a secondary (and largely useless) offensive to clear western Sicily. Patton deliberately refrained from informing Alexander that his clever ploy included the capture of both Palermo, the capital of Sicily, and eventually Messina. Patton's sudden willingness to challenge Alexander was a complete reversal of his earlier, almost stoical acceptance of the decision to give Montgomery the Vizzini highway. And, having permitted Montgomery free rein, Alexander was in no position to deny Patton's seemingly modest request. "He gave me permission to carry out my plan if I would assure him that the road net near Caltanissetta would be held. . . . If I do what I am going to do [capture Palermo], there is no need of holding anything, but 'it's a mean man who won't promise,' so I did."[41]

Lucas was one of the first to sense the strategic advantages of a concentrated drive on Messina and the north coast, and while Patton was in Tunis, Lucas had flown to Algiers to brief Eisenhower. After Lucas told him frankly that Patton must fight for what he thought was right, Eisenhower replied "he had never found an instance where the British had deliberately tried to put something over on us." The meeting concluded with an incredulous Lucas being directed by Eisenhower "to see that Patton was made to realize that he must stand up to Alexander and that he would not hesitate to relieve him from command if he did not do so."[42]

Patton quickly formed a Provisional Corps consisting of the 82d Airborne, 2d Armored, and 3d Infantry Divisions under the command of his trusted deputy, Maj. Gen. Geoffrey Keyes, and this force rapidly crushed all Italian resistance in its path. The 3d Division, whose troops Truscott had subjected to exceptionally harsh preinvasion training, accomplished the amazing feat of marching more than one hundred miles through formidable mountainous terrain in a little over seventy-two hours, most of it on foot. Although enemy opposition during the brief drive on Palermo had been scant, Truscott had trained his division to march at a pace of five miles per hour (rather than the accepted three) in what became known as the "Truscott Trot." Palermo fell on July 21, and the GIs entering the capital city were greeted by thousands of flag-waving, cheering Sicilians. When Truscott reported to an ebullient Patton the morning of July 23, he was greeted with the words: "Well, the Truscott Trot sure got us here in a damn hurry."[43]

As Patton toured Palermo on July 23 in his command car, he was cheered and heard shouts of "Down with Mussolini!" and "Long Live

America!" His new headquarters was the Royal Palace, where he was soon visited by a representative of the cardinal of Palermo and by other supplicants who came to be presented to Patton as they would have to a king in times gone by. Patton's junior aide, Stiller, suggested that he be permitted to find his boss "a nice modern house much better than this old dump!" However, the palace—where he could "eat K-rations on china marked with the cross of Saxony" in the huge state drawing room—well suited Patton's sense of history. Patton also "got quite a kick about using a toilet previously made malodorous by constipated royalty."[44]

On July 20 he decorated a number of soldiers with the Legion of Merit in a ceremony that was followed by a concert in which a band played the "Star-Spangled Banner" and "God Save the King":

> I could have been elected Pope right after the concert . . . all of the Dagos cheered. . . . I often wonder when one of them will try to kill me, but I think that [my] apparent lack of fear bluffs them and it is good for the troops to see my flags flying all over the front. One dies but once and I am on a high spot. A victorious memory may be better than to achieve success and be forgotten. However, I feel that I still have much to do, so probably won't get killed, but I do hate to get shot at just as much as I ever did.[45]

On July 24 Patton returned to Agrigento, where he gave an off-the-record press conference during which he exulted that his army's achievement exceeded anything the Germans had ever done. "Gentlemen," he proclaimed, "we had about 200 miles [sic] to go over crooked roads to get to Palermo." He also produced the statistics: 6,000 (Italians) killed or wounded, 44,000 captured, 190 enemy aircraft shot down and 67 guns seized. In his diary he proclaimed that "future students of the Command and General Staff School will study the campaign as a classic example of the use of tanks."[46]

Patton's exaggeration at a moment of great personal triumph that ended days of frustration was understandable. While the rapid movement of his beloved 2d Armored Division was indeed noteworthy, Patton might better have ascribed the American success to a combination of outstanding training and employment of *both* infantry and armor. Only a superbly trained unit like Truscott's 3d Division could have withstood the rigors of such a rapid advance in the heat, gagging dust, and hostile terrain of Sicily.

One of the many myths surrounding Patton was that he attempted to hog the credit for capturing Palermo, and even that he ordered that no Seventh Army troops be permitted to enter the city until he arrived to lead a triumphal entry personally, as the conquering hero. While the capture of Palermo certainly generated considerable publicity for Patton, he generously gave full credit to Geoffrey Keyes. After Patton's death, Beatrice revealed that "General Keyes told me that he went to see Georgie the night before he was to enter Palermo and that Georgie said, 'You took it. You enter and I will enter it after you.' "[47] In a *Life* magazine article a month later there are six photographs of Keyes and none of Patton.[48] Keyes later counted July 22, 1943, as the proudest day of his life.

The Palermo venture might have turned out differently had not Patton's loyal chief of staff, Hobart Gay, practiced one of the oldest tricks in the military book to protect his boss. On July 19 Alexander apparently realized that he had given Patton too loose a mandate. He sent a cable to Patton that appeared to alter significantly the terms under which the Seventh Army could drive on Palermo. Although Patton was present at his headquarters when the cable arrived on the morning of July 19, Gay decided he must not see the message, and he ordered dissemination of only the first portion but not the crucial second section, which would have obliged Patton to halt his drive.

The latter part of the cable disappeared into Gay's desk, and the Seventh Army signal section was directed to complain to the 15th Army Group that the message had been "garbled in transmission" and to request a duplicate. Gay's rationale was that the Seventh Army could hardly be expected to carry out an order it hadn't received. By the time all of this took place it no longer mattered, for the Seventh Army was about to enter Palermo. Patton only learned of Gay's subterfuge after the fact, and his reaction was never recorded. However, Gay's own diary suggests that Alexander was not reneging on his agreement with Patton but rather *confirming* the arrangement. The ineptness of relations between the 15th Army Group and the two armies is best illustrated by a remark made to Patton by a British liaison officer: "The Chief of Staff of the Eighth Army [Maj. Gen. Sir Francis "Freddie" de Guingand] said to pay no attention to any order from Alexander. Whether this is in good faith or as bait I did not and do not know. Nice people."[49] Several days later Patton was hugely amused when Alexander told him that "he was delighted with our ability to carry out his plans for the early capture of Palermo."[50]

The only senior officer in Sicily who found no joy in Patton's dash to Palermo was Omar Bradley, who thought it nothing short of insane. To begin with, Bradley totally misunderstood Patton's intentions when he flew to Tunis to see Alexander. He was certain that Patton had gone to persuade Alexander to allot a greater role for Seventh Army in the drive on *Messina*, and in particular, for a greater role for II Corps. When he learned the truth, Bradley was dismayed, not because the scheme apparently left Messina to the British and Palermo to the Americans, but that militarily he (rightly) considered it a vain and useless exercise that would merely exacerbate the problem of winning a decisive victory in Sicily. In his memoirs he wrote: "It was true that Palermo was to become essential to the logistical support of his Seventh Army but except for that single port there was little to be gained in the west. Certainly there was no glory in the capture of hills, docile peasants, and spiritless soldiers."[51]

To Bradley, Patton's venture smacked of grandstanding, "great theater" that generated the headlines he craved. Correctly surmising that Patton was never criticized for disobeying Alexander's [alleged] order, he noted caustically that "the victory and the headlines were too sweet. However meaningless in a strategic sense, it was our most dramatic and crowed-about 'success' to date. It made our soldiers, proud, lifted spirits at home, and impressed Alexander."[52]

Patton shrewdly saw Palermo as a means of restoring the confidence of American troops in themselves and, in the process, calling attention to their exploits in Sicily, which, up to then, had been largely reserved for Montgomery and the British. British historian Shelford Bidwell believes that "Patton went to Palermo because no one was in charge of the battle. Patton not only had to *be* a good general but *seen* to be a good general through his deeds. Newspaper headlines were one way to achieve this end. Patton instinctively knew what to do."[53]

Even though he had defied the odds and outfoxed Alexander, Patton was far from satisfied. He now cast his eye toward the real prize of Sicily: Messina, the only strategically significant objective of the Allies. There now began the most one-sided "race" of the war, as the final stage of the Sicily campaign saw Patton determined—by hook or crook—to beat Montgomery into Messina.

CHAPTER 34

From Triumph to Disaster
The Slapping Incidents

I was a damned fool.

—PATTON

The capture of Palermo by the Seventh Army marked a clear turning point in the Battle of Sicily. Not only were U.S. forces now very much full participants, but, thanks to both Patton and Montgomery, they were to dominate the final days of the campaign as the two Allied armies drove toward their final objective of Messina.

While the publicity focused on Palermo and the sweep into western Sicily, II Corps, with little fanfare, was thrusting toward the northern coast of the island. On July 23 the 45th Division cut Sicily in half. Patton lost no time ordering II Corps to begin a new offensive along the north coast. When they began to encounter stiff resistance from the 29th Panzer Grenadier Division, Patton sensed that the 45th Division needed a rest and replaced them with the 3d Division. To the south Allen's 1st Division also turned east and began a similar offensive along Highway 120, one of only two east-west interior roads in northern Sicily.

Patton was effusive in his praise for the 45th Division. "I hope you know how good you are," he told an infantry battalion, "for everyone else does. You are magnificent."[1] The only problem Patton ever had with the 45th Division was over Bill Mauldin, the famed cartoonist and war correspondent who got his start in Sicily in the *Forty-Fifth Division News* (the first division newspaper to appear during the war) by drawing cartoons of

Willie and Joe, two archetypal GI infantrymen with whom millions of American soldiers came to identify. Patton considered them "damned unsoldierly" and ordered the division commander, Maj. Gen. Troy Middleton, "to get rid of Mauldin and his cartoons." Middleton diplomatically replied: "Put your order in writing, George." No more was heard on the subject until Patton and Mauldin met during the Battle of the Bulge.[2]

At Montgomery's invitation Patton flew to Syracuse on July 25 to discuss a common strategy for the final phase of operations in Sicily. It is noteworthy that it was Montgomery, not Alexander, the Allied ground commander in chief, who took the initiative to make common cause with Patton. It was the first time the two generals had met since well before D day and the invitation had come from Montgomery, who cabled, "would be very honored if you will come over" so that "we can then discuss the capture of Messina."[3] The campaign was two weeks old and as yet there was still no master plan for the capture of Messina. Alexander continued to act without enthusiasm or firmness and seemed content to let his two army commanders run their own independent campaigns. However, by July 21 Montgomery had fully grasped not only the futility of his offensives to break the German defenses in the the Eighth Army sector, but had arrived at the cold realization that it was vital that he and Patton cooperate.

Montgomery's initiative might well have been classified as leadership by default. Alexander's style of leadership in Sicily varied not one iota from his command of British Middle East Forces, in which he had left the operations of the Eighth Army to Montgomery. He would later assert that he was well satisfied with the way the campaign was developing, in particular with Patton's conduct of the offensive into western Sicily. The timing of this operation, he told an official American historian, was about right. Alexander had wanted no setbacks to the Eighth Army before he "let Georgie go and exploit." In retrospect it seems clear that he could not have restrained Patton much longer than he did, in which case the Seventh Army commander would probably have said: "To hell with this," and gone ahead anyway. "But it never came to that," said Alexander after the war.[4]

Patton arrived fearing the worst. "I always feel like a little lamb on such occasions. . . . Some day I hope I can fight a nice war alone, but it is too much to hope for. Anyhow I love wars and am having a fine time." Their meeting began awkwardly as both made a show of hurrying to greet the other, and as they quickly got down to business around Monty's staff car, where a map of Sicily was spread across the hood, Patton fully expected there would be disagreement over priority on Sicily's sparse road network. The subject still rankled deeply with Patton, and he was determined that Montgomery was not going to rob him of the initiative he had gained by the capture of Palermo. To his amazement, Montgomery not only proposed that the Seventh Army rather than his own Eighth capture Messina, but he even gave Patton *carte blanche* to cross his boundary line if the situation were to develop favorably in the north.[5]

Deeply skeptical, Patton was certain that Montgomery had some ulterior motive. However, the truth was that Montgomery believed the British would never be properly positioned to capture Messina, and had decided that only Patton could end the campaign without further delay and unnecessary bloodshed. In postwar books and films, much has been made of the alleged rivalry between the two commanders for the bragging rights to Messina. The truth is that this misunderstood and historically distorted incident was largely a myth: The only rivalry was in the mind of Patton, who simply never accepted that Montgomery would willingly grant an American army the prize of Messina.

Yet Montgomery's reasons were quite uncomplicated: Not only was he fighting separate corps battles, but because of the large frontages and the harshness of the terrain, each of the Eighth Army divisions was also fighting its own independent battle. Instead of "hustling" the Germans, as he had claimed to Alexander, the four divisional offensives all developed unfavorably when the Germans savagely and successfully resisted all attempts to crack their defenses along the northern edge of the plain of Catania.

For all his outward arrogance, which so infuriated many Americans, Montgomery was first and foremost an honorable professional soldier, who had learned in the trenches of France during World War I to abhor the senseless waste of men so casually practiced by the British leadership. Contrary to some of the nonsense written about the so-called race for Messina, in Sicily Montgomery demonstrated that he would not jeopardize the Allied effort for the sake of personal glory.[6]

Patton's distrust was by no means unfounded. Ever since the D day landings, Montgomery had behaved as if the campaign were a British war, with the Americans present solely for his support. His tactics at Vizzini, where he imperiously usurped Bradley's boundary (earning him the everlasting enmity of the soft-spoken American), and his manipulation through a pliant Alexander, who granted his every wish, amounted to arrogant assertions that only the British could win the important battles in Sicily. However, what Patton never understood was that by July 25 this was no longer the same cocksure Montgomery who had only days earlier sensed a quick victory, or that the farthest thing from his mind at this moment was to claim Messina for the British. Thus his apparent, sweeping change of heart toward Patton was little more than a sobering awareness that the Sicily campaign could not be won unilaterally by the British.

The agreement between Patton and Montgomery was made before Alexander arrived. Patton recorded: "He looked a little mad, and, for him, was quite brusque. He told Monty to explain his plan. Monty said he and I had already decided what we were going to do, so Alex got madder and told

Monty to show him the plan. He did and then Alex asked for mine."[7] The meeting broke up on a sour note, with Patton complaining: "No one was offered any lunch and I thought that Monty was ill-bred both to Alexander and me. Monty gave me a 5¢ lighter. Some one must have sent him a box of them."[8]

Three days later Montgomery and his chief of staff flew to Palermo in a B-17 Flying Fortress to confer with Patton and very nearly lost their lives. The Palermo runway proved far too short to accommodate a B-17, and, had it not been for the extraordinary skill of the pilot, all aboard would undoubtedly have been killed. The B-17 was destroyed, and everyone emerged badly shaken, except Monty, who was quite unruffled.

Still piqued over his treatment at Syracuse, Patton deliberately failed to greet Montgomery in person at the Palermo airfield. However, when the party reached his new headquarters at the palace, a bevy of scout cars and motorcycles escorted Montgomery into the courtyard, where he was greeted by a band and an honor guard awaited his inspection. The ceremony was flawlessly executed in the best tradition of a Patton command. Montgomery's chief of staff later wrote: "Patton was a delightful host and was full of good stories about the campaign. One could sense his great love of the American fighting man."[9]

After an impressive lunch, both generals once again reviewed their plans to capture Messina, and for the second time Montgomery emphasized the importance of the American thrust. Despite nearly losing his life, Montgomery deemed his visit a grand success, even though Patton hoped that "Monty realized that I did this to show him up for doing nothing for me on the 25th." If Montgomery took notice, it did not appear in his diary. "We had a great reception. The Americans are most delightful people and are very easy to work with . . . their troops are quite first-class and I have a very great admiration for the way they fight."[10]

For the second time Patton remained wholly unconvinced of Montgomery's sincerity, and, unable to accept that he would champion a course of action more beneficial to American forces than British, continued to view their relationship as a contest of adversaries. A note sent to Troy Middleton reflected his skepticism: "This is a horse race in which the prestige of the US Army is at stake. We must take Messina before the British. Please use your best efforts to facilitate the success of our race."[11]

Why was Patton so mistrustful of Montgomery's proposal? The reasons lie in the months of disharmony and British distrust of the U.S. Army, which had left Patton, Bradley, Harmon, and others with a burning desire to make the British eat their words. Biographer Martin Blumenson believes that the condescending attitudes of Alexander, Montgomery, and other senior British commanders left Patton obsessed with reaching Messina first "not so much for his personal glory, although that was important, but rather to prove to the world that American soldiers were every bit as good as—indeed, better than—British troops."[12]

In Patton's mind the only way this could be overcome once and for all was by means of a great American triumph, which in Sicily meant the capture of Messina. American morale was scarcely uplifted by the bias of the BBC, which repeatedly played up the exploits of the Eighth Army at the expense of American forces, whose role was considered minor, with British troops fighting the bloodiest battles. The day after the 1st Division captured Enna, the BBC announced that the British had taken the city. But when the BBC snidely commented that "the Seventh Army had been lucky to be in western Sicily eating grapes," even Eisenhower had heard enough and penned a strongly worded letter to Churchill criticizing the BBC for undermining his efforts to create a truly unified Anglo-American command in the Mediterranean.[13]

The final two weeks of the Sicily campaign brought out the best and the worst in George S. Patton, and irrevocably altered what was left of his relations with Omar Bradley. By virtue of their temperaments, training, and lifestyles, it was inevitable that Bradley would eventually be at loggerheads with Patton. Although he seems to have concealed his true feelings masterfully in their official contacts, privately Bradley had grown increasingly disenchanted with his mercurial boss. Given his schoolmasterish personality, Bradley was a stickler for military protocol and was offended when its unwritten rules were violated. Since the invasion he had grown critical of Patton's direction of the American ground effort, and both the boundary-line incident and the Palermo operation had buttressed Bradley's conviction that Patton was turning out to be a poor commander.[14]

Bradley's dislike of Patton, both personally and professionally, came boiling to the surface in Sicily. Despite the fact that he was to become famous as a result of articles written by correspondent Ernie Pyle,[15] Bradley was uncomfortable with the outward trappings that accompany senior rank, especially the carnival atmosphere of Patton's movements as "he steamed about with great convoys of cars and great squads of cameramen."[16] Patton's profanity and theatrics irritated Bradley, who believed that Palermo, and now the fixation with Messina, were thinly disguised ploys for headlines and personal glory at the expense of *his* troops. While he has candidly admitted that he was just as anxious as Patton to seize Messina ahead of the British, Bradley deeply resented Patton's high-handed interference. "To George, tactics was simply a process of bulling ahead. He never seemed to think out a campaign. Seldom made a careful estimate of the situation. I thought him a shallow commander."[17] Bradley's sharp criticism was grudgingly moderated by the admission that, later, in northwest Europe in 1944–45, Patton had "a pretty good feeling of what the enemy could do and could not do. Had this instinct developed to a higher degree than any commander I knew."[18]

Bradley had his hands full guiding the difficult drive by II Corps toward Messina. The Germans had repelled one 3d Division attack after another, in terrain that was virtually impassable to vehicles except along the narrow coastal highway. Each day the American infantrymen fought to overcome the clever German defenses, in searing heat and thick dust that quickly sapped the energy of even the fittest of men. The lasting memory of one officer was "how damn tired we got just going day after day."[19]

Both generals agreed on the need to undertake small amphibious raids behind enemy lines. Bradley was annoyed by the constant reminders that he must gain Messina at all costs. One day, as the two conferred on the north coastal road, Patton "in a grandiose fashion, said 'I want you to get into Messina just as fast as you can. I don't want you to waste time on these maneuvers even if you've got to spend men to do it. I want to beat Monty into Messina.' I was very much shocked, and replied, 'I will take every step I can to get there as soon as I can.' To me the quickest way was through maneuver and that I continued to do."[20]

Patton made no attempt to interfere in the first amphibious "end run" worked out by Bradley and Truscott. Using navy LSTs based in Palermo, Truscott mounted Task Force Bernard (a reinforced battalion), which landed behind the San Fratello Line, at Sant'Agata di Minitello, and, after a furious fight, finally broke the back of the formidable German defenses along Mount San Fratello.[21]

The second amphibious operation took place twenty-five miles farther east, at Brolo, and was the object of bitter American controversy. After being forced from the defensive line at San Fratello Line, the 29th Panzer Grenadier Division established new defenses in the Cape Orlando Peninsula. By this time Patton was showing increasing signs of frustration at the sluggish advance by II Corps and kept up constant pressure on Truscott and Bradley to win the race for Messina. The prospect of another debilitating and costly battle like San Fratello led Patton on August 10 to summon Bradley and order Bernard's task force to land at Brolo the following morning.

The operation was being planned in conjunction with a flank attack against the main German positions around the village of Naso. Originally scheduled for August 10, the landing was delayed after a Luftwaffe air attack sank one of the LSTs. "Now Patton was in no mood for another postponement, and he left no doubt in Bradley's mind of this fact."[22] Unable to position his infantry and supporting artillery in time to launch the attack, and having been informed that control of the operation was entirely his, Truscott asked for a twenty-four-hour postponement. But when General Keyes visited him on the night of August 10, he warned Truscott that Patton was unlikely to approve any further postponement for any reason, especially since there would be correspondents accompanying the task force. According to Keyes, Patton wanted no criticism resulting from another delay. Both generals spoke on the telephone with Bradley, who emphatically approved a postponement. Keyes then called Patton, and the result was explosive. As Truscott relates:

> An hour later, General Patton came storming into my Command Post giving everybody hell from the Military Police at the entrance right on through until he came to me. He was screamingly angry as only he could be. "Goddammit, Lucian, what's the matter with you? Are you afraid to fight?" I bristled right back: "General, you know that's ridiculous and insulting. You have ordered the operation and it is now loading. If you don't think I can carry out orders, you can give the Division to anyone you please. But I will tell you one thing, you will not find anyone who can carry out orders they do not approve as well as I can." General Patton changed instantly, the anger all gone. Throwing his arm about my shoulder he said, "Dammit Lucian, I know that. Come on, let's have a drink— of your liquor." We did.[23]

Patton recorded his own version of their stormy confrontation after he arrived at the 3d Division command post at 8:45 P.M.:

> Truscott was walking up and down holding a map and looking futile. I said, "General Truscott, if your conscience will not let you conduct this operation I will relieve you and put someone in command who will." He replied, "General, it is your privilege to reduce me whenever you want to." I said, "I don't want to. I got you your DSM and recommended you for Major General, but your own ability really gained both honors. You are too old an athlete to believe it is possible to postpone a match." He said, "You are an old enough athlete to know that sometimes they are postponed." I said, "This one won't be. The ships have already started." Truscott replied, "This is a war of defile and there is a bottleneck delaying me in getting my guns up to support the infantry. They—the infantry—will be too far west to help the landing." I said, "Remember Frederick the Great: *L'audace, l'audace, toujours l'audace!* I know you will win and if there is a bottleneck you should be there and not here."[24]

Patton called Bradley to state that the operation would go on as scheduled, and that he accepted full responsibility for failure. He then strode from Truscott's command post and was not seen again during the battle for Brolo. "I am not going to the front today," he wrote, "as I feel it would show a lack of confidence in Truscott, and it is necessary to maintain the self-respect of generals in order to get the best out of them."[25]

Nor did Patton seem to have more than momentary regrets over the incident with Truscott: "I thought of Grant and Nelson and felt O.K. That is the value of history. . . . I may have been bull-headed but I truly feel that I did my exact and full duty and under rather heavy pressure and demonstrated that I am a great leader."[26] To Beatrice he wrote that he had "to get pretty tough" and claimed to have threatened to have both Bradley and Truscott reduced to colonel. "I have a sixth sense in war as I used to in fencing and besides, I can put myself inside the enemy's head, and also I am willing to take chances." When she annotated his war letters after his death, Beatrice noted that Patton had apparently even asked each to name his successor in case he were relieved.[27]

Bernard's 650-man task force was undetected in the early morning hours of a day that Truscott later said "I will never forget." Then a chance encounter with a passing German half-track led to the discovery of the landing and a swift and violent German reaction to this grave threat to their rear. By mid-morning Task Force Bernard was beating off the first of many counterattacks, and by noon, with the nearest 3d Division unit still some miles away, the situation quickly became extremely grave. As Truscott restlessly prowled the front, urging his troops to break through to Bernard, a steady stream of messages arrived pleading for air and naval support. The navy, believing its job done, had returned to Palermo only to be urgently recalled to help break up the German counterattacks.

The naval task force arrived off Brolo at 2:00 P.M. and delivered welcome supporting fire, and twice more was recalled to help break up counterattacks. In late afternoon a furious air-sea battle erupted when eight Focke-Wulf 190s attacked the fleet and all but one were shot down. Throughout the afternoon of August 11 there was an air of desperation as the 3d Division fought unsuccessfully to relieve their beleaguered comrades. When contact was finally established with the 3d Division early the next morning, the 29th Panzer Grenadier Division was long gone. Losses on both sides were heavy.[28] Patton's concern was expressed indirectly, when he arrived at Brolo in a half-track and was seen leaning over the side in conversation with Lieutenant Colonel Bernard. The battlefield was littered with dead Germans and smashed equipment, and some GIs were sitting among the corpses, eating. This proved too much even for Patton, who noted that it took a very strong stomach to observe such a gruesome scene.

The subject of death was never long out of Patton's mind, and on these occasions he inevitably chastised himself for even thinking of his dreaded fear of acting in a cowardly manner. "I was disgusted to find my pulse went up—I timed it—but I soon got myself in hand," he wrote after his command post was shelled on August 9. "Mostly I can have a shell hit or a mine go off quite close without winking or ducking. This is a great asset. . . . One must be an actor."[29] Near Cape Orlando, Patton's entourage was bombed by three Luftwaffe fighter bombers. "We could see the bomb bays open and the bombs drop, but as we were on a road with a wall on one side and a cliff on the other, we could no nothing about it." The bombs missed, but Patton realized he had once again evaded death. "I wish that there could have been more spectators present to see how officers conduct themselves in tight positions."[30]

On another occasion an artillery shell landed smoking but undetonated on a road only a few feet from Patton, who interpreted it as yet another validation of his inevitable destiny. "I knew then . . . for certain, that I was destined to live to do great things," he later told Frederick Ayer. "That dud shell was the second time I had been told. The first was in the Argonne in the First World War, Now I know I have been marked to do great things, but, of course, I don't know how much luck I may have left."[31]

No one in II Corps headquarters thought Patton was great. Bradley's staff was bitter and he was furious, and later wrote that Patton's decision had left him "more exasperated than I have ever been," but "as a subordinate commander of Patton's I had no alternative but to comply with his orders."[32]

According to the official historians, the Brolo operation accomplished very little except for compelling the 29th Panzer Grenadiers to abandon their positions a full day ahead of the German timetable. Despite the heavy losses and the acrimony over Patton's refusal to agree to further delay, the real mistake at Brolo was not his controversial decision but his failure to perceive that a larger, more coordinated operation would have garnered a prize worthy of the effort. A regimental-size force would undoubtedly have succeeded in trapping the 29th Panzer Grenadiers at Brolo, thus collapsing the northern anchor of the German defenses. The effects on other German units would have been equally calamitous; their escape routes would have been severed; and the eventual unsatisfactory ending to the Sicily campaign might have been averted.

Nor was Brolo the last American amphibious operation. Several days later Patton had formed another task force of the 45th Division to land on the beaches east of Cape Milazzo. Again, Truscott protested to Keyes, this time for the simple reason that the 3d Division had already advanced *beyond* the landing site. Mindful of the turmoil over Brolo, neither Keyes nor Truscott was willing to arouse Patton's ire again, and the operation took place with a predictably ludicrous ending when members of Truscott's staff greeted the first wave on the beach.[33]

The Brolo controversy failed to inhibit Patton, who continued to prod his subordinate commanders to push on to Messina, much to their outrage and Bradley's contempt. Calling him impetuous and utterly unpredictable, Bradley's bitterness was still palpable after the war: "I disliked the way he worked, upset tactical plans, interfered in my orders. His stubbornness on amphibious operations, parade plans into Messina sickened me and soured me on Patton. We learned how not to behave from Patton's Seventh Army."[34] There is little doubt that the confrontation on August 10 severely strained relations between Patton and Truscott, who was too good a soldier ever publicly to criticize him, even after his death.

Outwardly undeterred by the obvious distress of Bradley and Truscott, Patton continued turning up in the front lines without warning. A 3d Division infantry commander recalls his first encounter with Patton on the north coast of Sicily. His battalion was halted outside a village that was under heavy German artillery fire from a nearby ridge. Suddenly Patton appeared on the road above him:

> He stood up in his command car and looked down at me, and he said, "You, man, who are you?" And, I said, "I'm the 3d Battalion commander, sir, Lieutenant Colonel John A. Heintges." And he said, "Well, you son-of-a-bitch, why the—— damn hell aren't you moving?" I said, "Sir, we are temporarily halted here because the Germans have us under observation from that ridge about 600 yards in front of you." And I said, "You have no business being here. You must have gone through my guard on the other side of the village behind us." And, just then, the German artillery opened up and I have never seen anybody turn around on a narrow road as quickly as he did. His driver turned around on a dime and left nine cents change. And, that was Patton.[35]

Another incident that gained Patton notoriety occurred on a narrow bridge where a Sicilian farmer was unable to budge two obstinate mules blocking the road. Patton pulled out his pistol, shot the two animals through the head, and, as the farmer wailed in protest, ordered the animals thrown off the bridge. Patton's critics have cited this incident as an example of his brutal nature. But generals are paid to make difficult and often unpleasant decisions such as this one; and, as Patton later explained, he did not enjoy killing mules but preferred it to the alternative of having the Luftwaffe arrive to strafe the column and kill large numbers of his men.[36]

On the evening of 16 August, while a small British task force was still some miles south of Messina, a 3d Division patrol became the first Allied troops to enter the all but abandoned and ruined city. That same morning a British armored and commando force was landed ten miles south of Messina. The commandos were led by Lt. Col. J. M. T. F. Churchill (no relation to the prime minister but imbued with the same Churchillian spirit), who was determined to beat the Americans to Messina. In the back of his jeep were a large Scottish sword and a set of bagpipes for the occasion. To his intense dismay Churchill learned that American troops were already inside Messina.

At 8:00 A.M. on August 17, an Italian colonel reported to Truscott and offered to surrender. Truscott demurred, pending Patton's arrival at 10:00 A.M. to make his triumphal entry. On his arrival, Patton turned to the assemblage and said: "'What in hell are you all standing around for?" and proceeded to lead the American convoy into Messina to the accompaniment of harassing enemy artillery fire from Calabria.[37] Even this minor nuisance failed to deter Patton from his long-sought triumph. Behind Patton and Truscott came a second vehicle containing Keyes and Lucas, who recorded in his diary that the GIs "were tired and incredibly dirty. Many could hardly walk but were pushing on. These American boys of ours have remarkable stamina and are terrible in battle. I am glad I am on their side. . . . We entered the town about ten-thirty amid the wild applause of the people. . . . The city was completely and terribly demolished."[38] As the convoy proceeded through Messina, the third vehicle was hit and its occupants wounded by fragments from long-range artillery across the Strait of Messina. Patton's obsession with his destiny was enhanced by having once again emerged unscathed from enemy fire. "My luck still held," he wrote Beatrice almost apologetically.

Patton's command vehicle—its three-star pennants gleaming in the sunlight—rolled into the piazza in the center of Messina and met Colonel Churchill, who was soon joined by the British tanks. Correspondent Richard Tregaskis describes the joyful scene:

> Commandos, smiling and shouting, sprawled over the exterior of tanks, and the little parade was made festive with many colored flowers thrown by Sicilians. Some of the dirty-faced soldiers clutched huge bunches of grapes. Brigadier Currie [J. C. Currie, the tank force commander] dismounted from his tank . . . [as] General Patton, dazzling in his smart gabardines, stepped out and shook hands with the tall, lean brigadier. . . . "It was a jolly good race. I congratulate you."[39]

Patton mounted a pedestal to deliver a brief talk: "He was very nervous and he had a little Texas Ranger [Major Stiller] who was his bodyguard, who just adored him. God, he would look at him and old Georgie looked down at him and said, 'You little son-of-a-bitch, you're supposed to guard me, not look at me!'"[40]

In the film *Patton*, Montgomery is seen to lead a formal British parade down a street in Messina, only to find Patton there before him with a smirk on his face at having beaten his rival for the prize objective in Sicily. The film version is apocryphal fantasy; nevertheless, Patton thought that the token British presence in Messina was a plot by Montgomery to "steal the show."

Conspicuous by his absence that morning was Omar Bradley, who had failed to rendezvous with Patton outside Messina. "This is a great disappointment to me, as I had telephoned him, and he certainly deserved the pleasure of entering the town," Patton wrote in his diary that night, unaware of his subordinate's outrage. Like Montgomery, Bradley had little interest in victory marches, and his disgust with Patton had reached its zenith with Patton's order that no units were to enter Messina "until he could make triumphal entry. I was furious. I had to hold our troops in the hills instead of pursuing the fleeing Germans in an effort to get as many as we could. [The] British nearly beat him into Messina because of that." Momentarily Bradley was tempted to drive into Messina and greet Patton on a street corner when he arrived, which, he admitted, "would have been playing Georgie's game."[41]

In a reflective mood, Patton confessed: "I feel let down. The reaction from intense mental and physical activity to a state of inertia is very difficult. I feel that the Lord has been most generous. If I had to fight the campaign over, I would make no change in anything I did. Few generals in history have ever been able to say as much."[42] Just recovered from a fever transmitted by sand fleas, he was weak and despite feeling "awful" told his wife: "As I fought the perfect campaign and got a second DSC, I may have fulfilled my destiny, but hope not. Someone must win the war and also the peace."[43]

Although capturing Messina was a personal triumph for Patton after it appeared that the Seventh Army's fate was to act as a perpetual shield for the Eighth, he was quick to laud his army's accomplishments in a stirring order of the day. "I really am proud of the 7th Army. We have fought without a day's rest since the 10th of July. . . . 'Men live in deeds not years' and I have lived a long time in the last thirty days but I feel very humble it was the superior fighting ability of the American soldier, the wonderful efficiency of our mechanical transport, the work of Bradley, Keyes and the Army staff that did the trick. I just came along for the ride. . . . I certainly love war."[44]

Patton's triumph notwithstanding, the end of the fighting in Sicily was a dismal conclusion to what had been a campaign beset by controversy and indecision. For thirty-eight days the Allies fought some of the bitterest battles of the war in terrain every bit as harsh as they would soon find ahead of them in Italy and in northwest Europe. Yet their enemy had defied them to the end, and had added the final insult by pulling off one of the most dazzling strategic withdrawals in military history. At Dunkirk, British ingenuity and grit had saved the BEF, but not their precious equipment. In Sicily the Germans had saved themselves and virtually everything capable of being ferried to the mainland. Only those captured and the dead were left behind. And now that the campaign had ended, Patton found himself anxious "to get out of this infernal Island."[45]

                              *   *   *

In early August there began a chain of events that very nearly undid the great destiny Patton had predicted for himself. August 3 began auspiciously, when Patton learned that Eisenhower was to award him the DSC for "extraordinary heroism" on July 11 at Gela, an award Patton was delighted to receive, but that "I rather feel I did not deserve . . . but won't say so."[46] That afternoon he stopped off en route to II Corps to visit the 15th Evacuation Hospital outside Nicosia. Inside were many newly arrived wounded, most of them 1st Division troops. "All were brave and cheerful." Scenes of battle-wounded men always moved Patton and this occasion was no exception.

Then he encountered Pvt. Charles H. Kuhl of Company L, 26th Infantry Regiment (1st Division), who evinced no visible wounds. Patton asked him why he was being admitted, and the soldier replied he was not wounded: "I guess I can't take it." This reply instantly enraged Patton, who swore at Kuhl, called him a coward, and ordered him out of the tent. When the frightened soldier continued to sit motionless at attention Patton grew even more irate and, according to an eyewitness, "slapped his face with a glove, raised him to his feet by the collar of his shirt and pushed him out of the tent with a final 'kick in the rear.' "[47] That night Patton noted in his diary that he had met "the only arrant coward I have ever seen in this Army. . . . Companies should deal with such men, and if they shirk their duty, they should be tried for cowardice and shot."[48]

Two days later Patton issued a memo to all commanders in the Seventh Army:

> It has come to my attention that a very small number of soldiers are going to the hospital on the pretext that they are nervously incapable of combat. Such men are cowards and bring discredit on the army and disgrace to their comrades, whom they heartlessly leave to endure the dangers of battle while they, themselves, use the hospital as a means of escape. You will take measures to see that such cases are not sent to the hospital but are dealt with in their units. Those who are not willing to fight will be tried by court-martial for cowardice in the face of the enemy.[49]

A week later on August 10 Patton arrived unannounced at the 93d Evacuation Hospital. According to his diary he "saw another alleged nervous patient—really a coward. I told the doctor to return him to his company and he began to cry so I cursed him well and he shut up. I may have saved his soul if he had one."[50]

Firsthand accounts have recorded a far grimmer scene. In the receiving ward, Patton spoke to each soldier about his wounds, offering small talk and words of encouragement. The fourth man he stopped to speak to was Pvt. Paul G. Bennett, an artilleryman assigned to the 1st Battalion, 17th Field Artillery Regiment, who was shivering on a cot. In response to Patton's query as to what his problem was, Bennett replied: "It's my nerves." Patton, who believed that there was no such thing as "combat fatigue," and that those who claimed to suffer from it were there only to shirk combat duty, bellowed: "What did you say?" Bennett sobbed: "It's my nerves, I can't stand the shelling any more."

Patton was shaking with anger nearly as much as Bennett was shaking with fright. "Your nerves, Hell, you are just a goddamned coward, you yellow son of a bitch. Shut up that goddamned crying. I won't have these brave men here who have been shot seeing a yellow bastard sitting here crying." He turned to the receiving officer, Maj. Charles B. Etter, and ordered him not to admit this "yellow bastard." "You're a disgrace to the Army and you're going back to the front to fight, although that's too good for you. You ought to be lined up against a wall and shot. In fact, I ought to shoot you myself right now, God damn you!"

Patton then pulled his pistol from its holster and waved it in front of the terrified soldier's face. The tumult brought the hospital commander, Col. Donald E. Currier, on the run. Patton turned on Currier and ordered: "I want you to get that man out of here right away. I won't have these brave boys seeing such a bastard babied." Patton then slapped the quivering Bennett across the face while continuing to curse him. Then, turning to leave the tent, Patton saw the distraught soldier crying, and strode back to his cot and hit Bennett a second time, with such force that his helmet liner was knocked to the ground. The tumult brought other nurses and doctors to the receiving tent in time to witness the second slap. To prevent further mayhem, Colonel Currier interposed himself between Bennett and Patton.

As he continued his tour of the wards, Patton continued to talk about Bennett, once choking back a sob: "I can't help it, but it makes my blood boil to think of a yellow bastard being babied." Before departing Patton turned to Colonel Currier and said: "I meant what I said about getting that coward out of here. I won't have those cowardly bastards hanging around our hospitals. We'll probably have to shoot them some time anyway, or we'll raise a breed of morons."[51] In his wake Patton left a shocked and angry hospital staff.

The victim of Patton's second outburst was a twenty-one-year-old Regular Army soldier who had served in II Corps since March 1943. According to the official report, Bennett had exhibited extreme nervousness after a fellow soldier was wounded. Although nervous and distraught, Bennett apparently resisted evacuation and begged to stay with his unit, but was ordered to the 93d Evacuation Hospital by his battery surgeon.[52]

Patton went directly from the 93d Evacuation Hospital to the II Corps command post, where he exuberantly recounted the incident to a horrified Bradley. "Sorry to be late, Bradley. I stopped off at a hospital on the way up. There were a couple of malingerers there. I slapped one of them to make him mad and put some fight back in him."[53] According to Bradley:

> He was bragging how he had treated this man to snap him out of being a coward. Thought that if he made the man mad, he would be mad enough to fight. That men were showing a yellow streak. He didn't agree with me that every man has a breaking point. Some are low, some are high. We call the low points cowards. To George anyone who didn't want to fight was a coward. He honestly thought he was putting fight into these men. He was pleased with what he had done. He was bragging about the incident. Next day the surgeon of that hospital handed a written report to [Brig. Gen. William B.] Kean [the II Corps Chief of Staff]. Kean brought it to me. . . . After reading it, I told Kean to put it in a sealed envelope in the safe—only to be opened by Kean or me. I didn't forward the report to Ike because Patton was my Army commander—I couldn't go over Patton's head.[54]

Whatever his antipathy toward Patton, Bradley never hesitated in deciding that his conscience would not permit him to forward the report. However, the II Corps surgeon, Col. Richard T. Arnest, felt no such qualms and, taking matters into his own hands, sent it through medical channels to Eisenhower's chief surgeon, Brig. Gen. Frederick A. Blessé.*

The two slapping incidents quickly became common knowledge among American troops in Sicily. Even Alexander soon learned of them but refused to involve himself in what he considered a purely American problem, saying to Patton: "George, this is a family affair."[55]

One of the 93d Evacuation Hospital nurses told her boyfriend, a young captain in Public Affairs, who ensured that the news was immediately passed to the American correspondents attached to the Seventh Army. Four reporters, Demaree Bess of the *Saturday Evening Post*, Merrill Mueller of NBC, Al Newman of *Newsweek*, and John Charles Daly of CBS interviewed Major Etter and others, but otherwise made no attempt to file the story. Among themselves they decided that the matter must be brought to Eisenhower's attention. Bess, Mueller, and Quentin Reynolds of *Collier's*, flew to Algiers, and on August 19 a written summary prepared by Bess was presented to Bedell Smith. The Bess report noted that Patton had committed a court-martial offense by striking an enlisted man, and ended: "I am making

---

*"Wounded" or "injured" in French, "Blessé" surely ranks high among apt conjunctions of a man's name and career path.

this report to General Eisenhower in the hope of getting conditions corrected before more damage has been done."[56]

By then Eisenhower was already aware that he had a potentially explosive problem on his hands. Two days earlier, on August 16, the day before Patton had triumphantly entered Messina to end the Sicily campaign officially, General Blessé had presented him with Colonel Arnest's report. While shocked, Eisenhower did not immediately grasp the full implications of Patton's acts and merely remarked that he guessed it would be necessary to give Patton a "jacking up." Without further details, he was inclined to believe that the report might be exaggerated. However, the more he reflected on it, the more Eisenhower realized the implications. It would take far more than a mere "jacking up" to resolve what had become a red-hot crisis. He ordered Blessé to Sicily at once to investigate the incidents and also to carry a handwritten letter to Patton. "If this thing ever gets out, they'll be howling for Patton's scalp, and that will be the end of Georgie's service in this war. I simply cannot let that happen. Patton is *indispensable* to the war effort—one of the guarantors of our victory."[57]

Eisenhower's letter to Patton contained the strongest words of censure written to a senior American officer during World War II. While he clearly understood "that firm and drastic measures are at times necessary, it did not "excuse brutality, abuse of the sick, nor exhibition of uncontrollable temper in front of subordinates." If true, then "I must so seriously question your good judgment and your self-discipline, as to raise serious doubt in my mind as to your future usefulness." Eisenhower concluded: "No letter that I have been called upon to write in my military career has caused me the mental anguish of this one, not only because of my deep personal friendship for you but because of admiration for your military qualities; but I assure you that [such] conduct . . . will *not* be tolerated in this theater no matter who the offender may be." Contrary to popular belief, Eisenhower's letter did not require a personal apology to every soldier and unit in Seventh Army—only that "you make in the form of apology or otherwise such personal amends to the individuals concerned as may be within your power."[58]

The arrival of the three correspondents reinforced Eisenhower's awareness that he had a tiger by the tail. What they wanted was a deal: In return for killing the story they wanted Patton fired. Correspondent Reynolds summed up the strong anti-Patton bias within the press corps when he told Eisenhower that there were "at least 50,000 American soldiers on Sicily who would shoot Patton if they had the chance." John Charles Daly thought Patton had gone temporarily crazy.[59]

Eisenhower had no intention of submitting to an undisguised attempt to blackmail him into getting rid of Patton. Torn among loyalty to an old friend, the clear necessity that he must be disciplined, and the consequences of losing Patton altogether if the incidents became public, Eisenhower unhesitatingly decided that "Patton should be saved for service in the great battles still facing us in Europe, yet I had to devise ways and means to minimize the harm that would certainly come from his impulsive action and to assure myself that it was not repeated."[60] He summoned the three reporters and made his case for Patton, explaining that he had written a sharp letter of reprimand and had ordered him to apologize personally for his behavior.[61] While making it clear that there would not be censorship if they elected to ignore him, "They were flatly told to use their own judgment," Eisenhower said, hoping that they would see fit to keep the matter quiet in the interests of retaining a commander whose leadership he considered vital. More out of respect for Eisenhower than of compassion for Patton, the reporters entered into a gentleman's agreement not to publicize the story.[62]

Patton, who had been on top of the world after his victorious march on Messina, was severely jolted by the sudden fury emanating from Algiers. However, when General Blessé arrived with Eisenhower's letter of censure, Patton had yet to feel the full effects of his acts. His diary entry that day was more defiant than remorseful over the "nasty letter from Ike. . . . Evidently I acted precipitatly. . . . "[63] Patton finally began to discern just how much trouble he was in when a cable arrived from Algiers, ordering him to meet Lucas at the Palermo airfield that afternoon (August 20). Lucas would be carrying a personal message from Eisenhower, and he was to listen closely to what Lucas had to tell him.[64]

Lucas condoned the first slapping incident as Patton being Patton, believing that his impetuousness was the necessary companion to his intensive drive. But, after meeting Eisenhower the morning of August 20, Lucas realized that Patton was in deep trouble. "Ike has written him a letter and wants me to go back this afternoon and do what I can. He is in danger this time, I am afraid. . . . It seems too bad that a really brilliant victory should be marred by such things as this." Lucas's advice to Patton was "kindly but firm": Apologize to the soldiers he had slapped; apologize personally to every division in Seventh Army; and promise never to repeat the act.[65]

Thus it was on Lucas's advice rather than Eisenhower's order that Patton made amends to his army. On August 21 a colonel from Seventh Army had unexpectedly appeared at Bennett's battalion and ordered him fetched at once: "General Patton wants to see him." With barely time to button his shirt, Bennett was hustled off to Palermo. "On his face was the look of a man en route to the gallows."[66]

The inspector general noted that Bennett's morale was greatly raised and his mental improvement hastened by Patton's action, particularly his request "that if he cared, I would like to shake hands with him. We shook."[67] Bennett's brigade commander unwittingly strengthened Patton's conviction that he had done nothing seriously wrong by (apparently erroneously) stating that Bennett had been AWOL and had falsely represented his condition to his battcry surgeon. "It is rather a commentary on justice when an Army commander has to soft-soap a skulker to placate the timidity of those above," Patton wrote in his diary.[68]

On August 22 the doctors, nurses, and enlisted men from the two evacuation hospitals who witnessed the incidents were summoned to Palermo to hear Patton explain that during World War I a close friend had committed suicide owing to mental upset, but he believed that this officer might have been saved if stringent measures had been taken. Although he had overstepped himself, his actions at each of the hospitals had been guided by a sincere belief that the two soldiers might have been shocked back into reality. One doctor has written: "The General stated that he had always regarded cases of 'shell shock' as being most tragic. . . . He thought that if such men could at once be driven from their own self-concern—for example, by anger directed at someone else—they might be helped toward recovery."[69] Few were impressed by Patton's sudden compassion, and the evidence suggests that Patton's real regret was the problem his acts had brought upon himself. Dr. Currier in particular thought that Patton was feigning contriteness "in an attempt to justify what he had done," and that it was "no apology at all."[70]

Since World War II a great deal has been learned of the effects of combat on soldiers, but then "shell shock" or "battle fatigue" was equated by many with a pretext for malingering and cowardice. Patton simply could never concede that all men are not created with equal tolerance to combat conditions, or that in Bennett's case there was neither malingering nor cowardice.[71] Privately Patton continued to insist that "my motive was correct because one cannot permit skulking to exist. It is just like a communicable disease. I admit freely that my method was wrong and I shall make what amends I can. I regret the incident as I hate to make Ike mad when it is my earnest study to please him. . . . I feel very low."[72]

Patton also apologized to Private Kuhl, explaining that his intent had been to make Kuhl mad and say to himself: "I'll show that SOB Patton that I am not a yellow coward, become brave and redeem myself. But I see now that I used the wrong psychology. If you will shake my hand in forgiveness, I'll be much obliged to you." An eyewitness recorded that Kuhl's face "lit up with a broad grin. He grabbed and shook the general's hand. . . . It was very impressive and dramatic. I kept thinking, 'My Lord, here is a three-star general apologizing to a lowly private soldier.' I could not imagine anything similar happening in any other army."[73]

Kuhl's personal insight on the affair is instructive: "I think he was suffering a little battle fatigue himself." In 1970 an astute reporter wrote: "Unquestionably, Kuhl is correct in his assumption. Patton always drove himself as hard as he drove his men—to the outer limits of human endurance. Patton's difficulty was that he refused to acknowledge in himself the battle fatigue he deplored in his men. And so when the exhausted general met the exhausted private, the difference between the *slapper* and the *slappee* was only a matter of rank."[74] Although Patton would have scoffed at the notion, it is worth recalling that in the nearly three years since he was assigned to the 2d Armored at Fort Benning, he had been under tremendous strains, virtually without letup, that would have stressed the strongest of men. Patton may have thought of himself as a modern, invincible superman, but, as he had demonstrated in the aftermath of World War I, he was subject to the same human frailties as his soldiers. An argument can be made that, on the days he encountered Kuhl and Bennett, Patton may well have been suffering from a form of the very syndrome he deplored so strongly in them. As the Canadian coauthor of a landmark study of battle exhaustion has written: "the idea that soldiers might not always be capable of enduring stoically the extreme stress of battle is not easily accepted . . . [but] is as old as war itself. . . . During the Second World War . . . it was never entirely clear whether the psychological dysfunction was better left to the doctors or military policemen."[75]

What is indisputable is that Patton was clearly under considerable stress (much of it self-imposed) over his obsession with capturing Messina. The second slapping incident, on August 10, occurred only hours before Patton's contentious meeting with Truscott over the Brolo end run. In addition, on August 6, Patton had visited a field hospital and noted in his diary that he had seen two soldiers suffering from shell shock, one of whom kept going through the motions of crawling. There had been no complaint about these men, but he had witnessed several terrible injuries, including one man whose head was mostly blown off, who was soon to die. "He was a horrid bloody mess and not good to look at, or I might develop personal feelings about sending men into battle. That would be fatal for a General."

That same day the Seventh Army command post began coming under long-range German artillery fire. Patton admitted that the shelling bothered him. "I was worried for a few seconds and I feel ashamed of myself but I got over it. I have trained myself so that usually I can keep right on talking when an explosion occurs quite close. I take a sly pleasure in seeing others bat their eyes or look around."[76]

As was his longtime habit, Patton had been visiting field hospitals throughout the Sicily campaign and used these occasions to decorate his men. The day before the Kuhl slapping, Patton had visited a hospital where he pinned Purple Hearts on forty wounded soldiers, including one dying man who had an oxygen mask over his face. Removing his helmet, "I knelt down and pinned the Purple Heart on him and he seemed to understand although he could not speak."[77] Patton then arose and stood at attention. Codman accompanied him everywhere and later wrote that Patton also gave a rousing speech "to each ward, to the nurses, to the interns . . . Elementary if you like, but I swear there wasn't a dry eye in the house."[78] Afterward Patton spoke to a group of reporters and said: "I do a lot of human things people don't give me credit for and I'm not as big a [expletive] as a lot of people think. The commander of invading troops is under great tension and may do things he later regrets."[79]

With this scene fresh in his mind, the next day Patton encountered Kuhl. Could the severe strain of such events and the need to keep up the facade of his self-created tough-guy image been too much even for Patton?

Patton never understood or acknowledged that throughout the slapping affair Eisenhower had stood firmly behind him and refused to accept the possible loss of the leader who had triumphantly played a key role "in a campaign which will be a model for study in military schools for decades."[80] While Patton was complaining that Eisenhower had not congratulated him on his great victory, Eisenhower was agonizing over whether he might yet have to send him home in disgrace. Butcher recorded that "Ike is deeply concerned and has scarcely slept for several nights, trying to figure out the wisest method of handling this dilemma. The United Nations have not developed another battle leader as successful as Patton, Ike thinks."[81]

In fact Eisenhower was far more sympathetic than Patton ever knew. He found the very notion of sending Patton home in disgrace appalling:

> Ike makes a point that in any army one-third of the soldiers are natural fighters and brave; two-thirds inherently are cowards and skulkers. By making the two-thirds fear the possible public upbraiding such as Patton gave during the campaign, the skulkers are forced to fight. Ike said Patton's method was deplorable but his result was excellent. He cited history to show that great military leaders had practically gone crazy on the battlefield in their zeal to win the fight. Patton is like this. . . . Yet Ike feels that Patton is motivated by selfishness. He thinks Patton would prefer to have the war go on if it meant further aggrandizement for him. Neither does he mind sacrificing lives if by so doing he can gain greater fame. So Ike is in a tough spot; Patton is one of his best friends . . . but friendships must be brushed aside.[82]

Eisenhower usually kept Marshall abreast of everything he did, but in a long letter written on August 24, Eisenhower withheld details of the slappings, saying only that Patton's "brilliant successes" in the Sicily campaign

> must be attributed directly to his energy, determination and unflagging aggressiveness . . . in spite of all this—George Patton continues to exhibit

some of those unfortunate personal traits of which you and I have always
known and which during this campaign caused me some most uncomfortable days. His habit of impulsive bawling out of subordinates, extending
even to the personal abuse of individuals, was noted in at least two specific cases. I have had to take the most drastic steps; and if he is not cured
now, there is no hope for him. Personally, I believe that he is cured—not
only because of his great personal loyalty to you and to me but because
fundamentally he is so avid for recognition as a great military commander
that he will ruthlessly suppress any habit of his own that will tend to jeopardize it.[83]

A month later Eisenhower reiterated his belief in Patton as "a truly
aggressive commander and, moreover, one with sufficient brains to do his
work in splendid fashion."[84] However, had it been up to Bradley, Patton
would not have been let off so easily. "I would have relieved him instantly
and would have had nothing more to do with him. . . . He was colorful but
he was impetuous, full of temper, bluster, inclined to treat the troops and
subordinates as morons. His whole concept of command was opposite to
mine. He was primarily a showman. The show always seemed to come
first."[85]

As Patton made the rounds of every Seventh Army unit, his "apology"
generally took the form of an oblique reference of regret "for any occasions
when I may have harshly criticized individuals."[86] The inspector general's
report noted: "He usually referred to 'certain incidents that had better be forgotten.' He customarily used 'earthy language.' The effect of these talks differed with each individual. Many men were inspired to greater effort, others
were disgusted. The proportion of the latter is considered large enough to be
the cause of serious concern."[87]

The general reaction in most of the combat units was quiet indifference.
In Patton's former division, the 2d Armored, there was great enthusiasm and
a tendency to discount the rumors about Patton as untrue. In the 1st Division
the new commander, Maj. Gen. Clarence Huebner, worried that he bore
some of the responsibility for the slappings after remarking earlier in the
campaign to Patton that he believed there were some soldiers malingering in
the hospital to avoid combat. "Well, as luck would have it Patton went
straight to a hospital."[88] Patton's speech to the eighteen thousand men of the
"Big Red One," however, was greeted with stony silence, although one eyewitness recalls hearing a few scattered boos.[89] Huebner believed that this
was not so much a renunciation of Patton as an acknowledgement that he
was wrong and that the matter should be forgotten as quickly as possible.
Col. Monk Dickson and other members of Bradley's staff thought it was an
out-and-out repudiation of Patton.[90] Another officer recalls "that the 1st
Division had only just arrived in Licata and many of the troops had no idea

why Patton was there. Some asked why Patton had to do this. The division
was tired, dirty and had been in combat for thirty-eight days. Even I did not
know until three hours prior to the speech."[91]

Patton's appearance before the 60th Regiment of the 9th Infantry Division was another matter altogether. Even though the 9th Division had only
participated in the final two weeks of the campaign, its troops were nevertheless as familiar with Patton as those of the other divisions. The scene
described by an infantry major (later a four-star general) was one of the
most extraordinary moments of the war:

> We were assembled in a large . . . olive orchard [near Randazzo]. . . . Gen-
> eral Patton arrived in that famous command car of his with two metal flags
> on either side—three stars and the "Pyramid of Power," the Seventh Army
> emblem—a long, trailing cloud of dust, and MPs . . . we all stood at atten-
> tion and put on our helmets and the bugler sounded "Attention" and Gen-
> eral Patton mounted this . . . platform in front of these 3,000 [troops].
> . . . General Patton had a rather high, squeaky voice, and as he started to
> address the regiment he said, "Take seats," so we sat down on our hel-
> mets—it was a practice of those days, to keep us out of the mud or the dust
> . . . and General Patton started to give what we knew was to be his apol-
> ogy. But he never got past the first word, which was "Men!" And at that
> point the whole regiment erupted. It sounded like a football game—a
> touchdown had been scored because the helmets (steel pots) started flying
> through the air, coming down all over—raining steel helmets and the men
> just shouted "Georgie, Georgie"—a name which he detested. He was say-
> ing, we think he was saying—"At ease, take seats." Then he had the
> bugler sound "Attention" again, but nothing happened. Just all these
> cheers. So, finally General Patton was standing there and he was shaking
> his head and you could see the big tears streaming down his face and he
> said, or words to this effect, "The hell with it," and he walked off the plat-
> form. At this point the bugler sounded "Attention" and again everybody
> grabbed the nearest available steel helmet, put it on, being sure to button
> the chin strap (which was a favorite Patton quirk) and as he stepped into
> his command car and again went down the side of the regiment, dust
> swirling, everybody stood at attention and saluted to the right and General
> Patton stood up in his command car and saluted, crying. . . . He was our
> hero. We were on his side. We knew the problem. We knew what he had
> done and why he had done it. . . . He never came back.[92]

At another gathering of the 9th Division, there was a different ending.
The troops were gathered in a square around a stage and had no idea why
they were there. As one officer recalls, they soon learned why:

Patton was there to apologize for kicking ass at one of the Army's hospitals. It was difficult for the General to make the apology . . . and it took us a while to get the message. . . . General Patton was sweating hard. Suddenly, he stopped talking, reddened even more and stared out over our heads. He broke off and abruptly departed. What had broken his train of thought were dozens of condom-balloons, blown up by the rear ranks and wafted on the gentle breeze that came in behind us. It was a classic example of the soldier saying it all with the fewest possible words.[93]

Emotions also ran high when Patton appeared before the entire 3d Division outside Trapani. "We all loved General Patton regardless of what you hear. Sure, he cussed us out. He called us 'son-of-a-bitch' and 'bastards' . . . but there were a hell of a lot of tears shed when Patton had to apologize to my whole division," wrote a lieutenant colonel.[94] As Patton spoke there came unprompted shouts of "No! General, no no, no, General, no, no." Then, "Tears came to his eyes. The roar swelled in volume. . . . Choked with emotion, he left the speaker's stand abruptly, returned to his car, and drove away."[95]

Patton's most impassioned apology was reserved for his friend Dwight Eisenhower. In a lengthy letter Patton wrote: "I am at a loss to find words with which to express my chagrin and grief at having given you, a man to whom I owe everything and for whom I would gladly lay down my life, cause to be displeased with me."[96]

Although the whole "*l'affaire* Patton" was by the autumn of 1943 common knowledge to thousands, Eisenhower's gentleman's agreement with the press held until late November, when someone leaked the story to muckraking columnist Drew Pearson, who gleefully sensationalized the story on his weekly syndicated radio program. The resulting storm of criticism might have forced weaker men than Eisenhower, Marshall, and Stimson to sacrifice Patton on the altar of public opinion. Some congressmen and senators demanded Patton's dismissal, but Marshall and Stimson refused to bow to the pressure, while overseas Eisenhower defended his decision to retain Patton for Overlord.

Stimson sent a letter to the Senate defending Eisenhower's decision to retain Patton on the basis of the pressing need for his "aggressive, winning leadership in the bitter battles which are to come before final victory." Off the record he rebuked Patton, writing of "his disappointment that so brilliant an officer should so far have offended against his own traditions."[97]

But support for Patton came from many sources as well, including many of the mothers and fathers of his men. According to his aide-de-camp, Patton's personal mail was about 89 percent favorable.[98] Articles about the affair appeared in every conceivable publication, including the prison newspaper of the Atlanta penitentiary. Considering its sensational nature, the storm over Patton abated rather quickly, lending credence to the maxim that the public rarely remembers beyond yesterday's headlines. One of the saddest aspects of the slapping incidents was the dissolution of the long friendship between Patton and Gen. John J. Pershing, who privately became one of Patton's sternest critics. Although Patton wrote several letters to Pershing during the remainder of the war, the old general never replied.[99]

What actually went through Patton's mind during the two slapping incidents? Fortunately, a hitherto unpublished record exists in the form of a memoir written by his former cavalry commander, Maj. Gen. Kenyon A. Joyce, who recorded this account of what he called "as dramatic and soul searing a confession of an inexcusable act as I have ever heard."

In both instances, the action was inexcusable on my part. I am thoroughly conscious of that. There is no real extenuation but I will tell you what happened in one of the cases. It was at the peak of the campaign. I was under stress and nervous. I visited a hospital where there were a number of badly wounded and in one ward, there were many cases in desperate condition. [A reference to his visit to the 15th Evacuation Hospital on August 3.] I went along, visiting with the men and toward the end of the ward, I found a man who was bandaged in many places from head to foot. He had been hit by various fragments from a high explosive shell and was in critical condition. However, in reply to my greeting, he said, "General, I happened to be in the wrong place at the wrong time but the docs are fixing me up and I'll be O.K. and back with you soon." That man was great and had the fighting spirit that has made our country. It touched me deeply and I began emoting. You know me. I stammered something about "You are a fine soldier," and stepped to the next bunk.

That case was even worse. He was bandaged over one eye. He was in traction for a broken leg and broken arm and was a sight that would move anyone deeply. When I asked what had happened, he said he had stepped on a land mine and was lucky to be alive. I asked how he was getting along and he replied, "Oh, fine. They are patching me up in style," and then with a light shining in his unbandaged eye, "Save a place for me, General. I'll be back." I stammered something, I don't know what, as I was emoting all over the place by then. You know me when the tears get in the eyes, I'm sunk. I brushed them aside, however and moved to the last bunk in the ward.

At this point the trouble began, when Patton found

a man sitting on the side of his bunk, not bandaged or showing any signs of medical care that had been in such tragic evidence all down the ward. I approached him and said, "What happened to you, my man? Did you get wounded?" He looked like a cowardly rat and whined like one. Coming on top of the wonderful courage and great spirit I had seen displayed by other men in the ward, it caused a consuming revulsion to come over me.

I stepped closer to this supine creature, told him to get up on his feet and try to act like a soldier. I then asked him if the example of those brave men in the ward did not stir up something in him that would make him want to do his part. He said, "Ah, no, those guys don't mean nothin. I just can't take it."

With that, something burst in me. I said, "You rat" and slapped him across the face with my gloves, turned on my heel and walked away. It was inexcusable on my part. I was a damned fool but the contrast between those brave men of valor and this creeping thing did something to me. The other case [on August 10] was somewhat similar and in that I was equally a damned fool.[100]

Patton may have thought himself "a damned fool," but subsequent evidence leaves no doubt whatsoever that he was never willing to admit the existence of battle exhaustion. Six months later Patton would write in his diary that his order of August 5 not to admit such cases to any hospital "exactly expresses my feelings then and now."[101]

Patton's daughter Ruth Ellen had her own interpretation:

The answer is simple to those who knew him. He hated to see men killed; he honestly did not believe in battle fatigue—he called it cowardice—and he felt a coward has less of a chance to stay alive, because fear dulls intuition, and cowardice is catching—the ancient Greeks called it panic. He said, "One must fear, but one must conquer fear to continue with the scheme of God." He slapped that soldier [sic] as he had, years before, slapped hysterical daughters, because he thought the man needed to be snapped out of his collapse and restored to some pride. He afterwards wrote mother that he knew no officer should ever lay a hand on an enlisted man and he had been wrong and that he had apologized. That's all there was to it.[102]

The enormous publicity over the slappings ensured that the Germans also learned of Patton's problems. There was genuine perplexity over what the fuss was all about. In the German army it was not unusual for a soldier accused of malingering or cowardice to be summarily shot without recourse or trial. Why, they wondered, when success in battle was far more important than personal deportment, would the Americans even consider jeopardizing their own effort by punishing their most "thrustful" leader? Gen. Alfred Jodl, Hitler's operations chief of the OKW (army high command), thought the publicity about Patton nothing more than Allied propaganda to lull the Germans into believing Patton was to have no further role against them.

The indignity of an army commander apologizing to his men might have overwhelmed a lesser man, but Patton managed to retain not only his self-respect but even a sense of humor. At his first public appearance in September, the chairman of the American Red Cross introduced him to a large gathering of GIs. Patton announced that "I thought I'd stand here and let you fellows see if I am as big a son-of-a-bitch as you think I am." The assembled troops erupted in cheers.[103]

Had Patton been a lesser general his career would most certainly have ended after Sicily. That his superiors elected to retain him was best summarized by Assistant Secretary of War John J. McCloy, who told Eisenhower: "Lincoln's remark when they got after Grant comes to mind when I think of Patton—'I can't spare this man—he fights.'"[104] When President Roosevelt was asked about the incident at a press conference, he "recalled Lincoln's response when told that Grant drank too much, and only slightly misquoted: 'It must be a good brand of liquor.'"[105]

CHAPTER 35

Exile

In the doghouse . . . : In a state of disfavor or
repudiation.

—*WEBSTER'S THIRD NEW INTERNATIONAL DICTIONARY*

I am like a passenger floating on a river of destiny.

—PATTON

The Italian dictator Benito Mussolini and his fascist government were
deposed in late July. King Victor Emmanuel III immediately appointed Field
Marshal Pietro Badoglio to head the Italian government. Though he proclaimed that the war would go on as before, within weeks Badoglio began
secretly negotiating an armistice with the Allies, which led to Italy's secession from the Axis in September 1943. The Germans moved quickly to
revive earlier contingency plans: Rommel was hastily recalled from a special mission to Greece and given command of a newly formed Army Group
B, as the Germans began moving fresh troops into Italy under the pretense
of having to reinforce Kesselring in the event of an Allied invasion.

After months of indecision, the Allies had at last acted on Eisenhower's
recommendation that an invasion of Italy follow the fall of Sicily. At the
Quebec Conference in August, the Combined Chiefs of Staff brawled over
the scope of further moves in the Mediterranean. Churchill envisioned continued Allied operations into Italy and, if possible, an offensive that would
carry them into northern Italy, from where Allied aircraft could be employed
in a direct support role of the cross-Channel operation; while Marshall
remained adamant that nothing detract from Overlord. What was lacking

then or at any time during the Italian campaign was a statement of Allied grand strategy. But if the political goals were vague, even less clear were the aims of the forthcoming military operations there.

The Casablanca Conference never resolved whether Sicily was to be the stepping-stone to a larger objective in Italy, or merely an end in itself, but with Overlord deferred until May 1944, there was never the slightest doubt that the Allies would employ the massive force in Sicily and North Africa to invade Italy. Eisenhower was authorized to begin planning two operations to be directed against Italy as soon as Sicily fell. The Fifth Army was created in the spring of 1943 for the purpose of planning and executing operations in Italy, and for months Mark Clark's troops had been preparing for an amphibious invasion south of Naples, at Salerno. The Allied indecision and the fact that the command appointments for Overlord had yet to be made were to cloud Patton's future.

Within days of the fall of Messina the war moved to mainland Italy, with the Eighth Army invasion of Calabria, the Italian boot, and the Salerno landings on September 9, 1943. The Seventh Army was relegated to the bench, its future as yet undefined but apparently destined for disbandment. Its commander was unemployed and in the dog house.

In Sicily the U.S. Army came of age. Despite the tragic outcome of the airborne and glider landings, Ridgway's 82d Airborne and the British 1st Airborne Division proved there was future merit in this new weapon of warfare. American leadership at all levels was generally excellent. Troy Middleton led an untested National Guard division in the first of many campaigns, and Lucian Truscott whipped the 3d Division into one of the finest infantry divisions in the army. The veteran 1st Division upheld its stellar reputation at Gela and survived a change of leadership successfully, if grudgingly, when Terry Allen and Teddy Roosevelt were replaced in early August. At the higher level, Patton's deputy, Geoffrey Keyes, emerged as a future corps commander of considerable promise, and Omar Bradley drew universal praise for his calm, determined leadership, which had already made his name known to the public and would shortly carry him to the pinnacle of his profession.

Patton believed he could read this newfound maturity in the body language of his men. When he visited the 9th Division, which had received its baptism of fire in Tunisia, Patton noted: "The whole look of the men, especially their eyes and mouths, has changed. They have looked at death and laughed."[1]

Patton's drive on Palermo was a mixed blessing. Strategically Bradley was correct: It was unexceptional and of dubious military value. However, the very flamboyance of Patton's offensive brought both the Seventh Army and its commander to the attention of the world, and in the final days of the campaign Patton presided over the demise of the notion that American fighting ability was suspect. Despite the objections of Bradley and Truscott, Patton was correct to pursue the tactic of amphibious end runs rather than continue the slow and costly advance against stubborn German delaying actions, although they would have been far more successful if employed on a larger scale. The accomplishments of Truscott's infantry and the tanks of the 2d Armored Division, in terrain ill suited for mobile warfare, went largely unrecognized except by Montgomery, and it was not until the following summer in Normandy that Patton and his Third Army removed the last doubts with a stunning demonstration of the power and mobility of armor. The outstanding feature of the U.S. Army in Sicily was how quickly it overcame the setbacks of Tunisia and absorbed the lessons necessary to become a first-class army. A great deal of that credit properly belongs to Patton.

Contrary to popular belief, Montgomery greatly admired the accomplishments of the U.S. Army in Sicily. Just as the so-called Patton-Montgomery rivalry for the capture of Messina was a myth, so too was any notion that Monty ever denigrated Patton or the American GI. As his biographer points out: "Whatever he might feel about Patton's strategic judgment, Monty had been genuinely impressed by Seventh Army's mobility, speed, and on its eastern flank, rugged determination and professionalism." Moreover, Montgomery refused to permit anything derogatory about Patton to be printed in the newspapers of the Eighth Army. An article reporting that Patton was under investigation brought the editor a rare rebuff from Montgomery, who believed Patton to be a good man and that such criticism reflected badly on himself.[2]

For Patton, Sicily had been a roller-coaster ride that saw his career rise in dramatic fashion and then falter in a tide of criticism and bad publicity. In the months following the slapping incidents Patton resided in isolated splendor in the palace at Palermo, as Napoleon had during his exile on the island of Elba. Gradually the Seventh Army was stripped away to meet new commitments in Italy and England for Operation Overlord.

Of the many impulsive acts Patton was to commit during his lifetime, none was to prove more ruinous than the momentary acts of rage that overcame him in August 1943. At the very zenith of his career, just as he was being acclaimed as the hero and conqueror of Sicily and was on the verge of assuming a starring role in the forthcoming cross-Channel invasion, the slappings not only ruined him professionally but were indirectly to change the course of history. Important decisions were about to be made, and his name would certainly have led the list of those being considered to command the American invasion force in Overlord. Instead it took the intervention of Eisenhower and the steadfastness of Marshall and Stimson merely to keep from losing Patton's services altogether. And, while Eisenhower was adamant that Patton was essential to the defeat of Nazi Germany, it was equally true that he would not consider entrusting him with the command of all American ground forces. Patton had demonstrated, not once but twice, that his emotional stability was unreliable. Eisenhower was willing to tolerate his eccentricities but only under the right conditions. In practical terms it meant that for Patton to remain in any capacity, it would have to be under whoever was appointed senior American ground commander in Overlord.

Success in life is often the result of being in the right place at the right time, and the reality was that there were only three senior American generals who might have commanded the American invasion force: Jacob Devers, then the U.S. Army commander of all American forces in the United Kingdom; George S. Patton, and Omar Bradley. Devers had no combat experience and although in the right place, was clearly not the leading candidate for a command calling for an officer who had fought the Germans. Thus, in reality, the practical choice were between Patton and Bradley. Patton was the senior officer, and the command would have been his almost by default had the slappings in Sicily not occurred.

On August 29, at the end of a luncheon in Catania with Montgomery and Eisenhower, Patton handed his friend a letter of apology for the slappings. "He just put it in his pocket." By this time Patton had finally accepted how close he had come to disaster and was heartened when Lucas wrote that Marshall had received a glowing account of him from Eisenhower. "Well, that was a near thing but I feel much better. Ike ordered me to go to Algiers to see Mark Clark. He said that he may lose Bradley, Clark may be killed, and I will have to take over. I seem to be third choice but will end up on top."[3]

In early September Patton was stunned by Marshall's announcement that Omar Bradley had been selected to command the First United States Army—the designated American ground force for the cross-Channel invasion: "It is very heartbreaking. The only time I have felt worse was the night of December 9th, 1942, when Clark got the Fifth Army. . . . I feel like death but will survive. I always have." Bradley was delighted to be leaving Sicily, not only because of the promotion but most of all because it removed him from Patton's control. He took the principal staff officers of II Corps to England with him to form the new First Army staff. His successor, Patton's protégé Geoffrey Keyes, took over a bereft corps headquarters that had to be rebuilt from scratch.

When Bradley came to pay a final courtesy call, Patton swallowed his pride and brought out an honor guard and band for him:

We had quite a long talk and I told him a lot of my best ideas to tell General Marshall. I suppose I should have kept them to gain reputation by springing them myself, but I am not built that way. The sooner they are put into effect the better for our army. Bradley has a chance to help or hurt me with General Marshall. I hope he chooses the former course but I did not ask him to. My resilient nature worked all right and today I am almost back to normal. But I have to keep working on my belief in destiny and poor old destiny may have to put in some extra time to get me out of my present slump.[4]

Bradley was not fooled by Patton's bravado and thought he was "in a near-suicidal mental state. . . . This was an exceedingly awkward time for me personally," wrote Bradley. "This great proud warrior, my former boss, had been brought to his knees."[5]

Privately Patton foolishly continued to criticize Eisenhower, in the belief that he lacked the courage to support him. "Ike and Beetle [Smith] are not at all interested in me but simply in saving their own faces."[6] Patton's own judgment of himself in Sicily was colored by his misguided belief that he was being punished for "doing my plain duty to a couple of cowards. . . . Of course, I realize I did my duty in a very tactless way, but so long as my method pleased the God of Battles, I am content."[7] What he could not, and never did comprehend, was that he could not simply say "I'm sorry" and expect that he could carry on as if nothing had ever happened. His punishment was far more than exile in Palermo for the foreseeable future: It was to be the denial of an army group command in the decisive campaigns of World War II.

Patton's anger was sadly misplaced, as was his inability to comprehend that, had it not been for Eisenhower, he *would* have been summarily relieved and sent home in disgrace. One of Eisenhower's biographers draws the parallel between Eisenhower and Patton by quoting Lincoln on his temperamental commander of the Army of the Potomac: "All I want General McClellan to do is to win a victory. If he will do that, I will hold his horse."[8] In the autumn of 1943, Eisenhower was "holding Patton's horse" for the very same reason.

In late August Eisenhower wrote to Marshall to praise Patton for "quickly and magnificently" rehabilitating II Corps in Tunisia and for his leadership of the Seventh Army. He "has qualities that we cannot afford to lose unless he ruins himself. . . . His intense loyalty to you and to me makes it possible for me to treat him much more roughly than I could any other senior commander. . . . You have in him a truly aggressive commander . . . with sufficient brains to do his work in splendid fashion."[9]

Eisenhower was well aware that Patton's detractors were snidely twisting his vainglorious nickname by referring to him as "Our Blood, his Guts," and apparently privately agreed with them to some extent. "Ike feels Patton is motivated by selfishness," wrote Butcher. "He thinks Patton would prefer the war to go on if it meant further aggrandizement for him. Neither does he mind sacrificing lives if by so doing he can gain greater fame."[10] It was a reflection of Patton's obsession with carving his place in history that even his closest friend viewed him as intrinsically a glory hound. Or, as Eisenhower's biographer Piers Brendon has less charitably written: "Ike recognized that Patton's vicious and manic qualities were better calculated to win victories than the sober virtues of less inspired generals."[11]

Even after Drew Pearson's revelation, Eisenhower proclaimed: "I still feel my decision sound," and refused to retract it. But the slapping incidents left understandable wariness about the precise extent of Patton's future potential. Shortly after he was appointed Supreme Commander for Overlord in December 1943, Eisenhower made clear to Marshall: "In no event will I ever advance Patton beyond Army command."[12] With Patton in the doghouse, it never really became an issue in Marshall's mind of who should receive the coveted Overlord appointment. As for Bradley, Eisenhower praised him for "running absolutely true to form all the time. He has brains, a fine capacity for leadership and a thorough understanding of the requirements of modern battle."

The war had passed him by, and Patton's days became long, inconsequential, and increasingly boring. He drew some light relief during an evening with comedian Bob Hope, during which the two exchanged a lively banter. Hope said Patton was certainly "not the old blood and guts *I've* heard about." Then leaning over, Hope said, "Look, General, if you should ever be *out* of a job, I *believe* I can get you a solid week at Loew's State [theater]."[13]

Patton's loneliness was best exemplified when his friend Jimmy Doolittle decided on the spur of the moment to stop off in Sicily and announced to the airfield control tower he would like to land and pay his respects to Patton. "When I parked, there was Georgie in his famous jeep with the three-star flags flying, his helmet reflecting the sun gloriously . . . he rushed forward, threw his arms around me, and with great tears streaming down his face, said, 'Jimmy, I'm glad to see you. I didn't think anyone would ever call on a mean son of a bitch like me.' "[14]

Beatrice had been living with her sister, Kay, in Washington since August 1942, but in April 1943 she returned to the calmer atmosphere of Green Meadows, where she remained for the duration of the war. Her occasional radio broadcasts continued to emphasize the merits of patriotism, of soldiering, and of being the wife of a soldier. She continued to write long letters to her husband, filled with equal measures of family news, gossip, and advice. Her family usually demanded that she read aloud Patton's letters to her, which she did with what she describes as "judicious skipture," often to howls of laughter in response to one of his outrageous statements or brickbats.[15]

His letters remained a chronicle of his daily existence with as much detail as he could provide without incurring the wrath of the censors. For one of the rare times in his life, Patton did not share his woes with Beatrice. His only fleeting reference to his mounting difficulties was a vague comment on August 22: "As usual, I seem to have made Divine Destiny a little mad, but that will pass." It was only after Drew Pearson publicized the incidents that Patton revealed what had happened, noting lamely: "I did not write you about the incident, as I see no value in spreading my troubles and we all thought the whole thing was closed. It is too bad that the facts have never come out, as I was not much off base. I had just talked with hundreds of wounded heros and had no way of knowing that the two men in question were shell-shocked. No one told me." He later added: "I am morally certain he was not shocked but was just scared. . . . Can you blame me. We have all spanked babies." With this admission came the truth: "Of course I have not had a very happy time. I could not sleep except with pills and would wake up groaning, but that is all over. It has been a good experience and I am a better General as a result of it. . . . You know the motto of the 'Brave Rifles'—well, I too, have come out steel. I have a destiny and I shall live to fulfill it."[16]

Although he apologized to Beatrice in February 1944 for having "hurt you," he was barely contrite. "I know Damned well I did my duty and if more people did it the same way we would win a war instead of just fighting one. My little dictionary has not got Sycotpast [psychiatrist] in it but every division now has [one]."[17] It was a terrible time for his family; "his letters were filled with a monumental quiet despair. . . . His pain carried over to Ma, who knew him so closely, and it was a very stormy and unhappy time for both of them. Of course, she never lost faith," her morale sustained by advice her mother had once given her: that to bear misfortune nobly was itself good fortune. Beatrice prayed "they won't kick him to death while he is down. . . . The deed is done and the mistake made, and I'm sure Georgie is sorrier and has punished himself more than anyone could possibly realize."[18] To the end of her life, Bea Patton was conspicuously unrepentant over her husband's famous slaps, noting in the margin of one of his letters: "Too bad we have to support cowards," and decrying the fact that "their kind" were even permitted to propagate.

Patton's closest friends counseled patience, but as Everett Hughes wrote to Beatrice some months later: "Imagine George doing that [being patient] willingly. But he must and will. He knows that the Seventh Army did a good job, and he knows that the world knows he commanded it. This game is still young and he won't be on the bench long."[19] To Kenyon Joyce, Patton wrote wistfully: "At the moment I have not the faintest idea when, if ever, I shall be employed, but I hope for the best."[20]

In Italy all was not well. Mark Clark's inexperience and that of his Fifth Army staff had contributed at Salerno to a crisis so grave that the continued existence of the Allied beachhead was threatened. Prior to the invasion Patton had been appointed the reserve commander of the Fifth Army in the event Clark was killed or incapacitated, and when he was briefed on Salerno by the chief of staff, had unerringly pointed to a river marking the boundary (and a ten-mile gap) between the U.S. 36th Division and the British 56th Division as a problem the Germans could be counted on to exploit. "I told him that just as sure as God lives, the Germans will attack down that river," he wrote in his diary the night of September 2. When informed that a plan existed to deal with this contingency, Patton refrained from voicing his conviction that "Of course, plans never work, especially in a landing." Tactfully he suggested that possibility, "but it did not register. I can't see why people are so foolish. I have yet to be questioned by any planner concerning my experiences at 'Torch'" or the invasion of Sicily.[21]

When the situation at Salerno worsened and Patton's prediction came true, with Clark considering withdrawal from the American sector, Patton noted that "the only comfort I get out of it is the fact that my military judgment was proved correct. I hope they can stop them—a withdrawal would hurt our prestige and surely prolong the war. . . . We must attack or it will become a second Gallipoli . . . a Commander, once ashore, must conquer or die."[22]

Later, when Patton learned that Clark had relieved the invasion commander, Maj. Gen. Ernest J. Dawley, he declared that he would accept the command himself. "I would serve under the Devil to get a fight." Eisenhower declined, observing that he and Clark were hardly "soul mates." Eisenhower did hold out a carrot to his disgraced army commander, however: "I am to go to England and get an Army, probably under that victorious soldier, Jake Devers. Destiny had better get busy."[23] Patton was only partially mollified by Eisenhower's promise and continued to criticize him behind his back, scoffing at his friend's contention that he was always acting a part due to an inferiority complex. "The truth is that I have too little of such a complex—in fact I look down my nose at the world and too often let them know it." Nevertheless Eisenhower had hit the nail on the head.

Bradley's appointment seemed to suggest that the Seventh Army might not have a future. Patton summoned the heads of his staff sections and said: "I believe in destiny and that nothing can destroy the future of the Seventh Army. However, some of you may not believe in destiny, so if you can find a better job, get it and I will help you all I can. You may be backing the wrong horse." Patton wrote in his diary: "I feel that none of them will leave me," but also noted that, "apparently people have a very low opinion of my staff. Personally, I would not trade them for any other staff officers in Africa or the U.S." None accepted Patton's offer, and he would later write proudly to Beatrice: "They stuck like limpets."[24]

Patton returned from Algiers in early September 1943 complaining that the "so called Allied Headquarters is a British headquarters commanded by an American," and that there existed a deliberate attempt within AFHQ to denigrate the accomplishments of his army in Sicily. He was particularly irate that his recommendations for decorations were routinely disapproved by AFHQ or downgraded to a lesser award. Why, he wondered, was Omar Bradley good enough to be given command of an army but not good enough to receive the Distinguished Service Medal recommended by Patton? The answer seemed clear enough: He and his army were very much out of favor with AFHQ.[25]

When one assesses Patton's obsession with destiny and success, there is a tendency to overlook just how deeply he cared about his soldiers. In the aftermath of the slapping incidents, Codman's Harvard classmate, author John P. Marquand, visited Palermo. When Patton asked Marquand for his opinion of reaction in the United States to what American troops had accomplished in Sicily, Marquand replied that the newspapers left the impression it had been a cakewalk, and that the British had done the bulk of the fighting. Patton's response was deeply emotional:

> By God, don't they know we took on the Hermann Goering division? Don't they know about Troina? By God, we got moving instead of sitting down, and we had to keep moving every minute to keep them off balance or we'd be fighting yet—and what were they [the British] doing in front of Catania? They don't even know how to run around end. All they can do is make a frontal attack under the same barrage they used at Ypres.

Despite Patton's emotional outburst, Marquand believed Patton was not evincing anti-British sentiment as much as he was defending the U.S. Army. "He was speaking solely for his troops, aroused because their exploits had not been given proper attention." Patton went on to explain the virtues of the American GI, his endurance and his great sacrifices. "These things make me emotional, maybe too emotional sometimes." Patton pointed to a large stack of letters on his desk, noting that they were being sent to convey his personal appreciation to a number of Seventh Army soldiers for outstanding performance of duty: "I am glad to sign them . . . because usually in the Army you can expect loyalty from the bottom up, more than you can from the top down, and I ought to know."[26]

Marquand believed that Patton's appearance lived up to advance billing, but he found him taller and much more imposing than he had anticipated. Marquand was particularly impressed by Patton's hands, which he thought unusually "artistic for those of a combat general," and compared him to Sinclair Lewis.[27]

At times during his Sicilian exile Patton simply deluded himself that others were responsible for his plight. "Sometimes I almost believe that there is a deliberate campaign to hurt me; certainly it is hard to be victimized for winning a campaign. Hap thinks the cousins [the British] are back of it because I made a fool of Monty."[28] His report completed, Patton's days were filled with the inconsequential: tea with some citizens of Palermo ("women very fat"), attending services at the English church (where, he noted, there were more fleas than people); and dinners by the sea with yet other Sicilians, where "we danced to a cracked victrola and ate many horrible things. At least we killed off another day of uncertainty." On another occasion Patton took tea with "a very fat Bourbon princess with a black beard which she shaves" and her girlfriend "who should be a wrestler but is actually a famous pianist." Idleness led to mischievous thoughts and a raft of catty remarks about the virtues (or lack thereof) of others. Patton refused to grasp that the root of his unemployment was exclusively the slappings, and instead continued to believe the nonsense that the British were promoting Monty as a war hero at his expense. "That is why they are not too fond of me. One British general said to [an American general], 'George is such a pushing fellow that if we don't stop him he will have Monty surrounded.' I know I can outfight the little fart any time."[29]

Some of his remarks were both outrageous and racist, including his repetition of a rumor that Clark had been given high command as a concession to American Jews, and some pungent observations about black troops who were tried by court-martial for capital offenses. When three men were tried for rape: "I put two negro officers on the court. Although the men were guilty as hell, the colored officers would not vote death—a useless race."[30]

After visiting the Gela battlefield, he was amazed at the stupidity of the German tank attack without adequate artillery or infantry support, and came away more than ever convinced he had done the right thing by insisting on the 1st Division instead of the 36th. "I feel sure that a green division would have broken when the tanks came, but the veteran 1st Division stood its ground. . . . I had a hard time getting AFHQ to let me substitute the 1st. As usual God was on my side."[31]

The Germans rated Patton the most dangerous general on the Allied side and were keenly interested in his movements. As part of an Allied cover plan, Patton and several members of his staff were sent on a highly visible trip to Corsica to create the impression that he might be making preparations to lead an Allied force behind the German lines in northern Italy. Patton inspected many French and colonial formations and was ostentatiously seen inspecting a harbor and Napoleon's birthplace.[32]

Everett Hughes continued to counsel perseverance and advised him to stop worrying about what might happen and instead "go out and look at another monument, or if you have seen all of them, go build one of your own."[33]

When he toured the Italian front, Patton was dismayed to find troop morale low and the campaign floundering, and decided he wanted no part of the Fifth Army should Clark be sacked or killed.[34] Despite Eisenhower's promise to bring him to England, Patton might well have commanded the Fifth Army had it not been for the arrival of Jake Devers from England in early 1944 as the new deputy supreme commander and commander of all U.S. forces in the Mediterranean. Devers indicated that his choice was for Clark to continue commanding the Fifth Army and for Patton to lead Anvil, an invasion of the French Riviera planned for the summer of 1944.[35]

The Episcopalian Patton began attending Catholic churches, admittedly "largely for political reasons but also as a means of worshipping God, because I think He is quite impartial as to the form in which He is approached."[36] At Patton's invitation Eisenhower's driver, Kay Summersby, and his WAC private secretary, Ruth Briggs, visited him in Sicily. He personally escorted them on a sightseeing trip that ended at a medieval church outside Palermo. After delivering a brief lecture on medieval architecture, Patton "sank to his knees and prayed aloud for the success of his troops, for the health and happiness of his family and for a safe flight back for Ruth and me. He was completely unselfconscious." When Summersby recounted her experience, Eisenhower smiled and shook his head. "Georgie is one of the best generals I have, but he's just like a time bomb. You can never be sure when he's going to go off. All you can be sure of is that it will probably be at the wrong place at the wrong time."[37]

On his fifty-eighth birthday Patton was mired in the depths of despair. "One year ago today we took Casa. Now I command little more than my self-respect. General McNair wrote me, 'You are the Seventh Army.' He was more right than he knew." His mess sergeants presented him with an enormous birthday cake, in recognition of the Seventh Army's accomplishments in Sicily. Patton refused to cut it and instead insisted that Col. Harry A. "Paddy" Flint, another old friend, and a legendary infantryman, have that honor.

At a memorial service for American dead, Patton laid a wreath and said: "I consider it no sacrifice to die for my country. In my mind we came here to thank God that men like these have lived rather than to regret that they have died." At the behest of a chaplain, Patton wrote "A Soldier's Prayer," a paean to chivalry, duty, and honor:

God of our Fathers, who by land and by sea has ever led us on to victory, please continue Your inspiring guidance in this the greatest of our conflicts.
Strengthen my soul so that the weakening instinct of self-preservation, which besets all of us in battle, shall not blind me to my duty, to my own manhood, to the glory of my calling, and to my responsibility to my fellow soldiers.

Grant to our armed forces that disciplined valor and mutual confi-
dence which insures success in war.
Let me not mourn for the men who have died fighting, but rather let
me be glad that such heroes have lived.
If it be my lot to die, let me do so with courage and honor in a man-
ner which will bring the greatest harm to the enemy, and please, oh
Lord, protect and guide those I shall leave behind.
Give us the victory, Lord.[38]

In the end it took Drew Pearson's revelation in November 1943 to awaken Patton at last to the harsh realization that his career was in grave jeopardy. Nevertheless he managed to keep faith with the belief that he was "a passenger floating on a river of destiny." Although he had gone from commanding 240,000 men to less than 5,000, "pretty soon I will hit bottom and then bounce."[39]

By the autumn of 1943 the key Overlord appointments were still incomplete. At the Teheran Conference in November, Stalin put forth two blunt demands: When would the Allies open a Second Front and who would command it? Both Marshall and Brooke had aspirations of becoming the Allied supreme commander, but the betting in the autumn of 1943 was on Marshall, who had the support of both Stimson and Harry Hopkins. Marshall himself believed he would be appointed, but Roosevelt eventually concluded that he could not afford to lose Marshall, saying: "Well, I didn't feel I could sleep at ease if you were out of Washington."[40]

Instead, on December 6, 1943, the appointment went to Dwight Eisenhower, the only other American officer acceptable to Churchill. On Christmas Eve it was announced that Churchill had selected Montgomery to command Allied ground forces for the invasion.[41] The great command shake-up had begun. Conspicuously missing was the name of George S. Patton.

On December 8 Roosevelt stopped in Sicily en route home from Teheran and was greeted by Eisenhower and Patton at the Palermo airfield. Harry Hopkins took Patton aside and said: "Don't let anything that s.o.b. [Drew] Pearson said bother you." Even Eisenhower seemed affable, and Patton's diary records that "he felt sure I would soon get orders to go to the UK and command an Army."[42]

John J. McCloy found Patton "a bit downcast but soldierly." Now more uncertain than ever, Patton asked McCloy what his future held. McCloy replied that "he had never let friendship move him, but that he felt I had in my makeup certain chemicals no other General had; that I was a great fighter and an inspiring leader . . . and must be used." McCloy also told Patton that he looked and acted like a general, and that Marshall had said: "He will have an Army."[43]

McCloy's promise came a week after Eisenhower had reaffirmed his decision to stand behind Patton, in spite of his uncertainty that "we may yet have a lot of grief about it." Nevertheless, wrote Eisenhower: "I want you to know though that I think I took the right decision then and I stand by it. You don't need to be afraid of my weakening on the proposition in spite of the fact that, at the moment, I was more than a little annoyed with you."[44]

Stimson also offered encouragement:

> General Eisenhower's report on the matter, which I sent to the Senate and which was published by them in full, was admirable in form and temper. I believe that, like me, the great bulk of your fellow countrymen, while they deeply regret what you did, have not lost faith in your character and in your courage and skill as a battle leader. . . . Your wife, as usual, behaved with wonderful tact and devotion when she was interrogated by the press and her statement, I think, did much to forward the favorable result. But hereafter watch your step and above all, remember the difference between the battlefield and the hospital.[45]

Stimson was far too modest, for it was in no small part thanks to his active involvement in smoothing troubled waters in Washington that Patton had been saved from the ultimate disgrace of being relieved and sent home.

In December Patton made a well-publicized grand tour of the Mediterranean, designed to convince the Germans that something involving him was pending. He spent time with the British, where he was not disappointed to learn that many British Regular Army officers disliked Montgomery, and inspected the Polish II Corps, which would fight brilliantly in Italy as part of Alexander's army group. "The troops are the finest looking and best disciplined I have ever seen," wrote Patton, a high tribute indeed.[46] In Egypt he found the Pyramids "thrilling," but the Egyptian peasants "definitely lower than the Sicilian WHOM I have thought was the bottom."[47]

In June his son, George, had been "found" deficient in mathematics at the end of his second year at West Point. What would have been routine for any other cadet became the subject of correspondence with the chief of staff because he was the son of George S. Patton. His letters were sparse but welcomed by a son who had seen precious little of his father for so long. Patton advised him to take his setback in stride (as he himself had in 1905) and noted wryly: "As you have seen in the papers I am catching a little hell myself at the moment."[48]

To Beatrice he improvised on *The Mikado:*

To sit in gloomy silence
In a dim dank dock
Of a pestilential prison

With a life long lock
In sad anticipation
Of a short sharp shock
To be given by a chopper
On a big black block.[49]

On Christmas Day he was cheered by a great many presents and letters of support. For once he was ebullient and confident that his ordeal has been a test of his character and his will.

My men are crazy about me, and that is what makes me most angry with Drew Pearson. I will live to see him die. As a matter of fact, the ability to survive this has had a good effect on America, and on me. My destiny is sure and I am a fool and a coward ever to have doubted it. I don't any more. Some people are needed to do things and they have to be tempered by adversity as well as thrilled by success. I have had both. Now for some more success.

Yet his mood swings remained dangerously unpredictable and his anger toward Eisenhower irrational. One day he would write: "I am realy sorry to lose Ike," then, a day or two later, complain of "Divine Destiny's" alleged failure to take his side. "I wish to God Ike would leave and take Smith with him. They cramp my style. Better to rule in Hell than serve in Heaven."[50]

The new year, however, brought even more uncertainty, when Hughes alerted him that although the Seventh Army staff would plan Anvil, he was unlikely to command it and was destined for transfer to England. "I hope not," Patton wrote. "I would hate to play too far down in the team, also if I get to England . . . it means I go alone to pick up a new staff. I prefer my present one to any I have seen. They have stuck by me and I propose to stick by them. . . . There is no use worrying or guessing—soon we will know. Destiny will keep on floating me down the stream of fate."[51]

On New Year's Day he was officially relieved of command of the Seventh Army and ordered to report to Algiers "for further instructions." In addition to commanding Fifth Army, Mark Clark would also now be in caretaker command of Seventh Army. The uncertainty gnawed and was at times almost unbearable:

I feel very badly for myself but particularly for the staff and headquarters soldiers who have stood by me all the time in good weather and bad. I suppose that I am going to England to command another army but if I am sent there to simply train troops which I am not to command I shall resign. There is no appreciation that a staff is a living thing, not simply an animated table of organization. I cannot conceive of anything more stupid than to change staffs on a General, nor can I conceive of anything more inconsiderate than not to notify him where he is going. It is just one more thing to remember when the time comes to pay my debts. A Hell of a "Happy New Year."[52]

When he reported to Algiers on January 3, 1944, Patton was still fuming:

I cannot see how any normally intelligent person could inspire this fool change of staffs. It is unfair and insulting to me, but it is heartbreaking for the staff of Seventh Army who have been utterly loyal. . . . It is damnable. I hope I find out who did it and then get to a position so that I can really hurt him. . . . I have contemplated asking to be relieved but will stick it at least for the present.[53]

The following week he again visited the Italian front and liked nothing he saw, from the dismal stalemate at Cassino to Mark Clark, whom Patton found nervous and worried that his friend Johnny Lucas lacked the drive to carry out successfully the forthcoming Anzio landings. Clark and his chief of staff "were most condescending and treated me like an undertaker treats the family of the deceased. It was rather hard to take . . . but I had to be nice as I want my men promoted and decorated." At II Corps, Geoffrey Keyes took him to the front, where they visited a remote artillery observation post. At one point a salvo of four shells struck only thirty yards from where Patton had moments before stopped to take a photograph. Codman's helmet was peppered with shrapnel, and a piece buried itself in the ground just inches from his foot. "It *was* a fairly close thing" to being "curtains" for everyone, Codman wrote his wife.[54]

Instead of producing his usual stomach-churning knot, this latest near-death experience thrilled Patton. "Patton's luck still holds," he exulted to his son, and he wrote to Beatrice that, "The Lord had a perfect out for me and pulled his punch. You have no idea how much that near miss . . . cheered me up. I know I am needed!"[55]

At VI Corps, Patton told a worried Lucas: "John, there is no one in the Army I hate to see killed as much as you, but you can't get out of this alive," adding, "of course, you might only be wounded. No one ever blames a wounded general for anything." After instructing Lucas to "read the Bible when the going gets tough," Patton took one of his aides aside and said in all seriousness: "Look here, if things get too bad, shoot the old man in the back end, but don't you dare kill the old b—d." After his aide reported Patton's remarks, Lucas admitted he found himself "afraid to turn my back on him from D-day on."[56]

In Naples, Patton dined with some senior British officers, one of whom was the 1st Division commander, Maj. Gen. W. R. C. Penney. "Before lunch," wrote Penney, "we were all given a large tot of Scotch and George Patton gave two toasts. Looking at me he first said, 'Here's to the British,' and then he added a second toast, 'And here's to hell and damnation to any God damn commanding General who is ever found at his own C[ommand] P[ost].' "[57]

Although he no longer commanded even the Seventh Army, Patton returned to exile in Sicily. There were only so many Greek, Norman, Roman, and Carthaginian ruins and temples to visit, even in historic Sicily. There was ample time for reflection as he weighed his future. "The Catholic Church has been very much on my side, as have the veterans and the war mothers. I think I could run for office on the strength of my misdeeds. I am not the first General to catch hell; Wellington had plenty of it as did Grant, Sherman and countless others."

What especially concerned Patton was the attitude of some quite intelligent men who had told him that once the war ended: "We will have no more wars. When you ask them on what they base that forlorn hope they smile and reply, 'Things are different now.' Yet the avowed purpose of the Treaty of Vienna in 1814 was to see that this was the last war. . . . If we again think that wars are over, we will surely have another one and damned quick. 'Man is WAR' and we had better remember that." Later in 1944 Patton reacted with similar scorn when informed by a "visiting fireman of great eminence" that this would be the last war ever fought. "I told him that such statements since 2600 B.C. had signed the death warrant of millions of young men. . . . My God! Will they never learn??"[58]

He awoke the morning of January 18 in good spirits, which were soon dashed when Sergeant Meeks reported that he had heard on the radio the previous evening that Eisenhower had announced at his first press conference in London that Omar Bradley would command all American ground troops in the forthcoming cross-Channel invasion. Patton correctly noted that it meant Bradley would certainly command the U.S. army group to be established in France after the invasion. "I had thought that possibly I might get this command," Patton wrote with a combination of acute disillusionment and naïveté. "It is another disappointment, but so far in my life all the disappointments I have had finally worked out to my advantage. . . . If I am predestined, as I feel that I am, this too will eventually be to my advantage."

He mocked Bradley as "a man of great mediocrity," and documented his former subordinate's real and alleged sins as a commander, even blaming him for the airborne fiasco in Sicily:

On the other hand Bradley has many of the attributes which are considered desirable in a general. He wears glasses, has a strong jaw, talks profoundly and says little, and is a shooting companion of the Chief of Staff. Also a loyal man. I consider him among our better generals. I suppose that all that has happened is calculated to get my morale so that I will say "What the Hell! Stick it up your ass and I will go home," but I won't. I still believe.

This latest blow to his pride left Patton acutely aware that if he *was* eventually given command of an army in England, he would be subordinate to Bradley. "Well I have been under worse people and I will surely win."[59]

Patton's exile mercifully ended on January 22, 1944, when a cable arrived from Algiers ordering him to the United Kingdom. He cleaned out his desk and wrote to Beatrice: "This is the fourth time since I left home. I now feel much like we did when we quit [Fort] Sheridan for Myer in 1911—'Oh God how young we were,' and started into the unknown for the first time. Well, we have done it a lot of times since and have always ended up well known. I will do it again."[60]

He left Sicily for the last time, uncertain of what his future held. Nor was the usually well informed Everett Hughes able to enlighten him when Patton reported to Algiers. For the moment at least, all that seemed to matter was that he could take cold comfort in the fact that at least he had not been forgotten. He remained in Algiers for two days, saying good-bye to his Seventh Army staff "with sincere regret on both sides." On January 25, accompanied by Codman, Patton left Algiers in a C-54 cargo aircraft (carrying 250 extra pounds—his personal papers), bound for Prestwick, Scotland, and his last chance at reviving his shattered career.

His permanent legacy in Sicily was a bronze plaque that Patton had placed at his own expense in the British church in Palermo. It was made by several sailors from a broken propeller, for which he paid them twenty-five dollars. The engraving read:

TO THE GLORY OF GOD
In Memory of the heroic Americans of the Seventh Army and of the supporting units of the Navy and Air Force
who gave their lives for victory in the Sicilian campaign,
July 10–August 17, 1943.
—From their General [61]

PART IX

England

(1944)

We are fighting to defeat and wipe out the Nazis who
started all this goddamned son-of-a-bitchery.

—PATTON (ADDRESSING THIRD ARMY STAFF)

CHAPTER 36

Third Army Commander

Remember, you have not seen me. I am not here.

—PATTON

When Eisenhower secretly returned to the United States in early January 1944 before assuming command of the Overlord forces in England, the matter of Patton's future was still unresolved. Although Eisenhower seemed already to have settled on him to command an army, he and Marshall nevertheless debated Patton's fate. Marshall raised the question of perhaps letting Patton command Anvil, and although Eisenhower clearly favored him as important for Overlord, he did not object. However, as Marshall's official biographer writes, Eisenhower "noted that Devers and Patton were not congenial but on reflection decided that they were good enough soldiers not to let this factor interfere."[1] At one point Marshall had toyed with the idea of bringing Patton home for a short rest but decided against it in the belief that it would only refuel the unhealthy publicity over the slappings. When Marshall also questioned Patton's suitability for command, Eisenhower "assured the chief of staff that the volatile, offensive-minded Patton would always serve under the more even-handed Bradley."[2]

Ultimately it was Eisenhower's prerogative as the responsible commander to choose his subordinate commanders, and whatever his personal misgivings, Marshall fully supported Eisenhower's wishes, particularly since it was now clear that Patton was not to command the Anvil force. It is somewhat ironic that, as Forrest Pogue writes, while Patton continued to assume "that he was wandering in a wilderness, the officers able to decide his future were debating only which of two top assignments available he would handle best."[3] Thus, by the time Eisenhower returned to assume command of

Supreme Headquarters Allied Expeditionary Force (SHAEF), based in St. James's Square, London, Patton's fate had been decided.

Patton's aircraft landed at Prestwick, Scotland, the morning of January 26, 1944, in a proverbially British pea-soup fog. When Patton arrived at a military airfield outside London that afternoon, he was met by his West Point classmate Lt. Gen. J. C. H. Lee (Overlord logistical commander, deputy commander of U.S. forces, and an officer whom Patton detested), and the affable Commander Butcher, who was only there by coincidence to collect Ike's pet Scottie, Telek. There were no bands or honor guards to greet him this cold, dreary day. It was not lost on Patton that he was still something of an outcast in his own army.

At SHAEF Eisenhower greeted him with the news that he was to command the Third Army, which was still in the United States and would shortly begin arriving in the United Kingdom. Eisenhower also left Patton in no doubt where he stood with his boss by following his own advice and giving him "a severe bawling out" for his impulsive behavior. Butcher would later observe that Patton was "a master of flattery," who was able to turn any difference of opinion with Eisenhower "into a deferential acquiescence to the view of the Supreme Commander. . . . 'Ike, as you are now the most powerful man in the world, it is foolish to contest your views.'. . . Ike glumly and noncommittally passes off such flattery." Knowing that his comment would get back to Eisenhower, Patton told Butcher a few days later that Ike was "on the threshold of the becoming 'the greatest general of all time—including Napoleon.'"[4] Some months later Patton was delighted when someone related that General Wedemeyer had been overheard stoutly defending him in a heated conversation with Eisenhower: "Hell, get on to yourself, Ike; you didn't make him, he made you."[5]

Kay Summersby, the Irishwoman who had first been assigned to chauffeur Eisenhower when he arrived in England in 1942, was now a permanent member of the SHAEF staff and part of the supreme commander's inner circle. Detailed by Eisenhower to give Patton the "sixty-four-dollar tour" of the city, she thought he was the "most glamorous, dramatic general I'd ever met." Unlike other generals, Patton rode with Kay, "and his ramrod back never once unbent, never touched the seat." He was appalled by the magnitude of the bombing damage to London. "'Those sonsabitches,' he would mutter, 'those sonsabitches.' Then, he'd turn to me. 'I'm sorry, Miss Summersby. Excuse me, please.'" Summersby admitted that she had been around Americans long enough to learn how to curse, and had once yelled "Goddamn!" in the office. When Patton inquired from whom she had learned such language, Kay replied: "Dwight D. Eisenhower."

"I've heard Patton swear like a docker many times," she later wrote, "but I never felt actually embarrassed; he was a man's man, a real soldier, and yet he unfailingly treated women with an eighteenth century flourish." Next to Churchill and Eisenhower, no one she met impressed Kay Summersby more than this fascinating man. "Patton at once displayed that Old World gallantry which all his biographers seem to have missed. When he shook hands and bowed, everything was there but a Continental kiss of the hand. There was no hint of the expected American backslap or the wolfish eye. All he needed was a cavalier's cape and a sword."[6] However, as she noted, Patton's tough-guy image was marred by "the world's most unfortunate voice, a high-pitched womanish squeak."[7]

Although Patton sincerely liked Summersby, he was uncomfortable whenever she was present with Eisenhower and would instantly clam up, once saying: "I *do* have secrets from her." Patton usually laid out his best Myopia Hunt picnic spread for visiting women dignitaries, such as Clare Booth Luce or Margaret Bourke-White, but whenever Eisenhower arrived with Kay, Patton served coffee and sandwiches. [8]

Increasingly Eisenhower found that the informal evenings of conversation with Patton meant another sleepless night, for their conversations rarely ended before 3:00 A.M. Such evenings often began after dinner with off-color stories, and Patton's prudish side would lead him politely to remove Summersby from the room as if she were his daughter and should not hear such things. While Eisenhower winked at her, Patton would say: "You tell one of those handsome aides of mine to play Ping-Pong with you."

Patton emerged with mixed emotions from his first meeting with Eisenhower—obviously pleased to be getting an army command, but disappointed that the Third Army would not play a role until after the invasion. "All novices and in support of Bradley's First Army—not such a good job but better than nothing. . . . Well, I have an Army and it is up to me. 'God show the right.' As far as I can remember this is my twenty-seventh start from zero since entering the U.S. Army. Each time I have made a success of it, and this must be the biggest."[9] What Patton had heard from Eisenhower was nothing less than the fact that he was no longer the top American banana but henceforth merely a team player in the great Anglo-American endeavor called Overlord.

His pique was evident in his dismissal of the SHAEF chief of staff, Bedell Smith, recently promoted to lieutenant general, as an "s.o.b." Even so, in the weeks to come, Patton's evident distrust and wariness of Smith and the SHAEF staff led him at least once deliberately to "bootlick" Smith. After several such occasions a disgusted Patton wrote: "washed mouth out later.". . . After all the ass kissing I have to do, no wonder I have a sore lip," a reference to his badly infected lip.[10] Patton never seemed able to make up his mind whether he liked or despised the hard-bitten Smith, who jealously guarded his commander in chief's time and kept a tight rein on access to Eisenhower's office. Irascible and with a hair-trigger temper (exacerbated by a painful ulcer), Bedell Smith was the H. R. Haldeman of his time: He ruled by fear and was considered by most a terrifying martinet. "The public Smith was a truly formidable personality. Blunt, tense, and aggressive, Smith had no time for pleasantries or civilities. He assessed the quality of a subordinate's work on the spot, in exceedingly plain terms. One mistake could be condoned . . . but a second was not allowed. Smith frequently invoked the threat of relief, and seemingly took satisfaction in terrifying members of his staff."[11]

His unpleasant personality notwithstanding, Smith was an effective chief of staff, enjoyed Eisenhower's absolute confidence, and was a full-fledged member of his boss's brain trust. Known far and wide for his "Prussian" personality, Smith simply ran roughshod over anyone who gave or even hinted at the appearance of disturbing the harmony of SHAEF. Eisenhower said of him that "Smith was like a crutch to a one-legged man."[12] Patton was well aware of Smith's power and consequently understood that he must handle Smith with kid gloves if he was to avoid the irreparable harm that befell those who crossed him. When it came to a comparison between the two, however, the autocratic Bedell Smith made Patton resemble Little Red Riding Hood.

All the senior American generals were provided with flats for their use whenever they were in London. On the afternoon of February 3, when he arrived at his new quarters on Mount Street, in London's exclusive Mayfair district, Patton exploded with anger because it resembled a brothel and had been personally picked for him by his nemesis, J. C. H. Lee. Patton could hardly believe his eyes when he entered his bedroom. The floor was covered by a "white bear rug, the walls and curtains [were] of pink brocade," and there was an ornate dressing table and an enormous bed "lying low and lascivious under its embroidered silk coverlet. The silence was broken by a single highly charged exclamation: 'JESUS!'"

Joined for drinks by the new Seventh Army commander, Lt. Gen. Sandy Patch, Patton soon said: "Look, Sandy, I'd rather be shot than spend the evening sitting around this Anglican bordello." Codman procured tickets to a sold-out performance of *There Shall Be No Night* starring the American actors Alfred Lunt and his wife, Lynn Fontanne, and the two generals were smuggled into the Haymarket Theatre through a fire exit. During the performance a German buzz bomb crashed through the roof of a nearby building, but as bits of plaster fell down on the stage, the great Alfred Lunt did not so much as bat an eye. Patton's presence was soon noted by one and all, and it was questionable whether or not this was in compliance with Eisenhower's orders to keep a low profile. However, for what Eisenhower and the Allied planners had in mind, it was just as well that Patton was known to have been in England.

Patton and Patch dined with the Lunts at their Savoy Hotel flat after the performance, and when Patch was asked by Lynn Fontanne what he was fighting for, he replied: "To resist aggression, to defend our country, and to preserve our way of life." When Patton was asked the same question, he scoffed: "Sandy is talking through his hat." And, with a gleam in his eye, Patton said: "I, dear lady, have been fighting all my life and hope to continue indefinitely to do so for the simple reason that I love fighting."[13]

On at least one occasion buzz bombs landed near Patton's London flat: "I only woke up after fifteen minutes nothing big hit near us but as usual fragments fell on the roof." He could still be appalled by the carnage of war but thought the Germans were now getting the worst of it. Patton also took advantage of his time in London to visit a Savile Row tailor, where he said: "Damn the expense" as he was fitted for a new overcoat.[14]

Patton managed to persuade Smith and Eisenhower to approve the transfer to England of fifteen of his key Seventh Army staff officers, including Hap Gay, whom they urged him to get rid of. "But I'm not doing it."[15] Eventually, however, Patton succumbed to the pressure to replace Gay, and brought in Maj. Gen. Hugh J. Gaffey, who had commanded the 2d Armored Division in Sicily and was not at all thrilled about trading a coveted division command for a thankless staff job. Nevertheless Gaffey felt he owed a debt to Patton and became the newest member of his inner circle—the deputy chief of staff. Patton felt that the matter "was most distasteful. I really believe I would retain more self-respect if I resigned, but I am not quite that big-hearted. Gay was fine. . . . I told him the exact truth, that Ike ordered me to do it." The decision, he explained to Beatrice, was a matter of "self-preservation. . . . Not too noble I fear."[16]

Prior to Eisenhower's arrival in England, Omar Bradley had had no inkling that Patton was to be part of Overlord and was distressed to learn that he was to command the Third Army. Bradley candidly admitted that if Eisenhower had consulted him in advance, he would have counseled against Patton's appointment and argued that placing Patton under his former subordinate was too much to ask of anyone. However, Eisenhower thought differently and told Bradley: "All he wants is a chance to get back into the war." Bradley understood Patton's brilliant ability to gain ground, but it was not enough to offset his misgivings about his former commander, none of which he ever conveyed to Eisenhower. If the supreme commander wanted Patton, so be it. Nevertheless Bradley was relieved to know that until such time as the Third Army was committed to France the following summer, it was assigned directly under Eisenhower. Until then Patton would be Ike's problem, not his. Months later, when Patton did come under his command, Bradley deliberately tightened press censorship in an effort to keep Patton from creating yet another crisis as a result of intemperate speech. None of his subordinate commanders would be permitted to be quoted without Bradley's personal approval.[17]

It was no secret that the Allies intended to invade the continent of Europe at the first favorable opportunity in the spring or early summer of 1944. The most important secret of the war was *when* and *where* the cross-Channel invasion would take place. For the time being Patton's presence in England was kept secret, although Eisenhower made it clear that he had no intention of keeping Patton "under a blanket" for very long.

Lt. Gen. J. C. H. Lee, variously referred to as "Court House," "Jesus Christ" or "Jesus Christ Himself," was an officer who valued his creature comforts. Eisenhower inherited Lee and, despite the frequent urging of Bedell Smith, declined to get rid of him—advice he was later to rue when Lee and his entire organization moved into Paris after its liberation in August 1944, and took over the city's finest hotels for their personal use. As Smith bluntly said of him: "Lee was a stuffed shirt. He didn't know much about supply organization . . . [and] was one of the crosses we had to bear."[18] Although Eisenhower acted tough with Patton, he was entirely too soft on those, like Lee, who flaunted the system for their own benefit. Lee, a born-again Christian who saw no conflict between his piousness and a lavish lifestyle, was perhaps the most blatant example of one who violated the unwritten principle that an officer should endure the same hardships as his men. Patton would later wholeheartedly endorse a description of Lee as "a pompous little son-of-a-bitch only interested in self-advertisement."[19] Lee was the only general known to have worn his stars on both the front and the rear of his helmet.[20]

On January 27 Patton traveled aboard Lee's spacious personal train to Greenock, Scotland, where he welcomed the advance element of the Third Army, which arrived aboard the liner *Ile de France*. They were assembled in a nearby mess hall, assuming they were to be given a welcome address: "Even when Patton entered aglitter in gleaming boots and brass, they still suspected nothing. They thought he would do the spieling." Instead Patton merely said: "I am your new commander. I'm glad to see you. I hope it's mutual. There's a lot of work to be done and there's little time to do it. There's a special train waiting on the dock to take you to our CP. We will leave in an hour."[21] His most important point was that "I was still a secret and not to be mentioned."

In 1941 the Third Army was commanded by Lt. Gen. Walter Krueger, under whose tutelage it had become the preeminent training army in the United States during and after the Louisiana maneuvers. After Krueger left for the Pacific, the new commander was Courtney H. Hodges, who believed he was taking the Third to England in early 1944, but was rudely jolted when a letter addressed to "Lt. Gen. George S. Patton, CG, Third U.S. Army," was found in a bundle of mail with Beatrice's return address shortly before the headquarters embarked for England.

When the main body of Third Army headquarters arrived in England, the chief of staff and most of the general and special staff officers were immediately replaced with officers Patton knew and trusted, among them Harkins, Koch, Maddox, and Muller from his Seventh Army staff. Patton was still smarting over the criticism of his staff in Sicily, but his reply to his critics was simply to say: "See what they did." He thought part of the problem was that none had prewar reputations as staff officers. Nevertheless Koch was regarded by many as having the "the most penetrating mind in the U.S. Army's intelligence community." Muller has been praised as "the ablest quartermaster since Moses," and his signal officer, Elton Hammond, was extremely effective, as were many others. When the true mettle of the Third Army staff was proved by their masterful performance during the Battle of the Bulge, Omar Bradley would only later grudgingly concede: "Patton can get more good work out of a mediocre bunch of staff officers than anyone I ever saw."[22]

Third Army headquarters was established in Knutsford and in the nearby picturesque village of Peover (pronounced *pea*-ver), in the English Midlands, south of Manchester. Patton's quarters were in an eleventh-century manor house which, he wrote Beatrice, seemed to have been "last repaired in 1627 or thereabouts." His bedroom overlooked a small, ancient stone church (in which resided the sarcophagi of the lords of Peover), where he frequently meditated. Across the hall was the Third Army war room in which the war's closest-held secrets were discussed and plans laid. Since most GIs had a hard time pronouncing Peover, what Patton termed "the natural evolution of soldiers" resulted in the colloquial reference to the place as "Piss Over."

The day after the full Third Army staff arrived was cold and raw. Some one thousand officers and enlisted men assembled on the terrace in front of his headquarters to hear Patton, who was resplendent in a beautifully tailored uniform (on whose cap, shirt collar, and shoulders reposed a total of fifteen stars), gleaming cavalry boots with spurs, and his hand-tooled leather belt with its shiny brass buckle. In his hand Patton carried a riding crop, and by his side was his newly acquired English bullterrier, Willie.

One of his new staff officers wrote his family that most had never seen Patton in the flesh:

> When the drum ruffles and bugles sounded the General's march . . . we stood transfixed—there wasn't one square inch of flesh on 250 off
> and 750 enlisted men which was not covered in goose pimples

of the greatest thrills I shall ever know . . . that towering figure, impeccably attired froze you in place and electrified the very air. . . . In a somewhat boyish, shrill yet quiet voice he said, "At ease, gentlemen, I suppose you are all surprised to see me standing here in place of General Hodges—such are the fortunes of war. But I can assure you that the Third United States Army will be the greatest army in American history. We shall be in Berlin ahead of every one. To gain that end we must have perfect discipline. I shall drive you until hell won't have it. . . ." When he had finished, you felt as if you had been given a supercharge from some divine source. Here was a man for whom you *would* go to hell and back.[23]

Patton's speech was short, blunt and to the point:

I have been given command of Third Army. . . . I am here because of the confidence of two men: the President of the United States and the Theater Commander. They have confidence in me because they don't believe a lot of goddamned lies that have been printed about me and also because they know I mean business when I fight. I don't fight for fun and I won't tolerate anyone on my Staff who does. You are here to fight. . . . Ahead of you lies battle. That means just one thing. You can't afford to be a goddamned fool, because in battle fools mean dead men.

It is inevitable for men to be killed and wounded in battle. But there is no reason why such losses should be increased because of the incompetence and carelessness of some stupid son-of-a-bitch. I don't tolerate such men on my Staff.

Patton went on to describe how "some crazy German bastards decided they were supermen and decided it was their mission to rule the world. They've been pushing people around all over the world, looting, killing, and abusing millions of innocent men, women and children. They were getting set to do the same to us . . . we are fighting to defeat and wipe out the Nazis who started all this goddamned son-of-a-bitchery."

Men love to fight, he insisted, "always have and they always will. Some sophists and other crackpots deny that. They don't know what they're talking about. They are either goddammned fools or cowards, or both. Men like to fight, and if they don't, they're not real men. If you don't like to fight I don't want you around. You'd better get out before I kick you out." Patton reminded them that it would take more than guts to win a war, it also required men with brains. "A man with guts but no brains is only half a soldier. We licked the Germans in Africa and Sicily because we had brains as well as guts. We're going to lick them in Europe for the same reason. That's all, and good luck."[24]

Patton's speech was also laced with profane references to hitting "the goddamned Germans with a sock full of shit and when we wipe them out we will go over and get the purple-pissing Japs." One officer who knew little of Patton thought the speech "made an extremely bad impression on me, and, as we walked away, the officers I talked with all thought we had gotten the bottom of the barrel with this general." However, the officer was quick to note that "this first adverse impression did not last long, and I ended up having great admiration for him. By the time we were in combat, I would rather have served under him than any other general."[25] Pfc. David S. Terry, the bass drummer in the military band that had been brought in that day from a nearby replacement depot, remembers thinking: "I was eighteen years old and this was the first time I had ever heard anyone talk as I thought a warrior would. I thought, this man Patton is a warrior, and I'm glad we're on the same side."[26]

Patton left no one in any doubt as to who was in command of the Third Army. As units arrived and were indoctrinated into the Third Army way of doing business, everyone knew who their new army commander was. Formally, he was called to his face "the General" and privately, "Georgie." He was rarely referred to as "Old Blood and Guts." He was never idolized by his troops—his manner was, after all, too formal for that. He had few intimate friends and he scared hell out of most people. By design he was aloof from them. Nevertheless Patton inspired great loyalty in the troops of the Third Army. They came to identify with and to have confidence in him; they understood that a professional soldier and genuine fighting man was commanding them. And while most abhorred the war they were obliged to fight, they understood what it meant to have a commanding general like Patton. Gay's aide-de-camp, Maj. George Murnane, was around Patton from Indio in 1942 until September 1945, and was often entrusted with tasks that would cause Patton to say: "I ordered you to do it, Murnane. So if anything goes wrong, I'll take the blame." Patton's staff officers were often resentful of what they considered the unfair treatment accorded their boss by Eisenhower, but whenever Patton heard such criticism, "he tore the Hell out of us. He really simply would not tolerate such talk on our part. Of course, he'd argue hammer and tongs with Ike in person, but that was a different matter."[27]

Patton also stoutly defended his generals. In Normandy, Patton was informed that Eisenhower had assigned a certain general to command a Third Army division, and at once protested (in vain) that he did not want this incompetent so-and-so serving under him. Shortly thereafter Patton's worst fears were realized, when the officer made such a hash of things that Eisenhower directed his relief. "No way," countered Patton to his perplexed friend, who reminded him that he hadn't wanted the general in the first place. "True, but he was one of your spare generals then. Now he's one of my generals. I'll straighten him out myself;" and he did.

For his staff, working for Patton meant the freedom to express an opinion or make a recommendation and have it thoughtfully heard and often implemented. His war room contained not only maps of Europe but also updated maps of the other fronts and the Pacific war, about which he was briefed frequently. Some of his briefers were enlisted men, making Patton the only commanding general in the European Theater of Operations (ETO) to be so briefed. As Robert S. Allen notes: "Patton never launched a campaign without first thoroughly exploring it with his senior commanders. He never jammed an operation down their throats. It was his practice to assemble the corps commanders in the War Room, have the planning group outline a proposed operation, and then invite the former to 'work it over.' He encouraged free and frank discussion." However, once decided, the plan was to be carried out without further argument or discussion.[28]

During the campaign in France, the Third Army was frequently visited by civilian and military VIPs. Invariably Patton would call in members of his staff to meet them, and present their ideas and opinions. "He was proud of his official family and made no attempt to hide it," said Oscar Koch.[29] Patton fought hard for his staff and deeply resented the fact that his recommendations for the promotion of the senior section heads to brigadier general were routinely turned down, but just as routinely approved in other armies. Toward the end of the war, at a time when both Bradley and Eisenhower were absent, Patton initiated a letter to the War Department recommending the promotions of several section chiefs, favorably endorsed it as acting army group commander, and, in Eisenhower's name, sent it with a personal note to Stimson. To Patton's intense dismay, his recommendations were disapproved, but his staff knew he had tried his best.[30]

Col. Brenton G. Wallace, an assistant G-3, had extensive staff experience at all levels of command, including a British army, but he never worked on a staff that compared to Patton's Third Army. "Some headquarters are like merry-go-rounds. You feel as though you are going in circles, so many motions are superfluous." In the Third Army nothing was done to "make work." "Nothing was ever done there for show or appearance. Everything was practical and for a purpose. The 'Old Man' hated show and sham. He was interested in one thing only—efficiency; and his spirit permeated the whole organization. You had a feeling that Third Army was going in only one direction—forward."[31]

What counted in the Third Army was results, and "if you knew your job you were allowed to perform it in your own way and were never told how to do a thing, only requested in a quiet gentlemanly way to do it. The rest was up to you. . . . If you didn't, or if you were the cause of any friction . . . you were quickly and quietly gotten rid of, 'rolled' as we called it—sent to some other organization." From time to time, as a means of keeping morale high,

Patton would deliberately compliment them at his staff meetings. There were no prima donnas on the Third Army staff; they were all doers who understood what their commander wanted, and consequently got on with the job without the back-stabbing common in some headquarters.

The staff worked seven days a week, and outdoors steel helmets, side arms and neckties were *de rigueur*. Indeed, Third Army was the only organization where neckties were the rule rather than the exception. Patton was briefed twice daily, at 9:00 A.M. and 5:00 P.M., and demanded that current situation reports reach his headquarters an hour before each briefing. Later, in combat, the morning briefing would take place at 7:00, usually followed by a more formal one an hour later. However, when their work was done, there was no need to hang around merely as window dressing for the boss, and to be seen being "busy."

At the morning briefing the section chiefs were always present, and the current tactical situation was briefed by the G-3, the air situation by the Air Officer, and intelligence by the G-2, including a brief summary of the world situation and other fronts, culled from the BBC and U.S. radio newscasts. Briefings rarely lasted more than twenty minutes.

A Third Army historian has described Patton somewhat differently than he has been portrayed:

> He stood tall, as the saying goes, although he was never ramrod erect and actually walked with shoulders somewhat hunched forward. Nonetheless, he gave the impression of [greater] height . . . the only signs of age were a kind of greyness in the face and a crépy throat. In briefings he would sit and caress his throat with flashy rings on his fingers (one I remember was green), and if you concentrated on this, he was for all the world like a sultan viewing the dancing girls. The only GI bit of uniform I ever saw him wear was the service coat. His emphasis on a flashy uniform was very marked when you saw him with Bradley or Hodges. His riding britches were marvelous to behold. In repose he generally stood with his hands in the two front pouch pockets of the britches. The pearl-handled [*sic*] revolver or revolvers were used for trips to the front; normally he carried a small automatic. Although Hap Gay and others carried riding crops, Patton only carried his on ceremonial occasions. . . . With his own staff he never used obscenity but had a few blasphemous phrases. The obscenity was pretty largely for the GIs' edification.[32]

There was a separate room or tent complete with situation maps for the twenty-five or so liaison officers attached to the Third Army. Immediately after the morning briefing for Patton, a member of the staff would brief the liaison officers who, in turn, disseminated information to their respective headquarters, Thus, by noon each day, "every unit of Third Army had a summary of the complete situation and news from every other front."[33]

Patton's staff prepared a special operational map he used each day. Approximately ten by twenty inches and covered in acetate, it identified key towns and crossroads with secret code numbers so that Patton could refer to them over the radio when communicating with his headquarters. When the map was first presented to Patton during the Overlord planning, he looked it over and pronounced it fine. Then, with a twinkle in his eye, Patton said: "But it only goes as far east as Paris. I'm going to Berlin."[34]

In the Third Army, planning was three-dimensional. When Patton went to war there were always three campaigns in his head: the one he was fighting, the one he believed would follow—and the one beyond that.[35] He trained his staff to think ahead, and to plan accordingly to meet his needs.

There were other things that distinguished Patton's Third Army from the rest. He believed in recognition and insisted on prompt action to acknowledge bravery and success by means of awards and decorations, including the dreaded visits to hospitals to pin Purple Hearts on the wounded. Whenever Patton himself was decorated, he would tell his staff that he was merely being recognized for their accomplishments.[36]

All military units are identified by a code name and number. The number 6 is always reserved for the commander. Patton decided that the code name for Third Army would be "Lucky." Thus he became "Lucky 6" and in the field, his headquarters was called "Lucky Forward." He thought it an apt code name for what lay ahead.

His aide, Charles Codman, wrote to his wife that he thought she would like Patton, but warned that most of the time she would have to close her ears and consequently would miss his "amazing" language, which contained numerous and frequent "gems." Patton was, in the most literal sense, an *"enfant terrible—enfant* in his candor, intuitiveness, shrewdness, and unawareness; *terrible* in the intensity of his convictions, his self-discipline, and all the Spartan virtues."[37]

Patton encouraged frank and open discussion before he made a decision. However, once he made a decision, that was that. Orders were never complicated, normally never exceeded a single page, and often had a map sketched on the back. Patton's philosophy of command was: "Never tell people how to do things. Tell them what to do and they will surprise you with their ingenuity."[38]

As the units of Third Army began arriving in the United Kingdom over a period of weeks, Patton began visiting installations across England. However, to maintain the pretense that he was *not* commanding the Third Army, his message was always endlessly reiterated: "You have not seen me. Remember, I am not here!" Patton found it amusing but with only a vaguely defined mission for the Third Army, he fretted that destiny might yet elude him. He issued instructions to his subordinate commanders "on the art of slaughter" about which he had written so often and at such length that it had become almost routine to prepare a new one. Filled with Patton's principles of war, his instructions ended with his own personal reminder to them and to himself: "DO NOT TAKE COUNSEL OF YOUR FEARS."

As his way of doing business became known and, more important, understood, the Third Army began to take on a personality of its own. Patton was more than the usual remote, titular head that most senior commanders were: He was a commander who spoke to them in language they understood. Even those who disliked the profanity and posturing began to understand that what he had to say was comon sense and cut straight to the heart of why they were there and what was expected of them. For all Patton's bombast, his rules for success on the battlefield were wonderfully easy to grasp.[39]

Patton's junior aide, Al Stiller, was not particularly articulate, but a letter he sent to Beatrice in March 1944 became one of her most treasured possessions. Stiller wrote that she had "more reason to be proud of 'your man' than even you know. He is the real 'Leader' of this war. And before it is over he will again demonstrate it in no uncertain terms. I am very proud of the privilege to serve with him. I will never let him down. And will do my very all to return him to you. When the job is Finished. You can depend on it."[40]

Although he had been training units virtually his whole life, even Patton felt daunted by the task confronting him as the crucial day of the invasion of Normandy grew near. "This thing of imitating God and creating new worlds out of thin air is wearing but with the help of my luck and the Lord and the Staff I will do it." The state of preparedness of his new army varied widely, and he visited each division, inspecting, criticizing, praising, and ordering changes and improvements. He worried that too many of his troops lacked the killer instinct and were too willing to die for their country. Privately he frequently drew comparisons with the Fifth Army and its commander, Mark Clark, who "seems to take great pride in the number of men he gets killed . . . he is always boasting to me and making comparisons with the number I lost. I see no use in paying unnecessary casualties for victory."[41]

His troops were fresh from the United States and unfamiliar with him except by reputation. All were eventually the recipients of the "no bastard ever won a war by dying for his country, they won it by making the other poor dumb bastard die for his" speech that had become part and parcel of the Patton method. "All my life I have abhored speach makers and now I seem fated to make them all the time." To that end he acquired a truck with a loudspeaker mounted on it, that made him feel much like a politician but permitted him to "make inspirational talks at will." Eisenhower's admonishment about keeping still continued to rankle Patton, particularly if it were to suppress his speeches to his troops. "My talks make men fight," he defiantly retorted one evening in early April, pledging not to stop unless ordered to do so. "At the moment all the stress is on unostentatious men who are not criticized because they are colorless."[42]

He seemed pleased with his coterie of new corps and division commanders. His XX Corps commander was Maj. Gen. Walton H. Walker, whose nickname was "Bulldog" both for his build and for his fighting disposition. Patton sometimes referred to him as "fat Walker" and confessed a distrust for short, fat men, even though Walker was to prove an eminently effective corps commander.[43]

Among the skills he insisted that his troops practice was airplane recognition. In turn, this became an excuse to fly, and he exulted in May that he had flown 320 miles per hour in a two-seater. Patton's days were filled with the usual frenzy of activity, but in some of his rare leisure time he found himself "bitten by the golf bug," a sport at which he confessed to Beatrice that he did not exactly excel. To the contrary, he imparted a secret "you must not divulge"—namely that he had "harassed cows all over the pasture" practicing for a match in which a local Englishwoman "beat the hell out of me . . . what a comedown from polo! ! !" He also noted that "I could hunt a little but hate to take the chance of a bad fall with a fine fight in prospect—I fear I am loosing my nerve??"[44] Although unable to tell Beatrice precisely what lay ahead, he knew she would understand his oblique remark that: "The next party is realy something and I have a chance to play the same position I used to in football. If I do I may visit some of the places we saw when we lived with the ducks," a reference to an end-run and their sojourn at Saumur in the Loire Valley in 1912 and 1913.

In mid-February the Germans launched a ferocious counterattack against the Anzio beachhead, and the ensuing five-day battle became a fight to the death whose outcome was not decided until the last day, when the hard-pressed Anglo-American corps held by a thread. When it was over it was clear to both Alexander and Clark that John Lucas would have to be replaced. Alexander wrote to Brooke that, "What we require is a thruster like George Patton with a capable staff behind him. . . . Perhaps [Eisenhower] could advise on the best available American Commander."[45]

Eisenhower had been monitoring the situation in Italy with considerable interest. Late in the night of February 15, Butcher was awakened and informed there was an urgent message that Eisenhower must see at once. Although most of the blame for the desperate situation at Anzio was the fault of both Alexander and Clark, Eisenhower was startled to learn that Alexander was placing the blame for Anzio squarely on Lucas, and that he had not first consulted with either Clark or Devers, thus leaving the impression he was seeking a scapegoat. He telephoned Brooke and made it clear that "for Alexander to change from an American to a British Corps headquarters, or to change from one nationality to another during a crisis would be unwise, to say the least." He offered Patton for one month but only if Devers would personally request him.[46]

Several hours later Patton was awakened and ordered to London to report to Eisenhower on the morning of February 16. Ike's first words were: "I am afraid you will have to eat crow again for a little while." A perplexed Patton replied: "What have I done now?" When Eisenhower said: "You may have to take command of the beachhead in Italy and straighten things out," Patton responded that he considered it "a great compliment because I would be willing to command anything from a platoon up in order to fight."[47]

Patton was alerted to prepare to depart for Italy on short notice, and an aircraft was placed on standby in London. Despite what seemed a certain return to a combat role, he had decidedly mixed feelings about Anzio, uncertain he wanted any part of the mess there: "I hope I don't have to go back and straighten things out." However, when he learned of Alexander's compliment, "I have been skipping like a gazell [sic] ever since. I guess a real love of fighting belongs to but a few people." In his diary Patton wrote: "I told Ike that I was anxious to go. . . . I suppose I am the only person in the world who would be elated at the chance to commit personal and official suicide, but I am tickled to death and will make a go of it."[48]

The idea of employing Patton at Anzio evaporated as quickly as it had arisen. Without explanation he was told to return to his headquarters and never learned why he was not sent to Anzio. The principal reason, however, was the presence of Truscott, whom Eisenhower had earlier attempted without success to have transferred to England and placed in command of one of the two American assault corps for the Normandy landings. Patton regretted not having had the chance to command VI Corps, even though "it would have been very risky, but much honor could have been gained. No man can live forever."[49] Had Patton taken command at Anzio, it would have pleased senior officers on both sides, among them British Adm. Sir John Cunningham, who wrote to the First Sea Lord: "It's a thousand pities we didn't let Patton do the job!"[50]

Not long after his arrival in England Patton decided he wanted a dog, and assigned the task of locating a suitable bullterrier pup to Stiller and others, including Kay Summersby and Lady Leese, the wife of the new Eighth Army commander. In early March the great day arrived when an animal was located and brought to Peover, where it took to Patton "like a duck to water." Fifteen months old and originally named Punch, the dog had belonged to an RAF pilot who was lost during a bombing mission. He had taken the animal on six raids over Berlin, which may have accounted for why it loathed the sound of guns but loved to fly. As Patton described him, the dog was pure white except for "a little lemin on his tail which to a cur-

sory glance would seem to indicate that he had not used toilet paper."
Renamed Willie, the dog became a fixture at Patton's side. He also became
a damned nuisance for Sergeant Meeks, who was obliged to look after the
beast much as one would a small child. With his pink eyes and dour disposition, Willie was not every dog lover's notion of a faithful companion.

Despite his stern appearance, however, Willie was at heart good-
natured. In fact, Patton decided that he was not aggressive enough and
attempted unsuccessfully to train him to be more ferocious. Within days this
led to an encounter with an armored car in which Willie came off a distant
second best. Patton got the message and henceforth left him on his own.
Anything not nailed down was fair game for Willie, who once stole a letter
to Beatrice off Patton's desk and began to eat it. Willie's nights were generally spent tucked into Patton's bed, which he naturally considered was *his*,
not his master's. On one occasion when Willie howled at being left alone,
Meeks locked him in Patton's office, where he proceeded to eat an eraser
and was about to get to work on a box of thumbtacks when saved from disaster. Like many animal lovers, Patton could not resist sharing the foibles of
his latest companion with Beatrice and even sent her his footprint. Willie
could do no wrong, and the Third Army staff threw a birthday party for "Private Willie" one evening, complete with birthday cake. An amusing photograph depicts Patton cutting the cake while Willie sits with obvious enjoyment in a chair by the head of the table.

Patton and Willie were inseparable. The dog always accompanied Patton to his morning staff call and would curl up under his feet and fall asleep,
the only one present without a security clearance. At meals Willie was
always nearby, and Patton would take a piece of toast or bit of food and
flick it in his general direction. Willie usually did not even have to move but
merely opened his jaws and snapped them shut on whatever morsel his generous master had flung. Later, in France, to Patton's utter chagrin, Willie
would succumb to a serious case of cowardice and could generally be found
cowering under a chair at the first sound of shelling or bombs. Willie's other
bad habits included snoring and, as Blumenson notes, disgusting behavior
around women, with whom he was fond of "pushing his nose up their
dresses, clasping their legs, and rubbing his belly against them. Yet Patton
loved him because, as he said, Willie adored him . . . to Willie, Patton was
always right."[51]

Patton was prone to catch colds, and his doctors were always urging
him to cut down on his cigar smoking, which averaged a dozen per day. To
the dismay of his staff, Patton would occasionally swear off cigars, and they
would make bets on how long it would last. It was rarely more than two and
a half days, during which time he became grumpy. When he would reach
into his desk and grab a fistful of his beloved cigars, the staff would breathe
a sigh of relief that once again "Georgie" was fit to live with. Willie, too,

was happiest when Patton smoked: "When his master was off cigars, Willie was off his master. During abstinence periods, Patton was apt to be abrupt with Willie, so Willie kept his distance."[52]

Yet, in the space of fifteen short minutes in April 1944, it was Willie's master who would once again find himself in the doghouse—which, for the second time in less than a year, would nearly cost him the command of an army.

CHAPTER 37

In the Doghouse—Again
"The Knutsford Incident"

What would have passed as a local boner coming from anybody less than Patton had promptly exploded into a world crisis.

—BRADLEY

Although the Allies had squabbled endlessly over operations in the Mediterranean, there was no disagreement that the war would only be ended by means of a cross-Channel invasion in Normandy and the liberation of northwest Europe. The sticking point in 1943 had centered on the timing of the operation.

Hitler's so-called Fortress Europa was, except for the Pas de Calais, virtually nonexistent. In late 1943 Hitler appointed Rommel to prepare Germany's defenses against the expected second front. Shocked by what he found in Normandy, Rommel immediately began an all-out effort to plug its defensive weaknesses. As the commander of German Army Group B, Rommel controlled the Fifteenth Army defending the Pas de Calais, and the Seventh Army in Normandy. As the Allies raced the clock to finalize their plans for Overlord, so too did Rommel hasten the defenses of the Normandy coast.

Responsible for an area from Holland to Brittany, Rommel responded to his challenge in the manner that had made him a legend in North Africa. He tirelessly prodded Berlin for more troops, supplies, and construction equipment. Although he had no particular suspicion that Normandy was to be the

Allied target, the sector from Caen to the Cotentin Peninsula received special attention.

Montgomery's arrival in England as the acting Overlord ground commander in chief soon brought about major changes in the invasion plan. Originally conceived as a three-divisional assault on a thirty-mile front from near the mouth of the Orne River to Vierville-sur-Mer, the plan was exceedingly risky and the frontage dangerously narrow, making it easier for the Germans to locate and hold the assault force, and far more difficult for Allied forces to emerge rapidly and strike the enemy deep and hard. With Eisenhower's blessing Montgomery recast the Overlord plan into a five-division assault over a fifty-mile front. Two British divisions and one Canadian division would land east of Colleville, with the mission of capturing the strategically important city of Caen (the capital city of Normandy), and the Caen-Falaise Plain southeast of the city, from which the Allied air forces would establish tactical airfields.

In the American sector, the U.S. V Corps was to land on a six-thousand-yard beach code-named Omaha, which ran from Colleville in the east to Vierville in the west, above which were steep bluffs; they would then advance inland to secure a solid Allied beachhead. The difficult task would fall to the division Patton trusted the most, the veteran "Big Red One," reinforced by a regiment of the 29th Division, a first-class National Guard outfit. It was also deemed vital to seize the Cotentin Peninsula in western Normandy, not only to prevent the Germans from reinforcing and possibly annihilating the assault force but to facilitate the capture of the port city of Cherbourg, a key factor in fulfilling the insatiable logistical appetite for supplies, food, fuel, and ammunition of the Allied armies ashore. The assault on the Cotentin was to be carried out by the 4th Infantry Division of the U.S. VII Corps. The sea landings were to be preceded seven hours earlier by airborne and glider landings, beginning about midnight of the evening before and early morning hours of D Day.

The British 6th Airborne was to land by parachute and glider to seize the vital bridges across the Orne River and Canal, and protect the British sector from counter-attack from the area east of the river. In the Cotentin, the U.S. 82d and 101st Airborne Divisions would parachute in to protect the 4th Division and ensure a solid link-up with the other assault forces. The Allied forces ashore were to number some twenty divisions by D + 14, and by about D + 20 the Third Army would be committed to Normandy, with the probable mission of liberating the Brittany Peninsula. Overall the plan was both bold and risky, particularly given the notoriously unpredictable Channel weather (even in the late spring and early summer) and the enormous tides of the Normandy beaches. The perils were multiplied when Rommel ordered the installation of mines, metal stakes topped by antipersonnel

mines (called *Rommelspargel*—Rommel's asparagus), huge steel or concrete tetrahedron-shaped obstacles placed at the low-tide mark, and barbed
wire laced across the beaches. Inland, potential airborne landing zones were
similarly treated. Despite its risks, the revised plan issued on February 1,
1944, offered the best opportunity for a successful invasion. From Patton's
point of view the plan had one major drawback: The Third Army was not
destined to participate.

After the landings the assault forces were quickly to gain control of the
main centers of road communication, followed by deep thrusts inland by
armored formations to gain and control terrain for the Allied tactical airfields, and to provide blocking positions that would thwart German attempts
to counterattack and destroy the beachhead. The keys to success in Normandy were, in the British sector, the seizure of Caen and the equally
important Caen-Falaise Plain to its southeast. The American assault force
was Omar Bradley's First Army, whose primary tasks were to capture Cherbourg as quickly as possible, then to mount an offensive to secure the dry
ground south of Saint-Lô and Périers, from where the First Army would
advance toward Paris: and Patton's Third Army would be committed to
secure Brittany and potentially valuable ports. Until the preliminary operations were successfully carried out, Third Army would remain in England.

Toward the end of April 1944 Patton found himself in yet more trouble— of
such magnitude that it again threatened to remove him from the war only
weeks before the unleashing of its most decisive campaign. What later
became known as "the Knutsford Incident" began innocuously, when Patton
was invited to deliver a few remarks on the evening of April 25 at the opening of the Welcome Club for American GIs in Knutsford, a service club run
by local British volunteers. As Patton recounted the episode: "I had been
asked to open it but declined as I did not want to be too prominent. In fact, I
deliberately arrived 15 minutes late, but this did no good, as they were waiting for me. There were also some photographers in the yard, who took pictures but promised not to publish them, as I told them I was not there officially."

There were fifty or sixty people present, mostly local women, and when
the chairwoman introduced him, she said: "General Patton is not here officially and is speaking in a purely friendly way." On this particular evening
Patton seemed especially conscious of keeping his foot out of his mouth and
mentioned in his diary that "I was really trying to be careful." He also spoke
in the belief that there were no reporters present, misled by the chairwoman's introduction that he was not there "officially" as grounds for anything he said to have been "off the record." During his brief remarks Patton
spoke of how until that day "my only experience of welcoming has been to
welcome Germans and Italians to the 'Infernal Regions.' In this I have been

quite successful." Patton went on to note that "I feel that such clubs as this are a very real value, because I believe with Mr. Bernard Shaw . . . that the British and Americans are two people separated by a common language, and since it is the evident destiny of the British and Americans, and, of course, the Russians to rule the world, the better we know each other, the better job we will do."

Such organizations as the Welcome Club were "an ideal place for making acquaintances and promoting mutual understanding. Also, as soon as our soldiers meet and know the English ladies and write home and tell our women how truly lovely you are, the sooner the American ladies will get jealous and force this war to a quick conclusion, and I will get a chance to go and kill Japanese."[1]

To avoid even the remotest appearance of courting publicity, Patton declined an invitation to remain for dinner. To his dismay he soon learned that there *had* been a reporter present, who filed a story that appeared the next day in British newspapers. Although the British accounts were not headline makers, the wire services immediately picked up the story, which the censor somehow failed to catch and squelch, and which created a firestorm in the United States a day later. Some of the British accounts apparently failed to mention the Russians. In any event, the impression was left that Patton had predicted that Britain and the United States were destined to rule the postwar world.

In the wake of the slapping incidents, many American newspaper editors were eagerly awaiting a second opportunity to castigate Patton. His remarks apparently suggesting Anglo-American domination of the postwar world made terrific copy, and editors across the nation made the most of it. The *Washington Post* editorialized: "General Patton has progressed from simple assaults on individuals to collective assault on entire nationalities," while a prominent Republican, Karl Mundt of South Dakota, complained on the floor of the House of Representatives that Patton had managed to slap "the face of every one of the United Nations, except Great Britain." There were also howls of protest over Patton's references (possibly embellished by the press) to welcoming his enemies into hell and "English dames" and "American dames," which the *Washington Post* found "neither gracious nor amusing . . . we think that Lieutenant Generals . . . ought to talk with rather more dignity than this. When they do not they risk losing the respect of the men they command and the confidence of the public they serve. We think that this has happened to General Patton. Whatever his merits as a strategist or tactician he has revealed glaring defects as a leader of men." The editorial also recommended the disapproval of a permanent promotion list of Regular Army officers, then before the U.S. Senate for action.[2]

Although the SHAEF public relations office immediately began a damage-control effort by issuing a correction that included the Russians, it was too late. The damage had been well and truly done. This latest incident involving a Patton indiscretion could not have come at a worse moment for him personally and the U.S. Army professionally. As required by law, Marshall had recently submitted a list of officers to the Senate for approval. Now, with the uproar over Patton, the list was in sufficient peril that Marshall wrote: "I fear the harm has already been fatal to the confirmation of the permanent list."[3]

Once again the question was what to do about Patton. Marshall left Patton's fate entirely in Eisenhower's hands, noting: "Patton is the only available Army Commander for his present assignment who has had actual experience fighting Rommel and in extensive landing operations followed by a rapid campaign of exploitation."

With a multitude of problems consuming his time, absolutely the *last* thing Eisenhower needed was another Patton incident. Exasperated beyond measure at having to deal for the second time in less than a year with a major indiscretion, Eisenhower was angry and on the verge of sacking his friend. "I'm just about fed up," he told Bradley. "If I have to apologize publicly for George once more, I'm going to have to let him go, valuable as he is. I'm getting sick and tired of having to protect him."[4]

Eisenhower was especially irritated because "I had made a particular point of directing George to avoid press conferences and statements. He had a genius for explosive statements that rarely failed to startle his hearers. He had so long practiced the habit of attempting with fantastic pronouncements to astound his friends and associates that it had become second nature with him." The most recent warning had come less than three weeks earlier, when Eisenhower had warned him about talking too much, then backed off, saying: "Go ahead but watch yourself."[5]

Eisenhower was not alone in warning Patton to watch what he said. Ten days before the Knutsford incident both John J. McCloy and Gen. Joseph T. McNarney cautioned Patton. "McNarney kept saying that the thing for me was to keep out of trouble so that I could lead the men."[6]

Patton's first inkling of trouble came the day after his impromptu speech, when the SHAEF public relations office contacted Gay and revealed that Patton was being widely quoted in British papers, some of which made no mention of the Russians. It mattered little that a number of those present submitted statements to prove Patton had indeed included the Russians in his comments. On April 27 Bedell Smith called Patton and in Eisenhower's name verbally ordered him never again to talk in public without first submitting in writing his proposed remarks to the supreme commander for his personal censorship. Patton's reaction was predictable: outrage. "Every effort is made to show lack of confidence in my judgment." He blamed the British influence and complained bitterly that "Benedict Arnold was a piker compared with them," including "Ike and Beedle. . . . How sad." He was temporarily banned from speaking to four divisions, "a restriction that will surely cost lives, yet if I break it I will get relieved and that would mean defeat and a still larger loss . . . damn all reporters and gutless men."[7]

Although outwardly unruffled by this latest blow-up, Patton was sufficiently shaken that the following day he sent Gay to take the salute at a ceremony and retreated to a private automobile hidden from view on a side street. As the repercussions spread from Knutsford to Washington and across the United States, so too did Patton's realization that he was again in very serious trouble.

On April 29 a formal letter from Eisenhower severely rebuked Patton, warning that the incident

> is still filled with drastic potentialities regarding yourself. . . . I have warned you time and again against your impulsiveness . . . and have flatly instructed you to say nothing that could possibly be misinterpreted. . . . You first came into my command at my own insistence because I believed in your fighting qualities and your ability to lead troops in battle. At the same time I have always been fully aware of your habit of dramatizing yourself and of committing indiscretions for no other apparent purpose than of calling attention to yourself. I am thoroughly weary of your failure to control your tongue and have begun to doubt your all-around judgment, so essential in high military position. My decision in the present case will not become final until I have heard from the War Department. . . . I want to tell you officially and definitely that if you are again guilty of any indiscretion in speech or action . . . I will relieve you instantly from command.[8]

When Patton was summoned to London he had no idea what to expect when he faced Eisenhower. Everett Hughes had followed his boss from Algiers to London, and had told Patton that Eisenhower had drafted a cable to Marshall stating that he had no further use for Patton's services but then said, "Oh, hell," and tore it up.

But Eisenhower had *not* torn up the cable, which was sent to Marshall on April 30 and said: I HAVE SENT FOR PATTON TO ALLOW HIM OPPORTUNITY TO PRESENT HIS CASE PERSONALLY TO ME. ON ALL OF THE EVIDENCE NOW AVAILABLE I WILL RELIEVE HIM FROM COMMAND AND SEND HIM HOME UNLESS SOME NEW AND UNFORESEEN INFORMATION SHOULD BE DEVELOPED IN THE CASE.

There was no doubt as to Omar Bradley's opinion about Patton: "I fully concurred in Ike's decision to send Patton home. I, too, was fed up." It was Patton's good fortune that despite Eisenhower's request for Truscott's transfer to England, he was unavailable, having only recently succeeded Lucas in command of the VI Corps in the Anzio beachhead (and later as Fifth Army commander). Had this occurred there is little doubt that Eisenhower would have carried out the relief of Patton.[9]

On the morning of May 1, 1944, a nervous and contrite Patton reported to Eisenhower in utter uncertainty, knowing that he faced punishment ranging

> from a reprimand to a reduction. . . . These constant pickings are a little hard on the nerves, but great training. I feel that, if I get reduced and sent home, it might be quite important, as I would get into politics as an honest and straightspoken man would either be a great success or a dismal failure. . . . In spite of possible execution this morning I slept well and trust my destiny. God has never let me or the country down yet.

Patton's version continues that he was at once reassured by Eisenhower's cordial manner and that he was asked to sit down. Eisenhower began the conversation with: "George, you have gotten yourself into a very serious fix." Before he could continue Patton interrupted to state: "I want to say that your job is more important than mine, so if in trying to save me you are hurting yourself, throw me out."

Eisenhower did not mince words; he had even consulted Churchill, who had dismissed Knutsford as a tempest in a teapot. Patton insisted that if reduced to colonel he would demand the right to command an assault regiment, but Eisenhower said that he might yet need Patton's services to command an army at a later date and was not considering his reduction. However, he might be obliged to send him home, but had still to make up his mind.

Patton left crestfallen, his future more uncertain than ever:

> I feel like death, but I am not out yet. If they will let me fight, I will; but if not, I will resign so as to be able to talk; and then I will tell the truth, and possibly do the country more good. All the way home, 5 hours, I recited poetry to myself. . . . My final thought on the matter is that I am destined to achieve some great thing—what I don't know, but this last incident was so trivial in its nature, but so terrible in its effect, that it is not the result of an accident, but the work of God. His Will be done.[10]

Calling the incident the "unkindest cut of all," Patton wrote Bea "I had a pretty terrible day yesterday . . . it may be the end. . . . If I survive the next couple of days it will be O.K. . . . But still I get in a cold sweat when the phone rings. . . . Well, we ain't dead yet."[11]

In Eisenhower's very different (and possibly self-serving) version of their meeting, Patton remained silently at attention throughout, while Eisenhower explained that Patton had become a liability, and that it was now questionable if he ought to be retained in command of the Third Army. Patton promised to be a model of discretion and in a gesture of almost little-boy contriteness, he put his head on my shoulder. . . . This caused his helmet to fall off—a gleaming helmet I sometimes thought he wore in bed. As it rolled across the room I had the rather odd feeling that I was in the middle of a ridiculous situation . . . his helmet bounced across the floor into a corner. I prayed that no one would come in and see the scene. . . . Without apology and without embarrassment, he walked over, picked up his helmet, adjusted it, and said: "Sir, could I now go back to my headquarters?"[12]

Marshall again reaffirmed that Patton's future lay in Eisenhower's hands, cabling on May 2:

> THE DECISION IS EXCLUSIVELY YOURS. MY VIEW, AND IT IS MERELY THAT, IS THAT YOU SHOULD NOT WEAKEN YOUR HAND FOR OVER-LORD. IF YOU THINK THAT PATTON'S REMOVAL DOES WEAKEN YOUR PROSPECT, YOU SHOULD CONTINUE HIM IN COMMAND. IN ANY EVENT, I DO NOT WANT YOU AT THIS TIME TO BE BURDENED WITH THE RESPONSIBILITY OF REDUCING HIM IN RANK. SEND HIM HOME IF YOU SEE FIT. . . . CONSIDER ONLY OVERLORD AND YOUR OWN HEAVY BUR-DEN OF RESPONSIBILITY FOR ITS SUCCESS. EVERYTHING ELSE IS OF MINOR IMPORTANCE.[13]

Two days later Eisenhower sent Patton an "eyes only" cable: I AM ONCE MORE TAKING THE RESPONSIBILITY OF RETAINING YOU IN COMMAND IN SPITE OF DAMAGING REPERCUSSIONS RESULTING FROM A PERSONAL INDIS-CRETION. I DO THIS SOLELY BECAUSE OF MY FAITH IN YOU AS A BATTLE LEADER AND FROM NO OTHER MOTIVES.[14] What also seems plain is that Marshall was indirectly responsible for saving Patton. The evidence suggests that Eisenhower felt obliged to relieve Patton until he received the chief of staff's May 2 cable. Marshall not only took Eisenhower off the hook but cleverly suggested that he thought Patton ought to be retained for Overlord. Any other allusion would have meant the immediate and undoubtedly irrevocable death knell of Patton's career. As Eisenhower explained to Marshall on May 4, to relieve Patton would be enormously counterproductive, and "because your telegram leaves the decision exclusively in my hands, to be decided solely upon my convictions as to the effect upon OVERLORD, I have decided to retain him in command."[15]* Patton breathed a huge sigh of heartfelt relief: "Sometimes I am very fond of him and this is one of the

*There are conflicting versions of *when* Patton was told he would be retained. The official records make it clear that the date was May 3, although Eisenhower claims to have told him at their meeting on May 1. Given Patton's pitiful mood immediately afterward, and the fact that Marshall's important cable was not even sent until the following day, this seems unlikely.

times. When I read the wire, I called to Gay, 'The war is over, which I always say when I mean that trouble is over." Patton and his closest intimates celebrated with a toast that a junior aide thought callous—he had overheard Patton and thought he had been relieved of command.

Patton cabled Beatrice: "Everything is again O.K. because divine destiny came through in a big way, . . . I guess my trouble is that I don't realize that I am always news but you can bet I know it now. . . . The whole thing was so silly and started in a perfectly harmless informal talk. . . . Well the Lord came through again but I was realy badly frightened. . . . I have youthed thirty years since my last letter."[16]

Henry Stimson had about had it with Patton, too, and sent a blistering letter that he was "inexpressibly disappointed that . . . you could have made such an irresponsible speech; and my feeling is shared by literally everyone who has spoken to me about it. Your reported language was so inappropriate to the position which you hold as an American lieutenant general in the country of one of our important allies and among the civilized surroundings that I could not believe it at first that you had been accurately reported."

Stimson bluntly warned Patton that "practically the entire press of this country has formed an unfavorable opinion of you and are closely following your conduct in the expectation of playing up some future similar outburst on your part, to your final overthrow . . . the situation here is really very tense. . . . I am writing these unpleasant things in a spirit of the most honest friendship towards you." He praised Eisenhower for "exercising the utmost courage and fairness" in deciding to give Patton another chance. Stimson warned Patton in the bluntest language that "each time you have acted or spoken in this irresponsible, reckless, and arrogant way, you have laid an additional burden on his back. . . . The only way you can hereafter justify yourself and your commander is to keep your mouth absolutely shut until you have reached the beachhead and then, by successful drive and successful fighting, win your reputation back again as a soldier who can contain himself as well as conquer the enemy."[17] The secretary's letter left no doubt that he was supporting his longtime friend for perhaps the final time.

One evening when Patton was in residence in London, SHAEF public relations officer Col. "Jock" Lawrence arrived to inform Patton that Eisenhower had sent him to tell Patton personally that there were to be no more public statements by him or any member of his staff. Patton laughed heartily and demanded: "What did Ike *really* say?" Reluctantly Lawrence replied: "*He said you were not to open your goddamned mouth publicly until he said you could.*" Patton burst into uproarious laughter again.[18]

For Patton the end result of the Knutsford incident was what Fred Ayer calls "an almost desperate compulsion" to prove himself a great commander. As for his remarks about the Russians, Patton derisively remarked later: "Anybody, Freddy, who wants the Russians to rule any part of this world is a God-damned fool." Perhaps Omar Bradley summed up the Knutsford incident best when he observed: "What would have passed as a local boner coming from anybody less than Patton had promptly exploded into a world crisis."[19] However, it was Marshall who had the final word on Knutsford. A newly assigned public relations captain arrived at Third Army one day and asked to see Patton, who demanded to know what the hell he wanted. The captain replied: "Sir, General Marshall says you should shut up!" Patton at first blinked but then grinned and said: "Captain, I think you and I are going to get along all right."[20]

Shortly before D Day Patton was grumbling to McCloy, who told him to keep quiet about not playing a role in the forthcoming invasion. Ever the actor, Patton puffed himself to his full height and declared: "Your're taking a good deal of responsibility to come here on the eve of battle and destroy a man's confidence." McCloy retorted: "Listen, George, if I thought I could destroy your confidence by anything I might say, I would ask General Eisenhower to remove you," whereupon Patton immediately stopped emoting.[21]

Patton wrote to Everett Hughes:

You are probably damn fed up with me, but certainly my last alledged escapade smells strongly of having been a frame-up in view of the fact that I was told that nothing would be said, and that the thing was under the auspices of the Ministry of Information who was present. . . . You know what my ambition is—and that is to kill Germans and Japanese in the command of an Army. I cannot believe that anything I have done has in any way reduced my efficiency in this.

Although Knutsford became synonymous with Patton's near dismissal, the town's citizens were very kind to Patton and his men. One such occasion was a cocktail party Patton attended, accompanied only by M/Sgt. Mims, whose penchant for liquor led him to overindulge in the kitchen of the hostess. As Mims was unfit to drive, Patton chauffeured *him* on their return to Peover Hall. The following morning, not for the first or the last time, he found himself *Private* Mims. Eventually Patton, like an indulgent uncle, would feel sorry for him and restore his rank, under the pretense that it was hardly fitting for a three-star general to be driven by a mere private.[22]

The Third Army had no part in the cross-Channel invasion; however, Patton played an important role in Overlord to which only a handful of people were privy. There were only three sites by which the Allies could invade France: Brittany, which had no suitable landing beaches and was too far away from England to provide suitable air cover; Normandy, which could provide both;

and, most obvious of all, the Pas de Calais sector of northwest France, north of the River Seine to the Belgian border. Not only was this region the closest to Britain, it provided the most direct route of advance into Germany and afforded maximum air cover from British airfields in southern England. The Pas de Calais was such an obvious invasion site that the Germans had prepared and concentrated their heaviest defenses and the bulk of their troops in France there to repel the invasion.

In February 1944 the Overlord planners on the SHAEF staff had initiated the most daring deception operation of the war.[23] Code-named Fortitude South, its purpose was to convince Hitler and the German commanders in the West that the Normandy landings were merely a feint, and that the main Allied invasion was to be launched against the Pas de Calais by six Allied assault divisions to establish a bridgehead for a massive force of fifty divisions to follow.[24]

Patton's appointment as Third Army commander was deliberately kept secret. Publicly the Germans were permitted to learn that he was to command a fictitious First U.S. Army Group (FUSAG), created to convince the Germans that the Allies were planning to invade the Pas de Calais. FUSAG's principal forces were real: the First Canadian Army and Patton's own Third Army, both of which were destined to play follow-up roles in the Normandy campaign.

The architects of Fortitude South played to the German bias toward the Pas de Calais, and their belief that the Allies would invade there in order to eliminate the V-1 and V-2 rocket sites from which a German terror campaign against England was being carried out almost daily. In doing so they fulfilled the most essential ingredient of any successful deception operation: that there be sufficient elements of truth to reinforce an already existing belief. Moreover, the Germans had an exaggerated sense of the effect these weapons were having against the British populace. The result was, as the British official history records, the creation of "the most complex and successful deception operation in the entire history of war."[25]

The Germans had long feared Patton as the most able battlefield commander on the Allied side, and the most likely candidate to command the invasion force. There was no attempt to conceal the fact that Patton was in England, only the fact that he was commanding the Third Army. By April 1944 the secret decrypts of German message traffic, collectively known as Ultra, clearly showed that the Germans were convinced that the Allies fully intended to employ their best combat general to lead *Armeegruppe Patton*.

While the real Overlord force was assembling and training all over England, Scotland, and Northern Ireland, nonexistent troop units were created in East Anglia. Dummy troop concentrations were established, using cleverly designed wooden and rubber replicas of tanks, guns, boats, vehicles, fuel depots, hospitals, ammunition dumps, and troop cantonments, many of which were created by the wizards at the famed Shepperton Film Studios. Double agents working for Allied intelligence were fed information that confirmed the presence of a large invasion force in East Anglia.[26] To give even further credence to these fake installations, a signal network was established whose sole purpose was to transmit a steady stream of phony message traffic twenty-four hours a day. In this way the illusion was created that there were at least six divisions operating in East Anglia. The entire hoax was capped by the alleged presence of George S. Patton. Between the fake intelligence and the equally fictitious message traffic, the Germans soon built up the intended picture of an entire army group preparing to invade the Pas de Calais. By the spring of 1944 there was clear evidence that the Germans had fallen for Fortitude hook, line, and sinker.

Patton's name as the phantom army group commander was freely publicized. What could have been more obvious? Yet, had Eisenhower fired Patton in May 1944 and the Germans learned of it, the potential effect on the success of Overlord might have been disastrous. While his movements into southeast England were known, his inspections and visits to his Third Army units were always carried out in strictest secrecy. Again and again there was heard the refrain: "I am not here." Patton willingly embellished the fantasy by occasional visits to his "command" in East Anglia, even though he never enjoyed his role in Fortitude and died before ever learning the full extent of its success and the high regard in which he was held by Hitler and the German high command.

In mid-May German intelligence recorded the disposition of seventy-seven divisions and nineteen independent brigades in the United Kingdom, 50 percent more than actually existed. On May 28 the Allies intercepted a telegram sent by the Japanese ambassador in Berlin, in which he quoted Hitler to the effect that eighty Allied divisions were assembling in England. Only after diversionary operations were initiated in Scandinavia and an Allied bridgehead established either in Brittany or Normandy would the "real" invasion follow.[27]

Patton participated in every aspect of planning and training of the Third Army with his usual enthusiasm and keen eye for detail. Most of his pre-D-day speeches "stressed fighting and killing." He found some divisions better than others, but on the whole he was well satisfied with the progress his army was making toward the day when it would be committed to the battle. "The only worry I have about this show," he wrote, "is how I am going to get the Army across [the English Channel] and assembled on the other side. For the fighting I have no worry."[28]

On April 7, 1944, Good Friday, Montgomery assembled the senior Overlord commanders at his headquarters for a series of briefings by the ground, naval, and air commanders in chief. Those present included Eisenhower and, for part of the afternoon session, Churchill. Behind Montgomery, who spoke first for more than an hour, was a large outline map of Normandy bearing a series of colored phase lines depicting how he envisioned that the subsequent campaign would develop after the Allies secured a lodgement area in Normandy. Patton was an interested spectator and noted the clear improvement in interservice cooperation since Sicily. He was particularly impressed by Churchill's brief speech and very pleased when Montgomery referred to him by name.

The formal dress rehearsal for Overlord took place on May 15, a memorably frigid day in London, with the participants sitting huddled with their coats on to ward off the damp cold. The cast of characters who participated throughout this full day of briefings may have been the most glittering assemblage of Allied military leaders and senior officers ever brought together in one place: King George VI; Field Marshal Jan Smuts of South Africa; Churchill, Eisenhower, Tedder, Montgomery, Brooke, and the other British chiefs of staff; Adm. Sir Bertram Ramsay (the naval commander in chief), Marshal Sir Trafford Leigh-Mallory (the air commander in chief), Bradley, Lt. Gen. Miles Dempsey (the British Second Army commander), Lt. Gen. Courtney H. Hodges, Lt. Gen. H. D. G. Crerar (the Canadian First Army commander); the British, Canadian, and U.S. corps and division commanders; and a host of other senior officers from SHAEF and Whitehall.

The briefing was held in a lecture room of St. Paul's School, London, which Montgomery used as his headquarters and had once attended as a young man. In the front was a row of armchairs for the VIPs, but everyone else sat on hard wooden chairs formed in a semicircle, above the front of the room, on which resided a vast colored model of the invasion area. On the wall behind were smaller maps and charts.

Promptly at 9:00 A.M. Montgomery ordered the doors shut; thereafter no one was to be permitted to enter. Outside were stationed two large, fierce-looking American military policemen to enforce Montgomery's orders. As the conference was about to begin, there was an enormous hammering at the door, which caused Montgomery to look around angrily. The hammering continued, so he finally ordered the doors opened. They were flung wide and in marched George Patton. "Even Monty couldn't prevent his being late."[29]

Again Patton was largely a spectator at one of the most significant occasions of the war, about which few postwar accounts contain much specific detail. One by one those whose names would make history arose and delivered their plans for the great invasion and the campaign to follow. Montgomery opened the proceedings, which were largely dominated by his exposition of the Overlord plan. Before the lunch break King George VI delivered a brief speech; Patton thought it was "rather painful to watch the efforts he made not to stammer." Then came the commanders in chief, Smuts,

Churchill and, finally, Eisenhower—who closed out the day with the observation that "In half an hour Hitler will have missed his one and only chance
of destroying with a single well-aimed bomb the entire high command of
the Allied forces."[30]

At lunch Patton sat opposite Churchill, who wondered if Patton remembered him. "When I said I did, he immediately ordered me a glass of
whiskey." Patton was particularly impressed with the prime minister's "very
fine fighting speech."[31] The traumatic effects of the Knutsford incident were
as evident this day as King George's stutter, and Bradley was undoubtedly
not alone in observing Patton's "clam-like" demeanor. This day was clearly
not his, and he knew it.[32]

Senior British officers may have disdained Patton, but he seemed popular with the rank and file of the officer corps, who did not seem to take
offense at his blunt remarks the way their superiors did. Before D Day he
addressed a large group and expounded on his conviction that to dig in was
to die: "Of course, you British have had experience in landing on beaches,
in the last war, at Gallipoli. The trouble is, most of you are still there!" The
officers understood his message.[33]

A week before the invasion, Patton dined with his corps commanders,
and afterward he spoke informally of what lay ahead. "The only point I tried
to stress was that, in case of doubt, follow the old Confederate adage of
'Marching to the sound of guns.'"[34] The next day he formally greeted,
wined, and dined his friend and classmate, the tall and nearly bald William
H. Simpson (nicknamed "Big Simp"), whose Ninth Army was slated for
employment after the Normandy campaign. Simpson was a gallant, well-
liked officer who had quietly impressed Marshall and been advanced to
three-star rank after a solid military career that included chasing Pancho
Villa and winning a Silver Star commanding an infantry battalion in the
Meuse-Argonne campaign in 1918.[35]

Patton personally conducted the briefing for Simpson. "He was quite
impressed that I could also be my own G-3—of course, I always am," a reference to the fact that Patton always provided the guidance by which his G-3
developed the operational plans for the Third Army's employment. Some
commanders are content to let their G-3 develop the proposals, and then rule
on them, but Patton and Montgomery never let the tail wag the dog. Both
zealously regarded this as a commander's prime responsibility and the
essence of why they had been placed in command. Patton thought Simpson
relied too heavily on his staff, and deliberately conducted his own briefing in
the hope his friend would get his message.

The two were reunited later in Normandy as each awaited the commitment of his army. One evening they shared a cognac, and Patton said: "You
know, it is a funny thing. You and I are here now. We were at West Point
together—and here we are commanding armies. . . . You and Hodges and I

are older than either Eisenhower or Bradley, but we're going to do an awful lot of fighting for them . . . we older foxes are carrying the ball here."[36] Simpson's version is far earthier: "Isn't it peculiar that three old farts like us should be carrying the ball for those two sons-of-bitches?" which was later related by Simpson to Bradley.[37] Once when Simpson visited the Third Army in France, Patton gleefully introduced him to his staff by saying: "Gentlemen, this is General Simpson, When he isn't commanding Ninth Army, he acts as an advertisement for hair tonic." Simpson reputedly joined good-naturedly in the laughter.[38]

June 1, 1944, was an important date in Patton's career. Montgomery had moved his tactical headquarters to Southwick Park, near Portsmouth, on England's southern coast. That afternoon Bradley and Patton arrived to conduct a final review of the invasion plans: "General Montgomery, Bradley and I had tea and then we went to his office, and without the aid of any staff officers, went over the plans. Montgomery was especially interested in the operations of the Third Army and it was very important that two nights ago, I had rehearsed the whole thing for General Simpson, so I was very fluent."

Earlier that day Bradley had reviewed his final plans with Patton and told him that the First Army would conduct the break-out on the right flank of the Normandy bridgehead. Patton was keenly disillusioned by the news, believing that: "If everything moves as planned there will be nothing left for me to do."[39] The precise mission of the Third Army had deliberately been left vague and seemingly dependent on the turn of events on the battlefield. While Bradley would never publicly admit it, the evidence suggests that he would not have been unhappy had events unfolded in such a way that Patton's services could have been dispensed with. The First Army headquarters was heavily staffed with Bradley men brought from Sicily, and as Chester Hansen noted in his diary, Patton "[was] extremely unpopular in this headquarters. Most of our officers have carried with them the punctured legend from Sicily. When I told the Captain of the MPs to provide a motorcycle escort for Patton's arrival, he grinned and asked, [in sarcastic reference to the slapping incidents] 'Shall we have them wear boxing gloves?'"[40] Despite Bradley's antipathy toward Patton, however, both officers always behaved with scrupulous correctness, and this occasion was no different. "We gave Patton the red carpet treatment, including a motorcycle escort."[41]

Montgomery had no such qualms about Patton and firmly envisioned making full use of the Third Army in Normandy. During their June 1 meeting he twice stressed to Bradley the point that "Patton should take over for the Brittany, and possibly the Rennes [a city in the heartland of Brittany] operation."[42]

That evening Patton dined with Montgomery, Bradley, Dempsey, and Crerar in an informal atmosphere of comradeship and anticipation. He listened intently to their references to "if all went as planned," words Patton had learned never to trust. "It never goes as planned. In fact, no operation I have ever undertaken would have succeeded had the G-2 estimate of any enemy potentiality been [even] 50% correct."

Montgomery was well known for his fondness for making small wagers in a betting book kept by his aides. After dinner the book was duly produced, and Patton was asked whether or not England would be at war again ten years after World War II. Monty "bet she would not, therefore, to be a sport, I bet she would." Montgomery and Patton were the only two generals present who had commanded an army in combat. Port was passed around the table, and Montgomery lifted his glass in a toast to his army commanders. When no one moved to respond, Patton replied: "As the oldest Army Commander present, I would like to propose a toast to the health of General Montgomery and express our satisfaction in serving under him." The toast was genuine enough, even though Patton felt obliged to record in his diary that, "The lightning did not strike me." Nevertheless, at the conclusion of their evening together, Patton noted "I have a better impression of Monty than I had."[43]

When they parted the following morning, Montgomery said to Patton: "I had a good time and now we understand each other."[44] As future events would reveal, the remark was significant. While they did not always agree on matters of strategy, as long as the two maintained personal contact, they at least understood each other's point of view. Montgomery's official biographer has written: "Patton he regarded as a sabre-rattler, willfully ignorant of battle in its administrative [that is, logistical] dimension, and of army/air co-operation, but still the most aggressive 'thruster' in the Allied camp; Bradley he considered dull, conscientious, dependable and loyal."[45] Thus, while the two most high-strung, egotistical generals on the Allied side privately found fault with each other, there nevertheless existed between them a firmer basis for understanding the other than either was publicly willing to admit. Patton's distrust extended not simply to Montgomery but to the British army as a whole, and, for that matter, to any Allied general who threatened or even gave the appearance of thwarting or inhibiting his dreams of glory. It was one of the great ironies of their mercurial relationship that in Normandy only Patton would embrace the strategic concept articulated by Montgomery, which was summarily scorned by the other members of the Allied command team.

In other respects the two had a great deal in common. A British military biographer of Patton has written: "The aims of both Patton and Montgomery were in fact the same: both sought to convince their men that they had a great and noble object, the overthrow of the vile Nazi and Japanese tyrannies." Their commonality was also the essence of their success as generals:

Both played to win; both were equally contemptuous of convention; both had highly developed histrionic tendencies. Both were ardent Episcopalians; both were communicants; both avidly read the Bible including the blood-thirsty chapters of the Old Testament; both believed profoundly in the efficacy of prayer as an aid to victory. Their attitude to women was chivalrous. . . . Both captured the imagination of their own soldiers. . . . Patton's popularity even spread to the ranks of the British Army.[46]

Both were masters of manipulation, each influential in convincing their soldiers that what each of them did was important. The notion, perpetuated endlessly in the fifty years since the war by historians on both sides of the Atlantic—that Montgomery and Patton detested one another—is based not only on a misinterpretation of their private remarks about the other, but also on the passionate feelings of the members of the Anglo-American historical fraternity.

However, although a feud between the two most controversial Allied generals over the war's most contentious issues makes great copy, it does not always make good history. The plain fact is that the two generals held a grudging respect for each other, at least until the post-Normandy period, when events put them farther and farther from personal contact and generated bad feeling, primarily as the result of rumors and thirdhand information. The alleged Patton-Montgomery quarrels were always in Patton's obsessive, destiny-driven, tortured mind. When the Third Army was virtually stalemated near Metz in the autumn of 1944, it ought to have engendered in Patton a better understanding of (if not empathy for) what had happened to Montgomery during the first two months of the Normandy campaign.

Patton thought Montgomery was perhaps the best of the British generals but simply not bold enough to bring off Overlord, which, in Patton's opinion, required a commander with the courage of a riverboat gambler and the iron will of a Napoleon.

From his earliest involvement in the Overlord planning, Patton was convinced that once ashore in Normandy, the Allies would not have an easy time of it, and that the Second British Army would become stalled in the crucial left flank of the Normandy bridgehead around Caen (the ancestral home of William the Conqueror). His diary entry for February 18 reads: "Went to see Bradley at 0900 and told him that we should be prepared to land the Third Army at Calais if the First Army and the British become boxed up, as is highly possible." Nothing ever came of the idea. Not long afterward Patton prophetically wrote, "I fear that, after we get landed in France we will be boxed in a beachhead due to timidity and lack of drive, which is latent in Montgomery. I hope I am wrong."[47]

Patton was not part of the great gathering of Overlord commanders at Southwick House, in Portsmouth, before D Day, when Eisenhower took his momentous decision to launch the invasion force in the face of marginal weather that was expected to hold only long enough to land the assault divisions. In fact, for several days Patton was a spectator, as much in the dark about what was occurring as the average soldier cooped up aboard the thousands of ships awaiting the signal to proceed. On June 4 he restlessly attended church and dreamed he was leading the assault. D Day was originally planned for June 5, and Patton's diary merely notes: "Today might be D Day. I listened to the radio at 0600, but no news . . . there has been no news all day." Like the rest of the world, Patton learned of the Normandy landings on the BBC at 7:00 A.M., June 6, 1944.

"I hope I get in before it's all over. . . . I have horrible feelings that the fighting will be over before I get in but I know this is not so, as destiny means me to be in. . . . I started to pack up my clothes a little bit, always hoping, I suppose, that someone will get killed and I will have to go."[48] Perhaps characteristically, Patton's thoughts that day were mainly on himself, of his continued obsession with his destiny, and his recurring nightmare of not getting into the war before it ended: "It is Hell to be on the side lines and see all the glory eluding me but I guess there will be enough for all," he wrote wistfully. In times of personal tribulation Patton often turned to the Bible for solace, and he did so on June 6, 1944.[49]

He need not have worried; the opportunity not only to restore his reputation but to achieve his personal destiny was not far off. In a few short weeks Patton's successes would make worldwide headlines and reduce the unhappy year he had just endured to a distant memory.

CHAPTER 38

The Speech

I want those German bastards to . . . howl: "Jesus
Christ! It's the Goddamned Third Army and that sonofabitch Patton again!"

—PATTON

Before D Day Patton made the rounds of the Third Army to deliver his usual
prebattle speech. The routine was generally the same: Before his arrival the
site would always be buzzing with anticipation. Division-size audiences
were assembled around a platform and often up the sides of nearby hills,
where loudspeakers were sited to deliver Patton's message. Soon the hillside
would be a teeming mass of brown. His speeches became gala events in
themselves, with an honor guard and a band playing rousing marches.
Unlike Sicily and North Africa, where he generally roamed the front in a
half-track or jeep, in England, Patton used a black Mercedes driven by
Sergeant Mims.

The troops would nervously await his arrival, each wondering what this
general, whom most knew only by reputation, would say to them. Patton's
coming meant that the day when they would be committed to battle was not
far off. Most dreaded the thought and wondered how they would perform
under the terrible pressure of enemy fire. With his knack for understanding
what went through an ordinary soldier's mind, Patton would address that
very subject.

When his arrival was imminent, it would suddenly become quiet, and
then Patton would arrive, behind an MP escort, and emerge from the Mercedes. In his buff-and-dark-green uniform, helmet, and highly shined cavalry boots, he would march through their ranks to the front of the platform

and inspect the honor guard closely with his eyes, before mounting the platform with his escort.

A chaplain would deliver an invocation, and then Patton would be introduced by the corps or division commander; he would march stiffly to the microphone and then, as if satisfied all was well, command: "Be seated!" His Normandy speech was a considerable refinement on that presented in Sicily. What surprised all comers was Patton's voice, which a Third Army historian has described as "high but not squeaky or womanish. When he got angry, e.g., when he cursed Willie . . . his voice rose about an octave. In repose his face was nearly always set in hard lines, but he did have a very winning smile which he could use with great effect on visitors to the head-quarters."[1]

When he spoke to his troops, Patton always wore his "war face" fully set. Typically he began by saying:

Men, this stuff we hear about America wanting to stay out of the war—not wanting to fight—is a lot of bull-shit. Americans love to fight—traditionally! All real Americans love the sting and clash of battle. When you were kids, you all admired the champion marble player, the fastest runner, the big league ball players, the toughest boxers. Americans love a winner and will not tolerate a loser. Americans play to win all the time. I wouldn't give a hoot in hell for a man who lost and laughs. That's why Americans have never lost and will never lose a war, for the very thought of losing is hateful to an American.

If he isn't, he's a goddam liar! Some men are cowards, yes, but they fight just the same, or get the hell shamed out of them watching men fight who are just as scared. Some of them get over their fright in a minute under fire, some take an hour, and for some it takes days. But the real man never lets fear of death overpower his honor, his sense of duty to his country, and his innate manhood. All through your army career you men have bitched about what you call, "this chicken-shit drilling." That is all for a purpose—TO INSURE INSTANT OBEDIENCE TO ORDERS AND TO CREATE ALERTNESS. I don't give a damn for a man who is not always on his toes. You men are veterans, or you would not be here. You are ready! A man, to continue breathing, must be alert at all times. If not someone, sometime, some German sonofabitch, will sneak up behind him and beat him to death with a sockful of shit.

The troops roared their approval. "There are four hundred neatly marked graves somewhere in Sicily, all because ONE MAN went to sleep on his job," Patton roared back. He would pause as the men silently waited for him to continue. "But they are GERMAN graves, for we caught the bastard asleep before they did. We have the best food, the finest equipment, the

best spirit, and the best men in the world," he would shout, then lower his
head as if pensively in thought, and thunder: "Why, by God, I actually pity
those poor sons-of-bitches we are going up against. This individual heroic
stuff is a lot of crap. The bilious bastard who wrote that kind of stuff for the
Saturday Evening Post didn't know any more about real battle than he did
about fucking."

This generally brought howls of delight and clapping; some would gleefully slap their legs.

My men don't surrender. I don't want to hear of any soldier under my
command being captured unless he is hit. Even if you are hit, you can still
fight. That's not just bull-shit either. The kind of a man I want under me is
like the Lieutenant who, with a Luger against his chest, swept aside the
gun with his hand, jerked his helmet off with the other and busted hell out
of the Boche with the helmet. Then he picked up the gun and killed
another German. All the time this man had a bullet through his lung.
That's a man for you!

Most of Patton's speech later was recited by George C. Scott in the film,
but the following segment was not used:

All the real heroes are not storybook combat fighters either. Every single
man in the Army plays a vital part. Every little job is essential to the
whole scheme. What if every truck-driver suddenly decided that he didn't
like the whine of those shells and turned yellow and jumped headlong
into a ditch? He could say to himself, "They won't miss me—just one
guy in thousands." What if every man said that? Where in the hell would
we be now? No, thank God, Americans don't say that. Every man does
his job. Every man serves the whole. Every department, every unit, is
important to the vast scheme of things. The Ordnance is needed to supply
the guns, the Quartermaster is needed to bring up the food and clothes for
us—for where we are going there isn't a hell of a lot to steal! Every last
damn man in the mess hall, even the one who heats the water to keep us
from getting diarrhea, has a job to do. Even the Chaplain is important, for
if we get killed and he is not there to bury us we would all go to hell.
Each man must not only think of himself, but think of his buddy fighting
alongside him. We don't want yellow cowards in the Army. They should
be killed off like flies. If not, they will go back home after the war, goddam cowards, and breed more cowards. The brave men will breed more
brave men. One of the bravest men I saw in the African campaign was the
fellow I saw on a telegraph pole in the midst of furious fire. . . . I stopped
and asked him what the hell he was doing up there at that time. He
answered, "Fixing the wire, sir." "Isn't it a little unhealthy up there right

now?" I asked. "Yes, sir, but this goddam wire has got to be fixed." There was a real soldier . . . [and] you should have seen those trucks on the road to Gabès. The drivers were magnificent. All day they crawled along those sonofabitchin' roads, never stopping, never deviating from their course with shells bursting all around them. We got through on good old American guts. Many of the men drove over forty consecutive hours.

By now there would be dead silence. One account notes: "You could have heard a pin drop anywhere on that vast hillside." Then Patton would continue:

Don't forget, you don't know I'm here at all. No word of that fact is to be mentioned in any letter. The world is not supposed to know what the hell they did with me. I'm not supposed to be commanding this army. I'm not even supposed to be in England. Let the first bastards to find out be the goddam Germans. Some day I want them to raise up on their hind legs and howl: "Jesus Christ, it's that goddam Third Army and that sono-fabitch Patton again!"

The troops cheered and roared their approval. According to one observer: "This statement had real significance behind it—much more than met the eye—and the men instinctively sensed the fact and the telling mark that they themselves would play in world history because of it, for they were being told as much right now. Deep sincerity and and seriousness lay behind the General's colorful words, and well the men knew it, but they loved the way he put it as only he could do it."[2]

"We want to get the hell over there. We want to get over there and clean the goddam thing up. And then we'll have to take a little jaunt against the purple-pissing Japanese and clean their nest too, before the Marines get in and claim all the goddam credit!" They laughed in the knowledge that Patton was just jesting—at least they hoped it was in jest. In a quieter vein, he concluded his speech:

Sure we all want to go home. We want this thing over with. But you can't win a war lying down. The quickest way to get it over with is to get the bastards. The quicker they are whipped, the quicker we go home. The shortest way home is through Berlin! Why if a man is lying down in a shell-hole, if he just stays there all the day the Boche will get to him eventually, and probably get him first! There is no such thing as a fox-hole war any more. Foxholes only slow up the offensive. Keep moving! We will win this war, but we will win it only by fighting and by showing guts.

[Pause]

"There is one great thing you men will be able to say when you go home. You may all thank God for it. Thank God that, at least, thirty years from now, when you are sitting around the fireside with your grandson on your knee and he asks what you did in the great World War II, you won't have to say, 'I shoveled shit in Louisiana.' "[3]

Simplistic, profane, deeply offensive to some, his battle speeches were examples of Patton the actor at the peak of his form. The fact that Patton believed what he told his troops lent the essential air of authenticity to the carnival-like atmosphere he liked to create. Politicians use the same approach to whip up support for their cause. The "rah-rah" approach does not work for everyone, but it did for Patton, and by and large he achieved the desired result. What his critics have overlooked is that by reducing the terrible uncertainty of combat to the level of an endeavor shared by all, he not only raised morale but improved their chances of emerging from it with their lives intact.

Nevertheless some were greatly offended, including those who otherwise admired Patton. One was James H. Polk, whose fondness for Patton was only exceeded by his dislike of "those horrible speeches." When Polk assumed command of the 3d Cavalry Group in September 1944, Patton spoke to a small gathering of NCOs and had the good sense to modify his usual profane speech to a couple of sentences. Standing on a jeep, he said: "Now, this is my old regiment. You sons-of-bitches know exactly what you have to do in the next couple of days, and I expect you to do it. I'm going to be real proud of you when it is all over. Thank you very much." And drove off.[4]

Whether or not Patton's speeches were successful had a great deal to do with his audience. He delivered a never-to-be-forgotten pep talk in April 1944 in Armagh, Northern Ireland, where the U.S. 6th Cavalry Group was training. As the city fathers and dignitaries looked on, Patton said: "You men are not in the world's oldest profession; the women beat you to it." The frowns increased as Patton spewed forth invective, and nearby windows began to close as Armagh's citizens suddenly found a pressing need to be elsewhere. The troops, however, loved it.[5]

PART X

Normandy to the Rhine
(1944–1945)

Wars may be fought with weapons, but they are won by
men. It is the spirit of the men who follow and of the
man who leads that gains the victory.

—PATTON

I congratulate you and your heroic soldiers of the Third
Army. I commend you for the dashing and spectacular
victories which have played a great part in bringing
about this glorious day. The exploits of the Third Army
have been in the highest tradition of the armies that
have defended America throughout its history. You and
your gallant forces deserve the nation's homage.

—SECRETARY OF WAR HENRY L. STIMSON (MAY 8, 1945)

CHAPTER 39

The "Mighty Endeavor"

I'm proud to be here to fight beside you.

—PATTON

Roosevelt called the invasion of Normandy on June 6, 1944, "a mighty endeavor to preserve . . . our civilization and to set free a suffering humanity."[1] Although the D-Day landings succeeded after heavy opposition and a near disaster on bloody Omaha Beach, the strategy devised by Montgomery for securing the left flank was thwarted by the fanatical resistance of the enemy. Under the overall command of his longtime adversary, Rommel, the German army in Normandy managed to hold the vital city of Caen and the Caen-Falaise Plain. An opportunity to encircle the city from the west shortly after D Day was botched by one of Monty's veteran desert divisions, which he had personally hand-picked for an important role in Normandy. Repeated attempts to capture Caen through frontal attacks produced not only heavy losses but mounting criticism that Montgomery had lost control of the campaign. By the end of June the two sides were locked in a protracted stalemate outside Caen, and American airmen were making facetious remarks to their army counterparts about another Anzio.[2]

The situation was not helped by one of the most severe storms ever to hit Normandy, which sank a number of ships, wiped out the American Mulberry artificial harbor off Omaha Beach, and virtually shut down the war for five days. When Eisenhower went to see Churchill, the prime minister lamented: "They have no right to give us weather like this." Ike's newly commissioned son, John, was present and he remembers thinking: "In this respect, it seems that Churchill and General George Patton had an attitude in common. Dedicated to their respective missions, each seemed to feel that the Almighty had an obligation—a personal obligation—to render all necessary assistance to their accomplishments." By contrast, John thought his father, who did not share the notion that God owed him anything, more hard-headed than either Churchill or Patton.[3]

Nevertheless, Eisenhower too was growing increasingly disillusioned and moody because of the lack of progress, and he was being pressed by his deputy, Air Chief Marshal Sir Arthur Tedder, and others, to replace Montgomery as the ground commander in chief. Rumblings of discontent were heard in both London and Washington, and when, on June 27, Everett S. Hughes brought word of yet another delay, this time on the part of the First Army, Eisenhower was heard to say: "Sometimes I wish I had George Patton here."[4]

During the month of June Patton chafed while the war went on without him. "Time drags terribly," he lamented. "I am nervous as a cat . . . [and] full of gloom." Virtually every letter to Beatrice was filled with his apprehension that the war would end while he sat uselessly in Peover. His obsession with eventually playing a hero's role extended to wearing a shoulder holster after D Day, "to get myself into the spirit of the part. I suppose I am one of the few emotional soldiers who have to build up a role, but I have always hoped to be a hero, and now may be the time to attain my ambition." The Third Army staff was just as anxious to see the last of England, if only to eliminate the irritability of their commander.

On June 10 he addressed his corps and division public relations officers and said: "I want credit to go to the soldiers and junior officers." The following day he presented a Third Army flag to the vicar of Peover chapel and a bronze plate inscribed to commemorate that "the Commanding General and his Staff and Members of the American Third Army worshipped here during the Second World War, 1944."[5]

On D Day Patton wrote to his son:

> This group of unconquerable heroes whom I command are not in yet but we will be soon—I wish I were there now as it is a lovley sunny day for a battle, and I am fed up with just sitting. I have no immediate idea of being killed but one can never tell and none of us can live for ever so if I should go, dont worry but set your self to do better than I have.
>
> All men are timid on entering any fight. Whether it is the first fight or the last fight, all of us are timid. Cowards are those who let their timidity get the better of their manhood. You will never do that because of your blood lines on both sides. . . . What success I have had results from the fact that I have always been certain that my military reactions were correct. Many people do not agree with me; they are wrong. The unerring jury of history written long after both of us are dead will prove

me correct. . . . Soldiers, all men in fact, are natural hero worshippers. Officers with a flare for command realize this and emphasize in their conduct, dress, and deportment the qualities they seek to produce in their men . . . The influence one man can have on thousands is a never ending source of wonder to me.[6]

He also wrote to congratulate Bradley, assuring him that "he could not have done better." After nearly a year of inaction, Patton's spirits were raised by the receipt of a batch of fan mail, although he quickly disavowed one "damn fool" who wanted him to run for president. "I hope he keeps his ideas to himself." Beatrice was similarly buoyed by a letter from C. P. Summerall noting that Patton was "a general in the hearts of his soldiers . . . [who] stands alone in all the world in knowledge, ability and leadership."[7]

For some time after the landings Patton had only the vaguest notion of what was transpiring in Normandy, except that "apparently things are not going well and one gets the impression that people are satisfied to be holding on, rather than advancing." Patton's perceptions were correct. The Normandy bridgehead was in trouble, and the controversy over who was to blame would soon reach the boiling point.

Word filtered back from Normandy that the VII Corps commander, Maj. Gen. J. Lawton "Lightning Joe" Collins, had by mid-June already relieved two of his division commanders. "I doubt either the expediency or justice of such wholesale beheadings. It creates fear and lack of confidence," Patton penned in his diary. Citing Robert E. Lee's propensity to retain his generals and get the best out of them, Patton noted some weeks later:

Bradley and Collins are too prone to cut off heads. This will make division commanders lose their confidence. A man should not be damned for an initial failure with a new division. Had I done this with General Eddy of the 9th [Infantry Division] in Africa, the army would have lost a potential corps commander. Gay said that he read somewhere that when Napoleon was asked what sort of division commanders he had, he replied, "I don't know yet. They have never been defeated." I shall be more conservative in the removal of officers and have told the corps commanders so orally.[8]

Later Patton would twice refuse requests from the XX Corps commander, Walton Walker, to relieve one of his division commanders.

Not long after the Third Army was established in England, Patton summoned his G-2, Oscar Koch, to his office. Koch found him studying an ordinary Michelin road map of France.

The general straightened up and asked me to join him at the table. His finger continued to rest on the map."Koch," he said, "I want all of your G-2 planning directed here." . . . General Patton's finger rested deep in France. On Metz! Then, starting at Nantes on the Atlantic coast and sweeping his finger along the Loire River toward the east, [he] continued, "I do not intend to go south of the Loire unless it is necessary to avoid a right-angled turn."[9]

Before the invasion Patton had marked a similar map of Western Europe with the places where he expected to fight. "I did this before Sicily, and was correct." One of the places Patton marked as the site of his first big battle was Rennes, in Brittany. Of his propensity to study maps, Patton said: "If 'the greatest study of mankind is man,' surely the greatest study of war is the road net." To that end Patton prepared himself by reading *The Norman Conquest* to determine what roads William the Conqueror had utilized centuries earlier during his campaigns in Normandy and Brittany. The reason, Patton noted, was uncomplicated and invariably accurate: Roads in William's time had to be situated on passable terrain, and to determine what roads Third Army might use, the great Norman became his guide to bypassing the enemy on roads the Germans would not have thought to demolish (as was their custom).[10]

The map predicted with uncanny accuracy the swath that Third Army would make across France and into Germany, and years later his son would call attention to the annotations on the map in a note addressed: "To any biographer."[11] Critics have proclaimed that Patton would have made a poor army group commander, but this was an example of his tendency to think and plan both strategically and tactically.

In late June Patton spent a night with Eisenhower, whom he observed addressing troops before they deployed for France. He thought Ike's habit of gathering them around (as Montgomery did) quite clever, but could not help nitpicking Eisenhower's performance by remarking in his diary on too much use of the word "I," which he thought was the "style of an office seeker rather than that of a soldier. . . . I try to arouse fighting emotion—he tries [to get] votes—for what?"[12]

Frustrated by the growing stalemate in Normandy, Patton composed a paper that he had Hughes slip into Eisenhower's reading file, proposing to land a corps of two infantry and one armored division at Morlaix, on the north coast of Brittany. "We can make a rear attack on the Germans confronting the First U.S. Army, and then driving on to the line Alençon-Argentan, and thereafter on Evreux and Chartres, depending on circumstances, we will really pull a coup. On the other hand, if we play safe and keep on attacking with articulated lines driving to the south, we will die of old age before we finish."

Even though "I dressed my paper up with the names of Scharnhorst, Clausewitz and Moltke so as to catch Ike's eye . . . I have no ambition to be credited with the idea so long as I get the pleasure of executing it. It is a good paper."[13] Patton's hope was in vain, but it was yet another example of his boldness and ability to think ahead. It was also a forecast of the future employment of the Third Army.

His final act in England was to compose "A Soldier's Prayer"[14]— both a reflection of everything Patton had spent his life preparing for, and a means of mentally inspiring himself for what lay ahead: a visual reminder that he must not disgrace himself.

To preserve the fiction that Patton was still in England readying his army group for the "real" invasion in the Pas de Calais, the relocation of the Third Army from England to Normandy in early July was carried out in the greatest secrecy. Incredibly, the Germans were still taking the bait of Fortitude, with most of the Fifteenth Army still awaiting Patton to lead the main invasion there. Thus Patton's presence in Normandy had to be kept a closely guarded secret to avoid compromising the great deception.

On July 6 Patton, Codman, and Stiller drove to an airfield near Salisbury, where they boarded a C-47 transport, which was escorted to France by four P-47 fighters. Willie accompanied his master and soon curled up and went to sleep in Patton's jeep, which was also aboard. As the plane took off Patton glanced at his watch and announced: "Ten-twenty-five [A.M.]. Exactly a year ago to the minute we cast off from Algiers on the *Monrovia* for Sicily." Stiller interjected: "This time I doubt if we get our feet as wet." Patton's gloomy reply was: "I know. A hell of a way to make an amphibious landing." They flew over Cherbourg and the invasion beaches to a landing strip behind Omaha Beach, where they were met by Bradley's aide, Chester Hansen.

Patton looked at his watch again and announced: "Eleven-twenty-five. From Norfolk to Casablanca it took us eighteen days. From Algiers to Gela, Sicily, five days. And now France in one hour." Sighing, Patton directed, "Well, come on. Let's see if there is still a war going on." Suddenly he found himself surrounded by soldiers, sailors, and a few correspondents, whom he warned: "I was still a secret," as the shutters clicked in the cameras of the curious and fans who had never seen him in person. "News of his arrival in France spread like wildfire, and Army and Navy personnel rushed to see him." Patton wore but a single pistol and spoke briefly to them, saying: "I'm proud to be here to fight beside you. Now let's cut the guts out of those Krauts and get the hell on to Berlin. And when we get to Berlin, I am going to personally shoot that paper-hanging goddamned son of a bitch just like I would a snake."[15] Wherever he went Patton drew crowds of spectators. A navy lieutenant who later saw Patton on a beach observing the debarkation of Third Army tanks said: "And when you see General Patton . . . you get the same feeling as when you saw Babe Ruth striding up to the plate. Here's a big guy who's going to kick hell out of something."[16]

They remained overnight at Bradley's First Army headquarters to be briefed. Nearby artillery continually boomed so loudly that the briefing had to be sandwiched between the barrages fired by the howitzers, leading Stiller to observe: "I don't believe the General need worry about missing his war."[17]

After Bradley and Patton lunched with Monty the next day, Patton thought that

> Montgomery went to great length explaining why the British have done nothing. Caen was still their D-Day objective and they have not taken it yet. He tried to get General Bradley to state that the Third Army would not become operational until the [U.S.] VIII Corps had taken Avranches. Bradley refused to bite because he is using me as a means of getting out from under the 21st Army Group. I hope he succeeds.[18]

Quite mistakenly Patton believed that Monty was not anxious to see the Third Army operational, "as he fears I will steal the show, which I will."

Patton also wrongly believed that Montgomery was his rival for the headlines and the glory. To the contrary, as Ladislas Farago writes, Monty exercised command of the Allied ground armies "with exquisite discretion and tact, never interfering with Bradley's tactical arrangements and tacitly supporting his prudent directives. But at the same time he did nothing to restrain Patton and, in fact, gave him a free hand even when and where Bradley sought to stop him or slow him down." Monty also made it clear in his frequent praise of Patton to Brooke that he was pleased with the performance of the Third Army and its commander.[19]

As Patton would soon discover, despite what he may have thought, his problems in northwest Europe would prove to be less with Montgomery than with Eisenhower and Bradley. As had been the case in Sicily, what Patton never grasped about Montgomery was that he had much greater worries in Normandy than upstaging Patton. In Normandy he was now in temporary command of a far more dangerous and troubling campaign, with enormous and growing pressures. When the Third Army succeeded beyond expectations, no one was more pleased than Montgomery, who later cited the breakout as proof his master plan had worked to perfection—that while he had kept the Germans busy around Caen, Bradley and Patton were finally able to unleash the potent power of two American armies.

If there was a single, consistent Allied failure in Normandy that distinguished the campaign, it was that Eisenhower, Montgomery, and Bradley *all* seemed to have disregarded Napoleon's principle that battles are fought to

destroy the enemy, not for capturing terrain, an omission that was to manifest itself again in August. As one historian points out: "All three senior
commanders failed to grasp that the most important strategic objective of
the campaign was the elimination of German forces in Normandy."[20]

When Patton arrived, the Normandy campaign was a month old, and the
weather and the furious defense of Rommel's troops in the deadly
hedgerows of the Normandy bocage left the British unable to advance
beyond Caen. The First Army was fighting a debilitating battle in the
bocage, hedgerow by hedgerow. Until Cherbourg was captured in late June,
Bradley was unable to expand the American sector or carry out his intended
breakout toward Saint-Lô. There was no end in sight to the Battle of Normandy and, as Bradley later wrote: "By July 10, we faced a real danger of a
World War I–type stalemate."[21]

As criticism of the stalemate mounted, Montgomery insisted even more
obstinately (and for the rest of his life), that he was merely adhering to a
master plan to draw the bulk of the German forces and their panzers to the
Caen front, and away from Bradley's sector. However, what Montgomery's
concept of the Battle of Normandy *never* included was a protracted battle of
attrition for Caen. Later he sought to deflect criticism of his generalship by
suggesting that his enemies at SHAEF took advantage of the controversy to
discredit him. The outrage and frustration of the air commanders and many
American officers still rankled long after the war. They were among many
who failed to understand the reasons for the delay at Caen, and suspected
that U.S. forces were being used as a sacrificial lamb while the British dal-
lied in the east. One historian's 1978 account would dub the entire campaign
in northwest Europe "the War Between the Generals" and suggest that the
Allied brass were more interested in preserving their reputations than in
defeating the Germans.[22]

Into this cauldron of rivalry and disharmony stepped George S. Patton. The
roles of Bradley and Patton were now the exact reverse from Sicily, where
Bradley had been Patton's loyal but unhappy subordinate. Although Bradley
privately admitted he did not want Patton serving under his command, he
was soon to have ample reason to be pleased with his presence in Normandy. "My own feelings about George were mixed," said Bradley:

> He had not been my first choice for Army commander and I was still
> wary of the grace with which he would accept our reversal in roles. . . . I
> was apprehensive in having George join my command, for I feared that
> too much of my time would probably be spent in curbing his impetuous
> habits. But at the same time I knew that with Patton there would be no
> need for my whipping Third Army to keep it on the move. We had only to
> keep him pointed in the direction we wanted to go."

Patton's ability to mask his feelings convinced Bradley not only that there was no ill will but that Patton served him with "unbounded loyalty and eagerness."[23]

His reunion with Bradley left Patton in no doubt that he still considered himself on probation with his new boss. He told J. Lawton Collins, the VII Corps commander: "You know, Collins, you and I are the only people around here who seem to be enjoying this goddamned war." Patton paused and added: "But, I'm in the doghouse! I'm in the doghouse! I've got to do something spectacular!"[24]

When he learned of the attempted assassination of Hitler on July 20 by a group of German officers, Patton became so agitated that he went at once to Bradley's command post and pleaded: "For God's sake, Brad, you've got to get me into this fight before the war is over."[25] The smallest details nagged him. Early one morning shortly after arriving in Normandy, Codman found Patton pensively stroking his newly shaved jaw in front of a mirror. "Codman," he said: "I wish to hell I had a real fighting face." When Codman rejoined that he thought Patton's face was a reasonable facsimile, Patton replied: "No, no no! You are either born with a fighting face or you are not. There are a lot of them in Third Army. . . . Having practiced for hours in front of the mirror, I can work up a fairly ferocious expression, but I have not got, and never will have, a natural-born fighting face"—an opinion unequivocally not shared by the troops of his Third Army.[26]

One day Patton bounded into Bradley's command post with maps to expound another of his ideas, emerging an hour later somewhat subdued but with a twinkle in his eye despite not having gotten his way. Climbing into his jeep, he turned to Bradley and, waving his map case, said; "Between my screwy ideas and your brains, we certainly come up with some wonderful plans. Dammit, Brad, you and I make a wonderful team."[27] Bradley himself remembers Patton saying of his rejected plan: "You're right, Brad, goddammit, you're always right."[28]

Privately he thought his superior was too conservative and overcautious. Ideas and proposals to involve the Third Army more quickly and move the battle out of the hedgerows and swamps were flung at Bradley in profusion. "None of these ideas had any effect," Patton complained. "I am sometimes appalled at the density of human beings. I am also nauseated by the fact that Hodges and Bradley state that all human virtue depends on knowing infantry tactics. I know that no general officer and practically no colonel needs to know any tactics. The tactics belong to battalion commanders. If Generals knew less tactics, they would interfere less."

They were indeed an oddly matched pair: Bradley with his unfashionable glasses, modestly dressed in an unadorned uniform, and the elegantly dressed, flamboyant Patton. Eventually each managed to find common ground on which to coexist with the other, however uneasily at times. In

July 1944 patience was something George Patton held in short supply. Before too long Bradley himself was praising the "new" Patton, with whom he formed "as amiable and contented a team as existed in the senior command. No longer the martinet that had sometimes strutted in Sicily, George had now become a judicious, reasonable, and likeable commander." Bradley ascribed the change in Patton to the Sicily slappings, and thought that Private Kuhl, by being slapped, had contributed more than any private soldier in the U.S. Army to the winning of the war in Europe.[29] Perhaps, but the slapping incidents not only deprived Patton of army group command but left him so uncertain of himself and uncharacteristically skittish that at times he seemed unwilling or unable to assert himself with his superiors.

However, on the surface all was well. Patton made it a point not to disagree (at least not too strenuously) with Bradley, relying instead on flattery, cajolery, and a healthy dose of less-than-genuine humility that left Bradley convinced that Patton had fully recognized who was now the boss. Patton also conveniently absented himself from his headquarters at times when he was wheeling and dealing at the front and did not wish to receive any contrary orders from his superior.

Patton confined his misgivings about Bradley's generalship to his inner circle and the privacy of his diary and letters. Patton's eagerness to fight and demonstrate his prowess on the battlefield would sometimes overrule his normally sound common sense, leading to intemperate criticisms of his superiors when they failed to live up to his philosophy of waging war. By his own admission:

> Sometimes I get desperate over the future. Brad and Hodges are such nothings. Their one virtue is that they get along by doing nothing. I could break through in three days if I commanded. They try to push all along the front and have no power anywhere. All that is necessary now is to take chances by leading with armored divisions and covering their advance with air bursts. Such an attack would have to be made on a narrow sector, whereas at present we are trying to attack all along the line.[30]

Patton's personal assessments of his contemporaries may have been harshly self-righteous, but his observation of what was wrong on the battlefield was right on the mark. Eventually Bradley himself accepted the futility of attacking on a broad front in the Cotentin, which Montgomery tactfully pointed out on July 10 when he said: "'If I were you I think I should concentrate my forces a little more'—putting two fingers together on the map in his characteristic way."[31] Bradley would soon refine the idea into an operation he code-named Cobra.

The German army in Normandy fought with all the tenacity for which it is justly renowned, repeatedly counterattacking, even though vastly outnumbered and with only a shred of air support from the Luftwaffe. Moreover, for all the vast Allied preparations for D Day, it turns out that far too little consideration was ever given to the forthcoming Battle of Normandy, consequently the deadly fighting in the bocage came as a complete shock. Bradley's initial attempts to clear the Cotentin resulted in some forty thousand American casualties and exacerbated his frustration at his own inability to devise some way out of the Cotentin. The dirty, bloody business of advancing one hedgerow at a time was inexorably grinding down the soldiers of both sides, so much so that "a German corps commander called the struggle [in July 1944] nothing less than 'a monstrous bloodbath.'"[32]

Historian Max Hastings writes of the extensive high-level concern that some way be found to provide the Allied infantry divisions with the kind of impetus that seemed to be the exclusive domain of the more aggressive airborne infantry. "We were flabbergasted by the *bocage*," said General Quesada of IX Tactical Air Command, who was working daily alongside Bradley. "Our infantry had become paralyzed. It has never been adequately described how immobilized they were by the sound of small-arms fire among the hedges." A French general and friend of long standing reminded Patton of what he had told him in 1918: "The poorer the infantry the more artillery it needs; the American infantry needs all it can get." Patton was convinced that his friend "was right then, and still is."[33]

A week after his arrival in Normandy, Brig. Gen. Teddy Roosevelt died in his sleep of a heart attack. A man of immense courage who had served in the 1st Division in Sicily, Roosevelt was dynamic and never fitted the conventional mold of an infantry officer. His death came shortly before he was at last to have been given command of his own division. Patton mourned the loss of "one of the bravest men I ever knew." Not long after Patton attended Roosevelt's funeral, another close friend, Col. Paddy Flint, was killed in action. One of the U.S. Army's legendary regimental commanders, Flint was utterly fearless, and in Sicily had frequently deliberately exposed himself to enemy fire to inspire his men, usually noting disdainfully that the enemy couldn't hit the broad side of a barn. In Normandy he had tried this once too often. His death came as no surprise to Patton, who had written to Beatrice after seeing him in England a month earlier: "He expects to be killed and probably will be." Like Teddy Roosevelt, Paddy Flint was "a gallant soldier and a great friend. He died as he would have liked to, in battle. . . . I hope when it is my time to go I go as gallantly and as easily. . . . God rest his soul."[34]

"Lucky Forward" was established in an apple orchard not far from Cherbourg. Even as he prepared for war, Patton again indulged in his habit of recording his impressions of Normandy through a historian's eyes: "An arresting sight were the crucifixes at road intersections; these were used by

Signal personnel as supplementary telephone posts. While the crosses were in no way injured, I could not help thinking of the incongruity of the lethal messages passing over the wires."[35]

Patton had been in Normandy a mere ten days when he landed in hot water once again. He returned from escorting Secretary of War Stimson on a tour of the Normandy bridgehead[36] to learn that his public relations officer had briefed the war correspondents on Bradley's forthcoming Cobra operation. Several days earlier the G-3, Colonel Maddox, had briefed key members of the Third Army staff on what was afoot, after which Patton concluded with a stern warning that the operation was top secret and not to be discussed outside the G-3 tent. The resulting leak by Patton's public affairs officer was considered a monumentally serious breach of security, for which Patton, as the commander, was ultimately accountable. Patton was appalled and angry at what he considered his colonel's stupidity and breach of trust, particularly when he was already on thin ice with his superiors.

Patton immediately called Bradley to explain the situation, to apologize, and to assure him he would fire the offender. Bradley sighed, as much in exasperation as relief that he had elected to leave the matter in Patton's hands. Although Patton faithfully promised Bradley that he would sack the colonel for his breach of security, it was not until some time later that the officer was relieved, for reasons unrelated to the leak.[37]

This time Patton decided to take damage control into his own hands and went at once to speak to the correspondents: "I told them how dangerous this slip was, and that they had violated my trust in them by divulging the fact that they had been briefed to the other correspondents who were stationed with the First Army." There was an intense rivalry between correspondents of the two armies, and it was actually a First Army reporter who was responsible for the leak. Given the animosity toward Patton in the First Army, it was not surprising that someone had attempted to embarrass him. Fortunately any discussion of Cobra never went beyond the correspondents. Nevertheless the incident merely reinforced the belief of Patton's detractors that he had a lousy staff.[38]

(Patton was not the only one in trouble. Shortly before the Third Army became officially operational, Willie shamed his master when he carried on "a violent love affair with a French lady dog and also exhumed a recently buried German to the shame and disgrace of the military service."[39] When it came to Private Willie, George S. Patton, the stern disciplinarian, was, alas, a signal failure at maintaining order in his own household. When Sergeant Mims messed up, he was busted to private; when Willie misbehaved, his master merely shrugged. No one ever said life is fair.)

On July 3, the First Army launched an offensive that ground to an inexorable halt far short of its objectives in the dense bocage north of Saint-Lô.

Despite Montgomery's later claims of pinning down the German panzer divisions to his front, two of them had already slipped away to reinforce the German left flank and participate in a series of counterattacks against the First Army. In mid-July yet another breakthrough attempt failed when the British launched Operation Goodwood, a massive carpet-bombing to blast open a gap southeast of the city, through which three armored divisions charged in an attempt to seize the high ground of the Caen-Falaise Plain, achieving tactical success but strategic failure.[40]

Bradley's Cobra plan was similar to Goodwood: the employment of massive aerial carpet-bombing to blast open a corridor through which it was hoped that the First Army could finally break out from the confining hedgerow country west of Saint-Lô. Making the Saint-Lô–Périers road a boundary line, Bradley intended to use American fighter-bombers and strategic bombers to blast open gaps across the front, to be exploited by the massed forces of Collins's VII Corps. Once a gap of some three and a half miles had been opened by the bombers, Bradley intended to utilize two infantry divisions to hold open the shoulders of the bulge, while two armored divisions blitzed through toward Avranches and a motorized division drove on Coutances.

Patton had originated his own version of Cobra on July 2, but apparently never formally proposed it to either Bradley or Eisenhower. The plan suggested placing "one or two armored divisions abreast and going straight down the road, covering the leading elements with air bursts. I am sure that such a method, while probably expensive in tanks, due primarily to mines, would insure our breaking through to Avranches from our present position in not more than two days. This plan would have the advantage of not requiring setting up an amphibious operation. On the other hand, it is so bold that it would never be approved."[41]

No one could predict the outcome of Cobra. Thus Patton became Bradley's strategic reserve and his secret weapon to exploit whatever success the operation might bring. Although Bradley planned to employ one of Patton's corps, Troy Middleton's VIII, the American goal was a *breakthrough*, not a breakout. The popular conception is that Cobra was a brilliant masterstroke designed to achieve spectacular results. In reality Bradley simply (and somewhat desperately) searched for a suitable means of attaining a breakthrough that would enable the First and Third Armies to operate in terrain favorable to exploitation, which the bocage was not. In the end Bradley got both.

Unlike Goodwood, which began with such great promise and ended badly, Cobra began disastrously but ended in one of the greatest American triumphs of the war. The First Army captured the prize of Saint-Lô after days of savage fighting, and on July 25, fighter-bombers of the U.S. Ninth Air Force attacked the target area, followed immediately by fifteen hundred heavies of the Eighth Air Force, which disgorged some 3,400 tons of bombs. However, Cobra began badly, when a calamitous "short" bombing error by Allied bombers killed 111 GIs and wounded 490. Among the dead was Lt. Gen. Lesley J. McNair, commander of Army Ground Forces (and the architect of the training of the U.S. Army of World War II), who despite stern warnings had gone to the front to observe the bombing. For the third time that month Patton was summoned to bury a friend. Again he mourned, but complained that McNair had no business being so close to the front lines, conveniently forgetting that if he had been in McNair's place he would have done the same thing. McNair's death was kept secret, and his funeral was attended only by Patton, Bradley, and three others: "No band. A sad ending and useless sacrifice."[42]

For several days the operation appeared to have failed. Not even Bradley comprehended at the end of the first day of Cobra that what appeared to be continued stiff resistance was in fact a brittle facade which would soon crack under renewed pressure by VII Corps. What spelled the difference was a daring gamble by "Lightning Joe" Collins to deepen his penetration in the target area by committing two of his mobile armored columns before the time was ripe for an exploitation. VII Corps suddenly found their progress virtually unimpeded.

Panzer Lehr Division was one of the hardest hit of the German units defending along the edge of the target area. "It was hell," said its veteran division commander, who later recounted how Panzer Lehr was all but blown to pieces. As Cobra developed, Field Marshal Günther von Kluge, who had replaced Rommel after the latter was gravely wounded the afternoon of July 17, reported to Berlin: "It's a madhouse here. . . . You can't imagine what it's like." But when German resistance suddenly began to collapse, Bradley ordered VIII Corps to spearhead a thrust toward Avranches, the last remaining obstacle to a full-scale breakout into Brittany and the plains of southern Normandy, where the mobility of Patton's armored divisions could be exploited.[43] By July 28 Bradley recognized that it was time to turn Patton loose. The Third Army was to have become operational at noon on August 1, but in the interim he attached Middleton's VIII Corps to Patton.[44]

Patton's first stop was Middleton's headquarters, where he arranged to take the corps under the Third Army, doing so "very casually so as to not get people excited at the change." The next day, as Patton toured the front lines to assess the situation, he encountered Maj. Gen. Robert W. Grow, the commander of the 6th Armored Division, which was held up at a river. Patton demanded to know why a group of Grow's officers were busy studying a map instead of advancing across the shallow river. Unhappy with their reply, Patton went to the river and tested its depth, ignoring some Germans on the other side who did not bother to fire at him. He suggested to ex-cavalryman Grow that if he valued his job he had better move quickly.

   Patton also discovered a battalion of the 90th Division digging slit
trenches and told them to stop, reminding its officers that "it was stupid to
be afraid of a beaten enemy." Although he still questioned the ability and the
courage of his black troop units, Patton did not ignore them. On July 30 he
attended church at a black truck battalion, where "the colored preacher
preached the best sermon I have heard during this war."

   By July 29 there was a frenzy of movement and spectacular advances that
threatened the complete dismemberment of the German left flank in Normandy. But it was by no means a simple task; the advancing armor and
infantry fought a series of bloody battles with the retreating Germans. After
one particularly savage encounter between the 2d Armored Division and an
SS panzer force, one American officer described the carnage as "the most
Godless sight I have ever witnessed on any battlefield."[45]

   Patton was overjoyed that at last Bradley seemed to be acting aggressively and doing "just what I wanted to do. . . . I am very happy." Confident
that he could slash through the final bottleneck at Avranches-Pontaubault, an
exuberant Patton wrote in his diary the night of July 30: "I think we can
clear the Brest peninsula very fast."

   As he prepared to unleash the Third Army, Patton's intentions were
abundantly clear: "The thing to do is to rush them off their feet before they
get set." His morale went even higher when he was informed that a correspondent had quoted Marshall as saying to him: "Bradley will lead the invasion, but he is just a limited objective general. When we get moving, Patton
is the man with the drive and the imagination to do the dangerous things
fast." Patton thought the remark "very fine if true."[46]

   At VIII Corps on July 31 Patton found that Middleton had already carried out his orders to secure the line of the Sélune River but could not contact Bradley for further orders. Patton advised Middleton that military history taught that it was generally fatal not to cross a river. He also said that
while his takeover was not official until the next day, he had in reality been
in command since July 28, and VIII Corps was to get across the river
"now!" Then Patton learned that a bridge at Pontaubault had been secured
intact and that Maj. Gen. John S. Wood's 4th Armored Division had captured the nearby dams. He ordered Middleton to hustle 6th Armored across
the vital bridge. Once across and into Brittany proper, VIII Corps was to
employ one infantry and one armored division each toward Brest and
Rennes. Patton left, satisfied "this little kick" had had the desired effect.

   That night Patton wore one of his patented war faces when he delivered
a pep-talk to his staff, thanked them for their long endurance, saying that he
knew they would be as good fighting as they had been in planning. Part of
his speech was quoted by George C. Scott in the film. An ounce of sweat
was worth a gallon of blood, he reminded them, and the harder they pushed

the Germans and the more of them they killed, the fewer the American casualties:

> There's another thing I want you to remember. Forget this goddamn business of worrying about our flanks, but not to the extent we don't do anything else. Some Goddamned fool once said that flanks must be secured and since then sons of bitches all over the world have been going crazy guarding their flanks. We don't want any of that in the Third Army. Flanks are something for the enemy to worry about, not us. I don't want to get any messages saying that, 'We are holding our position.' We're not holding anything! Let the Hun do that. We are advancing constantly and we're not interested in holding on to anything except the enemy. We're going to hold on to him by the nose and we're going to kick him in the ass; we're going the kick the hell out of him all the time and we're going to go through him like crap through a goose . . . We have one motto, *"L'audace, l'audace, toujours l'audace!"* Remember that gentlemen.[47]

It had been one of his headiest days yet, and when Patton finally snatched some sleep after 1:00 A.M., August 1, it was at the start of a historic day for his Third Army.

When Bradley began to perceive what he had unleashed, he, for perhaps the first time during the war, exhibited both genuine relief at being free of the Cotentin and uncharacteristic fervor at what rewards apparently lay ahead. He thought that American forces might actually be in Rennes within two weeks—a notion Patton would have scoffed at. Bradley wrote to Eisenhower: "To say that personnel of the First Army Headquarters is riding high tonight is putting it mildly. Things on our front really look good. I told Middleton to continue tomorrow morning [with VIII Corps] toward Avranches and go as far as resistance will permit. As you can see we are feeling pretty cocky."[48]

Patton nursed the hope that the Third Army would eventually grow to nine infantry and three armored divisions. He thought that Eisenhower would strip that Third Army for the Ninth when it became operational. With rather typical Pattonesque logic he wrote: "This plan, if one must compromise, is all right." Then he added: "To hell with compromises."[49] Even Patton could never envision that by war's end he would have under his command, at one time or another, six corps, twenty-six infantry divisions, fourteen armored divisions, one French armored division, and one airborne division.[50]

The twin towns of Avranches and Pontaubault were the key to American success in Normandy. Situated at the neck of the Cotentin Peninsula, they were a "corner" that, once turned, would let the genie out of the bottle. By

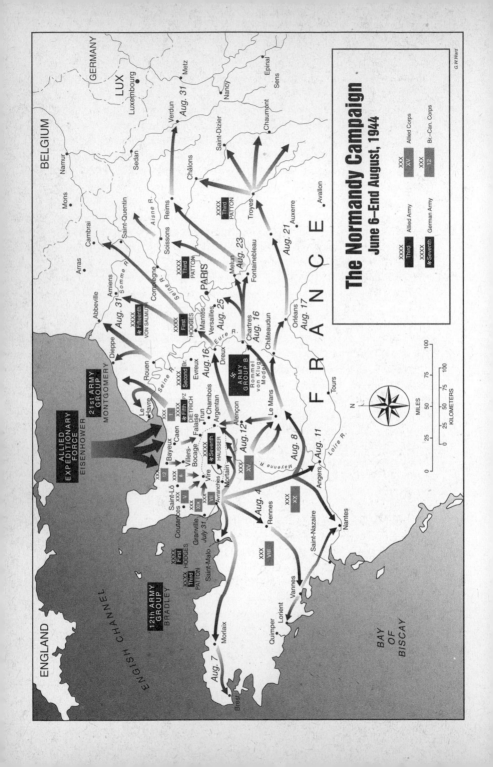

The Normandy Campaign
June 6–End August, 1944

July 30 Wood's 4th Armored, the VIII Corps spearhead, had driven clear to Avranches and, finding two bridges intact over the River Sée, had entered the city unopposed. As the gateway to an Allied advance into Brittany and southern Normandy, Avranches "in the summer of 1944 was a prize beyond compare." However, to ensure that this corridor to the south and west was held open, it was vital that the town of Pontaubault be captured as well. American good fortune held on July 31, when the bridge over the Sélune River, which Patton had been informed was still up, was also captured intact. When this last barrier fell, for the first time since June 6 the German defenses had not only been bent but broken by an Allied attack. Now, with the combined forces of VII and VIII Corps descending on Avranches like a torrent, the moment was at hand for the offensive into the Brittany Peninsula, and a pursuit of the remnants of the shattered German Seventh Army.

On July 1, when asked by Berlin what ought to be done next, an exasperated Commander in Chief West, Field Marshal Gerd von Rundstedt, had replied: "Make peace, you idiots! What else can you do?" Now, what von Kluge had described on July 31 as a *Riesensaurei*—"one hell of a mess"— was about to turn into a nightmare as the Allied general the Germans feared most, George S. Patton, moved in to fulfill in spectacular fashion the destiny he had predicted for himself. In August 1944 his name and that of Third Army would be emblazoned in headlines across the world.

CHAPTER 40

"A Damned Fine War!"

We are going so fast . . . my only worries are my relations not my enemies. . . . If I can keep on the way I want to go I will be quite a fellow.

—PATTON

August 1944 marked the turning point in Patton's fortunes and was the most electrifying and rewarding month of his career. At noon on August 1 the Third Army officially became operational. Patton was relieved to be back in command and eager to prove his mettle: "The waiting was pretty bad . . . but now we are in the biggest battle I have ever fought and it is going fine except at one town we have failed to take. . . . I am going there in a minute to kick someone's ass." He attempted to find something suitable to drink in celebration of the event, but the best his staff could come up with was a bottle whose contents purported to be brandy: "We tried to drink this, but gagged."

Patton was never happier; it was as if he had been freed from jail. There was a fresh cockiness in his stride that had not been evident for more than a year. Codman noted on August 8: "What a week." Patton had been like a man possessed, turning up seemingly everywhere to prod, cajole, exhort, and generally "rais[e] merry hell." Third Army was moving so quickly that it became necessary to relocate the command post each evening to a new location in what quickly came to be known as a series of one-night stands.[1]

Bradley became the commander of the 12th Army Group, which also became operational on August 1. Lt. Gen. Courtney Hodges, who had been understudying Bradley, assumed command of the First Army. Eisenhower did not make the announcement public, and his failure to do so became the basis for the opinion of many in the Third Army that it was part of an orchestrated plot at SHAEF to deny them and their commander credit. But when urged by his aide to publicize the spectacular arrival of Third Army into battle, Eisenhower replied: "Why should I tell the enemy?"[2]

A career infantryman who had entered West Point with Patton in 1904 but had flunked math, Hodges entered the army as a private soldier and earned a commission the hard way. A longtime friend of Bradley's from his Fort Benning days, Hodges was a good and decent man, schooled in the ways of the infantry. He and Patton could hardly have been more different. In his account of the U.S. Army in World War II, Geoffrey Perret writes that Chester Hansen thought that Hodges resembled "a small town banker in uniform." But Hodges exuded none of Bradley's exterior confidence and "seemed more worrier than warrior. . . . They were much alike in their working methods. When Hodges took over the First Army, the new broom swept nothing."[3]

The machinery was now in motion to finish what Cobra had started. Although Bradley later denigrated Patton's accomplishments during this crucial period of the war, he "put the Third Army on that one road south, pushing 200,000 men and 40,000 vehicles through what amounted to a straw" at Avranches:

> Every manual on road movement was ground into the dust. He and his staff did what the whole world knew couldn't be done: it was flat impossible to put a whole army out on a narrow two-lane road and move it at high speed. Everything was going to come to a screeching halt. He even intermingled units. Yet out of the other end of the straw came divisions, intact and ready to fight. If anybody else could have done it, no one ever got that man's name.[4]

Patton's audacity in driving his army through the Avranches-Pontaubault gap was the icing on the cake.

The mobility of the U.S. Army was now a reality, superbly supported by the XIX Tactical Air Force, commanded by Brig. Gen. Otto P. Weyland, a new friend and strong supporter of Patton's method of waging war. Patton's first order to Middleton was to thrust VIII Corps toward Brest (and the area of Quiberon Bay, where the Allies intended to construct a large port facility) and Rennes. By the afternoon of August 1, Wood's 4th Armored became the first American unit to enter Brittany; by early evening it had advanced forty miles, to the outskirts of Rennes. American units pouring through Avranches-Pontaubault became a traffic control problem. Patton's immediate priority was to pass his two corps—Middleton's VIII and Maj. Gen. Wade Haislip's XV—through this major bottleneck without delay. It was, said Patton later, something that could not be done, "but was." He gave the credit to the impressive accomplishments of experienced staff officers and his division commanders, some of whom personally directed vehicles through the incredible traffic jams. Nevertheless, Patton was all too aware of the terrific losses possible if the Luftwaffe caught his truck-borne infantry in one of them, and had to remind himself, Do not take counsel of your fears.[5]

When he found traffic hopelessly snarled in Avranches, Patton angrily leaped into an empty covered police box in the town center and began directing traffic, his arms waving furiously as he began untangling the mess. For the next hour and a half, Patton the commanding general became Patton the efficient traffic cop. Soon the vehicles began to move quickly through the bottleneck as amazed drivers recognized that it was not an MP vehemently waving them on but none other than Patton himself.[6]

"The operations for the first day are very satisfactory," he wrote. "Compared to war all human activities are futile, if you like war as I do." Everywhere he went there was abhorrent evidence of war: the dead, both men and horses; smoking and burning wreckage; all the horror that seemed both to repel and fascinate him. The roads were choked with a victorious army on the march, and the POWs of a defeated army. At a bridge Patton passed a German ambulance column that had been mistakenly attacked by Allied air and remarked how the entire area smelled of iodoform (a widely used antiseptic in 1944). Yet such scenes made him cringe inside.

With two armored divisions loose in Brittany—the 6th Armored to the north, driving on Saint-Malo and, ultimately, Brest; and the 4th Armored, preparing to assault the well-garrisoned Rennes—Patton had good reason to be pleased. In the days that followed these two divisions moved so rapidly that they outran their communications and became very difficult to control. VIII Corps and the Third Army actually lost track of their locations on several occasions.

On August 2 Patton caught up with a column of the 90th Division near Avranches. The division's record in Normandy was checkered, its performance apathetic. By the time it came under the Third Army, Bradley had removed the division's second commander, and it already had its third, a National Guard officer and former banker named Raymond McLain, a first-class commander who was beginning to orchestrate the remarkable turnaround of a heretofore poorly performing division. McLain had been Patton's personal choice, and he never had cause to regret insisting on his appointment.[7] Although he disliked changing division commanders in combat, even Patton would have agreed that Bradley had been right to do so. On this occasion, however, Patton was disgusted with their ragged appearance and their violation of his strict order never to remove or cover up their division patch and insignias of rank. He dragged a lieutenant from a ditch after an enemy aircraft had merely flown high overhead, and then began walking with the troops for nearly two miles, all the time talking and cajoling them to do better. Some were riding on artillery pieces: "I called them babies and they dismounted." When he encountered a GI with a horribly mangled leg, Patton stopped and personally administered morphine. Certain that the soldier would soon die, he remained to offer what little comfort he could until an ambulance arrived.

Later that afternoon Bradley innocuously revealed that he had taken it on himself to direct the 79th Division's mission in Brittany, saying that he knew Patton would concur. It was exactly what Patton had done to him at Gela on July 11, 1943. Loath to antagonize his superior but unwilling to let Bradley get away with it, Patton said that while he concurred, he chided him for succumbing to "the British complex of over-caution. It is is noteworthy that just about a year ago to the day I had to force him to conduct an attack in Sicily [Brolo]. I do not mean by this that he is avengeful, but he is naturally super-conservative."[8] In the days to come Patton would continue to exert his influence over his commanders to keep them moving, insisting that to attack, attack, attack would keep the enemy unbalanced. Without consulting Bradley, Patton ordered an element of the 5th Infantry Division to drive on Nantes, at the mouth of the Loire River. "I am sure he [Bradley] would think it too risky. It is slightly risky, but so is war."[9]

Patton was so eager to drive forward without letup that he could barely curb his impatience: "I could go faster but some think it is too risky." Others were "so damn slow," both mentally and physically, and lacked his self-confidence. "Am disgusted with human frailty," he noted in his diary on August 6. "However, the lambent flame of my own self-confidence burns even brighter." One of his critics later charged that as Grow's and Wood's armored divisions drove deeper into Brittany, Patton seemed so consumed with merely being in command again that he gave little thought to the fact that he might be going in the wrong direction.

Only when Brittany and the ports were secure could Patton turn eastward, "where the decisive battle of the European campaign would obviously be fought." But the success of Cobra had irrevocably altered the entire Allied concept of a deliberate, sweeping advance to the Seine. Neither the great breakout nor a pursuit across the plains of southern Normandy was ever in the Allied blueprint for the campaign. Patton was the first to grasp the immense possibilities open to the Third Army:

> What emerged was a concept quite different from that which had governed operations in the Cotentin. Patton saw his immediate objectives far in advance of the front, for his intent was to slash forward and exploit not only the mobility and striking power of his armored divisions but also the German disorganization. . . . There seemed little point in slowly reducing Brittany by carefully planned and thoroughly supervised operations unraveled in successive phases.[10]

However, as historian Russell F. Weigley contends, by "charging in all directions at the same time," Patton was guilty of

> distracting the Allies from their main offensive purpose, which should have been to pursue the German armies toward their destruction. But Patton's aggressiveness, also helped stimulate movement in the proper direction of opportunity and ultimate purpose. He himself merited the primary credit for hastening as many as seven divisions through the Avranches-Pontaubault bottleneck in only seventy-two hours despite stubborn German aerial attack.[11]

Bradley's gravest concern about Patton was that he would act too aggressively. Ignoring Patton's oft-repeated remark that in pursuit operations his flanks would take care of themselves, Bradley hedged his bets by ordering a division to seize Fougères to shield the First and Third Armies from a possible German reaction. In reality, as Weigley notes, Patton "was not so heedless in such matters as Bradley feared." Independently Patton had given the 5th Armored Division nearly identical orders. With more forces than he required to clear Brittany, Patton and Bradley both recognized that they had an offensive force capable of accomplishing far more than merely acting as a shield.[12]

However, during these crucial days, Patton was dealing from a position of weakness when it came to challenging the authority of Bradley and Eisenhower. For, as Blumenson points out, with Montgomery's image badly tarnished after Goodwood—which he had deliberately oversold in order to obtain the cooperation of the air chiefs to divert their bombers to a tactical role—Bradley's was growing after Cobra: "He was also the object of flattery from his fellow Americans, who rarely ceased to remind him of his preeminent place in the campaign. He was, after all, the senior American troop commander on the Continent. He had won his spurs by his victories, he was about to step up to army group, where he would be Montgomery's organizational equal."[13]

Patton well understood that he was quite fortunate to find himself again commanding anything, and he was both grateful and wary. More important, Patton was painfully aware that another serious misstep might be his last. What Patton would have done if he had been in Bradley's place and what he did as the Third Army commander were very different indeed, and he was determined to tread lightly. After Bradley turned down one of his proposals (which was often), Patton was occasionally successful in persuading him to adopt his course of action by convincing Bradley that it was *his* idea.

Patton seemed to view what he did in August 1944 as a race against Monty and the British. This time his goal was the Seine instead of Messina. To continue a major offensive in Brittany when there existed the answer to a combat commander's dream to the east—where both the First and Third Armies could be unleashed—was folly. When Wood found Rennes heavily garrisoned with German troops, he decided not to commit the 4th Armored to a potentially costly battle that would slow his advance, and instead bypassed the city so that he could quickly strike eastward into the unprotected German southern flank, not southwest to Vannes and Lorient, as his orders specified. On August 4 Hugh Gaffey, the Third Army chief of staff, perceived what Wood had done and immediately informed Patton, who in other circumstances might have been expected to take advantage of what Wood had orchestrated. On this occasion, instead, "Patton's response was categoric: comply with Bradley's 12th Army Group orders, he said."[14] The reason was simple enough. Patton was "desperately anxious not to become entangled in a wrangle with Bradley, when his own position remained that of a probationer."[15]

This meant putting a leash on his friend Wood, who was affectionately known to his troops as "Tiger Jack," a sobriquet given partly for fearlessly standing up to Patton. Wood was beside himself with disgust over being hamstrung at Rennes, believing that the 4th Armored was attacking in the wrong direction and should be turned loose to exploit to the east. Wood raged at Patton for failing to grasp the immense possibilities of an armored thrust through the virtually unprotected Orleans gap. One of the brightest, most aggressive, and brilliant of the World War II armored division commanders, John Shirley Wood has been called "the American Rommel." Like Patton, Wood was outspoken and did not suffer fools gladly, but, as the 4th Armored sat idle outside Rennes, Wood began to regard his friend and commander as one of them. When Middleton visited on August 4, Wood told him: "We're winning this war the wrong way, we ought to be going toward Paris." Wood "protested long, loud and violently" that he could have the 4th Armored in Chartres (175 miles northeast of Rennes) in two days. "But no! We were forced to adhere to the original plan—with the only armor available, and ready to cut the enemy to pieces. It was one of the most colossally stupid decisions of the war."[16] On August 5 Wood sarcastically signaled Patton: "Trust we can turn around and get headed in the right direction soon." When Patton later told him:"You almost got tried for that," Wood angrily shot back: "Someone should have been tried but it certainly was not I."[17]

By blazing a path for the Third Army, the 4th Armored would have opened the way to cutting off the Germans at the Seine. When Wood was finally permitted to turn 4th Armored to the east on August 15, it was too late. To his death Wood remained embittered over the lost opportunity, never knowing that under other circumstances Patton would have listened to him.

There now came the milestone for which generals are given stars: to decide whether the original Allied master plan to seize Brittany was still more important than taking advantage of the opportunity to drive to the Seine across the unprotected underbelly of Normandy. It was a mission tailor-made for Patton and his Third Army, but time was of the essence. Patton himself saw the opportunities, but his shaky status in the Allied hierarchy after Sicily and Knutsford left him unwilling to stick his neck out.

Resistance in Brittany was light and remained so until American units butted up against the fortress of Brest, which proved as tough a nut to crack as Cherbourg. As the First and Third Armies poured through the funnel at Avranches, Bradley, ever conservative, was not convinced that the German army in Normandy was beaten, and

> was determined to embark on no reckless adventures south-eastwards unless he was certain of holding the Avranches 'elbow' in their rear. "We can't risk a loose hinge" . . . he feared a German counter-attack north-westwards, breaking through to the coast and cutting off Patton's armored divisions from their fuel and supplies . . . and later accepted responsibility for the decision, for good or ill, to swing large American forces west into Brittany.[18]

Hitler had the same idea and several days later launched an ill-fated armored thrust at Mortain. Nevertheless, Normandy historian Max Hastings is correct when he criticizes Bradley's lack of imagination in failing to more quickly alter the Overlord plan to seize Brittany and the open door to the east. Even Montgomery only envisioned a single corps being diverted to Brittany, not the two that were employed there.

Patton's detractors, particularly those who served in the First Army, were resentful of the publicity accorded the Third at their expense, and have complained that he merely took advantage of Cobra—that the real credit belonged to J. Lawton Collins and VII Corps. "He did not break out. He walked out." As an old man, Bradley hypocritically took the final shot by criticizing Patton for what he himself had wrought: "Patton blazed through Brittany with armored divisions and motorized infantry. He conquered a lot of real estate and made big headlines, but the Brittany campaign failed to achieve its primary objectives."[19]

Why did Bradley fail to seize the moment earlier than he did? Bradley's defense was "the one overriding reason why I sent Patton and Middleton to Brittany: logistics . . . the lifeblood of the Allied armies in France," without which "we could not move, shoot, eat, land new troops or evacuate the wounded."[20] All true, but it should have been evident that even under the best of conditions the Brittany ports had only long-term logistical value.

Thus, although Patton had concluded that operations in Brittany could be minimized while the remainder of the Third Army drove toward the Seine through the Orléans gap—he felt unable to act aggressively, as he undoubtedly would have had his and Bradley's roles been reversed. The advantages were certainly not lost on Bradley, but not until August 3 did he make one of the most important strategic decisions of the war, which, as the U.S. official history records, altered the entire course of the campaign. Patton was to leave only "a minimum of forces" to secure Brittany. "The primary American mission was to go to the forces in Normandy who were to drive eastward and expand the continental lodgment area." Having noted how the spectacular success of Cobra had drastically changed the entire complexion of the Normandy campaign, Montgomery immediately concurred, and, to his credit, he resisted pressure from the logisticians to divert more forces to Brittany when Brest was found to be heavily defended, telling Brooke: "The main business lies to the east." Thus, within a few hours the original Allied strategy of Overlord had been scrapped.

The capture of Brest had been intended to relieve the logistical problem of resupplying what by the end of the Normandy campaign were more than two million Allied troops. Yet its vast distance from the front posed problems of its own, not the least of which was keeping resupplied the divisions fighting to capture the port. The SHAEF planners had decided before D Day that the Brittany ports were vital. The Normandy campaign had been mapped out with deliberation so that the logisticians could keep pace with the advancing armies. However, the great success of the Cobra breakout suddenly made the need to execute the plan highly questionable, and after Brest fell they realized that the port was simply too far away to be of much use. In the end Brittany became a monument to the notion that in war campaigns are supposed to unfold as planned.

Patton mistakenly envisioned a cheap victory in Brittany, but instead the operation drained badly needed supplies from his other corps and resulted in a huge military embarrassment to the Allies. One night Bradley candidly admitted to Patton why he persisted with the useless siege of Brest: "I would not say this to anyone but you, and have given different excuses to my staff and higher echelons, but we must take Brest in order to maintain the illusion . . . that the U.S. Army cannot be beaten." Patton thought that Bradley put more emotion into this statement than he believed Bradley had in him. "I fully concur in this view. Anytime we put our hand to a job we must finish it."[21]

Bradley's later claim that he had to capture Brest for logistical reasons was fatuous and untruthful. Although Middleton, Grow, and others suffered the wrath of Patton's ire when they did not overpower the Germans defending Brittany, the truth was that this failure of Allied generalship could be traced directly to both Bradley and Patton—Bradley for his lack of resolve in failing to recognize that he had been given an outdated and irrelevant mission; Patton for his unwillingness to tackle Bradley and, ultimately, Eisenhower, over the whole issue of Brittany. For all his newfound self-con-

fidence, Patton was simply unwilling to jeopardize his future as the Third
Army commander.

Patton's Achilles heel (which would be painfully evident later, in Lorraine) was that rather than cut his losses, he would attempt to storm his way
out of a bad situation in the name of prestige. One of his critics scornfully
notes that "the Third Army's wild rampage through Brittany obscured one
central fact—west was precisely the *wrong* direction. . . . Patton's greatest
deficiency as a tank commander: [was] his tendency to think as a traditional
cavalry tactician and to care little what direction he was attacking in, so long
as he was attacking."[22] Another biographer has written that Patton was "at
his best and most successful only where he could apply his brilliant loosed-
rein cavalry tactics against an already confused and mostly mediocre enemy.
This was to be the lesson of the Brittany campaign."[23]

While the Allied brass pondered their next course of action, in Avranches,
on highways, in rest areas and bivouacs, GIs everywhere learned what it
meant to be in the Third Army. Patton was enthralled by the danger he faced
each day. "We have been bombed, strafed, mortared, and shelled," wrote
Codman, who noted that Patton seemed to thrive as never before. The afternoon of August 7 Patton returned to his command post along a particularly
devastated stretch of highway choked with dust and littered with wrecked
and destroyed German vehicles and horribly blackened corpses. Fires were
burning in nearby fields, which were strewn with dead, swollen farm animals. Half turning in his seat, Patton shouted: "Just look at that, Codman.
Could anything be more magnificent?"

At that moment their vehicle passed an artillery battery firing an ear-
shattering salvo, and Codman leaned forward to hear Patton say: "Compared to war, all other forms of human endeavor shrink to insignificance."
His voice shaking with emotion, Patton concluded: "God, how I love it."[24] If
Patton worshiped in the house of Ares, it was for the personal satisfaction it
brought him as a successful general. Days when he flew above the battlefield produced more pensive reflections. Then Patton never ceased to marvel
at the incredible effort expended during World War I on the construction of
trenches, and to ponder what he called the "evils of war" that had resulted in
so much death. He thought them a wonderful subject for a sermon by a paci-
fist. The little white crosses that dotted the many military cemeteries
attested to human folly, "which has invariably resulted in more wars."[25]

As the Third Army's drive piled up the miles, Patton's mood was
euphoric. He wrote Beatrice that he was "quite tickled and not at all worried
although I have been skating on the thin ice of self-confidence for nine days.
I am the only one who realizes how little the enemy can do—he is finished
we may end this in ten days." His exuberance was boundless. "Get off your
butt and come up here. We're about to have a damned fine war!" he cabled

nephew Fred Ayer, who was assigned to SHAEF. Photographs taken in August 1944 show a different, smiling Patton from the one whose dour appearance glowered from those taken earlier. A week later he exulted: "In exactly two weeks the Third Army has advanced farther and faster than any army in the history of war. So far that is not known," a reference to the fact that it would be several days yet before the official announcement that Patton and the Third Army were spearheading the breakout. The Germans, however, needed no such reminders that Patton was now engaged against them.

After the war a rumor would persist that Patton disappeared for a day in mid-August, at the same time von Kluge was out of contact with his head-quarters. Von Kluge allegedly attempted to contact the Allies to discuss a surrender of German forces in the west. A paranoid Hitler, who had just escaped death at the hands of the German plotters, was inclined to smell a conspiracy and had von Kluge recalled to Berlin, but he committed suicide en route rather than face the tortures of the Gestapo. While there is some evidence that von Kluge may have entertained the notion of such discus-sions, Patton had no part in any meeting with the German commander, despite an assertion by a member of his staff that he claimed to have attempted to contact von Kluge.[26] The new German commander was Field Marshal Walter Model, who had earned the nickname of "Hitler's Fireman" on the Eastern Front. However, in the final days of the Normandy campaign, not even Model could reverse the defeat of the German army.

To others the business of war was a profession; to George S. Patton, the warrior, it was the fulfillment of his lifelong dream of leading a conquering army in a great battle. As the Third Army daily added glowing new statistics to its list of achievements, the adrenaline was flowing, and Patton was on the greatest "high" of his life. Yet Patton never lost sight of his mortality and of his fear that a stray bullet might cut short his run. The distances were now so vast (some four hundred miles between Brittany and the front lines in late August) that he often flew in an L-5 two-seater Piper Cub aircraft, its only distinctive feature a small three-star plate inserted between the struts. He rarely enjoyed it, confessing to Frederick Ayer: "You just wait to get hit." To Beatrice he wrote: "I dont like it. I feel like a clay pigeon." He also admitted that his fear of being strafed on the ground or even by his own air force "always scares me." Patton learned the consequences of habitually flying low the day a battery of 155 howitzers opened fire nearby. The shells whizzed past, blowing in the windows of the tiny aircraft, sending it buck-ing like a bronco. Thereafter Patton had the pilot fly higher.[27]

Patton also understood that the same gut-wrenching feelings existed in the pits of his own commanders' stomachs, and he made frequent trips to the front to reassure them; "they all get scared and then I appear and they feel better." Patton wryly noted the difference between Sicily, when he had "to put spurs on" and Normandy, where "I am being corked." On one visit to the 83d Division during the siege of Saint-Malo, Patton, perhaps lulled by the temporary quiet, wanted to go farther forward. He was restrained by the commander, Maj. Gen. Robert C. Macon, who said: "General, if you just move up there forty yards you will be in the enemy front line."[28]

Suddenly, in the press and on the radio, the star of the hour was Patton, and the media made the most of it, in the most glowing and often exaggerated manner. The transition from goat to hero was as instantaneous as the first headline. A United Press radio report began typically: "A fiction writer couldn't create him. History itself hasn't matched him. He's colorful, fabulous. He's dynamite. On a battlefield, he's a warring, roaring comet. No wonder the brass hats don't like Lieutenant General George Patton. When this war is ended . . . one of the vivid most human pages will be the saga of General Patton."[29] The sudden turnabout was worrisome. "I sincerely hope that these damn fools do not blow me up so high that I will burst like an over-inflated balloon."[30] His troops fought hard, took the inevitable casualties (despite the perception of the Third Army running amok, a high price was to be paid), and reveled in spearheading the breakout. One GI wrote tongue-in-cheek to his mother: "For God's sake, send some food over. All this man Patton orders is more tanks and more bullets!"[31]

Some, of course, scorned Patton's showmanship and his constant presence at the front as self-serving. Patton saw it differently, often saying: "If you want an army to fight and risk death, you've got to get up there and lead it. An army is like spaghetti. You can't push a piece of spaghetti, you've got to pull it."[32] Although the First Army was double the size of the Third, and actually had far more tanks, "it was Patton who came to symbolize the ultimate in pursuit. While Hodges was satisfied to measure progress in terms of Germans killed or captured, Patton measured his in miles."[33]

One night when the Third Army was still north of Avranches, the Luftwaffe launched a heavy bombing raid in a last-ditch attempt to stem the American advance. Bombs began falling nearby, and for many of the younger men of the Third Army staff it was their first taste of war. Most not on duty took shelter in nearby hedgerows. Patton sat in a deck chair in a nearby field, a cigar in his hand, cursing: "Those goddamned bastards, those rotten sons-of-bitches! We'll get them! We'll get them!" The word spread that if Patton was not afraid, there was no reason for anyone else to be either.[34]

When Maj. Gen. Manton S. Eddy took command of XII Corps, his 9th Division had been slugging it out in the bocage, where gains had been measured in yards. Within one hour of becoming a corps commander he was ordered to capture the town of Sens, fifty miles away, which was off his map. Eddy asked his new staff: "What the hell kind of war is this? I've been fighting for two months and have advanced five miles. Now in one day you want me to go 50 miles?" When he telephoned his new boss that night, Patton told him how pleased he was to have Eddy with him again. Eddy reminded Patton that XII Corps had only four divisions, and that there were an estimated ninety thousand Germans to his north and another eighty thousand to his south: "How much shall I worry about my flank?" Patton replied that it all depended on how nervous he was. "He had been thinking a mile a day good going. I told him to go fifty and he turned pale."

The next afternoon Eddy telephoned to proclaim: "General, I had a lovely drive. I'm in Sens. What's next?" Patton replied: "Hang up and keep going."[35] Although Eddy certainly understood Patton's methods, his infantry training had so indoctrinated him to the importance of his flanks that he never seemed comfortable with Patton's bold approach to war.

In fact, the notion of abandoning the doctrine of secure flanks was anathema to most of Patton's infantry generals. It ran counter to what they had been taught, and some never got used to the idea. Yet Patton's cavalier dismissal of his flanks was not as casual as it appeared. Between Ultra, the French Resistance fighters of the FFI (French Forces of the Interior) who helped to cover his exposed southern flank, and Weyland's airmen, Patton remained confident. As Robert S. Allen writes: "Patton never made a move without first consulting G-2 [Oscar Koch]. In planning, G-2 always had the first say . . . and then he acted on it. That explains why Third Army was never surprised and why it always smashed through vulnerable sectors in the enemy's lines."[36] Patton's secret weapon was his reliance on aerial reconnaissance and the close air support of Weyland's XIX Tactical Air Command.

Patton became a true believer in the value of air support in August 1944, when he saw for himself the carnage wreaked by Weyland's fighter bombers, who pounded the retreating Germans without letup, and heard the airmen's praises being sung by his tankers. Virtually every one of Patton's spearhead armored columns was supported by P-47 fighter bombers of XIX TAC, which roamed the skies "like hawks seeking prey."[37] Army Air Corps historian Geoffrey Perret writes that on August 1 Patton sent for Weyland and "greeted him with a full quart of bourbon. . . . By two in the morning the bottle was empty and the two generals were pals, swearing like boys to have no secrets from one another. Thereafter they attended briefings sitting in the front row, their arms around each other's shoulders."[38] One day, with Eisenhower in attendance, a briefer began: "Yesterday, XIX TAC continued to powerfully assist Third Army in making a greater wreck of the greater Reich." Eisenhower barely smiled, but Patton roared with delighted laughter.[39]

In Sicily, Patton had mistrusted and cursed the air forces; now, a year later, Weyland and his airmen could do no wrong. Each general felt that he had established a model relationship with the other, and indeed they had. Air-ground cooperation never worked better. Patton sang Weyland's praises to Hap Arnold, the boss of the Army Air Corps: "For about 250 miles I have seen the calling cards of the fighter-bombers," a reference to the burned-out hulks of German vehicles lining the roads. Their success, however, came at a high price in losses of both aircraft and airmen to help keep the Third Army rolling. So feared was the XIX TAC that twenty thousand Germans surrendered to a single infantry platoon near Orléans rather than face renewed attacks from the air.[40] Arnold and Spaatz once asked Patton if he was not worried about his flanks. "No worries," replied Patton. "The Air Force takes care of my flanks." Asked if he didn't worry about his supply situation, he replied: "Not a bit. I have a G-4. He worries about my logistics. I'll tell you though, he fainted three times on one day."[41] One of Patton's best-known pronouncements occurred when he turned to Weyland and said: "You guard the right flank. I can't be bothered. We can't fool with the damn thing, so everything south of the Loire River is yours. You hit it with air and watch it; we are going straight east."[42]

With Weyland's support Patton felt invincible. When Eddy persuaded the Third Army chief of staff to let him halt XII Corps short of the Meuse River because of dwindling fuel supplies, and seemed overly worried about the strength of his corps, Patton would have none of it. With typical Pattonesque bluntness, he ordered Eddy to "get off his ass . . . and to run till his engines stopped and then walk—that is how we took Commercy and found a million gallons of gas there—German gas. . . . We got 15 car loads of new 88 mm guns and 30 car loads of rations at Châlons—the lord will provide."[43]

The G-3 kept a chart (and another in Patton's van) reflecting the daily accomplishments of the Third Army, which was the first thing Patton looked at during his daily briefings. In less than a month of combat the results were spectacular. As of August 26 the score sheet read:

	Third Army	*German*
Killed	1,930(e)	16,000
Wounded	9,041(e)	55,500
Non-battle casualties	5,414	
Missing	1,854	
POWs		65,000
Totals	18,239	136,500

The Third Army captured or destroyed 4,353 German tanks, artillery, and vehicles, losing only 269 tanks, 74 guns, and 956 vehicles of its own. Patton's penchant for keeping accurate counts was the best of any commanding general's. Once, when Bradley asked for the number of enemy dead since D Day, only Third Army provided a prompt answer.[44]

As the Third Army advanced, it was cheered everywhere by the French, who lined the roads, throwing kisses and apples and offering wine to their liberators. Patton savored it and at the same time thought about the future. "It will be pretty grim after the war to drive ones self and not be cheered but one gets used to anything. . . . I used to wave back but now I just smile and incline my head—very royal."[45] His sister, Nita, wrote to praise him. "Well you are certainly getting your revenge on all the slimy jealous toads who tried to do you harm. . . . Papa [would be] proud of you these days, I know, and yearning over his fair haired boy. . . . Lordy I wish I could see you going down by the Arc de Triumph in your tank. You are a modern knight in shining armour. . . ."[46]

Patton learned from Eisenhower that Congress had at last acted to promote him in the Regular Army, and that "General Marshall has asked that you not spoil the record of a magnificent job by public statements." Patton was enjoying himself so hugely that he barely noticed, and professed not to "give a damn." On August 15 it was announced that his friend Sandy Patch had led the Seventh Army ashore in southern France, and that Patton was in command of the Third Army in France: "The first *I've* heard of it," he said with a grin.

One of Patton's gravest disappointments was being denied the liberation of Paris, an honor given to Jacques Leclerc and his Free French 2d Armored Division.[47] If Eisenhower had not transferred LeClerc's division to a First Army corps on August 17, after the change of strategy in the Falaise gap, that distinction would have been the Third Army's: "Patton was bitterly disappointed. His staff was stunned. It regarded the decision as the last straw in Eisenhower's increasingly apparent measures to exclude Patton from the glory road."[48]

On August 19 Codman flew to Vannes on the Brittany coast to fetch Patton's World War I French comrade, General Koechlin-Schwartz, who spent the night with him at the Third Army command post. Koechlin-Schwartz told Patton: "Had I taught 25 years ago what you are doing I should have been put in a madhouse, but when I heard that an armored division was headed for Brest I knew it was you."

Despite the daily headlines that screamed Patton's name, and the fact that the Third Army was running rampant across southern Normandy, the dark side of Patton still managed to intrude into his thoughts: "Civil life will be mighty dull—no cheering crowds, no flowers, no private airplanes. I am convinced that the best end for an officer is the last bullet of the war. Quelle vie."[49*]

Then, with little warning, Patton's triumphant march to the Seine was interrupted when a controversial decision by Bradley led to an Allied

---

*While his master was being hailed as a triumphant liberator, Willie was having a miserable war. One morning, as Patton's headquarters was preparing to move, someone shot a machine gun, and Willie, who was asleep in a nearby field, ran in terror and hid under Patton's desk. His concerned master had the Third Army surgeon check Willie's pulse, which was found to be very fast. Later that day Willie was attacked by wasps and became, in Patton's words, "realy sick."

attempt to envelop and strangle the German army in what has come to be known as the Battle of the Falaise Gap.

Although it took nearly three days too long for the Allied leadership to scrap its abruptly outmoded master plan to take advantage of Cobra's success, by August 3 Bradley finally acted to make official what Patton had already begun to orchestrate: a swing to the east by the Third Army. Patton was ordered to leave VIII Corps to secure Brittany, while the other three corps of the Third Army turned eastward toward the Seine.

Hitler believed that the outcome of the Normandy campaign would decide the destiny of Germany and, without consulting von Kluge, made one of the most monumentally stupid decisions of the war, by directing an immediate powerful armored counterattack between Mortain and Avranches, to recapture the neck of the Cotentin Peninsula and cut off all U.S. forces south of Avranches, and in Brittany.

The German attack on August 6–7 was repulsed at Mortain by the heroic stand of the U.S. 30th Division, and instead of splitting the First and Third U.S. Armies, Hitler's blunder left the German Seventh Army and elements of the Fifth Panzer Army exposed and in grave jeopardy of encirclement. It also became the *raison d'être* for turning Third Army loose across the plains of southern Normandy.

On August 7 Montgomery launched Totalize, a major offensive by Lt. Gen. H. D. G. Crerar's First Canadian Army to capture Falaise, while Sir Miles Dempsey's Second British Army swept the bocage on the right flank. Together the two armies would pivot to their left and drive east. Meanwhile, the First and Third Armies would thrust across the open southern flank and trap Army Group B (Fifth Panzer Army and Seventh Army) in a double envelopment along the River Seine. This was the so-called long envelopment or "long hook."

Mortain, however, suddenly changed the thinking of Eisenhower and Bradley. Haislip's XV Corps was spearheading the Third Army drive east and was already outside Le Mans. Bradley reasoned that if he could quickly swing XV Corps behind the Germans in the direction of Alençon and Argentan in a "short [left] hook," the Allies might be able to trap and annihilate the Seventh Army in the jaws of a pocket between XV Corps and the Canadians advancing south toward Falaise.

An exultant Bradley proclaimed it "an opportunity that comes to a commander not more than once in a century. We're about to destroy an entire hostile army. . . . We'll go all the way from here to the German border."[50] Eisenhower enthusiastically endorsed Bradley's change of plan, which would later spawn one of the greatest controversies of the war. Montgomery concurred and—although Argentan, Haislip's objective, was some twelve miles inside the British boundary— "welcomed the penetration."[51]

Even as he gave permission for Bradley to violate the interarmy group boundary, Montgomery was confident the Canadians would close the pocket at Argentan before XV Corps. Although the Canadians were nearly halfway to Falaise by August 9, German resistance was savage and the offensive soon slowed to a protracted, bloody battle. By the evening of August 12, XV Corps was in Alençon and Sées, prepared to drive north with four divisions to close the southern portion of the eighteen-mile gap. The Canadians, however, remained stalled north of Falaise.[52]

As XV Corps neared Argentan, Haislip reported to Patton that with additional forces he could effectively block the east-west roads in his sector. With the Canadians still well north of Falaise and making bitterly slow progress, Patton foresaw that they could not close the gap quickly, and on August 11 he aggressively told Haislip: "Pay no attention to Monty's God-damn boundaries. Be prepared to push even beyond Falaise if necessary. I'll give you the word."[53]

On August 12 he directed XV Corps to capture Argentan and then push slowly north toward Falaise. When Patton telephoned Bradley for permission to move north of Argentan, the 12th Army Group commander made one of the most controversial decisions of the war by ordering Haislip to halt at Argentan. Patton begged to be allowed to continue north, allegedly saying in half jest: "We now have elements in Argentan. Shall we continue and drive the British into the sea for another Dunkirk?" (The British, whose sensitivities over Dunkirk were still raw, soon learned of Patton's flippant remark and were not amused.)[54] Patton believed that XV Corps "could have gone on to Falaise and made contact with the Canadians northwest of that point, and definitely and positively closed the escape gap."[55] On the night of August 16, he wrote in his diary that Bradley's decision "will certainly become of historical importance." Although Patton thoroughly understood the significance of a decision that might further have cemented his place in history, his mild complaints reflect a distinct lack of his customary passion, as if his mind were already intent on his own preference of strangling the Germans by driving to the Seine and blocking its crossing points.

Bradley later defended his decision by arguing that he was fearful of a deadly collision with the Canadians. "In halting Patton at Argentan, however, I did not consult with Montgomery. The decision to stop Patton was mine alone. . . . I much preferred a solid shoulder at Argentan to the possibility of a broken neck at Falaise."[56] Patton thought he was merely "suffering from nerves."

The Canadians did not capture Falaise until late on August 16, which still left a fifteen-mile gap between the American and Canadian armies. It was not until August 19, when the Polish 1st Armored Division and the U.S. 90th Division joined forces farther east at Chambois, that the pocket was officially sealed. In the process, what remained of Army Group B absorbed terrible punishment. Most German units simply disappeared, as troops fled individually and in small groups toward the Seine. Of the fifty divisions in action in June, only ten could now even be called fighting units.[57]

The Allied triumph that ended the Normandy campaign ensured that Germany would ultimately lose the war. The magnitude of the German defeat was starkly visible around the villages of Trun, Saint-Lambert, and Chambois, which were littered with unburied dead, thousands of dead horses and cattle, and smashed and burning vehicles. Nevertheless, the final battles of the Normandy campaign have left a legacy of controversy and unanswered questions that have been the subject of an endless postwar debate that somehow the Allies had blundered in permitting between twenty to forty thousand German troops to escape across the Seine.[58] Yet these survivors retreated eastward in disarray, mostly on foot; their heavy weapons and armor remained behind, most of it destroyed in the savage battles around Trun and Chambois.[59]

Neither Patton nor Montgomery had any great enthusiasm for Bradley's so-called short hook. From the outset both clearly preferred a "long envelopment"—a drive to the Seine by the Third Army, the Canadian First Army, and the British Second Army to establish blocking forces to prevent the mass escape of Army Group B from Normandy. Even as late as August 14, both felt that a long envelopment still held promise. For Bradley, having elected to halt Patton at Argentan, other than sitting tight until XV Corps could be reinforced (an alternative he never considered), his remaining option was to strike quickly east toward the Seine while continuing to hold the Argentan shoulder temporarily with a smaller force: "George helped settle my doubts when on August 14 he called to ask that two of Haislip's four divisions on the Argentan shoulder be freed for a dash to the Seine."[60]

After persuading Bradley to permit the Third Army to strike east toward the Seine, Patton misquoted Shakespeare: "Oh, what a tangled web we weave when we first practice to deceive." Still: "I am happy and elated," he wrote in his diary.[61] Patton's divisions sliced across southern Normandy with little opposition. Le Mans, Orléans, Chartres, and Dreux all fell to the Third Army, while to the north Hodges's First Army began an offensive also aimed at blocking the Seine. When German resistance collapsed to the east of the Caen-Falaise Plain, the First Canadian Army and Dempsey's Second British Army began similar drives to barricade the Seine farther downriver. Henry L. Stimson euphorically observed "in delighted admiration" that Patton had "set his tanks to run around in France 'like bedbugs in a Georgetown kitchen.'"[62]

After the war Montgomery observed that "the battle of the Falaise pocket never should have taken place and was not meant to take place"— that is, the short envelopment was not his idea nor did it fit in with his concept of the battle.[63] In September 1944, when Patton was asked by a war correspondent if the Falaise encirclement had been part of the original Overlord master plan or an improvisation, he replied: "Improvisation by General Bradley. I thought we were going east and he told me to move north."[64]

Patton did not live long enough to participate in the postwar controversies and thus probably never knew that he and Montgomery had been in complete agreement over how the Normandy campaign ought to have concluded.

Although Montgomery still nominally commanded all Allied ground forces, the crucial decisions that took place in August 1944 were solely Omar Bradley's. He created the Falaise gap and, ultimately, he failed to close it. During these final, dramatic days of the Normandy campaign, Patton was little more than a spectator in what later became a great controversy.

Bradley arrived at Patton's headquarters on the morning of August 15 "fit to be tied." He told Patton that it had been decided at the highest level that the Third Army should not advance beyond Dreux and Chartres, or any farther toward the Seine because XV Corps might not be able to contain the Germans who had escaped from the pocket. Patton thought that Bradley had lost his nerve and, after being told not to advance any farther, replied "that he was already to the Seine River, in fact had pissed in the river that morning and had just come from there, what would [Bradley] want him to do— pull back?"[65]

One of the gravest mistakes of the war was the reversal of roles between Patton and Bradley. Had Patton commanded 12th Army Group—as indeed he would have had he not self-destructed his career in Sicily—the events of August 1944 would have been very different. Blumenson has concluded that "their problem of accommodating their distinctly different styles and outlooks was a contributory cause of why the Allies [later] faltered."[66] He presents a lucidly argued, scathing indictment of Bradley's indecisive generalship at the end of the Normandy campaign, noting that Bradley vacillated repeatedly over what he wanted to do, and then botched the decisions he did make. But, worst of all, Bradley "halted Third Army at the Seine and at Dreux, Chartres, and Orleans and thereby killed the pursuit at the very moment he ought to have been encouraging it. He permitted Patton to gain a bridgehead over the Seine at Mantes, then forbade him to reap the benefits of the action, nullifying Patton's wish to fashion a potent pincer. Finally, he let Patton charge off toward Germany even though the main work lay west of the Seine." Bradley's generalship was "troubled by doubt. He made instant decisions, then second-guessed himself. . . . He initiated potentially brilliant maneuvers, then aborted them because he lacked confidence in his ability to see them through to completion."[67]

Had Patton been in command there would not have been indecisiveness. Some armchair strategists will eternally complain that he would have taken dangerously misguided risks in order to win. However, Patton understood that the Allies had the initiative and was, Blumenson notes, "the single commander who grasped what needed to be done and how to do it." Moreover, the mistakes were not Patton's but attributable to Montgomery, Eisenhower, and Bradley. "All three were so intent on deciding where to execute the post-Overlord operations beyond the Seine River that they paid little attention to closing the jaws at Falaise or at the Seine."[68]

Patton and Montgomery might have accomplished together what Bradley and Montgomery did not. For all his private complaints about Montgomery, Patton was fully capable and willing to have worked with the British general. Even Blumenson, no admirer of Montgomery, concedes that Patton, as the top American ground commander, working with a friend of a quarter century, Eisenhower, would have been Montgomery's equal. "They would have worked closely and effectively together. For they respected each other. Their interests were professional and tied to the operational scene. Their strengths were complementary. The thrust of Patton and the balance of Montgomery would have produced a perfectly matched team . . . a less discordant Normandy campaign, a happier resolution of Overlord, a firm entrapment of the Germans west of the Seine, and a much earlier end of the war in Europe."[69]

Despite his mild-mannered facade, Bradley bore grudges longer and with far more vehemence than Patton. He never forgave Montgomery for trespassing on the II Corps boundary in Sicily, and he made no effort in Normandy to find an accommodation over the Falaise gap. Whether there had been the short hook to Argentan or the long envelopment to the Seine, from the outset Patton and Montgomery together would have combined to ensure that the remnants of the German army did not escape. Bradley did not, and the result was that as the Normandy campaign ended, the war of the generals was just heating up. At times Bradley almost gleefully criticized Montgomery for his real and alleged failures. However, although Montgomery remained in nominal control of the Allied armies until September 1 (when Eisenhower assumed command), for all practical purposes his tenure actually ended on August 1, when Bradley became 12th Army Group commander—and his equal.

Ahead lay further controversy over how to conclude the war that would polarize British and American commanders and make the dispute over Caen and Falaise seem trivial.

CHAPTER 41

"For God's Sake, Give Us Gas!"

If Ike . . . gives me the supplies, I'll go through the Siegfried line like shit through a goose.

—PATTON

At the end of August 1944 an unhealthy mood of overoptimism and self-deception swept through the ranks of the Allied high command as a result of the sudden, dramatic end to the Normandy campaign. So crushing was the victory, which was followed by Allied gains of fifty miles a day or more north of the Seine, that some began to perceive that the war was virtually over. How could it not be so with the German army in complete disarray and the Allies gobbling up territory at such a rate that it had long since outstripped its supply lines? German losses had been staggering and seemingly hopeless: 900,000 on the Eastern Front and another 450,000 in the west. Their quality and experience notwithstanding, what was overlooked was the fact that there still remained some 3.4 million troops in the army, of which well over a million were to be committed to the Western Front and the defense of the Reich.[1]

The mistake most consistently made by the Allied high command in northwest Europe was the failure to realize—despite repeated demonstrations to the contrary—the will and tenacity of the German army to resist against overwhelming odds and in the most appalling conditions. Sicily, Anzio, and Normandy were all examples that were soon to be repeated in Holland and in the Ardennes and Lorraine.[2]

The original Allied blueprint was predicated on an orderly German retreat across Normandy, and a solid defense of the Seine River line, not a rout such as had occurred as a result of Cobra and the entrapment of Army Group B. This pleasant turn of events did not include the luxury of a pause at the Seine while the Allied armies regrouped, advanced their logistical bases farther forward, and made plans to resume offensive operations toward Germany. With the remnants of the German army in full retreat toward the German border, it was time for pursuit, not consolidation. Unfortunately, however, SHAEF had failed to propose plans for dealing with success on such an epic scale.

Eisenhower stoutly maintained that the strategy he intended to employ was a "broad front" advance against Germany that would thrust Montgomery's 21st Army Group through Belgium toward the Ruhr industrial complex, while Bradley's 12th Army Group advanced to the south, covering the British right flank. The sudden collapse of German resistance in mid-August gave rise to a different proposal from Montgomery—for what he called a single, "full-blooded" thrust through Belgium and Holland toward the Ruhr by the two army groups abreast. This force, reasoned Montgomery (no longer the acting Allied ground commander now that Eisenhower had assumed command), would consist of some forty divisions and "would be so strong that it need fear nothing."[3]

Eisenhower refused, noting that there were inadequate sources of logistical support to give priority to 21st Army Group. He called the plan a "mere pencil-like thrust," inconsistent with his concept that the war would be fought and won by Allies advancing on a "broad front"—that is, with the two army groups advancing side by side from the North Sea to the Swiss border. Monty, however, repeatedly challenged the decision, to the point that Eisenhower privately questioned the loyalty of the newly promoted field marshal.

Bradley proposed his own plan, a thrust across central and southern France through the Frankfurt gap, and into the heart of the Third Reich, by both the First and Third Armies. Its essence was that the Third Army would advance into Lorraine and breach the Maginot and Siegfried Lines in the Saar. The First Army would advance on an axis to the north. Both routes, Bradley argued, were the most direct ones into the Reich. During the remainder of August, Bradley and Montgomery took turns lobbying Eisenhower.

Eisenhower rejected both plans, opting instead for his "broad front" advance: 12th Army Group to attack eastward toward the Saar (Germany's most important industrial region after the Ruhr) and the Frankfurt gap, Montgomery's army group to thrust northeast toward the Ruhr, but only after capturing Antwerp, the largest and best-equipped port in Europe, which Eisenhower correctly considered the vital key to supporting the Allied advance. To protect his right flank, Montgomery demanded (and received) the attachment of Hodges's First Army, over Bradley's strenuous objections. The Normandy beaches and ports were now too far from the front lines and increasingly incapable of supporting the enormous logistic appetite of the Allied armies. However, when the British captured Antwerp in early September, they failed to secure its access route to the sea—the Scheldt estuary—which remained in German hands, necessitating bloody combat by the Canadians to clear it, and delaying its opening until late November.

Thus the legacy of the great Allied victory in Normandy was rancor and controversy. So powerful were the differences between Eisenhower and Montgomery that their disagreements over strategy have since been the focus of more debate than any other aspect of the war in Europe.

In a single month the Third Army had rolled virtually unchecked across southern Normandy, and into Lorraine. Rheims and the great champagne vineyards, Verdun with its gruesome reminder of horror on an epic scale, and the dreadful Argonne, where Patton had been wounded in 1918: all fell to Third Army in late August or early September 1944. Patton's euphoria was that of a conquering hero. "To be Patton roaring across France, scattering a beaten foe, was to feel more like a god than a man. . . . To grab all three on the run and keep moving was to triumph over History."[4]

The Allied victory in Normandy was a logistician's worst nightmare. A German general once remarked that the blitzkrieg was paradise for the tactician and hell for quartermasters, but it was Ernie Pyle who described what followed as "a tactician's hell and a quartermaster's purgatory."[5] The Allies had literally outrun their supply lines, and the logisticians were obliged to maintain supply lines stretched to the breaking point, as four Allied armies dashed across France and into Belgium. On D + 100 days (September 14, 1944), the Allies had advanced to where the logisticians thought they would reach only in May 1945.[6] Thus, once beyond the Seine, logistics, not tactics, had become the dominant factor.[7]

That J. C. H. Lee was the chief Allied logistician was detrimental to an effort that required the hand of someone far more skilled in the school of logistical support than the general sneeringly referred to as "Jesus Christ Himself." Lee's narrow-mindedness, conventional peacetime quartermaster's mind, and his unwillingness to use every means at his disposal to improvise a logistic system tailored to the needs of the combat armies (the Red Ball Express* notwithstanding) produced what has aptly been called

* The Red Ball Express—in which six thousand trucks ferried supplies along a dedicated one-way highway system in an around-the-clock stopgap resupply from the rear area depots in Normandy to the troops in the rapidly expanding front—was created on August 25. Its object was to deliver 82,000 tons of supplies between August 25 and September 2, and it dispensed 89,000 tons. MPs directed traffic, and convoys rolled in what has been mythologized as the savior of the front. In fact the Red Ball provided only a small fraction of the supplies required and took such a heavy toll in trucks and drivers

"the tyranny of logistics" in the late summer and autumn of 1944. But, instead of concentrating on the problems of resupply, Lee seemed more interested in winning the race with SHAEF to claim the best hotels and facilities in Paris, and in indulging his mania for creature comforts.[8]

With insufficient supplies, particularly of gasoline and diesel fuel, to support both thrusts fully, Eisenhower allocated priority to Montgomery, to the great annoyance and dismay of Bradley and Patton, whose role was reduced to that of a minor thrust into Lorraine. Patton and Bradley aligned themselves against both Montgomery and Eisenhower, whom they believed had sold out the U.S. Army to the British. In frustration Patton raged to Bradley, when a convoy of trucks drove up bearing rations: "I'll shoot the next man who brings me food. Give us gasoline; we can eat our belts."[9] One British officer recalls a story making the rounds that when Patton's artillery was rationed to two and a half rounds per gun per day, he "once asked plaintively what he was supposed to do with half a round!"[10] Eisenhower's decision seemed to Patton proof positive of his pro-British bias, and in the coming months Patton frequently reminded the Third Army staff they now had two enemies to fight: the Germans and SHAEF, whose boss was the best general the British had.[11]

If the force had been composed of one nationality, Eisenhower might have agreed with Montgomery's plan. "But as things stood Eisenhower could not make his decisions solely on military grounds. He could not halt Patton in his tracks, relegate Bradley to a minor administrative role, and in effect, tell Marshall that the great army he had raised in the United States was not needed in Europe."[12]

The logistical pinch was felt immediately. The Normandy supply dumps were more than five hundred miles from Lorraine, and with the transportation system in western France all but destroyed during the pre–D Day campaign to isolate the Normandy battlefield, it could not be rebuilt overnight. (In the 1991 Gulf War, Gen. H. Norman Schwarzkopf's Desert Storm "Hail Mary" flanking maneuver during the hundred-hour ground campaign was considered logistically impossible but turned out to be "logistically brilliant," thanks to Lt. Gen. William G. Pagonis, the chief Allied logistician.[13]

---

that it could not have been sustained indefinitely. Severe overloading added to the wear and tear, creating a tire shortage and removing thousands of vehicles from service and into a maintenance system likewise incapable of keeping pace. This "non-stop conveyor belt of trucks . . . represented a calculated gamble that the war would end before the trucks broke down. . . . Red Ball Express itself consumed 300,000 gallons of precious fuel every day—nearly as much as a field army." The Red Ball continued in a modified form until the pursuit ended—a heroic but ultimately futile effort to keep the wheels from coming off the great Allied war machine, which had become a casualty of its own spectacular triumph. (From Christopher C. Gabel, *The Lorraine Campaign* [Fort Leavenworth, Kans.: Combat Studies Institute, 1985]).

Had Pagonis been another Lee, the maneuver could not have been undertaken. Where Schwarzkopf was blessed with Pagonis, Eisenhower was cursed with Lee.[14])

The crisis faced by the Allies in September 1944 had nothing to do with a shortage of supplies. Sufficient stocks of virtually all classes of supplies were already positioned in France to fill, for example, 90 to 95 percent of the fuel and supply needs of the forward armies. The dilemma was that they were in Normandy. The shortages facing the Allies at the end of August 1944 were not of fuel or ammunition, but transportation. As Daniel Yergin writes, the decisions made at this critical moment of the war were part of what Patton called the "unforgiving minute" of history, and once taken could not easily be reversed. "No one realizes [its] terrible value . . . except me," he seethed in indignation. "We got no gas because, to suit Monty, the First Army must get most of it. . . . Some way I will get on yet."[15]

The advancing Allied armies were consuming fuel at the rate of some 800,000 gallons per day, 350,000 of which were expended by the Third Army. At the end of August, Patton reckoned that the Third was short at least 140,000 gallons of fuel, and momentarily wondered if it was deliberate, before dismissing the notion of a conspiracy against him. His supplies on August 30 were barely a tenth of the Third Army's needs. The supply situation in the Allied armies was not enhanced by the diversion of a great many trucks to deliver food to the people of Paris, which had been liberated on August 25. Lee ostentatiously moved his hugely overstaffed headquarters of 8,000 officers and 21,000 enlisted men to Paris, where he took over 296 hotels, whose beds were still warm from their previous long-term guests, the Germans. In the process he required the use of an inordinate number of vehicles and consumed more than 25,000 gallons of gas that might have been put to better use than feeding his delusions of grandeur, which included routinely sending a bomber to North Africa to fetch fresh oranges for his breakfast table. The French rightly complained that the demands of the Americans in Paris were greater than those of the Germans.[16]

Lee's communications zone (COM Z) became an empire dedicated to his own aggrandizement. Instead of moving desperately needed winter clothing to France in September, Lee's priority was the shipment of tons of prefabricated housing for his officers and men. To the outrage of Bradley and Patton, Lee demanded thirteen infantry battalions to protect his supply dumps from black marketeers (Eisenhower gave him five, over Bradley's objections), and whenever disputes arose, he used his position as deputy U.S. theater commander to overrule even Bradley. Delivery was often so shoddy that the armies were obliged to send their own vehicles to pick up supplies that COM Z insolently claimed to have delivered.[17] Despite Patton's intense dislike of Lee, he and others were reluctant to cross him. When

Lee visited Third Army in October he was greeted by an honor guard, the Third Army band, and treated to an elegant lunch.

After Normandy, Patton was determined to attack without letup, not allowing the Germans to regroup and defend their borders. Before his gasoline and ammunition supplies were seriously cut, Patton envisioned rushing through the Siegfried Line. "I was certainly very full of hopes that day [in mid-September] and saw myself crossing the Rhine." Unlike the SHAEF G-2, the gullible Maj. Gen. Kenneth Strong, Col. Oscar Koch, Patton's own intelligence officer, was under no illusion that the German army was finished, and attempted to convince Patton. Initially Koch was rebuffed by Patton, who was convinced that the fabled *Westwall* (Siegfried Line) was an empty shell capable of being easily breached, but only if the Third Army was permitted to drive rapidly through Lorraine *before* the Germans could reinforce it. Patton persuaded a sympathetic Bradley to permit the Third Army to thrust toward the Moselle River, hoping that when it came to supply allocations, Eisenhower would be unable to ignore a *fait accompli.*

At a meeting with Eisenhower, Bradley, and Hodges at Verdun on September 2, Patton exaggerated the extent of Third Army's advance, noting that he had patrols already at the Moselle, (but untruthfully claiming one was also in the fortress city of Metz), and pleaded for his authorized share of the logistical spoils, arguing: "We could push on to the German frontier and rupture that goddamned Siegfried Line. I'm willing to stake my reputation on that." Eisenhower teased his friend: "Careful, George, that reputation of yours hasn't been worth very much." Patton grinned and shot back: "That reputation is pretty good now."[18] Eisenhower temporarily raised Patton's allowance and permitted an advance toward Mannheim and Frankfurt. When he learned of it, an irate Montgomery complained so bitterly to Eisenhower that priority should be given to the British advance that Ike put his hand on the Englishman's knee and said: "Steady, Monty, you can't talk to me like that. I'm your boss." When Patton learned what Montgomery had done, he became more determined than ever to get the Third Army even more deeply involved, and asked Bradley to avoid calling him for at least forty-eight hours.

In light of Lee's disgraceful performance, it is difficult to criticize the Third Army's brazen and unorthodox methods. Patton used every subterfuge in his bag of tricks to keep them rolling, including diversion of gasoline and ammunition destined for the First Army. Frequently Patton's troops obtained vital supplies by purporting to be from the First Army, and trucks of the Red Ball Express were mysteriously diverted to Third Army supply dumps.

Patton officially disavowed members of his army passing themselves off as First Army troops. Privately their initiative delighted him, particularly after a battalion of gas tankers were diverted to Third Army dumps by men disguised as First Army troops. "To reverse the statement made about the

Light Brigade," he wrote, "this is not war but is magnificent."[19] Although probably apocryphal, a story circulated that Patton himself had stood at a major intersection and ordered Red Ball trucks diverted to the Third Army. Even if untrue, it was nevertheless a tale Patton would have relished.

His confidence in his G-4, Col. Walter J. "Maud" Muller, never wavered, even during the gasoline crisis. Initially, within Third Army neither gasoline nor ammunition was rationed, as was the case in other armies. Patton instructed his forward units to keep rolling until they ran out of fuel, and not to worry about being immobilized. He had noticed the enormous quantities of ammunition left on the battlefields of North Africa and Sicily, and he told his artillerymen to shoot it up whenever they obtained it.[20] The Third Army temporarily overcame the shortage of ammunition; however, as the crisis deepened even Patton was forced to order vehicles immobilized in order to keep the remainder operational.

Muller and his supply officers resembled Ali Baba and the Forty Thieves. They became the masters at hijacking supplies and sending out raiding parties to relieve First Army supply dumps of whatever was not nailed down.[21] Light aircraft were sent on reconnaissance missions to locate fuel dumps suitable for "attack." Yet even Muller could pull just so many rabbits out of his hat, and he warned Patton that something would soon have to give.

Muller was a legendary scavenger, and his supply officers were regarded "somewhat as licensed pirates. They were understood to rove the rear areas as far back as the ports, where it was said that they not only misrepresented themselves as coming from other armies but brought with them truckloads of souvenirs—German flags and weapons—for purposes of barter, garnering thereby steaks and fresh eggs as well as gasoline and ammunition. For better or worse, Third Army was widely believed to live higher off the hog than the others."[22]

When the First Army complained of being ripped off by Third Army officers, Patton blandly replied: "I'm sorry to learn that First Army has lost some of its gas, very sorry indeed. But I know none of my officers would masquerade as First Army officers. They wouldn't stoop to that, not even to get gas SHAEF stole from us." It was coincidental, of course, that soon afterward several clever young quartermaster officers just happened to receive promotions.[23]

As a British historian of the Lorraine campaign has written: "A good G-4 required the hoarding instincts of a hamster and the rapacity of a bird of prey." Many complained about Muller's activities, but with Patton the complaints fell on deaf ears. Muller conveniently "forgot" to report captured stocks, which he used to augment the Third Army's dwindling stocks. On two occasions one hundred thousand gallons of precious gasoline were captured, which enabled the Third Army to drive as far as the Marne and later to the Meuse. Significant as these numbers may sound, one hundred thousand gallons was barely sufficient to keep a single armored division in combat for twenty-four hours, although with Patton's rigid constrictions on fuel consumption, that amount became a find of mixed blessings. When Capt. Jimmie Leach, the commander of B Company, 37th Tank Battalion (4th Armored), reported to his battalion commander, Lt. Col. Creighton W. Abrams, that he had gassed his tanks with captured German fuel, Abrams exploded. German gasoline was of a lower synthetic octane and tended to foul the engines of Sherman tanks accustomed to burning a better grade of fuel. Even godsends had their price.

Although the Third Army demonstrated a blatant disregard for the rules of the game, Patton believed that his army would never obtain its fair share unless unorthodox methods were employed. The plain fact is that combat troops couldn't care less about the source of their supplies. Patton's attitude was: "To hell with Hodges and Monty," and he bellowed to Bradley one day: "We'll win your goddam war if you'll keep Third Army going!" To the victor went the spoils, and Patton was not taking a backseat to anyone else in what had become the hottest competitive aspect of the war between the generals. To emphasize his shortage of gasoline, Patton's command vehicle regularly coasted into the 12th Army Group command post on empty, so that Sergeant Mims could gas up in Bradley's motor pool.

The Third Army's conspicuous indifference to playing by the rules was all too frequently counterproductive. Many "midnight requisitions" were self-destructive. For example, the trucks used to resupply the Third Army often ran on hijacked fuel. Shoddy record keeping throughout the system often resulted in delivery of supplies that were neither ordered nor required. Worst of all, with no time for maintenance, trucks were literally driven into the ground by the strain of long-distance logistical support. Considerable clothing was discarded by Third Army units and had to be retrieved by salvage units: when winter came it would be desperately needed. Most fuel was delivered in five-gallon jerry cans that became the heart of the system, yet by the end of August more than half the 22 million jerry cans in France had already been lost.[24] Nor did the Third Army help itself with its shoot-first attitude, that may have benefited its supply situation. "Tank crews found the shooting up of railway engines a rewarding target . . . [during] those halcyon days of pursuit."[25]

When he was not lobbying for help for his quartermasters, Patton spent most of his time taking the pulse of the Third Army. His popularity among his troops had never been higher, but Patton's paternalistic attitude toward the men of the Third Army managed to infuriate other American commanders. Hugh Cole, a member of the Third Army historical section and later the official U.S. Army historian of the Lorraine and Ardennes campaigns, notes that Patton

> looked upon the soldier as a man who not only had special obligations, but special rights as well. . . . He was never concerned about looting unless it was contrary to a specific order. As a result Third Army had a bad reputation, and once I was actually shaken down after spending a night in a 7th Army billet on the assumption that I must have stolen something. . . . [Patton] had troops hung [sic] for rape, just as he had troops shot for cowardice in front of the enemy (this despite all the later business about only one man [Pvt. Eddie Slovik] being executed in World War II). But he expected that the soldier would find feminine companionship (was entitled to it) and always kept the brothels open in any town we were in.[26]

Patton pragmatically believed that "in my experience it is futile to attempt to go against human nature. If we close these houses to our troops we will simply produce a flood of illicit prostitution which will be more difficult to control and which will cause greater infection." But when Patton wrote to Eisenhower to suggest that the new wonder drug, penicillin, be employed to help keep the brothels safe for his soldiers, he received a frosty reply pronouncing the idea "absolutely unacceptable to me."[27]

Although Patton was privy to the Ultra secret, until early August 1944 he had no idea that there existed a small, secret, special intelligence section attached to Third Army HQ which processed information passed from Bletchley Park. Thereafter the Ultra officer, Maj. M. C. Helfers, briefed Patton each morning. Helfers recalls the morning Patton was informed of a new order from SHAEF which directed there be no fraternization with Germans when the Allies reached Germany. Patton turned in his chair, grinned, and said, "You can tell the men of Third Army that so long as they keep their helmets on, they are not fraternizing."[28]

In October, Marshall visited Third Army where he decorated a soldier with the Silver Star for breaking up a German counterattack by destroying three Tiger tanks with a bazooka. After Marshall had finished, Patton asked if he might be permitted to make a presentation of his own. "On the spot Georgie composed an inspiring citation of his own and presented the soldier with the DSC which his aide quickly supplied. Can you wonder that the men of the Third Army were willing to follow Georgie to the end of the world?" wrote a brigadier general after the war.[29]

Patton's detractors allege that his actions were principally designed to enhance his own reputation. Few are aware of the small but important things he did virtually every day. After the capture of Metz, for example, Patton visited a hospital and stopped by the bed of a GI to inquire if he was aware that the city had fallen. When the soldier replied that he was, Patton grabbed his hand and smiled back: "Tomorrow, son, the headlines will read—'PAT-TON TOOK METZ!' which you, too, know is a damned lie, because you and your buddies are the ones who actually took Metz!"[30] When the Third Army captured fifty thousand cases of champagne, Patton had it distributed to his troops.

Although he was often uneasy mingling with his men, no one doubted his sincerity. No one can sniff out a pretender faster than a soldier, and no matter what the men of Third Army thought of him, most were pleased that he was their commanding general. Bradley was little known, and within First Army, few could even identify Courtney Hodges as their commanding general, much less be influenced or inspired by him. For, as historian Charles B. MacDonald (a decorated World War II infantry company commander) relates: "A front-line soldier was immensely well informed if he knew the name of his company commander, who had just arrived the day before to replace that other one who had lasted only a week."[31]

Patton's encounters with his soldiers were often memorable. On a highway near Metz, he stopped a lone GI marching down the road, and asked what the trouble was. The soldier replied that his unit had lost most of its officers and NCOs and the Pfcs were now running things. "What's wrong with that," asked Patton? "Plenty! Why, General, those sons-of-bitches are drunk with power."[32]

When Patton inquired of a lieutenant how his platoon had managed to make a successful assault crossing of the Moselle, he was told that they merely went to sleep the night before, awoke before dawn, and carried out their orders, thus making their fifth successful river assault, just as they had countless times during their training at Fort Benning. Patton barked to the division commander: "Relieve that lieutenant immediately! Never let him lead an assault crossing again, and goddamn it, promote him to captain today! How the hell can we learn about war when a hero like that makes it so simple!"[33]

Another time, when he learned that a crew had bailed out of their disabled B-17 bomber and were at a nearby railroad station, Patton immediately went there and decorated all nine on the spot with the Bronze Star. "You've done a swell job," he said before ordering Codman to escort them back to their base in England in the new C-47 sent to him by the Eighth Air Force commander, Tooey Spaatz. En route, one airman told Codman: "What a guy!" Another said: "I never thought I'd get to meet him. It was worth being shot down for."[34]

Evidence of his hand on the levers of command could be seen in a variety of ways. For example, Patton liked to leave the German dead on the battlefield, where fresh troops arriving in the Third Army for the first time would see them. Conversely, American dead were removed by graves registration teams as fast as humanly possible. He regularly called in company commanders to learn about their tactical experiences, and his letters of instructions issued in England before D Day were based partly on what he had learned from small-unit commanders after Sicily.[35]

And, of course, he continued to flirt with death. One night in Nancy the Germans began a heavy shelling, and one artillery shell struck the house across the street from Patton's quarters. He went to the rescue of a French couple trapped in the ruins. The man was freed, and while his wife screamed a French friend of Patton said: "I implore you, Madame, do not·derange yourself, be calm, be tranquil. Try to realize that the great General Patton is himself occupying himself with the removal of the bricks so that you, too, may be saved." Just then more shells arrived and deposited rocks on Patton, who said later, "I really believe that I was more frightened that night than in any time in my career."[36]

Patton still took unnecessary risks to appease his demons, which demanded that he continually test his courage. In late September, Patton, Codman, and Mims drove to the front for "a really forward look." Soon they were deep into "fairly fluid territory," and Mims began slowing down but was told to keep going, until the three finally halted in ominously quiet terrain well beyond the thinly held front lines. Patton seemed disappointed that there was nothing to see, and no action. Codman wondered why Patton did it. He could understand his need to set an example for his officers and men, but this particular foray made no sense. "I believe this type of chance-taking," he wrote his wife, "springs from a need, a compulsion on the General's part to compensate for the almost unbearable sense of frustration induced by enforced inaction and passivity."[37]

Both to see and be seen, Patton traveled in an open jeep that was equipped with a loud klaxon and three big stars affixed to the front and rear. It was common for him to return from a visit to the front and admit that he had been frightened. Yet in keeping with his inviolate rule that *no* general was ever permitted to exhibit fear, these admissions always came after the fact.[38] When he returned to his command post each evening, he made sure no one saw him leave. He also required an officer from each staff section in Third Army to visit the front each day.

One afternoon, after dismissing his MPs at a front-line regiment, Patton and Mims headed home in a jeep by themselves. At an intersection Patton misread his map, and they took a wrong turn and soon found themselves passing engineers and infantrymen who gave them strange looks. Patton finally said: "I think we turned wrong, Sarge." Just then rifle fire erupted all around them and Patton exclaimed: "By God, I know we have, turn this thing around," and they beat a hasty retreat. It was as close as Patton ever came to capture.[39]

*   *   *

It has been alleged that Patton condoned atrocities by the Third Army. He believed in the unwritten soldier's code that anything was fair in battle, but once captured, a soldier was entitled to be treated decently under the covenants of the Geneva Convention. If blatant atrocities were brought to his attention, Patton would take action to punish the offenders. However, when it came to the dross of the Third Reich, such as the Gestapo, Patton was fully prepared to turn a blind eye. In mid-August, for example, he wrote in his diary: "We captured a lot of Gestapo alive at Orleans which was a blunder."[40] In his eyes these men were not soldiers and were not entitled to humane treatment. Patton was always militarily correct when an officer was brought before him, but when the Third Army captured a paunchy SS general in Metz, he required the officer to stand at attention and deliberately used a Jewish lieutenant as his interpreter. Patton wore combat boots, carefully noting: "I always wear high boots when I talk to SS bastards. You will stand at attention when I speak to you and you will preface every answer to me with 'Sir.' I have captured a number of German generals but you are the most sordid son-of-a-bitch of them all. I ought to turn you over to the French." After excoriating the man for the crimes of the SS, he ordered: "Take the dirty bastard away."[41] Although Patton never lost his lifelong anti-Semitism and bias against blacks and other minorities, Hugh Cole points out that "when manpower got tight, Patton was the first—although this has now been forgotten—to integrate colored and white troops in the rifle companies."[42]

Patton's restraint toward his corps commanders was stronger than might be imagined. He rarely interfered, viewing his job as keeping the morale of his subordinate commanders and his troops high by means of personal appearances at the front, and in making the best use of his now legendary reputation as a fierce warrior. He used one gimmick to great effect. For example, if Robert Grow of the 6th Armored attended a briefing, Patton would deliberately praise Wood's 4th Armored ("I can still see the back of Grow's neck turning brick-red," writes Hugh Cole), and when Wood was in attendance, he would extol Grow's or some other armored division.[43] It usually worked effectively. Patton visited Verdun, which he found a "magnificent, though futile monument to heroism. You can see . . . where brave men died to maintain something they could have saved much more by attacking. To me [Fort] Douaumont epitomizes the folly of defensive warfare."[44] And on September 26 he returned to Bourg, the site of his tank training center in 1918. The first person he met on the street was a peasant who was "standing on the same manure pile on which he undoubtedly stood in 1918. I asked him if he had been there in the last war and he replied, 'Yes, General Patton, you were then here as a Colonel.'" Patton was delighted. "We had an Old Home Week. . . . I visited my old office, my billet, and chateau. . . . A day full of memories."[45] Ten days earlier he and Codman had visited the Saint-

Mihiel battlefield on the date the battle had ended twenty-six years earlier. Codman wondered aloud, if Patton had ever imagined in the heat of combat that he would be passing the same location in command of an army marching to the Rhine, what he would have said. "I would have said," replied Patton, "an intelligent and far-sighted prediction, because that is exactly what I aim to do."[46]

Despite his lapses of courage and the bad smells he gave off, Willie remained a centerpiece of Patton's life. In mid-November Eisenhower visited Third Army, and arrived driven by Kay Summersby, who later regretted having brought along Ike's Scottie, Telek. At lunch Willie was banished and Telek was under the table between the two generals, when suddenly war broke out beneath their feet. Willie had wandered in and found Telek occupying *his* place of honor. Kay wrote: "Patton let loose with every curse in his celebrated vocabulary. It was classic, that tirade. . . . I was terrified for Telek. It took four generals, the Theater's top brass, to separate Willie and Telek. And even then they had to throw water on the fighters." Patton banished Willie and profusely apologized, but Eisenhower insisted: "This is Willie's home. We should lock up Telek." Shaking his head, Patton replied, "No Sir! Telek outranks Willie, so Telek stays right here. Willie is confined to quarters, under arrest. That's Army protocol." Then, unable to contain himself, Patton bellowed, "But my Willie was chewing bejesus out of your Gawdamned little Scottie—rank or no rank!"[47]

Patton's earlier troubles left him nervous around the correspondents, and he opened a press conference on September 7 with: "Before starting the inquisition, I wish to reiterate that I am not quotable, and if you want to get me sent home, quote me, God damn it." To a question about supply, he said: "we cannot make five barley loaves and three small fishes expand like they used to."[48]

A bevy of USO entertainers did their best to buck up Third Army morale. Bob Hope, Bing Crosby, and Dinah Shore all passed through. In the eyes of the press, Patton was right up there with these elite when it came to entertainment. Although Patton's association with the press was tempestuous, they often found him hugely entertaining. On earlier occasions he had reduced Bob Hope, Jack Benny, and Al Jolson to the roles of straight men. Even though leery and prone to make remarks about their resemblance to inquisitors, at his September 23 weekly press conference Patton was in rare, colorful form. His dialogue was snappy, and after poking fun at Lee's Services of Supply, Patton turned to the subject of Monty and began to spoof the field-marshal's "lightning dagger-thrust" to the Ruhr. In a mock English voice, he aped Monty, saying, the Germans "will be off their guard, and I shall pop out at them like an angry rabbit." The reporters howled. Codman encountered one correspondent still weak with laughter and asked if he enjoyed the briefing. "Enjoy it? Listen, buddy, that guy in there, all by him-

self, without benefit of high-priced writers, music, or scenery—that guy is
EIGHTY-EIGHT ENTERTAINMENT."[49] Some weeks later at a press conference held in Luxembourg, Patton opened by pointing to the German positions on a map. "First of all, may I point out where these sons-of-bitches are
holding us up." Suddenly noting that there were two women correspondents
in his audience: Patton said, "I'm sorry ladies. Now let me show you where
these bastards are holding us up."[50]

Once, when Patton was asked about his plans in early September, Cornelius Ryan, then a *Daily Telegraph* correspondent, recalled, "In his high-
pitched voice and pounding the map, Patton declared that, 'Maybe there are
5,000, maybe 10,000 Nazi bastards in their concrete foxholes before the
Third Army. Now, if Ike stops holding Monty's hand and gives me the supplies, I'll go through the Siegfried line like shit through a goose.' "[51]

Marshall was well pleased by what he saw at the Third Army. Before
departing, he said "Well, George, you are doing a grand job. Now, is there
anything I can have sent over to you, or anything I can do for you?" Patton
grinned and replied, "Yes, there is one thing you can do for me and all this
fine staff of mine, General. After we have finished up over here, you can
send us all to the Pacific to clean up those little yellow bastards over
there!"[52]

Patton observing a battle near
Maknassy, Tunisia, late March
1943. *(Photo courtesy of the
Patton Museum)*

The Seventh Army
commander and his chief
of staff, Brig. Gen.
Hobart R. "Hap" Gay,
aboard the USS *Monrovia*,
off the coast of Sicily,
July 11, 1943. *(Official
U.S. Army photograph,
National Archives)*

Patton on the beach at Gela, Sicily, July 12, 1943. *(Photo courtesy of the Patton Museum)*

Although critics of Bernard Montgomery and George S. Patton have disparaged both generals as prima donnas intent on enhancing their own reputations, each was among the most consummate professional soldiers ever produced by their respective nations. Here Montgomery is visiting Patton's Seventh Army HQ in Palermo, July 25, 1943, to refine their mutually agreed strategy for capturing Messina and ending the Sicily campaign. *(Official U.S. Army photograph, National Archives)*

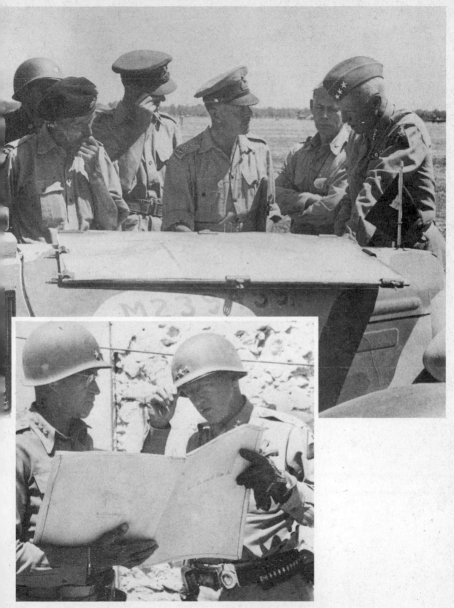

Maps are one of a commander's most important tools, and a great deal of Patton's day revolved around their use, as shown in the two examples here. *Top:* Montgomery explaining his strategy for the capture of Messina to Alexander, Bedell Smith, and Patton on July 25, 1943. *Bottom:* Patton and Bradley. *(Official U.S. Army photographs, Dwight D. Eisenhower Library)*

Patton's victorious entry into Messina on August 17, 1943, marked the end of the Sicily campaign. However, after it was learned that he had slapped two GIs in hospital wards earlier that month, Patton's triumph was short-lived. *(Official U.S. Army photograph, National Archives)*

Although Eisenhower did not specifically order it, Patton nevertheless apologized to every soldier in his army. He is shown here apologizing to the 1st Infantry Division, August 27, 1943. *(Official U.S. Army photograph, Dwight D. Eisenhower Library)*

Patton's lonely exile in Sicily in the autumn and winter of 1943 was briefly forgotten when Roosevelt made a stopover in Sicily in December. *(Official U.S. Army photograph, Dwight D. Eisenhower Library)*

While her husband was at war, Beatrice contributed on the home front by delivering radio broadcasts to reassure the families of military personnel of the importance of the service their loved ones were performing. Here Beatrice records a message to the women of France for broadcast by short-wave radio on November 8, 1943, the first anniversary of the Allied landings in French North Africa. *(Photo courtesy of the Patton Museum)*

With Maj. Gen. Geoffrey Keyes, Italy, January 1944. *(Photo courtesy of the Patton Museum)*

Eisenhower, Patton, Bradley, and Lt. Gen. Courtney H. Hodges in England before the invasion, 1944. *(Photo courtesy of the Patton Museum)*

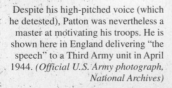

In 1944 Patton acquired Willie (his full name was William the Conqueror), a lovable but rather cowardly bull terrier who was his faithful companion until his death. *(Hulton Deutsch Collection)*

Despite his high-pitched voice (which he detested), Patton was nevertheless a master at motivating his troops. He is shown here in England delivering "the speech" to a Third Army unit in April 1944. *(Official U.S. Army photograph, National Archives)*

When Bradley's Operation Cobra unexpectedly
succeeded in late July 1944, Third Army poured
through the Avranches gap to carry out what
became one of the greatest pursuit operations
in American military history.
(Photo courtesy of the Patton Museum)

With the 4th Armored Division
commander, Maj. Gen. John S. Wood.
(Photo courtesy of the Patton Museum)

Checking German dispositions with
Maj. Gen. Hugh J. Gaffey during the
drive toward the Seine in August
1944. Patton required glasses for
reading but was rarely photographed
wearing them. *(Official U.S. Army
photograph, National Archives)*

Maj. Gen. John P. Lucas was a lifelong friend and confidant. He commanded VI Corps during the first weeks of the ill-fated Anzio campaign in 1944.

Maj. Gen. Manton S. Eddy commanded the 9th Division in North Africa and Sicily, and XII Corps in northwest Europe.

Maj. Gen. Troy H. Middleton commanded the 45th Division and VIII Corps under Patton. Gen. George C. Marshall rated Middleton the most outstanding regimental commander in France during World War I.

Field Marshal Sir Harold Alexander. *(Hulton Deutsch Collection)*

The Fifth Army commander and Patton's rival, Lt. Gen. Mark Clark.

(All Official U.S. Army photographs, National Archives, except for the photo at left)

With Bradley and
Brig. Gen. Otto P. Weyland,
commander XIX
Tactical Air Command.
Willie, as usual, is asleep.
*(Photo courtesy of the
Patton Museum)*

With members of the
French Resistance, August 1944.
*(Official U. S. Army photograph,
National Archives)*

With Sgt. John L. Mims in Normandy, 1944. As Patton's driver
from 1941–45, Mims safely
transported Patton thousands of
miles in the United States,
North Africa, Sicily, England,
and northwest Europe. He
departed Germany in the
autumn of 1945, shortly before
Patton's disastrous automobile
accident. *(Photo courtesy of the
Patton Museum)*

War correspondents were well treated in Third Army and kept apprised of military operations by briefings such as this one in France in 1944. *(Photo courtesy of the USMA Library)*

A visibly glum Patton with members of his staff and Maj. Gen. Walton H. Walker *(right)*, taken shortly after Third Army was ordered on the defensive along the Moselle River in Lorraine. Patton is flanked by Brig. Gen. Paul D. Harkins *(left)* and Brig. Gen. Otto P. Weyland. Willie awoke long enough to pose with his master. *(Official U.S. Army photograph, Dwight D. Eisenhower Library)*

Patton and Brig. Gen. Anthony C. McAuliffe, shortly after elements of Third Army linked up with the 101st Airborne Division to end the siege of Bastogne. When Patton learned McAuliffe had replied "Nuts" to a German surrender demand, he said: "Any man who is that eloquent deserves to be relieved [saved]. We shall go right away."*(Official U.S. Army photograph, National Archives)*

Bradley, Eisenhower, and Patton in the ruins of Bastogne, February 1945. *(Photo courtesy of the USMA Library)*

Patton congratulating engineer troops for bridging the Rhine, March 1945. *(Official U.S. Army photograph, National Archives)*

Patton urinated in the Rhine River at Oppenheim, March 24, 1945. "I have been looking forward to this for a long time," he said. *(Photo courtesy of the Patton Museum)*

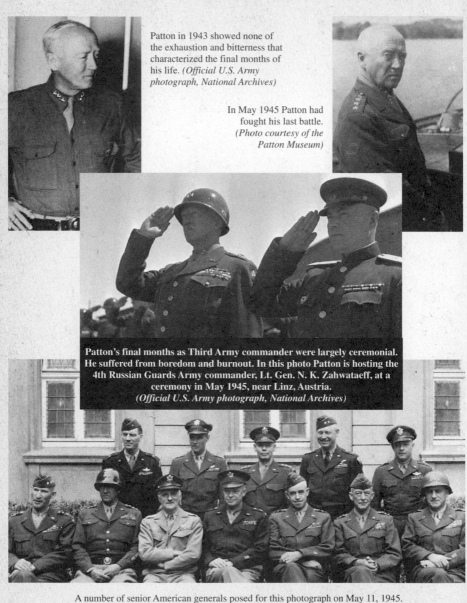

Patton in 1943 showed none of the exhaustion and bitterness that characterized the final months of his life. *(Official U.S. Army photograph, National Archives)*

In May 1945 Patton had fought his last battle. *(Photo courtesy of the Patton Museum)*

Patton's final months as Third Army commander were largely ceremonial. He suffered from boredom and burnout. In this photo Patton is hosting the 4th Russian Guards Army commander, Lt. Gen. N. K. Zahwataeff, at a ceremony in May 1945, near Linz, Austria.
(Official U.S. Army photograph, National Archives)

A number of senior American generals posed for this photograph on May 11, 1945.
Back row, left to right: Brig. Gen. Ralph Stearley, Lt. Gen. Hoyt Vandenberg, Lt. Gen. Walter Bedell Smith, Maj. Gen. Otto P. Weyland, and Brig. Gen. Richard Nugent. *Seated, left to right:*
Lt. Gen. William H. Simpson, Patton, Gen. Carl "Tooey" Spaatz, Eisenhower, Bradley, Hodges, and Lt. Gen. Leonard T. Gerow.*(Official U.S. Army photograph, Dwight D. Eisenhower Library)*

One of the last photos of the Pattons, June 1945. *(Photo courtesy of the USMA Library)*

Beatrice Patton greeting her husband at Hanscom Field, near Boston, June 1945. *(Photo courtesy of the Patton Museum)*

Patton during a ticker-tape parade in Los Angeles, June 1945. *(Official U.S. Army photograph, National Archives)*

Patton and Jimmy Doolittle at a press conference in Los Angeles, June 1945. *(Official U.S. Army photograph, National Archives)*

Patton was succeeded as Third Army commander by Lt. Gen. Lucian K. Truscott, an old friend and former subordinate in Sicily. *(Official U.S. Army photograph, National Archives)*

In Salzburg, Austria, Patton rode a magnificent Lippizaner stallion named Favory Africa, which Hitler himself had chosen as a gift for Japan's Emperor Hirohito. *(Photo courtesy of the Patton Museum)*

Patton's grim mood in the summer of 1945 was painfully evident in this photograph taken during his reunion in Los Angeles with his sister Nita *(center)*. *(Photo courtesy of the USMA Library)*

With grandchildren, summer 1945. *(Official U.S. Army photograph, National Archives)*

The last known photograph of Patton and Eisenhower, at an intra-army football game in Frankfurt, Germany, on October 15, 1945. Both men managed to conceal behind smiles the dissolution of their twenty-six-year friendship after Eisenhower relieved Patton from command of his beloved Third Army. *(Photo courtesy of Ernest C. Wilbur Jr.)*

In November 1945 Patton traveled to Sweden for a reunion with the Swedish Olympic team, against which he had competed in 1912. The trip included the inspection of an infantry regiment in Uppsala, where this photo was taken. *(Photo courtesy of the USMA Library)*

Patton's casket is carried from the railroad station in Luxembourg to its final resting place, the nearby American military cemetery at Hamm. Beatrice is escorted by Lt. Gen. Geoffrey Keyes. The pallbearer on the front left is Patton's orderly of eight years, M/Sgt. William George Meeks. *(Official U.S. Army photograph, National Archives)*

Patton's original grave site at Hamm, Luxembourg.
(Both photos this page are official U.S. Army photographs, National Archives)

Willie.

CHAPTER 42

A Sea of Mud and Blood
The Lorraine Campaign

> We roll across France in less time than it takes Monty to
> say "Regroup" and here we sit stuck in the mud of
> Lorraine.
>
> —PATTON

One of the most fought-over areas in military history, Lorraine was the shortest gateway to Germany and had for centuries been a traditional route of invading armies. After the pursuit across southern Normandy, the only logical employment of the Third Army was to advance into Lorraine. More important, it was in keeping with Eisenhower's blueprint to advance on a broad front, killing Germans and eliminating their fighting units this side of the Rhine, a strategy Bradley endorsed. With German forces in total disarray at the end of August, a virtually undefended Lorraine beckoned like the Rhine sirens of mythology.

When the Third Army (minus Middleton's VIII Corps, which was still embroiled in Brittany), rolled into Lorraine at the end of August, it was spearheaded by Walker's XX Corps in the north and Eddy's XII Corps in the south. Walker was to drive from Verdun toward Metz and gain a foothold across the Moselle, while Eddy launched a parallel advance on Nancy. Not until mid-September, when Eisenhower permitted Patton to carry out a reconnaissance in force and to realign his units along the Moselle, was either corps able to resume the offensive. Nancy, the provincial capital of Lorraine, had fallen, but the fortress city of Metz remained in

German hands, although it was virtually undefended at the end of August. During the Third Army's enforced pause in the first week of September, the Germans hastily assembled a force of infantry and panzers to defend the Moselle—in the words of the official U.S. Army historian, "the strongest of the German armies in the West."[1]

When the Third Army captured Verdun in early September, Third Army was barely thirty-five miles west of Metz, and a mere seventy miles from the temporarily unmanned Siegfried Line. On August 31, Third Army had received just 31,000 gallons of the 400,000 Patton had requested for his forward fuel dumps. Exasperated but still optimistic, he bounded into Bradley's CP and begged: "Dammit, Brad, just give me 400,000 gallons of gasoline and I'll put you inside Germany in two days." Bradley would have obliged, but even he lacked the power to produce the precious gasoline everyone prized more than gold.[2] Bradley later wrote that he, too, thought that given even a modest increase in fuel, the Third Army would certainly have gone farther, probably well into the Saar: "Three months and many casualties later we were to be forcefully reminded that in war, opportunity once forsaken is opportunity lost forever."[3] Patton was reduced to ordering his commanders to go as far as they could with what they had and then to get out and walk!

Patton raged at Montgomery for the loss of supplies and fuel to the British, seemingly never grasping that his destiny was actually in the hands of Eisenhower and the logisticians. Nor did he understand that Eisenhower felt a strong aversion to having to halt his friend, and when Antwerp fell on September 4, specifically noted the importance of getting Patton on the move again.

Aware of the inevitability of his advance being halted by Eisenhower, Patton resorted to what he called the "rock soup" method of advance through Lorraine, likening it to the example of a tramp who went to a house and asked for some boiling water with which to make rock soup. "The lady was interested and gave him the water, in which he placed two polished stones. He then asked if he might have some potatoes and carrots to put in the soup to flavor it a little, and finally ended up with some meat. In other words, in order to attack, we had first to pretend to reconnoiter, then reinforce the reconnaissance, and finally put on an attack—all depending on what gasoline and ammunition we could secure."[4] It was, Patton lamented, "a very sad method of making war." Although entirely sympathetic, Bradley was not optimistic that the Third Army's supply situation would improve.[5]

However, it did improve for a brief period beginning on September 5, after the Third Army had been all but immobilized for five days. Between August 26 and September 2, the Third Army averaged 202,382 gallons of gas per day. But on September 2 the well ran dry when the entire army received a mere 25,390 gallons. To resume the offensive east of the Moselle, at least 450,000 gallons a day would be needed. Between September 4–7 the drought ended in a sea of 1,637,000 gallons, and Patton was given the green light to mount a fresh drive toward the Siegfried Line and Frankfurt.[6] The fresh infusion of fuel lasted a mere ten days.

Whenever the Third Army seemed about to run out of gasoline, Patton would intervene to cajole and bargain in an effort to persuade Bradley to keep him going. However, on September 12 Bradley began applying the brakes, warning Patton that with the launching of Operation Market-Garden (the airborne invasion of Holland to seize a bridgehead over the Rhine at Arnhem) on September 17, the Third Army might well be stuck west of the Moselle indefinitely without ammunition or fuel. Patton pleaded: "Don't stop us now, Brad, but I'll make a deal with you. If I don't secure a couple of good bridgeheads east of the Moselle by the night of the 14th, I'll shut up and assume the mournful role of defender."[7] Bradley granted him two days, which proved ample time for Patton. And so it went. Whenever the final turn of the wheel seemed imminent, Patton would introduce some new scheme, until the day the taps were finally shut off and even Bradley could not get them turned back on. As late as mid-September, he continued to articulate orders indicating an advance by the 12th Army Group to the Rhine and the securing of a bridgehead between Mannheim and Cologne. Bradley also ordered an equal division of supplies between his armies, but it all came to naught when Eisenhower decreed that the priority of supply would go to Montgomery and to the U.S. First Army, whose drive on the British right flank was to screen the 21st Army Group drive into Arnhem in support of the airborne operation.

Eisenhower's only concession was to permit the Third Army to establish itself astride the Moselle (a move justified on defensive grounds—that Third Army's presence would help anchor the Allied right flank). As Eisenhower put it, Patton would have forty-eight hours "to become so heavily involved I might reconsider."[8] Later, when the supply situation grew desperate, Patton would continue to "edge" eastward in limited operations, rationalized by any reason he could conjure up to keep the Third Army moving: in other words, more "rock soup."[9] Eventually the Third Army simply sputtered to a halt in the inhospitable terrain of Lorraine. "Books will some day be written," he informed Beatrice, "on that 'pause which did not refresh' any one but the Germans."[10] Bradley did his best, repeatedly challenging Eisenhower on behalf of his armies, even warning that Market-Garden would fail.

Patton's dismay mounted at the end of September, when Eisenhower announced that he was inclined to make permanent the previous temporary attachment of Haislip's XV Corps to Jake Devers's 6th Army Group to

assist in the campaign in the Vosges Mountains of Alsace-Lorraine. Patton
was outraged to learn that he was to lose one of his favorite corps: "I am not
inclined to grumble or to think that the cards are stacked against me, but
sometimes I wish that someone would get committed to do something for
me. However, all my disappointments have turned out for the best. . . . I
should have more faith. If Jake Devers gets the XV Corps, I hope his plan
goes sour. The Lord is on my side, but he has a lot of getting even to do for
me. . . . Bradley and I are depressed. We would like to go to China and serve
under Admiral Nimitz."[11] When the transfer of XV Corps was later confirmed, Patton exploded: "May God rot his guts."[12]

The Allied failure at Arnhem and the lengthy battle to secure and open
the Scheldt estuary led to the so-called October pause, and to Patton's
lament to Everett Hughes: "At the moment I am being attacked on both
flanks, but not by the Germans." Others in the fifty years since the airborne
tragedy at Arnhem have speculated how differently Market-Garden might
have turned out had Patton commanded the land forces, which failed to link
up with the besieged airborne. Historian Rod Paschall writes: "Think what
might have happened if George Patton and not [Lt. Gen.] Brian Horrocks
had been at the helm of the [British] XXX Corps. Pity the Germans caught
between him and those British at Arnhem."[13]

As the weather worsened, so too did Patton's despondency and his dislike of Lorraine, "this nasty country where it rains every day and the whole
wealth of the people consists in assorted manure piles." The loss of XV
Corps was a heavy blow. "I decided to get drunk but could not. . . . If you
were here I would cry on your shoulder," he brooded to Beatrice. "Willie is
no use. Some times I almost think the fates are against me. . . . I took a long
walk alone—not even Willie[—and] put two officers in arrest for speeding
and felt better. If we had only Germans to fight war would be a joke but it
ain't! I love and miss you terribly."[14] His trips to the front helped cheer him
and once again nearly killed him when the Germans let loose a salvo of 150-
mm shells, one of which fell close enough to pitch dirt onto his jeep. Seconds later a dud shell landed only two feet from the running board.[15]

Field Marshal von Rundstedt was under no illusion where the greatest
danger came from. An astute student of warfare, Rundstedt "had found the
Americans more imaginative and daring in the use of armor, the British
superb with infantry. In each case, however, commanders made the difference. Thus von Rundstedt considered Patton a far more dangerous opponent
than Montgomery."[16]

Although desperately hard-pressed, the Germans well understood the
consequences if the Third Army were permitted to drive unchecked toward
the Saar. Two panzer grenadier divisions were hastily deployed from Italy
(one of which was Patton's nemesis from Sicily, the veteran 15th Panzer
Grenadiers) to defend the Moselle River at Nancy. The overall defense of

Lorraine was assigned to the Fifth Panzer Army, commanded by the able Gen. Hasso von Manteuffel, whose orders from Hitler were to counterattack and roll up the Third Army's vulnerable right flank. Fortunately for the Germans, the Third Army's gas tanks were running dry at the same moment as they were concentrating their defenses along the Moselle. The result was that by September 7 whatever opportunity existed to have driven uncontested to the Siegfried Line had passed, and both the Third Army's momentum and initiative were lost. Even the aggressive John S. Wood, whose 4th Armored had driven nearly four hundred miles since August 1, was reluctant to rush the Moselle, which turned out to be a formidable and stiffly defended terrain obstacle.

Nevertheless the river defenses were breached, and Nancy fell in mid-September. The heaviest battles were fought by Eddy's XII Corps just east of Nancy, around Lunéville and Arracourt in what Patton called "as bitter and protracted fighting as I have ever encountered. The Huns are desperate and are attacking at half a dozen places. . . . We have destroyed well over a hundred tanks and killed thousands. In one town . . . there were 1,600 dead [and] there was actually blood running down the gutter." At the forefront was the 4th Armored, which blunted a hastily mounted German counteroffensive to drive a fatal wedge into Patton's southern flank.

When Patton found Eddy discouraged and jumpy, he reminded him that Grant had once said: "In every battle there comes a time when both generals think themselves licked then he who is fool enough to keep on fighting wins. Manton you are a perfect fool so go in and win and he did. I told another general who had been kicked off a hill, 'take it by dark or be relieved'; he took it."[17] His tactics with his generals varied from cajolery or praise to threats. Occasionally Patton would tell an officer that he had such perfect confidence in him that to show it he was going home. When Eddy asked Patton to relieve one of his division commanders, Patton demurred, saying what was needed was to educate the general. It apparently worked, for Patton later praised him as one of his finest commanders.

The mounting problems between Eddy and the feisty Wood were another matter altogether. At Arracourt, the 4th Armored proved that it was as tough on the defensive and in counterattack as on the offensive, claiming the destruction of 281 German tanks, three thousand enemy killed, and another three thousand POWs. Unfortunately the 4th Armored's triumph sowed seeds of severe dissension between Wood and Eddy over the conduct of the battle.[18]

Virtually from the time the 4th Armored Division was assigned to XII Corps, a serious clash of both personalities and perspectives developed between Wood and the stolid, humorless Eddy. The volatile and imaginative Wood's penchant for bluntness did not sit well with his new corps commander. If ever there was an example of the philosophical and personal disparities between the aggressive armor generals and the Fort Benning–trained infantrymen, it was the appalling relations between these two men.

Their relationship had deteriorated to the point that, at Eddy's instigation, Patton was obliged to carry out the sad task of relieving one of his oldest and closest friends. Wood's relief came about because of two incidents, the first of which occured after he failed to carry out to Eddy's satisfaction an order to assist the 26th Division during an offensive in early November. Eddy complained to Patton, who warned Wood that he was verging on insubordination, and that he would have to relieve him unless Wood gave his assurance he would carry out both the letter and spirit of future orders.[19]

That Wood was an outstanding division commander whose genius burned brightly is beyond dispute. The problem was his obvious and outspoken antipathy toward Eddy, and his refusal to heed Patton's warning. After clearing it with Bradley, Patton relieved Wood. "I got P sent home on a 60 day detached service," he wrote his wife. "He is nearly nuts due to nerves and inability to sleep. I hope I can get him a job in the States. He is too hard to handle."

Patton appointed his chief of staff, Hugh Gaffey, to replace Wood, thus prompting one historian to charge wrongly that Patton was ambitious, jealous, and conspiring in his desire to promote Gaffey at Wood's expense. He was indeed jealous of his rivals, and no one was more ambitious or craved success more than Patton, but he did not conspire to remove one of his oldest and dearest friends, an act he personally found abhorrent but necessary.

Gaffey's appointment as the new commander came about because Patton considered him the best officer available. Among Gaffey's later accomplishments was to rouse Patton from bed in the early morning hours of December 26, 1944, to obtain permission for a tank charge that resulted in one of the great moments of the war: the liberation of Bastogne by Task Force Abrams.

There is ample evidence that Patton intended to return Wood to command. However, Marshall personally decided that Wood was more urgently needed training armored troops at Fort Knox than he was in Europe. A senior officer later revealed that Patton told Walton Walker: "Johnny, I don't give a damn what you think of 'P' Wood. He was one of the best division commanders I had and I want him back."[20]

By November 1944 the Lorraine campaign had degenerated into what Bradley has called "a ghastly war of attrition."[21] It was now an infantryman's war that made Sicily appear tame by comparison. Bradley wrote: "Patton attacked in grim weather—heavy, cold rain. The rivers were flooding. He had little or no air support. The Germans resisted every muddy inch of the way. . . . Patton slugged his way forward some thirty-five to forty miles to the Saar River, crossed it, but was stopped dead at the Siegfried

Line. His failure to break through deeper into Germany infuriated him. . . . In his frustration, he raged at me, Ike and Monty."[22] Patton cabled the War Department that he hoped "that in the final settlement of the war, you insist that the Germans retain Lorraine, because I can imagine no greater burden than to be the owner of this nasty country."[23]

The fog descended over Lorraine, it rained all the time, and the rivers and fields flooded, turning the ground into a gigantic sea of mud through which the Third Army was forced to slog. The roads resembled chocolate cream pie. The Germans left booby traps everywhere and time bombs planted in buildings, most timed to go off days or weeks later. Nonbattle casualties rose to alarming numbers: three thousand cases of trench foot in one division alone. At one point more than 18,000 of Patton's 220,000 troops were down with trenchfoot, the flu, and other assorted ailments, trench foot becoming a greater enemy than the Germans. Patton fumed at what he saw as a senseless waste of his men, and decried the poor quality of the GI combat boots. "The Germans have good boots," he complained "Why haven't we? Men unable to march cannot fight." He ordered that each soldier be required to change his socks once a day and threatened to relieve any commander who did not personally carry out his order and inspect his men's feet. "See that their socks fit" became as much a Third Army tenet as "*L'audace, l'audace.*" So did his constant exhortations: "Our mortars and artillery are superb weapons when firing. When silent, they are junk." A great deal of his time was spent teaching his officers how to deal with trench foot, saying: "This is more important for young officers to know than military tactics."[24] An adequate supply of quartermaster-issue socks now became as important and as precious as fuel for the Third Army's vehicles.

In early September Patton had breezily told Walker: "See you in Metz," but XX Corps had attacked for ten days along the Moselle and failed in the face of stiffening German resistance. Metz was protected by a series of thirty-five forts, anchored by Fort Driant (itself a chain of forts), which guarded the southern end of the approaches to the city, and had last been captured in A.D. 45 by Attila the Hun. Since then Metz had become a daunting obstacle to any general bent on its capture.[25] Ringed by artillery, moats, and automatic weapons, Driant was a formidable obstacle. Patton had originally hoped merely to outflank Metz and leave its defenders to wither for later mopping up. The enforced delay in early September made this impossible.

With VIII Corps now part of the First Army, Patton was obliged to maintain his "rock soup" tactics with two corps that were beginning to feel the pinch of a genuine shortage of infantry battalions, to sustain operations against an enemy that was being heavily reinforced at Metz and was counterattacking XII Corps in the Nancy bridgehead. Matters might have been righted had Patton been able to retain XV Corps, whose loss to Devers's 6th Army Group was grievous.

Three newly arrived infantry divisions in France, which might have been sent to the Third Army, were instead immobilized to support the Red Ball Express. The inevitable compromises, and the stop-and-start tactics brought about by the supply and fuel shortage, had a telling effect on the Third Army that inevitably calls into question Patton's decisions in the Lorraine campaign. For, as the author of an account of the battle for Metz writes:

> After the bitter fighting in September, it should have been obvious that to make a serious dent in the enemy line along the Moselle, a massive injection of fresh manpower was needed . . . limited operations merely cost American lives and achieved little. Either Patton should have been massively reinforced with infantry and unlimited air support, or he should have been ordered to stay put . . . political necessity was allowed to triumph over military common sense.[26]

Patton's years of study ought to have convinced him of the folly of siege warfare in the hostile environment of a place like Lorraine, which he knew well from firsthand experience in 1918. Between his frustration at the Third Army's dilemma and his insistence that he could win the war single-handedly if given the means, the attacks on Fortress Metz went forward. Yet he seemed reluctant to accept the fact that the great pursuit had ended and that he now faced circumstances of "too little gas and too many Germans, not enough ammo and more than enough rain."[27] Patton's later claims that in Lorraine he had held them by the nose and kicked them in rear end rang hollow.

The attack by XX Corps on October 3 on Fort Driant was firmly repulsed, and both Walker and Patton began to grasp that Metz would be a very tough nut to crack. Patton wanted to present Metz to Marshall as a trophy during his impending visit to the the Third Army and, as he had in Brittany, stubbornly insisted that Walker capture the fort even "if it took every man in XX Corps," saying he could not permit a Third Army attack to fail.[28]

But fail it did, and his diary records: "Things going very badly at Fort Driant. We may have to abandon the attack, since it is not worth the cost. I was over optimistic . . . but I hate to crush initiative." Later, however, when Walker, who admired Patton to the point of emulation, insisted on continuing the attacks, two of his division commanders rebelled and refused to continue. Patton backed them and ordered Walker to desist, although his affection for Walker was born of empathy for someone he was fond of referring to as "a fighting son of a bitch."

The attacks worried Bradley who, appealed to Patton: "For God's sake, George, lay off," promising him another chance later to pinch off Metz and take it from the rear. "Why bloody your nose with this pecking campaign." Although he had backed off, Patton renewed his attacks on Metz, this time with the justification of "blooding" new divisions in limited battalion-size attacks.[29]

Patton's frustration over Lorraine in general and Metz in particular made life hell for those around him. Even his ever-efficient aide, Charles Codman, felt Patton's wrath when Tooey Spaatz, Jimmy Doolittle, and Maj. Gen. Hoyt Vandenberg, commander of the Ninth Air Force, landed at a different airfield than expected. The three generals were delivered to the Third Army command post in the back of a truck, and Codman was informed in explicit terms that it did not matter "a good God damn" where the fault lay—he ought to have thought of covering the second airfield just in case.

One night in a French schoolhouse that had been commandeered as a corps command post, Patton was pacing the halls in the dark, lost in thought, when he stumbled over a sleeping soldier who angrily called him a "blockhead" and demanded: "Watch your step, can't you see I'm trying to sleep?" Patton roared in reply: "You're the first silly son-of-a-bitch around this place who knows what he's trying to do!"[30]

Patton's disposition grew as grim as the weather, and even Willie knew enough to stay on his best behavior around his master, particularly now that Patton had once again sworn off cigars. At dinner one night in late October, Patton's irritation spilled over when he swept toast from the table and bent a spoon double and flung it aside: "'How long, O Lord, how long?'* We roll across France in less time than it takes Monty to say 'Regroup' and here we are stuck in the mud of Lorraine. Why? Because somewhere up the line some so-and-so who never heard a shot fired in anger or missed a meal believes in higher priorities for pianos and ping-pong sets than for ammunition and gas." When his aide announced that Bradley was on a field telephone in the next room, Patton rose from his chair and, with Willie at his heels, went to the phone and said resignedly: "Hello, Brad, this is George. What does SHAEF want now?"[31]

As the weeks passed and the weather worsened, the Moselle reached flood stage, and still Metz remained a bone in Patton's throat. His engineers performed heroic feats. Patton stopped one day at a newly constructed bridge over a river: "Who in hell built this goddam bridge?" A petrified lieutenant replied: "I'm responsible for it, Sir." Patton snapped back: "All I got to say, lieutenant, it's a goddam good bridge."[32] In equal measures of jest and frustration, he wrote to Jimmy Doolittle:

---

*In his frustration Patton was quoting the last words George Bernard Shaw put in his *Saint Joan*'s mouth. In the epilogue of the play, she appears (after her canonization in 1920) to the spirits of some of the principal figures in her life, asking if she should return to them alive—only to be rejected again. (While Patton was making no claims to sainthood, it is interesting to note that Joan of Arc was, of course, from Lorraine.)

This is to inform you that those low bastards, the Germans, gave me my first bloody nose when they compelled us to abandon our attack on Fort Driant. . . . I have requested a revenge bombardment from the air to teach those sons-of-bitches that they cannot fool with Americans. I believe that this request will eventually get to you, and I am therefore asking that you see that the Patton-Doolittle combination is not shamed in the eyes of the world, and that you provide large bombs of the nastiest type, and as many as you can spare, to blow up this damn fort so that it becomes nothing but a hole.[33]

On another occasion, Patton unmercifully chewed out a junior staff officer until Gay tactfully pointed out that the officer had merely been carrying out an order Patton himself had neglected to rescind. Patton stood in silence contemplating the rug on the floor, and then ordered Stiller to find them two horses: "You and I are going riding up in the hills. I've been cooped up too long in this goddam office." He ordered: "Hap, have that officer back here at five o'clock. Also, everyone who was present when I bawled him out. I'm going to apologize."[34]

On the eve of the second anniversary of Torch, Patton was preparing to launch an armywide general offensive ordered by Bradley to capture the Saar-Moselle triangle and possibly break through to the Rhine, to open whenever the weather permitted. But the rains came down and the forecasts were not optimistic. Eddy and Grow arrived to request a postponement. After listening attentively Patton asked if they would "care to make recommendations as to your successor?" Escorting them to a map, he said: "This is what we are going to do," and for the next half hour "he poured and pumped the elixir of his own vitality." At the end Patton said: "And now, go back to your headquarters, have a big drink, and get some sleep."

"Don't worry, General, the attack will go on," they assured him. "Your're goddam right it will," replied Patton. Pausing at the foot of the stairs leading to his bedroom, Patton sighed and said to Codman: "I think this has been the longest day of my life. There is nothing I can do now but pray." He retired fully dressed to bed to read Rommel's *Infantry Attacks*, and was somewhat buoyed by a reference to how the Germans had managed in similar dreadful weather in 1914.[35] The next night he would observe that: "I have always maintained that there are more tired division commanders than there are tired divisions."

Patton observed from a hillside operational post as hundreds of heavy bombers and fighter-bombers attacked German positions in the most massive aerial bombardment since Cobra. "I'm almost sorry for those German bastards," he said. The attack went well and at dinner that night Patton "for the first time in days seemed relaxed and talkative."[36]

Patton's dream of capturing Fort Driant as his fifty-ninth birthday present was doomed to disappointment. "Armored Diesel" was served at an impromptu birthday party in the Third Army command post, but it was not a happy time. Patton had fallen victim to the very tactics he scorned in the infantry generals: He had failed to concentrate his forces for a decisive attack that might have taken Metz, then refused to accept that he had anything to do with that setback. A series of piecemeal attacks were nothing more than a return to the "penny-packet" warfare that had failed so badly in North Africa. In terrain unfit for tanks, such tactics played directly into the limited strength of the German defenders. In short, instead of the hoped-for triumph, Lorraine became Patton's bloodiest and least successful campaign. Patton and his corps commanders bore responsibility for failing to better employ their divisions. Often they were launched without adequate support or in tasks well beyond their capability. As a postwar U.S. Army study concludes: "Patton failed to concentrate Third Army's resources of the corps engaged in decisive operations. . . . The corps commanders were trapped between Patton, who continually urged aggressive action, and the grim realities of terrain, weather, and a determined enemy . . . [and] at times became preoccupied with local problems and lost sight of the broader issues," resulting in little cooperation between the corps and a disjointed campaign.[37]

Even when Metz finally fell on November 22, it was only after heavy fighting in the city to root out German defenders from their underground chambers room by room and tunnel by tunnel. Patton's senior opponent, Lt. Gen. Hermann Balck, the commander in chief of Army Group G, was scornfully critical. Balck freely admitted that his troops were "motley and badly equipped," and ascribed their success in defending Metz "mainly to the bad and timid leadership of the Americans."[38] It was the most scathing criticism ever leveled at Patton by one of his enemies.

Although Patton triumphantly entered Metz with sirens screaming, Fort Driant and a number of other smaller forts held out for another three weeks before some three thousand fanatics eventually surrendered when their food and their hope finally ran out. There were no further attacks against the forts, but they were subjected to heavy shelling with their own captured guns. The battle officially ended on December 13, when the last German defenders surrendered.

Patton reviewed the 5th Division, which had carried the brunt of the attacks and suffered heavy casualties. He awarded dozens of medals for valor and told the division: "I am very proud of you. Your country is proud of you. You are magnificent fighting men. Your deeds in the battle of Metz will fill the pages of history for a thousand years."[39] It was mostly a symbolic victory won at great cost by the brave men of the Third Army. In the midst of one the worst European winters in many years, the securing of the Moselle and the siege of Metz were distinguished as only the longest and bloodiest battles of an ugly campaign.[40]

*   *   *

By early December, spearheaded by the 4th Armored, XII Corps had advanced deep into the Saar Basin, astride the Sarre River, beyond which lay the *Westwall* and, a heartbeat away, German soil. As the official U.S. Army campaign historian has written: "The optimistic prediction by higher headquarters that Patton's troops would reach the Rhine by mid-December had been quietly forgotten. The Third Army commander himself had gradually abandoned the hope of a quick break-through to the Rhine; at this stage he seems to have been concerned simply with driving steadily forward, going as far as his strength and supplies would permit."[41]

To the north the First and Ninth Armies had encountered similar problems and were stalled near the Roer River. To break the winter stalemate and capture the Roer dams, Bradley and Hodges flung troops into the Hürtgen Forest in a series of futile attacks that made even Metz look tame, at a cost of thousands of casualties, in what is often referred to as America's Passchendaele. Official historian Charles B. MacDonald, who fought there, has called it "a misconceived and basically fruitless battle that could have, and should have, been avoided."[42]

Patton believed that the war could have been won in 1944, if only he had been given adequate fuel and ammunition, as did Liddell Hart, who has written that the war ought to have been won in 1944 at a great savings in lives. Patton cited August 29 as the critical date:

> Hereafter pages will be written on it, or rather on the events which produced it. It was evident that at that time there was no real threat against us as long as we did not stop ourselves or allow ourselves to be stopped by imaginary enemies. Everything seemed rosy when suddenly it was reported to me that the 140,000 gallons of gasoline which we were supposed to get for that day did not arrive. I presented my case for a rapid advance to the east for the purpose of cutting the Siegfried Line before it could be manned. It is my opinion that this was the momentous error of the war.[43]

Unfortunately the evidence weighs heavily against this notion. There was, of course, no conspiracy by Eisenhower, SHAEF, Montgomery, or Lee's supply services to keep the Third Army immobilized. Historian Anthony Kemp has drawn a compelling scenario for what would probably have happened if Patton had been given all the supplies he needed in the autumn of 1944. To begin with, the battle for Metz would never have occurred, inasmuch as the Third Army would have driven through the then undefended underbelly of Lorraine (the so-called Nancy gap) to the Siegfried Line, which could not have withstood a determined assault by Third Army. Then what? As Kemp notes:

> All the Allied strategists were in agreement that the prime objective was the Ruhr. Patton, however, with his forces in the Frankfurt-Mannheim

area, would have had to turn north and advance up the narrow Rhine valley. Simply advancing blindly into Germany would not have fulfilled any strategic purpose, and unless other Allied units had kept pace, the Third Army would have been out on a limb and liable to have its lines of communications cut. As we know that there were insufficient sources available to maintain an offensive along the whole front, Patton would have been on his own anyway. . . . Every mile advanced by Patton in September would have proportionately increased his lines of communication, requiring more and more trucks to keep up the flow of stores, trucks that were in desperately short supply. More divisions would have had to be immobilized in order to use their transport, which would have further aggravated the main background problem to the Lorraine campaign—shortage of manpower. In his grandiose dreams, Patton was simply flying in the face of a situation that had to be faced by all the army commanders in northwest Europe at the time.[44]

Even more convincing is the assertion by a senior engineer officer (responsible for airfield construction and maintenance) that close air support and aerial resupply beyond Lorraine would have been extremely limited, with few airfield sites between the *Westwall* and the Cologne-Maastricht Plain. "I don't doubt that we could have carried about two armored and one (motorized) divisions up to Cologne," he said. "But then where? Certainly not across the Rhine."[45]

Although he would not hear of it, the truth was that Patton was not a victim of Eisenhower or Montgomery, but of the broad-front strategy and a logistics system that was simply incapable of keeping pace with rapid, mobile warfare. The system was geared for the snail's pace of the Normandy bocage, not for what Cobra had wrought. Moreover, as Patton's earlier biographer Ladislas Farago rightly concludes: "Despite its apparent excellence and Patton's unbounded enthusiasm for it, his plan never had a chance to be accepted, or even to be taken seriously anywhere beyond General Bradley's command post. Not only was close air support vital if the plan was to succeed, but the Luftwaffe still retained the capability to interfere seriously with Patton's offensive. Among the other problems Patton faced but failed to acknowledge were the physical exhausion of the Third Army, which would have faced furious resistance from a German army defending its homeland for the first time. Indeed, from the moment the Third Army reached the Moselle, Patton had disregarded the reality that the conditions for a breakout and pursuit, which had carried it across the soft underbelly of Normandy, no longer existed in Lorraine. The war of movement that Bradley had ushered in with "Cobra" was already turning, even if imperceptibly as yet, into a war of position."[46]

Although German casualties may have numbered approximately 180,000, Lorraine was the Third Army's bloodiest battle. However, "to capture the province of Lorraine, a problem which involved an advance of only 40 to 60 air miles, Third Army required over 3 months and suffered 50,000 casualties, approximately one-third of the total number of casualties it sustained in the entire European war."[47]

It took Metz finally to convince Patton that his problems of supply and strategy were in the hands of others over whom he had little or no control. When the Normandy campaign ended there were more than 2 million men and 3.5 million tons of supplies in France. SHAEF had become a military Goliath with a gigantic logistical appetite that could not be satiated. But what ensured there could be no victory in 1944 could be found in the very nature of coalition warfare. If the United States had had a smaller and weaker coalition partner, Eisenhower might have felt justified in a single-thrust military operation in the autumn of 1944, but so long as Winston Churchill led Britain, such a decision was unimaginable.

Patton and the Third Army were not alone in their problems in Lorraine and the Saar. The failure of Market-Garden led to a general stalemate during the autumn and early winter of 1944–45. Although the three Allied army groups had advanced to the borders of the Third Reich, a hasty defense by German units and the advent of bad weather left them virtually immobilized in the mud, snow, and harsh terrain of the outer Ruhr, in the Ardennes forest, in Lorraine, the Saar, and in the Vosges. A series of battles of attrition gained little in the way of significant ground. Both sides suffered greatly from combat losses and the winter weather. Aachen and the Hürtgen joined Metz in the lexicon of Allied failures, making it plain that the war would not end in 1944.

December 16 was one of the most important dates of the war. In the Third Army it became apparent at Oscar Koch's 7:00 A.M. intelligence briefing that something unusual was afoot. For some time Koch had not only been keeping a close eye on German dispositions but reporting a buildup of panzer and infantry divisions and ammunition and gasoline dumps west of the Rhine in the Saar, and in the areas opposite the First Army from Aachen to the southern extremity of the Ardennes Forest. Massive rail movements, of increasing frequency and size, were noted, most north of the Moselle in the First Army sector. As the buildup continued, the possible ramifications became a matter of concern to Patton. On December 9 he was told that the Germans now had a two-to-one numerical advantage in the Ardennes sector guarded by Middleton's VIII Corps, then in the southwestern corner of the Ardennes in defensive positions, resting and reequipping after two of its divisions had been decimated in the Hürtgen.

Patton sat silently contemplating this latest news. At his direction the Third Army staff was planning a major new offensive to commence on

December 19, to breach the *Westwall* and drive to the Rhine. The operation was to be preceded by a massive three-thousand-plane bombardment by the air force, which Patton fondly noted would be the biggest blitz in the Third Army's history. Patton announced that until German intentions became clear, the ominous situation in the Ardennes would not interfere with the forthcoming Third Army offensive. Nevertheless planning was to be undertaken at once to counter any potential threat in the Ardennes. "We'll be in a position to meet whatever happens," Patton told his staff.[48]

What Patton did not know was that Eisenhower had already decided to transfer divisions from the Third Army to the northern group of armies to support a breaching of the Rhine and the main assault into the heartland of the Reich, "regardless of the results" of the Saar offensive.[49]

On December 16 Koch reported the Germans were in a state of radio silence, and when Patton asked him what it meant, the G-2 replied that when American units went under radio silence it invariably signaled an impending attack. "I believe the Germans are going to launch an attack, probably at Luxembourg," he predicted.

That evening after dinner the telephone rang and was answered by Colonel Harkins. On the other end was Leven C. Allen, Bradley's chief of staff, with orders immediately to detach the 10th Armored Division and send it north that night to reinforce VIII Corps.

Patton was irate to learn that Bradley was removing one of his key divisions at the most inopportune moment. He telephoned Bradley and outlined his reasons why detaching the 10th Armored was a bad idea. Bradley listened patiently and finally in exasperation reiterated that his order stood: The 10th Armored was to move *now*, "to help [Middleton's VIII Corps] repulse a rather strong German attack." Patton thought it was another case of Bradley taking counsel of his fears. When Patton persisted in his argument that to lose the 10th Armored would be to demean the price paid in blood by the Third Army in its advance to the *Westwall*, Bradley agreed with his logic but nevertheless: "I was compelled to give Patton a direct, unequivocal order to get the 10th Armored moving." He cut Patton short, saying: "I can't discuss this matter over the phone."

"This matter" to which Bradley referred cryptically was a massive German counteroffensive. All hell had broken loose in the Ardennes, in what would soon be christened the Battle of the Bulge. After disappointment and failure in bloody Lorraine, the Bulge was to become Patton's and Third Army's finest hour.[50]

CHAPTER 43

PATTON OF COURSE
The Battle of the Bulge

Every time I get a new star I get attacked.

—EISENHOWER

And every time you get attacked, Ike, I pull you out.

—PATTON

Seeking to repeat the success of 1940, Hitler gambled on a last-ditch attempt to split the Allied armies by a sudden, lightning thrust through the Ardennes to destroy all Allied forces north of a line running from Bastogne to Antwerp, naively believing that his armies could drive clear to Antwerp and compel the Allies to sue for peace. In launching a lightning attack in the Ardennes, the Germans were merely repeating what they had done twice before, in August 1914 and May 1940. In the midst of the worst winter in thirty-eight years, the Germans emerged from the fog and darkness to launch the deadliest and most desperate battle of the war in the West. Believing that there was little prospect of a major German attack in the poorly roaded, rugged, heavily forested Ardennes, Bradley had taken what he termed "a calculated risk" by lightly defending them with two newly arrived, inexperienced infantry divisions, and two battered veteran divisions absorbing replacements. Despite Germany's historical penchant for mounting counteroffensives when things looked darkest, no one believed that the Germans could pull off a major offensive in the Ardennes in secret. For

Hitler, the operation was a last gamble that he could yet take control of Germany's destiny. Massive in both scope and size, it was to be the first and only major German counteroffensive of the war in northwest Europe.

Three armies attacked: two panzer armies commanded by Generals Hasso von Manteuffel and Sepp Dietrich, and Gen. Erich Brandenberger's (largely infantry) Seventh Army, whose primary mission was to protect the flanks of the panzer armies from a counterattack by the Third Army. Time was of the essence if the attacking force was successfully to seize bridgeheads across the Meuse before the Allies could react. The road net in the Ardennes was sparse and to get to the Meuse the Germans had to pass through the towns of Saint-Vith and Bastogne where the main east-west roads converged.

What became known as the Battle of the Bulge began in heavy fog at 5:30 A.M. on December 16, when the two panzer armies (more than a quarter million men) attacked through the heavily wooded and most lightly defended sector held by the First Army. White-clad German troops advanced in waves, backed by panzers as the once-quiet Ardennes became bedlam. Dietrich's Sixth SS Panzer Army tore through the Losheim gap toward Liège, while to the south Manteuffel's Fifth Panzer Army drove west from the vicinity of Prüm through the Schnee Eifel toward Saint-Vith and Bastogne. Everywhere American units were caught flat-footed and became embroiled in desperate battles at Saint-Vith, Elsenborn Ridge, Houffalize, and later, Bastogne. The inexperienced U.S. 106th Division was quickly surrounded and nearly annihilated (seven thousand POWs alone), but even in defeat it managed to help buy time for others, such as Brig. Gen. Bruce C. Clarke, who began organizing the defense of Saint-Vith against a torrent that threatened to engulf everything in its path. Confusion reigned as the Allied commanders scrambled to assess whether this was merely a spoiling attack or something more serious. What they did not need their G-2s to tell them was that the Allied situation in the Ardennes had gone to hell in a hurry. As the German armies drove deeper into the Ardennes, the line defining the Allied front on the map took on the appearance of a large protrusion, or bulge, the name by which the coming battle would forever be known.

The German attack in the Ardennes revealed serious lapses in Allied intelligence, whose G-2s had failed to draw the right conclusions even though there was ample evidence of the German buildup. Postwar alibis later included an alleged lack of Ultra intercepts, an excuse thoroughly debunked when its existence was revealed in the late 1970s.[1]

A primary reason why the Third Army intelligence staff correctly assessed German intentions was attributable to Patton, who relied heavily on the daily judgments of his G-2, and lost no opportunity to probe anyone connected to the subject. The SHAEF G-2, Kenneth Strong, has written that

Patton demonstrated "an extraordinary desire for information of all kinds. He invariably came to see me when he was at Supreme Headquarters and would quiz me on details about the enemy, usually to satisfy himself that the risks he intended to undertake were justified."[2]

The much maligned Third Army staff excelled during the period prior to the German counteroffensive. Since mid-November, Oscar Koch closely followed the German buildup, reporting enormous German rail movements both east and west of the Rhine. He began to question its implications long before others jumped on the bandwagon. Their work was enhanced when Patton employed Col. Jimmy Polk's 3d Cavalry Group as a special source of intelligence by attaching units to each committed division and corps in the Third Army to augment and embellish the information on which he made crucial decisions from regular intelligence sources. Information tended to move through command channels at a snail's pace; his alternative source enabled information to be sent directly to the Third Army in a real time frame. Although unorthodox, the results established Patton and Montgomery (through his own personal liaison officers) as undoubtedly the best informed of the Allied senior commanders.[3]

On November 25 Patton noted that he believed "First Army is making a terrible mistake in leaving the VIII Corps static, it is highly probable that the Germans are building up east of them." Koch's gravest concern was the threat to the Third Army's northern (left) flank if the Germans attacked through the Ardennes, and at his instigation, Patton had obtained authorization to employ reconnaissance aircraft deep into the heavily forested Eifel region, north of the Ardennes, along Germany's western border with Belgium, and by December 9 "the situation north of the Moselle demanded special attention." He briefed Patton that the Germans were now fully capable of mounting "a large spoiling offensive" in the Ardennes, and during the next week continued to report the German buildup, his intelligence bolstered by the report of a high-ranking POW that a large-scale breakthrough was in the offing.[4]

Koch was the only Allied intelligence officer to anticipate trouble and plan how to deal with it. Thus, where other intelligence officers were lulling their commanders with false optimism and wishful thinking that nothing serious was imminent, the Third Army made plans to deal with what no one else believed would occur.[5] Where other intelligence staffs were usually defense-minded, the Third Army G-2 staff eagerly thought in offensive terms to suit their aggressive commander's needs and requirements: "It is striking that Patton was virtually alone among Ike's top commanders in rejecting the notion that the Germans, like the Americans themselves, were using the Ardennes as a quiet sector."[6]

At the briefing the morning of December 17, Patton mulled over the news of a German attack and stated that he believed "the thing in the north

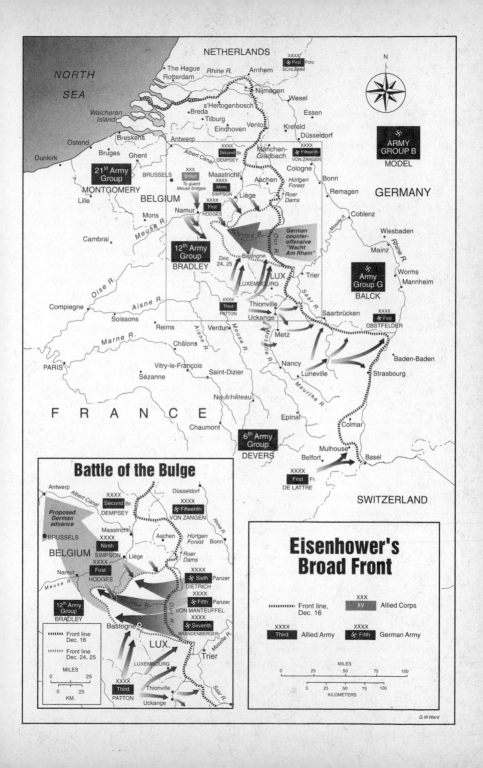

Eisenhower's Broad Front

Battle of the Bulge

is the real McCoy." The G-3, Col. Halley Maddox, believed that the Germans would have to commit their entire reserves in the Ardennes and called it a perfect setup for the Third Army. "If they will roll with the punch up north, we can pinwheel the enemy before he gets very far. In a week we could expose the whole German rear and trap their main forces west of the Rhine." Patton thought Maddox was right, but merely observed "that isn't the way those gentlemen up north fight. They aren't made that way. That's too daring for them. My guess is that our offensive will be called off and we will have to go up there and save their hides." Still hedging his bets, Patton ordered the immediate commitment of the 4th Armored in an effort to prevent losing it, as he had the 10th. Only later would he admit that even he had thoroughly underestimated at first the scope of the German attack.[7]

Bradley was in Versailles conferring with Eisenhower on December 16, when SHAEF first learned of the German attack, but dismissed it as merely a spoiling attack to hamper the Third Army offensive. Eisenhower disagreed, declaring: "That's no spoiling attack."[8] As the two generals weighed the initial Allied reaction, Bradley was openly apprehensive at the prospect of issuing orders that Patton would loudly and heatedly dispute, but Eisenhower growled: "Tell him that Ike is running this damn war."[9]

After summoning Patton to Luxembourg the following day, Bradley said: "George, I feel you won't like what we are going to do, but I fear it is necessary," displaying on the map how the German penetration was far deeper and more serious than Patton had previously thought. What could the Third Army do? Bradley asked. Patton replied that he would have two divisions on the move the next day, and a third in twenty-four hours, if necessary. With that the proposed Saar offensive was canceled. Bradley thought Patton was still discouraged at the necessity of abandoning it, but he now fully understood what was at stake and shrugged it off with the observation: "What the hell, we'll still be killing Krauts," grinning when Bradley assured him they would "hit this bastard hard."[10] Later that evening he telephoned that Patton was to be in Verdun the following morning to meet with Eisenhower and the other Allied commanders to work out a plan of action.

December 19, 1944, a historic day, began at 7:00 A.M., when Patton briefed his key staff officers and two of his corps commanders on the situation as he then knew it. An hour later he convened the full staff to explain his belief that the Third Army would be called on to come to the relief of the First, how and where to be decided at Verdun. The only certainty was that "while we were all accustomed to rapid movement, we would now have to prove that we could operate even faster. We then made a rough plan of operation." The plan assumed that Patton would be given control of VIII Corps, and that his newly arrived III Corps headquarters would take control of three divisions to relieve the First Army and move from the Saar via one or more of three possible routes. When Patton learned his mission, he would merely telephone Gay and announce the preestablished code word for whichever axis he was to employ. The arrangements complete, Patton, Harkins, and Codman departed for Verdun.

Eisenhower arrived at 11:00 A.M. in an armor-plated sedan escorted by MP jeeps with machine guns, "looking grave, almost ashen," his mood brightening only slightly when he encountered Patton and Bradley. Patton had arrived just moments earlier with his stern war face firmly in place. The meeting took place in a dismal second-floor room of a French barracks. Very little warmth emanated from a potbellied stove, and most kept their coats on to ward off the pervasive chill. The atmosphere was equally grim despite Eisenhower's fragile attempt at levity, when he opened by saying: "The present situation is to be regarded as one of opportunity for us and not of disaster. There will be only cheerful faces at this conference table." Patton immediately chimed in: "Hell, let's have the guts to let the sons of bitches go all the way to Paris. Then we'll really cut 'em up and chew 'em up." The smiles seemed forced.

Present were Eisenhower, Tedder, Bedell Smith, Bradley, Devers, Patton, and a handful of staff officers. Montgomery was absent but had sent his able chief of staff, Maj. Gen. Francis de Guingand. Most thought Monty's absence a calculated insult to both Eisenhower and themselves.

Eisenhower's tenseness was apparent when his G-2 (Kenneth Strong) and G-3 (Harold "Pinky" Bull) arrived a few minutes late: "Well," he snapped, "I knew my staff would get here; it was only a question of when."[11] If Eisenhower's mood was glum, he had good reason: Since Market-Garden not only had the Allies been stalled on the borders of the Third Reich, but there were increasing signs of plummeting morale, manifested by a rapidly rising desertion rate.[12]

Waffen-SS Lt. Col. Otto Skorzeny, who had daringly rescued Mussolini from captivity (after his arrest and confinement by the new Italian government, headed by Marshal Pietro Badoglio, overthrew the Italian dictator in July 1943) the previous year, led a commando group dressed in American uniforms behind Allied lines to seize bridges across the Meuse. A rumor that Skorzeny's real object was to assassinate Eisenhower created unprecedented panic at SHAEF headquarters in Versailles. Skorzeny's men roamed the Ardennes ambushing convoys, spreading alarm and confusion, altering or removing road signs, and generally raising hell as a guerrilla force.

The commanders quickly agreed to stop offensive action in all Allied sectors and concentrate on blunting the German drive. Eisenhower's strategy was to draw a stop line at the Meuse, beyond which there would be no further retreat. Once the German attacks had been contained, the Allies would counterattack. Eisenhower said: "George, I want you to command this move—under Brad's supervision of course—making a strong counterattack with at least six divisions. When can you start?"[13] Patton replied: "As soon as you're through with me," explaining that he had left three sets of instructions with his staff and by telephoning a given code word could put any plan in motion at once. "When can you attack?" Eisenhower asked. "The morning of December 21, with three divisions," Patton replied instantly.

Forty-eight hours! Eisenhower was not amused, wrongly assuming that Patton had once again picked a very inopportune moment to act publicly boastful. "Don't be fatuous, George," he retorted, in obvious disbelief. "If you try to go that early, you won't have all three divisions ready and you'll go piecemeal. You will start on the twenty-second and I want your initial blow to be a strong one! I'd even settle for the twenty-third if it takes that long to get three full divisions."

Eisenhower was dead wrong: It was not Patton the boastful but Patton the student of war at his absolute best. Where others at Verdun came with only vague ideas and without specific plans, Patton had devised *three* plans, each tailored to meet any contingency that his superiors might direct. "This was," writes Blumenson, "the sublime moment of his career." After more than thirty-four years, it was as if destiny had groomed him for this single, defining instant in which the fate of the war rested upon the right decisions being made and carried out by the men in that dingy room. While near panic existed elsewhere, in the Third Army there existed a belief in a magnificent opportunity to strike a killing blow. While others debated or waffled, Patton had understood the problem facing the Allies and had created a plan to counterattack the Germans and occupy Bastogne—which, although not yet surrounded, was clearly soon to be besieged. By contrast Bradley, whose army group had been attacked, "mostly observed" throughout the two-hour conference, "saying little, offering nothing." Even he realized that the only principal players were Eisenhower and Patton.[14]

Opinions vary, but certainly the reaction of some present that day was skepticism about yet another smug prediction by Patton that was quite out of place in this somber setting. Strong noted: "There was some laughter, especially from British officers, when Patton answered 'Forty-eight hours.'"[15] Codman witnessed "a stir, a shuffling of feet, as those present straightened up in their chairs. In some faces skepticism. But through the room the current of excitement leaped like a flame." John Eisenhower writes: "Witnesses to the occasion testify to the electric effect of this exchange. The prospect of relieving three divisions from the line, turning them north, and traveling over icy roads to Arlon to prepare for a major counterattack in less than seventy-two hours was astonishing, even to a group accustomed to flexibility in their military operations."

Suddenly to turn Third Army ninety degrees to the north along icy roads in terrible winter weather when its supply dumps were located to support the drive to the *Westwall* was a logistician's worst nightmare. It posed equally daunting challenges to Patton's subordinate commanders, but the months of training and experience in combat now paid off handsomely. "Altogether it was an operation only a master could think of executing," notes Blumenson. Moreover, only a commander with exceptional confidence in his subordinate commanders and in the professional skill of his fighting divisions could dare risk such a venture. Patton not only never hesitated but embraced the opportunity to turn a potential military debacle into a triumph.

Cigar in hand, Patton illustrated his intentions on the map by pointing to the obvious "bulge" in the Saint-Vith/Bastogne sector, and, speaking directly to Bradley, said: "Brad, the Kraut's stuck his head in a meatgrinder." Turning his fist in a grinding motion, he continued: "And this time I've got hold of the handle." He then replied to the inevitable questions with specific, well-rehearsed answers. "Patton would have liked to have seen the Germans drive some forty or fifty miles, then chop them off and destroy them, but he recognized that he would never muster support for that kind of daring."[16] Codman recorded: "Within an hour everything had been thrashed out—the divisions to be employed, objectives, new Army boundaries, the amount of our own front to be taken over by [Devers's] Sixth Army Group, and other matters—and virtually all of them settled on General Patton's terms." (Two of Patton's three corps were to be extricated for a counterattack into the Ardennes, with Patch's Seventh Army to take control of most of the Third Army sector in the Saar.) Simply put, it was perhaps the most remarkable hour of Patton's military career. Bradley later acknowledged that this was a "greatly matured Patton," and that the Third Army staff had pulled off "a brilliant effort."

With considerable understatement Patton wrote of this exceptionally important day: "When it is considered that Harkins, Codman, and I left for Verdun at 09:15 and that between 08:00 and that hour we [held] a Staff meeting, planned three possible lines of attack, and made a simple code in which I could telephone General Gay . . . it is evident that war is not so difficult as people think."[17]

If ever there was justification for Eisenhower to have saved Patton's career, it was now, and before they parted, Eisenhower, recently promoted to the five-star rank of General of the Army, said, "Funny thing, George, every time I get a new star I get attacked." Patton shot back affably: "And every time you get attacked, Ike, I pull you out" (a reference to the Kasserine Pass and the time Eisenhower attained his fourth star). Many years earlier Patton had said: "Ike, you will be the Lee of the next war, and I will be your Jackson." Whether or not Eisenhower qualified as Robert E. Lee, Patton was about to manifest a definite resemblance to Stonewall Jackson. The Third Army was poised to pull off one of the most remarkable feats of any combat army in history, "a maneuver that would make Stonewall Jackson's peregrinations in the valley campaign in Virginia and Gallieni's shift of troops in taxicabs to save Paris from the Kaiser look pale by comparison."[18]

After the meeting Patton began snapping out orders. To Harkins he commanded: "Telephone Gay. Give him the code number [*sic*], tell him to get started. . . . You know what to do. Codman, you come with me. Tell Mims we start in five minutes." Their destination was Luxembourg City, where Patton intended to relocate Lucky Forward. They stopped at XX Corps, where Walker persuaded Patton to remain overnight and lent him pajamas and toothpaste. Codman was dispatched to Nancy to coordinate the move of Patton's trailers to Luxembourg, and his journey became a nightmare spent dodging a more than one-hundred-mile-long column of fifteen hundred tanks and vehicles of the 4th Armored moving toward Bastogne.[19]

During the next three days Patton and Mims constituted Lucky Forward. With one pistol strapped to the outside of his parka and another tucked into his waistband, Patton sped from one division or corps to another. On December 20 "I visited seven divisions and regrouped an army alone," in a day spent conferring, issuing orders, wisecracking with GIs, and changing and fine-tuning dispositions. Like a cattle drover, he pushed, pulled, and exhorted everyone to keep moving, and to "Drive like hell" toward Bastogne. At the end of perhaps the most dynamic day of his life, Mims remarked: "General, the Government is wasting a lot of money hiring a whole General Staff. You and me has run the Third Army all day and done a better job than they do." Patton was pleased that he had earned his pay: "It was quite a day. . . . Destiny sent for me in a hurry when things got tight. Perhaps God saved me for this effort."[20]

Patton proved to be Omar Bradley's savior on this occasion. After the Verdun conference, Eisenhower returned to SHAEF and was strongly advised by his staff to split the Ardennes front in two until the situation could be brought under control, with Montgomery to be given temporary operational command of all Allied forces (principally the U.S. First and Ninth Armies) in the northern half of the Bulge, and Bradley to command only the southern flank (Third Army). Bradley's contact with Hodges was tenuous, and he was in no position to control the northern flank from Luxembourg. Calling it "an open and shut case," Bedell Smith (himself no admirer of Monty) backed the proposal. Eisenhower agreed, and telephoned Bradley of his decision. During the most serious confrontation between the two friends, Bradley protested so vehemently that Eisenhower felt compelled to say: "Well, Brad, those are my orders."[21] For the rest of his life Bradley bitterly (and erroneously) blamed Montgomery for inciting the order, and refused to admit that there seemed to be justification for SHAEF's (and later Montgomery's) loss of confidence in the exhausted, taciturn Hodges, who lacked Patton's flair "at a time when we needed Pattonesque bravado." Eisenhower attempted to soften the blow by recommending Bradley for promotion to four-star general, but Bradley never forgave him, and went to his grave convinced that his friend had betrayed him.

When the campaign ended Bradley informed Eisenhower he would never again serve under Montgomery and would ask to be sent home if ordered to do so, as "I will have lost the confidence of my command." A distraught Eisenhower observed: "I thought you were the one person I could count on for doing anything I asked you to." Bradley neglected to inform Eisenhower that Patton had given his assurance he too would resign.[22]

Now that they shared the same hotel, where their respective headquarters were located in Luxembourg City, Bradley and Patton saw a good deal of one another. Bradley needed an ally, and he found one in Patton, who commiserated with him over a decision he thought was inspired by Montgomery and Churchill: "General Eisenhower is unwilling, or unable, to command Montgomery," he grumbled.

Patton's first order to his troops read: "Everyone in this army must understand that we are not fighting this battle in any half-cocked manner. It's either root hog—or die! Shoot the works. If those Hun bastards want war in the raw then that's the way we'll give it to them!"[23] When he met with the staffs of three of his four corps in Luxembourg the night before the attack jumped off, Patton noted that their mood was one of doubt: "I always seem to be the ray of sunshine, and, by God, I always am. We can and will win, God helping. . . . I wish it were this time tomorrow night. When one attacks it is the enemy who has to worry. Give us the Victory, Lord."[24]

Bradley's role was largely reduced to that of an observer; the battle was Patton's to mastermind and control, and he uncharacteristically overcontrolled the 4th Armored, in his exuberance specifying tactics, the employment of tanks and artillery, how they would march on the road and fight, and generally interfering with Gaffey in a high-handed manner Patton would not for an instant have tolerated from others.[25] Until the Third Army could attack from the south, the original strategy had been to hold ground for as long as possible, retreat, blow up bridges, and delay again. Although it was against his principles to give up anything, Patton saw an opportunity for the Germans to become overextended before he struck their vulnerable left flank. As late as December 20, Patton contemplated ceding Bastogne to the Fifth Panzer Army, but admitted in his diary: "On Bradley's suggestion, in which Middleton strongly concurred, we decided to hang on to Bastogne, because it is a very important road net, and I do not believe the enemy would dare [by] pass it without reducing it."[26]

The afternoon of December 20 Patton met Middleton at Arlon and greeted him with the admonition: "Troy, of all the goddam crazy things I ever heard of, leaving the 101st Airborne to be surrounded in Bastogne is the worst!" A friend of long standing, Middleton was never in the least intimidated by Patton or his blustery remarks, and rejoined: "George, just look at that map with all the [six] roads converging on Bastogne. Bastogne is the hub of the wheel. If we let the Boche take it, they will be in the Meuse in a day." Patton understood the obvious need to hold Bastogne, and the two friends worked out an agreed axis of advance for the launching of an attack to relieve the beleaguered market town.[27]

In the forty-eight hours they had been given to get into position, Maj. Gen. John Millikin's III Corps had struggled for more than one hundred miles over icy, unfamiliar roads in snow and fog. But, as promised, at 6:00 A.M., across a twenty-mile front on the morning of December 22, some sixty-six hours after Patton's assurance at Verdun, three divisions launched the first Allied counterstroke of the Ardennes campaign.[28]

Bradley would later offer the highest praise of Patton he would ever accord:

> True to his boast at Verdun, Patton, having turned his Third Army ninety degrees, attacked on December 22. His generalship during this difficult maneuver was magnificent. One of the most brilliant performances by any commander on either side in World War II. It was absolutely his cup of tea—rapid, open warfare combined with noble purpose and difficult goals. He relished every minute of it, knowing full well that this mission, if nothing else, would guarantee him a place of high honor in the annals of the U.S. Army.[29]

The Third Army struggled against the weather and the Germans to liberate Bastogne, where the 101st Airborne and elements of his own 9th and 10th Armored Divisions were now surrounded. December 23 was the only day of fair weather, during which the air forces attacked the Germans and made more than two hundred supply drops into Bastogne, as its defenders repulsed a strong German attack. The 4th Armored spearheaded the drive toward Bastogne, but ran into increasing difficulty, now compounded by a fresh snowfall. "It is always hard to get an attack rolling," Patton observed, pleased that "the men are in good spirits and full of confidence."

Bastogne remained surrounded, and when the Germans demanded its surrender, the acting commander of the 101st Airborne, Brig. Gen. Anthony C. McAuliffe, replied with a defiant: "Nuts!" When it was repeated to Patton, he answered: "Any man who is that eloquent deserves to be relieved [rescued]. We shall go right away."

On Christmas Eve day Patton judged that "the German General Staff is running this attack and has staked all on this offensive to regain the initiative. They are far behind schedule and, I believe beaten. If this is true, the whole army may surrender." Patton was only partly right. Surrender was not an option for the Germans either. At that moment both combatants were at serious risk. Remembering history, Patton observed;, "On the other hand, in 1940 they attacked as at present. . . . They may repeat—but with what?" In response to pointed messages from the 101st expressing disappointment at not being rescued, Patton sent an ill-advised message to McAuliffe: "Xmas Eve present coming up. Hold on." Unfortunately the message was illusory: There was no chance of the 4th Armored reaching Bastogne by Christmas Eve or even Christmas Day.[30]

Patton well understood the extraordinary demands he had placed on his troops. When the Germans counterattacked violently across the III Corps line of advance to Bastogne, resulting in heavy tank losses and, in some cases, compelling temporary withdrawal, Patton accepted responsibility. "It is probably my fault, because I had been insisting on day and night attacks. This is all right on the first or second day of the battle and when we had the enemy surprised, but after that the men get too tired. Furthermore, in this bad weather, it is very difficult for armored outfits to operate at night." Once again he was impressed by "how long it takes to really learn how to fight a war."[31]

The legend of Patton's famous weather prayer has been erroneously por-trayed in the film as taking place during the drive to relieve Bastogne. In fact it occured in November in Lorraine, when Patton's frustration finally boiled over and he telephoned the Third Army chaplain, Msgr. (Col.) James H. O'Neill and announced, "This is General Patton. Do you have a good prayer for weather? We must do something about these rains if we are to win the war. We've got to get not only the chaplains but every man in Third Army to pray. We must ask God to stop these rains. These rains are the mar-gin that holds victory or defeat."[32] If O'Neill thought such a request odd— well, nothing was considered unusual to those who served under Patton, and he replied he would write something suitable within the hour. His prayer books contained nothing apropos, so O'Neill composed his own brief prayer. It read:

> Almighty and most merciful Father, we humbly beseech Thee, of Thy great goodness, to restrain these immoderate rains with which we have had to contend. Grant us fair weather for Battle. Graciously harken to us as soldiers who call upon Thee that, armed with Thy power, we may advance from victory to victory, and crush the oppression and wickedness of our enemies, and establish Thy justice among men and nations. Amen.[33]

O'Neill also brought Patton a proposed Christmas greeting to the Third Army, which read:

To each officer and soldier in the Third United States Army, I wish a
Merry Christmas. I have full confidence in your courage, devotion to
duty, and skill in battle. We march in our might to complete victory. May
God's blessing rest upon each of you on this Christmas Day.

G.S. Patton, Jr.

Lieutenant General

Commanding, Third United States Army[34]

Patton immediately approved both (which were printed on both sides of
250,000 wallet-size cards). O'Neill was amazed: "This was certainly doing
something about the weather in a big way." As a personal touch he suggested that Patton sign the Christmas greeting. Patton then explained why he
wanted them written. His rationale was as complex and thoughtful as the
man himself:

> Chaplain, I am a strong believer in prayer. There are three ways that men
> get what they want; by planning, by working, and by praying. Any great
> military operation takes careful planning or thinking. Then you must have
> well-trained troops to carry it out: that's working. But between the plan
> and the operation there is always an unknown. That unknown spells
> defeat or victory, success or failure. It is the reaction of the actors to the
> ordeal when it actually comes. Some people call that getting the breaks; I
> call it God.
>   God has His part, or margin in everything. That's where prayer
> comes in. Up to now, in the Third Army, God has been very good to us.
> We have never retreated; we have suffered no defeats, no famine, no epi-
> demics. This is because a lot of people back home have been praying for
> us. We were lucky in Africa, in Sicily, and in Italy, simply because people
> prayed. But we have to pray for ourselves too. A good soldier is not made
> merely by making him think and work. There is something in every sol-
> dier that goes deeper than thinking or working—it's his "'guts." It is
> something that he has built in there: it is a world of truth and power that is
> higher than himself. Great living is not all output of thought and work. A
> man has to have intake as well. I don't know what you call it, but I call it
> Religion, Prayer, or God.

Chaplain O'Neill's prayer and Christmas greeting are not to be confused with two other prayers purportedly written by Patton. The first, written on December 23, reads in part:

> Sir, this is Patton talking. The past fourteen days have been straight hell.
> Rain, snow, more rain, more snow—and I am beginning to wonder what's
> going on in your headquarters. Whose side are You on anyway?

For three years my chaplains have been explaining that this is a religious war. This, they tell me, is the Crusades all over again, except that we're riding tanks instead of chargers. They insist we are here to annihilate the German Army and the godless Hitler so that religious freedom may return to Europe. Up to now I have gone along with them, for You have given us Your unreserved cooperation. . . . But now, you've changed horses in midstream. You seem to have given von Rundstedt every break in the book and frankly, he's been beating the hell out of us. My army is neither trained nor equipped for winter warfare. And as You know this weather is more suitable for Eskimos than for southern cavalrymen.

But now, Sir, I can't help but feel that I have offended you in some way. That suddenly you have lost all sympathy for our cause. That You are throwing in with von Rundstedt and his paper-hanging God . . . our situation is desperate . . . my soldiers from the Meuse to Echternach are suffering tortures of the damned. . . . Damn it, Sir, I can't fight a shadow. Without Your cooperation from a weather standpoint I am deprived of accurate disposition of the German armies and how in hell can I be intelligent in my attack? All of this probably sounds unreasonable to You, but I have lost all patience with Your chaplains who insist that this is a typical Ardennes winter, and that I must have faith.

Faith and patience be damned! You have got to make up Your mind whose side You're on. You must come to my assistance, so that I may dispatch the entire German Army as a birthday present to Your prince of Peace.

Sir, I have never been an unreasonable man, I am not going to ask you for the impossible . . . all I request is four days of clear weather . . . so that my fighter-bombers can bomb and strafe, so that my reconnaissance may pick out targets for my magnificent artillery. Give me four days of sunshine to dry this blasted mud. . . . I need these four days to send von Rundstedt and his godless army to their Valhalla. I am sick of the unnecessary butchery of American youth, and in exchange for four days of fighting weather, I will deliver You enough Krauts to keep Your bookkeepers months behind in their work. Amen.

On December 27 there was another Patton address to God: "Sir, this is Patton again, and I beg to report complete progress. Sir, it seems to me that You have been much better informed about the situation than I was, because it was that awful weather which I cursed so much which made it possible for the German army to commit suicide. That, Sir, was a brilliant military move, and I bow humbly to a supreme military genius."

The provenance of these two Patton prayers, allegedly written during the Bulge, is less clear. Both were published after the war in a pamphlet issued by the national tourist office of Luxembourg.[35] When later published

in a regimental journal in Sweden, they drew howls of outrage. The chief of
Swedish army chaplains called them heresy. The ecclesiastics, of course,
thought Patton was mocking God. None realized that although it was partly
tongue in cheek, when it came to God, Patton was in dead earnest.

When the bad weather temporarily ended, Patton exulted, "Hot dog! I
guess I'll have another 100,000 of those prayers printed. . . . Get him up
here. I want to pin a medal on him."[36] When summoned to hear Patton praise
him, O'Neill became the only U.S. Army chaplain ever decorated with the
Bronze Star for writing a prayer. "Chaplain, you're the most popular man in
this headquarters. You sure stand in good with the Lord and soldiers." Then,
recalled O'Neill: "He cracked me on the side of my steel helmet with his
riding crop. That was his way of saying, 'Well done.' "[37]

International News Service correspondent Larry Newman, who covered
Third Army, wrote:

> Patton was never disheartened. In the midst of the battle—perhaps the
> most desperate a U.S. Army has ever had to fight—Patton called a confer-
> ence of all correspondents. As we filed into the room, the tenseness was
> depressing. But when Patton strode into the room, smiling, confident, the
> atmosphere changed within seconds. He asked "What the hell is all the
> mourning about? This is the end of the beginning. We've been batting our
> brains out trying to get the Hun out in the open. Now he is out. And with
> the help of God we'll finish him off this time—and for good."

Then Patton said, "I have a little Christmas card and prayer for all of you."
On one side was "a simple Christmas greeting with his signature." On the
rear was the weather prayer composed for him by Chaplain O'Neill.[38]

More than ever, Patton made it a point to be seen during the Bulge, always
riding in an open armored jeep. The cold was so intense that most soldiers
dressed in as many layers of clothing as they could manage, but Patton's
only concession to the glacial temperatures was a heavy winter parka or an
overcoat. He spent little time in his headquarters and most of each day on
the road, to see and be seen by his troops, and to endure the same wretched
conditions. Daily he prowled the roads of the Ardennes, sitting ramrod stiff,
often with his arms folded, his face unsmiling. More than once his face
froze. Word of his presence managed to filter through the amazing GI
grapevine with astonishing rapidity, as did his words of praise for his troops,
which were invariably reported down through the chain of command: "The
Old Man says . . ." or "Georgie says. . . ." However, Patton could not
indulge in his long-standing habit of driving to the front and flying home,
out of sight of the troops, and one cold, dark, miserable afternoon he
encountered a column of the 4th Armored moving toward Bastogne. Tanks

and vehicles were sliding off the road in the thick ice. Someone recognized Patton and let out a shout that began to roll down the column as soldiers in trucks and tanks began cheering. After the war a GI told Beatrice: "Oh, yes, I knew him, though I only saw him once. We was stuck in the snow and he come by in a jeep. His face was awful red and he must have been about froze riding in that open jeep. He yelled to us to get out and push, and first I knew, there was General Patton pushing right alongside of me. Sure, I knew him; he never asked a man to do what he wouldn't do himself."[39]

While the bureaucrats in the War Department and in the ETO dragged their heels, the soldiers who had to fight in the terrible winter weather did so with woefully inadequate uniforms and equipment. Arctic overshoes, parkas, and white camouflage uniforms were nonexistent, as was white paint for the Shermans. When Lee could not fill his needs, Patton took matters into his own hands and commissioned a French factory to manufacture ten thousand white capes per week for the Third Army.[40]

One evening correspondent Leland Stowe persuaded Patton to grant a private interview. Patton had his public face on, and was in rare form. He began by offering Stowe a large Havana cigar. Nearby lay his pistol belt and holsters. "Patton's squeaky falsetto soared nearly an octave, 'You see that pistol. . . . Take a good look. By God, its ivory-handled—not pearl! . . . All this cockeyed nonsense about me wearing a pearl-handled revolver . . . ! Just a bunch of goddamned ignoramuses. . . . Why, no real gunman would carry a pearl-handled pistol. It's bad luck. . . . Besides, I wear that particular gun because I killed my first man with it.' With that his grin expanded to Cheshire-cat dimensions."[41]

Stowe asked why the British Second Army had not attacked. Patton glowered at him "as if I were von Rundstedt. His falsetto crackled into full crescendo. 'Why in hell don't they attack? They're just being true to form. Afraid to take a chance . . . more afraid of losing a battle than anxious to win one. Well, that's Montgomery for you.'" He was interrupted by Bradley's G-2 with an undated intelligence report. Patton grinned and said: "I knew everything he told me at four o'clock this afternoon. . . . Of course they'll attack."

"Then he grabbed a phone and called one of his corps commanders. . . . All he said was: 'The Germans are preparing to attack at daybreak. Stop the bastards where they are!' Then he hung up, looking more pleased even than he had before." Asked about the remarkable swing of the Third Army to the north, Patton grinned and replied:

Yes, we broke all records moving up here. It was all done by the three of us . . . me, my chauffeur, and my chief of staff. All I did was to tell my division commanders where they'd got to be tomorrow. Then I let the others do it, adding modestly, "To tell you the truth, I didn't have anything much to do with it. All you need is confidence and good soldiers . . . if there is confidence at the top the soldiers all feel it. I know a lot of — soft-headed armchair generals accuse me of killing off my men. They don't know their fat behinds from a tommy gun. I don't waste men. I believe in saving my men's lives. And, by God! I've done it . . . again and again. More often than not the best way to save men's lives is to risk them . . . to take chances, and make your men fight better."

Passionately Patton continued:

I'll tell you one thing, Stowe. Maybe the G.I. hates discipline, but only until he learns that that's what makes a winning soldier. I'll put our god-dam, bitching, belching, bellyaching G.I.s up against any troops in the world. The Americans are sons of bitches of soldiers—thanks to their grandmothers! All you've got to do is to show them the value of disci-pline . . . give them the habit of obeying in a tight place. Yes. The Ameri-can is a hell of fine soldier.

Now in full stride, Patton joyfully declared that he did not give a damn what others thought about him or what he said. "You know what they can do," his smile "positively beatific. I've studied military history all my life. Georgie Patton knows more about military history than any living person in the United States Army today. With due conceit—and I've got no end of that—I can say that's true."[42]

Patton continued to raise hell with anyone he thought had let down his com-rades but to be soft on those who were "boys being boys." When soldiers from units overrun by the Germans began filtering back to friendly lines, Patton asked to see two dozen men to find out for himself what had occurred. As one was being interviewed, Patton suddenly shouted: "Chief of Staff, arrest this SOB!" For some minutes he blasted the soldier as a coward, and a "yellow SOB" for allegedly having deserted his friends. The soldier was turned over to the provost marshal, but later Patton ordered him released. The terrified man was escorted by two MPs to a senior staff officer and told he was free to go. "If ever a soldier looked like he was going to receive the death sentence, this man did. . . . I thought at first he was going to pass out," said an eyewitness.[43]

Patton took a fatherly, protective attitude toward the "Doughnut Dolly" Red Cross women assigned in the Third Army area, and was usually at the peak of showmanship around them. They saw the best and worst of him. One night a group overheard him give orders to a field commander on the phone. Turning to his attentive audience he said, "You have had the privi-lege of hearing the greatest general in Europe make a decision." Later, when they entertained him at a dinner in his honor he was persuaded to sing "Lilly from Piccadilly," but made up his own lyrics, which were unprintable. "At these moments, he was a very lovable elderly gentleman," said one. The next morning as he roared by, his motorcycle escort waved their Red Cross vehicle into a ditch, spilling their doughnuts and coffee everywhere.[44] On the other hand, he personally performed sentry duty while two Red Cross used the men's bathroom because there were none for women.

The Germans regularly shelled Luxembourg City at night with a large caliber gun. One night Patton telephoned Bradley to ask: "Brad, can't you send some Ninth Air Force planes to knock out that gun? Not that I mind, but it's driving Willie nuts."[45] When he had to relieve a senior officer with bleeding ulcers, Patton asked Beatrice to tell their son to be sure and "recognize" the officer's son, who was a plebe at West Point.

The defense of Saint-Vith, brilliantly orchestrated by Clarke, managed to stem the advance of Manteuffel's Fifth Panzer Army for eight days and was a key factor in the German failure in the Ardennes. One day during the battle, Clarke informed a sergeant manning a forward infantry outpost that he had heard that Patton's Third Army was attacking from the south. "The sergeant thought for a minute and said, 'That's good news. If Georgie's coming we have got it made.' I know of no other senior commander in Europe," said Clarke years afterward, "who could have brought forth such a response."[46]

On Christmas Eve, Patton, Bradley, and Codman attended a candlelight communion service at the Episcopal Church in Luxembourg City, sitting in a box once used by former Kaiser Wilhelm II. The dreadful weather notwithstanding, Patton ordered that every soldier in the Third Army was to have a hot turkey dinner on Christmas Day. To ensure that his orders were carried out, he and Mims spent the day driving from one unit to another. It was, wrote Patton in his diary, "a clear cold Christmas, lovely weather for killing Germans, which seems a bit queer, seeing Whose birthday it is." He found his troops cheerful, but "I am not, because we are not going fast enough." Mims recalled: "We left at six o'clock in the morning. We drove all day long, from one outfit to the other. He'd stop and talk to the troops; ask them did they get turkey, how was it, and all that." Patton was pleased that his men received a hot meal, for he sometimes had little faith in his mess sergeants, who, he complained, "couldn't qualify as goddam manure mixers. They take the best food Uncle Sam can buy and bugger it all up."[47] He also constantly checked for trench foot, and his troops inevitably heard the refrain, "Men we can get all kinds of equipment except we can't get more soldiers."

Patton again escaped death on Christmas Day. "As he approached our headquarters from the direction of Arlon a fighter plane (American without markings of any kind) strafed his small convoy of jeeps. As any other mortal, General Patton found refuge in the roadside ditch. After his departure and since only his dignity was harmed, we had a good laugh at his expense."[48] By the time he arrived at the 26th (Yankee) Division Patton was in a foul mood. The commanding general, Willard S. Paul, thought Patton had come to relieve him because of the slow progress of his division. Instead, Patton walked up to Paul, a much smaller man, threw him arm around his shoulder and said, "How's my little fighting sonofabitch?" Paul was so cheered not to have been relieved that he later said there was nothing he would not have done for Patton.[49]

By December 26, Patton was convinced that "the German has shot his wad," a judgment based on his observation of POWs who had not been fed in three to five days. "We should attack," he complained. "Why in hell the SHAEF thinkers hold the 11th Armored Division, 17th Airborne and 87th Infantry Divisions at Reims is beyond me. They should be attacking." He was unhappy that Eisenhower was not more aggressive, but saved his strongest criticism for Montgomery, whom he called "a tired little fart," and blamed him for virtually every problem in the Ardennes. When the SHAEF G-3 relayed a message that Eisenhower "is very anxious that I put every effort on securing Bastogne," Patton wrote in his diary, "What the hell does he think I've been doing for the last week?"[50]

The 4th Armored finally relieved Bastogne the afternoon of December 26, and although the town remained surrounded on three sides, Task Force Abrams had opened a corridor, thus ending only the first phase of one of the most dramatic battles of the war. In one of the few letters Patton managed to write to Beatrice during this period, he elatedly proclaimed: "The relief of Bastogne is the most brilliant operation we have thus far performed and is in my opinion the outstanding achievement of this war. Now the enemy must dance to our tune, not we to his. In the morning we are starting on a new series of attacks which may well be decisive . . . this is my biggest battle."[51]

The Battle of the Bulge, however, was far from over, and the bloodiest battles of the winter war in the Ardennes were yet to come. The relief of Bastogne by no means signaled the end of American troubles there. Indeed, it was not until January 12 (Beatrice's birthday) that the bloody three-week battle could be declared won. Three American divisions had at last broken the back of the siege and erased the final German salient, dooming some fifteen thousand of Hitler's best troops to capture. The previous night Patton had accurately predicted: "I believe that today ends the Bastogne operation. From now on it is simply a question of driving a defeated enemy. . . . I believe that the Bastogne operation is the biggest and best the Third Army has accomplished . . . and I hope the troops get the credit for their great work."[52]

When Patton visited Bastogne after it was liberated there were many

German corpses still frozen where they had fallen. "Finest battlefield I ever saw," he observed, but also "the most completely destroyed town that I have seen since the last war."[53] Later he told the correspondents: "It's a helluva lot easier to sit on your rear end and wait than it is to fight into a place like this. Try to remember that when you write your books about this campaign. Remember the men who drove up that bowling alley out there from Arlon."[54]

One of Patton's memorable confrontations during the battle was not with the Germans but with famed cartoonist and war correspondent Bill Mauldin, whose cartoons of the Willie and Joe dogface soldiers now appeared in *Stars and Stripes* and provoked Patton to fits of near apoplexy (he referred to it as a "scurrilous sheet" that subverted discipline). After Patton complained to SHAEF about Mauldin's "goddamned cartoons," Harry Butcher suggested a meeting between the two. After thundering that "if that little son of a bitch sets foot in Third Army I'll throw his ass in jail," Patton relented. A frightened Mauldin arrived, convinced he had been sent on "a suicide mission."

> There he sat, big as life even at that distance. His hair was silver, his face was pink, his collar and shoulders glittered with more stars than I could count, his fingers sparkled with rings, and an incredible mass of ribbons started around desktop level and spread upward in a flood over his chest to the very top of his shoulder, as if preparing to march down his back too. His face was rugged, with an odd, strangely shapeless outline; his eyes were pale, almost colorless, with a choleric bulge. His small, compressed mouth was sharply downturned at the corners, with a lower lip which suggested a pouting child as much as a no-nonsense martinet. It was a welcome, rather human touch. Beside him, lying in a big chair, was Willie, the bull terrier. If ever dog was suited to master this one was. Willie had his beloved boss's expression and lacked only the ribbons and stars. I stood in that door staring into the four meanest eyes I'd ever seen.

> They made quite a pair: the twenty-five-year-old baby-faced sergeant and the crusty general. Mauldin gave Patton his best parade ground salute, and Patton arose to offer his hand. Willie rose with his master—and fell off his chair. Patton told Mauldin to sit, and he took Willie's place. "The dog not only looked shocked now but offended. To hell with Willie. Butcher had been right. This was going to be O.K."

> Hardly. Mauldin's acid pen was unsparing, whether skewering the brass or the military police, and Patton launched into a tirade about "those god-awful things you call soldiers," insisting that Mauldin made American soldiers look "like goddamn bums." His ire rising, Patton demanded: "What are you trying to do, incite a goddamn mutiny?" The trouble was that "they

were not fighting the same war . . . what made Patton choleric was the way Willie and Joe *looked*. Scruffy. Disheveled. His own Third Army troops were clean-shaven, boots polished, neckties in place. Why, he sputtered to the Supreme Command, should *Stars and Stripes* star Willie and Joe?"

Patton then launched into a lengthy dissertation about armies and leaders of the past, of rank and its importance. "Patton was a master of his subject. . . . I felt truly privileged, as if I were hearing Michelangelo on painting. I had been too long enchanted by the army myself . . . to be anything but impressed by this magnificent old performer's monologue. Just as when I had first saluted him, I felt whatever martial spirit was left in me being lifted out and fanned into flame."

Patton's monologue encompassed four thousand years of military history and had barely reached the Hellenic wars when Mauldin unconsciously reached out to scratch Willie's ear, but he wisely stopped, later noting that it would have been his working hand, and convinced that with one snap of his jaws, Willie could have accomplished what Patton would have welcomed: to put him out of business. Waving a batch of Mauldin's cartoons, Patton raged that he was ruining morale: "The krauts ought to pin a medal on you." After nearly forty-five minutes of ranting, Mauldin had less than two minutes to offer his own views. When Patton had growled, Willie had bristled, ready to carry out his master's bidding. When the two parted, agreeing to disagree, Patton said: "You can't run an army like a mob. . . . All right, sergeant, I guess we understand each other now." Mauldin was thankful to escape from both Patton and Willie (who reclaimed his seat), and appreciative that he had not become Willie's lunch.

Word circulated that Patton had threatened to throw Mauldin in jail if he ever again set foot in the Third Army. Mauldin would later note that despite what Patton thought, he was neither antiofficer nor against discipline: "If you're a leader, you don't push wet spaghetti, you pull it. The U.S. Army still has to learn that. The British understand it. Patton understood it. I always admired Patton. Oh, sure, the stupid bastard was crazy. He was insane. He thought he was living in the Dark Ages. Soldiers were peasants to him. I didn't like that attitude, but I certainly respected his theories and the techniques he used to get his men out of their foxholes."

Mauldin had the last word after his great encounter with George S. Patton. In 1945 he won a Pulitzer Prize, and Willie (the soldier) appeared on the cover of *Time*. Patton was again left sputtering over Bill Mauldin's "goddamn bums."[55]

To celebrate the coming of 1945, Patton ordered a special greeting to be delivered to the Germans: "At the stroke of midnight on New Year's Eve every artillery piece in Third Army would fire on a likely target as a New Year's salute from Third Army to the Wehrmacht."[56]

The year 1945 thus opened with an earth-shattering roar, and forward observers reported hearing German screams ringing through the night. Patton wrote: "I hope this is true." His men, however, were more concerned about not freezing to death, one commenting: "This is a helluva way to start the new year."[57]

On January 3 Montgomery at last attacked in the north, while the Third Army fought to link up with First Army at Houffalize. The cold weather was now even more brutal; everyone suffered in temperatures that dropped off the thermometer. The stage had been reached of what GIs had nicknamed the "Bitter Battle for the Billets." At night they would hole up in whatever building, shelter, or basement they could find that would provide some measure of warmth, however small, and the following morning return to battle.

Freezing fog again grounded the air forces, the already icy roads became coated with sheets of sleet, and with progress bloody and painfully slow (a gain of two miles a day was considered a great accomplishment), the Germans, after expecting the worst, soon grasped that there was little danger of entrapment and fought even more desperately. The Allied attacks moved with all the sluggishness of a bulldozer, and "simply shoveled the enemy slowly back. . . . It did not, as Eisenhower wanted, destroy their effectiveness."[58] Throughout the remainder of January the pincers of the First and Third Armies mercilessly ground away in the forests and shattered towns of the Ardennes, turning the white snow blood red when the irresistible force met the immovable object.

Although it was clear by the end of December that Hitler's strategic aim of splitting the Allied front in half was doomed to failure, thus sealing Germany's ultimate fate, no one bothered to inform the soldiers of the German army who were "hungry, frozen, low on supplies and ammunition but high in morale."[59] To this might be added Patton's despondent observation in his diary on January 4 as the fighting raged around Bastogne: "We can still lose this war. The Germans are colder and hungrier than we are, but they fight better."

On January 8, 1945, Patton was on the road in an open jeep marked only with his three stars, his sole concession to the cold a lap robe and plastic sheeting along the passenger and driver's side to deflect the wind. As usual the roads were clogged with columns of vehicles stretching for miles. It was six degrees below zero. This particular column was filled with trucks carrying infantry of the 90th Division forward to battle; on the other side of the narrow highway were ambulances bringing the wounded to the rear. "When the men recognized Patton, they leaned out of the trucks, cheering wildly. The General's face broke into a smile. He waved. But he could hardly hold back the tears. Tomorrow many of those now cheering would be dead—because of his orders."[60] It was an incredible scene, wonderfully spontaneous and for Patton, "the most moving experience of my life, and the knowledge of what the ambulances contained made it still more poignant."[61]

"Radio th' old man we'll be late on account of a thousand mile detour."

After the pincers of the First and Third Armies at last closed the famous Bulge on January 16, Patton drove to the shattered town of Houffalize. While waiting for bulldozers to clear the highway, he said: "Little town of Houffalize, here you sit on bended knees. God bless your people and keep them safe—especially from the RAF!" (a reference to an order given him earlier that the Third Army was not permitted to attack the town until the RAF was able to bomb the Germans out of Houffalize.)[62] The devastation was worse than Bastogne, the absolute worst he had seen in this war.

Both moved and bitter, Patton composed a Christmas carol to the melody of "O, Little Town of Bethlehem":

> *Oh little town of Houffalize,*
> *How still we see thee lie;*
> *Above thy steep and battered streets*
> *The aeroplanes sail by.*
> *Yet in thy dark streets shineth*
> *Not any goddamned light;*
> *The hopes and fears of all thy years,*
> *Were blown to hell last night.*[63]

Finally even the pigheaded Hitler realized that stubborn American fighting men and the elements had combined to shatter his grandiose gamble. "Although Allied harmony heaved and strained and cracked a little under the pressure of his offensive, it did not give way. He did not get his new war."[64] Nor had he even bought Germany time, for the Allies would not in any case have launched their final offensive until the weather had improved in the early spring of 1945.

Patton's reputation soared as a result of a battle that gripped the minds of the public as had no other had since the Normandy landings, and the great breakout in August. To have fought in the horrific winter conditions that existed in the Ardennes in December 1944 was a feat beyond measure. Moreover, Patton was the only senior commander to emerge from the battle with his reputation intact. Those of Bradley, Hodges, and Montgomery were blemished. Bradley's for the perception that he had lost control, thus leading to Eisenhower's controversial decision to assign the northern half of the Bulge to Montgomery, who, in turn, lost credibility and spurred American ire when, at a press conference in January 1945, he appeared to patronize and belittle the American effort. Hodges, whose First Army actually bore the brunt of the German counteroffensive, was tarred with the same brush as Bradley for, among other things, failing to foresee the German attack. By the end of the battle, both Patton and Bradley were thoroughly soured on Montgomery, and where previously there had been a measure of mutual respect for the British general, there now existed only contempt.

The six-week campaign gnawed at the normally mild-mannered Bradley, who never recovered from what he considered Eisenhower's stinging rebuke and lack of confidence in him. He was particularly embittered when Eisenhower announced that Simpson's Ninth Army would remain under Montgomery's control, and that it had not been made clear that Montgomery had only been in temporary command of his two armies. Left with little to do, during January Bradley began interfering with increasing frequency in Patton's conduct of the battle.

Although the beleaguered, encircled 101st Airborne "Battered Bastards of Bastogne" and McAuliffe's famous reply when the Germans demanded his surrender have become the focus of the story of the Bulge, it was really Bruce Clarke's Combat Command B of the U.S. 7th Armored Division, and the hastily assembled troops from the 28th and 106th Divisions, that successfully defended Saint-Vith for eight days that averted disaster. Clarke's stubborn defense of Saint-Vith and the 1st, 2d, and 99th Infantry Divisions' equally valiant defense of Elsenborn Ridge frustrated the German advance long enough to permit the Allied commanders to rush reinforcements and plug the gaps. The 82d Airborne Division (commanded by James M. Gavin, now a major general) also fought tenaciously and contributed mightily to the ultimate German failure by denying Manteuffel's Fifth Panzer Army the rapid breakthrough that was the key to the success of the counteroffensive. And how close it was: The Germans never knew that "the biggest filling station in Europe" was located in the Ardennes, near Stavelot, and contained 2.5 million gallons, enough fuel to have kept Sepp Dietrich's fuel-starved panzers on the move and perhaps changed the outcome of this deadly campaign.[65]

If Patton and McAuliffe emerged as the best-known military figures of the Bulge, "the real victory in the Ardennes belonged to the American soldier," writes Charles B. MacDonald, "for he provided time to enable his commanders—for all their intelligence failure—to bring their mobility and their airpower into play. At that point the American soldier stopped everything the German Army threw at him."[66]

Winston Churchill hailed the Battle of the Bulge as "the greatest American battle of the war [that] will, I believe, be regarded as an ever famous American victory."[67] A British historian has written:

> The Battle of the Bulge has been the subject of as much bilious criticism as Waterloo, Gettysburg and Alamein. As a result only three reputations have emerged with almost unqualified acclaim—those of Eisenhower, the Allied air forces and the American soldier. Justice demands that to these should be added the name of Patton." Moreover, Eisenhower's "faint praise given [Patton] in *Crusade in Europe* for his part in the battle there is a hint of condescension which comes ill from the pen of one who despite all his talents and virtues was an unblooded soldier.[68]

The battles that raged for six weeks in the frozen hell of the Ardennes were among the bitterest and bloodiest of any fought in the West. Casualties on both sides were not only staggering, but "the Ardennes offensive was a rude awakening. The surprise lay not so much in the resurgence of German power as in the revelation of Allied weakness."[69] In such terrible conditions Patton was awed: "How men live, much less fight, is a marvel to me." The Third Army after-action report estimated 100,000 German casualties.[70] No official German casualty figures have ever been computed. However, the German High Command's own estimate of loss in the Bulge is 81,834, with 12,652 killed, 38,600 wounded and 30,582 missing.[71] Both sides lost enormous quantities of equipment; those of the Germans were irreplaceable.

The day the Bulge was closed Patton wrote to his son, George, that leadership is what wins battles. "I have it—but I'll be damned if I can define it." Of Omar Bradley he would soon write: "He is a good officer but utterly lacks 'it.' Too bad." He had again defied death when one of Hitler's new jets dropped a bomb nearby, scaring both Willie and his master. "They also shoot rockets at us, but one gets used to such things. It is like a thunderstorm. You are not apt to be in the way. And if you are, what the Hell, no more buttoning and unbuttoning."[72]

After being stymied and frustrated in Lorraine and the Saar, in the Ardennes Patton was presented with another opportunity not only to display his genius for war but to turn a wretched situation to his advantage, and that of the Allies. However, as with any military operation, there were flaws. Patton's maneuvering of Third Army to relieve Bastogne did not win the Battle of the Bulge. Indeed, as historians have pointed out, the relief of Bastogne was made in a sector occupied by inferior German formations, and the heaviest German attacks against Bastogne did not commence until December 26. They also note that credit must be given to the men of the First Army, who stubbornly held the northern shoulder against overwhelming odds.

Nor is Patton without critics of his generalship during this great battle. British historian Peter Elstob writes that, after a dramatic start, the disappointingly slow progress of the Third Army was due to more than the horrible weather conditions. Patton's problem was that he could not mount his kind of battle: "Instead of an all-out attack concentrated on a narrow front designed to cut through the base of the salient—he had to try to push a twenty-five mile wide front some twenty miles through country which seemed to consist of a series of natural defensive features. His attacking forces wore themselves down against Seventh Army's tough and elusive defenses. . . . Patton got his reinforcements late and committed them too soon."[73]

The advance on Houffalize was equally untidy. Patton committed two green divisions into battle without adequate reconnaissance, and seems to have launched a task force at night in order to claim Houffalize for the Third Army before the First Army and the British 30th Corps, a ploy reminiscent of the race for Messina. Nor did it help that the Third Army lost the crucial element of surprise during the relief of Bastogne, largely due to its notoriously poor radio discipline, which had alerted the Germans of their presence. Of the American armies, the Third Army's record in this sphere was abysmal.

One of the nastiest incidents of a thoroughly nasty campaign occurred in the northern sector when SS troops of Lt. Col. Joachim Peiper's Kampfgruppe Peiper brutally massacred 350 American troops and one hundred Belgian civilians at Malmédy and twelve other places. However disgraceful, such incidents seem to be the inevitable by-products of the ugly business of men killing each other on battlefields. In sports, or in war, tit-for-tat is usually the rule. The Ardennes appears to have been no different. Although the official U.S. Army history proclaims that there is no proof that American troops "took advantage of orders, implicit or explicit, to kill SS prisoners," John Toland has documented an example of some sixty POWs shot by members of the green U.S. 11th Armored Division, which had been rushed to the front and was engaged in its first combat. According to Toland: "word had been passed down to take no prisoners;" too late, it was learned that "Somebody fouled up. We're supposed to take prisoners."[74] Although Patton had no discernible role in this and other such incidents, it is clear he learned of the killings by the 11th Armored. His diary relates a visit on January 4, 1945, in which there were "some unfortunate incidents in the shooting of prisoners. (I hope we can conceal this.)" He added: "I can never get over the stupidity of our green troops."[75] His only reference to dead Germans in letters to Beatrice were gruesome descriptions of corpses he had found on the battlefield and photographed, once lamenting the absence of color film to record their "claret color."[76] Whatever his personal revulsion to atrocities, Patton seemed quite prepared to turn a blind eye to such incidents committed in the heat of battle. What is far more certain is that no such orders ever emanated from his headquarters. Nevertheless, such incidents tainted both sides.

If the entire Ardennes campaign resembled Wagnerian melodrama in the best German tradition, it was for Patton a western film: like the cavalry of yore to the rescue of homesteaders in a dramatic cliff-hanger, culminating with Third Army in the role of the cavalry, and Patton at its head, rallying his troops. No battle could have been more tailor-made for Patton's talents—and for his theatrics.

German losses in the Bulge were enormous and irreplaceable, thus setting the stage for the climactic battles of the war in the heartland of Germany. On December 30 the *Washington Post,* which had gleefully savaged him over Knutsford, now ran an editorial titled PATTON OF COURSE, noting that: "It has become a sort of unwritten rule in this war that when there is a fire to be put out, it is Patton who jumps into his boots, slides down the pole, and starts rolling."[77] Gay bitterly noted in his diary the fickleness of the media. "Seventh Army one year ago today was in disgrace; Headquarters of the Third Army, which is practically the same staff, today is being acclaimed by the world . . . [which] is calling on the same Commanding General . . . to stop the German thrust."[78] In a letter to a friend, Patton wrote: "Fortunately for my sanity, and possibly for my self-esteem, I do not see all the bullshit which is written in the home town papers about me."[79] In the aftermath of the greatest debacle of the war, America needed heroes, and in Patton and McAuliffe their need was fulfilled.

Patton remained as unpopular as ever with Courtney Hodges and the men of the First Army, who were increasingly irritated by articles in *Stars and Stripes* extolling Patton and the Third Army. The last straw occurred when Chaplain O'Neill's weather prayer appeared in a box on the front page. Hodges's staff howled in protest and demanded—and got—a separate edition printed for the First Army.[80] They were, of course, quite right. Neither Patton nor the Third Army won the Battle of the Bulge, but when it came to garnering publicity for his army, Patton was never apologetic that his staff proved more adroit in the publicity business than anyone else.

His enemy, Field Marshal von Rundstedt, said after his capture: "Patton, he is your best."[81] Everett Hughes gleefully repeated to Patton that Eisenhower had said: "Do you know, Everett, George is really a very great soldier and I must get Marshall to do something for him before the war is over."[82] Patton remained embittered that Eisenhower would never tell him so to his face. More than anything he wanted the approval and praise of his oldest friend. Few things were to disappoint him more deeply than his failure to obtain them.

Patton gave the credit to the soldiers who fought the battle. He rated them magnificent; they both moved and awed him. On January 29 he told the press: "We hit the sons of bitches on the flank and stopped them cold. Now that may sound like George Patton is a great genius. Actually he had damned little to do with it. All he did was to give orders." When a correspondent inquired why it had taken Montgomery so long to counterattack, Patton grinned and replied: "I am unable to comment on that. This is one time when I am not my brother's keeper."[83]

Third Army suffered 50,630 casualties (4,447 killed, 20,185 wounded, 3,228 missing, and 22,770 nonbattle casualties).[84] Summing it all up, Patton wrote: "During this operation the Third Army moved farther and faster and engaged more divisions in less time than any other army in the history of the United States—possibly in the history of the world. The results attained were made possible only by the superlative quality of American officers, American men, and American equipment. No country can stand against such an Army."[85]

What the Battle of the Bulge demonstrated is that, while possessed of tremendous vision—the ability to anticipate and react with impeccable foresight to an enemy move or countermove—Patton's greatest strength was not so much as a tactician but as an organizer, a mover and shaker. Gerald Astor, author of a recent account of the Bulge, is right on the mark when he writes that Patton's "true genius lay in his ability to put the show on the road, to move men and machines."[86] Patton understood that despite the terrible conditions under which his troops fought, it was equally difficult for the enemy, and that to attack, and to keep attacking, was what would win the battle. Patton himself rated determination to succeed higher than luck or genius and observed: "We have to push people beyond endurance in order to bring this war to an end."[87]

The Bulge also removed the uncertainty of a further grinding campaign in the Saar in favor of a battle that actually held greater promise. Instead of a winter campaign in the mud that at one point so frustrated Patton that he announced: "At the close of the war, I intend to remove my insignia and wristwatch, but will continue to wear my short coat so that everyone can kiss my ass," he was able to indulge in what he did best: fight on his own terms.[88]

Paul Harkins would later sum it all up: "The Bulge was an exhausting operation, filled with grim fighting, unimaginable situations, precise timing and movement, and a superhuman effort on the part of the American soldier." If George S. Patton had never before done or would never again do anything of significance, he had earned a place in history by his extraordinary generalship early in the Ardennes. It was a short, brutal campaign that not only solidified his reputation for generalship on the battlefield but left no doubt of the quality of the army on which he had put his imprint. It is the nature of war that perfection is never attainable, but this was Patton's and the Third Army's finest hour. No one else could have pulled off such a feat.

CHAPTER 44

Pissing in the Rhine

Dear SHAEF: I have just pissed in the Rhine. For God's
sake, send some gasoline.

—PATTON

Despite Hitler's failure in the Ardennes, the events of early 1945 were a
clear affirmation that the war was still far from over. Any doubts that there
would be an easy resolution were erased when, in early February, the Big
Three met at Yalta and reiterated their long-standing demand for nothing
less than unconditional surrender. This meant that the war would continue
until the Allied armies and the Red Army met somewhere in an annihilated
Nazi Germany.

Hitler also misread the Allied will to continue on the offensive after the
furious battles in the Ardennes, believing he would be given a respite that
would enable him to reinforce the crumbling Eastern Front. Eisenhower,
however, had no intention of letting up and was determined to pursue the
advantage by attacking across the broad front he had created the previous
August.

The arguments and debates after Normandy were merely a prelude to
1945, during which the contentiousness over the direction of the final campaigns of the war reached its zenith. Eisenhower was beset on both sides by
British and American generals with opinions contrary to those of the
supreme commander. It was, he said, like "trying to arrange the blankets
smoothly over several prima donnas in the same bed."[1]

While the generals fought over whose strategy would prevail, Brooke
and the British chiefs of staff, horrified at the notion of fighting major battles west of the Rhine, raised the stakes by pressuring Eisenhower to agree

to Montgomery's proposed "single, full blooded thrust" toward Berlin. They roundly condemned Eisenhower's strategy as akin to what someone sarcastically dubbed "Have a go, Joe," an expression commonly uttered by London prostitutes seeking business from American GIs.[2] To end this unseemly brawl, Marshall finally threatened the British by insinuating that Eisenhower would ask to be relieved as supreme commander unless they accepted his broad-front strategy.

Eisenhower, too, had studied history, and it had been his lifelong dream one day to emulate Hannibal's envelopment of the Romans at Cannae. SHAEF was turning into a Goliath that by March 1945 consisted of three army groups, totaling eighty-five Allied divisions, (twenty-three armored, five airborne, and fifty-seven infantry). To breach the great obstacle of the Rhine River, Eisenhower's strategy was to unfold in three phases, during which the Allied armies would invade Germany and overrun the Third Reich. While Montgomery made plans for Plunder, a massive airborne and amphibious operation in the Ruhr, to breach the lowlands of the Rhine in Holland with the First Canadian and Second British Armies, and the Ruhr with Simpson's Ninth U.S. Army, Bradley was to clear the Eifel and drive to the Moselle with the Seventh and Third Armies. Once across the Rhine, the armies of Hodges and Simpson would encircle the Ruhr and crush Model's Army Group B in the jaws of a huge Allied pincer: Eisenhower's Cannae. The Third Army and Devers's Sixth Army Group would advance toward Czechoslovakia and Austria and the Elbe River, where Eisenhower intended to halt the Allied advance and link up with the Red Army, which was crushing everything in its path during its advance from the east. Although Eisenhower seemed unhappy with limiting the two American army groups, he never altered his broad-front march through Germany.

In the greater Allied blueprint, American forces were again destined to play a secondary role in the advance toward Berlin. Patton believed that Eisenhower's plan was too timid and failed to attack von Rundstedt's defeated armies aggressively. Instead the American elements were confined to limited offensive actions, and only then if they conformed to the SHAEF plan. Patton fumed at the waste. "Active defense" was not his idea of how to win a war. He advised Beatrice that if she heard he was on the defensive: "It was not the enemy who put me there. I don't see much future for me in this war. There are too many 'safety first' people running it. However I have felt this way before and something has always turned up I will go to church and see what can be done."[3]

In frustration he returned to the "rock soup" approach he had first employed in Lorraine, taking advantage of whatever he thought Bradley or SHAEF would give him, buying time, deliberately remaining out of communication with Bradley, and taking objectives that would keep his units committed and therefore unavailable to be removed from his control. "I wonder if ever before in the history of the war, a winning general had to plead to be allowed to keep on winning," he mused. Shortly before an offensive against Bitburg, Patton was summoned to Bastogne by Eisenhower and was relieved to learn it was merely a photo opportunity instead of orders to halt his army. Whenever he could get away with it, Patton made it a practice to keep his superiors in the dark about his future actions. For the most part he was aided and abetted by Bradley, who cooperated where he could.

One of the myths about Patton is that he managed to hoodwink his superiors. It thrilled him to believe himself both audacious and disobedient of those who knew less about war than he did. But, as Bradley later revealed, the only person deluded was Patton himself. Both he and Eisenhower knew and understood exactly what Patton was doing. So long as Patton did not garner the limelight at Montgomery's expense, they generally approved of Third Army's operations. What Patton thought was insubordination was "purely make-believe, for without the knowledge of Patton's staff and mine, Eisenhower had agreed to it."[4] Indeed, neither was often fooled by any of his ploys, and it suited them to let Patton quietly believe that he was pulling a fast one. When he captured Trier in early March, Patton wrote to Beatrice that he had "fooled them again. . . . I had to beg lie and steal."

Patton was concerned when a congressman began a campaign to get him promoted to four-star general, fearing that Marshall would conclude that he himself was behind it. "They must promote Omar first or relieve him. We will all get four stars some day, and I never cared to be like the others." Patton later recounted that Eisenhower had apologized to him for not promoting him sooner, "but he has, however, good reasons—that is, you must maintain the hierarchy of command or else relieve them, and he had no reason for relieving them. . . . At the moment I am having so much fun fighting that I don't care what the rank is."[5]

As Patton happily admitted, during January and February 1945 "the supply situation, as well as the replacement situation, was the best it had ever been."[6] However, for American armies to be on the defensive and the Ninth Army assigned to the British for Montgomery's push to the Rhine left Patton "very bitter." With the British unable to jump off until February 10, Patton and Hodges were only permitted to continue attacking until the priority of resupply would again revert to the 21st Army Group. Only if their casualties and ammunition expenditure were within acceptable bounds would either army be permitted to continue offensively. Patton complained: "We felt it ignoble for the American armies to finish the war on the defensive. Another point which made us angry was the information that SHAEF was collecting a theater reserve" by taking divisions from the front-line armies.

The Ardennes counteroffensive had belatedly left Eisenhower and his staff determined never again to be caught with their pants down. Patton scorned the idea of a reserve as SHAEF's "new toy" and wondered: "Reserve against what? This seemed like locking the barn door after the horse was stolen. Certainly at this period of the war no reserve was needed—simply violent attacks everywhere with everything . . . the Germans do not have the resources to stop it."[7]

Eisenhower remained under substantial pressure from his three army group commanders, none of whom were prepared to sacrifice the priority of their operations to another. The extent of the divisiveness was never more evident than in Bradley's angry explosion when a senior SHAEF staff officer telephoned to inform him that the new theater reserve would necessitate the removal of several of his divisions. Patton, who had never seen Bradley lose his temper, observed with fascination as his superior reminded the caller that "I am goddam well incensed." Virtually every officer in the room stood up and applauded. Patton bellowed loudly enough to be heard clearly on the other end of the telephone: "Tell them to go to hell and all three of us will resign. I will lead the procession."[8]

From the Ardennes to V-E Day, Bradley's biggest ally was none other than Patton. This incident marked the pinnacle of their solidarity. But Bradley's favorable impression vanished when the derogatory, often scurrilous comments in Patton's wartime diary and letters to Beatrice were made public after Patton's death, and from that time forth Bradley's loathing of Patton knew few bounds.

Publicly Patton said of the stripping of his army: "We will comply promptly and without argument." Privately he told his staff: "It would be a foolish and ignoble way for the Americans to end the war by sitting on their asses. And, gentlemen, we aren't going to do anything foolish or ignoble like that—of course." The Third Army would therefore continue to attack under conditions of strictest secrecy: "Let the gentlemen up north learn what we're doing when they see it on their maps," he proclaimed, citing one of the most abused principles of warfare. "It may be of interest to future generals to realize that one makes plans to fit circumstances and does not try to create circumstances to fit plans. That way danger lies."[9]

In the Ardennes, Patton had overcontrolled his subordinates, but now tended to leave the operational details of war to his corps commanders, though he kept the heat on them to stay committed, thus avoiding enforced halts and the possible loss of divisions to the dreaded SHAEF reserve, wryly calling it "creeping defense." Nevertheless the famous Patton temper was still in evidence, as one corps commander learned the hard way, when two armored divisions were mistakenly directed at night along roads that crossed each other. A precarious situation ensued and an MP died directing traffic. A very angry

Patton drove up, stormed over to the corps commander, and shouted that he was damned if he would have any more of his MPs killed by such foolishness. "You made this mess; now you get out there and direct traffic yourself." For nine straight hours the general personally directed traffic through the intersection, doubtless never again making such a foolish mistake.[10]

Another corps commander briefed Patton on a forthcoming attack and presented an excellent plan. Asked when he could attack, the general replied: "Wednesday." Back came: "That's too late." When the general argued that Wednesday was the best he could do, Patton again said it was too late. "But general, you must remember that it takes time to do these things," he pleaded. "Then what the hell are you wasting your time here for?" The attack would take place Monday night. It did.[11]

The invasion of Hitler's Thousand-Year Reich turned out to be a deadly business. Although their cause was ultimately hopeless, the German army fought tenaciously in the dismal months of early 1945. They blew up the Roer dams and flooded the area, over which an offensive into the Reichswald (a dense state forest that proved exceptionally difficult to fight in) and the heart of the Siegfried Line cost six thousand British casualties. It was the same old story: They created every possible obstacle, fought stubbornly, and gave ground only when it could no longer be held. For both sides the cost was staggering. In six weeks the German army lost an estimated one-third of its total remaining strength defending the area west of the Rhine. Hitler decreed his army would fight to the death. His generals understood the limits of what could be asked of a soldier, and eventually even the fanatical Model disobeyed the Führer. The cost to the Germans of the broad-front thrust to the Rhine was 250,000 in POWs alone.

However, the Allies paid an equally high price. Tanks did not function well in the Eifel, where the Third Army was attacking, and, like Lorraine, it devolved into yet another series of bloody infantry battles fought by cold, weary GIs who amazed Patton with the ingenious ways they found to ward off the cold: "How human beings could endure this continuous fighting at sub-zero temperatures is still beyond my comprehension." The weather was so appalling that Patton actually gave Manton Eddy permission to halt an attack. Instead Eddy attacked even more vigorously. Still, occurrences of nonbattle casualties and self-inflicted wounds rose dramatically.

One of the most rugged areas of Germany, the Eifel sector of the famed Siegfried Line is bisected by the Moselle, Our, and Sauer Rivers. The so-called triangle area of the Eifel, and its principal city of Trier, is extremely hilly, heavily forested in places, dominated by fast-flowing rivers swollen by winter rains and snows, and the defenses of the Siegfried Line. The Eifel proved even grimmer than Lorraine—so difficult in fact, that it took the

Third Army a month of heavy fighting to reduce the bloody triangle in conditions fully as miserable as the Bulge. From Alsace to the Rhine estuary, across a 450-mile front, the Allied armies advanced with painful slowness in cold and mud and blood.

Patton correctly sensed that the time was ripe to attack and keep attacking, and was shocked when, on February 10, Bradley telephoned to report that Eisenhower was transferring divisions to Simpson's Ninth Army. How soon could the Third Army go on the defensive? Patton replied that he was the oldest serving general in both age and combat experience and he was goddamned if he would do so; he would resign first. Bradley replied that Patton owed too much to his troops to even consider such an act. Eisenhower permitted Bradley "to assume a posture of 'aggressive defense' . . . I chose to view it as an order to 'keep moving' toward the Rhine with a low profile."[12]

The ancient road to Trier was once tramped by Roman legions, a fact not lost on Patton, who likened his triumphal journey in 1945 to that of Caesar centuries earlier. He could "smell the coppery sweat of the legions" and was proud of following in Caesar's footsteps. He referred frequently to the fact that his current battles were modern-day versions of the Romans.' Many, he believed, would never understand him, once saying: "I'm going to be an awful irritation to the military historians, because I do things by sixth sense. They won't understand."[13] Where Bradley would laboriously work out his intended plans and strategy, Patton would matter-of-factly awaken with an idea already worked out in his head, which occurred on February 6, when the Palatinate campaign "popped into my head like Minerva. Whether ideas like this are the result of inspiration or insomnia, I don't know."[14]

The key to capturing Trier was for Patton to persuade Bradley to return the 10th Armored Division, sent to reinforce Bastogne in December. Both generals worried that SHAEF might overrule them, and Bradley agreed with Patton's suggestion that he stay away from his telephone for the next several days until it was too late to recall the division. After Trier fell, Patton received a message directing him to bypass the city as it would take four divisions to capture it, leading to his famous irreverent reply: "Have taken Trier with two divisions. Do you want me to give it back?"[15]

On a dirt road near Trier, Patton escaped death yet again when mortar shells began raining down from two German units firing blindly at each other. Patton turned to Gay and said: "Hap, in my considered judgment, the thing to do is get the hell out of here, drive home and have a drink." When he encountered a 155-mm gun tightly jammed under an overpass, Patton testily instructed a colonel that he could blow up the "goddam gun," blow up the "goddam bridge," or blow out his "goddam brains," he didn't much care which. Nearby, Patton spotted a man in civilian clothes, who he instinctively knew was a German officer, and ordered him taken to a POW camp. When

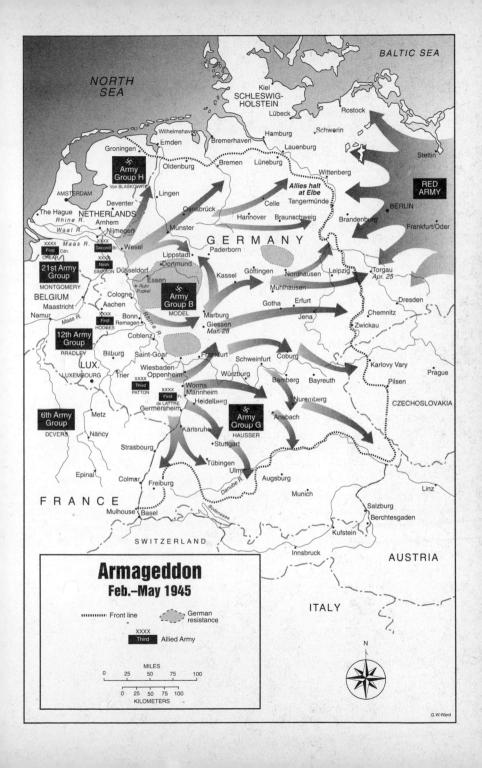

Armageddon
Feb.–May 1945

asked how he knew this, Patton said: "Goddammit . . . if after all these years I can't recognize a soldier when I see one, I should turn in my uniform."[16]

With the Third Army in limbo in mid-February, Patton took his first leave in thirty months, to Paris, where he visited Everett Hughes, drank champagne with an old French friend and her daughter, and attended the famed Folies-Bérgère. The manager's wife said that she hoped Patton would make the Folies his home whenever he was in Paris. "I can imagine no more restless place than the Follies, full of about one hundred practically naked women," he confided to his diary. The following day he went on a shoot with Bedell Smith and several others in the royal game preserve once used exclusively by the kings of France, but left early with a violent case of ptomaine poisoning, believing that someone was deliberately out to get him.[17] He later thought that the Germans were somehow tapping his telephones in the Third Army. From triumph in the Ardennes to near dismay in the Eifel, Patton's mood swings continued to be as fickle as ever.

On March 7 one of the great feats of the war occurred at the Ludendorff railway bridge across the Rhine at Remagen, which was captured intact by an American task force of the 9th Armored Division, commanded by Brig. Gen. William M. Hoge, and the First Army established a bridgehead east of the Rhine, before the bridge collapsed. The German response was to launch furious counterattacks, leaving sectors of the river to the south relatively unprotected. Both Bradley and Patton were determined to take advantage of the opportunity to beat Montgomery across the Rhine.

The final battles of the war were as much about prestige as any fought by the Allies. The rift between Montgomery and the American commanders in the wake of Montgomery's abysmal press conference left Bradley and Patton determined to prevent the 21st Army Group from reaping the victory headlines. Bradley admitted his own political naïveté and that of the other Americans who thought only in terms of military objectives. The unsophisticated Bradley had quickly learned how to play Montgomery's game, and in the process had made himself nearly as unpopular at SHAEF as the British field marshal. When Harold Bull (the SHAEF G-3) declared that the capture of the Remagen bridge did not fit in with "the plan," Bradley responded more like Patton: "What in hell do you want us to do, pull back and blow it up?" With Eisenhower's blessing, Bradley exploited Remagen, although he would gladly have strangled Bull, who seemed more concerned about "the plan" than he was about winning the war.

After capturing Trier, Patton sensed an opportunity to break loose from the confining terrain in which the Third Army had been fighting. Instead of advancing shoulder to shoulder with the 6th Army Group in the south, Patton saw an opening in the Palatinate and took it. Bradley persuaded Eisenhower to permit a new offensive whereby the Third Army would exploit its success by sweeping south from Trier to envelop the remaining German units defending the Siegfried Line, in conjunction with an attack by Patch's Seventh Army. Patton unleashed the 4th Armored in a successful dash to the Rhine, north of Coblenz, which the division reached on March 7, having thrust some fifty-five miles in less than forty-eight hours.

From March 11 to 25, the Third and Seventh Armies hammered the Germans in the Saar-Palatinate unmercifully. In the north, in the Moselle sector, Patton sent a corps through the Hunsrück Mountains as part of a two-pronged thrust to cut German Army Group G in two. The mountains were thought to be too rugged for tanks, but Patton was certain he could push Walker's XX corps through and did so, linking up with the 4th Armored. The two German armies defending the Saar-Palatinate triangle were no match for such attacks.

One account describes how "Patton's forces seemed to be everywhere at once—attacking the *Westwall*'s concrete casements from the rear, racing through the center of the Palatinate, sweeping southward along the Rhine itself. On March 19 alone the Third Army overran more than 950 square miles of territory. Coblenz was captured, and two more cities along the Rhine were within reach: Patton's tanks were just ten miles from Mainz, six miles from Worms."[18] The day the 4th Armored broke through to the Rhine, Tedder, the deputy SHAEF commander, noted the fresh arrow on the situation map and observed: "There goes Patton with another of his Phallic symbols."[19]

Eisenhower was sometimes known for his strange pronouncements, such as the day after the capture of the bridge at Remagen, when he told the Third Army staff: "The trouble with you people in Third Army is that you don't appreciate your own greatness. You're not cocky enough," an observation with which Courtney Hodges and the troops of the First Army would have taken serious issue. "Let the world know what you are doing, otherwise the American soldier will not be appreciated at his full value." When Eisenhower told Patton: "George, you are not only a good general, you are a lucky general, and, as you will remember, in a General, Napoleon prized luck above skill," Patton replied laughingly: "Well, that is the first compliment you have paid me since we served together."[20] When Gay wondered why Eisenhower had made his remark about the Third Army, Patton replied, "That's easy. Before long, Ike will be running for President. The Third Army represents a lot of votes." When those present began to smile, Patton continued: "You think I'm joking? I'm not. Just wait and see."[21]

The first elements of the Third Army reached the Rhine on March 9, two days after the capture of the bridge at Remagen by the First Army. Although his competitiveness usually bordered on obsession, Patton genuinely admired the achievement of Bill Hoge and his men. However, he chafed over when he would secure a bridgehead, and remained determined to beat Patch's Seventh Army across. "I am going to attack as soon as possible, because at this stage of the war, time is more important than coordination." He even apologized to Bradley for ceding the front pages to Hodges and the First Army. Bradley opined: "Well, even you have to regroup once in a while."[22]

Visualizing a Rhine crossing "as a glorious opportunity to score off Montgomery and steal some of the limelight from Hodges, whose Remagen coup had temporarily snatched the headlines for First Army," Patton had ordered his engineers forward and had collected large quantities of bridging material in anticipation of just such an opportunity.[23] His commanders would have preferred waiting a few days longer, but Patton had already selected sites for potential crossing and would not brook any delay.

The night of March 22 the first elements of the Third Army crossed the Rhine at Nierstein without opposition, about the same time as an infantry unit fought its way across farther south, at Oppenheim. Patton was not so foolish as to compromise his triumph, and telephoned Bradley on the morning of March 23: "Brad, don't tell anyone, but I'm across." Once again he had scored a coup at Montgomery's expense. "Well, I'll be damned. You mean across the Rhine?" replied Bradley. "Sure am. I sneaked a division over last night. But there are so few Krauts around there they don't know it yet. So don't make any announcement—we'll keep it a secret until we see how it goes." The presence of the Third Army did not remain a secret for more than a few hours, however. Throughout March 23 the Luftwaffe attacked the Rhine pontoon bridges in a desperate attempt to knock them out. In response Third Army gunners shot down thirty-three planes.[24]

That night Patton again telephoned his boss: "Brad . . . for God's sake tell the world we're across. . . . I want the world to know Third Army made it before Monty," who had launched his massive blitz of the Rhine, Operation Plunder, the same day.[25] Plunder was the largest military operation since D Day and, some believe, the most overrated of the war. Patton's liaison officer at Bradley's headquarters remarked: "Without benefit of aerial bombardment, ground smoke, artillery preparation or airborne assistance," Third Army had successfully breached the Rhine.[26] Patton wrote: "God be praised. . . . I am very grateful to the Lord for the great blessings he has heaped on me and the Third Army."[27]

March 24, 1945, was another banner day. With Eddy and Stiller, Patton arrived at the pontoon bridge over the Rhine at Nierstein. He stopped halfway across, said: "Time out for a short halt," and unzipped his fly "to take a piss in the Rhine. I have been looking forward to this for a long time."[28] On the eastern bank he deliberately stumbled, fell to one knee, grabbed two handfuls of dirt, stood, and as the soil wafted earthward through his fingers, exclaimed in a symbolic gesture: "Thus, William the Conqueror!"*

---

*Patton's symbolic act was a recreation of William's, when he reputedly stumbled and fell on his face emerging from his boat on the shores of England. "See," he said, "I

*   *   *

Generals savor the opportunity to needle one another and the final weeks of the war offered numerous opportunities. Leonard Gerow cabled Patton: CONGRATULATIONS ON YOUR BRILLIANT SURROUNDING AND CAPTURE OF THREE ARMIES, ONE OF THEM AMERICAN (a dig at the Seventh Army, which was pinched out by the Third Army's advance to the Rhine). Sandy Patch also cabled: CONGRATULATIONS ON YOUR BEING THE LAST TO REACH THE RHINE. Not to be outdone, Patton retorted: THANKS AND CONGRATULATIONS TO YOU FOR BEING THE FIRST TO BE KICKED OFF THE RHINE (referring to the withdrawal of a Seventh Army corps shortly after reaching the river).[29] The airmen loved the way the Third Army fought. The U.S. Ninth Air Force commander, Hoyt Vandenberg, cabled Patton: THAT IS THE WAY TO FIGHT A WAR, KEEP DRIVING. MY PILOTS WILL FLY THEIR HEARTS OUT IN A BATTLE LIKE THAT.[30]

On March 19 Patton wrote of his army: "We are the 8th wonder of the world. And I had to beg lie and steal to get started—now everyone says 'that is always what we wanted to do'?? I hope things keep smooth it seems too good to be true."

They didn't. As was often the case, triumph and folly seemed to go hand in hand. After his spectacular drive to the Rhine, and quietly beating Montgomery across, Patton was on a roll, writing confidently to Marshall that when the war ended: "I should like to be considered for any type of combat command from a division up against the Japanese. I am sure that my

---

have taken England with both hands." Lest it be thought that Patton's "piss in the Rhine" was the culmination of a childish fantasy, Churchill had long nourished a dream of urinating upon Hitler's vaunted Siegfried Line. He got his wish on March 3, 1945, during a visit to the front, near Jülich, in the company of Brooke, Simpson, and a bevy of VIPs and press. Brooke later wrote:

> As the photographers had all rushed up to secure good vantage points, he turned to them and said, 'This is one of the operations connected with this great war which must not be reproduced graphically.' To give them credit they obeyed their orders and, in doing so, missed a chance of publishing the greatest photographic catch of the war! I shall never forget the childish grin of intense satisfaction that spread all over his face as he looked down at the critical moment.

For years Churchill had looked forward with considerable anticipation to utilizing this means of symbolically displaying his utter contempt for Hitler, during which he was joined by some of the most famous personalities of the war. By contrast, a single (unpublished) photograph exists of Patton urinating in the Rhine. It was not until October that a delighted Patton learned from his son that Napoleon had once crossed the Rhine near Oppenheim. (Patton had also presented Willie with a bust of Hitler to urinate on at his leisure.) (Brooke quote from Martin Gilbert, *Road to Victory,* p. 1239. The author of the William the Conqueror quote was none other than Churchill, *The Birth of Britain* [New York, 1956]. The entire scene was recorded by Charles Codman in *Drive* [Boston: Little, Brown, 1957], p. 269. Napoleon is cited in MB-*PP* 2, p. 660.)

method of fighting would be successful. I also am of such an age that this is my last war, and I would therefore like to see it through to the end."[31]

To Beatrice he had written: "I am scared by my good luck. This operation is stupendous."[32] With the press he was in rare good humor. Shortly before the Rhine crossing he announced that the Third Army was about to capture its 230,000th POW. He had been denied permission to photograph the face of the 200,000th, but "this time we will take a picture of his ass." A week later the POW total reached three hundred thousand. He also asked the correspondents to ensure that the Germans knew that four of his armored divisions were pummeling them. It was "not for me—God knows I've got enough—I could go to heaven and St. Peter would recognize me right away—but it is for the officers and men." Patton also pointed out that "the Marines go to town by reporting the number [of their] killed. I always try to fight without getting [our] people killed."[33]

Within days of his Rhine triumph, however, a pall was cast over Patton's reputation that was eclipsed only by the slapping incidents. It began on March 23, when Patch met with Patton to discuss zones of responsibility east of the Rhine. They came to an amicable agreement, and it may have been during this meeting that Patch apprised Patton of an Allied POW camp for officers known as Oflag XIIIB. Located outside Hammelburg, near Schweinfurt, Oflag XIIIB contained some 3,000 Serbs of the Royal Yugoslav Army and about 800 Americans who had arrived in January, most of them captured during the Battle of the Bulge. American morale was terrible until early March, when another contingent of some 430 Americans arrived from Poland. The senior officer was Col. Paul Goode and his executive officer was Lt. Col. John Knight Waters, Patton's son-in-law, taken prisoner two years earlier. What ensued when he learned of the possible presence of Johnny Waters in Oflag XIIIB was the most controversial military decision of Patton's career, and one many would argue that ranked as his worst.

The events leading up to what has become known as the Hammelburg raid began innocently enough on March 25, when Patton's bodyguard, Maj. Al Stiller, arrived unannounced at the headquarters of the 4th Armored. Its new commander, William Hoge, was a fearless officer whose penchant for brusqueness had cost him promotions and blunted his career. However, when he received a strange order from his corps commander, Manton S. Eddy, on March 25, directing him to mount a special task force and liberate Oflag XIIIB, more than forty miles behind enemy lines, Hoge rebelled.

Stiller said little, merely that he was there "to go along" on the Hammelburg raid. However, generals' aides do not just "go along" on combat missions. Clearly Stiller seemed to be there for another purpose. For the second time a decidedly uneasy Hoge took his concerns to Eddy, who replied that he would "take care of Patton."[34]

When he learned that Eddy and Hoge were resisting, Patton went to XII Corps on the morning of March 26, but found Eddy away and ordered the chief of staff to get Hoge on the phone. "Tell him to cross the Main River and get over to Hammelburg." Despite insubordination by both Eddy and Hoge, Patton simply informed Hoge that he must carry out the order even if he had no men to spare. "I promise I'll replace every man and every vehicle you lose!" As historian John Toland reveals, "Hoge was embarrassed by the almost pleading tone in Patton's voice. With a baffled look he turned to Stiller, who had been listening. Stiller explained in a low voice that the 'Old Man' was absolutely determined to free the prisoners at Hammelburg—and revealed that John Waters, Patton's son-in-law, was one of the prisoners."[35]

The burden of furnishing the task force and mounting the raid on Hammelburg fell on Lt. Col. Creighton Abrams, the new commander of Combat Command B. To lead the mission, Abrams chose the operations officer (S-3), a young captain named Abraham Baum. If this mission had to be undertaken, both Hoge and Abrams wanted his entire combat command sent, but their request was denied by Eddy. Instead Task Force Baum was created, consisting of sixteen tanks, twenty-seven half-tracks, three 105-mm self-propelled guns, and 294 officers and men.[36]

Abrams had severe misgivings about the raid and believed it would take a miracle for Task Force Baum to survive. Stiller's presence was particularly troubling, and when questioned why he was risking his neck on a venture in which he had no apparent stake, Stiller allegedly replied, with a laugh, that he "was going for the thrills."[37]

The raid commenced on the night of March 26, so swiftly and unexpectedly that Task Force Baum encountered few problems during the first hours of its foray. En route the Americans liberated 700 Russians from a POW camp, who then ran amok, seizing weapons and booze and raising merry hell in the immediate area. After battling through Hammelburg itself, Baum's Sherman tanks crashed through the gates of the camp and liberated 5,000 POWs, 1,291 of whom were American, including Lieutenant Colonel Waters, who was gravely injured in the ensuing melee with a serious bullet wound in his left upper thigh, left buttock, and rectum. His life was saved by the camp's Serbian doctor who, although devoid of medical supplies, managed to drain the wound repeatedly, using bandages of paper, and a kitchen knife.

By now, however, the Germans knew of the presence of the American force, and were preparing to deal with it. Baum's remaining tanks and vehicles could only carry about 250 POWs; the remainder, including Waters, had to be left to fend for themselves. Then all contact with Task Force Baum was lost.

Patton spent a troubled day on March 27, writing in his diary: "I was quite nervous all morning over the 'task force' I sent to rescue the prisoners, as we could get no information concerning them. I do not believe there is anything in that part of Germany heavy enough to hurt them, but for some reason I was nervous—probably I had indigestion." He learned that the raid had turned into a bloody fiasco when Task Force Baum had been surrounded by elements of at least three German divisions and chopped to pieces in a series of desperate firefights. Scores were wounded, and most were captured and returned to Oflag XIIIB as POWs, among them Baum and Stiller. Of the 294 men who started the raid, all but one were listed as missing in action.[38]

On April 5 Patton wrote to Beatrice: "I feel terribly. I tried hard to save him and may be the cause of his death. Al Stiller was in the column and I fear he is dead. I don't know what you and [little] B[eatrice] will think. Don't tell her yet. . . . We have liberated a lot of PW camps but not the one I wanted." Patton later blamed Bradley. He also blamed Eddy and Hoge. In fact, he blamed everyone but himself.

On the same night (March 27) that he learned that Waters had been wounded and was expected to die, Eisenhower's son, John, happened to be visiting Patton, who said "that Waters had gotten his ass shot up. Patton then burst into tears. Later on, as we were leaving, he told me how much he owed my father. And he broke into tears again."[39]

On April 5 Patch's 14th Armored Division liberated Hammelburg. Stiller was not among the Americans there, but Waters was. Patton sent his personal physician, Dr. Charles B. Odom, to Hammelburg. Odom found Waters in very serious condition, and two cub aircraft were immediately sent by Patton to airlift him to Frankfurt. Some of the other wounded were resentful that Waters was clearly being given special treatment. An equally bitter Abe Baum thought that Odom's failure to ask about himself or Stiller was ample evidence of Patton's disinterest in their fate. When Patton met his son-in-law for the first time in more two years to decorate him with two Silver Stars (both of which had been recommended and approved by others), the first question Waters asked was: "Did you know I was there?" Patton replied: "I didn't know but thought you might be there."[40]

Patton later defended the raid on the grounds that the havoc raised by Task Force Baum behind German lines had diverted a number of enemy units to the Hammelburg area, thus easing the advance of Third Army during the final weeks of the war. The lingering question of Hammelburg is, Did Patton mount the raid solely because of his son-in-law, or were there other reasons? Circumstantial evidence suggests that Patton knew Waters was in Hammelburg at the time he ordered the raid. However, although he could have been fairly certain that Waters was no longer in Poland, Patton could not have known positively that he was in Hammelburg, merely that he might have been there. Perhaps this was sufficient basis for him to undertake the risk.

The only certainty in the entire tragic affair is that Bradley would have forbidden Patton to undertake the Hammelburg raid had he known of it in advance. Regardless of what Patton knew at the time, Hammelburg was the least defensible decision he ever made, and nearly as self-destructive as the slappings. His denials notwithstanding, the raid not only branded Patton a liar but tarnished the very fabric on which his fame rested—that his troops came first, and everything possible must be done to insure their survival. Instead, he had sent 307 men on a mission whose implicit purpose was the rescue of his own son-in-law.

Patton attempted to put on a brave face for an unjustifiable act. The camp would have been liberated in any event within a matter of days by forces that would not have been intimidated or destroyed. "It was a story that began as a wild goose chase and ended in tragedy," wrote Bradley, who viewed Hammelburg as yet another example of Patton's impetuosity: "I did not rebuke him for it. Failure itself was George's own worst reprimand."[41]

Patton perpetuated the fiction of Hammelburg at a press conference, in which he baldly deceived the correspondents by waving his personal and official diaries in the air and claiming he had known nothing of Waters's presence in Hammelburg until nine days after the raid. "We attempted to liberate the prison camp because we were afraid that the American prisoners might be murdered by the retreating Germans."[42] Those involved in the raid—Hoge, Baum, Stiller, and Creighton Abrams—were convinced that Patton had mounted Hammelburg to save his son-in-law. All, however, chose to remain silent in deference to Patton. Not until 1967, when he was himself a four-star general, did Abrams write that Stiller had admitted he made the trip "only because General Patton's son-in-law, Colonel Waters, was in the prison camp."[43]

Patton had had no news of Stiller until April 7, when he learned that his loyal aide had last been seen putting up a valiant fight before being captured and marched off toward Nuremberg, his fate unknown. Stiller was rescued unharmed but ten pounds lighter on May 1, when the Third Army liberated a POW camp at Moosburg. He was greeted by a tearful Patton, for whom an enormous burden of guilt had been removed.

Patton was fortunate that the press did not learn of the raid for ten days, and by then interest in Hammelburg had waned in the wake of Roosevelt's death. Initially Patton thought it likely that Hammelburg would be treated as another slapping incident, but then he decided: "What the hell! With the President's death you could execute buggery in the streets and get no farther than the fourth page."[44] The arrow pointing to Hammelburg on the Third Army operations map was abruptly removed when the camp was liberated. A reporter who happened by at that moment asked why this arrow was so far away from the remainder of Third Army and why it was being removed. No answer was forthcoming.[45]

Officially Hammelburg never happened. Baum suspected a cover-up when an uncharacteristically ill-at-ease Patton visited his bedside to award him the Distinguished Service Cross. Patton praised Baum: "You did one helluva job. I always knew you were one of the best." Baum, who greatly admired Patton, said in reply: "You know, sir, it's difficult for me to believe that you would have sent us on that mission just to rescue one man."

Patton replied: "That's right, Abe, I wouldn't." After promising to honor Baum's request that he be returned to the 4th Armored, Patton left. According to Baum, Patton's aide told him that the task force had been classified top secret, and that he was to use discretion when discussing it. Baum interpreted this to mean that his task force would get no recognition, and that he and his men had been "screwed again."[46]

Patton's letters to Beatrice written immediately after the event are highly suggestive. On April 5, 1945, he was deliberately vague, merely noting: "I heard that there was a prison camp at Hammelburg. Forty miles east of Frankfurt so I decided to liberate it. My first thought was to send a combat command (1/3 Div) but I was talked out of it by Omar and others. . . "

The most compelling evidence is contained in an earlier letter dated March 23 in which he reported: "We are headed right for Johns place and may get there before he is moved if he is moved he had better escape or he will end up in Bavaria. Frankly I think he was a fool not to do it when the others did of course I will not say so to [little] B."[47] A second letter dated four days later reveals: "Last night I sent an armored column to a place 40 miles east of Frankfurt where John and some 900 PW are said to be. I have been nervous as a cat all day as every one but me thought it was too great a risk I hope it works. Al Stiller went along. If I loose the column it will possibly be a new incident but I wont. . . ."[48]

Gay's diary entry for April 4—written well after the news of the raid had become public—offers the only evidence that Patton did *not* know of Waters's presence in Hammelburg. He recounted some of the action of Task Force Baum at the gates of Hammelburg: "This narrative is of particular interest to the Army Commander," wrote Gay, "because this was the first time he had news that his son-in-law, Colonel Waters, was one of the prisoners in the camp."[49] Even this was vague, and there can be two interpretations—that Gay was merely confirming what Patton hoped would be the presence of John K. Waters, or that he deliberately doctored the diary to deceive future historians.

In *War as I Knew It,* Patton blamed Eddy and Hoge: "There were two purposes in this expedition: first, to impress the Germans with the idea that we were moving due east, whereas we intended to move due north, and second, to release some nine hundred American prisoners of war who were at Hammelburg. I intended to send one combat command of the 4th Armored, but, unfortunately, was talked out of it by Eddy and Hoge . . . so I compro-

mised by sending one armored company and one of armored infantry."[50]
This version is as self-serving as his earlier complaint that Bradley had
denied him the necessary force to mount the raid.[51] Nevertheless the evidence is damning, and regardless of the full extent of his culpability, Hammelburg has become an enduring stain on Patton's reputation—one that his
critics have seized on as a perfect example of "Old Blood and Guts" in
action. To his death, Patton maintained that "throughout the campaign in
Europe I know of no error I made except that of failing to send a Combat
Command to take Hammelburg." His mea culpa was similar to Winston
Churchill's half-hearted denials that he bore no responsibility for the Allied
debacle at Anzio in 1944.[52]

In April 1945, XII Corps liberated the Merkers industrial salt mine. Inside
Manton Eddy discovered the entire German gold bullion reserve, hidden
some two thousand feet deep in a huge underground cavern. Patton ordered
a total embargo on the story until it could be confirmed. "His fury knew no
bounds" when a SHAEF censor permitted a leak of the story to pass unchal-
lenged. Although the officer was assigned to SHAEF, Patton ordered him
from the Third Army sector forthwith. When General Eddy informed Patton
that he still did not know for certain what lay inside a number of sealed
vaults belonging to the German Reichsbank, Patton again exploded. "Goddamnit, General Eddy. You blow open that fuckin' vault and see what's in
it."[53]

Eisenhower warned Marshall that some of Patton's "latest crackpot
actions may possibly get some publicity," a reference to the arbitrary relief
of the SHAEF censor, over whom he had no authority, for allowing the stories of Hammelburg and the great gold find at Merkers to leak to the press.
Several correspondents wrote bitterly critical articles about Patton, and
Butcher noted that "Ike had taken Patton's hide off. But I think Patton must
have as many hides as a cat has lives." Eisenhower was fully aware of Patton's "wild goose chase. . . . The story has now been released, and I hope the
newspapers do not make too much of it. . . . Patton is a problem child, but
he is a great fighting leader in pursuit and exploitation."[54]

Eisenhower and Bradley arrived to inspect this incredible cache, and as
the generals began the long descent in an ancient elevator held by a single,
none-too-secure cable in a pitch-black shaft, Patton quipped: "If that
clothesline should part, promotions in the United States Army would be
considerably stimulated." From the inky blackness an unamused Eisenhower said: "O.K., George, that's enough. No more cracks until we are
above ground again." All marveled at the sight of more than 4,500 twenty-
five-pound gold bars, worth an estimated $57,600,000, and millions more in
currency. There were also hundreds of paintings, and the few Patton examined he thought worth about $2.50 and best suited for saloons. An inter-

preter noted that the 3 billion Reichsmarks were badly needed to meet the army payroll. Bradley retorted that he doubted the Wehrmacht would be meeting many more payrolls. There were also thousands of gold and silver dental fillings, eyeglasses, and other items taken from those murdered in the concentration camps—a visible legacy of Hitler's loathsome Final Solution.[55]

At lunch Eisenhower asked why Patton had withheld the news of Merkers from the press, and what he would do with the money if the decision were his. Patton replied that there were two schools of thought in the Third Army: Half would like to see the gold made into medallions, "one for every son of a bitch in Third Army," while the remainder "wanted the loot hidden until Congress cracked down on peacetime military appropriations; then Third Army could drag out the money and buy new weapons." Eisenhower shook his head in amusement: "He's always got an answer!"[56]

The gallows humor of that morning was replaced in the afternoon by disgust and horror when the generals visited the Ohrdruf concentration camp, just liberated by the 4th Armored. In an effort to ensure that no one would testify to their crimes, the SS guards had murdered everyone before fleeing. Bodies were piled everywhere, along with macabre evidence of torture and butchery. Some had been set ablaze in huge funeral pyres. The stench was indescribable. It was evil incarnate, and it left Bradley speechless. Patton disappeared briefly and vomited against the side of a building. He refused to enter one room, containing the corpses of naked men starved to death. The generals had seen war in all its brutal forms, but none had ever seen anything like this. Eisenhower ordered that every American unit not in the front lines in Germany was to see Ohrdruf: "We are told the American soldier does not know what he is fighting for. Now, at least, he will know what he is fighting against."[57] Patton told Walton Walker: "You'll never believe how bastardly these Krauts can be, until you've seen this pesthole yourself."[58]

The stony expression on Patton's face in photographs taken at one of the camps liberated by the Third Army is testament to the utter disgust he and others felt. As an expression of his fury over Nazi bestiality, at Ohrdruf and other camps Patton compelled the local townspeople to dig graves and bury the corpses of the dead.[59]

Third Army also liberated the infamous Buchenwald death camp, where Patton was shown the system the Germans had perfected for disposing of six bodies at a time in the crematorium. Diplomat Robert Murphy later wrote that Patton "could not stomach the sights he saw at Buchenwald; he went off to a corner thoroughly sick. The inmates liberated by our forces were skeletons . . . many of the captives [had been] professional soldiers . . . and they pulled their wasted bodies into gallant salutes as Eisenhower, Patton and their staffs passed them. It was enough to make strong men weep—and some American officers did so unabashedly."[60]

After dinner the following night, Eisenhower revealed privately to Patton that he was soon to halt the First and Ninth Armies at the Elbe River to await the arrival of the Red Army. The Third Army would be given a new mission to drive southeast toward Czechoslovakia: "From a tactical point of view, it is highly inadvisable for the American Army to take Berlin and I hope political influence won't cause me to take the city. It has no tactical or strategic value and would place upon the American forces the burden of caring for thousands and thousands of Germans, displaced persons and Allied prisoners of war." Patton was the first senior officer to learn of Eisenhower's controversial decision. For a half century arguments have raged over whether the Allied armies were capable of capturing Berlin, and if an attempt should have been made. Among those utterly dismayed by Ike's decision was William Simpson. When told by Bradley that he must halt at the Elbe, he replied: "Where the hell did you get this?" Told "From Ike," Simpson could only obey his orders, convinced that it was a terrible mistake.

Patton's reaction to this pronouncement was similar incredulity: "Ike, I don't see how you figure that out. We had better take Berlin, and quick—and on to the Oder!" Later, in Gay's presence, Patton reiterated the need to drive on to Berlin, arguing that it certainly could be done in forty-eight hours by the Ninth Army. Eisenhower wondered aloud: "Well, who would want it?" Patton did not reply at once but, placing both hands on his friend's shoulders, said: "I think history will answer that question for you."[61]

Once again Montgomery and Patton found themselves in agreement. In his memoirs the field marshal wrote: "The Americans could not understand that it was of little avail to win the war strategically if we lost it politically. . . . It became obvious to me in the autumn of 1944 . . . we were going to 'muck it up.' I reckon we did."[62]

As Patton prepared for bed on the night of April 12, he noticed that he had not wound his watch. Turning on the BBC to obtain the correct time, he learned of Roosevelt's death in Warm Springs, Georgia, that afternoon. He immediately awakened Bradley and Eisenhower, who were spending the night at the Third Army. All agreed gloomily that Roosevelt's death was a tragedy and that he would be sorely missed at a critical moment in American history.[63]

Patton's first dig at the new president appeared in his diary for April 12: "It seems very unfortunate that in order to secure political preference, people are made Vice Presidents who were intended neither by the Party nor by the Lord to be Presidents."[64] Patton never knew that the little man with the big heart from Independence, Missouri, had been firing 75s in support of his tanks in the Argonne on September 26, 1918, but he did live long enough to learn that Harry S. Truman was a lot tougher than he looked.

Patton took an overnight leave in Paris in April, where he visited his son-in-law in a nearby hospital and had his teeth cleaned. Before flying back to the Third Army, he learned from *Stars and Stripes* that he had been promoted to four-star general. Bradley and Devers had been promoted ahead of him, despite Eisenhower's attempts to persuade Marshall to promote Patton and Hodges at the same time. There were compensations in the form of decorations from France (the croix de guerre and Grand Officer of the Legion of Honor) and Luxembourg (the croix de guerre), and, although he was happy to be a full general, Patton's pleasure was tempered by the conviction that he was an also-ran. Meeks heard the list read over Armed Forces Radio and came to Patton's room. "Good God, General, they are making [promoting] all the troop clerks." The ever-efficient Codman located the last set of four-star insignia in Paris, and a four-star flag.[65]

Tales of Patton—real and imagined—abounded in 1945. There was, for example, the day Patton was driving down a German autobahn, when he spied two GIs and two women and called out, "Don't you know there's a $65 fine for fraternizing with German women?" Fearfully one soldier replied, "Yes, sir. But these are liberated Russian women, sir, and we're just trying to to learn their language." Patton chortled: "Well, then you'd better get behind the bushes. It looks to me like they're teaching you to talk Russian with your hands."[66]

On the morning of March 7, Patton decided to review a new division coming into the line near Colmar. Mims was recovering from what Patton called his "quarterly atrocity" in the brig after a bender, and he was being driven this day in an open command car in an icy wind and rain by a corporal he had never seen before. Dressed in his "Eisenhower" jacket over a brown wool sweater (his only concession to the numbing cold), Patton noticed the soldier turning blue and asked if he had brought along a warm sweater. Informed that he had not, Patton removed his sweater and handed it to the shivering driver, saying: "Well, Corporal, you have one now."[67]

Patton lectured the division to be aggressive, for the Hun (his usual term of reference) would not expect such tactics. Always keep running forward and never stop firing your rifle against an enemy that will prove more scared than you: "The way most new soldiers use their rifles," he reminded them, "they are no more use than a pecker is to the Pope." When reminded that he had spoken excessively, Patton moaned: "Oh, my God, now I guess I'll have to write a letter of apology to the Vatican!" To the field-grade officers he said: "Sometimes getting hot food, or an extra pair of dry socks, up to every damned man in the lines is more important to victory than another ton of ammunition."[68]

At another division Patton was intentionally harsh, telling them that they had lost more men to nonbattle casualties than to combat, and more

POWs than in his entire military career—a disgrace they must erase. He left with a very public pat on the back for the commanding general, but not before privately informing him that "if conditions did not improve, he himself would become a non-battle casualty." Later that day Bradley informed Patton that the division was being transferred to the Ninth Army. "Damn this political war," he swore in his diary.

As Patton toured Germany by road and by air, he eagerly searched for a church, fortress, or castle to satisfy his craving to visit historic places. One night he spoke of a castle he had observed earlier that day. Codman hurriedly left the room, and when later asked why, replied: "I know the old man when he goes historical. He's going to go out and visit that place just as sure as can be in the morning." Codman had left to arrange for a team to depart at once to delouse the place. "Sometimes keeping up with your uncle," he told Patton's nephew, Fred Ayer, "is enough to kill off two men half my age."[69]

The 80th Division began what Patton dubbed the "Third Army War Memorial Project." Approaching an unsecured town, a unit would fire several projectiles containing a proclamation to the effect that the town must surrender. The burgermeister was directed to appear by a certain hour with a white flag, to offer his guarantee that there were no German troops inside to resist—or else. "The object," wrote Patton after the war, "was to let the inhabitants have something to show to future generations of Germans by way of proof that the Third Army had passed that way."[70] SHAEF had issued warnings of fanatical last stands and the possible existence of a so called National Redoubt in the mountains of Bavaria, which later turned out to be a myth. Still, even Patton took these threats seriously enough to carry a carbine to bed; and security around the army command post was augmented. When Gay was sniped at one night while walking outside his caravan, Patton ordered the closest two houses burned.[71]

Patton's penchant for making boastful remarks for their effect was never more in evidence than the day he and Weyland passed through a German village that had been left unscathed. Patton said: "Of course, we may have to come back here and create another Third Army memorial." His jocular mood ended moments later, when the two generals came on the shambles of what had once been a German supply column, now smoking wreckage, the dead bodies of humans and horses strewn everywhere. For some time they viewed the scene in silence before Patton finally spoke: "That is the greatest scene of carnage I have ever witnessed."[72]

In April Col. Robert S. Allen and several other members of the Third Army staff were ambushed and captured. Part of Allen's arm had to be amputated in a German hospital, and when he was liberated a short time later, Patton learned that the attack had been carried out by civilians. His diary recorded simply: "The town where it took place has been removed, together with, I hope, a number of the civilians."[73] After Ohrdruf and Buchenwald neither Patton nor anyone else who had seen the grisly death camps was in any mood to spare a moment's concern over the fate of towns or cities that had spawned the butchers who carried out such inhumane acts. Moreover, there were too many tales of atrocities, not by the German military but by civilians, perhaps an inevitable result of the terror bombing of Dresden and the enormous toll of noncombatants taken by Allied saturation bombing of German cities and towns. Patton's own son-in-law, John K. Waters, revealed that he and other American POWs were treated far more brutally when they were in civilian rather than in military hands.

Although it was against the Geneva Convention to deliberately kill civilians, the German army had defended towns from North Africa to their homeland. The Third Army's tactics, while perhaps not in strict consonance with the letter of the Geneva canons, was a practical means of avoiding unnecessary civilian casualties and the destruction of towns and cities. When Assistant Secretary of War McCloy visited Patton in April, the two discussed what Patton termed "the wanton and indiscriminate bombing of civilian cities . . . [which] we all feel . . . has no military value and is cruel and wasteful."[74]

As the end neared, the Third Army's statistics piled up, including during a twenty-four-day period four assault crossings of the Rhine, the capture of twenty-two major cities with a total population of more than 3 million; the liberation of 12,400 square miles of territory; the liberation of Ohrdruf and Buchenwald and 40,000 Allied POWs; seizure of the great gold bullion cache at Merkers; and the capture of 280,661 German POWs, with 14,300 more killed and 31,200 wounded. Weyland's airmen damaged or destroyed 10,774 motor vehicles, 663 armored vehicles and tanks, 446 locomotives, 5,685 railway cars, and 786 planes both in the air and on the ground.[75]

Before the war ended Patton experienced two more near brushes with death. Near Munich his aircraft was attacked by what was later identified as an Allied fighter piloted by a Polish RAF officer, who mistook his plane for a German Storch. Machine guns blazing, the Pole attacked; Patton's pilot managed to take evasive action and the fighter crashed, killing the pilot. Patton made light of the incident, noting that if he were to die anyway, "I might as well take some pictures of my impending demise." However, Patton was so nervous that he forgot to remove the cover from his camera lens, "and all I got were blanks." The second incident was documented in the film *Patton*. He was passing through a small German town in his jeep when a runaway oxcart came hurtling at him from a side street, its shaft narrowly missing Patton's head. Deeply shaken, he said: "God, what a fate that would have been! To have gone through all the war I've seen and been killed by an ox."[76]

On March 25 his son learned of the Rhine crossing and wrote his father. "Tonight I guess is one of the greatest moments in history and to think that

you're the guy that did it. Well I can't express myself. The Rhine bridged—
Hot damn! Not a casualty—I know you took a calculated risk, and won
again. I'll die happy, so help me God, if I can serve under you in combat for
one lonely month. I called up mother last night and she was right-square-
out-of-this-world with enthusiasm. . . . Damn I don't know how you do it, I
really don't. Well just stay healthy so the both of us can go to Asia. Hot
damn."[77]

Although Patton blamed the correspondents for Hammelburg, noting "how I
hate the press," he continued to take full advantage of them. When, on
March 24, *Life* magazine photographer Margaret Bourke-White arrived to
photograph Patton, she found him tangled up in a red silk parachute, wondering aloud what he might make from it for his family. He fussed over
what uniform he should wear to pose for her camera and complained:
"Things have been going too smoothly. It makes me jittery when things go
too well." With some prodding, he finally settled on his battle jacket and
steel helmet but insisted that White photograph his profile from the left side,
his favorite. "Don't show my jowls," he complained. "And don't show the
creases in my neck. Stop taking pictures of my teeth. Why are photographers always taking pictures of my teeth?" Tucking his head into his chest,
Patton groused: "This is the only angle at which the little hair I have will
show." As White worked, Patton bantered with Otto Weyland, and afterward
passed around bourbon, observing: "If I ran out of Bourbon and cigars, I'd
be a healthier man." Asked what he would do after the war, Patton replied:
"Go back to my yacht. She's a lovely boat." Patton begged Weyland to be
permitted to fly a mission the next morning over the front in a P-47 fighter.
Weyland emphatically denied the request. "You can't go on any mission,
General. That's one thing we won't let you do."

    Wherever White went, the men of the Third Army spoke highly of Pat-
ton. One described his visit to a hospital in Lorraine, where he left some in
tears when he walked to the middle of the tent and said: "I want to get you
boys home alive," and took off his helmet, saying: "I take my hat off to the
nation that produced such men as these." It was Patton the showman and
actor at his best. Where the acting left off and the real emotion began we
shall never know. When Margaret Bourke-White saw Patton for the last time
he was instructing a staff officer: "Tell those boys just to go hightailing
along. You see, I think this war is over now. I want my men to be up front,
to share the glory."[78]

    One of the highlights of Patton's life occurred two weeks later, also
courtesy of the detested press. His picture appeared on the cover of *Time,*
with the caption, "THIRD ARMY'S PATTON. The enemy has reason to
fear him." The correspondents loved to repeat stories of Patton, such as the
anecdote of an Allied officer's question to Eisenhower about his present

whereabouts: "Hell, I don't know. I haven't heard from him for three hours." Or the tale of Patton's "greeting to a black tank battalion arriving at his battle front—'I don't give a damn what color you are so long as you get out there and kill those sonsofbitches in the green suits.'"[79] The treatment of black troops by the American military during the Second World War was shameful. Patton had lost none of his prejudices; however, he was the first commander to not only employ black tank units but to personally welcome them to the Third Army. He addressed the men of the 761st Tank Battalion standing atop a half-track:

> Men, you are the first Negro tankers ever to fight in the American Army. I would never have asked for you if you weren't good. I have nothing but the best in my army. I don't care what color you are, so long as you go up there and kill those Kraut sonsabitches! Everyone has their eyes on you and is expecting great things from you. Most of all, your race is looking forward to you. Don't let them down and, damn you, don't let me down![80]

Historian Hugh Cole later pointed out that "when manpower got tight, Patton was the first—although this has now been forgotten—to integrate colored and white troops in the rifle companies."[81]

The softer, private side of Patton still appeared in letters home. He witnessed firsthand the pathos of civilians left displaced and destitute, and was particularly moved by a woman with all her worldly goods inside a perambulator, crying on a hillside; by an elderly man with three small children, wringing his hands; and by a woman with five children holding out a tin cup, also crying. "In hundreds of villages there is not a living thing. . . . They brought it on them selves but these poor pesants are not responsible. I am getting soft? I did most of it."[82]

As the final days of *Götterdämmerung* were played out in the ashes of the Third Reich, Patton was the most discussed and publicized American commander. Before being liberated by the 6th Armored Division, Americans in the Ziegenhain POW camp composed a song, sung to the tune of "The Battle Hymn of the Republic":

> *We're a bunch of Yankee soldiers*
> *living deep in Germany. We're eating*
> *soup and black bread, and a beverage*
> *they call tea. . . .[ending with] Come and*
> *get us, Georgie Patton, so we can come*
> *rambling home.*[83]

With Patton in command, the war was not over until the fat lady sang. On May 5 Maj. Gen. Clarence R. Huebner, the veteran commander of the

1st Division, now in command of V Corps, learned that his unit had been assigned to Third Army. He was just sitting down to dinner when his G-3 brought in the assignment orders, prompting Huebner to comment that he would give it twelve hours before Patton called with orders for V Corps to attack something. He had barely finished the soup course when the G-3 returned, a huge grin on his face, to summon Huebner to the telephone. On the other end was Patton, who asked where the hell he had been for so long. Huebner replied he had been "around" making a nuisance of himself, itself an interesting description of Omaha Beach and other places where the Big Red One had been fighting since the previous year. "Glad to have you back," said Patton. "I want you to attack Pilsen in the morning. Can you do it?" Of course Huebner could do it; one simply did not say "no" to such requests from Patton. "Fine, move fast now. We haven't got much time left in this war. I'll be up to see you. Goodbye." Huebner returned to dinner and admitted he had blown it. It had been twelve minutes, not twelve hours! [84]

Several weeks earlier Col. Jimmy Polk had received a similar call on his radio from Lucky Six. "Jimmy, are you going to get that bridge over the Danube at Regensburg intact for me today?" inquired Patton. "We will give it a good try," replied Polk. "Try hell, you get it and I'll make you a brigadier general; you fail and I'll sack your ass. Out." The Germans duly blew up the bridge, and no more was heard from Patton about Polk being sacked. It was merely motivation, Patton-style.

When and where the Third Army was to halt en route to Austria and Czechoslovakia became Eisenhower's primary challenge, and, ultimately, his decision. At Churchill's instigation, the British chiefs of staff exhorted the American Joint Chiefs to compel Eisenhower to liberate Prague and Czechoslovakia before the arrival of the Red Army. The State Department agreed that Czechoslovakia was a political prize that should be denied the Russians, and urged Truman's concurrence. Truman consulted Marshall, who passed the request back to Eisenhower, who replied that he thought the Red Army would liberate Prague before Patton could get there, and elected to halt Third Army at the pre-war border near Pilsen.

However, the Third Army had captured Nuremberg, advanced to the Danube, and been astride the Czech border for several weeks, primed to advance into both Czechoslovakia and Austria. Patton had begged for permission to push on but had been firmly restrained by a stop line beyond which the Third Army was not to advance without permission. Bradley thought that Prague could have been liberated within twenty-four hours. On May 4 Eisenhower finally authorized the Third Army to cross the Czech border, but there was to be no advance beyond Pilsen, which Huebner's V Corps easily occupied before halting. Bradley believed that Patton might ignore the new stop line and on May 6 excitedly telephoned to reaffirm

Eisenhower's order. "You hear me, George, goddamnit, *halt!*" Reluctantly, Patton complied, telling Gay: "We will probably never do this again, at least not in Germany."

Except for the formality of surrender, Patton's war was over.

The Czechs soon learned of the presence of Patton's troops in Pilsen, and the news brought rejoicing in the streets of Prague at what seemed imminent deliverance. Eisenhower, whose sole concern was to end the war as rapidly as possible, could see no strategic benefit to the Allied capture of Prague. No consideration was given to the political repercussions of ceding Czechoslovakia to the Russians. Instead Eisenhower left the decision to take Prague to the Red Army, which predictably requested no American advance beyond Pilsen. The ill-fated Yalta conference had resulted in Czechoslovakia being designated within the so-called Soviet sphere of interest.

An OSS team returned from Prague and briefed Patton that American entry into Czechoslovakia (spurred by an inaccurate report that the Third Army was a mere eighteen miles from the capital) had ignited an uprising against the SS. However, the situation was dire and the Czechs needed immediate help. While the Germans in Prague ruthlessly began suppressing the Resistance fighters in bloody battles in the streets, Third Army sat idle, a mere forty miles to the west. Patton again pleaded with Bradley for permission to advance beyond the Pilsen halt line. "For God's sake, Brad, those patriots in the city need our help. We have no time to lose." To take Bradley off the hook, Patton even suggested that he would remain incommunicado and would only report back to Bradley from a phone booth when Third Army was actually inside Prague. Bradley thought Patton's motivation was merely the headlines he would reap and was unwilling to authorize his advance to Prague without clearance from Eisenhower, who adamantly ordered Third Army halted. Under no circumstances was Patton to take Prague. Eisenhower wanted no international complications. Patton thought a powerful nation like the United States ought to let the other side worry about "complications."[85]

What is indisputable in the postwar debates over Eisenhower's decisions to halt at the Elbe and not to capture Austria and Czechoslovakia is that the decision brought about the repercussions that Churchill had correctly feared. The uprising by the Czechs against the German occupiers would be repeated in 1968, when a fleeting democracy would be ruthlessly suppressed by the same Red Army that liberated them in 1945. Although conceding that Eisenhower's reasons for halting at Pilsen were sound, Patton wrote shortly before his death: "I was very much chagrined, because I felt, and I still feel, that we should have gone on the Moldau River and, if the Russians didn't like it, let them go to hell."[86]

                    *   *   *

At 2:41 A.M., May 7, 1945, a German delegation signed documents of unconditional surrender at Rheims, and the following day the war in Europe officially ended. Although Eisenhower was exhausted, his famous grin reappeared at a historic moment, as he signaled a V for victory by holding aloft the two gold pens with which the German surrender document had been signed. Eisenhower awakened Bradley with the news; in turn, Bradley roused Patton from sleep to state: "Ike just called me, George. The Germans have surrendered."

Suddenly it was all over. The next morning Patton and his staff officers sat silently for several moments. Then Patton stood, nodded to Gay, and snapped his fingers at Willie, who obediently began following him toward the door. As the staff began to rise, Patton ordered, "Keep your seats." Gay began issuing instructions that beginning the following day, Lucky would no longer wear steel helmets, only helmet liners. From the doorway Patton interjected: "And make damn sure those liners are painted and smart-looking. I don't want any damn sloppy headgear around here." To his aide he remarked: "The best end for an old campaigner is a bullet at the last minute of the last battle." Nearly eight months later to the day, Patton would despair that this had not been his own fate.[87]

At the morning briefing Patton thanked his staff for their outstanding work: "He said that it was with regret, yet also with pleasure that he was forced to announce that this would be the last regular briefing in EUROPE for this headquarters."[88] "Most of them understood what I meant, and I added that I hope we will have other similar briefings in China. 2 1/2 years ago today we landed in Africa. During all that period until tonight at midnight we will have been practically in continuous battle. There is going to be a tremendous let down unless we watch ourselves."

Hap Gay wrote the final chapter of the Third Army's campaigns:

Not since 1806 has an army conquered Germany or crossed any large portion of it. I question if anyone a few years ago, even as late as the beginning of this war or as late as last year, felt the war would end with American forces having completely over-run Germany and passed into other countries. This time the Germans know what war at home means. The German nation has completely disintegrated, and even though they are an enemy nation, it is a saddening sight to witness.[89]

Patton had long since gotten the undivided attention of the Germans. What had begun in France in August was translated into genuine respect by 1945 after the Bulge. Patton's hatred of the Nazis and their odious regime was reciprocated in the autumn of 1944, when leaflets were dropped on a German POW camp in the Lorraine city of Toul, which, after urging escape, warned that Patton was "America's leading gangster and Public Enemy

Number One," who mistreated his prisoners. The Toul POWs meekly handed the propaganda leaflets to their keepers.[90] The Ardennes had made them conscious that the Third Army was not to be ignored. A captured officer spoke of the German fear of Patton. They considered him their gravest threat: "General Patton was always the main topic of military discussion. Where is he? When will he attack? Where? . . . How? With what? . . . [He was] the most feared general on all fronts. . . . The tactics of General Patton are daring and unpredictable. . . . He is the most modern general and the best commander of armored and infantry troops combined."[91]

The day after the armistice, Patton wrote: "The one honor which is mine and mine alone is that of having commanded such an incomparable group of Americans, the record of whose fortitude, audacity and valor will endure as long as history lasts. //s// G.S. Patton, Jr., General, 9 May 1945."

Defiant to the last shot, Patton baldly asserted:

> I can say this, that throughout the campaign in Europe I know of no error I made except that of failing to send a Combat Command to take Hammelburg. Otherwise, my operations were, to me, strictly satisfactory. In every case, practically throughout the campaign, I was under wraps from the High Command. This may have been a good thing, as perhaps I am too impetuous. However, I do not believe I was and feel that had I been permitted to go all out, the war would have ended sooner and more lives would have been saved. . . . As the Church says, "Here endeth the Second Lesson."[92]

Omar Bradley saw the end of the war rather differently and gave thanks that there would be no more desperate battles on beaches named Omaha or Utah or in places named Saint-Lô, Mortain, Bastogne, Saint-Vith, or Aachen. He remembered the 135,576 GIs who had given their lives in the battles to liberate Western Europe, and the memories haunted him. Sleepless and grim, Bradley "thanked God for victory."[93]

The most devastating event in human history had mercifully ended.

PART XI

An Unsoldierly Death
(1945)

He wasn't cut out to be an army of occupation
commander.

—THIRD ARMY ENGINEER OFFICER

The boundaries of an empire are the graves of her
soldiers.

—NAPOLEON

CHAPTER 45

Military Governor of Bavaria

Peace is going to be hell on me.

—PATTON
(MAY 1945)

Won't that old bastard ever get enough of war? He
wanted to fight in the Pacific, and I wish to God they'd
let him go.

—ANONYMOUS GI

The morning after the guns had fallen silent in Europe for the second time in
the twentieth century, Patton summoned the Third Army correspondents to
the war room of his headquarters in Regensburg. With the familiar figure of
Willie trailing behind him, Patton began by disclosing that the war would
end officially that morning, May 8, 1945. He then strode to the large war
map and pointed to Germany, Austria, Czechoslovakia, Poland, Bulgaria,
Hungary, and the smaller nations liberated by the Red Army. None of Patton's usual good humor was in evidence. Somberly he said:

> This [war] was stopped right where it started. Right in the Hun's back-
> yard which is now Hitler's graveyard. But that's not the end of the busi-
> ness by any means. What the tin-soldier politicians in Washington and
> Paris have managed to do today is another story you'll be writing for a
> long while if you live.
>
> They have allowed us to kick hell out of one bastard and at the same
> time forced us to help establish a second one as evil or more evil than the

first. We have won a series of battles, not a war for peace. We're headed
down another long road to losing another peace. This day we have missed
another date with our destiny, and this time we'll need Almighty God's
constant help if we're to live in the same world with Stalin and his mur-
dering cutthroats.

INS (International News Service) correspondent Larry Newman later
wrote that Patton's eyes brimmed with tears as he began to recall

the thousands from North Africa to the Channel, on the sea, in the air, in
the mud and filth, who gave their lives in what they believed was the final
fight in the cause of freedom. "I wonder how the dead will speak today
when they know that for the first time in centuries we have opened Central and Western Europe to the forces of Genghis Khan. I wonder how
they feel now that they know there will be no peace in our times and that
Americans, some not yet born, will have to fight the Russians tomorrow,
or ten, 15 or 20 years from tomorrow."
     We have spent the last months since the Battle of the Bulge and the
crossing of the Rhine stalling, waiting for Montgomery to get ready to
attack in the North; occupying useless real estate and killing a few lousy
Huns when we should have been in Berlin and Prague. And this Third
Army could have been. Today we should be telling the Russians to go to
hell instead of hearing them tell us to pull back. We should be telling
them if they didn't like it to go to hell and invite them to fight.
     Churchill was the only man in a position of power who knew what
we were walking into. He wanted to get into the Balkans and Central
Europe to keep the Russians at bay. He wanted to get into Berlin and
Prague and get to the Baltic coast on the North. Churchill had a sense of
history. Unfortunately, some of our leaders were just damn fools who had
no idea of Russian history. Hell, I doubt if they even knew Russia, just
less than 100 years ago, owned Finland, sucked the blood out of Poland,
and were using Siberia as a prison for their own people. How Stalin must
have sneered when he got through with them at all those phony conferences.[1]

Patton ended the press conference by noting he had not yet seen the
terms of surrender and didn't much care. "I catch them and somebody else
can cook them."
     Although he expressed a desire to help finish the war in the Pacific, Pat-
ton already understood that Marshall would never send him where there was
already a prima donna in command. He told the III Corps commander, Maj.
Gen. James van Fleet, "Jim, the war is all over. The S[O]Bs won't fight any
more. I would like to go to the Pacific theater but they won't let me. There

is already a star (MacArthur) in that theater and you can only have one star in a show."[2] Although he actively lobbied Secretary of War Stimson and others for an assignment in the Pacific, by mid-May Patton clearly understood there was scant chance of a transfer there, although he continued to imply to Beatrice it was yet possible.

Later that afternoon correspondents Newman and Cornelius Ryan went to Patton's trailer to pay their final respects before departing for Austria in search of fresh stories. They found Patton in his stocking feet, thumbing through a text of Napoleon's campaigns, a copy of Caesar's campaigns next to it. He asked if they remembered what he had said some weeks earlier in Trier: "I was sitting in my jeep under the arch built to commemorate Caesar's conquest of the city. One of you wise reporters asked me how I felt riding along the same battle route as Julius Caesar had more than 19 centuries ago? And I said: 'Hell, Caesar couldn't have been a supply sergeant in the Third Army.'" He laughed and then grew serious:

> This should be a day of rejoicing without a worry about the future and the peace we have all fought for. But thousands will sleep in strange and distant lands forever and still there is no peace. You cannot lay down with a diseased jackal. Neither can we ever do business with the Russians. Even the stupidest man thousands of miles away, back home, should have realized that by 1943 at the latest. The Russians really took us for suckers after we saved their hides. Now there'll be a lot of wining and dining with them and a lot of pinning of medals, but just wait until they send the fighting men back home and we get the postwar soldiers. Maybe then the tin soldiers on our side will see the handwriting on the wall. I hope I'm not here when that happens. I just couldn't stand being around and taking any lip from those S.O.B.'s"[3]

Despite his bitter tone, what Patton predicted before the ink had even dried on the German surrender documents was to prove a chillingly accurate prophecy. In the autumn of 1943 he had observed to a visiting American general: "It will be just as bad for us to have Russia win the war as it will be for Germany to do so. To be a success and to maintain world peace, the U.S. and the U.S. alone should destroy Germany and Japan and be ready to stop Russia."[4]

Patton made a similar pitch to visiting Undersecretary of War Robert Patterson, pleading for the great American armies to be left intact to counter the Russian threat. Let 30 percent of the trained troops remain in the United States if need be, he said, but "let us start training here, keeping our forces intact. Let's keep our boots polished, bayonets sharpened, and present a picture of force and strength to these people. This is the only language they understand and respect. If you fail to do this, then I would like to say to you

that we have had a victory over the Germans and have disarmed them, but
have lost the war." An exasperated Patterson replied that Patton did not
know the strength of the Russians and was unable to see the big picture.
"Mr. Secretary . . . for God's sake listen to what I am trying to tell you," he
pleaded. Asked what he would do, Patton replied he would keep the armies
intact, delineate the border between them, and if they did not return behind
it by a certain date, "push them back across it."

Unfortunately, as George C. Marshall had accurately predicted, the
United States has historically demonstrated little tolerance for a strong military establishment in times of peace. Consequently Patton's entreaties fell
on deaf ears. The nation and the world had endured the most horrendous
war in the history of the planet. With Germany beaten and the war in the
Pacific still raging, the last thing anyone would accept was the prospect of
further war in Europe, against an ally. Coming from Patton, the warnings
seemed like the rambling of a warmonger. Few listened; even fewer
believed what they heard, or if they did believe it, they were not willing to
pursue an unpopular course. The troops who fought World War II were
"proud of what they'd accomplished but in no mood to stand around admiring their handiwork. They'd approached the war like a job that had to be
done. And now that they'd done it, there was only one question left: Have I
got enough points to go home?"[5]

Yet what Patton was really advocating was better use of the *power* of
the United States and its massive army while it still existed: "We the Armed
Forces of the U.S.A. have put our government in the position to dictate the
peace. We did not come over here to acquire jurisdiction over either the people or their countries. We came to give them back the right to govern themselves. We must either finish the job now—while we are here and ready—or
later under less favorable circumstances."[6]

Patton's rationale was that the United States must avail itself of its awesome strength before demobilization decimated the ranks of its military
forces. By 1947 the armed forces of more than twelve million created by
George Marshall would have been reduced to a fraction of its wartime size.[7]
Unfortunately it was impossible to untangle Patton's astute warnings from
his own quest for glory. Now that his fighting days were over and his transfer to the Pacific little more than a fantasy, his reputation would forever rest
on his World War II exploits. His militancy was based in part on the conviction that, with the war over, "the excitement and the glory were already fad-
ing, and if an armed conflict was inevitable, as he believed, why not at once
when the Americans still had combat-effective forces in being?"[8] And, of
course, while he was still around to play an important role.

Moreover, Patton's credibility was hurt not for his prediction, but by his
call for war with the Red Army. Indeed, the folly of failing to heed the
warnings of Patton, Churchill and others in the aftermath of the war was

starkly demonstrated barely three years later during the Berlin blockade of 1948, which resulted in the creation of NATO the following year to counter the growing Russian threat in Europe.[9]

Patton's bellicosity has been dismissed as the ravings of a megalomaniac, yet of the major figures of the war, only Eisenhower was out of step with the prevailing attitude toward the Russians. His official biographer has written:

> Churchill wanted to get tough with the Russians; Montgomery wanted to (and did) stack the arms of the Germans surrendering to him in such a way that the Wehrmacht could be quickly rearmed if it came to a fight between the West and Russia; Patton was talking wildly about using the great American army in Europe, while it was still there, to drive the Red Army back to Moscow, in cooperation with the Wehrmacht. Herein lay the seeds of NATO, dominated by a German-American military alliance.

Only Eisenhower in 1945 took the approach that such an alliance was "irresponsible and self-defeating. He advocated, instead, making every effort to cooperate with the Russians to build a better world."[10]

Despite his sudden postwar outrage, during the war Patton had said little about the Russians, and one searches in vain to find in his writings the virulent anticommunism he began to display immediately after the German surrender. There is almost no evidence that he had harbored such wariness and hatred of the Red Army. To the contrary, as historian Hugh Cole points out:

> Although many officers in the Third Army, including Hap Gay, were always talking about how we should clean up the Communists now and not wait to do it later, Patton never gave any outward indication that I could see that he looked upon the Russians as our inevitable next enemy. In fact, during the last stages of the war when the press release would be read at the morning briefing, Patton would stand and initiate the applause whenever the Russians had made a good substantial gain.[11]

It is impossible to trace the evolution of Patton's change of heart about the Russians, but more than likely it began in April 1945, to the date when Eisenhower revealed that he would halt the Allied advance on the Oder River.

On May 9 Patton and Van Fleet were stopped at a crossroads in Czechoslovakia, where a group of Russians were being herded into prison camp. Patton sat silently and then said, pointing east with his swagger stick: "Jim, that's the way we should be going and take the rest of those Russians along with the Germans."[12]

As he had predicted, a number of ceremonies were held by the two armies, most of them rituals of one-upmanship. At one event hosted by the Russians in occupied Austria, Patton and his staff were entertained by a ballet troupe from Moscow and dined on stuffed pig, caviar, and seemingly endless rations of vodka. The Russians decorated an American female reporter as a symbol of American womanhood. Not to be outdone, Patton improvised a decoration on the spot for the prima ballerina, which was efficiently produced by a member of his staff.

Patton's first brush with the Russians occured shortly after V-E Day. He was smoking a cigar and working quietly in his new headquarters in Bad Tölz, when Gay entered and said there was a Russian brigadier general outside who insisted on seeing him. "What the hell does the son-of-a-bitch want?" Informed that it had to do with some Germans in river craft on the Danube who had surrendered to Third Army, Patton commanded: "Bring the bastard in, and you and Harkins come with him."

The English-speaking Russian demanded on behalf of his army commander that the river craft be turned over to the Red Army. Patton calmly reached into a drawer and pulled out a pistol, slamming it on his desk. Then he stood and with sudden rage shouted: "Gay, goddamnit! Get this son-of-a-bitch out of here! Who in hell let him in? Don't let any more Russian bastards into this headquarters. Harkins! Alert the 4th and 11th Armored and 65th Division for an attack to the east." The Russian was shaking and had turned white as he was escorted from Patton's office. After carrying out their commander's orders, Gay and Harkins returned to find a smiling Patton calmly smoking his cigar. "How was that?" he asked. "Sometimes you have to put on an act, and I'm not going to let any Russian marshal, general or private, tell me what I have to do. Harkins, call off the alert of the divisions. That's the last we'll hear from those bastards." And it was.[13]

Although he admired their fighting qualities, Patton thought Russian soldiers little better than robots, and described the hero of the Red Army, Marshal Georgi Zhukov, as apelike, "comic opera, covered with medals." On May 13 he had hosted and decorated the commander of the 4th Russian Guards Army in Linz. At lunch the whiskey flowed, and the Russian attempted to drink Patton under the table. In his diary he smugly noted that he had "walked out under my own steam," while the Russian general "went out cold." Patton had watered down his bourbon and seemed pleased when the Russian general's aide said he had a cast-iron stomach. When Patton indulged in the Russian tradition of breaking his glass after a toast, the Russian embraced him. Patton's reaction to the event: "They are a scurvy race and simply savages. We could beat hell out of them."

The next day Patton and his staff were feted by a different Russian army commander, whom he described as "a very inferior man" who "sweated profusely." He conceded that "they certainly put on a tremendous show," noting that the fifteen miles of road they passed over "had actually been swept" for his entourage. "After lunch they had a very splendid show which unquestionably had been flown in from Moscow. They did their best to get us drunk, but we had taken the precaution of drinking two ounces of mineral oil before starting on the expedition. We were also very careful of what we drank."

Patton was both impressed and distressed at the severe discipline of the Russians: "The officers with few exceptions give the appearance of recently civilized Mongolian bandits. The men passed in review with a very good imitation of the goose step. They give me the impression of something that is to be feared in future world political reorganization." Gay's diary recorded: "Everything they did impressed one with the idea of virility and cruelty."[14] Patton's verdict was that "in addition to his other amiable characteristics, the Russian has no regard for human life and is an all out son of a bitch, barbarian, and chronic drunk."[15]

Patton's hatred of the Russians did not dim with the passage of time. In early September he attended a large military review in Berlin hosted by Marshal Zhukov. His new aide, Maj. Van S. Merle-Smith overheard this exchange when some enormous Russian tanks passed in review. Zhukov: "My dear General Patton, you see that tank, it carries a cannon which can throw a shell seven miles." Patton: "Indeed? Well, my dear Marshal Zhukov, let me tell you this, if any of my gunners started firing at your people before they had closed to less than seven hundred yards I'd have them court-martialed for cowardice." Major Merle-Smith observed: "It was the first time I saw a Russian commander stunned into silence."[16]

In other aspects of postwar debate, Patton was both rational and insightful. He was, for example, a lifelong advocate of universal military training, firmly believing it was a citizen's duty to serve his nation, that democracy was a shared burden, thus joining the select company of Leonard Wood, Elihu Root (Theodore Roosevelt's secretary of state, 1905–9), Henry Stimson, and George Catlett Marshall. Patton's prediction regarding the compelling need for universal military training after World War II proved chillingly accurate: "You just wait and see," he said. "The lily-livered bastards in Washington will demobilize. They'll say they've made the world safe for democracy again. The Russians are not such damned fools. They'll rebuild; and with modern weapons. But if we have compulsory universal military service, if we vote for it first at the polls, or in Congress, then in years to come the rest of the world will know we mean what we say."[17] Most American church leaders were opposed to the concept, and when he informed the War Department of his intent to discuss the matter with his longtime friend Cardinal Spellman of New York City and with Archbishop Richard Cushing of Boston, Patton was firmly enjoined to "keep my mouth shut and that I'm a warmonger."[18]

Had he seen it, Patton certainly would have agreed with Marshall's 1945 report to the War Department, which said in part: "We have tried since the birth of our nation to promote our love of peace by a display of weakness. This course has failed us utterly."[19]

On May 10 one of the most unlikely high-level meetings of the war took place. Eisenhower summoned Hodges, Simpson, Patch, Patton, and their air officers to deliver a lecture on cooperation and to extract a promise not to criticize publicly the campaigns fought in Europe:

> General Eisenhower talked to us very confidentially on the necessity for solidarity in the event that any of us are called before a Congressional Committee . . . he then made a speech which had to me the symptoms of political aspirations, on cooperation with the British, Russians, and the Chinese, but particularly with the British. It is my opinion that this talking cooperation is for the purpose of covering up probable criticism of strategical blunders which he unquestionably committed during the campaign. Whether or not these were his own or due to too much cooperation with the British, I don't know. I am inclined to think that he over-cooperated.[20]

To his staff Patton merely said: "Eisenhower is running for President."[21]

Except for Patton's opinions and his conspiratorial interpretation, no biographer of any other participant has ever acknowledged this meeting or offered an interpretation of its intent. The controversial British historian David Irving has asserted "the big cover-up is beginning. There is to be no criticism of the strategic blunders that Eisenhower has unquestionably, in Patton's view, committed during the campaign."[22] Until the day he died Patton believed Eisenhower's failure to take Berlin to be a monumental blunder, once writing: "I believe historians will consider [it] a horrid crime, and a great loss of prestige in letting the Russians take the two leading capitals of Europe [Berlin and Prague]."[23]

The trouble with Irving's wild accusation of an Allied cover-up is not only that Patton's criticism varied little from his earlier criticism of Eisenhower's generalship, but that he could not even make up his mind whether Ike's alleged blunders were made on his own or as a result of "too much cooperation with the British." Moreover, Patton's assertions presume that Eisenhower had acknowledged that he had *indeed* made serious blunders during the northwest Europe campaign—a point that Eisenhower himself never conceded.[24]

On May 23, 1945, the Third Army headquarters moved from Regensburg to a former Waffen-SS barracks in Bad Tölz on the Isar River in the foothills of the Bavarian Alps, some twenty miles from Munich. The headquarters was housed in an enormous nine-hundred-room building, renamed Flint Kaserne in honor of Paddy Flint, whom he loved and admired. Patton's palatial residence on a nearby lake included an indoor swimming pool and a bowling alley and was once owned by the wealthy publisher of Hitler's *Mein Kampf.*

When it was reported that a desk reputedly belonging to Rommel had been located in Stuttgart, Codman was sent to appropriate it for Patton. Whether or not it was actually Rommel's, the "desk" turned out to be a small, rather feminine and dainty table that was hardly in keeping with Rommel's (or Patton's) image. A disappointed Patton kept the desk for its symbolic value. [25]

In Nuremberg on Memorial Day Patton visited a temporary military cemetery. The permanent military cemeteries, whose green lawns and simple snow-white crosses—now so well known to Americans—had yet to be established. In the aftermath of the war the cemeteries were mostly open fields with hundreds of graves marked by temporary wooden crosses. With a large entourage of reporters and photographers in his wake, Patton arrived resplendent in his dress uniform, his boots gleaming, his pistols strapped to his waist. He was handed some flowers and at random chose a grave, at which he knelt, placed the flowers against the cross, and remained in silent prayer for some moments. The cemetery was a quagmire of deep, sticky mud, but this did not deter Patton, whose legs were covered with muck when he rose. As the press moved toward him, the military police attempted to block their advance, but were waved away by Patton, who turned and pointed to the grave. In his high-pitched voice he issued one of those pronouncements, at once profound and profane, that only he was capable of: "That man died so bastards like you can continue to breed!" [26]

Patton dealt with a myriad of problems resulting from the war. The liberation of the infamous Dachau concentration camp in April 1945 provoked a fifty-year argument over which American division was the first on the scene, with increasingly acrimonious exchanges occurring between veterans of the U.S. 42d and 45th Infantry Divisions. A task force of the 45th Division, commanded by Lt. Col. Felix L. Sparks, one of the heroes of the Anzio beachhead the previous year, [27] arrived on the morning of April 29, and after a brief firefight, eliminated the SS troops guarding the camp. Shortly thereafter a brigadier general named Linden, the assistant division commander of the 42d Division (a Seventh Army unit), arrived and demanded entry. Sparks refused, citing his orders to permit no one in or out until advised by his regimental commander. Harsh words were exchanged, and Sparks drew his pistol to signal that he intended to carry out his orders to the letter. The general left, threatening Sparks with court-martial charges, and later filed a complaint with the Seventh Army inspector general, who investigated the incident.

The Seventh Army drew up court-martial charges, and Sparks was placed under arrest several weeks later. The charges were referred to Patton in his new capacity as the occupation commander of Bavaria, and Sparks reported to Patton who said: "Colonel, I have some serious court-martial charges against you and some of your men here on my desk." Sparks explained what had happened; Patton paused and then said: "I have already had these charges investigated, and they are a bunch of crap. I'm going to tear up these goddamn papers on you and your men." Patton ripped them up, and tossed them into his wastebasket. "You have been a damn fine soldier. Now go home." Sparks saluted and left. The entire interview had lasted a mere three minutes. Patton knew a good soldier when he saw one, and Felix Sparks was one of the U.S. Army's most outstanding officers, whose exploits as a rifle company commander in the Anzio beachhead were legendary.[28] Patton was soft on those who achieved great deeds in battle, and the petty charges against Sparks fell under the heading of "chickenshit." Unless their crimes were heinous, such men who served under him received medals and praise, not court-martials.

Another problem brought Patton credit he barely deserved: the rescue of the magnificent white Lipizzaner horses of Austria. Bred since 1580 in Austria, Germany, Hungary, and other neighboring countries, the Lipizzaners became the riding and carriage horses of the Austrian Imperial Court, and for two hundred years had been trained at Vienna's famed Spanish Riding School, which the Nazis permitted to stay open during the war.

When Vienna was under bombardment in the spring of 1945, the school's head, Col. Alois Podhajsky, a former Austrian Olympic rider, somehow managed to smuggle the animals to Sankt Martin im Innkreis, a small village in Lower Austria, near Linz, where it was hoped they would be safe from the Red Army. However, fodder was in short supply, and their future was uncertain. In April, Podhajsky sought American help from the XX Corps commander, Walton Walker. Although unable to speak to Walker himself, he did manage to get the attention of a major who pleaded Podhajsky's case to Walker: If the U.S. Army did not move quickly to save the Lipizzaners, they would soon be claimed as German war booty and would be performing in Moscow for Stalin and the Russians, lost forever to the world. At last convinced of the plight of the Lipizzaners, Walker decided to involve Patton by inviting him to a demonstration, explaining that "in that way we can get more stars in back of this than I've got."[29]

With Patton observing, even the horses were nervous and seemed to sense the vital importance of their performance. After a superb exhibition of riding and horsemanship, Podhajsky proudly rode toward Patton who stood to greet him. "We ask your protection," said Podhajsky. A stone-faced Patton merely nodded and replied: "Magnificent! These horses will be wards of the U.S. Army until they can be returned to the new Austria." In reply "Pod-hajsky and his riders slowly raised their hats. The Lipizzaners stood at attention. No one who was there will ever forget the scene."[30] The Lipizzaner story was later the subject of a Walt Disney film, which depicts Patton as the horses' generous benefactor.

The reality, as biographer Farago writes, was that "actually, they disgusted him."[31] Patton's diary for that day notes:

> It struck me as rather strange that, in the midst of a world at war, some twenty young and middle-aged men in great physical condition, together with about thirty grooms, had spent their entire time teaching a group of horses to wriggle their butts and raise their feet in consonance with certain signals from the heels and reins. Much as I like horses, this seemed to me wasted energy. On the other hand, it is probably wrong to permit any highly developed art, no matter how fatuous, to perish from the earth— and which arts are fatuous depends on the point of view. To me the high-schooling of horses is certainly more interesting than either painting or music.[32]

With the war over it was inevitable that at some point Patton would be permitted a home leave for the first time in nearly three years. Although Beatrice had cabled her husband how proud she was of him (he replied that her "modest estimate of me means more than the opinion of the rest of the world"), Patton had mixed feelings about his eventual reunion with his wife of thirty-five years.[33] The reason had to do with Beatrice's half-niece, Jean Gordon, the same young woman with whom he had apparently become involved with in Hawaii nine years earlier.

In the summer of 1944 Jean had suddenly arrived in England as a Red Cross volunteer. Although Patton's diaries make only oblique and innocuous references to Jean, his West Point classmate, Everett Hughes, recorded some very indiscreet remarks. Blumenson writes: "Jean Gordon breathlessly telephoned Patton from London early in July," and he apparently visited her shortly before departing for Normandy. When "Hughes wondered about their relationship . . . Patton told him, with some exaggeration: 'She's been mine for twelve years.'"[34] If this comment was true, it meant that Patton and Jean had been involved from the time she was seventeen and a frequent guest of Ruth Ellen. When David Irving's controversial *The War Between the Generals* was published in 1981, the alleged affair became the subject of an article in *Parade* magazine, in which Ruth Ellen angrily replied to the allegations: "Jean Gordon was my best friend. Her father died when she was 7. She was a bridesmaid at my wedding and at my sister's wedding. She was one of the family. Daddy looked upon her as one of his own daughters. To accuse him of having an affair with Jean is a disgusting lie."[35]

However, Robert Patton's 1994 family memoir reveals that Beatrice believed that Jean's reappearance signaled a resumption of the affair she suspected her husband had had in Hawaii in 1936. It was a tense and undoubtedly painful time for Beatrice, who was obliged to wait at home with "tight-lipped consternation . . . [she] can only have felt helpless and perhaps even foolish to complain about a matter that seemed trivial in comparison to her husband's headline-making exploits in Europe. She loved him; she was his most ardent and patient advocate. To think he'd had an affair with Jean in 1936 hurt Beatrice terribly. For Jean now to be with him in his moment of triumph mixed that pain with fury."[36]

In late July 1944 Beatrice wrote to her husband that she was aware of Jean's presence, prompting a cavalier dismissal, and a denial that he had even seen her. "The first I knew about Jean's being here was in your letter," he wrote on August 3. "We are in the middle of a battle so don't meet people so don't worry."[37]

However, Jean arrived in France in early September, having contrived an assignment to the Third Army as a Red Cross "doughnut dolly." The Red Cross contingent was supervised by a woman named Betty South, who has steadfastly denied Jean's involvement with Patton. "I knew her very well. Jean Gordon was an extremely sensitive, well-bred girl from a wealthy family. She adored General Patton, but strictly in a father-daughter relationship." According to South:

> The man she truly loved was a handsome young captain in the Fourth Armored. Unfortunately, he was married. But he told Jean right off that if ever he got out of the war alive, he was going back to his wife and son in San Antonio. Those were the ground rules when they fell in love. Jean honestly felt she could abide by them. But she couldn't. . . . They had one final fling [in September 1945], then he went back to the States. Jean tried to drive the memory of him out of her mind. But she couldn't. She grew steadily more depressed and morose."[38]

Nevertheless, if Everett Hughes is to be believed, the affair with Patton did continue in England and Europe, even as Beatrice pressed him about her presence in the Third Army, eliciting this reply on March 31, 1945: "Don't worry about Jean. I wrote you months ago that she was in this Army . . . [but the] letters must have been lost. I have seen her in the company of other Red + several times, but I am not a fool, so quit worrying."[39] His curt dismissal of Jean's presence as of little consequence probably did little to calm Beatrice.[40]

Hughes's diary contains a number of references to Patton and Jean Gordon, alleging, for example, that they had quarreled a week or so before Patton began a ten-day leave in England in mid-May, and that they apparently renewed their liaison in London during the few days prior to a summons from Eisenhower ordering Patton to return at once to Germany. At a farewell dinner at the Ritz, Patton admitted that "he was scared to death of going back home to America."[41]

Hughes's observations and commentary notwithstanding, Patton himself left no indication or acknowledgment of an affair with Jean Gordon. His behavior suggests that in both 1936 and 1944–45, the presence of the young and attractive Jean was a means of assuaging the anxieties of a middle-aged man troubled over his virility and a fear of aging.[42]

Although his troops may not have known or appreciated it, Patton took a *laissez-faire* attitude toward Eisenhower's policy of nonfraternization, openly proclaiming: "Anything my men have fought to capture, they are, by God, entitled to." He meant women, and deliberately turned a blind eye to such activities, while always insisting there be an adequate supply of the new wonder drug penicillin available. The result was a low rate of venereal disease in the Third Army. Patton often discoursed on the care and management of women, advising, for example, that immediately on returning from a long absence, a man should immediately take his wife or lover to the nearest hotel for several days, away from family and children—advice he would fail to act on during his own home leave in June.[43]

Patton's leave in England was cut short so that his superiors could take advantage of his fearsome reputation. The Yugoslav leader, Tito, was attempting to annex Trieste, a political dilemma the Allies tried to solve by threatening Allied military intervention against him; Tito eventually compromised after a tense confrontation.[44] Patton thought he detected sinister implications of Russian treachery behind every fresh problem in occupied Europe. In this instance he thought that Tito might be acting as a puppet of the Russians, and viewed it as yet another occasion to fight them sooner rather than later.[45]

Patton's increasing anxiety was a sign that he was as ill-prepared for peace in 1945 as he had been in 1919. He seemed unable to accept that the war was over, his final battle had been fought, and that the remainder of his life would inevitably have to take a different direction. The same feelings of depression, of overwhelming letdown that existed after World War I now returned. However, unlike in 1919, his relationship with Jean Gordon had so severely strained his marriage that he could not bear to face his long-suffering wife. Robert Patton writes: "For Georgie, the shock of peacetime that many veterans experience was intensified because he was happy in war, his happiness a sort of steady state in which his swings between feeling good and feeling bad were only incidental anomalies in a prevailing condition of total engagement in life. . . . Without a world war to give them a vast and appropriately awesome arena in which to play out, he could not expect to survive very contentedly or very long."[46]

Patton had barely settled in at Bad Tölz when he learned he was to return to the United States for thirty days to participate in "a goddamn bond-raising tour." Despite an annoying case of strep throat, Patton made a triumphal return to the United States on June 7, 1945. Escorted by fighters and B-17s flying in formation, his plane landed at Hanscom Field, near Boston. Patton's entourage included some fifty officers and men of the Third Army. When the engines shut down, there was a roar from the crowed, followed by a sudden hush as the plane's door opened.

A bevy of VIPs awaited Patton on the tarmac. To one side stood his son, on summer leave from West Point; Ruth Ellen; Little Bee; and a much older-looking Bea, whose hair had begun to show gray, perhaps in defiance of her husband's long-ago admonition that she was never to grow old. Without waiting for the ramp to be put in place, Patton jumped from the side door and attempted to get to Beatrice, but he was deterred by the dictates of military protocol. A visibly annoyed Beatrice had been told where she and the family would stand and had conveyed her displeasure. They were obliged to wait while Patton was greeted by an honor guard and a seventeen-gun salute. Massachusetts governor Maurice J. Tobin doffed his hat in respect, and Patton returned the honor by removing his gleaming helmet with its four stars and the emblems of his three commands: I Armored Corps, Seventh Army, and Third Army. Finally Patton was free to greet Beatrice, and the two embraced for the first time in nearly three years.

For some twenty-five miles through the suburbs of Boston, flags fluttered in tribute. As young women tossed flowers at him, Patton stood proudly (alternating from smiles to grimaces) in an open car, his helmet clutched to his chest or in his right hand, waving, as an enormous crowd, estimated at one million, cheered his triumphal return. Many wept openly. In downtown Boston a snowstorm of confetti and paper descended on him. Addressing a crowd of thirty thousand (some accounts put the number at nearly fifty thousand) from the Hatch Shell on the Esplanade of the Charles River, Patton said, "My name is merely a hook to hang the honors on. This great ovation by Boston is not for Patton the general, but Patton as a symbol of the Third Army." He spoke of those who would never return home, buried in cemeteries from Normandy to Austria, and of the thousands of wounded in hospitals, and of the great work of the Army Medical Corps, which he credited with saving many lives. He also managed once again to put his foot in his mouth. The fact that a soldier was killed in action often made him a fool rather than a hero, he told the audience. Some four hundred wounded soldiers of the Third Army were present, to whom Patton pointed and said: "These men are the heroes." He ended his speech by saluting them.[47] Massachusetts and Boston had never seen anything like it.

At a state dinner held in his honor at the Copley Plaza Hotel (the first tendered anyone since Marshal Foch, the World War I Allied supreme commander), Patton entered to the strains of the Armored Force March, composed several years earlier by Beatrice. One newspaper account recorded that "Patton lost his composure in the face of all the tributes. He bowed his head and tears streamed down his cheeks. It was some time before he dared to trust his voice, and even then it shook with feeling. Three times as he spoke of the men of the Third Army in terms of glowing praise he choked up and stopped to regain his self composure." Few noticed Beatrice lean over and pat him encouragingly on the side as he did so. "At the end, as he resumed his seat amid cheers and applause, he took out his handkerchief and wiped the tears from his eyes." He had spoken movingly of the sacrifices of his men, saying: "When we mourn for such men who have died we are wrong because we should thank God that such men were born." When Governor Tobin praised Beatrice as "a model of the soldier's wife," Patton rose and jumped up and down, his hands clasped over his head. In his speech he referred to Beatrice and his family as "my generals." It was one of the most emotional days of Beatrice's life. "I can hardly speak, I'm so overcome," she declared. "This has been a proud and wonderful day." Calling her husband's return a miracle, she told a reporter that Georgie had said he had expected to die, "and I had prepared myself for whatever might happen."

The next morning the *Boston Daily Record* headlined in two-inch black letters: GEN. PATTON IN TEARS AT HUB TRIBUTE [48] A photographer captured an emotional Patton with a handkerchief to his face. Later, however, there were howls of protest at his perceived insensitivity from the Gold Star parents of those killed in action. Telegrams and letters poured into the War Department, addressed to Marshall and Stimson. An attempt at damage control by the Pentagon public relations staff failed to nullify the harm wrought by Patton's latest indiscretion. Once again, his intent had been misread: he had meant to honor these men, not demean them. The private anger of some notwithstanding, the reaction of Boston's newspapers to his speech was universal praise. To Bostonians, Patton was a hero whom they could claim as their own.

It was after midnight when the exhausted couple returned to Green Meadows to snatch a few brief hours of sleep. Early the next morning they flew first to Denver, where he delivered a speech, and then Southern California, for similar ceremonies in Los Angeles and Pasadena. Son George returned to West Point, never to see his father again. [49]

The Los Angeles Coliseum was filled to its one-hundred-thousand capacity to cheer Patton and witness a mock tank battle. [50] Again he wept at the public's acclaim and in a resounding speech vowed the defeat of Japan. Many of Hollywood's elite jockeyed to become part of this event. Pointing to two women who seemed to know him, he was told they were gossip columnists Hedda Hopper and Louella Parsons. Patton grumbled that he had never heard of them. The Army public relations personnel assigned to watch over Patton had installed a switch on the microphones so he could be instantly cut off if his words became too intemperate. However, someone had neglectfully left a microphone "live," over which was heard: "Magnificent. Almost as good as the real thing. And God help me, I love it [war]." As Charles Codman writes: "The General's phrasing was clear, distinct, and coast to coast."[51]

With their wives sitting on the stage, Patton and Jimmy Doolittle delivered a series of speeches. At the Hollywood Bowl, Doolittle spoke first and ended by saying: "If General Patton and I have achieved any success in fighting the war, these two lovely ladies are responsible for that success." Their wives stood to deafening applause. Once again failing to realize the mike was open, Patton said "with glorious amplification" to his friend: "You son of a bitch, I wish I'd said that."[52]

Patton and Doolittle were greeted by huge crowds estimated at one million. They wore their ribbons, which, Patton proclaimed, "our dead won." As Doolittle later recalled: "Georgie was his usual blunt self in public. He could not be muzzled, and he made off-the-cuff remarks about our Allies, the Soviets, politicians, and even the next war . . . it was always a typical fire and brimstone Patton speech."[53]

He greeted nearly five hundred of his near and distant friends and relatives at a gala gathering of the Patton clan, including the venerable Mary Scally, who was now ninety-five. In Boston, Patton had promised that once in California he would attend the Church of Our Savior in San Gabriel, where he had been baptized and confirmed: "God has been very good to me. I would like to go there and give Him my thanks." On his first Sunday home that debt was paid. At the graves of his beloved parents he prayed in silence and left a wreath.[54] He also left a wreath in memory of his aide, Dick Jenson. Patton never forgot those whom he loved and respected.

His brief sojourn in Southern California reunited Patton with Nita, who had filled the rooms of her home in San Gabriel with memorabilia of her brother. From childhood their kinship had remained unbroken. Nita shared her brother's triumphs and heartbreaks as if they were her own, and her bitterness at those who disparaged George Patton never wavered. She dedicated much of her life to advancing her brother's cause, exulting in Georgie's triumphal return in June 1945 as sweet revenge on those who had attempted to damage his reputation."[55]

To the dismay of his detractors, and in spite of his verbal lapses, Patton had indeed become an authentic American folk hero, his popularity exceeded only by Eisenhower's.[56] The perception of Patton as "Old Blood and Guts" was vastly different from the personality of this enigmatic man when he voluntarily emerged from behind his protective facade. An example occurred when he and Doolittle were feted at a victory parade in their honor. As their motorcade neared the speaker's platform, a saxophone player in the band set down his instrument and stood with a camera in his hand to photograph Patton, who spotted him clicking away. He began gesturing futilely to gain the young man's attention, then stood up, got out of the car, and pushed through the crowd, headed not for the speaker's platform but toward the musician. A reporter observing this spectacle, wrote: "Grinning broadly, the bemedalled general, scourge of Germany, the professional soldier whose salty language was legend wherever GIs gather . . . whose name was a household word in those troubled times—held up the parade and ceremony in his honor while he plucked a lens cap from the man's camera. 'Better take this off,' he chuckled at the embarrassed photographer. 'You'll get a better picture.' Then, as the crowd waited, General Patton posed for the saxophone player while he focussed and snapped his shutter. Then, he turned and mounted the stairs to the platform to begin his speech."[57]

In San Marino, Patton was given a gala welcome as a hometown hero and agreed to speak to the assembled citizens. He was unnerved when a Cub Scout presented him with a bouquet of flowers, and he handled it more like a hand grenade about to explode. Finally unable to contain himself, Patton put it down and announced: "I know I've made a lot of widows during my life, but this is the first time I've had to stand around looking like a damned bride."[58]

In Washington daughter Ruth Ellen had been working as an occupational therapist in the amputee ward of Walter Reed Army Hospital. She insisted he visit the ward, whose patients had begged her to persuade Patton to do so. He reluctantly agreed. With Beatrice, doctors, VIPs, and press trailing in his wake, Patton arrived in full uniform, his trademark swagger stick in a gloved left hand. Before entering the ward, he rounded on the reporters: "I'll bet you goddam buzzards are just following me to see if I'll slap another soldier, aren't you? You're all hoping I will!" As Ruth Ellen recalled:

> I had worked in that ward for two years, and I guess for my own protection I had ceased to see how truly tragic and horrible it was. . . . But that ward was a shock to Ma and Georgie. He strode down to about the center of the rows of beds. All of the patients were looking at him with their hearts in their eyes. Suddenly, he whipped out a large white handkerchief and burst into tears. He looked around and said, "Goddammit, if I had been a better general, most of you would not be here." He turned on his heel and walked rapidly out the door with the crowd of officials scrambling after him. The men cheered as he went by. One of them said to me, "Mrs. Totten, did you hear what your father said to *me*?" He had gotten across to them as no one else could have done. They did not want sympathy.[59]

Yet, in the minds of his critics, the ghosts of Sicily lingered unceasingly. No matter what he did, Patton would always be the general who had inexcusably slapped two soldiers.

Before departing Washington, George and Beatrice were invited to lunch at the home of daughter Bee. While Beatrice was upstairs, Patton somberly turned to Bee and Ruth Ellen and said in a conversational tone: "'Well, I guess this is goodbye. I won't be seeing you again. Take care of your little brother, and tell John and Jim to take care of you." Pausing, he said: "I think I'll see your mother again."

Shocked at this outburst, both women began talking at once and Ruth Ellen cried: "Oh, come on Daddy, it's crazy! The war is over." Patton replied, "Well, my luck has run out. Every shell that has struck near me, struck closer each time. Front-line infantrymen use up their luck a lot faster than a rear line cook. You are born with a certain amount of luck, like money in the bank, and you spend it and it's gone. It's too damned bad I wasn't killed before the fighting stopped, but I wasn't. So be it."[60] It was the last time his children ever saw him.

Throughout the remainder of his leave Patton continued to say the politically incorrect thing, often using allegory that went over the heads of most. An increasingly lonely man, Patton felt that his was the voice of truth in a wilderness of indifference. Only Beatrice understood the "word, verse, chapter and book" of his life: "She agonized with him." He spoke of retiring and traveling around the world in the *When and If*. "He said he was not going to write his memoirs—that the truth was never known about a war until one hundred years after it was over, until everyone who had fought in it was dead. He said that most military memoirs were 'apologia pro vita sua.'"[61]

In Washington, Patton learned that he would not be sent to the Pacific and complained: "I am persona non grata in the Pacific." Yet he was equally unhappy at the prospect of returning to Germany. "He was very upset as he knew that his talents were not suited to such an assignment. He said he was a military leader, not a politician."[62] Patton's lobbying efforts in Washington had included a meeting with President Truman, who declined to intercede on his behalf with Marshall and Stimson over a Pacific assignment.

Protests over his Boston speech continued to draw fresh howls of outrage and dismay from those who had lost sons and now demanded yet another apology, this time for defaming those who had died for their country. Before Patton arrived in Washington, Marshall's executive officer (who twenty-five years later would produce the film *Patton*), Col. Frank McCarthy, sent a memo to his boss that Patton's presence in Washington would "call for some special attention." McCarthy thought that Marshall ought to talk with him as soon as he got to Washington, before Patton's mouth could again get himself and the army into further trouble. Unperturbed, Marshall declined, noting that he would see Patton in due course.[63]

Still, Marshall's staff were scrambling to deflect the criticism of the Boston speech, thus seeming to lend credence to critics like Walter Bedell Smith, who tartly observed: "George's mouth doesn't always carry out the orders of his brain."[64] This crisis eventually passed, but Bedell Smith had the final say when he noted: "Patton acts on the theory that it is better to be damned than say nothing—that some publicity is better than none."[65] Yet Patton's resounding oratory had achieved the primary aim of his month-long trip to the United States, and elicited a note of thanks from Treasury Secretary Henry Morgenthau for helping to sell millions of war bonds for the climactic days of the war against Japan.[66]

Still, both Marshall and Stimson were sufficiently concerned over what Patton might say in Washington that the secretary personally chaired his press conference to ensure that there would be no further public faux pas. Indeed, so thoroughly did Stimson seize control of the June 14 press conference that Patton was confined to innocuous but colorful and quotable comments about the Germans and his beloved Third Army, on whose troops he continued to heap praise. Stimson artfully explained away Patton's recent indiscretion, and when reporters asked where he would go next, the secretary replied that Patton would return to Germany to complete "all the unfinished business there." His Boston blunder notwithstanding, Patton's popularity remained at an all-time high.[67]

Before the atomic bomb was dropped on Hiroshima and Nagasaki, an invasion of Japan had not been ruled out.[68] Patton's name was on a list of six generals submitted by the War Department to MacArthur, who (predictably) wanted no part of him.[69] Patton lobbied everyone in Washington he knew who might help him, but neither Marshall nor Stimson was willing to cross MacArthur by assigning Patton to his command without his consent. Mark Clark, now the American occupation commander in Austria, had written to Eisenhower that he did not want that "sonofabitch" near him or in any way involved in corraling the troublesome Tito. Patton was perceived as a loose cannon whom no one wanted. Even Eisenhower had written to Marshall that Patton was a "mentally unbalanced officer." The chief of staff even briefly considered sending a psychiatrist to observe Patton's behavior at his press conference. Personal feelings of friendship and respect notwithstanding, by the summer of 1945 Patton's superiors saw him as an increasingly troublesome burden, his usefulness having evaporated with the war's end and his skills as a combat leader now superfluous.[70] Unwilling or perhaps simply unable to recognize the obvious perils of sending him back to Germany, they made no effort to transfer Patton to a post in the United States or to retire him while his dignity and pride remained intact. He was not the first warrior to find discontent with peace. Patton may have been an American hero to millions, but in the corridors of power he had become an outcast.

*   *   *

Patton dreaded returning to a job he no longer relished, but even this apparently seemed preferable to the strains on his marriage from having to explain Jean Gordon. Little is known of what transpired during the thirty days George and Beatrice Patton were together, or if they even discussed what had clearly become a festering problem, but given Beatrice's evident bitterness over the affair, it seems inevitable that the subject of Jean must have arisen. It does not appear to have been resolved. Nonetheless. photographs with his grandchildren in his arms show a smiling Patton.

When he returned to the Third Army via Paris in early July, Patton seemed visibly relieved, confessing to Hughes that "Beatrice gave me hell. I'm glad to be in Europe." His homecoming had clearly troubled him. During the airplane trip, Patton had readily agreed with an observation by Sergeant Meeks that their month in the United States had been like a thirty-day jail sentence.[71] Although reassured to find that he was "still a fairly good sized [social] lion," Patton found Paris a gloomy place and hastened to return to the more comfortable surroundings of the Third Army. In Bavaria, Patton was given a royal welcome that included a fighter escort, and a band and honor guard at the airfield. Troops and tanks lined the road to Bad Tölz. It was exactly what Patton needed to restore his sagging spirits. "It gave me a very warm feeling in my heart to be back among soldiers," he wrote in his diary that night.

The task of administering war-torn Bavaria was daunting. Patton was expected to reinstate a semblance of normalcy by restoring basic services while at the same time denazifying and demilitarizing the local government in compliance with Eisenhower's expressed goal of ridding Germany of all vestiges of the former Third Reich. However, as the Americans learned when the Third Army captured the Nazi party archives in Munich, 75 percent of Germany's sixty million population were members of the party.[72] To achieve Eisenhower's goal seemingly required the skills of a Houdini.

While Patton was in the United States, problems had begun piling up. From the dreadful conditions in the displaced person (DP) camps (whose occupants Patton scorned as "too worthless to even cut wood to keep themselves warm"), to filling important appointments of civil positions in Bavaria, there seemed no end to the myriad problems facing the new military governor.

For all his intellectual prowess, Patton was utterly unprepared for his peacetime role. It can be argued that being a warrior was too deeply ingrained in him to allow him ever again to function effectively in peace. His biographer Roger Nye has noted that "Patton's extensive study of American history [did not] seem to sensitize him to the strains of isolationism and war weariness of the American populace."[73] Yet, Patton's vision of postwar

Europe was remarkably prescient. He believed that the time had come to look beyond the evil that Hitler had wrought and embark on restoring Germany to a strong, viable nation that would eventually become an American ally. "Only the Germans, it seemed to him, could emerge from the ashes of defeat and contribute to the stability of Europe. Only they could stand as a bulwark against the Russians and Communism."[74]

Patton visited the 80th Division to present Presidential Unit Citations to two of its units at a parade. The division was superbly turned out, and Patton appeared in full regalia.

> Even from across the parade ground you could tell which one he was. (Do his detractors ever think of that?) And he gave us the kind of talk we expected to hear. As brutal as Patton could be to anyone he thought was a malingerer, he was always lavish in praising his veterans. He told us what great soldiers we were and how honored he was to command us. He said that he knew how much we wanted to go home. . . . He asked us to maintain an interest in military preparedness and to keep America strong because . . . "You never know when some crazy sons of bitches like Hitler and Mussolini will come along and start another war." He then climbed into his armored car, trooped the line with flags flying and sirens blaring, and returned to the reviewing stand to salute us as we passed in review. It was all vintage Patton, and we loved it.[75]

Patton relished such ceremonial visits, the training and discipline of his units, and duck and stag hunting with Ernie Harmon, who found him utterly disinterested in the governance of Bavaria. Patton would curtly brush off Harmon's "stupid questions" when all he wanted to do was hunt and not be reminded of his responsibilities. Conversely, Harmon seems to have helped fuel Patton's hatred of the Russians and his forbearance of Nazis in high places.[76] When not hunting, Patton preferred riding with Gay in the beautiful Bavarian countryside on two "liberated" French horses. He added a squash court to his magnificent villa, complaining that he would have gone mad without this diversion. When not actively exercising, Patton's spare time was devoted to reading. His vices were few and his routine lackluster. He continued his lifelong habit of smoking cigars, only occasionally heeding Dr. Odom's advice that he ought to quit. Once, when he developed a sore on his lip, Patton walked into Harkins's office bellowing:"Goddamnit, Paul, I'm stopping smo-*goddamn*-king for good. You keep these ci-*fuckin'*-gars." Harkins left the box of cigars in his desk drawer, and as the days passed noticed more and more of them missing. Patton would furtively slip into his office and remove them one at a time, much as he did with the centers of chocolates.[77]

Today Patton would be described as suffering from a severe case of burnout, and it left him simply unwilling to become seriously committed to solving the increasingly complex day-to-day details of military government. As the weeks passed, his pronouncements became more virulent, and fre-. quently anti-Semitic. While genuinely concerned over the plight of the average German, who faced exceptional hardship in the coming winter, Patton continued to make public pronouncements that undermined Eisenhower's policy and were completely out of step with American public opinion. Among his comments was that it was "no more possible for a man to be a civil servant in Germany and not have paid lip service to Naziism than it is for a man to be a postmaster in America and not have paid at least lip service to the Democratic Party or Republican Party when it is in power."[78]

Even the ceremonial appearances had begun to ring hollow. His speeches had the primary aim of reassuring and motivating his troops. Although Patton had apparently lost none of his zeal, with no war to fight in the summer of 1945, there was little he could do to motivate homesick soldiers whose primary focus was the number of points they had accumulated toward rotation. He had begun to realize that his efforts were in support of a cause for which even he had no appetite. His pronouncements lacked purpose, and the only enemy left seemed to be the "lily-livered civilians" in the United States who were screwing up the peace. As a Third Army colonel observed: "Instead of killing Germans, the job was now to govern them," which required Patton to "guide his late enemies into paths of sweetness and light." The soldiers of the Third Army "were willing enough to have their Army commander remind them of the glories of their service under him, but as for the future—well, they had come to a parting of the ways with Patton and all that he stood for. . . . Patton was forever preaching a gospel of warfare that was somehow alien and antipathetic to young Americans who had reluctantly and only temporarily suffered themselves to become soldiers." About the only fighting Patton could appeal for was an exhortation that "if any goddamned civilian ever tries to make fun of your uniform, you are to knock the son of a bitch down."[79]

In July 1945 a soldier being tried for killing POWs once again cited the so-called Patton defense, claiming that he was only carrying out his commanding general's order. Queried by a judge advocate, Patton responded that while he had encouraged his men to kill Germans in battle, he had also insisted that once captured, a POW be treated humanely according to the Rules of Land Warfare (part of the Geneva Convention). "I have no apologies or excuses to make for any statement which I have made to troops in combat," he declared.[80]

Truman and Patton met for the last time in July 1945, when the new president arrived in Europe to attend the Potsdam Conference. As his plane flew toward Berlin, Truman could see for himself the frightful devastation wrought by the war and could well appreciate Patton's admonishment in a speech: "You who have not seen it do not know what hell looks like from the top."

In Berlin, Truman presided over the raising of the American flag in the new U.S. sector of the now-divided former German capital. Among those present was Patton, a man with whom Truman was wholly incompatible. Patton, the flashy dresser who "seemed to glow from head to foot," and his antithesis, the short Missourian in a plain business suit and panama hat. Of Patton, Truman's biographer observes: "There were stars on his shoulders, stars on his sleeves, more stars than Truman had ever seen on one human being. He counted twenty-eight." Soon after assuming the presidency, Truman noted in his diary: "Don't see how a country can produce such men as Robert E. Lee, John J. Pershing, Eisenhower, and Bradley and at the same time produce Custers, Pattons and MacArthurs."[81] Truman's dislike of Patton was in stark contrast to Franklin Roosevelt, who once said of him: "He is our greatest fighting general, and sheer joy."[82] Only Marshall seemed "most friendly almost gushing." As Marshall's biographer observes, even as an acclaimed four-star general, "Patton still craved his old friend's approval."[83]

Patton's letters from Berlin were gloomy. To Nita he observed: "I am very much afraid that Europe is going to go Bolshevik, which, if it does, may eventually spread to our country," while to Codman he wrote: "One cannot help but feel that Berlin marks the final epitaph of what should have been a great race." To Beatrice he suggested that "all thinking men who are not running for office" were gloomy over Europe's uncertain future.

Despite the bitter years of war and the gruesome revelations of the Final Solution, in his fervent condemnation of his new enemy, the Russians, Patton seems to have dismissed German aggression and atrocities. "Berlin gives me the blues," he wrote Beatrice from Potsdam. "We have destroyed what could have been a good race" and replaced them with "Mongolian savages . . . all of Europe will be communist."[84] His growing anti-Semitism coupled with despair of the fate of Germany became frequent, rambling topics in his diary. The dissolution of Germany was all a plot by American Jewish leaders. He accused Treasury Secretary Morgenthau and Bernard Baruch of "Semetic revenge against Germany," and characterized the Jews who survived the death camps (most of whom were DPs, whom he despised with equal fervor) as "lower than animals." His rambling vitriol inevitably led to the hated press, whom Patton thought lackeys of

Semetic influence . . . the attitude of the American people as evinced by the press and the radio is such that I am inclined to think I made a great mistake in serving them for nearly forty years, although I had a very good time doing it. . . . If we let Germany and the German people be completely disintegrated and starved, they will certainly fall for Communism, and the fall of Germany for Communism will write the epitaph of Democracy in the United States. The more I see of people, the more I regret that I survived the war.

It distressed him that the great wealth of experience was being dissipated by the great drawdown of the army: "The tactics of the next war will be written by someone who never fought. . . . I console myself with the thought that I have, insofar as the ability lay within me, done my damndest. But it seems very queer that we invariably entrust the writing of our regulations for the next war to men totally devoid of any but theoretical knowledge."[85]

As the summer passed, Patton's melancholy grew. When the war ended with the surrender of Japan, he wrote to Beatrice: "Now the horrors of peace, pacafism, and unions will have unlimited sway. I wish I were young enough to fight in the next one. It would be real fun killing Mongols. . . . It is hell to be old and passé and know it." In his diary that same night he wrote: "Another war has ended and with it my usefulness to the world. It is for me personally a very sad thought. Now all that is left is to sit around and await the arrival of the undertaker and posthumous immortality. Fortunately I also have to occupy myself with the de-Nazification . . . of Bavaria."[86]

It has been conjectured that in the final months of his life Patton was suffering the long-term effects of too many head injuries from a lifetime of falls from horses, and from road accidents. Various explanations have been offered to substantiate his increasingly peculiar behavior, including the fact that his prejudices became more open and virulent. While it is perfectly true that Patton suffered repeated and potentially serious injuries to his head (the worst of which was the injury in Hawaii in 1936 that resulted in his two-day blackout), and seemed to give the appearance of being punch-drunk, like a boxer who has been hit in the head too often, medical evidence to support the notion of a subdural hematoma is wholly lacking.

Nonetheless it seems virtually inevitable, as evidenced by his amnesia in Hawaii, that Patton experienced some type of brain damage from too many head injuries. The extent to which it influenced Patton's behavior will never be known. The only certain means of determining the degree, if any, of brain damage would have been an autopsy after his death, but Beatrice adamantly refused to allow what she regarded as the desecration of the remains of her beloved Georgie, despite a request from the army.[87]

Yet Patton's lifelong struggle to achieve and maintain the role of a macho, tough-guy, antihero, coupled with an ambition that bordered on the maniacal, does not in itself justify a determination of a subdural hematoma or any other mental condition. Even the fact that Patton was a Type A overachiever who consistently "overextended his physical, mental, and emotional capacities . . . [or that] bitterness, resentment, and jealousy marred the latter months of his life . . . in part a consequence of the waning of his own powers," is little more than evidence of a lifetime of acting out the role he had created for himself more than forty years earlier.[88] To the contrary, it more closely resembled the depression and enormous letdown of 1919. Patton began to feel alone and unappreciated, particularly by his friend Eisenhower, from whom he craved praise and recognition. His fragile ego and low self-esteem required constant bolstering, and with Beatrice far away, and their marriage facing an uncertain future in the wake of Jean Gordon, Patton seemed lost in a postwar world that was wholly foreign to his own personal beliefs.

At the age of nearly sixty, George S. Patton had spent virtually his entire life in training for and participating in war. He existed to fight; every fiber of his being was directed toward the quest for a goal as unobtainable as the Holy Grail. Unlike the heroes of mythology and legend, he sought but could never quite attain the godlike status for which he hungered. Had Patton attained the universal public praise of his peers and his superiors, he might have been content to realize that he had attained his boyhood goals of becoming a famous warrior. It was hardly surprising that he found himself—again—shortchanging himself, doubting his place in history, and craving yet one more opportunity to prove himself on a field of battle. In his delicate, even delusional state of mind, he seemed unable to grasp the fact that his military career was effectively over, that there was nowhere for him to go except downhill. He continued as if his army still had a wartime mission, instead of administering peacetime military government.

In the government's haste to demobilize its citizen soldiers, the armed forces of the United States were already well on the way to achieving the dubious distinction of unpreparedness that characterized the interwar years. And in the process, Patton, the warrior on whom so many had relied, had become one of its victims, an anachronism in an allegedly new era of peace.

Patton had no wish to become the military governor of Bavaria, an assignment for which he proved singularly ill suited. Patton would willingly have settled for an assignment more suited to his talents. And for their failure to find a suitable position for Patton (or barring this, retiring him), both Eisenhower and Marshall bore responsibility. Both men knew Patton well enough to realize that to leave him in his present position was to invite self-destruction and the relief from command, which finally came about in September 1945. Patton's previous behavior during the war could lead to no other conclusion except that his penchant for saying the wrong thing would eventually land him in hot water once again. Moreover, to suggest that two five-star generals (Marshall and Eisenhower, who would shortly succeed him as chief of staff), the most powerful men in the U.S. Army, could not, between them, have found suitable employment for Patton is absurd. Yet neither general seems to have given the problem of Patton's continued employment in the ETO anything more than cursory thought.

Patton's great friend, Jimmy Doolittle, best summed up how Eisenhower used and misused Patton. "To Ike, Patton was one of those rare, indispensable leaders who won wars; unfortunately, men like Georgie were too often unfit for peacetime assignments. I have often thought Ike used Georgie as one would use a pit bulldog. When there was a fight, he would tell Georgie to 'sic 'em.' But when the fight was won, he would have to put him in isolation somewhere until the next scrap."[89]

One of Patton's senior-staff officers points out the extent of the mess that was postwar Germany:

> The roads were blown up; the bridges were blown up; the waterworks didn't work in the towns; the sewers didn't work . . . but the order was that you were to deal with no Nazis . . . We had a million soldiers, a million POWs, and a million displaced persons in the Third Army area, which meant that we were going to have to house them in the wintertime and feed 'em . . . it was important that we get the roads open again and the people in the local economy working and the agricultural operations going so they could help us with the problems of feeding.[90]

Patton dragged his heels over purging Nazis. "My soldiers are fighting men and if I dismiss the sewer cleaners and the clerks my soldiers will have to take over those jobs. They'd have to run the telephone exchanges, the power facilities, the street cars, and that's not what soldiers are for."[91] In March 1944 Patton had already dismissed future antifraternization as hopelessly unenforceable, telling Eisenhower's political adviser, who "was terribly worried about how to prevent fraternization between Germans and Americans during the first seven months of our occupation . . . [that] he was crossing a bridge before he got to it and that, anyway, nothing could stop fraternization."[92]

Patton's penchant for getting the job done the best way he knew how led him to turn a blind eye to a German's political persuasions under the Hitler regime so long as he could function:

> General Patton felt that it was more important to get things going with people that knew how to repair things. So his order was, "Goddamn it, let's get these things fixed up." And the only way you could do it was to deal with the people that knew how, and if they're Nazis you go ahead and do it. . . . And so we went to work trying to get the cities operating, the sewer plants going, the water works going, the hospitals rehabilitated

so they could take care of themselves and probably the displaced persons. And that didn't go over too well with the Supreme Command, and Patton was in trouble for dealing with the Nazis.[93]

In August a reporter interviewing Patton asked what his future plans were. He replied he would stay in the army until there was no chance of further fighting, "and then he supposes he will get out." Asked about a congressman's attempt to have him named secretary of war, Patton scoffed that he did not want the job. He was also horrified when some prominent right-wing Republicans proposed that he become a candidate for president, but as an absolutist. He told Everett Hughes: "I am like Sherman. I would not run if nominated, nor serve if elected! I intend to remove my insignia and wrist-watch, but will continue to wear my short coat so that everyone can kiss my ass." Other propositions emerged but were rebuffed. Patton may have sounded like a darling of the Right, but he never seriously contemplated running for public office nor was he swayed by the various offers floated by those who would have misused him as their own mouthpiece.

The rejuvenation of war-ravaged Germany was a long-term task that would require a great deal of time and patience. Patton was not the man for that job. As a soldier he was used to a highly structured military system in which he could efficiently and effectively train and command troops responsive to his leadership. In postwar Germany, Patton was not intellectually or temperamentally suited to play a major role in the long-term reconstruction of a shattered nation. Patton was a man of action whose direct approach to problem solving was simply incompatible with being a military governor. Although too stubborn to acknowledge it, Patton was an exhausted warrior playing a role he neither wanted nor was willing to give his best effort to fulfill.[94]

In short, Patton was like a duck out of water. Stubborn and still jealous of Ike (and equally resentful over his friend's seeming lack of recognition and appreciation of his accomplishments), Patton seemed hell-bent on a collision course that would inevitably result in his removal.

CHAPTER 46

"All Good Things Must Come to an End"
Patton's Relief

Instead of killing Germans, what he knew best,
Patton is asked to govern them, what he knows
least. It won't work.

—COL. GEORGE FISHER

The true reason for my relief is lack of intestinal fortitude
on the part of the person we used to refer to as "Divine
Destiny."

—PATTON

If Patton's performance in Bavaria was under a cloud, it was not apparent when Bedell Smith telephoned on August 21 to praise the Third Army's efforts.[1] And, after Henry Stimson returned from Germany, he reported to Beatrice that Bavaria under Patton was the best-governed area in the ETO, a judgment echoed two weeks later by John J. McCloy.[2] That same month Eisenhower even recommended Patton to Marshall as one of his possible successors when he returned to the United States. Patton's "difficulty," wrote Eisenhower, "would be his habit of talking at injudicious moments, but he would have the loyalty of a very competent staff and a fine battle prestige."[3]

Yet the winds of discontent with Patton were growing, fueled in part by the same Bedell Smith, who, only days before his relief, had telephoned, saying: "You're my best friend, George. Probably that is the reason why you always give me my worst headaches." Unmoved by Smith's references to their avowed friendship, Patton turned to Harkins afterward and pronounced him "a goddamn snake."[4]

SHAEF had been disbanded, and the U.S. element in occupied Germany commanded by Eisenhower was located in Frankfurt and now called U.S. Forces European Theater (USFET). However, Patton's real enemy in USFET was not Walter Bedell Smith but a civilian whom he had never even met, Professor Walter L. Dorn, a civilian assistant to the civil affairs officer (G-5), Brig. Gen. Clarence L. Adcock. A history professor on leave from Ohio State University, Dorn could trace his Germanic roots to the time of Martin Luther. In the tradition of his liberal forebears, Dorn was single-mindedly determined to restore Germany to its former glory by eradicating Nazism in all its repellent forms. When they first met on September 28, 1945, Patton was distrustful and perceived Dorn to be a "smooth, smart-ass academic type" whom he thought was "very probably a Communist in disguise." Patton was not alone; Robert Murphy, Eisenhower's diplomatic adviser, considered Dorn "a wild-eyed intellectual" whom he gravely mistrusted.[5]

When Dorn began to look into the denazification program in Bavaria, he did not like what he found. The object of his displeasure was the minister president, Dr. Fritz Schaeffer, and his administration. Although Schaeffer had been appointed before the Third Army took over responsibility for Bavaria from the Seventh, Patton had uncharacteristically paid so little attention to Schaeffer that he later could not even recall meeting him. Without informing anyone in the Third Army, the USFET G-5 began a secret investigation of the Bavarian government, and when he learned that the Schaeffer administration was riddled with former Nazis, Dorn took direct aim at Patton with righteous indignation.[6]

POW Camp 8 in the Bavarian resort city of Garmisch-Partenkirchen was filled with nearly five thousand hard-core, dangerous, and brutal Nazis who were scornfully referred to by their captors as "the scum and the brutes" of the late Third Reich. On September 8 Patton inspected the American garrison there and appeared utterly disinterested and bored until boldly approached outside the POW compound by the *Lagerkommandant* (the camp commander of the POWs), a former German captain, who complained that some Germans were being interned without justification as political prisoners. Patton said to the American officers accompanying him in a voice that the POWs could hear: "It's sheer madness to intern these people." An appalled American Jewish officer reported the incident to USFET, where it and others soon landed on General Adcock's desk.

Adcock briefed Bedell Smith, who said: "There is no rational explanation for what General Patton is doing. I don't doubt any longer that old George has lost his marbles." In Frankfurt, Patton's actions were viewed as deliberate insubordination. Eisenhower was vacationing on the Riviera when a courier arrived bearing the news that he was to become chief of staff when Marshall retired at the end of 1945. However, an accompanying letter from Bedell Smith attached to Adcock's damning report put a sudden end to his vacation. It warned that Patton was out of control in Bavaria and hinted that Eisenhower ought to return to Germany at once to deal with the problem before it got further out of hand, and possibly affected his forthcoming appointment.[7]

Eisenhower immediately traveled to Bavaria, where he spent the night as Patton's guest on September 16. On the surface all seemed normal between the two friends, who talked until 3:00 A.M. When Eisenhower explained that his likely successor was his deputy, Gen. Joseph McNarney, Patton acidly replied that he did not care to serve under him. "I thought it unseemly for a man with my combat record to serve under a man who had never heard a gun go off. I stated that there were only two jobs in the United States which I felt I could take. One was President of the Army War College . . . and the other was Commanding General of the Army Ground Forces," a post now held by his nemesis, Jake Devers, whom Eisenhower noted he would be empowered to remove when he became chief of staff. Eisenhower also stated that the War College was to become a joint army–navy–air force institution, and its first president would be a navy admiral. This was not the case; in fact Eisenhower's close friend Leonard Gerow had already received the appointment, another bitter pill for Patton, who thought it "a joke. He was the poorest corps commander in France. . . . I guess there is nothing left for me but the undertaker. However . . . I can always resign."[8] There is no record of any discussion of Patton's administration of Bavaria.

Although Patton had first briefly mentioned resignation a month earlier, this was the first time he seems seriously to have considered quitting, and noted in his diary that "the only thing I can do is go home and retire. However, Eisenhower asked me to remain at least three months after he left so as to get things running quietly. I tentatively agreed to this."[9]

However, fresh reports of conditions in the DP camps and accounts of Patton's behavior and rampant anti-Semitic remarks compelled Eisenhower to return to Bavaria a week later. At one enormous DP camp the supreme commander was horrified to find that the guards were Germans, some of them former SS men. When Patton complained that the camps had been clean and decent before the arrival of their present occupants, who were "pissing and crapping all over the place," an exasperated Eisenhower angrily retorted: "Shut up, George." Patton defiantly continued his diatribe: "I've never seen a group of people who seem to be more lacking in intelligence and spirit. There's a German village not far from here, deserted. I'm planning to make it into a concentration camp for some of these goddamn Jews." Patton's anti-Semitism reached its zenith during the summer of 1945. Time after time he seized virtually any opportunity to curse Jews and Jewry and their alleged pernicious influence on everything from the peace to his own troubles as Bavarian occupation commander.

Conditions were indeed dreadful, but Patton refused to accept the explanation of an UNRRA representative that their experience in the German concentration camps had been so terrible that as DPs they were having to relearn basic human sanitation and the very rudiments of living. Patton was unmoved, and "marvelled that beings alleged to be made in the form of God can look the way they do, or act the way they act." On the road to Bad Tölz the memory of the smells and sights of the camp were too much for Patton, who ordered the staff car halted so that he could vomit his lunch along the roadside. He suggested to Eisenhower that they go fishing "to get the smell of shit out of our lungs," and that evening took a hot bath to cleanse himself of the stench and the memory.[10] According to an eyewitness, Patton "accepted with true penitence Eisenhower's scolding and his stern injunction that all the DP camps be cleaned up. . . . Eisenhower treated Patton as though he were his 'beloved naughty boy.' "[11]

Patton's bizarre behavior triggered equally outlandish reactions from his enemies. Undeterred and unaware he was under constant scrutiny, Patton no longer seemed to care what he said or to whom. It was as if he were playing his own peculiar game of Russian roulette. Not surprisingly Patton's sanity was being widely questioned, particularly by Dorn and Adcock, who sent a psychiatrist disguised as a supply officer to the Third Army in an attempt to determine if Patton was indeed mad. Adcock also had the Signal Corps secretly tap Patton's telephone and place microphones in his quarters. During a tapped telephone conversation with McNarney (depicted in the film *Patton* as taking place with Bedell Smith), Patton exploded in irrational anger at the Russians, demanding to know why anyone gave a damn what they thought: "Hell. . . . We are going to have to fight them sooner or later. . . . Why not do it now while our Army is intact and we can have their hind end kicked back into Russia in three months? We can do it easily with the help of the German troops we have, if we just arm them and take them with us. They hate the bastards."[12]

Unaware the line had been tapped by the U.S. Army, McNarney thought that the Russians might have done so, and replied in utter exasperation: "Shut up, Georgie, you fool!" But Patton was on a roll and blustered that he would like to find some way of starting a war with Russia: "You don't have to get mixed up in it at all if you are so damn soft about it and scared of your rank—just let me handle it down here. In ten days I can have enough incidents happen to have us at war with those sons of bitches and make it look like their fault." McNarney did the only thing he could: He hung up.[13]

While Patton was still at Bad Tölz, biographer Farago alleges that Nazi fanatics conspired to install a former SS colonel of cavalry and an "unreformed Nazi," Baron von Wangenheim, as Patton's groom. A gold medal winner in equestrian riding in the 1936 Olympic Games, von Wangenheim had much in common with Patton. During leisurely rides in the countryside von Wangenheim is thought to have poisoned Patton's mind with anti-Russian propaganda. Patton frequently professed that he held no grudges against the professional soldiers against whom he had fought, and he found von Wangenheim a charming and stimulating companion, never realizing the German officer's ulterior motive.[14]

Patton's flip attitude and apparent indifference to denazification were conspicuously on display at a conference held in Frankfurt on August 27, when he spoke out against the Russians and signed a letter proposing the release of some Nazi internees. His remarks deeply offended Eisenhower, who finally exploded, saying: "I demand that you get off your bloody ass and carry out the denazification program as you're told instead of mollycoddling the goddamn Nazis."[15] He wrote to Beatrice that night: "If what we are doing is 'Liberty, then give me death.' I can't see how Americans can sink so low. It is semitic, and I am sure of it." Two days later Patton was vigorously complaining about a directive that gave special consideration to Jews. "If for Jews, why not Catholics, Mormons, etc? . . . I am going to quit the army when I leave here or so I think now. . . . The stuff in the papers about fraternization" is being written "by Jews to get revenge. Actually the Germans are the only decent people left in Europe. It's a choice between them and the Russians. I prefer the Germans."[16]

Patton had become a very lonely man, with few friends. His closest associates—those on whom he had relied and who, in turn, had kept him out of trouble, had all departed. Codman, who understood Patton so well, had been discharged due to an illness in the family; Stimson, his great benefactor since 1912, had just retired, and even Omar Bradley, who, despite their differences, had helped protect him, was now in Washington directing the new Veterans Administration. Muller and Koch were gone, and of his inner circle, only Gay and Harkins remained. There was a huge void in his life that left Patton adrift on a sea of indifference. The death of fellow aristocrat President Franklin Roosevelt had distressed him deeply. In short, the backbone of his existence and his success was all gone, and the result was not only an increasing seclusion but a growing inability (and desire) to cope with the daily routine of military occupation. Col. George Fisher summed up Patton's dilemma: "Instead of killing Germans, what he knew best, Patton is asked to govern them, what he knows least. It won't work."[17]

*   *   *

By September, Eisenhower had begun to lose what little patience he had left on the Nazi issue, and he wrote Patton a letter that said in part: "Make particularly sure that all your subordinate commanders realize the discussional stage of this question is long past, and that any expressed opposition to the faithful execution of the order cannot be regarded leniently by me."[18] It also appeared, as Col. Barney Oldfield (erstwhile First Army press relations officer) has written, that "Some of the old war correspondents, now rooting through the rubble of the Reich for political stories, felt that General Patton's stewardship in Bavaria might be out of line with the occupation statutes and the wishes of General Eisenhower."[19]

Patton's days were already numbered when the spark that led to his dismissal was lit on September 22. He was asked to hold a press conference at Bad Tölz. In fact, it was not even a regular press conference but a brief, routine affair that lasted barely fifteen minutes. Patton was clearly on edge that morning and anxious to be elsewhere, and rather than hold a press conference after his normal morning briefing by the staff, he permitted the correspondents to sit in on the staff meeting. Colonel Fisher relates that after the final presentation

> Patton jumped to his feet and made as if to dash off. At the same instant, up popped the journalists with a few questions. It did seem that what they wanted to know was reasonable enough. And they wanted the information from the Army commander himself instead of from his staff. And they bore credentials from Washington as well as from Frankfurt. And they did evidence a degree of deference toward the Old Man that was notable. But Patton was just not in the mood. Something was gnawing his guts that morning. His own people knew enough to let him alone at such moments. But of course the press could not wait

The result was that Patton "barked out the first things that happened to come into his head."[20]

A correspondent attending his first session with Patton asked a calculatedly loaded question: Why were Nazis being retained in governmental positions in Bavaria? Previous questions had already clearly established that the correspondent was hostile. In the background Hap Gay animatedly shook his head back and forth in an unmistakable signal for Patton *not* to answer the question. Patton observed Gay's alarm but declined to heed his advice, and answered by saying:

> I despise and abhor Nazis and Hitlerism as much as anyone. My record on that is clear and unchallengeable. It is to be found on battlefields from Morocco to Bad Tölz. In supervising the functioning of the Bavarian government, which is my mission, the first thing that happened was that the

outs accused the ins of being Nazis. Now, more than half the German people were Nazis and we would be in a hell of a fix if we removed all Nazi party members from office.

Then came the words that were to seal his fate:

The way I see it, this Nazi question is very much like a Democratic and Republican election fight. To get things done in Bavaria, after the complete disorganization and disruption of four years of war, we had to compromise with the devil a little. We had no alternative but to turn to the people who knew what to do and how to do it. So, for the time being we are compromising with the devil. . . . It's regrettable, but a very urgent and vital job has to be done to put this shattered economy back on its feet again. . . . I don't like the Nazis any more than you do. I despise them. In the past three years I did my utmost to kill as many of them as possible. Now we are using them for lack of anyone better until we can get better people.[21]

To this point Patton had merely articulated what he and every other Allied occupation commander was doing. The problem was that the others kept a low profile about their activities. Too stubborn to grasp that he was being goaded, Patton had lost his temper and uttered the damning words that would soon cost him his command.

There were immediate repercussions the following day when Carl Levin of the *New York Herald Tribune* quoted Patton's reference to "this Nazi thing . . . [as] just like a Democrat-Republican [*sic*] election fight." To make matters worse, the GI paper, *Stars and Stripes,* was printed in the *Herald Tribune* building in Paris and thus picked up the story at once and ran with it. "The papers charged that 'at least twenty Nazis were still in high positions' in the Bavarian government, and there were generous implications that Patton was unfit for occupation command." Similar sinister stories appeared in the *New York Times*, by correspondent Raymond Daniell, and by Edward Morgan in the *Chicago Daily News*.[22]

Other stateside papers quickly seized on the story. As in Sicily and Knutsford, Patton was again the center of controversy. Some newspapers editorialized that it was time for Patton to go; others were more forceful, suggesting he be relieved forthwith. An editorial in *PM* (published in New York City) demanded Patton's immediate dismissal. "This is not the kind of job Gen. Patton is prepared to tackle. . . . He ought, for the good of his country and for his own, [to] be brought home before his reputation as a warrior is tarnished by an ignominious failure as a man of peace." Conversely, a Boston newspaper columnist named Bill Cunningham thought Patton was a scapegoat and, "if he wants his day in court . . . this column is one place he can have it."[23]

Although the first report by Daniell was buried in the back pages of the *Times*, its principal focus was the words "the Nazi thing is just like a Democrat and Republican election fight," and what, in retrospect, seems a clear attempt to portray Patton in a bad light. However, Daniell's strongest remarks appeared on September 27, when he wrote: "Eisenhower's headquarters cracked down hard on Gen. George S. Patton, Jr. today . . . " On September 28 he cabled:

ALL INDICATIONS POINTED TODAY TO A SHOWDOWN BETWEEN GEN. GEORGE S. PATTON, JR. AND GEN. DWIGHT D. EISENHOWER. . . . IT IS QUITE CLEAR HIS FUTURE ROLE IN THE OCCUPATION OF GERMANY WILL DEPEND ON HIS ABILITY TO CONVINCE GENERAL EISENHOWER THAT HE IS TEMPERAMENTALLY, EMOTIONALLY AND INTELLECTUALLY CAPABLE OF CARRYING OUT THE DIRECTIVE IN SPIRIT AS WELL AS IN LETTER.[24]

When Eisenhower learned of Patton's indiscreet political analogy, his famous temper "erupted in the granddaddy of all tempers: General Patton had made his worst and final mistake."[25] In a last-ditch effort to save Patton from himself, Eisenhower accepted Patton's statement that he had been misquoted. Bedell Smith telephoned to suggest Patton agree to Ike's suggestion that he hold another press conference to "set the record straight." However, Patton was barely contrite and gave the appearance of again condemning Allied policy, thus turning loose the press in yet another chorus of criticism. As the storm swirled, Patton wrote to Beatrice to revile Jews and Communists. "Clearly," observes Blumenson, "he had become delusional."[26]

Patton recorded his belief that those present that morning represented "the ragtag and bobtail remnants of the great U.S. press. . . . Today there was very apparent hostility, not against me personally but against the Army in general . . . [it] made me mad, which is what I think they wanted. There is a very apparent Semitic influence in the press. . . . They have utterly lost the Anglo-Saxon conception of justice and feel that a man can be kicked out without the successful proof of guilt before a court of law."[27]

On September 25 Eisenhower cabled Patton that he must have been incorrectly quoted. Would Patton "please take the first opportunity to fly up here on a good weather day and see me for an hour." For the moment Eisenhower again gave Patton the benefit of the doubt. The next day Bedell Smith held a press conference of his own, during which he was aggressively questioned about Patton's actions. When Smith was asked point-blank "whether you think that [Eisenhower's denazification] program can be carried out by people who are temperamentally and emotionally in disagreement with it," he denied that this assessment applied to Patton. However, like hounds baying after a fox, the reporters bombarded Smith with endless questions about Patton's fitness to continue in command. The transcript leaves little doubt that some of the reporters exhibited a clear bias against Patton, attacking his ignorance of what was going on in Bavaria and, in one instance, wringing from Smith the concession: "I am not prepared to state that it [denazification] has been carried out effectively in every case." Smith went on to suggest that within a week or so they might see a "marked improvement," and weakly defended Patton by stating: "General Patton is a soldier and will carry out his orders."[28]

Friends and supporters of Patton believe that Smith's press conference demonstrated "a definite scheme to undermine and discredit General Patton" by reporters Levin, Morgan, and Daniell. Frank Mason, a former INS correspondent, newspaper executive, and special assistant to Secretary of the Navy Frank Knox, who was freelancing in Germany, believed the denazification program was "bullshit" and that Patton had been railroaded.

Mason's own investigation produced the following version of events. A notice that correspondents could attend the Third Army staff meeting on September 22, after which Patton would attempt to answer any questions, was posted on the press camp bulletin board four days in advance. Reporters Daniell, Levin, and Morgan arrived the night before. The following morning

the team of Levin, Daniels [*sic*] and Morgan started to work out on Patton. Morgan's part was to stand close to Patton with a pipe in his mouth and insolently to puff smoke into Patton's face. This was related to me by several newspapermen who were present, and who were humiliated by the scene. Levin and Daniels, using a sneering tone, with the attitude of a criminal prosecutor interrogating a hardened criminal, and both talking at once at times, went after Patton for permitting a German named Schaeffer to hold public office in Munich. . . . The other correspondents stood back and not one of them got into the cross fire of Levin and Daniels attacking Patton. Every one agrees that before Patton could answer a question, the other would start up with another question, and as I was told "crossed Patton up." Such remarks were made by Levin and Daniels in a condescending tone as: "Don't you know that the directives are that all Nazis should be removed from office?" Levin was so disagreeable in his tone that Patton once said, "You are so smart. You know everything. Why do you ask me?"

The exchange became so heated that Patton demanded to know if Levin and Daniell were calling him a liar:

Levin and Daniell jumped at this chance. . . . Obviously they were trying to get Patton to lose his temper. Some of the witnesses told that despite what the Levin-Morgan-Daniels clique wrote, that Patton was really quite cagey in his answers, and that if there had been a shorthand transcript, which there was not, that Patton did not do so badly as they made it sound.

Levin and Daniels would make long involved statements of policy, and then ask Patton to confirm or deny that such-and-such was the policy. Patton said to Daniels, "Don"t you put words in my mouth.". . . One of the regular correspondents . . . who is an old timer and thoroughly reliable told me that he had heard the three, Levin, Morgan and Daniels, plotting at the breakfast table how they would get Patton . . . [they] did considerable exulting, I am told, at the successful outcome of their plan to get Patton.[29]

Mason concluded that a definite conspiracy existed on the part of three "radical journalists" whose object seemed to be to eliminate any German officials unacceptable to the Russians and to warn other American generals of the consequences of opposing the appointment of anyone unacceptable to the Russians, resulting in "the formation finally of governments in German cities in the British and American zones which will be in effect . . . puppets of Russia."[30]

Although the conspiracy theory expounded by Mason has never been proved, there is certainly sufficient evidence to conclude that there were a handful of anti-Patton journalists who sought (and on September 22, 1945, found) a means to either demean (or, better yet, destroy) Patton and—perhaps more important—exploit his strongly anti-Russian views by embarrassing his superiors and arousing the American press and public to demand his removal. In this aim they succeeded brilliantly against a foe who barely fought back. Carl McArdle, a reporter for the *Philadelphia Bulletin*, later disparaged the uproar caused by Levin, Morgan, and Daniell as a story not even worth filing. No one in Germany who had known the circumstances or who knew Patton had "paid the slightest attention."[31]

For a warrior of Patton's fearsome reputation, his reaction to the events swirling around him in the summer of 1945 was curiously subdued. Instead of fighting back, he seemed almost content to curse his enemies as pro-Communist, and leave it at that. It was as if the fight had gone out of him. His press secretary, Maj. Ernest C. Deane, was outraged and later blamed himself for being pressured by the press into persuading Patton to agree to meet with them. However, the original intent was to keep the local correspondents informed, and only when Daniell (the Berlin Bureau chief of the *New York Times)* and others learned of it did it include such outside reporters. Deane could have known neither of this nor of their intent to attack Patton, and other than taking the precaution of recording all of Patton's public utterances, could have done little or nothing to prevent it. Deane thought Patton's behavior was "deliberately self-destructive" and that he had "put himself into the hands of his enemies."[32]

After Patton's death a reporter who had been present at the September 22 press conference visited Beatrice to express regret at what had taken place that fateful morning. An unforgiving Beatrice dismissed the apology as too late to matter. Raymond Daniell later denied that it was he who had been to see Beatrice (who could not remember the man's name). He also disclaimed any notion that he had been part of a conspiracy to "get" Patton, but did admit there were some who had sought to entrap Patton into saying something embarrassing that would blacken his name.[33]

When Mason interviewed Patton several days after his ill-fated press conference, he wrote in confidence to Roy Howard (president of Scripps-Howard Newspapers) that Patton said: "We fought the revolutionary war to establish the RIGHTS OF MAN: we fought the civil war to end slavery; we fought this war only to lose the two things we won for humanity in the two earlier wars."

Mason concluded: "Roy, he is sound. He is American. He is getting hell for daring to block the radicals in running interference for a red government for Germany. He has not said this to me. I do not even know whether he even knows it. Maybe he thinks that they are just going after George Patton."[34]

It was a tight-lipped, exhausted, and grim Patton who reported to Eisenhower's spacious office in the IG Farben Building late in the afternoon of September 28 after a seven-and-a-half-hour journey by auto in heavy rains. In anticipation of what was certain to be another tongue-lashing from Ike, Patton was austerely dressed in an ordinary uniform, his swagger stick and pistols deliberately left in his staff car. Although he understood that he was again in trouble, Patton thought that he could once again unleash his charm and flattery on Eisenhower to talk his way out of this latest controversy.[35]

Although Ike was grinning, the seriousness of the summons quickly became evident. It was a familiar theme: Patton had again failed to keep his mouth shut, and in so doing had embarrassed himself, Eisenhower, and the U.S. Army. Patton defended himself by insisting that his words had been deliberately and maliciously distorted. Kay Summersby thought Eisenhower had "aged ten years in reaching the decision, which was inevitable in the light of Patton's past mistakes and the universal furor over this one."[36] She has described a three-way conversation between Ike, Patton, and Bedell Smith as "long and acid," and "one of the stormiest sessions ever staged in our headquarters. It was the first time I ever heard General Eisenhower really raise his voice." Patton's diary records: "General Eisenhower was more excited than I have ever seen him." Four days earlier Eisenhower had written in exasperation to Mamie that "George Patton has broken into print again in a big way. That man is yet going to drive me to drink. He misses more good opportunities to keep his mouth shut than almost anyone I ever knew."[37]

Finally Eisenhower gingerly broached the real subject of the meeting: Patton's relief. Under the guise of an idea that had just come to him, Eisenhower suggested that since the Fifteenth Army (a small headquarters created solely for the purpose of preparing a history of the war in Europe) would soon require a commanding general, it might be "a good idea" if Patton was transferred to this allegedly important job. Patton wasn't buying this "idea" and had a different solution: "I told him in my opinion I should be simply relieved, but he said he did not intend to do that and had had no pressure from the States to that effect. I said then that I thought I should be allowed to continue the command of the Third Army and the government of Bavaria."[38] The two generals discussed Patton as if he were a third party. Patton said he thought his greatest virtue and gravest fault were both his honesty and lack of ulterior motive. Eisenhower shot back that Patton's greatest virtue and fault was his audacity.

Soon after Patton arrived, Eisenhower summoned Adcock to his office to report on their investigation of the situation in Bavaria. Adcock brought along Dorn, whom Eisenhower had never laid eyes on before that day and testily demanded to know who the hell he was. Dorn skillfully and surgically destroyed Patton by painting Minister President Schaeffer, an ultraconservative, longtime Catholic politician, as an intriguer whose administration was rife with former Nazis. It was, said Dorn, "little better than the Nazis it replaced," and in deliberate defiance of Eisenhower's proscription.

Patton was torn between immediate resignation and acceptance of a demeaning "paper" command: "If I am kicked upstairs to the Fifteenth Army should I accept or should I ask for relief and put in my resignation? By adopting the latter course I would save my self-respect at the expense of my reputation but . . . would become a martyr too soon." In the end:

> I accepted the job with the Fifteenth Army because I was reluctant, in fact unwilling, to be a party to the destruction of Germany under the pretense of de-Nazification . . . the utterly un-American and almost Gestapo methods of de-Nazification were so abhorrent to my Anglo-Saxon mind as to be practically indigestible. . . . I believe Germany should not be destroyed, but rather rebuilt as a buffer against the real danger which is Bolshevism from Russia.[39]

During Dorn's damning assessment of Patton, he remained polite, took notes, and at one point called Harkins on the telephone and gave orders that Schaeffer and the other ex-Nazi ministers were to be fired immediately, writing in his notes: "regardless of the setback it would give to the administration of Bavaria, and the cold and hunger it would produce."[40]

It was usual after such rebukes of Patton for Eisenhower to invite his friend for dinner, conversation, and entertainment, but this night there was none of the old camaraderie: "If you are spending the night of course you will stay with me, but since I feel you should get back to Bad Tölz as rapidly as possible, I have my train set up to take you and it leaves at 7:00

o'clock." Looking pointedly at his watch, Eisenhower said without compassion: "It's leaving in half an hour." There was to be no reprieve, no friendly dinner. It was over: their friendship, and Patton's command of the Third U.S. Army. George S. Patton had been relieved of command, and Eisenhower's none too subtle hint left him little choice; "I took the train."[41]

The press had been camped in the hallway outside Eisenhower's office and knew perfectly well that something important had transpired. When the two generals appeared, Eisenhower curtly declined their demand for a statement of what he and Patton had discussed saying only: "I have conferences with my Army Commanders whenever I feel like it—period."[42]

If Dorn thought himself victorious, however, he was sadly mistaken, and his triumph was short lived. Eisenhower found Dorn too smug for his liking and seemed as wary of him as Patton was. He ordered Adcock to "get rid of that goddamn professor." Dorn was sent to Berlin to work for Gen. Lucius Clay, Eisenhower's deputy commander, but Clay eventually sacked the self-righteous professor.

Eisenhower's only brief reference to the affair noted

> [his] final brush with George Patton's impulsiveness. It was an example of a strong-minded man's tendency to oversimplify history. Whenever we met, George would say, 'I hope you know, Ike, that I'm keeping my mouth shut. I'm a clam' . . . George Patton was aware that a principal purpose of our occupation mission was to cleanse the continent of Nazi control and influence. Now his words could have been misinterpreted by those who wanted an audible spokesman for a soft policy toward Nazi leaders. This was senseless. I ordered George to my headquarters and said, 'The war's over and I don't want to hurt you—but I can't let you be making such ridiculous statements. I'm going to give you a new job.' . . . From then on he had no occasion to meet the press and I had no occasion to criticize George for indiscretions."[43]

Eisenhower's son, John, remembers his father telling him with sadness the night of September 28: "'I had to relieve George Patton from Third Army today. . . . We could survive this tempest. Actually I'm not moving George for what he's done—just for what he's going to do next.' No personal animosity was involved whatsoever. The two generals went to a football game together in Frankfurt a couple of days after the switch-over." Patton was far too proud ever to reveal his disappointment and dismay to his friend; his bitterness remained cloaked, and Eisenhower never grasped the terrible hurt he had done Patton in relieving him of his beloved Third Army.[44]

His aide Major Merle-Smith accompanied Patton during the trip home and found him "very calm and very humble." During dinner Patton wondered aloud what there was left for him to do? "I've obeyed orders and done my best; and now there's nothing left. I think that I'd like to resign from the Army so that I could go home and say what I have to say." Merle-Smith suggested that to resign was tantamount to admitting his enemies were right. Patton agreed.[45]

Of his September 22 press conference, Patton wrote:

> I was intentionally direct, because I believed that it was then time for people to know what was going on. My language was not particularly politic, but I have yet to find where politic language produces successful government. One thing which I could not say then, and cannot yet say, is that my chief interest in establishing order in Germany was to prevent Germany from going communistic. I am afraid that our foolish and utterly stupid policy in regard to Germany will certainly cause them to join the Russians and thereby insure a communistic state throughout Western Europe. It is rather sad to me to think that my last opportunity for earning my pay has passed. At least, I have done my best as God gave me the chance."[46]

General Clay thought that all army commanders in Germany, not just Patton, were unsuited for the role of military governor, which he viewed as more political than military and wanted separated from the army chain of command. When Clay, whose standing with Eisenhower was high, was approached for his recommendation after Patton's press conference, he endorsed his reassignment. "It wasn't just [Patton's statement] . . . it was my firm belief that military government should not be under the Army commanders."[47]

Virtually no one who knew Patton believed he was fitted to carry out an important occupation role. Among them was his longtime friend and five-star aviation pioneer, Gen. Hap Arnold, who had known him since his cadet days at West Point. In his memoirs Arnold observed that the army periodically produced nonconformists who, "cannot get along in the regulation-controlled, unexciting peacetime Army. Yet in war they stand out among their fellows. . . . [Patton] was a man of self-determination—he had to be different—a crusader. . . . I often thought that he would have great difficulty in adjusting himself to the restraints and restrictions of postwar service in Europe."[48]

Only one of Eisenhower's many biographers attempts to explain his role in the relief of Patton, and he is scathingly critical of those who were responsible for retaining Patton in command of the Third Army. "To attempt to make a military governor out of Patton . . . was an egregious blunder, one that must be charged against the highest levels of the War Department."[49] He might have added that his own subject was equally culpable for not having recognized his friend's unsuitability while there was still time to have removed him with face-saving dignity.

Ladislas Farago notes that Marshall attempted to find a suitable assignment for Patton, but also that one of Patton's greatest admirers, the financier and presidential adviser Bernard Baruch (ironically, one of the purveyors of the "Semetic" influence Patton so despised), was unsuccessful when he lobbied Marshall to find a suitable new job for Patton in the summer of 1945. "The decision to leave Patton in Europe was justified with the explanation that he was too fast for the Japanese," but, in reality, it merely meant that General Douglas MacArthur wanted no part of George S. Patton Jr.[50]

When he relieved Patton, Eisenhower claimed that it was partly his fault for not having found him more suitable employment, but rationalized that there had been no vacancy in the United States for him. More to the point is that a suitable position *could* have been created for Patton had either Marshall or Eisenhower elected to do so. If Patton was singularly unsuited for his occupation role, his superiors were as much a part of the problem as they were of its solution. The post Patton was given as Fifteenth Army commander was previously held by Leonard T. Gerow, who had been appointed commandant of the War College in August. Surely Patton could have been given this posting, even as a four-star general, which would have permitted him to do what he did best when not fighting: teach.

Patton claimed that Eisenhower admitted his own culpability:

He stated he was at fault as much as I was in that knowing my strength and weakness as he did, he should not have put me in as Military Governor. I told him it was my considered opinion that Bavaria was the best governed state in Germany. . . . Ike said that had he possessed any adequate command for me at the time, he would have given it to me rather than have me act as Military Governor of Bavaria.

In the war's aftermath Patton had asked Robert Murphy if he thought he had fought his last battle. "He inquired, with a gleam in his eye, whether there was any chance of going on to Moscow, which he said he could reach in thirty days, instead of waiting for the Russians to attack the United States when we were weak and reduced to two divisions." On September 29 Patton invited Murphy to lunch at Bad Tölz, but their conversation was interrupted by an urgent telephone call from Bedell Smith who stated he had been directed to read a letter from Eisenhower to Patton. "There was no love lost between Smith and Patton," wrote Murphy, "and the latter suspected trouble. Pointing to an extension telephone, he said to me, 'Listen to what the lying SOB will say.' I did not know that the decision had been taken to relieve Patton of his command until I heard Smith performing his duty as tactfully as possible. Patton vigorously pantomimed for my benefit his scornful reactions to Smith's placatory remarks."[51] Eisenhower's letter attempted to soften the blow by stating that it was not meant to reflect on

Patton's performance but "my belief that your particular talents will be better employed in the new job and that the planned arrangement visualizes the best possible utilization of available personnel."[52] He was directed to assume command of the Fifteenth Army on October 8. Patton was too old a soldier to buy into such nonsense. It was plain and simple: He had been fired. "The last paragraph of the letter was full of soap which meant a little less, probably, than the paper used to receive it. So another die has been cast and probably for the best, as I am sure hell is going to pop soon in this government business. . . . Truscott will get the Third Army and I am sure will do as good a job as he can in an impossible situation."[53]

Although Eisenhower insisted there would be no public announcement until the change of command took place on October 8, it was impossible to contain the scuttlebutt of Patton's sacking within the press corps. Someone leaked the story to a reporter, thus compelling Eisenhower to make an official announcement on October 2. The story made headlines, and even the *Stars and Stripes* trumpeted: PATTON FIRED. For Bill Mauldin's compatriots on the GI newspaper it was belated but sweet revenge.

Eisenhower's private report to Marshall reflected his frustration at Patton's verbosity after repeated verbal and written reminders to guard his tongue: "One thing that always provokes me is that so many of my troubles [with Patton] need never have arisen—they come up from some completely useless cause . . . being Patton he cannot keep his mouth shut either to his own subordinates or in public."[54]

There was considerable sympathy for Patton in some U.S. newspapers, where his dismissal was viewed not only as disgraceful but as a harbinger of the success of communism in discrediting a genuine American hero. The *New York Daily News* editorialized that it must have been part of a high-level plot "from someone above Eisenhower" to get rid of him, and demanded: PATTON SHOULD ASK A TRIAL. The *New York Times* more calmly noted:

> Patton has passed from current controversy into history. There he will have an honored place. Perhaps he himself will share the sense of relief his countrymen feel at so safe and quiet a transfer. He was obviously in a post which he was unsuited by temperament, training or experience to fill. It was a mistake to suppose a free-swinging fighter could acquire overnight the capacities of a wise administrator. His removal by General Eisenhower was an acknowledgment of that mistake.[55]

Although Daniell, Morgan, and Levin have defended their stories and denied the allegations that there was a conspiracy, both Patton and the members of his family have always thought that his relief stemmed from a deliberate effort to "get" him. However, after exhaustive investigation, Ladislas

Farago concluded that "Daniell and his colleagues were tough reporters" whose only transgression was "lack of taste and good manners, not participation in some sinister plot." The only conspiracy to get Patton existed not in the press but in SHAEF headquarters "Patton was toppled from his pedestal, not by those 'itinerant' corespondents and the philistines," but by Bedell Smith; the G-3, Harold Bull; and the G-5, Clarence Adcock, who "had concluded that Patton was unfit to govern the Germans."[56]

Yet Patton remained blinded by his biases and his anger. "An investigation would be futile," he wrote Beatrice. "The noise against me is the only means by which the Jews and Communists are attempting and with good success to implement a further dismemberment of Germany . . . were it not for the fact that it [the 15th Army Group] will be a kick up stairs I would like it much better than being a sort of executioner to the best race in europe."[57]

Of his decision to accept a meaningless job, Patton mused: "Am I weak and a coward? Am I putting my posthumous reputation above my present honor? God how I wish I knew. . . . No one gives a damn how well Bavaria is run. All they are interested in is how well it is ruined."[58]

One of Patton's last official acts as Third Army commander was to decorate his faithful driver, M/Sgt. John L. Mims, with a Silver Star during a private farewell ceremony shortly before Mims rotated. When Mims elected to return to the United States after more than four years as Patton's driver, Patton had raged: "Hell, no, he isn't going home," Then Stiller bluntly told him: "General, I want to tell you . . . there isn't a God damn thing you can do about it. . . . He has performed his duty over here and he's entitled to go. He has a young wife back home." As usual, when presented with the irrefutable, Patton conceded: "You're right. Tell him to get ready to go."[59]

Patton and Truscott, the two old cavalrymen who had once been adversaries in Sicily, spent two days together before the change of command. "The only sign he ever gave of the blow which he must have felt came when he said: 'Lucian, if you have no objection, I want to have a big formal ceremony and turn over the command to you with considerable fanfare and publicity. I don't want Ike or anyone else to get the idea that I am leaving here with my tail between my legs.'"[60]

It was raining on the morning of October 7, and the change of command ceremony was held in a gymnasium. Patton's farewell speech was brief:

> All good things must come to an end. The best thing that has ever happened to me thus far is the honor and privilege of having commanded the Third Army. . . . Please accept my heartfelt congratulations on your valor and devotion to duty, and my fervent gratitude for your unwavering loyalty. . . . A man of General Truscott's achievements needs no introduction. His deeds speak for themselves. I know that you will not fail him. Goodbye and God bless you."[61]

The band played "Auld Lang Syne," Patton passed the Third Army flag to Truscott, who made a brief speech, and then it was over: The Third Army had a new commander. The two generals left to the strains of the Third Army march and "For He's a Jolly Good Fellow." After a luncheon in his honor, Patton boarded the Third Army train and was gone. His friend Otto Weyland wrote movingly: "I feel that the Third Army has died. To me, the Third Army meant Patton. When you left it, it ceased to be a thing alive." Blumenson thinks Patton would have agreed: "Divisions, corps and, armies, he was sure, had souls."[62]

It can be disguised as reassignment or retirement or by some other means, but the truth is that nothing is more hurtful to an officer than to be relieved of command. There is simply no greater feeling of emptiness or humiliation. Although Patton wrote to Gen. Thomas T. Handy at the War Department in Washington, D.C., that his "new assignment is more in keeping with my natural academic tendencies than is that of governing Bavaria," the plain truth was that it was window dressing to conceal his feeling of desolation at losing his beloved Third Army.[63] Patton's devastation can only be imagined. "Nothing in his dress or bearing reflected the torture of his soul," noted a colleague.[64]

The distress of those devoted members of his staff still assigned to Third Army was palpable. Robert S. Allen writes:

Patton took the heartbreaking blow in silence. But while outwardly subdued, inwardly he seethed in anguished fury and searing despair. He was baffled as to just where and how he had erred. He could not understand the rancorous storm of abuse and castigation. He had done no more than other Allied commanders administering occupied areas! So why pick him? If he was wrong, why weren't they? Particularly he was cut to the quick by Eisenhower's stinging rebuke [of Patton as Bavarian occupation commander] in their talk. Patton felt that not only was that wholly uncalled for, but grossly ungrateful and unfair. From Africa to the ETO he had given Eisenhower the utmost in loyal, unstinting and peerless service. . . . Surely simple gratitude alone warranted more than a humiliating verbal spanking in private and degrading condemnation in public. For that reason alone, Patton indignantly felt, he should have been treated with more consideration and appreciation. To him the whole debasing affair confirmed strongly a suspicion . . . that malicious and envious forces in and out of the Army were determinedly bent on destroying him and discrediting his matchless record as a battle commander.[65]

It would hardly have assuaged Patton's hurt had he known that his relief occurred because his oldest friend had simply lost patience with him. Institutions are too large and too impersonal ever to admit to mistakes; when mistakes are made it is individuals who pay the price of failure. In the early autumn of 1945 the U.S. Army was no exception. Its leadership had bungled by leaving Patton in command of Third Army when the war ended. Now he paid the price by being ignominiously dismissed and relegated to the command of a so-called army that had no troops, no equipment, and no mission save that of executing a task that any competent colonel or brigadier general could have performed equally well. To her death, Beatrice Patton believed that her husband had been betrayed by the army he had served so well, and, in particular, by Dwight Eisenhower's "cowardice toward the press. Crucified and thrown to the wolves. Not one voice was raised in his defense."[66]

Patton arrived at Bad Nauheim aboard the Third Army train in the early hours of October 8 and was greeted by the acting commander, Bradley's former chief of staff, Maj. Gen. Leven C. Allen. Patton was alone in his private railroad car, dressed in jodhpurs and tapping his leg with a riding crop, when Allen arrived. After a tense moment Patton said: "Well, you know damn well I didn't ask for this job, don't you?" The headquarters was in an old hotel and all went well until lunch, when Patton descended the stairs into the main lobby, now the officers' mess. Willie preceded him and, sniffing the food odors emanating from within, trotted ahead, followed by Patton, who shouted in his most commanding voice: "Willie!" Some one hundred officers in the mess came immediately to attention. Seeing the grim-faced Patton stride into their midst, those present expected the worst from this fearsome warrior. Eyeing them in silence, he seemed to soften visibly and said: "There are occasions when I can truthfully say that I am not as much of a son of a bitch as I may think I am. This is one of them." Colonel Allen later wrote: "The relieved staff roared with surprised delight. From then on, it was as wholeheartedly for him as the Third Army staff had been."[67]

For some days the Fifteenth Army had been rife with rumors about Patton and what strict new edicts he would impose upon his arrival. John Eisenhower was one of those assigned to Patton's new headquarters, and, even though he was a family friend, also awaited his arrival with trepidation:

> I was unsure of what his attitude would be. Nevertheless, his generous nature showed through, and as the officers of the headquarters went through the receiving line, he pulled me aside for a brief and congenial chat. In his typical impish manner, however, Patton had to jar his audience in his remarks that evening. He started out by saying, "I have been here and have studied your work today." Then, his voice rising to a pitch, he shouted, "I have been SHOCKED"—and then in a lower tone—"by the excellence of your work."

Patton announced that he intended to return to the United States by March 1946, "and we were by God going to have all our reports finished by that time! Everyone became frantic . . . the final report . . . showed the fine hand of the new commander. Even though I felt that First Army had contributed more to victory than had the Third, I noticed that Patton's army was mentioned about three times as often as any other in the theater."[68]

Not long afterward Patton encountered a GI lolling in a chair in the hallway, stupidly ignoring the new commanding general, even though he could clearly see him approaching. As the soldier belatedly snapped to attention, Patton delivered a memorable tongue-lashing and then walked on, continuing an earlier conversation about Ping-Pong as if nothing at all had occurred.[69]

For weeks he chafed at what he believed was the manifest unfairness of his relief for having done his job no differently than any other commander in occupied Germany. His days were scholarly, filled with work on his forthcoming memoir, and in composing a series of articles dealing with the war and the subjects closest to his heart: leadership, tactics, combat, and horsemanship. With tongue in cheek, he tentatively titled his book *War as I Knew It: Or, Helpful Hints to Hopeful Heroes*. His essays on leadership and tactics—an exposition of the art of command—eventually appeared as the final two chapters of *War as I Knew It*. To this day they remain an important part of Patton's legacy, and are models of clarity. When the new secretary of war, Robert Patterson, learned of Patton's intentions, "he expressed the hope that I would notify all the people who received [mss.] copies of 'War As I Knew It' that it must be kept a dark and deadly secret."[70]

There was ample evidence of Patton's intense distress and that his life had seemingly reached a vital crossroads. The day after his arrival, Patton told a reporter that he had done his duty as best he could, and that his conscience was clear.[71] But, as Robert S. Allen writes: "As the days passed, Patton became increasingly tense and restless. He took long drives by himself, and at times nervously paced the floor of his office. At dinner, he said little and went to his quarters early. He smoked more cigars than usual. It was obvious he was undergoing deep and gnawing turmoil."[72]

Although Paul Harkins was a soft-spoken, taciturn New Englander, he understood Patton well, and for his sixtieth birthday on November 11, Harkins turned the grand ballroom of Bad Nauheim's world-renowned spa-hotel into an approximation of its prewar glory. Patton's friends from near and far attended a gala evening that included cigars which he had once again given up. The event was a total surprise and delighted Patton, who savored the pomp, and most of all, of being in the limelight again, even if for only an evening.

Patton had released his faithful aide, Charlie Codman, to return to civilian life. In a series of bitter letters in September and October 1945, Patton wrote Codman: "This thing is exactly like a decomposing corpse and as

Kipling says in his poem, 'Gentleman Ranks [*sic*]': 'We are dropping down the ladder rung by rung.' I think you were very lucky not to come back and see our magnificent Army sloughing away."

By mid-October Patton had made up his mind to resign his commission rather than retire: "My present plan is to finish this job, which is a purely academic one, about the first of the year," he wrote Codman, "and then submit my resignation after which I can do all the talking I feel like and may write the book you suggest, 'War and Peace as I Knew It.'. . . . My private opinion is that practically everyone but myself is a pusillanimous son of a bitch and that by continued association with them I may develop the same attributes. . . . The true reason for my relief is lack of intestinal fortitude on the part of the person we used to refer to as 'Divine Destiny.'"[73]

Patton continued to lash out at Jews, and dismissed the assault on him as the "scurrilous attacks made on me by the non-Aryan press. . . . I am really very fearful of repercussions which will occur this winter and I am certain we are being completely hood-winked by the degenerate descendents of Genghis Khan. People who talk about peace should visit Europe where, as I believe the Lord said, I bring you not peace but the sword. The envy, hatred, malice and uncharitableness in Europe passes belief."

During his brief sojourn in Massachusetts, Patton had been exposed to the strongly anti-Semitic views of Beatrice's half-brother, Charles, and the husband of a half-sister, both of whom may have poisoned his mind with intemperate suggestions that Jews were responsible for everything from the war to their own plight as the victims of Hitler's Final Solution. The mess that was postwar Germany, Patton's own diminishing role in a postwar world he had trouble understanding, and his increasingly irrational tendency to lash out at anyone suspected of creating his present situation, inevitably led him to blame those he understood least. Russians and Jews became the targets of his vitriol. It did not seem to matter that some of those he trusted and most respected were Jewish, among them his own G-2, Oscar Koch.

After some weeks of contemplation Patton finally revealed his decision to Gay one evening at dinner:

> I have given this a great deal of thought. I am going to resign from the Army. Quit outright, not retire. That's the only way I can be free to live my own way of life. That's the only way I can and will live from now on. For the years that are left to me I am determined to be free to live as I want to and to say what I want to. . . . It has occupied my mind almost completely the last two months, and I am fully convinced this is the only honorable and proper course to take."

Gay counseled caution and urged Patton to talk it over first with Beatrice and with Frederick Ayer and other close friends: "These people are part of your life and you don't want to make a decision as momentous as this and which will affect them as much as it will you, without discussing it with them."

Patton, however, was adamant that he was doing the right thing, that he would be better off "outside the tent pissing in, than inside the tent pissing out."[74] His last letter to Beatrice, written on December 5, announced that he was to depart for the United States on December 14 aboard the battleship *New York,* and would be home for Christmas: "I have a months leave but don't intend to go back to Europe. If I get a realy good job I will stay, otherwise I will retire. . . . I hate to think of leaving the Army but what is there?" To Frederick Ayer he wrote: "The main thing is that I wish to have sufficient money to be very extravagant for the next fifteen years."

Patton's melancholy was broken in late November when his old friends in Sweden extended an invitation to a reunion with the members of the Swedish Modern Pentathlon team against whom he had competed in the 1912 Olympic Games. He left on November 27 aboard a special train whose opulent cars had once been used by Field Marshal von Hindenburg. In Sweden, Patton was greeted by an entourage of VIPs that included the chief of the army and eight former members of the Pentathlon team. He breakfasted in a suite at the Grand Hotel with Count Folke Bernadotte and the following morning met King Gustavus Aldolphus V of Sweden, and later the crown prince. At the Olympic Stadium a special ice carnival and hockey game were staged for Patton's benefit. In a nearby cave the pistol competition of 1912 was reenacted, with Patton placing second—"13 points better than I made in 1912." The days that followed were a steady round of exhibitions and demonstrations; visits to schools, cathedrals, museums, a university, and military units; and a great deal of comradeship at the dinner table and at the bar.

Ostensibly Patton had been invited to Sweden to address the Swedish-American Society. But the visit turned out to be much more, and for the first time in many months Patton felt at home, not among his own people but with men he had not seen in thirty-three years. Here he could let his hair down, knowing that his every utterance was not being scrutinized for its potential political worth. When he left to return to Germany, Patton noted in his diary: "It is my considered opinion that anyone who can survive a trip to Metz [which Patton had recently visited] and one to Sweden within a week is apt to live forever or die of a stroke."[75]

Patton's final diary entry was dated December 3, and his description of a luncheon hosted by Bedell Smith for Eisenhower's replacement, Gen. Joseph McNarney, was filled with vindictiveness and deep-seated indignation. "I have rarely seen assembled a greater bunch of sons-of-bitches," he wrote:

It is certainly quite a criticism of our form of Military Government to find that . . . General Clay, and . . . General McNarney, have never com-

manded anything, including their own self-respect. . . . The whole luncheon party reminded me of a meeting of the Rotary Club in Hawaii
where everyone slaps every one else's back while looking for an appropriate place to thrust the knife. I admit I was guilty of this practice,
although at the moment I have no appropriate weapon.

Eisenhower returned to the United States, and it was soon confirmed
that he would not return to Europe. "There is great activity among the rats to
get out," Patton wrote Beatrice. During his absence Patton—as the senior
officer in the ETO—became the acting theater commander. He kept his distance from Bedell Smith, who continued to run the USFET, often asking
that papers requiring his signature be sent to him at Bad Nauheim. His bitterness deepened. He noted: "I finally after a fight of three years got the
DSM for all my people ten in all. I think it is amusing that no one trys to get
any for me. I got nothing for Tunisia, nothing for Sicily and nothing for the
Bulge. Brad and Courtney were both decorated for their failures in that
operation."[76]

As Patton's mood sank, Gay's concern rose in direct proportion. It hurt him
deeply to view the obvious torment of his longtime friend and boss. He
began casting about for some diversion to occupy Patton's mind, even if
only for a few hours. Finally he decided that what Patton needed was a
pheasant shoot. Casually he broached the subject on the evening of December 8. "You haven't done any hunting for quite a while," Gay told Patton.
"You could stand a little relaxation before you take off for home. . . . I know
exactly where to go for some good hunting." Several days earlier Patton had
intended to hunt wild boar, but the snows kept him from the hunt. Gay now
proposed that they hunt pheasant in the Rhine Palatinate west of Speyer, an
area known to be rich in game.

Patton's interest was piqued, and he seemed to brighten at the thought:
"You've got something there, Hap. Doing a little bird-shooting would be
good. . . . Yes, let's do it. You arrange to have the car and guns on hand early
tomorrow and we'll see how many birds we can bag."[77] It would be Patton's
last day, and a festive occasion. The next day he was scheduled to fly to
England aboard Eisenhower's personal plane, and would sail aboard the
USS *New York* to New York from Southampton on December 10.

Sunday, December 9, was typically raw, cold, and dismal. Patton, Gay,
and their driver, Pfc. Horace L. Woodring, left Bad Nauheim shortly after
dawn in the general's Cadillac sedan. It was a journey from which Patton
would never return.

CHAPTER 47

"A Helluva Way to Die"

We had all been brought up being told he was going to die in a foreign land, and he was going to be buried there.

—RUTH ELLEN TOTTEN

This is a helluva way to die.

—PATTON

The morning of December 9 was clear and very cold. At approximately seven o'clock the intercom sounded in the quarters occupied by Patton's personal staff. On the other end was M/Sgt. Meeks with orders for Patton's new driver, Pfc. Horace Woodring, to ready the general's 1939 Model 75 Cadillac, the last of a special, elegant series designed and built to withstand Europe's primitive roads. After the departure of M/Sgt. Mims in late May, Patton had been chauffeured by a series of different drivers picked from the duty roster in the Third Army's motor pool, and within a few weeks he had been slightly injured in a road accident, the latest in a lifetime of equestrian and motor-vehicle mishaps.

At Bad Nauheim, Patton again had a regular driver in Horace Woodring, a young man who, in retrospect, was hardly an ideal choice. The nineteen-year-old Woodring's "whole life seemed to have been wrapped up in automobiles." Already demoted three times for infractions of Eisenhower's nonfraternization edict, Woodring could scarcely believe his good fortune at being able to drink and wench at night with German fräuleins and drive like hell for Patton in the daytime. Not surprisingly Woodring appealed to Patton, who thought him a breath of fresh air after the conservative Mims, who had nevertheless managed to keep his master alive and accident-free during the five years he had driven him many thousands of miles across some of the most forbidding and dangerous terrain on earth. "Woodring is the fastest and the mostest. He's better than the best Piper Cub to get you there *ahead* of time," boasted Patton, referring to a recent trip to Liège, Belgium, during which Woodring had driven some 150 miles in less than two hours.[1]

According to Gay, Patton's two-vehicle convoy left Bad Nauheim about 9:00 A.M. and consisted of himself, Patton, and Woodring in the Cadillac, with T/Sgt. Joe Spruce trailing in a jeep that carried the guns and a hunting dog (*not* Willie, whose only claim to be a hunting dog was his proficiency at sniffing out a potential meal and adroitly snatching food scraps tossed his way). They began their journey along the Kassel–Frankfurt–Mannheim autobahn, but soon detoured near Bad Homburg to enable Patton to indulge his passion for ancient places by visiting the ruins of a restored Roman outpost in the nearby Taunus Mountains. By about 11:30 A.M. they had left the autobahn and were on National Route 38 in the northern outskirts of the industrial city of Mannheim. Despite the propensity of both Patton and Woodring for high-speed driving, Patton had earlier commented on the minor car trouble that had resulted in two stops during their journey south. "This is a very careful driver," noted Patton. "He seems to sense when there is something wrong with the car."

Sometime after 11:00 A.M. they encountered a military police checkpoint on Route 38, but Woodring believed that Patton's four-star pennant on the front bumper of the Cadillac entitled him to pass without stopping. However, a young MP thought otherwise, flagged down the Cadillac, and demanded that the occupants identify themselves. Patton had thoroughly soaked his boots and socks exploring the Roman outpost and ever since had been riding in the front seat with Woodring to take advantage of the heater in an attempt to dry out. At the checkpoint Patton got out and walked over to the shivering MP, "patted him on the back and told him in a pleased paternal voice: 'You are a good soldier, son. I'll see to it that your C.O. is told what a fine MP you make.'"[2]

It was so cold that Patton complained that the hunting dog would freeze to death in Spruce's "goddamn truck" and ordered him transferred to the front seat of the Cadillac next to Woodring. Patton resumed his former place in the right rear of the roomy backseat.[3] Both sides of the road were littered with wrecked vehicles, and, never content to merely sit back and relax, Patton sat forward on the edge of the seat to better observe the sights they passed.

In the northern Mannheim suburb of Käfertal, Woodring slowed the big Cadillac to five to ten miles per hour and then momentarily stopped at a railroad crossing as a train passed. When the gates went up, the Cadillac crossed the tracks and then sped up to about thirty miles per hour, as Spruce passed them to lead the way (Woodring did not know the route to the hunting site). From the opposite direction a two-and-a-half-ton truck driven by T/5 Robert L. Thompson suddenly turned left to enter a nearby quartermaster depot.[4]

According to Farago's meticulous investigation, Thompson was out joyriding after a night of boozing with some enlisted buddies, and had no business even being on the road that Sunday morning. What later helped fuel the fires of the conspiracy theorists was that Thompson "was allowed to vanish before long to become the mystery man of the incident."

It was now approximately 11:45 A.M., and at virtually the same instant as Thompson began making his left turn, Patton exclaimed: "Look at all the derelict vehicles! How awful war is. Think of the waste." Momentarily distracted, Woodring took his eyes off the road, and in split seconds Thompson's vehicle suddenly appeared directly across his path, less than twenty feet away. Woodring slammed on the brakes and wrenched the steering wheel to the left in an attempt to avoid hitting the truck. "I had time to say 'Sit tight," said Gay, recalling that he thought that "one should relax like when falling from a horse; then we crashed." The Cadillac and the truck collided at approximately a ninety-degree angle. The right front bumper of the big truck struck the right front side of the Cadillac, smashing the radiator and the front fender. Neither driver was injured.

According to Gay, Patton

> apparently was thrown forward and then backward, because at my next recollection of the accident (I was unhurt except for slight bruises) I was sitting in the back seat on the left side, half-faced to the right with my right arm around General Patton's shoulders. His head was to the left and I was practically supporting him on my right shoulder in a semi-upright position. He was bleeding profusely from wounds of the forehead and scalp. He was conscious.

Although Patton was instantly aware that he had been gravely injured, his first concern was for Gay and Woodring, whom he asked if they were hurt. Both replied no. Patton said: "I believe I am paralyzed. I am having trouble in breathing. Work my fingers for me. Take and rub my arms and shoulders and rub them hard." Gay did as he was told, and then Patton demanded: "Damn it, rub them." Gay immediately understood that something was seriously wrong, and told him they must await help, as it would be inadvisable to attempt to move him. As they waited Patton plaintively said, "This is a helluva way to die."[5]

Why was Patton so severely injured while Gay and Woodring escaped with only minor cuts and bruises? Both Woodring and Gay anticipated the collision and were braced for the impact while Patton, who had been sitting on the edge of his seat, gazing out the window, was totally oblivious. It was nearly six feet between the front and rear seat partitions of the spacious Cadillac, and with Patton sitting unbraced, he became a flying missile when the automobile slammed into the truck. Patton either struck his head on a steel frame that held the glass partition separating the front and rear seats in place in the "closed" position, or on the clock mounted on the partition—probably both.[6] According to Woodring the partition was open at the time of the accident. "It took all the skin from the General's forehead for approximately three inches above his eyebrows and three inches across, partially scalping him and completely separating his spinal column."[7] It is also likely that some of the damage to Patton's face and scalp occurred when his head struck the overhead interior light mounted in the ceiling.[8]

Gay sat holding his injured friend for approximately five minutes until assistance arrived in the person of an MP officer, Lt. Peter K. Babalas, who opened the rear door and asked if there was anything he could do. Gay instructed him to summon a doctor and an ambulance. Lieutenant Babalas "asked me who the injured officer was and I told him, 'General Patton.'" Gay continued to support Patton until two medics arrived and took charge of the general, first bandaging his torn scalp. About fifteen minutes later an ambulance arrived bearing two medical officers, who very carefully transferred Patton to a stretcher and placed him in the back of the ambulance. Although there was a military hospital in Mannheim, the obvious seriousness of Patton's injury led to an immediate decision to send him with a military police escort to the better-equipped and -staffed 130th U.S. Army Station Hospital in Heidelberg, twenty-five miles away. Gay followed in the staff car of a brigadier general. Throughout his ordeal Patton remained conscious, but had said only in a whisper while still in the backseat of the Cadillac: "I think I'm paralyzed."

Thompson's truck had been moving at barely ten miles per hour, and it was estimated that Patton's Cadillac was traveling at approximately thirty miles per hour. Both Gay and the military police later exonerated both drivers of any fault, even though Woodring and Gay stated that Thompson had never at any time signaled his intention to turn left.[9] The official conclusion was that—although clearly preventable—it had been nothing more than a tragic accident.[10]

Unfortunately Patton's accident was poorly documented at the time and remains so to this day. Later investigations have pinpointed certain previously vague details,[11] but, as with other public figures, his accident and subsequent death have spawned a myriad of myths, gossip, speculation, and outright fabrications by opportunists who would profit from such an event. Among them have been a film called *The Brass Target*, starring George Kennedy as an overweight Patton, and a novel, both of which hypothesize that the accident was in fact a well-planned assassination.[12] In 1987 a former GI asserted that he was the first on the scene of the accident (alleged to have occurred not in Mannheim-Käfertal, but near Heidelberg, only a few scant miles from the hospital where Patton was taken), and that Eisenhower had shaken his hand after Patton's funeral and commended him, an interesting statement given the fact that Eisenhower had returned permanently to the United States on November 11 and did *not* return to Germany to attend Patton's funeral.[13] Yet another has proclaimed that he was hired by the head of the OSS, William J. Donovan, to assassinate Patton, but that someone else did the job on December 9 using a specially designed Czech weapon that fired a piece of metal making his injuries appear to have resulted from the accident. When Patton did not die, the assassin allegedly finished the job by slipping into the hospital and administering cyanide.[14]

Those who suggest that Patton was somehow murdered have failed to provide the slightest evidence of how anyone could have planned such a caper or ensured that Patton's Cadillac would be momentarily stopped for the passage of a train at the crossing just down the street from the scene of the accident. Other than a handful of men on his personal staff, no one even knew where Patton would be, what route he would follow, or what time he would arrive at his destination.

The perfunctory investigation by the army and the failure to hold a full-scale formal inquiry opened the door to the vultures and the publicity hounds, who saw profit in conspiracies, lies, and half-truths. The army's failure to investigate the accident thoroughly was as incomprehensible as it was inexcusable. Until Ladislas Farago undertook to do so, it remained shrouded in contradictions and what he quite rightly described as an appallingly inadequate inquiry. Lieutenant Babalas filed the preliminary report and concluded that both drivers were careless. Some years later, however, when Babalas, then a Virginia state senator, wrote to the Pentagon seeking a copy of his report of the accident, none could be located.

The ambulance arrived at the 130th Station Hospital at 12:45 P.M., and Patton's litter was positioned on an operating table in the emergency room, which soon swarmed with medical personnel.[15] The first to treat him were the hospital commander, Col. Lawrence C. Ball, and the chief surgeon, Lt. Col. Paul S. Hill, who found Patton conscious but suffering from severe traumatic shock, with a barely readable pulse of 45, and a barely obtainable blood pressure reading of 86/60. There were other clear signs of shock: Patton's face was ashen, and his extremities were cold and damp, but his only complaint was of pain in his neck.

Patton's face and scalp were a bloody mess from cuts that had opened his skull clear to the bone. "There was a long, deep Y-shaped laceration from the bridge of his nose up over his scalp. It was obvious he had neither

sensory nor motor function below his neck. He had lost quite a bit of blood
. . . [and] was beginning to develop surgical shock." Hill recognized there
would have to be surgery to fix the wound properly and for the moment
only cleaned and dressed it before administering a tetanus shot to ward off
infection.

Patton's litter was then placed on wooden sawhorses, with his head
lower than his feet to help control the shock. A corpsman bent over to
remove Patton's uniform and said: "General, I'm afraid that we'll have to
cut off your uniform. Will that be all right with you?" Patton replied: "Hell,
yes, Sergeant. Cut away. It's been done before." His holster and revolvers
were removed and safely locked up; however, to avoid any damage to his
spine, Patton's bloodstained clothing was cut away from his body but not
removed until later. Once his arms were freed of his uniform, blood and
plasma transfusions were begun through veins in both arms. Later, his uniform, shoes, and underclothing quickly disappeared into the hands of souvenir hunters and were never seen again.

During the next hour or so, Patton's mood fluctuated from despair to
gallows humor. He muttered: "If there's any doubt in any of your Goddamn
minds that I'm going to be paralyzed the rest of my life, let's cut out all this
horse-shit right now and let me die." Then his mood quickly improved, and
he chuckled: "Relax, gentlemen, I'm in no condition to be a terror now."
Still later Patton exclaimed: "Jesus Christ, what a way to start a leave."
When the 130th Station Hospital chaplain was permitted at his side to say
several prayers, Patton said: "Well, let him get started. I guess I need it."[16]

The hospital had no elevators and Patton could not be moved upstairs
for X rays. A doctor and several technicians had to wrestle a six-hundred-
pound portable machine down the stairs from the floor above. The X rays
revealed the worst: Patton was paralyzed from the neck down.[17]

As Dr. Hill began repairing the damage to Patton's head, he heard Patton utter: "Seventy-two." Asked what he meant, Patton replied it was his
seventy-second stitch, and described in colorful language how he had
acquired the dozens of injuries and scars from polo, horseback riding, and
automobile accidents.[18] The scar on his right buttock, now a permanent
crease, brought forth the observation: "It may be symbolic for something or
another that the only permanent memento of my historic service in the First
World War is this goddamn scar on my ass."[19] After his face and scalp were
stitched, Patton was moved to a small room on the first floor.[20]

Bad news always spreads quickly, and the corridors of the hospital were
soon flooded with army brass. Unable to visit Patton and without evident
purpose other than the moral support of being near at hand, they aimlessly
and, in the minds of some of the medical staff, uselessly crowded the hallways. One of the orthopedists required splints and, when asked who ought
to be detailed to bring them, replied testily: "Hell, man, send a brigadier

general. They seem to be a dime a dozen right now." The doctor was nearly correct: At least eleven generals and a number of lesser ranks were present, some of whom overheard his remark. The doctor blushed and was given some very disapproving looks. Order was quickly restored with the arrival from Frankfurt of the theater surgeon, Patton's longtime friend, Maj. Gen. Albert W. Kenner, who had been one of the first officers notified of Patton's accident. Another of Patton's oldest and dearest friends was his former deputy commander in Sicily, Geoffrey Keyes, who had been promoted to lieutenant general and had succeeded Sandy Patch as commander of the Seventh Army. Keyes, whose headquarters were in a nearby *Kaserne* (barracks), efficiently took charge of the nonmedical administrative details.

When Kenner learned of the seriousness of Patton's condition, he immediately put in motion a series of steps that led to securing of the services of several renowned neurosurgeons. He first telephoned the U.S. Army surgeon general in Washington, who advised him to contact a noted British neurosurgeon, Brig. Hugh Cairns of Oxford University. Cairns readily agreed to assist, and Kenner sent a plane to London to fly him and another orthopedic specialist to Heidelberg.

On the evening of December 9 skeletal traction was applied to Patton's head, which was immobilized by a pair of what were called Crutchfield tongs. More closely resembling a medieval torture apparatus, these so-called ice tongs were clamped to Patton's head to provide traction by stretching his injured spine. However, in addition to being painful, they tended to slip.

Dr. Cairns and his colleague arrived on the morning of December 10. Both examined Patton and pronounced themselves satisfied with the steps taken to immobilize his neck. They advised some changes that turned out to be an equally dreadful new method of traction. "Ordinary fish hooks were purchased at a local German sporting goods store, and a metal fish hook was inserted just below the bony process beneath Patton's eyes." Each hook was then attached to a tensile strut, or tong, to maintain constant tension. Whether or not due to this makeshift device, Patton's condition did begin to stabilize,[21] though he stoically endured almost unbearable pain. After several days the tissue surrounding the hooks began to die. Finally, after nine days of agony, the tongs were mercifully removed, and traction was maintained by encasing Patton's neck and shoulders in a special plaster jacket.

Patton spent a restless first night filled with the dread that accompanies any trauma, particularly one of this extreme seriousness. Not until Beatrice's arrival the afternoon of December 11 did his spirits show improvement. In the days that followed, a parade of different consultants and specialists examined Patton. None were even remotely hopeful that there was any realistic prospect of his ever regaining the use of his limbs.

*   *   *

In Washington the surgeon general set in motion steps to secure the services of an American neurosurgeon. Col. R. Glen Spurling had served in the ETO as chief neurosurgical consultant and, after the war, in Washington in a simi- lar capacity for the army surgeon general. Spurling was to have been dis- charged from the army on December 20, and on the afternoon of December 10 was aboard a train bound from Louisville, Kentucky, to Washington, D.C., when he was paged by the conductor, who conveyed a message that Spurling was to leave the train in Cincinnati to deal with an emergency. Deeply worried that some catastrophe had occurred in his family, Spurling was hustled from the train and handed a message by the stationmaster, which read: "You will abandon the train in Cincinnati, proceed to the airport . . . where an Army plane will fly you to Washington, hence to Germany." It was signed by the adjutant general, the army's highest-ranking administra- tive officer. A perplexed Spurling asked if the stationmaster had any idea of it was all about. "Haven't you been listening to the radio? General Patton has been seriously injured in Germany and I have a suspicion this has some- thing to do with it."[22]

At the airport a C-47 transport awaited him on the tarmac, its engines already warmed up. In Washington he was met by Lt. Col. Walter T. Ker- win, a Pentagon staff officer sent by Marshall, and Beatrice Patton, who explained the situation. Within minutes the aide, Spurling, and Beatrice were airborne to Germany, via Newfoundland and the Azores, in freezing discomfort. In the Azores the plane was grounded due to the dreadful weather conditions in Europe, but an hour later they left for Paris after a decision had been made to risk the bad weather. Spurling wondered why the lives of ten people (three passengers and a crew of seven) were being risked "in order to get to the bedside of a man who probably had no chance to sur- vive." After three attempts that nearly caused the plane to crash in foul weather at a Paris airfield, it was diverted to Marseilles, where it landed with its gas tanks nearly dry. The party was met by Patton's classmate and nemesis, J. C. H. Lee, and transferred to Lee's personal plane for the final leg of the long and tiring journey. There was another close call as the plane landed in frightful weather in Mannheim, not far from the site of the acci- dent. The navigator told Spurling it had been the closest call he had ever experienced: "Thank God I'm going home next week! I'll never fly another airplane."

Beatrice optimistically told a large group of reporters on the afternoon of December 11: "Well, I have seen Georgie in these scrapes before and he always came out of it all right." Although Patton's vital signs were some- what improved, and he seemed in good spirits, his first words were: "I am afraid, Bea, this may be the last time we see each other." Their reunion lasted half an hour, and it left Beatrice visibly shaken, her public confidence replaced by a gnawing private dread. Nevertheless, during the many hours she spent at Patton's bedside in the last days of his life, Beatrice managed to exhibit the same outward confidence she had shown the press.[23]

Spurling's initial examination of Patton filled him with gloom. There was not only complete paralysis from his neck down, but he was having difficulty breathing. Of equal concern was the fact that both his bowels and bladder were likewise paralyzed. Patton's official medical record on December 11 read: "Prognosis for recovery increasingly grave."[24] An operation to relieve the pressure on his badly damaged spinal cord was impossible, and would not have eliminated the paralysis.

Patton awoke to find Kenner, Cairns, and Spurling all at his bedside. "He was fully conscious, in fact almost jovial. . . . After the usual greetings, he spoke to me, 'Colonel Spurling, I apologize for getting you out on this wild goose chase, and I am particularly sorry since it probably means that you won't be home with your family for Christmas. This is an ironical thing to have happen to me—after the best of the Germans have shot at me, then to get hurt in an automobile accident going pheasant hunting.'"

Later Patton summoned Spurling to his bedside for a confidential chat. "I knew what was coming," wrote Spurling. "The General wanted the truth. After some casual pleasantries, Patton said, 'Now, Colonel, we've known each other during the fighting and I want you to talk to me as man to man. What chance have I to recover?'" Spurling's reply was evasive: Everything depended on his progress in the coming days. Dissatisfied, Patton bluntly demanded to know: "What chance have I to ride a horse again?" This time the reply was unequivocal: "None."

"In other words," said Patton, "the best I could hope for would be semi-invalidism." The reply was yes. Patton thought for a moment and then said: "Thank you, Colonel, for being honest."

Then in a flash he returned to his old jovial mood. He said, "Colonel, you're surrounded by an awful lot of brass around here—there are more generals than privates. . . . I just want you to know that you're the boss—whatever you say goes." They discussed visitors, and after Spurling recommended barring everyone but Beatrice, Keyes, and the medical staff, Patton readily concurred: "I think that is a good decision. . . . After all, it's kind of hard for me to see my old friends when I'm lying here paralyzed all over." Spurling spelled out Patton's medical problems frankly and noted the need for him to conserve his strength. "He grasped them immediately without question," saying, 'I'll try to be a good patient.'" In fact, throughout his terrible ordeal Patton was a model patient who never complained and was unfailingly gracious and considerate of his doctors and nurses. He was generally cheerful, never once referring to his condition, preferring instead to reminisce with his doctors about his dream of again sailing the *When and If*, and the forthcoming publication of his autobiography.

No effort was spared to ensure that he received the best possible care. In all Patton was treated by some fourteen physicians of varying specialties. One was William Duane Jr., a professor of neurosurgery, who remained at Patton's bedside each night. Duane got to know both Beatrice and his famous patient well and was fascinated by the interplay between them. At Patton's request a number of books from his personal library in Bad Nauheim were sent to the hospital. Duane observed as Beatrice read from the English translation of the memoirs of Napoleon's confidant, ambassador, foreign minister, and general, Armand de Caulaincourt. A sure sign that Patton's spirits had improved was his observations about Napoleon's prophecies and his destiny. Although Patton had modeled himself on John J. Pershing, it was Napoleon who became his idol, not only for his dash and élan, but particularly for his great sense of destiny.[25] While Beatrice helped to sustain Patton's morale, his doctors could do little more than keep him in traction and on medication. Dr. Hill had noted an increasing quantity of fluid on the spinal cord and was deeply worried that Patton's survival was now in question.

News of the accident was flashed worldwide by the wire services, and by December 10 there began what soon became a veritable flood of cards, letters, and cables from well-wishers. They came from ordinary soldiers and citizens, as well as from American and foreign senior officers and politicians, and included a telegram and a letter from Eisenhower. One of the messages of encouragement sent to Patton was from Harry S. Truman: "I am distressed at the painful accident which you have suffered and want you to know that I am thinking of you at this time. You have won many a tough fight and I know that faith and courage will not fail you in this one. I am thankful that Mrs. Patton will be at your side to strengthen and sustain you."[26]

Nearly thirty reporters descended upon the hospital, many of whom had left the Nuremberg war crimes trials to cover the story and loudly clamored for news. Most were cooperative, but a few resorted to unsavory tactics to obtain news of Patton's condition. Unable to get past MPs detailed to guard the hallways and Patton's room, several reporters dressed up in hospital garb and attempted to find someone who would talk. One even attempted to bribe the German cook, who prepared and delivered Patton's meals, with cigarettes, nylons, and Hershey bars, the standard payoffs in postwar Germany.[27]

Some reported Patton near death, while others confidently predicted his eventual recovery. One reporter learned from a night nurse that Dr. Spurling had ordered a glass of whiskey for his famous patient each night before dinner in the hope of stimulating his flagging appetite. Patton usually left it half empty, but headlines the following day announced that he was following his usual custom of drinking his whiskey straight. As Spurling relates: "Patton never asked for a drink throughout his illness. . . . Each night before his dinner, when the whiskey would appear, he would invite all of the doctors on duty into his room for a drink with him. We would always wait for the General . . . and his usual words were, 'Cheerio, boys.'"

In an attempt to head off press speculations, Kcyes held a press conference each morning at 11:00 A.M., at which Spurling provided an update of the previous twenty-four hours:

We had decided to make our reports very noncommittal for we couldn't do much else and still stick to the truth. The way some reporters would 'doctor' up the stories, though, would exasperate us beyond measure. They just kept playing up the fact that the general was going to get well, that he would be back again at his old duties. These statements were pure falsifications. General Marshall . . . requested that a confidential medical report be sent to him daily. These confidential bulletins always told the stark truth and I am sure there was never any question in Washington but what General Patton was done for.[28]

In addition to briefing the press, Spurling also had to cope with well-meaning friends and acquaintances of Patton. "His many friends either came personally to the hospital or called by phone many times each day. Since we had decided there would be no visitors in the sickroom, it fell to my lot to interview many of them. . . . Some of the Army generals were particularly difficult to handle   [and were] outraged by being kept on the sidelines." After Bedell Smith insisted on visiting Patton, Spurling bluntly informed him that he was not welcome, and broke his own rule of not bothering Patton by casually mentioning his problem with Smith. Patton's reaction was that he did not want to see the "old so-and-so. He's never been a particular friend of mine. Colonel, it's up to you to keep him out." Beatrice was even more adamant: "Under no circumstances" was Smith to gain entry to Patton's hospital room.[29]

Spurling knew Patton only vaguely from casual contact earlier in the war, and it was through his budding friendship with Geoffrey Keyes that he learned a great deal about Patton: "General Keyes felt that General Patton was the military genius of our age." Of Patton's legion of friends, few loved him more than Keyes, who could not bear the thought that his friend might not survive. Despite an increasingly grave prognosis, Keyes remained in total denial, writing to his wife on December 16 that "the doctors are very optimistic now and say, barring unforeseen complications, Gen Patton is out of danger as far as saving his life is concerned . . . each day now there is improvement and real cause for encouragement." Sadly, Keyes's optimism was totally unfounded and based mainly on Spurling's attempt to hold out some hope rather than Dr. Hill's franker, far gloomier prognosis.[30] Yet for several days there were hopeful but misleading signs that the crisis might indeed be over. Patton's vital signs stabilized, and there were slight indications of motor movement in his paralyzed extremities.

Despite the uncertainty of whether her husband would even survive his ordeal, Beatrice constantly reassured friends and colleagues, answered the piles of letters arriving each day, and kept a vigil over her beloved Georgie. Dr. Spurling was particularly impressed by the woman whom he described as possessing

> an immense amount of vitality. She isn't pretty by ordinary standards, yet her personality radiates like a brilliant gem. Within five minutes after I met her in Washington, we were friends. She is quiet and unassuming; she speaks slowly, softly and correctly. Her eyes are alive and they punctuate every thought. She is as different from George Patton as night from day. Yet, her devotion to him is one of the most beautiful human relationships I have ever known. 'Georgie' was her boy—a bad boy at times but a very good one most of the time. She had lived her life for him. They had worked together and played together. . . . Throughout these trying days she kept her equanimity; nothing ever seemed to ruffle her. . . . Viewed from any angle, Mrs. Patton is a grand lady.[31]

As Ladislas Farago has eloquently described her, Beatrice was "at her best when the chips were down. . . . Having managed his life for so long, she [was] now arranging his death. It was an act of superb courage that required an extraordinary character, and Beatrice Patton had both in abundance . . . for she was an Ayer and a Patton, and a woman who lived by the rule of noblesse oblige."[32]

One of Patton's few visitors was Paul Harkins, who found him in fine spirits. When Patton asked how he looked, Harkins was unable to hide his distress, and replied he looked fine. Patton broke into laughter and replied: "Paul, goddammit. You're a lying sonofabitch." Harkins left in tears, never to see Patton again.[33]

From the moment of her arrival in Heidelberg, Beatrice's primary concern was that Patton might develop another embolism similar to the one that nearly killed him in 1937. Each day she would say to Spurling: "If he just doesn't get an embolism I think he may pull through." (Embolisms, sudden arterial blockages by a clot carried to the site by the bloodstream, are frequently caused by immobilization that restricts normal blood circulation.) On December 19 Patton was unable to expel mucus from his bronchi, which were under pressure from his crushed spine.

On the afternoon of December 20, Beatrice had been reading to him, when Patton suddenly said: "I feel like I can't get my breath." A doctor was summoned and found his skin bluish, and oxygen was quickly applied to restore his breathing. X rays taken the following morning confirmed that there had indeed been a small pulmonary embolism in his lung (which, if not removed, deprives the brain of oxygen and results in death) that had apparently obstructed his right upper lung.[34]

By the morning of December 21 Patton was cheerful, and his condition appeared to have improved. But it was illusory: His vital signs betrayed continued pulmonary distress. Privately his doctors thought that he had only about forty-eight hours to live. Various accounts of Patton's hospitalization suggest that he said to Beatrice that morning: "It's too dark. . . . I mean too late."

She spent most of the afternoon reading to him, but Patton was drowsy and often fell asleep. His only visible sign of distress was a bout of coughing that was alleviated by turning him on his side. He was asleep at 5:15 P.M., when Beatrice, Spurling, and Hill left for dinner in the hospital mess across the street. At about 6:00 P.M., Dr. Duane suddenly appeared to summon them back to Patton's bedside. By the time they arrived, Patton had died in his sleep.*

The new commanding general of U.S. forces in Germany, Joseph T. McNarney, made the official announcement: "It is my painful duty to announce the death of a great fighter and a great man. General George S. Patton, Jr. died peacefully at 5:30 p.m. [sic] tonight at the 130th Station Hospital of injuries suffered in an automobile accident last December 9. His injuries were grave but his fight to overcome them was gallant. He went down fighting. General Patton could have died no other way."

Despite Patton's marvelous physical condition, which was the primary reason he had survived for twelve days after the accident, in his own mind he seems to have sensed that his death was imminent, and welcomed it. Several times on December 21, Patton said he was going to die, the last time to his favorite nurse, Lt. Margery Rondell. "I am going to die. Today," he told her shortly after 3:00 P.M. Physically and mentally exhausted beyond measure, Patton was being kept alive by the largely artificial means of medications and stimulants, such as pure protein. The night of December 20–21 a weary Patton lamented to the duty nurse: "Why don't they just let me die?"[35] The false hopes held by Beatrice and Keyes were not shared by those who attended Patton each day, and understood that it was merely a matter of when, not if, Patton would die.

Whether the press was deceived or merely got it wrong, it was widely reported that Beatrice had been at Patton's bedside when he "died peacefully." Both Dr. Duane and Lieutenant Rondell were on duty, and the official medical notes for December 21 indicate that since 4:00 P.M. Patton had been

*His comment three years earlier, before the landings in North Africa, that Pershing might outlive him proved chillingly accurate. Black Jack died in 1948.

very drowsy, and his pulse irregular. At 5:00 and again at 5:30 P.M. Rondell recorded that he was sleeping. At 5:55 P.M., with no outward sound, Patton's heart had stopped beating. He died quietly, just as he had predicted. There was nothing overt to signal his passing; Lieutenant Rondell had merely sensed that Patton had died. While Dr. Duane ran from the room to notify Beatrice, Spurling, and Hill, nurse Rondell sat silently awaiting their return. The vigil had ended.

The army officially listed the cause of death as "pulmonary edema and congestive heart failure." The final entry in Patton's medical log read simply and starkly: "1755 hours: Expired."[36]

On December 21 Patton's daughters both experienced vivid premonitions of his death. Ruth Ellen was asleep but awoke with a start to find her bedroom bathed in light, and her father stretched out on the seat in front of a bay window. "He had his head propped on his hand. He was in full uniform and looking at me fixedly. I sat up in bed—I could see him plainly. When he saw I was looking at him, he gave me the sweetest smile I have ever seen. It was loving and reassuring and his very own. Then, he was gone. . . . I felt as if some burden had been lifted."

When Ruth Ellen telephoned her sister early the following morning she was not surprised to learn that Bee had been about to call her. She, too, had been asleep, when the phone on her bedside table rang. "She picked it up and then there was a lot of static, as if it were an overseas call, and she heard Georgie's voice say: 'Little Bee, are you all right?' Then it was cut off." When she called the overseas operator she was informed there had been no calls whatsoever from Germany that night. One of the sisters finally articulated what both already knew: "I guess he's dead then. Poor Ma."[37]

Ruth Ellen was certain that Patton believed that he would not die in the manner he wanted to. He had strived, prayed, and hoped God and Lady Luck would grant him a hero's death that was a fitting end of a lifetime spent attempting to re-create the perfect life of a warrior. Such warriors did not die in bed, they died heroically on the battlefield. George S. Patton, however, was not fated to die a hero's death, but in a hospital bed as an incurable quadriplegic, hooked to life-sustaining devices.

One of Patton's unusual traits was his fearless disregard of death, of which he had often said: "Did you ever stop to think how much more exciting death can be than life?"[38] But he could not abide helplessness in himself, and except for the blood clots, might well have willed his own death to avoid such a terrible fate. Shortly before he died, as Beatrice sat at Patton's bedside, reading to ease his mind, he suddenly said: "I guess I wasn't good enough." Beatrice interpreted his remark to mean his disillusionment at not having died in battle like Hugh Mercer, Waller Tazewell Patton, and the first George Patton.[39] However, had he ever been willing to admit it, there was little heroic about the deaths of his grandfather and great-uncle, from horribly painful wounds that brought about slow and agonizing death.[40]

One of his premonitions did come true: "Georgie had a lifelong conviction that he would die in a foreign land," wrote Ruth Ellen many years later.[41] Since Patton had prophesied his own death while on leave in June 1945, Ruth Ellen was neither surprised nor shocked by her vision. "We were all prepared, my sister and I, but [son] George hadn't been prepared. The three of us had been brought up being told he was going to die in a foreign land, and he was going to be buried there. We grew up with that. When you are a kid and your father says something like that, you believe it."[42]

With Christmas rapidly approaching, there were important decisions to be made about Patton's funeral and where he would be buried. It was also deemed important that all traces of his death pallor be removed for the viewing and his final earthly journey. One of the more bizarre idiosyncratic rituals of contemporary Western culture is the employment of undertakers to restore the dead to a lifelike appearance one final time. A casket was flown in from London, but there was no undertaker available to prepare Patton's body for burial. There was a momentary quandary over locating an American mortician until Colonel Ball recalled there was an enlisted man stationed nearby who had once worked in a funeral home. He was hastily summoned to Heidelberg.[43]

To avoid press snooping, Patton's body was removed to the basement directly under the room where he had spent the final twelve days of his life. There it temporarily resided on a table in one of the cubicles that had once been stalls occupied by horses of the German cavalry. (The hospital was formerly the site of a German cavalry school.) Sergeant Meeks arrived with Patton's personal four-star storm flag and tearfully handed it to Dr. Hill, who, aided by Dr. Ball, gently draped it over his body. After the mortician completed his work, Patton's casket was removed to the stately Villa Reiner, high on a mountain overlooking the Neckar River and the charming city of Heidelberg, where, enveloped with flowers, it lay in state.

Although there was no trained pathologist immediately available, Spurling nevertheless urged Beatrice to permit an autopsy, but she politely declined. Even if there had been a pathologist present, it seems doubtful she would have authorized anyone to carve up the corpse of her beloved husband. Thus, the extent of the damage done to his body during a lifetime of injuries will forever remain conjectural. Far more sensitive, however, was the question of where Patton would be buried. Apparently the matter had never been resolved, inasmuch as Beatrice's original intention was reflected in cables exchanged between the Seventh Army and the War Department concerning arrangements for a special plane to return Patton to the United States for burial at West Point.

A distressed Geoffrey Keyes took Dr. Spurling aside and explained that since the beginning of the war no American officer or enlisted man had been sent home for burial. "He was fearful of adverse reaction on the part of American mothers whose boys had been buried overseas. Yet, General Keyes felt he could not put the matter straight to Mrs. Patton because of his long friendship with both of them." Moreover, it was well known that once Patton was buried, Beatrice was strongly against ever moving him for reburial elsewhere. Spurling spoke "bluntly" to Beatrice, and a potentially explosive problem was hastily resolved when she replied: "Of course he must be buried here! Why didn't I think of it? Furthermore, I know George would want to lie beside the men of his army who have fallen."[44]

Her decision clearly reflected Patton's wishes. Prior to the Sicily landings in July 1943, Patton had written to Frederick Ayer: "If I should conk, I do not wish to be disinterred after the war. It would be far more pleasant to my ghostly future to lie among my soldiers than to rest in the sanctimonious precincts of a civilian cemetery."[45] In fact Patton was adamant that he did not wish to be buried in the United States, insisting that his family promise to bury him overseas: "In God's name don't bring my body home." Perhaps he was thinking of a remark by Napoleon, which he often quoted: "The boundaries of a nation's empire are marked by the graves of her soldiers." He also spoke of a soldier's death: "Why goddammit, to bury a soldier anywhere else is simply to cater to a bunch of snivelling sob sisters retained by those carrion-eating ghouls, the coffin makers and undertakers."[46]

From a choice of three large military cemeteries where Patton might have been buried among other Third Army men, Beatrice selected the cemetery at Hamm, Luxembourg, near the site of his most famous battle. Not all of his friends agreed with her choice. A distraught Everett Hughes blurted out that if he were chief of staff he would unhesitatingly have ordered Patton's body brought home—the obvious and proper thing to do, he believed.

On December 22 Beatrice and her brother, Fred, who had arrived from Boston several days earlier, journeyed to Bad Nauheim. In addition to viewing Patton's final place of duty, one of Beatrice's purposes was to speak with the sons of friends and colleagues of Patton whom she had come to know well as a military wife. Among them was Lt. John Eisenhower, who remembers admiringly: "In the midst of her grief she was thoughtful enough to call us in, one at a time, for a short visit so she could report back to our families that we were well. She said little about the general. . . . I have rarely been witness to such a gesture of fortitude and kindness."[47]

Patton's personal effects were prepared for shipment to Green Meadows, and perhaps the most poignant photograph in the vast Patton collection is that of the ever-faithful Willie, forlornly lying amid Patton's baggage, mourning his missing master, whose absence he can sense in his soul but cannot comprehend.

Christmas 1945 was an emotionally painful time for Cadet George S. Patton, who was in his final year at West Point. Young Patton was on Christmas leave and en route to spend Christmas with his sister Bee in Washington when he first read a headline framed in black in a New York newspaper: GENERAL PATTON DIES IN SLEEP. His father's death had come as no surprise, for the doctors at the West Point hospital had warned him that the injuries were so serious he would be paralyzed *if* he survived. What made the news even more grievously painful was Beatrice's decision that he would not be permitted to fly to Europe to attend his father's funeral, which was to be held on his birthday. "I was right in the middle of some very tough examinations. . . . The decision was made by her that I should not go. In retrospect, I don't agree . . . ," he recalled during an interview in 1985, with a mixture of regret and bitterness that the passage of forty years had done little to abate.[48]

Despite never knowing his father as well as he would have liked, his son has fond memories: "He was a first-class father. He taught me all kinds of things: to shoot, ride, and we worked on ship models together. . . . When I was a little boy I was scared to death of him. Then when I got to be about 11, I just decided one day I wasn't going to be scared to death of him. He never laid a hand on me, not once. He locked me in my room once. That was the worst thing he ever did."[49]

Patton's favorite nephew, Fred Ayer Jr., was in Boston with two marine officers on December 21 when he learned of his uncle's death: "There was a moment of dreadful silence. I saw tears in the eyes of one very tough man. Then he shook his head, lifted his glass, 'There died the best god-damned Marine the Army ever bred.'"[50]

The headlines in the December 22 edition of the *Stars and Stripes* blared: PATTON DIES. In stateside papers his death was generally treated in a more subdued and respectful fashion. Cards, notes, letters, and telegrams began pouring in to Heidelberg from all over the world; many were quite lengthy. They came from politicians, statesmen (among them Truman and Clement Attlee, the new British prime minister), but particularly from ordinary soldiers who had served under Patton. The French National Assembly, in a rare tribute, sent a special message of condolence to the "President of the United States and the noble American nation," describing him as "a great soldier who was one of the liberators of France."[51]

There were as yet only a handful of newspapers being published in Germany, and the event thus drew scant attention in the German press. An unnamed GI on occupation duty wrote to his parents:

> Last night one of the greatest men that ever lived died. That was Patton. The rest of the world thinks of him as just another guy with stars on his shoulders. The men that served under him know him as a soldier's leader. I am proud to say that I have served under him in the Third Army. . . . All the flags are at half-mast. We are making every Heinie that passes stop and take off his hat. They can't understand our feelings for him. I don't know whether or not you can understand them either.[52]

Among the first to cable his condolences was Omar Bradley, and at VMI, where the Pattons were revered, the flag flew at half staff in honor of the former cadet they still claimed as one of their own. Even one of the two soldiers he had slapped in Sicily in August 1943, Pvt. Charles Kuhl, was quoted as saying that he "felt deep regret" on learning of Patton's death.

For two days Patton lay in state in the Villa Reiner. Behind the flag-draped casket, which was surrounded with flowers, were flags denoting the armies he had commanded. Throughout the room candles burned, and four MPs with rifles mounted a round-the-clock honor guard as a steady procession of soldiers paid their final respects to the controversial general. Beatrice, who had said her final good-bye and could not bear to pay a final public visit, sent her brother in her place.

The afternoon of December 23 the casket was sealed and then conveyed on the back of a half-track, surrounded by his pallbearers and escorted by a platoon of the 15th Cavalry (the unit in which Patton had begun his military career in 1910 at Fort Sheridan, Illinois) in jeeps and armored cars to the Protestant Christ Church in Heidelberg. Behind the cortege dozens of generals, headed by Joseph McNarney, marched from the Villa Reiner to the church, where an Episcopalian service was conducted by two U.S. Army chaplains.

Outside, the massed bands of several divisions played as the casket arrived and departed the church. A photograph taken by an Associated Press photographer reveals that the ceremony was sparsely attended, mainly by army officers and representatives of twelve countries.[53]

None of Patton's three children attended the funeral, and while there were justifiable reasons, which included the short time between his death and burial, grandson Robert believes that Beatrice simply wanted her last minutes with her husband of thirty-five years to be unshared with anyone— including her own children.[54]

Beatrice, Fred Ayer, and Keyes sat in the front row within a few feet of the casket. Behind it someone had placed a nearly bare Christmas tree on which hung a few meager strips of tinsel. The ceremony lasted a mere twenty-two minutes, during which the hymn "The Son of God Goes Forth to War" was sung, and his two favorite psalms, the Sixty-third and the Nineti-eth, were read by the chaplains. A thirty-six-voice soldiers' chorus sang "The Strife Is O'er." "There were no eulogies," reported the AP.

From the church Patton's casket was again transported aboard the half-track for the mile-long journey to the railroad station, where two funeral trains awaited. Few Germans were on the streets of Heidelberg late in that cold, dark afternoon, and those that were stood stony-faced as the cortege passed along streets lined by a massive honor guard of some six thousand half-frozen GIs. At the station the guns of a 1st Armored Division artillery battery boomed a seventeen-gun salute, and then the flag-draped casket and thousands of flowers were placed in the baggage car of the lead train, in which rode Beatrice and her small entourage. Everyone else rode in the sec-ond train. At 4:30 P.M. the trains began the journey to Luxembourg.[55]

Among those accompanying Beatrice was Lucian Truscott. When Truscott took command of Third Army, Patton thought he was upset at hav-ing to take command under such circumstances. On the funeral train Truscott was afraid that Beatrice would not speak to him, even though he was an old family friend. "When I offered him my hands, he threw his arms around me and burst into tears," wrote Beatrice.[56] Conspicuously missing in the assemblage of Army brass at Patton's funeral was Walter Bedell Smith. Before he died Patton had told Beatrice emphatically that he did not want either Eisenhower or Bedell Smith attending his funeral. After his relief Pat-ton was particularly bitter toward Smith, telling Eisenhower at a dinner in October: "In light of what happened, I cannot hereafter eat at the same table with Beetle Smith."[57] The original list of Patton's pallbearers included "Red" Muller and Paul Harkins, plus a number of other high-ranking offi-cers, among them Smith. At Beatrice's request Smith's name was removed and replaced with Sergeant William Meeks, who had served Patton faith-fully for ten years as his personal batman. Neither Bedell Smith's nor Dwight Eisenhower's name appeared on the final list of his honorary pall-bearers. Patton's final judgment of his longtime friend, Eisenhower, was equally harsh: "I hope he makes a better President than he was a General."[58]

Another who did not attend the funeral was Dr. Glen Spurling. Beatrice thought he had given up enough, and encouraged him to leave at once so that he might spend Christmas with his family. However, the foul weather grounded all aircraft in Europe for several days, and Spurling did not arrive back in the United States until December 27. "This time I took no chances about getting out of the Army," and hand-carried his discharge papers to the Pentagon. "After all, I didn't want to take a chance on another general's breaking his neck."[59]

It was after midnight when the trains arrived in Luxembourg City. At numerous points along its route they had stopped so that Patton might be honored. During the brief passage through France there were a series of spontaneous and touching ceremonies. The trains made six stops, and despite the darkness and torrential rains, honor guards and bands had turned out smartly to pay homage to Patton. French officers who had known and respected him placed wreaths of flowers aboard the train.

George S. Patton was buried at midmorning on December 24, during a twenty-five minute ceremony in the unfinished American military cemetery (now called the Luxembourg American Cemetery and Memorial) situated on a hill at Hamm, a suburb of Luxembourg City. Appropriately, M/Sgt. Meeks was given the honor of driving the half-track used to transport his casket to the cemetery. As it had been in Heidelberg, the road to Hamm was lined for more than three miles with American, Belgian, French, and Luxembourgian troops.

Patton's grave in the thick, reddish clay of the Ardennes was dug by German POWs, and was next to that of a Third Army soldier from Detroit who had been killed in action during the Battle of the Bulge.[60] An account written by a United Press correspondent captures the essence of the somber ceremony that took place in a driving rain, attended by dozens of generals, soldiers, and Luxembourgers.

> Patton was buried in what he himself once called "damned poor tank country and damned bad weather." But he was buried in precision-like military ceremony, touched by pomp and tendered by grief. Big generals and little soldiers were there, as were the royalty and the commoners of this tiny country from which Patton drove the Germans in that crucial battle last Christmastide.
>
> But the focal figure, standing there under the dark sky against the background of green hills, was Patton's widow. A raw wind, whirling across the top of the bluff on which the cemetery is located, ruffled Mrs. Patton's veil as she watched the ceremony at the grave . . . Her eyes were red, but for the rest she was the same good soldier her husband had been. In the final minute of the ceremony Master Sgt. William G. Meeks, the Negro from Junction City, Kan., who had served Patton faithfully as his orderly . . . presented the general's widow with the flag that had draped the coffin. There were tears in Meeks' eyes. His face was screwed up with the strain. He bowed slowly, and handed the flag to Mrs. Patton. Then he saluted stiffly to her. For an instant their eyes met and held.
>
> Meeks turned away, a 12-man firing squad raised its rifles and a three-round volley of salutes echoed into the Luxembourg hills. The bugler played the soft, sad notes of "Taps."[61]

*   *   *

Among those most grieved by Patton's death was "Bulldog" Walker, who flew from Texas at his own expense to be present for the interment, only to be unable to land in the overcast that shrouded Luxembourg. All the frustrated Walker could do was fly above the long cortège. Beatrice could hear his plane but did not see him until both were reunited in Paris later that day. As the commander of the U.S. Eighth Army during the Korean War, Walker, too, met a tragic death in a jeep accident on December 23, 1950, five years and one day after the death of his idol, George S. Patton.[62]

After the ceremony the prince of Luxembourg promised Beatrice that he would make it his personal business to look after Patton's grave.

Larry Newman later wrote that a correspondent friend had bitterly said on the day Patton was buried in Hamm:

> Georgie Patton didn't die from an automobile accident. He died of a broken heart when they took away his army. . . . They wouldn't let him speak his mind about the reds and what they had in store for the United States and the free world. I guess it's just as well he died over here. The apologists, the peace at any price cowards, the friends of the Soviet Union, always hated him. And brother, did he hate them! At least he's buried alongside his beloved soldiers who died so those people back home could attack the things patriots love.[63]

The most eloquent of the public tributes to Patton appeared the day after his death

The New York Times
Saturday, December 22, 1945

GENERAL GEORGE S. PATTON
History has reached out and embraced General George Patton. His place is secure. He will be ranked in the forefront of America's great military leaders. The enemy who reached their judgment the hard way, so ranked him. This country, which he served so well, will honor him no less.

George Patton had a premonition he would die in battle. It is a wonder he did not, for he took chances in the heat of the fight that made even his hard-bitten soldiers shudder. . . .

Long before the war ended, Patton was a legend. Spectacular, swaggering, pistol-packing, deeply religious and violently profane, easily moved to anger because he was first of all a fighting man, easily moved to tears, because underneath all his mannered irascibility he had a kind heart, he was a strange combination of fire and ice. Hot in battle and ruthless too, he was icy in his inflexibility of purpose. He was no mere hell-for-leather tank commander but a profound and thoughtful military student. He has been compared with Jeb Stuart, Nathan Bedford Forrest and Phil Sheridan, but he fought his battles in a bigger field than any of them. He was not a man of peace. Perhaps he would have preferred to die at the height of his fame, when his men, whom he loved, were following him with devotion. His nation will accord his memory a full measure of that devotion.

The grandest of the memorial services for Patton took place on January 20, 1946. The twelve hundred attendees in the Washington Cathedral included Henry L. Stimson and Dwight Eisenhower. Some came from overseas to attend the service, others from distant cities. At Beatrice's request the principal eulogy was delivered by Harry Semmes, who extolled his friend as having had "a full life dangerously lived." President Truman was expected to attend but elected to send his military aide. His absence was duly noted and roundly criticized in an editorial in the *Washington Times Herald* for having found ample time for sailing on the Potomac, but no time to honor one of America's war heroes.[64]

The outer wall of the chapel at Hamm is inscribed with the words: "1941–1945. In proud remembrance of the achievements of her sons and in humble tribute to their sacrifices this memorial has been erected by the United States of America." There is a wide disparity between generals and private soldiers on the battlefield, but it is military tradition that in death all soldiers are equal. Patton's headstone at Hamm is thus the same, simple, dignified white cross (or Star of David for Jewish dead) that appears over the graves of everyone buried in an American military cemetery. His grandson has written: "His burial at Hamm with hundreds of his men served to transcend the egotistical bluster that had distanced him from the common soldier. It made him one of them. A casualty of war. Another dead American son."[65]

Until it was eventually moved to what is now its permanent location in the shadow of the memorial that overlooks the cemetery, so many people visited Patton's grave that the area surrounding it was trampled. In this position Patton is alone and again at the head of his men, this time commanding the 5,076 who lie permanently in Hamm.[66]

GEORGE S. PATTON, JR.
GENERAL THIRD ARMY
CALIFORNIA DEC 21 1945

George S. Patton's legacy is that he will lie for eternity in the midst of men who died in what he regarded as a noble calling. Were we able to ask him, Patton would undoubtedly proclaim: "I'm damn glad to be here."

EPILOGUE

After burying George S. Patton, Beatrice returned to the United States on Christmas Day, 1945. She put on a gay, brave face in order not to upset her grandchildren, but Ruth Ellen thought her mother looked "very small and somehow old." Her manner reminded Ruth Ellen of the false gaiety displayed by Ellie Ayer the day after Grandfather Ayer had died in 1918:

> The heart that truly loves never forgets
> But as truly loves on to the close
> As the sunflower turns to her God when He sets
> The same look that she turned when He rose.[1]

Her grief remained private and deep. Beatrice kept what she called a thought book, and three weeks after Patton's death she wrote: "the star is gone . . . the accompanist is left behind. There must be a reason, but I pray for understanding. . . . The tears run down behind my eyes."[2]

Back at Green Meadows she devoted herself to organizing her husband's massive collection of papers, which were later deposited in the Library of Congress. Beatrice had always been Patton's most articulate and passionate benefactor, and now she traveled widely, always with the object of doing all she could to perpetuate his memory. At a seemingly endless series of ceremonies to dedicate statues, buildings, streets, and parks named after him in both Europe and the United States, Beatrice became adept at extolling his accomplishments as she accepted in his name honors and awards from grateful cities, towns, and organizations. [3]

Patton's World War II diary was purged of nearly all its earthy and outrageous language and, with the assistance of his former deputy chief of
staff, Col. (later Gen.) Paul D. Harkins, it was published in 1947 as *War as I
Knew It*. Noted Civil War historian Douglas Southall Freeman, whom Patton greatly admired, wrote the introduction, and, at Beatrice's instigation,
would have become her husband's official biographer had he not died soon
thereafter. Some twenty years passed before the Patton family invited the
noted historian Martin Blumenson to edit Patton's papers, which were published in two volumes in 1972 and 1974 as *The Patton Papers*. Others,
including Harry Semmes, wrote memoirs of Patton, and Beatrice's nephew
Fred Ayer produced the most revealing account of Patton the man until the
publication of grandson Robert's 1994 memoir.

War as I Knew It became a bestseller, but stripped of Patton's candid
language, which would have made it unique, the book was merely politically correct and only mildly controversial, never approximating the true
character of its author. Nevertheless it served to achieve Beatrice's aim of
keeping his memory alive. *War as I Knew It* remains in print and widely
read by those curious about Patton.

However, all this lay in the future. Beatrice had a more urgent item of unfinished business to conclude. Without telling him why, she asked her brother,
Fred, to arrange for a room at a Boston hotel, where she would like to meet
with Jean Gordon, who had returned from Europe the previous November.
Beatrice, the last to arrive, entered the room quietly. Neither Fred nor Jean
had the slightest hint of what was to ensue. After removing her hat and coat,
Beatrice suddenly pointed her finger at Jean and recited the deadliest curse
known to Hawaiians: "May the Great Worm gnaw your vitals and may your
bones rot joint by little joint." Beatrice had learned it many years earlier in
Hawaii, and Jean's face suddenly turned "from rose to pearl to gray." The
cold, hostile expression on Beatrice's face so appalled her brother that he
fled from the room. "Fred said that there was so much malevolence in the
room that he jumped up and grabbed his hat and ran out, and only slowed
down when he reached the street."

Whatever else Beatrice may have said to Jean Gordon, both women took
with them to the grave. On January 8, 1946, only days after the confrontation
with Beatrice, Jean stuck her head in a gas oven in the apartment of a friend
in New York, a suicide only days shy of her thirty-first birthday. For years an
unsubstantiated rumor circulated in the Patton family that Jean had left a
defiant note stating: "I will be with Uncle Georgie in heaven and have him
all to myself before Beatrice arrives." If such a note ever existed, it has never
surfaced, and Jean's suicide remains as mysterious as it was tragic. [4]

Beatrice never told anyone what she had done, and it was not until years

after her death that Fred Ayer unburdened himself to Ruth Ellen, who thereafter referred to Jean in her unpublished memoir as "the Faithless Friend."
As her family and others who ran afoul of her knew only too well, Beatrice
Ayer Patton was a formidable woman whom one did not provoke lightly,
and her memory for those who wronged her or a member of her family was
infinite. Ruth Ellen once described her mother as "a good hater." However,
in the aftermath of Jean's death, Beatrice became a tower of strength to her
grieving half-sister.

Beatrice's jealousy of Jean Gordon was that of an older woman for a
young and attractive mistress who has stolen her husband's interest. Jean
was, writes Robert Patton, "educated, bright, and amusing," with a fine wit
and an interest in history, sailing, and horses. Jean told a friend that she
thought that with the war now over, perhaps Patton's death had been a blessing in disguise. "Jean had an understanding of him that was insightful and
not frivolous, ample reason for his wife to deem her a serious rival. Beatrice, out of love, could forgive Georgie's indiscretion; but Jean she was
determined to punish."[5]

In 1952 tragedy again struck the Pattons when Little Bee died suddenly at the
age of forty-one of heart failure. Beatrice was devastated, and Ruth Ellen
thought "she looked completely defenseless."[6] The following year Beatrice's
doctor warned her that she had an aneurysm in her aorta that might kill her at
any time. He suggested that she stop driving and horseback riding, but otherwise it wouldn't make any real difference what she did. The aneurysm would
eventually kill her, in a day, a week, or perhaps even a year, regardless of
whether she led a normal life or took to her bed. Beatrice accepted the danger
of possibly killing someone while behind the wheel of an automobile, and
quit driving, but she adamantly refused to give up riding. She had a horror of
the living death of old age in a nursing home, or of being infirm, hooked up
to IVs, reliant on others. Secretly, she made her son-in-law, Jimmy Totten,
promise to bring her a bottle of pills if she ever found herself in such a condition. When her children remarked that she did not seem to be driving her
automobile anymore, Beatrice shrugged it off by saying her eyesight was
failing. To the end she remained a very private person.

On September 30, 1953, she was riding in a drag hunt at the Myopia
Hunt Club near Green Meadows with her brother, when, without warning,
she fell off her horse seconds before it was to have jumped a fence. By the
time she hit the ground Beatrice was dead of the aneurysm. It was widely
reported that she died from the fall, but the truth was that the fall was the
result of her instant death, in the manner she would most have wanted. She
had once spoken of the sweetness of death and of awakening in God's arms.
During her lifetime she did not express fear of dying, believing reincarnation was her ultimate destiny.[7]

It was perhaps fitting that the press she so roundly mistrusted managed to get her death wrong, attributing it to the fall from her horse. The family did not bother issuing a correction, no doubt in the belief it was no one's business.[8]

Beatrice Ayer Patton's grave site is under an elm at Green Meadows, but it is now empty. Her real desire had always been to be buried with her beloved Georgie. Beatrice regarded it as unthinkable to have Patton reburied in the United States and after his death she had written to old friends in high places in an effort to gain permission to be buried in the same grave in Hamm. Military regulations forbade this, however, and the army gently but firmly informed her it would not be possible. But her children knew of her wish, and in 1957 Ruth Ellen, her son Michael, and Patton's son, George, visited his grave in Luxembourg. On a wall nearby, the family beheld a cat, which in Egyptian mythology was Bast, the goddess of happy death. Robert Patton writes that Ruth Ellen opened her purse, removed an envelope, and sprinkled Beatrice's ashes onto her father's grave: "He knows I never failed him," Beatrice wrote the day after he died. Nor had her children failed her.[9]

From the time of their courtship those many years before, Beatrice Patton had proved to be a tower of strength that fueled her husband's fragile ego and helped carry him through two world wars and the dreadful and trying years between, when it seemed that Patton had no future. Through thick and thin, Beatrice stood by George S. Patton, even at times when most others would have given up. Her apparent fragile physical demeanor masked a woman of truly remarkable determination, as the portly colonel who had insulted Georgie in 1920, and paid the price, would willingly have testified. Daughter Ruth Ellen has left this tribute to her parents from 2 Sam. 1:20: They "were lovely and pleasant in their lives, and in their death they were not divided: they were swifter than eagles; they were stronger than lions."[10] The most touching tribute came in 1946 from Little Bee, her older daughter, who wrote shortly after Patton's death: "I'm no good at saying stuff—but Ma, you have given all of us something to shoot at. You are as magnificent in adversity as you are in success. . . . God bless you."[11]

Although Patton's sister, Nita, never married, she raised two adopted sons of whom she was immensely proud. Both became career officers, one in the navy, the other in the army. She died in 1971 at the age of eighty-four.[12]*

After his death, Beatrice was determined that a statue of her late husband be erected at West Point. She commissioned renowned sculptor James E. Frazier, who thoroughly studied Patton, and even had his son pose in the

---

*Willie was eventually relocated from Germany to Green Meadows to spend the remainder of his life in his master's home, which would have been Patton's wish.

clothes he intended to use in the statue. Through the efforts of Patton's friends a memorial fund was established, and the resulting bronze statue—a marvelous likeness—was dedicated in August 1950. (The hands were cast from a mold into which Beatrice had thrown his four silver stars.) It faces the West Point Library, a location Patton would wholly have approved. At the well-attended, star-studded ceremony, the principal speeches were delivered by Douglas Southall Freeman and Beatrice, who spoke movingly of her husband's heartaches and triumphs as a cadet. "Life is a book," she said, "in which each chapter is different but every chapter is built on those that have gone before. Glory with honor never grown old. May this statue stand for Duty—Honor—America."[13]

Some fifty yards away there is now a statue of General of the Army Dwight David Eisenhower. It faces the Plain of West Point, where the Corps of Cadets parades numerous times each year. It is a bittersweet irony that the statues of these two lifelong comrades should have their backs turned to each other.

In the Episcopal Church of Our Savior, in San Gabriel, California, whose bricks were fired in the ovens of Benjamin Davis Wilson's Lake Vineyard ranch, there is now a General George S. Patton memorial stained-glass window, dedicated in October 1946. In the lower right-hand corner Patton is pictured mounted in a tank, an Armored Force patch on his left shoulder. The centerpiece of the window depicts Saint George on a horse, slaying the dragon. The insignia of the many divisions, corps, and the two armies he commanded also appear. Saint George's shield is emblazoned with the *A* of the Third Army patch. Inscribed around the charger are these words from 2 Tim. 4:7:

I have fought a good fight
I have kept the faith.
I have finished my course.

Across the bottom is the simple inscription: "In loving memory, General George S. Patton, Jr. 1885–1945."[14]

POSTSCRIPT

Patton's Legacy
A Genius for War

If the art of war consisted in not taking risks glory would be at the mercy of very mediocre talent.

—NAPOLEON[1]

He was one of those men born to be a soldier.

—DWIGHT D. EISENHOWER

Whether one admired or despised this extraordinary man, what is indisputable is that George Smith Patton Jr. was one of the unique Americans of this or any other century. He represented the individuality and passion for life of a man who was so thoroughly a product of America. Born out of conflict, the United States has to this day found itself torn between its democratic ideals and the wars that have rent the planet since the dawn of history. From the time the first Patton arrived in Virginia in the pre-Revolutionary period, to the wars of George S. Patton Jr. in the twentieth century, the Pattons were a family deeply affected by war. That the great-nephew of one Confederate hero and the grandson of another should follow in the footsteps of his forebears is hardly surprising.

Patton's deeply religious nature was wholly inconsistent with his public image, yet he never wavered in his deeply held belief that God had chosen him to fight a great war for the United States and, as one of his staff officers wrote with chilling accuracy: "So long as God chose to keep him in this capacity, he would live. And when God concluded that he had served his purpose, he would die. And this was good enough for George Patton."[2]

Opinions of his place in history have varied from outright condemnation to oversimplification and unadorned adoration. Lucian Truscott, the hard-boiled cavalryman who clashed with Patton in Sicily believed: "He was perhaps the most colorful, as he was certainly the most outstanding, battle leader of World War II."[3]

His old-fashioned virtues included the genuinely held belief that part of a citizen's responsibilities was to fight for his country when called on. To Patton the "Duty, Honor, Country" of the West Point creed was a living thing. He regarded war as a noble calling engaged in by honorable men for a just purpose, and never ceased in his attempts to convince his troops that men ought to tell their sons proudly that they had once been soldiers.[4]

As a tactician, Patton ranged from superb to average in situations beyond his control. However, what separated him from his peers and cemented his reputation was his daring, freewheeling approach to modern warfare. His grasp of the capabilities of the weapons and equipment at his disposal was unexcelled. No one, for example, made better use of tactical air support. As historian Hugh Cole concluded: "Patton's tactics were daring. He had no scruples about getting an armored combat command across a river by itself with no infantry anywhere in sight. He always sought opportunities to make raid-type penetrations and was always looking for a flank to turn. He knew the value of combined arms, and . . . he had a cavalryman's eye for ground."[5]

In his postwar recollection of Patton, Leland Stowe wrote: "Every wakeful minute, he was playing the indomitable, inimitable, and incomparable professional warrior. He played the part so boldly that much of it became a reality. . . . You were not going to be permitted to forget Georgie—no more than the Nazi generals ever could—Georgie, by his own opinion, was an obstreperous, fighting, cantankerous bastard—and proud of it!"[6]

Patton was a complete student of military history who should be remembered as well for his achievements as a trainer of troops as for his tactical innovations and military writings. As British military biographer H. Essame reminds us, not only was Patton willing to use innovative—and at the time, daring—tactics to achieve results on the battlefield, but when war came he was one of the few American commanders prepared to fight the German army: "With the advent of armor he saw that chances of flank attack could be taken more readily than with infantry."[7] What Patton and Rommel practiced during World War II is, fifty years later, an accepted means of successfully waging war. The success of Israel's tank victories in the desert during the various Arab-Israeli wars, and those of the Allies in the 1991 Gulf War, fully attest to their legacy.

Essame believed that

Patton was unquestionably the outstanding exponent of armored warfare produced by the Allies in the Second World War. In terms of blood and iron he personified the national genius which had raised the United States from humble beginnings to world power: the eagerness to seize opportunities and to exploit them to the full, the ruthless overriding of opposition, the love of the unconventional, the ingenious and the unorthodox, the will to win whatever the cost and, above all, in the shortest possible time.

However, Essame also observed that Patton "had even less political sense than Montgomery and that is saying much."[8]

Roger Nye, who so ably chronicled Patton's intellectual side, points out what his critics have ignored: "The Patton mind that emerged from the crucible of private study was capable of creating a kind of warfare that was so fast and so destructive of the enemy that the battle could be won with a minimum of friendly casualties and expenditure of materiel. The Patton mind also envisioned a warfare that placed limits on the amount of violence to be visited on innocent bystanders, and anticipated peace treaties without vengeance or retribution."[9]

Patton had many detractors, and one of his severest was literary critic and social commentator Dwight Macdonald, who loathed him as "brutal and hysterical, coarse and affected, violent and empty," and pronounced him "an extreme case of militarist hysteria." To Macdonald, Patton's utterances were "atrocities of the mind," indicative of what war had done to his personality. Another, writing in 1964, called Patton

a swaggering bigmouth, a Fascist-minded aristocrat . . . the last of our generals to call the Germans "the Hun." His horizons were limited; he was born for war, as he freely confessed. . . . As a very young man, safely attached to the headquarters of a less valiant army than his, I knew that I feared and despised him. If you drove in the Third Army sector without steel helmet, sidearms, necktie, dogtags, everything arranged according to some forgotten manual, Patton's fiercely loyal M.P. gorillas would grab you. You could protest, but say one word against their pigheaded general?—I never had the nerve . . . there is no doubting his sincerity, and no doubt that compared to the dreary run of us, General Patton was quite mad.[10]

While George S. Patton was by no means universally loved, even those who disliked him took a certain perverse pride in being able to state that they had served in the Third Army. When asked what outfit they had served in, most Third Army veterans replied simply: "I was with Patton." The tank driver who had once complained: "Won't that old bastard ever get enough of war" was later interviewed and practically came to attention, saying with evident pride: "I'm one of Patton's men."[11]

Among Patton's most prominent detractors is CBS's *60 Minutes* commentator Andy Rooney, who was a reporter for the *Stars and Stripes* during the war: "I detested Patton and everything about the way he was. It was because we had so few soldiers like him that we won the war. . . . Patton was the kind of officer that our wartime enlisted men were smarter than. It was the independent action of the average GI that made our Army so successful . . . not the result of the kind of blind, thoughtless devotion to the next higher authority that Patton demanded." After Rooney criticized Patton on a CBS documentary in 1992, he received this brief letter from Ruth Ellen: "Dear Mr. Rooney, My father wouldn't have liked you either. Sincerely, Mrs. James W. Totten."[12]

By contrast well-known New York radio personality Don Imus employs a George C. Scott–like voice of Patton to satirize events and public personalities in the news on his syndicated *Imus in the Morning* program, often with hilarious results.[13]

What made Patton different? Historian Eric Larrabee suggests that he was not merely a man thoroughly schooled in his profession:

A commander must be able to "think like" the unit he commands. All its weapons and their capabilities, the terrain on which it is disposed, the state of its supply and of its training and morale—in short, what it can reasonably be asked to do . . . all must be an extension of his own mind. To think like a platoon over a few hundred yards is no great trick; to think like a division over a score of square miles is difficult, but the gene pool seems to cast up an adequate number of men who can manage it. To think like a corps is an unusual gift . . . and to think like an army is a rarity itself. Patton could think like an army. When the occasion demanded it he could turn it on a dime, and this raises him high above the rest of the Allied generals. "Perhaps of them all," writes R. W. Thompson, "only George Patton was truly capable of commanding an army."[14]

Tales of those who served under Patton will endure. More than one unsuspecting soldier or civilian has paid the price of a black eye or broken nose for having insulted Patton or those who served in the Third Army. Ask any soldier what his outfit was, and invariably he will reply, such and such a division or regiment. Ask anyone who served in the Third Army and he will answer: "Third Army" or "I was with Patton." Men who ranged from bank presidents and corporate lawyers to truck drivers and car salesmen had a

common bond of having been Patton men. They may not—most would
not—have condoned the slappings, but it did not change the unique sway he
held over them, one made permanent by his untimely death.

Patton was a far too complex human being to be dismissed as merely a first-
class SOB, or as a hero without flaw. What seems indisputable is that
George S. Patton was not cheated of a full life. In his sixty years his accomplishments and his experiences were rich beyond measure. Just when one is
prepared to believe an understanding of the man has been achieved, he
would invariably act in a manner that would confuse or confound. In short,
there is no niche into which one can conveniently place Patton.

Gen. Paul Harkins has said of Patton: "He liked to fight, he'd rather
fight than eat."[15] Despite the war faces he wore, Patton never deceived those
who really knew him, and few understood Patton better than his longtime
aide Charles Codman, who noted that despite Patton's facade of the gruff,
tough, profane swashbuckler, he never heard a dirty or sacrilegious tale in
more than two years of constant companionship. As for the tough-guy
image, the rages (which were mostly feigned) never for a single moment
fooled his personal staff or those who knew him intimately. Not even Willie,
at whom he tended to thunder in public but in private would croon baby-
talk. Above all, Codman noted, for the most part, the press had only a one-
dimensional view of the man. They were not present when he displayed the
human side of the real Patton. "Compassionate" is not a word one would
expect to apply to him, but Codman is correct when he insists that this quality is what set George S. Patton apart from others.

There were no correspondents present when Patton fulfilled perhaps the
most difficult duty of a commander: consoling the wounded whom he had
sent into battle. Virtually every day Patton could be found visiting a field
hospital or aid station, where he often knelt by the cot of a seriously
wounded GI, sometimes in silence, other times offering words of praise and
encouragement. Only those closest to him knew of Patton's lifelong aversion to the suffering of others. "It is this quality—so difficult, nay impossible, to square with the business of war making—which sheds light upon
some of the contradictions and anomalies of the General's character."[16]

Nor was Patton the warmonger that his own self-created image has suggested. Gen. Isaac D. White, who knew Patton as well as anyone, believes:
"He loved the opportunity that war presented to use the skill, the leadership
and the courage his profession required, as a surgeon loves his profession,
but not disease, illness and injury."[17]

Indeed, what set Patton apart from others in his profession was the least
obvious virtue, his humanity. As Codman notes: "To him the concepts of
duty, patriotism, fame, honor, glory are not mere abstractions, nor the shop-
worn ingredients of Memorial Day speeches. They are basic realities—self-

evident, controlling. Bravery is the highest virtue, cowardice the deadliest sin."[18]

One of Patton's former Third Army staff officers has pointed out: "Most of the things he said that drew criticism could have been said by any other general officer and little attention would have been paid to them, but in Patton's colorful phraseology they got wide circulation."[19] The public rudeness and bombast was quietly replaced in private by the Patton who would bring in an exhausted division commander for a quiet night or two of rest away from the pressures of command; the man who wept unashamedly over what he had wrought on the battlefield; the man who could one moment curse another in the vilest language, only to recant in private for hurting his feelings. Where arrogant men like J. C. H. Lee put themselves first, Patton lived a simple life as commander of the Third Army. He cared so little about what he ate that he consumed whatever was put in front of him (frequently overdone or burned), often disdaining lunch as a waste of time, to the chagrin of his aides who were reduced to sneaking dry rations out of his sight, during their travels.[20]

After the war Gen. Fritz Bayerlein, the able commander of the Panzer Lehr Division and a veteran of North Africa, assessed the escape of Rommel's Panzer Armee Afrika after Alamein. "I do not think General Patton would have let us get away so easily," said Bayerlein, comparing Patton with Guderian and Montgomery with von Rundstedt.[21] Grudging admiration of Patton was even expressed by Adolf Hitler, who referred to him as "that crazy cowboy general."[22]

Historian S. L. A. Marshall never admired Patton and thought the Third Army nothing special, while conceding: "He was in many ways, a fair-minded man. . . . I'd put him in the same category as Orde Wingate and Stonewall Jackson. Many of his idiosyncrasies were more than that. . . . There was a lot of the little boy in Patton. . . . I think he was about half mad. Any man who thinks he is the reincarnation of Hannibal or some such isn't quite possessed of all his buttons." Marshall also thought that Patton's profanity was forced: "It didn't come over as something that was natural with him."[23]

Even his biographer Ladislas Farago once wrote privately: "I am convinced that he was, if not actually mad, at least highly neurotic, for reasons (primarily libidinal) that a Freudian psychoanalyst would have no trouble in explaining." [24]

Eisenhower's son, John, had a different view of Patton:

There was something a bit scary about Patton. . . . To pretend to love war like he did, there had to have been a screw loose somewhere. For some people the Army was a vocation—a paycheck—but Patton was one of those for whom it was an *avocation*.. . . [but] One thing that I noticed about Patton, which was never portrayed in the George C. Scott movie about him, was that he was a very generous man, with courtly manners . . . when he was away from the front. At the front he was very aware that he was playing a certain role—urinating in the Rhine River and things like that, while journalists took pictures. He was very excitable, though. Very high-strung. When I was with him at dinner one night he broke into tears twice, listening to what he himself had just said.[25]

We can only begin to perceive the complex man that was George S. Patton. Daughter Ruth Ellen has said: "I don't think anybody really knew Georgie, not even my mother. He was such a stratified person, so many pressures from so many traditions and impossible comparisons to the great and beloved dead heroes. He would have very much liked the fact that the movie could not strip him to the soul, for the world to remark." One of his favorite poems could have been written for him. It is called "The Soul Speaks":

> Here is Honor, the dying knight
> And here is Truth, the snuffed out light
> And here is Faith, the broken staff,
> And there is Knowledge, the throttled laugh,
> And here is Fame, the lost surprise,
> Virtue, the uncontested prize,
> And Sacrifice, the suicide
> And there, the wilted flower, Pride,
> Under the crust of things that die
> Living, unfathomed, here am I.[26]

His family successfully fought the making of a film about Patton for nearly twenty years, but when they finally consented, they were immensely pleased with Frank McCarthy's production of *Patton,* starring George C. Scott. Ruth Ellen observed:

Of course, in the movie, you only saw what he called his "war face." As a young man, and even as an old man, in repose, he was almost painfully beautiful, curly blonde hair, big blue eyes with long dark lashes, a gorgeous aristocratic nose, in other words, "the works." He spent hours practicing his war face in front of mirrors. I can well remember him doing it—and once, I will never forget it, he was reciting

> In peace, there's nothing so becomes a man
> As modest stillness and humility
> But when the blast of war blows in our ears

Then imitate the action of a tiger
Stiffen the sinews, summon up the blood
Then lend the eye a terrible aspect
Now set the teeth and stretch the nostril wide
Hold hard the breath and bend up every spirit
To his full height! On, On you noblest English!

It was years before I found out . . . that it was from Act III of Shakespeare's Henry V, but my sister and I always shouted, "On, On, you noblest English," to our horses going over jumps in horse-shows—it was very inspiring.[27]

But, most of all, what Ruth Ellen remembered about her father was that

Georgie was the kindest man that ever lived, and he spent a great deal of his life trying not to show it—he thought it was a sign of weakness. Animals, dogs, old ladies and little children adored him. His mother-in-law had his picture on her dressing table—I have it now on mine—and across the back of it is written in her hand, "The bravest are the tenderest; the loving are the daring." When one of his family got hurt, as we frequently did because we spent most of our waking moments on the backs of his horses, he was always there to cut off the riding boot, grab off his shirt and stuff it under the broken collarbone, or apply one of his huge linen handkerchiefs to whatever was bleeding.

You cannot sum up a human being. Georgie Patton was the result of his life long training, sitting beside his Aunt Nannie in front of the fire, at the ranch, hearing of the fearless trek of his grandfather across the plains to California . . . hearing from his own father how he was taken in front of his father's saddle to see the body of the great Confederate hero, Jeb Stuart, lying on the billiard table at the Yellow Tavern, like a statue of a God, his form outlined beneath a damp white sheet, his great bronze beard spilling over his silent breast, and the lines of weeping Confederate soldiers filing by; hearing in his mind the dying words of General Mercer, at Princeton, and of Tazewell Patton at Gettysburg—and behind them the parade of mighty heroes, Alexander, and Achilles, Hannibal crossing the Alps, Harold at Hastings, Napoleon, Nathan Hale. All this combined in George S. Patton, Jr. to make him the man he was. He always wanted to be among the heroes. All his life he summoned unto himself a company of heroes, alive and dead, and in his death he lies among heroes to whom he was a hero. As a man thinketh, so is he.[28]

One of the tragedies of Patton's untimely death is that he never completed the memoir or autobiography that he would inevitably have come to write. *War as I Knew It* is based mainly on his military actions, as taken from his diary. The posthumous edition was a mild version of what might have been. As Geoffrey Keyes has noted, "The accident that killed General Patton destroyed what could have been the greatest book to come out of World War II."[29]

Leland Stowe writes: "Looking back, there's one thing that seems a great and everlasting pity. There will never be a completely unexpurgated version of the vocal Blood-and-Guts. No Americans who never met him or served with him will ever remotely conceive how Georgie Patton really talked. Not even a Norman Mailer novel can give you more than a faint approximation. I hereby apologize to the rollicking, mocking ghost of Lt. Gen. George S. Patton, Jr. But he could never have got his conversation on the printed page himself. How can anyone else hope to do it? It's too bad, Georgie. The history books will never do right by you. But given the limitations, they'll do well enough. And what a helluva lot of fun you must be having in Valhalla!"[30]

There was this final tribute by Eisenhower: "He was one of those men born to be a soldier, an ideal combat leader whose gallantry and dramatic personality inspired all he commanded to great deeds of valor. His presence gave me the certainty that the boldest plan would be even more daringly executed. It is no exaggeration to say that Patton's name struck terror at the hearts of the enemy."[31]

Yet, when, in 1946, Eisenhower was sent the proofs of a book titled *Patton and His Third Army*, he wrote in pencil what his biographer, Merle Miller, calls "his frankest appraisal of the late Georgie." Describing Patton as "one of my oldest, dearest and most intimate friends covering a span of 26 years," Eisenhower thought the book gave too much credit to Patton and would be ruined unless revised to give credit chiefly to Bradley. "George Patton was the most brilliant commander of an army in the open field that our or any other service produced. But his army was part of a whole organization and his operations part of a great campaign. Consequently, in those instances where Patton obeyed orders, the story only hurts itself by assuming that Patton conceived, planned and directed operations in which he was in fact—the brilliant executor." It was Bradley, not Patton who directed the campaigns in which Patton fought so brilliantly, said Eisenhower.[32]

In the end, each paid a debt to the other. Eisenhower rescued Patton from certain career-ending ruination several times and, in return, Patton gave him victories on the battlefield. It was a pretty fair exchange.

Whatever his claims to the contrary, Omar Bradley was no friend of Patton, and, in fact, was the primary culprit in popularizing the "our blood, his guts" label hung on Patton. Consider this statement, which appears in *A Soldier's Story*:

Canny a showman though George was, he failed to grasp the psychology of the combat soldier. For a man who lives each day with death tugging him at the elbow lives in a world of dread and fear. He becomes reproachful of those who enjoy rear-echelon security and safety. . . . George irritated them by flaunting the pageantry of his command . . . [his] exhibitions did not awe the troops as perhaps Patton believed. Instead, they offended the men as they trudged through the clouds of dust left in the wake of that procession. In Sicily Patton, the man, bore little resemblance to Patton the legend.[33]

A former Third Army lieutenant who was a combat veteran of three campaigns offers a counterpoint to those who, like Bradley, claim that Patton's melodramatics had no impact:

Without a solid record of success, Patton's theatrics would have been sheer buffoonery. Dizzy Dean is reputed to have said, "If you done it, it ain't bragging." Patton "done it" over and over again . . . Moreover, he never squandered the lives of his troops needlessly. Patton's sense of what was possible on the battlefield was unequalled. Fiascoes like the Fifth Army's attack on the Rapido River or the First Army's attack in the Hürtgen Forest never occurred under his command. In the Third Army we knew what General Patton expected us to do, and we believed that if we did it we would win. That's what generalship is about.[34]

Some, however, would argue that the Hammelburg raid was a clear example of an instance in which Patton did indeed squander needlessly the lives of soldiers, and it remains an unfortunate stain that has justifiably reinforced the contentions of his critics.

Biographer Farago is right when he notes of Patton:

The picture of the man ran to extremes. At one end was General Semmes' effusive characterization of Patton as "the symbol of America.". . . At the other is the vulgar presentation of the General as the swashbuckling "Blood and Guts Patton.". . . In the popular mind he survives as a great captain of war, to be sure, but mostly as the general who had slapped an enlisted man, then redeemed himself by leading a dashing and dramatic campaign at the head of a competent and romantic army. But Patton was not so simple as that. He was not so evil as the man seemed to be who had struck a nerve-racked soldier. And he was not quite so good as the legend. This is part of the Patton mystique.[35]

When Patton self-destructed his career in Sicily, he may have altered the course of the war. The slappings deprived the Allies of the one commander who would have reacted very differently in two of the decisive battles of the campaign in northwest Europe. As the U.S. army group commander, would he have waffled over the so-called Falaise gap as did Bradley? Instead, Patton would have ensured the entrapment of the German army in the west at the Seine. Blumenson is not alone in believing the war might have ended (or at least have had its end significantly hastened) in the autumn of 1944 if the German army in Normandy had been annihilated west of the Seine.[36]

With Patton in command would there have been a Battle of the Bulge? At the very least, it might have been a vastly different battle. It seems likely from what we know that Patton would not simply have ignored the mounting evidence of an impending German offensive, and would not have waited for the Germans to attack before taking his own countermeasures.

How Patton would want to be remembered we can only guess, however. There is little doubt that he would have been proud of the words of his final message on May 9, 1945, to the officers and men of his beloved Third Army:

> The one honor which is mine and mine alone is that of having commanded such an incomparable group of Americans the record of whose fortitude, audacity and valor will endure as long as history lasts.

One of George S. Patton's fundamental beliefs was that he would one day be reborn to lead men in battle once again. Most people would, I believe, prefer to think that there was and ever will be but one Patton. He was deeply proud to have been a soldier, and no place was this reflected more eloquently than in his favorite quote, taken from John Bunyan's *Pilgrim's Progress,* which Beatrice Patton selected as the dedication for *War as I Knew It*:

> My sword I give to him that shall succeed me in my pilgrimage, and my courage and skill to him that can get it. My works and scars I carry with me, to be a witness for me that I have fought His battles who now will be my rewarder.
> So he passed over and all the trumpets sounded for him on the other side. [37]

Patton Family Genealogy

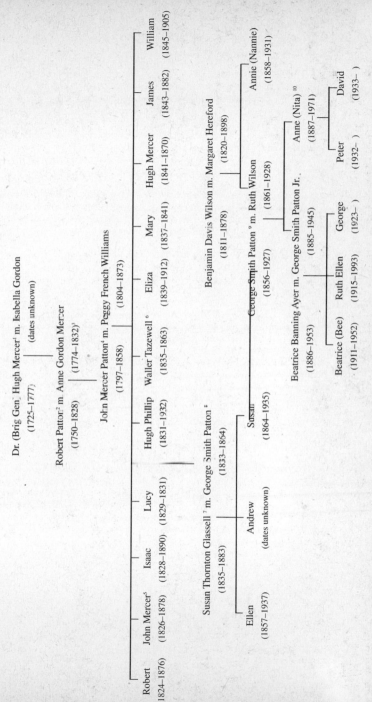

Dr. (Brig. Gen.) Hugh Mercer[1] m. Isabella Gordon
(1725–1777)   (dates unknown)

Robert Patton[2] m. Anne Gordon Mercer
(1750–1828)   (1774–1832)[3]

John Mercer Patton[4] m. Peggy French Williams
(1797–1858)   (1804–1873)

Robert   John Mercer[5]   Isaac   Lucy   Hugh Phillip   Waller Tazewell[6]   Eliza   Mary   Hugh Mercer   James   William
(1824–1876)   (1826–1878)   (1828–1890)   (1829–1831)   (1831–1932)   (1835–1863)   (1839–1912)   (1837–1841)   (1841–1870)   (1843–1882)   (1845–1905)

Susan Thornton Glassell[7] m. George Smith Patton[8]
(1835–1883)   (1833–1864)

Ellen   Andrew   Susan
(1857–1937)   (dates unknown) (1864–1935)

Benjamin Davis Wilson m. Margaret Hereford
(1811–1878)   (1820–1898)

George Smith Patton[9] m. Ruth Wilson
(1856–1927)   (1861–1928)

Beatrice Banning Ayer m. George Smith Patton Jr.   Annie (Nannie)
(1886–1953)   (1885–1945)   (1858–1931)

Anne (Nita)[10]
(1887–1971)

Beatrice (Bee)   Ruth Ellen   George
(1911–1952)   (1915–1993)   (1923–)

Peter   David
(1932–)   (1933–)

1. Born in 1725 and educated at Marischal College, University of Aberdeen, Hugh Mercer became a physician in 1744. Mercer was an adventurous sort who fought in the ill-fated Jacobite Rebellion but was obliged to flee Scotland to escape death at the hands of the Duke of Cumberland when the cause of the Young Pretender, Bonnie Prince Charlie, was crushed at Culloden in April 1746. Mercer emigrated to Pennsylvania in the autumn of 1746 and, after practicing on the frontier for ten years, became a soldier and joined Gen. William Braddock's expedition against Fort Duquesne in 1755 as its surgeon. Seriously wounded and left behind, Mercer survived a perilous trek through the wilderness to rejoin Braddock. During the French and Indian War, Mercer began a lifelong friendship with George Washington. He later served in the Continental Army, beginning in 1775 when he was elected a brigadier general by the Continental Congress. Serving under his long-time friend Washington, Mercer was in command of a brigade on the morning of January 3, 1777, during the Battle of Princeton, when "his horse was shot from under him. As he was attempting to rally his brigade on foot, he was surrounded by Redcoats, clubbed on the head with the breech of a musket, forced to the ground despite his efforts to defend himself with his sword, and bayonetted in seven places. After the battle he was carried to a neighboring farmhouse where he died [on January 12, 1777]. . . . The Continental Congress voted to erect a monument in his honor. . . . Washington repeatedly lauded his judgment and experience." (Malone, *Dictionary of American Biography*, pp. 542–43.)

2. It was previously believed that Robert Patton did not marry Anne Gordon Mercer until 1793, however the (Fredericksburg) *Virginia Herald* mentions their wedding in the October 18, 1792, edition. Another probably apocryphal tale has it that Robert returned from an Indian campaign to find his sweetheart, Anne, in the process of marrying another man, and that "Mr. Patton rode into the church and carried off the lady, some say that in so doing he was forced to pistol [whip] the prospective groom." However, a brief account of the wedding by a noted Fredericksburg historian suggests no such dramatics, merely that Miss Mercer, dressed in white satin and muslin and looking "infinitely lovely . . . gave her hand to the delighted Mr. Patton."

Members of Robert Patton's family were later buried in the Masonic Cemetery in Fredericksburg, but the burial location of Robert and Anne Gordon Mercer Patton is unknown. The logical place for Robert Patton's grave would have been in the cemetery of St. George's Episcopal Church in Fredericksburg. Shortly after Robert Patton's death in 1828, the Episcopal Church in Virginia went through what Fredericksburg historian John T. Goolrick describes as "a critical period," when, "for various reasons many of the St. George's congregation had become dissatisfied. . . [and] brought about the subsequent rapid growth of Presbyterianism. . ." According to Goolrick: "The present [Presbyterian] brick edifice on George Street was erected in 1833, the ground having been donated by Mrs. Robert Patton, the daughter of General Hugh Mercer." Robert Patton's mansion, White Plains, survived until 1946 or 1947, when it was torn down to make way for a Dairy Queen ice cream shop. The last known occupant of the house was a Miss Amelia Brulle. (John T. Goolrick, *Historic Fredericksburg*, Fredericksburg, 1921, p. 175,, and letter to the author from Barbara P. Willis, Virginiana Librarian, CRRL, April 17, 1992.) See also the microfilm records of the (Fredericksburg) *Virginia Herald,* edition of Wednesday, Nov. 5, 1828, CRRL. George S. Patton's great-great grandmother, Anne Gordon Mercer Patton, died in Richmond in 1832. (Source: *Richmond Daily Whig*, May 16, 1832, and *Bulletin of the Virginia State Library* 14 , Oct. 1921.)

3. Accounts conflict whether Anne Gordon Mercer Patton was fifty-eight or fifty-nine when she died in 1832. In *Death Notices from Richmond, Virginia, Newspapers, 1821–1840*, the *Richmond Constitutional Whig* and the *Richmond Enquirer* gave her age at death as fifty-nine. Her date of birth may well have been 1773.

Some accounts suggest that Robert and Anne Mercer Patton had eight children, others six. For example, the genealogical tables in Goolrick's, *The Life of General Hugh Mercer*, record only six: Robert, Hugh Mercer, John Mercer, Eleanor Anne, William Fairlie, and Margaretta L. Patton. The correct number appears to be seven; Dora C. Jett in *Minor Sketches of Major Folk and Where They Sleep* (p. 89) records the death in Oct. 1801 of George Weedon Patton, an infant of seven months.

4. Source: Goolrick, *The Life of General Hugh Mercer*, pp. 109–11. Six of John Mercer Patton's surviving sons fought in the Civil War:

- Col. Waller Tazewell Patton—commanded 7th Virginia—killed at Gettysburg, July 1863
- Col. George S. Patton—22d Virginia—killed at Winchester, September 1864
- Col. Isaac Williams Patton—commanded 20th Louisiana, captured Vicksburg
- Col. John Mercer Patton—commanded 21st Virginia and a brigade under Stonewall Jackson during the first batttle of Winchester, 1862
- Lt. Hugh Mercer Patton—wounded at Second Manassas, 1862
- Lt. James French—wounded at Cold Harbor
- Cadet Sgt. William M. Patton, VMI

Other Patton kin serving the Confederacy were: Brigadier General Hugh W. Mercer—commanded Georgia troops (grandson of the first Hugh Mercer), and Colonel George Hugh Smith—commanded 25th Virginia & 62d Virginia—wounded at McDowell, May 1862; wounded and captured at Second Manassas; fought in the battle of New Market, May 1864. (Sources: Halsey, "Ancestral Gray Cloud Over Patton"; Walker, *Biographical Sketches*, p. 570; Horace Edwin Hayden, *A Genealogy of the Glassell Family of Scotland and Virginia* [published in Pennsylvania in 1891, reprinted 1979 by the Genealogical Publishing Co. of Baltimore], and various Patton family papers.)

5. Although somewhat pompous in his later years, John Patton is remembered for his good nature and his fondness for good company, good food and, especially, conversation. On one occasion he said to a group of departing guests, "Come again soon, gentlemen, for you have given me a most delightful day." One of the guests, a noted scholar, drinker, and talker, replied: "I reckon we have; you have done all the talking!" The Chesapeake and Ohio Railway was granted a right of way through his property in Albemarle County but only on the condition that all passenger trains must stop at his house if signaled. When he desired to avail himself of the train, he would have one of his servants flag it down and make it wait while leisurely strolling through his garden, pipe in mouth, while his servants watched. It was said that although both Colonel Patton and God Almighty were of equal rank, to his servants, the good colonel was nearer at hand. One Confederate officer described John Mercer Patton as, "but a pigeon-headed fellow as narrow as any king's that ever tormented mankind." (Source: "The Meadows," Box 31, PP-LC, and Halsey, "Ancestral Gray Cloud Over Patton.")

In 1862, during one of the early skirmishes of the war, Colonel John Mercer Patton reported to Stonewall Jackson that his 21st Virginia had killed or captured all but one of an attacking Federal cavalry force. "He regretted, said Patton, to see the bluecoats shot down. . . Jackson asked quietly, 'Colonel, why do you say that you saw those Federal soldiers fall with regret?' In surprise, Patton answered that the Union troopers had shown so much more courage and valor than the enemy usually had that he had sympathy with the brave men and wished that their lives might be spared. 'No,' said Jackson in a dry tone, 'shoot them all; *I do not wish them to be brave.*'" (Quoted in Douglas Southall Freeman, *Lee's Lieutenants*, Vol 1, "Manassas to Malvern Hill" [New York, 1942], p. 424.)

6. Profile of Colonel Waller Tazewell Patton, 7th Virginia Infantry in Walker, *Biographical Sketches,* pp. 425–26. Patton's letter said: "My sufferings and hardships during two weeks that I was kept out in the field-hospital were very great. I assure you that it was the greatest consolation, whilst lying in pain on the cold, damp ground, to look up to that God to whom you so constantly directed my thoughts in infancy and feel that I was His son by adoption. When friends are far away, and you are in sickness and sorrow, how delightful to be able to contemplate the wonderful salvation unfolded in the Bible! Whilst I have been very far from being a consistent Christian, I have never let go my hope in Jesus, and find it inexpressibly dear now. I write these things to show you my spiritual condition, and to ask your prayers continually for me."

Waller Tazewell Patton was a deeply Christian man whose solace during his intense suffering was his abiding faith in God. A relative who remained at his bedside in the Union hospital in Gettysburg during the two weeks he survived after being horribly wounded on July 3 later wrote:

He was aware of the approach of death, and met it as became a soldier and a Christian. . . .When he became too weak to write, he tried to repeat the hymn "Rock of Ages, cleft for me." His friend read the hymn, and he tried to repeat it after him. He then called upon the chaplain. . . to read the 14th chapter of St. John. After prayer, he called us to his bedside, and shook hands with us, one by one. He retained to the last the utmost patience under his sufferings, and expressed his gratitude for every little service rendered him, taking us by the hand. And thus he fell asleep in Jesus, amid the tears of all around him. . . .His soul enjoys perpetual peace. God grant that we who survive may so live that we may meet him in that better land where there is no war, and where God will wipe away all our tears!

Waller Tazewell Patton's twenty-ninth birthday was on July 15, 1864, only a week before his death in Gettysburg Hospital. Patton was one of eleven VMI graduates to command regiments in Pickett's division on July 3, 1864. Every single one was either killed (6) or wounded (5). Patton's two roommates at VMI were both killed at the head of their regiments during Pickett's Charge. (Source: B. David Mann, *They Were Heard From: VMI Alumni in the Civil War* [Lexington, 1986], p. 9.) The death of their commander brought sadness throughout his brigade, prompting an officer of another regiment to compose a poem in his memory:

Patton farewell!
Let history tell
Thy grand heroic story
Wreathe with growing fame
Around thy name
Of great undying glory.

(Source: Riggs, *7th Virginia Infantry*, p. 28.) As young George S. Patton II later recalled of his uncle: "There was great excitement as to whether or not Virginia would secede and my father had determined to go south if she failed to do so. Finally, the convention was called at Richmond to vote on the Ordnance of Secession. Uncle Tazewell . . . was a member of this convention. We all went to Richmond and . . . I remember distinctly that on the day the Ordnance was passed Uncle Tazewell held me up in his arms so that I could see and, pointing to the platform, told me to always remember that I had seen this historic ordnance ratified. That night there was a torch light procession through the streets and all the windows were illuminated with candles. I was greatly impressed with the grandeur of the scene." ("A Child's Memory.")

7. Susan Thornon Glassell Patton's family was of Scottish descent and also traced its lineage to King Edward I of England and King Phillip III of France. Some sixteen barons were reputed to have signed the Magna Carta, which, as Martin Blumenson writes, "in feudal fashion, the Pattons felt that their blood conveyed the birthright of leadership, together with a sense of honor and an obligation to responsibility." (MB-PP 1, p. 24.)

Married Colonel George Hugh Smith, first cousin of the late Colonel George Smith Patton in 1870. Two children from the marriage: Anne Ophelia and Eltinge Hugh (1876–1887). Susan Thornton Patton's brother was William Glassell. Glassell had been a U.S. Navy officer but resigned his commission rather than take an oath to the Union. In October 1863 he commanded the first Confederate torpedo boat *David*, which attacked the USS *New Ironsides* which was blockading Charleston harbor. Patton family lore has it that Glassell disdained a sneak attack and instead hailed the enemy ship and announced to an officer that this was, "the Confederate States steamship *David* come out to attack you." When the officer attempted to sound the alarm, Glassell killed him with a blast from both barrels of a shotgun. Only later did he learn that he had killed his roommate and best friend at the Naval Academy. However, according to the official history of the Confederate Navy, the *David* was discovered by a Union sentinel aboard the *New Ironsides*, and, "without making any reply to his hail, Glassell kept on and fired with a shotgun at the officer of the deck . . . who fell mortally wounded."

The attack failed when the *New Ironsides* was rammed by the *David* and was crippled but did not sink. The *David* was swamped and sank from the fires aboard the *New Ironsides*. Under musket fire from the *New Ironsides*, Glassell and his men abandoned ship and after an hour in the water, he was taken prisoner, and later paroled. His last official act as a Confederate naval officer was to fire the fleet in the James River before the fall of Richmond in April 1865. (Sources: "My Father as I Knew Him," and J. Thomas Scharf, *History of the Confederate States Navy* [New York, 1878], pp. 758–59.)

8. One of the myths about Colonel George S. Patton I is that at the time of his death he had been approved for promotion to brigadier general but his commission failed to reach him in time. However, as the historian of the 22d Virginia has written, "There is some sketchy evidence to support this claim but no valid documentation has ever been found, although he was recommended for the position a number of times during the war." Although his brother John had lobbied hard for his promotion, "George was against his brother doing this and preferred for his actions on the field of battle to speak for themselves. . . ." (Source: Lowry, *22d Virginia Infantry*, p. 75.)

9. Born George William Patton. Changed his middle name to Smith in honor of both his own father and his stepfather, George Hugh Smith. Some accounts lists the date of birth of George Patton II as 1855, however his son's memoir of him states that he was nine years old when the Civil War ended in 1865, and other documents, such as his obituary in the VMI Alumni News and other records in the VMI archives, confirm his date of birth as September 30, 1856.

10. Peter and David Patton are the adopted sons of Nita Patton.

NOTES

Abbreviations

AEF	Allied Expeditionary Force
BA	Beatrice Ayer
BAP	Beatrice Ayer Patton
C&GSC	U.S. Army Command & General Staff College, Fort Leavenworth, Kansas
CMH	Center of Military History, U.S. Army, Washington, D.C.
COM Z	Communications Zone (Lt. Gen. J.C.H. Lee's command)
CRR	Charles R. Codman, *Drive* (Boston: Little, Brown, 1957)
CRRL	Central Rappahannock Regional Library, Fredericksburg, Virginia
CSI	Combat Studies Institute, U.S. Army Command & General Staff College, Fort Leavenworth, Kansas
DDE	Dwight David Eisenhower
EL	Dwight D. Eisenhower Library, Abilene, Kansas
EP	Dwight D. Eisenhower Papers, Eisenhower Library, Abilene, Kansas
EPP	Dwight D. Eisenhower Presidential Papers, Eisenhower Library
ETO	European Theater of Operations
FA	Fred Ayer, *Before the Colors Fade* (London: Cassell, 1964)
FM	Frank McCarthy Papers, George C. Marshall Library, Lexington, Virginia
GCM	George C. Marshall
GCML	George C. Marshall Library, Lexington, Virginia
GCM-P	George C. Marshall Papers, Marshall Library, Lexington, VA
GSP	George Smith Patton, Jr.
GSP-*W*	George S. Patton, Jr. *War as I Knew It* (New York: Bantam, 1980)
GSP II	George Smith Patton II (Patton's father)
HS	Harry Semmes, *Portrait of Patton* (New York: Paperback Library, 1970).
IWM	Imperial War Museum, London
JJP-P	General John J. Pershing Papers, Manuscript Division, Library of Congress, Washington, D.C.
LF	Ladislas Farago, *Patton: Ordeal and Triumph* (New York: Ivan Obolensky, 1963)
LF-*LDP*	Ladislas Farago, *The Last Days of Patton* (New York: McGraw-Hill, 1981)
LHC	Liddell Hart Center for Military Archives, King's College, London
MB	Martin Blumenson, *Patton, The Man Behind the Legend, 1885–1945* (New York: Morrow, 1985)
MMB-NA	Modern Military Branch, National Archives
MB-*PP*	Martin Blumenson, *The Patton Papers,* vols. 1 and 2 (Boston: Houghton Mifflin, 1972, 1974)

NA   National Archives, Washington, D.C.
NYT   *New York Times*
ONB   Gen. Omar Nelson Bradley
ONB-*SS*   Omar N. Bradley, *A Soldier's Story* (New York: Holt, 1951)
ONB-*AGL*   Omar N. Bradley, with Clay Blair, *A General's Life* (New York: Simon & Schuster, 1981)
PP-LC   George S. Patton Papers, Manuscript Division, Library of Congress, Washington, D.C.
PP-USMA   Patton Papers, United States Military Academy, West Point, New York
RET   Ruth Ellen (Patton) Totten
RET-M   Ruth Ellen (Patton) Totten Mss. ("MA: A Button Box Biography")
RHP   Robert H. Patton, *The Pattons: The Personal History of an American Family* (New York: Crown, 1994)
SHAEF   Supreme Headquarters, Allied Expeditionary Force
USAMHI   U.S. Army Military History Institute, Carlisle Barracks, Pa.
USMA   United States Military Academy, West Point, N.Y.
USMA-SC   Special Collections, USMA Library, West Point, N.Y.
VMI   Virginia Military Institute, Lexington, Va.
VMI-A   Virginia Military Institute Archives, Lexington, Va.
WP   *Washington Post*

Prologue   Who Was George S. Patton?

1. Charles Whiting, *Patton's Last Battle* (New York, 1987), p. 269.
2. ONB-*SS*.
3. The papers of Frank McCarthy, the producer of *Patton*, detail the participation and financial remuneration of General of the Army Omar N. Bradley.
4. John E. Fitzgerald, "The Paradoxical Patton," n.d., n.p. [1970], Box 3, PP-LC.
5. Ibid., and FA, p. 100.
6. Quoted in Brooks Clark, "Gunning for Glory," *Pursuits* (Spring 1989).
7. FA, p. 116.
8. George C. Scott, "Why 'Patton,'" *20th Century-Fox Informational Guide to the film* Patton, 1970, FM.
9. FA, p. 5. Patton's daughter, the late Ruth Ellen Patton Totten, inherited not only her father's love of poetry but his extraordinary memory and a prolific ability to quote verse.
10. Ibid., pp. 5–6.
11. Gerald Clarke, "Biography Comes of Age," *Time,* July 2, 1979.
12. Interview of Frank McCarthy, Box 17, FM.

Chapter 1   The Pattons of Virginia

1. Ashley Halsey, "Ancestral Gray Cloud Over Patton," *American History Illustrated*, Mar. 1984.
2. MB, p. 20, and LF, p. 49.
3. There is no record of any governor of Bermuda having died under mysterious circumstances in the eighteenth century. Robert apparently did stop in Antigua, British West Indies, en route to Virginia from Scotland circa 1769–70. (Various histories of Bermuda in the Library of Congress; "My Father as I Knew Him and of Him from Memory and Legend" [Box 5, PP-LC], hereafter cited as "My Father as I Knew Him"; and RHP, p. 18.)
4. Robert Patton appears to have been in Culpeper as "merchant and factor, res. Glasgow, pre-1776," per entry in David Dobson, *The Original Scots Colonists of Early America, 1612–1783,* n.d., p. 270, extract furnished by genealogist Julie Poole, who researched Scottish sources for the author.
5. In the period of the 1760s, the Scottish tobacco trade from the American colonies

was dominated by three giant syndicates, of which William Cunninghame was one. The capitalization of William Cunninghame was £72,000. The British pound in 1774 was worth $37.86 in 1973 dollars, thus making William Cunninghame one of the first multimillion-dollar firms to do business in the American colonies. (T. M. Devine, ed., *A Scottish firm in Virginia, 1767–1777: W. Cunninghame and Co.* [Edinburgh, 1984], p. x., and *Statistical History of the U.S.,* data furnished by the Boston Public Library.)

6. *A Scottish Firm in Virginia, 1767–1777,* contains numerous references to Robert's activities. In 1773 he was referred to as Robert Paton, but by Aug. 1775 the spelling had changed to "Robert Patton."

7. The usual period of indenture was five years, thus it seems likely that Robert arrived in Virginia about 1769 and left their employ in 1774 or 1775. More certain is that by 1776 he was an independent merchant in Fredericksburg.

8. See the family genealogy.

9. *Virginia Herald,* Oct. 18, 1792.

10. Extract from the Virginia Legislative Papers in *Virginia Historical Magazine* 18 (1910), pp. 25–26. It is thought that the sloop *Speedwell* was confiscated by the Spotsylva-nia Committee of Safety in 1776 and its master, Capt. John Lindsay, permitted to keep his wages of £42, which, in turn, he assigned to Robert Patton.

11. John Mercer Patton, entry in Malone, *The Dictionary of American Biography,* pp. 316–17.

12. *Virginia Herald,* Apr. 12, 1792, and Nov. 7, 1793, microfilm records in the Virgini-ana Room, CRRL. Patton frequently advertised these and similar items for sale through his merchant business. Other traces of Robert Patton include that of a founding stockholder and commissioner of the Bank of Richmond in 1792–93; service as a Fredericksburg council-man in 1793; trustee of the local charity school and the Fredericksburg Academy; officer of the Mutual Insurance Fire Company. Patton also sold lottery tickets and did other charitable work, such as organizing a Fourth of July celebration and aiding a local citizen whose home had burned down. The latter was in repayment for aid given him when his own house burned in 1792.

13. *Virginia Herald,* Sept. 3, 1805.

14. Robert Patton's White Plains was a magnificent two-and-one-half-story brick house of some fifteen rooms, with three chimneys and floors and large front columns of marble, surrounded by shade trees, flowers, and shrubs on what is now Princess Anne Street, the original Fredericksburg–Washington highway. (N. M. Deaderick, "White Plains," *Works Progress Administration of Virginia, Historical Inventory* [*of Spotsylvania County homes*], July 13, 1937. Copy in CRRL.)

15. *Virginia Herald,* passim; Carrol H. Quenzel, *The History and Background of St. George's Episcopal Church, Fredericksburg, Virginia* (Richmond, 1951), p. 92; William W. Hening, *Laws of Virginia,* vol. 13 (Philadelphia, 1823), p. 599; and Edward Alvey Jr., *History of the Presbyterian Church of Fredericksburg, Virginia, 1808–1976* (Fredericksburg, Va., 1976), pp. 8–11. Between 1806 and 1808 the Episcopalian church had no rector and those serving between 1808 and 1812 failed to stem the rush of local citizens to the new Presbyterian church, whose minister was a dynamic speaker. "As soon as notice that Mr. Wilson [the minister] would preach, the room was thronged with Episcopalians." (All references in CRRL.)

16. Pension records of the Revolutionary War in CRRL, and RHP, p. 13.

17. *A Scottish Firm in Virginia, 1767–1777,* pp. 233–34.

18. Deaderick, "White Plains," and Dora M. Jett, *Minor Sketches of Major Folk and Where They Sleep* (Richmond, 1928), p. 90. According to Robert H. Patton, his marriage to Anne Mercer was an unhappy one, and a journal found after his death revealed that Robert's true love was named Nelly Davenport, who died of consumption in 1790 in Antigua, where he had taken her in hopes that the weather might help her regain her health. (See RHP, chap. 2, passim.)

19. The myth that the name "Patton" was an alias was based on the false assumption that he had fled Scotland after the Revolution. Inasmuch as Robert Patton was not born

until five years *after* the Revolution, there is nothing to suggest that he was anything more than one of the many Scots who emigrated to North America.

20. Deaderick, "White Plains."

21. See the family genealogy.

22. The Virginia Assembly had passed a resolution condemning the president's action. Patton was a strong supporter of Jackson, and when the governor sent him a copy of the resolution in an obvious attempt to change Rep. Patton's vote, the young congressman refused to budge. (Malone, *The Dictionary of American Biography,* pp. 316–17.)

23. Ibid. John Patton was also an acknowledged leader of the Richmond bar and, although never a practicing physician, he was elected president of the Board of Visitors of the Medical College of Virginia from 1854 until his death in 1858.

24. It has long been believed that the number of Patton sons sired by John Mercer Patton was either seven or eight. In fact, there were nine brothers and three daughters. See the family genealogy.

25. Formerly a state arsenal created in 1818, the Virginia Military Institute was chartered by the Commonwealth of Virginia on July 11, 1839, and generally modeled on West Point. Its object was to provide a three-year course of instruction to young Virginia men in order to prepare them "for the varied work of civil life. . . . The military feature, though essential to its discipline, is not primary in its scheme of education." Its creators saw the benefits of a system that was military in nature but also provided "for their personal benefit, and for the advantage of Virginia, an education which . . . fulfilled the practical pursuits of life." (Colonel William Couper, *One Hundred Years at V.M.I.,* vol. 1, [Richmond, 1939], passim.)

26. Academic reports for the years 1850–1852, G. S. Patton I file, VMI-A. The course of instruction at VMI was then of three years' duration and in his first year young Patton was rated ninth in his class, and in his second, 1851, fifteenth.

27. It seems clear from the many letters in the VMI archives that Francis Smith had formed a close attachment to both George and Waller Tazewell Patton and did all he could to bolster their careers after graduation. (Cadet files, VMI-A.)

28. See the family genealogy.

29. MB-*PP* 1, p. 24.

30. Terry Lowry, *22d Virginia Infantry* (Lynchburg, Va., 1988), p. 2.

31. Charles D. Walker, *Biographical Sketches of the Graduates and Elèves of the Virginia Military Institute* (Philadelphia, 1875), p. 424. An earlier photograph of young Patton as a VMI cadet is not flattering.

32. Steve Cohen, *Kanawha County Images—A Bicentennial History, 1788–1988,* p. 83. Patton also designed the uniforms worn by the Kanawha Riflemen. His grandson would follow in his footsteps in the two world wars of the twentieth century when he devised uniforms for his tank corps. During their early years the Riflemen functioned mainly as a ceremonial unit at social affairs and parades in their flashy uniforms, to, as one member described it, proudly guard "the fried chicken and lemonade." (Lowry, *22d Virginia Infantry,* p. 3.)

33. A meeting of the Kanawha Riflemen on April 19, 1861, produced an official statement: "Whereas an unjust and unnecessary war has been forced on the country by the administration at Washington, in which our state may be required to take part; we the Kanawha Riflemen, hereby declare it to be our fixed purpose never to use our arms against the State of Virginia, or any other Southern State. . . we hold ourselves ready to respond to every call that may be made on us to defend our State." (Cohen, *Kanawha County Images,* p. 83.)

34. "A Child's Memory of the Civil War," memoir dictated to GSP by his father in Jan. 1927, Box 5, PP-LC, hereafter cited as "A Child's Memory."

35. Ibid. Like many Southern families, some of the Pattons had slaves. Patton's father notes that at Spring Farm, his grandparents' estate, there were "lines of slave cabins behind the detached kitchen." Patton's misspelled words when he transcribed his father's oral reminiscences were the result of dyslexia, about which more will be said in later chapters.

36. "My Father as I Knew Him."

37. RHP, pp. 34 and 71; and Robert's obituary in the *Culpeper Observer,* May 19, 1876, microfilm in Culpeper (Va.) Public Library.

38. Halsey, "Ancestral Gray Cloud Over Patton."

39. James McPherson, *Battle Cry of Freedom* (New York, 1988), p. 662.

40. Quoted in David F. Riggs, *7th Virginia Infantry,* (Lynchburg, Va., 1982), p. 24. Waller Tazewell Patton was but one of the seven thousand wounded Lee was obliged to leave behind when his battered army abandoned Gettysburg for the sanctuary of Virginia.

41. "My Father as I Knew Him." Maj. Henry T. Lee was a Union lieutenant and an aide-de-camp to Maj. Gen. Abner Doubleday, commander of a division defending Union positions to the left of Cemetery Ridge during the third day of the Battle of Gettysburg. After the war Lee settled in Southern California, where he became a friend of George S. Patton II, the father of George S. Patton Jr. Lee related that he witnessed the mortal wounding of Waller Tazewell Patton.

42. Profile of Col. Waller Tazewell Patton, in Walker, *Biographical Sketches,* pp. 425–26.

43. Ibid.

44. "A Child's Memory," and "My Father as I Knew Him." The treatment employed by a young medical student to save Patton's arm was bizarre. "He fixed a pulley in the ceiling and passing a cord through it, fastened one end to the wrist and put a brick on the other to hold the arm vertical. He then arranged a tin bucket with a small hole so that water from the pump would drip constantly on it. The arm was saved." Colonel Patton's young son saw a good deal of the horror of war. In 1862 he contracted scarlet fever and nearly died. When the family lived in Richmond, the family home was directly across the street from a field hospital. "From the window of my room I could see them bringing in the wounded soldiers. Most of the operations must have been amputations, for I recall seeing cart loads of arms and legs being taken away." One day he encountered the young slave of a Confederate general carrying the officer's amputated leg wrapped in a blanket.

45. Patton's dislike of the Union was made clear in a letter written in early 1862, which he closed by noting how much "he hated Yanks." (Lowry, *22d Virginia Infantry,* p. 25.)

46. "My Father as I Knew Him." In a slightly different version of this incident in the other memoir, "A Child's Memory," the story was apparently told to George S. Patton II after the war by Brig. Gen. Henry Heth: "The young colonel doubled over from the impact of a minie bullet. 'I am hit in the belly,' he cried out. 'It is all over.' Nearby . . . Heth leaned over the fallen man and probed the wound with his finger. 'I think not, George,' he said. Heth held up a small bent object, a $20 gold piece from Patton's pocket vest. The bullet spent itself against the coin."

Colonel Patton's life had indirectly been saved by his wife. Before the battle her brother sent her some gold coins, and she had insisted her husband carry some of them at all times on a money belt she had made for him in the event he was ever again captured.

47. Gen. William Woods Averell, USMA, 1855, commanded a Union cavalry division that operated in western Virginia from 1862 to 1864. Averell and Patton had been well acquainted before the war, and in response to his friend's request, Averell sent provisions and an armed detail to protect the house where the Pattons were staying. When he came to call that evening, he remarked on the bad cold he had caught. Their hostess said she had a fine remedy that would cure him. Young Patton thought that she would surely have poisoned the concoction and was disappointed when Averell failed to drop dead on the spot. "The sergeant on guard at the house once stood me on the table and asked me if I was a Rebel. I swallowed my Adam's apple and weakly said that I was. Whereupon he put his cap on my head and said, 'Now you are a Yankee.' When I told this story to my mother my head was vigorously scrubbed." ("A Child's Memory.")

48. "A Child's Memory."

49. Halsey, "Ancestral Gray Cloud Over Patton." Cadet John R. Patton was left behind in Lexington to guard the VMI buildings. (Walker, *Biographical Sketches,* p. 571.)

50. "A Child's Memory."

51. Lt. Gen. Jubal A. Early, *A Memoir of the Last Year of the War for Independence in the Confederate States of America* (Lynchburg, Va., 1867), pp. 90–91. Early incorrectly cited Patton's middle initial as "W" instead of "S."

52. Ibid., p. 91. Patton's death also drew the attention of the Union commander, Sheridan, who noted in a letter shortly after the battle that "Col. George S. Patton was mortally wounded at the battle of Opequon September 19, 1864 and died . . . in Winchester shortly afterwards." (Gen. Philip Sheridan to Maj. Gen. E. O. C. Ord, 27 Oct. 1864, Box 37, PP-LC. Sheridan always referred to the events of Sept. 19, 1864, as "The battle of Opequon," so-named for a creek east of Winchester. Jubal Early scornfully noted in his memoir of the Civil War: "I know no claim it has to that title, unless it be the fact that, after his repulse in the fore part of the day, some of his troops ran back across that stream.")

53. Profile of Colonel George S. Patton, in Walker, *Biographical Sketches,* p. 422.

54. "A Child's Memory." His faithful black batman-slave Peter took Patton's horse, saddle, and saber and, riding only at night, managed to elude Union pickets and foragers and arrived safely at the home of John Mercer Patton Jr. In recognition of this act of faith and devotion, the Pattons supported Peter for the rest of his life. Col. Patton was apparently wounded in the hip by a shell fragment and at first it appeared as if he would recover. ("My Father as I Knew Him.")

55. "My Father as I Knew Him." After the Civil War it was illegal to wear a Confederate uniform, therefore the reburial took place at night in secret.

56. Walker, *Biographical Sketches,* pp. 423–24.

57. "A Child's Memory."

58. James I. Robertson Jr. , *Civil War Virginia* (Charolettesville, 1994), pp. 174–75.

59. RET-M.

Chapter 2   Don Benito Wilson

1. Her son, George, remembers how his mother expected to die and at the height of her illness a great calm came over her and with it a removal of all fear of death or dying. ("My Father as I Knew Him.")

2. Midge Sherwood, *Days of Vintage, Years of Vision,* vol. 2 (San Marino, 1987), p. 152.

3. Ibid., p. 153.

4. Born in Madison County, Virginia, Andrew Glassell and his family moved to Alabama when he was a child. In 1854 he moved to California. Although Glassell did not participate in the Civil War, for its duration he retired from the practicing law in lieu of signing a mandatory oath to the Union. In 1878 Glassell became the first president of the Los Angeles County Bar Association. Until 1889 Los Angeles County included what is now Orange County; thus Glassell also had the distinction of being the first president of the Orange County bar. Glassell is also credited with founding the town of Orange in what was then called the Santa Ana Valley. (See RET-M, and letter, C. E. Parker [chairman of the Orange County Bicentennial Committee] to Mr. Jere M. H. Willis Jr., Jan. 29, 1976, copy in CRRL.)

5. Sherwood, *Days of Vintage,* vol. 2, p. 156.

6. Ibid., p. 157.

7. Ibid., p. 159.

8. Ibid., chap. 9, passim. In an attempt to overcome the effects of her illness Susan Glassell Patton drank whiskey three times a day.

9. Patton tells of being forbidden to drink water because of typhoid. One day, while alone in bed feverish and thirsty, he spied a pitcher of water on a washstand. So weak that he could only crawl, he went to get a drink but spilled the water on himself. When he awoke his fever was gone. Patton's recovery was undoubtedly aided by his uncle's cow, which produced an ample source of milk. (See "My Father as I Knew Him," and Sherwood, *Days of Vintage,* vol. 2, p. 154.)

10. "My Father as I Knew Him."

11. Sherwood, *Days of Vintage,* vol. 2, p. 168.

12. Quoted in ibid., p. 160. In previous biographies of Patton it has been assumed that the change of name was to honor Patton's stepfather, Col. George Hugh Smith, and no doubt this eventually served as a double honor for both his late father and Smith, the man whom he came to revere and adore. However, as Sherwood's account makes clear, the decision to change his middle name occurred well before Susan Glassell Patton and George Hugh Smith were married.

13. Record of George Hugh Smith, VMI archives. Smith's maternal grandparents were Isaac Hite Williams and Lucy Coleman Slaughter, whose sister married John Mercer Patton, the grandparents of Col. George S. Patton.

14. Many of Smith's reminiscences of his experiences in the Civil War, particularly the Battle of Newmarket in May 1864, are in Box 37, PP-LC.

15. Author interview of Ruth Ellen Totten, Aug. 26, 1991.

16. RET-M.

17. Ibid.

18. Sherwood, *Days of Vintage,* vol. 2, p. 166.

19. "My Father as I Knew Him."

20. Ibid.

21. Col. R. A. Marr to GSP II, Jan. 26, 1904, Box 5, PP-LC.

22. He was not alone in the impoverished postwar years of the 1870s. The VMI cadet companies were obliged to attend in rotation one of four churches in Lexington each Sunday. At one the minister complained that none of the cadets ever put money in the collection plate. He was told that it was impossible inasmuch as their uniforms contained no pockets in which to carry coins. The minister foolishly replied that if they were righteously minded they would find a way. The next time the company attended this church every cadet brought a copper coin in his mouth. As the collection plate was passed, each solemnly spat his coin into it, thus ending any further attempts to collect donations from the Corps of Cadets. (See "My Father as I Knew Him.")

23. Record of George Smith Patton II, VMI, and MB, p. 24.

24. Sherwood, *Days of Vintage,* vol. 2, p. 169.

25. Quoted in ibid., p. 182. Another paper, the *Arizona Citizen,* trumpeted: "The rash young orator in Los Angeles who proposed making the streets of Washington run deep with blood, if necessary, to inaugurate Grover Cleveland as President, has been overtaken by a terrible retribution. He has got married."

26. *History of Los Angeles County, California* (1880; reprint, Berkeley, 1959), p. 36.

27. The expedition to California followed the Old Spanish Trail which ran north of the Grand Canyon, much of the time in the company of a regular caravan of New Mexico traders. (John Walton Caughey, "Don Benito Wilson: An Average Southern Californian," *Huntington Library Quarterly* [Apr. 1939], and *History of Los Angeles County, California,* p. 36).

28. Robert Glass Cleland, *From Wilderness to Empire: A History of California* (New York, 1969), p. 74.

29. Caughey, "Don Benito Wilson."

30. Harold D. Carew, *History of Pasadena and the San Gabriel Valley,* vol. 1 (Pasadena, Calif. 1930), p. 249, and Sherwood, *Days of Vintage,* vol. 1 (San Marino, 1982), chap. 3, passim. Wilson's marriage to Yorba also produced a son, John, who was murdered in his youth.

31. *History of Los Angeles County, California,* p. 37.

32. FA, pp. 12–13.

33. Ibid., and *History of Los Angeles County, California,* p. 37.

34. Caughey, "Don Benito Wilson."

35. FA, p. 13.

36. Sherwood, *Days of Vintage,* vol. 1, p. 70, and Cleland, *From Wilderness to Empire,* chap. 11, passim.

37. *History of Los Angeles County, California,* p. 37.

38. MB, p. 26. Wilson's landholdings were immense and in addition to the Jurupa Ranch, included part of the great Rancho Santa Ana and numerous other rancheros in the Los Angeles area. However, of his many landholdings, it was Lake Vineyard that Don Benito came to prize as the true Wilson homestead. (A detailed account of the Southern California land boom of which Wilson was so much a part is in Glenn S. Demke, *The Boom of the Eighties in Southern California* [San Marino, 1944].)

39. Benjamin Cummings Truman, publisher of the *Los Angeles Star,* article of Apr. 1874, quoted in Sherwood, vol. 1, p. 332. The "jackrabbit" quote on p. 310 refers to nearby land that a friend and business associate of Wilson's, Dr. John Strother Griffin, sold to a group of colonists from Indiana. As Sherwood notes: "The Yankees had the last laugh. The 2,576.35 acres they chose for their new community . . . [was] later named Pasadena."

40. Sherwood, *Days of Vintage,* vol. 2, p. 125. A California journalist has described the combined Wilson and Shorb estates as "a constant succession of ever varying enchanting scenes . . . equal in picturesque and irregular beauty to any through which I have ever wandered, and I am tolerably familiar with the scenery of California, Arizona and Nevada, to say nothing of fifteen or twenty other states." (Sherwood, *Days of Vintage,* vol. 1, p. 333.) According to Sherwood, the San Gabriel Wine Company could crush 250 tons per day in a structure that was built of more than 1,200,000 bricks. In all, the winery consisted of a large complex of modern buildings and equipment linked to the outside world by a spur line to the Southern Pacific Railroad. (Sherwood, vol. 2, p. 125.)

41. RET-M.

42. Ibid.

43. Sherwood, *Days of Vintage,* vol. 2, p. 211.

Chapter 3 "The Boy"

The principal sources are Ruth Ellen Totten's unpublished memoir and "My Father as I Knew Him." Essays written at the Classical School for Boys are in Box 5, PP-LC. All others are cited below.

1. MB, p. 16, and RHP, p. 82.

2. MB, p. 27.

3. The skunks were not only tolerated but encouraged in the belief that they cut down on the rats and mice. Patton's father showed his son how to catch ground squirrels in box traps. They were considered pests and for each squirrel young George trapped he earned ten cents. Unfortunately one of the squirrels turned out to be a skunk, which he "did not find out about until too late." (Box 5, PP-LC).

4. Essay written at the Classical School for Boys, Feb. 6, 1902.

5. RHP, p. 81.

6. RET-M.

7. FA, pp. 18–19.

8. RHP, p. 88.

9. Patton's West Point letters mention of the necessity that someone always accompany Aunt Nannie on her travels. See Boxes 5 and 6, PP-LC.

10. RHP, p. 95.

11. Ibid., p. 106.

12. MB, p. 29.

13. RHP, pp. 88–89.

14. Ibid., p. 89.

15. Ibid., p. 90.

16. FA, p. 201.

17. MB, p. 30, and McPherson, *The Battle Cry of Freedom,* p. 737.

18. Author interview of Ruth Ellen Totten, Aug. 26, 1991.

19. MB, pp. 24–25.

20. RHP, p. 96. As Robert Patton observes, no Patton was more unsure of himself than Papa Patton, whose inability to follow in his own father's military footsteps produced a

lifelong sense of frustration and inadequacy that was overcompensated for by his exaltation of the Pattons at the expense of the Wilsons.

21. Ibid., p. 97; and MB, pp. 24–25.

22. *Los Angeles Star*, Mar. 14, 1878, quoted in Sherwood, *Days of Vintage*, vol. 1, p. 460. During Patton's lifetime a number of landmarks in Southern California bore Wilson's name, including Mount Wilson and later, the famous Mount Wilson Observatory, named not for the president but for the California pioneer.

23. RHP, pp. 93–94.

24. FA, p. 18.

25. Ibid., p. 19.

26. Harold N. Levinson, M.D., *Smart But Feeling Dumb* (New York, 1984), p. v.

27. Ibid., p. 1. Based on years of research and clinical studies of some ten thousand dyslexics, Dr. Levinson has concluded that the source of the disorder is due to a dysfunction of the inner ear and can be present at birth or acquired in childhood because of accident, allergy, or disease. A recent study conducted at Harvard Medical School "suggests that people with dyslexia have an inborn defect in the brain's visual system that is part of the cause of the language and reading disorder that often causes severe learning disabilities." ("Study Finds Visual Defect in Brain May Cause Dyslexia," *Boston Globe*, Sept. 17, 1991.)

28. Ibid., p. 1.

29. Ibid., passim.

30. Ibid., p. 3.

31. Quoted in Steve Dietrich, "The Professional Reading of General George S. Patton, Jr.," *Journal of Military History* (Oct. 1989), and Box 5, PP-LC.

32. FA, p. 4.

33. Box 5, PP-LC, and MB, p. 36.

34. "Notes Before World War I," FM.

35. MB-*PP* 1, p. 36. A sampling of Patton's notes shows that his reading consisted of such works and subjects as fifteen decisive battles of the world, *Myer's Greek History*, social life in Greece and Persia in the *Manual of Ancient History*, and the *History of Greece* (Box 5, PP-LC).

36. One of Patton's assignments was to write about how he spent his weekend. After apologizing for the letter being so "I" oriented, he wrote: "And now I must close this letter, which I fear is very stupid."

37. Letter of Jan. 29, 1902.

38. MB-*PP* 1, p. 42.

39. Letter of Jan. 31, 1902.

Chapter 4   The Belle of Boston

The principal source is Ruth Ellen Totten's unpublished memoir. All others are cited below.

1. RET-M, and MB, p. 35.

2. FA, p. 39. After the terrible fire of Oct. 1871 destroyed most of the city, Ayer made an interest-free loan to the Chicago city fathers in return only for title to one of the downtown business sections after it was rebuilt.

3. In all, Ayer produced seven children, four by Cornelia: Ellen (1859), Jamie (1862), Charles "Chilly" (1865), and Louise (1876).

4. Beatrice never got used to the copious amounts of wine that flowed at Lake Vineyard. Although the ladies apparently rarely indulged other than at dinner, there were several decanters from which to choose placed on the sideboard in the dining room. Mrs. Patton told her that when she was growing up there had always been a bucket of wine and a silver-handled gourd dipper placed in the front hall, from which the men would slake their thirst upon entering the house after a hard day on the estate hunting or riding.

5. FA, p. 40.

6. Ruth Ellen Totten has described Avalon as having "stucco walls and bas reliefs of

cherubs over the French windows. The circular staircase runs up from and around a marble paved hall, lit by a giant skylight. The enormous living room runs the length of the front of the house, with great tall windows framed in wisteria, and there is a musician's balcony over the grand piano. The whole downstairs always smell of lilies and roses, the ghosts of happy summers."

7. MB, p. 38.

8. Letter of Nov. 14, 1902, in RET-M.

9. MB, p. 38, and Box 5, PP-LC. GSP's letter of Jan. 10, 1903, was his first ever to Beatrice.

Chapter 5   A Father's Influence

Unless otherwise cited, all correspondence is from Box 6, PP-LC.

1. GSP to BAP, Jan. 16, 1903.

2. *Official Register of the USMA*, "Information Relative to the Appointment and Admission of Cadets to the United States Military Academy," 1904, Box 45, PP-LC.

3. Bard was born in Pennsylvania in 1841. During the Civil War he was a volunteer Union scout and in 1864 moved to California where he became involved in Republican politics. First elected to the U.S. Senate in Feb. 1900, Bard was defeated for reelection in 1904.

4. Entry for Sen. Thomas R. Bard, *Biographical Directory of the American Congress, 1774–1971* (Washington, D.C., 1971), p. 552, and Sherwood, *Days of Vintage*, vol. 1, p. 39.

5. MB-*PP* 1, chap. 3, passim, and Box 5, PP-LC.

6. Quoted in ibid., p. 55.

7. Frances C. Woodman to GSP II, Feb. 26, 1903. Had it not been for the influence exerted by the son of the Episcopal bishop of Los Angeles, Woodman would not even have considered Mr. Patton's request.

8. MB-*PP* 1, p. 57.

9. Ibid., p. 58.

10. "My Father as I Knew Him."

11. Author interview of Ruth Ellen Totten, Aug. 26, 1991.

12. MB, p. 43. There is nothing to indicate that Patton paid much attention to Aunt Nannie during her protective sojourn in Lexington as the family watchdog over her beloved nephew.

13. "My Father as I Knew Him."

14. Since its founding in 1839, 1,762 of 1,925 graduates, excluding the 109 who had died prior to 1861. (Mann, *They Were Heard From: VMI Alumni in the Civil War*, p. 2.)

15. Semiannual examination report of Cadet G. S. Patton, Feb. 15, 1904, Box 45, PP-LC.

16. "My Father as I Knew Him."

17. William Bancroft Mellor, *General Patton: The Last Cavalier* (New York, 1971), p. 33.

18. Col. William Couper to William Bancroft Mellor, Feb. 13, 1945, Patton file, VMI-A.

19. GSP to BA, Oct. 21, 1909.

20. GSP to his GSP II, Nov. 28, 1903.

21. "Patton Once Scrub 'Tackle' on VMI Team," *Richmond News Leader*, Dec. 19, 1944, and GSP to H. H. Spence, Jan. 31, 1945, Patton file, VMI-A.

22. L. Harvie Strother to GSP II, Jan. 30, 1904. Patton narrowly avoided serious trouble the previous month when one of his friends rigged tin cans over the door of one of the faculty member's quarters. One night Patton knocked on his door just as the officer arrived. "I saluted and stood aside to let him pass but when he opened the door to enter a great number of tin cans, carefully arranged by Dikeman (another cadet) fell with a great crash. The inference is obvious. I was standing in the door [and] he came and surprised me at my work, then I pretended I had been waiting to see him about lessons. Ha! Only he did *not* think this." Which, for George Patton was fortunate, for with such shenanigans

on his record, a favorable recommendation from the superintendent might well have been denied.

23. GSP to GSP II, Dec. 13, 1903.
24. Ibid., Jan. 10, 1904.
25. GSP II to GSP, Jan. 19, 1904.
26. GSP II to Strother, Jan. 20, 1904.
27. Certificate of Hammond P. Howardy, Surgeon, VMI, n.d., but circa, late Jan. 1904.
28. Certificate, Classical School for Boys, Feb. 1, 1904.
29. Prof. Philip Bard to BAP, Aug. 26, 1947, Box 34, PP-LC.
30. GSP II to GSP, both letters dated March 4, 1904.
31. GSP to Bard, March 6, 1904.
32. Extracted from GSP's letters of March 13 and 18, 1904, Box 34, PP-LC.
33. Mellor, *General Patton,* p. 34.
34. War Department to GSP, May 24, 1904.
35. "My Father as I Knew Him."
36. GSP to GSP II, March 13, 1904.

Chapter 6   "The Military School of America"

Patton's correspondence and academic reports are in Boxes 5 and 6, PP–LC.

1. USMA, *Register of Graduates,* p. 242. Jefferson's motives were pragmatic and rooted in the "the possibility that officers trained for military engineering could also build bridges, dams, harbors, and roads for the civilian population. An institution that could produce trained military officers who also possessed their-rare engineering skills satisfied those who wanted a permanent officer corps and those who considered civilian projects paramount." (Simpson, *Officers and Gentlemen,* pp. 6–7.)

2. Ibid., p. 247. e.g., the class of 1818 admitted 105 cadets, but only 22 graduated in 1818.

3. Jeffrey Simpson, *Officers and Gentlemen* (Tarrytown, 1982), p. 25, and USMA, *Register of Graduates,* passim.

4. Simpson, p. 59.

5. Ibid., p. 150. During the era after the Civil War the first blacks were admitted to West Point and the first graduate was Henry O. Flipper, in 1877. Between 1865 and 1915, thirteen black cadets were admitted but only three graduated. One became a full colonel but all were relegated to service with black regiments. As lamentable as this record was, the Naval Academy's was even worse: It failed to admit a single black during the same period. Blacks were shamefully treated at West Point, not by hazing but by being ostracized and, as historian Stephen Ambrose notes, "prejudice against Negroes was neither higher nor lower at West Point and in the army than it was throughout the nation—which meant that it was high."

6. Ibid., pp. 100–101, and Stephen E. Ambrose, *Duty, Honor, Country* (Baltimore, 1966), p. 222.

7. Ambrose, *Duty, Honor, Country,* p. 222.

8. GSP to his GSP II, Aug. 15, 1904. Patton loved parades, but found them oppressive in the hot weather, and loathed the heavy dress hats worn by the Corps of Cadets.

9. RHP, p. 124.

10. GSP to his mother, June 21, and GSP II, July 24, 1904.

11. GSP to GSP II, Aug. 21, 1904. The corporal who came to Patton's rescue that night was Charles Tillman Harris Jr. (1884–1961), who graduated fourteenth in the class of 1907. The most famous cadet of the class of 1907 was Henry H. "Hap" Arnold, who later became a pioneer aviator, the commander in chief of the Army Air Forces in World War II, and a five-star general.

12. GSP to GSP II, July 3, 1904.

13. Ibid.

14. RHP, p. 128.

15. GSP to GSP II, Aug. 15, 1904.

16. Ibid., July 17 and 31, 1904.

17. Unpublished ms. of Col. George S. Pappas, "The Class of 1891 a Hundred Years Later."

18. GSP to BA, July 1904.

19. GSP to GSP II, July 24, 1904.

20. Ibid., July 31, 1904.

21. Ibid., Aug. 21, 1904. Patton also noted that discipline was "equally good in small things such as keeping eyes to the front and talking in ranks . . . there is never a question about obeying orders."

22. Ibid., July 17 and Aug. 21, 1904.

23. GSP to BA, July 10, 1904.

24. GSP to GSP II, July 17, 1904.

25. Ibid., Aug. 28, 1904.

26. Pappas, "The Class of 1891 a Hundred Years Later."

27. As a result of a $6 million congressional appropriation, West Point's antiquated physical plant was completely rebuilt during the decade between 1903 and 1913. (See Simpson, *Officers and Gentlemen,* p. 164.)

28. GSP to GSP II, Sept. 4, 1904.

29. Pappas, "The Class of 1891 a Hundred Years Later."

30. GSP to GSP II, May 1905.

31. Ibid.

32. Both Robert Patton and Martin Blumenson have written of how Patton reinvented his own persona. (See RHP, MB, and MB-*PP* 1, passim.)

33. MB, p. 31.

34. An example occurred in the spring of 1905, when Mr. Patton asked him to attempt to locate a certain species of flower. Patton wrote back apologetically: "I wish that I might have been able to help you find those flowers. It would be great to wander in the woods in the mornings and look for flowers I thought I would try it this morning in the hills back of West Point but all I could find was a little turtle about an inch long. And I felt so poetically sad that I carried the beast back to its pond and let it go." (GSP to GSP II, dated May 1905.)

35. GSP to GSP II, Sept. 18, 1904.

36. Ibid.

37. Ibid, Sept. 11, 1904.

38. See letter, Joseph D. Neikirk to David J. Winton, Dec. 19, 1963, Box 3, FM.

39. RET-M. Mrs. Totten notes: "As far as I can read into the hundreds of letters everyone wrote to everyone else, Aunt Nannie and/or Mrs. Patton would return to California to regain their strength and their personal identity whenever Georgie was on leave and staying with the Ayers."

40. Mellor, *General Patton,* pp. 49–50.

41. GSP to GSP II, Sept. 27, 1904.

42. Ibid, Oct. 5, 1904.

43. Ibid., dated 1904, and written sometime in Nov. or Dec.

44. GSP to BA, Oct. 15, 1904.

45. GSP to GSP II, Oct. 15, 1904.

46. Ibid.

47. Academic and conduct report, Oct. 1904.

48. GSP to his mother, Nov. 8, 1904.

49. GSP to GSP II, Nov. 12, 1904.

50. GSP to BA, Dec. 11, 1904.

51. Academic and conduct report, Nov. 1904

52. GSP to GSP II, dated 1904.

53. RHP, p. 132.

54. GSP to GSP II, Dec. 10, 1904.

55. GSP to BA, Dec. 11, 1904.

56. Pappas, "The Class of 1891 a Hundred Years Later."

57. GSP to GSP II, Nov. 25, 1904.

58. Ibid., Dec. 26, 1904.

59. RHP, p. 117.

60. Ibid., pp. 117–18.

61. GSP to BA, Dec. 18, 1904.

62. Ibid., March 30, 1905, in "Excerpts from Cadet Letters of GSP to Father," Binder 1, PP-USMA.

63. GSP to GSP II, May 14, 1905.

64. Ibid., May 21, 1905.

65. Ibid., March 18, 1905.

66. GSP to BA, March 7, 1905.

67. GSP to GSP II, March 18, 1905. Many of Patton's classmates returned to West Point with monumental hangovers; he came back nauseated from eating too much candy.

68. Ibid., April 19, 1905.

69. Ibid., April 9 and 19, and academic and conduct report, March 1905.

70. Ibid, April 9 and 24, 1905.

71. Ibid., June 3, 1905.

72. GSP II to GSP, June 10, 1905.

73. Several days later Patton received a brief letter from the War Department that read, "Upon the recommendation of the Academic Board Cadet George S. Patton, Jr., Fourth Class, United States Military Academy, is turned back to join the next fourth class."

74. Levinson, *Smart but Feeling Dumb*, p. 3.

Chapter 7   "If at First You Don't Succeed"

Letters and USMA papers and notebooks are from Box 6, PP-LC, or in "Excerpts from Cadet Letters of GSP to Father," PP-USMA.

1. GSP to BA, July 10, 1905.

2. "My Father as I Knew Him."

3. GSP to BA, July 10, 1905.

4. GSP to GSP II, Aug. 28, 1905.

5. Ibid., May 13, 1906. Even when in high spirits, as he obviously was when this letter was written, Patton still ended by noting: "I am quite well and stupid."

6. Ibid.

7. RET-M.

8. The opening date of Patton's small black notebook is Aug. 1905. There were actually two notebooks, a small one begun in 1905 (hereafter referred to as "1905 Notebook"), and a second cadet notebook begun in 1906 (referrred to as "Cadet Notebook"). He inscribed inside the cover of the "1905 Notebook" an offer of a five-dollar reward that "will be paied to any person finding this book if he returns it." In the "Cadet Notebook" Patton later wrote at the top of the opening page: "This is the notebook kept by George S. Patton, Jr. while a cadet at the usma. Many of the ideas are good many bad. It shows however interest in the profession of arms. //s// G.S. Patton, Jr." The "Cadet Notebook" dealt mainly with copies of letters and essays about tactics and the lessons of war that interested Patton, along with some ideas he derived from his reading. The "1905 Notebook" was shorter and the more interesting of the two. It recorded Patton's early poetry, ideas, and thoughts on a variety of subjects from 1905 to 1909.

9. Ibid., Patton's principles were: (1) cut line[s] of communication; (2) cause [the] enemy to form front to flank; (3) operate on internal lines; (4) separate bodies of enemy and fight in detail; and (5) direct attack. This observation also appeared in his notes: "A Saxon can die without a murmer. A French man can die laughing. But only a Norseman can laugh as he kills."

10. GSP to BA, Aug. 9, 1905.

11. GSP to GSP II, Aug. 28, 1905.

12. GSP to BA, Sept. 1, 1905.

13. Ibid., Oct. 23, 1905.

14. MB, p. 54.

15. GSP to GSP II, Nov. 5, 1905. Patton wrote this poem for the West Point yearbook, *Howitzer,* but decided not to submit it for the usual reason that he did not believe it was good enough.

16. Academic and conduct report, Oct. 1905.

17. GSP to BA, Sept. 1, 1905.

18. GSP to his mother, Nov. 11, 1905.

19. GSP to GSP II, Dec. 4, 1905.

20. Ibid, Dec. 24, 1905. Patton remained first in drill regulations, twenty-second in math, and thirtieth in English.

21. Academic and conduct report, Dec. 1905.

22. GSP to BA, dated Jan. 1906.

23. GSP to GSP II, Jan. 6, 1906.

24. GSP to BA, Jan. 21, 1906.

25. GSP to GSP II, March 24, 1906.

26. Patton was topped for first corporal by Cadet Edwin St. John Greble Jr., one of the outstanding members of the Class of 1909, and a cadet against whom he competed through-out his tenure at West Point.

27. GSP to BA, July 9, 1906.

28. Ibid., Aug. 28, 1906. Later Patton wrote: "A lot of cadets have told me that I dropped from 2 to 6 [corporal] because I was too military and the cadet officers thought I was sort of reflecting on them. However that may be, I am certain that I am the best officer of my class and that there is not another man who can control it when marching by itself except me. Besides, there is many a change between now and first class June so I have not given up being adjutant yet, though at first I felt pretty bad."

29. RHP, pp. 128–29.

30. "1905 Notebook."

31. Entries in "1905 Notebook," late Aug. 1906.

32. GSP to GSP II, Nov. 11, 1906.

33. Ibid., Sept. 23, 1906.

34. Ibid., Oct. 7, 1906.

35. Ibid., March 17, 1907,

36. "My Father as I Knew Him."

37. Entry in "1905 Notebook," Nov. 27, 1907.

38. Ibid. Entries are from early 1909.

39. GSP to GSP II, Feb. 5, 1908. To Beatrice he lamented that he lacked the ability to excel in anything, including peeling potatoes. "It is awful to see other people do not only some things but all things better than you do."

40. A note in Patton's handwriting indicates that these words were first written in 1903 and later copied into his "1905 Notebook."

41. GSP to BA, March 23, 1908.

42. Ibid., dated March 1908.

43. BA, quoted in RET-M.

44. MB, p. 144.

45. Ibid., p. 55.

46. GSP to BA, dated Feb. 1908.

47. MB-*PP* 1, pp. 140–41.

Chapter 8 Love and Marriage

The primary sources are Ruth Ellen Totten's memoir and Box 6, PP-LC.

1. GSP to his mother, Jan. 4, 1909, quoted in RET-M. Beatrice also wrote to Ruth Patton that "we love each other very much indeed and since we have found our tongues to

tell one another, you are next in our hearts. Papa and Mama and Aunt Nannie are the only others."

2. GSP to BA, Jan. 6, 1909.

3. GSP to GSP II, Jan. 10, 1909.

4. RHP, p. 134.

5. GSP to Frederick Ayer, Jan. 3, 1909.

6. Ayer to GSP, Jan. 10, 1909.

7. GSP to Frederick Ayer, Jan. 18, 1909.

8. GSP to BA, Jan. 17, 1909.

9. Ibid., Jan. 27, 1909. In his notebook Patton quoted a Dr. Canfield: "A passion for money may bring peace, but cannot bring honor."

10. GSP to GSP II, Jan. 31, 1909.

11. RHP, pp. 136–37.

12. GSP to BA, March 26, 1909.

13. Ibid., March 29, 1909.

14. GSP to GSP II, April 5, 1909.

15. GSP to BA April 6, 1909. About this time Patton began appending to his closing small circles with plus marks inside, which were his version of an "I love you" heart.

16. MB-*PP* 1, p. 147.

17. 1909 USMA *Howitzer*.

18. Ibid.

19. Colonel Roger H. Nye, USA-Ret., "The Patton Library Comes to West Point," in *Friends of the West Point Library Newsletter* 12 (Mar. 1988), and lecture, "The Professional Reading of General George S. Patton, Jr.," April 1988. Lt. Col. G. J. Fieberger, a USMA professor of engineering, wrote *Elements of Strategy*, the standard cadet text of Patton's period.

20. Thomas J. Fleming, *West Point* (New York, 1969), p. 288. Patton had in fact written his own rules for "quilling" in his notebook, called "Essentials of Quill." Designed for himself as rules he attempted to follow, they later became tenets by which Patton enforced his own standard of discipline:

1. Start the day you enter. 2. Do every thing possible to attract attention. 3. Brace hard and at all times. The less you are spoken to the more you should brace. 4. Always be very neat and when you get any new clothes let every one know it. 5. Do with all the snap and power you possess whatever you do. 6. When ordered to do a thing carry out the spirit. 7. Brace through publication of orders and when 'at ease' and never stop quilling.

21. GSP to BA, Sept. 1, 1908.

22. MB, p. 58.

23. MB-*PP* 1, p. 176.

24. 1909 USMA *Howitzer*. Patton's entry in the Furlough Book for the Class of 1909 read: "Quill Georgie"—"To get back, so as to be near that dear skin book," and "He stands erect, right martial is his air, His form and movements."

25. GSP to GSP II, Nov. 1908.

26. George S. Patton IV, quoted in FA, p. 54.

27. GSP to BA, June 3, 1909.

28. Patton's final class standing in his last academic report were: Civil and Military Engineering—37; Law—80; Ordnance and Science of Gunnery—62; Drill Regulations—5; Practical Military Engineering—10; Conduct, First Class—16. He had accumulated 42 demerits for the year 1908/09. Dickinson had been appointed Secretary of War three months previously by President William Howard Taft. (*Official Register of the Officers and Cadets of the USMA*, June 1909, USMA-SC, and MB-*PP* 1, p. 175.)

29. *USMA Register of Graduates*.

30. "My Father as I Knew Him."

31. In "My Father as I Knew Him," Patton described it as a stopwatch repeater that was marked down from $600 to $350 because it was thicker than the current style. His Aunt Nannie bought him a chain for it, and Beatrice a locket.

32. GSP to Ellen Banning Ayer, Aug. 1909.

33. FA, pp. 9–10. Ayer records this event as 1910, however, inasmuch as George and Beatrice were married in May 1910, it probably occurred in 1909, the year Patton graduated from West Point.

34. BA to Aunt Nannie, Sept. 11, 1909.

35. GSP to GSP II, Sept. 23, 1909.

36. MB-*PP* 1, p. 185.

37. Russell F. Weigley, *History of the United States Army* (New York, 1967), p. 561. For 1909 the figure was $192,487,000. By comparison, the cost of the Civil War between 1862 and 1865 was $2.7 billion.

38. GSP to BAP, Sept. 13, 1910.

39. Weigley, *History of the United States Army,* passim.

40. Ibid., p. 328.

41. Ibid., p. 334. About one-quarter of the frontier forts were abandoned in the late 1880s and early 1890s, but too many remained, some of which were virtually condemned by the surgeon general in 1897. (See Edward M. Coffman, *The Old Army* [New York 1986], p. 346.)

42. GSP to BA, Sept. 13, 1909.

43. Ibid., Sept. 25, 1909.

44. Ibid., Oct. 3, 1909.

45. MB-*PP* 1, p. 190.

46. GSP to BA, Dec. 13, 1909.

47. RHP, p. 126.

48. GSP to BA, Oct. 1, 1909.

49. Lucian K. Truscott Jr., *The Twilight of the U.S. Cavalry: Life in the Old Army, 1917–1942* (Lawrence, Kans.,1989), p. 4. Although Truscott did not enter the army until 1917, the conditions he describes were virtually identical to those of the period 1909–1917 experienced by Patton.

50. GSP to GSP II, Jan. 17, 1910.

51. GSP to BAP, Jan. 23, 1910.

52. Frederick Ayer to GSP, Mar. 7, 1910.

53. Ibid.

54. GSP to GSP II, Apr. 19, 1910.

55. GSP to BA, Apr. 25, 1910.

56. GSP to Ellen Banning Ayer, Apr. 20, 1910.

57. GSP II to BA, Apr. 25, 1910.

58. GSP to BA, May 12, 1910.

59. Ibid., May 16, 1910.

60. Ibid., May 22, 1910.

61. RET-M.

62. The bridesmaids were Patton's sister, Nita, Kay Ayer, Rosalind Wood and Helen Longyear. The ushers were Patton's cousin, Lt. George Patton Brown, U.S. Navy, and four West Point classmates (1909): Lts. James A. Brice, Francis G. Delano, Philip S. Gage, and Stanley S. Rumbough. (MB-*PP* 1, pp. 206–7, and *Register of Graduates*.)

63. RET-M, and MB-*PP* 1, pp. 206–7.

64. Quoted from an unnamed newspaper account reproduced in RET-M.

65. Ibid., and MB-*PP* 1, pp. 206–7.

66. RET-M.

67. Patton apparently intended to keep a brief record of their travels in a small black notebook, titled: "My Trip Abroad." However, after only three days the entries abruptly ended. Other than brief descriptions of places they visited in Cornwall, very little is revealed about their trip or honeymoon. One of Patton's first books was Carl von Clausewitz's famous *On War*. (See MB, p. 68, and Box 7, PP-LC.)

Chapter 9   "And Baby Makes Three"

Primary source for the Fort Sheridan period is Ruth Ellen Totten's memoir. Correspondence is in Box 7, PP-LC.

1. Beatrice's parents knew the "Wild West" quite well; Ellie had grown up in St. Paul, and Frederick Ayer had traveled extensively selling his patent medicines.

2. BAP to Aunt Nannie, Aug. 1, 1910.

3. MB-*PP* 1, p. 209.

4. Ibid., p. 210. Capt. Marshall's first efficiency report rated Lt. Patton "a young officer of especial promise" and "the most enthusiastic soldier of my acquaintance."

5. The Patton family and Aunt Nannie took rooms in Chicago but spent most of their visit at Fort Sheridan.

6. GSP to Aunt Nannie, Mar. 12, 1911.

7. BAP to ibid., July 11, 1911.

8. GSP to ibid., July 2, 1911.

9. Quotes are from Lafcadio Hearn in *Lafcadio Hearn: Life and Letters,* vol. 2, and Elizabeth I. Adamson, in *The Treasury of Modern Humor.* Both quotes appear in Edward F. Murphy, *The Macmillan Treasury of Relevant Quotations* (New York, 1978), pp. 70–71.

10. GSP to BAP, March 29, 1918, PP-USMA.

11. GSP to Aunt Nannie, Mar. 12, 1911.

12. Ibid., July 22, 1911.

13. See the "Small Accounts Book" in Box 7, PP-LC.

14. MB-*PP* 1, pp. 212–13. The type of auto the Pattons purchased is not known, but after attending an auto show in Chicago Patton regretted that he had not spent more for a Stevens-Dourly, which cost the princely sum of $1,750, telling his father that it was probably a better value.

15. GSP to GSP II, Apr. 23, 1911.

16. Ibid., Apr. 12, 1911.

17. MB-*PP* 1, p. 216.

18. RET-M. The new Lake Vineyard was "the most perfect house ever built. Each room had its own fireplace with hand-carved mantelpieces. . . . Aunt Nannie's large bedroom was next to George's and one of the doors from her bathroom opened into his room. "Someone, probably Bama [Patton's mother], had tactfully put a large immovable arm chair against it-on Georgie's side of the door."

19. Capt. F. C. Marshall to GSP, June 26, 1911.

20. GSP to BAP, July 2, 1911.

21. GSP to Aunt Nannie, Sept. 18, 1911.

22. The source is unknown, but suffices to note that Patton regarded it as a positive omen of the merits of using one's influence.

Chapter 10   "A Young Man on the Make"

Unless otherwise cited, the source for correspondence is Box 7, PP-LC.

1. William Gardner Bell, *Quarters One* (Washington, D.C.: 1988), p. 10.

2. Ibid., p. 11.

3. MB-*PP* 1, pp. 223–24.

4. Ibid., p. 224.

5. Ibid., p. 226.

6. Box 7, PP-LC.

7. RET-M.

8. Mellor, *General Patton,* p. 56.

9. MB-*PP* 1, p. 226.

10. Ibid.

11. More than one hundred nations now vie for the prized gold, silver, and bronze medals and often for the lucrative financial endorsements offered many of the medal winners in high-profile events.

12. The original Pentathlon had consisted of a javelin and discus throw, a mile run, a standing broad jump, and some form of wrestling, all of which were designed to demonstrate the abilities of the perfect Greek soldier. While Patton was the only American officer represented in the Pentathlon, four other U.S. Army officers were teammates who competed in the jumping and riding competitions. They included Patton's friend Guy V. Henry Jr., and Ben Lear, later a lieutenant general. (GSP, "Report on the Olympic Games [to the adjutant general of the army]," Sept. 19, 1912, Box 7, PP-LC.)

13. Ibid.

14. RET-M.

15. FA, p. 61.

16. GSP, "Report on the Olympic Games"; RET-M.; MB-*PP* 1, p. 229, and *Mellor, General Patton,* p. 60.

17. "My Father as I Knew Him."

18. Sixty-eight men were entered in the Modern Pentathlon of which only forty-two actually showed up to compete. (MB-*PP* 1, p. 229.)

19. RET-M.

20. After two practice sighting shots, each competitor fired 20 rounds for record in groups of five each. Patton scored a total of 169 out of a possible 200. (GSP, "Report on the Olympic Games.")

21. MB-*PP* 1, p. 230. Patton described the fencing competition as "a great strain on all participants."

22. Mellor, *General Patton,* p. 61, and Louis J. Gulliver, "Gen. Patton Olympic Athlete," *Army and Navy Register,* Sept. 16, 1944.

23. Simply defined, a steeplechase is a race in which horse and rider must jump a series of obstacles, usually in the form of fences and ditches. The Stockholm course consisted of twenty-five jumps and some fifty other impediments, in the form of single and double drainage ditches and fallen logs, over rock-strewn hills and forests north of the city. The ditches were from three to four feet deep and about two feet wide and were covered with grass, thus rendering them almost invisible. The riders were permitted to walk the course three days before the event and had to memorize the layout. Horse and rider only saw the course for the first time together during the actual event, a practice that continues to this day. One Dane broke most of his ribs in a fall and seven others fell at one or more of the double ditches, which were situated a mere fourteen feet apart. (*The Complete Book of Riding* [New York, 1989], passim; GSP, "Report on the Olympic Games," and MB-*PP* 1, p. 230.)

24. MB-*PP* 1, p. 231; Mellor, *General Patton,* p. 62; and undated paper by Patton that described each of the Swedish participants titled "Swedish modern pentatlonmen [*sic*] 1912."

25. "My Father as I Knew Him."

26. Gulliver, "Gen. Patton, Olympic Athlete."

27. Quoted in *The Twentieth Century,* vol. 1, *The Progressive Era and the First World War (1900–1919)* (New York, 1992), p. 110.

28. The modern cynicism that blurs any distinction between amateur and professional athletes did not exist in 1912.

29. GSP, "Report on the Olympic Games." Patton was also ahead of his time when he recommended that elimination contests be held at least nine months prior to future games, "with a view to selecting those best qualified and that the final choice of a team should be the result of a competition contest."

30. MB-*PP* 1, p. 232.

31. FA, p. 59, and MB-*PP* 2, p. 811.

32. Los Angeles *Times,* Wed., Aug. 7, 1912, Box 66, PP-LC. The absence of international telephone links meant that news from Stockholm took nearly a month to reach Los Angeles. According to the *Times,* Patton fell once during the steeplechase, one of several apparent errors in their story, the other being that Patton had taken honors in shooting.

33. MB-*PP* 1, p. 233. As Blumenson notes, the rank of adjutant is peculiar to the French army and is "somewhere between the grades of warrant officer and sergeant major."

34. GSP to BAP, Sept. 7, 1912.

35. Ibid., Sept. 2, 1912.

36. Ibid., April 8, 1915, Binder 7, PP-USMA.

37. GSP to GSP II, Sept. 22, 1912.

38. GSP to Frederick Ayer, Sept. 14, 1912.

39. BAP to Aunt Nannie, Sept. 12, 1912.

40. Weigley, *History of the United States Army*, chaps. 14 and 15, passim.

41. G.S.P. Report on the Olympic Games, and MB, p. 74.

42. MB-*PP* 1, p. 248.

43. BAP to Aunt Nannie, March 20, 1913.

44. Ibid., March 1913, Box 19, PP-LC.

45. GSP to BAP, June 10, 1913, ibid.

46. FA, p. 62.

47. Ibid., p. 63.

48. Ibid., p. 62. Rearing is dangerous, and the use of water balloons to break the habit is also still commonly practiced today. The objective of sitting on a downed horse's head is to teach the animal that the rider is the master, and is to be obeyed or, translated into human terms: "Thou shalt not rear."

49. Ibid., p. 23.

50. MB-*PP* 1, chap. 12, passim.

51. U.S. Bureau of Labor Statistics, Consumer Price Index (CPI). The CPI for 1913 was 9.9 vs. 136.2 for 1991; that is, $1.00 in 1913 = $13.75 in 1991. Barbara Conathan, research librarian, Falmouth (Mass.) Public Library, tracked down the C.P.I. information from the U.S. Bureau of Labor Statistics. The $300.00 cost is noted in MB-*PP* 1, p. 260.

52. MB-*PP* 1, p. 260.

53. RET-M.

54. Mellor, *General Patton*, p. 64.

55. Lt. Hassler to GSP, quoted in MB-*PP* 1, p. 262.

56. Quoted in FA, p. 8.

57. Ibid.

58. RET-M.

59. MB-*PP* 1, p. 262.

Chapter 11 ". . . A Home Where the Buffalo Roam"

Unless cited otherwise, Patton correspondence is in Box 7, PP-LC.

1. Truscott, *Twilight of the U.S. Cavalry,* p. 76.

2. Secretary of the Treasury Albert Gallatin, quoted in Coffman, *The Old Army,* p. 38.

3. Truscott, *Twilight of the U.S. Cavalry,* p. xx. Although he was by then a retired four-star general of the Infantry, Lucian Truscott proudly signed the preface to his memoirs, "L. K. Truscott, Jr., Cavalryman."

4. Ibid., p. 75.

5. GSP to GSP II, Oct. 16, 1913.

6. GSP to BAP, Sept. 25, 1913.

7. Ibid., Oct. 2, 1913.

8. HS, p. 37.

9. Box 19, PP-LC, and MB-*PP* 1, pp. 269–71. "Pluck, determination, quick eye to see and will to take a fleeting chance . . . confidence . . . are instilled by racing. We talk a great deal about bold riding, and, outside the club, do very little . . . bold riders . . . will reap the reward . . . the training a man's nerve gets, the confidence it gives him in his horse, are worth the effort."

10. RET-M.

11. The Adjutant General to 2d Lt., G. S. Patton, Jr., subj: Efficiency Record, May 8, 1914.

12. Maj. Gen. Hugh L. Scott to GSP, July 13, 1914, Box 3, PP-USMA.

13. James E. Sullivan, Secretary, American Committee, Sixth Olympiad to GSP, June 24, 1914, MB-*PP* 1, p. 241.

14. Patton returned with an expensive new mare during a whirlwind trip that covered more than five hundred miles in twenty-six hours. Its cost was $400 or in 1991 money, the princely sum of $5,500. (U.S. Bureau of Labor Statistics, Consumer Price Index [CPI], in *Statistical Abstract of the United States.*)

15. "My Father as I Knew Him."

16. BAP to GSP II, Sept. 25, 1914.

17. FA, p. 8. At Saumur the previous summer, Patton and Lieutenant Houdemon had both expressed their belief that war with Germany was imminent. If the United States remained neutral, Patton had been assured he would receive a commission in Houdemon's cavalry regiment.

18. GSP to Leonard Wood, Aug. 3, 1914.

19. Leonard Wood to GSP, Aug. 6, 1914.

20. Patton greatly admired Rhodes, who would later become the commanding general of the 82d and 34th Divisions in the AEF during World War I.

21. GSP to GSP II, Feb. 11, 1915, Binder 7, PP-USMA.

22. GSP to BAP, Mar. 1, 1915, ibid.

23. GSP to GSP II, Feb. 11, 1915, ibid.

24. Ibid., May 16, 1915, Binder 7, PP-USMA.

25. GSP to BAP, Mar. 9, 1915, ibid.

26. Box 19, PP-LC.

27. GSP to BAP, Apr. 8, 1915, Binder 7, PP-USMA.

28. MB-*PP* 1, p. 287.

29. FA, pp. 65–66.

30. When she learned that the Pattons expected to accompany the regiment, his mother-in-law, Ellen Banning Ayer, reacted in her usual dramatic fashion. It would be a "cruel parting," but "My 'B's' [Beatrice and Little Bee] are ever with me—and away you will all fly soon. God help us all." Ellen Ayer's letters were as eccentric as her personality, referring in the letter cited here to southern California as "your native land," but her fondness for Patton was always evident in her infrequent correspondence with him. (Ellen Banning Ayer to GSP, June 5, 1915, Box 7, PP-LC.)

31. GSP to BAP, June 21, 1915, Binder 7, PP-USMA.

32. MB-*PP* 1, p. 292.

33. Truscott, *The Twilight of the U.S. Cavalry,* p. 90.

34. "My Father as I Knew Him." The horses soon multiplied and "were a great nuisance to Papa but he never complained."

35. MB, pp. 77–78. Patton's new commander was Maj. George Taylor Langhorne, USMA 1889. Hickok was a member of the West Point class of 1892, and in 1918 would become a brigadier general and the chief of the staff of the 5th Division of the AEF.

36. MB-*PP* 1, chap. 15, passim.

37. Ibid., p. 297.

38. Frank E. Vandiver, *Black Jack: The Life and Times of John J. Pershing,* vol. 1 (College Station, Tex., 1977), p. 582.

39. GSP to BAP, Sept. 30, 1915, Binder 7, PP-USMA.

Chapter 12   Pershing and the Punitive Expedition

1. Herbert Molloy Mason Jr., *The Great Pursuit* (Garden City, N.Y., 1961), p. 38. Pancho Villa was born Doroteo Arango, but changed it to Francisco Villa as a young member of a *bandito* gang.

2. Ibid., pp. 39–40.

3. Vernon L. Williams, "Lieutenant George S. Patton, Jr. and the American Army: On the Texas Frontier in Mexico, 1915-16," *Military History of Texas and the Southwest,* vol. 17, 1982. (Also published as *Lieutenant Patton and the American Army*

in the Mexican Punitive Expedition, 1915–1916 [Austin, Tex. 1983]). Hereafter "Lt. G. S. Patton, Jr."

4. Vandiver, *Black Jack,* vol. 1, p. 583.
5. Ibid., p. 171.
6. Ibid., p. 583.
7. Ibid., pp. 592–93, and vol. 2, p. 598.
8. Williams, "Lt. G. S. Patton, Jr."
9. GSP to BAP, Oct. 20, 1915, Binder 7, PP-USMA.
10. Ibid., Oct. 26, 1915.
11. Ibid.
12. Ibid., Nov. 11, 1915.
13. Ibid.
14. GSP to Aunt Nannie, Nov. 1, 1915. At a remote ranch he encountered another character, who was the most profane man he had ever met. The rancher despised Woodrow Wilson and as he puffed on his pipe, said, "with the deepest feeling God Damn the Dirty S—— B——. I have never heard one say a thing with so much emphasis" (GSP to Frederick Ayer, Nov. 11, 1915).
15. MB-*PP* 1, p. 306.
16. GSP to GSP II, Nov. 27, 1915, Box 7, PP-LC.
17. GSP to Ellen Banning Ayer, Nov. 28, 1915, Binder 7, PP-USMA.
18. Mason, *The Great Pursuit,* p. 185.
19. RET-M.
20. Ibid.
21. Vandiver, *Black Jack,* vol. 2, p. 606.
22. Williams, "Lt. G. S. Patton, Jr."
23. Ibid. Although the numbers are in dispute and virtually impossible to verify with accuracy, Villa apparently lost more than one hundred killed and 23 wounded. One of the heroes of the battle was a young West Point graduate named John Porter Lucas, whose destiny during World War II would converge with that of his friend, George S. Patton. Lucas and another lieutenant helped save Columbus from being overrun by Villa's irregulars, who would have slaughtered the entire population of the town and its garrison.
24. Leonard Wood to GSP, Mar. 13, 1916, Box 8, PP-LC.
25. GSP to GSP II, Mar. 12, 1916, Box 8, PP-LC.
26. "Personal Glimpses of General Pershing," Box 49, PP-LC. Essay written in 1924. Also in Vandiver, *Black Jack,* vol. 2, p. 607.
27. "Personal Glimpses of General Pershing."
28. GSP to GSP II, Mar. 12, 1916, Box 8, PP-LC.
29. Williams, "Lt. G. S. Patton, Jr."
30. Vandiver, *Black Jack,* vol. 2, p. 665.
31. GSP to GSP II, Apr. 6, 1916, Box 3, PP-USMA.
32. "Personal Glimpses of General Pershing."
33. Quoted from a small black notebook kept by Patton, dated Jan. 8, 1917, "Inspection by Gen. J.J. Pershing," Box 8, PP-LC. Also cited by Vandiver in *Black Jack,* vol. 2, p. 666. The notebook was a small but concise treatise on how to conduct an inspection and emulated exactly what Pershing himself had done recently in the Punitive Expeditionary Force. In today's terms it was Pershing's version of a combination Command Maintenance and IG inspection.
34. World War II diary of George S. Patton, Oct. 21, 1942, Box 4, PP-LC, hereafter "Diary."
35. Vandiver, *Black Jack,* vol. 2, pp. 608–9.
36. Ibid., pp. 609–10.
37. Ibid., chap. 17, passim.
38. Mason, *The Great Pursuit,* p. 106. Even Pancho Villa had recognized the potential of air power and had attempted, without success, to form his own small air force. Between 1908 and 1913 Germany had spent a whopping $28 million on aviation, while

the United States had spent less than 2 percent of this figure, or a miniscule $453,000 (pp. 106–7).

39. Richard O'Connor, *Black Jack Pershing* (Garden City, N.Y., 1961), p. 123.

40. Ibid., chap. 6, passim, and Weigley, *History of the United States Army,* p. 351.

41. The plight of the gallant aviators who flew these deathtraps was so scandalous that at least one risked court-martial by giving an interview to the *New York World*, which trumpeted their dilemma in bold headlines.

42. "Personal Glimpses of General Pershing."

43. Col. George A. Dodd was a sixty-three-year-old, cigar-chomping, veteran Indian fighter with a deserved reputation as one the cavalry's most reliable and aggressive regimental commanders. It was no coincidence that Pershing chose Dodd to command the flying columns of the Punitive Expedition. (Mason, *The Great Pursuit,* p. 95.)

44. "Personal Glimpses of General Pershing."

45. Col. H.A. Toulmin Jr., *With Pershing in Mexico* (Harrisburg, Pa.,1935), p. 53.

46. O'Connor, *Black Jack Pershing,* p. 125.

47. "Diary Kept by George S. Patton, Jr., while on duty with the Headquarters Punitive Expedition into Mexico," March 15, 1916, Box 1, PP-LC, hereafter "Patton Diary, Punitive Expedition."

Chapter 13   The Bandit Killer

1. "Personal Glimpses of Pershing," Box 49, PP-LC.

2. Milton F. Perry and Barbara W. Parke, *Patton and his Pistols* (Harrisburg, Pa., 1957), p. 12.

3. GSP to GSP II, Apr. 28, 1916, Binder 7, PP-USMA.

4. GSP to BAP, Apr. 13, 1916, Box 123, PP-LC.

5. David McCullough, *Truman* (New York, 1992), p. 588. Despite the presidential edict in 1948, the armed forces were slow to comply with integration, notably the army, which dragged its feet. The army fighting in Korea was not integrated until Gen. Matthew B. Ridgway assumed command of the Eighth Army in 1951, and U.S. forces in Europe did not begin to integrate until 1953. see RHP, pp 124–25, for Papa's racial bias.

6. *The Twentieth Century,* vol. 1, p. 22.

7. 1st Lt. Innis Palmer Swift, USMA, 1905, later a major general commanding a task force to recapture the Admiralty Islands in 1944, and the commanding general of a corps in the U.S. Sixth Army.

8. GSP to GSP II, Apr. 17, 1916.

9. Villa had supported slain President Francisco Madero. A cousin was married to the daughter of the wealthy landowner of the San Miguelito ranch, and it is thought that part of their debt for his support was to offer hospitality and protection to Villa and members of his band.

10. The exact number of men in Patton's band has never been precisely established. His official account written for Pershing states there were fifteen, but his description indicates the actual total should have been sixteen: Patton, two civilian guides, ten soldiers and three drivers. ("Report on the Death of Col. Cardenas," undated, but written immediately afterward. Box 48, PP-LC.)

11. Williams, "Lt. G. S. Patton, Jr."

12. Ibid.

13. Correspondent Frank B. Elser's account of the shoot-out in *NYT,* May 23, 1916, is based on his interviews of Patton and the other participants.

14. Mason, *The Great Pursuit,* p. 187, and Elser, *NYT,* May 23, 1916.

15. Elser, Patton's "Report on the Death of Col. Cardenas"; GSP to BAP, May 14, 1916, Box 8, PP-LC; MB-*PP* 1, chap. 16, passim, and Mason, *The Great Pursuit,* pp. 185–87.

16. Perry and Parke, *Patton and his Pistols,* p. 4. The authors note that the Colt Peacemaker had since its introduction "seen use in the hands of soldiers, cowboys, police, and

gunmen, and had gained a place in song and story equalled by no other weapon. Famous marshals and outlaws preferred it; generals and privates had proved it in combat. Adopted by the Army . . . in 1873, it was rivalled by no other revolver. . . [and was] as much a part of the American Tradition and Legend as the 'Kentucky Rifle' and 'The Old West.'"

17. FA, p. 73. Ayer was never entirely certain whether or not this tale told him by Patton in the late 1920s was true or not, but he tended to believe it.

18. Vandiver, *Black Jack,* vol. 2, p. 646.

19. Elser, *NYT,* May 23, 1916, and MB-*PP* 1, p. 337.

20. GSP to BAP, May 14, 1916, Box 8, PP-LC.

21. Ibid., May 17, 1916.

22. Mason, *The Great Pursuit,* p. 221.

23. GSP to BAP, July 29, 1916, Box 3, PP-USMA. Patton wrote nearly a dozen poems in 1916. Many are in Box 60, PP-LC.

24. Box 60, PP-LC, and Carmine Prioli, ed., *The Poems of General George S. Patton, Jr., Lines of Fire* (Lewiston, N.Y., 1991), pp. 41–42.

25. Patton Diary, Punitive Expedition, Aug. 19, 1916.

26. A "paper" attachment is one in which a soldier, NCO, or officer is assigned or attached to a unit that is authorized his rank and position but actually serves elsewhere. In this case Patton was attached to the 10th Cavalry but never actually served a single day with the unit.

27. GSP to BAP, July 29, 1916, Box 3, PP-USMA. In 1992, the Buffalo Soldiers of the 9th and 10th Cavalries received long-overdue recognition in the form of the Buffalo Soldier Memorial, a thirteen-foot bronze statue dedicated at their home station of Fort Leavenworth, Kansas, by the officer who inspired its creation, Gen. Colin Powell, the chairman of the Joint Chiefs of Staff. Maj. Charles Young (1865–1922) graduated from West Point in 1889 and retired as a full colonel in 1917.

28. GSP to Frederick Ayer, June 27, 1916, Box 8, PP-LC.

29. See correspondence between GSP and Capt. R. F. McReynolds (the collector of internal revenue in Los Angeles) and with his father, both July 6, 1916. (Box 3, PP-USMA, and Box 8, PP-LC.)

30. GSP to GSP II, July 12, 1916. Patton could hardly have been more critical of Wilson than in this letter, but in September wrote to Beatrice that, "I wish Pa was out of politics so I could say what I think about Wilson." Considering what he *did* say, Patton could hardly have been more scathing.

31. Ibid.

32. Weigley, *History of the United States Army,* p. 350. Between 1914–17, there were numerous skirmishes between those who saw the urgent need for a fundamental reform and forces determined to limit the powers of the General Staff and the role of the military. At the heart of the problem were Wilson's utopian ideas about the role of the military in a constitutional government, ideas (however noble), that had little to do with the reality that the United States in 1916 was on the brink of entering World War I.

33. Vandiver, *Black Jack,* vol. 2, p. 658.

34. GSP to GSP II, Sept. 7, 1916, Box 8, PP-LC.

35. GSP to BAP, Oct. 7, 1916, Box 3, PP-USMA.

36. RET-M.

37. MB-*PP* 1, p. 355.

38. RET-M.

39. Virtually unknown outside Los Angeles, Mr. Patton was decisively tied by his opponent to the vast Huntington railroad and land empire, and was never seriously in the running against the highly popular Johnson, who had the priceless advantage of already being the governor of California.

40. "My Father as I Knew Him." The evidence suggests that Wilson would have won the state with or without the support of Patton's father. (See MB-*PP* 1, p. 357.)

41. JJP to GSP, Oct. 16, 1916, Box 4, PP-USMA.

42. GSP to BAP, Nov. 26, 1916, ibid.

43. Ibid., Nov. 28 1916.

44. Ibid., Dec. 1, 1916. Patton refused a $2.50 payment from the *Cavalry Journal* for an article about the saber, noting: "I will not take money for defending the saber. It would be sacrelidge."

45. Ibid., Dec. 20, 1916.

46. Quoted in RET-M. see also Prioli, *Lines of Fire*, pp. 37–38.

47. GSP II to BAP, May 17, 1917, Box 8, PP-LC.

48. RET-M.

49. Extracted from GSP's letters to BAP, Jan. 14 and 15, 1917, Box 8, PP-LC.

50. Mason, *The Great Pursuit,* pp. 231–32.

Chapter 14   "Over There: The Yanks Are Coming"

Unless otherwise cited, all letters to and from Patton are from Box 8 and diary entries Box 1, PP-LC.

1. James W. Rainey, "The Questionable Training of the AEF in World War I," *Parameters* (the journal of the U.S. Army War College) (Winter 1992–93).

2. British casualties at Gallipoli were 41,000 killed and missing and 78,500 wounded. The French lost 9,000 killed and 13,000, and together the British and French sustained another 100,000 nonbattle losses due to sickness. The Turks lost 66,000 killed and 152,000 wounded. (Brig. Peter Young, ed., *A Dictionary of Battles [1816–1976]* [New York, 1977], p. 358.)

3. Michael Howard, *War in European History* (London, 1976), p. 112. As Sir Michael points out: "The Napoleonic principles on which soldiers had been raised for a hundred years—*Niederwerfungstrategie,* as the Germans termed it, the strategy of overthrow—were no longer valid."

4. Two of the British lieutenant colonels serving in the front lines were a prospective prime minister, Winston Churchill, and a field marshal in the next world war, Lt. Col. Bernard Law Montgomery, who was severely wounded in the chest during the First Battle of Ypres in 1914, where he won the Distinguished Service Order (DSO).

5. Weigley, *History of the United States Army,* p. 353.

6. Ibid., p. 356. By April 1917 the Regular Army consisted of 5,800 officers and 133,000 men, while the National Guard numbered 3,200 officers and 67,000 men. (Rainey, "The Questionable Training of the AEF in World War I.")

7. Ibid., p. 353.

8. Edward M. Coffman, *The War to End All Wars* (New York, 1968), chap. 2, passim.

9. Ibid., p. 29.

10. Ibid., p. 43.

11. Army Times, eds., *The Yanks Are Coming* (New York, 1960), p. 59; Coffman, *The War to End All Wars,* p. 40, and Laurence Stallings, *The Doughboys* (New York, 1963), p. 25.

12. *The Yanks Are Coming,* p. 64.

13. "The Present Saber—Its Form and the Use for Which It Was Designed," *Cavalry Journal* (Apr. 1917).

14. MB-*PP* 1, p. 380. See also Microfilm Reel 1, PP-LC.1

15. The other two were from the commander of Fort Bliss and his erstwhile squadron commander, Major Edmund M. Leary, USMA 1893 who was killed in Sept. 1919 in an air crash.

16. GSP to JJP, Apr. 11, 1917.

17. Ibid., Apr. 23, 1917. Beatrice had become ill as well, and with sister Kay about to be married, Patton complained: "If I can possibly get away from this combined Hospital and matrimonial establishment, I am going up to Washington . . . and see if I cant get into some regiment near here." Eventually all recovered.

18. GSP II to GSP, Apr. 30, 1917, Box 8, PP-LC. The unmailed letter was later found among his papers.

19. Several days earlier the adjutant general's office, unaware that Pershing would be claiming his services, had decided to reassign him to their office, with provisional duty at nearby Front Royal, Virginia, where he would be in charge of procuring horses for the U.S. Army. When he learned what the adjutant general intended, Pershing orchestrated Patton's immediate assignment to his new headquarters.

20. Everett Strait Hughes was a member of Patton's original West Point class of 1908. Once a field artilleryman and later an ordnance officer, Hughes had also participated in the Punitive Expedition. More will be seen of Hughes during the World War II chapters.

21. MB-*PP* 1, p. 388.

22. "My Father as I Knew Him."

23. MB-*PP* 1, p. 389.

24. RET-M.

25. "My Father as I Knew Him." Patton has written that Papa, "kept all the letters I ever wrote to him and had those I wrote from France filed separately They are in his safe. I never kept letters [*sic*] so have none of his I could never think he would die before me."

26. "Diary of U.S. Expedition to France," various entries between June 9 and 13, 1917, Box 1, PP-LC, hereafter "WW I Diary."

27. Coffman, *The War to End All Wars,* p. 123.

28. GSP to BAP, June 14, 1917.

29. Ibid., June 24, 1917.

30. WW I Diary, June 24, 1917.

31. Ibid., June 24, 1917; GSP to BAP, June 25, 1917, and Vandiver, *Black Jack,* vol. 2, p. 737.

32. GSP to BAP, July 16, 1917.

33. Ibid., July 17, 1917.

34. BAP to Aunt Nannie, Aug. 16, 1917. Beatrice was soon her old, plucky self. Writing to Aunt Nannie a week later, "I guess. . . the best way to minimize our own troubles & disappointments—do something for somebody else. Well, I am going to do my best; my very best. But I'm a poor loser."(Letter of Aug. 22, 1917.)

35. RET-M.

36. GSP to BAP, Aug. 24, 1917.

37. GSP to BAP; GSP to GSP II, and Secretary Franklin K. Lane to GSP II, all dated Sept. 25, 1917. Because of his unique relationship with Pershing, Patton continued to believe he could risk Black Jack's displeasure, but as Lane pointed out to Mr. Patton: "There is real danger of hurting your son if you do [intervene with Wilson on his son's behalf]. The pressure for exceptions to this rule is great, and I have known no man to profit by desiring that an exception be made to a rule for him." Several days later Patton wrote to Beatrice, "I suppose the idea of taking the bull by the horns scared Pa to death." (Letter of Sept. 27, 1917.)

38. GSP to BAP, June 28, 1917.

39. Ibid., July 29, 1917.

40. Ibid., Aug. 27, 1917.

41. Overall, the peak strength of the US Army in World War I was 3,703,273. (Coffman, *The War to End All Wars,* p. 357.)

42. GSP to BAP, Aug. 27, 1917.

43. MB-*PP* 1, p. 405.

44. GSP to BAP, July 23 and Aug. 8, 1917.

45. Ibid., July 14, 1917.

46. Vandiver, *Black Jack,* II, pp. 743-44.

47. GSP to BAP, July 24, 1917, Binder 7, PP-USMA. Patton rated Haig, a former cavalryman, "more of a charger than I am and was very nice to me. He is a fine looking man . . . [who] plays polo and hunts."

48. Diary of Field Marshal Sir Douglas Haig, July 20, 1917, quoted by Vandiver in *Black Jack,* vol. 2, p. 748.

49. GSP to Mrs. C. G. Rice, July 24, 1917.

50. Vandiver, *Black Jack,* vol. 2, p. 751.

51. George C. Marshall, *Memoirs of My Services in the World War, 1917–1918* (Boston, 1976), p. 16. Patton's diary on July 13 also records their brief meeting.

52. GSP to BAP, Aug. 26, 1917.

53. GSP to Frederick Ayer, Aug. 26, 1917, Box 4, PP-USMA.

54. JJP to GSP II, Aug. 17, 1917, ibid.

55. GSP to BAP, Aug. 20, 1917.

56. Situated on the borders of Champagne and Burgundy, five hours from Paris by automobile, Chaumont was famous as the site a century earlier where Francis II of Austria, Alexander I of Russia, and Frederick William of Prussia met to sign a treaty in March 1814 that formally committed them to an alliance against Napoleon shortly before he was finally vanquished and exiled to Elba. (See O'Connor, *Black Jack Pershing,* p. 198.)

Chapter 15   Tank Officer

1. Stallings, *The Doughboys,* pp. 23–24. Patton later observed:

No adulation could persuade him to countenance the placing of American men in French and British units. It is to his iron resolution to form an American Army that we owe the great heritage of victorious America, victorious in her own right, and by her own means. But for him her man power would have been bled white to fill the depleted ranks of allied units, where their valor would have been unmarked and their achievements unheralded. ("Personal Glimpses of General Pershing.")

2. Ibid., p. 25.

3. Coffman, *The War to End All Wars,* p. 137.

4. Quoted in Vandiver, *Black Jack,* vol. 2, p. 725.

5. Among the criticism was that some of the AEF's best officers were stolen from the divisions that badly required their leadership to fill the ranks of the cadre in its many schools. Units were constantly sending their officers to this or that school, thus stripping units at the time they should have been training as a team. Others complained that there was simply too much repetitive drilling over and over, to the point where units lost their edge. A discussion of Pershing's training methods is in Rainey, "The Questionable Training of the AEF in World War I."

6. *The Yanks Are Coming*, p. 87.

7. GSP to BAP, Sept. 19; and WW I Diary, Sept. 7, 1917.

8. GSP to GSP II, Sept. 4, 1917. The furnishings in the house were "in the damdest taste you ever saw. All gilt and bras[s] and quite new. The man wanted to be thought a sport so he bought all sorts of game heads which are growling at one every where. There is a huge crocodile fifteen feet long with gaping jaws who appears ready to spring on you when you enter the front hall . . . a wild boar threatens you when you eat." (GSP to BAP, Sept. 2, 1917.)

9. GSP to BAP, Oct. 5, 1917.

10. Ibid., Sept. 22, 1917.

11. Ibid., Oct. 10, 1917.

12. GSP to Frederick Ayer, Oct. 1, 1917, PP-USMA.

13. GSP to BAP, Sept. 27, 1917.

14. Ibid., Oct. 5, 1917.

15. Ibid., Sept.19, 1917.

16. F. Mitchell, *Tank Warfare: The Story of the Tanks in the Great War* (London, 1933), pp. 4–5. As J. F. C. Fuller points out in *Tanks in the Great War, 1914–1918* (London, 1920), chap. 1, the first tanks were living ones: the medieval knights in armor. Ideas for battle cars and carts date to the early fifteenth century.

17. Quoted in Mitchell, *Tank Warfare,* p. 7.

18. So named for German Crown Prince William, who had been derisively nicknamed "Little Willie" by British troops. The "Little Willie" prototype tank proved impractical and eventually evolved into the first operational models, which were called "Big Willies." (Ibid., pp. 8–9.)

19. To preserve what was now becoming a prized secret from the Germans, the responsibility for the development and production of the land-cruiser was given to a small group called the Executive Supply Committee. The cover name given this group was "Tank Supply Committee" and soon what had previously been called "land-cruiser," "landship," and "caterpillar machine-gun destroyer" was shortened to "tank." J. F. C. Fuller asserts that this was the first use of the word "tank" in "the history of the machine." (*Tanks in the Great War, 1914–1918*), p. 29.

20. Mitchell, *Tank Warfare,* p. 34.

21. Dale E. Wilson, *Treat 'Em Rough!: The Birth of American Armor, 1917–20* (Novato, Calif., 1990), pp. 2–3. There are two editions of this book, the original 1989 edition published by Presidio Press and a 1990 edition published by the Military Book Club. The pagination is different; all citations that follow are from the 1990 book club edition.

22. Ibid., p. 9.

23. Ibid., p. 10. Wilson notes that the Mark VI was a prototype heavy tank, of twenty-seven to thirty tons, that was never produced.

24. GSP to BAP, Oct. 19, 1917, and lecture, "Tanks Past and Future," Feb. 20, 1928, Box 50, PP-LC.

25. Appended to letter GSP to BAP, Oct. 24, 1917.

26. Ibid., Oct. 26, 1917.

27. Ibid., Oct. 28, 1917.

28. Col. (later Brig. Gen.) LeRoy Eltinge, USMA Class of 1895, was Patton's former troop commander in the 15th Cavalry at Fort Myer in 1912. Later, at the Mounted Service School, Eltinge had been one of those who had impressed Patton. Eltinge had tutored Patton at Fort Bliss as he prepared for his promotion examination to first lieutenant in 1915, and was one of those to whom he owed a great debt.

29. WW I Diary, Nov. 3, 1917.

30. "Tanks Past and Future."

31. WW I Diary, undated.

32. GSP to GSP II, Sept. 25, 1917, and to BAP, Sept. 23, 1917.

33. GSP to JJP, Oct. 3, 1917, Box 20, PP-LC.

34. GSP to GSP II, Nov. 6, 1917, Box 20, PP-LC. Patton had appended a P.S., which read: "Don't talk about T. it is a secret," In a letter to his father-in-law several months later, Patton admitted that Mr. Ayer had also influenced his decision to join the Tank Corps: "In November I got a letter from you saying that war was so wasteful of life and that some means ought to be devised of reducing the killing. At that very moment I was talking with Col. Conner about whether I should go into the Tank service. Your letter decided me and next morning I asked for the Tanks. They save life as two men in a tank are as good as ten out of one." Mr. Ayer was so enthralled by Patton's letter that he carried it in his wallet, where Ellie Ayer later found it after his death. Beatrice kept it as a permanent part of her husband's papers. (Letter of Jan. 20, 1918, PP-USMA, and MB-*PP* 1, p. 474.)

35. "Tanks Past and Future."

36. AEF special order of Nov. 19, 1917, quoted in "History of the 304th (1st) Brigade, Tank Corps" n.d. but written by Patton after the end of World War I. Box 60, PP-LC, hereafter "History of the 304th."

37. GSP to BAP, Nov. 3, 1917.

38. Quoted in RET-M.

39. GSP to BAP, Dec. 2, 1917.

40. Ibid., Nov. 9, 1917. Opinions were divided among Patton's superiors, with Colonels Eltinge, Malone (the head of all AEF schools), McCoy, Shallenberger, and Hugh Drum in favor and Gen. J. G. Harbord, Pershing's chief of staff, and Col. Fox Conner

opposed. "They said it was a gamble while the infantry is sure. All said I was right in getting out of here."

41. GSP to JJP, Nov. 12, 1917, memo titled, "Military Appearance and Saluting."

42. WW I Diary, Nov. 18, 1917.

43. GSP to BAP, Nov. 20, 1917. His orderly was a French sergeant named Count Everest de Pas who, was "very attentive and calls me either lieutenant or general as the humor takes him."

44. MB-*PP* 1, p. 444.

45. Ibid., p. 446, and Wilson, *Treat 'Em Rough!* p. 16. Both officers absorbed their lessons so well that they left the French with several recommendations for minor modifications that were later accepted and implemented.

46. *The American Heritage History of World War I* (New York, 1964), pp. 220–21.

47. GSP to BAP, Nov. 26, 1917.

48. WW I Diary, Dec. 1, 1917. His diary, unlike the one he kept during World War II, was austere and reveals only that from his discussions with Fuller he "got interesting data."

49. Memo, "Light Tanks," dated Dec. 12, 1917, Box 9, PP-LC.

50. Letters, GSP to BAP and Aunt Nannie, both dated Dec. 14, 1917.

51. Ibid., GSP's second letter of Dec. 14 to Beatrice.

52. Ibid., Dec. 15, 1917.

53. WW I Diary, Dec. 25, 1917.

Chapter 16 "Great Oaks from Little Acorns Grow"

1. Brig. Gen. Samuel D. Rockenbach, "Operations of the Tank Corps, A.E.F., with the 1st American Army," Dec. 1918, quoted in Wilson, *Treat 'Em Rough!,* p. 22, and GSP to BAP, Dec. 23, 1917, Box 8, PP-LC.

2. Wilson, *Treat 'Em Rough!* pp. 25–26.

3. RET-M.

4. GSP to BAP, Jan. 30, 1918.

5. MB-*PP* 1, p. 464.

6. GSP to BAP, Aug. 10, 1918.

7. LF, p. 53.

8. RET-M. Shortly after moving into the hotel Patton heard the voices of a woman and of a man attempting to comfort her. The real estate agent was aghast: "Mais, mon Colonel, that is impossible! There is no one in the city of Langres that would come upstairs in this hôtel!" Patton was then told the story of the Count d'Aulan, who had returned from his Crusade only to find his wife in bed with a page. He is alleged to have bricked them both into an upstairs room and to have gone back to war, from which he did not return.

9. Ibid.

10. GSP to BAP, Dec. 19, 1917.

11. Ibid., Dec. 19, 1917, Box 8, PP-LC.

12. Ibid., Dec. 20, 1917. He advised Beatrice: "Don't lose my poems. They may be priceless some day and I never keep copies."

13. Ibid., Dec. 23, 1917.

14. Ibid., Feb. 22, 1918.

15. Ibid., Jan. 10, 1918, and MB-*PP* 1, chap. 23, passim.

16. Ibid., Jan. 12, 1918.

17. Ibid., Jan. 14, 1918.

18. Memorandum No. 1, Army Tank School, Jan. 27, 1918, Box 9, PP-LC, signed personally by GSP. The memo was vintage Patton. Standards were very high and by action and deed he was making it clear that on his watch nothing less would be tolerated.

19. GSP to BAP, Jan. 20, 1918.

20. Ibid., Jan. 25, 1918. Although Patton became notorious for cursing officers, he is not known for having cursed a noncommissioned officer.

21. MB-*PP* 1, pp. 473–74. Patton's sojourn in the haunted, dungeonlike château-hotel turned out to be temporary, and he moved several times. Each time his nemesis attempted to have him kicked out of his new billet.

22. 2d Lt. Will G. Robinson, Tank Corps, quoted in Wilson, *Treat 'Em Rough!* p. 35.

23. Ibid.

24. Wilson, *Treat 'Em Rough!* p. 36. For his enlisted men it was a circular brass disk containing the outline of a British Big Willie surrounded by a wreath; minus the disk, the officer insignia was similar.

25. "Notes on Co-operation between tanks and the other arms especially infantry," Box 48, PP-LC. Patton urged:

Cooperation between the tanks and infantry must be of the closest. . . [and] worked out in the most minute detail . . . this mutual understanding . . . must be more than a mere matter of routine. Tank and infantry officers must associate and by conversation at mess and elsewhere interchange ideas and so become thoroughly conversant with all the difficulties which beset their respective arms. A failure to do this has—in my opinion—been the cause of much waste in time, money and blood.

26. GSP to BAP, Feb. 26, 1918.

27. Ibid., Jan. 20 and Feb. 8, 1918.

28. Ibid., Apr. 11, 1918

29. Ibid., Jan. 21 and 23, 1918.

30. Ibid., Jan. 25, 1918.

31. Ibid., Jan. 31, 1918.

32. Wilson, *Treat 'Em Rough!,* p. 30, and GSP to BAP, Feb. 17, 1918.

33. GSP to BAP, March 4, 1918.

34. Ibid., Feb. 2 and 24, 1918, and MB-*PP* 1, p. 492. (The woman's real name was Madame de Vaux de la Porquière).

35. Stimson received a commission in the field artillery and for a brief period was in temporary command of a regiment before becoming a student at the General Staff School at Langres. (See Godfrey Hodgson, *The Colonel* [New York, 1990], pp. 84–85, and Stimson's memoirs, *On Active Service in Peace and War* [New York, 1948], pp. 96–97.)

36. GSP to BAP, Feb. 18, 1918.

37. Ibid., Feb. 22, 1918.

38. Patton, "History of Army Tank School, A.E.F.," Nov. 22, 1918, RG 120, Folder 229, MMB-NA.

39. GSP to BAP, Mar. 12, 1918.

40. Ibid., Feb. 27, 1918.

41. Ibid., Mar. 13, 1918. Beatrice later scribbled in the margin: "Nonsense! I didn't mean it that way at all."

42. Ibid., Apr. 10, 1918.

43. "History of Army Tank School, A.E.F." As Martin Blumenson observes, "It was all quite typically Patton, detailed and thorough" (MB-*PP* 1, p. 487).

44. Extracts from "Lecture on Discipline," Box 48, PP-LC.

45. HS, p. 48.

46. GSP to BAP, Apr. 25, 1918.

47. Both quotes in Wilson, *Treat 'Em Rough!* p. 37.

48. MB-*PP* 1, p. 490, and GSP to BAP, Feb. 23, 1918.

49. GSP, "Mechanized Forces," *Cavalry Journal,* Sept.-Oct. 1933.

50. Ibid.

51. MB-*PP* 1, p. 484.

52. Patton's horse Sylvia Green thought that his "tanks are some new sort of racing animal. She is not afraid of them but snorts in contempt when she passes them as if in derision of their lack of speed," wrote Patton to his father on April 19, 1918.

53. GSP to BAP, Mar. 28, 1918.

54. "History of the 304th."

55. GSP to BAP, Mar. 27, 1918.

56. Ibid., Mar. 20, 1918.

57. Ibid., Feb. 23, 1918.

58. Ibid., Mar. 13, 1918.

59. Ibid., Mar. 19, 1918.

60. Quoted in Wilson, *Treat 'Em Rough!*, p. 32.

61. GSP to BAP, Mar. 20, 1918.

62. HS, pp. 45–46, and LF, p. 54. There is another version in his diary for March 23 that Patton drove seven of the ten tanks off the train and one of his lieutenants the other three. In a letter the following day, Patton alludes to having had to back all ten tanks off the train. He also makes the same claim in his paper, "History of the 304th." (See also MB-*PP* 1, pp. 508–9.)

63. Patton would soon inform Beatrice that he was becoming more and more like Henry Ford than he was a soldier, except that he produced tankers instead of automobiles. "I think I might have been a manufacturer if I had put my mind on it but I am glad I did not. Because you might never have married me if it had not been for my brass buttons?" (Letter of Apr. 13, 1918.)

64. HS, p. 46. The Renault light tank weighed 7.4 tons and was powered by a water-cooled 39-horsepower, 4-cylinder engine. It carried 24 gallons of gasoline and had a range of around 24 miles. The main armament was either a 37-mm gun or an 8-mm Hotchkiss machine gun and 4,800 rounds of ammunition. The armor plating on the Renault was an almost microscopic 0.3 to 0.6 inches. The tank itself was 16 feet, 5 inches long; 5 feet, 8 inches wide; and 7 feet 6 ½ inches high. (Wilson, *Treat 'Em Rough!* Appendix A, p. 223.)

65. GSP to BAP, Apr. 25, 1918.

66. Ibid., June 15, 1918. The officer was Col. Dawson Olmstead, USMA 1906, who was serving in the Office of the Inspector General, GHQ, AEF.

67. "Tanks Past and Future."

68. At this early stage, Patton still envisioned the employment of tanks much as a choreographer would a dance, with each movement prescribed and carried out as taught and practiced.

69. GSP to BAP, June 9, 1918.

70. Wilson, *Treat 'Em Rough!* pp. 34–35; MB-*PP* 1, pp. 523–24, GSP to BAP, Apr. 22, 1918.

71. GSP to BAP, Apr. 3, 1918.

72. Ibid., Mar. 19, 1918, one of two letters sent to BAP on this date.

Chapter 17 Baptism of Fire

Unless otherwise cited, all letters from GSP to his family in 1918 are in Binder 8, PP-USMA.

1. Box 60, PP-LC. Written in the period of uncertainty shortly before he established the tank center, Patton's poem is a reminder to himself to persevere. An extract reads:

> *In wondrous cat like ability*
> *It is not in the intricate planning*
> *Nor yet in regretting the past*
> *To land on my feet with agility*
> *No one is greater than I. . . .*

> *For these gifts Oh God I thank you*
> *Prey let me continue the same*
> *Since by doing things well which are nearest*
> *Perhaps I shall yet rise to fame. . . .*

So seize I the things which are nearest
And studiously fall on my feet
Do ever in all things my damndest.
And never oh never retreat! (dated Dec. 20, 1917)

2. Quotation by a World War II veteran in John Mortimer's novel *Dunster* (New York, 1993), p. 71.

3. GSP to BAP, June 13, 1918.

4. Joseph Wadsworth Viner (1888–1983) was a cavalry officer and a 1913 graduate of West Point who participated in Pershing's Punitive Expedition. Viner was typical of the outstanding caliber of officer Patton attracted to the Tank Corps. By the end of World War I Viner was a lieutenant colonel and the chief of staff of the AEF Tank Corps.

5. GSP to BAP, May 20, 1918. His letters to Papa were no more reassuring than those to his wife.

6. Quoted in MB-*PP* 1, p. 534.

7. GSP to BAP, May 25, 1918, Box 10, PP-LC.

8. Ibid., May 29, 1918.

9. Ibid., May 26, 1918.

10. Ibid., July 7, 1918.

11. Ibid., June 3, 1918, Box 10, PP-LC.

12. Ibid.

13. Wilson, *Treat 'Em Rough!*, pp. 41–42. Such mischief helped to keep legions of staff officers in the general headquarters of the AEF, and in the War Department gainfully employed. In this instance the War Department (whose knowledge of tanks could be measured by the eyedropper) elected to overrule Rockenbach and the AEF who had devised a perfectly sensible numbering scheme. However, it is one of the unwritten commandments of any bureaucracy that: "If it ain't broke, fix it anyway." Patton's two battalions became the 344th Tank Battalion, commanded by Maj. Serano Brett, and the 345th by Capt. Ranulf Compton. In November 1918, the 1st Tank Brigade was redesignated the 304th Tank Brigade. To avoid confusion, it is here referred to only as the 1st Tank Brigade. (See *Treat 'Em Rough!* pp. 116–17, n. 53.)

14. Ibid., p. 41. The French had told Patton it would be necessary to completely overhaul the Renault tank after every fifty hours of operation. However, "Due to the care and efficiency of our repair men, our tanks often went 500 hours without complete overhaul." (Patton, "History of the 304th [1st] Brigade, Tank Corps," Box 60, PP-LC.)

15. Ibid., p. 43.

16. GSP to BAP, June 27, 1918.

17. Wilson, *Treat 'Em Rough!* pp. 42–43.

18. Another classmate at Langres was Maj. Harold R. Bull, USMA 1914, who in World War II would serve as Dwight D. Eisenhower's G-3 on the SHAEF staff. Bull was no friend to Patton, and the two rarely agreed on anything.

19. MB-*PP* 1, p. 544.

20. GSP to BAP, July 18, 1918. Box 10, PP-LC.

21. Quoted in MB-*PP* 1, p. 562, and "History of the 304th." Before departing for Chaumont he hastily scribbled a note to Beatrice that left no doubt as to where he would soon be going: "I will wire you before the 15 of Sept that I am all right so as to relieve your mind on this letter." An even more morose note was dispatched to his parents in which Patton spoke of possibly not returning.

22. Coffman, *The War to End All Wars*, p. 217.

23. Description in James G. Harbord, *The American Army in France, 1917–1919* (Boston, 1936), p. 421. Maj. Gen. Harbord was a friend of Patton's and Pershing's chief of staff. As Harbord points out, most Army officers who were graduates of the army service schools were familiar with the topography of the Saint-Mihiel salient, based on map studies

of this region of France while students. In addition, US divisions had had ample time to study and become familiar with the Saint-Mihiel sector.

24. Wilson, *Treat 'Em Rough!* p. 92, and Robert E. Rogge, "The 304th Tank Brigade," *Armor* (July-Aug. 1988. The numbers were: 150 heavy (British) tanks; three French brigades totaling 225 light Renault tanks; Patton's 144 Renaults, including the three attached French battalions; plus another 36 St. Chamond and Schneider heavy tanks.

25. Ibid., pp. 93 and 96. The Schneider was a larger and heavier version of the Renault that weighed 14.9 tons and mounted a 75-mm gun and two machine guns. (See *Treat 'Em Rough!* pp. 225–26.)

26. GSP to BAP, Sept. 1, 1918.

27. Patton, "The Report of Personal Experiences in the Tank Corps," Dec. 16, 1918, Box 47, PP-LC.

28. Rogge, "The 304th Tank Brigade."

29. WW I Diary, Sept. 8, 1918, and Rogge, "The 304th Tank Brigade."

30. William R. Kraft Jr., "The Saga of the Five of Hearts," *Armor* (July-Aug. 1988). The Five of Hearts went on to become one of the most famous tanks of the AEF for its exploits during the Meuse-Argonne offensive. After the war it was emplaced in front of post headquarters at Camp (later Fort) Meade, Maryland, the postwar home of the U.S. Army Tank Corps.

31. "History of the 304th," and Rogge, "The 304th Tank Brigade." Lt. Brown was assigned to Harry Semmes's Company A, 344th Tank Battalion.

32. Patton's "Special Instructions for the 326th Bn., and 327th Bn.," Sept. 8, 1918, Box 10, PP-LC.

33. GSP to BAP, Sept. 16, 1918, and GSP to his father, Sept. 20, 1918, both Box 12, PP-LC, and John Toland, *No Man's Land: The Story of 1918* (London, 1980), p. 418.

34. Brig. Gen. Samuel D. Rockenbach, "Operations of the Tank Corps A.E.F. at St. Mihiel, in the Meuse-Argonne Operation, and with the British E.F.," Dec. 27, 1918, USAMHI. Copy furnished the author courtesy of Dale Wilson.

35. *The American Heritage History of World War I,* p. 332.

36. Patton set up a system of flag signals to send instructions to other tanks. But, as Harry Semmes writes, "when the flags were poked through the hole in the turret of the tank, machine gun bullets shot off the flag as fast as it appeared." The only other means of communication was by means of homing pigeons and this too failed. Each tank carried two pigeons in a wicker basket. Unfortunately, they had to be kept on the floor of the turret where there was already insufficient space. The gunner was forced to stand on the baskets which ended up flattened within five minutes, its occupants meeting an unforeseen and unhappy end. In an effort to protect the pigeons from gas each was issued a "gas mask" in the form of a Bull Durham (tobacco) sack. However, as Semmes wryly notes: "It would have taken Houdini to have put one on a pigeon in the dark and crowded turret of a tank in combat." Major Brett tells the story of having managed to save one pigeon, which he duly dispatched with a message to Rockenbach's headquarters. Apparently confused by the rough ride inside a tank, it flew only as far as a nearby tree, where Brett threw sticks at it to induce it to move. "It finally took off and flew toward Germany." (HS, pp. 62–63.)

37. GSP to GSP II, Sept. 20, 1918.

38. Ibid.

39. Ibid.

40. William Manchester, *American Caesar* (Boston, 1978), p. 115. MacArthur's memory was faulty, with the result that he and his biographer incorrectly identify Patton as a major.

41. John F. Wukovits, "Best-Case Scenario Exceeded," *Military History* (Dec. 1992).

42. "The Report of Personal Experiences in the Tank Corps." Daughter Ruth Ellen has written of the incident at the Essey bridge:

> Georgie volunteered to walk across it, taking a chance at tripping any wires, before he would allow his precious tanks to be threatened. He wrote Ma that he

knew instinctively that it was mined, and that every step he took might be his last. As he put his foot on the bridge, he was suddenly separated from the little uniformed figure he could see walking stiffly and carefully across the bridge, lifting its feet and putting them down with exquisite precision. After a moment, he knew he was watching himself. He saw the figure step off the bridge at the far side, duck down under it, and then come up with something in its hands; and then he was standing there with the broken connections, ordering his troops to cross. The bridge had been mined at the far side, and the trip wires were not visible from the approach. (RET-M.)

The author has been unable to locate the letter in the Patton Papers.

43. Manchester, *American Caesar,* pp. 115–16.

44. RET-M.

45. Ibid.

46. Quoted in ibid.

47. Douglas MacArthur, *Reminiscences* (New York, 1964), p. 63. MacArthur does not mention crossing the Essey bridge with Patton, whom he misidentifies as the squadron commander of the tanks supporting his brigade. He also refers to Patton as "an old friend," which is strange, for there is no indication they had ever met before that day at the Essey bridge, or afterward.

48. Toland, *No Man's Land,* p. 423.

49. "Report of Personal Experiences in the Tank Corps."

50. Ibid. Patton wrote to his father that, "At last the bright thought occured to me that I could move across the front in an oblique direction and not appear to run yet at the same time get back. This I did listening for the machine guns with all my ears, and laying down in a great hurry when I heard them, in this manner I hoped to beat the bullets to me." (Letter of Sept. 20, 1918.)

51. Ibid.

52. "History of the 304th."

53. GSP to GSP II, Sept. 20, 1918.

54. Ibid.

55. Quoted in Wilson, *Treat 'Em Rough!* p. 104.

56. "History of the 304th," and *Treat 'Em Rough!* p. 104.

57. Both quotes by Rockenbach in "V.M.I. Smoker on the [57th] Anniversary of the Battle of New Market," held in Washington, D.C., on May 14, 1922. Copy in PP-USMA. This paper appears to be the minutes of the reunion, which Patton probably attended. Rockenbach was the main after-dinner speaker of the evening.

58. MB-*PP* 1, chap. 29, passim.

59. Ibid., p. 596.

60. Ibid., p. 597.

61. "History of the 304th;" HS, p. 55; "Report of Personal Experiences in the Tank Corps"; and Rogge, "The 304th Tank Brigade."

62. According to Dale Wilson, the term originated at the tank training center at Camp Colt, Pennsylvania, which was commanded by a young captain (soon to be a lieutenant colonel) named Dwight D. Eisenhower. The Camp Colt newspaper was called *Treat 'Em Rough*, and eventually the name became the motto of the men of the Tank Corps. (*Treat 'Em Rough!* chap. 4.)

63. "History of the 304th."

64. HS, p. 65.

65. Ibid., and lecture, "Tanks Past and Future," delivered in Hawaii, Feb. 27, 1928, Box 50, PP-LC.

66. "Tanks Past and Future."

67. JJP, "Report of the First Army: Organization and Operations," 1923, quoted in William A. Morgan, "Invasion on the Ether: Radio Intelligence at the Battle of St. Mihiel, September 1918," *Military Affairs* (Apr. 1987).

68. Quoted in Paul F. Braim, *The Test of Battle: The American Expeditionary Forces in the Meuse-Argonne Campaign* (Newark, Del., 1987), p. 85. Before the Saint-Mihiel offensive, the soldiers of the US First Army had no idea that the objective was merely to eliminate the salient and the pet story that made the rounds of the troops went as follows: "First doughboy: 'Pershing says he'll take Metz if it costs a million men.' Second doughboy: 'Ain't he the generous son-of-a-bitch.'" (Quoted in *The American Heritage History of World War I,* p. 329.)

69. Gen. von Gallwitz, quoted in Toland, *No Man's Land,* p. 424.

70. *The American Heritage History of World War I,* p. 332.

71. Rockenbach, "Operations of the Tank Corps A.E.F.," and GSP to BAP, Sept. 16, 1918. Only one of the officers and three of the enlisted men were actually wounded while in a tank. Losses recorded are from Sept. 12–16, 1918.

72. "History of the 304th."

73. Rockenbach, "Operations of the Tank Corps A.E.F."

74. Wilson, *Treat 'Em Rough!,* p. 113.

75. The numbers bear out the magnitude of Saint-Mihiel: 40,000 tons of ammunition stored in forward supply dumps; 120 water plants dispensing 1,200,000 gallons per day; engineer stockpiles to build 300 miles of railway line; 100,000 tons of crushed rock for building roadways; 2,000 French trucks to haul men and supplies and more than 2,000 hospital beds. (Braim, *The Test of Battle,* p. 83.)

Chapter 18 Valor Before Dishonor

Unless otherwise cited, all letters from Patton to his family in 1918 are from Binder 8, PP-USMA.

1. George C. Marshall, *Memoirs of My Services in the World War, 1917–1918* (Boston, 1976), p. 149. As Marshall pointed out, the seventy-two guns in each division occupied fifteen kilometers of road space. To move a single division from Saint-Mihiel to the Argonne required nine hundred trucks for the troops alone.

2. Toland, *No Man's Land,* p. 429.

3. Wilson, *Treat 'Em Rough!* pp. 122–24.

4. Description and quote from Coffman, *The War to End All Wars,* p. 300.

5. HS, pp. 56–57.

6. Coffman, *The War to End All Wars,* p. 301.

7. Toland, *No Man's Land,* p. 430.

8. HS, p. 56.

9. GSP to BAP, Sept. 20, 1918.

10. MB-*PP* 1, chap. 30, passim.

11. "History of the 304th," and *Treat 'Em Rough!,* pp. 128–29.

12. GSP to BAP, Sept. 22, 1918.

13. Ibid., Sept. 19, 1918.

14. Ibid., Sept. 20, 1918.

15. Ibid., Sept. 25, 1918.

16. MB-*PP* 1, p. 610.

17. Maj. Gen. Robert Lee Bullard, the US III Corps commander, quoted in Braim, *The Test of Battle,* p. 97.

18. Toland, *No Man's Land,* pp. 431–32, and Braim, *The Test of Battle,* p. 97.

19. McCullough, *Truman,* p. 129.

20. "The Report of Personal Experiences in the Tank Corps," and HS, p. 59.

21. Ibid.

22. GSP to BAP, Sept. 28, 1918.

23. In 1918 there existed a narrow-gauge railway spur line that ran from Clermont to Varennes.

24. "History of the 304th."

25. Wilson, *Treat 'Em Rough!* p. 135.

26. As is common in attempting to recount the chaos of battle after the event, the timing of these events is in dispute. One eyewitness has Patton reaching the crossroads south of Cheppy at about 8:00 A.M. Patton's orderly, Pfc. Angelo, records the time as 9:00 A.M., and Capt. M. K. Knowles 10:00 A.M. Knowles's appears to be the most accurate version if for no other reason than the fact that the fog would have not lifted as early as 8:00 or even 9:00. According to Paul Braim, the sun began to break through the fog at 9:30. (Braim, *The Test of Battle,* p. 98, and the statements of various officers and enlisted men contained in memorandum by Captain. Knowles, "Report of Circumstances of wounds received by Col. G.S. Patton, Jr.," Dec. 17, 1918, Box 11, PP-LC. Patton's letter to Beatrice, dated Sept. 28, notes the time of arrival south of Cheppy as 9:30 A.M. It is reasonable that they would have remained there until at least 10:00 A.M., while Patton composed and sent his pigeon message.

27. GSP to BAP, Sept. 28, 1918.

28. Ibid.

29. "History of the 304th."

30. Statement by 1st Lt. Paul S. Edwards, "Gallant and Exemplary Conduct of Col. Geo. S. Patton, Jr. and the circumstances leading up to his being wounded in the Argonne attack—Sept. 26, 1918," dated Nov. 27, 1918, Box 45, PP-LC.

31. GSP to BAP, Sept. 28, 1918. According to Pfc. Joseph T. Angelo: "Then he and Capt. English chained three tanks together and we three stood on the parapet to give the hand signals to the tanks. The machine gun fire was heavy, also the shells and several of the men digging were hit but the colonel, Capt. English and myself stood on the parapet through it all and were not hit." Angelo later made two official statements describing Patton's actions on Sept. 26. One was dated Nov. 21, 1918, given to Capt. Edmund N. Hebert, the 1st Tank Brigade adjutant; the other was sworn to before Captain Knowles on Dec. 23, 1918. Both in Box 11, PP-LC.

32. RET-M.

33. Statement of 1st Lieutenant Edwards.

34. The quote above has combined extracts from the after-action statements of Angelo and Edwards. For the sake of clarity, the author has added quotation marks around the phrase "who comes with me."

35. GSP to BAP, Sept. 28, 1918.

36. MB-*PP* 1, p. 613.

37. "My Father as I Knew Him."

38. From Shakespeare's *Measure for Measure,* 1.4.78.

39. GSP to BAP, Sept. 28, 1918.

40. Ibid.

41. One of Patton's entourage, Sgt. Edgar W. Fausler, later helped to carry him to an aid station. Fausler cites the time and place as approximately 10:30 A.M., at a point about eight hundred meters southwest of Cheppy. (Fausler statement, Nov. 19, 1918, Box 11, PP-LC.)

42. GSP to Nita, Oct. 26, 1918.

43. MB-*PP* 1, p. 614.

44. Wilson, *Treat 'Em Rough!,* p. 138.

45. MB-*PP* 1, p. 615.

46. RET-M.

47. FA, p. 98.

48. Wilson, *Treat 'Em Rough!* pp. 138–39; MB-*PP* 1, pp. 614–15; WW I Diary, Sept. 26, 1918; Statement of 1st Lt. Edwards; and GSP to BAP, Sept. 28, 1918.

49. McCullough, *Truman,* p. 130. As Fred Ayer later wrote, Patton believed in a god of battles, who protected those who properly served him and remained faithful to a just cause, which Patton interpreted to mean *his* side and *his* cause. (FA, p. 100.)

50. GSP to BAP, Oct. 24, 1918.

51. Wilson, *Treat 'Em Rough!* pp. 131 and 133. The concussive effects of Semmes's head wound caused him to fantasize that he was being embalmed. Semmes later had no recollection of this incident but admitted it had been the object of considerable gallows humor among his friends in the Tank Corps. (HS, pp. 64–65.)

52. GSP to BAP, Oct. 24, 1918. "Fate" and the god of battles notwithstanding, had the wound hemorrhaged, Patton would certainly have bled to death in the shell hole.

53. GSP to Nita, Oct. 26, 1918.

54. GSP to BAP, Oct. 2, 1918. "It is strange," he observed, but "the 'gentlemen' make less noise over their wounds than the others. But there is little howling even on the train I heard hardly any noise."

55. Ibid., Oct. 4, 1918.

56. HS, p. 65, and GSP to BAP, Oct. 4, 1918.

57. GSP to BAP, Oct. 12, 1918.

58. Ibid., Oct. 15, 1918.

59. Ibid., Oct. 16, 1918. Patton broke out with a severe itch over most of his body. When the itch was finally cleared up, he was left with two boils on his back and two more on his chest that caused him more discomfort than his wound.

60. Ibid., Oct. 17, 1918.

61. GSP to Aunt Nannie, Oct. 19, 1918. When some of his friends criticized his promotion, an angry Patton observed: "If they had wanted to risk their skins in fighting instead of looking for staff jobs they might have been promoted too."

62. GSP to BAP, Oct. 19, 1918.

63. Ibid., Oct. 20, 1918. Many historians have credited Patton and Semmes with going AWOL from the hospital in order to return to duty at the front. The source is Harry Semmes's memoir, and it is patently untrue. Why the usually reliable Semmes recounts such a fantasy remains a mystery, but the truth is that with Patton's wound still open, he was unfit for duty. As badly as he desired to play a further role in the war, there is no evidence that he even considered leaving the hospital near Dijon. (In *Portrait of Patton*, p. 65, Semmes asserts: "As soon as Patton's wounds and mine were sufficiently healed to permit them to be sewn up, Patton got a car and together we deserted from the hospital to return to troops. Together we were forgiven by Pershing upon Patton's personal report to him.")

64. Rod Paschall, *The Defeat of Imperial Germany, 1917–1918* (Chapel Hill, 1989), pp. 186–87.

65. Coffman, *The War to End All Wars,* p. 299, and GSP to Aunt Nannie, Oct. 19, 1918.

66. HS, p. 59, and *Treat 'Em Rough!,* pp. 139–40.

67. GSP to BAP, Oct. 10 and 12, 1918.

68. Account of 1st Lt. Harvey L. Harris, quoted in *Treat 'Em Rough!* p. 147.

69. MB-*PP* 1, p. 636.

70. WW I Diary, Nov. 11, 1918.

71. RET-M, and Stanley Weintraub, *A Stillness Heard Round the World* (New York, 1985), pp. 17–22.

72. *The American Heritage History of World War I,* p. 344.

Chapter 19 Bitter Aftermath

1. Siegfried Sassoon (1886–1967) served as a junior officer in the Royal Welch Fusiliers in France during World War I, where he earned a Military Cross and the nickname "Mad Jack" for his bravery under fire. While recuperating Sassoon began writing poetry that angrily attacked those who ran the war. Quote is from *Dreamers,* written in 1918.

2. In the post-Vietnam era such behavior is often thought to result from what has been labeled posttraumatic stress disorder. Among its symptoms are nightmares, sleep disturbances, constant reliving of the war or specific events, alienation from others, lack of zest for life, and an inability to experience any sort of profound emotional response. The behavior of many Vietnam War veterans is similar to the experiences felt (but never defined) by their World War I, World War II, and Korean War brethren.

3. Harbord, *The American Army in France,* p. 436; italics in original.

4. From Euripides, *Iphigenia in Tauris,* quoted in John Bartlett, *Familiar Quotations,* 16th ed. (Boston, 1992), p. 68.

5. Box 60, PP-LC, written sometime in 1920. In all, Patton wrote some eighty-six poems during the period 1903–45.

6. James Wellard, *General George S. Patton, Jr.: Man Under Mars* (New York, 1946), p. 43. Virtually everything in this quote about what Patton allegedly did on September 26 is fantasy.

7. Stallings, *The Doughboys,* pp. 366 and 368.

8. Quoted in McCullough, *Truman,* p. 132.

9. Box 60, PP-LC.

10. Box 11, ibid.

11. Pershing is thought to have had a brief reunion with Nita in London during an official visit in late June 1919 but failed to invite her to the forthcoming ceremony. (MB-*PP* 1, p. 707, states that Pershing wrote Patton he had seen his sister in London, and Vandiver, *Black Jack,* vol. 2, pp. 1016–21, discusses Pershing's trip, but makes no mention of any stopover to see Nita Patton.)

12. Veterans like Patton were scornful of those who did not serve, including his own brother-in-law, Keith Merrill, a diplomat serving in the U.S. Embassy: "Had he not produced an infant I should not have believed him a man at all as it is I think his manhood does not reach above his belt." (GSP to BAP, Nov. 25, 1918.)

13. When he visited Pershing on Feb. 10, 1919, Patton attempted but failed to act as an intermediary between the two lovers.

14. Vandiver, *Black Jack,* vol. 2, p. 1008.

15. RET-M.

16. Ibid.

17. Ibid.

18. In 1922 Nita nearly married a man named Brain, but backed out at the last minute: "I am fated to be free. . . . So, you'll have me back on your hands for keeps. And the worst of it is I'm glad. Love is not for such as I." Notes her niece, Ruth Ellen: "She spent the rest of her life taking care of everyone." (Nita to Aunt Nannie, June 25, 1922, in RET-M.)

19. RET-M.

20. GSP to BAP, Nov. 16, 1918.

21. "History of the 304th."

22. GSP to BAP, Nov. 24, 1918.

23. Ibid., Nov. 26, 1918.

24. Ibid. Although it was not the same thing as combat, Patton did express satisfaction at what he had accomplished. "Realy it is almost as much fun working out manuvers as it is working out fights only the denouement is not so nice as in the case of a fight."

25. Brett to GSP, circa Oct. 23, 1918, MB-*PP* 1, p. 627. Patton learned from Brett that he had been only a hundred yards away attempting to round up some tanks to attack the German machine guns when he was wounded. "Of course, I didn't know you were there. But it was the hottest little hell I have ever burned in, and believe me I wouldn't have given three cents for my future pay vouchers for a while."

26. GSP to Brett, Nov. 25, 1918. On Dec. 1, the two officers returned to Cheppy, where they retraced their actions of Sept. 26 and Patton took photographs of the shell hole where he lay wounded. (WW I Diary, Dec. 1, 1918.)

27. GSP to BAP, Nov. 18, 1918. The "A" is a reference to the highly coveted letter awarded to varsity athletes at West Point.

28. Ibid. Marked "Nov 18, Night," it was the second letter sent on this date to Beatrice.

29. WW I Diary, Nov. 26, 1918, and Wilson, *Treat 'Em Rough!* p. 227.

30. MB-*PP* 1, p. 651. Blumenson believes the basis for disapproval was inadequate documentation of Patton's feat on Sept. 26. The firsthand observations of Knowles, Edwards, and Angelo proved to have been too sketchy, and only later was this oversight corrected when new, more detailed statements were taken from all concerned in Dec. 1918. In fact, it was Rockenbach's eagerness and haste to see his subordinate decorated that ultimately undermined the recommendation.

31. Microfilm Reel 1, PP-LC. Award recommendations and officer efficiency reports then, as well as now, were treated as separate matters. Thus Patton's excellent report was of no particular value in the decision regarding the DSC.

32. Ibid.

33. GSP to BAP, Nov. 22, 1918.

34. Ibid., Nov. 25, 1918.

35. GSP to BAP, Dec. 16, 1918. This is one of the few letters Beatrice wrote that survived burning in her fireplace after George Patton's death. Like a romantic schoolboy, Patton noted his reaction to her compliment in his diary: "I am glad she likes me."

36. GSP to BAP, Nov. 22, 1918.

37. Copy of citation in Box 3, PP-USMA.

38. GSP to Rockenbach, Dec. 5, 1918, Patton file, VMI-A.

39. GSP to BAP, Dec. 8, 1918, and McCullough, *Truman,* p. 135.

40. Ibid., Dec. 11, 1918. The dog cost Patton the extravagant sum of $200.

41. GSP to Aunt Nannie, Dec. 29, 1918.

42. Microfilm Reel No. 1, and WW I Diary, Dec. 31, 1918.

43. Coffman, *The War to End All Wars,* chap. 11, passim.

44. Ibid., p. 359.

45. GSP to GSP II, Jan. 28, 1919.

46. GSP to BAP, Feb. 9, 1919.

47. Ibid., Feb. 10, 1919. Patton had gone to Chaumont to present Pershing with several shell casings that had been made into tobacco and cigar cases.

48. Ibid., Feb. 18, 1919, and *Truman,* p. 136. (Among those in the 35th Division to shake Pershing's hand was Harry S. Truman, who was told "he had a fine-looking bunch of men.")

49. GSP to JJP, Feb. 23, 1919. Similar good-bye/thank-you letters were sent to a number of generals and colonels who had helped or befriended him.

50. GSP II to GSP, Feb. 20, 1919. It seems likely that Mr. Patton never really got over the obscene poem his son had written in Mexico, called "The Turds of the Scouts." (See chap. 13.)

51. MB-*PP* 1, p. 689.

52. As his biographer, W. A. Swanberg, writes, press baron William Randolph Hearst "fought for American neutrality at a time when neutrality was considered little short of treason. He defended the Germans. He assailed the British. He was called pro-German and anti-English." Hearst's highly unpopular anti-war stand was constantly elucidated in his newspapers and angered virtually everyone in uniform. Hearst was accused of being "the spokesman of the Kaiser," and it was this approbation that led to the revolt against Hearst's presence at the welcome ceremonies. (See *Citizen Hearst* [New York, 1961], passim.

53. *NYT,* Mar. 18, 1919. Most of the article was devoted to Patton and his exploits on Sept. 26, 1918.

54. *New York Herald,* Mar. 18, 1919.

55. GSP to GSP II, Apr. 1, 1919. Box 11, PP-LC, and MB-*PP* I, p. 694.

56. *New York Herald*, Mar. 18, 1919.

57. All press clippings are in Box 11, PP-LC.

58. The letter was dated Mar. 16, 1919, ibid.

59. RET-M.

Chapter 20   Eisenhower, Patton, and the Demise of the Tank Corps

Unless otherwise noted, all letters are from Box 11, PP-LC, and descriptions of Patton family life are in the Totten memoir.

1. MB-*PP* 1, p. 705, and Wilson, *Treat 'Em Rough!,* p. 210.

2. GSP to GSP II, June 2, 1919, Box 9, PP-USMA.

3. Microfilm Reel 1, PP-LC.

4. Quoted in MB-*PP* 1, p. 710, and GSP to GSP II, July 17, 1919.

5. "My Father as I Knew Him," and MB-*PP* I, p. 711. The local newspapers also greeted his return with headlines in the manner of a conquering hero.

6. RET-M; Merle Miller, *Ike the Soldier* (New York, 1987), p. 182, and GSP to GSP II, Apr. 1, 1919, Box 9, PP-USMA. Patton's daughters were not particularly enamored of eating at home. They preferred the mess hall where "none of the food had been 'good for you.'" Bee and Ruth Ellen also made "great friends" with the prisoners and their guards. Most were young men serving short sentences for going AWOL, drunkenness, or other offenses not serious enough to rate a stay at the mother of all military prisons, the infamous U.S. Disciplinary Barracks, at Fort Leavenworth, Kansas.

7. GSP to Rockenbach, Mar. 27, 1919, Patton file, VMI-A.

8. Quoted in RET-M.

9. Ibid. Some years later at another dinner party, Beatrice found herself seated next to the same officer, who somewhat desperately whispered that he was sorry she had to be seated next to him and would have switched place cards had he known in advance. "I don't suppose you will speak to me." No, agreed Beatrice, she would not, but in order not to embarrass her hostess, she decreed that when conversation was called for they would recite the multiplication tables to one another.

10. MB-*PP* 1, p. 705.

11. Wilson, *Treat 'Em Rough!* p. 60.

12. No class in the history of West Point ever produced more successful generals than did the class of 1915, whose members, in addition to Eisenhower, included: General of the Army Omar Bradley, Joseph McNarney, James Van Fleet, and George Stratemeyer. Of the 164-man graduating class, 59 members later became general officers, and a virtual "who's who" of the division and corps commanders who served in the Mediterranean and in the ETO. (USMA *Register of Graduates*.)

13. Miller, *Ike the Soldier*, p. 183.

14. Dwight D. Eisenhower, *At Ease: Stories I Tell to My Friends* (New York, 1967), p. 169.

15. Miller, *Ike the Soldier*, p. 182. Another biographer has said: "Tact became a way of life to Ike, whereas Patton (to paraphrase Winston Churchill on John Foster Dulles) was a bull who always carried his own china shop around." (Piers Brendon, *Ike: His Life and Times* [New York, 1986], p. 49.)

16. John S. D. Eisenhower, quoted in Miller, *Ike the Soldier*, p. 186.

17. Brendon, *Ike,* p. 49, and RET-M.

18. Quoted in Stephen E. Ambrose, *Eisenhower: Soldier, General of the Army, President-Elect, 1890–1952,* vol. 1 (New York, 1983), p. 65.

19. Brendon, *Ike,* pp. 42-43. Their son, Dwight Doud, whom they called by the nickname Icky, spent a good deal of his time in the Patton household. Shortly before Christmas 1920 Icky became ill with scarlet fever and died at the age of three. Icky's death, Eisenhower later said, "was the greatest disappointment and disaster of my life, the one I have never been able to forget completely." (Quoted in Eisenhower, *At Ease,* p. 181.) The children of Camp Meade loved to fish in a nearby ditch, using bent-pin hooks, and grasshoppers for bait. One day Icky returned to the Patton quarters with a fish he had just caught and asked Beatrice to cook it for him. By then the fish "was pretty run down. Ma told us to go and play. Pretty soon she came out with Icky's fish on a dish; a piece of buttered toast under it; garnished with a sprig of parsley and a lemon wedge. Icky ate it in ecstasy. Ma told us later that she had opened a can of sardines and dressed one up for the occasion."

20. Robert Leckie, *Delivered from Evil* (New York, 1987), p. 11.

21. D. Clayton James, *The Years of MacArthur, 1880–1941,* vol. 1 (Boston, 1970), p. 261.

22. Weigley, *History of the United States Army,* p. 396.

23. Eisenhower, *At Ease,* and Miller, *Ike the Soldier,* pp. 200 and 202.

24. "The U.S. Army Between World Wars I and II," Association of the U.S. Army Background Brief, No. 28, Mar. 1992. A further seven thousand troops served in the Philippine Scouts.

25. Ibid.

26. Time-Life editors, *This Fabulous Century, 1920–1930,* vol. 3 (New York, 1969), p. 25.

27. Ambrose, *Eisenhower,* vol. 1, p. 73. Even after a lifetime of public service, during which he met the most famous men of the twentieth century, Eisenhower would still repeat: "Fox Conner was the ablest man I ever knew." A wealthy, soft-spoken Southerner from Mississippi and an 1898 graduate of West Point, Conner was blessed with an insightful mind and the charm that inevitably drew men like Patton and Eisenhower into his circle of friends and admirers.

28. Ambrose, *Eisenhower,* vol. 1, p. 73.

29. Eisenhower, *At Ease,* p. 170.

30. Ibid., p. 171.

31. Ibid., p. 173. Shortly before Christmas 1919, Patton's routine was interrupted by the latest accident, one that was both painful and embarrassing. He was riding his horse on the target range at Meade when the animal spooked and bucked him forward onto the pommel of his western saddle, severely bruising his genitals. Still in severe discomfort a month later, he was placed on a month's leave of absence by the Camp Meade surgeon, who advised "absolute rest in bed."

32. GSP to the Secretary of the Naval War College, Sept. 24, 1919.

33. The Christie tank was designed with tracks that could be removed and replaced with wheels. The convertible tank was a revolutionary development that offered great promise by eliminating the need for tank carriers. The prototype was little more than a mobile platform and bore scant resemblance to a tank. Among Christie's unique accomplishments was the design of the turret track used on battleships. (See MB-*PP* 1, chap. 36, and George F. Hofmann, "The Demise of the U.S. Tank Corps and Medium Tank Development Program," *Military Affairs* [Feb. 1973].)

34. Quoted in Eric Larrabee, *Commander in Chief* (New York, 1987), p. 416.

35. Ladislas Farago suggests that Patton loaned money to help Christie's tank project, while Martin Blumenson merely notes that he may have privately subsidized him. (LF, p. 101, and MB, p. 121.)

36. Mildred Hanson Gillie, *Forging the Thunderbolt* (Harrisburg, 1947), pp. 274–75. The tank demonstrated for Rockenbach was the second of Christie's three prototype amphibious tanks. The vehicle weighed seven tons and could travel seven and a half miles per hour in the water. Christie died in poverty in Jan. 1944.

37. The DD (duplex-drive) tanks were medium M4 Sherman tanks modified to float and propel themselves with twin propellers. Their potential was never fully tested on D Day because of the bad weather on June 6. Most floundered and were lost in the heavy seas offshore. (For further discussion of the DD tank, see Gordon A. Harrison, *Cross-Channel Attack* (Washington, D.C., 1951), p. 192, n. 123, and pp. 304, 309, and 315.)

38. Larrabee, *Commander in Chief,* p. 416.

39. Because of the dearth of material being written in the aftermath of World War I, some articles written for the *Cavalry Journal* appeared in the *Infantry Journal.* An example was GSP's 1920 article, "Tanks in Future Wars."

40. Eisenhower, "A Tank Discussion," *Infantry Journal* 17 (Nov. 1920).

41. *Infantry Journal* (May 1920).

42. Fuller, *Tanks in the Great War,* p. 276.

43. Eisenhower, *At Ease,* p. 173.

44. Ibid.

45. Miller, *Ike the Soldier,* p. 186. In 1922 the name was formally changed as part of an effort to streamline the army school system.

46. Roger H. Nye, *The Patton Mind* (Garden City Park, N.Y., 1993), pp. 50–51, and Eisenhower, *At Ease,* p. 176. The author was a general named Morrison who headed the Department of Military Art at Fort Leavenworth, whom Patton believed lacked the knowledge and common sense required in a successful tactician.

47. Stephen E. Ambrose, "A Fateful Friendship," *American Heritage,* Apr. 1969.

48. Most of Eisenhower's correspondence with Patton during the interwar years disap-

peared when one of his steamer trunks was lost enroute from the Philippines to the U.S. in 1939. The handful of letters from Patton that survive in the EL are dated after 1939.

49. Farago, quoting William Bancroft Mellor, in LF, p. 103.

50. "Comments on Cavalry Tanks," *Cavalry Journal,* Vol 30, July 1921.

51. Quoted in Hofmann, "The Demise of the U.S. Tank Corps and Medium Tank Development Program."

52. Ambrose, *Eisenhower,* vol. 1, pp. 71–72.

53. MB-*PP* 1, pp. 716–17.

54. Hofmann, "The Demise of the U.S. Tank Corps and Medium Tank Development Program."

55. Ibid.

56. Truscott, *The Twilight of the U.S. Cavalry,* pp. 156–57.

57. The army system of promotions resulted in most officers holding two separate ranks. Regular officers held what was known as a temporary active duty grade, usually one or two ranks higher than their permanent one. Permanent grades were established by Congress, while the temporary grades were those needed to fill the needs of the army during any particular year. During World War I, for example, Patton held the temporary grade of colonel and the permanent grade of captain. Thus, when Congress ordered massive personnel reductions after World War I, officers who, like Patton, were retained, reverted to their permanent ranks.

58. MB-*PP* 1, p. 739. Rockenbach also kept Mitchell in command of the 305th Tank Brigade, Patton's sister unit in the AEF.

59. Nye, *The Patton Mind,* p. 50.

60. GSP to Nita, Oct. 19, 1919.

61. MB-*PP* 1, p. 742.

62. Ibid., pp. 703 and 738. The modern-day version of the crude preference statement filled out by Patton in early 1919 is often referred to as a "dream sheet" in which an officer records in order of preference where he/she would like to serve, and in what capacity.

63. Microfilm Reel No. 1, PP-LC, and MB-*PP* 1, p. 738.

64. GSP to Rockenbach, Sept. 28, 1919.

65. GSP speech to the 304th Tank Brigade, Sept. 28, 1919.

Chapter 21 "If You Want to Have a Good Time, Jine the Cavalry"

The primary source for anecdotal material about Patton family life at Fort Myer is Ruth Ellen Totten's memoir.

1. From "The Cavalryman," lecture delivered to his officers in 1921; Quoted in Nye, *The Patton Mind,* p. 53.

2. LF, p. 104, and Truscott, *The Twilight of the U.S. Cavalry,* p. 1.

3. Gregory J. W. Urwin, *The United States Cavalry* (New York, 1983), p. 182.

4. Maj. Gen. Robert W. Grow, "The Ten Lean Years: From the Mechanized Force (1930) To the Armored Force (1940)," part 1, *Armor* (Jan.-Feb. 1987). Grow was one of the participants during the turbulent era of the 1930s when mechanization finally came about in spite of the blindness of the cavalry and the obstinacy of the infantry. During World War II Grow commanded an armored division under Patton.

5. Nye, *The Patton Mind,* p. 53.

6. Ibid.

7. MB-*PP* 1, pp. 749–50.

8. Patton field notebook, 1921–22, Box 59, PP-LC. Quotes are not necessarily in the order in which they appear in the notebook but are grouped for relevance.

9. Dorothy Brandon, *Mamie Doud Eisenhower* (New York, 1954), pp. 116-17. Throughout the ride Beatrice was seated on the lap of a very large sergeant, who kept his arms firmly wrapped around her as the tank jolted and shuddered through the demonstration.

10. LF, p. 102. Farago states that the demonstration was for the prototype Christie tank and that Beatrice actually drove the machine. That may well have been part of the program;

however, it is doubtful that Patton would have chanced her actually driving a tank, much less this crude prototype. Dorothy Brandon's version in *Mamie Doud Eisenhower* is more probable.

11. Brandon, *Mamie Doud Eisenhower,* p. 117.

12. When she recovered from laughing at the incident, Mrs. Rivers told Beatrice not to worry. "We don't have any children of our own, and I dislike children very much, but I know your little girl didn't ruin my tulips to be mean."

13. RET-M.

14. Microfilm Reel No. 1, PP-LC.

15. GSP to Col. J.R. Lindsey, Oct. 10, 1921, Box 49, PP-LC.

16. Gen. Hamilton H. Howze, "'Upon the Fields of Friendly Strife,'" *Army* (Aug. 1987). The title of the article is from a quote by Douglas MacArthur about polo that reads: "On the friendly fields of strife are sown the seeds which, in other days and other ways, will bear the fruits of victory." During World War II, Howze led a combat command of the 1st Armored Division at Anzio, and it was his Task Force Howze that was poised to capture Rome during the breakout from the Anzio beachhead in May 1944, when Gen. Mark Clark made his controversial decision to defy Gen. Sir Harold Alexander (the Allied army group commander in Italy) and attack Rome from another direction, at great cost. (See D'Este, *Fatal Decision,* chaps. 21 and 22.)

17. The following year Patton wrote two papers on the subject, titled "Polo in the Army" and "Army Polo (No. 2)" in which he delivered his usual cogent arguments as to how the problems of the army team might be solved.

18. Truscott, *The Twilight of the U.S. Cavalry,* p. xv.

19. Howze, " 'Upon the Fields of Friendly Strife.' "

20. FA, p. 65.

21. Both anecdotes in FA, p. 71.

22. Ibid., pp. 75–76.

23. "George C. Marshall Interviews and Reminiscences for Forrest C. Pogue," p. 546.

24. Patton's poems are in Box 60, PP-LC. Underneath "The Soul in Battle," Patton wrote, "The idea is here but failed to hatch."

25. Ibid.

26. LF, p. 106.

27. Ibid.

28. Ibid., p. 110.

29. Observation on discipline and Patton's use of marginal notes is in Dietrich, "The Professional Reading of General George S. Patton, Jr."

30. Ibid. As Dietrich writes, GSP's notes (in Bernhardi's 1921 book) show special interest in "political and economic issues, national preparedness for war, the use of aircraft, tactics, cavalry, and the strategic threat presented by Japan. Patton's appreciation for combined arms, integrating the effects of infantry, cavalry, artillery, and air power, flows through his comments."

31. Nye, *The Patton Mind,* p. 54.

32. GSP's extensive notes are in Box 60, PP-LC.

33. Nye, *The Patton Mind,* p. 58.

Chapter 22   Past and Future Warrior Reincarnate

1. Field Marshal Earl Alexander of Tunis, *The Alexander Memoirs* (New York, 1962), p. 44, and Miller, *Ike the Soldier,* p. 184.

2. FA, pp. 102–3, and CRC, p. 271.

3. Other famous men who believed in reincarnation were the composers Gustav Mahler and Richard Wagner, and auto magnate Henry Ford. The Reverend Billy Graham has also articulated that one's soul has "conscience, memory, intelligence, and consciousness. . . . Your body soon goes to the grave but your soul lives on." (Quoted in Susy Smith, *Reincarnation* [Los Angeles, 1967], p. 14.)

4. Ruth Ellen Totten, lecture to the Topsfield (Massachusetts) Historical Society, 1974, quoted in George Forty, *Patton's Third Army at War* (London, 1978), pp. 58 and 60.

5. Ibid., pp. 60–61.

6. FA, pp. 94–95.

7. Ibid., p. 95.

8. Quoted in HS, p. 16.

9. RET-M.

10. Smith, *Reincarnation,* p. 75.

11. Nye, *The Patton Mind,* p. 65.

12. Ibid.

13. Ibid., p. 76. As author Smith notes, were Patton's ancestors "destined to follow his continual fighting career throughout history, always keeping an eye on him, and yet not have the opportunity for rebirth themselves?" More to the point, was this a form of intellectual selfishness on Patton's part?

14. RHP, pp. 239-40.

15. Nye, *The Patton Mind,* p. 6, and author interview of Ruth Ellen Totten, Aug. 26, 1991. See Nye, chap. 28, for Patton's vision of death as a Viking during his hospitalization in 1937.

16. RHP, p. 240.

17. Nye, *The Patton Mind,* p. 6. It is equally true that Patton's varied reading would have given him ample fuel for his memories of his past lives.

18. Ibid., p. 7.

19. Author interview of Aug. 26, 1991.

20. Roger Nye, *The Patton Mind,* p. 162. Bad Wimpfen is a fortified Roman town that in the thirteenth century was the imperial residence of the Hohenstaufens during the Holy Roman Empire. It was typical that Patton was drawn to a medieval town and church of which none of his contemporaries would have had the slightest knowledge or interest.

21. Quoted by Nye in *The Patton Mind,* p. 9.

22. Ibid., p. 65, and "Through a Glass Darkly," signed GSP, May 27, 1922, Box 60, PP-LC.

23. "Through a Glass Darkly."

24. Quoted in Nye, *The Patton Mind,* p. 6. Patton also noted that in ancient Bohemia the wagon was the first armored vehicle (p. 7).

25. HS, pp. 44–45.

26. RET-M.

27. FA, p. 81.

28. Ruth Ellen Totten, letter of Apr. 16, 1987, quoted in Carmine A. Prioli, "King Arthur in Khaki: The Medievalism of General George S. Patton, Jr.," *Studies in Popular Culture* 10:1 (1987).

29. RET-M. These two episodes are detailed in chapter 23.

30. Smith, *Reincarnation,* pp. 14–15, and author interview of Ruth Ellen Totten, Aug. 26, 1991. Patton's daughters also experienced psychic dreams. (See chap. 47.)

31. RET-M.

32. Prioli, "King Arthur in Khaki."

33. Nye, *The Patton Mind,* p. 7.

34. RET-M.

35. Prioli, quoting from the film *Patton* in "King Arthur in Khaki." Although the conversation in the film is apocryphal, it would have been typical of Patton to have spoken these words.

Chapter 23   Student Days, Boston Baked Beans, and Hawaiian Leis

Anecdotal material on Patton family life is from the Totten memoir.

1. Brig. Gen. Paul Robinett, "George Smith Patton, Jr.," Box 16, Robinett Papers, GCML.

2. MB-*PP* 1, p. 772.

3. The original French fortress was located somewhere near the present site, but its exact whereabouts have never been determined. (See John W. Partin, ed., *A Brief History of Fort Leavenworth, 1827–1983* [C&GSC, 1983].)

4. Lt. Col. Phillip W. Childress, "Manifest Destiny, 1837–1858," in ibid. Fort Leavenworth also served as the primary western supply depot for the army.

5. The first school established at Fort Leavenworth was the School of Application for Infantry and Cavalry (1881), later the U.S. Infantry and Cavalry School (1886). At the turn of the century Secretary of War Elihu Root established four schools for junior officers, two of which were the School of the Line (formerly the Infantry and Cavalry School), and the General Staff School, to train staff officers. When the Army War College in Washington, D.C., took over the functions of the latter, what emerged was the Command and General Staff School, whose mission was "to train officers in the use of combined arms in the division and corps and the command and staff functions for division and corps as they related to tactics and logistics." (Jonathan M. House, "The Fort and the New School, 1881–1916," in ibid.)

6. A notable exception was Pershing, who failed to receive an appointment in 1889, at a time when the prestige of the school had yet to be fully established and before it became a college for future staff officers. (See House, and Vandiver, *Black Jack,* vol. 1, pp. 77–78.)

7. Ibid., p. 50. The curriculum covered a wide variety of subjects ranging from tactics and logistics to every aspect of operations at the division, corps, and army levels. (Sources on Fort Leavenworth and the curriculum of Patton's day are Timothy K. Nenninger, *The Leavenworth Schools and the Old Army,* [Westport, Conn./London, England, 1978]; Boyd L. Dastrup, *The US Army Command and General Staff College: A Centennial History* [Fort Leavenworth, 1982]; Elvid Hunt, *History of Fort Leavenworth, 1827–1937* [Fort Leavenworth, Kans.], and Partin, *A Brief History of Fort Leavenworth*).

8. One of the lone exceptions was a future friend and subordinate of Patton, a maverick cavalryman, Terry de la Mesa Allen, who attended Leavenworth the same year as Dwight Eisenhower (1925–26). Allen managed to start off on the wrong foot the very first day by skipping class in order to keep a tennis date with a senior officer's attractive blond daughter. Not surprisingly the commandant referred to Allen as the "most indifferent student ever enrolled there." (Carlo D'Este, *Bitter Victory: The Battle for Sicily, 1943* [New York, 1988], p. 269.)

9. GSP to BAP, n.d. (but circa late 1923, shortly before their son was born), Box 12, PP-LC.

10. Ambrose, *Eisenhower,* vol. 1, p. 80. When Eisenhower attended in 1925–26 he refused to succumb to the pressure and decided a fresh mind each morning was preferable to the usual exhaustion of the average student. He usually studied only two and one-half hours and was in bed by 9:30 P.M.

11. Partin, *A Brief History of Fort Leavenworth,* p. 52.

12. "My Father as I Knew Him."

13. RET-M. Their mischief included attempting to hit one of the highly polished black-and-white marble squares in the foyer by spitting from the second-floor balcony of the Merrill homestead. They evaded detection until one day when the Italian butler came from the pantry with a tray laden with tea and hit the wet square. Years later Ruth Ellen would chortle as she related: "The sights and sounds were worth the punishment that followed."

14. Joseph J. Angelo to GSP, Feb. 3, 1924, Box 12, PP-LC, and MB-*PP* 1, p. 776.

15. Another honor graduate was one the army's finest soldiers, Troy Middleton, who became one of Patton's corps commanders in World War II. According to the article in the local *Leavenworth Times* describing the graduation of Patton's class: "Regulations do not permit the publication of class standing or rank in class. They do permit publication of the honor graduates, arranged alphabetically." When Eisenhower attended two years later he became the top student in the class by two-tenths of a point over his friend Leonard T. Gerow. Either the regulation was changed or class standings (at least of the top twenty-five)

were revealed despite the prohibition. (*Leavenworth Times,* June 20, 1924, Microfilm Box 847, Leavenworth Public Library.)

16. Ibid., June 1, 1924.

17. Patton asked Eisenhower to return the notes, which were an inch thick, and later wrote in the margin of the title page: "Prepared by Major G. S. Patton, Jr., GSC (Cav), Honor Graduate, C&GS, 1924. Every user of these notes has graduated from the Command and General Staff School in either the honor or Distinguished Group—G.S.P., Jr." (Nye, *The Patton Mind,* p. 61, and Box 59, PP-LC.)

18. GSP to DDE, July 9, 1926, EP. Patton's letter is frequently quoted by, among others, Ambrose in *Eisenhower,* vol. 1, p. 82, and Maj. Mark C. Bender, *Watershed at Leavenworth: Dwight D. Eisenhower and the Command and General Staff School* [C&GSC, 1990].

19. Bender, *Watershed at Leavenworth,* p. 47. The notebook seems to have clearly helped but, as he would during World War II, Patton tended to underestimate the brilliance of Dwight Eisenhower.

20. GSP to DDE, July 9, 1926, EP, and Ambrose, *Eisenhower,* vol. 1, p. 82.

21. Ibid., and Nye, *The Patton Mind,* pp. 74–76. Patton's reflections about those whom he called "skulkers" is from p. 76 and a lecture, "Why Men Fight," given in 1927, Box 50, PP-LC. He advocated placing officers behind the front lines with orders to shoot anyone who attempted to desert. Patton's strong feelings about skulking are in keeping with his own personal code of honor and would become manifest in August 1943 in Sicily.

22. Brig. Gen. Harry A. Smith to Maj. Gen. Robert L. Howze, Mar. 22, 1924, Box 121, PP-LC.

23. GSP to JJP, Sept. 22, 1924, Box 12, PP-LC, and *Boston Evening Transcript,* Sept. 13, 1924, microfilm in Boston Public Library. Pershing was always flattered by the attention and praise heaped on him by Patton, and wrote that it was "particularly pleasant and somewhat rare to hear such good things of oneself while still alive. . . . I want you to know that I am particularly grateful for the very loyal and efficient support you have always given me." (Letter of Sept. 30, 1924, ibid.)

24. Having qualified at Leavenworth as a General Staff officer, Patton fully expected to be assigned next to the War Department General Staff and was surprised to receive orders to Hawaii.

25. MB-*PP* 1, p. 780; Nye, *The Patton Mind,* p. 63, and "Armored Cars with Cavalry," *Cavalry Journal* (Jan. 1924), copy in Box 49, PP-LC.

26. Essay, "Cavalry in the Next War," Feb. 4, 1930, Box 50, PP-LC.

27. Lecture to 11th Cavalry, Boston, Nov. 7, 1924, Box 49, PP-LC; MB-*PP* 1, p. 783, and Nye, *The Patton Mind,* p. 63.

28. Italics in original; Patton managed to save a number of books, which were later carefully rebound. Virtually everything else had been reduced to waterlogged junk. (RET-M, and letters of Mar. 27, Apr. 7 and 10, 1925, PP-USMA)

29. GSP to BAP, Apr. 2, and 10, 1925. He soon learned that he had been assigned to Hawaii only because his predecessor was reputed to be "the meanest man in the Army," and his superiors had successfully schemed to get rid of him.

30. Microfilm Reel 1, and letter, Brig. Gen. H. A. Smith to GSP, May 19, 1925, Box 21, PP-LC. Smith warned Patton not to become too accomplished a staff officer and thus be assigned exclusively to these duties in the future. Until his untimely death in 1929, Smith continued to be one of Patton's greatest friends and boosters.

31. General Harbord was unaware of Patton's interest and, having already pledged his support to another officer, was unable to help.

32. *Boston Evening Transcript,* Wed., Mar. 24, 1926, Box 12, PP-LC. Beatrice was also motivated to find any means possible to help keep up her husband's flagging morale during the interwar years.

33. Hawaii was a memorable experience for Patton's daughters. At her mother's insistence, Ruth Ellen finally learned to swim. Her swimming efforts had begun on a bad note seven years earlier at Avalon, when she was three and Patton announced that little animals

all swam naturally. To prove his point he threw baby Ruth Ellen into Salem Harbor. After she sank for the third time, Beatrice managed to make the point to her stubborn husband that "here was one little animal who did not swim naturally. He had to dive in to get me, and ruined a new pair of of white flannel trousers: a loss he brooded over for years."

34. Ruth Ellen was awake in the next room when Patton returned to their beach cottage and the safe haven of his wife's presence. The woman was described as "a lean and scrawny woman, stylishly dressed. . . but so old! So old! I was appalled at the thought that anyone who appeared so juiceless could be contemplating the magic of love—and in the nude!" Henceforth Patton would never return home when the woman was giving Beatrice a lesson.

35. In addition to being proficient on the piano and the guitar, Beatrice also learned to play the mandolin, the harpsichord, and the musical saw.

36. Quoted in MB-*PP* 2, p. 6.

37. "My Father as I Knew Him."

38. Ibid.

39. Beatrice was equally devastated by the loss of the gentle man who had been more like a father to her than a father-in-law: "She had known the Pattons her whole life long and was deeply devoted to them . . . they were her other parents."

40. RET-M, and RHP, pp. 209–10. After a lifetime living with the Pattons at Lake Vineyard, Aunt Nannie moved out on her own soon after Mr. Patton's death.

41. It has always been thought that Patton went to California by himself; however, a letter to his lifelong friend, Capt. (later Maj. Gen.) Floyd Parks, from Beatrice indicates that Parks looked after the Patton children during their absence in California. (BAP to Floyd Parks, n.d., but clearly written on Lake Vineyard stationery shortly after Mr. Patton's death, Parks Papers, EL.)

42. "My Father as I Knew Him."

43. Ibid. He signed it "Your devoted son, G. S. Patton, Jr, July 9, 1927."

44. MB-*PP* 1, p. 811. There is a thin line between a senior staff officer acting in the name of a commander and the usurpation of the commanding general's prerogative to discipline a subordinate commander. In this instance Patton clearly crossed the line. As Blumenson notes: "For a major to 'correct' a brigadier general was inadmissible."

45. Ibid., p. 813.

46. Ibid., p. 816.

47. Harry Smith wrote in response to a letter: "When I read Mrs. Smith the sentence, 'May we soon have a war, so that you can exercise command,' she said, 'You needn't read any more. I know George Patton wrote that.' " (Smith to GSP, Oct. 26, 1926, Box 21, PP-LC).

48. Microfilm Reel 1, PP-LC. Also quoted in Nye, *The Patton Mind*, p. 82.

49. GSP to BAP, Nov. 7, 1926, PP-USMA.

50. During his brief leave, Patton's main interest was the magnificent stable that came equipped with every modern convenience. But it was Beatrice who turned Green Meadows into their permanent home. (RET-M.)

Chapter 24   The Washington Years

Anecdotal material about the Pattons is in the Totten memoir.

1. MB, p. 130. Death or serious injury from equestrian-related activities, particularly steeplechasing and show jumping, is not only common but, short of war, ranks very high in the lexicon of dangerous pursuits. A five-foot jump is so dangerous that it can be undertaken with relative assurance of success by only the most experienced horseman.

2. John Knight Waters (1906–1989) graduated 148th of 296 in the USMA class of 1931.

3. MB-*PP* 1, p. 840. An example was a 1929 paper called "The Value of Cavalry" in which Patton waffled that "the effectiveness of Cavalry is in no way reduced by the advent of mechanical units. . . [the] mobility of cavalry and its universal adaptability are unal-

tered." The paper ended up in the War Department G-3 staff, where he alleged that his close friend, Maj. Adna R. Chaffee, had it copied and submitted to the G-3 as having been written by his own staff. The G-3 took the paper to Patton's boss, Major General Crosby, saying it was a great paper and why couldn't someone in Crosby's office have written such a fine paper. Crosby then showed the G-3 the original paper with Patton's name as the author. Patton was incensed that the G-3 and Chaffee, who would soon become the driving force behind the creation of the armored force, never acknowledged he had authored the paper. (Paper is dated Aug. 15, 1929, Box 50, PP-LC.)

4. William J. Woolley, "Patton and the Concept of Mechanized Warfare," *Parameters* (Autumn 1986). This article is the most incisive and informative of the writings about Patton and mechanization during the interwar years.

5. "Mechanization and Cavalry," cowritten by Maj. C. C. Benson, *Cavalry Journal* 39 (Apr. 1930). Also quoted in Nye, p. 87.

6. Patton's tour of duty in Washington coincided with the publication of a number of books about cavalry and tanks by British, French, and German authors, each of which he avidly studied. He also spent considerable time reflecting on the larger subject of future wars, a theme to which he returned to time and again. His last paper in Hawaii suggested that whenever the next war came it would be fought with large conscript armies similar to those of World War I, rather than by small professional forces. ("Modern Cavalry," lecture to the Marine Corps School, Quantico, Va., Jan. 1931, quoted in Nye, *The Patton Mind*, p. 87.)

7. Nye, *The Patton Mind*, p. 89. Patton's favorite biography was G. F. R. Henderson's *Stonewall Jackson and the American Civil War*, which his father had first read to him at the age of twelve.

8. Ibid., p. 90.

9. Ibid., p. 85.

10. John S. D. Eisenhower, *Strictly Personal* (Garden City, N.Y. 1974), pp. 8–9.

11. Woolley, "Patton and the Concept of Mechanized Warfare."

12. Ibid.

13. Ibid. Readers interested in delving further into the evolution of mechanization in the U.S. Army will find this article and Grow's four part article in *Armor*, "The Ten Lean Years: From the Mechanized Force (1930) to the Armored Force (1940)," the most instructive.

14. For a synopsis see Patrick J. Cooney, "U.S. Armor Between the Wars," *Armor* (Mar.-Apr. 1990).

15. "Mechanized Forces," *Cavalry Journal* (Sept.-Oct., 1933), based on a lecture given several times by Patton, who attempted to make the leap from rejecting the need for armor to that of courage as the prime ingredient for winning a battle. His attempt to connect the two was specious, and even Patton would have undoubtedly acknowledged that while leadership and courage are the heart of an army, a cavalryman with a rifle would not have fared well against an enemy tank, particularly the new family of tanks being created by the German army.

16. MB-*PP* 1, p. 844.

17. Jon Clemens, "Waking Up from the Dream: The Crisis of Cavalry in the 1930s," *Armor* (May-June 1990). During his tour of duty at Fort Myer from 1932 to 1935, Patton's commanding officer, Col. Kenyon Joyce, assisted RKO pictures in producing a western film called *Keep 'Em Rolling*, starring Walter Huston. The film premiered at Fort Myer with Stimson in attendance. It was a typical melodramatic Western of that era, in which an old cavalry soldier deserts rather than permit his beloved but aged horse to be destroyed in accordance with an order from on high. As Joyce writes: "To mention but one of those who brushed a tear during the showing was one of the most hard-bitten soldiers of all time, George Patton." (Unpublished memoir, Joyce Papers, USAMHI.)

18. In 1931–32 the War College mission was defined as (1) "to train officers for the conduct of field operations of the Army and higher echelons; (2) to instruct officers in War Department General Staff duties; (3) to train officers for joint operations of the Army and Navy; (4) to instruct officers in the strategy, tactics and logistics of large operations in past

wars with special reference to the [first] world war." (Extracted from "Course at the Army War College, 1931–1932," USAMHI.)

19. Simon Bolivar Buckner, Jr., a member of Patton's original class of 1908, became a lieutenant general in World War II and was killed in action on the island of Okinawa in June 1945.

20. The paper was cumbersomely titled "The Probable Characteristics of the Next War and the Organization, Tactics, and Equipment Necessary to Meet Them." (Box 52, PP-LC.)

21. Ibid. Nothing is known of what eventually happened to the paper.

22. Clemens, "Waking Up From the Dream." One of the final directives issued by Patton's friend and mentor, Gen. C. P. Summerall, before leaving office in 1930, was an order: "Assemble that mechanized force now. Station it at Fort Eustis [Virginia]. Make it permanent, not temporary." MacArthur directed the individual branches to fund mechanization from their own pockets, but none were deep enough during the Great Depression to finance anything more than a token armored force, which, at Chaffee's instigation, was moved in 1931 to Fort Knox, Kentucky (now the permanent home of armor and the Armor School). Among the first to answer the call was Maj. Serano Brett.

23. Quoted in MB-*PP* 1, p. 889.

24. GSP to his mother, Nov. 30, 1931.

25. The artist was Donald Gordon Squier. The caption reads: "Col. George S. Patton, Jr., Student at the Army War College, 1932." The reference to Patton as a full colonel was obviously in deference to his wartime rank, for in 1931–32 he was still a major and would remain so until his promotion to lieutenant colonel in March 1934.

26. Microfilm Reel 1, PP-LC.

27. Clemens, "Waking Up From the Dream."

28. Ibid. In June 1932 Patton was awarded the Purple Heart and a wound stripe (worn on the uniform sleeve) for his wound at Cheppy.

29. The bonus was in the form of "Adjusted Service Certificates"—that is, a bond. An excellent description of the Bonus March and Patton's role in the affair is Truscott, *The Twilight of the U.S. Cavalry,* chap. 5.

30. "Joe Angelo Pleads for Veterans," date and publication unk., circa 1931, copy in Box 12, PP-LC. During his testimony Angelo extravagantly embellished his own role in ways for which there is no evidence whatsoever. Beatrice was so upset by the implications of Angelo's remarks about her husband that she termed him "a catspaw—a pathetic type." Nevertheless, she took her children to meet Angelo, who recounted the events of Sept. 26, 1918, which had become a part of Patton family lore. (See RHP, pp. 212–13.)

31. Quoted in Gene Smith, *The Shattered Dream* (New York, 1970), p. 142.

32. The 3d Cavalry was trained to perform riot duty, and in the use of tear gas. The afternoon of July 28, Hurley delivered to MacArthur an order to enter and clear downtown Washington, where a number of veterans were occupying condemned buildings, "without delay." Hoover's original orders to Hurley had been to round up and incarcerate every single bonus marcher, so that each could be identified and the ringleaders tried in court. After consulting MacArthur, Hurley, believing the order was not only impossible to carry out but potentially disastrous, amended the order to that of clearing the city. (Smith, *The Shattered Dream,* pp. 157–58.)

33. MB, p. 134.

34. Smith, *The Shattered Dream,* p. 161.

35. James, *The Years of MacArthur,* vol. 1, p. 401.

36. Brendon, *Ike,* p. 63. MacArthur's belief that this was the start of a Communist uprising throughout the country had absolutely no basis. (See also Smith, *The Shattered Dream,* p. 158.)

37. Smith, *The Shattered Dream,* p. 164. Smith's source is Matthew Josephson, *Infidel in the Temple* (New York, 1967), pp. 99–100, who wrote that Patton thought the melee far more serious than portrayed. Patton's 1934 recollection was borne out by newsreel film depicting some veterans fighting back with considerable fervor. "After all, those veterans were only four blocks from the Capitol. It might have been a bad thing if they had got in

there. I myself was hit on the head and had to be taken to the rear." According to Joseph-son's account: "A heavy, well-aimed brickbat dropped Patton from his horse, rendering him unconscious."

38. Quoted in Truscott, *The Twilight of the U.S. Cavalry,* p. 129. Truscott's account appears to dispute another dubious version in which Joe Angelo was one of those who fled into the night, "flapping from his shirt . . . the DSC given him in 1918 for saving George Patton's life." (Smith, *The Shattered Dream,* p. 166, whose source is Waters, *B.E.F.,* p. 121.) Patton had recommended Angelo for the Medal of Honor for saving his life and had presented him with an engraved watch, but there is no indication the Pattons ever paid to establish a business for him. Patton's grandson believes it was an unfair dis-missal of Angelo because he had "unwittingly brought embarassment on him." (See RHP, pp. 212–13.)

39. "Federal Troops in Domestic Disturbances," Box 12, PP-LC. The twenty-three-page paper was written with typical Patton thoroughness, providing both an historical and legal basis for federal intervention in domestic disturbances.

40. Roger Daniels, *The Bonus March* (Westport, Conn., 1971), pp. 344–45, n. 30. Daniels writes of several alleged atrocities committed by the army: "Part of the story may have come from the boasting of the then Major George S. Patton, the swashbuckling armored commander of World War II, who rode with the cavalry that day and told many yarns about his heroics."

41. Josephson, *Infidel in the Temple,* p. 275.

42. Lecture, Nov. 11, 1932, Box 52, PP-LC. Patton said:

Personally I believe that our form of preparation lays in the maintenance of an adequate and immediately available regular army and navy with which to get to the fire quickly and hold it in check until our national man power in smaller num-bers and with better training becomes available. . . . When our time comes to sleep in Arlington [National Cemetery] we can lie down with quiet minds content in the knowledge that as soldiers and citizens we have done our full duty.

43. Mellor, *General Patton: The Last Cavalier,* pp. 98–99.

44. Source and quote from ibid., p. 97.

45. Nina Carter Tabb, article about Cobbler Hunt, Oct. 20, 1934, newspaper unknown, clipping in Box 4, PP-LC.

46. Col. Charles L. Scott to GSP, Apr. 8, 1934, Box 2, PP-LC; italics in original. Scott was a cavalry officer and a 1905 graduate of West Point.

Chapter 25   War Clouds

Unless otherwise cited, the source for Patton's family life is the Totten memoir.

1. Beatrice Ayer Patton's *Blood of the Shark:* "A Romance of Early Hawaii," was pub-lished in Honolulu and few copies seem to have survived. (One is in the Library of Congress.) The book was a considerable success in Hawaii where it eventually went through four printings, the first of which sold out the first week. (RHP, p. 234.)

2. Peggy O'Connell Parker, "The *When and If*—Patton's Own," *Army* (Feb. 1978).

3. RHP, p. 233.

4. Ibid., pp. 233–34.

5. On another occasion, when Beatrice remained home with bronchitis, Patton and Ruth Ellen sailed to Maui to attend a five-day social event. Patton realized the hostess had an eye for him and before retiring to the guest house told his daughter, whose bedroom was next door, that if she heard him call out and ask if she was all right, she was to quickly come to his room. Later Patton yelled, and Ruth Ellen dutifully ran to his room in time to see the hostess scuttling out the door in a state of semiundress clearly visible in the moonlight. Pat-ton roared with laughter until the tears ran down his face. At breakfast the hostess glared at Ruth Ellen with loathing and said: "What a hell of a little spoil-sport you turned out to be!"

6. RHP, p. 235. The hurt Patton inflicted on his wife was no less painful for Ruth Ellen, who witnessed firsthand the near ruin of her parents' marriage by her father's brazen dalliance with her best friend.

7. Ibid., pp. 232–33.

8. MB-*PP* 1, p. 910.

9. Robert Calvert Jr., " 'Drum, Drum, I Wish He Would Stop Beating His Own Drum,' " *Army* (Sept. 1989).

10. Ibid., and RET-M. There is no evidence to support the allegation by biographer Ladislas Farago that a vindictive Drum savaged Patton in his efficiency report and plotted further revenge. As the endorsing officer, Drum had the right to enter any comments he chose on Patton's efficiency report, but in his next report, written shortly after the polo incident, he wrote: "Heretofore I have noted on this officer's Efficiency Reports a weakness in 'Tact.' In the last year he has overcome this weakness in a satisfactory manner. Colonel Patton has those qualities so essential to a superior combat leader." If Drum previously wrote anything unfavorable about Patton, it does not appear in any official record. The reports Drum endorsed in May 1936 and June 1937 were concurrences with the rating officer, who cited Patton as "Superior" and "of very high general value to the service." (Patton's efficiency reports are in Microfilm Reel 1, PP-LC.)

11. Edward S. Miller, *War Plan Orange* (Annapolis, 1991), p. 47.

12. "Surprise," paper submitted to the Chief of Staff, Hawaiian Division on June 3, 1937, copy in Box 53, PP-LC. Also see Nye, *The Patton Mind,* pp. 104–7. At the conclusion of his prophetic paper, Patton had written: "It is realized that the events above enumerated are not likely of occurrence. On the other hand the vital necessity of Japan of a short war and of the possession at its termination of land areas for bargaining purposes may impel her to take drastic measures. It is the duty of the military to foresee and prepare against the worst possible eventuality."

13. Miller, *War Plan Orange,* p. 51. One of the main topics of study at the Army War College during the 1930s was the army's role in War Plan Orange (war between the United States [code-named Blue] and Japan [Orange] without benefit of allies). In the later Rainbow plans the United States was part of a larger military alliance. (See also Henry G. Gole, "War Planning at the War College in the Mid-1930s," *Parameters* [Spring 1985].)

14. Parker, " 'The *When and If*'—Patton's Own." The boat was left to be sold and the family headed for Green Meadows. Actress Greta Garbo was apparently set to purchase the schooner, but later changed her mind after considerable publicity over the sale.

15. FA, p. 104.

16. RET-M. There is no direct evidence that the convening of a board of officers was authorized by army regulations of the time, and the affair seemed to possess all the elements of a witch-hunt in which the object was to pin blame on Patton for his accident.

17. Patton's injury was a violent trauma in which there was serious risk of complications. As orthopedic surgeon Dr. Thomas W. Brooks Jr. points out, medical knowledge and treatment of thrombophlebitis in 1937 was very limited and the risk of complications, including death, was high. Among Patton's problems were a chronic draining infection (that today would be treated by antibiotics) and the possible need to amputate his leg if it failed to heal. Dr. Brooks also notes that there was nothing miraculous about his survival and that the time it took to heal was about average for someone with this type of injury. That Patton successfully recovered can be attributed to blind luck. (Telephone interview with Dr. Thomas W. Brooks, Jr., June 14, 1993.)

18. FA, p. 105.

19. Ibid., p. 106.

20. RHP, p. 241.

21. Ibid.

22. In addition to serving on the faculty of the Cavalry School, Patton also was the executive officer of the Academic Division, and of the 9th Cavalry Regiment.

23. A full portrait of Terry Allen is in D'Este, *Bitter Victory,* chapter 15. See also Col. Bryce F. Denno, "Allen and Huebner: Contrast in Command," *Army* [June 1984].)

24. Col. C. L. Scott had earlier complimented Patton for not complaining when a judge's decision had gone against him at the Cobbler Hunt in 1934.

25. Efficiency reports of June 30, 1934, and May 14, 1935, Microfilm Reel 1, PP-LC.

26. LF, p. 128.

27. GSP to BAP, Aug. 4, 1938, PP-USMA.

28. Ibid., Aug. 23, 1938.

29. Ibid., Aug. 27, 1938.

30. Wainwright was the highest-ranking officer captured by the Japanese in the Philippines after the fall of Corregidor in early 1942.

31. Quoted in LF, p. 133.

32. Patton file, CMH.

33. RET-M. The Wilde quote is from *The Ballad of Reading Gaol.*

34. Gen. James H., Army Polk, "Patton, 'You Might as Well Die a Hero (Dec. 1973).'"

35. HS, p. 78.

36. Brig. Gen. Paul Robinett, "George Smith Patton, Jr," Robinett Papers, GCML.

37. The first Totten was the tenth graduate of West Point in 1805, and Ruth Ellen's fiancée was merely the latest. As engineers and coast artillerymen, Tottens had helped build the Panama Railroad, and Fort Totten, New York, was named for another ancestor.

38. Joanne Holbrook Patton's maternal great-great-grandfather was Brig. Gen. Eli DuBose Hoyle, who graduated in 1812 and later returned as superintendent from 1833–1838. Fort DeRussy, Hawaii (site of the Armed Forces Recreation Center) was named in his honor.

39. Even after their engagement, Patton would still drop shoes on the floor overhead as a signal that it was time for Lieutenant Totten to leave. Sometimes it took as many as five shoes before his message could no longer be ignored.

40. FA, pp. 109–10.

Chapter 26   Division Commander

Unless otherwise cited, letters from GSP to BAP are in Box 13, PP-LC.

1. Quoted in LF, p. 128, written in Sept. 1939.

2. Ibid., chap. 4. Marshall's standing with the president was enhanced by the support of FDR's powerful adviser, Harry Hopkins, who also greatly admired him. See also Thomas Parrish, *Roosevelt and Marshall* (New York, 1989), pp. 98. Quote from Parrish, p. 519.

3. Forrest C. Pogue, *George C. Marshall: Organizer of Victory, 1943–1945* (New York, 1973), vol. 3 of his superb four-volume biography of Marshall.

4. Maj. Gen. Elwood Quesada, oral history transcript, May 23, 1975, USAMHI.

5. GSP to BAP, July 27, 1939, PP-USMA. Patton's standard of living had clearly not slipped when he asked Beatrice to send him a check for $5,000, "as I am getting pretty low." He was not quite as generous with Joe Angelo, who was on relief, sustained only by a menial job with the WPA. In a letter to an ex-captain who had written him of Angelo's plight, Patton replied with a check for $25.00. "My mother and I helped him considerably, but due to changed conditions I am not able to do as much for him now as then." Patton seemed to have tired of anything to do with his former batman.

6. Parrish, *Roosevelt and Marshall,* p. 117. Patton never understood that what precious little free time Marshall had was reserved for his family, not for talking "shop," which is what Patton's presence would have entailed.

7. "George C. Marshall Interviews," p. 510.

8. Maj. Gen. John K. Herr to GCM, July 25, 1940, Herr Collection, in the possession of Fanny de Russy, Washington, D.C. Among the few officers over sixty who survived were Ben Lear, and John L. DeWitt, who were given key posts in the War Department.

9. McNair memo to GCM, "Higher Commanders," Oct. 7, 1941, PP-USMA. Forty-eight years later Patton's son would justifiably scoff that "McNair's predictions were not too hot."

10. Quoted in Leonard Mosley, *Marshall: Organizer of Victory* (London, 1982), p. 189. Although Patton failed to recall it, Marshall had written to him in [Sept. 29] 1936 and noted that "if he [Marshall] ever reached a position of high command he would want his services, since Patton was the type of man who would go through hell and high water." (Forrest C. Pogue, *George C. Marshall: Ordeal and Hope, 1939–1942* [New York, 1966], p. 103.)

11. Quoted in LF, p. 129. Another Marshall favorite was the maverick Terry Allen, whose behavior and idiosyncrasies were far more outlandish than Patton's. However, behind the facade of Allen's devil-may-care attitude was an astute mind, and an outstanding leader. In 1940 Marshall promoted Allen from lieutenant colonel to brigadier general and in early 1942 gave him the coveted command of the 1st Infantry Division, the Big Red One.

12. Ibid., p. 130.

13. Ibid., Chap. 8.

14. MB, p. 146.

15. GCM to GSP, Sept. 23, 1939, MB-*PP* 1, p. 944. Similar pairs of stars were sent to his friend and mentor, Kenyon Joyce, who had recently been promoted to major general. As Blumenson notes, about this time there began a lessening of the close relationship the two officers had enjoyed for many years. Joyce criticized Patton's penchant for wide flanking movements when his cavalry "defeated" a mechanized force during maneuvers in 1939. Both men attempted to paper over their growing philosophical differences, but future correspondence would only exacerbate the obvious disparity between the dedicated horse cavalryman and the proponent of mechanization.

16. Christopher R. Gabel, *The U.S. Army GHQ Maneuvers of 1941* (Washington, D.C.,1991), pp. 23–24. Present at the maneuvers were the chiefs of the Infantry and Cavalry who, as Gabel notes, "Significantly . . . were not invited to the meeting." Chaffee deliberately termed the proposed new divisions "armored" to distinguish them from the few mechanized tank units previously under the infantry and cavalry branches. Although there is no record of Patton's contribution, his presence can be attributed to Chaffee. Gabel's monograph is the best account written of the important Louisiana Maneuvers.

17. Ibid., p. 189. The first meeting in Washington to work out the details of Marshall's directive was tense and acrimonious. An historian of the armored force has written that there was "powerful opposition to removal of mechanized units from the control of the branches. . . . The creation of a separate armored corps was termed fantastic, and the whole plan denounced as a conspiracy to grab power. . . . It was 'the old order in military circles holding out to the bitter, reactionary end.' " (See Gillie, *Forging the Thunderbolt,* p. 166.)

18. Herr waited too long to propose a meaningful role for the cavalry by converting horse units to mechanized. As Robert W. Grow would later write: "He staunchly refused to give up a [single] horse unit. So he lost it all." The new armored force was created at Fort Knox, Kentucky, under Adna Chaffee. (See Grow, "The Ten Lean Years" part 4, *Armor* [July-Aug. 1987.])

19. Nye, *The Patton Mind,* p. 120.

20. MB-PP 1, p. 952.

21. Commanded by Chaffee and designated the I Armored Corps.

22. Chaffee to GSP, quoted in MB-PP 1, pp. 953-54.

23. GCM to GSP, July 19, 1940, GCM-P.

24. Nye, *The Patton Mind,* p. 113.

25. Described in the *Washington Evening Star,* July 24, 1940.

26. GSP to BAP, Sept. 12, 1940, PP-USMA.

27. Ibid., Aug. 31, 1940.

28. Ibid., Sept. 3, 1940. In place of the usual intimate closing, it was signed simply "George."

29. Ibid., Aug. 27, 1940.

30. See the photographs of the Pattons in costume dress in MB, pp. 148–49.

31. RET-M; and Martin Blumenson review, *Army,* Nov. 1995.

32. Charles Whiting, *48 Hours to Hammelburg* (London, 1970), p. 197.

33. Donald E. Houston, *Hell on Wheels: The 2d Armored Division* (San Rafael, Calif., 1977), p. 38.

34. Quoted in LF, p. 140. Nevertheless, Patton believed that his new troops were good material to train. In a letter to Pershing written soon after his assignment to the 2d Armored, Patton wrote that his men were "all southern boys and it seems to me that over seventy percent of them have light hair and eyes—the old fighting breed: not [the] sub way soldiers as one gets in northern recruits especially from New York and Penn." (GSP to JJP, Sept. 24, 1940, Box 155, JJPP.)

35. GSP to BAP, Sept. 3, 1940, PP-USMA.

36. HS, p. 23.

37. Houston, *Hell on Wheels,* chap. 2, and LF, chap. 9.

38. "Armored Operations in Poland," a seventeen-page lecture to the officers of the 2d AD on Sept. 3, 1940, Box 12, PP-LC. Also quoted in Nye, *The Patton Mind,* p. 122. Patton wrote to Beatrice that he spent a week to write the speech, which he had memorized and delivered without notes. He "was the first one to get clapped so it must have been good." (Letter of Sept. 3, 1940, PP-USMA.)

39. Quoted by Houston in *Hell on Wheels,* p. 43.

40. GSP to Terry Allen, Sept. 29 and 30, 1940, Box 12, PP-LC.

41. Ed Cray, *General of the Army: George C. Marshall, Soldier and Statesman* (New York, 1990), p. 176,

42. The Patton Papers contain dozens of such letters. Many are moving and recalled the heady days of the First World War. Patton's aides were kept busy, but most requests had to be compassionately turned down or referred to the Adjutant General for action—and inevitable rejection Man Mountain Dean was one of the most popular figures in America, but after Pearl Harbor he was prepared to forsake his career in order to join Patton. (Letter of Jan. 5, 1942, Box 16, PP-LC.)

43. Houston, *Hell on Wheels,* p. 43.

44. Quoted in HS, p. 82. In Apr. 1941 Patton wrote to retired chief of staff Gen. Malin Craig, that he was

> probably the most unpopular man, not only in the Second Armored Division, but in the Army, as I got very tired of being the only person in this outfit who makes any corrections. So, today, I had the regimental commanders in and told them that from now on I would first write them a letter of admonition, and second relieve them from command if any units under them fail to carry out standing orders. I hope this meets with your approval in doing this. I assure you I did not use any profanity while making this statement. (Letter of Apr. 24, 1941, Box 22, PP-LC.)

45. Oscar W. Koch, *G-2: Intelligence for Patton* (Philadelphia, 1971), p. 159.

46. Gen. I. D. White oral history, Oct. 29, 1977, White Papers, USAMHI. White is thinly disguised as "General Black" in the Brotherhood of War series of novels by best-selling author W. E. B. Griffin.

47. White oral history.

48. HS, pp. 16–17.

49. Address to the 2d Armored Division, July 8, 1941, Box 54, PP-LC.

50. "General George S. Patton, Jr.," reflections of Lt. Gen. Raymond S. McLain, sent to BAP on Feb. 21, 1952, Box 17, PP-LC.

51. MB, p. 153.

52. McLain, "General George S. Patton, Jr."

53. MB, p. 153.

54. Quoted in HS, p. 22.

55. McLain, "General George S. Patton, Jr."

56. Quoted by Houston in *Hell on Wheels,* p. 40.

57. LF, p. 141. Yet another version by syndicated columnist Robert S. Allen, who served in the Third Army G-2 section, is that a reporter coined it in the spring of 1942 during tank

maneuvers at the Desert Training Center. Harry Semmes also ascribes it to an evening during one the series of lectures given the division officers on leadership and armored tactics, Patton expounded in colorful terms on what the coming war would be like. Patton habitually emphasized the importance of strong leadership, suddenly exclaiming: "War will be won by blood and guts alone." The entire division soon knew what Patton had said. (HS, pp. 12–13.)

58. Ibid., pp. 144–45.

59. Quoted in speech, "Farewell to Members of the 2d Armored Division," Apr. 4, 1941, PP-USMA. The 3d Armored and 4th Armored Divisions were being formed, and a number of 2d Armored troops were being reassigned to active duty the new 3d Armored Division at Fort Polk, Louisiana. Patton's complaint at their loss brought a blast from General Scott, who noted that the armored force could not be expanded rapidly, as the War Department had decreed, unless trained men were used to form the cadre of new units. "How many experienced men did you have in your tank center overseas [in 1918]?" No more was heard from Patton on this subject. Scott had made his point. (Houston, *Hell on Wheels,* p. 53.)

60. Author interview of Gen. I. D. White, July 13, 1985.

61. Lt. Gen. William W. Quinn, *Buffalo Bill Remembers* (Fowlerville, Mich., 1991), p. 428; and FA, p. 111. Quinn, whose nickname was Buffalo Bill, was the 4th Division military police officer and the one who reported the incident to Fredendall.

62. Quinn, pp. 428–30.

63. Memoir of M/Sgt. John L. Mims, USA (Ret.), "I Knew General Patton Personally," Box 66, PP-LC. On another occasion, at a lecture for division officers, Patton publicly apologized for mistakenly reprimanding two young reserve officers, noting that he had not been in possession of the facts when he had done so. (HS, p. 82.)

64. On one occasion Patton inspected the MP mess, stood in line and ate from a mess kit like his soldiers. What Patton never knew is that Mims had previously called the commanding officer to warn him of the general's impending visit. In return, the MPs would take care of Mims whenever he got drunk.

65. Mims memoir, "I Knew General Patton Personally."

66. Interview of M/Sgt. John L. Mims, USA (Ret.), by Dr. Herbert P. LePore, Mar. 20, 1987, Mims File, USAMHI. Mims was typical of the career soldiers of his era. He usually got drunk on payday and spent his paycheck, often ending up in jail after a binge. He was demoted more times than he could remember, more than once by Patton himself, who always forgave Mims and restored his rank and privileges. Patton would say: "Sergeant I hate to bust you but I can't let that make any difference." Often in as little as a week, back would come Mims, as if he had never left.

67. Mims memoir, "I Knew General Patton Personally."

68. Gen. I. D. White to the author, Sept. 2, 1985. Patton began keeping a brief diary when he reported to Fort Benning, and his entry for Nov. 27 notes that "the girls entertaining became so vulgar that I left without saying good night." The following day he noted his distress: "I told them that I felt that the actions of the entertainers was [*sic*] as much a shock and surprise to them as it was to me. . . . I warned them that such things must not happen again." (GSP diary, 1940–41, Nov. 28, 1940, Box 45, PP-LC.)

69. *Life,* July 7, 1941.

70. GSP to DDE, Oct. 1, 1940, EP.

71. Quoted in Ambrose, *Eisenhower,* vol. 1, pp. 124–25.

72. Ibid., p. 125. Eisenhower's desire to serve under Patton was revealed to his friend, Mark Clark, then stationed in Washington, whom he asked to intercede with the chief of infantry to avoid his being assigned an infantry division chief of staff and thus losing the chance for a transfer to the new Armored Corps.

73. GSP to DDE, Nov. 1, 1940, EP.

Chapter 27   The 1941 Tennessee, Louisiana, and Carolina Maneuvers

1. Quoted in Pogue, *Marshall: Ordeal and Hope,* p. 89, and Gabel, *The GHQ Maneuvers,* p. 64.

2. Extracted from *The Times Atlas of the Second World War* (New York, 1989). With Britain solely dependent on resupply by sea, German naval strategy was to force its surrender by sinking at least 750,000 tons of shipping each month for a year, thus equalling the sixty million tons a year the British required for survival.

3. Parrish, *Roosevelt and Marshall,* p. 114. With less than 190,000 men in 1939, the army was well below its authorized strength of 210,000.

4. Gabel, *The GHQ Maneuvers,* chap. 1.

5. McNair, quoted in ibid., p. 8.

6. Ibid., p. 5.

7. "Address of Officers and Men of the Second Armored Division, Ft. Benning, Ga," May 17, 1941, Box 54, PP-LC.

8. Houston, *Hell on Wheels,* p. 63. Houston's excellent history of the 2d Armored is the only known published source to describe the Tennessee maneuvers in detail, and, along with Gabel, *The GHQ Maneuvers,* is the principal reference.

9. Ibid., pp. 64–65. During the maneuver, while attempting to avoid a truck, one of Patton's tanks drove right into the town hall of a small municipality, and was buried in bricks. When queried by reporters, Patton replied that it was not the tank crew's fault "because the damn city hall was not on the map."

10. Ibid., p. 69.

11. Patton file, CMH.

12. MB, p. 157.

13. A similar exercise was held in Texas in June 1941 by Krueger's Third Army.

14. Houston, *Hell on Wheels,* chap. 5, and Geoffrey Perret, *There's a War to Be Won* (New York, 1991), pp. 43–44

15. See Gabel, *The GHQ Maneuvers,* p. 111.

16. Houston, *Hell on Wheels,* p. 88.

17. Among the reporters covering the maneuvers were Hanson Baldwin of the *NYT* and Eric Sevareid of CBS. Told Eisenhower was the man to see, the correspondents began converging in his tent for informal bull sessions. All were impressed by Dwight Eisenhower, whose praises they sang in newspaper columns across the United States, even though many misspelled his name. Others who wrote favorably about Eisenhower were syndicated columnists Robert S. Allen and Drew Pearson, whose "Washington Merry-Go-Round" column was read by millions. Another who emerged with enhanced credentials was McNair's chief planner, Brig. Gen. Mark Wayne Clark, who wrote the maneuver scenario. (See Ambrose, *Eisenhower,* vol. 1, p. 130.)

18. Gabel, *The GHQ Maneuvers,* p. 187.

19. Ibid., p. 194. Adna Chaffee's death in Aug. 1941 was a grievous loss to the U.S. Army, but ultimately enhanced Patton's career. Had Chaffee lived he would likely have been given the command of I Armored Corps in North Africa.

20. Quoted in ibid., p. 121.

21. Ibid., p. 134. Among Drum's infractions, which he later attempted to cover up, were the illegal positioning of troops outside the designated maneuver area before the exercise opened and signal lines laid across the Pee Dee River (the site of the future battle).

22. Porter B. Williamson, *Patton's Principles* (New York, 1979), chap. 1. A young reserve officer, Williamson was assigned to headquarters, I Armored Corps, until Aug. 1942. After the war he found that the lessons Patton had taught him had application to the business world. His book, while about Patton, is also a primer of good business principles. "I served with General George S. Patton, Jr. No man ever served *under* Gen. Patton; he was always serving *with* us." Williamson's account offers one of the few depictions of Patton's command of the I Armored Corps in 1942 prior to its deployment to North Africa.

23. Ibid. Patton was impressed with Williamson's map and asked him where he got it. "From *National Geographic,*" he replied. Patton observed:"Your map is a hell of a lot better than mine! I have been using gas station maps! All of my military maps are out of date."

24. Ibid. When Patton's command tank stalled, he hitched a ride with Williamson but never once issued an order to the lieutenant's driver. Instead, he would politely request,

"Would you ask your driver to turn at the next road." Always notoriously distrustful of Reserve officers, Patton gave Williamson the ultimate compliment when he said: "You sure taught me that you reserve officers can solve problems." Patton got his gas, and a bond was forged between the general and the lieutenant. Although Drum accused him of buying gas for his tanks from local gas stations out of his own pocket, Williamson states that he did not do so during the Carolina maneuvers.

25. AP correspondent Hal Boyle, in "Patton Makes 'em Mad So They Fight Harder," *Roanoke World News,* July 13, 1943, PP-USMA.

26. There are embellished accounts of Drum's capture, including that he was captured the day before the exercise opened. Another was that Patton's troops had roared into the First Army ccommand post with sirens blaring and lights flashing. Houston's account of Drum's capture at a roadblock is corroborated by a *NYT* account on Nov. 17, 1941.

27. Houston, *Hell on Wheels,* p. 98, and Williamson, *Patton's Principles,* p. 15.

28. Marshall offered Drum a post in China as the American military advisor to Chiang Kai-shek, but when Drum insisted on a more active role, an angry Marshall withdrew his offer and instead appointed Vinegar Joe Stilwell. Once again Drum proved to be his own worst enemy and wrecked what remained of his military career. Patton is usually thought of as the prime example of an officer whose mouth consistently harmed his career, but Hugh Drum's intemperate acts were fatal. Moreover, he had violated the cardinal unwritten rule that an officer is expected to answer a call to serve without regard to his personal desires. (See Calvert, "Drum, Drum.")

29. Pogue, *George C. Marshall: Ordeal and Hope,* p. 208.

30. Gabel, *The GHQ Maneuvers,* pp. 187–88.

31. Patton warned that "waffle ass" was fatal in reconnaissance units. His advice was typical of his efforts to save the lives of his men:

> When any of you gets to a place where your experience tells you there is apt to be an anti-tank gun or a mine or some other devilish contrivance of the enemy, don't ride up in your scout car or tank like a fat lady going shopping, stop your vehicle, take a walk or a crawl and get a look but remember that in walking or crawling you must not go straight up the road, you must go well off to a flank probably as much as one thousand yards. Then when you have gotten where you can see the probable location of the enemy, use your field glasses and find out what is there. This walking is hard work but it is very much easier than getting killed and getting killed is what will happen to you in battle unless you use proper precautions in reconnaissance. (From Patton's critique of the Tennessee maneuvers, delivered to the entire 2d Armored Division, on July 8, 1941, Box 54, PP-LC.)

32. GSP to Floyd Parks, Aug. 1, 1941, EP.

33. HS, p. 73. Patton's confidence in the mobility of the 2d Armored was again tested after the Carolina maneuvers when the division's three-hundred-mile road march back to Fort Benning was accomplished with only one halt en route.

34. LF, p. 159.

35. Gabel, *The GHQ Maneuvers,* p. 188–89.

36. GSP speech to 2d Armored Div., Dec. 1941, Box 54, PP-LC.

37. Ibid.

38. Patton was replaced by Maj. Gen. Willis D. Crittenberger who later commanded a corps during the Italian campaign. In 1975 in what was certainly one of the proudest moments of his life, Patton's son assumed command of the 2d Armored at Fort Hood, Texas.

39. Charles Scott was unfairly criticized for the performance of the corps during the Carolina maneuvers and was shunted off into a dead-end advisory job, his career in the armored force at an end. Oscar Koch relates that after Patton thanked Marshall for the assignment to command I Armored Corps, the chief of staff said: "I had nothing to do with

your selection; 2nd Armored selected you." Overcome by emotion when he related the anecdote, Patton choked up with tears of gratitude. (See I. D. White to W. E. Butterworth, Nov. 5, 1975, White Papers, USAMHI.)

40. Oral history No. 2 of Gen. Jacob L. Devers, Nov. 18, 1974, EL.

41. Quoted in Williamson, *Patton's Principles*, chap. 2, which is the only known account of GSP's colorful assumption of command of the I Armored Corps.

42. Ibid.

43. GSP to GCM, Jan. 22, 1942, and GCM to GSP, Jan. 26, 1942, both in Box 79, GCM-P.

44. The relevant letters are in Box 24, PP-LC.

45. Quoted in HS, p. 83, and RET-M. To a wife who had spoken to him of her husband's challenging new assignment, Patton replied tactlessly: "He's got a good job, but he's going to get shot!"

Chapter 28   Countdown to War

1. Eventually more than one million troops trained at the Desert Training Center.

2. "The Desert Training Center and C-AMA," Study No. 15, prepared by the Historical Section, Army Ground Forces, 1946, Combined Arms Library, Fort Leavenworth, Kansas. Hereafter referred to as DTC, Study No. 15. Patton's companions included Cols. Hugh Gaffey and Hobart R. Gay, and Lt. Col. Walter J. Muller.

3. Patton, "The Desert Training Corps," *Cavalry Journal* (Sept.-Oct. 1942).

4. "DTC, Study No. 15."

5. MB, p. 160.

6. "DTC, Study No. 15."

7. Ibid.

8. Quote supplied by the Friends of the General Patton Memorial Museum, Chiriaco Summit, California.

9. Williamson, *Patton's Principles*, chap. 2. Patton also sent Williamson to the Sears store in San Bernardino with orders to purchase as many washbasins as he could locate. Williamson asked for one hundred thousand. The astounded manager asked who would pay, and was told without hesitation—that either the U.S. government or General Patton would take care of it. The manager said he would take Patton's check (*Patton's Principles*, pp. 69–70).

10. Ibid., pp. 121–22. When informed that a politically appointed National Guard colonel was living with a shapely woman in the desert in an air-conditioned trailer, and with his wife in Indio, Patton vowed to "retire that SOB." His military police seized the trailer, arrested the officer and his mistress, and with red lights and sirens blaring, hauled them and the trailer to Indio, where they were handed over to his wife. The colonel soon retired from a newly discovered "disability."

11. "DTC, Study No. 15." While such "prettying up" as painting rocks and creating "eyewash" for VIPs was not part of Patton's regimen, neatness was. A messy base camp was evidence of sloppy leadership, and was not tolerated.

12. Williamson, *Patton's Principles*, pp. 100–101.

13. "DTC, Study No. 15." One of the sergeants was the leviathan Man Mountain Dean, whose physical presence was said to make men gasp.

14. Ibid.

15. Williamson, *Patton's Principles*, p. 36.

16. "DTC, Study No. 15."

17. Williamson, *Patton's Principles*, p. 40. Patton embellished the location of his wound to get his point across.

18. Williamson, *Patton's Principles*, passim.

19. Ibid. One of dozens of fantasy tales about Patton that circulated during the war had him dying and ringing the bell outside the gates of heaven to summon Saint Peter, who eventually appeared, dressed in toga and sandals. An indignant Patton lashed into the hap-

less saint for failing his duty by not being at the gate twenty-four hours a day and obviously not having read the Bible. As for his appearance, a scruffy beard and lack of a necktie and helmet were scorned as Patton visited his wrath on the unsuspecting Peter. (Ibid., pp. 159–60.)

20. Ibid., pp. 86–87.

21. "DTC, Study No. 15."

22. Williamson, *Patton's Principles,* pp. 86–7.

23. Mims, "I Knew General Patton Personally;" and GSP to James A. Ulio, Apr. 2, 1942, Box 46, PP-LC.

24. GSP to Malin Craig, May 6, 1942, Box 16, PP-LC.

25. Williamson, *Patton's Principles,* pp. 41–43.

26. GSP to McNair, May 2, 1942, Box 12, PP-LC.

27. Ibid., May 20, 1942.

28. GSP to Devers, July 14, 1942, quoted in MB-PP 2, p. 71.

29. Ibid.

30. "Operation 'Symbol,'" the unpublished diary of Brig. (later Lt. Gen. Sir] Ian Jacob. Copy furnished the author by the late Sir Ian Jacob.

31. LF, pp. 174–75; and "George C. Marshall Interviews," p. 510. Years later Marshall again grinned when he related to Dr. Forrest C. Pogue that his decision to send Patton packing had "scared him half to death."

32. Cray, *General of the Army,* p. 323.

33. In July 1942 Eisenhower wrote from London that he hoped his friend would become involved in his European endeavor, as he considered Patton the epitome of a battlefield commander.

34. *American Experience*: "Dwight David Eisenhower," PBS television documentary, 1993.

35. GSP to son George, July 13, 1942, PP-USMA.

36. RHP, p. 249.

37. Pogue, *George C. Marshall: Ordeal and Hope,* pp. 406–7, and Perret, *There's a War to Be Won,* p. 138.

38. Gay replaced Hugh Gaffey, who had been promoted and reassigned to the 2d Armored Division. Gaffey would return in 1944 as chief of staff of Patton's Third Army.

39. Lt.Gen Hobart R. Gay, oral history, USAMHI, and taped interview with Lt. Col. Roger Cirillo.

40. MB-*PP* 2, p. 757, and CRC, p. 12.

41. Quoted in Pogue, *George C. Marshall: Ordeal and Hope,* p. 404. Whether Marshall picked Patton, whose assignment was approved by Eisenhower, or vice versa, is a moot point. Both officers wanted Patton for a command role in Torch.

42. World War II diary of Gen. George S. Patton Jr., Aug. 9, 1942, Box 3, PP-LC, hereafter referred to as "Diary." In the summer of 1942 Patton began keeping a detailed diary that encompassed the entire war and is a major source of knowledge of his war years. After Patton's death Beatrice with the assistance of Col. (later Gen.) Paul D. Harkins, a longtime friend and Third Army deputy chief of staff, began the arduous task of organizing and transcribing the Patton diaries. Harkins wrote in a memo dated Jan. 2, 1953, "The original longhand diary was dictated to a stenographer [by Beatrice]." The typed diary "was in many cases identical with the longhand diary, but contained some changes. Where this occurred, both versions were included in a compilation by Mrs. Patton. This compilation was then reviewed and where the texts were not identical, one or the other was crossed out. . . . A third edition was typed . . . [and] is a consolidation of the longhand and Patton dictated diary complete after eliminating duplications and selecting one or the other when they did not quite agree." This third version is generally used by most researchers, including the author. Beatrice Patton was scrupulous in transcribing Patton's papers and made no attempt to doctor or alter for history her husband's words. Patton's letters, for example, were not only transcribed and typed exactly as written, but included his numerous dyslexic misspellings and mistakes of punctuation.

· 43. Ibid., Aug.11, 1942.

44. Lucian K. Truscott, *Command Missions* (New York, 1954), p. 59. During the evening Patton and an RAF group captain engaged in a spirited discussion not of the war but English nobility.

45. Brendon, *Ike,* p. 86.

46. Truscott, *Command Missions,* pp. 60–61.

47. En route to the United States his plane hit such heavy headwinds that the pilot had to decide whether to continue or return to the U.K. Patton's plane landed in Gander with less than one hour's fuel.

48. Diary, Sept. 24, 1942.

49. Patton and Clark quoted in letter, DDE to GCM, Aug. 17, 1942, in Joseph P. Hobbs, *Dear General: Eisenhower's Wartime Letters to Marshall* (Baltimore, 1971), p. 34.

50. MB, p. 217. After Torch Lambert was replaced as G-3 by Col. Halley G. Maddox.

51. Brenton G. Wallace, *Patton and His Third Army* (Nashville, Tenn., 1981), p. 17. Patton's first G-2 and G-3 were replaced after Torch by Oscar Koch and Muller.

52. LF, p. 96.

53. Ibid., pp. 96–97, and Pogue, *George C. Marshall: Ordeal and Hope,* p. 405.

54. Beatrice Patton, personal notes, dated Dec. 7, 1949, Box 36, PP-LC. A reference to the storm the Pattons had survived aboard the *Arcturus* in 1937.

55. Quoted in Ernest N. Harmon, *Combat Commander* (Englewood Cliffs, N.J., 1970), p. 77; and Norman Gelb, *Desperate Venture: The Story of Operation Torch, the Allied Invasion of North Africa* (New York, 1992), p. 150.

56. Samuel Eliot Morison, *Operations in North African Waters, Oct. 1942–June 1943,* vol. 2, *History of United States Naval Operations in World War II* (Boston, 1947), pp. 32–33. As Morison notes, "British and American high commanders remembered very well that the Dardenelles campaign, the principal amphibious operation in World War I, failed because Gen. Sir Ian Hamilton insisted on a postponement of several months in order to gain time for training." Truscott had later been Patton's personal representative to the Torch planners in London.

57. Harmon, *Combat Commander,* p. 69.

58. Ibid.

59. Koch, *G-2: Intelligence for Patton,* p. 153.

60. Brig. Gen. (Ret.) Edwin H. Randle, "The General and the Movie," *Army* (Sept. 1971). Randle's first formal meeting with Patton occurred several weeks later as his regiment was loading at Newport News, Virginia. Introduced by Harmon, Patton put out his hand and said, "Well, Randle, when you get on the beaches over there don't get messed up with any feline fecal matter. He didn't use those fancy words, but the short, vulgar ones which meant the same thing. His grin was derisive, but charming, too. I was startled and a little shocked. I had expected more serious advice. I said, 'Yes, sir. I will be careful.' He laughed and moved on."

61. Quoted in Gelb, *Desperate Venture,* p. 151.

62. Ibid., p. 133.

63. Ibid., pp. 132-33.

64. Diary, Oct. 21, 1942. When Patton said he was carrying the same revolver he had used at Rubio, Pershing said, " 'I hope you kill some Germans with it.' "

65. BAP, personal notes, dated Dec. 7, 1949.

66. Diary, Oct. 21, 1942.

67. GSP to Frederick Ayer, Oct. 20, 1942, Box 12, PP-LC. Patton's brief note said that he found it impossible to tell Beatrice how much she had meant to him since they were sixteen: "Your confidence in me was the only sure thing in a world of dreadful uncertainty." (RHP, p. 252.)

68. MB-*PP* 2, p. 91. Patton also wrote letters to each of his children and a number of other intimate family friends, such as Guy Henry, and the widow of his first company commander, Mrs. Francis C. Marshall. He asked General Harbord to look after his son, George.

69. Larrabee, *Commander in Chief,* p. 486.

70. Diary, Oct. 21, 1942.

71. LF, pp. 191–92, and Morison, *Operations in North African Waters,* n. 67, p. 42.

72. BAP, personal notes, dated Dec. 7, 1949.

73. RHP, p. 251.

74. Diary, Oct. 23, 1942. Somehow Patton found time before sailing to write a paper for Beatrice on the care of horses.

75. Morison, *Operations in North African Waters,* pp. 41–42.

76. Oral History of Walter Amsel, Sept. 24, 1974, extract furnished the author by Dr. Robert W. Love, Jr., Dept. of History, USNA.

77. BAP, personal notes, dated Dec. 7, 1949.

78. Ibid. Although each letter was similar in content, Patton's letter to Harmon not only spelled out what he wanted the 2d Armored Division to accomplish, but reminded him of what U.S. Grant once said: "In every battle there comes a time when both commanders consider themselves beaten. Then he who has the hardihood to continue the attack, wins. We must conquer or die. Skulkers or malingerers will be shot. For myself, I pledge that I shall only leave Africa as a conqueror or as a corpse."

79. Nye, *The Patton Mind,* pp. 128 and 129. Before he left California, Patton published the twenty-one-page "Notes on Tactics and Techniques of Desert Warfare," which summarized what he had learned from trial and error. His message was urgent: "The Notes emphasize those features of desert operations which differ from the traditional technique of tactics and which are vital in the successful prosecution of any campaign and which must be followed if the command is to avoid disaster." (Box 3, PP-USMA.)

80. Truscott, *Command Missions,* p. 89.

81. Diary, Oct. 23, 1942. On Oct. 18 Patton and Stimson bade each other a tearful farewell. Stimson also wrote to remind Patton he had been chosen for his leadership, courage and for his unique fighting ability. Stimson limited his advice to a plea that Patton not sacrifice himself, as it was certain to be a long, difficult war in which his services would be badly needed.

Chapter 29   The "Torch" Landings

1. Keith Sainsbury, *The North African Landings, 1942* (Newark, Del., 1976), p. 153.

2. Morison, *Operations in North African Waters,* p. 115, and Gelb, *Desperate Venture,* p. 130. Source for the missions of the Western Task Force is Morison, pp. 33–34.

3. Perret, *There's a War to Be Won,* p. 137. Two weeks earlier Clark landed secretly by submarine in Algeria where he and diplomat Robert Murphy negotiated a precarious arrangement with the French army commanders.

4. GSP to Maj Gen A.D. Surles, Nov. 6, 1942, Box 24, PP-LC.

5. LF, p. 17, and Morison, *Operations in North African Waters,* chap. 2.

6. Randle, "The General and the Movie."

7. Maj. Gen. George S. Patton, USA (Ret.), quoted in Jeffrey St. John, "Reflections on a Fighting Father," *The New American,* Dec. 16, 1985. Patton regarded reading the Koran as a necessary preparation for his forthcoming role there.

8. Diary, Oct. 28, 1942.

9. BAP, personal notes, Dec. 7, 1949. At 6:00 A.M. each morning the *Augusta*'s loudspeakers announced: "General Quarters all hands. Man your battle stations." Patton's duty station was his cabin where he remained and performed exercises to keep himself fit. He used a rowing machine loaned by the *Augusta*'s captain and ran in place holding on to the dresser in his cabin.

10. Ibid., Nov. 3, 1942.

11. Diary, Nov. 3, 1942. It was the only occasion in which Patton publicly uttered the dreaded "S"-word—"surrender"—in any context during the war.

12. LF, pp. 18–19.

13. Quoted in ibid., p. 19.

14. Roosevelt's broadcast at 1:30 A.M. (coincidental with the Oran and Algiers land-

ings) was made over Patton's objections to Eisenhower. He and others believed the broadcast would compromise Torch and eliminate any chance of surprise. (See LF, pp. 21–22.)

15. Morison, *Operations in North African Waters,* chap. 4.

16. Captain Selwyn H. Graham, USNR, "We Knew General George Patton, Jr., When—," *Shipmate,* Oct. 1988.

17. Perret, *There's a War to Be Won,* pp. 139–41, and Gelb, *Desperate Venture,* chap. 15. See Truscott, *Command Missions,* for a complete description of the Mehdia landings and their problems.

18. Gelb, *Desperate Venture.*

19. Ibid., pp. 97–99.

20. LF, p. 98.

21. Quoted in CRC, p. 21.

22. Diary, Nov. 9, 1942. Another version is that the exhausted soldier was sleeping when Patton prodded him with his own rifle, ordered him to get up, then gently put his hand on the man's shoulder and said: "I know you're tired. We're all tired. That makes no difference. The next beach you land on will be defended by Germans. I don't want one of them coming up behind you and hitting you over the head with a sockful of shit." (Quoted in CRC, p. 22.)

23. Morison, *Operations in North African Waters,* p. 158.

24. Diary, Nov. 9, 1942; LF, p. 195, and Morison, *Operations in North African Waters,* p. 80. Truscott's three-day fight for Port Lyautey was the most vicious, prolonged, and difficult of the invasion battles.

25. MB-*PP* 2, p. 109.

26. GSP to DDE, Nov. 15, 1942, Box 24, PP-LC.

27. Harmon, *Combat Commander,* pp. 97–98. Patton also praised Harmon's "tremendous drive, leadership, and willingness to take risks." Eisenhower, who barely knew Harmon, would undoubtedly recall Patton's praise in early 1943 when he summoned him to Tunisia.

28. Diary, Nov. 18, 1942.

29. George F. Howe, *Northwest Africa: Seizing the Initiative in the West* (Washington, D.C., 1957), p. 173.

30. CRC, p. 46.

31. HS, p. 135.

32. GSP to Stimson, Dec. 7, 1942, Box 13, PP-LC.

33. Diary, Nov. 11, 1942. For a comprehensive account of the Casablanca landings see Howe, *Northwest Africa,* the official U.S. Army history.

34. Howe, *Northwest Africa,* p. 173. Figures include both U.S. Army and Navy personnel.

35. HS, p. 136.

36. BAP to GSP, Nov. 8, 1942, PP-USMA. The trait of gossiping about others was always a part of the Pattons' wartime letters. Beatrice dutifully reported anything of interest she learned in her many contacts with friends in the Washington area. She also shared her husband's disdain for certain of his rivals, such as Mark Clark.

37. Ibid., Dec. 2, 1942.

38. Quoted in ibid., Dec. 5, 1942. Although she anticipated a joyless Christmas without him, Beatrice wrote: "I try all the time to think only of how you are doing that which you have longed for and trained for all your life, and how glad I am that you are having your chance to show what you really are." On what would have been her father's 120th birthday, she wrote of the conversation she had had with him in 1918: "'My little daughter,' he said, 'you must be sure he has what he needs. George is bringing in much glory to our family and we must always see that he is as comfortable as we can make him.' That was the last thing he ever said to me," wrote Beatrice. Pa always loved you and trusted you." (Letter of Dec. 7, 1942.)

39. GSP to BAP, Dec. 2, 1942.

40. Walter F. Dillingham to GSP, Nov. 5, 1942, Box 12, PP-LC. Dillingham also

observed that, "If you have picked, as you must have, a staff of officers who are qualified on the polo field you will be supported by men who are quick in making decisions and direct and intelligent in their method of attack."

41. Morison, *Operations in North African Waters,* p. 177.

42. GSP to Stimson, Dec. 7, 1942, Box 24, PP-LC.

43. Martin Blumenson, "Patton as Diplomat," *Army* (July 1973). Eisenhower's naval aide, Capt. Harry C. Butcher, accompanied Patton on an inspection trip through Casablanca. As they passed along the harbor, filled with sunken French warships, Patton permitted only the quickest of glimpses and refused to let the driver stop for fear of offending French pride. Butcher wrote in his diary that night: "This illustrates the sensitive feeling General Patton has shown to the French." (Diary of Harry C. Butcher, Jan. 19, 1943, EL.)

44. GSP to the sultan of Morocco, Nov. 10, 1942, Box 12, PP-LC.

45. The description of Patton's visit to Rabat on Nov. 16 is based on his diary, a paper written later that day, "Description of the Visit of the Commanding General and Staff to General Nogues and the Sultan of Morocco," Box 12, PP-LC; and letters to BAP written soon after the event, in Box 36, PP-LC.

46. GSP to BAP, Nov. 19, 1942, Box 36, PP-LC.

47. Ibid.

48. "Visit to Marrakech and Ouarzazate," Box 3, PP-USMA.

49. Diary, Nov. 17, 1942, GSP to BAP, Nov. 19, 1942, Box 36, PP-LC.

50. Ibid., Nov. 20, 1942.

51. Ibid., Nov. 24, 1942.

52. John S.D. Eisenhower, *Allies* (New York, 1982), p. 203.

53. Diary, Dec. 1, 1942.

54. Ibid., Feb. 3, 1943.

55. Ibid., Feb. 18, 1943. Paget was commander in chief of British Home Forces.

56. Ibid., Feb. 5, 1943, and MB-*PP* 2, p. 168.

57. GSP to DDE, undated, Box 13, PP-LC, and MB-*PP* 2, p. 169.

58. "Account of General Patton's Visit to the Tunisian Front," Dec., 1942, Box 3, PP-USMA.

59. Ibid.

60. Brig. Gen. Paul M. Robinett, *Armor Command* (Washington, D.C., 1958), p. 110.

61. Robinett, "George Smith Patton, Jr."

62. Miller, *Ike the Soldier,* pp. 440–41.

63. Ibid., p. 439, and Robert H. Ferrell, ed., *The Eisenhower Diaries* (New York, 1981), p. 84. Entry dated Dec. 10, 1942.

64. "Account of General Patton's Visit to the Tunisian Front."

65. Diary, Feb. 19, 1943.

66. Brig. Gen. William H. Hobson, "Gen. George Smith Patton, Jr., U.S.A.—A Tribute," 1947, Box 34, PP-LC.

67. DDE to Maj. Gen. Thomas T. Handy, Dec. 7, 1942, EP.

68. A full account of Churchill's Mediterranean strategy is in Martin Gilbert, *Winston S. Churchill,* vol. 6, *Road to Victory: 1941–1945* (New York), 1986, passim.

69. The term "Combined Chiefs of Staff (CCOS)" was given to the U.S. and British Chiefs of Staff operating together to formulate strategic policy and issue command guidance to the Allied commanders-in-chief in the field, such as Eisenhower. In reality the Casablanca Conference was a series of informal meetings of the CCOS.

70. Soon to be given the code name Operation Overlord by Churchill. (See Martin Gilbert, *Winston S. Churchill,* vol. 6, *Finest Hour, 1939–1941* (London, 1983), p. 966.

71. David Fraser, *Alanbrooke* (New York, 1982), p. 314.

72. Ibid.

73. Arthur Bryant, *The Turn of the Tide* (London, 1957), p. 559.

74. Diary, Jan. 9, 1943.

75. Ibid., Jan. 12, 1943.

76. HS, p. 139.

77. Doris Kearns Goodwin, *No Ordinary Time: Franklin and Eleanor Roosevelt: The Home Front in World War II* (New York, 1994), pp. 402–3.

78. Diary, Jan. 18 and 19, 1943. Whether or not he was serious, Harry Hopkins noted Patton's accomplishments in French Morocco and asked if he would like to be an ambassador. Patton said he would resign if given such a job. (Diary, Jan. 18, 1943.)

79. Gerald Pawle, *The War and Colonel Warden* (London, 1963), pp. 223–24, based on the war recollections of Cdr. C. R. Thompson, personal assistant to Churchill, 1940–45. Churchill, the inveterate storyteller and night owl, did not leave Patton's villa until nearly 3:00 A.M. and insisted on walking back alone to his own quarters, but was challenged by one of Patton's sentries, who summoned the corporal of the guard, saying: "I have a feller down here who claims he is the Prime Minister of Great Britain. I think he is a goddamn liar." The incident amused Churchill, who repeated it on numerous occasions. (HS, pp. 139–40.)

80. Frank McCarthy interview, Box 17, FM.

81. Hastings Ismay, *The Memoirs of General the Lord Ismay* (London, 1960), p. 289.

82. Alanbrooke Diary and "Notes on My Life," Alanbrooke Papers, LHC.

83. Until deciphered by British historian David Irving's assistant in the 1970s, the Hughes diary (written in an incomprehensible scribble) in the Library of Congress remained largely overlooked. Hughes was a notorious gossip with many personal axes to grind, and the diary is a source a historian must approach with care. Nevertheless, the diary and papers are instructive in assessing Patton's state of mind.

84. Devers to BAP, Feb. 12, 1943, Box 13, PP-LC.

85. Diary, Jan. 16, 1943.

86. Quoted in Pogue, *George C. Marshall: Organizer of Victory,* p. 182.

87. Quoted in Peter Lyon, *Eisenhower: Portrait of the Hero* (Boston, 1974), p. 192.

88. Diary, Jan. 28, 1943.

89. D'Este, *Bitter Victory,* p. 68, and Diary, Feb. 14, 1943.

90. Ibid., and Jan. 28, 1943.

91. BAP to GSP, Dec. 26, 1942. Although he wrote after Christmas that he was "having a lot of fun and am still young enough to learn," a sure sign of his boredom was that although he still continued to raise hell about saluting, Patton no longer cussed out offenders but merely jotted down the names of their commanding officers who were required to report to him personally. "This is the only place in Africa where any one salutes at all."

92. Diary, Feb. 2, 1943. Patton candidly admitted that hitting the boar in the eye was pure luck.

93. CRC, pp. 84–85.

94. BAP to GSP Jan. 10, 1943. When the tale was told of how a eunuch had saved his royal lady from an attacking panther with two shots, someone commented: "General Patton would certainly have dispatched the matter with a single shot." (CRC, p. 86.)

95. GSP to BAP, Jan. 11, 1943.

96. Ibid., Feb, 1943. Among Patton's administrative headaches was his own driver, M/Sgt. Mims, who had once again indulged in his well-known penchant for drink and this time had "run over a Frenchman. He will have to take his medicine," which traditionally meant a reduction to private, a fine, and some form of restriction to headquarters.

97. Diary., Feb. 19, 1943. Patton never made anything but tongue-in-cheek references to a future in politics, and his passing comments were little more than flights of fancy.

98. Box 13, PP-LC, Mar. 3, 1943.

99. Ibid.

100. BAP to GSP, Mar. 2, 1943.

101. Note by BAP appended to GSP's letter of Apr. 8, 1943. Patton never told his daughter that Waters was due to have been sent home, or that his division commander, Orlando Ward, had deliberately sent him to the rear on a mission he expected would take at least five days and probably keep him out of battle. Waters, however, apparently did the job in three days and returned to his unit in time for the battle of Sidi-Bou-Zid. Patton also thought his letter to Little Bee a failure, as "I am not too good a liar." (GSP to BAP, Mar. 2, 1943.)

Chapter 30   A Summons to Battle

Unless otherwise cited, letters from GSP to BAP in 1943 are in Box 13, PP-LC.

1. An American colonel who once made the fatal mistake of referring to his counterpart as "that *British* bastard," was relieved and sent back to the United States. (See David Schoenbrun, *America Inside Out* [New York, 1984], p. 92.)

2. Brendon, *Ike*, p. 99.

3. Alexander's assignment was a clever ploy by the British at Casablanca to take *de facto* control of the Allied campaign in North Africa from Eisenhower, whom they regarded as inept. The Combined Chiefs of Staff approved a restructuring of the Allied command that provided for three deputy commanders, one each for air, ground, and sea—all British. While Eisenhower would continue to command in name only, the actual operational control of Allied forces would be in the hands of *British* commanders. Alexander was officially Eisenhower's deputy commander for land forces; in reality, his principal function was to command the 18th Army Group.

4. ONB-*SS*, p. 25.

5. Robinett, *Armor Command*, p. 190.

6. The British practice of employing "penny packet" teams of infantry, artillery, and armor to fight independently of one another had already proved ruinous because they violated one of the fundamental principles of war—concentration of force. When Montgomery took command of the Eighth Army, one of his first acts was to disband these "jock columns" and "brigade groups," as they were known. (See C. E. Lucas Phillips, *Alamein* [London, 1962], pp. 50–52.)

7. ONB-*SS*, p. 27.

8. Butcher diary, Feb. 23, 1943.

9. Martin Blumenson, *Kasserine Pass* (Boston, 1967), pp. 3–4.

10. Robinett, *Armor Command*, p. 198.

11. Ambrose, *Eisenhower*, vol. 1, p. 228. The prevailing belief among Ambrose and others is that Eisenhower was reluctant to remove Fredendall because he was Marshall's hand-picked choice. However, Fredendall's appointment was actually sponsored by McNair.

12. Ibid., p. 231.

13. B. H. Liddell, ed., *The Rommel Papers* (London, 1953), p. 407.

14. Robinett, *Armor Command*, p. 137.

15. Miller, *Ike the Soldier*, p. 472.

16. H. Essame, *Patton the Commander* (London, 1974), p. 67. See also William R. Betson, "Sidi Bou Zid—A Case History of Failure," *Armor* (Nov.-Dec. 1982). Betson excoriates Fredendall's cavalier command of II Corps, and his misuse of the 1st Armored Division.

17. Quoted in Eisenhower, *Crusade in Europe* (London, 1948), pp. 159–60.

18. George F. Howe interview of Maj. Gen. Ernest N. Harmon, Sept. 15, 1952, USAMHI, and GSP diary, Mar. 2, 1943.

19. Nigel Nicolson, *Alex: The Life of Field Marshal Alexander of Tunis* (London, 1973), p. 212.

20. Partly from British fear of furthering the embarrassment of American failure in the first ground combat of the war against the Germans, Fredendall's removal was disguised to appear as if he were returning home to employ his troop training skills. He was given a hero's welcome, the command of a training army, and an undeserved promotion to lieutenant general. One Washington newspaper complained that Fredendall was the scapegoat for Eisenhower's negligence in what was termed "one of the Army's worst scandals" for which Fredendall "is paying the penalty." (Ray Tucker, "National Whirligig" column, date unknown. Copy in PP-USMA.)

21. Diary, Mar. 4, 1943.

22. Box 13, PP-LC. Forty-six members of the I Armored Corps staff signed the letter-petition. In addition to his aides only four officers accompanied Patton: his chief of staff, Hugh Gaffey; the G-3, Colonel Kent Lambert; Colonel Oscar Koch (then the assistant G-2);

and the assistant G-4. The rest remained behind under the temporary command of Geoffrey Keyes to continue planning Husky.

23. Butcher, *Three Years with Eisenhower* (London, 1946), p. 235. Eisenhower also instructed Patton to be "cold-blooded" about getting rid of inefficient officers, which, given Patton's reputation, was like carrying coals to Newcastle. More than anything else, what finally seems to have inspired Eisenhower to act were reports of Fredendall's anti-British remarks and his lack of cooperation.

24. Diary, Mar. 5, 1943.

25. Bradley Commentaries, USAMHI.

26. Diary, Mar. 11, 1943.

27. Alexander to Brooke, Apr. 3, 1943, Alanbrooke Papers.

28. Alexander's claim to Brooke that he would do everything in his power to help reverse the situation was undoubtedly sincere, but as subsequent events would demonstrate, he never seemed able to overcome his innate distrust of the American soldier. He told the CIGS: "I have only the American 2d Corps. There are millions of them elsewhere who must be living in a fool's paradise. If this handful of Divisions here are their best, the value of the remainder may be imagined." (Ibid.)

29. GSP to BAP, Mar. 13, 1943.

30. ONB-*SS*, pp. 44–45.

31. AP correspondent Hal Boyle, *Roanoke World News,* July 13, 1943.

32. Robinett, *Armor Command,* p. 198.

33. Ibid., p. 199, and William Tecumseh Sherman, *Memoirs* (New York, 1875), p. 408. Robinett had a not untypical love-hate relationship with Patton, whom he thought severely exaggerated the state of discipline within II Corps and the extent of his miracle in restoring it. Like so many, Robinett could never reconcile his admiration for the success of Patton's means with his revulsion for the negative aspect of his ends, complaining that "intemperate and vile language, vanity, lack of judgment, and excessive ambition are not commendable in anyone." (Ibid.)

34. Diary, Mar. 12, 1943. Patton soon modified his criticism of the II Corps staff, calling them "better than what I had been told."

35. HS, p. 142.

36. Diary, Mar. 13, 1943.

37. GSP to BAP, Mar. 6 and 11, 1943.

38. Diary, Mar. 12, 1943.

39. GSP to BAP, Mar. 13, 1943.

40. Bradley Commentaries. After the war, Marshall told how Patton's profanity offended a bishop who complained that Patton had cursed on the radio in Los Angeles and couldn't something be done to shut him up? Marshall replied that George Washington had forbidden profanity in the Continental Army and had failed miserably. "If George couldn't do it with his small force, think of our problem with eight million men. Look at your constituents around the table.... I have never heard more profanity than I have here." The bishop replied: "I see your point," and no more was heard of muzzling Patton. (Pogue Interviews, p. 512.)

41. ONB-*SS*, p. 52.

42. Oral history of Lt. Gen. John A. Heintges, 1974, USAMHI.

43. Interview of William S. Jackson, Jan. 15, 1994. Jackson was a close friend of the late Bill Donley, who related the incident to him. Telephone conversations between Patton and Allen were often heated and frequently surreptitiously listened to by members of the 1st Division signal section. The air would turn blue as the two friends cursed each other in the most obscene terms. Patton may have outranked Allen, but in all but their public exchanges, Allen cursed his friend and superior as an equal.

44. ONB-*AGL*, p. 140.

45. Bradley Commentaries.

46. GSP to Bradley, Apr. 23, 1943.

47. Clay Blair to the author, Mar. 11, 1986. As Blair notes, during his research he dis-

covered that no heavy drinkers ever got anywhere with Bradley, with the lone exception of Col. B. A. "Monk" Dickson, the II Corps G-2, whom he greatly admired but who was the only principal staff officer on Bradley's First Army staff who failed to be promoted to brigadier general.

48. Martin Blumenson, "Bradley-Patton: World War II's 'Odd Couple,'" *Army* (Dec. 1985).

49. S. L. A. Marshall oral history, Marshall Papers, USAMHI.

50. Charles B. MacDonald, *The Mighty Endeavor* (New York, 1969), pp. 354–56. An official U.S. Army historian after the war, MacDonald is best known for his acclaimed account of the life of an infantryman, *Company Commander*. MacDonald all but calls Hodges a butcher for feeding one division after another in brutal frontal attacks in the Hürt-gen where it was impossible to win because the Germans had flooded the nearby Roer River dams. Both MacDonald and Lt. Gen. James M. Gavin refer to the sorry tale of the Hürtgen as an American Passchendaele. (See Gavin's memoir, *On to Berlin* [New York, 1978], p. 268.)

51. Lt. Gen. James M. Gavin to the author, Oct. 8, 1985.

52. Gavin, *On to Berlin*, p. 232. Although Bradley is not mentioned by name, Gavin clearly meant to single him out, a point emphasized privately in the author's conversations with General Gavin.

53. GSP to BAP, Mar. 30, 1943.

54. Quoted in MB-*PP* 2, p. 206.

55. GSP to BAP, Mar. 11, 1943.

56. Ibid.

57. Ibid., Mar. 25, 1943.

58. Ibid., Mar. 15, 1943.

59. Diary, Mar. 21, 1943, and news clipping, date ands source unknown, but probably the *Los Angeles Times,* circa late Mar. 1943, copy in PP-USMA.

60. Ibid., Mar. 21, 1943.

61. Chester B. Hansen diary, USAMHI.

Chapter 31   Allies

1. ONB-*SS*, p. 52.

2. Diary, Mar. 15 and 16, 1943.

3. ONB-*SS*, p. 52.

4. Lt. Col. Henry Gerard Phillips, *El Guettar* (N.p., 1991), p. 71.

5. Diary, Apr. 9, 1943.

6. Drew Middleton, quoted in part 13 of *The World at War* television series, 1976.

7. Although Alexander's intent was clearly not to thrust II Corps into a pivotal role, had Patton succeeded in cutting the Axis in half, Alexander would nevertheless have been hard pressed to have denied II Corps a larger role.

8. GSP to BAP, Mar. 30, 1943, Box 12, PP-LC.

9. Ibid.

10. Bradley Commentaries.

11. Diary, Mar. 24 and 25, 1943.

12. Ibid., Mar. 28, 1943. As he had shown at Indio, Patton disliked chaplains and often treated them very badly, usually cursing or demeaning them. An example occurred during a visit to the 47th Regiment when Patton encountered the regimental chaplain and cursed him for not being in the front lines with his troops. An eyewitness noted that, "When I say cussed, I mean he used every bit of invective in his expansive vocabulary. Frankly, I thought it was uncalled for and very demeaning for a clergyman." (Hansen diary and Phillips, *El Guettar,* pp. 57–58.)

13. Diary, Mar. 31, 1943.

14. MB-*PP* 2, p. 196.

15. Diary, Apr. 4, 1943. Alexander's recommendation that Eisenhower sack Ward was

the only time he took the extraordinary step of recommending the relief of a senior American commander: "In my opinion," he wrote, "General Ward is not the best man to command the American 1st Armoured Division. I am sure he has many fine qualities but the fact remains that the splendid force he commands is not the power in the field it should be." Although relief from command was generally career-ending, there was considerable compassion for Ward, who returned to the United States to command the Tank Destroyer Center at Camp Hood, Texas. In 1945 he was given command of another armored division that fought in northwest Europe. (MB-*PP* 2, p. 211; and Alexander to GSP, Apr. 1, 1943, EP.)

16. See D'Este, *Fatal Decision*, chap. 9. Eisenhower wrote to Marshall: "I completely agree with General Patton as to the necessity for this action." (See Miller, *Ike the General*, p. 498.)

17. Quoted by John P. Marquand in his introduction to CRC, p. xvii.

18. Diary, Mar. 23, 1943.

19. GSP to BAP, Dec. 26, 1942, Mar. 25, and Apr. 13, 1943, Box 12, PP-LC. Although he had railed against women in a combat zone in the past, Patton was proud of the "swell job" his front-line nurses were doing.

20. Phillips, *El Guettar*, p. 9.

21. Capt. C. P. Brownley, 9th Division, quoted in Henry G. Phillips, *Sedjenane: The Pay-off Battle* (N.p., 993), p. 20. Captain Brownley thought Bradley "the very antithesis of his predecessor in every way. . . Patton's entourage [tied] up traffic for miles when he passed . . . I can recall coming across Bradley parked at the side of the road waiting patiently for a supply convoy to move through."

22. Ibid., passim. Another tale of Patton in the 9th Division is of the day he arrived at an infantry outpost and, pointing to a distant German tank, ordered the company commander to knock it out. "I didn't have to tell him not to wave his map around. A German mortarman did that for me," remembers the officer.

23. Randle, "The General and the Movie." The next day Randle encountered Patton on a road crowded with tanks and vehicles moving forward. When four soldiers of the 1st Armored passed him in a jeep wearing knitted caps Patton yelled, "Stop, goddammit!" and demanded: "Where are your goddam helmets?" They hastily donned helmets from the floor of the jeep and Patton ordered them to hand over "those goddam caps." Then in a far softer tone he ordered them to "get out of here and keep those goddam helmets on *all* the time." Looking at Randle, Patton grinned. "He was enjoying every minute of it."

24. Phillips, *Sedjenane*, p. 66.

25. Witnessed by Bradley, whose own aide, Chester Hansen, was in the trench next to Jenson and was unhurt, it was Bradley's closest-ever brush with death. (See ONB-SS, pp. 61–62.)

26. Diary, Apr. 1, 1943.

27. MB-*PP* 2, p. 206.

28. Air Vice-Marshal Arthur Coningham commanded the Northwest African Tactical Air Force, and in that role was the Allied tactical air commander. A New Zealander with a Patton-like tendency to put his foot in his mouth, Coningham was a pioneer aviator who was proud of the nickname "Maori," which eventually became simply "Mary." In World War I he won the DSO, MC, and DFC as a pilot in the Royal Flying Corps, and in 1925 set a long-distance record by leading a flight from Cairo to Nigeria, which inspired the opening of the African air routes. Coningham was a decisive commander and a brilliant strategist who was quick to grasp a situation and react with sound judgment. Unfortunately, not long after assuming command of the Desert Air Force in support of the Eighth Army, the strong-willed Coningham clashed with Montgomery over what the airman believed was a deliberate denial of credit for the feats of the RAF. It was one of the bitterest quarrels of the war. (See D'Este, *Bitter Victory*, passim, and Vincent Orange, *Coningham* [London, 1990], pp. 146–50.)

29. Miller, *Ike the Soldier*, p. 496.

30. Lord Tedder, *With Prejudice* (London, 1966), p. 411; Ambrose, *Eisenhower*, vol. 1, p. 232; Orange, *Coningham*, p. 149, and ONB-AGL, p. 148.

31. Gen. Lawrence S. Kuter, "Goddammit, Georgie!" *Air Force* [Feb. 1973.]

32. Diary, Apr. 3, 1943; Kuter, "Goddammit, Georgie!"; Tedder, *With Prejudice,* p. 411; and ONB-*SS*, pp. 63–64. Orange, *Coningham,* p. 147, suggests there was only *one* Luftwaffe aircraft and that the story has been embellished in the telling, however, everyone present agrees that there were four aircraft, except Tedder, who records it as three Focke-Wulf 190s.

33. Ibid. The actual incident is well portrayed in the film *Patton.* Tedder left feeling: "Taking the long view, I thought the effect might be good. Patton was now a friend of ours and I thought that the chance of bellyaching signals from the Army would be greatly reduced." No one present has ever admitted diving for cover but it may be readily assumed.

34. This account of the Coningham-Patton confrontation on Apr. 4 is reconstructed from Patton's diary, Coningham's letter to Air Vice Marshal H. E. P Wigglesworth, dated Apr. 5, 1943 (AIR 23/7439, PRO), and Kuter, "Goddammit, Georgie!" Kuter notes that immediately on his return, Coningham provided him a full account of the meeting, which he immediately recorded in his diary.

35. Coningham to Wigglesworth, Apr. 5, 1943. If Coningham's intention was to single out only a handful of II Corps personnel, his signal certainly conveyed a very different impression: "If Sitrep is in earnest and balanced against . . . facts, it can only be assumed that II Corps personnel concerned are not battle-worthy in terms of present operations."

36. Ibid.

37. Diary, Apr. 4, 1943, which includes complete copies of the various cables and letters sent by Patton during this incident.

38. Copy in diary, Apr. 5, 1943.

39. DDE to GSP, Apr. 5, 1943. The full text of Eisenhower's two-page letter is in Alfred D. Chandler Jr., ed., *The Papers of Dwight David Eisenhower: The War Years,* vol. 2 (Baltimore, 1970), pp. 1073–74 (hereafter *The Eisenhower Papers*). Although Eisenhower noted that he did not intend the letter as a lecture, its tone clearly conveyed this meaning.

40. Miller, *Ike the Soldier,* p. 497.

41. Diary, Apr. 7, 1943.

42. The British blamed the 34th Division for failing to carry out a mission while attached to one of their corps. The assessment was wholly unwarranted after the corps commander, Lt. Gen. J. T. Crocker, had decreed the tactics to be used by the 34th Division commander, Maj. Gen. Charles W. "Doc" Ryder, who had a better plan which he was not permitted to carry out.

43. Diary, Apr. 11 and 12, 1943.

44. GSP to Alexander, Apr. 11 and Apr. 12, 1943, copies in diary. In *A Soldier's Story* Bradley leaves the impression that saving the 34th Division was purely his doing, when, in fact, both deserve credit for challenging Alexander. In his later memoir, *A General's Life,* Bradley acknowledged Patton's role. (ONB-*SS*, pp. 67–68; ONB-*AGL*, p. 150; and Diary, Apr. 10–13, 1943.)

45. Ibid. Under Ryder the 34th Division would indeed vindicate itself, not only in Tunisia, but later in Italy during the bloody battles at Cassino. After the war a British historian would write that their exploits "must rank with the finest feats of arms carried out by any soldiers during the war." (Fred Majdalany, *The Battle of Cassino* [London, 1975], p. 99.)

46. Diary, Apr. 12, 1943.

47. In *A Soldier's Story* Bradley also took credit for Eisenhower's eventual change of heart in assigning II Corps a larger role, and conveniently ignored Patton's role in standing up to Alexander.

48. Liddell Hart, *The Rommel Papers,* p. 523.

49. ONB-*AGL*, p. 151.

50. Eisenhower, *Crusade in Europe,* p. 166.

51. "Tunisia," CMH pamphlet, 1993.

52. ONB-*AGL*, p. 155.

53. Diary, Apr. 14, 1943.

54. ONB-*AGL*, p. 154.

55. Ibid., p. 151. After the publication of *A General's Life* (which was completed by coauthor Clay Blair after Bradley's death), there was some criticism that the overall tone of book was uncharacteristically harsh. Acutely aware of his dilemma, Blair not only had a number of distinguished military historians read the manuscript, but, armed with thousands of pages of Bradley's transcribed words, believes that what he wrote was not only accurate and verifiable but would have met with Bradley's approval had he lived (p. 11).

56. Quoted in Nicolson, *Alex*, p. 229. Readers interested in a full account of this campaign may wish to consult the official histories: I. S. O. Playfair and C. J. C. Molony, *The Mediterranean and Middle East*, vol. 4 (London, 1966), and Howe, *Northwest Africa*. A useful unofficial account of combat operations is Kenneth Macksey, *Crucible of Power: The Fight for Tunisia 1942–1943* (London, 1969).

57. An incident between Anderson and Harmon during the Bizerte offensive is illustrative. When Harmon briefed Anderson on the plan he and Bradley had devised to employ the 1st Armored, "Anderson waved his swagger stick vaguely and commented, 'Just a childish fantasy, just a childish fantasy.' With that he stalked out of the tent . . . under my breath I muttered to myself, I'll make that son-of-a-bitch eat those words." (See Harmon, *Combat Commander*, p. 130.)

58. ONB-*AGL*, p. 159.

59. U.S. losses also included 8,978 wounded and 6,528 missing in action. Overall Allied losses were 70,341: French—16,180; British—First Army, 23,545; Eighth Army, 12,395; from Nov. 12, 1942, to May 13, 1943.

60. Diary, Apr. 16, 1943.

61. Ibid., Apr. 15, 1943.

62. Ibid.

63. Ibid., Apr. 14, 1943. Bradley later admitted that he too never trusted Mark Clark, writing that he lacked experience to command Fifth Army. "I had serious reservations about him personally. He seemed false somehow, too eager to impress, too hungry for the limelight, promotions and personal publicity." (ONB-*AGL*, pp. 203–4.)

64. Butcher diary, Apr. 17, 1943.

65. Diary, Apr. 16, 1943.

66. Ibid., Apr. 17, 1943.

67. Commander, 1st Derbyshire Yeomanry to GSP, Apr. 15, 1943, Box 13, PP-LC.

68. Butcher diary, Apr. 17, 1943.

69. Diary, Apr. 15, 1943, and Mrs. Richard Jenson to GSP, same date. She agreed with Patton's suggestion that her husband's grave site remain in North Africa.

70. Butcher diary, Apr. 17, 1943.

71. Diary, Apr. 17, 1943.

Chapter 32 "A Dog's Breakfast"

1. The Canadian 1st Division and the U.S. 45th Division were being staged directly from Britain and the United States. London and Washington thus became two of *five* separate centers of planning. The inevitable result was considerable confusion, compounded by the fact that Cunningham, Tedder, and Alexander each eventually established their operational headquarters in separate locations.

2. Diary, Apr. 29, 1943.

3. Ibid., May 3, 1943.

4. "Remarks Made at Conference in Algiers on 2 May 1943 by Gen. Montgomery," Butcher diary.

5. Quoted in LF, p. 273.

6. Samuel Eliot Morison, *Sicily-Salerno-Anzio* (Boston, 1954), p. 20.

7. Diary, May 7, 1943, and GSP to BAP, May 7, 1943, Box 13, PP-LC.

8. Ibid., May 22, 1943.

9. LF, p. 273; and Alexander Despatch, "The Conquest of Sicily," Oct. 9, 1945, Alexander Papers, PRO (W.O. 214/68).

10. Diary, May 5, 1943.

11. Ibid., May 16 and 17, 1943. Patton also wrote on May 9 to congratulate Bradley on "your magnificent victory."

12. Ibid., June 10, 1943. In fact, Patton sent Lambert (now in the 1st Armored Division) a scorching letter to "stop being a goddam fool! You know very well that you have no business sending [such] letters home." (GSP to Lambert, May 12, 1943, Hughes Papers, in Irving Microfilm, USAMHI.)

13. Eisenhower unpublished ms., copy furnished by John S. D. Eisenhower; and Larry I. Bland and Sharon Ritenour Stevens, eds., *The Papers of George Catlett Marshall*, vol. 3, *"The Right Man for the Job," December 7, 1941–May 31, 1943* (Baltimore, 1991), p. 700.

14. ONB-SS, p. 119.

15. Diary, June 2, 1943.

16. Quoted in Nigel Hamilton, *Monty: The Making of a General, 1887–1942* (London, 1981), p. 490. When he was not emoting for the benefit of his troops and the public, Patton repeatedly carried out small but important courtesies to others. For example, shortly before Husky, one of Patton's teeth required an inlay. After the dentist finished, Patton asked him to write down his name, rank, and serial number. Shaken, the dentist asked if he was going to prefer charges. Patton smiled and replied: "You know, when someone does something nice for me, there is only one way I can repay him." Several days later the dentist's commanding officer received a letter of commendation. (Norman Wahl, "Appointment in Mostaganem," *American Journal of Orthodontics* [Oct.1990].)

17. FA, p. 3.

18. Diary, June 14, 1943. Yet, British officers who met Patton found him cordial and gentlemanly. Montgomery's liaison officer, Maj. Harry Llewellyn, recalls his first encounter with Patton who assured him he did not need permission to visit American units. "Just say, I am a friend of the General's." As Llewellyn observes: "I believe there were two Pattons, the really tough, thrusting cavalry officer and the soft-spoken gentle man—perhaps I was lucky only to see the latter side." (See Harry Llewellyn, *Passport to Life* [London, 1980], pp. 149–50.)

19. Ibid., May 20, 1943.

20. "Description of Victory Parade Held at Tunis, May 20, 1943," Box 13, PP-LC. Patton was critical of the marching of American troops, who passed in review without flags or even guidons. "In spite of their magnificent appearance, our men do not put up a good show in reviews. I think that we still lack pride in soldiers and we must develop it."

21. ONB-SS, p. 109, and ONB-*AGL*, p. 170; GSP, "Description of Victory Parade"; Miller, *Ike the Soldier,* p. 505, and LF, p. 256;

22. Even Patton's attempts at humor generally tended to fall flat; such as the day Eisenhower telephoned that he was to meet "my American boss" (Marshall) the following day. "I asked him, 'when did Mamie arrive?'" Eisenhower was not amused. (Ibid., June 1, 1943.)

23. Diary, May 8, 1943. Patton told Eisenhower the story of Grant at Vicksburg in which he said, "'Don't let my personal reputation interfere with winning the war.' I feel the same way and told Ike the story. He needs a few loyal and unselfish men around him, even if he is too weak a character to be worthy of us."

24. GSP to BAP, May 17, 1943.

25. Ibid., June 1, 1943.

26. GSP to Stimson, May 10, 1943, copy in Diary. Patton was thrilled to receive letters from both Marshall and Roosevelt, praising him for his work in Tunisia. Marshall had written: "You did a masterful job in reorganizing II Corps." "Very nice," noted Patton, "as he seldom praises."

27. Diary, May 15, 1943.

28. Ibid., May 28, and June 2, 1943.

29. Ibid., May 5, 1943.

30. Bradley Commentaries.

31. Diary. May 7, 1943. Patton had observed the confusion and their "reasons why

things cannot be done. I straightened them out by the simple method of showing a confidence which I don't feel, although I am pretty confident. I believe in my fate, and, to fulfill it, this show must be a success."

32. Koch, *G-2: Intelligence for Patton,* pp. 156–57.

33. Diary, June 22, 1943.

34. See MB-*PP* 2, p. 243–44.

35. Eisenhower memo, June 11, 1943, EP.

36. Ibid.

37. Diary, June 1, 1943.

38. Ibid., May 17, 1943.

39. GSP to Frederick Ayer, May 5, 1943, PP-USMA.

40. Diary, July 4, 1943.

41. GSP to Malin Craig, May 10, 1943, Box 24, PP-LC.

42. Diary, July 5, 1943. As Patton completed in longhand the notebooks that comprised his diary he sent them to Beatrice with instructions to place them in a safe deposit box in a Boston bank: "They may be of great value some day." Throughout the war Eisenhower remained blissfully ignorant of Patton's increasingly shrill complaints about his generalship recorded in his diary and in letters to Beatrice and other confidants. It was not until 1947 that Eisenhower learned their full extent, when he read an excerpt of Patton's posthumous memoir, *War as I Knew It.* His reaction to Patton's accusation that he had unnecessarily prolonged the war and cost thousands of additional lives by his broad front strategy was: "I am beginning to think that crackpot history is going to guide the future student in his study of the late conflict." (See Miller, *Ike the Soldier,* p. 439.)

43. Ibid., June 29, 1943.

44. GSP to Frederick Ayer and BAP, both July 5, 1943, Boxes 12 and 123, PP-LC.

Chapter 33 "Born at Sea, Baptized in Blood"

1. Quoted in Morison, *Sicily-Salerno-Anzio,* p. 64.

2. See *Bitter Victory,* p. 221. The full text of the letter to Seventh Army, dated June 27, is in Patton's diary.

3. Diary, July 9, 1943.

4. The problems of air-ground and air-naval cooperation are discussed in *Bitter Victory,* chap. 8.

5. Quoted in Morison, *Sicily-Salerno-Anzio,* p. 22.

6. During this engagement Darby won the Distinguished Service Cross for personally destroying an Italian tank. Several days later Patton offered Darby a regiment in the 45th Division and a promotion to full colonel. Darby declined, saying he preferred to remain with his beloved Rangers. An amazed Patton wrote in his diary that this was the first time he had ever seen a man turn down a promotion: "Darby is a really great soldier."

7. Early in the Husky planning it had been intended to use the inexperienced, untested 36th Infantry Division (a Texas National Guard unit) as one of the U.S. invasion forces. Patton already had a similar unit in the 45th Division, and wisely did not want two-thirds of his invasion force consisting of untested divisions.

8. LF, p. 291.

9. Quoted in Albert N. Garland and Howard McGaw Smyth, *Sicily and the Surrender of Italy* (Washington, D.C., 1965), p. 170; and LF, p. 291.

10. Bradley Commentaries and ONB-*AGL*, p. 183.

11. Diary, July 11, 1943.

12. ONB-*SS*, p. 130.

13. *New York Herald Tribune* July 15, 1943, and *Los Angeles Evening Herald-Express* July 14, 1943, copies in PP-USMA. The correspondent was British reporter for the London *Daily Mail* and had not set foot in Sicily. His account was based on what he had been told while aboard a Royal Navy destroyer offshore at Gela. An article in August noted Patton had gone through Sicily "like a squirrel up a greased stump."

14. Albert C. Wedemeyer, *Wedemeyer Reports!* (New York, 1958), p. 226.

15. LF, pp. 396–97.

16. "I had a hell of a lot of casualties because of these ruses they pulled on us," a battalion commander later recalled. (See Oral history of Lt. Gen. John H. Heintges, USAMHI.)

17. Diary, July 15, 1943. Patton's description of Bradley as "a most loyal man" was tinged with sarcasm and seems reflective of their deteriorating relations.

18. Chester Hansen Papers, USAMHI. According to Hansen, during the campaign Patton raised the question of whether or not an enemy soldier who shoots until his ammunition is exhausted and then gives up loses his right to surrender. There is no evidence this was anything more than a rhetorical question or that it was ever said in the presence of any of his troops.

19. James J. Weingartner, "Massacre at Biscari: Patton and an American War Crime," *The Historian* 52 (1989).

20. Ibid.

21. Diary, Apr. 5, 1944.

22. D'Este, *Bitter Victory,* p. 319.

23. GSP to BAP, Box 123, PP-LC.

24. See Carlo D'Este, *Decision in Normandy,* p. 507. There were examples of such behavior in every theater of war.

25. Butcher diary, July 13, 1943, and diary of Maj. Gen. John P. Lucas, USAMHI.

26. Butcher diary, ibid. Patton noted that Eisenhower seemed disinterested in his situation briefing and more interesting in nit-picking the alleged inadequacy of his reports. Patton was also criticized for being "too prompt in my replies and should hesitate more, the way he does before replying," an apparent reference to the Coningham incident. "I think he means well, but it is most upsetting to get only piddling criticism when one knows one has done a good job."(Diary, July 12, 1943.)

27. Lucas diary. On July 12, Patton moved ashore to a house in Gela, complete with bedbugs, and his first night there dined on a can of cheese and champagne "captured" by Stiller.

28. Diary, July 13, 1943.

29. Ridgway quoted in ONB-*AGL*, p. 186.

30. Ibid., p. 189.

31. Dominick Graham and Shelford Bidwell, *Coalitions, Politicians and Generals* (New York, 1993), p. 219.

32. Unpublished memoirs of General Sir Oliver Leese, copy supplied the author by Nigel Hamilton.

33. This often cited quote appears in numerous sources, including ONB-*SS*, p. 138.

34. Garland and Smyth, *Sicily and the Surrender of Italy,* p. 422.

35. ONB-*AGL*, p. 188.

36. Ibid., p. 189.

37. Diary, July 13, 1943.

38. Truscott, *Command Missions,* p. 218. Alexander, during his July 13 visit, had made a point of remarking to Patton that if Agrigento could be captured "through the use of limited forces in the nature of a reconnaissance in force, he had no objections." (Diary, July 13, 1943.)

39. GSP to BAP, letters dated July 19 and 20, 1943, Box 123, PP-LC.

40. Diary, July 17, 1943.

41. Ibid.

42. Lucas diary, July 20, 1943.

43. Truscott, *Command Missions,* p. 227.

44. GSP to BAP, July 27, 1943.

45. Diary, July 20, 1943.

46. Ibid., July 24, 1943; GSP-*W*, pp. 61–62; and LF, p. 300. The distance traveled to Palermo was slightly more than one hundred miles.

47. Quoted in HS, pp. 160–61.

48. "The Surrender of Palermo," *Life*, Aug. 23, 1943. Patton's picture was on the covers of both *Time* and *Newsweek* on July 26.

49. Diary, July 24, 1943, based on a conversation held on July 16.

50. Ibid., Aug. 2, 1943.

51. ONB-*SS*, p. 140.

52. ONB-*AGL*, p. 193.

53. Shelford Bidwell, Conversation with the author, Aug. 9, 1984.

Chapter 34   From Triumph to Disaster

A portion of this chapter originally appeared in chapter 29 of *Bitter Victory*, and has been updated and supplemented with additional research material.

1. Diary, July 30, 1943. The 45th Division fought in eight campaigns, including Salerno and Anzio, suffered 62,563 battle and nonbattle casualties, and distinguished itself as the most combat-experienced division in the U.S. Army.

2. Frank James Price, *Troy H. Middleton* (Baton Rouge, La., 1974), p. 160.

3. Cable, Montgomery to GSP, July 23, 1943, quoted in *Bitter Victory*, p. 443.

4. George F. Howe, the Alexander interviews, USAMHI.

5. GSP to BAP, July 24, 1943, Box 123, PP-LC. On July 13 Alexander had given the Eighth Army the exclusive use of the four roads leading to Messina, which effectively banned the Seventh Army from any participation in its capture.

6. A recent example is Leonard D. Ash and Martin Hill, "Allied Enemies," *The Retired Officer* (July 1993).

7. Diary, July 25, 1943.

8. Ibid. By validating the Monty-Patton agreement, Alexander tacitly acknowledged that the Seventh Army was no longer the stepchild of the 15th Army Group, and had attained full equality with Eighth Army, with which it would now share the final battles for Messina.

9. Maj. Gen. Sir Francis de Guingand, *Operation Victory* (London, 1947), p. 301.

10. Montgomery diary, July 28, 1943, quoted in *Bitter Victory*, p. 449.

11. GSP to Middleton, July 28, 1943, quoted in ibid., p. 449.

12. MB-*PP* 2, p. 307.

13. See *Bitter Victory*, Chap. 27.

14. Bradley Commentaries.

15. Ernie Pyle coined Bradley's nickname, "The GI General."

16. Bradley Commentaries. While he did nothing to stop the practice, even Patton finally noted that "Sgt. Mims has gotten so self-important that he blows my siren all the time."

17. Ibid.

18. Ibid. Bradley's claim that "during the Sicily planning George seldom got into the details of the planning," is refuted by the evidence of his considerable involvement before D-Day.

19. Gen. Ben Harrell (G-3 of the 3d Division in Sicily), 1971, oral history, USAMHI. On one grim day the 15th Infantry Regiment took 103 casualties without gaining any ground. In his memoirs Truscott recalls that, "In ordinary times, the distance to Messina would be a few hours drive of great scenic beauty. Now it was an area in which a determined enemy had every advantage for defense and delay." (Truscott, *Command Missions*, p. 230.) The use of mines continued to be the most devilishly effective German tactic in Sicily. Seemingly possessed of an inexhaustible supply, the Germans planted mines in defense of their positions and along the Allied route of march as they retreated to Messina.

20. Ibid.

21. Garland and Smyth, *Sicily and the Surrender of Italy*, p. 357. Task Force Bernard was Lt. Col. Lyle W. Bernard's 2d Battalion, 30th Infantry, supported by artillery, tanks, and combat engineers.

22. Ibid., p. 389.

23. Truscott, *Command Missions,* p. 235.

24. Diary, Aug. 10, 1943. The naval representative also favored postponement because the operation had begun an hour late. (See Morison, *Sicily-Salerno-Anzio,* chap. 10.)

25. Ibid., Aug. 11, 1943.

26. Ibid., Aug. 10, 1943.

27. GSP to BAP, Aug. 11, 1943. "Last night I remembered Frederick's 'L'audace, toujours, l'audace' and Nelson putting the glass to his blind eye and saying, 'Mark well, gentlemen, I have searched diligently and see no signal to withdraw. Fly the signal to continue the action.'"

28. Task Force Bernard lost 171 men, plus most of its supporting tanks and artillery; German losses were equally severe.

29. GSP to BAP, Aug. 9, 1943.

30. Col. Robert H. Fechtman, "An Infantry Lieutenant's Tale: 'We Walked All Over Sicily,'" *Army* (Aug. 1993); and Diary, Aug. 12, 1943.

31. FA, p. 139.

32. ONB-*SS*, p. 158.

33. Truscott, *Command Missions,* pp. 242–43, and Morison, *Sicily-Salerno-Anzio,* pp. 413–15. Patton was fully aware that the 157th RCT landing might accomplish little, but apparently considered their presence might help the 3d Division capture Messina more quickly, and believed it was the quickest way to move the regiment forward by avoiding the congested coastal highway. Once again Bradley's protests had failed to sway Patton.

34. Bradley Commentaries. Bradley was equally critical of Patton's staff and what his G-2 sarcastically called "that great silence we call Army." Lucas was told Seventh Army never maintained telephone lines into II Corps (an established military doctrine) and that no army staff officer ever visited him during the campaign (Lucas diary, Aug. 14, 1943.) Bradley was equally critical of the Seventh Army's failure to maintain adequate stocks of ammunition in their rear depots. Although Lucas defended him, Bradley was adamant that Patton's known dislike of detail—particularly supply matters—contributed to the breakdown of essential supplies at a critical moment in the campaign: "We were always short on supply largely because Seventh Army showed a complete lack of understanding of the fundamentals of supply." (In British and U.S. military doctrine the corps is strictly a tactical command, responsible for orchestrating the combat operations of a [flexible] number of divisions and other attached units. Supply, transport, and communications are army responsibilities.)

35. Heintges oral history, USAMHI.

36. Frank McCarthy interview, FM.

37. Quoted in Garland and Smyth, *Sicily and the Surrender of Italy,* p. 416. Returning from Messina, Patton's chief of staff, Brig. Gen."Hap" Gay, encountered a staff car parked along the roadside.: "In front of the car stood Lt. George Murnane [Gay's aide]—looking utterly disgusted. In the car sat a brigadier general . . . he appeared more disgusted than did Murnane—down the embankment crouched a high-ranking staff officer from General Eisenhower's headquarters—hiding or seeking protection from artillery fire—actually our own, not the enemy's." When Patton's arrived a short time later, Gay wrote: "I have never seen Patton so completely abashed—ashamed of an American." The officer was the AFHQ Chief of Staff, Maj. Gen. Walter Bedell Smith. Patton was later told that when a nearby artillery battery opened up, "Smith thought it was enemy shells arriving and jumped from the car into the ditch in one leap, and refused to leave it, even when told it was quite safe. When I got back in, he was still pale and shaky." Smith, for whom Patton had no great love, and to whom he sometimes referred as an "S.O.B.," was taken away in Patton's command vehicle. (Gay diary, Aug. 17, 1943, USAMHI, and Diary, same date.)

38. Lucas diary, Aug. 17, 1943. The awful destruction in Messina affected Patton who said, "I do not believe the indiscriminate bombing of towns is worth the ammunition, and it is unnecessarily cruel to civilians."

39. Richard Tregaskis, *Invasion Diary* (New York, 1943), p. 89.

40. Harrell oral history. Patton well understood Stiller's limitations, noting in his diary on May 28, 1943: "I will have to get rid of Stiller. He is too dumb and too crude. He tries hard but never clicks." Of course, Patton did no such thing. Stiller filled an important role in Patton's daily life.

41. Diary, Aug. 17, 1943, and Bradley Commentaries. According to Bradley's aide, Chester Hansen, his boss apparently was not notified of the rendezvous and was "quite dis gusted." Patton later apologized to Bradley for his failure to ensure that he was notified. Even if he had, it seems doubtful if Bradley would have put in an appearance in Messina. Neither of Bradley's two memoirs clarifies this point. (See also Hansen diary.)

42. Diary, ibid.

43. GSP to BAP, Aug. 18, 1943, Box 13, PP-LC

44. Ibid., Aug. 9, 1943, .

45. GSP to Walter Dillingham, Aug. 18, 1943, Box 24, PP-LC.

46. GSP to BAP, Aug. 18, 1943. The initiative for this decoration came from Lucas, who prepared and delivered the recommended citation to Eisenhower.

47. Report of Col. F. Y. Leaver, MC, CO, 15th Evacuation Hospital, Aug. 4, 1943, EP. Lucas accompanied Patton and recorded a scene of "brave, hurt bewildered boys. All but one . . . [who] said he was nervous and couldn't take it. Anyone who knows him can realize what it would do to George. The ward sister was really nervous when he got through." (Lucas diary, Aug. 3, 1943.) After the war Lucas commented that "Patton slapped him on the cheek with his folded gloves and ordered him from the tent." Kuhl was about to undergo treatment for the third time in Sicily for "exhaustion" and was later diagnosed as "high strung." After the slapping incident his medical file indicated that Kuhl was suffering from diarrhea and mild malaria.

48. Diary, Aug. 3, 1943. After the incident Kuhl wrote to his parents that Patton had slapped him, but urged them to "just forget about it."

49. Patton memo, dated Aug. 5, 1943, copy in diary, Mar. 16, 1944.

50. Ibid., Aug. 10, 1943.

51. This account of Patton's actions on Aug. 10 is based on a composite of the official reports of investigation (EP), Patton's diary, the account of correspondent Demaree Bess of the *Saturday Evening Post*, Bradley's *A Soldier's Story* and *A General's Life*, Eisenhower's *Crusade in Euro*pe, and Garland and Smyth, *Sicily and the Surrender of Italy*, chap. 21.

52. Report of Lt. Col. Perrin H. Long, MC, to the Surgeon, North African Theater of Operations, U.S. Army (NATOUSA), EP.

53. ONB-SS, p. 160.

54. Bradley Commentaries.

55. Quoted in LF, p. 329.

56. Bess Report, EP.

57. Quoted in Ambrose, *The Supreme Commander*, p. 229.

58. DDE to GSP, Aug. 17, 1943, copy in *The Eisenhower Papers*, vol. 2, pp. 1340–41.

59. Garland and Smyth, *Sicily and the Surrender of Italy*, p. 427; LF, p. 335; and Butcher diary, Aug. 20, 1943.

60. Eisenhower, *Crusade in Europe*, p. 199; Garland and Smyth, *Sicily and the Surrender of Italy*, p. 431; and Ambrose, *The Supreme Commander*, p. 230.

61. The reprimand was unofficial and thus never became a part of GSP's official personnel (201) file.

62. Eisenhower, *Crusade in Europe*, p. 200, and Phillip Knightley, *The First Casualty*, rev. ed. (London, 1982), p. 304. Bedell Smith successfully implored the press pool in Algiers to keep mum about the affair, something that would be unheard of in the 1990s.

63. Diary, Aug. 20, 1943.

64. Lucas enjoyed the full confidence of both generals and was the ideal officer to convey Eisenhower's extreme concern to Patton. He was also ordered to investigate the incidents from the soldiers' point of view. Eisenhower later sent two colonels from the NATOUSA inspector general's office to investigate the incidents and a theater medical consultant, Lt. Col. Perrin H. Long, MC, who was ordered to submit an "eyes only" report to

Eisenhower. Ironically Dr. Long was from Patton's adopted hometown of Hamilton, Massachusetts.

65. Lucas diary, entries of Aug. 20–21, 1943.

66. Account of Bennett's battalion commander, Lt. Col. Joseph R. Couch (1st Battalion, 17th Field Artillery), in the "The Day Gen. Patton Slapped a Soldier," *Washington Post*, June 3, 1979. When he learned of the incident Couch agonized over what action he ought to take: "As battalion commander I could hardly ignore it, but as a junior lieutenant colonel I could not confront a lieutenant general." His older, wiser group commander advised Couch to do nothing: "I bit my lip and took his advice." (Patton file, CMH.)

67. Diary, Aug. 21, 1943; and NATOUSA inspector general's report.

68. Diary, ibid.

69. Statement of Capt. H. A. Carr, MC, 15th Evacuation Hospital, EP.

70. Quoted in Miller, *Ike the Soldier*, p. 536.

71. NATOUSA IG Report, Sept. 18, 1943, EP.

72. Diary, Aug. 20, 1943.

73. Account of Dr. Daniel Franklin (one of the investigating officers) in *Infantry*, circa 1975, Patton file, CMH.

74. "A Matter of Rank," *Richmond News Leader*, May 30, 1970. For the rest of his life, Kuhl (who later participated in the Normandy invasion but was again evacuated with combat fatigue), was haunted by the incident.

75. Bill McAndrew, "War's Mental Casualties," *Ottawa Citizen* Jan. 9, 1991. McAndrew and Terry Copp have coauthored *Battle Exhaustion: Soldiers and Psychiatrists in the Canadian Army, 1939–1945* (Montreal, 1991).

76. Diary, Aug. 6, 1943.

77. Ibid., Aug. 2, 1943.

78. CRC, p. 111. When one soldier told him he had been wounded in the chest, Patton replied: "Well, it may interest you to know that the last German I saw had no chest and no head either." Codman wrote of the day when Patton awarded the forty Purple Hearts: "God, he did it well."

79. Allan Carpenter, *George Smith Patton, Jr., The Lost Romantic* (Vero Beach, Fla., 1987), p. 79.

80. Butcher, *Three Years with Eisenhower*, p. 338.

81. Ibid., pp. 338–39.

82. Butcher diary, Aug. 21, 1943. Butcher also noted: "After I had retired last night Ike sat in my room for half an hour debating the question. Not so much with me as with himself."

83. DDE to GCM, Aug. 24, 1943, copy in Butcher diary.

84. Ibid., Sept. 26, 1943.

85. Bradley Commentaries.

86. Quoted in MB-*PP* 2, p. 338.

87. NATOUSA inspector general's report.

88. Quoted in MB-*PP* 2, p. 333.

89. Telephone interview of Malcolm Marshall, Oct. 15, 1993.

90. MB-*PP* 2, p. 340; and ONB-AGL, p. 205.

91. Interview with Gen. Robert W. Porter Jr. USA (Ret.), Sept. 15, 1985. At the time Porter was the 1st Division intelligence officer (G-2).

92. Oral history interview of Gen. Theodore J. Conway, Sept. 1977, USAMHI. Also published in *Parameters* (Dec. 1980). Conway served in the 60th Infantry Regiment in Sicily.

93. Unpublished account of Lt. Col. Henry G. Phillips, 1985.

94. Heintges oral history, USAMHI.

95. MB-*PP* 2, pp. 341–42.

96. Quoted in Eisenhower, *Crusade in Europe*, p. 201. Patton reiterated what he had told the medical personnel about the alleged World War I incident and concluded with the observation: "After each incident I stated to officers with me that I had probably saved an immortal soul."

97. Stimson to Sen. Robert R. Reynolds, Chairman, Senate Military Affairs Committee, reprinted in the *Army-Navy Register*, Dec. 4, 1943; and Henry L. Stimson and McGeorge Bundy, *On Active Service in Peace and War* (New York, 1947), p. 499.

98. CRC, p. 134.

99. Patton's letters to Pershing are in Box 155, JJ-P.

100. Joyce manuscript, USAMHI. Patton later told Eisenhower's driver, Kay Summersby, "I always get in trouble with my gawdamned mouth!" (Quoted in Kay Summersby, *Eisenhower Was My Boss* [New York, 1948], p. 81.)

101. Diary, Mar. 16, 1944.

102. Ruth Ellen Totten, address to the Topsfield, Massachusetts, Historical Society, 1974. Patton's son, although he personally believed them wrong and unjustified, nevertheless empathized with his father: "I personally and professionally believe he should not have done it," said George S. Patton in 1985. "At the time of the incident . . . he was pretty well worn out, and extremely high strung, trying to get to Messina ahead of the British. . . . I believe if he were sitting here today, he would say he shouldn't have done it. You just don't strike an enlisted man. But I could understand why he did it. I might have done the same thing under the same circumstance s. . . [with] very, very seriously wounded, perhaps dying men, and this enlisted man was a shirker . . . it was too much for the Ol' Man." (Maj. Gen. George S. Patton, USA [Ret.], quoted in "Reflections on a Fighting Father.")

103. Quoted in LF, p. 338.

104. Quoted in McCloy memorandum to Eisenhower, Dec. 13, 1943, EP.

105. Eric Larrabee, *Commander in Chief* (New York, 1987), p. 486.

Chapter 35   Exile

All letters to Beatrice Patton are in Box 123, PP-LC, unless otherwise cited.

1. Diary, Oct. 3, 1943.

2. Hamilton, *Monty: Master of the Battlefield, 1942–1944*, p. 374.

3. Diary, Aug. 29, 1943. The letter to Eisenhower was remorseful and thanked him for "your fairness and generous consideration for making the matter personal instead of official."

4. Ibid., Sept. 7, 1943.

5. ONB-*AGL*, p. 208. Eisenhower had cabled Patton of his intention to disband the Seventh Army.

6. Diary, Nov. 28, 1943.

7. Ibid., Sept. 21, 1943.

8. Miller, *Ike the Soldier,* p. 543.

9. DDE to GCM, Aug. 24 and Sept. 6, 1943, Box 53, GCM-P, and Butcher diary.

10. Butcher diary, Aug. 21, 1943.

11. Brendon, *Ike,* p. 116.

12. Both quotes in Garland and Smyth, *Sicily and the Surrender of Italy,* p. 431.

13. Quoted in CRC, p. 115.

14. Gen. James "Jimmy" Doolittle, *I Could Never Be So Lucky Again* (New York, 1991), p. 363.

15. BAP to GSP, July 30, 1943, PP-USMA.

16. GSP to BAP, Aug. 22 and Dec. 21 & 23, 1943. Beatrice did not learn of her husband's troubles until she heard Drew Pearson's broadcast. Kay Summersby records that Eisenhower lost a good many nights' sleep agonizing over what to do about his friend.

17. Ibid., Feb. 6, 1944.

18. RET-M.

19. Everett S. Hughes to BAP, circa Jan. 20, 1944, Box 14, PP-LC.

20. GSP to Kenyon Joyce, Sept. 26, 1943, Box 25, PP-LC.

21. Diary, Sept. 2, 1943. Torch was more difficult than Husky, which was far larger in all respects.

22. Ibid., Sept. 15, 1943.

23. Ibid., Sept. 17 1943.

24. Both quotes in diary, Sept. 6, 1943.

25. In other respects Patton found himself out of step with the mainstream of military protocol. To his friend, A. D. Surles, chief of War Department information, he wrote of what he believed was the sheer stupidity, for security reasons, of failing to identify publicly units fighting the Axis: "The Bible states that, 'Man does not live for bread alone,' and I am as sure as God that he does not fight for pay alone. He fights for glory and fame, which is publicity, and we do our damdest to see that he gets none at all." (GSP to A. D. Surles, Oct. 13, 1943, Box 3, PP-USMA.)

26. CRC, pp. xix–xx. Marquand wrote the introduction to Codman's memoir. Patton later asked Marquand not to publicize any intemperate references to his distinguished colleague, Bernard Montgomery: "I seem to be in a lot of hot water lately, and suppose we just forget the episode." Paddy Flint commanded a regiment of the 9th Division attached to the 1st Division during the bloody battle for Troina.

27. Ibid., p. xiv.

28. Diary, Sept. 10, 1943.

29. Ibid., Sept. 29, 1943.

30. Ibid., Nov. 8, 1943.

31. Ibid., Oct. 14, 1943.

32. Lt. Gen. Garrison H. Davidson, "Grandpa Gar: The Saga of One Soldier as Told to his Children," 1974 ms., USMA-SC. When he was shown the sofa upon which the emperor was born, Patton "closed his hand loosely and rubbed his fingernails over the covering"; then, with a twinkle in his eye, blew on them, "Just for good luck."

33. Everett S. Hughes to GSP, Nov. 17, 1943, Box 12, PP-LC.

34. According to M/Sgt. Mims, "Everywhere he went they would stand up and salute him and then they would see who it was and they would throw up their hats and everywhere they would start shouting. . . ." (Mims interview, USAMHI.)

35. MB-*PP* 2, p. 399. Anvil (later renamed Dragoon) was the invasion of southern France, and a subsidiary operation to Overlord, designed to draw off German formations from northern France.

36. GSP to Frederick Ayer, Sept. 26, 1943, Box 12, PP-LC.

37. Kay Summersby Morgan, *Past Forgetting* (New York, 1976), pp. 165–66.

38. "A Soldier's Prayer," Box 123, PP-LC.

39. Diary, Nov. 17, 1943.

40. Quoted in Pogue, *George C. Marshall, Organizer of Victory, 1943–1945,* p. 321. Prior to Quebec it had been assumed that supreme commander would be British, partly in return for Eisenhower's selection in the Mediterranean in 1942, but mainly because Overlord would be mounted from Britain and British forces would play a dominant role, at least initially. In July 1943 Churchill had told Brooke he wanted him to take the invasion command, only to change his mind at Quebec for the same reasons FDR did over Marshall.

41. For an account of the selection of the Overlord commanders see *Decision in Normandy*, chap. 4.

42. Diary, Dec. 8, 1943.

43. Ibid., Dec. 7, 1943. "I should have a group of armies," wrote Patton, "but that will come."

44. DDE to GSP, Dec. 1, 1943, Box 14, PP-LC.

45. Stimson to GSP, Dec. 18, 1943, ibid.

46. "The Flight into Egypt," Box 3, PP-LC, is a lengthy account of Patton's trip. His admiration for the Poles and their commander, Lt. Gen. Wladyslaw Anders, was unbounded.

47. Diary, Dec. 13, 1943.

48. GSP to son George, Dec. 11, 1943, Box 34, PP-LC.

49. GSP to BAP, Dec. 2, 1943, Box 14, PP-LC.

50. Ibid., entries written between Dec. 21 and 29, 1943. Eisenhower rated him "Superior" on his efficiency report but also noted that he would only recommend him for army

command. He was "outstanding as a leader of an assault force. Impulsive and almost flamboyant in manner. Should always serve under a strong but understanding commander." (Report, July 1–Dec. 31, 1943, Microfilm Reel 1, PP-LC.)

51. Diary, Dec. 31, 1943.

52. Ibid., Jan. 1, 1944. He proudly wrote to Frederick Ayer: "Not a man quit. I had to loan some of them recently and they all begged me to be sure and get them back, and two of them cried. So did I." (Letter of Jan. 14, 1944, Box 12, PP-LC.)

53. Ibid., Jan. 3, 1944. Patton also took note that Beatrice had sent him a copy of a biography of Wellington, whose staff also changed on several occasions. "Fate," he wrote.

54. Ibid., Jan. 9, 1944, and CRC, p. 135.

55. Letters dated Jan. 11, 12, and 14, 1944, Box 14, PP-LC. Patton also promised that when he did get back into battle, "I have a lot of swell new ideas to kill Huns with."

56. Lucas diary, Jan. 10, 1944. He had also hoped to make contact with son-in-law Lt. Col. James Totten, who was commanding an artillery battalion, and who would be gravely wounded at Anzio. They never saw each other.

57. Maj. Gen. W. R. C. Penney to Brig. Douglas Renny, Feb. 8, 1956, Penney Papers, LHC.

58. GSP to Frederick Ayer, Jan. 14, and to BAP, Mar. 3, 1944, Box 12, PP-LC and PP-USMA.

59. Diary, Jan. 18, 1944, and letter to BAP, Jan. 12, 1944.

60. GSP to BAP, Jan. 23, 1944.

61. Diary, Mar. 16, 1944. The sailors who made the plaque for Patton only reluctantly accepted a $25.00 payment.

Chapter 36   Third Army Commander

Unless otherwise cited, Patton's 1944 correspondence is in Box 14, PP-LC, or in the West Point Collection.

1. Pogue, *George C. Marshall: Organizer of Victory, 1943–1945*, p. 376.

2. Cray, *General of the Army*, p. 444.

3. Pogue, *George C. Marshall: Organizer of Victory, 1943–1945*, p. 376.

4. Butcher diary, Jan. 27 and Feb. 11, 1944. Patton's diary records no mention of his session with Eisenhower, only that at dinner the same evening, "Ike very nasty and show-offish," which he blamed on the presence of Eisenhower's driver, confidante, and alleged lover, Kay Summersby.

5. Ibid., Apr. 5, 1944.

6. Summersby, *Eisenhower Was My Boss* (New York, 1948), pp. 21–22. On one occasion Patton pulled one of his revolvers and waving it under her nose said: "Look at that. People always talk about my pearl-handled revolvers. That's not pearl. That's ivory. Good solid ivory. From an elephant." (Morgan, *Past Forgetting*, p. 162.)

7. Morgan, *Past Forgetting*, p. 162.

8. Miller, *Ike the Soldier*, p. 639.

9. Diary, Jan. 26, 1944.

10. Ibid., Feb. 18, Jan. 31, and Mar. 2, 1944. Several months later Patton noted the practice of general officers exchanging signed photos of one another before D Day. He received a touching one from Bedell Smith the same day Smith had called to silence him after the Knutsford Incident. He told Beatrice that the photo reminded him of the Judas trees that grew in Virginia. (Diary, May 24, 1944.)

11. William P. Snyder, "Walter Bedell Smith: Eisenhower's Chief of Staff," *Military Affairs* (Jan. 1984).

12. D. K. R. Crosswell, *The Chief of Staff: The Military Career of General Walter Bedell Smith* (New York, 1991), p. 300.

13. CRC, chap. 7, passim. The play was by Robert E. Sherwood.

14. GSP to BAP, Feb. 5, 1944. p. 19.

15. Diary, Jan. 27, 1944.

16. Ibid., Mar. 6, 1944; and letter to BAP, Mar. 16, 1944. Patton seemed relieved when Gay and Gaffey hit it off well and thought it vindication of a difficult and personally unpleasant decision. "He is a perfectly unselfish man," Patton wrote of Gay when the subject was first broached by Eisenhower.

17. ONB-SS, pp. 229–32.

18. Interview of Lt. Gen. Walter Bedell Smith by Forrest C. Pogue, 1947, USAMHI.

19. Diary, Aug. 7, 1944.

20. Russell F. Weigley, *Eisenhower's Lieutenants* (Bloomington, Ind., 1981), p. 84.

21. Robert S. Allen, *Lucky Forward* (New York, 1947), p. 16. Allen, who in civilian life was a veteran Washington newspaperman and political columnist, became a Third Army assistant G-2, under Colonel Oscar Koch.

22. ONB-*SS*, p. 473, and H. Essame, *Patton: The Comander*, p. 122. See also Lt. Col. Paul G. Munch, "Patton's Staff and the Battle of the Bulge," *Military Review* (May 1990). The following observation is attributed to Patton, who wrote tongue in cheek: "The typical staff officer is a man past middle age, spare, wrinkled, intelligent, cold, passive, noncommittal, with eyes like a codfish, polite in contact, but at the same time unresponsive, cool, calm, and as damnably composed as a concrete post . . . a human petrification with a heart of feldspar and without charm or the friendly germ; minus bowels, passions or a sense of humor. Happily they never reproduce and all of them finally go to hell."

23. Letter, Third Army staff officer, to his family, Aug. 16, 1944, copy in Box 14, PP-LC.

24. Allen, *Lucky Forward*, pp. 19–20.

25. Col. C. Cabanné Smith, *My War Years: 1940–1946: Service on Gen. Patton's Third Army Staff* (N. p., 1989), p. 34.

26. Interview with David S. Terry, Mar. 31, 1994. Before the ceremony Patton came to look them over and pronounced himself satisfied. The Third Army band had not yet arrived and the band used that day had come from the 10th Replacement Depot.

27. George F. Murnane Jr., quoted in FA, p. 149.

28. Allen, *Lucky Forward*, chap. 2, passim. Bradley and Truscott would have disagreed and argued that Brolo had been shoved down their throats. But Patton too learned from experience and although he never commented further on this particular operation, the evidence suggests it concerned him more deeply than he was ever prepared to admit.

29. Koch, *G-2: Intelligence for Patton,* p. 156.

30. Ibid., p. 157.

31. Wallace, *Patton and His Third Army,* pp. 16–17.

32. Dr. Hugh M. Cole, "George Patton," memoir in the S. L. A. Marshall Military History Collection, University of Texas at El Paso.

33. Wallace, *Patton and His Third Army*, p. 21.

34. Koch, *G-2: Intelligence for Patton,* p. 61.

35. Alden Hatch, *George Patton: General in Spurs* (New York, 1950), p. 199.

36. On Mar. 27, 1944, Patton traveled to SHAEF headquarters, where he and several other American officers were decorated on behalf of the king by Field Marshal Sir Alan Brooke, the Chief of the Imperial General Staff. When Patton was presented with the Companion of the Order of Bath (CB), Brooke said: "Don't wince, Patton. I shan't kiss you." Patton also wrote: "He also said that I had earned the decoration more than any other American. He probably said the same thing to each one." (Diary.)

37. CRC, p. 144.

38. Munch, "Patton's Staff and the Battle of the Bulge," and Gen. John W. Foss, "Command," *Military Review* (May 1990).

39. MB-*PP* 2, pp. 424–25.

40. Albert C. Stiller to BAP, Mar. 6, 1944.

41. Diary, Mar. 25, 1944.

42. Ibid., Apr. 16, 1944.

43. Ibid., Apr. 20, 1944.

44. Ibid., Feb. 26, 1944.

45. Cable, Alexander-Brooke, Feb. 15, 1944 (PREM 3-248/4), PRO.

46. Butcher diary, Feb. 17, 1944, and D'Este, *Fatal Decision*.

47. Diary, Feb. 16, 1944.

48. Ibid. As much as he would have relished the assignment, Patton knew he would not only have been walking into a mess at Anzio, but would have found Jake Devers "a damned nuance [nuisance] to work under."

49. Ibid., Feb. 17, 1944.

50. Adm. Sir John Cunningham to Adm. Sir Andrew Cunningham, Feb. 11, 1944, the Papers of Admiral of the Fleet Viscount [Andrew] Cunningham of Hyndhope, Dept. of Manuscripts, the British Library.

51. MB, p. 217.

52. Allen, *Lucky Forward*, p. 43.

Chapter 37   In the Doghouse—Again

1. The documents and sources pertaining to the so-called Knutsford incident are in Eisenhower, *Crusade in Europe*, pp. 246–47; Pogue, *George C. Marshall: Organizer of Victory, 1943–1945*, chap. 19; Cray, *General of the Army*, chap. 26; Miller, *Ike the General*, pp. 593–96; Chandler, *The Eisenhower Papers*, vol. 3, pp. 1837–41; the SHAEF papers, NA, and Patton's diary. The period in question is between Apr. 27 and approximately May 10, 1944.

2. Cable, GCM to DDE, Apr. 29, 1944, SHAEF Papers. Through Everett S. Hughes, Patton offered that same day to authorize having his name withdrawn from the list in order not to jeopardize the promotions of others.

3. Ibid. Patton's exact words have been the subject of some debate but as Eisenhower reported to Marshall, he apparently said: "Since it seems to be the destiny of America, Great Britain and Russia to rule the world, the better we know each other the better off we will be."

4. Quoted in ONB-SS, p. 231.

5. Diary, Apr. 7, 1944. Patton had previously noted in his diary that "I shall certainly attempt to say nothing which can be quoted."

6. Ibid., Apr. 18, 1944.

7. Ibid., Apr. 27, 1944.

8. DDE to GSP, Apr. 29, 1944, in Chandler, *The Eisenhower Papers*, vol. 1, pp. 1839–40.

9. ONB-AGL, p. 222; and cable, DDE to GCM, Apr. 30, 1944, SHAEF Papers. According to Bradley's collaborator, historian Clay Blair, Bradley was considerably miffed to learn that Eisenhower had not consulted him in reaching his decision to retain Patton. Bradley attributed Eisenhower's decision partly to the fact that there was no hope of getting Truscott from Italy, and that Churchill had dismissed Knutsford as trifling.

10. Diary, May 1, 1944.

11. GSP to BAP, May 2, 1944.

12. Eisenhower, *At Ease*, pp. 270–71. In his conservatively written *Crusade in Europe* (p. 247), Eisenhower referred to Patton's contriteness and concludes: "I laughingly told him: 'You owe us some victories; pay off and the world will deem me a wise man.'" A slightly different version is in the unpublished memoir of Col. Justus "Jock" Lawrence, USMA-SC. According to Lawrence, Eisenhower told him that when Patton "started to cry all over again . . . I could no longer stand it. This was too much for me! I stretched out on the couch in my office and burst into laughter, which I now regret for it was, in retrospect, cruel. General Patton stood at strict attention, not even looking at me lying on the couch, laughing." After Patton departed: "I had to tell someone, so I called in Beetle [Smith] and told him what had happened. It is probably the only time in the all the years of my long experience with Smith that I saw Beetle really lose himself in laughter!"

13. Cable, GCM to DDE, May 2, 1944. Marshall who seems to have been pointing Eisenhower toward giving Patton yet another chance, noted that none of the editorials in the United States, however, caustic, had actually called for his relief from command.

14. Cable, DDE to GSP, May 3, 1944, in Chandler, *The Eisenhower Papers*, vol. 3, pp. 1846–47. Eisenhower indicated that what principally saved Patton was the fact that he had resisted making a speech well before the meeting, and had spoken in the [mistaken] belief his remarks were off the record.

15. DDE to GCM, May 3, 1944, ibid., p. 1846.

16. GSP to BAP, May 3, 1944.

17. Stimson to GSP, May 5, 1944, copy in Irving Microfilm, USAMHI. The same day Stimson wrote to praise Eisenhower for "the great courage you have shown in making this decision . . . has filled me with even greater respect and admiration than I had for you before." Eisenhower replied that he believed that "from the viewpoint of OVERLORD the decision was a correct one."

18. Lawrence memoir. The "keep your goddamned mouth shut" order was apparently delivered after Patton's meeting with Eisenhower on May 1.

19. FA, p. 141 and ONB-SS, p. 230.

20. Don Lawson, *The United States in World War II* (New York, 1963), p. 164.

21. Walter Isaacson and Evan Thomas, *The Wise Men* (New York, 1986), p. 196.

22. George Forty, *Patton's Third Army at War* (London, 1978), p. 25.

23. F. H. Hinsley et al., *British Intelligence in the Second World War*, vol. 3, part 2 (London, 1988), p. 47.

24. There were actually two Fortitude operations, Fortitude North, a simulated threat to Scandinavia by both Anglo-American forces and the Red Army, and Fortitude South, the Pas de Calais operation. The original deception plan was drawn up by Lt. Gen. Sir Frederick E. Morgan's COSSAC (chief of staff to the Supreme Allied Commander) planning staff in Sept. 1943. The initial invasion force against Normandy was inadequate in both size and scope and it was considered essential that some means be found to ease the pressure on the invasion force both before and after Operation Overlord by pinning down as many German forces as possible in the Pas de Calais. Eventually COSSAC's deception plan evolved into Fortitude in 1944.

25. Michael Howard, *British Intelligence in the Second World War*, vol. 5, *Strategic Deception* (London, 1990), p. 105.

26. Daniel Wyatt, article on Operation Fortitude in *World War II*, May 1994.

27. Howard, *British Intelligence*, p. 131.

28. Diary, May 18, 1944.

29. Thames Television interview of Goronwy Rees for *The World at War*. Rees was a member of the 21st Army Group staff. A more detailed version of the May 15 Overlord briefings at St. Paul's School is in *Decision in Normandy*, chap. 6.

30. Diary, May 15, 1944; and *Decision in Normandy*, p. 88.

31. Ibid., May 15, 1944.

32. ONB-AGL, p. 241.

33. Wallace, *Patton and His Third Army*, p. 204.

34. Diary, May 29, 1944.

35. Perret, *There's a War to be Won*, p. 343.

36. David Irving, interview of Gen. W. H. Simpson, Mar. 15, 1972, Irving Microfilm, USAMHI.

37. ONB-AGL, p. 341, based on a postwar interview by Clay Blair with Simpson.

38. Allen, *Lucky Forward*, n., p. 159.

39. Diary, June 1, 1944.

40. Chester B. Hansen diary, June 2, 1944.

41. ONB-AGL, p. 243.

42. Diary, June 1, 1944.

43. Ibid.

44. Ibid., June 2, 1944.

45. Hamilton, *Monty: Master of the Battlefield*, p. 602.

46. Essame, *Patton the Commander*, passim.

47. Diary, Feb. 26, 1944.

48. Ibid., June 4–6, 1944.
49. GSP to BAP, June 6, 1944.

Chapter 38 The Speech

1. Cole, "George Patton."
2. There is more than one version of Patton's famous pre D-Day speeches. The version cited here is primarily from a speech titled "A General Talks to His Army," copy in the Patton file, VMI archives; and another titled "General Patton Talks to His Troops," which is alleged to have been taken down verbatim by one of Patton's officers, who was also a former court reporter. There are some minor variations in other versions of "The Speech," including one published in 1963 by an admirer named C. E. Dornbusch, titled *Speech of General George S. Patton, Jr., to his Third Army on the Eve of the Normandy Invasion* (Cornwallville, N.Y., 1963).
3. "A General Talks to His Army." Patton believed that various articles in the *Saturday Evening Post* unnecessarily stressed the heroics of individuals and gave a false impression of how battles were won. The real message of his "speech" was that training and teamwork win battles.
4. Gen. James H. Polk, oral history, Polk Papers, USAMHI.
5. Brig. Gen. Robert W. Williams, "Moving Information: The Third Imperative," *Army* (Apr. 1975).

Chapter 39 The "Mighty Endeavor"

1. Roosevelt, quoted in "Normandy," U.S. Army campaign pamphlet, 1994, CMH.
2. Before departing for Normandy Patton found Eisenhower discouraged by Montgomery's perceived inaction, but inclined to (at a later date) attach an American army to the British army group. Patton complained in his diary: "Why an American army has to go with Montgomery, I do not see, except to save the face of the little monkey."
3. John S. D. Eisenhower, *Strictly Personal*, p. 67.
4. Everett S. Hughes diary, quoted in Max Hastings, *Overlord: D-Day and the Battle for Normandy* 1944, (London, 1984), p. 197.
5. Diary, June 11, 1944. At a small ceremony in front of Peover Hall on June 17, the Third Army colors were presented and blessed, but the flag and plate were not be installed until the existence of Third Army was published in August.
6. GSP to his son, George, June 6, 1944, Box 5, PP-USMA.
7. GSP to BAP, and Summerall to BAP, both June 9, 1944.
8. Diary, June 17 and July 7, 1944.
9. Koch, *G-2: Intelligence for Patton*, p. 53.
10. GSP-W, p. 89.
11. The memo appended to his father's map reads: "To any biographer—This is a historic document since GSP, Jr. marked it well before the campaign plans had developed. He was very close in his prediction. //ss// G. S. Patton, LTC, USA, July 1962." (Box 67, PP-LC.)
12. Diary, June 26, 1944. Eisenhower's efficiency report (Jan. 26 to June 30, 1944) on Patton cited him as "Superior," with a rating of second as an army commander of twenty-six lieutenant generals "of his grade known to me personally," but eighth in "general rating." Patton was "a brilliant fighter and leader. Impulsive and quick tempered. Likely to speak in public in an ill-considered fashion." (Reel 1, GSP microfilm.)
13. Ibid., July 2, 1944. Whether or not an operation and the logistics to support it could have been mounted on such short notice is problematical.
14. Quoted in diary entry of July 5, 1944.
15. MB-*PP* 2, p. 477.
16. George Creel, "Patton at the Pay-Off," *Collier's*, Jan. 13, 1945.
17. CRC, pp. 151–53. The endless earth-shaking noise was a potent reminder of just

how long Patton had been idle. It was "the most infernal artillery preparation I have ever heard," he wrote in *War as I Knew It,* p. 93.

18. Diary, July 7, 1944.

19. LF, pp. 492–93.

20. Cole C. Kingseed, "When the Allies Forfeited a Golden Opportunity," *Army* (Mar. 1994).

21. ONB-AGL, p. 272.

22. David Irving, *The War Between the Generals* (New York, 1981).

23. ONB-SS, p. 355. After the publication of his war memoirs, when Patton's private remarks became public, Bradley's attitude would harden noticeably, and in his second volume (*A General's Life*) there was little charity toward Patton.

24. J. Lawton Collins, *Lightning Joe* (Baton Rouge, 1979), pp. 247-48.

25. ONB-SS, p. 357.

26. CRC, p. 275. Among those Patton singled out who had "fighting faces" were his own aide, Al Stiller, and his old friend Paddy Flint.

27. "General Bradley as Seen Close Up," *New York Times Magazine,* Nov. 30, 1947.

28. ONB-SS, p. 356.

29. Ibid., Chap. 18, passim.

30. Diary, July 14, 1944.

31. Lt. Gen. Sir Miles Dempsey, quoted in *Decision in Normandy,* p. 333.

32. Quoted in Martin Blumenson, *The Battle of the Generals* (New York, 1994), p. 121.

33. Diary, Aug. 19, 1944, and Hastings, *Overlord,* p. 289.

34. GSP to BAP, June 17, 1944, Box 14, PP-LC, and diary, July 25, 1944. Teddy Roosevelt was later posthumously awarded the Medal of Honor for exceptional gallantry on D Day. Bradley has said of him: "He braved death with an indifference that destroyed its terror for thousands upon thousands of younger men. I have never known a braver man nor a more devoted soldier."

35. GSP-*W*, p. 94.

36. Stimson had again warned Patton to keep his mouth shut, not to criticize others, and to "let my actions speak for me."

37. ONB-*SS*, p. 356. The correspondents all signed a letter to Patton which attempted to exonerate the PRO and place the blame on one of their own.

38. Patton's diary for July 18 records that the officer had done it "from dumbness, and from a misplaced sense of loyalty to me. I think he is honest but stupid." Patton's own justification for not firing the officer at once was that to do so might cause another leak. It was equally true that for all his outward bravado, Patton hated to sack his officers. Two days later Patton learned that the Third Army Civil Affairs section had distributed two secret papers that contained criticism of the First Army. He had the papers hastily recalled, but his faith in a close member of his staff was shaken.

39. GSP-*W*, p. 95.

40. Despite Montgomery's insistent claims to the contrary both then after the war, there is solid documentary evidence that the ultimate objective of Goodwood was Falaise. (Discussion and sources in D'Este, *Decision in Normandy,* p. 399, paperback editions only. Important additional evidence about Goodwood, previously unavailable, was discovered only after publication of the original British and U.S. editions in 1983.)

41. Diary, July 2, 1944.

42. Ibid., July 26, 1944.

43. Troy Middleton's VIII Corps was then under First Army operational control, but to take advantage of Cobra's apparent success, Bradley elected to place the corps under Patton's operational control prior to Aug. 1 (Martin Blumenson, *Breakout and Pursuit* [Washington, D.C., 1961], p. 37, and chap. 3, passim.)

44. Patton suggested the immediate employment of airborne troops to land behind enemy lines and capture the vital bridges and dams over the Sélune River near Avranches, but received a "not very congenial" response: "It takes so long to get paratroops moving

that the war would be over before we could use them." All of which would have been news to Matthew Ridgway and James Gavin of the 82d Airborne Division, who had conducted a successful night emergency airborne drop into the besieged Salerno beachhead with only a few hours warning.

45. Weigley, *Eisenhower's Lieutenants,* p. 160.

46. Diary, July 27, 1944.

47. The versions in the film *Patton* and LF, p. 447, are virtually identical. The only difference is that in the film, Patton is speaking to his soldiers when, in reality, these words were spoken to the Third Army staff the night before it became operational, either late on July 31 or in the early hours of Aug. 1, 1944.

48. Bradley to DDE, July 28, 1944, EP, EL.

49. Diary, July 30, 1944.

50. Appendixes G and H, GSP-*W*.

Chapter 40   "A Damned Fine War!"

1. CRC, p. 158.

2. Miller, *Ike the Soldier,* p. 669.

3. Perret, *There's a War to Be Won,* p. 345. Bradley has given Patton only token credit for his accomplishments with Third Army, and has reserved his strongest praise not for Patton, but for Hodges, whose record as an army commander was uninspired, particularly later in the war when he and Bradley orchestrated a suicidal series of attacks into the deadly Hürtgen Forest. Bradley has described him as greatly underrated, anonymous because of his soft-spoken, retiring manner, and "one of the most skilled craftsmen of my entire command. Yet as a general's general his stature among our U.S commanders was rivaled only by that of Simpson . . . Hodges successfully blended dexterity and common sense. . . .Of all my army commanders he required the least supervision." Conspicuous by his absence was any mention of Patton, who uncharitably referred to Hodges as "realy a moron." (ONB-*SS*, p. 226, and Diary, Aug. 18, 1944.)

4. Ibid., p. 340.

5. GSP-*W*, p. 96.

6. CRC, pp. 158–59; and Paul D. Harkins, *When the Third Cracked Europe* (Harrisburg, Pa., 1969), p. 18.

7. See chap. 26.

8. Diary, Aug. 2, 1944.

9. Ibid., Aug. 7, 1944.

10. Blumenson, *Breakout and Pursuit,* p. 349.

11. Weigley, *Eisenhower's Lieutenants,* p. 189.

12. Ibid.

13. Blumenson, *The Battle of the Generals,* p. 147.

14. Ibid.

15. Hastings, *Overlord,* p. 331.

16. Ibid. Wood was also known as "P" (for "professor"), a nickname earned at West Point for his tutoring of other cadets. When ordered by Patton to resume his original mission of capturing Vannes, Wood complained to Middleton: "The high command are winning the war in the wrong way." (Price, *Troy H. Middleton,* p. 188.)

17. Quoted in Caleb Carr, "The American Rommel," *MHQ* (Summer 1992). Patton complained in his diary on Aug. 4 that Wood had gotten bull-headed, turned east, and had to turn back toward his objectives of Vannes and Lorient.

18. Hastings, *Overlord,* p. 329.

19. Ibid., and ONB-*AGL*, p. 285.

20. ONB-*AGL*, p. 285. Bradley has also argued that he was concerned about the potential trouble the 50,000 German troops in Brittany might have had on U.S. forces if not isolated and captured.

21. MB-*PP* 2, p. 532.

22. Carr, "The American Rommel."

23. Charles Whiting, *Patton* (New York, 1970), p. 61.

24. CRC, p. 159.

25. GSP-*W*, p. 125.

26. Richard Lamb, "Kluge" in Correlli Barnett, ed., *Hitler's Generals* (London, 1989), p. 407. Lamb writes that Lt. Col. George R. Pfann, Hobart Gay's senior aide, was the author of the claim.

27. Mellor, *General Patton: The Last Cavalier,* p. 153.

28. Hatch, *George Patton,* p.140.

29. Copy in Box 14, PP-LC.

30. GSP to Beatrice's half-brother, Charles F. Ayer, Sept. 1, 1944, in MB-*PP* 2, p. 536.

31. Mellor, *General Patton: The Last Cavalier,* p. 151.

32. Wallace, *Patton and His Third Army,* pp. 206–7.

33. Perret, *There's a War to Be Won,* p. 366.

34. Wallace, *Patton and His Third Army,* pp. 200–201.

35. Harkins, *When the Third Cracked Europe,* p. 24, and GSP to BAP, Aug. 20, 1944.

36. Allen, *Lucky Forward,* p. 68. Patton wrote to Pershing, who was ill in Walter Reed Army Hospital: "I have consistently refused to worry about what the enemy could do to me, and have spent my time trying to decide what I could do to the enemy . . . we are still attacking—although most people continue to tell me that I am crazy." There was no reply from the old general. (Letter of Aug. 24, 1944, Box 155, Pershing Papers, LC.)

37. Perret, *There's a War to Be Won,* p. 367.

38. Geoffrey Perret, *Winged Victory: The Army Air Forces in World War II* (New York, 1993), pp. 312–13.

39. Allen, *Lucky Forward,* p. 69.

40. Perret, *There's a War to Be Won,* p. 367.

41. H. H. Arnold, *Global Mission* (New York, 1949), p. 543.

42. Polk oral history, USAMHI. Polk was present when the order was issued to Weyland.

43. GSP to BAP, Sept. 1, 1944, Box 15, PP-LC.

44. Diary, Aug. 28, 1944.

45. GSP to BAP, Aug. 20, 1944.

46. Nita to GSP, Aug. 19, 1944, MB-*PP* 2, p. 519.

47. On Aug. 16 the volatile LeClerc, whose 2d Armored Division was then attached to Haislip's XV Corps, came to see Patton complaining that his division had not been included in the Third Army thrust to the Seine. LeClerc also sent Patton a letter that if he was not permitted the honor of liberating Paris he would resign. Patton had assigned the French division to guard the southern flank of the Falaise gap for military reasons, and although he had no desire to hurt French feelings, had no intention of being intimidated. "I told him in my best French that he was a baby, and I would not have division commanders tell me where they would fight, and that anyway, I had left him in the most dangerous place. We parted friends." Patton turned to "P" Wood, who witnessed the encounter and said: "You see, John, he's a bigger pain in the butt than you are." (LF, p. 536.)

48. LF, p. 534.

49. Diary, Aug.19, 1944.

50. ONB-*SS*, pp. 375–76.

51. Ibid., p. 375.

52. Montgomery mistakenly believed that the Fifth Panzer Army would react furiously to hold open the southern perimeter of the gap until they could make good their escape to the east.

53. Quoted in LF, p. 518.

54. The real source of the quote apparently was a cryptic telephone conversation between Patton and Gay on the afternoon of Aug. 17, and had nothing to do with closing the gap between Argentan and Falaise, but rather an attack toward Trun by a provisional corps he had formed under Hugh Gaffey. Patton had warned Gay that he would likely call

with instructions whether Gaffey's corps or Gerow's V Corps would carry out the attack. When Gay asked for a clarification of Patton's instructions that it was to be Gerow's corps, he replied: "Another Dunkirk." There was no intent to denigrate the Canadians who were operating in the area. As Martin Blumenson states, Patton "meant that if Gerow's corps reached Trun and found no Canadians there, Gerow was to continue as far as he could go." (Gay diary, Aug. 17, 1944, USAMHI, and MB-*PP* 2, p. 515.)

55. Diary, Aug. 16, 1944.

56. ONB-*SS*, p. 377.

57. D'Este, *Decision in Normandy*, n. 4, p. 430, and pp. 430–31.

58. A discussion of this controversy is in ibid., chap. 26, and D'Este "Falaise: The Trap Not Sprung," *MHQ* (Spring 1994).

59. Quoted in *Decision in Normandy*, p. 456. The most compelling evidence of Allied success is the assessments of many senior German officers. Allied investigating teams later found a staggering array of equipment on the battlefield that included over 9,000 tanks, self-propelled guns, artillery pieces, armored cars and other military vehicles. Fewer than 120 armored fighting vehicles were thought to have made it across the Seine. In one instance a column of some 3,000 vehicles was destroyed. Some 10,000 German soldiers perished in the Falaise and Trun-Chambois pockets and an estimated 50,000 were taken prisoner.

60. ONB-*SS*, pp. 378–79.

61. Diary, Aug.14, 1944.

62. Stimson and Bundy, *On Active Service in Peace and War*, pp. 659–60.

63. Ibid., p. 447. The historian of the 21st Army Group wrote in 1954 that Montgomery originally envisioned "the rolling up of the German left flank [at Caen] not [the] right . . . He did not envisage—and could never have envisaged—the double envelopment battle that resulted in the dissolution of the German armies in Normandy."

64. MB-*PP* 2, p. 542.

65. Diary, Aug. 16, 1944. Notes of the meeting were recorded by Stiller. Patton sarcastically observed that Bradley's motto seemed to be: "In case of doubt, halt. I am complying with the order, and by tomorrow I can probably persuade him to let me advance. I wish I were Supreme Commander."

66. Blumenson, "Bradley-Patton: World War II's 'Odd Couple.' " *Army* Dec. 1985.

67. Blumenson, *The Battle of the Generals*, pp. 273–74.

68. Blumenson, *Parameters* (Winter 1994–95). Patton mistakenly believed Bradley was destined for the army group command before the slapping incidents even occurred. (Letter to BAP, Aug. 18, 1944.)

69. Blumenson, *The Battle of the Generals*, p. 279.

Chapter 41   "For God's Sake, Give Us Gas!"

1. Leckie, *Delivered from Evil*, pp. 767–68.

2. The euphoria even reached the United States, where the War Production Board rashly canceled some contracts in the mistaken belief that the Germans were beaten and the war was nearly over.

3. Montgomery, *Memoirs* (London, 1958), p. 266.

4. Perret, *There's a War to Be Won*, p. 366.

5. Quoted in Roland G. Ruppenthal, *Logistical Support of the Armies*, vol. 1 (Washington, D.C., 1953), p. 489.

6. Perret, *There's a War to Be Won*, p. 369.

7. A point that has been conclusively reiterated in the 1993 PBS eight-part documentary on the history of oil, *The Prize*.

8. See Martin van Creveld, *Supplying War: Logistics from Wallenstein to Patton* (Cambridge, England, 1977), and a controversial new book by Dominick Graham and Shelford Bidwell, *Coalitions, Politicians and Generals* (New York, 1993).

9. Hatch, *George Patton*, p. 184. Patton's outbursts were for the benefit of his superi-

ors; he placed just as high a priority on ensuring that his men were properly fed and clothed as he did on fuel and ammunition.

10. Brig. (Ret.) H. B. C. Watkins to the author, Nov. 5, 1993.

11. Such criticism soon reached Eisenhower's ears, and he sent Butcher to the Third Army to investigate claims that junior officers were repeating the story. Patton denied it, and Butcher, who genuinely liked Patton, attempted to defuse the situation with Eisenhower.

12. Ambrose, *The Supreme Commander*, p. 513.

13. See Bernard Trainor, "Schwarzkopf the General," Naval Institute *Proceedings* (May 1994). Trainor points out, Schwarzkopf's plan was "tactically pedestrian," but that "future students of warfare will study the U.S. Army's swing through the desert not for its strategic or operational merits, but for its logistical accomplishments."

14. There were some very able logisticians in the ETO who did an outstanding job under the most trying conditions. But the fact remains that Lee, as the chief logistician, was more politician than commander. He lived like an emperor, rewarded his friends with the finest black market goodies, punished anyone who dared to criticize him, and lived in splendor while not far away men were bleeding and dying at the front. (See Graham and Bidwell, *Coalitions, Politicians and Generals*, for a scathing but fair assessment of Lee.)

15. Diary, Aug. 30, 1944.

16. Brendon, *Ike*, p. 161, and Perret, *There's a War to Be Won*, p. 372. Although Eisenhower eventually directed Lee to vacate his headquarters in Paris, he failed to remove Lee from command, despite Bedell Smith's repeated urgings, and the recommendation of Assistant Secretary of War Patterson. At the last minute, however, Eisenhower changed his mind, possibly in gratitude for Lee's performance in organizing the Overlord logistics, and decided to stick with him for the remainder of the war. (Jean Edward Smith, *Lucius D. Clay: An American Life* [New York, 1990], pp. 181–82.)

17. ONB-SS, p. 428, and Perret, *There's a War to Be Won*, p. 372. By his own later admission, Lee had sufficient assets within COM Z to have established and maintained his own security.

18. Leckie, *Delivered from Evil*, p 766.

19. GSP-W, p. 121. One can visualize the grin on Patton's face as he wrote to Beatrice about the "horrid rumor" circulating that he and his men would stoop to such Machiavellian behavior. In Oct. 1944 Patton admitted to Frederick Ayer, "I've already stolen enough gas to put me in jail for life, but it's nowhere enough to keep us rolling. Someday I may even steal a whole damned division of armor and bust the hell out of here."

20. Cole, "George Patton." At Thionville, an artillery commander seized six 90-mm guns emplaced in the hills in 1871 and used its World War I ammunition (so slow its trajectory could be followed with the naked eye) to fire at the Germans. At one point the Third Army operated its own captured German ammunition plant to help alleviate the growing shortage.

21. Anthony Kemp, *The Unknown Battle: Metz, 1944* (New York, 1981), p. 22. Some things never change. The author observed an incident in Vietnam in 1965, in which a field-grade officer heading a scavenging party attempted to "appropriate" a new generator belonging to another unit. When the officer refused to heed the warnings of the NCO responsible for the generator, the sergeant was obliged to draw his pistol. The scavenging party quietly withdrew.

22. Larrabee, *Commander in Chief*, p. 487.

23. Allen, *Lucky Forward*, p. 47.

24. Van Creveld, *Supplying War*, p. 221.

25. Ibid., and Essame, *Patton the Commander*, p. 192.

26. Cole, "George Patton." A distinction must be drawn between the military offense of desertion for which Private Slovik was executed, and other capital offenses such as rape and murder, for which there were numerous death sentences ordered and carried out during the Second World War.

27. DDE to GSP, Oct. 19, 1944, Box 15, PP-LC.

28. Lt. Col. M.C. Helfers, "My Personal Experiences with High Level Intelligence," Helfers Collection, Citadel College archives, Charleston, S.C.

29. Brig. Gen. William H. Hobson, USA, Ret., "General George Smith Patton, Jr., U.S.A.—A Tribute," 1947, Box 34, PP-LC. Codman always carried a bagful of Bronze Stars and other medals for immediate use whenever Patton decided on the spur of the moment to decorate someone.

30. Ibid.

31. Charles B. MacDonald, *A Time for Trumpets* (New York, 1985), p. 614.

32. Quoted in Allen, *Lucky Forward,* p. 41.

33. Harkins, *When the Third Cracked Europe,* pp. 51–52.

34. CRC, pp. 183–84, and Doolittle, *I Could Never Be So Lucky Again,* p. 420.

35. Cole, "George Patton."

36. GSP-*W*, p. 151.

37. CRC, pp. 186–87.

38. Allen, *Lucky Forward,* pp. 34–35.

39. Mims interview, 1987. When Everett Hughes visited Patton in September, he was taken on the ride of his life to the front at speeds rarely slower than sixty miles per hour, with Willie perched next to him, both choking from the dust. (Irving, *The War Between the Generals,* p. 278.)

40. Diary, Aug. 18, 1944.

41. Allen, *Lucky Forward,* pp. 179–80, and FA, p. 196.

42. Patton's reprehensible attitude was not unusual in light of recent revelations of anti-Semitism in the United States during World War II. As the 1994 PBS documentary "America and the Holocaust: Deceit and Indifference" recounted, the United States was rife with anti-Semitism in the 1940s. In 1942 an opinion poll disclosed that, after Germans and Japanese, Jews were regarded as the gravest menace to the country. Worst of all, the United States government, particularly the patrician State Department, clearly knew of the Holocaust but attempted at every turn to bar Jewish refugees from the United States and deliberately failed to publicize the terrible events occurring at Auschwitz and the other death camps.

43. Cole, "George Patton."

44. GSP-*W*, p. 127. On another occasion Bradley asked him to escort visiting Supreme Court Justice James Byrnes on a tour of World War I battlefields. Patton provided Brynes with vivid descriptions of his battles. When they stopped to ask directions from a French farmer, he asked if they had seen the nearby American military cemetery. "Seen it!" said Patton, "I was damn near in it." (Henry Lee Munson, "My Most Unforgettable Character," FM.)

45. Diary, Sept. 26, 1944.

46. CRC, p. 181.

47. Summersby, *Eisenhower Was My Boss,* pp. 197–98. Willie's bad behavior did not end with Telek. In early September, the head of the War Mobilization and Reconversion Office, Anna Rosenberg, arrived at the Third Army dressed in khaki pants that offended Willie, who promptly embedded his teeth in the tiny woman's leg. (LF, p. 586.)

48. Press conferences of Sept. 7 and 23, 1944. "I always feel at these meetings we should have on black hoods over our heads as in the Inquisition," he told the correspondents.

49. CRC, pp. 182–83.

50. Jim McNamara interview, FM.

51. Cornelius Ryan, *A Bridge Too Far* (London, 1974), p. 78.

52. Quoted in Wallace, *Patton and His Third Army,* p. 210.

Chapter 42   A Sea of Mud and Blood

1. Hugh Cole, quoted in Kemp, *The Unknown Battle,* p. 32.

2. ONB-*SS*, pp. 402–3.

3. Ibid., p. 403.

4. Quoted in D'Este, *Decision in Normandy,* p.468.

5. Diary, Sept. 3, 1944.

6. Hugh M. Cole, *The Lorraine Campaign* (Washington, D.C., 1950), pp. 52–53.

7. Quoted in Kemp, *The Unknown Battle,,* p. 83.

8. David Eisenhower, *Eisenhower at War, 1943–1945* (New York, 1986), p. 466.

9. Kemp, *The Unknown Battle,* pp. 83–84;

10. GSP to BAP, Sept. 10, 1944, Box 15, PP-LC.

11. Diary, Sept. 23, 1944.

12. Ibid., Sept. 27, 1944. On Nov. 24, when Eisenhower and Bradley visited 6th Army Group, Devers learned that he was to lose two divisions to Third Army for the forthcoming attack against the Siegfried Line. Devers protested that his force was ready to advance to the Rhine in what he and the Seventh Army commander, Lt. Gen. Alexander M. "Sandy" Patch, were certain would be "a clean breakthrough," which was now to be disrupted by Eisenhower's decision. "This, in my opinion, was a mistake," Devers wrote in his diary. Despite their mutual rivalry Devers emerged from a long talk with Patton on Dec. 5 in an upbeat mood: "George has the right idea and I believe that between us we can crack this front and get on to the Rhine, probably faster than they can in the north. I wish he were under my command. Then I would be sure." (Diary of Gen. Jacob L. Devers, copy in CMH.)

13. Rod Paschall, "The Failure of Market-Garden," *MHQ* (Spring 1994.)

14. GSP to BAP, Sept. 27, 1944.

15. GSP-W, p. 142.

16. Ryan, *A Bridge Too Far,* p. 61. See also Caleb Carr, "The Black Knight," *MHQ* (Spring 1994).

17. GSP to BAP, Sept. 21, 1944.

18. Christopher R. Gabel, *The 4th Armored Division in the Encirclement of Nancy* (U.S. Army monograph, CSI, 1986.)

19. See A. Harding Ganz, "Patton's Relief of General Wood," *Journal of Military History* (July 1989). Dr. Ganz's article is the most thorough investigation yet written about Wood's relief.

20. Ibid. An *MHQ* reader has noted that "both Patton and Eddy had little choice, faced with Wood's insubordination. But rather than put a permanent blot on his military record, they attributed the relief solely to combat fatigue." Under the best of circumstances the relief of a senior commander is unpleasant, but Patton's handling of the Wood affair showed restraint, not opportunism. When a corps commander recommends to his army commander the relief of a division commander, and by his future actions that commander demonstrates the instability shown by Wood, the choice—no matter how distasteful—is clear.

21. ONB-*AGL*, p. 343.

22. Ibid., p. 342.

23. Gabel, *The Lorraine Campaign.*

24. Hatch, *George Patton,* p. 147, and LF, p. 635.

25. Whiting, *Patton,* p. 72.

26. Kemp, *The Unknown Battle,* p. 84.

27. Perret, *There's a War to Be Won,* p. 368.

28. Whiting, *Patton,* p. 75, and MB-*PP* 2, pp. 562–63.

29. ONB-*SS*, p. 427

30. The date and place of this well-known tale is generally thought to have been in Lorraine or the Saar in the late autumn of 1944, but John S. D. Eisenhower's account of the Battle of the Bulge, *The Bitter Woods* (New York, 1969), dates it to Dec. 30, 1944, when Patton was at the command post of the 11th Armored Division, and stumbled over Troy Middleton's sleeping driver.

31. CRC, p. 203.

32. Jim McNamara interview, FM.

33. GSP to Doolittle, Oct. 19, 1944, Doolittle Papers, LC.

34. CRC, p. 228.

35. LF, pp. 635–36.

36. CRC, chap. 11, passim. When Spaatz and Doolittle visited Third Army Patton effusively praised them. "Individuality plays a much greater part in warfare than most people realize. Frankly, I look on yesterday's bomber missions as so many expressions of your personal friendship . . . for me."

37. Gabel, *The Lorraine Campaign*.

38. Quoted in Whiting, *Patton*, p. 83.

39. Kemp, *The Unknown Battle*, p. 219.

40. The only permanent reminder of the siege of Metz is a single street named *Rue du XX Corps Américain*.

41. Cole, *The Lorraine Campaign*, p. 520.

42. Charles B. MacDonald, *The Battle of the Huertgen Forest* (New York, 1984), p. 205.

43. Charles M. Province, *The Unknown Patton* (New York, 1983), p. 173.

44. Kemp, *The Unknown Battle*, p. 228.

45. Ibid., pp. 228–29, and LF, p. 599.

46. LF, pp. 598–99.

47. Gabel, *The Lorraine Campaign*.

48. Koch, *G-2: Intelligence for Patton*, pp. 86–87.

49. Hugh M. Cole, *The Ardennes: Battle of the Bulge* (Washington, D.C., 1965), pp. 485–86.

50. ONB-*AGL*, p. 356; ONB-*SS*, p. 465; HS, p. 218; GSP-*W*, pp. 188–89; Whiting, *Patton*, p. 85; MB-*PP* 2, p. 595, and diary, Dec. 16, 1944. Bradley's accounts suggests he initiated the call to Third Army the night of Dec.16; however, the evidence indicates that the first call came from Leven Allen, and was followed by Patton's call of protest to Bradley.

Chapter 43 PATTON OF COURSE

1. Ralph Bennett, *Ultra in the West* (London, 1979), and Jacques Nobécourt, *Hitler's Last Gamble* (New York, 1967), p. 132. Although the Germans had imposed a rigid radio silence, there were sufficient intercepts before the attack to have at least aroused Allied suspicions.

2. Maj. Gen. Kenneth Strong, quoted in Harold C. Deutsch, "Commanding Generals and the Uses of Intelligence," *Intelligence and National Security* (July 1988). Deutsch notes that "Patton ranks as among the most enthusiastic and successful users of intelligence among the Allied leaders, especially with respect to Ultra."

3. R. W. Williams, "Moving Information: The Third Army Imperative," *Army* (Apr. 1975)."

4. MacDonald, *A Time For Trumpets*, pp. 68–69; Koch, *G-2: Intelligence for Patton*, chap. 7, passim; Deutsch, "Commanding Generals," and diary, Nov. 25, 1944.

5. The First Army G-2, Colonel B. A. "Monk" Dickson, who allegedly identified the coming offensive, was a pretentious opportunist whose intelligence estimates before the German offensive are replete with equivocations—one moment seeming to stop just short of predicting an attack, only to hint at an imminent German collapse. The best intelligence work in the First Army G-2 was being accomplished by the Ultra officer, Adolf G. Rosengarten, who predicted the offensive, but was ultimately ignored by Dickson, who "went off to Paris on leave on 15 December—scarcely the act of a man who knows that the army he serves is about to take the brunt of a panzer assault." (See Ronald Lewin, *Ultra Goes to War* [London, 1978], p. 356.)

6. Deutsch, "Commanding Generals." Where other commanders tended to underuse Ultra, for Patton it became "the *pièce de résistance* of every intelligence briefing" (although of necessity given separately, to preserve secrecy). Whatever his feelings about

an attack in the Ardennes, Patton did not share his misgivings with either Bradley or Hodges.

7. LF, pp. 673–74.

8. Deutsch, "Commanding Generals."

9. Eisenhower, *The Bitter Woods,* p. 215.

10. ONB-*SS*, p. 469.

11. Principal sources for this account of the Verdun meeting are: Weigley, *Eisenhower's Lieutenants,* pp. 499–501; Patton, *War As I Knew It,* pp. 190–92; MacDonald, *A Time For Trumpets,* pp. 420–21; Nobécourt, *Hitler's Last Gamble,* pp. 219–22; Bradley, *A Soldier's Story,* pp. 470–73, and *A General's Life,* pp. 358–59; Eisenhower, *Crusade in Europe,* pp. 382–85; Blumenson, *The Patton Papers,* vol. 2, pp. 599–601; Miller, *Ike the Soldier,* pp. 727–28; Ambrose, *The Supreme Commander,* pp. 558–59; John S. D. Eisenhower, *The Bitter Woods,* pp. 256–57; David Eisenhower, *Eisenhower at War, 1943–1945,* pp. 566–69; Codman, *Drive,* pp. 231–33; Maj. Gen. Sir Kenneth Strong, *Intelligence at the Top* (London, 1968), pp. 161–63; Jacob L. Devers's diary; GSP diary for Dec. 19, and his letter of Dec. 21 to BAP.

12. The desertion rate had become so serious that Eisenhower denied an appeal of the death sentence given by a court-martial to Pvt. Eddie Slovik. In an attempt to set an example that would gain attention, Eisenhower became the first president since Lincoln in the Civil War to order an American soldier executed for desertion. (See William Bradford Huie, *The Execution of Private Slovik* [New York, 1954].)

13. Patton noted sarcastically in his diary: "The fact that three of these divisions exist only on paper did not even enter his head," a reference to Middleton's three divisions badly battered in the Hürtgen and unfit for further combat.

14. Montgomery's official biographer has written: "What stunned Eisenhower was not the speed with which Patton reoriented his divisions, but the alacrity with which . . . Patton abandoned his planned Saar offensive. Eisenhower was bewildered. Was this the same commander who had railed against Monty's call for concentration of the main Allied land forces since August, and had made it a matter of American honor that his Saar offensive not be closed down?" (Nigel Hamilton, *Monty: The Battles of Field Marshal Montgomery* [New York, 1994], p. 481.)

15. Strong, *Intelligence at the Top,* p. 163.

16. MacDonald, *A Time For Trumpets,* p. 421. Nor would such a strategy have been workable, for that deep into the Allied front numerous American supporting units would have been wiped out.

17. GSP-*W*, p. 181.

18. MacDonald, *A Time For Trumpets,* p. 421. Lt. Gen. Jake Devers was the unhappiest participant at the Verdun conference with his new role of taking over much of Third Army's sector. It meant Seventh Army would have to cease offensive operations at a time when Devers believed that 6th Army Group was being called upon to bail out 12th Army Group "just as we are about to crack the Siegfried Line." Although he understood fully the necessity for his new role, his views were never made public until the publication of the official history in 1993. It was, he said, a "tragedy" because SHAEF "has not seen fit to reinforce success on this flank." Devers was also privately critical of Montgomery and Patton, whom he thought guilty of "wild and inaccurate statements." (Devers diary, Dec. 19, 1944; Jeffrey J. Clark and Robert Ross Smith, *Riviera to the Rhine* [Washington, D.C., 1993], p. 491, and Michael A. Markey, "Quartermaster to Victory," *Army* [Aug. 1994].)

19. Lt. Col. Steve E. Dietrich, *Patton's Greatest Battle,* unpublished monograph. Patton's quarters in the field consisted of two converted trucks; a mobile office and another that served as his private bedroom, parlor, and bath. It contained a two-way radio that Patton often listened to, but almost never used. One of his two telephones had a green receiver and was connected directly to Eisenhower and Bradley. It also had a scrambler that frustrated Patton, who reserved for it some of his best profanity, when it seemed to scramble his words ever before he could even utter them. The corrugated stairs proved hazardous to Willie, who lost several toenails before it was covered by wooden boards. (GSP-W, p. 275, n. 1.)

20. Diary Dec. 20, 1944; GSP to BAP, Dec. 21, and Dietrich, *Patton's Greatest Battle.*

21. ONB-AGL, p. 363; and Crosswell, *The Chief of Staff,* pp. 284–87.

22. Toland, *Battle,* pp. 347–48.

23. Whiting, *Patton,* p. 94.

24. Diary, Dec. 21, 1944.

25. Eisenhower, *The Bitter Woods,* p. 336.

26. Diary, Dec. 20, 1944.

27. Eisenhower, *The Bitter Woods,* p. 334.

28. Toland, *Battle,* p. 207.

29. ONB-*AGL,* p. 367; and Hansen diary, Dec. 24, 1944. Hansen's diary recorded that Patton "is boisterous and noisy, feeling good in the middle of a fight."

30. MacDonald, *A Time for Trumpets,* p. 525.

31. Diary, Dec. 24, 1944.

32. O'Neill, "The Story Behind Patton's Prayer." Patton also instructed O'Neill to produce a training letter to the other 486 chaplains assigned to Third Army on the subject of prayer. The account of Paul Harkins suggests a clash between the commanding general and his chief chaplain, in which O'Neill allegedly challenged the notion of men of his profession praying for good weather in order more effectively to kill other men. Harkins suggests that the prayer was ordered and written shortly before Christmas, and gives the date as December 14. O'Neill states it was December 8. (See Harkins, *When the Third Cracked Europe,* p. 44. Farago's account in *Patton: Ordeal and Triumph* gets the date right, but uses Harkins's version, pp. 660–61.)

33. Msgr. James H. O'Neill, "The Story Behind Patton's Prayer," in *Our Sunday Visitor,* published in Huntington, Indiana, Aug. 15, 1971, copy in PP-LC; and Paul Williams, "1944 Christmas Card Memorable One," newspaper article, date and publication unknown, FM.

34. Williams, "1944 Christmas Card."

35. Quoted in Danny S. Parker, *The Battle of the Bulge* (Philadelphia, 1991), p. 194; and "Battleground Luxembourg," published by the Luxembourg National Tourist Office.

36. Harkins, *When the Third Cracked Europe,* p. 44.

37. Account of Harkins in revised edition of GSP-W, p. 177, and O'Neill, "The Story Behind Patton's Prayer." O'Neill's recollection is that this event occured in late Jan. 1945 in Luxembourg.

38. "Larry Newman Sees 'Blood and Guts' at the Battle of the Bulge," in Louis L. Snyder, ed., *A Treasury of Great Reporting* (New York, 1949), pp. 669–70—based on Newman's dispatch published in various U.S. newspapers in late Dec. 1944.

39. Postwar annotation by Beatrice Patton to GSP diary entry of Jan. 21, 1945. Also quoted in Hatch, *George Patton,* p. 195.

40. Ibid., pp. 193–94; and Ruppenthal, *Logistical Support of the Armies,* vol. 2, pp. 228–33.

41. Leland Stowe, "Old Blood and Guts Off the Record," postwar date of writing unknown, copy in Combined Arms Library, C&GSC.

42. Ibid.

43. S. L. A. Marshall Collection, USAMHI.

44. Ibid. A second such incident took place some months later in Germany, when their vehicle was forced into a culvert as Patton sped by. "The General hadn't even cast us a glance."

45. Henry Lee Munson, "The Most Unforgettable Character," FM.

46. Quoted in Forty, *Patton's Third Army at War,* p. 128.

47. FA, p. 199.

48. Ibid., p. 140. Quote by Lt. Col. Hal Pattison of the 4th Armored.

49. HS, p. 224.

50. Diary, Dec. 26 and 27, 1944.

51. GSP to BAP, Dec. 29, 1944, PP-USMA.

52. Diary, Jan. 11 and 12, 1945.

53. John S. D. Eisenhower, *The Bitter Woods.* n. p. 329; and Diary, Jan. 30, 1945.

54. "Larry Newman Sees 'Blood and Guts' at the Battle of the Bulge."

55. Bill Mauldin, *The Brass Ring* (New York, 1971), chap. 15, passim; Nancy Cald-well Sorel, "George S. Patton and Bill Mauldin," *Atlantic Monthly,* Mar. 1988; David Lamb, "Bill, Willie, and Joe," *MHQ* (Summer 1989); and *Time,* June 18, 1945.

56. Forty, *Patton's Third Army at War,* p. 140.

57. Toland, *Battle,* p. 320.

58. Janice Holt Giles, *Those Damned Engineers* (Boston, 1975), p. 364.

59. Ibid., p. 328.

60. Toland, *Battle,* p. 342.

61. Diary, Jan. 8, 1945, and GSP-*W*, pp. 215–16.

62. Harkins, *When the Third Cracked Europe,* p. 46.

63. Prioli, ed., *Lines of Fire*, p. 155.

64. Giles, *Those Damned Engineers,* p. 371.

65. Yergin, *The Prize,* pp. 347–48.

66. MacDonald, *A Time for Trumpets,* p. 618. Although Patton declined to get into "a pissing match" about it, he was privately critical that (in his opinion) the 4th Armored deserved a good deal more credit than it got at the expense of the 101st Division. (GSP to Gen. Thomas T. Handy, Feb. 6, 1945, Box 1, Handy Papers, GCML.)

67. Quoted in MacDonald, *A Time for Trumpets,* p. 614.

68. Essame, *Patton the Commander,* pp. 232–33.

69. Nobécourt, *Hitler's Last Gamble,* p. 17.

70. Eisenhower, *The Bitter Woods,* p. 430. A 1994 U.S. Army historical study esti-mates German losses at between 81,000 and 103,000. (See *Ardennes-Alsace* [Washington, D.C., 1994], p. 53.) However, in *War As I Knew It,* pp. 192 and 217, German losses between Dec. 22, 1944, and Jan. 29, 1945, are shown as 135,300: 29,700 killed, 82,800 wounded, 22,800 POWs, figures that later data apparently do not substantiate.

71. Parker, *Battle of the Bulge,* pp. 291–93. With Luftwaffe (air and ground) troop losses, the totals are believed to have been closer to 100,000.

72. GSP to son George, Jan. 16, 1945, quoted in MB-*PP* 2, p. 625; and diary, Jan. 26, 1945.

73. Peter Elstob, *Hitler's Last Offensive* (London, 1971), pp. 381–82.

74. Cole, *The Ardennes,* p. 264, and Toland, *Battle,* pp. 320–23. Although never pub-licly documented, there is believed to have been at least one other incident involving the shooting of German POWs by another division in the Third Army. Details of the SS mas-sacres in Malmédy and elsewhere are documented by Cole, pp. 260–68. There are numer-ous examples of tit-for-tat retaliation, including Sicily, Italy, and Normandy, where the 3d Canadian Division and the 12th SS Panzer Division (the Hitler Jugend) each began shoot-ing prisoners after nearly forty Canadians were lined up and shot dead against a wall of the Château Audrieu. (See *Decision in Normandy,* p. 507.)

75. Diary, Jan. 4, 1945. The reference to "the stupidity of our green troops" clearly meant their performance in battle, but may have also alluded to the shooting incident. So rapidly do things change in war that, within a month, Patton would praise the men of the 11th Armored as "A-1."

76. GSP to BAP, Jan. 22 and 24, 1945, PP-USMA.

77. Quoted in RHP, p. 269.

78. Gay diary, Jan. 1, 1945.

79. Quoted in MB-*PP* 2, p. 626.

80. T. Michael Booth and Duncan Spencer, *Paratrooper: The Life of Gen. James M. Gavin* (New York, 1994), p. 270.

81. Quoted in FA, p. 181.

82. Diary, Jan. 18, 1945.

83. Quoted in Allen, *Lucky Forward,* p. 280.

Third Army casualty figures are from GSP-*W*, pp. 192 and 217 and represent Dec. 22, 1944 and Jan. 29, 1945. MacDonald calculates overall U.S. casual-including 19,000 killed and 15,000 captured.

85. Box 55, PP-LC.

86. Gerald Astor, *A Blood-Dimmed Tide* (New York, 1992), p. 226.

87. MB-*PP* 2, pp. 593, 595.

88. MB, p. 245.

Chapter 44   Pissing in the Rhine

1. Quoted in Franklin M. Davis Jr., *Across the Rhine* (Alexandria, Va., 1980), p. 22.

2. Ibid., p. 23.

3. GSP to BAP, Feb. 4, 1945, Box 15, PP-LC.

4. ONB-*AGL*, p. 394; and ONB-*SS*, p. 501.

5. GSP to BAP, Feb. 25 and March 16, 1945, Box 15, PP-LC. See also MB-*PP* 2, p. 656.

6. GSP-*W*, p. 219.

7. Ibid., p. 223; and Diary, Feb. 2 and 22, 1945.

8. Gay diary, Jan. 24, 1945. The third person referred to was Courtney Hodges.

9. Diary, Feb. 26, 1945, and LF-*LDP*, p. 33.

10. Hatch, *George Patton,* pp. 202–3.

11. Koch, *G-2: Intelligence for Patton,* p. 162.

12. ONB-*AGL*, p. 392. Patton chose to interpret the directive "in the bitterest possible terms," cursing and damning this "foolish and ignoble political war." (Diary, Feb. 2, 1945.)

13. Hatch, *George Patton,* p. 204.

14. Diary, Feb. 6, 1945.

15. Quoted in HS, p. 240, and various other sources.

16. FA, pp. 192–93.

17. Diary, Feb. 17, 1945.

18. Davis, *Across the Rhine,* p. 77.

19. Diary, Mar. 13, 1945.

20. CRC, p. 264.

21. John Toland, *The Last 100 Days* (New York, 1965), p. 237.

22. MB-*PP* 2, p. 653; and diary, Mar. 9, 1945. In Apr someone complained to Bradley that a Third Army correspondent had written an article impugning First Army. Patton's attitude was that Third Army did not need to defend itself by criticizing another army, and he instructed his public information officer never to release such a story.

23. Alexander McKee, *The Race for the Rhine Bridges* (London, 1971), p. 213.

24. Davis, *Across the Rhine,* p. 83.

25. Quoted in ONB-*AGL*, p. 412.

26. Forty, *Patton's Third Army at War,* p. 160.

27. Diary, Mar. 23, 1945.

28. Ibid., Mar. 24, 1945. In mid-April Patton returned to open a railway bridge across the Rhine, built in nine days by a West Point classmate. He was to cut red tape with a pair of scissors but instead growled for someone to give him "a goddamn bayonet."

29. CRC, pp. 265–66.

30. Gay diary, Mar. 25, 1945. Vandenberg was quoting one of his air commanders as exemplifying "the feeling of admiration for the Third Army throughout the Ninth Air Force."

31. GSP to GCM, Mar. 13, 1945, in Davis, *Across the Rhine,* p. 74.

32. GSP to BAP, Mar. 23, 1945, Box 15, PP-LC.

33. Box 55, PP-LC.

34. Toland, *The Last 100 Days,* pp. 286–87.

35. Ibid., p. 287. According to Toland, the head of the U.S. Military Mission to Russia, Maj. Gen. John Deane, had learned that Waters and the other American POWs had been removed from their camp in Poland and were being forcibly marched west, where they eventually ended up in Oflag XIIIB at Hammelburg. Deane's information was cabled to Eisenhower, who informed Patton.

36. Interviewed after the raid, Baum stated he had 307 men; other accounts put the

total as approximately 294. A later account co-authored by Baum was also 294. (See Richard Baron, Abe Baum, and Richard Goldhurst, *Raid!: The Untold Story of Patton's Secret Mission* [New York, 1981], p. 270.)

37. Interview of Capt. Abraham J. Baum, Apr. 10, 1945 (part of the debriefings conducted by the 4th Armored Division after the raid), Irving Microfilm, USAMHI.

38. Baron, Baum, and Goldhurst, *Raid!* p. 298. The final tally was 9 killed and 16 whose fate was never determined, but who were presumed dead.

39. Interview of John S. D. Eisenhower, *World War II*, May 1994.

40. Charles B. Odom, *General George S. Patton and Eisenhower* (New Orleans, 1985), pp. 70–71; and Baron, Baum, and Goldhurst, *Raid!* pp. 252–54.

41. ONB-*SS*, pp. 542–43.

42. Toland, *The Last 100 Days,* p. 299.

43. Frederick E. Oldinsky, "Patton and the Hammelburg Mission," *Armor* (July-Aug. 1976).

44. Perret, *There's A War To Be Won* n., p. 445.

45. Baron, Baum, and Goldhurst, *Raid!* p. 250.

46. Ibid., pp. 263–66. Baum believed that if he had been recommended for a Medal of Honor the raid would have been investigated by Congress, but that a DSC and the classification of the mission would ensure its cover-up.

47. GSP to BAP, Mar. 23, 1945, Box 15, PP-LC. Two weeks earlier he had written that he would ask Eisenhower for a DSC for Waters, but doubted he would succeed. "I think between our selves that J. leaned over backwards in not escaping[. O]ne owes something to ones self and family." (Letter of Mar. 8.)

48. Ibid., Mar. 27, 1945.

49. Gay diary, Apr. 4, 1945.

50. Box 55, PP-LC.

51. Patton clearly attempted to incriminate Bradley when he wrote in his diary on Mar. 31: "So far I have only made one mistake, and that was when I lost two companies of the 4th Armored Division in making the attack on Hammelburg. I made it with only two companies on account of the strenuous objections of General Bradley to making any at all. Had I sent a combat command as I had first intended to do, this mistake would not have occurred." The diary entry appears to be an attempt unfairly to cast blame on Bradley where none existed.

52. Box 55, PP-LC. See also *Fatal Decision pp.* 400–401.

53. LF-*LDP*, p. 45.

54. Butcher, *Three Years with Eisenhower,* pp. 672–73; and DDE-GCM, Apr. 15, 1945, in Chandler, *The Eisenhower Papers*, vol. 4, pp. 2616–17.

55. LF-*LDP*, p. 46.

56. Toland, *The Last 100 Days,* p. 370. The total gold in Merkers has been estimated at 100 tons; however, other sources state that 4,500 twenty-five pound gold bars were found (or approximately 56 tons). Gold in 1945 was valued at $32 per ounce. Thus gold coins obviously added to the overall value of the cache.

57. CRC, pp. 281–82; and Butcher, *Three Years with Eisenhower,* p. 672.

58. Toland, *The Last 100 Days,* p. 370, and Allen, *Lucky Forward,* pp. 369–71. A 4th Armored combat commander ordered Ohrdruf's civilian officials (who predictably disclaimed any knowledge of what had occurred in the grisly concentration camp in their midst) to attend a memorial service for the victims and to view the corpses in the enormous open-pit crematorium. That afternoon the burgomeister and his wife committed suicide.

59. Cole, "George S. Patton."

60. Robert Murphy, *Diplomat Among Warriors* (New York, 1964), p. 255.

61. Gay diary, Apr. 12, 1945, and Toland, *The Last 100 Days,* p. 371. Patton, who usually entered a full account of his day in his diary did not do so this night, and instead made only an oblique reference to "a proposed stop line." Stephen E. Ambrose in *Eisenhower and Berlin, 1945,* rev. ed. (New York, 1986), p. 97, observes: "Patton's romanticism, the influence of the Civil War on his mentality, was never more evident."

62. Montgomery of Alamein, *Memoirs* (London, 1958), p. 332. Montgomery believed Third Army ought to have been permitted to liberate Czechoslovakia, and never understood why it was halted on its western frontier in late April.

63. Toland, *The Last 100 Days,* p. 377.

64. Diary, Apr. 14, 1945, and Toland, *The Last 100 Days,* p. 377.

65. Diary, Apr. 18, 1945; GSP-*W,* p. 288; and LF-LDP, pp. 82–83. Patton was also presented with a large medallion by the grateful citizens of Metz.

66. Betty South, "We Called Him 'Uncle Georgie,'" *Quartermaster Review* (Jan.-Feb. 1954).

67. FA, p. 213.

68. Ibid., pp. 214–15.

69. Ibid., p. 217.

70. GSP-*W,* p. 278.

71. Butcher, *Three Years with Eisenhower,* p. 667.

72. CRC, pp. 267–68.

73. Diary, Apr. 13, 1945.

74. Ibid., Apr. 8, 1945.

75. Allen, *Lucky Forward,* p. 379. Third Army losses during the same period were 1,757 killed, 5,885 wounded, 782 missing, and the loss of 49 pilots of the XIX TAC.

76. Allen, *Lucky Forward,* pp. 381–82.

77. Cadet G. S. Patton to GSP, Mar. 25, 1945, PP-USMA.

78. Margaret Bourke, White, *Dear Fatherland, Rest Quietly* (New York, 1946), Chap. 3, passim.

79. *Time,* Apr. 9, 1945.

80. Quoted in Trezzvant W. Anderson, *Come Out Fighting: The Epic Story of the 761st Tank Battalion, 1942–1945* (Salzburg, Austria, 1945), p. 21. Whiting, *Patton's Last Battle,* p. 35, citing M. Motley, *The Invisible Soldier* (Detroit, 1977), states that Patton added: "They say it is patriotic to die for your country, well let's see how many patriots we can make out of those German motherfuckers!"

81. Cole, "George S. Patton."

82. GSP to BAP, Mar. 23, 1945, Box 15, PP-LC.

83. Quoted in *Time,* Apr. 9, 1945.

84. GSP-*W,* fn., p. 307.

85. LF-*LDP,* pp. 53–4 and 63; Toland, *The Last 100 Days,* chap. 32, passim; ONB-*SS,* pp. 547 and 549; ONB-AGL, p. 433; and Ambrose, *Eisenhower and Berlin, 1945,* pp. 83–87.

86. GSP-W, p. 309. There were Third Army reconnaissance units in the vicinity of Prague, which may have ignited the rumor that American liberation of the city was imminent.

87. Allen, *Lucky Forward,* pp. 401–02.

88. Gay diary, May 8, 1945.

89. Ibid.

90. Wellard, *General George S. Patton, Jr.,* pp. 209–10. In 1994, a former German staff officer of the 4th Panzer Army would praise Patton for having saved him and him comrades from capture and reprisals by the Red Army in 1945. After fifty years he still felt a debt of gratitude to Patton, which he hoped to repay in the form of a published article. (Letter, Dr. Joseph Pointner to the U.S. Ambassador to Austria, Mar. 1994.)

91. Lt. Col. Freiherr von Wangenheim, Box 16, PP-LC., CMH files

92. Box 55, PP-LC. Patton's General Order for VE-Day also noted:

During the 281 days of incessant and victorious combat, your penetrations have advanced farther in less time than any other army in history, you have fought your way across 24 major rivers and innumerable lesser streams. You have liberated or conquered more than 82,000 square miles of territory, including 1,500 cities and towns, and some 12,000 inhabited places. Prior to the termination of active hostilities, you had captured in battle 956,000 enemy soldiers and killed or wounded at least 500,000 others. France, Belgium, Luxembourg, Germany, Austria and Czechoslovakia bear witness to your exploits. All men and women of the six corps and 39 divisions that have at different times been members of the Army have done their duty. . . . In proudly contemplating our achievements let us never forget our heroic dead whose graves mark the course of our victorious advances, nor our wounded whose sacrifices aided so much our success. . . . During the course of this war I have received promotion and decorations far above and beyond my individual merit. You won them: I as your representative wear them."

93. ONB-*AGL*, p. 436.

Chapter 45   Military Governor of Bavaria

1. Larry G. Newman, "Gen. Patton's Premonition," *American Legion Magazine*, July 1962. When Patton was asked why the Third Army had not taken Prague, the correspondents were expectantly poised to write the answers in their notebooks. To appreciative laughter, Patton cannily replied that it was because he had not been ordered to do so. (GSP-*W*, p. 312.)

2. Quoted by the III Corps chief of staff in S. L. A. Marshall Papers, USAMHI.

3. Patton also once complimented the Third Army staff by observing: "You know, Julius Caesar would have a tough time being a brigadier general in this army!" (Koch, *G-2: Intelligence for Patton,* p. 156.)

4. Diary, Nov. 14, 1943.

5. Perret, *There's a War to be Won,* p. 543.

6. Recollections of Lt. Col Albert C. Stiller of meeting between GSP and Patterson, May 7, 1945.

7. In 1945 the army numbered 8,267,958, but by 1947 it had been reduced to 991,285, and by the advent of the Korean War in 1950, a mere 593,167. (Weigley, *History of the United States Army,* p. 569.)

8. MB-*PP* 2, p. 699.

9. Weigley, *History of the United States Army,* p. 501.

10. Ambrose, *Eisenhower,* 1, p. 400.

11. Cole, "George S. Patton."

12. Jim McNamara anecdotes, FM.

13. Harkins, *When the Third Cracked Europe,* p. 65.

14. Diary, May 14, 1945; and Gay diary, same date.

15. Ibid., Aug. 8, 1945.

16. FA, p. 245. Smith was a relative of George Hugh Smith.

17. Ibid., 240.

18. Ibid., pp. 240–41.

19. Quoted by Stallings in *The Doughboys,* p. 234.

20. Diary, May 10, 1945.

21. Odom, *General George S. Patton and Eisenhower,* p. 73.

22. Irving, *The War Between the Generals,* p. 2.

23. Diary, Aug. 1, 1945.

24. In addition Irving incorrectly states that the meeting was held around a great oak table at Eisenhower's Frankfurt headquarters. The meeting was actually held at 12th Army Group, even though Bradley was on leave that day and thus not present. Irving claims that only five senior generals were present, "their faces [bearing] the congratulatory grins of men who have done a job they are proud of." If Eisenhower's intent was to initiate a cover-up he would hardly have included the airmen, most of whom were brigadier generals.

Moreover, there *could* not have been a cover-up without the willing participation of Omar Bradley. (*The War Between the Generals,* p. 1; and Diary, May 10, 1945.)

25. LF-*LDP*, pp. 70–1.

26. Anecdote in S. L. A. Marshall Papers, USAMHI.

27. See the author's *Fatal Decision*, passim.

28. Emajean Buechner, *Sparks: The Combat Diary of a Battalion Commander (Rifle) WW II* (Metairie, La., 1991), p. 147. Patton was fully within his authority to order the charges dismissed. Sparks later retired as a brigadier general and went on to a distinguished career as a Colorado Supreme Court justice.

29. J. J. Hanlin, "The General and the Horses," *American Legion Magazine,* Feb. 1963.

30. Frederic Sondern Jr., "The Wonderful White Stallions of Vienna," *Reader's Digest,* Apr. 1963.

31. LF-*LDP*, p. 60.

32. Diary, May 7, 1945. There still remained the problem of the Lipizzaner mares and foals, which were located on a stud farm at Hostau, near Pilsen. In addition to the Lipizzaners, the Germans had assembled some of the finest breeding herds in Europe. Col. Charles H. Reed, commander of the 2d Cavalry Group, whose unit had been spearheading the Third Army advance, learned of their location on Apr. 26. Believing that the Russians might well destroy the stud, Reed was determined to save it.

He selected Capt. Tom Stewart to arrange for the surrender of the guardians of the farm. Liberated were some 400 horses, 247 of them Lipizzaners, which were removed to Germany and eventually sent to the Riding School's new home in the American sector of occupied Austria. One of the better-known photographs of Patton shows him riding a white Lipizzaner once chosen by Hitler for presentation to Japan's Emperor Hirohito. Although one unsubstantiated account credits him with authorizing Reed to mount the rescue, Reed undoubtedly acted on his own to save the Hostau stud. Patton was probably informed of the plans, but otherwise played no major role. (The Lipizzaner story is compiled from: a letter from artist Don Stivers to the author [Sept. 11, 1992]; Paula Hauser, "The Great Escape"; *Equine Images* [1992]; "Hostau Reminiscences," unpublished account of Capt. Tom Stewart; and H. H. Douglas, "The Lipizzaners," parts 1 and 2, *Armored Cavalry Journal* [Sept.-Oct., Nov.-Dec. 1947]; Frederic Sondern Jr., "The Wonderful White Stallions of Vienna," and J. J. Hanlin, "The General and the Horses.")

33. GSP to BAP, May 9, 1945, PP-USMA. Beatrice had apparently commented on praise for Patton by John K. Waters during a telephone call to Little Bee.

34. MB, p. 228. Patton seems to have exaggerated in his reference to a twelve-year affair. There is no indication of any contact or infidelity with Jean Gordon prior to her arrival in Hawaii in 1936.

35. Quoted in "General Patton and his Niece," *Parade,* 1981. In her unpublished memoir, Ruth Ellen Totten describes with evident pain her father's infatuation with Jean in Hawaii. Robert Patton notes that "to maintain appearances of family propriety, Beatrice had stonily endured Jean's presence as a bridesmaid at Ruth Ellen's wedding in 1940." (RHP, p. 270.)

36. RHP, p. 271.

37. GSP to BAP, Box 14, PP-LC.

38. Betty South, quoted in *Parade,* 1981. Betty South's version is colored by the fact that she was protective of both Patton's reputation and Jean's. See, for example, South, "We Called Him Uncle Georgie."

39. GSP to BAP, Mar. 31, 1945, PP-USMA.

40. RHP, p. 271.

41. Irving, *The War Between the Generals,* p. 407. Irving's source is the Hughes diary.

42. Patton was fond of proclaiming that "a man who won't fuck, won't fight." He once toured a château with a major he was considering for the position of junior aide. The master bedroom was replete with mirrors along one wall and on the ceiling. The naive young officer turned to Patton and asked what the mirrors were for. Patton decided that

anyone that ignorant could never succeed as one of his aides and did not offer him the post. (FA, pp. 207–8.)

43. FA, pp. 208–9.

44. Weinberg, *A World at Arms,* p. 833.

45. Diary, May 18, 1945.

46. RHP, p. 271.

47. MB, p. 271.

48. See various accounts of Patton's homecoming in the Boston *Globe*, *Herald Traveler*, *Daily Record*, and *Post*, copies in the PP-USMA.

49. In Denver, after being praised as "one of the greatest generals of all time," Patton replied that while the compliment was appreciated: "I should tell you that I am really a better poet than a general. Publishers haven't recognized that as yet, but it's the truth just the same." (Robert S. Allen, "The Day Patton Quit," *Army* [June 1971].)

50. Hatch, *George Patton,* p. 218.

51. CRC, p. 319.

52. Doolittle, *I Could Never Be So Lucky Again,* p. 443. Doolittle never saw his friend again, and often wondered how he would have tolerated peace had he lived.

53. Ibid., pp. 442–43.

54. MB, p. 278.

55. LF-*LDP,* p. 89.

56. Ibid., p. 87.

57. Robert P. Studer, "The Day That General Patton Held Up the Parade" (ca. 1970). The Studer article is in Binder 17, PP-USMA.

58. FA, pp. 243–44.

59. RET-M.

60. Ruth Ellen Totten, quoted in Jeffrey St. John, "Reflections on a Fighting Father"; and RET-M.

61. RET-M.

62. Odom, *General George S. Patton and Eisenhower,* p. 74.

63. Memo, Frank McCarthy to GCM, June 12, 1945, GCM-P.

64. LF-*LDP,* p. 85.

65. Ibid., p. 86.

66. Ibid., p. 89.

67. Ibid., pp. 91–93; and MB-*PP* 2, pp. 721–22.

68. Diary, Aug. 18, 1945. Patton thought the dropping of the atomic bomb on Hiroshima "a most unfortunate" act, "because now it gives a lot of vocal but ill-informed people—mostly fascists, communists, and s.o.b.'s assorted—an opportunity to state that the Army, Navy, and Air Forces are no longer necessary as this bomb will either prevent war or destroy the human race." He also noted, "Actually, the bomb is no more revolutionary than the first throwing-stick or javelin or the first cannon or the first submarine. It is simply, as I have often written, a new instrument added to the orchestra of death which is war." (See also Maxwell D. Taylor, *The Uncertain Trumpet* [New York, 1959], pp. 2–3.)

69. Allen, "The Day Patton Quit."

70. LF-*LDP,* chaps. 6 and 7, passim.

71. RHP, p. 271.

72. LF-*LDP,* p. 67.

73. Nye, *The Patton Mind,* p. 162.

74. MB-*PP*, 2, p. 709.

75. John S. Ingles, "Patton Revisited," *VMI Alumni Review* (Summer 1990).

76. Harmon, *Combat Commander,* pp. 274–75, and LF-*LDP*, p. 162.

77. LF-*LDP,* pp. 131–32. Patton would swear between syllables only when he was genuinely upset.

78. MB-*PP* 2, p. 738.

79. Colonel George Fisher, Third Army Chemical Warfare Officer, interview in *Combat Forces Journal* (May 1951).

80. Ibid., p. 732.

81. McCullough, *Truman*, chaps. 9 and 10, passim.

82. LF-*LDP*, p. 51.

83. Cray, *General of the Army*, p. 545.

84. GSP to BAP, July 21, 1945, PP-USMA.

85. Various diary entries, July-mid-Sept. 1945, passim.

86. GSP to BAP, July 21, 1945, and diary, Aug. 10, 1945.

87. Various medical texts describe the condition of a subdural hematoma. For example, a recent study in Arthur K. Asbury, M.D., ed., *Diseases of the Nervous System: Clinical Neurobiology*, vol. 2, (Philadelphia, 1992), notes that mental confusion, fluctuating states of drowsiness, and headache are common features. That such conditions do not appear to have existed in Patton, nor epilepsy nor even coma—other results of a serious head injury—all are likely grounds for ruling out this diagnosis.

88. MB-*PP* 2, p. 843.

89. Doolittle, *I Could Never Be So Lucky Again,* p. 363. When Doolittle once addressed a letter to "Dear General Patton," he received a two-sentence reply that read: "Why the goddam hell do you get so formal as to address me as General Patton? It would seem to me that two men of our low mental and moral characteristics should be more informal."

90. Brig. Gen. R. Fraser, "Engineer Memoirs," Office of Engineer History, Fort Belvoir, Va. Fraser was chief of engineering construction for Third Army in 1945.

91. Frank McCarthy interview, FM.

92. Diary, Mar. 27, 1944.

93. Frasier, "Engineer Memoirs."

94. MB-*PP* 2, p. 706

Chapter 46   "All Good Things Must Come to an End"

1. MB-*PP* 2, pp. 740–41.

2. Diary, Sept. 29, 1945.

3. DDE to GCM, Aug. 22, 1945, copy in Irving Microfilm. Eisenhower's first choice to succeed him was Bedell Smith.

4. LF-*LDP*, p. 211

5. Ibid., pp. 198–99.

6. Patton's contentious tenure as procounsel of Bavaria and his subsequent relief from the command of Third Army are intricate subjects, themselves worthy of an entire book. Such an account was published in 1981 by Patton's earlier biographer, Ladislas Farago. *The Last Days of Patton,* his revealing account of the last six months of Patton's life, should be consulted by readers interested in a more detailed enumeration of Patton's relief.

7. LF-*LDP*, chaps. 12 and 13, passim.

8. GSP to BAP, Aug. 18, 1945, PP-USMA.

9. Diary, Sept. 16, 1945.

10. Ibid., Sept. 17, 1945; Lyon, *Eisenhower,* p. 360, and LF- *LDP*, Chap. 13, passim.

11. Lyon, *Eisenhower,* p. 360.

12. LF-*LDP*, p. 207.

13. Ibid., p. 207.

14. Ibid., p. 133. Whatever von Wangenheim's influence, Patton's diary is largely devoted to an anti-Russian diatribe.

15. Ibid., p. 146.

16. Diary, Aug. 29 and GSP to BAP, Aug. 31, 1945.

17. Fisher interview in *Combat Forces Journal*, PP-USMA.

18. Barney Oldfield, *Never a Shot in Anger* (New York, 1956), p. 295.

19. Ibid.

20. Fisher interview.

21. Allen, "The Day Patton Quit." Allen maintains Gay was present on Sept. 22, as does Blumenson who cites Gay's own journal. Farago, however, is incorrect when he

alleges that Gay was on leave in the United States and when he returned Patton all but embraced him, noting that he would not have gotten into trouble if Gay had been present on the day of the news conference. (See also LF-*LDP*, pp. 196–97; Gay diary, Sept. 22, 1945, and MB-*PP* 2, pp. 761–62.)

22. Oldfield, *Never a Shot in Anger,* pp. 295-96.

23. Victor Bernstein in *PM,* Sept. 24, 1945, copy in PP-USMA, and journalist Bill Cunningham in unnamed Boston newspaper, Box 29, PP-LC.

24. *NYT,* Sept. 28, 1945.

25. Summersby, *Eisenhower Was My Boss,* p. 278.

26. MB, p. 288.

27. Diary, Sept. 22, 1945.

28. Transcript of Bedell Smith's press conference, Sept. 26, 1945, Box 17, PP-LC.

29. Frank E. Mason to Roy Howard, Sept. 26, 1945, Box 17, PP-LC.

30. Ibid.

31. FA, pp. 248–49.

32. Quoted in RHP, p. 279.

33. FA, pp. 249–50.

34. Mason, letter of Sept. 26, 1945. Later, Mason tried unsuccessfully to set the record straight, but after Patton's death was unable to interest anyone except his biographers Martin Blumenson and Ladislas Farago in the matter, both of whom have recounted the story of how a small cabal of reporters took advantage of Patton. Given Patton's previous problems with the press, his staff seems to have been negligent in failing to ensure that *all* his press conferences were recorded by a stenographer, to prevent just such an occurrence as that of Sept. 22.

35. LF-*LDP,* p. 210.

36. Summersby, *Eisenhower Was My Boss,* p. 278. The dates of Patton's meetings with Eisenhower are wrong, but otherwise this is one of the few memoirs even to comment on the stormy session that ended their long friendship.

37. Dwight D. Eisenhower, *Letters to Mamie* (New York, 1978), p. 272; letter of Sept. 24, 1945.

38. Diary, Sept. 29, 1945.

39. Ibid.

40. LF-*LDP,* p. 216.

41. Ibid., pp. 216–17, and Diary, Sept. 29, 1945.

42. Diary, Sept. 29, 1945.

43. Eisenhower, *At Ease,* pp. 307–8.

44. John S. D. Eisenhower, *Strictly Personal,* p. 114.

45. FA, p. 255.

46. Box 55, PP-LC.

47. Smith, *Lucius D. Clay,* p. 227.

48. Arnold, *Global Mission,* pp. 546–47.

49. Lyon, *Eisenhower,* p. 361. Lyon criticizes but never elaborates on Patton's alleged failures in Morocco, but presumably they encompass his inexperience and naïveté in dealing with French general Auguste Noguès.

50. LF-*LDP,* pp. 95–97.

51. Robert Murphy, *Diplomat Among Warriors* (Garden City, N.Y., 1964,) pp. 294–95.

52. DDE to GSP, Sept. 29, 1945, EP.

53. Diary, Sept. 30, 1945.

54. DDE to GCM, ibid., GCM-P.

55. Copies in Box 29, PP-LC, and in PP-USMA.

56. LF-*LDP,* pp. 190–91. In the spring of 1946 Major Deane revealed that at the time of Patton's fateful press conference he had "a pipeline into the press center in Frankfurt," and learned through it that "Beetle Smith had a key hand in sending the press to Third Army to dig up dirt about the the DP situation . . . the Old Man was in trouble in Washington because some of the boys at Frankfurt were surely taking advantage of the

opportunity to finish him off." (Quotes in letter, Ernest C. Deane to Mason, Feb. 22, 1950, Box 17, PP-LC.)

57. GSP to BAP, Sept. 29, 1945, PP-USMA.

58. Ibid.

59. Mims interview, 1987. When he learned that Mims had asked for a pistol to take home with him as a souvenir, Patton held up the large convoy he was to have ridden in while an Ordnance officer located and delivered one to him.

60. Truscott, *Command Missions,* p. 508.

61. Diary, Oct. 7, 1945. One unsubstantiated account alleges that Patton hosted a reception in Truscott's honor that included singer Paul Robeson, and that by the time of the change of command ceremony, "Patton was feeling no pain and neither was Truscott . . . the two of them were trooping the line holding each other up in the quadrangle at Bad Tolz and [then] Patton took off." The account does not agree with those of Truscott and Patton, whose diaries for Oct. 7 note that the change of command ceremony took place at noon, and that he left promptly at 2:30 P.M. to board the Third Army train which departed at 3:00. Moreover, it was raining that day and the ceremony was moved indoors. (Fraser, "Engineer Memoirs.")

62. MB, p. 288.

63. LF-*LDP,* p. 12.

64. RHP, p. 279.

65. Allen, "The Day Patton Quit."

66. RHP, p. 279.

67. Allen, "The Day Patton Quit."

68. John S.D. Eisenhower, *Strictly Personal,* pp. 114–15.

69. Munson, "The Most Unforgettable Character."

70. GSP to Codman, Sept. 21, 1945, Box 31, PP-LC.

71. MB, p. 289.

72. Allen, "The Day Patton Quit."

73. GSP to Codman, letters written between Sept. 21 and Nov. 1, 1945, copies in Box 31, PP-LC.

74. LF-*LDP,* p. 237.

75. Diary, Dec. 2, 1945.

76. GSP to BAP, Nov. 24, 1945, Box 17, PP-LC.

77. Allen, "The Day Patton Quit."

Chapter 47   "A Helluva a Way to Die!"

Primary sources of Patton's accident, hospitalization, and death are: Gay diary and papers, USAMHI; Farago, *The Last Days of Patton,* chaps. 18 and 19, *After the Battle* 7 (1975); various articles and documents in the archives of the historian, U.S. Army Europe (USAREUR), Heidelberg, Germany; and PP-LC and PP-USMA.

1. Description of Woodring from LF-*LDP,* pp. 239–40.

2. Ibid., p. 242.

3. Ibid., pp. 241–42.

4. Professor Denver Fugate, who has thoroughly investigated Patton's accident, notes that before the war the depot Thompson attempted to enter was formerly the Mannheim Sanitation Depot, and being used as a QM facility. It was returned to the Germans in 1947. (Letter, Fugate to the author, Jan. 20, 1995.)

5. Account of Maj. Gen. Hobart R. Gay, Box 17, PP-LC. The death quote is ascribed to Allen, "The Day Patton Quit," and to FA, p. 263. Ayer, who traveled to Heidelberg to comfort his sister and brother-in-law, describes the statement as occurring shortly before Patton's death. Patton may well have uttered these words both at the time of the accident and from his hospital bed.

6. It is probable that *both* the clock and the overhead ceiling light of the Cadillac caused Patton's injuries.

7. Statement of Pfc. Horace Woodring. One of the first medical officers to later attend Patton thought he had hit his head on a metal rod in the roof of the Cadillac. (MB-*PP* 2, pp. 818–19.) Doctors at the 130th Station Hospital later treated Gay for minor contusions and a small cut on his left knee.

8. MB-*PP* 2, p. 819.

9. Ibid., and LF-*LDP*, pp. 244–45. In exonerating Thompson, it is not clear if the military police were aware that he was operating his truck in violation of several military regulations the morning of Dec. 9, 1945. What is now clear is that the report of MP Lieutenant Peter Babalas ought to have triggered a full inquiry into the facts and circumstances surrounding the accident. (Reel No. 2, GSP microfilm.)

10. Woodring later wrote an official statement of the events of Sunday, Dec. 9, 1945, and Hap Gay recorded his version in an unpublished account. Woodring, when interviewed in 1979 by Farago, did not recall ever having made a statement which was allegedly ascribed to him in 1952, and quoted at length in MB-*PP* 2, pp. 817–19. While Woodring's statement does contain references to Patton's injuries and death "that would presuppose medical expertise which Woodring conspicuously lacked in 1945, in 1952, or at any time . . . ," there seems little doubt it was made by him, as it contained information no one else could have known. (See LF-LDP, pp. 298–99.)

11. For example, *After the Battle* magazine investigated the accident and photographed the scene in 1975; a medical officer from the U.S. Army Medical Command (based in Germany) also investigated the accident and examined German police reports. The exact site of the accident was on Mannheimer Strasse in Käfertal, a suburb of Mannheim, only a few yards from the Mannheim city limits, where the street name changes to Käfertaler Strasse. (USAREUR Historical Office.)

12. Frederick Nolan, *The Oshawa Project* (London, 1974), in which Patton is absurdly fictionalized as a mere brigadier general named Robinson Campion Jr. by a British thriller writer.

13. John Enigl, "Death of Patton," *Military* (Aug. 1987). This article includes the assertion by someone named Earl Staats, who in 1945 was a young enlisted man stationed near Heidelberg, that he was the first outsider to reach the Cadillac after the accident. According to Staats, Patton stopped at Handschuhsheim (a small town on Route 3 just north of Heidelberg, which was nowhere near the route taken by Patton, and was, in fact, some twenty miles southeast) to visit his commanding officer on the morning of December 9, and that the accident occured a few blocks from his barracks. Staats also alleges that Gay was not present when he arrived on the scene. (Source of all Staats's allegations is the article by Enigl.)

14. The claims of Douglas Bazata were published in a Baltimore paper called *Spotlight*, on Jan. 4, 1993.

15. Although a message that Patton was enroute to Heidelberg was radioed by the military police, it was either ignored or mishandled at the 130th Station Hospital. Not until the actual moment of his arrival was anyone in authority aware that the hospital was about to receive Patton.

16. Dr. Paul S. Hill to Historical Unit, Walter Reed Army Medical Center, Oct. 21, 1964. Copy in USAREUR Historical Office; Warren A. Lapp, M.D., "An Ex-Medical Officer's Recollections of the Last Days of General George S. Patton, Jr., *MSCK* (Jan. 1987); and MB-*PP* 2, p. 820.

17. The official medical diagnosis of Patton was "a fracture of the third cervical vertebra, with a posterior dislocation of the fourth cervical vertebra. . . . Whether or not the spinal cord had been transected or merely traumatized was a matter for conjecture. . . ." (Lapp and Hill accounts, and GSP's official medical record.)

18. Hill account.

19. LF-*LDP*, p. 253.

20. Patton remained in the same first-floor room until he died. Later a plaque was installed to commemorate the place where he was treated and died. However, in early 1980s the hospital underwent renovations and the plaque was removed and sent to the hos-

pital commander's office for safekeeping. By the time the renovations were completed several years later there was no one still assigned to the hospital who knew where the plaque properly belonged. It remained hanging in the commanding officer's office until the USAREUR historian, Billy A. Arthur, informed the hospital staff it was in the wrong building. Nevertheless it took two more years to persuade the stubborn hospital brass that they were mistaken and even then the plaque was not reinstalled in its rightful location for an additional two years. Finally, in 1986, it was replaced outside the door of Room 110, the X-ray facility. (*After the Battle* 7 [1975], and Janet D'Agostino, "Patton: Death Throes of a General," *Stars and Stripes*, Dec. 20. 1989.)

21. Lapp account. An iron lung was brought to Heidelberg from Belgium but never utilized.

22. Dr. R. Glen Spurling, "The Patton Episode," a memoir sent to BAP on Aug. 10, 1950, Box 17, PP-LC

23. LF, p. 790.

24. Reel No. 2, GSP microfilm, PP-LC.

25. LF-*LDP*, pp. 276–77.

26. DDE-GSP, Dec. 10, 1945, EP; and Hill letter. Hill had jotted the text of the message on a scrap of paper which he kept. Eisenhower's letter included a remark that he had found a new (unspecified) post he planned to offer Patton in the United States.

27. LF-*LDP*, pp. 273–74.

28. Spurling, "The Patton Episode." Bedell Smith cabled Eisenhower on Dec. 20 that Patton's death was believed imminent within the next forty-eight hours. (See Chandler, *The Eisenhower Papers*, vol. 6, pp. 673–74.)

29. Ibid., and LF-*LDP*, p. 275.

30. Letter of Dec. 16, quoted in MB-*PP* 2, p. 828, and LF-*LDP*, p. 282.

31. Spurling, "The Patton Episode."

32. LF-*LDP*, pp. 285–86.

33. Ibid., p. 288.

34. Before his condition worsened, Spurling held out hope that if he remained stable, it might be possible to transfer Patton to the Beverly, Massachusetts, hospital near his home where, coincidentally, his life had been saved in 1937. President Truman's personal plane was to be placed at the Army's disposal to transport Patton. (Hill letter.)

35. Ibid., and LF-LDP, p. 292. Patton's statements of his impending death were recorded in the medical log in Dr. Duane's handwriting.

36. War Department Form 52-1, Reel No. 2, GSP microfilm; and LF-*LDP*, pp. 292–93. Dr. Duane had written on Patton's chart: "Died at 1755, 21 December 1945 with sudden stopping of heart." There is little doubt, however, that the real cause of death was another blood clot that starved his body of oxygen and plugged circulation, causing his heart to shut down, and his death.

37. RET-M, and author interview with Ruth Ellen Totten.

38. Odom, *General George S. Patton and Eisenhower,* p. 79.

39. LF-*LDP*, p. 283.

40. Totten lecture to the Topsfield Historical Society, 1974.

41. RET-M.

42. Ibid.

43. LF-*LDP*, p. 300, and Hill letter.

44. Notes of Frederick Ayer, Dec. 27, 1945, PP-USMA; and Spurling, "The Patton Episode."

45. GSP to Frederick Ayer, June 9, 1943. Patton had also added: "However, don't worry about this because I have no intention of departing this life."

46. Ibid.

47. John S. D. Eisenhower, *Strictly Personal,* p. 115. Other evidence of Beatrice's distinctiveness came in the form of a case of wine donated to the hospital mess in appreciation of their kindness.

48. Maj. Gen. George S. Patton, USA (Ret.), quoted in St. John, "Reflections on a

Fighting Father." St. John writes: "Time has not allowed the hurt to heal. Reporters have not helped," recounting how Patton's son had been hounded for years by the media on the subject of his father's death.

49. Ibid.

50. FA, p. 264.

51. *NYT,* Dec. 22, 1945.

52. Although papers concerning GSP's death are in a number of different boxes in the Library of Congress and West Point, most of the tributes are in Box 33, PP-LC.

53. Despite claims by an AP reporter that six hundred people attended the service, the photograph that was widely reproduced in American newspapers suggests the figure was considerably lower.

54. RHP, p. 282.

55. Sources for Patton's funeral ceremonies in Heidelberg and Luxembourg are: MB-*PP* 2, pp. 834–35; LF-*LDP*, pp. 302–3; Forty, *Patton's Third Army at War,* p. 188; *After the Battle* 7; Frederick Ayer's unpublished account, and AP reports and other newspaper accounts.

56. Note by BAP, appended to diary entry, Oct. 7, 1945.

57. LF-*LDP*, p. 305.

58. Odom, *General George S. Patton and Eisenhower,* pp. 79–80. When Smith offered to visit Patton's bedside, Beatrice politely declined with the excuse that having visitors would be too stressful. (Letter of Dec. 14, 1945, Box 10, W. B. Smith Papers, EL.)

59. Spurling, "The Patton Episode."

60. Oldfield, *Never a Shot in Anger* fn., p. 296.

61. UP account, Dec. 24, 1945, copy in PP-USMA.

62. FA, p. 192.

63. Newman, "General Patton's Premonition."

64. *Washington Times Herald*, Jan. 21, 1946. The editorial also concluded that no matter what his personal feelings toward Patton, Truman's duties as president outweighed them. In a nation already beginning to become edgy over the menace posed by the Soviet Union, Truman's refusal to honor Patton was also viewed as playing directly into the hands of the Communists.

65. RHP, pp. 286–87.

66. "Luxembourg American Cemetery and Memorial," American Battle Monuments Commission pamphlet.

Epilogue

1. RET-M.

2. RHP, p. 285. In another entry Beatrice wrote directly to her late husband that although unable then to dream of him, she felt compelled to articulate what she felt so deeply in her heart.

3. Ibid., p. 287.

4. RET-M, and ibid., p. 288. Jean Gordon's date of death is confirmed in Scott C. Steward, *The Sarsaparilla Kings* (Cambridge, Mass., 1993), p. 92, a history and genealogy of the Ayers.

5. Ibid., pp. 288–89.

6. RET-M. The day her sister died, Ruth Ellen had a premonition that left her in tears that "Something terrible is happening."

7. A "drag" hunt is one in which the hounds are not following a fox but instead a trail of scent created by dragging a sack coated with anise over the hunt course.

8. The *New York Herald Tribune* and the *Roanoke* (Virginia) *Times* were typical, both telling their readers that Beatrice had been killed by the fall, probably from internal injuries. Even *Time* wrongly ascribed her death to acts of her horse.

9. p. 293.

10. RET-M.

11. Beatrice ("Little Bee") to BAP, Jan. 26, 1946, Box 34, PP-LC.

12. Interview with Robert Patton, Jan. 17, 1995.

13. BAP's speech dated Aug. 19, 1950, Box 123, PP-LC.

14. From photo in Box 5, PP-USMA.

Postscript Patton's Legacy: A Genius for War

1. Quoted by Essame in *Patton the Commander,* p. 258.

2. Lt. Col. Paul W. Philips, "George Patton, A Memoir," address delivered in Fort Wayne, Indiana, 1966. Copy in Fort Wayne–Allen County Public Library, furnished by Carl E. Hudson, Jr.

3. Truscott, *Command Missions,* p. 509.

4. Cole, "George S. Patton." What Patton would have made of the Vietnam war can only be conjectured.

5. Ibid.

6. Leland Stowe, "Old Blood and Guts Off the Record."

7. Essame, *Patton the Commander,* p. 258.

8. Ibid., pp. 258–59.

9. Nye, *The Patton Mind,* p. 162.

10. LF-LDP, p. 86; and John Phillips, "The Ordeal of George Patton," *New York Review of Books*, Dec. 31, 1964 (review of Farago's *Patton: Ordeal and Triumph*).

11. See chap. 45, and quote in a Boston newspaper, Sept. 1945, copy in Box 29, PP-LC.

12. Andy Rooney to the author, May 28, 1994.

13. WFAN-AM 66.0, New York City. Syndicated in various other U.S. cities.

14. Larrabee, *Commander in Chief,* p. 487. The R.W. Thompson quote is from *The Battle for the Rhineland* (London, 1958), p. xii.

15. Oral history of Gen. Paul D. Harkins, USAMHI.

16. CRC, p. 272.

17. Gen. I. D. White, "Patton—The Man and the Film," *New Hampshire Sunday News*, 1970.

18. CRC, p. 271.

19. Wallace, *Patton and His Third Army,* p. 209.

20. It was not until the autumn of 1944 that a first-class quartermaster officer whom Patton had known before the war was brought in to revamp his mess and bring about other improvements in food service in the Third Army. (CRC, p. 274.)

21. Desmond Young, *Rommel* (London, 1950), p. 177.

22. Leckie, *Delivered from Evil,* p. 767.

23. S.L.A. Marshall oral history, Marshall Papers, USAMHI.

24. Letter, Ladislas Farago to Frank McCarthy, Nov. 21, 1968, Box 3, FM.

25. Interview of John S. D. Eisenhower, *World War II,* May 1994.

26. RET, lecture to the Topsfield Historical Society, 1974.

27. Ibid.

28. Ibid.

29. LF-*LDP*, p. 13.

30. Stowe, "Old Blood and Guts Off the Record."

31. Quoted in Wallace, *Patton and His Third Army,* pp. 205–6.

32. Miller, *Ike the Soldier,* p. 658. Two resolutions were introduced into the 82d Congress in 1951 by Massachusetts representatives to posthumously promote Patton to the rank of five-star general, thus placing him alongside both Eisenhower and Bradley. Neither passed and both were opposed by the Pentagon on grounds that it was against policy ever again to promote officers to five-star rank. (See H.J. Res. 72, and H.R. 174, Reel 2, GSP microfilm, PP-LC.)

34. ONB-*SS*, pp. 159–60.

35. John S. Ingles, "Patton Revisited." Ingles records the only exception to his statement that Patton never squandered lives was the Hammelburg raid.

35. LF, p. 796.

36. Blumenson, in *Parameters* (Winter 1994–95).

37. Also quoted in FA, p. 266.

SOURCES AND
SELECT BIBLIOGRAPHY

Unpublished Sources

The two most important principal collections are the George S. Patton Papers in the Manuscript Division of the Library of Congress, and the Patton Collection in the Special Collections Division United States, Military Academy, West Point. The Library of Congress collection consists of nearly 41,000 items and occupies more than thirty-seven linear feet of shelf space in the Manuscript Division. However, there is a wealth of information about Patton in the archival collections in the following institutions:

U.S. Army Military History Institute (USAMHI), Carlisle Barracks, Pennsylvania

The papers of General of the Army Omar N. Bradley,* including the papers and diary of Lt. Col Chester B. Hansen; the diary and papers of Maj. Gen. John P. Lucas, interview of M/Sgt. John L. Mims; the papers of Gen. James H. Polk; Gen. I. D. White, Lt. Gen. Hobart R. Gay; Lt. Gen. Garrison H. Davidson; Maj. Gen. Kenyon Joyce; Maj. Gen. Terry Allen; Brig. Gen. Guy V. Henry; Brig. Gen. Oscar Koch, Brig. Gen. Halley G. Maddox, and S. L. A. Marshall. Miscellaneous "Vignettes of Military History" and OCMH collections and interviews at the U.S. Army Center of Military History.

Oral histories: Gen. Theodore J. Conway; Gen. Paul D. Harkins; Gen. Ben Harrell; Gen. James H. Polk; Gen. John K. Waters; Gen. I. D. White; Lt. Gen. Hobart R. Gay; Lt. Gen. John A. Heintges; Lt. Gen. Elwood R. Quesada, and S. L. A. Marshall.

Library: David Irving microfilm collection; Patton bibliography; bibliography and papers about the Punitive Expedition; Col. George J. Fisher, "The Boss of Lucky Forward"; Notes on Patton's death; various monographs; GSP's prayer for good weather during the Battle of the Bulge, December 1944; "Essays in Some Dimensions of Military History"; "Speech of Gen. George S. Patton, Jr., to his Third Army on the eve of the Normandy Invasion"; GSP's "Notes on Bastogne Operation"; interview of Capt. Abraham J. Baum; press conference of Secretary of War with GSP, Washington, D.C, June 14, 1945; and "An Historical Sidelight on Armored Warfare."

---

*The Bradley papers are deposited both at USAMHI, Carlisle and West Point. As collaborator Clay Blair notes in *A General's Life*, there is some duplication but those at Carlisle deal mainly with the Second World War, while those at West Point are personal.

Liddell Hart Centre for Military Archives, King's College, London
The papers of Sir Basil Liddell Hart (notes of 1944 conversations with Patton); and the papers and diary of Field Marshal Lord Alanbrooke.

Manuscript Division, Library of Congress, Washington, D.C.
The papers of Gen. George S. Patton Jr.; the diary and papers of Maj. Gen. Everett S. Hughes; the papers of General of the Armies John J. Pershing; the papers of Gen. Carl A. Spaatz; and the papers of Gen. James H. Doolittle.

Special Collections Division, United States Military Academy Library, West Point, New York
George S. Patton Jr. collection; the papers of General of the Army Omar N. Bradley; Col. B. A. Dickson collection; and various miscellaneous collections, histories, and registers of the Corps of Cadets, USMA.

The Citadel Archives, Charleston, South Carolina
Melvin C. Helfers papers; Friedman Cryptologic collection; the papers of Gen. Mark W. Clark.

Dwight D. Eisenhower Library, Abilene, Kansas
The (prepresidential) papers of Dwight D. Eisenhower; the papers and diary of Capt. Harry C. Butcher, USN; the papers of Lt. Gen. Walter Bedell Smith; the papers of Lt. Gen. Floyd Parks; oral histories of Gen. Jacob L. Devers; Gen. Elwood Quesada; Gen. William H. Simpson; Lt. Gen. LeRoy Lutes; various official documents and miscellaneous papers; and various photographs.

Combined Arms Research Library, U.S. Army Command & General Staff College, Fort Leavenworth, Kansas
Various monographs and secondary sources (Civil War, World Wars I and II).

Leavenworth, Kansas, Public Library
Microfilm records of the *Leavenworth Times*, 1923–24.

Public Record Office, Kew, London
The personal papers of Field Marshal Earl Alexander of Tunis (WO 214); the Churchill papers (PREM 3); Allied Force Headquarters (AFHQ papers (WO 204); SHAEF papers (WO 219); 21st Army Group papers (WO 205); correspondence of Air Marshal Sir Arthur Coningham, Tunisian campaign; and various campaign and historical files, Mediterranean and northwest Europe.

Imperial War Museum, London
Various published and unpublished articles, letters, photographs, and papers in the Department of Printed Books, Department of Documents; and Department of Photographs. Oral history transcripts from the Thames Television series *The World at War, 1939–1945.*

National Archives, Suitland Annex, Suitland, Maryland
Diary of Lt. Gen. Hobart R. Gay.

Kreitzberg Library, Norwich University, Northfield, Vermont
The papers of Gen. I. D. White and Maj. Gen. Ernest N. Harmon.

Sterling Library, Yale University, New Haven, Connecticut
The papers of Secretary of War Henry L. Stimson (microfilm).

Virginia Military Institute, Lexington, Virginia
Papers pertaining to the Patton family in the VMI archives: Waller Tazewell Patton, George Smith Patton I, George Smith Patton II, George Smith Patton III, George H. Smith, and John Mercer Patton. Correspondence to and from George S. Patton; newspaper and magazine articles about Patton and his family; and photographs of Patton family.

George C. Marshall Research Library, Virginia Military Institute, Lexington, Virginia
The papers of General of the Army George C. Marshall; the diary and papers of Gen. Lucian K. Truscott Jr.; the papers of Lt. Gen. Thomas T. Handy; George F. Howe collection; the papers of Brig. Gen. Frank McCarthy; the papers of Brig. Gen. Paul M. Robinett; and miscellaneous magazine and newspaper articles.

S. L. A. Marshall Military History Collection, Special Collections Division, Library, University of Texas at El Paso
Memoir of Patton by Dr. Hugh M. Cole; and miscellaneous correspondence and papers about GSP.

Fort Wayne–Allen County Public Library, Fort Wayne, Indiana
"Gen. George Patton: A Memoir," copy of speech by Lt. Col Paul W. Philips, contributed by Carl E. Hudson Jr.

Center of Military History, U.S. Army, Washington, D.C.
Published and unpublished articles and papers, Patton file, CMH archives. Various monographs about World War II published by CMH.

Still Photo Branch, National Archives, Washington, D.C.
World War II photographs; photographs of George S. Patton.

Special Collections Division, Mugar Library, Boston University, Boston, Massachusetts
The papers of Ladislas Farago.

Library, U.S. Naval War College, Newport, Rhode Island
Oral histories: Admiral John L. Hall; Admiral H. Kent Hewitt; and Admiral Alan G. Kirk. Monograph: "From Algiers to Anzio," the unpublished diary and recollections of Major Gen. John P. Lucas.

United States Navy Historical Center, Washington, D.C.
The papers of Samuel Eliot Morison.

Central Rappahannock Regional Library, Fredericksburg, Virginia
Microfilm records and primary and secondary source documents and books in the Virginiana Room pertaining to the Patton family, particularly Robert Patton's emigration to Virginia and his career and life in Fredericksburg.

Office of History, U.S. Army Corps of Engineers, Alexandria, Virginia
Oral histories of Gen. Bruce C. Clarke and Brig. Gen. Harvey R. Fraser.

Patton Museum of Cavalry and Armor, Fort Knox, Kentucky
Photographs.

Gen. Phineas Banning Residence Museum, Wilmington, California
Photograph of the Patton, Ayer, and Banning families (1902).

Miscellaneous Sources

"Operation Symbol," the late Lt. Gen. Sir Ian Jacob's unpublished personal diary of the Casablanca Conference; the unpublished manuscript of General of the Army Dwight D. Eisenhower.

Published Sources

BOOKS

Alexander, Field Marshal Earl of Tunis. *The Alexander Memoirs, 1940–1945*. New York: McGraw-Hill, 1962.

Allen, Robert S. *Drive to Victory*. New York: Berkley Publishers, 1947.

———. *Lucky Forward*. New York: Vanguard Press, 1964.

Ambrose, Stephen E. *Duty, Honor, Country: A History of West Point*. Baltimore: Johns Hopkins Press, 1966

———. *The Supreme Commander*. London: Cassell, 1970.

———. *Eisenhower: Soldier, General of the Army, President-Elect, 1890–1952*. New York: Simon & Schuster, 1983.

Anders, Curt. *Fighting Generals*. New York: Putnam, 1965.

Army Times Editors. *The Yanks Are Coming*. New York: Putnam, 1960.

—. *Warrior: The Story of General George S. Patton*. New York: Putnam, 1967.

Atkinson, Rick. *The Long Gray Line*. New York: Pocket Books, 1991.

Ayer, Fred. *Before the Colors Fade*. Boston: Houghton Mifflin, 1964.

Ayer, Frederick. *The Reminiscences of Frederick Ayer*. Boston: n.p., 1923.

Baron, Richard. *Raid! The Untold Story of Patton's Secret Mission*. New York: Putnam, 1981.

Bell, William Gardner. *Quarters One*. Washington, D.C., U.S. Government Printing Office, 1988.

Bender, Mark C. *Watershed at Leavenworth: Dwight D. Eisenhower and the Command and General Staff School*. Fort Leavenworth, Kans.: C&GSC, 1990.

Blair, Clay. *Ridgway's Paratroopers*. New York: Dial Press, 1985.

Bland, Larry I., and Sharon Ritenour Stevens, eds.. *The Papers of George Catlett Marshall*, vol. 1, *"The Soldierly Spirit": December 1880–June 1939*. Baltimore: Johns Hopkins University Press, 1981.

———. Vol. 2, *"We Cannot Delay": July 1, 1939–December 6, 1941*. Baltimore: Johns Hopkins University Press, 1986.

———. Vol. 3, *"The Right Man for the Job": December 7, 1941–May 31, 1943*. Baltimore: Johns Hopkins University Press, 1991.

Blumenson, Martin. *Kasserine Pass*. Boston: Houghton Mifflin, 1967.

———. *The Patton Papers, 1885–1940*. Boston: Houghton Mifflin, 1972.

———. *The Patton Papers, 1940–1945*. Boston: Houghton Mifflin, 1974.

———. *Patton: The Man Behind the Legend, 1885–1945*. New York: Morrow, 1985.

———. *The Battle of the Generals*. New York: Morrow, 1994.

Bourke-White, Margaret. *Dear Fatherland, Rest Quietly*. New York: Simon & Schuster, 1946.

Braim, Paul F. *The Test of Battle*. Newark: University of Delaware Press, 1987.

Bradley, Omar N. *A Soldier's Story*. New York: Henry Holt, 1951.

Bradley, Omar N., with Clay Blair. *A General's Life*. New York: Simon & Schuster, 1983.

Brereton, Lewis H. *The Brereton Diaries*. New York: Morrow, 1946.

A Brief History of Fort Leavenworth, 1827–1983. Fort Leavenworth, Kans.: U.S. Army Combined Arms Center and Fort Leavenworth, Kansas.

Brooks, Stephen, ed. *Montgomery and the Eighth Army*. London: The Bodley Head, 1991.

Brown, Vernon. *The Emperor's White Horses*. New York: McKay, 1956.

Bullard, Robert L. *Personalities and Reminiscences of the War*. Garden City, N.Y.: Doubleday, Page & Co., 1925.

Butcher, Harry C. *Three Years with Eisenhower.* London: Heinemann, 1946.

Carew, Harold C. *History of Pasadena and the San Gabriel Valley.* 3 vols. Pasadena, Calif.: S.J. Clarke Publishing Co., 1930.

Carpenter, Allan. *George Smith Patton, Jr.* Vero Beach, Fla. Rourke Publications, 1987.

Cary, James. *Tanks and Armor in Modern Warfare.* New York: Franklin Watts, 1966.

Center of Military History, U.S. Army. *American Armies and Battlefields in Europe.* Washington, D.C., U.S. Government Printing Office, 1992.

C&GSC. *A Military History of the United States Army Command and Gen. Staff School, 1881–1963.* N.d.

Chambers, Lewis. *Stonewall Jackson and the Virginia Military Institute: The Lexington Years,* Lexington, Va.: Historic Lexington Foundation, 1982.

Chandler, Alfred D., ed. *The Papers of Dwight David Eisenhower: The War Years.* Vols. 1–5. Baltimore: Johns Hopkins University Press, 1970.

Chandler, Alfred D., and Louis Galambos, eds. *The Papers of Dwight David Eisenhower: Occupation, 1945,* vol. 6, and *The Chief of Staff,* vols. 7–9, 1978.

Charlton, James, ed. *The Military Quotation Book,* New York: St. Martin's Press, 1990.

Churchill, Winston S. *The Great War.* Vol. 1. London: George Newnes, 1933.

Clarke, Jeffrey J., and Robert Ross Smith. *Riviera to the Rhine.* Washington, D.C.: CMH, 1993

Cleland, Robert Glass. *From Wilderness to Empire: A History of California.* New York: Alfred A. Knopf, 1969.

Codman, Charles R. *Drive.* Boston: Little, Brown, 1957.

Coffman, Edward M. *The War to End All Wars: The American Military Experience in World War I.* New York: Oxford University Press, 1968.

———. *The Old Army.* New York: Oxford University Press, 1986.

Cole, Hugh M. *The Lorraine Campaign.* Washington, D.C.: U.S. Government Printing Office, 1950.

———. *The Ardennes: Battle of the Bulge.* Washington, D.C.: U.S. Government Printing Office, 1965.

Collins, J. Lawton. *Lightning Joe.* Baton Rouge, La.: Louisiana State University Press, 1979.

Commandants, Staff, Faculty, and Graduates of the Command and Gen. Staff School Fort Leavenworth, Kansas, 1881–1939. Fort Leavenworth, Kans.: Command & Gen. Staff School Press, 1939.

Couper, Colonel William. *One Hundred Years at V.M.I.* 4 vols. Richmond, Va.: Garnett & Massie, 1939.

Cray, Ed. *General of the Army: George C. Marshall, Soldier and Statesman.* New York: Touchstone Books, 1990.

Creveld, Martin van. *Supplying War: Logistics from Wallenstein to Patton.* Cambridge University Press, 1977.

Crosswell, D. K. R. *The Chief of Staff: The Military Career of General Walter Bedell Smith.* Westport, Conn.: Greenwood Press, 1991.

Crozier, William Armstrong, ed. *Virginia County Records,* vol. 1, *Spotsylvania County, 1721–1800.* Baltimore: Genealogical Publishing Co. Inc., 1978.

Cullum, George W. *Biographical Register of the Officers and Graduates of the U.S. Military Academy at West Point: Supplement Vol. VI-B, 1910–1920.* Saginaw, Mich.: Seeman & Peters Printers, 1929.

Cunningham, Admiral of the Fleet Viscount. *A Sailor's Odyssey.* London: Hutchinson, 1951.

Daniels, Roger. *The Bonus March.* Westport, Conn.: Greenwood Press, 1971.

Dastrup, Boyd L. *The US Army Command and General Staff College: A Centennial History,* Fort Leavenworth and Manhattan, Kans.: J.H. Johnston and Sunflower University Press, 1982.

Davis, Franklin M. *Across the Rhine.* Alexandria, Va.: Time-Life Books, 1980.

Davis, William C. *The Battle of New Market.* Baton Rouge, La.: Louisiana State University Press, 1975.

Demke, Glenn S. *The Boom of the Eighties in Southern California.* San Marino, Calif.: Huntington Library, 1944.

D'Este, Carlo. *Decision in Normandy.* New York: E.P. Dutton, 1988.

———. *Bitter Victory: Th e Battle for Sicily, 1943.* New York: E.P. Dutton, 1988.

———. *Fatal Decision: Anzio and the Battle for Rome.* New York: HarperCollins, 1991.

Doolittle, General James H. *I Could Never Be So Lucky Again.* New York: Bantam, 1991.

Dornbush, Charles E. *Speech of General George S. Patton, Jr., to His Third Army on the Eve of the Normandy Invasion.* Cornwallville, N.Y.: Hope Farms Press, 1980.

Dumke, Glenn S. *The Boom of the Eighties in Southern California.* San Marino, Calif.: Huntington Library, 1944.

Dupuy, R. Ernest. *Men of West Point.* New York: William Sloane Assoc., 1951.

Dupuy, Trevor N. *Hitler's Last Gamble.* New York: HarperCollins, 1994.

Dyer, George. *XII Corps: Spearhead of Patton's Third Army.* Baton Rouge, La.: Army Navy Publishing Co., 1947.

Eisenhower, David. *Eisenhower at War, 1943–1945.* London: Collins, 1986.

Eisenhower, Dwight D. *Crusade in Europe.* Garden City, N.Y.: Doubleday, 1948.

———. *At Ease: Stories I-Tell My Friends.* New York: Doubleday, 1967.

Eisenhower, John S. D. *The Bitter Woods.* New York: Putnam, 1969.

———. *Strictly Personal.* Garden City, N.Y.: Doubleday, 1974.

———. *Allies: Pearl Harbor to D-Day.* Garden City, N.Y.: Doubleday, 1982.

Eisenhower Foundation. *D-Day: The Normandy Invasion in Retrospect.* Abilene, Kans., 1971.

Elman, Robert. *Fired in Anger.* New York: Doubleday, 1968.

Embrey, Alvin T. *History of Fredericksburg, Virginia.* Richmond, Va.: Old Dominion Press, 1937.

Essame, H. *Patton: The Commander,* London: B.T. Batsford, 1974.

Evans, Clement A. *Confederate Military History Extended Edition.* Vol. 3, *West Virginia.* Vol. 8, *Alabama.* 1899. Reprint: Wilmington, N.C., 1987.

Farago, Ladislas. *Patton: Ordeal and Triumph.* New York: Ivan Obolensky, 1963.

———. *The Last Days of Patton.* New York: McGraw-Hill, 1981.

Ferrell, Robert H., ed. *The Eisenhower Diaries.* New York: W. W. Norton, 1981.

Fleming, Thomas J. *West Point: The Men and Times of the United States Military Academy.* New York: Morrow, 1969.

Forty, George. *Patton's Third Army at War.* New York: Scribner's, 1978.

Foster, Gaines. *Ghosts of the Confederacy.* Baton Rouge: Louisiana State University Press, 1987.

Frankel, Nat. *Patton's Best: An Informal History of the 4th Armored Division.* New York: Hawthorn Books, 1978.

Freeman, Douglas Southall. *Lee's Lieutenants,* vol. 1, *Manassas to Malvern Hill.* New York: Scribner's, 1942; reprint, 1970.

Fuller, J. F. C. *Tanks in the Great War, 1914–1918.* London: John Murray, 1920.

———. *Generalship: Its Diseases and Their Cure.* Harrisburg, Pa.: Military Service Publishing Co., 1936.

Gabel, Christopher R. *The U.S. Army GHQ Maneuvers of 1941.* Washington, D.C.: CMH, 1991.

Garland, Albert N., and Howard McGaw Smyth. *Sicily and the Surrender of Italy.* Washington, D.C.: U.S. Government Printing Office, 1965.

Gavin, James M. *On to Berlin.* New York: Viking, 1978.

Gelb, Norman. *Desperate Venture: The Story of Operation Torch, the Allied Invasion of North Africa.* New York: Morrow, 1992.

Gillie, Mildred Hanse. *Forging the Thunderbolt: A History of the Development of the Armored Force.* Harrisburg, Pa.: Military Service Publishing Co., 1947.

Goodwin, Doris Kearns. *No Ordinary Time: Franklin and Eleanor Roosevelt: The Home Front in World War II.* New York: Simon & Schuster, 1994.

Goolrick, John T. *The Life of Gen. Hugh Mercer.* New York & Washington, D.C.: Neale Publishing Co., 1906.

———. *Historic Fredericksburg.* Fredericksburg, Va.: 1921.

Greenfield, Kent Roberts, ed. *Command Decisions*. London: Methuen, 1960.

Hamilton, Nigel. *Monty: Master of the Battlefield, 1942–1944*. London: Hamish Hamilton, 1983.

———. *Monty: The Field Marshal, 1944–1976*. London: Hamish Hamilton, 1986.

———. *Monty: The Battles of Field Marshal Bernard Montgomery*. New York, Random House, 1994.

Harkins, Paul D. *When the Third Cracked Europe*. Harrisburg, Pa.: Stackpole, 1969.

Harmon, Ernest N. *Combat Commander*. Englewood Cliffs, N.J.: Prentice-Hall, 1970.

Hastings, Max. *Overlord: D-Day and the Battle for Normandy. 1944*. London: Pan Books, 1985.

Hatch, Alden. *George Patton, General in Spurs,* New York: Julian Messner, 1950.

Hayden, Rev. Horace Edwin. *A Genealogy of the Glassell Family of Scotland and Virginia*. Baltimore: Genealogical Publishing, 1979.

Headley, Robert K. Jr. *Genealogical Abstracts from 18th-Century Newspapers*. Baltimore: Genealogical Publishing Co., 1987.

Herr, John K., and Edward S. Wallace. *The Story of the U.S. Cavalry*. Boston: Little, Brown, 1953.

History of Los Angeles County, California. 1880. Reprint, Berkeley: Howell-North, 1959.

Hobbs, Joseph P. *Dear General: Eisenhower's Wartime Letters to Marshall*. Baltimore: Johns Hopkins University Press, 1971.

Hodgson, Godfrey. *The Colonel*. New York: Alfred A. Knopf, 1990.

Horne, Alastair, with Montgomery, David. *Monty: The Lonely Leader, 1944–1945*. New York: HarperCollins, 1994.

Houston, Donald E. *Hell On Wheels: The 2d Armored Division*. San Rafael, Calif.: Presidio Press, 1977.

Howe, Henry. *Historical Collections of Virginia*. 1845. Baltimore: Regional Publishing Co., 1969.

Howe, George F. *Northwest Africa: Seizing the Initiative in the West*. Washington, D.C.: U.S. Government Printing Office, 1957.

Hunt, Elvid. *History of Fort Leavenworth, 1827–1937*. Fort Leavenworth: the Command and Gen. Staff School Press, 1937.

Ingersoll, Ralph. *Top Secret*. New York: Harcourt, Brace, 1946.

Irving, David. *The War Between the Generals*. New York: Congdon & Lattès, 1981.

Isaacson, Walter and Evan Thomas. *The Wise Men*. New York: Simon & Schuster, 1986.

Jacob, John N. *George C. Marshall Papers, 1932–1960: A Guide*. Lexington, Va.: George C. Marshall Foundation, 1987.

Jett, Dora C. *Minor Sketches of Major Folk and Where They Sleep*. Richmond: Old Dominion Press, 1928.

Josephson, Matthew. *Infidel in the Temple*. New York: Alfred A. Knopf, 1967.

Kemp, Anthony. *The Unknown Battle: Metz, 1944*. New York: Stein & Day, 1985.

Kenamore, Clair. *From Vauquois Hill to Exermont: A History of the Thirty-Fifth Division of the United States Army*. St. Louis, Mo.: Guard Publishing, 1919.

Kent, Gerald T. *A Doctor's Memoirs of World War II*. Cleveland, Ohio: Cobham and Hatherton Press, 1989.

Knightley, Phillip. *The First Casualty*. London: Quartet Books, 1982.

Koch, Oscar W. *G-2 Intelligence for Patton*. Philadelphia: Army Times/Whitmore, 1971.

Larrabee, Eric. *Commander in Chief*. New York: Harper & Row, 1987.

Lawrence, Joseph Douglas. *Fighting Soldier: The AEF in 1918*. Boulder, Colo.: Associated University Press, 1985 .

Lewis, Thomas A. *The Guns of Cedar Creek*. New York: Harper & Row, 1988.

———. *The Shenandoah in Flames: The Valley Campaign of 1864*. Alexandria, Va.: Time-Life Books, 1987.

Liddell Hart, Basil H., ed. *The Rommel Papers*. New York: Harcourt, Brace, 1953.

Liddell Hart, Basil H. *History of the Second World War.* London: Cassell, 1970.

———. *The Other Side of the Hill.* London: Pan Books, 1978.

Linethal, Edward T. *Changing Images of the Warrior Hero in America: A History of Popular Symbolism.* New York: Edwin Mellen Press, 1982.

Llewellyn, Harry. *Passports to Life.* London: Hutchinson/Stanley Paul, 1980.

Love, Robert W., Jr. *History of the United States Navy, 1775–1941.* Vol. 1. Harrisburg, Pa.: Stackpole, 1992.

Lowry, Terry. *22d Virginia Infantry.* Lynchburg, Va.: H.E. Howard, Inc., 1988.

Lyle, Royster, Jr. *The Architecture of Historic Lexington.* Charlottesville: University Press of Virginia, 1977.

Lyon, Peter. *Eisenhower: Portrait of the Hero.* Boston: Little, Brown, 1974.

MacDonald, Charles B. *A Time for Trumpets: The Untold Story of the Battle of the Bulge.* New York: Morrow, 1985.

———. *The Siegfried Line Campaign.* Washington, D.C.: Office of the Chief of Military History, 1963.

Mann, B. David. *VMI Alumni in the Civil War.* Lexington, Va.: VMI Museum, 1986.

Marshall, George C. *Memoirs of My Service in the World War: 1917–1918.* Boston: Houghton Mifflin, 1976.

Mason, Herbert Malloy, Jr. *The Great Pursuit.* New York: Random House, 1970.

Matheny, E. A. *History of the Third United States Army, 1919–1962.* Fort McPherson, Ga., 1967.

Mauldin, Bill. *The Brass Ring.* New York: W.W. Norton, 1971.

McCullough, David. *Truman.* New York: Simon & Schuster, 1992.

McKee, Alexander. *The Race for the Rhine Bridges.* London: Pan Books, 1974.

Mellenthin, Maj. Gen. F. W. von. *Panzer Battles.* Norman: University of Oklahoma Press, 1956.

Mellor, William Bancroft. *Patton, Fighting Man.* New York: Putnam, 1946.

———. *General Patton: The Last Cavalier.* New York: Putnam, 1971.

A Military History of the US Army Command & General Staff College, 1881–1963. Fort Leavenworth, Kans.: C&GS.

Mitchell, F. *Tank Warfare: The Story of Tanks in the Great War.* London: Thomas Nelson & Sons, 1933.

Miller, Merle. *Ike the Soldier: As They Knew Him.* New York: Putnam,1986.

Mission Accomplished: Third Army Occupation of Germany, 9 May 1945–15 Feb 1947. N.p. 1947.

Mitchall, Ralph M. *The 101st Airborne Division's Defense of Bastogne.* Fort Leavenworth, Kans.: C&GSC, Combat Studies Institute, 1986.

Mitcham, Samuel W., Jr., and Friedrich von Stauffenberg. *The Battle of Sicily.* New York: Orion Books, 1991.

Montgomery, Field Marshal Viscount. *Memoirs.* London: Collins, 1958.

Morgan, Kay Summersby. *Past Forgetting.* New York: Simon & Schuster, 1976.

Morrison, James L., Jr. *The Memoirs of Henry Heth.* Westport, Conn.: Greenwood Press, 1974.

Morison, Samuel Eliot. *Operations in North African Waters: October 1942–June 1943.* Boston: Little, Brown, 1947.

———. *Sicily-Salerno-Anzio.* Boston: Little, Brown, 1954.

Mosley, Leonard. *Marshall.* London: Methuen, 1982.

Nadeau, Remi. *City-Makers: The Story of Southern California's First Boom, 1868–76.* Costa Mesa, Calif.: Trans-Anglo Books, 1965.

Nenninger, Timothy. *The Leavenworth Schools and the Old Army.* Westport, Conn.: Greenwood Press, 1978.

Newmark, Harris. *Sixty Years in Southern California, 1853–1913.* Boston: Houghton Mifflin Co., 1930.

O'Connor, Richard. *Black Jack Pershing.* Garden City, N.Y.: Doubleday, 1961.

Odom, Charles B. *General George S. Patton and Eisenhower.* New Orleans: Word Picture Productions, 1985.

Official Register of the Officers and Cadets of the United States Military Academy. West Point, N.Y., 1909.

Ogorkiewicz, Richard M. *Armor: A History of Mechanized Forces*. New York: Praeger, 1960.

Oldfield, Col. Barney. *Never a Shot in Anger*. New York: Duell, Sloane & Pearce, 1956.

Orange, Vincent. *Coningham*. London: Methuen, 1990.

Palmer, Dave Richard. *The River and the Rock: The History of Fortress West Point, 1775–1783*. New York: Greenwood Publishing Co., 1969.

Parrish, Thomas. *Roosevelt and Marshall: Partners in Politics and War*. New York: Morrow, 1989.

Paschall, Rod. *The Defeat of Imperial Germany, 1917–1918*. Chapel Hill, N.C.: Algonquin Books, 1989.

Patton, Beatrice Ayer. *The Reminiscences of Frederick Ayer*. London: Redwood Press, 1971.

———. *Blood of the Shark*. Honolulu: Paradise of the Pacific Press, 1936.

Patton, George S., Jr. *War as I Knew It*. New York: Bantam, 1980.

——— *Diary of the Instructor in Swordsmanship*. Fort Riley, Kans.: Mounted Service School Press, 1915.

Pearl, Jack. *Blood and Guts Patton*. New York: Monarch Books, 1961.

Peifer, Charles, Jr. *Soldier of Destiny*. Minneapolis: Dillon Press, 1989.

Perley, Sidney. *Ayer Genealogy*. Newburyport, Mass.: Parker River Researchers, 1986.

Perret, Geoffrey. *There's a War to Be Won*. New York: Random House, 1991.

——— *Winged Victory: The Army Air Corps in World War II*. New York: Random House, 1993.

Perry, Milton F., and Barbara W. Parke. *Patton and his Pistols*. Harrisburg, Pa.: Stackpole Co., 1957.

Pershing, Gen. John J. *My Experience in the World War*. New York: Stokes, 1931.

Phillips, Henry Gerard. *El Guettar*. N.d., 1991.

Phillips, Robert F. *To Save Bastogne*. New York: Stein & Day, 1983.

Pogue, Forrest C. *George C. Marshall: Ordeal and Hope 1939–1942*. New York: Viking, 1966.

———. *George C. Marshall: Organizer of Victory, 1943–1945*. New York: Viking, 1973.

Prioli, Carmine A., ed. *Lines of Fire: The Poems of General George S. Patton, Jr.* Lewiston, N.Y.: Edwin Mellen Press, 1991.

Province, Charles M. *The Unknown Patton*. New York: Bonanza Books, 1983.

Pullen, John J. *Patriotism in America: A Study of Changing Devotions, 1770–1970*. New York: American Heritage Press, 1971.

Puryear, Edgar F. *Nineteen Stars*. Novato, Calif.: Presidio Press, 1971.

Pyle, Ernie. *Brave Men*. New York: Henry Holt & Co., 1944.

Quinn, Lt. Gen. William W. *Buffalo Bill Remembers*. Fowlerville, Miss.: Wilderness Adventure Books, 1991.

Reeder, Red. *Heroes and Leaders of West Point*. New York: Thomas Nelson, 1970.

Register of Graduates and Former Cadets of the United States Military Academy 1802–1990. West Point, N.Y., 1990.

Ridge, Warren J. *Follow Me!* New York: Amacom, 1989.

Riggs, David F. *7th Virginia Infantry*. Lynchburg, Va.: H.E. Howard, Inc., 1982.

Robertson, James I., Jr. *Civil War Virginia*. University Press of Virginia, 1991.

Robinett, Paul M. *Armor Command*. Washington, D.C.: McGregor & Werner, 1959.

Rohmer, Richard. *Patton's Gap*. London: Arms and Armour Press, 1981.

Sainsbury, Keith. *The North African Landings, 1942*. Newark: University of Delaware Press, 1979.

Scharf, J. Thomas. *History of the Confederate States Navy*. New York: Rogers & Sherwood, 1887.

Semmes, Harry H. *Portrait of Patton*. New York: Appleton-Century-Crofts, 1955.

Shaara, Michael. *The Killer Angels*. New York: Ballantine, 1974.

Shapiro, Milton. *Tank Command: Patton's 4th Armored Division*. New York: David McKay, 1979.

Sherwood, Midge. *Days of Vintage, Years of Vision.* Vol. 1. San Marino, Calif.: Orizaba Publications, 1982.

————. *Days of Vintage, Years of Vision.* Vol. 2. San Marino, Calif.: Orizaba Publications, 1987.

Shipley, Thomas. *The History of the AEF.* New York: George H. Doran & Co., 1920.

Smith, Francis G. *History of the Third Army.* Washington, D.C.: Historical Section, Army Ground Forces, 1946.

Smith, Gene. *The Shattered Dream.* New York: Morrow, 1970.

Smith, Jean Edward. *Lucius D. Clay: An American Life.* New York: Henry Holt & Co., 1990.

Smith, Susy. *Reincarnation.* Los Angeles: Sherbourne Press, 1967.

Smythe, Donald. *Pershing: General of the Armies.* Bloomington: Indiana University Press, 1986.

Soldiers and Statesmen. Military History Symposium, U.S. Air Force Academy, Washington, D.C.: 1973.

Sorley, Lewis. *Thunderbolt: From the Battle of the Bulge to Vietnam and Beyond: General Creighton Abrams and the Army of His Times.* New York: Simon & Schuster, 1992.

Speidel, Hans. *We Defended Normandy.* London: Jenkins, 1951.

Stallings, Lawrence. *The Doughboys: The Story of the AEF, 1917–1918.* New York: Harper & Row, 1963.

Steward, Scott C. *The Sarsaparilla Kings.* Cambridge, Mass.: N.p., 1993.

Stewart, George R. *Pickett's Charge.* Boston: Houghton Mifflin, 1959.

Summersby, Kay. *Eisenhower Was My Boss.* New York: Prentice-Hall, 1948.

Surnames in the United States Census of 1790. Baltimore: Genealogical Publishing Co., 1971.

Taylor, Maxwell D. *The Uncertain Trumpet.* New York: Harper, 1959.

Tedder, Lord. *With Prejudice.* London: Cassell, 1966.

Thayer, Charles. *Bears in the Caviar.* Philadelphia: J.B. Lippincott, 1950.

Thomas, Emory M. *The Confederacy as a Revolutionary Experience.* University of South Carolina Press, 1971.

Toland, John. *The Last 100 Days.* New York: Random House, 1967.

————. *No Man's Land: The Story of 1918.* New York: Doubleday, 1980.

————. *Battle: The Story of the Bulge.* New York: Random House, 1959.

Tompkins, Frank. *Chasing Villa.* Harrisburg, Pa.: Military Service Publishing Co., 1934.

Toulmin, H. A., Jr. *With Pershing in Mexico.* Harrisburg, Pa.: Stackpole, 1935.

Truscott, Lucian K. *Command Missions.* New York: E.P. Dutton, 1954.

—. *The Twilight of the U.S. Cavalry.* Manhattan: Kansas University Press, 1990.

U.S. Third Army Headquarters. *In Memoriam, George S. Patton, Jr., Gen. U.S. Army.* Bad Tölz: 1946.

Vandiver, Frank E. *Black Jack: The Life and Times of John J. Pershing.* College Station, Tex.: A&M University Press, Vols I and Vol II, 1977.

Walker, Charles D. *Biographical Sketches of the Graduates and Elèves of the Virginia Military Institute,* Philadelphia: J.B. Lippincott & Co.,1875.

Wallace, Brenton G. *Patton and His Third Army.* Nashville: the Battery Press, 1981, (reprint of 1946 edition).

Waters, W. W. *B.E.F.: The Whole Story of the Bonus Army.* New York: John Day, 1933.

Wedemeyer, Albert C. *Wedemeyer Reports!* New York: Henry Holt & Co., 1958.

Weigley, Russell F. *History of the United States Army.,* New York: Macmillan, 1967

————. *The American Way of War: A History of United States Military Strategy and Policy.,* Bloomington: Indiana University Press, 1973.

————. *Eisenhower's Lieutenants.* Bloomington: Indiana University Press, 1981.

Wellard, James. *Gen. George S. Patton, Jr.: Man Under Mars.* New York: Dodd, Mead & Co., 1946.

Wert, Jeffrey D. *From Winchester to Cedar Creek.* New York: Touchstone, 1987.

Whiting, Charles. *48 Hours to Hammelburg.* 1970. Reprint, New York: PJB Books, 1982.

————— . *Bounce the Rhine.* New York: Stein & Day, 1986.
————— . *Patton's Last Battle.* New York: Stein & Day, 1987.
————— . *Patton.* New York: Ballantine, 1970.
Williams, Mary H. *Chronology, 1941–1945.* Washington, D.C.: Center of Military History, 1989.
Williams, Vernon L. *Lt. Patton and the American Army in the Mexican Punitive Expedition, 1915–1916.* Austin, Tex.: Presidial Press, 1983.
Williamson, Porter B. *Patton's Principles.* New York: Touchstone Books, 1982.
Wilmot, Chester. *The Struggle for Europe.* New York: Harper & Row, 1952.
Wilson, Dale E. *Treat 'Em Rough!: The Birth of American Armor, 1917–20.* Novato, Calif.: Presidio Press, 1990.
Woollcott, Alexander. *The Command of Forward.* New York: Century Co., 1919.
Wyant, William K. *Sandy Patch: A Biography of Lt. Gen. Alexander M. Patch,* New York: Praeger, 1991.
Yergin, Daniel. *The Prize.* Simon & Schuster, 1991
Young, Desmond. *Rommel.* London: Collins, 1950.

ARTICLES

"A Merry Christmas from Gen. Patton." *The Static Line,* Dec. 1981.
Allen, Robert S. "The Day Patton Quit." *Army* (June 1971).
Ambrose, Stephen E. "A Fateful Friendship." *American Heritage,* Apr. 1969.
"American Light Tank Brigade at St. Mihiel." *Infantry Journal* (Mar. 1928).
Bailey, Tom. "The Patton Story." *Soldiers* (July 1971).
"The Battle of Bridges." *Cavalry Journal* (1942).
Bell, W. G., and Martin Blumenson. "Patton the Soldier." *Ordnance* (Jan.-Feb. 1959).
Betson, William R. "Sidi Bou Zid: A Case History of Failure." *Armor* (Nov.-Dec. 1982).
Blumenson, Martin. "George S. Patton's Student Days at the Army War College." *Parameters* 5, no. 2 (1976).
————— . "Patton and Montgomery: Alike or Different?" *Army* (June 1972).
————— . "Patton as Diplomat." *Army* (July 1973).
————— . "Patton in Mexico." *American History Illustrated,* Oct. 1977.
————— . "Patton and the Press." *Army* (May 1982).
————— . "Bradley-Patton: World War II's 'Odd Couple.' " *Army* (Dec. 1985).
————— . "Eisenhower Then and Now: Fireside Reflections." *Parameters* 12, no. 2 (summer 1991).
————— . "The Hammelburg Affair." *Army* (Oct. 1965).
————— . "Disaster at Kasserine Pass." *Army* (Feb. 1993).
————— . "Essence of Command: Competence, Iron Soul." *Army* (Mar. 1993).
Brown, Kent Masterson. "The Heroic Agonies of the Colonels Patton." *Virginia Country* 1 (Aug. 1983).
Calvert, Robert. " 'Drum, Drum, I Wish He Would Stop Beating His Own Drum.' " *Army* (Sept. 1989).
Cariff, Frank. "The Patton Paradox." *Reserve Officer,* Apr. 1946.
"The Cavalry Maneuvers at Fort Riley, Kansas, 1934." *Cavalry Journal* (1934).
Chynoweth, Bradford G. "Cavalry Tanks." *Cavalry Journal* (July 1921).
Clark, Brooks. "Gunning for Glory." *Pursuits* (Spring 1989).
Clemens, John. "Waking Up from the Dream: The Crisis of Cavalry in the 1930s." *Armor* (May-June 1990).
"Command and Gen. Staff School." *Cavalry Journal* 25 (1926).
Conniff, Frank. "The Patton Paradox." *Reserve Officer,* Apr. 1946.
Cooney, Major Patrick J. "U.S. Armor Between the Wars." *Armor* (Mar.-Apr. 1990).
Creel, George. "Patton at the Payoff." *Collier's,* Jan. 13, 1945.
Cromie, Robert. "'Me? I Was with Patton,' " *TV Guide,* Sept. 6, 1986.
"The Death of George S. Patton." *After the Battle* 7 (1975).
Denove, Joseph P. "Division Review." *Cavalry Journal* (May-June 1941).

DeRome, Jeff. "George C. Scott Gives Patton a Fair Shake." *The American West* (Nov.-Dec. 1985).

Dietrich, Steve E. "The Professional Reading of Gen. George S. Patton, Jr." *Journal of Military History* (Oct. 1989).

Douglas, H. H. "The Lipizzaners: Royal Horses of Austria." Part 1, *Cavalry Journal* (July-Aug. 1947).

———. "The Lipizzaners: Royal Horses of Austria." Part 2, *Cavalry Journal* (Nov.-Dec. 1947).

"Draughts of Old Bourbon Address to Third Army." *American Mercury,* Jan. 1951.

Dupuy, Colonel R. Ernest. "Love at First Sight." *Army-Navy-Air Force Register,* Sept. 23, 1961.

Eisenhower, Dwight D. "A Tank Discussion." *Infantry Journal* (Nov. 1920).

Eliot, George Fielding. "To Get the Best." *Combat Forces Journal* (Jan. 1951).

Falls, Cyril. "Aftermath of War: War as Gen. Patton Knew It." *Illustrated London News,* Oct. 2, 1946.

Fisher, Col. George J. "The Boss of Lucky Forward." *Combat Forces Journal* (May 1951).

Fleming, Thomas. "West Point in Review." *American Heritage,* Apr. 1988.

Foley, Maj. Gen. Thomas C. "Desert Shield Deployment Rivals Patton's Rush to the Bulge," *Armor* (Jan.-Feb. 1991).

Ganz, A. Harding. "Patton's Relief of Gen. Wood." *Journal of Military History* (July 1989).

Garland, Albert N. "They Had Charisma." *Army* (May 1971).

Gavin, James M. "Two Fighting Generals." *Atlantic Monthly,* Feb. 1965.

"Gen. Patton at Knutsford." *After the Battle* 7 (1975).

"Gen. Patton's War Letters." *Atlantic Monthly,* Dec. 1947.

"Gen. Patton Incident." *Army and Navy Register,* Dec. 4, 1943.

"Gen. Patton's Indiscretions." *Christian Century,* 3 Dec. 1943.

"George S. Patton: A Personality Profile." *American History* (July 1966).

Gillette, Jean. "Romancing the Game." *Spur,* Jan.-Feb. 1991.

Graham, Capt. Selwyn H., USNR. "We Knew General; George Patton, Jr. When—." *Shipmate,* Oct. 1988.

Grant, Brig. Gen. Frederick D. "West Point Our Great American Military University." *Metropolitan Magazine,* May 1905.

Greenberg, Lawrence M. "Rolling Advance Stymied." *World War II* (July 1991).

Grow, Maj. Gen. Robert W. "Ten Lean Years: From the Mechanized Force (1930) to the Armored Force (1940)." Parts 1–4, *Armor* (Jan.-Feb., Mar.-Apr., May-June, July-Aug. 1987).

Hammond, Colonel Elton F. "Signals for Patton." *Signals,* Sept.-Oct. 1947.

Halsey, Ashley. "Ancestral Gray Cloud over Patton." *American History Illustrated,* March 1984.

Hanlin, J. J. "The Gen. and the Horses." *American Legion Magazine,* Feb. 1963.

Hawkins, Brig. Gen. Hamilton S. "Cavalry and Mechanized Force." *Cavalry Journal* 40 (Sept.-Oct. 1931).

"'Hell on Wheels' Rolled from Africa to Berlin." *Armor* (July-Aug. 1990).

"Highlights from Gen. Patton's Own Story." *Saturday Evening Post,* Nov. 1, 1947.

Hofmann, George F. "The Demise of the U.S. Tank Corps and Medium Tank Development Program." *Military Affairs* (Feb. 1973).

Irzyk, Brig. Gen. Albin F. "Bastogne: A Fascinating Obscure Vignette." *Armor* (Mar.-Apr. 1986).

—. "The 'Name Enough' Division." *Armor* (July-Aug. 1987).

Jennys, David R. "Defenders Reluctantly Engaged." *World War II* (Nov. 1992).

Johnson, G. W. "Prometheus Patton." *Virginia Quarterly Review* (Apr. 1945).

Karl, Dennis R. "Drive for Berlin: The Debate over Strategy in the Invasion of Germany." *American History Illustrated,* June 1985.

Keating, Susan Kating. "Issue of POWs and War Crimes Is an Old One." *Insight,* Feb. 11, 1991.

Koch, Oscar W. "2d Armored Division Maneuvers in Tennessee." *Cavalry Journal* (Nov.-Dec. 1941).

Koyen, K. "Gen. Patton's Mistake: Third Army, Fourth Armored Division and the Hammelburg Affair." *Saturday Evening Post,* May 1, 1948.

Kraft, William R., Jr. "The Saga of the Five of Hearts." *Armor* (July-Aug. 1988).

Kuter, Gen. Lawrence S. "Goddammit, Georgie." *Air Force Magazine,* Feb. 1973.

Lapp, Warren A., M.D. "An Ex-Medical Officer's Recollections of the Last Days of Gen. George S. Patton, Jr." *MSCK,* Jan. 1987.

Linderman, Gerald F. "Military Leadership and the American Experience." *Military Review* (Apr. 1990).

Maginnis, Gerald F. "A2 Leader Styles: Preparing for the Airland Battle." *Military Review* (Dec. 1985).

"Mechanized Force Becomes Cavalry." *Cavalry Journal* (1931).

Munch, Lt. Col. Paul G. "Patton's Staff and the Battle of the Bulge." *Military Review* (May 1990).

Nenninger, Timothy K. "The World War I Experience." *Armor* (Jan.-Feb. 1969) .

———. "Tactical Dysfunction in the AEF, 1917–1918." *Military Affairs* (Oct. 1987).

Newman, Larry G. "Gen. Patton's Premonition." *American Legion Magazine,* July 1962.

Noyer, William L. "The Final Voyage." *The Retired Officer,* March 1989.

Nye, Colonel Roger H. "The Patton Library Comes to West Point." *Friends of the West Point Library Newsletter* (March 1988).

———. "Whence Patton's Miltary Genius?" *Parameters* (Winter 1991–92).

Oberbeck, S. K. "Total Warrior." *Newsweek,* Oct. 7, 1974.

"Official Text of Letter and Report on Patton Case." *Army and Navy Journal,* Dec. 4, 1943.

Ogan, Ronald A. "Patton's Peacemaker." *American Rifleman,* May 1986.

Oldinsky, Frederick E. "Patton and the Hammelburg Mission." *Armor* (July-Aug. 1976).

O'Neill, Msgr. James H. "The Story Behind Patton's Prayer." *Our Sunday Visitor,* Aug. 15, 1971.

"The Original George S. Patton—A Profile." *Civil War Times,* June 1961.

Orsini, Eric A. "The Great Gen.'s Only Mistake." *Army Logistician* (Mar.-Apr. 1986).

Painton, Frederick C. "Old Man of Battle." *Reader's Digest,* Sept. 1943.

Parker, Peggy O'Connell. "The *When and If*—Patton's Own." *Army* (Feb. 1978).

Patton, Beatrice Ayer. "A Soldier's Reading." *Armor* (Nov.-Dec. 1952).

Patton, George S., Jr. "The Army at the National Horse Show." *Cavalry Journal* (Jan. 1923).

———. "The 1929 Cavalry Division Maneuvers." *Cavalry Journal* (Jan. 1930).

———. "Cavalry Work of the Punitive Expedition." *Cavalry Journal* (Jan. 1917).

——— . "Comments on Cavalry Tanks." *Cavalry Journal* (July 1921).

———. "Defense of the Saber." *Cavalry Journal* (July 1916).

———. "The Desert Training Corps." *Cavalry Journal* (Sept.-Oct. 1942).

———. "The Effect of Weapons on War." *Infantry Journal* (Nov. 1930).

———. "Form and Use of the Saber." *Cavalry Journal* (Mar. 1913).

——— "Gen. Patton Writes Home." *Los Angeles Examiner,* Sept. 27, 1943.

———. "Mechanized Forces: A Lecture." *Cavalry Journal* (Sept.-Oct. 1933).

———. "Motorization and Mechanization in the Cavalry." *Cavalry Journal* (July 1930).

———. "Present Saber: Its Form and Use for Which It Was Designed." *Cavalry Journal* (Apr. 1917).

———. "Report of Operations of the Army Polo Team of 1922." *Cavalry Journal* (Apr. 1923).

———. "Success in War." *Cavalry Journal* (Jan. 1931).

———. "Tanks in Future Wars." *Infantry Journal* (May 1920).

———. "What the World War Did for Cavalry." *Cavalry Journal* (Apr. 1922).

———. "Mechanization and Cavalry." *Cavalry Journal* (Apr.1930) (written with Maj. C. C. Benson).

"The Patton Affair." *New Republic,* Dec. 6, 1943.

"Patton at the Pay-Off." *Collier's,* Jan. 13, 1945.

"Patton of the Armored Force." *Life,* July 7, 1941.

"Patton's Vehicles." *After the Battle* 7 (1975).

"Patton's Slap." *Newsweek,* Dec. 6, 1943.

"Patton Rebuked." *Scholastic,* Oct. 22, 1945.

"Patton—The Great Soldier." *80th Division Service Magazine,* [1946].

Pew, Thomas W., Jr. "On the Way to War." *American West* (Nov.-Dec. 1985).

Phillips, John. "The Ordeal of George Patton." *New York Review of Books,* Dec. 31, 1964.

"The Photography of Patton." *After the Battle* 7 (1975).

Polk, Gen. James H. "Patton: 'You Might as Well Die a Hero.'" *Army,* Dec. 1975.

Prioli, Carmine A. "King Arthur in Khaki: The Medievalism of Gen. George S. Patton, Jr." *Studies in Popular Culture* 10 (1987).

———. "The Poetry of Gen. George Smith Patton, Jr." *Journal of American Culture* (Winter 1975).

Randle, Brig. Gen. Edwin H. "The Gen. and the Movie." *Army,* Sept. 1971.

Rarey, Captain C. H. "Lessons from the use of Tanks by the American Army." *Infantry Journal* 23 (1928).

Reid, Brian Holden. "Major Gen. J. F. C. Fuller and the Problem of Military Movement." *Armor* (July-Aug. 1991).

"Remove Patton From Germany." *New Republic,* Oct. 1, 1945.

"Report on Gen. Patton." *Army and Navy Journal,* Nov. 27, 1943.

Roberts, Joseph B., Jr. "American Warriors and Their Arms." *American Rifleman* (Apr. 1987).

Rogge, Robert E. "The 304th Tank Brigade: Its Formation and First Two Actions." *Armor* (July-Aug. 1988).

Root, E. Merrill. "George Patton." *American Opinion* (Feb. 1972).

Schneider, James J. "An Open Letter to Gen. George S. Patton." *Military Review* (June 1986).

Schreier, Konrad F., Jr. "American Tanks Meet the Test." *Armor* (Sept.-Oct. 1990).

"The Second Armored Division Grows Up." *Cavalry Journal* (1941).

Semmes, Brig. Gen. Harry H. "General George S. Patton, Jr.'s Psychology of Leadership." *Armor* (May-June 1955).

Shane, T. "These are the Generals." *Saturday Evening Post,* Feb. 6, 1943.

Sheehan, Vincent. "The Patton Legend–And Patton as Is." *Saturday Evening Post,* June 23, 1945.

Sondern, Frederic, Jr. "The Wonderful White Stallions of Vienna." *Reader's Digest,* April 1963.

Sorel, Nancy Caldwell. "George S. Patton and Bill Mauldin." *Atlantic Monthly,* Mar. 1988.

South, Betty. "We Called Him 'Uncle Georgie.' " *Quartermaster Review* (Jan.-Feb. 1954).

Strobridge, Truman R. "Old Blood and Guts and the Desert Fox." *Military Review* (June 1984).

Wahl, Norman, DDS. "Appointment in Mostaganem." *American Journal of Orthodontics* (Oct. 1990).

Weigley, Russell F. "From the Normandy Beaches to Falaise-Argentan Pocket." *Military Review* (Sept. 1990).

"Where Patton Prepared for North Africa in Southern California." *Sunset,* Feb. 1989.

White, I. D. "Patton—The Man and the Film." *Military Affairs* (Dec. 1970).

Williams, Brig. Gen. Robert W. "Moving Information: The Third Army Imperative." *Army,* Apr. 1975.

"With Major Gen. Patton." *Christian Science Monitor,* Jan. 9, 1943.

Woolley, William J. "Patton and the Concept of Mechanized Warfare." *Parameters* (Autumn 1986).

Wukovits, John F. "Best-Case Scenario Exceeded." *Military History* (Dec. 1992).

MONOGRAPHS

American Battle Monuments Commission. *Luxembourg American Cemetery and Memorial.* Hamm, Luxembourg: 1961.

Berlin, Dr. Robert H. *U.S. Army World War II Corps Commanders: A Composite Biography.* Fort Leavenworth, Kans.: Combat Studies Institute, 1989.

Gabel, Christopher R. *The Lorraine Campaign: An Overview, September–December 1944.* Fort Leavenworth, Kans.: Combat Studies Institute, 1985.

—. *The 4th Armored Division in the Encirclement of Nancy.* Fort Leavenworth, Kans.: Combat Studies Institute, 1986.

Gardner, Major Gregory C. *Generalship in War: The Principles of Operational Command.* Fort Leavenworth, Kans.: School of Advanced Military Studies, 1987.

Higgins, George A. *The Operational Tenets of Generals Heinz Guderian and George S. Patton, Jr.* Fort Leavenworth, Kans.: C&GSC, 1985.

Hodgson, James B. *Historical Example: Third Army Advance to the West Wall.* 1953.

Mellor, Sgt. Sidney L. *The Desert Training Center and C-AMA.* Historical Section, Army Ground Forces, 1946.

Partin, John W., ed. *A Brief History of Fort Leavenworth, 1827–1983.* Fort Leavenworth, Kans.: Combat Studies Institute, 1983.

Patton, George S. *Saber Exercise, 1914.* Washington, D.C.: U.S. Government Printing Office, 1914.

Plummer, Oscar E. *Gen. Patton's Purple Heart.* N.p., 1981.

Stowe, Leland. *Old Blood-and-Guts Off the Record.* Fort Leavenworth, Kans.: C&GSC Library.

Vermillion, Maj. John M. *The Main Pillars of Generalship: A Different View.* Fort Leavenworth, Kans.: School of Advanced Military Studies, 1986.

ACKNOWLEDGMENTS

One of the most difficult but rewarding aspects of such a major undertaking is to find adequate means of thanking the many people who have helped make this biography possible. To begin with I record my immense debt to the eminent historian, and Patton's official biographer, Martin Blumenson. Since our first association nearly fifteen years ago Martin has not only enthusiastically supported my work but has encouraged me to investigate areas that he himself had already written about. Where others might jealously guard their domain, Martin has offered stimulation and encouragement. His writings about Patton, and a host of other subjects, are models of clarity and have been both a source of inspiration and an important reference in the writing of *Patton: A Genius for War*.

Martin Blumenson is one of a unique group of military historians who emerged during and after World War II to chronicle the most devastating conflict in history. Other than Charles B. MacDonald, who was a decorated infantry company commander in some of the bloodiest battles in the ETO, most served in some capacity as staff historians in the field. All were in a singularly uncommon position to observe firsthand and later record the battles and campaigns of the war. In the postwar years they served in the Office of the Chief of Military History to write the official histories of the U.S. Army in World War II, the famous "Green Books."

Charlie MacDonald wrote *Company Commander* and several other notable works that have become important references for anyone studying or writing about World War II. Others in this group include Gordon A. Harrison, Dr. Hugh Cole, Maurice Matloff, Roland Ruppenthal, and George F. Howe.

Although their contributions are in some instances indirect, each has played a part in the shaping of this book through his influence on the author. These men set the high standards for the writing of military history, and their impact upon the successive generations of historians who followed them has been, I believe, considerable. They were role models for those of us who believe that to undertake the writing of military history and biography is an awesome responsibility.

In June 1992 the U.S. Army Center of Military History brought together many of these distinguished men at a conference of military historians. A number of those in attendance were later heard to observe that its highlight was the reminiscences of this exceptional group. What seems certain is that we will never see their like again, and I thank each of them for furthering our knowledge of the tumultuous events about which they reported.

I also take this opportunity to acknowledge the assistance and inspiration of Forrest Pogue, who was so helpful many years ago when I was searching for the focus of what eventually became *Decision in Normandy*. Since then I have learned how many countless others Dr. Pogue has unselfishly assisted, always with no other motive except to share his formidable knowledge. Pogue has written the history of SHAEF and a monumental four-volume official biography of one of the greatest Americans of the twentieth century, Gen. George C. Marshall. Forrest Pogue's fame may rest on his books, but to those of us whom he has helped in his kindly manner, he is quite simply a jewel we are all deeply proud to call a friend and mentor.

The late Ruth Ellen [Patton] Totten was instrumental not only in supporting my decision to write about her father but in imparting a wealth of background about the Pattons. She also graciously lent the author a copy of her exceptional unpublished memoir of her mother, "MA: A Button Box Biography." The memoir is especially valuable for its revelation of the life of Beatrice Ayer Patton, whose extraordinary role in the life of George S. Patton has been too little known. Mrs. Totten died in 1993, and I welcome this opportunity gratefully to acknowledge her important contributions to this biography of her father.

Mrs. Totten has written of the unusual title of her memoir:

> Thinking about our lives and the way they were changed by the fact of being the children of Gen. George S. Patton, Jr. and Beatrice Banning Ayer is like looking through my button box. I can pick up one button and remember that it came off a blue woolen reefer that belonged to one of the children . . . then another button, from our daughter's first-day-at-school dress and another button, a pink quartz turtle, bought in a Japanese store in Honolulu during my green and salad days—all my life brought back to me by buttons!
>
> Being our father's children was a special influence in all of our lives, but the greatest, most pervasive, and most interesting influence in my life was Ma. She was not famous. But from the time I can remember, she made her Georgie famous to us. He was the Apollo at Delphi—the king of the castle—the perfect knight. When we read of Achilles, we knew that Georgie looked like him. Beowulf was in his image; Roland was Georgie; and Siegfried, with his sword and his blond curls, was the very picture of Georgie. He is a twice-told tale, and Ma is a tale that has never been told.

My sincere appreciation to Col. Michael W. Totten, USA (Ret.), James P. Totten, and Beatrice Totten Britton for permission to quote from their mother's unpublished manuscript.

I am also deeply grateful to Maj. Gen. George S. Patton, USA (Ret.), not only for his assistance but for his kind permission to use and quote from the Patton Papers. At the outset Gen. Patton thought there might be only limited interest in his father in the 1990s, and it is my sincere hope that *Patton: A Genius for War* will serve to mitigate that doubt.

No work of this scope can be undertaken without the advice and assistance of a great many others. What has proved so rewarding during the lengthy research for this book is how graciously so many have given of their time and resources to provide information about Patton and the turbulent times during which he lived. Individuals and institutions alike have invariably cooperated in ways large and small that have resulted in this portrait of George S. Patton, Jr. To each I record my appreciation.

The assistance of the following institutions is gratefully acknowledged:

In the United Kingdom
The staffs of the Imperial War Museum and the Keeper and Staff of the Public Record Office, Kew, London; Miss Patricia Methven, archivist of the Liddell Hart Centre for Mili-

tary Archives, King's College, London; the director and staff of the Imperial War Museum, London: Department of Documents, Department of Printed Books, and the Department of Photographs; and the Hulton Picture Library.

In the United States

The staff of the Special Collections Division, Boston University; the staff of the Boston Public Library, who responded efficiently to various requests for assistance, particularly Mary Francis O'Brien; Mrs. Barbara P. Willis; Sue Willis and John Johnson of the Central Rappahannock Regional Library, Fredericksburg, Virginia; Karen Duffy, reference librarian, Mary Washington College, Fredericksburg, Virginia; the Virginia Historical Society; the Library of Congress: the always helpful and efficient staff of the Manuscript Division, and the Local History & Genealogy Branch, particularly Anne Toohey; the staff of the Combined Arms Library, U.S. Army Command & Gen. Staff College, Fort Leavenworth, Kansas, especially reference librarians Carol Ellis Ramkey and Elizabeth Snoke; the Library of the Naval War College, Newport, Rhode Island; and Jane Yates, acting director, Archives-Museum, The Citadel, Charleston, South Carolina. Also, Mrs. Jacqueline Painter, archivist emeritus, and Paul Heller, librarian, Kreitzberg Library, Norwich University; Thomas F. Burdett, curator, S. L. A. Marshall Collection, University of Texas at El Paso, and Dr. John Votaw, Director of Cantigny: the 1st Division Museum, Wheaton, Illinois, and archivist Eric Gillespie.

A special word of appreciation to Alan C. Aimone, chief of the Special Collections Division, USMA Library, and Ms. Gladys Calvetti, for their assistance with the West Point collections, particular the Patton papers; and to Mrs. Charlyn Richardson for assistance in tracking down and processing the photographs from the USMA collection. The staff of the George C. Marshall Research Library, Lexington, Virginia: Larry I. Bland, Ronald R. Marryott, president, Marshall Foundation, and Royster Lyle Jr. At the Virginia Military Institute: archivist Diane Jacob, Prof. Norman S. Stevens, Keith Gibson, and my friend, Dr. Bruce C. Vandervort, who went out of his way to ensure that my research trip to VMI was productive. I am greatly indebted to Katie Talbot, librarian of the Patton Museum of Cavalry and Armor, Fort Knox, Kentucky, who responded to my request for assistance with many of the photographs that appear in this book. Also Jan Losi, Director, Gen. Phineas Banning Residence Museum, Wilmington, California; and Martin K. Gordon, Historian-Archivist, Office of History, U.S. Government Printing Office Army Corps of Engineers, Alexandria, Virginia. The Dwight D. Eisenhower Library, Abilene, Kansas, provided some of the photographs, and archivist Thomas W. Branigar ensured that my research was rewarding.

Brig. Gen. Harold W. Nelson, USA (Ret.), former chief of the U.S. Army Center of Military History, has been an ardent benefactor to whom I owe an enormous debt of thanks for his support and assistance. At CMH I thank Hannah Zeidlick, Lt. Col. Steve E. Dietrich, Dr. John Greenwood, Billy A. Arthur, and Lt. Col. Roger Cirillo.

I make special acknowledgment to the following who deserve mention for their contributions: Robert H. Patton, the grandson of George S. Patton Jr., has written a superb memoir of his family and kindly made available photographs of the early Pattons that have enriched the visual presentation of this book. I sincerely thank him for his helpfulness and cooperation.

Cadet James Brown (class of 1993, Massachusetts Maritime Academy) assisted immeasurably in enabling me to understand dyslexia and the problems faced daily by dyslexics. Renowned dyslexia expert and author, Dr. Harold E. Levinson, kindly read portions of the manuscript dealing with dyslexia.

T. M Pickens, of the Gen. Patton Memorial Museum, Chiriaco Summit, California, and Richard E. Fagan, of the Needles Resource Area, Bureau of Land Management, U.S. Department of the Interior, furnished information about the Desert Training Center.

Charles M. Province, founder and president of the George S. Patton Historical Society, provided useful bibliographical material, and Lou Reda and Sammy Jackson of Lou Reda Productions several of the photographs.

Dr. Robert J. T. Joy, (Col., MC, USA, Ret.), of the F. Edward Hébert School of Medicine, Uniformed Services School of the Health Sciences, Bethesda, Maryland, provided a wealth of medical information, advice, and counsel that greatly aided my understanding of Patton's various medical conditions. Orthopedic surgeon Dr. Thomas W. Brooks Jr. advised on Patton's 1937 accident; and my friend and physician Dr. William E. Litterer, of Falmouth, Massachusetts, reviewed Patton's medical condition in December 1945, thus enabling me to explain the causes of his death in lay terms.

Robert K. Krick, chief historian, Fredericksburg & Spotsylvania National Military Park, Virginia, provided useful references to the early Pattons, and Dr. Vincent Orange, Department of History, University of Canterbury, Christchurch, New Zealand, and the biographer of Air Marshal Sir Arthur Coningham, kindly furnished information about the Tunisian incident. Bruce H. Siemon, command historian of U.S. Army Europe & Seventh Army, Heidelberg, Germany, made available USAREUR's archival records of Patton's accident and death.

World War II cartoonist and author Bill Mauldin kindly permitted reproduction of his evocative cartoon of Willie and Joe on their "thousand-mile detour" of Third Army.

My longtime friend Nigel Hamilton assisted in many ways large and small, and his advice and encouragement helped bridge the long period between the conception of this book and its completion.

Richard F. Reidy Jr. has been a friend of nearly thirty years, whose enthusiastic encouragement of my pursuit of this project and frequent hospitality were welcome and important aspects of numerous research trips to Washington, D.C. Author and friend James F. Murphy Jr. likewise proved a sounding board for ideas and an infectious source of Irish humor and motivation, for which he is justly noted.

For nearly fifteen years the Falmouth (Massachusetts) Public Library has been an incomparable source of information and reference material, without which this book could not have been written. Toni Robertson and Gail Rose of the Inter-Library Loan Department have once again cheerfully and efficiently fulfilled my requests for books, articles, microfilm, and photocopies. Very special thanks to reference librarians Barbara Conathan, Jill Erickson, and Kathy Mortenson, all of whom proved conclusively on numerous occasions that one need not be a magician to pull rabbits from hats—and who responded unerringly to my questions and guided me to sources of information, no matter how obscure.

I offer my grateful thanks for his assistance and support to Col. Roger Nye, USA (Ret.), former deputy head of the Department of History at West Point and author of *The Patton Mind*, a superb biography of Patton as an intellectual. Colonel Nye assisted in locating several of the photographs and shared his expertise in a variety of helpful ways that were of immeasurable value to me.

Historian Kent Masterson Brown furnished material about Waller Tazewell Patton and the Battle of Gettysburg, and read an early draft of the Civil War chapter.

I am grateful to the late Col. G. Gordon Bartlett, USA (Ret.), for his advice and assistance during the writing of this and an earlier book on Anzio. John S. D. Eisenhower provided a copy of his father's unpublished manuscript and offered commentary on the long friendship between Dwight Eisenhower and George S. Patton.

Genealogist Julie Poole of St. Andrews, Scotland, researched Robert Patton and his emigration to Virginia on my behalf.

Special thanks to artist Don Stivers for furnishing articles and accounts of the Lipizzaners, and to my friend, historian Edward M. Coffman, for assistance with the World War I chapters.

The U.S. Army Military History Institute, Carlisle Barracks, Pennsylvania, is a national treasure whose library and archives contain rich sources of our military history from the Revolutionary War to the present. The dedication of the MHI staff is exceptional, and I offer my heartfelt thanks to former director, Col. Rod Paschall, the present director, Col. Stephen L. Bowman, archivist-historian Dr. Richard J. Sommers, and assistant archivist David A. Keough, to whom I make a special acknowledgment. As always, John J. Slonaker and his library staff insured that no stones were left unturned in my research. To everyone at this superb facility, my deep appreciation for your assistance, your guidance, and your friendship.

I owe a very special debt of gratitude to two superb military historians whom I am indeed fortunate to count as friends, Dale E. Wilson and Lt. Col. Roger Cirillo. Dale, the author of *Treat 'Em Rough!: The Birth of American Armor, 1917–20*, contributed his expertise as an editor and historian by reading and critiquing the manuscript and offering sound advice on a variety of subjects. Roger not only provided the hospitality of his home, but read portions of the manuscript and shared without reservation his encyclopedic knowledge of military history in general and World War II in particular. Roger also supplied the rare photo of Patton and Pershing taken in 1917. I am pleased to record my heartfelt thanks to both.

For the author photograph I gratefully acknowledge my friend and neighbor, David Marlin, whose photographic beat was New England during a distinguished career with CBS.

Others whom I thank for their assistance: Dr. Dean C. Allard; Thomas Allsopp; Gerald Astor; Brig. Gen. Raymond E. Bell Jr., USAR; David Bennett; Brig. Shelford Bidwell, OBE; James R. Bird; Clay Blair; Robert H. Cowley; Gerald B. Forrette; William P. Frank; Professor Denver Fugate; Christopher R. Gabel; Ralph J. Garcia; the late Lt. Gen. James M. Gavin, USA (Ret.); Dominick Graham; Tallmadge Hamilton, Jr.; Professor David K. Hart; Professor James E. Herget; Warren E. Huguelet; Carl E. Hudson, Jr.; the late Lt. Gen. Sir Ian Jacob; Brooks Kleber; Col. Jimmie Leach, USA (Ret.); John Lehman; Dr. William J. McAndrew; the late Gen. Andrew P. O'Meara, USA (Ret.); Col. George Pappas, USA (Ret.); Lt. Col. USA (Ret.) Henry G. "Red" Phillips; T. M. Pickens; Craig Prediger; Charles Price; David F. Riggs; Don W. Rightmyer; Andrew A. Rooney; G. L. Seligmann; Midge Sherwood; C. Cabanné Smith; Prof. Norman S. Stevens; Scott Steward; David S. Terry; Donald C. Van Roosen; William Walton; Brig. H. B. C. Watkins, British Army (Ret.); the late Gen. I.D. White, USA (Ret.); Ernest C. Wilbur Jr., and G. Kenneth Yates.

No author could be more blessed than to have as editors Stuart Proffitt (HarperCollins London) and M. S. "Buz" Wyeth (HarperCollins New York). Buz recognized the potential and importance of writing the fascinating saga of George S. Patton and enthusiastically backed the project from start to finish. He also edited the original manuscript, thereby confirming that the art of editing is indeed alive and well. To both I offer my gratitude for their counsel, criticism, and, above all, patience for the more than five years during which the book evolved from conception into reality. I also thank Susan H. Llewellyn of Harper-Collins, New York, who marked our second collaboration with another first-class job of copyediting, and Arabella Quin of HarperCollins, London. Special thanks to my longtime New York agent Julie Fallowfield (McIntosh & Otis) and my British agent, Sheila Watson (Watson, Little), who helped make it all happen.

Last, to my family—particularly my wife, Shirley Ann—who have endured the presence of George S. Patton in *their* lives for so many years, my inestimable gratitude for your love and support—and for your ideas for making this a better book.

Although I have been able to call on a vast array of distinguished and knowledgeable people during the research and writing of *Patton: A Genius for War*, any mistakes of fact or interpretation are my sole responsibility.

INDEX